ENCYCLOPEDIA OF PHYSICS

EDITED BY

S. FLÜGGE

VOLUME XXII

GAS DISCHARGES II

WITH 337 FIGURES

SPRINGER-VERLAG
BERLIN · GÖTTINGEN · HEIDELBERG
1956

HANDBUCH DER PHYSIK

HERAUSGEGEBEN VON

S. FLÜGGE

BAND XXII

GASENTLADUNGEN II

MIT 337 FIGUREN

SPRINGER-VERLAG
BERLIN · GÖTTINGEN · HEIDELBERG
1956

ISBN-13: 978-3-642-45849-1 e-ISBN-13: 978-3-642-45847-7
DOI: 10.1007/978-3-642-45847-7

SOFTCOVER REPRINT OF THE HARDCOVER 1ST EDITION 1956

Inhaltsverzeichnis.

Seite

Ionization Growth and Breakdown. By Professor Dr. F. LLEWELLYN JONES, M. A., Department of Physics, University College of Swansea Swansea/Wales (Great Britain). (With 35 Figures) 1

A. Spatial growth of ionization in static uniform fields at low gas pressures 2
B. The similarity principle 17
C. Growth of ionization in non-uniform fields 21
D. Spatial growth of ionization in static fields at high pressures 27
E. The temporal growth of ionization 40

The Glow Discharge at Low Pressure. By Dr. GORDON FRANCIS, Clarendon Laboratory, Oxford (Great Britain). (With 96 Figures) 53

I. Introduction 53
II. General description 53
III. Methods of measurement in discharges 61
IV. Similarity laws 70
V. The cathode region 81
a) Experiment 81
b) Theory 101
VI. The negative glow and FARADAY dark space 110
VII. The positive column 114
a) Main types 114
b) The uniform column in atomic gases: main experimental results . . . 115
c) The uniform column in molecular gases: main experimental results . . 118
d) Gradient in very wide tubes 120
e) Theory 120
f) Extensions of the theory 128
g) Comparison of theory and experiment 131
h) The striated column 137
VIII. The anode region 145
IX. Cathode sputtering 154
X. Special types of discharge 161
XI. Miscellaneous effects 195

References 203
General references 203
Special references 203

Radiation from Low Pressure Discharges. By Professor Dr. RICHARD G. FOWLER, Department of Physics, University of Oklahoma, Norman/Oklahoma (USA). (With 17 Figures) 209

Glossary of special usages 209
A. General aspects of gas radiance 209
B. Excitation 218
C. Depopulating processes 238
D. Special discharges 240
Major Reference Sources 253

Seite

Elektrische Bögen und thermisches Plasma. Von Professor Dr. W. FINKELNBURG und Privatdozent Dr. H. MAECKER, Siemens-Schuckert-Werke AG., Forschungslaboratorium, Erlangen. (Mit 104 Figuren) . . . 254

A. Allgemeine Lichtbogenphysik . . . 255

B. Lichtbogen-Phänomenologie . . . 266

I. Die Abhängigkeit der Bogenerscheinungen von den äußeren Bedingungen . 266
II. Allgemeine Eigenschaften stationärer und nichtstationärer Lichtbögen . . 276
III. Übersicht über die wichtigsten Lichtbogentypen . . . 291

C. Theorie des thermischen Plasmas . . . 305

I. Das thermische Gleichgewicht und Quasineutralität im Bogenplasma . . 305
II. Einführung in die Thermodynamik irreversibler Prozesse . . . 315
III. Diffusions- und Energieströme in reagierenden Gasen . . . 323
IV. Eigenschaften des reinen Plasmas . . . 325
V. Irreversible Prozesse im System mit drei Komponenten . . . 329
VI. Diskussion der Plasmagleichungen . . . 335
VII. Berechnung der thermodynamischen Zustandsgrößen . . . 340
VIII. Rückführung der Koeffizienten auf die Stoßquerschnitte . . . 344
IX. Strahlung des Plasmas . . . 353

D. Die Bogensäule . . . 366

I. Allgemeines . . . 366
II. Messung der Zustandsgrößen in der Säule . . . 369
III. Die Bogensäule als Ganzes . . . 383

E. Kathodische Erscheinungen und Kathodenmechanismus von Lichtbögen 403

I. Kathodische Erscheinungen und empirische Daten . . . 403
II. Kathodenfalltheorien . . . 412

F. Anodische Erscheinungen, anodisch bestimmte Sonderformen von Lichtbögen und Anodenfallmechanismen . . . 422

I. Anodische Erscheinungen und empirische Daten . . . 422
II. Anodisch bestimmte Lichtbögen und ihre Eigenschaften . . . 427
III. Anodenfalltheorien . . . 434

Bibliographie . . . 439
Bücher über elektrische Lichtbögen . . . 439
Spezielle Literatur . . . 440

Electrical Breakdown of Gases with Steady or Direct Current Impulse Potentials. By Professor Dr. LEONARD B. LOEB, Department of Physics, University of California, Berkeley, California (USA). (With 9 Figures) . . . 445

Foreword . . . 445

I. Introduction. Basic definitions . . . 445
II. The growth of the TOWNSEND discharge . . . 457
III. Properties of the threshold equation . . . 462
IV. Statistical fluctuations of TOWNSEND or other breakdowns and the various time lags . . . 469
V. Breakdown in non-uniform fields and near highly stressed anodes. Photoionization in gases. Streamers and burst pulses . . . 480
VI. The role of the streamer in uniform field breakdown and the long time lags of sparking . . . 490
a) Application of MEEK's theory to sparks in air . . . 490
b) Search for transition from streamer to TOWNSEND spark . . . 491
VII. Breakdown in asymmetrical fields at steady potential: The coronas, glows and sparks . . . 506
VIII. Breakdown in non-uniform but symmetrical fields at steady potentials . . 525
IX. Impulse breakdown in non-uniform fields . . . 526
X. Sparks in very short gaps, at high pressures and in vacuum . . . 527

References . . . 528

Seite

Breakdown in Gases: Alternating and High-Frequency Fields. By Professor Dr. SANBORN C. BROWN, Research Laboratory of Electronics, Massachusetts Institute of Technology, Cambridge, Mass./(USA). (With 33 Figures) 531

Table of symbols 531

A. Diffusion controlled breakdown 531

I. Behavior of the average electron 531

II. Theories based on the BOLTZMANN transport equation 539

a) High-frequency electric fields 539

b) The effects of nonuniform fields 552

c) Effects of a superimposed d.c. field 558

d) Effects of a superimposed magnetic field 559

B. Electron attachment controlled breakdown 566

C. Mobility-controlled breakdown 568

D. Secondary electron resonance breakdown 570

E. Summary of high-frequency breakdown mechanisms 574

The Lightning Discharge. By Professor Dr. B. F. J. SCHONLAND, F.R.S., Atomic Energy Research Establishment, Harwell, Didcot/Berks. (Great Britain). (With 43 Figures) 576

A. Introduction 576

B. General features of ground and air discharges 577

C. Photographic studies of luminosity in the discharge 580

I. Methods 580

II. Results obtained by the photographic method 585

a) Flashes to ground 585

b) Air discharges 596

c) Cloud discharges 597

d) The apparent diameters of streamer channels 597

D. Electrical Observations; methods and results 598

I. Methods of low time resolution 598

II. Methods of high and intermediate time-resolution 601

III. Currents and current-time variations 605

IV. Discharges to very tall structures 607

V. Fast electrical field-change studies 608

VI. Radar studies of interstroke streamers in the cloud 615

VII. Heights of the regions initiating successive strokes 616

VIII. Cloud and air discharges 617

E. The mechanism of breakdown and of streamer propagation 618

Major references and notes 627

Sachverzeichnis (Deutsch-Englisch) 629

Subject Index (English-German) 641

Ionization Growth and Breakdown.

By

F. LLEWELLYN JONES.

With 35 Figures.

1. Introduction. Early work[1] in this field was mainly carried out at the CAVENDISH Laboratory, Cambridge, by J. J. THOMSON and J. S. TOWNSEND, who investigated how ionization in a gas develops to lead finally to the breakdown of the insulating properties of the gas. When a potential difference is established between electrodes in a gas, the gas behaves as an insulator unless the potential exceeds a certain definite value, which is called the *sparking* or *breakdown* potential. In practice, this potential V_s is sharply defined; the gas insulates well at a potential only a few volts less than V_s. When the applied potential V is greater than V_s, even if only by a very small amount, the insulation breaks down, and a large current can pass if the circuit conditions permit, i.e. if the source of potential has a low impedence. The physical problem is to account for the development of the ionization in an electric field from a small number of initiatory electrons up to any specified final current, in terms of fundamental electronic, atomic and ionic collision processes in the gas and at electrode surfaces.

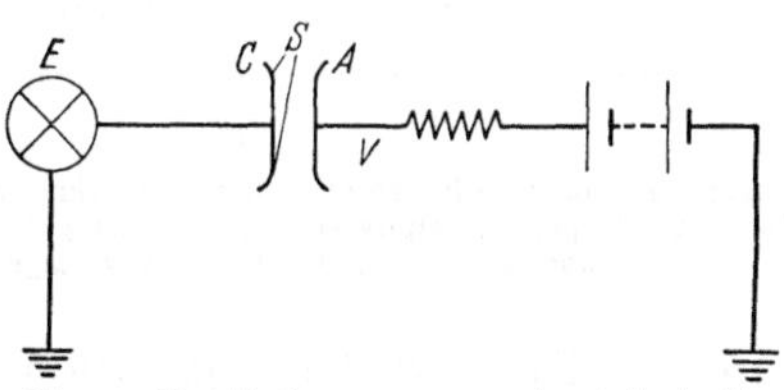

Fig. 1. Circuit for measurement of ionization currents.

Consider a practical case of a simple parallel-plate gap in air as shown in Fig. 1; V is a source of steady potential, A the anode, C the cathode irradiated from a source S, and E an electrometer to measure currents $\sim 10^{-10}$ A. As the anode potential is increased, the gas in the space remains almost a perfect insulator until a certain potential V_s is applied, then suddenly a flash appears across the gap. A high current can pass, depending upon the gas pressure, distance between the plates, and resistance of the source of potential; the electrodes can become heated and a complete short circuit of the source can be produced. The appearance of the discharge depends upon the conditions. At low pressures, the spark flash has the form of a wide glow; but at high pressures, the flash is narrow and in the form of a luminous streak, then naturally referred to as a *ribbon, kanal* or *streamer*[2], as for example in the lightning flash. With the small initial current, the potential across a gap remains at V_s, but as the current increases a glow is produced and the potential falls, and can reach the arc value if the electrodes

[1] J. J. THOMSON: Conduction of Electricity through Gases, 2nd edit. Cambridge 1906. — J. J. and G. P. THOMSON: Conduction of Electricity through Gases, 3rd edit. Cambridge 1933. — J. S. TOWNSEND: MARX's Handbuch der Radiologie, Bd. 1. Leipzig: Akademische Verlagsgesellschaft 1916. — Electricity in Gases. Oxford: Clarendon Press 1915. — Theory of Ionization of Gases by Collision. London: Constable 1910. — R. SEELIGER and G. MIERDEL: Handbuch der Experimentalphysik, Vol. 13, Part 3. Leipzig: Akademische Verlagsgesellschaft 1929. — W. O. SCHUMANN: Elektrische Durchbruchfeldstärke von Gasen. Berlin: Springer 1923.

[2] J. M. MEEK and J. D. CRAGGS: Electrical Breakdown of Gases. London: Clarendon Press 1953.

start thermal or field emission of electrons. The time sequence of events[1] is indicated by Fig. 2, and it is seen that the times involved can be very short.

It is possible to divide the complete breakdown process, from insulating gas to highly conducting arc, into three phases as the voltage falls and the current rises, the phases corresponding to the low self-maintained (TOWNSEND) *spark current*, the *glow*, and the *arc*. The glow phase, however, can in many cases be of very short duration, thus giving two main phases of the complete breakdown process. In seeking an explanation of the complete flash in terms of the fundamental physics of atomic collisions, it is important first to consider how the growth of ionization produces the transition of the gas from its completely insulating state to one in which the current can rise to a value which depends on the external

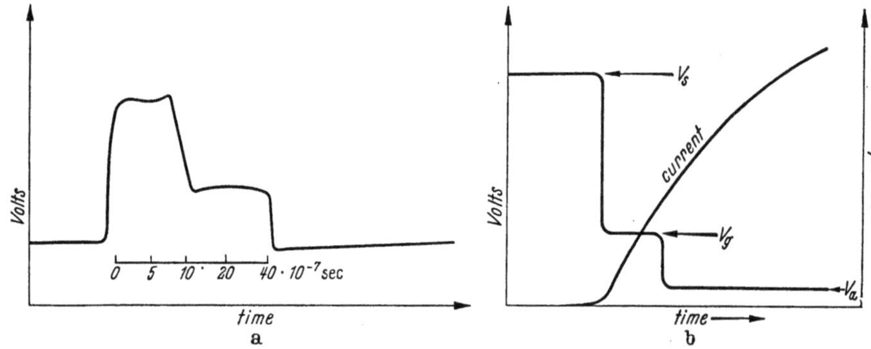

Fig. 2a and b. (a) Change of voltage with time in the development of breakdown. (SEELIGER and MIERDEL: Handbuch der Experimentalphysik, vol. 13, part 3, p. 39, Fig. 28. 1929.) (b) Schematic indication of increase of current and corresponding fall of gap voltage with time; breakdown voltage is applied at time origin.

circuit. The added complication due to the rapid increase in current density leading to a low-voltage arc can be considered afterwards as a separate problem of established electrical discharges governed by space charges[2].

In this article, ionization growth will be considered both for uniform and non-uniform fields, and for impulse as well as static applied potentials. The physical phenomena in the various phases appear to differ so much, that from time to time different theories have been proposed to account for them. It is now interesting to consider how far a consistent theory of the mechanism is at present possible.

A. Spatial growth of ionization in static uniform fields at low gas pressures.

($p d < 150$ mm. Hg × cm.).

2. Experimental curves. Consider the development of a current I from a small initial photo-electric current I_0 ($\sim 10^{-12}$ A) under a uniform electric field E in a gas at pressure p between parallel plane electrodes, distance d apart as in Fig. 1. When electrons and ions move through the gas, their mean energies attain equilibrium values dependent on the ratio E/p. Thus, in order to investigate how the current growth depends on the space through which the electrons move, or the nature of the electrode system, it is essential to maintain the mean energies of the electrons and ions constant and to measure the current I as a function of gap distance d *when the ratio E/p is maintained constant.* The constancy of

[1] R. SEELIGER and G. MIERDEL: Handbuch der Experimentalphysik, Vol. 13, Part 3, p. 38 (Fig. 28). 1929.

[2] F. LLEWELLYN JONES: Electrical Discharges. Rep. Progr. Phys. Lond. **16** (1953). — M. J. DRUYVESTEYN and F. M. PENNING: Rev. Mod. Phys. New York **12** (1940).

the ratio E/p is the basis of this type of experimental investigation; it is therefore necessary that the applied potential should be sufficiently steady and that the current should not distort the field. Consequently, this work is usually carried out with current densities in the range 10^{-15} to 10^{-8} A/cm.^{2}.

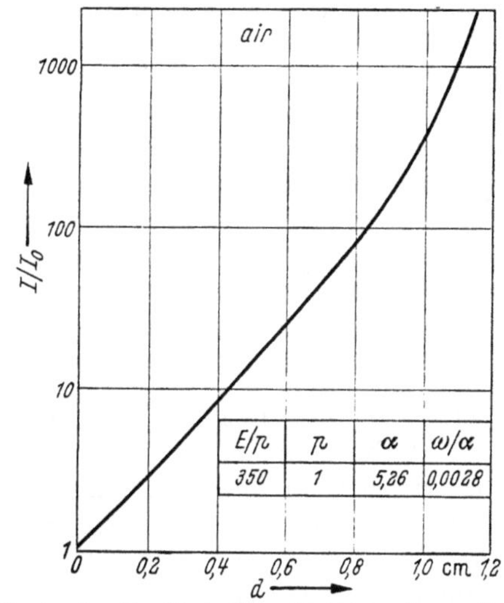

Fig. 3. Growth of ionization with distance in a uniform field in air. (TOWNSEND: Electricity in Gases, p. 278.)

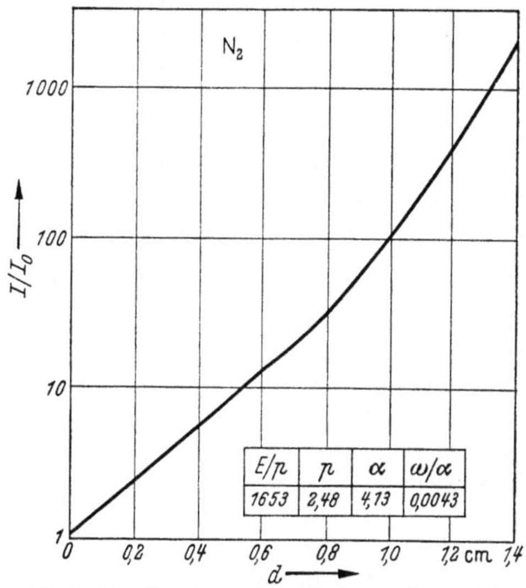

Fig. 4. Growth of ionization with distance in a uniform field in nitrogen. [AYRES: Phil. Mag. 45, 353 (1923).]

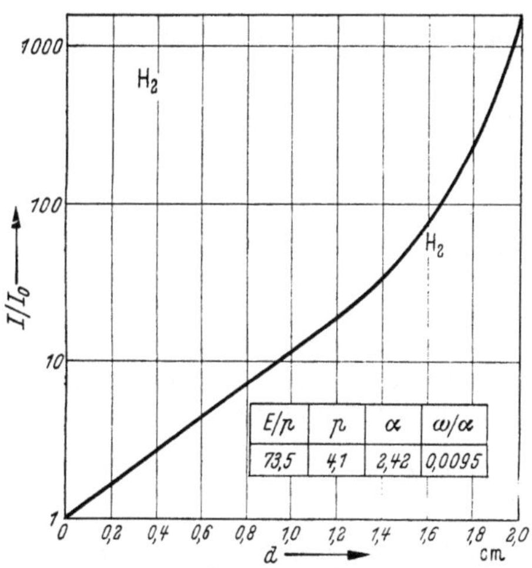

Fig. 5. Growth of ionization with distance in a uniform field in hydrogen. [AYRES: Phil. Mag. 45, 353 (1923).]

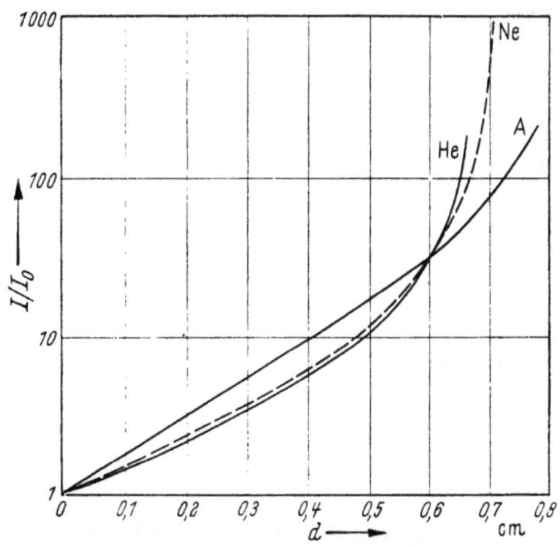

Fig. 6. Growth of ionization in uniform fields in monatomic gases. [He: TOWNSEND and McCALLUM, Phil. Mag. 17, 693 (1934); Ne: TOWNSEND and McCALLUM, Phil. Mag. 6, 870 (1928); A: KRUITHOF and PENNING, Physica, Haag 3, 526 (1936).]

With E/p constant, and with an initiatory photo-electric cathode emission of I_0 produced by irradiation from a constant source of ultraviolet light, comprehensive measurements of the ionization current I have been made in a large number of gases[1]; and, using modern vacuum and gas purification techniques, further measurements have been made very recently[2]. Typical curves obtained are given in Figs. 3, 4 and 5 for air, nitrogen and hydrogen respectively, and in Fig. 6

[1] J. S. TOWNSEND: Electricity in Gases. 1915.
[2] D. J. ROSE: Conf. on gaseous electronics. New York 1954.

for helium, neon and argon, where I/I_0 is plotted on a logarithmic scale against d for a given constant value of E/p.

The initial stage of current growth is influenced by back-scattering of the electrons by the gas molecules just outside the cathode surface, and also by the increase of mean energy with distance travelled from the cathode. Let d_1 represent the distance ($\sim$ some free paths) required for the mean energy of the electrons to build up to the equilibrium value corresponding to the value of E/p in the gap. For values of gap distance greater than d_1 (which is often negligible) the curve of the $\log I$ *vs.* d graph is linear if d is not too large, and the slope α is a constant for the gas at that value of E/p. Thus the observed current growth is given by the relation

$$I = I_0 \, e^{\alpha d}.$$

With different values of E/p similar curves are obtained but with different slopes, and for a given gas α/p is a function of E/p.

The significance[1] *of* α. In moving through the gas, electrons can ionize the gas molecules, and the probability of their doing so depends on the cross-section for ionization, which is a function of the electron energy. Let n_0 electrons be emitted per second from the cathode at $x=0$: at the point x the number has increased to n_x by ionizing collisions. Let these n_x electrons in moving through a lamina dx at x in the direction of E generate by collisions $\alpha n_x dx$ new electrons, so that $dn = \alpha n_x dx$. The number n_d of electrons which strike the anode at $x=d$ is then $n_0 e^{\alpha d}$. Such a group of electrons is called an *avalanche*. It follows that

$$I = I_0 \, e^{\alpha d} \quad \text{or} \quad \log I/I_0 = \alpha d, \tag{2.1}$$

where α is the TOWNSEND coefficient of primary ionization which can be related to the fundamental atomic properties as follows.

Let the electron velocities u be distributed about their mean according to $dn = n f(u)\, du$, and let $\sigma(u)$ be the atomic cross-section for ionization by an electron of velocity u. The total number of ionizations per second made by a group n of electrons is

$$N n \int f(u)\, u\, \sigma(u)\, du;$$

i.e.

$$N n \int \left(f(u)\, u\, \sigma(u)\, du\right) \frac{dx}{W_-}$$

ionizing collisions are made while the group traverses a lamina dx of gas, where N is the number of gas atoms per unit volume, the drift velocity[2]

$$W_- = e E \, \frac{\overline{(2l/u + dl/du)}}{3m}$$

and l is the mean free path of electrons with velocity u. It is, of course, less than $1/N\sigma$ unless every collision ionizes.

There is no difficulty in principle, therefore, in relating α to fundamental atomic properties, and many attempts have been made. The earliest attempt was made by TOWNSEND[3] who, ignoring the distribution and ionization probability functions, and assuming that all electrons which exceeded V_i in energy ionized obtained the simple formula:

$$\frac{\alpha}{p} = A \, e^{\frac{-B}{E/p}}.$$

---

[1] See the article by A. H. VON ENGEL on Ionization by collision, Vol. XXI of this Encyclopedia.

[2] P. M. DAVIDSON: Proc. Phys. Soc. Lond. B **67**, 159 (1954).

[3] J. S. TOWNSEND: Theory of Ionization by Collision. 1910.

When $A=5$, $B=13$, this formula agrees with the values of α in hydrogen measured by AYRES for $100<E/p<600$ V/cm. per mm. Hg. More recent calculations have been made for other gases[1]. The main difficulty in practice in calculating α exactly is that of knowing sufficiently accurately the energy distribution function and the cross-sectional terms as a function of energy. Thus α is a statistical quantity specified when E and p are constant or the effects of grad E or dE/dt negligible.

3. The spatial growth of current in static fields. The graphs in Figs. 3 to 6 are practically straight initially but curve upwards at the larger values of d, finally becoming vertical at $d=d_s$. It is observed that just when d attains these values of d_s, the breakdown flash occurs. The form of these curves is given within the experimental error by the relation

$$I=I_0\frac{e^{\alpha d}}{1-C(e^{\alpha d}-1)}, \tag{3.1}$$

where α and C are constant when E/p is constant at a given p. Further, I will be the same if E/p and pd are the same.

Table 1 gives measured values of the current I obtained in hydrogen at the two pressures 1 mm. Hg. and 2 mm. Hg for various distances d. When $p=1$ mm. Hg $E=350$ V/cm., $\alpha=5.25$ and $C=0.0027$; when $p=2$ mm. Hg $E=700$ V/cm., i.e. E/p was the same as when $p=1$ mm. Hg. It is seen that the measured currents are the same in both cases. The third row gives the current calculated from equation (3.1). These results confirm that α/p and C are functions of E/p.

Table 1.

pd	0.2	0.4	0.6	0.8	1.0
I ($p=1$ mm.) .	2.86	8.3	24.2	81	373
I ($p=2$ mm.) .	2.9	8.3	23.8	80	374
Equation (3.1) .	2.87	8.3	24.6	80	380

It is reasonable to assume that C represents some process of ionization in addition to the α process. C is the second TOWNSEND, or secondary, ionization coefficient.

When $d<d_s$, I is proportional to I_0; and if the original irradiation of the gap is cut off, I_0 becomes zero, and the total current I also ceases. From the form of the equation (3.1) it follows that when

$$1-C(e^{\alpha d}-1)=0 \tag{3.2}$$

the value of d makes I tend to increase without limit. This d is d_s. For any value of C, however small, it is always possible to find a value of d_s which satisfies this relationship. When d_s is measured, then $V_s(=E d_s)$ can be determined.

α) *The sparking potential and breakdown criterion.* The value V_s is thus obtained by finding α and C directly from measurements of the growth of ionization current and inserting these values in the criterion (3.2). The values of V_s so obtained can then be compared with the potential at which the breakdown is observed to occur. This has been done for many gases and good agreement found[2]; the data for air and hydrogen, for example, are illustrated in Fig. 7.

Equation (3.2) is TOWNSEND's criterion for breakdown in uniform static fields. The agreement of the observed breakdown potentials of various gases with those calculated from the criterion (3.2), is thus in accordance with the view that the breakdown criterion is set by the mechanism of the growth of ionization

[1] H. S. W. MASSEY and E. H. S. BURHOP: Electronic and Ionic Impact Phenomena. Oxford: Clarendon Press 1953.

[2] J. S. TOWNSEND: Electricity in Gases, Chap. IX. 1915.

current prior to breakdown, of which growth breakdown is a natural development. It also follows that at the *breakdown condition* specified by equation (3.2) the effects of space charge are still insignificant, although the current can begin to increase rapidly. Full understanding of breakdown now requires the elucidation of the secondary processes represented by the coefficient C.

β) Possible secondary processes.

1. Secondary emission due to incidence of positive ions on the cathode. $n_x \alpha\, dx$ electrons produced in the gas at distance x from the cathode involve the production of $n_x \alpha\, dx$ positive ions, which at the cathode can produce $\gamma\, n_x \alpha\, dx$ electrons; γ is the number of electrons liberated from the cathode per ionizing collision in the gas[1], and it is therefore not dependent on the value of the current.

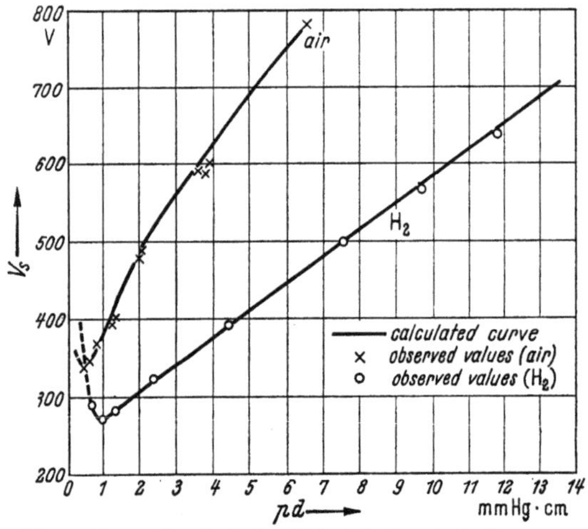

Fig. 7. Observed and calculated V_s, pd values for hydrogen and air. (H_2: TOWNSEND, Electricity in Gases, p. 324, 1915; Air: TOWNSEND, The Theory of Ionization of Gases, p. 56, 1910.)

2. Cathode emission due to incidence of photons. When n_x electrons pass through a lamina dx of gas they excite molecules and produce $\alpha' n_x\, dx$ photons, a fraction, ϑ of which will fall on the cathode and with a photo-electric efficiency f will produce approximately

$$f \vartheta\, e^{-\mu x} \alpha' n_x\, dx = \delta n_x\, dx$$

electrons: the absorption $e^{-\mu x}$ in the gas may be neglected at low pressures. The number of photoelectrons emitted from the cathode[2] per ionizing collision in the gas is thus δ/α. The total electron emission from the cathode due to positive ions and photons per ionizing collision in the gas is $\delta/\alpha + \gamma$. The interception of the photons by the cathode depends on the geometry of the electrode system; and this consideration is important when considering asymmetrical systems such as a wire and coaxial cylinder.

3. Cathode emission due to incidence of excited atoms. Excited atoms return to the ground state in times $\sim 10^{-8}$ sec. Thus many do not survive motion through the gas; those which do reach the cathode can produce secondary electrons. A metastable atom which is a particular case of an excited atom, has a relatively long life ($\sim 10^{-1}$ sec.) unless forced to revert to the ground state by a suitable collision (PENNING effect)[3]; metastable atoms are more likely to reach the cathode and there produce electrons. This action is similar to the photo-electric effect, except that the transport of energy to the cathode is slower, and the process can be represented by a coefficient ε/α of same dimensions as δ/α.

4. Gas ionization by positive ions. Another form of ionization has also been considered, namely that due to the direct collisional ionization of neutral atoms by positive ions[2]. This process can be designated by a coefficient β, where

$$dn_x = \beta\, p_x\, dx,$$

p_x being the concentration of positive ions at x. There is a dearth of reliable experimental data on this process[4].

[1] J. J. THOMSON: Conductions by Electricity through Gases. 1906.
[2] J. S. TOWNSEND: Phil. Mag. **45**, 444 (1923).
[3] M. J. DRUYVESTEYN and F. M. PENNING: Rev. Mod. Phys. **12**, 113 (1940).
[4] H. S. W. MASSEY and E. H. S. BURHOP: Electronic and Ionic Impact Phenomena. 1953.

5. *Step-wise ionization.* This occurs when either electrons or photons ionize gas atoms which are themselves already excited: electrons or photons of energy given by

$$\tfrac{1}{2} m u^2 \sim h\nu > e V_i - e V_{ex}$$

are necessary. The stepwise ionization rate depends on the number of excited neutral atoms and on the number of colliding electrons or photons. Spontaneous rapid return of excited atoms to the ground state makes the population of excited atoms generally negligible. The process in general depends on the square of the electron concentration, and this would make the secondary coefficient C depend on the value of the current.

6. *Photo-ionization.* The experimentally determined values of the cross-sections[1] indicate that at low pressures, the influence of the photo-ionization on the growth of current is negligible. At higher pressures ~ 760 mm. Hg the process has to be quantitatively examined, and will be discussed below.

γ) Growth of current when the secondary processes: β, γ, δ and ε act. Let I_- refer to electrons, and I_+ to ions; and assume that the initial electrons from the cathode quickly attain the mean energies corresponding to the value of E/p in the gap, i.e. that the coefficients α, β, γ, δ and ε are each constant from $x=0$ to $x=d$. The coefficient δ can, in the steady state, include ε due to diffusion of metastable atoms. The normally excited atom density will be low.

$$\frac{dI_-}{dx} = -\frac{dI_+}{dx} = \alpha I_- + \beta I_+,$$

$$I_- + I_+ = \text{total current } I = \text{constant},$$

Hence,

$$\frac{dI_-}{dx} = \alpha I_- + \beta(I - I_-) = \beta I + (\alpha - \beta) I_-$$

and therefore

$$\beta I + (\alpha - \beta) I_- = D\, e^{(\alpha-\beta)x}.$$

At the anode

$$I_+(d) = 0 = I - I_-(d) = \frac{\alpha I - D\, e^{(\alpha-\beta)d}}{(\alpha - \beta)},$$

hence

$$D = \alpha I\, e^{-(\alpha-\beta)d}$$

and

$$I_- = I\, \frac{\alpha\, e^{-(\alpha-\beta)(d-x)} - \beta}{\alpha - \beta}.$$

But

$$I_-(0) = I_0 + \gamma I_+(0) + \delta \int_0^d I_- \, dx.$$

Combining this with the previous expression it follows that

$$I = \frac{I_0 \left(1 - \frac{\beta}{\alpha}\right) e^{\alpha d}}{e^{\beta d}\left[1 - \frac{\beta}{\alpha} + \frac{\beta\delta}{\alpha} d - \left(e^{(\alpha-\beta)d} - 1\right)\left(\gamma + \frac{\delta}{\alpha - \beta} + \frac{\beta}{\alpha}(1 - \delta d)\right)\right]}.$$

Experimental data show that, in general, β/α, and βd, and δd are $\ll 1$. Thus, at all currents up to the sparking criterion this expression reduces to

$$I = I_0 \frac{e^{\alpha d}}{1 - \frac{\omega}{\alpha}\left(e^{\alpha d} - 1\right)} \tag{3.3}$$

[1] See article by G. L. Weissler, vol. XXI of this Encyclopedia. Sect. 3. 1956.

where, neglecting second order terms,

$$\omega = \beta + \alpha\gamma + \delta + \varepsilon. \tag{3.4}$$

The relation (3.3) is precisely the same as that found experimentally and given in (3.1), when the only secondary process is that of positive ions producing secondary electron emission from the cathode, then

$$I = I_0 \frac{e^{\alpha d}}{1 - \gamma (e^{\alpha d} - 1)}, \tag{3.5}$$

and similarly for the other separate secondary processes: β/α, δ/α, etc.

δ) Shape of growth curve. Each of the different and distinct secondary processes discussed above thus gives practically the same analytical form for the expression for the current growth, so that the processes cannot easily be distinguished from the shape of the curve giving the observed growth of current. Other investigations must therefore be made to distinguish between the processes, the two most significant of which are the γ and δ processes.

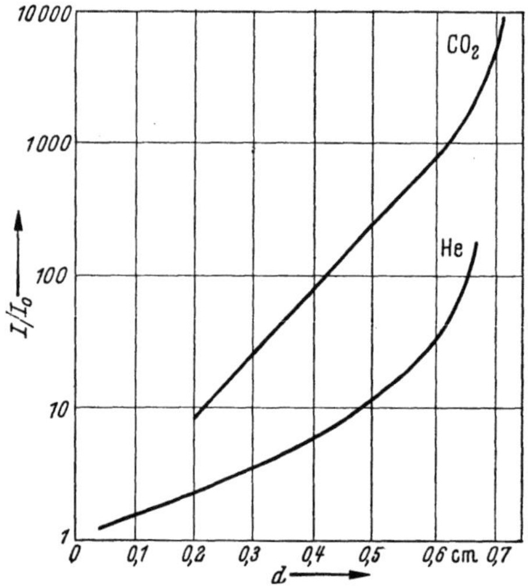

Fig. 8. Comparison of growth curves in CO_2 and He. [CO_2: TOWNSEND, Electricity in Gases, p. 319, 1915; He: TOWNSEND and McCALLUM, Phil. Mag. 17, 693 (1934).]

The general shape of the $\log I$ *vs.* d graph is determined by the value of ω/α; e.g. in helium[1] ω/α is high, and there is practically no linear section in the graph corresponding to the equation $I = I_0 e^{\alpha x}$. In CO_2, on the other hand, ω/α is small, and the graph has a long linear section as shown in Fig. 8. As the pressure increases, and E/p for breakdown diminishes, the ratio ω/α also diminishes, and the part of the graph where the upcurving occurs becomes smaller. The above derivation of equation (3.3) assumes that the distance d_1 through which the initial electrons from the cathode must travel in order to acquire the constant mean energy is negligible compared with the gap d. At low values of E/p, however, this is not the case and it is found experimentally that, for instance, the growth of ionization currents in monatomic gases, such as neon, helium[1] and argon[2] at low pressures, is described by putting $(d - d_1)$ for d in equation (3.3), i.e.

$$I = I_0 \frac{e^{\alpha (d - d_1)}}{1 - \frac{\omega}{\alpha} (e^{\alpha (d - d_1)} - 1)}.$$

The coefficients α and ω are functions of the field and the above derivations of (3.3) also assume that the field in the gap is uniform and that the space charge of the current produces no significant changes in α or ω. However, if I_0 and therefore I is sufficiently high, α can be increased by field distortion due to space charge, and TOWNSEND[3] attributed a lowering of V_s at high I_0 to this cause.

[1] J. S. TOWNSEND and S. P. MACCALLUM: Phil. Mag. **6**, 857 (1928) (Ne); **17**, 678 (1934) (He). — J. S. TOWNSEND: Electricity in Gases, p. 318. 1915.

[2] A. A. KRUITHOFF and F. M. PENNING: Physica, Haag **3**, 515 (1936).

[3] J. S. TOWNSEND: Theory of Ionization by Collision. 1910.

It has also been assumed that no electron emission from the cathode occurs other than that due to external radiation or to ions or photons from the electron avalanche. This assumption is justified for fields $< 10^4$ V/cm., but may not be justified for higher fields, which can produce emission from electrodes at room temperature[1]. At high pressures of some atmospheres in small gaps, the electric field V/d can exceed 10^4 V/cm., and electron emission can be produced from the electrodes at room temperature due to the action of the field. This results in an enhancement of the secondary coefficient ω/α, and consequent lowering of the sparking potential[2].

4. Physical significance of the breakdown criterion. *α) The criterion.* The current cannot continually increase without limit when $V = V_s$, because no practical source can supply the power. The physical interpretation of the breakdown criterion is as follows.

If d_s satisfies the relationship (3.2), it follows from (3.3) that I can be finite even when I_0 is zero. Thus, this condition (3.2) is really a mathematical statement of a replacement condition; viz. that the ionization processes in the gas are such that they replace, by generation in the gas or at the cathode, the original I_0 initiatory electrons. Consequently, the current in the gap is self-maintained. This is a physical condition which can be realised in the laboratory. Thus, the mathematical criterion represented by equation (3.2) represents a condition, the physical significance of which is that when $d = d_s$, then I_0 could be reduced to zero, and I could then remain finite and V would remain equal to V_s and be independent of I when this is small. On this view, the sparking, or breakdown potential V_s can be defined as the potential, independent of the current, which maintains a small current: this physically specifies the breakdown criterion, and is unrelated to the attainment of infinitely large currents.

The problem of what happens when the current is allowed to increase rapidly can be regarded as a problem of space charge conditioned electric discharges[3]; and it is thus convenient to separate the range of phenomena from insulating gas to final arc or glow discharge into two parts. The one part, when the processes represented by α and ω/α bring the gas to the state given by $d = d_s$, $V = V_s$; and the other part which occurs when any given current, I, already established in the gas, is allowed to increase finally passing through the well-known phases involving the emission of light and heat and which may also involve various electrode effects.

Consider now the practical realization of this criterion (3.2). A saturated diode introduced in series with the gap can restrict the current to any specified value, enabling V to be measured for the various values of I, thus giving the functional dependence of I on V. Fig. 9 gives an example of such a curve obtained in hydrogen by C. G. MORGAN and E. JONES at Swansea, showing that I is independent of V over a wide range; the range itself depending on p and d.

The experimental determination of such curves is, however, not easy. In the first place, the diode will not be completely saturated. Secondly, the nature of the gap cathode, in general, determines the maintenance condition through the

---

[1] F. LLEWELLYN JONES and E. T. DE LA PERRELLE: Proc. Roy. Soc. Lond., Ser. A **216**, 167 (1953). — F. LLEWELLYN JONES and C. G. MORGAN: Proc. Roy. Soc. Lond., Ser. A **218**, 88 (1953). — C. G. MORGAN and D. HARCOMBE: Proc. Phys. Soc. Lond. B **66**, 665 (1953).

[2] F. LLEWELLYN JONES and C. G. MORGAN: Phys. Rev. **82**, 970 (1951). — W. S. BOYLE and P. KISLIUK: Phys. Rev. **97**, 255 (1955).

[3] F. LLEWELLYN JONES: Electrical Discharges. Rep. Progr. Physics (London: The Physical Society) **16**, 216 (1953). — R. SEELIGER and G. MIERDEL: Handbuch der Experimentalphysik, Vol. 13, Part 3, p. 38. — J. S. TOWNSEND: Phil. Mag. **8**, 750 (1904).

action of cathode effects; the micro-geometry and micro-state of the cathode then greatly influence the stability of the current, and these can change while current passes unless the cathode surface is clean, smooth and de-gassed. This effect becomes more important at higher pressures because of the localisation of the current due to various causes such as the diminution of lateral diffusion. This makes it extremely difficult to obtain static V, I characteristic curves at high gas pressures.

For the higher values of the current the potential V falls but only to a certain limit V_g, at which a glow appears near the anode[1]. This phase at low pressures is a glow discharge. Thus, by interposing a controlling device for the current, it

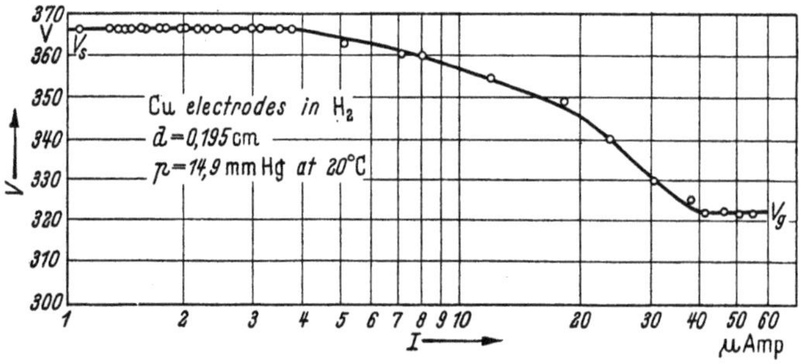

Fig. 9. V, I characteristics at breakdown for hydrogen.

is possible to examine slowly the phenomena which can occur almost instantaneously ($<10^{-6}$ sec.) when the current is not so controlled.

β) PASCHEN'S *Law.* The dependence of the breakdown potential V_s on the product pd, was first established by DE LA RUE and MULLER[2]. Later, PASCHEN[3] concluded, from an extensive study of air, CO_2 and H_2 over a range of values of pd, that V_s is a function $\Psi\,(pd)$ of the product pd only; this result is known as PASCHEN'S Law. CARR[4] confirmed its validity for a number of different gases for values of pd from 0.1 mm. Hg · cm. to 15 mm. Hg · cm. Typical PASCHEN curves are given in Fig. 10 from which it can be seen that E/p diminishes as pd increases.

PASCHEN'S Law follows analytically from the TOWNSEND criterion for breakdown because theory and experiment show that α/p and ω/p are functions of E/p. Putting $\alpha/p = f(E/p)$, and $\omega/p = F(E/p)$ in equation (3.2), it follows that

$$1 - \frac{F(V_s/p\,d_s)}{f(V_s/p\,d_s)}\left\{\exp f\left(\frac{V_s}{p\,d_s}\right)\cdot p\,d_s - 1\right\} = 0$$

i.e.

$$V_s = \Psi(p\,d_s) \tag{4.1}$$

which is PASCHEN'S Law.

The above derivation assumes that the field E is always uniform when pd is changed, but this is only the case when the ratio L/d of the linear dimension L of the electrodes to the gap distance d is sufficiently great, the value depending on the gas[5]. At low pressures especially, the charges on the walls of the containing

[1] R. SEELIGER and G. MIERDEL: Handbuch der Experimentalphysik, Vol. **13**, Part 3, p. 38.
[2] W. DE LA RUE and H. W. MULLER: Phil. Trans. Roy. Soc. Lond. **171**, 109 (1880).
[3] F. PASCHEN: Wied. Ann. **37**, 69 (1889).
[4] W. R. CARR: Phil. Trans. Roy. Soc. Lond., Ser. A **201**, 403 (1903).
[5] S. P. MCCALLUM and L. KLATZOW: Phil. Mag. **17**, 291 (1931). — F. M. PENNING: Proc. Acad. Sci. Amst. **34**, 1305 (1931).

glass envelope can influence the ionization in the gap unless L/d is sufficiently high.

Early measurements by CARR of V_s for about 50 different metals at low pressures could distinguish no important change in V_s for different electrode materials. This appeared to indicate that the cathode surface played no part in the current growth, and therefore that secondary ionization was due to a process such as β entirely in the gas[1]. However, no conclusive experimental evidence of the action of this process under these conditions has so far been obtained, and as a result further work using modern vacuum techniques was carried out to investigate the role of the cathode surface.

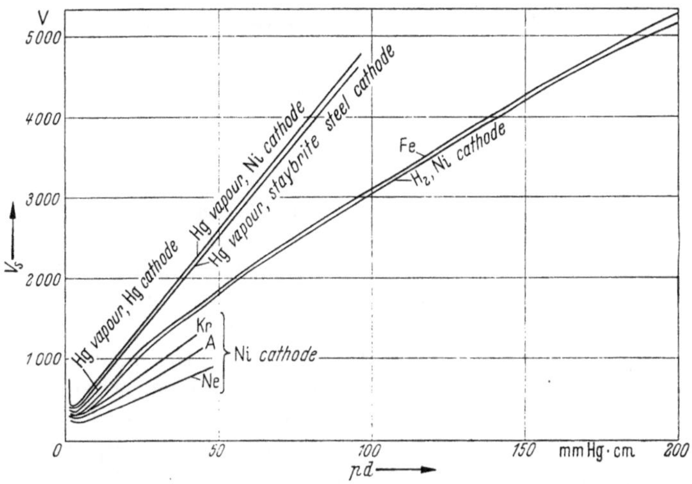

Fig. 10. PASCHEN curves for various gases. [Ne: A, Kr, and Hg: LLEWELLYN JONES and GALLOWAY, Proc. Phys. Soc. Lond. 50, 207 (1938); H_2: LLEWELLYN JONES and HENDERSON, Phil. Mag. 28, 185 (1939).]

γ) Cathode effects. Putting $\alpha = p f(E_s/p)$ in (3.2), then, approximately[2],

$$dV_s = \frac{d(\omega/\alpha)}{\omega/\alpha} \Big/ f'\left(\frac{E_s}{p}\right).$$

The change in V_s due to a change in ω/α thus depends on E_s/p, so that dV_s is constant when E_s/p is constant. It is characteristic of discharges in gases that E_s/p becomes very nearly constant for high gas pressures, and in hydrogen E_s/p only falls from 32 to 28 V/cm. per mm. Hg when pd is increased from 90 to 200 mm. Hg · cm.; for higher pressures E_s/p is nearly 20. Consequently, the sparking potential curves for cathodes with different values of ω/α should be nearly parallel at high pressures, and the curves obtained experimentally[2] and given in Fig. 10 show this to be approximately the case.

The fractional change in V_s is given by the equation

$$\frac{dV_s}{V_s} = \frac{d(\omega/\alpha)}{\omega/\alpha} \Big/ V_s \cdot f'\left(\frac{E_s}{p}\right).$$

For high values of pd the ratio E/p is practically constant so that dV_s/V_s is nearly inversely proportional to V_s, and it follows that dV_s/V_s can become small enough to fall within the experimental error. This accounts for the fact

[1] J. S. TOWNSEND: Electricity in Gases, p. 332. 1915.
[2] F. LLEWELLYN JONES: Phil. Mag. 28, 192 (1939).

that, at high pressures ($pd > 90$), the sparking potential curves of both hydrogen[1] and deuterium[2] for the different cathode materials are barely distinguishable. The sparking potential between parallel plane electrodes has been measured[1,2] throughout a range of gas pressures from 1 to 520 mm. Hg for six different de-gassed cathodes: clean and slightly oxidized Staybrite steel, pure and commercial aluminium, nickel and copper, and the values given in Fig. 11.

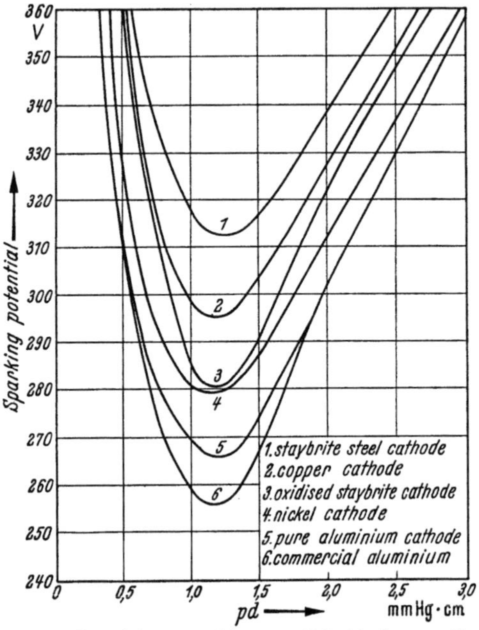

Fig. 11. The minimum sparking potential of hydrogen with different cathodes.

Marked dependence of the sparking potential on the cathode material was found, especially at the minimum values. The change in sparking potential due to a change of cathode at any given pressure was practically constant throughout the pressure range. Consequently, at 520 mm. Hg the fractional change in potential was very small; and at atmospheric pressure, the material of the cathode would appear to have a very small effect on the sparking potential of hydrogen even when the secondary ionization processes are entirely cathode emission. In the lower pressure range ($pd < 200$ mm. Hg · cm.) the change of the cathode material only affects V_s if the material is clean and de-gassed: otherwise, the surface, having an adsorbed gas layer is practically the same for different cathodes.

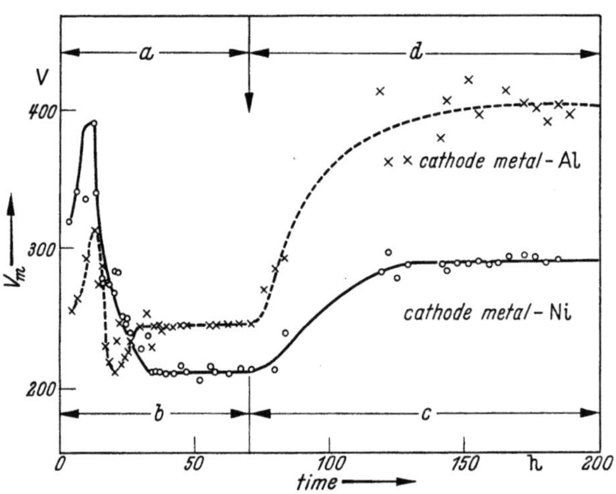

Fig. 12. Variation of V_m with time. (a) Cleaning of Al cathode; (b) deposition of Al film on Ni cathode; (c) cleaning of Ni cathode; and (d) deposition of Ni film on Al cathode. [LLEWELLYN JONES and DAVIES: Proc. Phys. Soc. Lond. B 64, 401 (1951).]

A comprehensive investigation of the effect of the state of the cathode has been made[3] by measuring the minimum sparking potential V_m as one metal was gradually deposited on a cathode of different metal. Large changes in V_m were recorded and shown in Fig. 12. With tarnished or oxidised electrodes, it is not easy to obtain accurate values of V_s, the current often becoming intermittent, presumably due to local changes in the cathode surface. Prolonged heating of the electrodes at 400° C either *in vacuo* or in hydrogen appeared to be ineffective in removing the last traces of oxide on

[1] F. LLEWELLYN JONES and J. P. HENDERSON: Phil. Mag. **28**, 185 (1939).
[2] F. LLEWELLYN JONES: Phil. Mag. **28**, 328 (1939).
[3] F. LLEWELLYN JONES and D. E. DAVIES: Proc. Phys. Soc. Lond. B **64**, 397, 579 (1951).

cathode surfaces; on the other hand, bombardment of the cathode by positive ions of hydrogen (including some protons) removed oxide films in a comparatively short time. When the oxide film had been removed, the onset of sparking was easily determined, and regular and reproducible PASCHEN curves were obtained.

A thin film of alien metal deposited on a cathode as base altered V_m, but the V_m obtained was characteristic neither of the underlying metal nor of the deposited metal in bulk. In determining the value of V_m (or the cathode emission) characteristic of a pure metal it is desirable to have both electrodes made of the same metal, otherwise the surface of each electrode is liable to become contaminated by deposition of atoms due to sputtering from the other electrode during the outgassing process.

Values of the cathode secondary emission coefficient ω/α for pure and mercury-contaminated air were calculated from the sparking criterion (3.2) and are given in Fig. 13. For the copper cathode covered with amalgam (curve a) there is evidence for a strong increase in ω/α at $E/p = 100$ V/cm. per mm. Hg and also for a noticeable increase in ω/α for the cases of Hg films on Ni (curve b) and on Al (curve *c*). No increase is exhibited for Al and Ni without Hg films (curves d and e). Further data are given in Table 2 in which AYRES' values of α were used to calculate ω/α in hydrogen. The experiments clearly demonstrate the great dependence of V_m on the nature of the cathode surface.

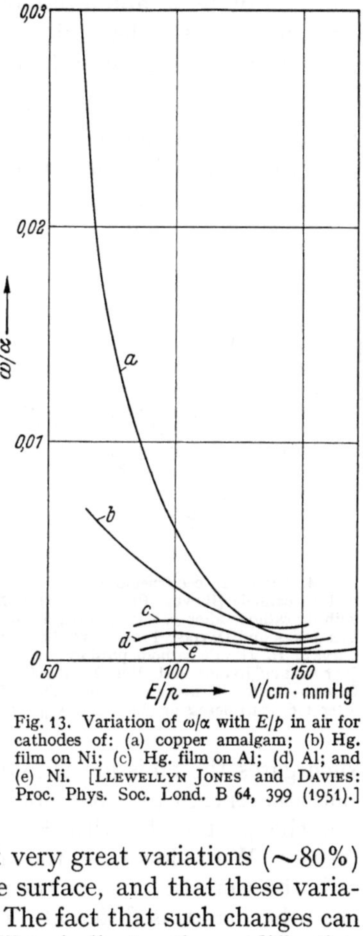

Fig. 13. Variation of ω/α with E/p in air for cathodes of: (a) copper amalgam; (b) Hg. film on Ni; (c) Hg. film on Al; (d) Al; and (e) Ni. [LLEWELLYN JONES and DAVIES: Proc. Phys. Soc. Lond. B 64, 399 (1951).]

It can be seen from Table 2 and Fig. 14 that very great variations (~80%) in V_m can be produced by changes in the cathode surface, and that these variations are greater than those previously recorded. The fact that such changes can be brought about by deposition of thin metallic films indicates the modification

Table 2. *Effect of nature of cathode surface on minimum sparking potential.*

Gas	Cathode	V_m	E/p	ω/α
Air + Hg vapour	Copper amalgam	460	720	0.004
	Hg film on aluminium	390	885	0.014
	Hg film on nickel	390	885	0.014
	Hg film on Staybrite Steel	390	585	0.006
Air	Oxidized aluminium	416	905	0.01
	Oxidized nickel	421	957	0.01
Hydrogen (surfaces treated by glow discharge)	Aluminium	243	200	0.04
	Aluminium deposited on nickel	212	200	0.06
	Nickel	289	180	0.02
	Nickel deposited on aluminium	390	245	0.0085
	Commercial aluminium	225	200	0.125
	Aluminium on Staybrite steel	205	210	0.07
	Staybrite steel	274	190	0.025
	Steel deposited on aluminium	282	190	0.023

of the effective work function of the surface by these films. The deposition of a film of aluminium either on nickel or on Staybrite steel doubled the electron emission, while the deposition of a film of nickel on pure aluminium reduced the emission by 85%, and a film of iron on commercial aluminium reduced the emission by 50%. The deposition of metallic films on the cathode can be important in measurements of the growth of current in metallic vapours such as that of mercury[1].

δ) Distinguishing the secondary processes. These results demonstrate the predominating influence of the cathode processes, such as γ or δ, over the purely gas processes such as β or photo-ionization[2]. Evidence of the nature of the predominating cathode processes can be obtained from the form of the variation of the generalized secondary coefficient ω/α with E/p. Secondary emission due to the impact of positive ions on the cathode is mainly a resonance effect and depends on the energy released by the ions at the electrode surface; this energy is made up partly of potential energy eV_i, and partly of kinetic energy. The change of kinetic energy $\frac{1}{2}MU^2$ with E/p depends on the nature of the gas, and in monatomic gases, for example, is likely to be small because of charge transfer effects; so that in such gases the emission is mainly due to the potential energy of the ions. The potential energy eV_i does not depend on E/p, so that in many cases γ should not change greatly with E/p other than to show a small increase as E/p increases. With slow ions, the energy of the ejected electron is $(eV_i - 2\varphi)$, where φ is the work function of the metal surface. The case of photo-electric action, on the other hand, is different. The number of photons emitted in a lamina of gas dx depends on the number of excitations to a given level, and this can be indicated by a term such as $n_x \alpha' dx$. Thus, the number of photons falling on the cathode, and therefore the electron emission, per ionizing collision in the gas is proportional to $n_x \alpha' dx / n_x \alpha dx$. Now this ratio depends on the shape of the electron energy distribution function, and its magnitude depends on the value of the mean electron energy, α'/α *increasing* as the mean energy or E/p *diminishes*. Consequently, an increase in ω/α as E/p diminishes indicates the action of a photoelectric effect.

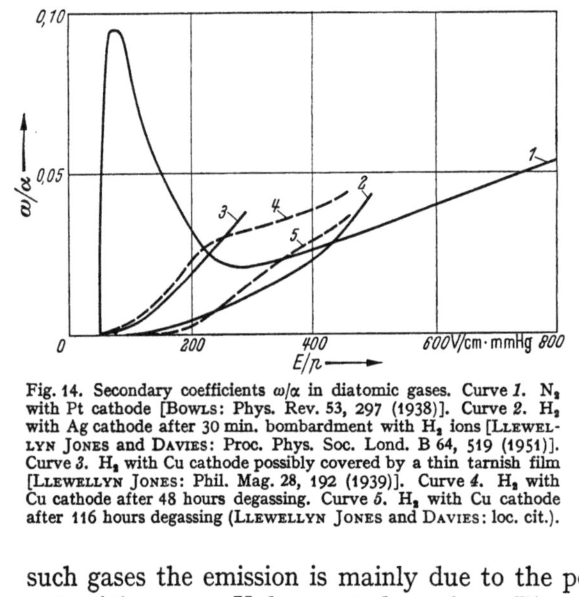

Fig. 14. Secondary coefficients ω/α in diatomic gases. Curve *1*. N_2 with Pt cathode [BOWLS: Phys. Rev. 53, 297 (1938)]. Curve *2*. H_2 with Ag cathode after 30 min. bombardment with H_2 ions [LLEWELLYN JONES and DAVIES: Proc. Phys. Soc. Lond. B 64, 519 (1951)]. Curve *3*. H_2 with Cu cathode possibly covered by a thin tarnish film [LLEWELLYN JONES: Phil. Mag. 28, 192 (1939)]. Curve *4*. H_2 with Cu cathode after 48 hours degassing. Curve *5*. H_2 with Cu cathode after 116 hours degassing (LLEWELLYN JONES and DAVIES: loc. cit.).

Examples of the variation of ω/α with E/p, are given in Fig. 13 for air, in Fig. 14 for diatomic gases[3], and in Fig. 15 for the monatomic gases[4]. Fig. 15 shows that

[1] F. LLEWELLYN JONES and W. R. GALLOWAY: Proc. Phys. Soc. Lond. **50**, 207 (1938). — R. GRIGOROVICI: Z. Physik **111**, 596 (1939).

[2] Recent work on relative significance of β and γ has been reported by H. FETZ: Brit. J. Appl. Phys. **6**, 288 (1955).

[3] F. LLEWELLYN JONES: Phil. Mag. **28**, 192 (1939) (H_2). — F. LLEWELLYN JONES and D. E. DAVIES: Proc. Phys. Soc. Lond. B **64**, 579 (1951)(H_2). — W. E. BOWLS: Phys. Rev. **53**, 297 (1938) (N_2).

[4] A. A. KRUITHOFF and F. M. PENNING: Physica, Haag **3**, 529 (1936) (Argon). — A. A. KRUITHOFF: Physica, Haag **7**, 530 (1940) (Neon). — J. S. TOWNSEND and S. P. MCCALLUM: Phil. Mag. **6**, 892 (1928) (Ne); **17**, 693 (1934) (He).

with pure, clean, de-gassed cathode surfaces in monatomic gases the variation of ω/α with E/p is small and generally increases with E/p, thus indicating the predominating influence of positive ion action at the cathode. On the other hand, with undegassed electrodes, or with electrodes having thin tarnish layers, or with surfaces of comparatively low work function, an increase of ω/α as E/p diminishes is sometimes found at low values of E/p, for example in air and nitrogen. This result indicates the influence of the photoelectric effect. With non-uniform fields much higher values of the ratio E/p are possible, and the proportion of γ to δ effects may well be different.

There is considerable direct evidence[1] to show that positive ions are able to eject electrons from the cathode, but data are not at all consistent. However, recent determinations of γ for He^+, Ne^+, Ar^+, Kr^+ and Xe^+ have been made by HAGSTRUM[2], and the values of γ for He^+ *in vacuo* are comparable with those found by TOWNSEND and MACCALLUM from measurements of the growth of ionization in helium when back-scattering of electrons at the cathode surface is taken into consideration.

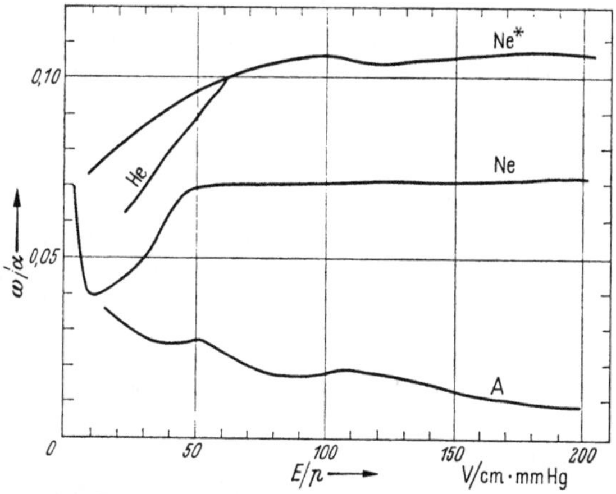

Fig. 15. Secondary coefficients in monoatomic gases. Ne with Ni cathode [TOWNSEND and McCALLUM: Phil. Mag. 6, 872 (1928)]. He with Ni cathode [TOWNSEND and McCALLUM: Phil. Mag. 17, 693 (1934)]. Ne with Cu cathode [KRUITHOFF: Physica, Haag 7, 530 (1940)]. A with Cu cathode [KRUITHOFF and PENNING: Physica, Haag 3, 529 (1936)].

Analyses of the components of the total secondary ionization have been made using pulse techniques by HORNBECK[3], MOLNAR[4] and by ENGSTROM and HUXFORD[5], who were able to separate the effects due to metastable atoms from those due to photons or ions.

ε) The V, I characteristic. The following argument has been used to indicate approximately the nature of the discharge after space charge effects have become important. Since the velocities of the positive ions are $\sim 10^{-2}$ those of the electrons, the space charge effects are mainly due to them. They cause the electric field to have its greatest value in the region of the cathode[6]; in fact ultimately the field is small except in a range from the cathode to a plane distant, say, d_c from it. Practically all the ionization then takes place in this range and roughly speaking $p d_c$ is related by the PASCHEN curve to the potential fall across this range, say V_c, which is called the cathode fall and is less then V. With increasing current d_c diminishes, and V_c falls along the PASCHEN curve until d_c becomes $d_{\min}$, and $V_c \sim V_{\min}$. As the current increases further, the space charge which previously had not occupied the full width of the system, spreads laterally, but d_c remains constant. At this stage an increase in current is obtained

[1] H. S. W. MASSEY and E. H. S. BURHOP: Electronic and Ionic Impact Phenomena. 1953.
[2] H. D. HAGSTRUM: Phys. Rev. **96**, 325, 336 (1954).
[3] J. A. HORNBECK: Phys. Rev. **83**, 374 (1951).
[4] J. P. MOLNAR: Phys. Rev. **83**, 933, 940 (1951).
[5] R. W. ENGSTROM and W. S. HUXFORD: Phys. Rev. **58**, 67 (1940).
[6] J. S. TOWNSEND: Phil. Mag. **8**, 750 (1904). — A. SCHUSTER: Proc. Roy. Soc. Lond. **47**, 541 (1890).

without an increase in potential across the cathode fall. The current density is constant and the potential across the gap V_g is independent of I; this condition is called the normal glow discharge. The form of the V, I characteristic is thus as given in Fig. 9, up to the establishment of the normal glow discharge.

When the discharge has covered the electrodes, further lateral spread is not possible; the current density then increases and further concentration of the space charge reduces pd further and the voltage rises corresponding to values of $pd < pd_{\min}$. On the other hand, when the pressure is suitable the increase of current density may introduce further ionization processes, such as thermal ionization in the gas, as well as intensified bombardment of the cathode to produce thermionic cathode emission. Thermal effects are unstable — because the ionization increases with the temperature. This latter stage, the transition from the glow to arc, is not so easily established in low pressure cases as at high pressures, partly because of the lower current density. The cross section of the current stream will be largely determined by the lateral diffusion of the electrons as they move across the gap, the diffusion becoming ambi-polar at the higher current densities owing to space charge[1, 2, 3].

The fact that, after the static breakdown criterion is satisfied ($V = V_s$), the increase of spark current I_D causes a fall in the potential V_D across the gap, led FUCKS[4] to propose the empirical relations between V_D, pd and the pre-current density I

$$V_I = V_s\{1 - \Phi(p, d)\sqrt{I}\}$$

and

$$I_D = \Psi(p, d)\sqrt{I_0}$$

where I_0 is the pre-current density passing before the breakdown is produced.

It can be shown[5] theoretically for rare gases that

$$V_s - V_D = k_1 I_D^2,$$

but experimentally it has been found that

$$V_s - V_D = k_2 I_D.$$

Explanations have been attempted on the basis of the variation of α and ω/α with E.

Irradiation of the cathode by an external source can also affect the value of V_s, and the fall can be appreciable[5] with $I_0 > 10^{-8}$ A/cm.^{2}. It is found experimentally that

$$V_{s(0)} - V_{s(I)} = C_1\sqrt{I_0}$$

where $V_{s(0)}$ and $V_{s(I)}$ are the potentials with and without radiation, and

$$I_D = C_2\sqrt{I_0}.$$

---

[1] W. SCHOTTKY: Phys. Z. **25**;, 342, 635 (1924).
[2] J. S. TOWNSEND: C. R. Acad. Sci. Paris **186**, 55 (1928).
[3] F. LLEWELLYN JONES: Brit. J. Appl. Phys. **1**, 60 (1949).
[4] W. FUCKS: Arch. Elektrotechn. **40**, 1, 16 (1950).
[5] M. J. DRUYVESTEYN and F. M. PENNING: Rev. Mod. Phys. **12**, 115, 119 (1940).

B. The similarity principle.

5. Similarity relationships. TOWNSEND[1] showed that PASCHEN'S Law was a special case of a more general similarity theorem which applied not only to breakdown in uniform fields, but also to breakdown which was dependent on ionization by collision in non-uniform fields. This theorem was later extended by HOLM[2] to account for the maintenance of current between geometrically similar electrodes, and its relation to fundamental processes has been discussed by VON ENGEL and STEENBECK[3]. More recently, MARGENAU[4] showed theoretically how the principle might be extended for high frequency alternating fields, and this has also been demonstrated experimentally[5].

Consider two geometrically similar systems I and II in which a leading dimension a_2 in system II is $K \times a_1$, where a_1 is the corresponding dimension in system I. $A_1 B_1$ are electrodes in I and $A_2 B_2$ are the electrodes in II. The potential V_1 at any point P_1 in the inter-electrode space is $\int\limits_{A_1 B_1} \sigma_1 d S_1/r_1$ where $d S_1$ is an elementary surface area on A_1 or B_1 at a distance r_1 from P_1, and σ_1 is the surface charge density. The potential V_2 at the corresponding point P_2 in II is given by

$$\int\limits_{A_2 B_2} \sigma_2 \frac{d S_2}{r_2}$$

and is equal to V_1 when

$$\sigma_2 = \frac{\sigma_1}{K}$$

and then $E_2 = E_1/K$. If p_2 is also made equal to p_1/K, the ratio E/p will be the same in the two systems. Thus, all quantities which are functions of E/p, are the same in the two systems. The current I is proportional to

$$\exp \int \alpha\, d r = \exp \int p f\left(\frac{E}{p}\right) d r$$

when the externally generated current is the same, and it follows that

$$I_1 = I_2 .$$

The current density

$$J_1 = \frac{I_1}{d S_1} = K^2 \frac{I_2}{d S_2} = K^2 J_2,$$

so that

$$\frac{J_1}{p_1^2} = \frac{J_2}{p_2^2} = \text{constant.}$$

Also, the space charge densities set up by the currents satisfy

$$\varrho_1 = K^2 \varrho_2,$$

since, as will be shown below, the drift velocities of electrons or ions will be the same in the two systems. Because $\varrho = -\nabla^2 V/4\pi$, it follows that space charge will not destroy the similarity of potential distribution. Further, in system I a group of electrons travels from the cathode to any given place in time T_1, but a group in system II will only reach an homologous place in the time $T_2 = K T_1$,

[1] J. S. TOWNSEND: Electricity in Gases, p. 365. 1915.
[2] R. HOLM: Phys. Z. **25**, 497 (1924).
[3] A. H. v. ENGEL and M. STEENBECK: Elektrische Gasentladungen, Vol. II. Berlin: Springer 1934.
[4] H. MARGENAU: Phys. Rev. **73**, 197 (1948).
[5] F. LLEWELLYN JONES and G. D. MORGAN: Proc. Phys. Soc. Lond. B **64**, 560 (1951).

since the drift velocities are the same. Table 3 gives the transformation relation of quantities in homologous regions.

Table 3.

	System	
	I	II
Linear dimension	a	Ka
Gas density or pressure	p	p/K
Potential across tube	V	V
Field strength	E	E/K
Surface charge density	σ	σ/K
Space charge density	ϱ	ϱ/K^2
Total mass of gas between plates	M	MK^2
Total electric charge in gas	e	eK
Ratio of concentration of ionized to neutral molecules	n_i/n_n	$n_i/n_n K$
Electric current	I	I
Current density	J	J/K^2
Voltage current characteristic	V, I	V, I
Temporal-growth rate of current	$\partial I/\partial t$	$\partial I/K\partial t$

In considering similarity for high frequency fields the parameter is no longer E/p only, but involves also a time factor which must be changed by the factor K This parameter must contain a term f/p (f is the frequency of the field) as well as E/p.

6. Physical significance of the similarity principle. The similarity principle is a consequence of the nature of the significant fundamental gas processes. In similar systems E/p is constant. The mean energy $(\frac{1}{2}mu^2)$ of a cloud of electrons in a gas is dependent upon E/p only; therefore, the velocity of agitation u is a function of (E/p); the electron drift velocity W_- of electrons is $\sim\left(\frac{E\,e\,l}{m\,u}\right)$ and is therefore a function of E/p. The drift velocity W_+ of positive ions is $\sim\left(\frac{E\,e\,L}{M\,U}\right)$, where U is the mean thermal velocity and L is the ion mean free path, is also a function of E/p. The probabilities of ionization and excitation are functions of the energies of the colliding electrons, and therefore are functions of E/p. The diffusion coefficient D of electrons $(\frac{1}{3}\,l\bar{u})$ is a function of $(E/p)\cdot(1/p)$.

Let n_1 be the concentration of carriers in system I. dn_1/dt_1 is the rate of increase in system I, and dn_2/dt_2 is the rate of increase in system II. Now

$$\frac{dn_1}{dt_1} = \frac{d(K^2 n_2)}{dt_2/K} = K^3\,\frac{dn_2}{dt_2}.$$

Therefore corresponding homologous rates per unit volume should have this ratio $1:K^3$; and wall current densities should have the ratio $1:K^2$ in similar discharge systems.

Consider processes by which the ion concentration n is altered in a discharge: the total change is due to the following processes of gain or loss.

$$\frac{dn}{dt} = \frac{dn}{dt}\,\text{(electron collisions)} + \frac{dn}{dt}\,\text{(photons)}$$
$$+\frac{dn}{dt}\,\text{(ion collisions)} + \frac{dn}{dt}\,\text{(diffusion)}$$
$$+\frac{dn}{dt}\,\text{(drift)} + \frac{dn}{dt}\,\text{(recombination)}$$
$$+\frac{dn}{dt}\,\text{(electron attachment)}.$$

Thus, when individual collision processes occur in II at a rate which is $(1/K^3)$ that in I, the similarity relationship holds, and, alternatively, if in two cases the similarity relationship is obeyed the corresponding collision rates will transform according to the above ratio.

Various fundamental processes will now be considered to see whether they satisfy the similarity relationship.

α) Ionization by electron collision. The ionization rate per unit time per unit volume is

$$n \frac{u}{l} P(u) = \Phi(u) \cdot n \cdot p.$$

The ionization probability $P(u)$ and velocity u are the same when E/p is the same, so that the number of ionizing collisions per unit time per unit volume will transform as $\alpha_1 n p$ where α_1 (and, later, $\alpha_2, \alpha_3, \ldots$) is a coefficient which is the same in the two systems. Therefore this ionization process will transform as $1/K^2 \cdot 1/K = 1:K^3$, which is in accordance with the similarity principle. In fact, the discussion of Sect. 5 above depended on this.

β) The drift motion of electrons or ions in an electric field.

$$\frac{dn}{dt} + \operatorname{div}(n W) = 0.$$

Now

$$\operatorname{div}(n W) = \frac{\partial}{\partial x}(n W_x) + \frac{\partial}{\partial y}(n W_y) + \frac{\partial}{\partial z}(n W_z),$$

and transforms as $1/K^2 \cdot 1/K = 1/K^3$ since W is the same; so that the motion of the electrons in the field in homologous volumes will transform in accordance with the similarity principle.

γ) Electron attachment to neutral molecules. The attachment coefficient ξ is defined by the relationship

$$\frac{dn}{dt} = \xi \frac{n u}{l} = \alpha_2 n p$$

and both sides transform as $1/K^3$, so that electron attachment conforms to the similarity principle.

δ) Recombination in the gas. Let recombination be defined by

$$\frac{dn}{dt} = R n_e n_p$$

where n_e and n_p are the concentrations of electrons and ions.

There are many ways in which recombination can take place. When the recombination coefficient R is proportional to $1/p$, then $R \cdot n_e n_p$ transforms as $(1/K) \cdot (1/K^4)$, i.e. as $1:K^5$, and therefore does not satisfy similarity. When R varies as $\sqrt{p}$ the transformation follows $K^{-\frac{9}{2}}$, again not conforming to similarity. Similarity also fails when R is independent of p.

ε) Step-wise ionization. Another important process is step-wise ionization given by

$$\frac{dn}{dt} = \alpha_3 n_e n_{ex},$$

where n_{ex} is the excited atom concentration. But

$$n_{ex} \text{ is proportional to } n_e n \quad \text{and} \quad \frac{dn}{dt} = \alpha_3 n_e^2 n$$

and transforms as $1/K^5$, therefore not conforming to similarity. Thus, step-wise ionization of this form cannot occur to a marked extent in similar discharges.

ζ) Photoionization. Assume photons are generated in the gas by radiative recombination. The rate of generation of the photons then depends on the product of ion and electron concentration, on the number of gas atoms on which radiation falls, and on the length a of the discharge. Therefore

$$\frac{dn}{dt} = \alpha_4 n_e p a n_p$$

does not transform as $1:K^3$.

When radiation density i comes from an outside source, the generation is a process described by

$$\frac{dn}{dt} = \alpha_5 i n,$$

and therefore transforms as $1:K$ if i is constant. Similarity would not hold unless $i \propto 1/K^2$.

η) Photo-electric emission from the cathode.

a) External radiation i falling per unit area on the cathode will produce a photo-electric current density J given by

$$J = \alpha_5 i$$

thus the similarity relation will be satisfield when i varies as $1/K^2$.

b) When photons are generated by single excitation of neutral molecules, the current density J produced by their action on the cathode transforms as $1:K^2$, and conforms to similarity.

c) For photons generated by recombination,

$$J = \alpha_6 n_e n_p a,$$

and this transforms as $1:K^3$, not conforming to similarity.

7. Application of the similarity principle. In some pure gases metastable atoms are formed by single electron impact and this process transforms as does the α process. The predictions of the similarity principle depend upon the assumptions made concerning the relative importance of the various processes of destruction. In a pure gas and with small ionization currents, destruction takes place mainly at the walls of the tube to which there is lateral motion by diffusion. With sufficient impurity, however, a metastable atom is bound to ionize in a collision of the second kind with an impurity molecule, to an extent which makes this the principal destructive process. In each of these cases similarity holds. In an intermediate stage, however, when the two destructive processes are of comparable importance the more complex equation is not compatible with similarity unless the percentage of impurity is the same in all the systems.

These considerations are in accordance with the observations[1] of sparking potentials with helium containing traces of impurity. When the helium was purified continually, it was found that in the last stages similarity was at first upset but finally re-established. The similarity principle has been applied to observations of low pressure discharges in air, hydrogen and helium both for breakdown and for maintained discharges, both with high frequency and with static fields[1]. Another process which does not conform to the similarity relation is the emission of electrons from electrodes under the action of electric fields $\gtrsim 10^4$ V/cm., as has

[1] F. Llewellyn Jones and G. C. Williams: Proc. Phys. Soc. Lond. B **66**, 345 (1953). — F. Llewellyn Jones and G. D. Morgan: Proc. Phys. Soc. Lond. B **64**, 560, 574 (1951).

been observed with tarnished surfaces[1]. This accounts for the fact that for breakdown in gases at very high pressures, PASCHEN'S Law fails[2,3]. Such high fields can also be obtained even at low pressures by the action of positive ions in the vicinity of tarnished electrode surfaces, and the resulting electron emission will be dependent on E, and will not conform to the similarity relation.

C. Growth of ionization in non-uniform fields.

8. General considerations. In non-uniform fields the value of E/p may well be different at one electrode from that at the other, and a polarity effect is to be expected independently of any effect due to difference in the electrode materials. The ionization coefficients are continuous functions of the field, and it is to be expected that the same general mechanism of ionization current growth occurs in the non-uniform field as in uniform fields. The transition from the uniform to the non-uniform field can be regarded as being produced by a continuous change in the field distribution; and the difference between the two cases would then be due to the difference in the relative values of the various ionization coefficients on account of the difference in the electric fields. Consequently, it should be possible to set up a quantitative expression for the growth of current for the non-uniform cases, just as has been done for uniform fields.

However, there are certain qualifications[4] to the theory of ionization growth in uniform fields given above in Sect. A which might be necessary in non-uniform fields. Firstly, the gradient of the electric field must not be too great, otherwise the value of α in any given elementary region will not be the same as that corresponding to the value of the electric field in that region if the field were uniform in the gap. This question has been examined experimentally for a highly divergent field[5]. Secondly, there must be a restriction to the magnitude of the current density. For example, if the cathode were a sharply pointed electrode, the ion current density at the point could then be so high that the space charge would distort the field to an extent not easily assessable. These two qualifications modify the values of the ionization coefficients which correspond to the value of the field in slowly diverging fields and negligible space charge, and make quantitative investigation difficult. It is for this reason that it is not easy to draw quantitative conclusions about the fundamental processes from observations on electric discharges from sharp points at high gas pressures, when the electric field is highly divergent and space charge distortion occurs. Consequently, in attempting a quantitative treatment of non-uniform fields, it is desirable to work with slowly diverging fields for which experimental data concerning ionization coefficients in uniform fields can be applied, and with low current densities with which the electric field in the gap is determined only by the potentials of the electrodes.

Further, ionization processes in non-uniform fields may act differently from those in uniform fields. For example metastable atoms by drifting between the electrodes may eventually produce secondary electrons in regions where the electric field is very different from that where the metastable atoms were first produced.

---

[1] F. LLEWELLYN JONES and E. T. DE LA PERRELLE: Proc. Roy. Soc. Lond., Ser. A **216**, 267 (1953).

[2] F. LLEWELLYN JONES and C. G. MORGAN: Phys. Rev. **82**, 51 (1951).

[3] W. S. BOYLE and P. KISLIUK: Phys. Rev. **97**, 255 (1955).

[4] J. S. TOWNSEND: Theory of Ionization of Gases by Collision, Chap. IV. 1910.

[5] L. H. FISHER and G. L. WEISSLER: Phys. Rev. **66**, 95 (1944).— P. L. MORTON: Phys. Rev. **70**, 358 (1946). — G. W. JOHNSON: Phys. Rev. **73**, 284 (1948).

9. General maintenance condition. Consider two electrodes of any shape and distance apart, the field varying from point to point in the gap. Assume that a small current passes in a steady state, and for simplicity, consider that its maintenance is due to a general cathode action (δ and γ effects) and a β positive ion collision process in the gas; at low gas pressures, photo-ionization of the gas may be neglected, but photo-ionization must be considered quantitatively for the case of a very remote or very pointed cathode, especially at high pressures.

The continuity equations are:

$$n_- W_- + n_+ W_+ = \frac{I}{e},$$

$$\frac{\partial n_-}{\partial t} + \operatorname{div}(n_- W_-) = \alpha\, n_- W_- + \beta\, n_+ W_+,$$

$$\frac{\partial n_+}{\partial t} - \operatorname{div}(n_+ W_+) = \alpha\, n_- W_- + \beta\, n_+ W_+.$$

Arguments similar to those used above in Chapter A for uniform fields lead to the condition for the self-maintenance of a small ionization current (the breakdown criterion), and when β is negligible

$$0 = 1 - \int_0^d \left(\frac{\omega}{\alpha}\right) \left\{\alpha \exp\left(\int_0^x \alpha\, dx\right)\right\} dx, \tag{9.1}$$

where

$$\omega = \alpha\gamma + \delta.$$

For quantitative discussion of the consequences of this equation, the case of coaxial cylinders is of special interest. This is partly because it permits simple quantitative treatment, and partly because of its practical importance, (for example, in GEIGER-counters with low gas pressures, and also in relation to discharges from high tension power lines at high gas pressures). It is thus of interest to consider whether any correlation exists between the mechanisms of ionization growth in non-uniform and in uniform fields, in order to find whether a general theory can apply to both sets of conditions. Correlation can be investigated by comparing the values of ω/α found with different gases in uniform fields, with those obtained in the non-uniform fields of a simple coaxial electrode system; and the properties of such a system will now be discussed.

Wire and Coaxial Cylinder. With a wire of radius a and a coaxial cylinder of radius b, the field at the cathode depends on the polarity of the wire. The field E_a at the wire can be high ($\sim$300 V/cm.), while at remote regions and near the cylinder the field E_b can be low ($\sim$30 V/cm.). Hence the coefficient of the cathode ionization processes can be found for different values of E/p (i.e. et E_a/p and at E_b/p) simply by investigating the breakdown criterion for the different polarities of the wire.

Assuming that the number ω/α of secondary electrons produced for each primary (α) electron depends on the field at the cathode and not on the field where the primary electron (and also the operating photon or ion) was produced, it follows that the maintenance criterion can be written

$$1 - \frac{\omega}{\alpha}\left\{\exp\int_a^b \alpha\, d r - 1\right\} = 0 \tag{9.2}$$

and the field E_r at any radial distance r when (9.2) is satisfied is related to the potential V_s by the equation

$$E_r = \frac{V_s}{r \log (b/a)} \tag{9.3}$$

when α/p and ω/α are functions of E/p, it follows that

$$\frac{\omega/\alpha}{1+\omega/\alpha} \sim \frac{\omega}{\alpha} = \exp\left\{\frac{V_s}{\log(b/a)} \int\limits_{E_a/p}^{E_b/p} \frac{(\alpha/p)\, d(E/p)}{(E/p)^2}\right\} \tag{9.4}$$

which gives the dependence of V_s on ω/α.

Equation (9.3) can be written as a similarity relation:

$$V_s = a\, E_a \log\left(\frac{b}{a}\right) = F(p\,\Lambda) \tag{9.5}$$

where Λ is a characteristic length. When the wire radius is small, or when that of the cylinder is large, the region of ionization is confined to the wire; E_b/p can be negligible and collision processes at the cylinder can have no influence on the maintenance process, and breakdown is then determined by the value of the field E_a at the wire. The potential of the cylinder is then merely that required to set up the required field at the wire. In two cylindrical systems with a wire of same radius a but different cylinder radii b_1 and b_2 then the corresponding potentials V_1 and V_2 are given by the corona relation

$$V_1 = V_2 \frac{\log(b_1/a)}{\log(b_2/a)}\,. \tag{9.6}$$

It is now interesting to consider experimental data.

10. Experimental data. *α) PASCHEN curves.* PASCHEN curves[1], for air, hydrogen and helium, representing a gas mixture, simple gas, and a monatomic gas, for three tubes A, B and C, with wire diameters ($2a$) of 0.159, 0.318 and 0.160 cm. and cylinder diameters ($2b$) of 3.945, 3.950 and 2.080 cm. are given in Fig. 16, 17 and 18 which show to what extent the above conclusions are verified. These data are in general agreement with those of HUXLEY[2] (nitrogen), and BRUCE[3] (hydrogen) and of BOULIND[4] (air).

β) Secondary processes. The values of ω/α for the three cases of air, hydrogen and helium given in Figs. 16, 17 and 18 are calculated from equation (9.4) and are given in Figs. 19, 20 and 21.

1. Air. For low values of pa photo-ionization is negligible. Fig. 19 shows that in air at low values of E/p, ω/α decreased with increasing E/p.. Consequently the emission from the cylinder, when cathode, was partly photo-electric; on the other hand at high values of E/p (>100), ω/α increased with E/p so that emission from the wire, when cathode, was due to positive ions. It follows that the generalized secondary coefficient ω/α for a positive wire may be approximately represented by the sum of two coefficients; i.e.

$$\left(\frac{\omega}{\alpha}\right)_+ = \gamma_i + \left(\frac{\delta}{\alpha}\right)_+$$

where $(\delta/\alpha)_+$ represents secondary emission due to incidence of photons on the cylinder (in these circumstances E/p was low at the cylinder) and the coefficient γ_i is the electron emission coefficient due to the incidence of low energy positive

[1] F. LLEWELLYN JONES and G. C. WILLIAMS: Proc. Phys. Soc. Lond. B **66**, 345 (1953).
[2] L. G. H. HUXLEY: Phil. Mag. **5**, 721 (1928); **10**, 185 (1930).
[3] J. H. BRUCE: Phil. Mag. **10**, 476 (1930).
[4] H. F. BOULIND: Phil. Mag. **18**, 909 (1934).

ions on the cylinder. When the wire was negative, on the other hand, the liberation of secondary electrons was due almost entirely to the impact of high energy positive ions, because most of the radiation liberated in the gas still fell on the cylinder, which was now the anode. The photo-electric yield from the wire was thus low in comparison with that due to the impact of high energy positive ions.

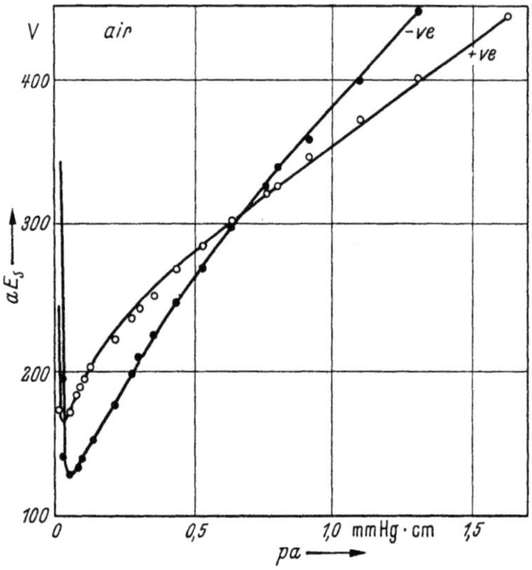

Fig. 16.

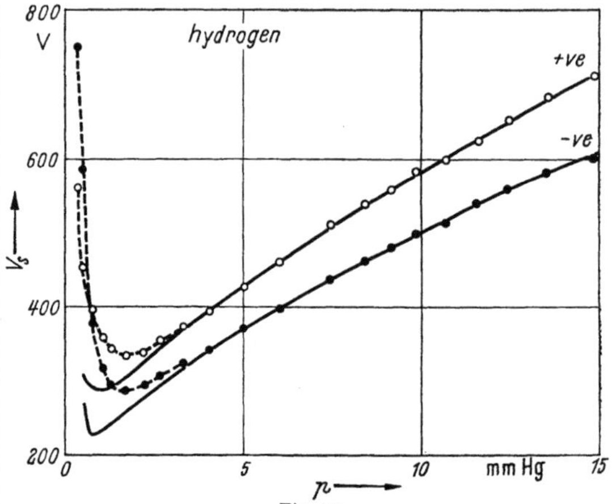

Fig. 17.

Fig. 16. aE_s and pa curves for positive and negative discharges for air. The full curves give values of aE_s for tube C for positive and negative discharges. Points (white and black) give values of aE_s for positive and negative discharges in the geometrically similar tube B. The similarity relationship is obeyed. E_s is the value of E_a at breakdown.

Fig. 17. Measured values of V_s and calculated values using the corona relationship equation. The measured values (white and black circles) of V_s for tube C are compared with those (curves) calculated from the measured values for tube A using the corona relationship equation (9.6).

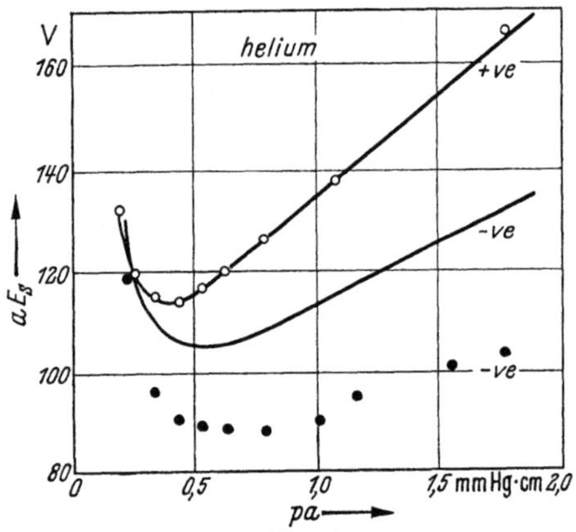

Fig. 18.

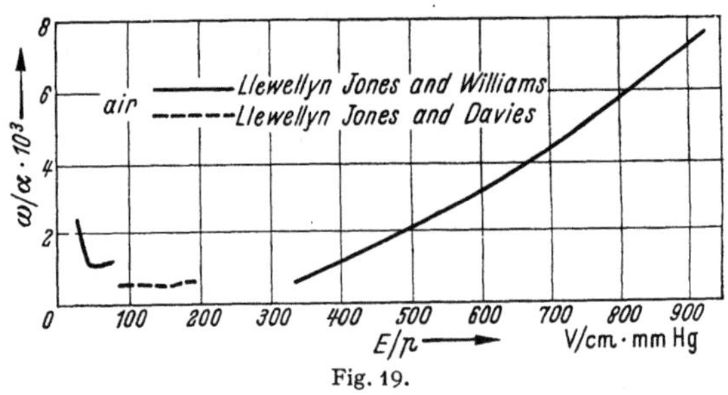

Fig. 19.

Fig. 18. aE_s and pa curves for positive and negative discharges for helium. The full curves give values of aE_s for tube C for positive and negative discharges. Points (white and black circles) give values of aE_s for positive and negative discharges in the geometrically similar tube B. The similarity theorem is obeyed for positive breakdown but not for negative.

Fig. 19. $(\omega/\alpha,\ E/p)$ curves for air for cathodes of nickel.

The generalized secondary ionization coefficient for negative breakdown $(\omega/\alpha)_-$ is then given by

$$\left(\frac{\omega}{\alpha}\right)_- = \gamma_h$$

where γ_h was the value of γ at the high values of E/p found at the wire.

It follows therefore that whether $(\omega/\alpha)_-$ will be greater or less than $(\omega/\alpha)_+$ in any given case depends on the relative magnitudes of δ/α, γ_h and γ_l. Since ω/α is related to the breakdown potential V_s by expression (9.4), the relative magnitudes of the breakdown potentials for positive and negative breakdown depend on the relative magnitudes of ω/α, γ_h and γ_l, which depend on the nature of the cathode.

2. *Hydrogen.* Compared with air hydrogen is a simple non-active gas. In air a clean metal surface is not obtainable, but this is not the case in hydrogen; thus the influence of surface tarnish layers on V_s and ω/α can be more readily investigated in hydrogen than in air. When the cylinder was the cathode, secondary emission could be due to the incidence of both photons and positive ions; and Fig. 20 shows that in hydrogen at the low values of E/p, the coefficient ω/α increased rapidly with E/p from a value of about 10^{-3} at 15 V/cm. per mm. Hg to a value of about 6×10^{-2} at 150 V/cm. per mm. Hg. When the wire was negative, photo-electric emission could not play any predominant role, and Fig. 20 shows

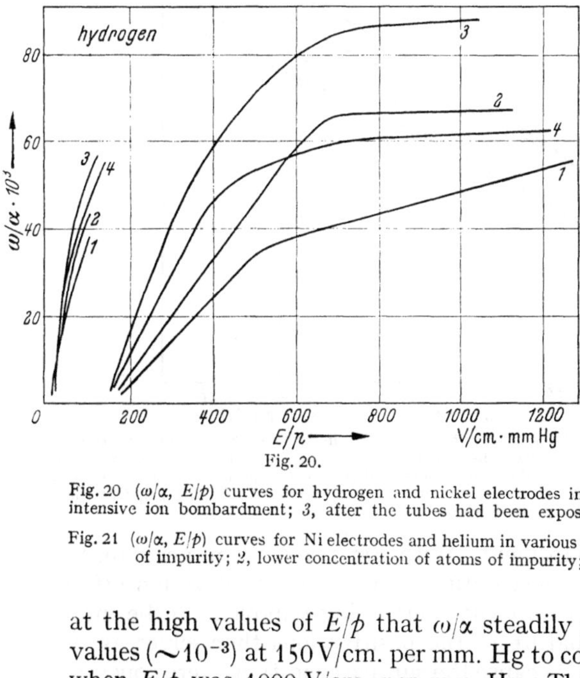

Fig. 20.

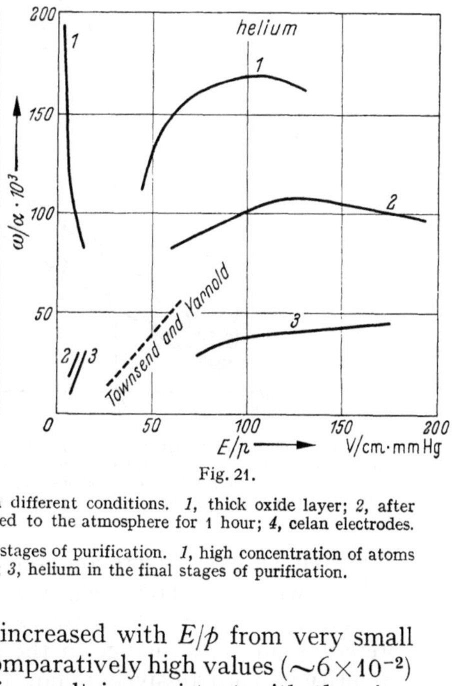

Fig. 21.

Fig. 20 (ω/α, E/p) curves for hydrogen and nickel electrodes in different conditions. *1*, thick oxide layer; *2*, after intensive ion bombardment; *3*, after the tubes had been exposed to the atmosphere for 1 hour; *4*, celan electrodes.

Fig. 21 (ω/α, E/p) curves for Ni electrodes and helium in various stages of purification. *1*, high concentration of atoms of impurity; *2*, lower concentration of atoms of impurity; *3*, helium in the final stages of purification.

at the high values of E/p that ω/α steadily increased with E/p from very small values ($\sim10^{-3}$) at 150 V/cm. per mm. Hg to comparatively high values ($\sim6\times10^{-2}$) when E/p was 1000 V/cm. per mm. Hg. This result is consistent with the view that the emission was caused by positive ions of hydrogen with energies which steadily increased with E/p at these high values of E/p.

When the electrodes were covered with oxide layers and $E/p<35$ V/cm. per mm. Hg. Fig. 22 shows that the similarity relationship did not hold exactly with a positive wire; for higher values of E/p, however, and also for all cases investigated when the wire was negative, the similarity relationship held within the experimental error. An electrode process which can account for this is the enhanced emission[1] from a cathode due to the high electric field set up by positive ions on thin surface insulating layers; such a process depending on E and not on E/p.

3. *Helium.* Results are given in Fig. 21.

These (ω/α, E/p) curves for helium have the same general features as those obtained for air and hydrogen in that, at low values of E/p when the cylinder was the cathode, the curve exhibits an increase indicating the preponderance of the photo-electric coefficient δ/α, in the generalized coefficient ω/α. Purification

[1] F. Llewellyn Jones and G. C. Williams: Proc. Phys. Soc. Lond. B **66**, 345 (1953).

generally lowered the values of ω/α but the sharp rise at relatively low values of $E/p < 20$ V/cm. per mm. Hg, when the wire was positive indicated the influence of the photo-electric effect. It is, however, not possible to make highly accurate computations in this range of E/p. On the other hand, when the wire was negative, and for high values of $E/p > 50$ V/cm. per mm. Hg, ω/α did not greatly change with E/p.

An important feature of the curves is the considerable decrease in ω/α produced by the progressive removal of the final traces of impurities. For these measurements the electrode surfaces could be regarded as clean, especially the wire,

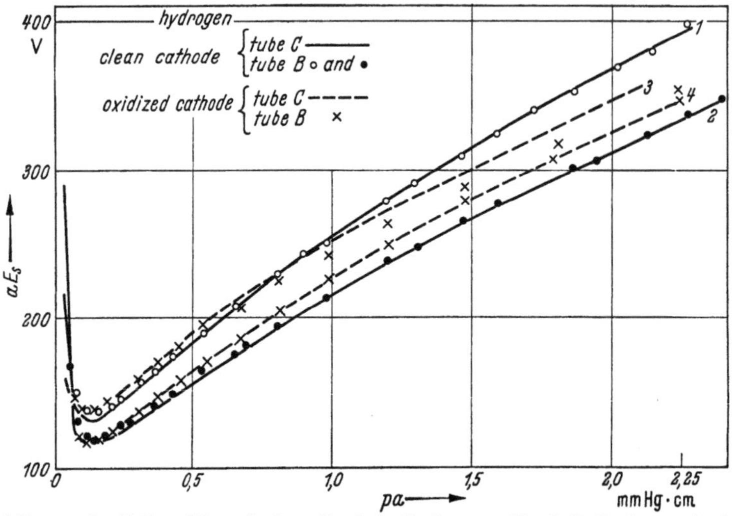

Fig. 22. The influence of oxidation of the cathode on the similarity theorem. The similarity theorem is obeyed for both positive and negative discharges when the electrodes are clean, because the points ○ and • for tube *B* coincide with the full curve for tube *C*. When the electrodes are oxidized the theorem is not obeyed exactly for a positive discharge because the broken curve for tube *C* does not correspond with the crossed curve for positive wire in tube *B*. *2*, *4* negative wire; *1*, *3* positive wire.

which had suffered severe ion bombardment. With $E/p > 100$ V/cm. per mm. Hg and the wire as cathode, ω/α was almost entirely due to the positive ion effect. The fall in ω/α as the final traces of impurity were being removed can be interpreted as due to a change in the nature of the positive ions themselves as the impurity was progressively removed, because the surface conditions were constant. Such a change can be produced by the action of metastable atoms which ionize atoms of the impurity in collisions of the second kind to produce positive ions of the impurity molecules. The similarity theorem did not hold for a negative wire; while, on the contrary, at low values of E/p with the purest specimens of helium, the theorem held.

11. Conclusions. The observed values of the breakdown potentials for positive and negative discharges in non-uniform fields are determined by the relative importance of the various possible secondary processes, which itself is determined by the polarity of the discharge. These cathode secondary processes differ greatly in efficiency, so that the relative values of V_s for the positive and negative discharges depend on the nature of the cathode surface. The mechanism of ionization growth to breakdown in non-uniform fields is thus of the same general nature as that for uniform fields. Further, a correlation has been established between the actual values of the secondary ionization coefficient for breakdown in non-uniform fields and those for uniform fields for the three types of gases, viz. air (a gas mixture), hydrogen (a pure diatomic gas) and helium (a monatomic gas).

D. Spatial growth of ionization in static fields at high pressures.

($150 < pd < 1000$ mm. Hg cm.; $V \leq V_s$.)

12. The problem. In the early work on the TOWNSEND theory of current growth, carried out half a century ago in the *Cavendish* Laboratory, the steady voltage supply was usually a bank of accumulator cells; the question of voltage instability did not arise, and the measurements were restricted to the lower values of $pd \lesssim 150$ mm. Hg·cm. Because of the lack of evidence to the contrary, it was assumed that the same theory, when generalised to take account of later knowledge of atomic collision processes, applied also to the breakdown criterion at higher pressures. With the introduction of the high-speed cathode ray oscilloscope and of high voltage sources, further breakdown phenomena could be investigated at high pressures in normal gaps. However, it must be emphasised that a highly stable source is necessary if the ionization coefficients are to be determined at high values of pd from measurement of the growth of ionization current; it is essential that E/p should be constant, otherwise the shape of the growth curve has no significance.

In 1926, ROGOWSKI[1] at Aachen, applied a sharply peaked impulse of e.m.f. to a cm. gap and found that at high pressures the voltage collapsed in times $\sim 10^{-6}$ or 10^{-7} sec. This result appeared to introduce a considerable difficulty in the explanation of spark breakdown on the basis of the TOWNSEND theory; the occurrence of a γ secondary process would appear to require times longer at least by a factor 10^2 than those observed owing to the long transit time ($d/W_+ \sim 10^{-5}$ sec.) of the ions. It was then concluded from this result that positive ions would have no time to cross the gap before the completion of breakdown and this was further taken to suggest that no secondary process took part in the build up of ionization (because the photo-electric effect was ignored). It was considered that the TOWNSEND mechanism required several avalanche generations, and was therefore comparatively slow. However, it is pertinent to point out here, that, at that time, no rigorous mathematical investigation of the rate of growth of ionization due to primary α and secondary ω/α processes in fields $E > V_s/d$ had been made, so that the real significance of the observed times of voltage collapse with an overvolted gap could not then be assessed.

The possibility that secondary processes did not occur in the growth of ionization currents could readily be tested by measuring the spatial growth of photoelectric currents in the gas under a static electric field with $V < V_s$ and then examining the shape of the $\log I$ *vs.* d curve obtained, in precisely the same way as had been done at the lower gas pressures, and described in part A above. The action of secondary processes must give a curve following equation (3.1), while the action of only a primary process would follow equation (2.1), i.e. give a straight line for the $\log I$ *vs.* d graph. Experimental investigations[2] of this question were carried out by MASCH, PAAVOLA, SANDERS, POSIN, and HOCHBERG and SANDBERG. Typical[3] $\log I$ *vs.* d graphs obtained in air were apparently linear up to the value of d at which a breakdown flash occurred. When an upcurving of the $\log I$ *vs.* d graph is recorded the cause of breakdown is readily explicable in terms of the action of both primary and secondary processes. On the other

[1] W. ROGOWSKI: Arch. Elektrotechn. **16**, 496 (1926).

[2] K. MASCH: Arch. Elektrotechn. **26**, 589 (1932). — M. PAAVOLA: Arch. Elektrotechn. **22**, 443 (1929). — F. H. SANDERS: Phys. Rev. **41**, 667 (1932). — D. Q. POSIN: Phys. Rev. **50**, 650 (1936). — B. HOCHBERG and E. SANDBERG: J. techn. Phys. USSR. **12**, 65 (1942).

[3] L. B. LOEB: Fundamental Process of Electrical Discharges in Gases, p. 371, New York: J. Wiley & Co. 1939.

hand, with linear growth curves indicating nothing but the action of an α process, breakdown apparently presents a problem.

There were also other aspects of the breakdown problem at high pressures to be considered. No definite evidence had been recorded to indicate any influence due to the cathode on breakdown at high pressure and this had been taken to indicate that under these conditions the γ and δ effects were negligible[1]. It should be noted, however, that it can be deduced[2] from equation (3.1) that any changes in electrode-dependent γ or δ would have a small effect on V_s at high values of pd. Again, it was considered that the comparatively narrow spark track found at high pressures, rather than the more diffuse glow obtained at low pressures, indicated a different mechanism from the TOWNSEND process. It should be noted, however, that high pressure naturally reduces the lateral diffusion of the avalanche, so that a wide diffuse track would not be expected at high gas pressures.

Further, interesting photographs of the propagation of tracks of the spark channels, have been obtained by RAETHER[3] using a cloud-chamber technique with overvoltages $\gtrsim 10$ to 20%. The rates of *propagation of ion density* in these tracks were measured and found to be $\sim 10^7$ cm./sec., and it was considered that the mechanism of current growth based on the continuous development of ionization by primary and secondary processes would involve many transits of ions, and thus would be too slow to account for this luminosity propagation.

13. Kanal and streamer theories. The interpretation of all the above phenomena as indicating that, at high values of pd (<150 mm. Hg·cm), spark breakdown can occur without the influence of any ω/α process led to the proposal of the canal theory by RAETHER[3], and the streamer theory by LOEB[1], for breakdown with static or with impulse fields, in place of the generalised TOWNSEND type mechanism based on the continuous development of ionization by primary and secondary processes. It was considered then that the TOWNSEND criterion equation (3.2) did not apply to static breakdown when $pd \sim 760$ mm. Hg · cm. corresponding to a 1 cm. gap in the atmosphere.

On both canal and streamer theories of avalanche propagation, additional electrons are considered to be generated in the gap by photo-ionization due to photons produced in the primary electron avalanche at a rate which becomes very rapid because of the distortion of the field due to a positive ion space charge at the head of the avalanche. Such new electrons initiate additional avalanches producing extremely rapid effective propagation of the avalanche across the gap; and it was maintained that breakdown times of about 10^{-7} sec. become feasible.

The nature of the generation of the requisite photons and the way in which such photons could produce the extra ionization and contribute to the increase in current was not quantitatively explained. These theories were applied to air; it is not easy to see how the mechanism dependent only on photoionization could occur in pure gases and especially in monatomic gases. However, the streamer theory was originally considered[1] as applying satisfactorily to breakdown in a 1 cm. gap in the atmosphere.

Both these views are qualitative, and a quantitative treatment of the *kanal* or streamer mechanism relating the sparking potential under static fields to ionization coefficients (for ions and photons) which can be measured in any gas has not been given, but two criteria have been proposed.

[1] L. B. LOEB and J. M. MEEK: Mechanism of the Electric Spark. Stanford: University Press 1941.

[2] See Sect. 4γ above.

[3] H. RAETHER: Z. Physik **112**, 464 (1939); **117**, 375, 524 (1941). — Ergebn. exakt. Naturw. **22**, 73 (1949).

α) Raether's criterion. An electron from the cathode travels towards the anode producing a Townsend avalanche, and a positive space charge is considered as distorting the applied field in the gap, especially at the head of the avalanche. The space charge is dependent on the number of ions in the avalanche, and Raether[1], from his cloud chamber studies with gaps over-volted by 10 to 20%, suggested that when this number exceeds a certain value N, the field distortion is sufficient to produce a rapidly increasing ionization by which the avalanche builds up into an anode-directed *kanal*, and then a cathode-directed *kanal* which bridges the gap. Raether's criterion can thus be expressed as a critical amplification:

$$e^{\alpha d} = N = \text{constant},$$

and indeed by suitable empirical choice of N with known values of α it is possible to predict values of d_s which are approximately the same as those observed in the static case; the value of N when $\alpha d = 20$ was at first suggested as the critical value.

β) Meek's criterion. Meek[2] suggested a criterion for Loeb's streamer theory based on the following further considerations. The local space charge field in the avalanche was assumed powerful enough to attract the photo-electrons (produced by the rapid absorption of the photons emitted by the avalanche), from all directions radially to the avalanche head, thus extending the ionization throughout the gap. To find a criterion it is then *postulated* that conduction, i.e. complete breakdown, will occur soon after the field of the space charge is of the same order as the applied field. Assuming photo-ionization still negligible the order of magnitude of the field produced by a spherical space charge was calculated, and Meek related this to the value of the previously undistorted field E by a constant of proportionality k, and found the empirical relation for air when $k = 1$:

$$\alpha d + \log\frac{\alpha}{p} = 14.46 + \log\frac{E}{p} - \frac{1}{2}\log(p\,d) + \log d.$$

This is the Meek spark criterion. When α/p, d and p and V_s are known, it is possible to find k; and the equation was considered then able to give V_s for other values of pd. Values deduced in this way were approximately the same as those previously published at high pd.

γ) Zeleny's comments. These criteria do not take into consideration the actual mechanism of ionization development or the processes postulated as essential for breakdown, that is, the production of photons and the way they contribute to the current. In analysing the derivation of this equation, Zeleny[3] pointed out that the space charge field would not act in the way proposed, and that since suitable values of the important factor k had to be assumed, the equation is really empirical, giving values of V_s which differ very little from those deduced by the expression

$$e^{\alpha d} = N = \text{constant},$$

as previously discused by Schumann[4]. Thus, both the Raether and Meek criteria reduce to the same form, which itself is deducible from the Townsend

[1] W. Kohrmann and H. Raether: Z. angew. Phys. 6, 211 (1954). — H. Raether: Z. Physik **112**, 464 (1939).

[2] J. M. Meek and J. D. Craggs: Electrical Breakdown of Gases 1953 — L. B. Loeb and J. M. Meek: Mechanism of the Electric Spark. 1941.

[3] J. Zeleny: J. appl. Phys. **13**, 444 (1942). See also C. E. R. Bruce, J. Inst. Elect. Eng. **89**, 352 (1942) and L. H. Fisher, Phys. Rev. **69**, 530 (1946).

[4] W. O. Schumann: Elektrische Durchbruchfeldstärke in Gasen. Berlin: Springer 1923.

criterion for static breakdown. For, it has been shown above, that at the high values of pd, V_s does not vary greatly with ω/α, and for calculating V_s the TOWNSEND criterion can be approximately expressed as

$$e^{\alpha d} - 1 = \text{constant} \sim N. \tag{13.1}$$

Thus these three criteria appear very similar.

However, both *kanal* and streamer theories when proposed for static breakdown envisaged criteria controlled by sudden space charge distortion and a growth of pre-breakdown ionization which should give a linear log I *vs.* d graph[1]. Consequently, it is necessary to examine the published data on the growth of currents which had been assumed as showing that no secondary ionization occurred at high values of pd, in order to find, whether, if a secondary effect did in fact exist, those particular measurements would have disclosed it.

14. Recent work on ionization growth. α) *Previous experimental data.* Such examination[2] showed that the range of investigation was, in fact, not sufficient to indicate the action of a secondary process if it had existed; and further, that the stability of the electric fields used was not sufficiently great to permit the necessary range to be investigated. If E/p fluctuates the growth of current measured could actually be that corresponding to a higher value of E/p than was being assumed in the measurements; thus, the breakdown would appear to occur for a smaller value of d than expected on the assumed value of E/p. Calculation shows that if the growth of photo-electric currents in air is to be measured in the region where $E/p \sim 41.6$ V/cm. per mm. Hg then distances must be used which are within 20% of the sparking distance d_s. To do this, certain experimental conditions must be fulfilled. Gas pressure, electrode distances and gap voltages must be measured with an accuracy within 0.1%, and recent work[2,3] in this field will now be described.

β) *Growth in air.* An ionization chamber (illustrated in Fig. 23) capable of operating at 100 kV with potentials steady to within 0.05% with electrode separations up to 5 cm. was used at Swansea to measure the growth of small initial photo-electric currents I_0 ($\sim 10^{-14}$ amp. from an area of 0.3 cm.² on the cathode) in air over a range of E/p from 39 to 45 V/cm. per mm. Hg at a pressure of 200 mm. Hg, for values of pd up to about 760 mm. Hg · cm., thus including the parameter value for a 1 cm. spark in the atmosphere.

Apart from the experimental difficulties connected with the maintenance of the stability of high voltages there were certain operating difficulties due to the use of air. Oxygen can produce progressive changes in the cathode surface due to the formation of oxides and tarnish layers. Such layers can have a profound effect upon the electron emissive properties of the surface owing to the action of positive ions near the surface in the presence of the electric field[4]. Thus, the maintenance of even a small ionization current in air can not only change the photoelectric sensitivity of the cathode, but also change the secondary emission under

[1] L. B. LOEB: Fundamental Processes of electrical discharges in gases, p. 371 and 407. 1939.

[2] F. LLEWELLYN JONES and A. B. PARKER: Proc. Roy. Soc. Lond., Ser. A **213**, 185 (1952).

[3] F. LLEWELLYN JONES and A. B. PARKER: Nature, Lond. **165**, 960 (1950).

[4] F. LLEWELLYN Jones and E. T. DE LA PERRELLE: Proc. Roy. Soc. Lond., Ser. A **216**, 267 (1953). — F. LLEWELLYN JONES and C. G. MORGAN: Proc. Roy. Soc. Lond., Ser. A **218**, 88 (1953). — C. G. MORGAN and D. HARCOMBE: Proc. Phys. Soc. Lond. B **66**, 665 (1953).

positive ion bombardment. Care has therefore to be taken in the treatment of the surface and the measurement of the current. It is important, however, to investigate the electrical properties of air and the criterion which determines its

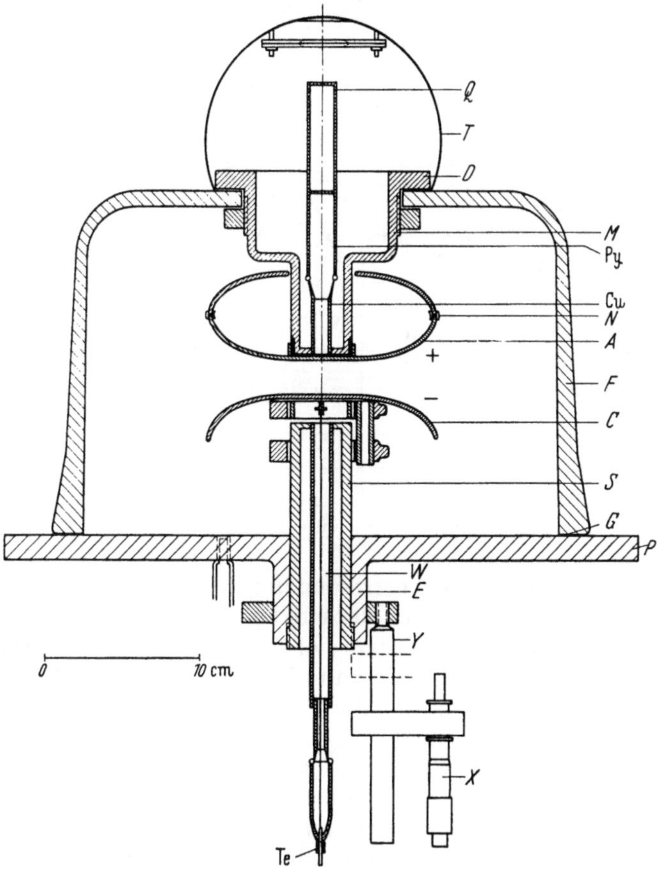

Fig. 23. Ionization chamber used for measurement of growth of ionization current at high values of pd, from Proc. Roy. Soc. Lond., Ser. A **213**, 192 (1952).

breakdown, because considerable previous work has been done with that gas and also because of its technological importance. The experimental results are given below for air filtered, dried and passed through liquid air traps.

Table 4. *Experimental values of I_0, α/p, ω/α and calculated values of d_s for various values of E/p when $p = 200$ mm. Hg.*

E/p V/cm. per mm. Hg.	$10^{15}\ I_0$ A	α/p (cm. mm. Hg.)$^{-1}$	$10^6\ (\omega/\alpha)$	d_s cm.
39	1	0.0161	8	3.73
40	6	0.0181	23	2.94
40	20	0.0181	$\geqq 150$	—
41	6	0.0196	40	2.53
42	6	0.0224	46	2.22
43	5	0.0252	113	1.83
44	3	0.0295	105	1.57
45	6	0.0345	84	1.37

A typical set of results obtained in the investigation at Swansea is given in Fig. 24 as graphs of log I *vs.* d. The curves show that for distances up to about 1.25 cm. the relation between log I and d was linear for all values of E/p from

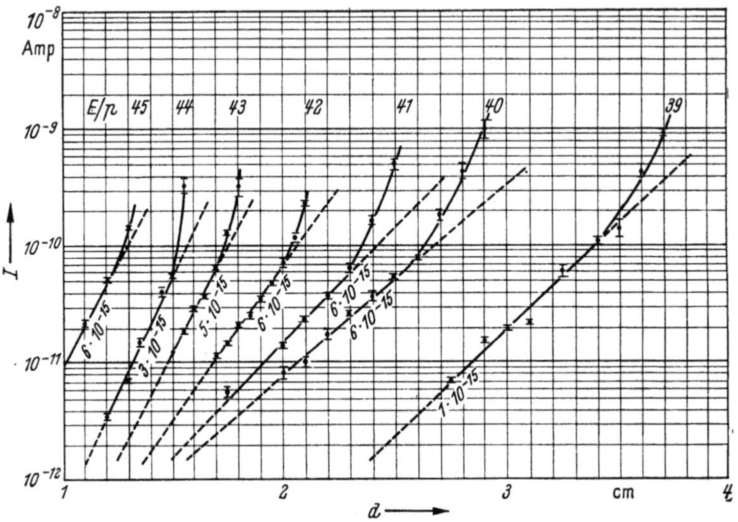

Fig. 24. Log I *vs.* d relations for air for different values of E/p. Values of I_0 are indicated on the curves.

40 to 45, which include the case corresponding to a cm. gap in the atmosphere. For larger values of d the curves departed from linearity, giving measurable values of ω/α, and showed the existence of secondary processes. The values of α were obtained from the slope of the linear portions and values of ω/α from the upcurving sections and are given in Table 4 as functions of E/p. Observations were made as close as possible to the sparking distance, but in no case was a spark obtained *before* the curve indicated the presence of an ω/α process. These values of ω/α and α when substituted in equation (3.2) give the theoretical value of V_s. After each set of current growth measurement, a single but careful measurement of V_s was made and these are shown in Fig. 25 which shows agreement in the range 10 to 30 kV between theoretical and measured values.

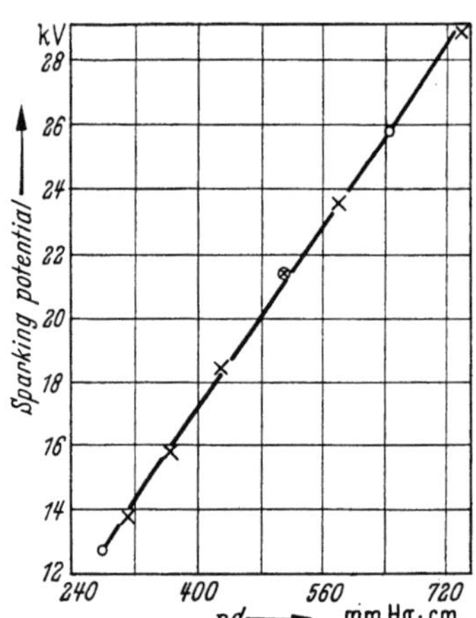

Fig. 25. Calculated (×) and observed (o) sparking potentials in air.

γ) The mechanism of spatial ionization growth in air in uniform static fields. The following conclusions are drawn from the experimental results.

1. Even at high values of pd ($\sim$ corresponding to 1 cm. spark in air) equation (3.1) based on the action of primary and secondary processes still describes the growth of ionization currents in uniform fields, as is the case at low values of pd.

2. The mechanism setting the static breakdown criterion is that which causes the growth of ionization in the pre-breakdown regime when $V < V_s$. The criterion in fact followed equation (3.2).

3. No evidence was obtained for the sudden appearance of any new mechanism setting the spark criterion when $d = d_s$.

4. The secondary ionization largely involved cathode emission due to the incidence of ions and photons.

5. Since the values of α and ω/α which satisfy the breakdown criterion were those obtained from the growth of a small current at values of $d < d_s$ it would appear that the space charge of the current in the vicinity of $d = d_s$ had not developed significantly to alter α and ω/α.

KOHRMANN and RAETHER[1] also consider that, when $e^{\alpha d} < e^{20}$, the electron avalanche develops in terms of primary α and secondary processes in a negligible space charge field and in times $\sim$ several gap transit times. When $e^{\alpha d} > e^{20}$, however, they consider that the field distortion due to space charge will have a controlling influence, and the development will become extremely rapid. For the 1 cm. gap in the atmosphere and with static fields αd is less than 20, so that the breakdown should be set by the criterion (3.2), as indeed is found experimentally.

δ) Growth in nitrogen and hydrogen. It is now necessary to consider the case of nitrogen in order to find (1) whether an upcurving of the log I *vs.* d graph occurred in nitrogen as well as in air, in order to see whether breakdown in air (a mixture) in which photo-ionizations can occur is fundamentally different from that in a simple gas; (2) whether the breakdown potential or the secondary coefficients depended upon the current, i.e. whether double ionization processes or photon production by recombination processes occurred; and (3) whether space charge was significant in setting a breakdown criterion.

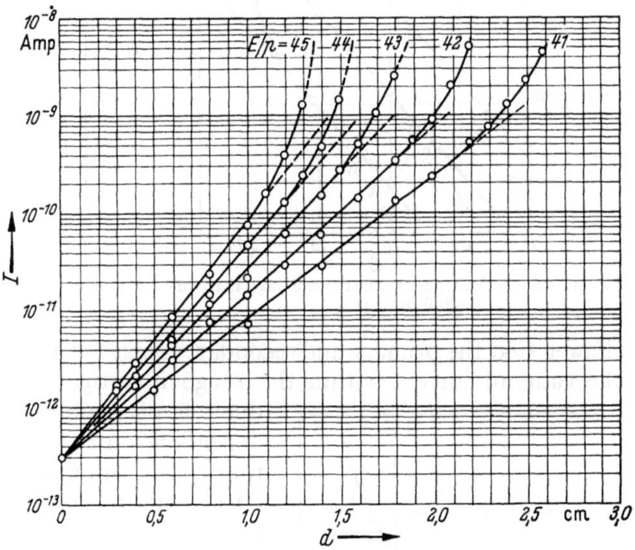

Fig. 26. Log I *vs.* d curves in nitrogen for different values of E/p at a pressure of 300 mm. Hg.

The use of nitrogen avoids difficulties which result from the progressive oxidation of the cathode. A careful investigation of the photo-electric efficiency of the cathode was made, and after treatment in a glow discharge in nitrogen the photo-electric yield became constant. The use of nitrogen, however, involved further experimental difficulties because of the necessity of increasing the sensitivity of the current measuring apparatus compared with that required for air. The anode voltages (10 to 40 kV) were stabilized to 0.05% and the growth of currents was measured for tank nitrogen passed through liquid air traps, using initial photo-electric currents in the range 6×10^{-13} A to 6×10^{-11} A from an area of 0.3 cm.[2].

Typical results[2] are given in Fig. 26. These show that nitrogen gives an upcurving in the log I *vs.* d graphs as did air. Fig. 27 gives curves obtained at the same E/p up to the breakdown distance d_s for three different initial currents,

[1] W. KOHRMANN and H. RAETHER: Naturwiss. **17**, 400 (1954).

[2] J. DUTTON, S. C. HAYDON and F. LLEWELLYN JONES: Proc. Roy. Soc. Lond., Ser. A **213**, 203 (1952).

and, within experimental error, these have the same slope, indicating the same value of α, and the same value of d_s or V_s, and the same value of (ω/α). This

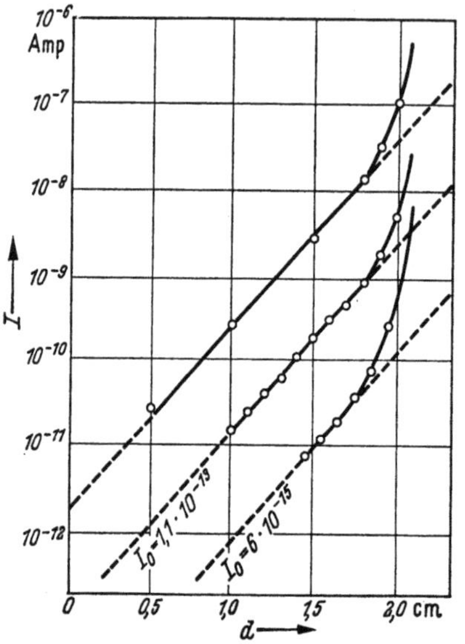

Fig. 27. Log I *vs.* d graphs obtained over a large range of initial currents I_0

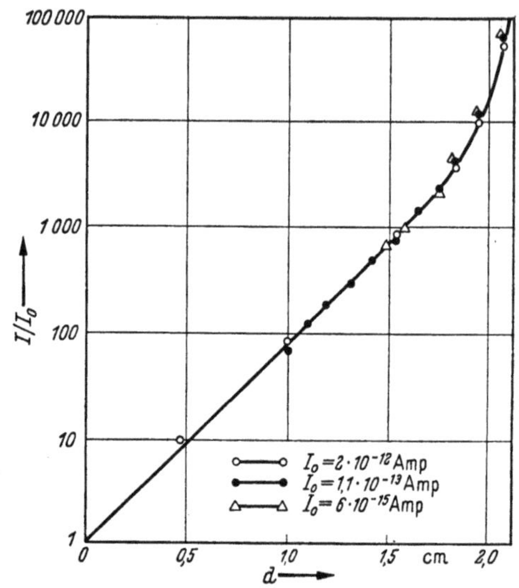

Fig. 28. Data of Fig. 27 superimposed, showing sparking distance is independent of I_0.

showed that the primary and secondary ionization coefficients did not depend upon the value of current within the range investigated, and that therefore the production of photons by recombination processes depending on the square of the current did not take place appreciably. Further, no significant effects due to space charges occurred in setting the conditions for V_s. Calculations by Crowe, Bragg and Thomas[1] of space charge effects for the present range of current densities are in agreement with this conclusion. The points of Fig. 27 when plotted as log I/I_0 *vs.* d curves, all fall on the same curve as given in Fig. 28. Previous measurements of Posin[2] gave different curves for different currents and had been taken to indicate the effects of space charge[3]; but such variations could be expected if the cathode surface had not been treated to give it stable emissive properties. After carefully treating the electrode surfaces by glow discharges in hydrogen, growth curves were measured in that gas by Crompton, Dutton and Haydon[4], and a typical curve is given in Fig. 29. Similar curves have been obtained by Wilkes, Hopwood and Peacock[5].

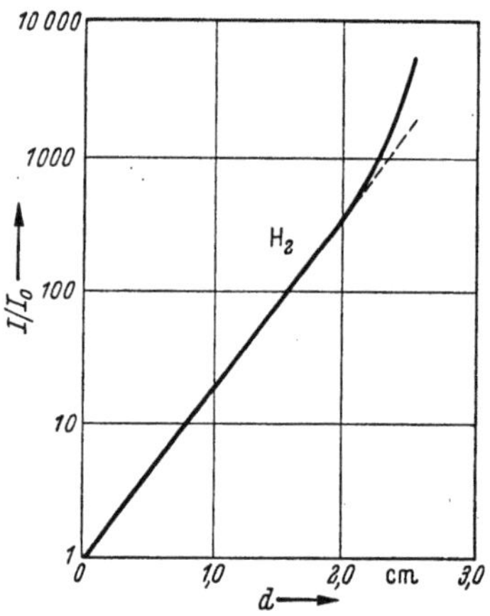

Fig. 29. Growth of ionization with distance in a uniform field in hydrogen at high pressure. $E/p = 23$ V/cm. mm. Hg.; $p = 150$ mm. Hg.

Comparison of Figs. 24, 26 and 29, giving growth curves in air, nitrogen and hydrogen at high pressure, with Figs. 3, 4 and 5 obtained for the same gases at

[1] R. W. Crowe, J. K. Bragg and V. G. Thomas: Phys. Rev. **96**. 10 (1954).
[2] D. Q. Posin: Phys. Rev. **50**, 650 (1936).
[3] L. B. Loeb: Fundamental Processes of Electrical Discharges in Gases, p. 386.
[4] R. W. Crompton, J. Dutton and S. C. Haydon: Nature, Lond. **176**, 1079 (1955). — Proc. Phys. Soc. B **69**, 2 (1956). — Appl. Sci. Res. B **5**, 111 (1955).
[5] A. Wilkes, W. Hopwood and N. J. Peacock: Nature, Lond. **176**, 837 (1955).

low pressure, support the view that the growth of current and the general mechanism setting the breakdown criterion is the same at low and high pressures.

ε) Conclusion. The above experimental data support the view that the generalized mechanism which sets the criterion for breakdown in a uniform field is based on the continuous development of ionization by primary and secondary processes and is applicable over the wide range of conditions in the different gases which have so far been quantitatively investigated, i.e. at least up to $pd = 800$ mm. Hg·cm. and including the case of a 1 cm. gap in the atmosphere. The results also indicate that the criterion for breakdown is set before space charge significantly alters the ionization coefficients corresponding to the applied field. Further, in this mechanism the development of the avalanches is mainly maintained from behind by a secondary ionization agency which is effective when it acts in regions nearer and nearer the cathode, and most efficient when acting at the cathode itself. The relative importance of the various possible secondary processes is influenced by the nature and geometry of the cathode. When $V > V_s$ the complete breakdown will involve high current densities, intense ionization, and the transition to an arc. The rate of development during these high current stages will depend on the external circuit and space charge effects.

Before discussing temporal growth it is of interest first to consider photo-ionization as a secondary mechanism, when the requisite photons are being generated by collision processes in the electron avalanche.

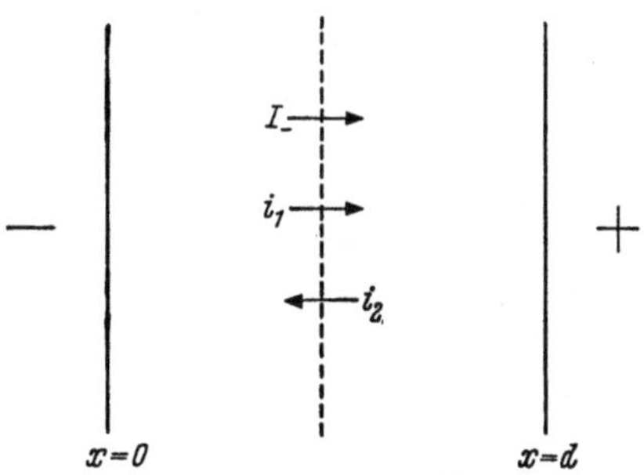

Fig. 30. The electron current and light intensities at a distance x from the cathode of a plane parallel gap.

15. Photo-ionization in the electron avalanche in uniform fields. *α) Introduction.* Little was known of this process when the theory of the spark was first investigated, but photo-ionization has been strongly advocated in recent years by RAETHER for his ***kanal*** theory and by LOEB for the streamer view, albeit on qualitative grounds, as an essential process in the electrical breakdown of gases at the higher pressures. The importance of photo-ionization lies in the fact that it is a secondary process taking place entirely in the gas, not dependent on any electrode effects.

In view of the experimental confirmation of the spatial growth equation (3.1) *at these high values of $pd \sim 1000$ mm. Hg·cm.*, it is now important to investigate whether photo-ionization, as a secondary process, can predict the growth of the important pre-breakdown ionization currents in accordance with the curve observed experimentally, and thus be included in the group of possible secondary processes which satisfy equation (3.2) the criterion for spark breakdown. The following analysis[1] makes a quantitative theoretical investigation of the role of photo-ionization in the development of ionization in gases under uniform electric fields; and then, on the basis of present knowledge of the relevant molecular cross-sections, the results are applied to the important case of air.

Consider, for simplicity, an electron avalanche in a plane-parallel gap d, and the flux of radiation as indicated in Fig. 30. I_0 electrons leave the cathode per second, and collide with gas molecules which they ionize or excite in producing an avalanche; suppose that photons with sufficiently high energies for ionization

[1] J. DUTTON, S. C. HAYDON and F. LLEWELLYN JONES: Proc. Roy. Soc. Lond., Ser. A **218**, 206 (1953) (with mathematical appendix by P. M. DAVIDSON).

are produced as a result of the excitations. These photons will travel in all directions: some will be absorbed by the gas atoms; and additional ionization and excitation of gas atoms will occur. Of the atoms excited, a fraction will fluoresce and re-emit photons of the same wavelength. These can be given off in any direction, and so will cause a redistribution of the radiation intensity. The rest of the excited atoms will degenerate by returning to states other than the ground state, emitting photons of longer wavelengths which will not be capable of ionizing. The new electrons produced by the ionizing absorptions will give rise to subsidiary electron avalanches which will contribute to the total current I_-.

The problems are, firstly, to find whether the processes can lead to a criterion such as equation (3.2) from which the sparking distance and breakdown potential can be predicted; and secondly, to find whether these processes can lead to spatial growth of the initial current I_0 in accordance with the equation (3.1), because these are experimental facts which must be accounted for.

It is necessary first to consider whether the essential high energy photons can in fact be produced in these conditions. In pure gases, photons with energies sufficient to produce ionization are only likely to be produced by the excitation of an ion or by recombination of ions and electrons. These processes will depend in the square of the current. If they were the predominant processes causing breakdown in pure gases, a marked dependence of V_s on the value of I_0 should be observed. That no such dependence was observed in nitrogen supports the view that in simple gases the requisite photons are not produced in sufficient quantities and that photo-ionization plays no significant role in pure gases. In air, a mixture of gases, on the other hand, photo-ionization becomes important. Electrons can excite a nitrogen molecule to levels of energy greater than 15 eV, and the high-energy photons produced can subsequently ionize an oxygen molecule, which has an ionization potential lower than 15 eV. This process, not being dependent on the square of the current, would conform to the observations on the growth of currents. It has long been known[1] that radiation capable of ionizing air is emitted from spark discharges in air, but the evidence[2] appeared to indicate that the effect is negligible with small ionization currents. In recent years[3] experimental measurements of the absorption coefficients of such radiations have been made. Thus it may be considered that air is a gas in which photo-ionization can occur, and it is now necessary to investigate whether the process is important in comparison with the other ionization processes in the breakdown mechanism in uniform fields in air.

β) Solution of the continuity equations for ions, electrons and radiation. The rigorous theoretical treatment of the growth of ionization due only to photo-ionization and the α process involves a solution of the following continuity equations

$$\frac{dI_-}{dx} = \alpha I_- + \mu_\iota (i_1 + i_2),$$

$$\frac{di_1}{dx} = \frac{1}{2}\vartheta I_- - \mu_i i_1 - \mu_e i_1 + \frac{1}{2}\mu_r (i_1 + i_2),$$

$$-\frac{di_2}{dx} = \frac{1}{2}\vartheta I_- - \mu_i i_2 - \mu_e i_2 + \frac{1}{2}\mu_r (i_1 + i_2).$$

[1] E. Wiedemann: Z. Elektrochem. p. 159 (1895). — Hoffmann: Wied. Ann. **60**, 269; **61**, 767 (1897). — J. J. Thomson: Proc. Cambridge Phil. Soc. **10**, 74 (1899).

[2] H. A. Wilson: Phil. Mag. **6**, 180 (1903).

[3] Vide G. L. Weissler's article on photo-ionization in vol. XXI of this Encyclopedia.

The coefficient ϑ is the number of photons capable of ionization produced on the average by an electron in travelling unit distance along the field, and $N\mu_i\,dx$ is the number of atoms ionized in distance dx by N such photons. The photons, however, also produce excitation, specified by a coefficient μ_e which is defined in analogy with μ_i, and of the photons absorbed in this way a fraction μ_r/μ_e will be re-emitted as fluorescence with energy greater than 15 eV. The coefficient of reflection of these photons at the electrodes will be taken as zero.

The above equations relate the electrons current I_-, with the photon currents in either direction i_1 and i_2, at the plane distant x with the various ionization and absorption cross sections. The solution due to DAVIDSON for the electron current $I_-(x)$ is

$$I_-(x) = \sum_{1,2,3} A_n \exp(\alpha_n x), \tag{15.1}$$

where the three values of α_n have the same dimensions as the primary ionization coefficient α and are the three roots of the cubic equation:

$$\alpha_n^3 - \alpha_n^2 n - \mu(\mu - \mu_r)\,\alpha_n + \mu\,\mu_i\left[\vartheta + \frac{\alpha(\mu - \mu_r)}{\mu_i}\right] = 0,$$

where $\mu = \mu_i + \mu_e$. Inserting the boundary conditions in these equations the three values of the coefficient A_n are obtained.

$$\begin{aligned} A_1 + \Psi A_2 + \Phi A_3 &= 0, \\ A_1 + \Psi_1 A_2 + \Phi_1 A_3 &= 0, \\ A_1 + \quad A_2 + \quad A_3 &= I_0, \end{aligned}$$

in which the Ψ's and Φ's are expressions in terms of the coefficients and d.

Solving these gives the solutions for A_1, A_2 and A_3 which all have the same denominator. For certain values of the coefficients and at a certain plate separation d_s this denominator vanishes. The condition is that

$$\Phi\Psi_1 - \Phi_1\Psi + \Phi_1 - \Phi + \Psi - \Psi_1 = 0.$$

This condition is the breakdown criterion. Expressed fully in terms of the various coefficients, the expression becomes

$$\left.\begin{aligned} &\frac{\varrho^2 - \alpha_1^2}{\mu + \alpha_1} \cdot \frac{\mu + \alpha_2}{\varrho^2 - \alpha_2^2}\left[\frac{\mu - \alpha_3}{\varrho^2 - \alpha_3^2} \cdot \frac{\varrho^2 - \alpha_1^2}{\mu - \alpha_1} - 1\right] e^{(\alpha_2 - \alpha_1) d_s} \\ &+ \frac{\varrho^2 - \alpha_1^2}{\mu + \alpha_1} \cdot \frac{\mu + \alpha_3}{\varrho^2 - \alpha_3^2}\left[1 - \frac{\mu - \alpha_2}{\varrho^2 - \alpha_2^2} \cdot \frac{\varrho^2 - \alpha_1^2}{\mu - \alpha_1}\right] e^{(\alpha_3 - \alpha_1) d_s} \\ &+ \frac{\varrho^2 - \alpha_1^2}{\mu - \alpha_1}\left[\frac{\mu - \alpha_2}{\varrho^2 - \alpha_2^2} - \frac{\mu - \alpha_3}{\varrho^2 - \alpha_3^2}\right] = 0 \end{aligned}\right\} \tag{15.2}$$

where

$$\varrho^2 = \mu(\mu - \mu_r).$$

γ) Significance of the absorption coefficient. Equation (15.2) is the breakdown criterion based on the assumption that photo-ionization is the only operative secondary mechanism. This compares with the much simpler criterion (3.2) which assumes other secondary processes. From this criterion (15.2) it is now possible to calculate the distance d_s at which breakdown will occur for any given set of conditions on the basis of known cross-sectional data. In order to see whether known values of the constants can lead to a physically significant solution the criterion will now be applied to the particular case of air for which experimental data are available. As, however, some of the cross sectional data are only known within a range of values, they have here been treated as parameters, and the computations carried out over a range of such parameters.

A typical result for a simple case is given in Fig. 31 in which fluorescence and degeneracy are for simplicity ignored, and a simple absorption coefficient μ is taken; d_s is the distance at which breakdown will occur for any given value of μ. It readily follows that:

1. for the case of air in uniform fields, there is only a limited range of absorption coefficient $0<\mu<4.1$ over which breakdown is at all possible on the basis of photo-ionization; negative values of d_s have no significance;

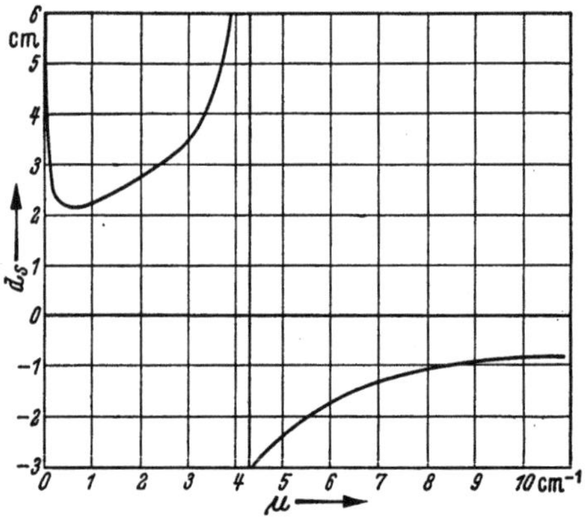

Fig. 31.

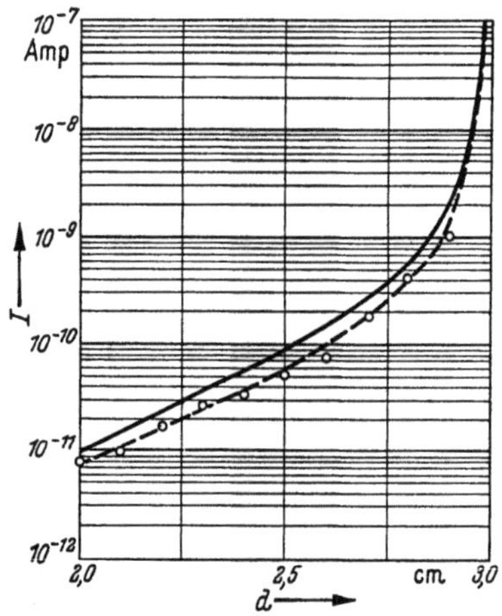

Fig. 32.

Fig. 31. Typical graph showing the variation of the sparking distance d_s with absorption coefficient μ (for conditions of no resonance fluorescence or degeneracy: $\vartheta=0.042$; $k=1.5\times10^{-8}$).

Fig. 32. Log I *vs.* d curve comparing the measured values (0) of the ionization currents I for air at $E/p=40$ V/cm. mm. Hg with: (a) those obtained theoretically (———) assuming photo-ionization as the only secondary mechanism; and (b) those obtained assuming a curve of the TOWNSEND type (– – –), represented by equation (3.1) ($I_0=6\times10^{-15}$ A; $d_s=2.94$ cm.).

2. that as μ approaches the limits of this range the sparking distance becomes infinite; and

3. within this range there are two values of μ which give the same breakdown distance.

The curves which have been computed for the more complicated cases which include resonance fluorescence and degeneracy are similar, and in all these cases the values of the various coefficients of absorption necessary to give the observed sparking distance d_s are found to be not inconsistent with the published experimental data on absorption ceofficients.

The fact that the breakdown distance calculated above is in agreement with that observed, though a necessary condition, is itself not sufficient to establish that photo-ionization is significant in breakdown. The process must not only account for the observed breakdown distance, but it must also account for the observed growth ot the ionization current which produces the breakdown. Consequently, it is necessary to evaluate the expression for the current $I_-(d)$ from the growth equation (15.1) for various values of d, and the resulting curve showing the development of the current across the gap is given in Fig. 32 by the full line curve.

The experimental data are represented by the points, and it can be seen that although there is general agreement there is a definite difference. On the other hand, it is seen that the spatial growth expression (3.1), which is shown by the dotted line, actually passes through the experimental points. It should be noted that both the theoretical curves [the full curve representing equation (15.1) and the dotted curve representing equation (3.1)] have been calculated using the same end-points, i.e. the same value of current at the cathode I_0 and the same sparking distance d_s. The full line curve was evaluated for a wide range of such parameters and it has been found that a very wide range of such values have an insignificant effect on that curve. Thus the difference between the full curve and the experimental points is not due to lack of knowledge of cross-sectional data.

δ) The role of photo-ionization in uniform fields. It is concluded from this analysis that:

1. on the basis of present knowledge of atomic cross-sections photo-ionization can account approximately for the observed sparking distances in air, on the other hand photo-ionization is not a probable secondary process in pure gases, because there is no process (a linear function of the current) which will give a sufficient number of high energy photons;

2. the photo-ionization process can, under certain restricted conditions, lead to a growth of pre-breakdown ionization currents in air, but the form predicted does not completely agree with that measured;

3. consequently, for the breakdown in air in uniform fields cathode processes must be invoked; a conclusion which has been confirmed by the examination of electrode effects.

Further interesting conclusions concerning the way in which photons can act in the breakdown mechanism may be drawn from this analysis.

When photo-ionization was originally proposed[1] as an essential factor in the static breakdown mechanism in uniform fields, it was stated on qualitative grounds that the absorption, and therefore the ionization, must take place *near* the head of the electron avalanche as in that way a narrow filamentary spark channel would be produced. The present calculations show, that, on the contrary, the amplification of the current by photo-ionization requires ionization *remote* from the avalanche head; this follows at once from the restrictions of the values of the absorptions coefficients illustrated in Fig. 31. The physical reason for this is the necessity of providing a long distance of travel for the electrons in the electric field to produce sufficient amplification of the photo-electrons. A further interesting fact about photo-ionization as a secondary process which arises from the restriction in the μ's is that proof of the existence of the process is no indication of its significance in breakdown.

In those cases, when owing to the absence, for example, of any cathode effect and of ionization by collision in the gas by positive ions, photoionization is the only feasible secondary process, these calculations show that the general mechanism of breakdown would then be still the same as that involving other secondary processes, in that the effective photo-ionization tends to take place at points remote from the avalanche head, so regenerating the avalanche from behind; in fact, in just the same way as the more efficient process of photo-electric emission from the cathode. Thus, when some photo-ionization occurs together with the other secondary processes, photo-ionization contributes to the generalised secondary coefficient ω by adding a further linear term η so that approximately

$$\omega = \beta + \alpha\gamma + \delta + \varepsilon + \eta .$$

[1] L. B. Loeb and J. M. Meek: Mechanism of the Electric Spark, Chap. II. 1941.

ε) Non-uniform fields. In practice breakdown in a non-uniform fields is of great interest, and an extreme example of this is the lightning flash. It has been shown that qualitative considerations of the possible action of photo-ionization in breakdown can be misleading; this is especially so in the more complicated conditions of a non-uniform field. To investigate whether photo-ionization is an effective process in any particular case of non-uniform field breakdown, the complete quantitative calculations must be carried out for that field configuration as has been done above for uniform fields. The above analysis carried out for a uniform field indicates the kind of data which are required to make these further investigations; in particular precise knowledge is required of the dependence of the various coefficients on the electric field, especially α. This conclusion has particular relevance to an attempted explanation of the lightning flash[1] which so far has only received general qualitative treatment. In this connection, important information still needed, apart from the dependence of α on the electric field, is that concerning the distribution of the electric field in the neighbourhood of the thundercloud, in the important phase *just before the flash*.

The above calculations indicate that for air in uniform fields the photo-ionization term η is generally a small fraction of the total generalised coefficient ω, thus η can be made to predominate only in conditions when the γ or δ terms are small. This is the case at high pressures in uniform fields if the cathode surface is inactive; but it is also the case in non-uniform fields either when the cathode is pointed so that the δ effect is negligible, or when the anode is pointed and the cathode very remote. For the positive discharge from a point only the β and η secondary effects can act. There is, at present, insufficient evidence, especially of the β process, on which to assess the relative importance of the two effects in any given case. Accurate measurements of the growth of ionization currents are needed for the case of non-uniform fields which must be known accurately as a function of distance, so that computations may be made using measured photoionization cross-sections, on the lines of those given above for uniform fields. Similarly, the long gap is in general a particular case of a non-uniform field. Further experimental data are required for uniform parallel plates, for $pd \gg 1000$ mm. Hg·cm. and also in non-uniform fields[2].

E. The temporal growth of ionization.

16. Fundamental principles. In the account given above of the development of ionization under static electric fields with an initial steady supply I_0 of electrons at the cathode, the current becomes constant in time. When the applied potential V is less than the potential V_s given by the TOWNSEND breakdown criterion equation (3.2), the resulting current I is proportional to I_0; when I_0 ceases, then I ultimately ceases. It is important now to consider the current which can flow when $V = V_s$. When the replacement condition is satisfied, a current I becomes self-maintained even when I_0 is zero. When I_0 is not zero and $V = V_s$, the current will increase with time at a steady rate, and its value can be readily calculated for the conditions when I is still insufficient to distort the field and so does not affect the coefficients α and ω/α. The time required for the growth of current to any given magnitude is dependent on the gap transit time of the ions or electrons involved in the processes represented by ω.

The condition, in which I flows with V exactly equal to V_s is, however, extremely difficult to realise in practice. This is because very small fluctuations

[1] See article on Lightning by B. F. J. SCHONLAND, in this volume.

[2] See H. NORINDER and O. SALKA: Ask. Fysik **3**, 347 (1951); **5**, 493 (1952).

in V at times make $V > V_s$, and it is therefore important to consider how ionization grows in the gas when the applied potential V exceeds V_s, even if the excess is small. It is clear that when the current is small, the coefficients α and ω/α will have the values corresponding to the applied field V/d, and that these values will be greater than the values obtaining at the replacement condition when $V = V_s$ and the field (V_s/d) lower than V/d. It therefore follows that with the higher field the current will increase at an increasing rate.

In recent years experimental techniques have been developed which make it possible to apply suddenly to a gap accurately defined potentials which can be maintained constant thereafter for predetermined periods, provided the current is not excessive. It is clear that in these conditions the current will increase at a rate which is dependent on the field in the gap, and will continue to increase as long as the field is maintained. In practice the gap is regarded as having broken down when the current attains a magnitude sufficiently large to produce a collapse of the applied voltage. The interval t elapsing from the instant of application of the field to the time when the current attains this magnitude is called the formative time lag. As has been discussed previously in Sect. 4, for the static case, the later stages of the current growth will depend on how the current density and its space charge influence ionization processes.

It is important, therefore, to investigate for $V > V_s$ the temporal growth of an ionization current on the basis of the same primary and secondary ionization processes which are known to be operative under uniform static fields. This is a justifiable procedure because the ionization coefficients are themselves continuous functions of V, and their values do not appear to be affected by the currents which can satisfy the equations (3.1) and (3.2). By using the values of the ionization coefficients α and ω/α obtained from measurement of the growth of currents in static fields when $V < V_s$, the time of growth of current to any given value can be calculated in any given case of higher applied fields as a function of the field, and this can then be compared with the growth times determined experimentally. Such comparisons can thus provide important information about the ionization processes in the vicinity of the value $V = V_s$, both when $V < V_s$ and when $V > V_s$ for cases when the lag $t \gg l/u$.

The time t required for the current growth is particularly sensitive to the nature and relative importance of the individual secondary processes represented by the generalised coefficient ω. This is readily seen because the transit times of the positive ions involved in the γ process are $\sim 10^2$ times greater than those for the electrons produced by the photon δ process. Thus when V is very near V_s, and a number of transits are required to produce a final given value of current, the times involved in growth based on a γ process are greater than those based only on a δ process by a factor $\sim 10^2$. For a 1 cm. gap in the atmosphere the formative time with a γ process should be $\gtrsim 10^{-5}$ sec. while that for a δ process should be $\gtrsim 10^{-7}$ sec. It should be emphasised here, however, that this conclusion applies only to breakdown with uniform fields when V is not very much greater than V_s, and the times may be quite different in a highly overvolted gap, when successive complete transits may not be necessary and the ionization produced by primary and secondary processes can grow exceedingly rapidly in time throughout the gap space; this aspect will be discussed below.

17. The temporal growth of ionization in uniform fields. *α) General considerations.* There have been a number of investigations of the growth of ionization in uniform fields and early work is described by R. SEELIGER and G. MIERDEL[1].

[1] R. SEELIGER and G. MIERDEL: Handbuch der Experimentalphysik, Vol. 13, 2. 1929.

ROGOWSKI[1] considered the case of highly overvolted gaps, and for these conditions the calculations are necessarily approximate because the current increase and consequent field changes are exceedingly rapid. Of particular interest, however, are the more rigorous calculations of SCHADE[2] and STEENBECK[3] relating to cases of lower overvoltages $\lesssim 20\%$. Later, BARTHOLOMEYCZYCK[4] investigated the temporal growth from a consideration of the continuity equations

$$\frac{\partial I_-/\partial t}{W_-} = -\frac{\partial I_-}{\partial x} + \alpha I_-,$$

$$\frac{\partial I_+/\partial t}{W_+} = +\frac{\partial I_+}{\partial x} + \alpha I_-,$$

with the boundary conditions

$$I_-(0,t) = \gamma I_+(0,t) + \delta \int_0^d I_-(x,t)\, e^{-\mu x}\, dx,$$

$$I_+(d,t) = 0;$$

and gave an approximate solution

$$I_-(x,t)\, e^{-\alpha x} = C \exp\left[\lambda\left(t - \frac{x}{W_-}\right)\right]$$

for calculating the current as a function of time and distance. C is a constant, and λ is the real root of $F(d)=0$, where

$$F(x) = 1 - \frac{\alpha\gamma}{\varphi}(e^{\varphi x} - 1) - \frac{\delta}{\psi}(e^{\psi x} - 1),$$

$$\varphi = \alpha - \frac{\lambda}{W}; \qquad \psi = \alpha - \mu - \frac{\lambda}{W_-},$$

$$\frac{1}{W} = \frac{1}{W_-} + \frac{1}{W_+}.$$

GUGELBERG[5] noted the omission of the term I_0 from the $I_-(0,t)$ boundary condition, and DAVIDSON[6] pointed out that moreover BARTHOLOMEYCZYCK'S formula does not satisfy the initial conditions, but can be altered so as to satisfy the continuity equations and the correct boundary conditions (but still not the correct initial conditions) by using the following procedure. If λ is given BARTHOLOMEYCZYK'S value, and if

$$P = 1 - \gamma(e^{\alpha d} - 1) - \frac{\delta}{\alpha - \mu}\{e^{(\alpha-\mu)d} - 1\},$$

then a short calculation leads to DAVIDSON'S approximate expression for the growth equation

$$I_-(x,t)\, e^{-\alpha x} = \frac{I_0}{P} + C \exp\left\{\lambda\left(t - \frac{x}{W_-}\right)\right\}, \tag{17.1}$$

which, with an accompanying expression for $I_+(x,t)$, satisfies the continuity equations and the corrected boundary conditions. The constant C may be given a value which gives approximate agreement with the initial conditions; for example, if the initial value of $I_-(0, t)$ is c, the solution can be made to have that initial value of $I_-(0,t)$ by taking C as $c - I_0/P$.

DAVIDSON[6] also gave an accurate solution of the continuity equations in which the solution not only satisfies the continuity equations and the correct boundary

[1] W. ROGOWSKI: Arch. Elektrotechn. **20**, 99 (1928).
[2] R. SCHADE: Z. Physik **104**, 487 (1937).
[3] M. STEENBECK: Wiss. Veröff. Siemens-Konz. **9**, 42 (1930).
[4] W. BARTHOLOMEYCZYCK: Z. Physik **116**, 235 (1940).
[5] H. L. VON GUGELBERG: Helv. phys. Acta **20**, 250, 307 (1947).
[6] P. M. DAVIDSON: Brit. J. Appl. Phys. **4**, 170 (1953).

conditions, but also satisfies the initial conditions completely; giving any prescribed initial distribution of positive and negative charge in the whole range $x=0$ to d at $t=0$. DAVIDSON's analysis will now be given.

β) DAVIDSON's exact expression for ionization growth. The equation $F(d)=0$ has, in addition to the real root, an infinite number of complex roots; the last term in the growth equation (17.1) must be replaced by a summation containing these various λ's and with C's determined by the initial conditions. The mathematical problem itself is of little interest here, and it will be sufficient to state the final expressions obtained for I_- and I_+. This latter expression is of great importance in determining when the effects of space charge become significant.

It is convenient to think of the expression for, say, $I_-(x,t)$ as consisting of two parts: (i) the value which it would have if the initial charge distribution had been absent; and (ii) the value which it would have if the constant generation I_0 had been absent. The quantitiy $I_-(x,t)$ which will be present in the experiment is the sum of the two parts. The same remarks apply to $I_+(x,t)$.

In the part (ii), $I_-(x,t)$ and $I_+(x,t)$ are conveniently calculated from the initial distribution, say $\varrho_- = f_-(x)$ and $\varrho_+ = f_+(x)$, by means of four g's (GREEN's functions), of which, for example, $g_\pm\,(x\,t\,x_1)$ is the $I_-(x,t)$ due to the presence at time zero of unit positive quantity in the region of $x=x_1$. Thus the required $I_-(x,t)$ of part (ii) is

$$\int_0^d \left[g_-^- f_-(x_1) + g_-^+ f_+(x_1)\right] dx_1$$

and a corresponding expression gives $I_+(x,t)$.

At values of t less than x/W_- the values of g_-^+ and g_+^-, and also of the I_- and I_+ of part (i), are of a simple nature, readily seen on visualizing the motion. For example, g_-^- is zero in this range of time if $x<x_1$. If $x>x_1$ there is a sudden pulse at $t=(x-x_1)/W_-$, conveying a quantity $\exp\alpha(x-x_1)$. At all later times the series expressions given below for these quantities are valid. Similarly the values of g_-^+ and g_+^+ are readily seen at times up to $(x_1/W_+) = (x/W_-)$ after which the series are to be used.

The series expressions are:

Part (i)

$$\left.\begin{aligned}
\frac{I_-}{I_0} &= \frac{e^{\alpha x}}{P} + \sum Q,\\
\frac{I_+}{I_0} &= \frac{e^{\alpha d} - e^{\alpha x}}{P} + \sum R\,Q,\\
\text{where}\quad Q &= (\lambda D)^{-1}\exp\left\{\left[\alpha - \frac{\lambda}{W_-}\right]x + \lambda t\right\},\\
R &= \frac{\alpha}{\varphi}\left[e^{\varphi(d-x)} - 1\right],\\
D &= \frac{\gamma\alpha}{\varphi^2 W}\left[1 - (1-\varphi d)\,e^{\varphi d}\right] + \\
&\quad + \frac{\delta}{\psi^2 W_-}\left[1 - (1-\psi d)\,e^{\psi d}\right].\\
\text{Part (ii)}\quad g_-^- &= \sum G_-^- \qquad g_+^- = \sum R\,G_-^-,\\
g_-^+ &= \sum G_-^+ \qquad g_+^+ = \sum R\,G_-^+,\\
G_-^- &= \lambda\,Q\,F(x_1)\exp\left\{\left[\frac{\lambda}{W_-} - \alpha\right]x_1\right\},\\
G_-^+ &= \gamma\,\lambda\,Q\exp\left[-\frac{\lambda x_1}{W_+}\right].
\end{aligned}\right\}\qquad(17.2)$$

From these expression it is possible to calculate the electron or ion density at any point x in the gap at any time t after the initial electrons are produced. The expressions (17.2) thus give the spatio-temporal growth of ionization in the gap, and this is of considerable importance in calculating not only formative time lags in overvolted gaps, but also the spatio-temporal development of the luminosity produced during the later stages of ionization growth. DAVIDSON[1] has also given the exact expression for the growth in an alternative form, which is particularly convenient for considering cases of high over-voltage.

18. Experimental investications at low pressures ($pd < 150$ mm. Hg · cm.). TANK and GRAF[2] were the first to demonstrate experimentally that the time of growth in low pressure discharges could be observed in reproducible experiments. VON GUGELBURG[3] measured the current flowing through a resistance in series with a discharge gap. Ionization currents between 10^{-11} to 10^{-4} A were observed in H_2, He, Ar, Kr, N_2, and admixture of Ne and Ar for values of $pd < 100$ mm. Hg × cm. with uniform fields and also non-uniform fields. The results were compared with the times calculated on the basis of his analysis, and estimates of the relative contribution (δ/ω) of the photo-electric effect to the total secondary emission then made. In Ne and He there appeared to be no perceptible photoeffect, whereas in Xe the contribution was 20%, and in H_2 even greater.

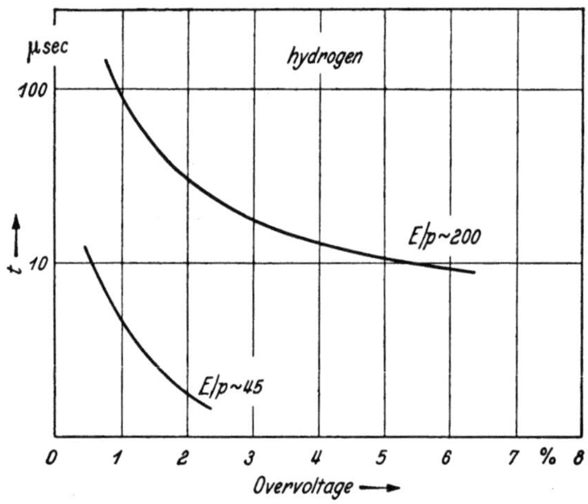

Fig. 33. Formative time lags in hydrogen.

More recently, MORGAN[4] has made measurements of the growth of current in highly stable uniform electric fields as a function of the overvoltage ΔV over a wide range of the parameter E/p corresponding to conditions in low pressure discharges in hydrogen. For low values of E/p (< 50 V/cm. per mm. Hg) the formative time lags were of the order of 5 μsec. at overvoltages ~1%; while at larger values of E/p (~200 V/cm. per mm. Hg) t was of the order of 100 μsec. A typical result is given in Fig. 33. The ratio γ/δ increased with E/p.

The form of this curve is in general agreement with previous experimental work of SCHADE[5] and of BARTHOLOMEYCZYK[6]. Curves relating the growth time t to the overvoltage ΔV were calculated from DAVIDSON's growth equation (17.1) and comparison of experimental curves similar to those in Fig. 33 with the theoretical curves enable the values of the individual coefficients δ/α and γ composing ω/α to be found for the electrode surface used. There appears to be a change of the relative importance of the two secondary ionization processes, the γ effect becoming increasingly important as E/p increases.

[1] P. M. DAVIDSON: Phys. Rev. **99**, 1072 (1955).
[2] F. TANK and K. GRAF: Helv. phys. Acta **2**, 33 (1929).
[3] H. L. v. GUGELBURG: Helv. phys. Acta **20**, 250, 307 (1947).
[4] C. G. MORGAN: Conference on Ionized Gases at Birmingham University (1954). Appl. Sci. Res. B **5**, 1 (1955).
[5] R. SCHADE: Z. Physik **104**, 487 (1937).
[6] W. BARTHOLOMEYCZYK: Z. Physik **116**, 235 (1940).

One of the most significant results of this investigation of the formative time lags at low pressures, is, firstly, that even when the applied field is undistorted by space charge, the rate of growth of ionization in the gap can be very high, yielding formative time lags $\sim 10^{-6}$ sec. with overvoltages as low as 2%. Thus, the assumption that the general mechanism of ionization growth dependent on primary α and secondary ω processes in a uniform field is too slow to give the times $\sim 10^{-6}$ sec. observed is unfounded: at higher overvoltage the growth times can be even shorter, eg. $\sim 10^{-7}$ sec. Secondly, when the current begins to develop rapidly, the rate is so great that the formative time lag of breakdown does not greatly depend on the *magnitude* of the final current taken to indicate complete breakdown.

These data on ionization growth at the lower pressures with impulse fields when $V > V_s$ are complementary to those obtained in static fields when $V \leq V_s$. They are also consistent with the view that the breakdown criterion equation (3.2) is set by the attainment of the replacement condition through the action of primary and secondary processes before the space charge due to the current significantly distorts the field: distortion only occurs at a later stage in the ionization growth.

19. Experimental data at higher pressures ~ 760 mm. Hg. Experimental investigations of the formative time lag in uniform fields in air have been made by FISHER and BEDERSON[1]. The method used was to apply an approach voltage V_A less than the sparking potential V_s to the gap and then suddenly to apply a voltage pulse V_P to make the total voltage $V_P + V_A > V_s$. The time between the application of the pulse and the collapse of the voltage across the gap is taken as the formative time lag t. The experiments were carried out for a range of pressure from about 300—760 mm. Hg and a range of gap distances from 0.3 to 1.4 cm. These results are given in Fig. 34 together with theoretical curves[2] based on the ionization growth equation (17.1). While the theoretical and experimental curves are of the same form the agreement is not perfect, but better agreement can be obtained, by choosing different proportions of photon and positive ion effect.

For numerical computations[2] of formative time lags in air at atmospheric pressure $(V_s/p d_s)$ was taken as 41.6 V/cm. mm. Hg and ω/α was calculated from the static sparking potential V_s using measured values of α.

The dependence of the shape of the time lag curve on the value of the cathode current $I_-(0, t)$ taken, is illustrated in Fig. 35 for the three cases when $\delta/\omega = 1$, 0.5 and 0 respectively. Because the value of $I_-(0, t)$ given by equation (17.1) depends on the value assumed for the initial photo-electric current I_0, the curves plotted in Fig. 35 show also the dependence of the times on the value of I_0 for a given final value of $I_-(0, t)$. The dependence of the value of t on I_0 is more marked the smaller is δ/ω (i.e. the greater the contribution of electrons liberated at the cathode by positive ions to the total secondary ionization), and the smaller the overvoltage.

These calculations lead to the conclusion that the action of primary α and secondary ω/α processes at high pressures can lead to short formative time lags; for example, when δ/ω is 0.5, t is about 10^{-6} sec. with an overvoltage as low as 2%. Further, the time lag is not greatly dependent on the assumed value of the final current, assuming the field remains constant. When the current attains high values producing significant field distortion, the rate of the growth of ionization

[1] L. H. FISHER and B. BEDERSON: Phys. Rev. **81**, 109 (1951).

[2] J. DUTTON, S. C. HAYDON and F. LLEWELLYN JONES: Brit. J. Appl. Phys. **4**, 170 (1953).

can remain high, even if the over-all field V/d diminishes as V collapses, if arc-like conditions prevail.

Measurements of the current growth in uniform fields using un-degassed brass electrodes in air, have been made by BANDEL[1]. The range investigated was 10^{-6} A up to 10^{-2} A in times between 10 and 100 μsec.; the theoretical computations were based on DAVIDSON's approximate growth equation which wa found to give fair agreement up to a stage when space charge effects becams appreciable and a rapid increase accurred. BANDEL also concluded that owine to the action of δ effect in the experimental conditions, the ionization curreng was initially spread over the whole electrode surface.

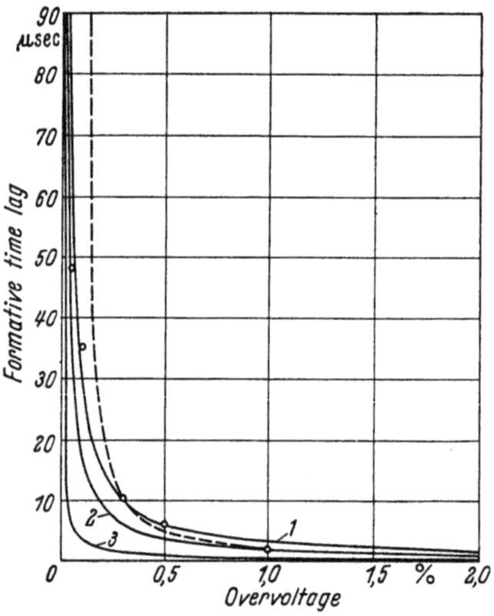

Fig. 34. Comparison of the theoretical curves of a formative time lag against percentage overvoltage (assuming various values for $I_-(0,t)$ and δ/ω) with the measurements of FISHER and BEDERSON.

	δ/ω	$I_-(0,t)$
1	1.0	10^{6} A
2	1.0	10^{-1} A
3	1.0	10^{-9} A
---	0.9	10^{-1} A

o = measured values of FISHER and BEDERSON.

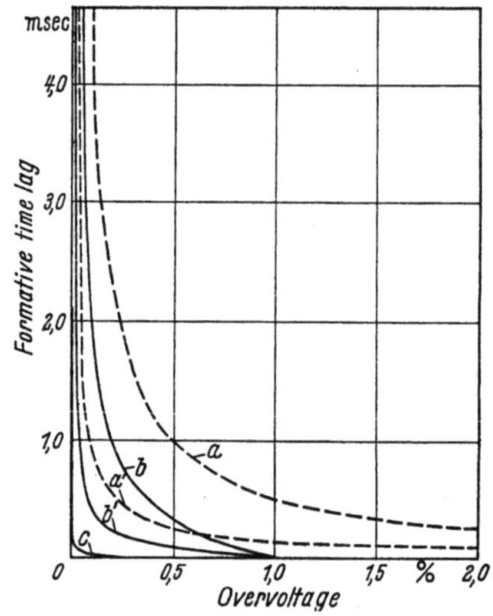

Fig. 35. Theoretical curve showing the variation of the formative time lag with percentage overvoltage for various values of δ/ω and $I_-(0,t)$.

	δ/ω	$I_-(0,t)$
a	0	10^{-1} A
a'	0	10^{-9} A
b	0.5	10^{-1} A
b'	0.5	10^{-9} A
c	1.0	10^{-1} A

Time lag data for breakdown in air in a uniform field with high voltages up to 150 kV has recently been obtained by AKED, BRUCE and TEDFORD[2]. For small electrode spacings the data agree with those of FISHER and BEDERSON for a similar range. At low overvoltages ΔV long time lags were obtained which appeared to increase indefinitely as ΔV approached zero; t decreased rapidly at high overvoltages, and were of the order of 1 μsec. at 2% overvoltage. The authors point out that these values are of the nature and order predicted on the basis at the growth equation (17.1).

It is difficult to give a quantitative treatment of the growth of ionization in a highly non-uniform field[3] for the reasons previously given, so that a full quantitative discussion of experiments on breakdown in non-uniform impulse fields

[1] H. W. BANDEL: Phys. Rev. **95**, 1117 (1954).
[2] A. AKED, F. M. BRUCE and D. J. TEDFORD: Brit. J. Appl. Phys. **6**, 233 (1955).
[3] An account of published work in this field is given by MEEK and CRAGGS, Electrical Breakdown of Gases, Chap. 3.

has not yet been given. Nevertheless, the results of GÄNGER[1] obtained with a sphere-to-plane configuration are of interest. Measurements of the formative time lag over a wider range of pressure in air from atmospheric down to 25 mm. Hg with potentials up to 50 kV were made. GÄNGER investigated whether any transition occurred from a TOWNSEND type of ionization growth mechanism in an undistorted field to a fast *kanal* or streamer type. No discontinuities were, however, found in the measured $(t, \Delta V/V)$ curves over the pressure range investigated. Furthermore he observed long time lags at low over-voltages. He concluded that for static breakdown, the TOWNSEND mechanism operated, in which the β effect was feasible, the *kanal* mechanism only occuring with over-voltages, with which, too, there is a small electrode polarity effect.

FISHER and his colleagues[2] have also made investigations of the formative time lag for positive point corona in air and for uniform field breakdown in nitrogen, hydrogen, oxygen, and argon, for ranges of pressure and distance similar to those used in air. In uniform fields the results obtained for nitrogen and hydrogen were found to be very similar to those obtained for air. The lags decreased rapidly with overvoltage and there was no observed pressure dependence. In oxygen and argon, on the other hand, the lags were much larger for a given overvoltage than those measured for air, and there was a dependence of time lag on pressure. From the results in argon it was concluded that the TOWNSEND pre-breakdown ionization growth mechanism involved a delayed photon δ effect; and such a delayed action has been found in discharges in argon GEIGER counters and attributed by COLLI[3] to a long lived molecular state. Further investigations of this effect have been made by MENES[4] who recorded oscillographically the current growth and the light emission after the application of the field. The results are in fair agreement with COLLI's data at atmospheric pressures, but deviate appreciably at the lower pressures, and further work is needed.

20. Later stages of ionization growth—the spark diameter. In the previous sections discussion of the ionization growth has been carried quantitatively to the stage when the space charge begins to become significant. For reasons previously discussed above in Sect. A, the current, if permitted, will finally distort the field, especially at the cathode and this will enhance the ionization rate and reduce the over-all voltage across the gap. The process is unstable in so far as increase of current density increases the ionization, and the resulting collapse is sudden. At high pressures the final current density is high, and it is interesting to estimate theoretically the diameter of the *kanal*, or streamer, on the basis of the avalanches in which the positive ion field is significant.

It can be shown[5] that for these conditions the continuity equation for a particle density n is

$$\frac{\partial n}{\partial t} = \alpha W_- n + D_- \nabla^2 n - \operatorname{div} n \{W_-(E') - W_-(E)\}, \tag{20.1}$$

where E' is the intensity of the electric field due to the charges on the electrodes and the space charge in the gas and E is the applied electric field V/d due only to the charges on the electrodes. Because of the presence of ions, space charge

[1] B. GÄNGER: Arch. Elektrotech. **39**, 508 (1949).

[2] L. H. FISHER and M. MENES: Phys. Rev. **86**, 134 (1951); **94**, 1 (1954). — L. H. FISHER and G. A. KACHIKAS: Phys. Rev. **79**, 232 (1950); **88**, 878 (1952); **91**, 775 (1953). — L. H. FISHER and I. LESSIN: Phys. Rev. **93**, 649 (1954).

[3] L. COLLI: Phys. Rev. **95**, 892 (1954).

[4] M. MENES: Gaseous Electronics Conference, New York University, October 1954.

[5] F. LLEWELLYN JONES: Brit. J. Appl. Phys. **1**, 60 (1950).

in the electron avalanche may be ignored, E may be put equal to E'. It is seen from equation (20.1) that a quantity $n = n_0 \exp(\alpha W_- t)$ satisfies

$$\frac{\partial n_0}{\partial t} = D \nabla^2 n_0 \tag{20.2}$$

which is the diffusion equation which the electrons would satisfy if there were no internal generation (by ionization), i.e. if the total number remained n_0. If, therefore, the spatial distribution of n_0 is calculated at any time t by solving equation (20.2), then, since $n = n_0 (\alpha W_- t)$, it is clear that the spatial distribution of n, the number of electrons including those internally generated, is the same as that of n_0, i.e. it is also given by equation (20.2). It follows that the cross section A_1 of the electron avalanche at the anode is given approximately by

$$\pi \overline{R}^2 = 4\pi D_- t,$$

i.e.

$$A_1 = 4\pi D_- \frac{d}{W_-} = 4\pi d^2 \frac{E_1}{V}, \tag{20.3}$$

where E_1 is the mean energy of the particles.

The positive ion concentration originally at the anode will move to the cathode where the electron concentrations is low, and space charge cannot therefore be neglected. The equation of motion of the positive ion is obtained from (20.1) by putting $\alpha = 0$ giving

$$\frac{\partial n}{\partial t} = D_+ \nabla^2 n - \operatorname{div} n \{W_+(E') - W_+(E)\}.$$

But $W_+ = \mu E$ where μ is the positive ion mobility. Putting $E'' = E' - E$,

$$\frac{\partial n}{\partial t} = D_+ \nabla^2 n - \operatorname{div} n\, \mu E'',$$

where E'' is now the field due to the space charge and its multiple images in the electrodes. Now $\operatorname{div} E'' = 4\pi n e$ hence,

$$\frac{\partial n}{\partial t} = D_+ \nabla^2 n - 4\pi \mu e n^2 - \mu (E'' \operatorname{grad} n).$$

In overvolted gaps in air at atmospheric pressure E/p is in the range 40 to 100 V/cm. per mm. Hg; $E_1 \sim 4$ to 10 eV; and until the gas becomes greatly heated E_2 for ions is ~ 0.1 eV. Equation (20.3) then gives the ratio A_1/A_2 of the cross-section for electrons and ions as 0.025, thus showing that the diffusion of the ions may be neglected, so that dispersal must be due to their space charge. It may be assumed that the main components of the space charge fields E' and the concentration gradient grad n are both radial and perpendicular to the axial field E. Thus the radius r of a cylindrical surface enclosing a fixed number N of ions per cm. expands at the rate $dr/dt = \mu E' = \mu 2Ne/r$.

Integration gives

i.e.,

$$\left.\begin{aligned} r_t^2 - r_0^2 &= 4e\mu N t \\ (A_2)_t - (A_2)_0 &= 4\pi e d^2 \alpha (\exp \alpha d)/2V. \end{aligned}\right\} \tag{20.4}$$

It is assumed that the avalanche was generated by the emission of a single electron from the cathode, and if N is given the approximate value $\alpha \exp(\alpha d)$ and $(A_2)_0$ the value of A_1 given by equation (20.3), and setting $t = d^2/\mu V$, the time for a positive ion to cross the gap, then the area $(A_2)_t$ given by equation (20.4) may be regarded as the area of the resulting cathode hot-spot, and $r_1 = \sqrt{A_2 \pi}$ its radius.

On the basis of these simplifying assumptions, the area A_2 through which the current strikes the anode in a 1 cm. gap in air (with $V=32$ kV, $E_1=3.5$ eV, $\alpha=17$, $A_1=1.4\times10^{-3}$ cm.^{2}), is given by $(A_2)_t-(A_2)_0=1.0\times10^{-2}$ cm.^{2}, giving $r_t=6.0\times10^{-2}$ cm. Such comparatively small areas are observed in practice, and high current densities can occur at the cathode at which intense local heating and evaporation are produced[1]. Thus a filamentary *kanals* can be expected at high pressures when the growth development is due to the action of primary and secondary processes in which positive ions play an important part.

21. The growth concept of complete breakdown of an overvolted gap. *α) Spatio-temporal growth.* The formal account, given above in the preceding sections, of the growth of ionization produced by primary (α) and secondary (ω) processes, emphasizes the spatio-temporal character of the growth. At any point along the avalanche path the ionization increases rapidly in time at a rate which can be calculated from DAVIDSON's expressions (17.1) and (17.2) for the growth equation, when the appropriate coefficients α and ω and the field E are all known. This rate is highly sensitive to the value of the overvoltage $V-V_s$, and calculation shows that equation (17.1) predicts extremely high rates of ionization growth when $(V-V_s)/V_s\sim10\%$, because of the increase of α and ω with the field: high electron and ion densities are attainable in a very short time.

β) Physical significance. When short formative time lags $\sim10^{-6}$ sec. were first observed[2], they were interpreted as being incompatible with ionization growth in a uniform field by primary and secondary processes, especially those occurring at the cathode, because, it was considered, the build-up of current would involve a large number of transits of the gap by avalanches, requiring a comparatively long time ($\gtrsim10^{-4}$ sec.). However, the short time lags had been observed with *highly overvolted* gaps, for which conditions the growth equation shows that short time lags are to be expected.

With uniform fields, when $V<V_s$ the initial current I_0 from the cathode is reinforced by emission $f I_0$ due to secondary processes in the avalanche as it crosses the gap, where $f<1$. When $V=V_s$, then $f=1$, and the current becomes self-maintained, being capable of continuing by virtue of the fact that secondary processes occurring as the avalanche crosses the gap just maintain the initial electron production even when I_0 is zero: the time involved in this process is necessarily long. With an overvoltage, however, the value of the coefficient α will be enhanced and the secondary ionization produced while an avalanche crosses the gap will now be more than adequate to replenish the necessary electrons, since $f>1$. Traversal of the full gap distance d is not necessary to provide the regeneration necessary to give $f=1$, which is the condition for continued growth when I_0 is finite; very rapid ionization growth is possible even with moderate over-voltages. The occurrence with high over-voltages of formative times (measured to the collapse of gap voltage) which are less than the estimated avalanche transit time d/W_- does, however, require further consideration. Such short times have, in fact, been recorded, and these must now be considered.

γ) Highly over-volted gaps. Using small sphere gaps, formative times of 2.5×10^{-8} sec. were observed by WHITE[3], and shorter times of 5×10^{-9} sec. by WILSON[4]. These times were, however, still greater than the avalanche transit

[1] F. LLEWELLYN JONES: Brit. J. Appl. Phys. **1**, 60 (1950).

[2] R. SEELIGER and G. MIERDEL: Handbuch der Experimentalphysik, Vol. 13, Part 3, § 1.

[3] H. J. WHITE: Phys. Rev. **49**, 507 (1936).

[4] R. R. WILSON: Phys. Rev. **50**, 1082 (1936).

times estimated for the 1 mm. gap used with over-voltages of about 50%, assuming the drift velocity of electrons to[1] be $0.224 \times 10^7 \sqrt{E/1000}$ cm./sec. Later, NEWMAN[2] found times as low as 5×10^{-9} sec. with a 2.57 mm. sphere gap using an over-voltage of 120%, which time was about the same as the transit time. FLETCHER[3] observed even shorter formative times (measured from the application of the pulse of potential to the instant of collapse of the gap voltage) as low as 0.57×10^{-9} sec., which was much less than the estimated avalanche transit time of about 4×10^{-9} sec. with the 1 mm. gap used. This result has been taken to indicate that the TOWNSEND mechanism of ionization growth in an undistorted field could not produce collapse of the gap voltage in those cases, and that some different and faster mechanism was required in those cases of highly over-volted gaps[4,5]. To account for such times, FLETCHER adopted the streamer concept, in which the space charge field of the initial avalanche sets the criterion for the subsequent almost instantaneous complete ionization of the gap by some fast mechanism not quantitatively specified. The space charge field is assumed to attain the value V/d of the applied field in the gap (the criterion postulated by MEEK and by RAETHER) when the avalanche attains the critical size $N = \exp \alpha x$; and the formative time t is then practically given by

$$t = \frac{\ln N}{\alpha W_-}.$$

Using values of α given by SANDERS[6] and values of W_- estimated by RAETHER[1], the times were calculated in this way, and were found to be in agreement with those observed for the highest over-voltages when the times were less than about 10^{-8} sec., but not for the lower over-voltages.

Such an explanation, however, neglects to take into account the effect which the charge in the avalanche has upon the potential difference across the gap. This effect must now be considered. Because of the high over-voltage, a large avalanche can grow in the gap within the transit time, and the movement of this charge can lower the potential of the anode before any charge from the gap in fact reaches it. This fall in potential, recorded on an oscilloscope, could be interpreted as the breakdown of the gap occurring in a time less than the time taken for an avalanche to cross the gap. The rate of fall of potential V due to the movement of charges (positive and negative), such as $n_x e$ at x in the gap, is given by the following relation[7,8]

$$C \frac{dV}{dt} = -\frac{\sum n_x e \dot{x}}{d} + I_{\text{ext}} = -\frac{N_e W_-}{d} + I_{\text{ext}},$$

where e is the magnitude of the electronic charge, I_{ext} is the current in the external circuit, C the capacity of the gap, and N the total number of electrons between the plates. Using this relation, calculations for the case of the experimental arrangements used by FLETCHER have been made by DICKEY[7], who found that the calculated time intervals to the voltage collapse were about the same as those observed. There was thus no particular need to introduce

[1] H. RAETHER: Z. Physik **107**, 91 (1937).
[2] M. NEWMAN: Phys. Rev. **52**, 652 (1937).
[3] R. C. FLETCHER: Phys. Rev. **76**, 1501 (1949).
[4] W. KOHRMANN and H. RAETHER: Naturwiss. **17**, 400 (1954).
[5] J. M. MEEK and J. D. CRAGGS: Electrical Breakdown of Gases, Chap. 6, pp. 274–279.
[6] F. H. SANDERS: Phys. Rev. **44**, 102 (1933).
[7] F. R. DICKEY: J. Appl. Phys. **23**, 1336 (1952).
[8] K. J. SCHMIDT: Z. Physik **139**, 251 (1954).

some unspecified fast process (*kanal* or streamer) for these cases of highly over-volted gaps recorded by FLETCHER. DICKEY further considered that such streamer space charge action would not speed up the already extremely fast process of ionization growth. His calculations shows the experimental results are, in fact, satisfactorily accounted for by the growth of the avalanche in a practically undisturbed field.

Accurate assessment of the case when the over-voltage is lower ($\lesssim 20\%$), when the formative time (to voltage collapse) is of the order of a few estimated avalanche transit times requires more accurate values of the mobilites than are at present available. A theoretical expression for the (V, t) characteristic of an over-volted gap is given by the relation, neglecting the current in the external circuit,

$$V(t) = V_0 - \frac{1}{C} \int_0^t I(t)\, dt,$$

where $I(t)$ is the spatially averaged current in the gap at time t obtained from the expression for $I(x, t)$, given in equation (17.2). For times of the order of a few transit times the alternative form given by DAVIDSON[1] of that general expression for $I(x, t)$ is more convenient, as explained in Sect. 17β above.

δ) *Propagation of luminosity across the gap.* There are further interesting consequencies of the rapid spatio-temporal growth of ionization when $V > V_s$. The rate of excitation of the gas molecules depends on the electron density, and when this attains a certain value, say N, the resulting light emission associated with the growth of current then becomes sufficient to be recorded, either by eye or photographic plate. It is now interesting to see at what time, and where, this threshold density is attained in the gap; for this the growth equations (17.2) based on the action of primary α and secondary ω processes can be used.

Simple considerations show that a given concentration during the flash growth first appears in a position in the gap determined by the overvoltage, and that this value of concentration will be reached later at points nearer to the anode to which the avalanche, which is growing in time, is also travelling; then at subsequent times the given ions density N will appear at positions in the gap nearer and nearer the cathode. Thus the luminosity of the growth will appear as a growing filament which may be at first anode directed (depending on the overvoltage and the length of the gap), and later be cathode directed[2]. This is just what has been observed[3]. The *speed of transit of luminosity* is, on this view, determined by the *spatio-temporal growth of ionization in the gap* and not entirely by the transit speed of the primary avalanche. The theory indicates that extremely high speeds of luminosity transit can be obtained even with low overvoltages. Thus, the observed high speeds of luminosity transit are not inconsistent with the continuous growth of ionization by primary and secondary processes, the same processes which when $V < V_s$ lead to the well known *spatial* growth of steady pre-breakdown current.

The above considerations show that the growth of ionization up to the attainment of the breakdown criterion in uniform electric fields for values of the parameter pd up to about 1500 mm. Hg · cm. can be described in terms of a general theory which is based upon the action of both primary and various secondary ionization processes. Which secondary process predominates in any given case

[1] P. M. DAVIDSON: Phys. Rev. **99**, 1072 (1955).
[2] F. LLEWELLYN JONES: Brit. J. Appl. Phys. **5**, 49 (1954).
[3] W. KOHRMANN and H. RAETHER: Z. angew. Phys. **6**, 211 (1953).

depends, it has been shown, upon the nature and geometry of the electrodes and upon the value of parameter E/p. In this range of pd experimental data are consistent with the view that the breakdown criterion is set before the space charge of the ionization current becomes large enough significantly to distort the field. The same ionization processes also determine the early stages of current growth when V exceeds V_s; the field remains undistorted until the current increases sufficiently, at which stage the primary and secondary processes are enhanced and the current build-up accelerated so producing a voltage collapse. Further work is required in the investigation of long gaps and for higher values of pd, for which cases it is possible that the electric field can become non-uniform not only because of the geometry of the system but also because of the current. Further work is also required on the elucidation of the final discharge phase of the breakdown process including the problem of glow to arc transition when the breakdown flash acquires the properties of an arc[1].

[1] A. v. HIPPEL: Dielectrics and Waves. New York: J. Wiley & Co. 1953. — F. LLEWELLYN JONES: Electrical Discharges. Rep. Progr. Phys. **6**, 252 (1953). (London: The Physical Society.) M. J. DRUYVESTYN and F. M. PENNING: Rev. Mod. Phys. **12**, 140 (1940).

The Glow Discharge at Low Pressure.

By

GORDON FRANCIS.

With 96 figures.

I. Introduction.

1. The low pressure glow discharge is one of the most well known phenomena in physics, and has been closely studied for a century. For fifty years or more the results were almost entirely qualitative, and many interesting and complicated types of behaviour were discovered. During the last thirty or forty years we have learnt to describe quantitatively many of the atomic processes which cause ionization in gases, and the results have been applied successfully to explain some aspects of the glow discharge.

These are, however, still some important properties of these discharges which are not fully understood, notably in the cathode region, and in the striated positive column. The radiation from the discharge, which for many years was thought to be of interest only for practical applications and for spectroscopy, is now believed to play a vital part in maintaining the discharge itself.

This article deals mainly with discharges at pressures below a few millimetres of mercury, and surveys the literature from about 1920 to the middle of 1955. The earlier work can be found in the previous Handbook of GEIGER and SCHEEL. Properties which are treated in other articles, for example radiation, chemical effects and fundamental atomic processes, are mentioned only briefly, where it is essential for a proper understanding of the mechanism of the discharge.

II. General description.

2. Formation of a glow discharge. Consider a cylindrical glass tube with two plane electrodes at its ends filled with a gas at low pressure ($<$ about 10 mm. Hg): the shape of the vessel is specified only for convenience and will affect only certain details of the discussion but not the principal points. A d.c. potential is applied across the electrodes from a source in series with a variable resistance to control the current flowing through the circuit, which we measure with a sensitive instrument, e.g. a galvanometer. As the potential V across the electrodes is increased very slowly we observe first, if the instrument is sensitive enough, very small currents ($\sim 10^{-18}$ amps) occurring in random bursts: the size and frequency of these bursts increases with increasing voltage. A somewhat larger steady current, of about 10^{-12} or 10^{-13} amps would be observed if the cathode were illuminated with ultra violet light the current becoming larger with increased intensity of illumination. Removal of the light causes the current to revert to random bursts. Such currents are due to electrons multiplying in the gas by making ionising collisions with gas molecules; the original electrons come from ionisation by cosmic rays, giving random bursts of current, or when the light shines on the cathode, from photoelectric emission. These are shown in Fig. 1, region AB.

If now with the cathode still illuminated the potential V is further increased the current begins to increase rapidly, and within a certain very small range of voltage suddenly rises to a much larger value determined by the series resistance in the circuit. The current does not change if the light shining on the cathode is removed: it is independent of any external source of ionisation and this state of affairs constitutes a self sustained discharge. The lowest voltage at which this occurs is called the breakdown voltage V_B [1].

The explanation is as follows. Each electron leaving the cathode moves along the tube and by virtue of its random motion produces ions and also excited

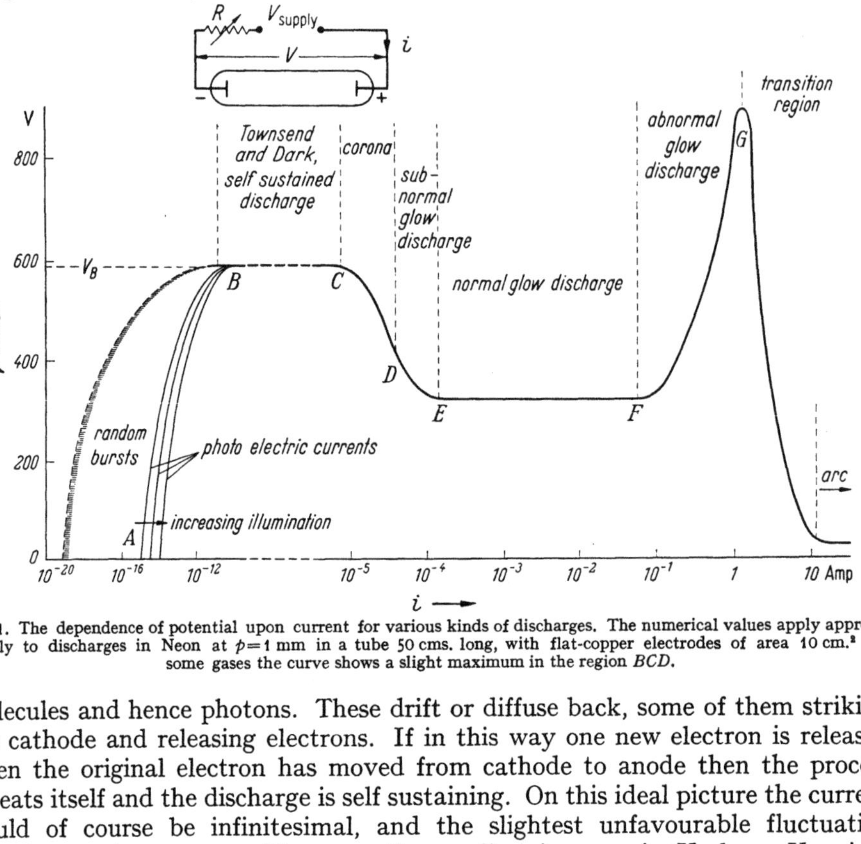

Fig. 1. The dependence of potential upon current for various kinds of discharges. The numerical values apply approximately to discharges in Neon at $p=1$ mm in a tube 50 cms. long, with flat-copper electrodes of area 10 cm.² In some gases the curve shows a slight maximum in the region *BCD*.

molecules and hence photons. These drift or diffuse back, some of them striking the cathode and releasing electrons. If in this way one new electron is released when the original electron has moved from cathode to anode then the process repeats itself and the discharge is self sustaining. On this ideal picture the current would of course be infinitesimal, and the slightest unfavourable fluctuation would cause it to cease. However, the smallest increase in V above V_B, gives increased multiplication in the gas and gives an enormously increased current, so that in practice we observe this abrupt change at a voltage slightly greater than V_B, but by an amount too small to measure.

Self sustained discharges of this kind (region BC, Fig. 1) are usually called *dark discharges*. By adjusting the external resistance the current can be allowed to rise to nearly 10^{-6} amps with hardly any change in V. These currents are so small that the light emitted and the space charges are negligible, the field being given only by V and the geometry. Although most of the charges flow to the electrodes, some, especially electrons, flow to the walls. Here their charge is only partially neutralised by a similar flow of positive ions to the walls, and there results a radial electric field. The wall charges act as an electrostatic lens and tend to make the field more uniform, and thus more concentrated near the cathode. The effect of wall charges and lateral motion of ions on the starting of a glow

[1] There are other definitions differing only slightly from this: see M. J. Druyvesteyn and F. M. Penning. Rev. Mod. Phys. **12**, 87 (1940)

discharge have been treated in detail, mainly by SEELIGER and OLLENDORF [277] and also the properties of a glow discharge in metal tubes[1,2].

If now the current is allowed to increase further (up to about 10^{-4} amps) two striking effects are observed. Firstly the discharge becomes faintly visible with light and dark spaces arranged along the tube in a characteristic manner, which will be described later. Secondly the potential across the electrodes drops considerably, until it reaches a constant value at a current of order 10^{-4} amps This region is shown as *CDE* in Fig. 1. The discharge is often intermittent here. Evidently space charges have now begun to play a part in the mechanism of the discharge, distorting the field along the tube. The fall in voltage shows that this rearrangement results in more efficient processes, making ionisation easier or reducing losses or both. The current is now determined not only by the external resistance but by the discharge as well. In these circumstances, when the properties of the discharge are first determined by space charges, a *glow discharge* is said to have been established.

In practice it is not necessary to go through all the process just described in order to obtain a glow discharge. If $V > V_B$ is applied across the electrodes the first random electron produced in a favourable place (near the cathode) will start a self sustained discharge and the whole process described by the curve $A \to B \to C \to D$ will take place in a fraction of a second, stopping finally when the current rises to the value satisfying

$$V + iR = V_{\text{supply}}. \tag{2.1}$$

When the cathode is intensely irradiated so that large photoelectric currents are produced the region *BCD* may be entirely absent, and even the properties of the glow discharge slightly changed, notably the cathode fall[3]. The starting potential is clearly considerably greater than the working potential: glow discharges are often difficult to start due to the presence of patches of charge on the walls of the vessel, perhaps remaining over from a previous discharge and prevented from dispersing by their mutual attractions or surface conditions. This gives rise to capricious behaviour in starting, although of course it does not affect the discharge once it has been established. When this happens the discharge may be encouraged to start by running a rod or the hand over the outside of the tube.

3. Current-voltage relationships: the normal, subnormal and abnormal discharge (see Fig. 1). When the glow discharge develops from the dark discharge as just described, the potential falls as the current is allowed to increase. This is the region of the *subnormal* discharge (*DE*).

If the current is allowed to increase further the voltage remains constant, even when the current varies over 2 to 3 orders of magnitude (e.g. 10^{-4} to about 0.1 amp). This is the region of the *normal* discharge (*EF*) and here we observe visibly that the discharge covers only a part of the cathode, the area covered being proportional to the current and the current density remaining constant. This is true whatever the shape of the cathode. For example if the cathode is rod-shaped then a discharge carrying a small current covers only the tip; with increasing current the discharge envelops an increasing length of the rod, the increase in the length covered being proportional to the increase in current.

When the whole cathode is covered and the current still further increased the potential across the discharge rises considerably: this is an *abnormal* discharge (*FG*). Still further increase in current causes the potential to rise to a maximum and at large currents to fall to a very low value; we have then an *arc* discharge.

[1] J. ZELENY: J. Franklin Inst. **226**, 35 (1938).

[2] G. MIERDEL and M. STEENBECK: Z. Physik **106**, 311 (1937).

[3] W. KLUGE: Z. angew. Phys. **5**, 173 (1953); also with A. SCHULZE, Z. Physik **142**, 83 (1955)

Here many processes which do not contribute to the glow discharge become important, notably heating of the gas.

The values given in Fig. 1 are typical for the kind of discharges ordinarily found, i.e. for tubes of diameter a few cm. and length about 50 cm., gas pressure ~1 mm. Hg. The distance between the electrodes affects only slightly the voltage required to pass a given current through the discharge provided it is not obstructed, although appreciably larger breakdown voltages are required in longer tubes. It is the current density rather than the current which determines the properties of the discharge, and so in wider tubes with large electrodes the currents in the regions *EFG* (Fig. 1) might all be increased, say by a factor 10.

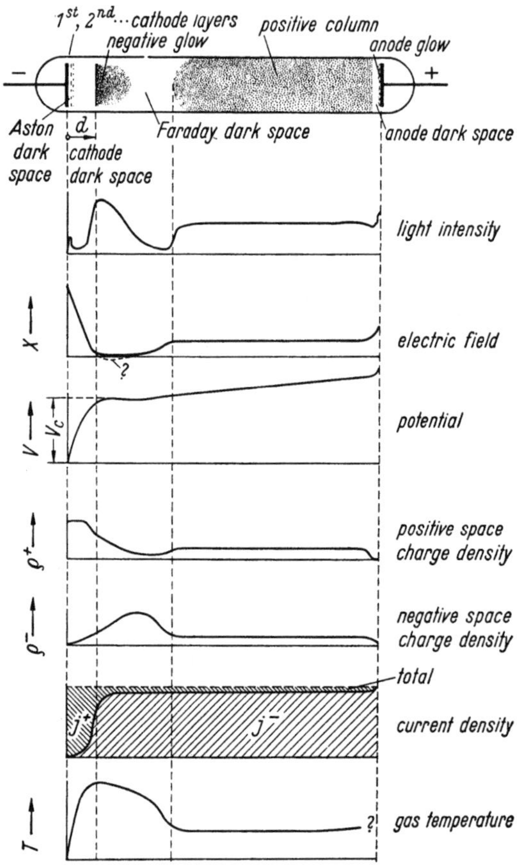

Fig. 2. Variation of discharge parameters along the length of the discharge (schematic).

4. Regions of the discharge. The glow discharge has a distinctive arrangement of luminous and dark regions between the cathode and anode. The actual position of the various regions, and indeed the occurrence of a few of them depends upon the kind and pressure of the gas, and upon the current, or the voltage across the discharge. Most regions are, however, common to all.

Consider as a typical example a normal discharge in a long cylindrical tube a few cm. wide, filled with neon, at a pressure of 1 mm., and carrying a small current, e.g. $\sim 10^{-4}$ A. Then starting from the cathode we have the following regions (see Fig. 2).

(1) The ASTON or primary dark space: a very thin totally dark layer bounded by: —

(2) The first cathode layer—a thin layer of feeble light.

(3) This pattern may be repeated, another very thin dark space followed be the second cathode layer etc. This will be very much fainter than the first, and the third cathode layer much fainter still so that rarely are more than two such layers seen.

(4) The cathode dark space (sometimes called the CROOKES or HITTORF dark space). This is usually taken to be the entire space between the cathode and the boundary of the next region, the negative glow, and thus it includes the cathode layers and dark spaces. Except for the ASTON dark space it is not totally dark, but appears so to the eye in comparison with the luminous parts of the discharge.

(5) The negative glow. This is the most luminous zone in the discharge. It has a sharp boundary with the cathode dark space, but is diffuse on its anode side. The light increases to a maximum near the middle of the glow and diminishes towards the anode where it merges into the FARADAY dark space. The

length of the glow is comparable with that of the cathode dark space: its colour is characteristic of the gas.

(6) The FARADAY dark space. This again is only relatively dark, and in pure neon does not exist at all the space being filled with red light (aureole). A small trace of impurity produces a dark space, however, and it seems to exist in all other gases. Its length depends on the radius of the tube, being large in wide tubes, and it is usually somewhat longer than the negative glow.

(7) The positive column is a long luminous column filling the rest of the tube almost to the anode. Its colour is characteristic of the gas but is not the same as that of the negative glow, nor is it as bright. It is often uniform, but sometimes, more especially in molecular gases, striated, i.e. it consists of a regular pattern of lighter and darker regions distributed along its length. The striations may be stationary or moving; sometimes they may move so fast that the column appears uniform to the eye. The boundary with the FARADAY dark space (called the "head" of the column) is diffuse.

(8) The anode dark space and glow. Often a thin dark space can be seen at the end of the positive column, and a glow on the surface of the anode.

The curvature of the boundaries between these regions is significant. Speaking always of the boundary on the cathode side we see that the boundaries of all the negative zones are parallel to the cathode surface, except perhaps near the walls where then are concave towards the cathode, while that of the positive column is convex. If we suppose that the shape of a boundary is connected with the local electric fields this suggests that near the cathode the walls have a positive charge and near the positive column a negative one.

Now we discuss how the appearance of the discharge changes under different conditions, as we vary one at a time the following factors: —

α) Pressure: An increase of pressure causes all the negative zones (cathode dark space, negative glow and FARADAY dark space) to be compressed towards the cathode. The positive column extends to fill remainder of the tube, except for the short anode region. As the pressure increases the compressed negative zones become progressively more difficult to separate visually, until when $p \sim 100$ mm. one can distinguish only the FARADAY dark space, the negative glow apparently being on the surface of the cathode. At about this pressure the positive column also contracts radially, no longer filling the whole width of the tube.

A decrease of pressure naturally causes the reverse effect: the negative zones expand and their boundaries become more diffuse. The positive column is driven into the anode and disappears altogether when the pressure is sufficiently low. Further reduction in pressure causes the FARADAY dark space and the negative glow successively to disappear into the anode. As the FARADAY dark space moves in, the anode glow suddenly disappears: as the last traces of the negative glow disappear the discharge goes out, unless the applied voltage is considerably increased. The discharge is then said to be *obstructed*. Further decrease in pressure causes the negative glow to disappear entirely and requires still further increase in potential to maintain the discharge. The distance between cathode and anode is now less than the normal length of the dark space. An extreme case of this is the *canal ray discharge* for which a very low pressure ($\sim 10^{-3}$ mm.) and very high voltages (~ 2 to 50 kV) are employed. Here the ions moving from anode to cathode have great energy and make very few collisions, thus possessing beam properties. If they are allowed to pass through a small hole in the cathode they emerge as a well defined pencil, called *canal rays*.

β) Distance between the electrodes: If the anode is moved further apart from the cathode a slightly larger voltage is required to maintain the discharge, and the positive column extends to occupy the additional length. There is no effect on the negative zones. Conversely as the anode is moved towards the cathode the positive column becomes shorter and slightly less voltage is needed. However, as the anode moves through the FARADAY dark space the anode glow disappears, accompanied by a slight fall (10 to 15 volts) in the voltage necessary to sustain the discharge; as it moves near the cathode edge of the negative glow the discharge becomes obstructed and the voltage necessary rises rapidly.

γ) Potential and current: If the potential is raised the length of the cathode dark space decreases, whereas the length of the negative glow increases slightly. There is also an increase in the current through the discharge and consequently a general increase in the brightness of the luminous parts, since more energy is now being dissipated in the gas and some of this appears as additional radiation. At higher potentials a new luminous layer, the cathode glow, appears on the surface of the cathode, considerably brighter and different in colour from the first and second cathode layers which are no longer visible. Also at higher potentials the boundaries of the light zones are much more sharply defined. An increase in the current causes only a general increase in luminosity when the discharge is normal (V = constant), except of course that the regions close to the cathode expand radially until the negative glow covers the whole cathode surface. In the abnormal discharge the increased potential causes the effects just described.

δ) The gas: Each gas gives a discharge of a characteristic colour, given in the table below:

Gas	First cathode layer	Negative glow	Positive column
Air	pink	blue	
Hydrogen	brownish-red	pale blue	pink
Nitrogen	pink	blue	red
Oxygen	red	yellowish-white	pale yellow with pink centre
Helium	red	green	red to violet
Argon	pink	dark blue	dark red
Neon	yellow	orange	brick red
Krypton		green	
Xenon		olive-green	
Bromine		yellowish-green	reddish
Chlorine		yellowish-green	whitish-green
Iodine		buff	reddish-blue
Lithium	red	light red	
Sodium	pink to orange	whitish	yellow
Potassium	green	pale blue	green
Rubidium	pink	blue	pinkish-red
Cesium	pink	milky green	yellowish-brown
Mercury	green	green	greenish
Calcium	violet-blue	violet-red	
Magnesium		green	green
Aluminium	violet-blue	violet-blue	
Thallium		green	whitish-blue
Cadmium		red	bluish green
Arsenic		bluish	greenish
Silver		pink	bluish-green
Lead		yellowish-red	violet
Zinc	violet-red	blue to violet	red
$SnCl_4$		green	sky blue
CCl_4		pale green	whitish-green
HCl		green	pink
NO		bluish-white	
NO_2		bluish-white	
NH_3		yellowish-green	

The length of the cathode dark space also varies with the gas; in gases which are easily ionised, i.e. have a large ionisation coefficient and low ionisation potential, the dark space is short.

ε) Material of the cathode. The cathode has a marked effect upon the voltage necessary to maintain the discharge. Lower voltages are needed when the cathode is a good emitter of electrons under bombardment by positive ions or photons. The colour of a certain cathode light, when it appears (see Sect. 28) depends on the cathode substance. There is sometimes a small change in the length of the cathode dark space but otherwise the cathode material does not influence the appearance of the discharge.

5. Field, charge, and current distribution[1]. The distributions of longitudinal electric field, positive and negative space charge and current density in a typical glow discharge are shown in Fig. 2. The important points to note are that the cathode dark space is a region of high field due to the intense positive space charge which lies in front of the cathode, and that most of the potential across the whole discharge appears across this dark space. The field is almost zero in the negative glow and FARADAY dark space, but is small and constant along the positive column.

6. Rays of charged particles in the discharge: shadow effects. The production of canal rays in a very low pressure discharge has already been explained: they are a beam of positive ions which have shot through a hole in the cathode into the space behind. There they can be detected by the faint light they produce in their path through the gas, or if the space is evacuated by the fluorescence produced when they hit the glass wall of the vessel. Since they are beams of moving charged particles, they can be bent by electric and magnetic fields, and were used in this way by Sir J. J. THOMSON in his classic experiments by which he measured the charge to mass ratio of the ions. This was the earliest mass spectrograph.

Similar beams of charged particles, electrons, move at high speed from cathode to anode: these are *cathode rays*. They have properties similar to canal rays, producing faint light in their path, and fluorescence when they hit the tube wall, and can be bent by electric and magnetic fields. It was this last property of cathode rays which led to the discovery of the electron. Any solid object placed in the path of a beam of cathode rays directed at the tube wall will cast a clear shadow on the fluorescent patch, showing that the rays travel in straight lines[2].

It is easy to understand how the electron and ion streams have these beam properties at such low pressures ($\sim 10^{-3}$ mm.), but experiments show that they persist in the cathode dark space up to quite high pressures. For example WEHNELT[3] found that a small obstacle placed in the cathode dark space cast a shadow both on the cathode and the negative glow. The shadow was rather larger than the obstacle suggesting either divergent beams or more probably some random motion of the particles. The experiments also suggested that the two beams are interdependent, since apparently no electrons leave the cathode from the shadowed area where no ions hit it. Further evidence of the beam nature of the streams of charged particles near the cathode can be found. If a plane cathode is moved about in a large spherical bulb the negative zones remain attached to the cathode, while the positive column fills the rest of the vessel to the anode.

---

[1] Wall charges are positive near the cathode, negative elsewhere [*94*], Sect. 44ff.

[2] Narrow constrictions in tubes also act as virtual cathodes, from which cathode rays emerge [*10*], p. 262.

[3] WEHNELT: Wied. Ann. **67**, 421 (1899).

7. General mechanism of the discharge. Electrons leave the cathode with little energy (a few eV) and are rapidly accelerated by the strong field. Within a very short distance they gain energy corresponding to the maximum of the excitation function, and the first cathode layer appears. Spectral lines of lowest energy lie nearest to the cathode, as would be expected. The layer is sharply defined, due to the beam like properties of the electron stream, and very thin because the electron energy rises rapidly with distance from the cathode. Electrons which have excited, however, and lost nearly all their energy in the collision are accelerated again and produce a much fainter second cathode layer. Further from the cathode the electrons are moving too fast to produce any appreciable excitation, but they make ionising collisions and the electrons so produced are themselves accelerated. The ionisation thus increases with distance from the cathode, while the electric field decreases almost to zero at the boundary of the glow.

Thus we have two main groups of electrons crossing the boundary into the negative glow; a large number of slow ones produced by ionisation in the cathode dark space, and rather fewer fast ones (the primaries) coming straight from the cathode and having almost the full energy of the cathode fall. In both groups their speed is now decreasing with distance. The slow electrons which are easily able to excite but hardly able to ionise produce the intense light which is the cathode end of the glow. Here spectral lines of greatest energy appear nearest the cathode (SEELIGER's rule). The fast electrons dissipate their energy similarly in many inelastic collisions but penetrate further from the cathode, so that the total light intensity in the glow rises to a maximum and then diminishes as we move away from its cathode edge. Photons and sometimes ions from the glow, and ions from the dark space hit the cathode and produce enough secondary electrons to sustain the discharge.

All beam properties are lost in the glow, the motion of the electrons becoming quite random. Probably little recombination occurs, except at the walls, all the light emitted being due to simple excitation. A net negative space charge develops due to an excess of electrons at the anode end of the glow and a small electric field extends from there to the anode.

The FARADAY dark space may be regarded as a modified repetition of the cathode dark space. Electrons are pulled out of the glow by this weak field and after moving a considerable distance gain sufficient *random* energy to excite and ionise; this point marks the head of the positive column. Again spectral lines of lowest energy appear nearest the cathode. The boundary is vague because the electrons have no beam properties, and also because they start from various points in the glow with various initial energies. The points where they gain ionising and exciting energy are thus distributed in space. There is also this fundamental difference between the two dark spaces: in the cathode dark space there is little excitation because the electrons move too fast, in the FARADAY dark space they are too slow. (In pure Ne the FARADAY dark space is filled with red light due to the diffusion of resonance radiation from the glow and also metastable atoms which can be excited by slow electrons.)

Conditions are usually uniform along the length of the positive column, except at the ends. The weak field sustains a small rate of ionisation due to the random motion of the electrons. The concentrations of ions and electrons ($\sim 10^{10}$ to 10^{13} per cm^3.) are equal, and the charges flow radially by an ambipolar process to the walls where they recombine at a rate equal to their rate of production. The current along the axis of the column is sustained by electrons which flow in from the FARADAY dark space, travel along with a small drift velocity superimposed on their random motion, and are collected at the anode. The attraction of electrons and the repulsion of positive ions by the anode results in a negative space charge and an enhanced field in front of its surface. This produces the anode fall in potential, and in this short region additional ionisation occurs so that ions are fed into the positive column in numbers just sufficient to balance those that flow out of the column into the FARADAY dark space.

The situation is more complicated in the striated column, although the same general principles apply. The few stationary striations which sometimes occur at the beginning of the column in rare gases are formed in a manner analagous to the cathode layers. The average gradients in striated and uniform columns are almost equal, but it is quite wrong to deduce that striations are merely small fluctuations in the uniform column. There are large variations both in electric field and light intensity.

III. Methods of measurement in discharges.

8. Current, potential, electric field. The total current through the discharge and the total potential between anode and cathode are measured with normal d.c. instruments, i.e. an ammeter or microammeter and a voltmeter. The distribution of current over the surface of plane cathodes and anodes has been measured by dividing them into concentric rings and measuring the current to each part [*11*], *12*], [*13*][1]. The current carried by ions to the cathode, and its radial distribution have been measured by placing a collector at negative potential behind a small movable hole drilled in the cathode [*14*] to [*17*]. Errors can arise here due to ionisation caused by the ion stream passing through the gas and by secondary electron emission when it hits the collector. Both errors have been eliminated by placing a grid at a suitable potential near the mouth of the collector.

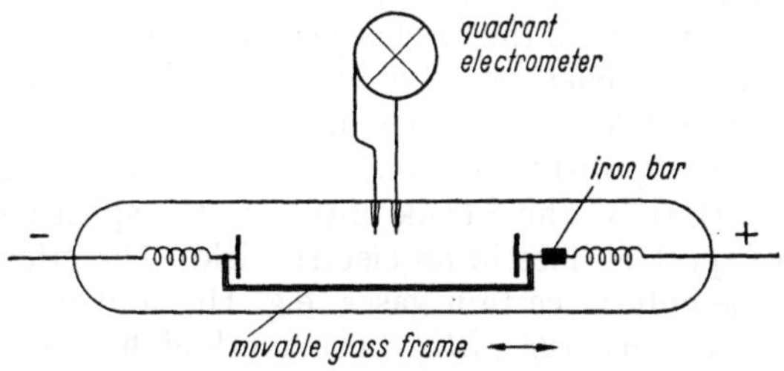

Fig. 3. Schematic diagram of H. A. WILSON's method of measuring field and potential difference. The electrode system is moved by an external magnet.

The potential at points in the discharge (space potential) is nearly always measured by using a LANGMUIR probe (see Sect. 9). In earlier work, before about 1925, the potential acquired by an isolated probe (thin wire) placed in the discharge was measured by electrostatic methods, and this was assumed to be equal to the space potential. These measurements were probably not more than a volt or two in error in regions like the positive column where there are large numbers of positive and negative charges: the probe acquires a small negative potential sufficient to equalise the random ion and electron currents flowing to it. Much less error arises from using two such probes to measure potential differences between near points in the same uniform region of the discharge, since the errors of each probe cancel[2,3] (Fig. 3). Large errors occur, however, when an isolated probe is placed in a region, e.g. cathode dark space where charges of

[1] See also H. A. WILSON: Phil. Mag. **4**, 608 (1902). — F. W. ASTON: Proc. Roy. Soc. Lond., Ser. A **79**, 80 (1907); **86**, 168 (1912).

[2] H. A. WILSON: Phil. Mag. **49**, 505 (1900).

[3] J. M. HOLM: Phil. Mag. **11**, 194 (1931). HOLM uses a condenser in place of the electrometer in Fig. 3, to avoid the disturbance of the leads. The charge on it is measured with a ballistic galvanometer. — WILSON and HOLM's papers are typical examples: many earlier references will be found in [*1*], [*2*], [*3*].

one sign predominate, especially if the charged particles travel fast: the probe, being bombarded by these charges gains a potential appreciably different from the adjacent space. The space potential can sometimes be inferred from optical properties, e.g. at the first cathode layer, but his requires some knowledge of the processes in the discharge.

The electric field is measured in a number of ways, namely:

(I) By the difference in space potential ΔV_s between two points Δx apart, measured as in Fig. 3 or more commonly now by LANGMUIR probes: $X = \Delta V_s / \Delta x$. This is the best method in regions of low field strength. Hot probes have also been used (see Sect. 11).

(II) By the deflection of an electron beam shot transversely across the discharge, a method suggested by J. J. THOMSON and first used by ASTON [*3*], [*18*] to [*22*], [*307*]. The deflection of the beam is proportional to the field, but due to edge effects the numerical relation is best derived by calibrating the deflection in a known uniform field. The contribution of edge effects has been considered theoretically[1]. This method has recently been improved in sensitivity and, with careful design at the places where the beam enters and leaves the discharge chamber, can give accurate results. Such electron beams, carrying very small currents, do not disturb the discharge and can be used at pressures up to about 1 mm. Hg, for beam energies of order 10^4 eV [*23*]. The method is applicable in relatively strong fields ($X > 30$ v/cm.) and has been used in the cathode dark space, in the striated positive column and in the anode region.

(III) By the STARK effect i.e. the splitting of atomic energy levels and hence of spectral lines in an electric field. The effect is appreciable and easily measurable only in certain gases, e.g. He, and in strong fields, that is in the cathode dark space [*24*], [*25*]. The method has been applied in these conditions and has given consistent results, except in narrow tubes where the presence of wall charges seriously distorts the shape of the field. In the simplest case the splitting $\delta\nu$ in a spectral line of frequency ν is given by $\delta\nu \propto X^2$.

(IV) By the mechanical force on an electrode (usually the cathode). This method gives only the field X_c at the cathode surface; the force per unit area $P = X_c^2/8\pi$. One has to correct for the momentum given to the cathode by the ions. The method has given X_c of the right order[2].

9. LANGMUIR probe measurements: space potential, electron concentration and energy. A particularly powerful method of using probes was introduced in 1923 by LANGMUIR and MOTT SMITH [*26*], and has since been developed and widely used[3]. The probe is suitable for use in regions of low field where the density of both positive and negative charges is relatively high ($N > 10^8$ per cm.^{3}); and the method consists in observing the current drawn to the probe as its potential is varied. The probe used in this way has been an indispensable tool for the study of discharges, but naturally it has limitations which were well known to those who introduced it but not always appreciated by later users. For this reason the results of probe measurements should often be regarded with caution. We give now only the simplest theory of the plane probe, to show the principle of the method. The plasma is assumed to contain only electrons and positive ions (i.e. no negative ions), in large and nearly equal concentrations.

---

[1] P. F. LITTLE: D. Phil. Thesis, Oxford 1952.

[2] W. BRAUNBECK: Z. Physik **21**, 204 (1924).

[3] There are many papers on probe theory, and references will be found in LOEB's book, p. 256, [*7*]. Here we quote only the recent papers and the more important older ones. See reviews by G. MIERDEL, Arch. tech. Messen **105**, V 31—7 (1940), and F. WENZL, Z. angew. Phys. **2**, 59 (1950).

The basic circuit is shown in Fig. 4: the probe is shielded on one side so that only one face of it is presented to the plasma. The area of the probe must be large compared with the mean free path of the ions and electrons. There are three potentials in the circuit namely that of the probe itself, the space just in front of the probe and some fixed reference potential, usually the anode. Current flows plasma → probe → anode → plasma (or the other way around). For the operation of the probe we shall refer to its potential with respect to the space potential e.g. a "negative probe" means $V_{\text{probe}} - V_{\text{space}} < 0$.

When the probe is strongly negative all electrons are repelled, and ions are attracted: a layer, or sheath, of ions forms on the surface of the probe and can sometimes be observed as a dark zone separating it from the plasma. The current to the probe, consisting entirely of positive ions is limited by their space charge, and is constant. The current density is given by:

$$\left.\begin{aligned} j_+ &= \frac{1}{9\pi}\sqrt{\frac{2e}{M}}\,\frac{V^{\frac{3}{2}}}{d^2} \\ &= \text{constant} \end{aligned}\right\} \tag{9.1}$$

where d is the thickness of the sheath, V the potential across it ($= V_{\text{space}} - V_{\text{probe}}$), and M the mass of the ions. The thickness d of the sheath is an important quantity whose relation to the electron mean free path λ_e determines the limit of application of the method.

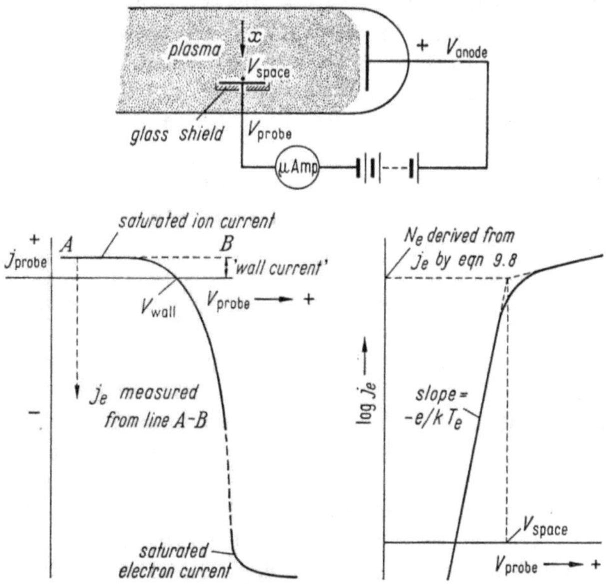

Fig. 4. Basic circuit and ideal current-voltage curves of the plane probe.

We assume that the electrons in the surrounding space are distributed in velocity and move in all directions. If the probe is now made less negative some of the faster electrons can reach the probe surface, against the retarding potential V across the sheath—namely those which have a velocity v_x in the x direction, perpendicular to the probe such that

$$\tfrac{1}{2} m v_x^2 \geq e V. \tag{9.2}$$

An electron current as well as an ion current thus flows to the probe, reducing the total current, which now varies with the probe potential. Provided the probe is always negative i.e. draws a constant ion current, the velocity distribution of the electrons can in principle be derived from the variation of the probe current with potential. This variation has a special form when the distribution is MAXWELLian, i.e. for the x component only:

$$\frac{dN_{v_x \to v_x + dv_x}}{N} = f(v_x)\,dv_x = \frac{1}{\sqrt{\pi}\,w}\cdot e^{-\frac{v_x^2}{w^2}}\cdot dv_x \tag{9.3}$$

where w is the most probable velocity. We define an electron temperature T_e in the usual way[1]

$$k T_e = \tfrac{2}{3}\cdot\left(\tfrac{1}{2} m \bar{v}^2\right) = \tfrac{1}{2} m w^2 = \tfrac{2}{3} e \bar{V} \tag{9.4}$$

[1] Note that the concept of temperature applies *only* to a MAXWELLian distribution: every distribution has, however, a mean energy.

where $\overline{v^2}$ is the mean square velocity, $e\overline{V}$ is the mean energy. Then it can easily be shown that provided the probe is always negative the electron current density j_e is

$$j_e = e N_e \sqrt{\frac{k T_e}{2\pi m}} \cdot e^{-\frac{eV}{kT_e}}, \tag{9.5}$$

$$\log j_e = \text{const} - \frac{eV}{kT_e} \tag{9.6}$$

hence

$$\frac{d}{dV}(\log j_e) = \frac{d}{dV_{\text{probe}}}(\log j_e) = -\frac{e}{kT_e}. \tag{9.7}$$

The logarithm of the probe current density plotted against the probe potential gives a straight line in this section, whose slope gives T_e. The effective area of the probe must be known; for the plane probe this is nearly equal to the area of the probe itself, but for cylindrical and spherical probes the thickness of the sheath increases the effective area, and this has been tabulated for probes of various radii [*4*]. Equation (9.7) above applies equally to these probes, except that we use i_e the current per unit length for the cylinder. Fig. 4 illustrates the curves obtained; j_e must be calculated by subtracting the constant ion component j_+ from the total current.

At a certain potential, when the probe is still negative, the total current to it is zero, the ion and electron components being equal. This is called the *wall potential*, or *floating potential* because it is that potential which an insulated wall would acquire in order to equalise the currents of ions and electrons moving to it by their own diffusion. The current due to each component is then given by equation (9.1), and this is sometimes loosely termed the "wall current".

Now if the probe potential is made equal to the space potential the sheath of positive ions disappears, and at more positive potentials is replaced by a much thicker sheath of electrons. The positive ion current drops quickly to zero leaving only a saturated electron current to the probe: in this region the $\log j_e - V$ curve shows a marked change in slope and the point $V_{\text{probe}} = V_{\text{space}}$ is found from the intersection of the extrapolated linear parts (Fig. 4). Then $V = 0$, and from eq. (9.5)

$$j_e = e N_e \sqrt{\frac{k T_e}{2\pi m}}. \tag{9.8}$$

T_e is already known from the slope of the curve, hence the electron concentration N_e is found.

Attempts have been made, assuming $N^+ \approx N_e$ in the plasma, to calculate the ion temperature T_+ by equating the saturated ion current density j_+, eq. (9.1), to the random flow $\frac{1}{4} N^+ \bar{v}_+$ from the plasma into the sheath[1]. These have given absurdly high values of T_+, which on general grounds should not be much different from the gas temperature. The reason is that, in forming the sheath, the field of the probe alters the energy distribution of the ions until they satisfy the condition[2]:

$$\frac{2\int\limits_0^\infty \sqrt{V} f(V) \cdot dV}{\int\limits_0^\infty \frac{1}{\sqrt{V}} \cdot f(V)\, dV} > \frac{k T_e}{e}. \tag{9.9}$$

where eV is the ion energy.

---

[1] I. Langmuir and L. Tonks: Phys. Rev. **34**, 876 (1929).

[2] R. L. F. Boyd: Nature **165**, 228 (1950); Proc. Roy. Soc. Lond., Ser. A **201**, 329 (1950), who refers to earlier calculations by Langmuir.

10. Limitations of probe measurements, and disturbing factors. The probe must cause the least possible disturbance to the discharge; the space occupied by its sheath and the current which it draws must be small compared with the volume and current of the discharge itself. These conditions give lower limits to the gas pressures, ion densities and discharge currents in which it may be used, which are about $p \sim 10^{-4}$ mm.; N^+, $N_e > 10^8$ per cm.^{3}, $j \sim 10^{-7}$ to 10^{-6} A/cm.^{2}. The theory of the probe also assumes that electrons entering the ion sheath reach the probe without making collisions with gas molecules: for this to be true the gas pressure must not be greater than about 1 mm. A non-MAXWELLian electron energy distribution gives a non linear $V - \log j_e$ curve from which no temperature can be deduced (see also Sect. 11), but groups of electrons each having a MAXWELLian distribution with a different mean energy will give several linear portions on the curve[1].

Apart from the inevitable disturbance of the plasma itself by the presence of the probe, which can never be entirely eliminated many effects can occur which falsify the results obtained by probe measurements. Excited gas molecules, especially metastables, photons, and fast electrons all cause secondary electrons to be emitted from the probe surface; electrons and ions are also reflected from the surface and all these effects give probe currents which are not described by the equations of Sect. 9. Adsorbed gas layers cause contact potentials such that the actual potential of the collecting surface is different from that applied to the probe from the circuit; chemical reactions (e.g. in Hg discharges) and impurities on the surface resulting in a changing or inhomogeneous work function have a similar effect.

Changes in the probe surface were first studied systematically by DÜRRWANG[2], and by VAN BERKEL[3], the latter noting that the probe curves depended on the rate at which observations were made. He deduced that the work function φ changed with time t according to

$$\varphi_0 - \varphi_t \propto e^{-\frac{t}{\tau}}.$$

The effect could be eliminated by applying a large alternating potential to the probe, and displaying the $V - i$ curve on an oscilloscope; by increasing the frequency of the potential the whole curve could be plotted too quickly for φ to change. Irregular readings due to the contamination of probe surfaces by adsorbed oxygen layers had also been noticed by several workers[4,5]. SPENCER-SMITH [*34*] measured the change in surface resistance of a probe as its potential was altered. Recently other workers have found similar results. In Hg discharges the probe curves obtained can depend on whether they are measured with an increasing or decreasing potential. The reason is that when the probe is strongly positive a layer of amalgam is destroyed by electron bombardment; readings taken soon afterwards before the layer reforms are reliable. Deposits of evaporated material from the cathode (especially oxide coated cathodes) have a strong influence on probe measurements. Various authors[6,7,8] suggest that probe characteristics should be measured only after the surface has been cleaned by

---

[1] See for example: N. L. OLESON and C. G. FOUND: J. Appl. Phys. **20**, 416 (1949); and references [*31*], [*81*], [*115*] to [*118*], also Sect. 39.

[2] J. DÜRRWANG: Helv. phys. Acta **8**, 333 (1935).

[3] W. P. J. VAN BERKEL: Physica, Haag **5**, 230 (1938).

[4] E. S. LAMAR and K. T. COMPTON: Phys. Rev. **9**, 1069 (1931).

[5] A. L. REIMANN: Phil. Mag. **20**, 594 (1935).

[6] T. A. ANDERSON: Phil. Mag. **38**, 179 (1947).

[7] M. A. EASLEY: J. Appl. Phys. **22**, 590 (1951).

[8] G. WEHNER and G. MEDICUS: J. Appl. Phys. **23**, 1035 (1952).

heating, or electron or ion bombardment, and have devised electronic methods for plotting the curves quickly. Results obtained in this way do not agree with some earlier measurements. Heating the probe may, however, disturb the local gas density.

When small cylindrical probes are used consisting of a short length of wire protruding from an insulating stem the ion sheath formed around the stem covers part of the probe itself, and seriously reduces its effective area, which can, however, be deduced by surrounding the probe with a small metal tube whose potential is varied[1].

The presence of plasma oscillations causes electrons periodically to acquire high drift velocities, and changes their energy distribution. Probe curves are then seriously disturbed.

11. Extensions of probe theory and technique. *α) Energy distribution.* When the electron velocity distribution is not MAXWELLian the probe characteristic (i.e. $V - \log j_e$ curve) is nonlinear. The actual velocity distribution $f(v)$ can be found from its curvature (if it is certain that this is not due to any of the disturbing factors mentioned in Sect. 10). From a relation due to DRUYVESTEYN [27]:

$$\frac{1}{N}\frac{dN}{dv} = f(v) = f\left(\sqrt{\frac{2eV}{m}}\right) = \frac{4m}{Ae^2}\,V\,\frac{d^2 i}{dV^2} \tag{11.1}$$

where A is the area of the probe, $V = V_{\text{space}} - V_{\text{probe}}$, i is the total current to the probe, and dN the number of electrons in a velocity internal dv.

In terms of energy, expressed in volts:

$$\frac{1}{N}\frac{dN}{dV} = F(V) = \frac{1}{A}\sqrt{\frac{8m}{e^3}}\cdot\sqrt{V}\,\frac{d^2 i}{dV^2}\,. \tag{11.2}$$

The second derivative d^2i/dV^2 can be found in several ways:

(I) By graphical differentiation, performed twice, starting from the $i - V$ curve, a laborious and inaccurate procedure.

(II) By superimposing a small alternating potential $\delta V \sin \omega t$ on to the probe potential. If $i = f(V)$ for normal d.c. probe measurements, then $i = f(V + \delta V \sin\omega t)$ when the oscillation is superimposed, and an expansion of this expression by TAYLOR's theorem shows that an additional d.c. current δi flows given by

$$\delta i \approx \left(\frac{\delta V}{2}\right)^2 \cdot \frac{d^2 i}{dV^2} \quad \text{for small } \delta V. \tag{11.3}$$

Hence by measuring the increase in d.c. current when the a.c. is superimposed d^2i/dV^2 can be evaluated directly, at any chosen value of probe potential[2].

(III) By connecting the probe in a bridge circuit, devised by VAN GORCUM, and using a small a.c. voltage superimposed. The method eliminates one graphical differentiation and increases the accuracy of the remaining one [28].

(VI) By applying a saw-tooth voltage wave form to the probe making $V \propto t$, so that $\frac{d^2 i}{dV^2} \propto \frac{d^2 i}{dt^2}$; with suitable circuits the second derivative can be displayed directly on an oscilloscope[3]. Recently BOYD and TWIDDY [29] have developed an electronic method of multiplying $\sqrt{V}$ and d^2i/dV^2 and displaying $F(V)$ directly on an oscilloscope.

---

[1] M. FIEBICH: Acta Phys. Austriaca **4**, 170 (1950).

[2] SLOANE and MCGREGOR: Phil. Mag. **18**, 193 (1934). See also LANDALE Proc. Cambridge Phil. Soc. **25**, 355 (1929).

[3] KAGAN, FEDOROV, MALYOKEV and GAVALLAS: Dokl. Acad. Nauk SSSR. **76**, 215 (1951).

β) Probes in special conditions (high pressure, magnetic field etc.). Probes of general size have been discussed by DAVYDOV and ZIMANOVSKAYA[1]; and the use of probes at rather higher pressures, when electron collisions occuring within the sheath affect the current collected, by BORGNIS[2]. It is known that when a probe is used in a magnetic field there is no linear part in the characteristic. The field causes the electrons to move in spiral paths so that they no longer move straight to the probe surface when they have crossed the boundary of the sheath: a theory has been given by SPIWAK and REICHRUDEL[3], and very recently by BICKERTON and VON ENGEL [*308*]. WEIZEL and OLMESDAHL have considered the interpretation of the curves obtained when a probe is placed in a plasma and at the same time exposed to a beam of electrons [*30*].

EMELEUS and his associates quite early found an anomalous kink in the $V-i$ curves of probes placed in a positive space charge, whose presence can therefore be detected by this peculiarity [*31*].

γ) Special types of probes. The *hot probe* (or electron emitting probe) was developed by LANGMUIR: the probe is heated to incandescence and its potential measured when it begins to emit electrons [*32*]. Its presence changes the gas density locally[4].

Screened probes have been developed recently by BOYD [*33*], although earlie attempts were made by LANGMUIR and SENA. The probe itself is a fine wire grid, behind which is placed a plane collector which can be biased positively or negatively with respect to the grid. In this way the electron and ion components of the probe current have been separated, and residual currents due to metastable atoms etc. identified. The electron energy distribution can also be measured directly for relatively high energies (eg. $>15\ eV$).

Mass spectrometer probes have been used to identify the types of ion occurring in a discharge, particularly in electronegative gases. The ions are drawn to the probe, pass through a small hole, and are accelerated by an electric field. Their masses (strictly e/M) are then analysed either by a magnetic field [*34*], [*35*], [*36*], or by a series of electrical shutters opened and closed by a radio frequency electric field [*37*].

The constant current probe[5] is a device used to investigate sudden or transient changes in a plasma. Two identical floating probes (i.e. not connected to any fixed potential) are placed at points having the same space potential. They are made to operate at different fixed values of current, kept constant by a pentode circuit, but both on the linear part of their respective characteristics. Then the difference $V_1 - V_2$ between the potentials they acquire is proportional to T_e. With an oscillograph circuit a continuous record of T_e as a function of time can be obtained, and a modified method of use gives also the variation of the field in the discharge.

The floating double probe[6] consists of two identical probes placed a short distance apart in the plasma: a small variable and reversible potential V_d is applied *between* them, but the whole system is unconnected to any fixed source of potential. One measures the current flowing in the circuit as the differential voltage V_d is varied. The theory is rather different from that of the LANGMUIR

[1] B. DAVYDOV and L. ZIMANOVSKAYA: J. techn. Phys. USSR. **3**, 715 (1936).

[2] F. BORGNIS: Ann. Phys. **31**, 497 (1938).

[3] G. SPIWAK and E. REICHRUDEL: J. techn. Phys. USSR. **5**, 715 (1938).

[4] The development of an ion layer around an incandescent probe is treated by V. POLIN and S. GVOSDOVER: Phys. Z. Sowjet. **13**, 47 (1938).

[5] B. T. BARNES and S. EROS: J. Appl. Phys. **21**, 1275 (1950).

[6] E. O. JOHNSON and L. MALTER: Phys. Rev. **80**, 58 (1950).

probe, and depends on KIRCHHOFF'S current law which requires that the total net current $i_e + i^+$ flowing to the circuit from the plasma is zero. This arrangement has two great virtues:

(I) The total current drawn from the discharge cannot exceed the ion current to one probe (this is about $\frac{1}{100}$ the current drawn by a conventional probe).

(II) Since its potential is floating it can be used to measure the space potential of a decaying plasma; this is not possible with a conventional probe which would itself affect the plasma potential.

12. Microwave measurements of electron concentration. It is well known from the theory of radio propagation in the ionosphere that space containing N electrons per cm.^{3} (we omit here the suffix in N_e for convenience) behaves as a medium of dielectric constant

$$K = 1 - \frac{4\pi N e^2}{m \omega^2} \tag{12.1}$$

to an electric field of angular frequency ω. MARGENAU and others have shown that when electrons have a MAXWELLian energy distribution in a gas at low pressure, the imaginary part of the conductivity σ_i is

$$\sigma_i = -\frac{N e^2}{m\omega}\left[1 - k_1\left(\frac{\nu_c}{\omega}\right)^2\right] \tag{12.2}$$

provided $\omega > \nu_c$, the collision frequency of the electrons. k_1 is a numerical constant.

These conditions are usually satisfied in discharges, and the term $k_1(\nu_c/\omega)^2$ is only a correction factor[1].

It is clear that N can be measured by the effect the ionised medium has on the passage of an electric field of suitable frequency ω. This principle was first applied in 1919 by VAN DER POL[2], who placed a discharge between LECHER wires, and compared its conductivity with that of an electrolytic solution. Modern methods of producing very high frequencies have facilitated more accurate measurements. In the first recent experiments the effect of terminating a coaxial line in a region of a discharge was observed [*38*], [*39*], but this did not give very satisfactory results. In later improved methods, N has been determined from:

(I) Transmission or reflection (i.e. absorption and phase shift) of the wave in a wave guide, part of which is filled with the discharge [*40*], [*41*].

(II) The change $\Delta\omega_0$ in resonant frequency ω_0 of a cavity when a discharge tube is inserted [*42*], [*43*], [*44*][3]

$$\frac{\Delta\omega_0}{\omega_0} = \frac{2\pi N e^2}{m\omega^2} \times G \text{ (geometrical factor)}. \tag{12.3}$$

It is a fundamental assumption of these measurements that N is constant over the region explored: the results give only $\overline{N}$. The method is applicable for $N = 10^6$ to 10^{10} electrons/cm.^{3}, and has given quite accurate results in good agreement with probe measurements. It has the great advantage that no solid object has to be introduced into the discharge.

In some measurements the effect of a localised microwave field on the discharge current has been measured, but it is not known what the results mean[4].

---

[1] J. SLATER: Rev. Mod. Phys. **18**, 473 (1946). — H. MARGENAU: Phys. Rev. **69**, 508 (1946). — ROSE, KERR, BIONDI, EVERHART and BROWN: M.I.T. Techn. Report 140, Oct. 1949. — M. A. BIONDI: Rev. Sci. Instrum. **22**, 500 (1951). More references may be found here.

[2] B. VAN DER POL: Phil. Mag. **38**, 352 (1919).

[3] See also W. ECKHARDT, R. HONERJÄGER and E. SCHULZ-DU BOIS: Z. angew. Phys. **6**, 246 (1954).

[4] M. A. LAMPERT and A. D. WHITE: Electr. Comm. **30**, 124 (1953).

13. Electron temperature from electrical noise measurements. It was observed several years ago that discharges produce noise in the microwave region. If it is assumed that the power radiated in a frequency interval Δf is $kT_e \cdot \Delta f$, the values of T_e calculated from power measurements agree well with those found by probes[1,2]. Several empirical laws analogous to the laws of black body radiation, in particular WIEN's displacement law, have been noted[3], and there is evidence that the discharge emits radiation like a black body at temperature T_N ("noise temperature"). Measurements by EASLEY and MUMFORD[4] in Hg–A discharges show that T_N and T_e vary in the same way with changes in discharge current and pressure. PARZEN and GOLDSTEIN[5] find that the noise power can be accounted for by fluctuations in the electron current in the plasma. By calculating the probability that electrons of a given velocity will have a certain number of collisions per second they calculate the power P_f radiated in the frequency interval $f, f+df$:

$$P_f = \left[kT_e + \frac{P_0}{N\nu}\left(2 + \frac{\nu^2 - 4\pi^2 f^2}{\nu^2 + 4\pi^2 f^2}\right)\right] df \qquad (13.1)$$

where P_0 is the total power dissipated in the discharge, ν the collision frequency of the electrons. Usually the second term is small and hence $T_e \approx T_N$.

14. Measurements of gas temperature. The gas temperature influences the discharge only through the associated change in gas density: there are no processes which depend on the actual thermal motion of the gas atoms. Although we refer always to the gas pressure (e.g. in factors like j/p^2, pd) we really mean the number of atoms per cm.^{3} i.e. the gas density, for it is this factor which controls the number of collisions. Gas temperatures are usually low in glow discharges (although it was at one time believed that the light emitted was due to incandescence of the gas[6]): appreciable increase of temperature occurs only in strongly abnormal discharges (large V_c) carrying large currents ($\sim$1 A); at high pressure, and in the positive column of discharges in narrow tubes.

For the general methods of temperature measurement in gases we refer to earlier reviews[7,8], and mention here only those commonly used. The temperature is deduced by measuring:

(I) The temperature of a thin thermocouple or bolometer wire, inserted into the discharge. This is determined by the thermal equilibrium between the wire, the discharge and the walls, and in general will not be equal to the temperature of the gas. However, for small temperature differences, low gas pressures and very thin wires, the discrepancy is small. Allowance must be made for the energy delivered to the wire by currents flowing to it from the discharge.

(II) The change in gas pressure. A simple application of the gas laws $pV = nRT$ is necessary. When the discharge occupies the entire volume of the vessel, except for the negligible volume of the pressure gauge $T_1/T_0 = p_1/p_0$ where p_0 and T_0 are the initial values. If the discharge occupies a volume V_1, the remaining (unheated) volume being V_2, then

$$\frac{p_0(V_1 + V_2)}{T_0} = \frac{p_1 V_1}{T_1} + \frac{p_1 V_2}{T_0}. \qquad (14.1)$$

This gives only an average temperature in the whole discharge.

[1] G. C. SOUTHWORTH: J. Franklin Inst. **239**, 285 (1945).
[2] L. GOLDSTEIN and N. L. COHEN: Phys. Rev. **73**, 83 (1948).
[3] K. S. KNOL: Philips Res. Rep. **6**, 288 (1951).
[4] M. A. EASLEY and W. W. MUMFORD: J. Appl. Phys. **22**, 846 (1951).
[5] P. PARZEN and L. GOLDSTEIN: Phys. Rev. **79**, 190 (1950).
[6] See for example R. W. WOOD: Wied. Ann. **59**, 238 (1896).
[7] A. v. ENGEL and M. STEENBECK: Arch. techn. Messen **5**, 2163—1 (1933).
[8] M. PIRANI and R. ROMPE: Trans. Amer. Electrochem. Soc. **69**, 417 (1936).

(III) The change in gas density ϱ, usually by an optical method making use of the associated change in refractive index μ. For relatively small increases in temperature we assume $\delta\mu \propto \delta\varrho$. An interference method is usually used, the discharge tube being inserted in the path of one beam of a suitable interferometer (e.g. JAMIN's). The movement of fringes is counted when the discharge is switched on. This is a most sensitive method, and the shape of the fringes can be made to indicate directly the radial temperature distribution [*45*].

Absorption methods (e.g. absorption of X-rays and α-rays) have also been used to estimate a change in density.

The temperature is deduced from simple kinetic theory, since $p = \frac{1}{3}\varrho\,\overline{c^2} \propto \varrho\, T$

$$\frac{\varrho_1}{\varrho_0} = \frac{T_0}{T_1}. \tag{14.2}$$

(IV) Optical effects due to the actual motion of the gas atoms or molecules. The DOPPLER effect causes a broadening of any observed spectrum line, from which the speed of the atom and hence the gas temperature can be deduced. The relation is not simple because the width of a line depends on a number of other factors. Rotational band spectra have been observed in molecular gases, and the temperature deduced from the relative intensity of the various bands. It is assumed that molecules are excited only by collisions between themselves, and that the concentration of excited states of energy eV obeys the BOLTZMANN relation

$$n = n_0\, e^{-\frac{eV}{kT}} \tag{14.3}$$

where T is the gas temperature. It is, however, a matter of controversy whether the band spectra are excited by electron collisions, when at least some of the levels would be distributed according to $\exp(-eV/kT_e)$ where T_e is the electron temperature. Hence too high a gas temperature would be deduced (see [*45*]).

IV. Similarity laws.

15. The purpose of similarity relations. Any chosen property of a discharge, for example the potential V, depends on a number of others, the current, gas pressure, size of the vessel and electrodes. The dependence is often complicated, and sometimes cannot be expressed analytically at all but only in the form of curves, whether theoretical or experimental, linking various parameters. Clearly it would require an infinite number of such curves to describe all possible states of the discharge given by every combination of dimensions, i, p etc. We look therefore for some simplification by grouping together these parameters. To give an example: The starting potential V_s of a gas between plane parallel electrodes depends both on their distance d apart, and on the pressure p. It is found experimentally that V_s does not change if the product $p\,d$ is kept constant (except at the highest pressure) i.e. $V_s = f(p\,d)$. $p\,d$ is thus one of the combinations we look for, because for the purpose of discussing breakdown it describes completely the behaviour of the gas. Any particular $V_s = f(p\,d)$ refers of course only to one combination of gas and electrode material. The importance of such grouping was first realised by DE LA RUE and MÜLLER[1], later by PASCHEN[2], TOWNSEND[3] and others who noted that $X\lambda$ represents the energy

[1] W. DE LA RUE and W. H. MÜLLER: Phil. Trans. Roy. Soc. Lond. **171**, 65 (1880).
[2] F. PASCHEN: Wied. Ann. **33**, 69 (1889).
[3] J. S. TOWNSEND: Electrician **71**, 348 (1913).

gained by an electron moving along a free path λ, and $p \cdot d$ is proportional to the total number of molecules between two electrodes of given size. If $X\lambda$ (or X/p) and pd are kept constant the multiplication of electrons across the gap is thus fixed. These groups of parameters and others, can be deduced more generally as shown below, partly following the treatments of HOLM[1], VON ENGEL and STEENBECK[2] [*4*], and DÄLLENBACH[3].

Similarity relations are not confined to gas discharges but can be found in many branches of physics. For example, in hydrodynamics the transition from laminar to turbulent flow is governed by the parameter $\eta/\varrho a$ where η is the viscosity, ϱ the density of the liquid, a the radius of the tube: the period of a magnet oscillating in a magnetic field is governed by I/MH. Thus $\eta/\varrho a$, I/MH may be regarded as "similarity" parameters.

16. Derivation of the relations. *α) General.* The object of the treatment is to define two discharges identical in only one or two chosen properties and different in others, and to derive groups of parameters which are invariant and can therefore be used to describe those properties identical in both discharges.

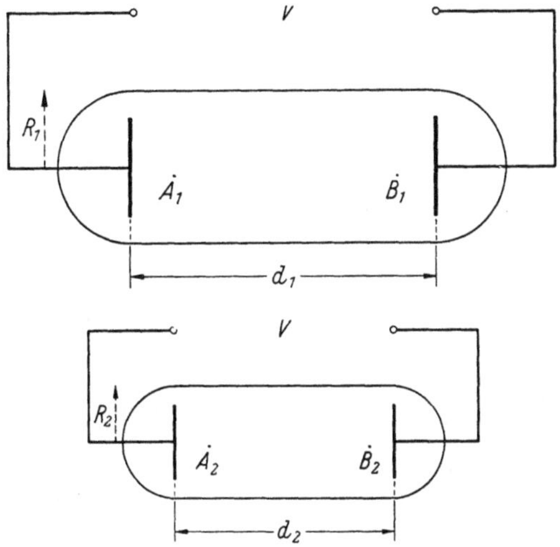

Fig. 5. Similar Discharges.

Consider two discharges in the same gas with the same electrode material, in which all the corresponding linear dimensions are different by a factor a (see Fig. 5). This includes the vessel, electrodes and the properties of the gas, e.g. mean distance between molecules. We assume that the gas temperature is the same in both, and that corresponding points (e.g. A_1, A_2; B_1, B_2) have the same potential difference between them; the potential across the electrodes is V. Such discharges are called similar[4].

Then the following relations hold for corresponding points in the two discharges:

For the linear dimensions of vessel and electrodes:

$$\left.\begin{aligned} d_1 &= a\, d_2, \\ r_1 &= a\, r_2, \\ R_1 &= a\, R_2, \end{aligned}\right\} \tag{16.1}$$

where r is any radially measured distance.

For elements of area:

$$d A_1 = a^2\, d A_2. \tag{16.1a}$$

[1] R. HOLM: Phys. Z. **25**, 497 (1924); **26**, 412 (1925).

[2] M. STEENBECK: Wiss. Veröff. Siemens-Werk **11** (2), 36 (1932). — The most extensive account.

[3] W. DÄLLENBACH: Phys. Z. **26**, 483 (1925).

[4] Sometimes it is defined that the currents also shall be the same at corresponding points in both discharges. This additional assumption can be partly avoided, and is discussed later. For this reason the presentation given here is somewhat different from the usual.

For the mean free paths of electrons, ions or molecules in the gas:

$$\lambda_1 = a\,\lambda_2. \tag{16.2}$$

Hence the gas density (number of molecules per cm.[3])

$$n_1 = \frac{n_2}{a}. \tag{16.3}$$

Since by definition the temperature is the same in both, pressure p is $\propto n$, hence for the pressure

$$p_1 = \frac{p_2}{a}. \tag{16.4}$$

The potential

$$V_1 = V_2, \quad \text{by definition.} \tag{16.5}$$

The electric field $X_x = -\frac{\partial V}{\partial x}$.

Hence

$$X_1 = \frac{X_2}{a}. \tag{16.6}$$

Surface and volume charge densities σ, ϱ_{net} can be derived from

$$\left.\begin{aligned} X_{\text{surface}} &= 4\pi\,\sigma: \\ \frac{dX}{dx} &= 4\pi(\varrho^+ - \varrho^-) \end{aligned}\right\} \tag{16.7}$$

whence

$$\sigma_1 = \frac{\sigma_2}{a} \tag{16.8}$$

i.e.

$$(\varrho_1^+ - \varrho_1^-) = \frac{1}{a^2}(\varrho_2^+ - \varrho_2^-). \tag{16.9}$$

From these we can derive

$$X_1\,\lambda_1 = X_2\,\lambda_2, \tag{16.10}$$

$$\frac{X_1}{p_1} = \frac{X_2}{p_2}. \tag{16.11}$$

Under certain circumstances the energy of the electrons and ions is determined by X/p alone. Consider the average net gain in energy, $\delta\varepsilon$ of an electron in one free path: the electron gains energy from the electric field and loses a fraction k of its total energy in collision, at the end of its path. Then

$$\delta\varepsilon = e\,X\,\lambda_e - k\,\varepsilon. \tag{16.12}$$

In the steady state $\delta\varepsilon = 0$. If λ_e, k are constants the solution is $\varepsilon = f(X\lambda)$ i.e. $= F(X/p)$. The same is true if k, λ_e vary only with the energy corresponding respectively to inelastic collisions, and the RAMSAUER effect. But suppose that electrons collide also with already excited atoms; the average energy lost in all collisions then depends both on ε and on the relative concentration N^*/n of excited atoms. We could write

$$k = f\left(\varepsilon, \frac{N^*}{n}\right). \tag{16.13}$$

To find N^*/n we equate the rates of production and loss of excited states. The production is due, we suppose, to direct collisions between electrons (concentration N_e) and gas molecules, giving

$$\left(\frac{dN^*}{dt}\right)_{\text{prod}} \propto N_e \cdot n. \tag{16.14}$$

The rate of loss is given by

$$\left(\frac{dN^*}{dt}\right)_{\text{loss}} \propto N^* \tag{16.15}$$

for molecules which simply radiate or $\propto N^* n$ for metastable states destroyed by colliding with gas molecules. This gives

$$\frac{N^*}{n} \propto N_e \quad \text{or} \quad \propto \frac{N_e}{n}. \tag{16.16}$$

Since N_e is not normally dependent only upon X/p but also upon the gas pressure it follows that when two-stage processes occur the mean energy of the electrons is *not* defined simply by X/p.

This qualitative argument has been presented here firstly because it shows that the Ramsauer effect does not upset similarity relationships, and secondly because it shows that even if X/p is the same in both discharges the velocities of the charged particles are equal *only when two-stage processes are absent.* This point is often ignored at this stage, although demonstrated later. However, we derive:

β) Assuming only single stage processes. Drift or random velocity

$$v_1 = v_2. \tag{16.17}$$

Mean energy (which can be defined for any general energy distribution)

$$\varepsilon_1 = \varepsilon_2. \tag{16.18}$$

If the particles have Maxwellian distributions; the mean energy can be defined in terms of temperature, giving

$$(T_{e,+,-})_1 = (T_{e,+,-})_2 \tag{16.19}$$

for electron, positive and negative ion temperatures respectively.

For processes involving time the relation can be found from $v = dx/dt$, giving:

For time intervals

$$dt_1 = a\, dt_2. \tag{16.20}$$

For frequencies,

$$\text{e.g. collision frequency } \nu_1 = \nu_2/a. \tag{16.21}$$

Applied frequency of the field in a.c. discharges $f_1 = f_2/a$.

The relation between magnetic fields can be derived from the radius of curvature (r') of the path of a moving charge. Since $Hev = \frac{m v^2}{r'}$, $H \propto v/r'$, hence:

$$\text{For magnetic fields } H_1 = \frac{H_2}{a}. \tag{16.22}$$

γ) Relations involving the current. We have already found relations between the net space charge densities

$$(\varrho^+ - \varrho^-)_1 = \frac{1}{a^2}(\varrho^+ - \varrho^-)_2 \tag{16.9}$$

ϱ^- includes both electrons and negative ions.

It does *not* necessarily follow (as is often assumed) that ϱ^+, ϱ^- are each separately $1/a^2$ times the corresponding densities in the second discharge, except in regions where there are charges of only one sign, i.e. either $\varrho_1^+, \varrho_2^+ = 0$ or

$\varrho_1^-, \varrho_2^- = 0$. However, we cannot assume this proportionality as a general law. The general expression for the total current density is:

$$j = \varrho^+ v^+ + \varrho^- v^-. \tag{16.23}$$

Now this can be written

$$j = (\varrho^+ - \varrho^-)(v^+ - v^-) + \varrho^- v^+ + \varrho^+ v^-. \tag{16.24}$$

To find the relation between j_1 and j_2 we write j_1 in this form, use equation (16.9) and note that for single stage processes only, $v_1 = v_2$, then we get:

$$\left.\begin{aligned} j_1 &= (\varrho_1^+ - \varrho_1^-)(v_1^+ - v_1^-) + \varrho_1^- v_1^+ + \varrho_1^+ v_1^- \\ &= \frac{1}{a^2}(\varrho_2^+ - \varrho_2^-)(v_2^+ - v_2^-) + \varrho_1^- v_2^+ + \varrho_1^+ v_2^- \end{aligned}\right\} \tag{16.25}$$

and we write

$$j_2 = (\varrho_2^+ - \varrho_2^-)(v_2^+ - v_2^-) + \varrho_2^- v_2^+ + \varrho_2^+ v_2^-. \tag{16.26}$$

Now it is clear that a simple proportionality between j_1 and j_2 can arise only when

$$\varrho_1^- v_2^+ + \varrho_1^+ v_2^- = \frac{1}{a^2}(\varrho_2^- v_2^+ + \varrho_2^+ v_2^-) \tag{16.27}$$

which together with equation (16.9) results in

$$\varrho_1^+ = \frac{1}{a^2}\varrho_2^+, \qquad \varrho_1^- = \frac{1}{a^2}\varrho_2^-. \tag{16.28}$$

Since elements of area in the two discharges are related by $dA_1 = a^2 dA_2$, the total currents across corresponding elements are

$$i_1 = j_1 dA_1 = \frac{1}{a^2} j_2 a^2 dA_2 = i_2. \tag{16.29}$$

This equality of current in the two discharges has been used by some writers as part of the definition of similar discharges, i.e. two discharges are defined as similar when the potential *and* current at corresponding points are equal and all linear dimensions are different by a factor a. The derivations are then much simpler, for we write:

$$\left.\begin{aligned} i_1 = j_1 dA_1 = i_2 = j_2 dA_2, \\ \therefore j_1 = \frac{1}{a^2} j_2, \end{aligned}\right\} \tag{16.29}$$

$$\therefore (\varrho_1^+ v_1^+ + \varrho_1^- v_1^-) = \frac{1}{a^2}(\varrho_2^+ v_2^+ + \varrho_2^- v_2^-). \tag{16.30}$$

Equation (16.9) and $v_1^+ = v_2^+$, $v_1^- = v_2^-$ (excluding 2-stage processes) gives immediately

$$\varrho_1^+ = \frac{1}{a^2}\varrho_2^+ \quad \text{and} \quad \varrho_1^- = \frac{1}{a^2}\varrho_2^- \tag{16.28}$$

and hence for the densities of charged particles

$$N_1^+ = \frac{1}{a^2} N_2^+, \qquad N_1^- = \frac{1}{a^2} N_2^-. \tag{16.31}$$

The first derivation, not assuming that the currents in the two discharges are equal, requires a little more complicated algebra, but it has the advantage of making fewer assumptions and proving that the only possible *simple* relation between i_1 and i_2 is their equality.

Now we assume that the current and voltage are the same in both discharges and examine the processes which can occur. It is not possible to *prove* that certain processes will lead to concentrations of charged particles differing in the two discharges by the factor $1/a^2$ but we can show that some combinations of processes are consistent with the assumption of this $1/a^2$ relation, while other combinations are not. Such processes we distinguish as "permitted" or "forbidden"—terms purely of convenience, having nothing to do with what can happen physically.

17. Permitted and forbidden gas processes, in the steady-state discharge. The concentration N of any charged particle at any point in a steady state discharge can be derived by equating the rates of production and loss at that point. Since N_1 must equal $(1/a^2)\, N_2$, and $dt_1 = a\, dt_2$ then

$$\left(\frac{dN}{dt}\right)_1 = \frac{1}{a^3}\left(\frac{dN}{dt}\right)_2 \tag{17.1}$$

dN/dt is the sum of the rates of production by a number of elementary processes, e.g.

$$\left(\frac{dN}{dt}\right)_{\text{total}} = \left(\frac{\partial N}{\partial t}\right)_{\text{electron collision}} + \left(\frac{\partial N}{\partial t}\right)_{\text{drift}} + \left(\frac{\partial N}{\partial t}\right)_{\text{diffusion}} + \left(\frac{\partial N}{\partial t}\right)_{\text{photonisation}} + \cdots. \quad (17.$$

Each process, whether of gain or loss, must transform by the factor $1/a^3$ in order that eq. (17.1) shall be satisfied. Considering the processes separately, we refer always to corresponding points or small volumes in the two discharges.

α) Ionisation by single collisions of electrons with gas molecules. The rate of ionisation depends on the concentration N_e of electrons and their energy, and on the concentration n of gas molecules. Assuming that the electron energy is the same in both discharges, the rate of production of ions and electrons is

$$\left(\frac{\partial N}{\partial t}\right)_1 = C\, N_{e_1} n_1 = C\,\frac{N_{e_2}}{a^2}\,\frac{n_2}{a} = \frac{1}{a^3}\left(\frac{\partial N}{\partial t}\right)_2. \tag{17.3}$$

The process is therefore permitted.

The constant C, and in the following equations $C_1, C_2, \ldots$ merely indicate proportionality.

β) Stepwise ionisation, collisions of the second kind. Ionisation by these processes can occur when the electron hits an already excited molecule, when two excited molecules collide, or when an excited (metastable) molecule of one gas hits and ionises a neutral molecule of another (PENNING effect). This last process can occur only when the excitation potential of the one gas exceeds the ionisation potential of the other.

The equilibrium concentration N^* of excited molecules is given by equating rates of production and loss, assuming that electrons produce, by single collisions, excited states having an average life τ:

$$\left(\frac{\partial N^*}{\partial t}\right)_{\text{production}} \propto N_e\, n, \tag{17.4}$$

$$\left(\frac{\partial N^*}{\partial t}\right)_{\text{loss}} \propto \frac{N^*}{\tau}. \tag{17.5}$$

Hence

$$\frac{N^*}{\tau} \propto N_e\, n. \tag{17.6}$$

For excited molecules which simply radiate within about 10^{-8} secs., and whose radiation is not absorbed in the gas, $\tau \ll$ time between collisions and is therefore a constant. This gives $N \propto N_e n$, i.e.

$$N_1^* = \frac{1}{a^3} N_2^* . \tag{17.7}$$

For metastable atoms or molecules (concentration N_m^*) destroyed by collision with gas molecules, or on the walls $\tau \propto 1/n$ or $\tau \propto R$ hence

$$(N_m^*)_1 = \frac{1}{a^2} (N_m^*)_2 . \tag{17.8}$$

This is true provided that no metastables are produced by electrons falling from higher levels.

In conditions where resonance radiation is appreciably absorbed and re-emitted in travelling through the gas the rate of production of resonance states (N_r^*) in any given volume is determined by the net flow of resonance photons through this volume, as well as by eq. (17.4). These photons come from other parts of the discharge, and the net flow in any one direction (e.g. x) is proportional to $\frac{\partial}{\partial x}\left(\frac{N_r^*}{\tau}\right)$. Hence for resonance states:

$$\frac{N_r^*}{\tau} \propto N_e n + \text{const} \frac{\partial}{\partial x}\left(\frac{N_r^*}{\tau}\right) \tag{17.9}$$

and taking $\tau = \text{constant}$

$$N_r^* \propto N_e n + \text{const} \frac{\partial N_r^*}{\partial x} . \tag{17.10}$$

From this it follows that N_r^* does not transform with any simple factor of a, in particular

$$(N_r^*)_1 \neq \frac{1}{a^2} (N_r^*)_2 . \tag{17.11}$$

The rate of ionisation caused by electrons hitting these excited molecules is proportional to the concentrations of both. We use relations (17.7), (17.8), (17.11) to derive the following. For ionisation by:

$$\left.\begin{aligned}
&\text{Electrons—excited molecules} && \left(\frac{\partial N}{\partial t}\right)_1 = C_1 N_e N_1^* = \frac{1}{a^5}\left(\frac{\partial N}{\partial t}\right)_2 ,\\
&\text{Electrons—metastable molecules} && \left(\frac{\partial N}{\partial t}\right)_1 = C_2 N_e (N_m^*)_1 = \frac{1}{a^4}\left(\frac{\partial N}{\partial t}\right)_2 ,\\
&\text{Electrons—resonance states} && \left(\frac{\partial N}{\partial t}\right)_1 = C_3 N_e (N_r^*)_1 \neq \frac{1}{a^3}\left(\frac{\partial N}{\partial t}\right)_3 .
\end{aligned}\right\} \tag{17.12}$$

All these processes are therefore forbidden. The rate of ionisation by the collision of two excited molecules is proportional to the concentrations of both:

$$\text{Two excited molecules:} \quad \left(\frac{\partial N}{\partial t}\right)_1 = C_4 N_A^* N_B^* \neq \frac{1}{a^3}\left(\frac{\partial N}{\partial t}\right)_2 \tag{17.13}$$

where $N_A^* N_B^*$ is any combination of N^*, N_m^*, N_r^*, hence all possible processes of this kind are forbidden. Ionisation by the PENNING effect in two gases X, Y (Y having $V_i < V_{\text{exc}}$ of X) gives:

$$\text{Metastable } X\text{—neutral } Y: \quad \left(\frac{\partial N}{\partial t}\right)_1 = C_5 \cdot (N_m^*)_X n_Y = \frac{1}{a^3}\left(\frac{\partial N}{\partial t}\right)_2 . \tag{17.14}$$

This is a permitted process.

γ) Electron attachment and detachment: electrons attach to neutral molecules at a rate proportional to the concentration of both, forming negative ions. Exactly as for ionisation by single collisions it follows that

$$\left(\frac{\partial N^-}{\partial t}\right)_1 = \frac{1}{a^3}\left(\frac{\partial N^-}{\partial t}\right)_2 \tag{17.15}$$

and the process is permitted. Detachment of electrons from negative ions due to collisions with gas molecules follows a similar law, since N^- transforms with $1/a^2$, n with $1/a$. This also is permitted.

δ) Photoionisation, proceeds at a rate proportional to the concentration of gas molecules and of photons, N_ν, (usually from excited molecules of another gas) able to ionise them. N_ν is $\propto \frac{\partial N^*}{\partial t} \propto \frac{N^*}{\tau}$. Using either non-resonance [eq. (17.6)] or resonance photons [eq. (17.10)] gives

$$\left(\frac{\partial N}{\partial t}\right)_1 \propto N^* n \quad \text{or} \quad \propto N_\nu^* n \neq \frac{1}{a^3}\left(\frac{\partial N}{\partial t}\right)_2. \tag{17.16}$$

The process is forbidden.

ε) Drift (mobility motion), and diffusion can be regarded as processes of gain or loss depending on whether the net flow of particles is in or out of the volume considered. *Self repulsion* is a special case of drift in the field, in a region where charges of only one sign are present.

The rate of production or loss in a small element $dx\,dy\,dz$ due to drift of charged particles in the electric field is given by

$$\frac{\partial N}{\partial t} = \frac{\partial}{\partial x}(N v) + \frac{\partial}{\partial y}(N v) + \frac{\partial}{\partial z}(N v) \tag{17.17}$$

where v is the drift velocity, N represents generally the concentration of any type of charged particle. Taking for example the x component we get for electrons, and ions

$$\left(\frac{\partial N}{\partial t}\right)_1 = \frac{\partial}{\partial x_1}(N_1 v_1) = \frac{\partial}{a\cdot\partial x_2}\left(\frac{1}{a^2} N_2 v_2\right) = \frac{1}{a^3}\left(\frac{\partial N}{\partial t}\right)_2 \tag{17.18}$$

and similarly for the y, z components.

The rate of change of concentration in a similar small element due to diffusion is

$$\frac{\partial N}{\partial t} = -D\left(\frac{\partial^2 N}{\partial x^2} + \frac{\partial^2 N}{\partial y^2} + \frac{\partial^2 N}{\partial z^2}\right). \tag{17.19}$$

Now D is $\propto \lambda v$ where v is the random velocity. D could refer to any type of diffusion, e.g. ambipolar. Taking one component we get for ions, electrons

$$\left(\frac{\partial N}{\partial t}\right)_1 = C_6 \lambda_1 v_1 \frac{\partial^2 N_1}{\partial x_1^2} = C_6 a \lambda_2 v_2 \frac{\partial^2 \frac{1}{a^2} N_2}{a^2 \partial x_2^2} = \frac{1}{a^3}\left(\frac{\partial N}{\partial t}\right)_2 \tag{17.20}$$

and similarly for y, z. Both processes are permitted.

ζ) Charge transfer is of special interest in discharges when fast ions hitting neutral molecules exchange their charge but do not share kinetic energy, the collision resulting in a slow ion and a fast neutral molecule. The rate of production of such molecules is proportional to the concentration of ions and gas molecules, i.e. $\propto N^+ n$ and hence transforms to $1/a^3$. Thus n^0 the concentration of fast neutral molecules arising from charge transfer is related by $n_1^0 = n_2^0/a^2$. This process is permitted.

The reverse process can occur when a fast neutral molecule picks up a charge from a slow ion and itself becomes a fast ion. The rate of production of such fast ions[1] is proportional to n^0 and to N^+ giving

$$\left(\frac{\partial N_{\text{fast}}^+}{\partial t}\right)_1 = C\, n_1^0\, N_1^+ = C\,\frac{1}{a^2}\, n_2^0\,\frac{1}{a^2}\, N_2^+ = \frac{1}{a^4}\left(\frac{\partial N_{\text{fast}}^+}{\partial t}\right)_2 \tag{17.21}$$

and hence

$$N_{1\,\text{fast}}^+ = \frac{1}{a^3}\cdot N_{2\,\text{fast}}^+ . \tag{17.22}$$

This process is forbidden.

η) Recombination between positive and negative ions follows the law

$$\left(\frac{\partial N}{\partial t}\right) = -\,\alpha\, N^+ N^- \tag{17.23}$$

where α is the recombination coefficient. At high pressure $\alpha \propto 1/p$, and so transforms with the factor a, while N^+, N^- each transform with $1/a^2$, thus finally $\partial N/\partial t$ transforms with $1/a^3$. The process is then permitted. At low pressure $\alpha \propto p$, $\partial N/\partial t$ transforms with $1/a^5$ and the process is forbidden.

For recombination between electrons and positive ions α is never $\propto 1/p$, but rather $\propto \sqrt{p}$, hence $\partial N/\partial t$ never transforms with $1/a^3$ and the process is always forbidden.

ϑ) Some implicit assumptions. It is quite obvious in the treatment of diffusion and drift that in taking $v_1 = v_2$ we assumed the same electron energy distribution in both discharges, implying the absence of step-wise ionisation. This same assumption is implicit in the derivations for all those processes which we have found to transform correctly, because we have assumed that the constants of proportionality $(C, C_1, C_2, \ldots)$ do not change from discharge 1 to discharge 2.

It follows that when 2-stage ionisation occurs in a discharge not only does that process itself violate the similarity laws (this is the most pronounced effect) but to a smaller extent it upsets the behaviour of normally permitted processes.

18. Permitted and forbidden secondary processes. If we denote the electron current density emitted from a wall or electrode by some secondary process as j_e', then the process is consistent with equal currents flowing in both discharges if

$$(j_e')_1 = \frac{(j_e')_2}{a^2} .$$

We shall now apply the results of the last section using particularly the relations between the concentrations of the various particles. Secondary electron emission due to the impact of:

α) Electrons or ions (excluding fast ions formed by a double charge transfer process) is permitted, since

$$(j_e')_1 = \delta\,(j_e)_1 = \delta\,\frac{1}{a^2}\,(j_e)_2 = \frac{1}{a^2}\,(j_e')_2 , \tag{18.1}$$

$$(j_e')_1 = \gamma_i\, j_1^+ = \gamma_i\,\frac{1}{a^2}\, j_2^+ = \frac{1}{a^2}\,(j_e')_2 , \tag{18.2}$$

where γ_i, δ are the usual secondary emission coefficients.

[1] We assume here that the fast neutral molecules are those already formed by the process of charge transfer at some other point in the discharge. If we considered simply the fast molecules of the MAXWELLian distribution their concentration would be related as $1/a$ and this process would be permitted. However, the number of sufficiently fast molecules normally present in a gas at ordinary temperatures is negligible.

β) Fast ions, formed by a double charge transfer process [see (17.22)] is forbidden, since

$$(j'_e)_1 = \gamma'_i N^+_{1\,\mathrm{fast}}\, v_1 = \gamma'_i \frac{1}{a^3} N^+_{2\,\mathrm{fast}}\, v_2 = \frac{1}{a^3} (j'_e)_2 . \tag{18.3}$$

γ) Fast neutral molecules, formed by charge transfer is permitted

$$(j'_e)_1 = \gamma_n n^0_1 v_1 = \gamma_n \frac{1}{a^2} n^0_2 v_2 = \frac{1}{a^2} (j'_e)_2 . \tag{18.4}$$

δ) Metastable atoms is permitted, since N^*_m transforms as $1/a^2$

$$j'_e = \gamma_m \tfrac{1}{4} N^*_m v, \tag{18.5}$$

and v is invariant.

ε) Photons is usually permitted.

The cathode subtends the same solid angle ω at corresponding small volumes $\delta x\, \delta y\, \delta z$ in the two discharges. Non-resonance photons fall directly on to the cathode giving a total secondary current i_e:

$$(i_e)_1 = \gamma_p \frac{\omega}{4\pi} \frac{N^*_1}{\tau} (\delta x\, \delta y\, \delta z)_1 = \gamma_p \frac{\omega}{4\pi} \frac{1}{a^3} \frac{N^*_2}{\tau} \cdot a^3 (\delta x\, \delta y\, \delta z)_2 = (i_e)_2 . \tag{18.6}$$

This photoelectric effect is therefore permitted.

Resonance photons diffuse and give $j'_e \propto \frac{\partial}{\partial x}(N^* v)$ but since N^*_r does not transform as $1/a^2$ neither does j'_e. This potoelectric effect is forbidden.

Finally, secondary emission by

ζ) Field Emission is forbidden since $j'_e = A\, V^2 e^{-B/V}$. V and hence j'_e are the same in both discharges.

19. Summary and conclusions to be drawn from the similarity laws. The results we have obtained are summarised in Tables 1 and 2, and a list given of those groups of parameters which remain invariant in the two discharges. It should be noted that the invariance of some groups (e.g. j/p^2) depends upon more assumptions than that of others (e.g. X/p). It should also be noted that although the entire treatment assumes the motion of charges to be determined by collisions with gas molecules, any high vacuum process in which charged particles gain energy in free fall through a potential drop $(\frac{1}{2} m v^2 = e(V_1 - V_2))$ also obeys similarity laws.

The greatest care is needed in attempting to infer anything about physical processes when a given discharge does or does not obey similarity laws. The derivations given prove only consistency; they do not prove (for example) that single stage ionisation causes $N^+_1/N^+_2 = 1/a^2$.

Therefore logically we can deduce only that if the similarity laws are obeyed then all the forbidden processes can be excluded[1]; we cannot deduce that forbidden processes are necessarily present if the laws are broken. However, it is often reasonable to infer that if a discharge first obeys and then progressively deviates from the laws as some parameter (e.g. discharge current) is varied, then forbidden processes are appearing, or the gas temperature is changing.

Similarity relations in discharges with externally produced ionisation have also been discussed[2].

[1] Even this statement needs qualifying: we have no calculations to show what *proportion* of any forbidden process would cause a noticeable deviation from the laws.

[2] W. Fuchs: Naturwiss. **35**, 283 (1948).

Table 1.

Parameters	Relations	Invariant groups of parameters derived "Similarity parameters"
Potential All linear dimensions Current Gas temperature	$V_1=V_2$ $d_1=a d_2$ $R_1=a R_2$ $i_1=i_2$ $T_1=T_2$ } Basic assumptions[1]	V, i, T
Mean Free Path of any particle Gas density Gas pressure Electric field Wall charge density Volume charge density	$\lambda_1=a\lambda_2$ $n_1=n_2/a$ $p_1=p_2/a$ $X_1=X_2/a$ $\sigma_1=\sigma_2/a$ $\varrho_1^+, \varrho_1^- = \frac{1}{a^2}\varrho_2^+, \frac{1}{a^2}\varrho_2^-$ } Derived from the basic assumptions	X/p, $p d$, pR
Total and partial current density Particle density: Electrons, $+ve$ and $-ve$ ions, metastables[2], fast neutrals (by charge transfer) Excited states, except resonance Velocity, energy of charged particles Electron, ion, temperature Time interval Collision frequency Applied frequency Magnetic field	$(j, j^+, j_e)_1 = \frac{1}{a^2}(j, j^+, j_e)_2$ $(N_e, N^+, N^-, N_m^*, n^0)_1 = \frac{1}{a_2}(N_e \ldots)_2$ } Derived from above assuming also no stepwise ionisation except the PENNING effect $N_1^* = N_2^*/a^3$ $v_1=v_2$ $\varepsilon_1=\varepsilon_2$ $(T_e, T_+)_1=(T_e, T_+)_2$ $dt_1=a\,dt_2$ $\nu_1=\nu_2/a$ $f_1=f_2/a$ $H_1=H_2/a$	T_e, T_+ H/p j/p^2 N_e/p^2 and for a.c. discharges fd, fR, f/p

Table 2. *Permitted and forbidden processes.*

	Permitted	Forbidden
In the gas	Ionisation by single collision PENNING effect, Electron attachment and detachment Drift, Self Repulsion, Diffusion } of charges Charge transfer ion→fast neutral Recombination ion—ion at high p	All stepwise ionisation, collisions of 2nd kind except PENNING effect Photoionisation, Charge transfer: fast neutral→ion[3] All recombination except ion—ion at high pressure
On the walls	Secondary emission by impact of: Electrons (δ) Ions (γ_i) Non-resonance photons (γ_p) Metastables (γ_m) Fast neutrals (γ_n)	Electrons emitted by fast ions formed by charge transfer[3]. Photoelectric effect by diffusing resonance photons Field emission.

[1] But see Sect. 16 for a discussion on the assumption $i_1=i_2$.

[2] Provided metastables are produced only by direct excitation, and not by electrons falling from higher levels.

[3] See footnote in Sect. 17.

V. The cathode region.

a) Experiment.

20. Description and basic mechanism. The cathode region is the most important part of the discharge. Most of the potential drop between the electrodes occurs in the cathode dark space, that region of short length d between the cathode and the negative glow. The properties of this space are almost independent of the rest of the discharge: a discharge can exist without a positive column, FARADAY dark space or negative glow, but not without a cathode dark space. In particular its length d, the potential across it V_c, called the *cathode fall*, and the current density j do not depend on the length of the discharge provided that the anode does not approach very close to the edge of the dark space; nor upon the diameter of the tube provided it is not too narrow. If the cathode is moved, or its plane rotated, the cathode dark space and negative glow remain attached to it, their boundaries being parallel to the cathode surface, and the current flowing normally across them: the positive column bends towards the anode. This leads us to believe that the cathode dark space is the place where the processes essential to the maintenance of the discharge occur. Before a discharge is established the electric field is given simply by the applied potential and the geometry of the electrodes and tube, but when appreciable ionisation takes place in the gas a large positive space charge grows close to the cathode. Its size and position determine the field, and hence the ionisation in this region. Light emitted here is partially polarised, indicating a high field.

The essential processes in the cathode region are as follows. Electrons leaving the cathode are accelerated across the dark space, excite and ionise the gas, and are multiplied in number by a factor M. Positive ions, photons, excited and neutral molecules hit the cathode and release new electrons. Suppose that for each of the $(M-1)$ electrons formed in the gas γ electrons are released at the cathode by these various mechanisms. Then for the discharge to be self sustaining the usual condition must be satisfied:

$$\left.\begin{aligned} \gamma(M-1) &= 1, \\ \text{or} \qquad 1+\frac{1}{\gamma} &= M. \end{aligned}\right\} \tag{20.1}$$

Here we shall not specify either the exact secondary mechanism at the cathode (γ), nor the expression for M. In earlier literature it has usually been assumed that electrons are released from the cathode only by ions ($\gamma=\gamma_i$), and M is expressed as $e^{\int_0^d \alpha\,dx}$ where α is TOWNSEND'S ionisation coefficient. Recent work has shown that the photoelectric effect at the cathode is comparable with that due to ion bombardment, and also that the use of α leads to wrong results in non uniform electric fields such as exist in the cathode dark space. It has also been shown that the negative glow plays some part in the mechanism of the dark space. When this contribution becauses important the above condition becauses somewhat modified in form, through not in principle (see Sect. 37).

However, it is clear that the multiplication M depends upon the nature and pressure of the gas; so also does the secondary coefficient γ which is influenced by the size and energy of the particles or photons hitting the cathode. γ also depends on the cathode material and the state of its surface. We therefore expect the properties of the cathode dark space to depend markedly upon these factors, and the results of many measurements can be understood from this brief outline without anticipating the exact theories which will be discussed later.

21. The cathode fall, V_c. The large potential drop between the cathode and the negative glow is one of the most important quantities describing the glow discharge. Although the length of the dark space may be defined in slightly different ways the uncertainties lie in a region of very low field strength and the resulting differences in V_c are trivial. For a given gas and pressure V_c remains constant when the current through the discharge is altered, in a certain range even by several orders of magnitude. HITTORF first showed this to be true provided that the discharge covers only part of the cathode. The current density at the cathode remains constant (HEHL'S law) and the area of the cathode covered by the discharge is proportional to the total current. A discharge with these properties is said to be a *normal discharge*, and to have a normal cathode fall V_n, and current density j_n. The normal cathode fall is a characteristic constant for each combination of gas and cathode material. If the similarity principle applies, V_c should be a function of j/p^2 and this is often found to be so. Corresponding to V_n there are characteristic constant values, j_n/p^2 and $p d_n$. The cathode fall ceases to remain constant: —

(I) when, with increasing current, the discharge covers all the cathode. V_c and j then increase as the current i increases. This is the *abnormal discharge*, having an abnormal cathode fall $V_c > V_n$.

(II) When the current is so small that the discharge covers a region of the cathode of diameter comparable with the length of the dark space. Then a considerable fraction of the ions in the dark space is lost by lateral motion. An increased cathode fall is necessary to sustain the discharge, so here V_c increases with decreasing current. This is the *subnormal* discharge.

These regions are marked on Fig. 1.

It was at one time believed that V_n was closely related to the minimum sparking potential for the given combination of gas and cathode, and probably equal to it. Accurate measurements have shown that this is not true. V_n is always less (often considerably less) than $(V_s)_{\min}$, probably because in a discharge the maintenance condition is satisfied with the help of ions, photons etc. from the negative glow.

α) Measurements. V_c has been defined and measured in a number of ways: the earlier workers placed a small wire at some point in the negative glow and using an electrostatic voltmeter measured its potential with respect to the acthode[1]. This method suffered from the inaccuracies mentioned in the section on probes. A wire placed at the cathode edge of the glow and standing in a stream of electrons would acquire a potential appreciably negative with respect to the space; if placed at the anode end of the glow it would be nearer space potential but the measurement would then give the potential drop across cathode dark space and glow.

Modern measurements are made by the following methods:

(I) *Probes.* The minimum space potential in the glow is measured with a LANGMUIR probe, the cathode fall being defined as the difference between this potential and that of the cathode. The method suffers from the possible errors, and the limitations (of pressure, current density) discussed in Sect. 10.

(II) *Electric field measurements.* The field at every point in the dark space can be measured by the deflection of an electron beam. A linear fall is found,

[1] See general references [*3*], [*10*].

the field becoming practically zero at the edge of the glow. The integrated value of the field to this point gives the field V_c. The method seems to vary in accuracy, depending greatly on the design of the apparatus, but some workers in highly abnormal discharges have found results agreeing with other measurements. It is probably best to regard this kind of measurement as a useful check for large V_c.

(III) *Electrode potentials.* The potential difference across the electrodes is measured as the anode is moved towards the cathode, while the discharge current is kept constant. Fig. 6 shows how it varies. At first there is a slow linear fall as the anode moves through the positive column; when it is pushed into the FARADAY dark space a sudden drop in potential occurs (10 to 25 volts depending on the gas), which marks the disappearance of the anode fall and coincides with the visual disappearance of the anode glow. Further movement of the anode causes hardly any change until it reaches nearly the edge of the cathode dark space, when at a critical point the potential necessary to maintain the constant current increases rapidly. V_c is then defined as the potential between the electrodes, when the anode fall has just disappeared.

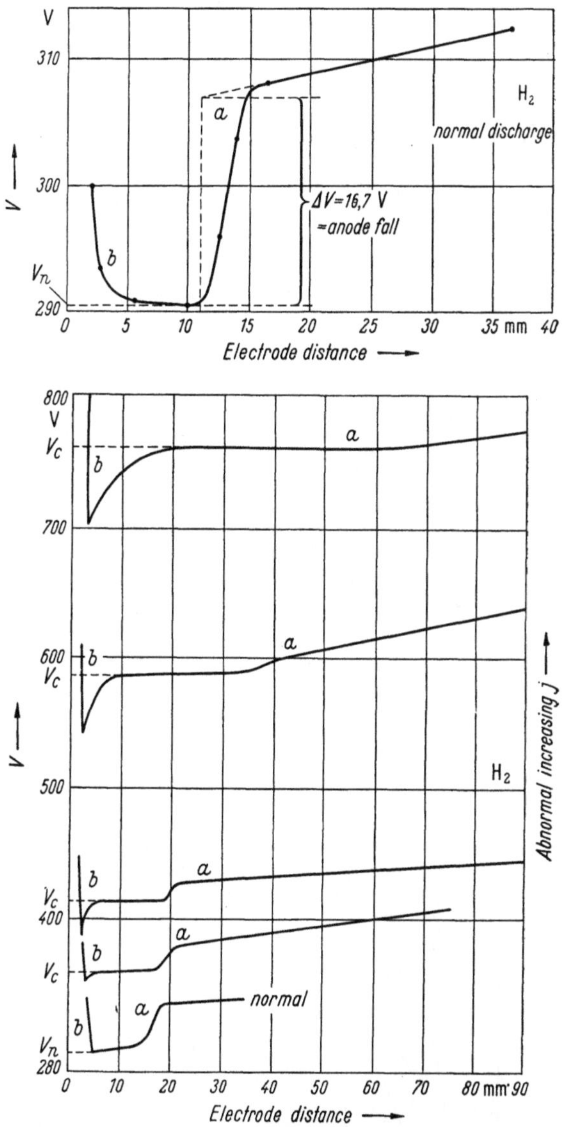

Fig. 6. Voltage V, between electrodes as the anode is moved towards the cathode. The anode glow and anode fall begin to disappear at a; the discharge becomes obstructed to the left of b. The minimum just to the right of b (at high current density) is discussed in Sect. 84.

This method can be used even without a movable anode, by so adjusting the gas pressure that the discharge spreads out from the cathode until the anode becomes immersed in the FARADAY dark space or negative glow.

All the modern values of the cathode fall have been obtained in this way, and there is little doubt that this is the most accurate type of measurement.

Each method has certain inherent errors, those due to probes have been mentioned, while the insertion of the anode into the negative glow must necessarily disturb that region of the discharge and possibly to some extent the mechanism of the dark space. It is also true that the definition of V_c depends strictly upon the method used to measure it. However, these criticisms should

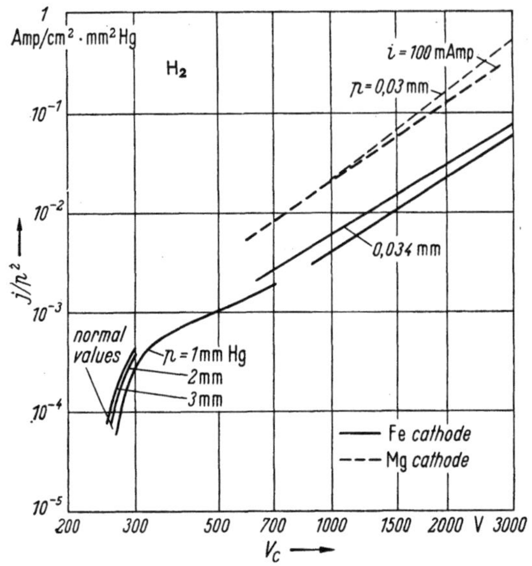

Fig. 7. j/p^2 as $f(V_c)$ for abnormal discharge in H_2. (GÜNTHERSCHULZE's results, collected by DRUYVESTEYN and PENNING), showing the differences in the values found under different conditions of current and pressure.

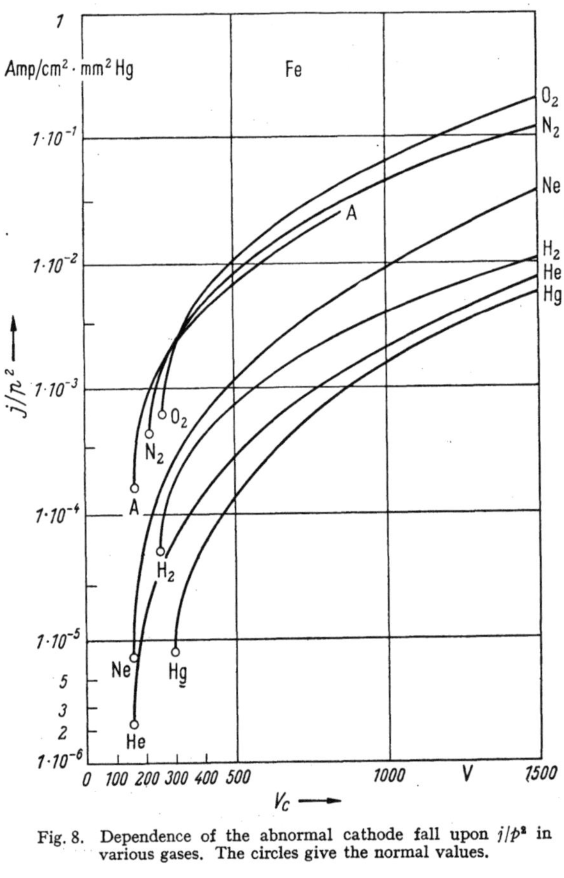

Fig. 8. Dependence of the abnormal cathode fall upon j/p^2 in various gases. The circles give the normal values.

not be exaggerated; such errors are not likely to be more than a few per cent, and the results given by different methods are mutually consistent. Much greater errors arise from impurities in the gas and on the cathode surface, and from heating of the gas and cathode which causes local variations in the gas density.

β) Abnormal cathode fall. The results of many measurements by GÜNTHERSCHULZE and associates [46] to [49] are shown in Figs. 7 and 8. The abnormal V_c has been measured for various currents and current densities greater than the normal. It has been found rather easier to obtain consistent and reproducible values of the abnormal than the normal cathode fall for reasons which are only now being appreciated. These are discussed below. In strongly abnormal discharges the cathode can become very hot and alter the gas density close to it. To avoid this effect massive cathodes must be used, and low gas pressure. Several empirical relations have been found: e.g.

$$\left.\begin{aligned}\frac{j}{p^2} &= A\,V_c^B\\ &\text{(for } V_c \text{ 1000 to 3000 volts).}\end{aligned}\right\}\qquad(21.1)$$

A and B are constants for H_2, N_2, which obey the similarity laws; but not for O_2, He, and the rare gases. ASTON [12], [13] has also found that molecular gases obey:

$$\left.\begin{aligned}V_c &= E + F\,\frac{\sqrt{j}}{p},\\ \text{or}\qquad \frac{j}{p^2} &= k\,(V_c - E)^2.\end{aligned}\right\}\qquad(21.2)$$

where E, F, k are constants.

22. Normal cathode fall, V_n. For a proper measurement of V_n edge effects at the cathode and heating of the gas have to be eliminated, although heating effects are usually not serious here.

Ideally, guard rings should be used around the cathode and small currents employed, provided that the radius of that part of the cathode covered by the discharge is kept much greater than d. In spite of the most careful control of purity both of the gas and the cathode surfaces many experimenters have obtained results not agreeing with each other, and often not reproducible in their own apparatus. Differences of up to 30 volts were observed, for example, in a series of measurements by PIKE, who used films of sputtered metal as cathodes [*50*].

We now believe, through the work of PENNING and his associates, that the conditions of purity of both gas and cathode required for reproducible measurements are much more stringent than was previously thought. The standard methods of baking the glass vessels under vacuum in a furnace at about 400 to 600° C, and heating metal parts by high frequency eddy currents are often not adequate.

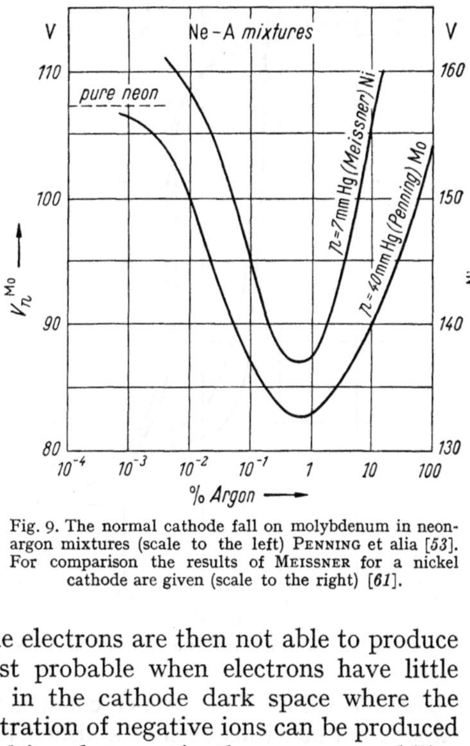

Fig. 9. The normal cathode fall on molybdenum in neon-argon mixtures (scale to the left) PENNING et alia [*53*]. For comparison the results of MEISSNER for a nickel cathode are given (scale to the right) [*61*].

Impurities in small amounts act chiefly in three ways:

(a) By collisions of the second kind. A gas molecule X in an excited state, colliding with a molecule of the impurity Y can ionise it if the excitation potential of X is greater than the ionisation potential of Y. The probability is greatest if these two energies are nearly equal and if the excited state is metastable. This process increases the ionisation, and is often called the "PENNING effect".

(b) By electron attachment. Electronegative gases (e.g. Cl_2, I_2, O_2) attach electrons to form heavy negative ions; the electrons are then not able to produce further ionisations. This process is most probable when electrons have little energy (0.1 to 1 eV), and rarely occurs in the cathode dark space where the electrons are fast. An appreciable concentration of negative ions can be produced however in the negative glow. The resulting decrease in the average mobility of the negative carriers is similar to that resulting from an increase in gas pressure, and the length of the dark space decreases (see Sect. 89). The resulting change in space charge distribution alters the cathode fall, but not much is known precisely about the magnitude of the effect.

(c) By chemical action at the cathode. Oxygen in particular forms an extremely stable and adhesive layer of oxide on many metals (e.g. Mo, W, Al). It is usually a monatomic layer and can thus be formed when the smallest quantity of oxygen is present in the gas. This layer can be removed only by the most intense heating at about 1800 to 2000° C (usually not practicable) or by sustained bombardment with fast positive ions (energy $\sim$500 eV).

The effect (a) was clearly demonstrated by PENNING: the addition of 1 part of argon to 10^4 parts pure neon reduced the normal current density j_n by a factor 4, and showed an equally large effect upon the starting potential [*51*], [*55*]. Accurate modern measurements are shown in Fig. 9. The effect here is large because $(V_{exc})_{Ne} \cong (V_i)_A$.

Recent experiments by PENNING, MOUBIS and JURRIAANSE [53] have shown the importance of the cathode surface, and have led to a technique for accurate and reproducible measurements of V_n. They observed that a glow discharge in neon between two carefully cleaned molybdenum electrodes which originally covered most of the cathode surface (diameter 1 cm) began after some time, to contract radially. The process became very rapid as the glow contracted to a spot (diameter 1 mm.) and it was accompanied by a large drop in the voltage across the electrodes[1] (see Fig. 10). This they attributed to the removal, by ion bombardement, of a thin layer of oxide on the cathode, the discharge concentrating on that part where the pure metal is first exposed, and where γ is greatest. This was confirmed by noting that if the cathode were previously deliberately oxidised no contractions of the discharge took place even after several hours. On the other hand increasing the current density between the original "clean" electrodes caused the contraction to occur much sooner. The process was accompained by sputtering of the cathode metal on to the walls of the tube. After the contraction had taken place the whole cathode could be cleaned by steadily increasing the discharge current, until the discharge covered the whole surface. The area covered by the discharge was then observed to be proportional to i, and the voltage across the electrodes was of constant and reproducible value. The proper conditions for the measurement of the normal cathode fall had thus been established, and each point on the cathode was electrically as well as geometrically equivalent with respect to the anode[2].

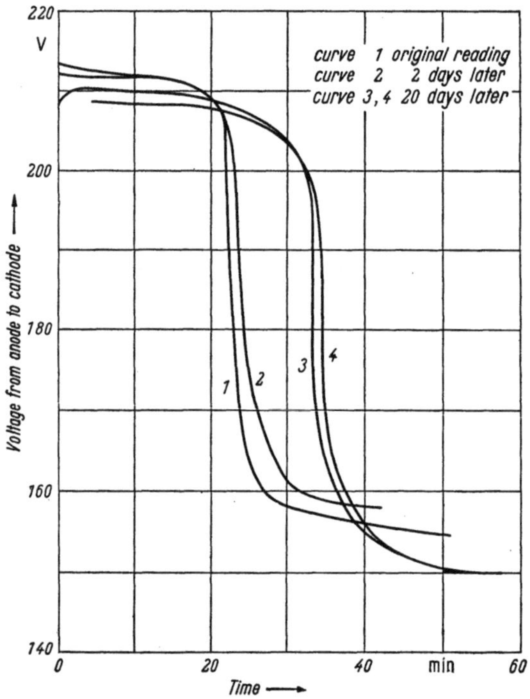

Fig. 10. Decrease of cathode fall with time: Ne with Mo cathode [53].

The material sputtered on to the walls also cleans the gas by trapping any small amounts of oxygen. It has been found that a thick unbroken layer of sputtered Mo on all the walls of the apparatus is necessary for accurate measurement, and that this layer performs three essential functions, namely:

(a) It acts as a getter absorbing impurities left when the system is first evacuated [55]. A sputtered layer is a much better absorber than one formed by evaporating in vacuum [56].

(b) It absorbs gas coming from the walls and prevents it reaching the cathode (see also CAMPBELL [57]).

[1] A similar change of cathode fall with time as a result of changing surface conditions was found by SEELIGER and BARTHOLOMEYCZYK: Ann. Physik 1, 241 (1947).

[2] Extremely clean glass surfaces have been produced by treatment in a glow discharge, the purity confirmed by electron microscope studies, by the uniform condensation of moisture, and by a very high coefficient of dry friction: KARASEV and IZMAILOVA: J. techn. Phys. USSR. 24, 871 (1954).

(c) It screens the glass walls, preventing the liberation of gas by excited gas atoms.

(The action of metastable Neon atoms in removing O_2 from glass has been shown by LANGMUIR and also DRUYVESTEYN [58], [59].)

Even with these precautions it was found extremely difficult to reproduce results with a carbon cathode, and a modified technique was required for Cu and Ag. Values for the normal cathode fall derived from these measurements are given in Table 3. In some of them a rod (i.e. solid cylinder) was used instead

Table 3. *Values of the minimum discharge voltage* $V_{\min}$ *and the normal cathode fall* V_n *(reduced to plate cathode and 40 mm. pressure) in neon* (after PENNING and MOUBIS [54]).

Atomic number	Material	Cathode		p (mm.)	$V_{\min}$ (volts)	Experimental Error ΔV (volts)	Corrections (volts)		V_n (volts) (40 mm., plate)	Work function φ (volts)
		shape	$2r$ (mm.)				shape	pressure		
6	C			20	200	± 20			(195)	4,36
14	Si	rod	1,8	40	142	± 5	−5		137	3,59
22	Ti	plate		20	116	± 2		−3	113	4.16
23	V	plate		40	118	± 3			118	4.11
24	Cr	plate		40	123	± 5			123	4.45
26	Fe	rod	1,5	20	137	+10, −2	−4	−4	129	4.63
28	Ni	rod	2	20	145	± 2	−5	−4	136	4.91
29	Cu	rod	3.7	20	156	± 5	−3	−5	148	4.48
40	Zr	plate		40	104	± 3			104	4.12
41	Nb	plate		40	106	± 3			106	4.00
42	Mo	plate		40	107	± 2			107	4.15
45	Rh	plate		40	124	± 3			124	4.58
47	Ag	rod	2.3	20	152	± 5	−5	−4	143	4.74
72	Hf	rod	2	40	108	± 2		−3	105	3.53
73	Ta	plate		20	109	± 3		−3	106	4.12
74	W	plate		20	114	± 2		−3	111	4.54
90	Th	plate		20	115	± 3		−3	112	3.38

of a plane cathode, and a geometrical correction is made. All these results are reduced to a pressure of 40 mm., since a slight variation of V_n with p was found in Neon.

Older values of V_n [6], [10], [46], [68] to [71] are given in Table 4 and also the normal current density, expressed as j_n/p^2, in Table 5.

The normal cathode fall has also been measured on glass, and on single crystal surfaces as the cathode and found to depend on the particular surface chosen. Other measurements have been made at the melting points of Sn, Pb, Hg and Bi[1]. The actual values obtained, by probe measurements are not very accurate due to the changing density of the gas, but when the cathode is allowed to cool V_n changes suddenly by a few volts as the metal solidifies. It is thought that this may be connected with the change in work function φ, but since the values of φ are not known under these conditions this cannot be decided.

Much more sputtering occurs in the abnormal than in the normal discharge, and this may explain why consistent values of the abnormal cathode fall have generally been easier to obtain.

---

[1] H. KURZKE: Ann. Phys. **25**, 688 (1936). A. WEHNELT and H. KURZKE: Ann. Phys. **20**, 919 (1934).

α) Dependence of V_n on the cathode material. It has often been hoped to find some simple relation between the normal cathode fall and a fundamental property of the cathode material, for example the work function φ. There is a general rule that materials having the lowest work function give the lowest cathode falls. From the earlier measurements it seemed that, for some metals, $V_n \propto \varphi$ [*60*], [*50*]. There were, however, many wide deviations from this law, and it seems that the different states of the cathode surfaces may have masked any changes due to differences in φ. Modern values show no proportionality.

Table 4. *Older values of the normal cathode fall V_n in volts.* PIKE's *values* [*50*] *are given in brackets. Full references are in* [*10*], [*10a*], [*6*]. *For composite cathodes see* [*3*], *vol. 2, p. 332.*

Cathode material	Work function φ (volts)	Gas									other gases
		He	Ne	A	Air	H_2	N_2	O_2	Hg	Cl_2	
Na . .	2.3	80	75		200	185	178				Na 115
K. . .	2.3	59	68	64	180	94	170				K 80
Cu . .	4.0	177 (204)	220 (200)	130 (196)	370	214	208		447		CO 484, CO_2 460, NH_3 385
Ag . .	4.1	162 (184)	150 (170)	130	280	216	233		318		
Au . .	4.7	165 (180)	158 (184)	130 (196)	285	247	233				
Mg . .	2.7	125	94	119	224	153	188	310			
Ca . .	4.0	86	86	93			157				
Sr . .	2.7	86		93			157				
Ba . .	2.5	86		93			157				
Zn . .	4.3	143		119	277	184	216	354			CO 480, CO_2 410, NH_3 330
Cd . .	4.0	167	160	119 (123)	266	200	213				
Hg . .	4.5	143				337	226		340[3]		
Al . .	3.0	140	120	100	229	170	180	311	245	280	
Th . .	4.0		125								
C . . .	4.4					280			475[3]		
Sn . .	4.4			124	266	226	216				
Pb . .	4.0	177	172	124	207	223	210				
Ta . .	4.1	(171)	(158)	(156)							
Sb . .	4.6			136	269	252	225				
Bi[1] . .	4.3	137		136[1]	272	240	210				
Mo . .	4.2	(171)	115 (160)	(145)					353		
W . .	4.5	(155)	125 (153)	(~140)					305		
Fe . .	4.6	150	150	165	269	250	215	290	298[3]		Kr 215, X 306, Na 115, K 80, Rb 395, Cs 340
Co . .	4.3				380						
Ni . .	4.9	158	140	131	226	211	197		275		
Pd . .	5.0				421						
Ir. . .	4.6				380						
Pt . .	5.4	165 (194)	152 (178)	131	277	276	216	364	340	275	I_2 377, Br_2 335, NH_3 370, CO 490, CO_2 475, NO 365
Glass[2]					310	260					

Also on Pt cathode: NO_2 387, HCl 330, HBr 350, $HgCl_2$, $HgBr_2$, HgI_2 370 to 430.

It may well be that no simple relation does exist between V_n and φ. The essential factor which determines V_n is γ, not φ, and γ depends both on the cathode and on the ions and photons hitting it, i.e. on the gas. Quite apart from any possible chemical action at the cathode we do not know enough about the mechanism of secondary emission to assert that different emissions are due only to different work functions. However, some of the values of V_n and atomic number do fall into a regular pattern, as the graph (Fig. 11) by PENNING shows.

β) Dependence of V_n on the nature, pressure and temperature of the gas. Since V_n is a quantity characteristic of each combination of cathode material and

[1] Also on single crystal faces in A. 111 plane 150, $11\bar{1}$ plane 153, H. KURZKE: Z. Physik **104**, 735 (1937); see also K. G. EMELEUS and J. W. BECK: Phil. Mag. **8**, 121 (1929).

[2] K. WOLF: Z. Physik **112**, 96 (1939) thin soft glass discs heated to 300° C.

[3] Much lower values (~250 v) were found by N. L. MINDLIN: J. techn. Phys. USSR. **20**, 175 (1950).

gas it clearly depends markedly on the latter. Here again no simple relation is found between V_n and any one property of the gas. We find, however, as would be expected from the basic mechanism, that gases having the lowest ionisation potentials and the greatest probabilities of ionisation by electron collision give the lowest cathode falls. An example showing how V_n varies with the rare gases using the same cathode Mo is given in Fig. 12, and similar trends[1] can be seen in Table 4.

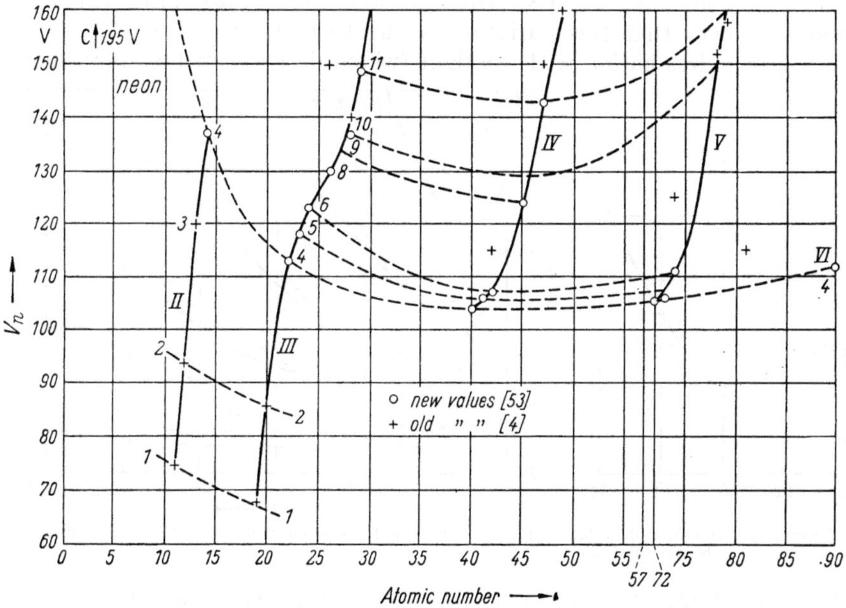

Fig. 11. Values of the normal cathode fall V_n in neon as a function of the atomic number of the cathode material, the rare earths being omitted. The points for elements in the same row of the periodic system (Roman figures) are joined by solid curves, those for elements in the same column (Arabic figures) by broken curves. In this way a rather regular pattern appears. The point for carbon (atomic number 6) lies off the graph.

According to the similarity principle V_n ought not to vary with pressure. A small variation has, however, been found by JURRIAANSE, and earlier by MEISSNER although he thought that the increase of V_n with decreasing pressure at low p is due to edge effects at the cathode [*61*]. In JURRIAANSE's experiments a $V-i$ curve was taken at each pressure, a flat portion (see Fig. 13) indicating that the measured voltage was genuinely V_n. Similar results were found for

Table 5. j_n/p^2 in 10^{-6} A/sq. cm. (mm. Hg).

Cathode material	He	Ne	A	Kr	Xe	H_2	N_2	Air	Hg
Cu						64		240	15
Au						110		570	
Mg	3	5	20						
Zn						80			
Al						90		330	4
Fe (also Ni)	2	6	160	43	16	72	400		8
Pt	5	18	150			90	380	550	
Glass						~80		~ 40	

[1] See also A. GÜNTHERSCHULZE: Z. Physik **34**, 549 (1924).

neon and argon. All $V_n - p$ curves have the same inclination and can be represented approximately by

$$V_n = 12\log\left(\frac{1}{p}\right) + C \tag{22.1}$$

where $C \approx 124$ volts (p in mm. Hg).

MEISSNER's experiments showed also that j_n/p^2 is not constant in neon but varies according to:

$$j_n = c \cdot p^b \tag{22.2}$$

where c, b are constants, $b = 1.5$. This was confirmed by the other experiments. MEISSNER believes that this variation is due to metastable Ne atoms ionising impurities, since irradiation with Ne light (which destroys metastables) increases

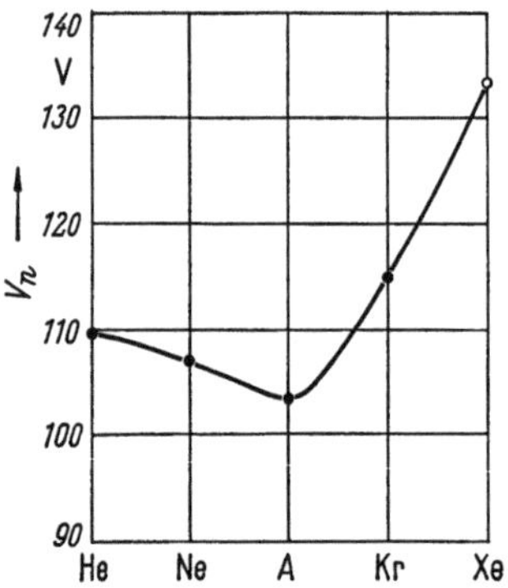

Fig. 12. The value of the normal cathode fall V_n on molybdenum in the rare gases [53].

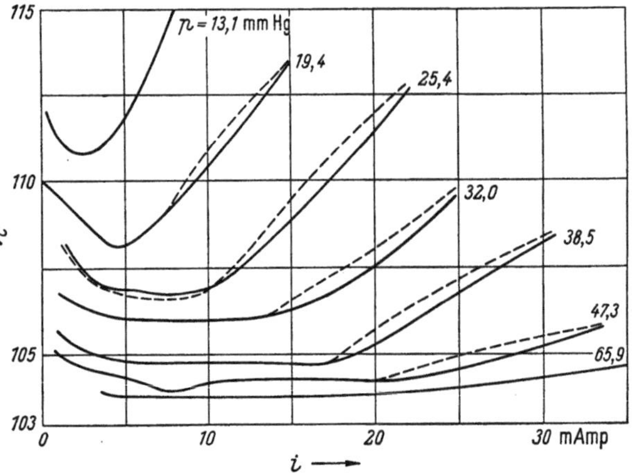

Fig. 13. Cathode fall as function of the current for various pressures [53].

b to 1.85. JURRIAANSE, however, in his experiments tested the purity of his gas, observing that the starting potential was hardly changed by similar irradiations. V_n and j_n were influenced by the light and it seems more probable that the effect is genuine in the pure gas and is due to a 2-stage ionisation process in the cathode fall. This would of course violate the similarity laws.

A variation of the normal cathode fall with average gas temperature T has also been found [53], dV_n/dT being about -16×10^{-3} V/° C for neon, -4×10^{-3} V/° C for argon at a temperature of 25° C, varying slightly with the current. It has been shown however, by varying the wall temperature and applying the gas laws, that this small effect is due entirely to changes in the density of the gas. The cathode temperature also affects V_n and j[1].

23. Discharges without a cathode fall. The cathode fall would not be necessary if there existed at the cathode some mechanism giving a sufficient supply of electrons without the action of fast ions or energetic photons impinging on it. Two such types of glow discharge are known:

(a) The hot cathode discharge. In this electrons are released by thermionic emission, the cathode often being in the form of a heated tungsten or oxide coated filament.

(b) the spray discharge. This has been found to occur when a carbon cathode is covered with an insulating layer (e.g. of Al_2O_3 powder), and when Al is covered with a coherent layer of oxide. Here positive ions charge up the outer insulating surface and electrons are accelerated by the resulting field through the insulating layer (see Sect. 87).

[1] H. JACOBS and J. MARTIN: J. Appl. Phys. **21**, 681 (1950).

24. The electric field in the dark space. It is not possible to use a LANGMUIR probe in the dark space because the electrons have almost beam-like properties and their energy distribution is far removed from MAXWELLian. ERNST has, however, measured the distribution of potential in the space, using an electron-emitting probe, although his results are not in good agreement with earlier probe measurements by BROWN and THOMPSON [*62*], [*63*]. The most reliable method seems to be the measurement of the deflection of an electron beam fired across the discharge, a method first used by ASTON and J. J. THOMSON, later by GEDDES and in a more sensitive form by LITTLE and VON ENGEL, and by STEIN [*18*] to [*21*]. These show a linear decrease of the field with distance from the cathode (see Fig. 14), a result confirmed by STEUBING who used the STARK effect [*25*]. (Earlier work by BROSE [*24*] using this method had given different results, but this was probably due to a distortion of the electric field caused by the use of narrow tubes)[1]. Most earlier probe measurements were in normal discharges (e.g. ERNST, in A and H_2), but only STEIN has measured in the normal cathode fall with an electron beam, all other results by this method having been obtained in abnormal or hollow cathode discharges. The beam method is most accurate in regions of high field strength: there is some discrepancy between the results of various authors on the low fields near or in the glow. STEIN, for example, sometimes finds a small negative field in the glow, but he points out that there were systematic errors in his apparatus due to the electron beam not being

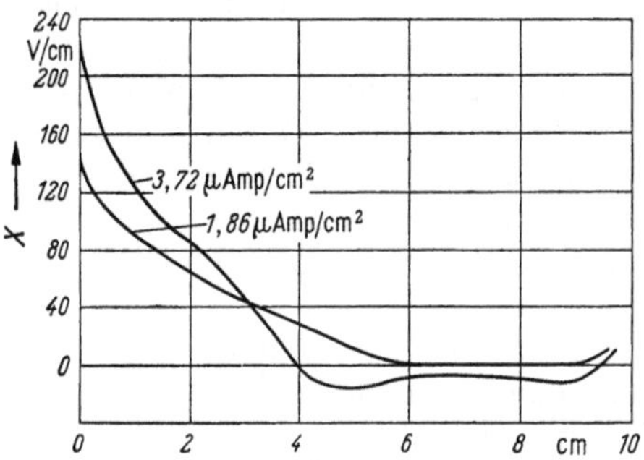

(a) N_2 $p = 55 \cdot 10^{-3}$ mm Hg

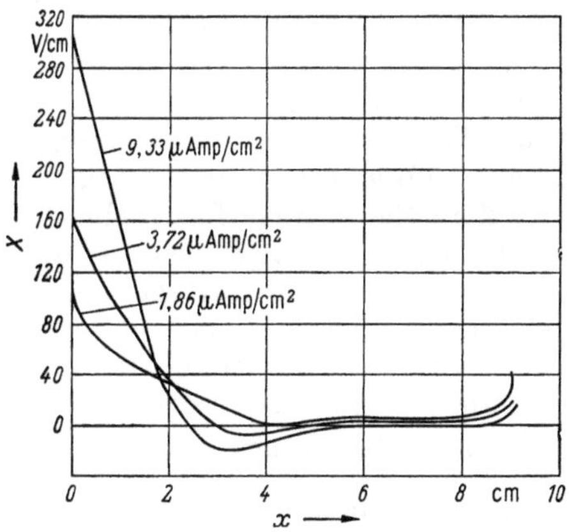

(b) N_2 $p = 9 \cdot 10^{-2}$ mm Hg.

Fig. 14a and b. Field in cathode dark space: STEIN [*20*]. x is the distance from the cathode.

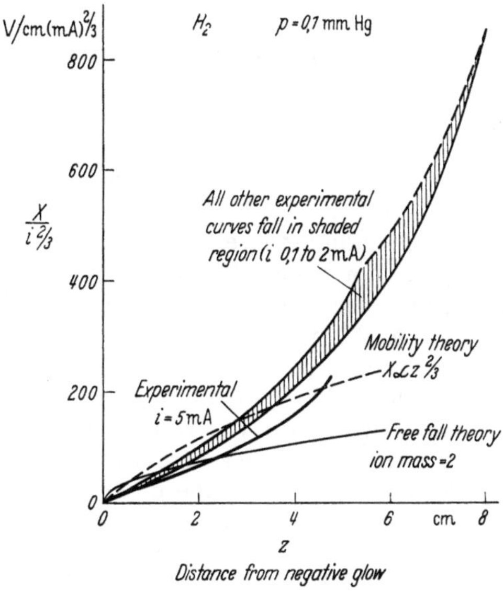

Fig. 14c. Field X in the cathode dark space, at low pressure at various discharge currents i, as a function of distance z from the cathode edge of the negative glow. WARREN [*307*].

---

[1] The effect of the tube width is discussed by FOSTER, who was able to reproduce varying results of earlier workers by altering the diameter of his tubes. S. J. FOSTER: Nature, Lond. **117**, 197 (1936).

exactly parallel to the plane of the cathode. This resulted in the integrated field being not quite equal to the measured cathode fall, and a correction was applied. However, it appears that there may have been slight errors in the measurements of the large field near the cathode due to curvature of the field lines, and these would affect the value of V_c obtained by integration, and hence the corrected values of X. Other beam measurements have not shown this discrepancy between $\int X\,dx$ and V_c, nor indicated negative fields in the glow. Probe measurements by EMELEUS and coworkers were at one time thought to show the existence of a negative field, but this was later regarded as a mistake (Sect. 39). The weight of evidence points to the field decreasing linearly through the dark space to nearly zero in the negative glow. It is, however, possible that a negative field in the glow may exist periodically due to striations[1]: this could give a spread-out deflection of the electron beam. At low pressure the decrease of field is slightly less linear (see Fig. 14a). The point where the extrapolated field becomes zero can be taken as the edge of the dark space.

Very recently a refined electron beam technique has been developed by WARREN [*307*] who has made extensive measurements of the field in the cathode dark space of abnormal discharges in air, N_2, H_2, He, A, to an accuracy of 2%. These all show an almost linear decrease of field to zero with distance from the cathode, except at the lowest pressures used (about 0.1 mm Hg). No negative fields were found in the negative glow. It should be noted that the fields measured by electron beam deflections are average fields across the diameter of the discharge, including the effects of the walls. If the pressure is not too low wall effects do not penetrate far into discharge, but WARREN finds that at $p \sim 0.03$ to 0.1 mm. Hg the field along the axis is about 10% greater near the negative glow, and about 10% less near the cathode, than the average field. Many of his results can be grouped in the form shown in Fig. 14c, $X/i^{\frac{2}{3}}$ being plotted against the distance $z\,(=-x)$ from the negative glow. In this way they can be compared with two theories of the cathode region (Sects. 33, 35) one which assumes that ions move across the dark space as though in vacuum (free fall), and predicts $X/i^{\frac{2}{3}} \propto z^{\frac{1}{3}}$; the other which assumes that ions have a drift velocity $v^+ \propto \sqrt{X}$, and predicts $X \propto z^{\frac{2}{3}}$. The latter gives rather better agreement.

25. Length of the cathode dark space. d may sometimes be measured visually with great accuracy, e.g. to a fraction of a millimetre in O_2. Such a sharp boundary between cathode dark space and glow occurs only in electro-negative gases which easily form negative ions thus preventing the diffusion of electrons. In other gases, and generally at low pressures the boundary is much less sharp and can sometimes be defined by electrical methods. The effect of moving the anode towards the cathode while keeping the current constant has already been described (Sect. 21), the point where the voltage rises abruptly can in some circumstances be used to define the edge of the dark space (see Fig. 6). This is only true for low values of V_c: in abnormal discharges the voltage begins to rise in the glow. Field measurements are also used to define d, the point where the extrapolated field becomes zero being taken as the boundary. The results of the three types of measurement agree well. Values obtained by GÜNTHERSCHULZE in H_2 are shown in Fig. 15, and for other gases in Fig. 16 [*46*], [*48*], [*49*], [*64*] to [*71*].

By the similarity theorems the essential parameter to describe the length of the dark space is not d but $p \cdot d$. For a given value of cathode fall V_c, $p d$

[1] See for example: J. J. THOMSON: Phil. Mag. **8**, 1 (1929), Ref. [*194*] and the literature in Sect. 65 to 68.

should be constant. Fig. 15 shows that this is not quite true, although when p changes by a factor 100, pd changes by only a factor 2. Deviations from the similarity laws are here probably due to heating of the gas. For the normal cathode fall there is, of course, a characteristic value pd_n corresponding to V_n for each combination of gas and cathode. These are given in Table 6.

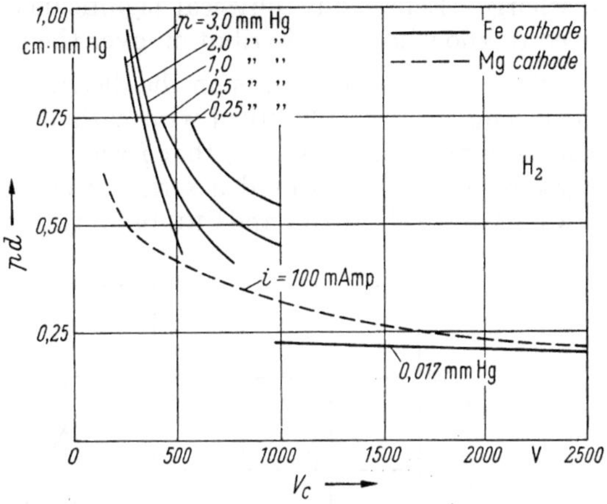

Fig. 15. Length of cathode dark space as $f(V_c)$: results of measurements in H_2 by GÜNTHERSCHULZE, collected by DRUYVESTEYN and PENNING.

Several empirical relations between d and other parameters have been found, e.g.

$$pd = A + \frac{Bp}{\sqrt{j}} \tag{25.1}$$

and

$$d\sqrt{j} = C + D(V - V_0) \tag{25.2}$$

both due to ASTON. Here A, B, C, D, V_0 are constants and j is the current density at the cathode[1]. ASTON also found empirically

$$j \propto \frac{V^{\frac{3}{2}}}{d^2} \tag{25.3}$$

which has the form given by a space charge limited current [72], [73].

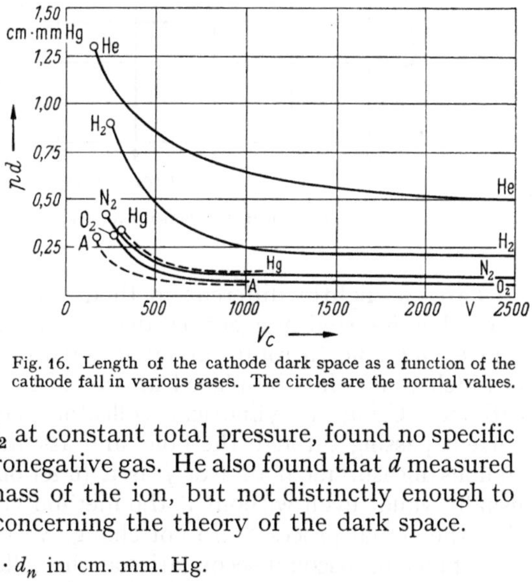

Fig. 16. Length of the cathode dark space as a function of the cathode fall in various gases. The circles are the normal values.

SPENCER SMITH [34] and, more recently, SEELIGER [74] have studied the effect on the thickness of the dark space when two gases are mixed in varying proportions. SPENCER SMITH noted that I_2 when added to He caused d to decrease, and thought that the effect was due to the replacement of He_2^+ by I_2^+ rather than the formation of negative ions. SEELIGER, using various binary mixtures of He, A, Xe, Hg, Cl_2 at constant total pressure, found no specific effect due to the presence of an electronegative gas. He also found that d measured visually appears to vary with the mass of the ion, but not distinctly enough to enable us to draw any conclusions concerning the theory of the dark space.

Table 6. $p \cdot d_n$ in cm. mm. Hg.

Cathode	He	Ne	A	H_2	N_2	Air	Hg	Other Gases
Cu				0.8		0.23	0.6	
Mg	1.45			0.61	0.35			O_2: 0.25
Hg				0.9				
Al	1.32	1.64	0.29	0.72	0.31	0.25	0.33	O_2: 0.24
C				0.9			0.69	
Fe	1.3	0.72	0.33	0.9	0.42	0.52		O_2: 0.31; Kr: 0.26 Xe: 0.23
Glass				0.8		0.3		

Also H_2 on cathode of Zn 0.8, Cd 0.87, Pb 0.84, Ni 0.9, Pt 1.0.

[1] V is the total potential across the discharge, which here is nearly equal to V_c.

26. The current density, and its distribution over the cathode surface. The average current density j has been derived by measuring the total current flowing to the cathode, using a normal, d.c. instrument (ammeter, microammeter or galvanometer), and dividing by the area of the cathode covered by the discharge. Before any physical significance can be given to the value found we need to know how its distribution over the cathode surface varies with V_c, p, etc.

The earliest experiments on the distribution of current were made by WILSON [*75*], who used a cylindrical wire cathode, and ASTON who used a plane disc with a guard ring, assuming that j was constant over the inner electrode. Later work by CHILDS who used a series of concentric rings showed that this is not always true [*11*]. ASTON and CHILDS worked at relatively low voltages (< 1000 V).

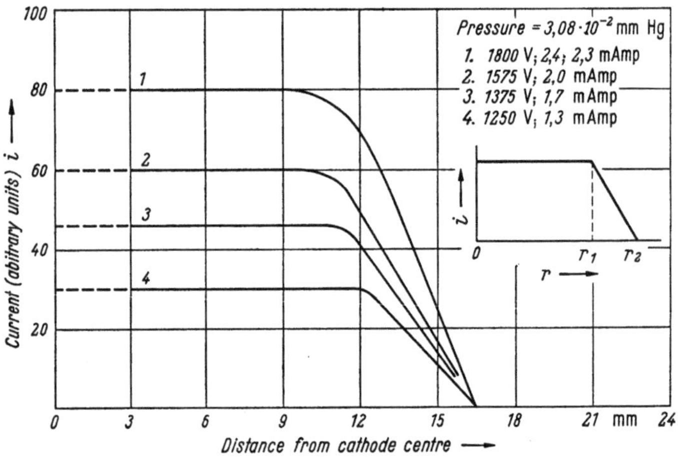

Fig. 17. Current distribution over the cathode surface at low pressure [*14*].

Later a slightly different method was developed by CHAUDHRI and BAQUI [*14*]: they inserted in a normal disc-shaped cathode a circular plug mounted off-centre, but able to rotate about its own centre. The plug had a small hole in it which could thus be placed at any distance from the centre of the complete cathode. Using a cylindrical collector placed behind, they measured the ion current passing through the hole, and assuming that $i \propto i_+$ thus obtained a continuous measurement of i or j as a function of the radius. The assumption is usually valid; even without it the method gives relative measurements provided that the actual processes do not change with the radius. No special precautions were taken to discount secondaries at the collector, but again this does not matter for relative measurements. Some results are given in Figs. 17 and 18. BREWER and MILLER find slightly different results at low pressure [*17a*].

These experiments show that there is always a central portion of uniform current density: the area of this depends only on the pressure, except at the lowest pressure when it decreases as V and i increase. The total area covered by the discharge varies likewise. Except at the lowest pressure ($\sim 10^{-2}$ mm.) the total area at the edge where j varies is too small to affect the average value calculated by dividing the total current by the total area. These measurements have been extended to higher V and lower p in studies on the canal ray discharge, which is discussed later. Each curve of i as $f(r)$ can be approximated by a constant part and a linear fall at the edge, as shown in Fig. 17. The uniform current density in the middle is given by

$$j = \frac{i}{\pi r_1^2 + \frac{\pi}{3}(r_2 + 2r_1)(r_2 - r_1)} \tag{26.1}$$

where i is the total discharge current. CHAUDHRI and BAQUI also confirmed ASTON's empirical law

$$V_c = E + F\frac{\sqrt{j}}{p} \tag{26.2}$$

and showed that it is true even at low pressure (<0.3 mm.) and in strongly abnormal discharges. It is valid only for that part of the cathode over which j is uniform.

These results show that in discharges with $V < 500$ volts and $p > 0.1$ mm. there is negligible error in j by neglecting the variation of the current at the edge of the discharge.

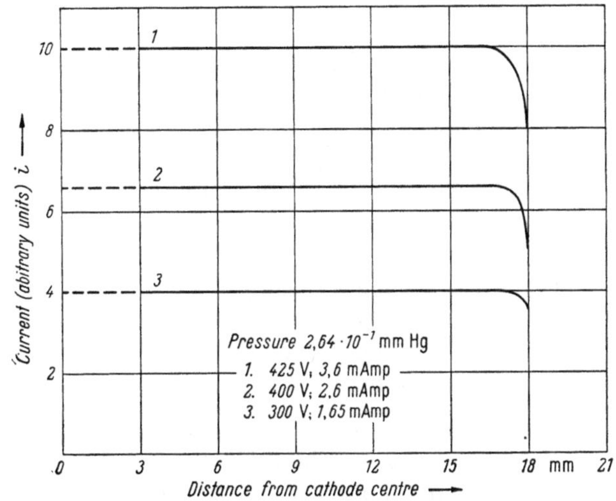

Fig. 18. Current distribution over the cathode surface at rather higher pressure [14].

27. Energy distribution of ions and electrons in the dark space. The current carried to the cathode by positive ions has been measured by making a small hole in the cathode and observing the current to a collector placed behind it: by magnetic deflection or by retarding potentials in front of the collector the energy distribution of the ions can also be measured. Unless proper precautions are taken the result can be falsified by two sources of error:

(a) secondary electron emission from the collector;

(b) ionisation in the gas caused by fast ions colliding with gas molecules.

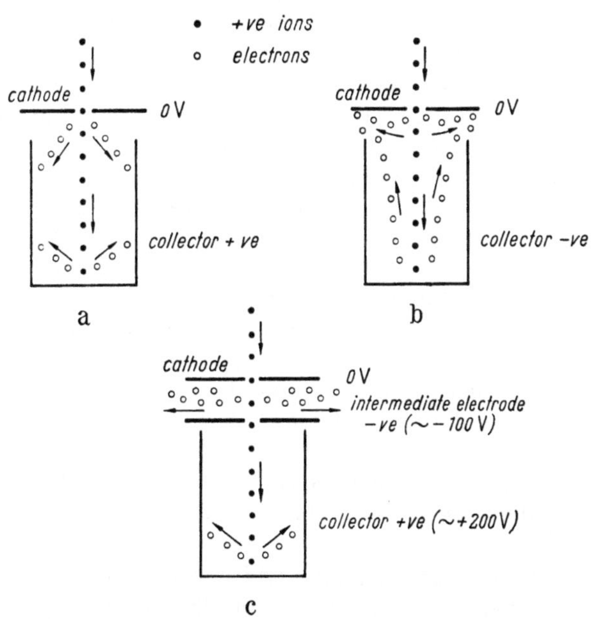

Fig. 19. Measurement of positive ion current to the cathode.

Both errors increase with increasing ion energy, but they can be eliminated by placing a grid at an intermediate potential at the entrance of the collector (see Fig. 19). The values of the relative potentials given there are typical.

Reliable experiments of this kind were done by CAMPAN [*16*], and CHAUDHRI and OLIPHANT [*17*] (see also [*17a*]). CAMPAN found that in nearly normal discharges in H_2, O_2, i_e/i_+ was between 6 and 16% (i_e, i_+ are the currents at the cathode). BREWER and MILLER [*17a*] found $i_+/i \propto 1/p$ between 0.5 and 10^{-2} mm. Hg, the fraction also becoming larger in heavier gases.

CHAUDHRI [*76*] has by a similar method also measured the energy distribution of the electrons in the dark space. Here the electrons are allowed to flow through a small hole in the *anode* into an evacuated space, where their energies are analysed. The general conclusions of all these measurements are:

(a) At low p, high V i.e. a strongly abnormal discharge, the electrons are almost homogeneous having energy $= eV_c$.

(b) At higher p lower V they have a wide range of energies. The form of the distribution curve is roughly independent of i and V, and the maximum of the curve always lies at the maximum energy $\approx eV_c$.

(c) At constant pressure more electrons possess the full energy when the current is increased.

(d) With increasing pressure relatively more electrons have low energies, and the fraction of current carried to the *anode* by the electrons decreases.

It therefore appears that, except at high pressure, nearly all the electrons originate at the cathode.

28. Optical properties of the dark space: first cathode layer, cathode glow, and ASTON dark space. The cathode dark space is not completely dark; it emits a faint light but the intensity is very much smaller than that from the negative glow. The light is mainly due to excitation of gas molecules by electron collisions [*77*]. Apart from this general faint luminosity there are bright layers close to the cathode, of three distinct types although these are often confused under the general term "cathode glow". The three kinds are:

I. The *first cathode layer*, separated from the cathode by the thin ASTON *dark space*[1], length d_a. This is observed only at low pressure at low values of V_c, and in the rare gases, although it is thought to be a general phenomenon [*80*]. The ASTON dark space is completely dark, and the first cathode layer appears at the point where the electrons have first acquired enough energy to excite the gas[2]. A little further from the cathode a second and third layer can be seen, much fainter than the first (see Fig. 2). These layers were observed by HOLST and OOSTERHUIS in a TOWNSEND discharge, and explained in this way [*82*], [*114*].

II. The *cathode glow*, which appears at higher pressure and larger values of V_c. This looks like a velvety coating on the surface of the cathode and is brighter than the first cathode layer; no dark space exists between the cathode and this light. The cathode glow is due to excitation of gas molecules by positive ions from the dark space, or by fast neutral atoms formed by charge transfer [*83*] to [*85*]. If a hole is drilled in the cathode the light can be seen emanating from the path of the beam of ions which penetrates behind the cathode (canal rays).

III. The *cathode light*[3], which occurs only when the cathode is covered with a thin metallic or oxide layer of alkali or alkaline earth (MgO, Na_2O, Li_2O) [*64*]. The light comes from excited atoms of these metals sputtered from the surface, and consists of their strongest spectral lines. The sputtered atoms have energies of about 5 eV.

---

[1] Sometimes called the primary dark space [*78*], [*79*].

[2] Probe measurements (not very accurate in this region) have shown that the space potential here lies between the resonance and ionisation potentials of the gas [*81*].

[3] This name suggested by DRUYVESTEYN and PENNING. For references see [*8*], p. 130.

Length, d_a, of the ASTON *dark space.*

The thickness of the ASTON dark space and its dependence on the cathode fall have been measured in several pure gases (He, Ne, Kr, Xe, A, H_2, N_2, O_2) and some mixtures (He+Na, Li, Cs, Tl) by GÜNTHERSCHULZE and KELLER [*64*]. d_a decreases more strongly than d (cathode dark space length) with increasing V_c (see Fig. 20). The product $p d_a$ obeys the similarity law well: for example in He a change in p by a factor of 100 causes $p d_a$ to change by only 20%. This would be expected since the first cathode layer is produced by single collisions, electrons exciting the gas molecules to their lowest excited level. Since these energy levels are known it would seem that the electric field in the dark space could be calculated. There have been attempts to do so by GÜNTHERSCHULZE and KELLER, and also by KOSSEL, but the results have not been satisfactory [*64*], [*79*], [*78*].

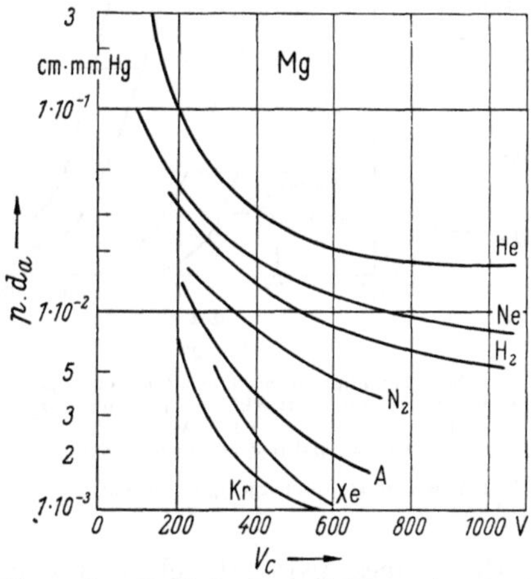

Fig. 20. Length of the ASTON dark space (expressed as $p d_a$) as a function of the cathode fall. For different cathode materials the ordinate ($p d_a$) should be multiplied by: Mg×1, Al×0.88, Cu×1.1, Zn×1.02, Fe×1.35. The curves for Ne, A, H_2, N_2 are partly extrapolated.

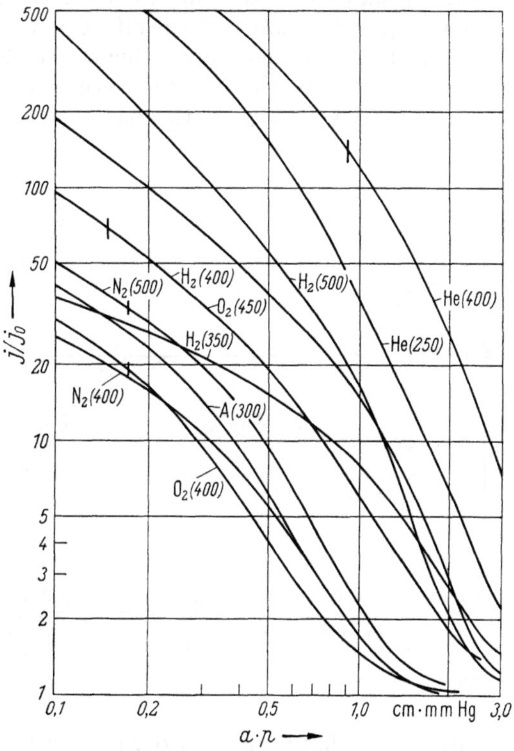

Fig. 21. The hollow cathode effect. j/j_0 as a function of the distance a between the electrodes, the cathode fall being kept constant (its value in volts indicated in brackets). The position of the vertical lines gives the length of the cathode dark space ($d \cdot p$), and j_0 is the current density in a conventional discharge.

29. Effect of cathode geometry; the hollow cathode discharge; concave and convex cathodes. Experiments already described have shown that electrons in the cathode dark space behave very much like a beam directed normally away from the cathode surface. If two plane cathodes were to be placed facing each other the beams from each would be in opposite directions and the negative glows would coalesce. We would expect a discharge with such a cathode arrangement to have unusual properties which would depend on the distance apart of the cathodes. A cathode in the form of a hollow cylinder would be expected to produce essentially similar results, the only difference being rather more complicated geometry and the dependence of the properties on the diameter, and to a smaller degree, the length of the cylinder.

A discharge with such electrode arrangements is called a *hollow cathode,* or sometimes a SCHÜLER, discharge. Its main properties are a current density orders of magnitude larger than in a conventional discharge for the same cathode fall, and a negative glow giving most intense light. For this reason it has been widely used as a spectroscopic source; the low cathode fall energy results in a low gas temperature and so there is little line broadening due to the DOPPLER effect[1]. These unusual properties are, of course, more marked when the cathodes are closer together. The effect occurs in rare and molecular gases, and has been studied by PASCHEN, SCHÜLER, VAN VOORHIS and SHENSTONE, GÜNTHERSCHULZE, LOMPE, SEELIGER, and WOLTER, and LITTLE and VON ENGEL [*86*] to [*91*], [*19*]. Fig. 21 shows how the current density increases with decreasing cathode separation a (the "hollow cathode effect"); the optical properties are discussed in another article. Recently LITTLE and VON ENGEL [*19*] have measured the fields in the cathode dark spaces of the hollow

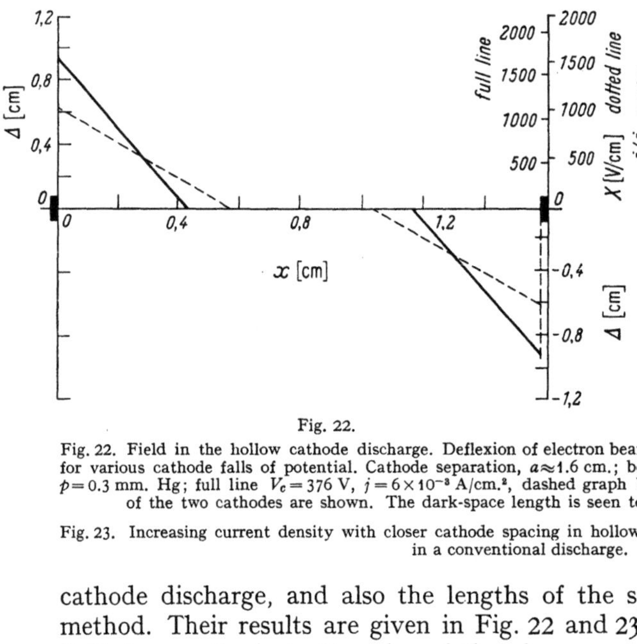

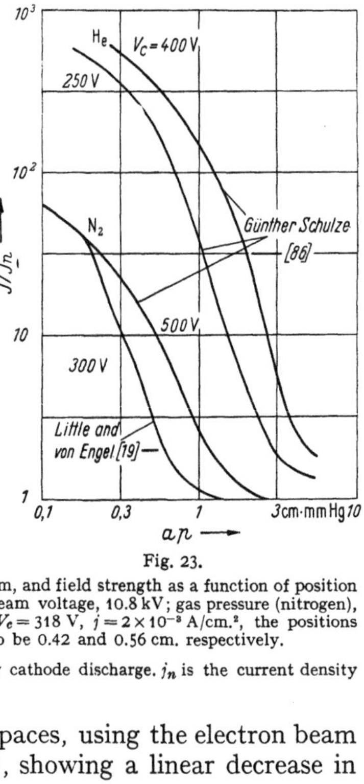

Fig. 22.

Fig. 23.

Fig. 22. Field in the hollow cathode discharge. Deflexion of electron beam, and field strength as a function of position for various cathode falls of potential. Cathode separation, $a \approx 1.6$ cm.; beam voltage, 10.8 kV; gas pressure (nitrogen), $p = 0.3$ mm. Hg; full line $V_c = 376$ V, $j = 6 \times 10^{-3}$ A/cm.², dashed graph $V_c = 318$ V, $j = 2 \times 10^{-3}$ A/cm.², the positions of the two cathodes are shown. The dark-space length is seen to be 0.42 and 0.56 cm. respectively.

Fig. 23. Increasing current density with closer cathode spacing in hollow cathode discharge. j_n is the current density in a conventional discharge.

cathode discharge, and also the lengths of the spaces, using the electron beam method. Their results are given in Fig. 22 and 23, showing a linear decrease in field from cathode to the edge of the glow.

The theory of the hollow cathode discharge is well understood physically, but good quantitative agreement with experiment is not possible until more detailed knowledge of certain constants is known. This is discussed in Sect. 37 where a recent theory believed to be applicable both to conventional and hollow cathode discharges is described. Physically it is clear that by arranging the cathodes in this way many of the products of ionisation and excitation which in an ordinary discharge are lost to the walls will now fall upon one of the cathodes and by secondary processes contribute to the current. A smaller separation between the cathodes reduces the length of the dark space, which at constant V_c increases the field and hence the density of ions. This also causes the ions to strike the cathode with larger velocity and the current density rises rapidly. The intense light from the glow also produces copious photoelectric emission from the cathodes.

[1] See for example measurements of temperature in a hollow cathode discharge (N_2, N_2—He mixtures), from the band spectra of N_2^+ and the broadening of He lines. Compare Sect. 14. R. SCHOEN and J. R. HOLMES: J. Opt. Soc. Amer. **44**, 502 (1954).

Some empirical laws result from the measurements mentioned: firstly Fig. 23 shows that all the curves of j/j_n plotted against ap are similar in shape except for the upper parts. This divergence is probably due to heating of the gas at large current densities, causing a reduction in gas density, or equivalent pressure. Hence the true values of ap are smaller and of j/p^2 are larger. No heating occurs at low currents and no such correction is needed for j_n, so the net result is that j/j_n should be larger than the uncorrected value. Similarity rules were found to hold in these measurements.

The parameter ap is not fundamental since it includes the lengths of dark spaces and glow. However, LITTLE and VON ENGEL, who measured d, find:

$$\frac{j}{p^2} \propto \frac{1}{(pd)^{\frac{5}{2}}} \quad \text{approximately} \qquad (29.1)$$

also for small values of ap i.e. small cathode separation,

$$\frac{j}{p^2} \propto \frac{1}{(ap)^{\frac{5}{2}}} \quad \text{approximately} \qquad (29.2)$$

hence in this range $a \propto d$.

At large values of ap two independent glows appear. Also it seems probable that

$$\frac{j}{p^2} \propto \frac{V_c^3}{(pd)^{\frac{5}{2}}} \qquad (29.3)$$

although measurements were made over only a small range of V_c because of the difficulty of keeping the gas cool. The value of j used here is corrected for its non-uniform distribution over the cathode surfaces, it being greater near the anode side.

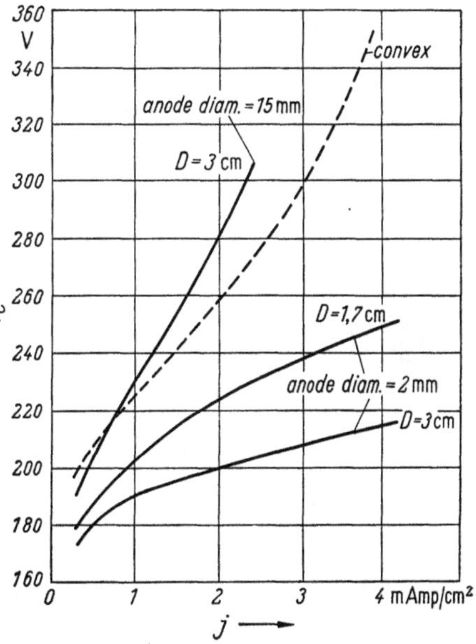

Fig. 24. Current density as a function of V_c for hollow cathodes of various diameters D (LOMPE and SEELIGER). $p = 2$ mm., length of cathode 13 cms.

Experiments on concave cathodes were performed many years ago by CAMPBELL SWINTON and later by ASTON, showing that a focussing effect is produced and the ionisation is localised [*92*], [*93*]. Later experiments have shown how the cathode fall and current density vary with the curvature of the cathode (Fig. 24). Convex cathodes (e.g. any cathode in the form of a rod) produce the opposite effect [*109*], the current density being reduced for the same cathode fall.

30. Energy balance at the cathode. The power generated in the cathode region iV_c is dissipated mainly in the heating of the cathode and the gas: radiation can be neglected and also heating of the walls and anode if they are far enough away [*94*]. The heating of the cathode is due to the impact of positive ions; we regard them as falling through an average potential drop $\overline{V}$, then the power P_c delivered at the cathode is

$$P_c = i_+ \overline{V} = k i V_c \qquad (30.1)$$

where k is a fraction < 1. We neglect the ionisation energy eV_i.

Now in a strongly abnormal discharge $\overline{V} = V_c$ very nearly: the ions are so fast that they are hardly deflected in their few collisions with gas molecules, and the gas molecules which are hit by the ions probably arrive at the cathode also. Ions which lose their charge by charge transfer arrive at the cathode as neutral molecules but still with almost the full kinetic energy[1]. Putting $\overline{V}=V_c$, and remembering that $i=i_+ + i_e = i_+(1+\gamma)$ we get

$$k(1+\gamma)=1, \quad \gamma=\frac{1-k}{k}. \tag{30.2}$$

Hence by measuring k, γ can be found. Here γ is a general factor, defined by $\gamma=i_e/i_+$, and not necessarily the coefficient for the emission of electrons by the ions themselves. GÜNTHERSCHULZE and his coworkers [*95*], [*96*] measured the heating of the cathode at low pressures, and derived for large cathode falls only,

$$\gamma = A(V_c - V_0) \tag{30.3}$$

where A, V_0 are constants, A depending on the cathode material, V_0 being between 100 and 400 volts. This is valid when $V_c \gg V_0$, corresponding to $\gamma > 0.1$.

At higher pressures (1 to 8 mm. Hg) and lower cathode falls ($V_c \simeq V_n$) as much as 40% of the power goes into heating the gas [*97*], [*98*].

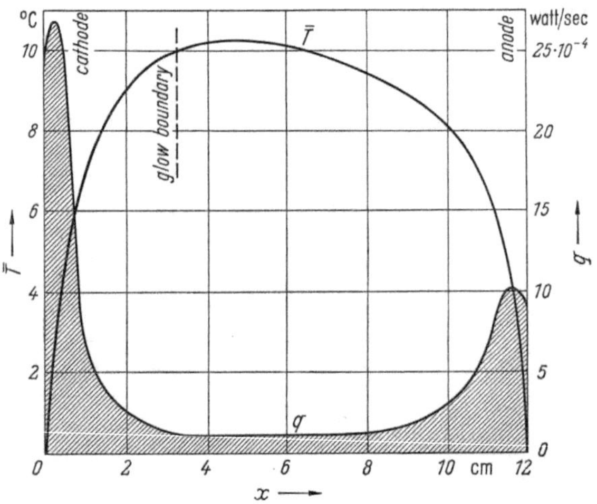

Fig. 25. Distribution of mean temperature $\overline{T}$ and heat production q along a discharge in H_2. x is the distance from the cathode [*98*]. The discharge has no positive column.

31. Temperature distribution along the discharge. Measurements made by the methods described in Sect. 14 have shown that the temperature is highest in the cathode dark space, decreases through the negative glow to a minimum in the FARADAY dark space and rises to a small constant value in the positive column (e.g. WOOD 1896, bolometer measurements). Fig. 25 shows a modern measurement in a discharge without a positive column. This distribution is what we would expect, since most of the electrical power is expended in the cathode dark space ($V_c \sim V_{\text{tube}}$). Not all of the energy dissipated in this region is used in heating the gas, however; if it were, a temperature $T \sim 2000°$ C would be attained in an abnormal discharge with $p \sim 0.1$ mm., whereas $T > 200°$ C is rarely found [*96*]. Most of the energy goes into heating the cathode. Very much higher gas temperatures could be expected at rather higher pressures (Sect. 96).

Band spectra measurements[2] in the positive column of N_2 in capillary tubes have confirmed the parabolic radial distribution of temperature which is to be expected if the production of heat is uniform or varies as the zero order BESSEL function across the tube (see Sect. 62). Here the gradient was high (probably ~100 V/cm.) and axial temperatures of order 500° C were found. Lower temperatures were found by interference methods, however [*45*].

32. Fluctuations (striations) in the cathode fall. Until recently it has been thought that striations occur only in the positive column, but the work of DONAHUE and DIEKE [*194*] has suggested that the striations observed there travel in a modified and less pronounced form, through the negative glow and dark space to the

[1] I. RUNGE: Z. Physik **61**, 174 (1930).
[2] T. HAMANDA: Electrotechn. J. **1**, 57 (1937). — ORNSTEIN, BRINKMAN and HAMANDA: Proc. Roy. Acad. Amsterd. **39**, 50, 315, 484 (1936).

cathode. The striations themselves cannot be followed through the cathode dark space, but weak oscillations in the light from the cathode glow have been seen. LOEB thinks that fluctuations in the cathode region due to rapidly changing surface conditions may indeed be the cause of the striations in the positive column: this is discussed later. Striations have also been found to affect measurements of field strength made by the STARK effect (footnote 1, Sect. 24).

b) Theory.

33. Theory of the cathode fall. The object of any theory of the cathode fall is to obtain relations between V_c, j, and d, and to calculate them from the pressure and known constants of the gas and cathode material (e.g. α, or the ionisation function of the gas, and γ for the cathode). The dependences are usually expressed in terms of the similarity parameters V_c, j/p^2 and pd. The theory should also give the variation of the field X in the dark space, and demonstrate the existence of a minimum value of $V_c = V_n$, the normal cathode fall.

We must distinguish now between theories which regard the cathode dark space as an isolated independent unit, and those which consider the dark space and the negative glow together as the unit whose processes sustain the discharge. All the earlier theories belong to the first group; here the boundary of the negative glow is regarded simply as another electrode, sometimes also as a source of ions, to allow for those which flow into the dark space from the glow. The theory consists essentially of solving the fundamental equation of maintenance (see Sect. 20)

$$\gamma(M-1) = 1 \tag{33.1}$$

although not necessarily in this form.

Theories of the second group, including the effect of the negative glow (more than merely as a source of ions) give a slightly different maintenance equation.

In some early theories the cathode dark space was regarded as a plane space charge layer, of thickness d, between two parallel plates, the field being zero at the edge of the glow. The ions were assumed not to hit any gas molecules in their flight across the space, and the space charge limited current density was given by

$$j = \frac{\sqrt{2}}{9\pi} \cdot \sqrt{\frac{e}{M}} \cdot \frac{V_c^{\frac{3}{2}}}{d^2} \tag{33.2}$$

where M is the mass of the ion.

This equation was applied by GÜNTHERSCHULZE to the normal, and independently by RYDE to the abnormal, cathode fall, where the assumptions are more justified [*99*], [*100*]. RYDE's results are compared in Fig. 26 with the experimental values of ASTON and WATSON [*73*], who found the empirical law

$$j \propto \frac{(V-G)^2}{d^2} \qquad (G = \text{constant}). \tag{33.3}$$

The agreement is remarkably good in view of the assumptions made, and becomes better when the secondary emission due to ions (coefficient γ_i) is considered. Putting $j = j^+(1+\gamma_i)$ gives

$$j = \frac{\sqrt{2}}{9\pi} \sqrt{\frac{e}{M}} (1+\gamma_i) \frac{V_c^{\frac{3}{2}}}{d^2}. \tag{33.4}$$

The theory gives too large values of j, probably due to the neglect of ion collisions.

A major fault in these theories is their prediction that the field X should vary as $z^{\frac{1}{3}}$, where z is the distance from the negative glow boundary (see Sect. 24) whereas ASTON had found a linear variation: also it is not explained how ionisation occurs, since one must assume that electrons also do not hit gas molecules.

J. J. and G. P. THOMSON ([3], vol. 2) outline a qualitative theory, inserting expressions for M and γ in eq. (33.1), showing that the cathode fall should depend on the cathode material, ionisation potential and ionisation function of the gas.

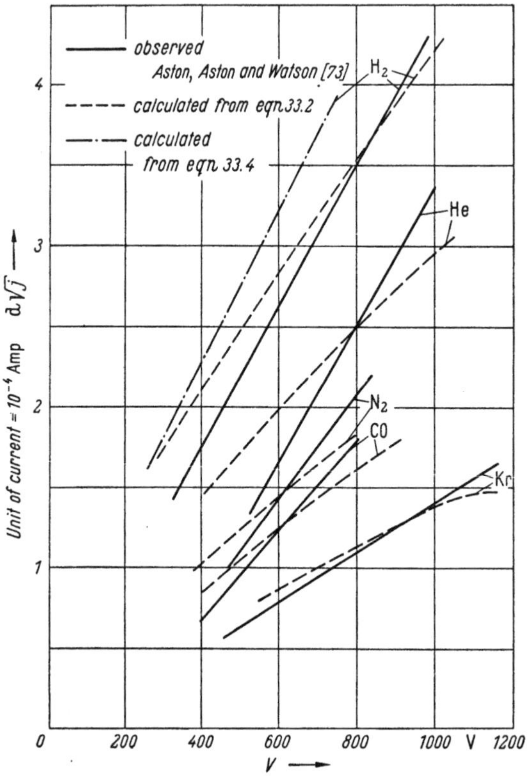

Fig. 26. Comparison of RYDE's theory with experimental results.

Many attempts have been made at a more general solution of the cathode fall, and the following section gives an outline of the methods employed. This refers only to theories of the first type.

34. General theory, neglecting radiation from the negative glow.

The physical picture is as follows. Electrons are released from a plane cathode, one electron for every γ ions that hit it. These electrons starting with a small energy (~ 1 eV) are accelerated towards the negative glow, ionise gas molecules and produce ions at a rate which increases towards the boundary of the glow. The space charge of the ions and electrons determines the field which largely controls their rate of generation and removal. At the cathode there are few electrons and the current is carried mainly by ions: at the boundary of the glow, where the field is very small, the numbers are more equal but because of their greater mobility the electrons carry the current. The cathode fall V_c is defined as the potential between the cathode and a plane near the boundary of the glow such that small changes in the position of this plane make little difference to V_c. The equations describing this picture are now given, x being measured from the cathode.

The total current density j is related to the secondary process at the cathode:

$$j = j_c^+ (1 + \gamma) = e N_c^+ v_c^+ (1 + \gamma) \tag{34.1}$$

where j_c^+ is the ion current density at the cathode, N_c^+, v_c^+ the corresponding ion density and drift velocity.

The multiplication M in the gas requires a complete knowledge of the electric field in the dark space, and the ionisation coefficient or ionisation function of the gas. This can be described by 3 sets of equations (taking here only the x component i.e. along the axis of the discharge).

I. POISSON's equation:

$$\frac{dX}{dx} = -4\pi e (N^+ - N^-) \approx -4\pi e N^+ \tag{34.2}$$

the contribution of the electrons (N^-) usually being neglected on account of their high speeds.

II. Continuity: in a steady state the rates of production and loss of ions and electrons are equal in any elementary volume. Suppose z is the number of ionisations per cm.³ per sec. at some point x, where the field is X, then

$$\left.\begin{aligned} &\text{For ions: } z = f(X) = \frac{d}{dx}(N^+ v^+), \\ &\qquad\qquad \text{diffusion of ions being usually neglected.} \end{aligned}\right\} \quad (34.3)$$

$$\text{For electrons } z = f(X) = \frac{d}{dx}\left\{N^- v^- + D\frac{dN^-}{dx}\right\}. \quad (34.4)$$

Some authors have taken z as determined by the potential rather than the field, but this does not affect the formal development here. $f(X)$ is of course a known function.

III. Mobility:

$$\text{For ions } v^+ = F(X) \text{ usually} = b_+ X \text{ (constant mobility) or } \propto \sqrt{X}, \quad (34.5)$$

$$\text{For electrons similarly } v^- = b^- X \quad \text{or } \propto \sqrt{X}. \quad (34.6)$$

These three sets of equations suffice, in principle, to determine X, N^+, N^- each as $f(x)$. A boundary condition $X=0$ when $x=d$ introduces the length of the dark space. Since N^+, v^+ are both known, j_c^+ $(= eN^+v^+$ at $x=0)$ can be calculated.

Finally

$$V_c = \int_0^d X\,dx \quad (34.7)$$

gives sufficient equations to determine V_c, j and d.

This treatment uses all the equations of balance and therefore *includes* the fundamental equation of maintenance (33.1) which does not have to be introduced separately.

The equations are also quite general and with suitable modification (e.g. the introduction of recombination, electron mobility and radial diffusion terms) could be made to apply to any part of the discharge. Several authors [*101*], [*102*], [*103*] have attempted in this way to derive expressions describing the whole discharge[1], but the solutions depend upon the boundary conditions chosen. This requires that a great deal must be known about the discharge already, and the various regions of the discharge have to be treated separately. Thus the solutions obtained are not as general as they appear at first sight.

35. Actual theories: comparison with experiment. Various theories have been derived from a selection of the equations given. Most authors use TOWNSEND'S α for the rate of ionisation, and write $e^{\int_0^d \alpha\,dx}$ for the multiplication M. They make different assumptions about v^+ and α as $f(X)$, and about the boundary condition giving the ion current at the edge of the glow. Others have preferred to omit some of the equations, using instead the experimentally determined distribution of the field in the dark space; these are the only theories which

[1] These will not be treated here, except as they deal with the cathode fall. We refer to the review of M. J. DRUYVESTEYN and F. M. PENNING: Rev. Mod. Phys. **12**, 87 (1940).

have been extensively compared with experiment. A further variation is the introduction of a minimum energy principle to account for the normal cathode fall.

α) COMPTON and MORSE (1927) [*104*] invoked this minimum principle, saying that the potential distribution which arises is that most favourable to the passage of the current, subject to the limitations of POISSON's equation[1]. They neglect recombination and diffusion, and find the maximum value of the multiplication $M = e^{\int_0^d \alpha\,dx}$ for a given $V (= V_n)$. α is assumed to be given by TOWNSEND's relation

$$\frac{\alpha}{p} = A\,e^{-\frac{B}{X/p}} \tag{35.1}$$

where A and B are constants. Recognising that the field changes in space, they use here instead of the actual field X_x, its value half a mean free path nearer the cathode: α thus depends on X and dX/dx. With a somewhat arbitrary, but reasonable, definition of d_n they obtain

$$V_x = 1 \cdot 2\,V_n \left(1 - e^{-\frac{7}{4} \cdot \frac{p x}{p d_n}}\right), \tag{35.2}$$

$$p\,d_n = 0.85\,\frac{V_n}{B}\left(1 + \frac{B}{0.97 A V_n} + \text{smaller terms}\right), \tag{35.3}$$

$$\frac{j_n}{p^2} = \text{const}\,\sqrt{\frac{\lambda_+}{M}}\,\frac{V_n^{\frac{3}{2}}}{1 - \frac{N_0^+}{N_d^+}} \tag{35.4}$$

where N_0^+, N_d^+ are the ion concentrations at $x = 0$, d; their ratio being calculated from POISSON's equation and mobility. The results were not in very good agreement with experiment, except for the corollary of eq. (35.3) which shows that $\frac{p d_n}{V_n}$ is almost a constant depending only on the gas and not on the cathode material.

β) MORSE (1928) [*105*] abandoned the minimum principle and used the general equations given, assuming $v^+ \propto \sqrt{X}$. He also assumed:

$$\frac{D_e}{b_e} = \frac{D_+}{b^+} = \text{const} \tag{35.5}$$

which is most dubious, since it is equivalent to assuming $T_+ = T_e$. He derived

$$X = C_1\,e^{-A' p x} + C_0 \tag{35.6}$$

where A' depends on the gas. This is of the same form as that found by using the minimum principle, and of course does not agree with the measured variation. Due to the difficulty of deriving suitable boundary conditions, the expressions for j/p^2, pd and V_c involve an arbitrary constant: this has the value zero for the normal cathode fall. Comparison with experiment was possible only from the ratios of V_n obtained with various combinations of gas and cathode; better agreement was found than in the earlier theory. This theory also shows that a constant V_n must exist, a result previously assumed.

γ) VON ENGEL and STEENBECK (1934) ([4], vol. 2) omitted equations of continuity and used instead the known linear distribution of the field

$$\frac{X}{X_c} = 1 - \frac{x}{d} \tag{35.7}$$

[1] See also A. v. HIPPEL: Z. Physik **76**, 1 (1932). — W. ROGOWSKI: Z. Physik **82**, 473 (1933).

where X_c is the field at the cathode. Assuming $v^+ \propto X$, α given by the eq. (35.1) and taking $j_+ = 0$ at the boundary of the glow ($x = d$) they derived:

$$j = \frac{V_c^2}{d^3} \frac{b^+}{\pi} (1 + \gamma), \tag{35.8}$$

$$\left(C_2 \frac{j}{p^2}\right)^{\frac{2}{3}} = (C_1 V_c)^{\frac{1}{3}} \cdot f\left(C_1 V_c \cdot C_2 \frac{j}{p^2}\right)^{\frac{1}{3}} \tag{35.9}$$

where C_1, C_2 are constants involving γ and properties of the gas,

$$f(x) = \int_0^x e^{-\frac{1}{y}} dy. \tag{35.10}$$

The length of the dark space is derived from the basic eq. (33.1) written in the form

$$\int_0^d \alpha\, dx = \log\left(1 + \frac{1}{\gamma}\right). \tag{35.11}$$

These equations obey similarity laws, and the second one has a form independent of the gas and cathode material. As j/p^2 is varied V_c passes through a minimum, which corresponds to the normal cathode fall V_n; we then put $j = j_n$, $d = d_n$. When this point is reached (j having decreased to j_n) further decrease of the discharge current causes no change in the current density, which remains constant at j_n, while the discharge contracts to cover a smaller area of the cathode. V_c does not follow the dotted part of the curve (Fig. 27) but remains equal to V_n. The theory can be compared with experiment over a wide range of conditions (Figs. 28 to 30), and with the exception of He gives quite good agreement[1]. Wide discrepancy occurs in He also for other theories and one reason is probably that the curve of α as $f(X)$ in this gas is not the S-shape described by Townsend's eq. (35.1). Smaller variations are found for argon.

δ) Rogowski (1932, 1939), Weizel, Rompe, and Schön (1939) [*101*], [*103*] have derived theories based on the general equations of Sect. 34, the later treatment being an extension of Rogowski's. Diffusion is neglected, α is taken as a constant, equal to its maximum value at the cathode surface, v^+ and v^- are both assumed $\propto \sqrt{X}$ and a small ion current is assumed to flow from the glow, its value being estimated from the ion density in the negative glow. The properties of the glow are a separate part of the theory. They derive the equation

$$\frac{d}{dx}\left(\sqrt{X} \frac{dX}{dx}\right) = -\alpha \sqrt{X} \frac{dX}{dx} + \frac{4\pi j a}{b^+} \tag{35.12}$$

and with the boundary conditions:

$$x = 0, \quad j_e = \gamma j_+; \qquad x = -d, \; X = 0 \text{ and } j_+ = j_0 \tag{35.13}$$

(they measure x negative moving away from the cathode), the solutions are

$$V_c = \left(\frac{6\pi}{b^+ p} \cdot \frac{j}{p^2}\right)^{\frac{2}{3}} \left(\frac{\alpha}{p}\right)^{\frac{5}{3}} \cdot f(y_0, y_d) \tag{35.14}$$

where $y_0 = \frac{1}{1+\gamma}$, $y_d = \frac{j_0}{j}$ and f is a complicated function.

$$p\, d = \frac{1}{(\alpha/p)} \log \frac{\left(1 - \frac{j_0}{j}\right)}{\gamma}. \tag{35.15}$$

[1] See M. J. Druyvesteyn: Physica, Haag **5**, 875 (1938), for an extensive test of the theory in abnormal discharges.

Comparison with experiment is difficult due to lack of precise data for j_0, the ion current from the glow; but with the meagre data available γ calculated from (35.15) rises from 2 to $6 \cdot 10^{-2}$ as V_c increases from 50 to 200 eV, in agreement with experiment. V_c is implicit in eq. (35.15) through the value of α which depends on X/p at the cathode.

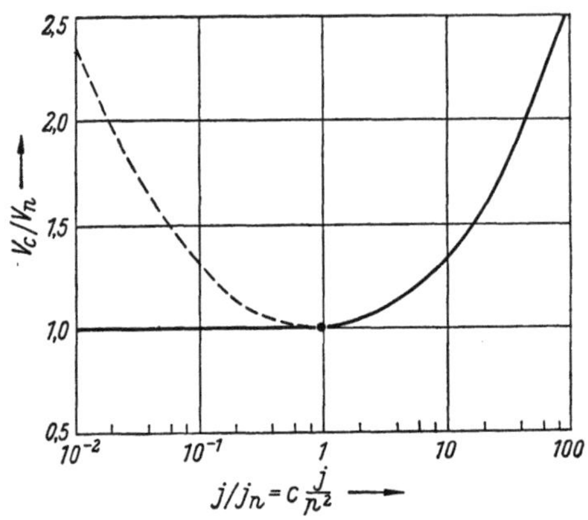

Fig. 27.

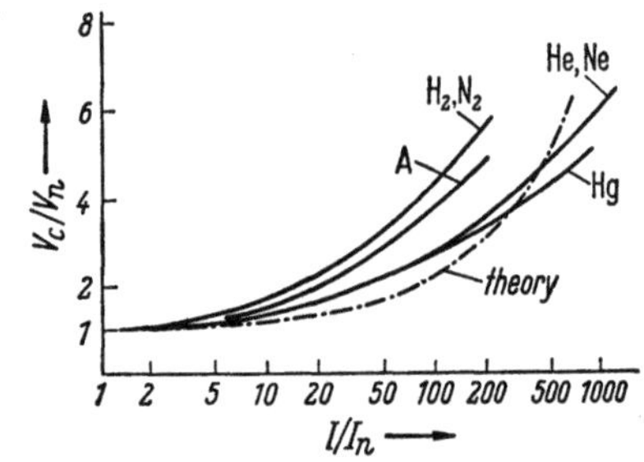

Fig. 28.

Fig. 27. Calculated values of V_c/V_n as a function of j/j_n (V. ENGEL and STEENBECK).

Fig. 28. Cathode fall (iron cathode) as a function of the current density from GÜNTHERSCHULZE's experiments [46], [47] compared with theory (v. ENGEL and STEENBECK).

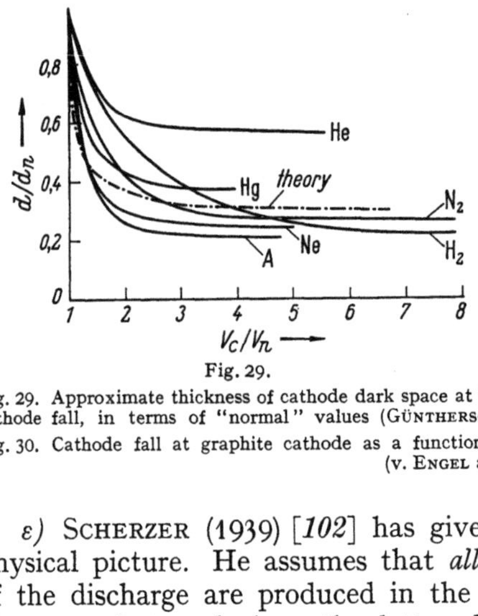

Fig. 29.

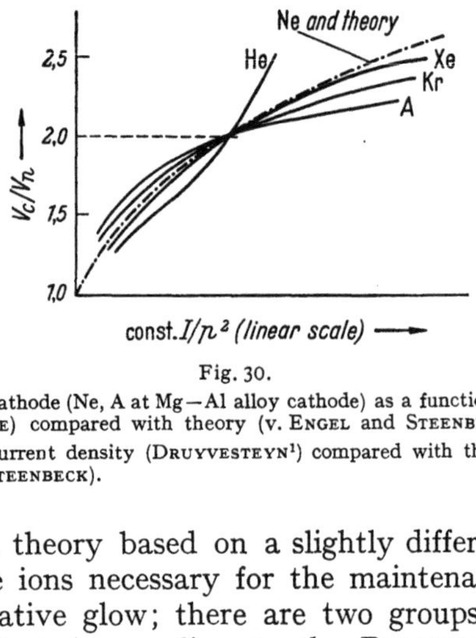

Fig. 30.

Fig. 29. Approximate thickness of cathode dark space at iron cathode (Ne, A at Mg—Al alloy cathode) as a function of cathode fall, in terms of "normal" values (GÜNTHERSCHULZE) compared with theory (v. ENGEL and STEENBECK).

Fig. 30. Cathode fall at graphite cathode as a function of current density (DRUYVESTEYN[1]) compared with theory (v. ENGEL and STEENBECK).

ε) SCHERZER (1939) [*102*] has given a theory based on a slightly different physical picture. He assumes that *all* the ions necessary for the maintenance of the discharge are produced in the negative glow; there are two groups of electrons, fast and slow, the latter distributed according to the BOLTZMANN relation

$$N^- = a \cdot e^{-\frac{eV}{kT_e}}. \tag{35.16}$$

a is determined by the electron concentration in the glow. The ion velocity is taken as $v^+ \propto X = b^+ X$, where b^+ is not the normal mobility equation but a modified form due to the effects of charge transfer

$$b^+ = \frac{e\lambda_+}{\sqrt{3.5\, M\left(V_r + \frac{e X \lambda_+}{14}\right)}} \tag{35.17}$$

(V_r = least excitation potential of the gas).

[1] M. J. DRUYVESTEYN: Physica, Haag **5**, 975 (1938).

α is assumed to be negligibly small in the dark space, and constant in the negative glow: the equilibrium of charged particles is maintained by recombination described by a constant coefficient r. The basic equation derived (omitting numerical constants) is of the form:

$$\frac{d^2}{dx^2}(b^+ X^2 + b^+ k T_e \cdot N^-) = -r e N^- \frac{dX}{dx} - r(e N^-)^2 + \alpha j_s \tag{35.18}$$

where j_s is the current density carried across the dark space by fast electrons.

No complete solution can be found, but approximations can be made by dividing the discharge into dark space, ionisation and recombination zones. The boundary conditions are defined at the cathode and the anode, which lies near the far edge of the glow. The results show a linear potential drop across the dark space, and its length is given by

$$d \propto \frac{V^{\frac{1}{3}}}{\alpha_0^{\frac{2}{3}}} \tag{35.19}$$

where V is now the total potential between cathode and anode (in the conditions specified i.e. anode in the glow); α_0 is the value of α in the glow.

As far as is known no extensive comparison with experiment has been made.

ζ) SEELIGER [*106*] has pointed out a connection between the last two theories, in that WEIZEL, ROMPE, and SCHÖN consider the simultaneous motion of charges due to the electric field and diffusion, which for small drift velocities is equivalent to the BOLTZMANN distribution as used by SCHERZER. The properties of the glow are also discussed.

Defects of the theories[1].

None of the theories has yet been able to account for the observed linear change of field in the dark space. This requires a uniform positive ion space charge, which could be produced by a uniform generation of ions[2], provided $v^+ \propto X$, but no physical reason has been found why this should be so. In non uniform fields it is known that $\alpha(x)$ is not determined by $X(x)$, but that the ionisation is considerably greater than can be accounted for in this way [*107*]. COMPTON and MORSE attempted to make allowance for this in their theory. SEELIGER[3] has used eq. (35.8) in a modified form to examine what kind of function $\alpha/p = f(X/p)$ is necessary to account for the main experimental results, which can be expressed:

$$\frac{\partial d}{\partial j} < 0, \quad \frac{\partial d}{\partial V_c} < 0, \quad \frac{d V_c}{d j} > 0 \quad \text{and} \quad \frac{\partial V_n}{\partial j} = 0. \tag{35.20}$$

He finds that only a function such that α/p rises more slowly than linearly with X/p can satisfy these conditions. This may partly account for the disagreement in helium (see Figs. 29, 30). However, it seems that ionisation in non-uniform fields requires a new and more fundamental treatment before this problem can be solved [*108*]. Similar objections can be raised against the use of mobility equations, which apply only when the ions are in equilibrium with the field. Charge transfer, neglected in most theories, is likely to have a great effect on the motion of the ions, and hence indirectly on the space charge in the dark space.

WARREN [*307*] has compared the field distribution in the cathode dark space of abnormal discharges with that predicted by several theories. The assumptions

[1] For a more detailed description and criticism refer to the review by A. v. ENGEL: Rep. Progr. Phys. **8**, 338 (1941).

[2] J. J. and G. P. THOMSON: Vol. 2, p. 300.

[3] R. SEELIGER: Z. Physik **122**, 209 (1944).

that ions move freely, or with a drift velocity $v^+ \propto \sqrt{X}$ give results not agreeing well with experiment. It appears also that in strongly abnormal discharges relatively little ionisation occurs in the dark space, but that many ions flow in from the negative glow.

36. The normal cathode fall: constancy of current density. There have been many attempts to explain why the cathode fall and current density remain constant provided the discharge does not cover the entire cathode, nor is confined to a small area in the middle. Most authors invoke a minimum energy principle, assuming that the normal cathode fall is the least potential which can drive the necessary current through the discharge, and the constant current density provides the space charge necessary for this potential drop. No physical descriptions of the processes by which this constancy is maintained are given[1].

LOEB [7] has suggested that the effect is due to changing conditions at the cathode surface, and hence changing γ; this argument may have some relevance at the edges of the discharge but does not explain why with decreasing current the discharge contracts uniformly towards the centre.

It appears more probable that the effect is connected with radial losses due to radial electric fields and diffusion. An increased current will give more ionisation, increasing the space charge of ions and hence their outward flow by diffusion. It is clear that the discharge will tend to spread outwards, but no simple physical picture which gives $i \propto \pi r^2$ can yet be given [71].

It is clear that the distribution of ions at the edge of the discharge (i.e. radial edge) controls the losses by diffusion; the constancy of j_n refers probably only to the central uniform region. Little error has arisen in the past by neglecting the edge effects because the region of varying ion density is small (see Sect. 26). It may be noted, however, that on spherical cathodes the transition to an abnormal discharge defined by complete coverage of the cathode does not quite correspond to that defined by an increase in V_c [109].

37. Theory including radiation from the glow: theory of the hollow cathode discharge. LITTLE and VON ENGEL [19] have recently included in the theory of the cathode fall the important secondary effects at the cathode due to radiation produced both in the dark space itself and more particularly in the glow. The theory also applies, with a simple modification, to the hollow cathode discharge.

In accordance with the experimental results for both ordinary and hollow cathode discharges the field is assumed to fall linearly with distance from the cathode, becoming zero at the edge of the glow [eq. (35.7)]. The electron current density at the cathode $(j_e)_c$ consists of three groups, the secondary electrons released by ions and photons from the dark space, and by photons from the glow.

$$(j_e)_c = \gamma_i j_c^+ + f_d \gamma_p n_d (j_e)_c + f_g \gamma_p n_g (j_e)_g \tag{37.1}$$

where j_c^+ is the ion current density at the cathode, f_d is the fraction of photons from the dark space which hit the cathode, n_d the number of photons having $h\nu > \varphi$ (the work function), produced by each primary electron and its secondaries in the dark space; and similarly for the photons produced in the glow.

γ_i, γ_p are the secondary coefficients for ions and photons. Since the whole current in the glow is carried by electrons $(j_e)_g = j$, and since $j = (j_e)_c + j_c^+$ we get

$$\frac{j_c^+}{j} = \frac{1 - D - G}{1 - D + \gamma_i} \tag{37.2}$$

[1] Arguments which regard the current as being carried by a number of discharges in parallel, each operating at the same V_n are naive, since they assume the "parallel discharges" are independent, which in practice they are not.

where $D = f_d \gamma_p n_d$, $G = f_g \gamma_p n_g$. When photons have no effect $D = G = 0$ and we get the familiar form $j = j^+ (1 + \gamma_i)$.

One can derive

$$\frac{j}{p^2} = \frac{V_c}{2\pi (p d)^2} v_c^+ \frac{1 - D + \gamma_i}{1 - D - G} \tag{37.3}$$

v_c^+, the velocity of the ions at the cathode is treated in some detail, including the effects of charge transfer, and forward scattering, giving, for ions of mass M

$$v_c^+ = \sqrt{\frac{2 e V_c}{M} \cdot \left\{ \frac{2\lambda_0}{p d} - \left(\frac{\lambda_0}{p d} \right)^2 \right\}} \tag{37.4}$$

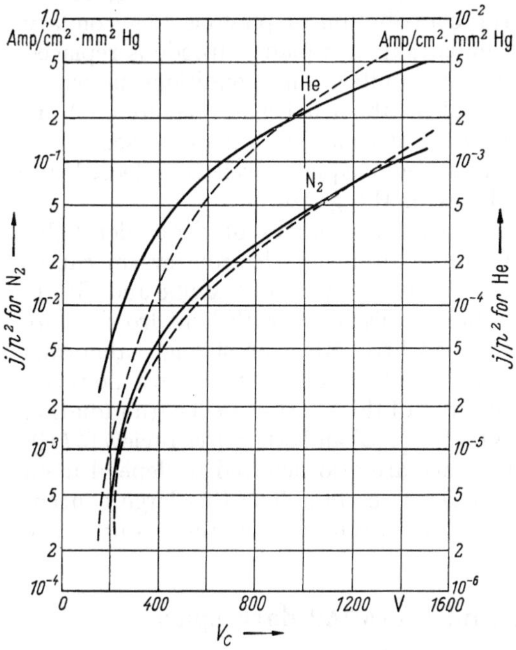

Fig. 31. j/p^2 as a function of the cathode fall for a conventional glow discharge with iron cathode. — — — Experiment (GÜNTHERSCHULZE); —— theory.

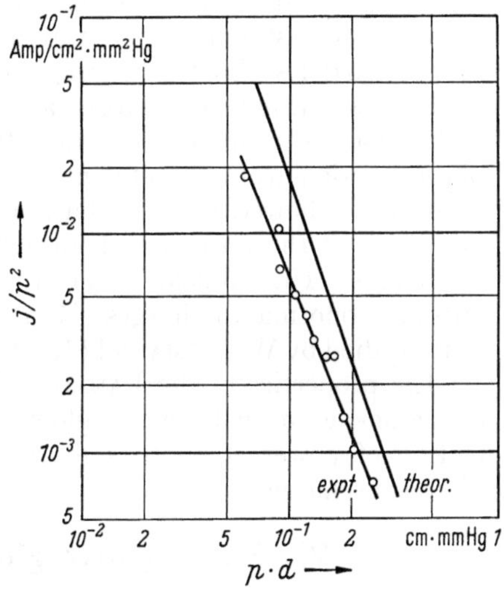

Fig. 32. Current density j/p^2 as a function of the cathode dark space length $(p \cdot d)$ for a hollow-cathode discharge in nitrogen with aluminium cathodes. $V_c = 300$ V.

where λ_0 is the mean free path of the ions *for charge transfer*. The term in brackets is derived from the parabolic distribution of potential which arises from the linear distribution of field. D and G have to be calculated from geometrical factors, photoelectric yields (which have recently been measured) and the excitation produced by the electrons, which can be calculated approximately. The final result is

$$\frac{j}{p^2} = \sqrt{\frac{e \lambda_0}{\pi^2 M}} \cdot \frac{V^{\frac{3}{2}}}{(p d)^{\frac{5}{2}}} \sqrt{1 - \frac{\lambda_0}{2 p d}} \cdot S(\gamma) \tag{37.5}$$

where $S(\gamma)$, the secondary process term, is a complicated function including D, G, γ_i, $p d$, V_c and an average ionisation coefficient.

In a normal glow discharge $S(\gamma)$ is only a correction term, and j_n/p^2 should not depend much on the cathode material. In the abnormal discharge the influence of $S(\gamma)$ is rather stronger; the geometrical factors increase as the glow moves closer to the cathode. In the hollow cathode discharge the photon contribution is very marked because f_d, f_g which in other discharges are $< \frac{1}{2}$, are here ≈ 1, and j/p^2 rises rapidly. On the simplest assumptions, considering only the effect of photons from the glow, the relation between pd, j/p^2 and V_c has been calculated and compared with experiment (see Figs. 31 and 32). The agreement is quite good.

The results are also in agreement with experiments on the obstructed discharge. In abnormal conditions (large V_c) the negative glow plays a more important part in maintaining the discharge. When such a discharge is obstructed the potential necessary to maintain the current begins to increase before the glow has been completely removed. This can also partly be explained by the greater flow of ions across the boundary into the dark space from the glow at higher V_c[1].

It has often been supposed that the contribution of photons to secondary processes must be very small on the grounds that, if a plane cathode is replaced by a gauze, no change in discharge current results[2]. The conditions in which this experiment were done (whether normal or abnormal, gas pressure etc.) are not known, and the result cannot be regarded as conclusive. Also the secondary emission by fast ions or neutral molecules could equally well be affected due to the particles shooting through the holes in the gauze.

Hollow cathode discharges are widely used as sources of ultraviolet light, and it has been shown that radiation emitted from an electron swarm causes photoemission from a metal plate [*110*]. Many years ago DEMBER and GEHLHOFF [*111*] showed that irradiation of a helium discharge with light from an arc lamp halved the normal cathode fall, but had little effect on strongly abnormal discharges. This indicates that the internal irradiation must have a powerful effect in abnormal discharges. Many properties of this short wave length radiation were studied by WIEDEMANN [*112*], DAUVILLIER [*113*] and others (see review [*311*]).

The properties of the FARADAY dark space are also believed to depend upon energetic radiation from the glow, and it seems certain that this large concentration of powerful photons must play an important role in the maintenance of the discharge.

VI. The negative glow and FARADAY dark space.

38. Description. These two regions, although they appear separately as a very bright layer followed by a relatively dark space, are best treated together since the properties of the FARADAY dark space are inseparable from those of the glow. To the eye, the light intensity rises rapidly to a maximum from the edge of the cathode dark space (see Fig. 2), decreases more slowly where the glow merges diffusely into the FARADAY dark space, and rises fairly sharply again at the head of the positive column. In pure rare gases the FARADAY dark space is filled with light (aureole), but can be distinguished from the glow by its redder colour. The FARADAY dark space is usually considerable longer than the negative glow. The distance from the cathode to the head of the positive column increases as p decreases (with i constant) and also as i increases (with p constant).

The processes which take place in these regions are qualitatively well known, but due to their complexity no satisfactory quantitative theory exists, although several attempts have been made. Our information comes entirely from electrical measurements made with LANGMUIR probes, and from spectral observations.

39. Electrical measurements. The older measurements are reviewed in references [*3*] and [*10*]. Probes cannot be used near the boundary of the cathode dark space where they are exposed to a beam of relatively high speed electrons

[1] See DRUYVESTEYN and PENNING: Rev. Mod. Phys. **12**, 87 (1940), p. 135, also WARREN [*307*].
[2] J. J. and G. P. THOMSON: Vol. 2, p. 296.

coming from the cathode. Attempts to use them there have given anomalous probe curves usually associated with the presence of a positive space charge [*115*] (see also Sect. 11). Many investigations by EMELEUS and associates[1] [*31*], [*81*], [*115*] to [*118*] using probes at the brightest part of the glow or beyond have given the following picture. The field in the glow and FARADAY dark space is almost zero, except just in front of the positive column. Small negative fields have sometimes been reported in the glow (e.g. [*20*]) but the evidence for their existence is rather dubious. In discharges with an almost normal cathode fall there are three distinct groups of electrons in the middle of the glow, having mean energies of about 25 eV, 7 eV and 0.5 to 2 eV and called respectively primary, secondary and ultimate electrons [*119*]. The middle group are thought to be electrons ejected from atoms when they are ionised[2]. Recent measurements in helium, with modern screened probes have confirmed the properties of these three groups [*276*].

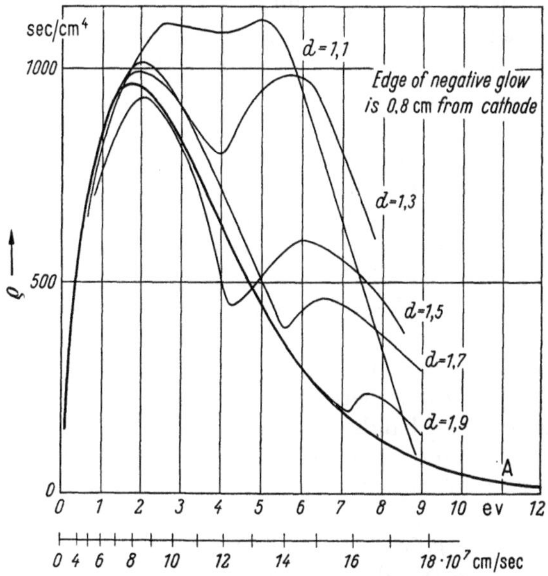

Fig. 33. Velocity distribution of electrons in the negative glow and FARADAY dark space of a neon discharge [*28*]. $dN = \varrho(v)\,dv$ d = distance from cathode in cm. Discharge current = 1.8 A, tube diam. = 2.8 cm. p = 4.7 mm. Hg. Curve A is the MAXWELLian distribution found in the positive column ($\overline{V} = 2.65$ volts, $N_e = 9.2 \cdot 10^{10}$).

The term "primary" for the fast group is rather confusing here since this usually applies to the electrons entering the glow at the boundary of the cathode dark space. It seems very likely from spectral observations and direct measurements of electron energies that many electrons arrive at the glow with energies much greater than 25 eV. However, these probe measurements were made in the middle of the glow, where the original electrons from the cathode dark space would have lost their directed motion and some of their energy; also with a normal cathode fall fewer electrons have the full energy eV_n.

There are many more slow electrons than fast; e.g. in He, $p = 1$ mm. $j = 2 \cdot 10^{-5}$ A/sq cm., $N_e \approx 5 \cdot 10^6$ (25 eV), $5 \cdot 10^7$ (7 · 3 eV), $4 \cdot 10^9$ (0.6 eV) electrons per cm.[3]. Towards the FARADAY dark space all the electrons become slow ("ultimate"). The mean energy of this slowest group increases as p and j decrease. No ionisation occurs in the FARADAY dark space, the current being carried by diffusion of electrons and ions from the glow due to their own concentration gradients. At the far end, however, the field rises and electrons are accelerated into the positive column. Direct measurements of the electron energy distribution have shown how it changes through the glow, tending to MAXWELLian at the edge of the positive column (see Fig. 33).

A few measurements have been made with large probes set in the glow [*120*], [*121*], [*122*]. It has been found that variation of the probe potential over a

[1] Some of the earlier probe measurements were misinterpreted because the properties of probes were at that time not fully understood. Corrected results were published later [*31*].

[2] The energy of these electrons does not depend much upon the energy of the incident ionising electron. See MOTT and MASSEY: Theory of atomic collisions, 2nd ed. Oxford U.P. 1949.

large range has little effect upon the voltage necessary to sustain the discharge, and also that the current to a negative probe can sometimes exceed the total discharge current. It is not certain that this implies appreciable recombination in the glow, for which there is little other evidence[1]. The anode when placed well into the glow acquires a potential lower than that of the space; these effects have been attributed entirely to diffusion processes and are not compatible with an early theory by ROGOWSKI. Later experiments[2] have shown that the negative glow can be cut into two or more independent parts by a grid held at a suitable potential; the potentials of the several glows may be quite different and they are not necessarily connected by any exchange of charge carriers.

40. Spectral measurements[3] (*125* to *130, 132* to *136*). The spectral distribution falls into four regions. Nearest the cathode, in the brightest part of the glow the spectrum consists of arc and spark lines, i.e. lines of excited atoms and excited ions. These are produced by simultaneous excitation and ionisation by fast electrons coming from the cathode dark space. DRUYVESTEYN found that neon spark lines appear only when eV_c exceeds the energy required to excite the ion in a single impact: DAUVILLIER [*131*], studying a normal discharge in A, observed the L radiation, showing that electrons of at least 240 eV were present near the cathode edge of the glow. Many other processes also occur here: in helium for example He_2 bands have been seen. These are thought to be due, not to recombination but to three body collisions between two normal and a metastable atom: metastable molecules result, which are excited by slow electrons. If nitrogen is added these metastables are destroyed by colliding with N_2 which itself dissociates; He_2 bands are then replaced by the lines of N atoms [*130*].

Photometric measurements on the *total* light from the negative glow in Ne show that most of the excitation is probably due to electrons with a mean energy of about 25 eV, which agrees with the probe measurements. When $V_c < 285$ volts the total light emission from the glow excluding the aureole is independent of V_c, proportional to i and not much dependent on p.

In the second region, the transition from the glow to the FARADAY dark space, direct excitation by electron impact becomes of less importance, compared with collisions of the second kind involving metastable atoms, and the absorption of resonance radiation from the main glow.

In the FARADAY dark space itself the latter processes alone occur, although knowledge of the actual details is vague. Metastable atoms are known to diffuse in appreciable concentration from the negative glow, and can emit light when they are hit by neutral atoms or slow electrons. (Only slow electrons are present in this region.) The aureole found in the rare gases disappears if the metastables are destroyed by a small amount of impurity, e.g. Ne + a trace of A. It has been suggested that short wave radiation from the glow, including resonance radiation, plays an important part both in the production and in the excitation of these metastable atoms.

The existence of such radiation was proved by J. J. THOMSON[4] and DAUVILLIER [*113*], [*131*] and investigated in a vacuum spectrograph by KRUGER[5]. Some

[1] F. L. MOHLER: Bur. Stand. J. Res. **19**, 447, 559 (1937), from afterglow studies finds only a small amount of recombination.

[2] W. WEIZEL and W. OLMESDAHL: Z. Physik **114**, 719 (1939).

[3] The important existence is so scattered amongst the papers that this section would be unreadable if references were given in the text. Many previous measurements are mentioned in the papers of EMELEUS; evidence for simultaneous ionisation and excitation is mostly in [*125*] and for the role of metastables and resonance radiation in [*133*] to [*135*].

[4] J. J. THOMSOM: Phil. Mag. **49**, 761 (1925).

[5] G. KRUGER: Phys. Rev. **34**, 1122 (1929).

of it does not diffuse but travels in straight lines [1] through the gas, and experiments on its method of propagation have been made in discharge tubes of various shapes with obstacles placed in them. Before this was recognised it was thought that the FARADAY dark space was swept by beams of cathode rays: this is true at low pressure [2,3] ($p \sim 10^{-2}$ mm.) but at higher pressure the range of the fast electrons from the cathode is too small. No numerical estimate of the effect of this radiation upon the light emission or electrical properties of the FARADAY dark space has been possible. Recently fundamental measurements on the absorption of light in gases have provided numerical data from which it might be possible to elucidate the processes in this region [*312*].

The transition distance from the FARADAY dark space to the positive column emits very little light. Here the field increases, and the properties of this region have been treated theoretically by J. J. and G. P. THOMSON (ref. [*3*], vol. 2, p. 361) and have shown that its length should be proportional to p/j provided the pressure is not too low. (These authors regard this region alone, where the field begins to rise, as constituting the FARADAY dark space.)

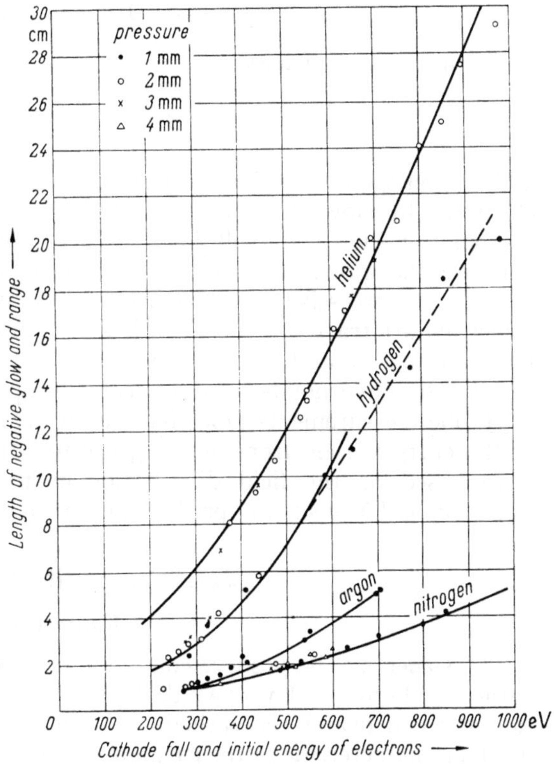

Fig. 34. Range of electrons (lines) and length of negative glow (points) for various values of initial electron energy (in eV) or cathode fall (in volts) respectively.

41. Length of the negative glow, l_g. BREWER and WESTHAVER [*137*] have measured the length of the negative glow in several gases, for various values of the cathode fall, and have found it equal to the range of electrons measured independently with an ionisation chamber [4]. Their results are given in Fig. 34. l_g increases with increasing V_c and decreasing pressure.

42. Theory of the negative glow. It is known that a beam of fast electrons, having energies up to the full cathode fall, shoots across the boundary from the cathode dark space into the glow. There the electrons dissipate all their energy, first losing large amounts of energy in exciting and ionising the gas, sometimes doing both simultaneously; later losing smaller amounts in exciting collisions only, and finally, in the FARADAY dark space having very little energy (< 1 eV) and being able only to excite metastable atoms.

[1] LANGMUIR and FOUND: Phys. Rev. **36**, 604 (1930); **39**, 237 (1931).

[2] GOLDSTEIN: Wied. Ann. **67**, 84 (1899). The rays were observed by the light they produced. See also J. J. and G. P. THOMSON: Vol. 2, p. 340.

[3] TAIT, SMITH and WHIDDINGTON: Proc. Leeds Phil. Soc. **5**, 707 (1949).

[4] J. F. LEHMANN: Proc. Roy. Soc. Lond., Ser. A **115**, 624, (1927). — BREWER and WESTHAVER altered V_c by changing i and thus do not give $p\,l_g$ as $f(V_c)$ and $F(i)$ separately.

This picture has been adopted in the theory of WEIZEL, ROMPE, and SCHÖN [*103*]; a somewhat similar treatment is due to SCHERZER [*102*] and an earlier theory was given by ROGOWSKI [*101*] WEIZEL, ROMPE, and SCHÖN regard the glow as a nearly neutral plasma crossed by an electron beam, and derive the basic equations:

Electron current:

$$j_e = j_1 + e D_e \frac{d N_e}{d x} + e N_e b_e X. \tag{42.1}$$

Ion current

$$j_+ = - e D^+ \cdot \frac{d N^+}{d x} + e N^+ b^+ X. \tag{42.2}$$

Gain and loss of ions

$$\frac{d j_+}{d x} = \alpha |j_e| + \beta |j_+| N_e - \varrho e N_e N^+ - \tau e N^+ N_e^2 \tag{42.3}$$

and an analogous equation for electrons. The terms represent ionisation by fast electrons, by slow electrons (two stage), recombination (coefficient ϱ) and recombination by 3-body collisions. j_1 is the current carried by the electron beam entering the glow. Together with POISSON'S equation these are soluble by a method of successive approximation. The boundary conditions are naturally somewhat arbitrary and this point has been discussed by SEELIGER who finds that when the anode is in the glow, the field can be described by an elliptic function [*106*]. The results have not been numerically compared with experiment.

Probe measurements [*138*] have been made to check some early calculations of the energy of electrons entering the negative glow [*139*]. These calculations were based on the field distribution in the cathode dark space derived by COMPTON and MORSE, and predicted electron energies lower than those measured.

VII. The positive column.

a) Main types.

43. General. The positive column is that region of luminosity bounded on its cathode side by the FARADAY dark space and very near the anode by the anode fall: the end nearer the cathode is termed the "head" of the column. It has a colour characteristic of the gas, not so bright as the negative glow, and emitting fewer blue lines. Its properties are independent of the length of the column, provided that this is not too short, and it is usually studied in long cylindrical tubes. At pressures less than a few mm. Hg the column spreads laterally to the walls, filling the whole cross section of a tube; at higher pressures (a few cm.) it begins to contract radially. All the properties of the column show radial symmetry about the axis of the tube, except certain types of contraction.

There are three main types of column:

α) Uniform. This is the most common type; the luminosity and electrical properties are constant along the length of the column, except near its ends. Uniform columns are found especially at low pressures ($<$ about 1 mm.) and in pure rare gases. It should be noted, however, that the appearance of the column may be very deceptive: many years ago WÜLLNER, SPOTTISWOODE, and later ASTON and KIKUCHI [*140*], [*141*], [*142*] showed by rotating mirror experiments, that apparently uniform columns contain very rapidly moving striations (velocity $\sim 10^5$ cm./sec.). Recent work by DONAHUE and DIEKE has shown that such conditions are very much more common than has been realised.

β) Column with stationary striations. Here the luminosity varies in a periodic or nearly periodic manner, regions of bright light alternating with darker spaces. This effect has been found in limited ranges of current and gas pressure. In pure rare gases only a few striations occur, at the head of the column: that nearest the cathode is the sharpest, they become progressively more diffuse and the remainder of the column is uniform. In molecular gases, or any mixture containing a molecular gas an unlimited number of striations can occur: these show no general tendency to become more diffuse nearer to the anode. The pattern of light and dark regions is regularly repeated along the length of the column; the striations are not necessarily all identical, sometimes narrow and wide, or very sharp and rather diffuse striations alternate, or more complicated patterns are seen. Striations of this kind are favoured by relatively high pressure (>1 mm.) and by the presence of impurities whose effect is especially noticeable in rare gases.

γ) Column with moving striations. These have been observed mainly in the rare gases, where striations move from anode to cathode with a speed of order 10^4 cm./sec. Recent work has shown that they occur also in H_2, air and Hg vapour, and that these striations are accompanied by fainter ones moving from cathode to anode. They sometimes move much faster ($\sim 10^5$ cm./sec.), causing the column to appear uniform, and like stationary striations require a pressure of about 1 mm. or more for their formation.

It was known a long time ago that the field along the axis of the column (the gradient) depends upon the gas pressure, the diameter of the tube and sometimes (in rare gases) on the current. These main experimental results, and also LANGMUIR probe measurements showing the dependence of the electron temperature will be given first; other more detailed measurements will be described after the necessary theory has been given.

b) The uniform column in atomic gases: main experimental results.

44. The field X (Gradient). The longitudinal field has been measured in various ways: in the older measurements by inserting two or more floating probes into the column and measuring the difference between the potentials they acquire either with an electrostatic voltmeter or by applying externally a compensating potential so that no current flows between the probes. Other workers have used a movable anode and, keeping the discharge current constant, measured the extra potential needed to sustain a short additional length of column. Most measurements since about 1930 have been made with LANGMUIR probes spaced along the column, and used to deduce the space potential, electron temperature and concentration. All methods give, of course, the average field, $X = \Delta V/\Delta x$ where Δx is usually a few cm.[1].

The results of many experiments agree well, [*143*] to [*156*], [*161*] and are summarised in Figs. 35, 36, and 37. The field as a function of the gas density (expressed as usual in terms of pressure) shows usually a maximum between $p = 10^{-2}$ and 10^{-1} mm. Hg; the rise in X when $p \sim 1$ mm. Hg is followed by the contraction of the column when p exceeds a few cm. Hg. In general, gases with the lowest ionisation potentials give the lowest fields, although there are exceptions (cf. K, $V_i = 4.3$ volts, Hg, $V_i = 10.4$ volts, and also at $p > 0.4$ mm., Hg and Kr, $V_i = 13.3$ volts) (Fig. 35). However, it is true over the whole pressure range considered for gases of similar atomic structure, e.g. the rare gases.

[1] These methods are accurate only provided the column has a certain minimum length. See W. HOLTZ and G. KÖMMNICK: Z. Physik **99**, 252 (1936).

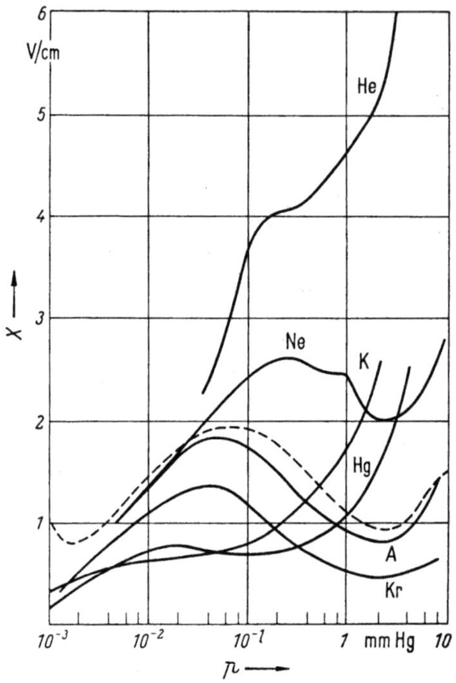

Fig. 35. Potential gradient in the column as a function of pressure (KLARFELD [150], diam 20 mm., $i = 0.3$ A). Dotted line, GROOS [148] Argon, diam 19.5 mm., $i = 0.1$ A.

X is much less dependent upon the current, especially at lower pressures ($< 10^{-2}$ mm.). However, at low current density ($<$ about 10^{-6} amp./sq. cm.) the field rises rapidly with decreasing current. This corresponds also to the subnormal discharge, the positive column of which has been discussed recently by EMELEUS and BURNS [157]. (It is, of course, the processes in the *cathode* region which determine whether a discharge is subnormal or not, but here the change in field in the positive column also adds a little to the $V-i$ curve of the whole discharge.) At higher currents and in narrow tubes the curves have a positive slope in some regions, which is most pronounced at low pressures in He and Ne (Fig. 38), but also occurs in A and Kr[1].

The diameter of the tube has a marked influence on the gradient which decreases as the tube becomes wider: the decrease is less rapid at higher pressures. At pressures above 1 mm. the

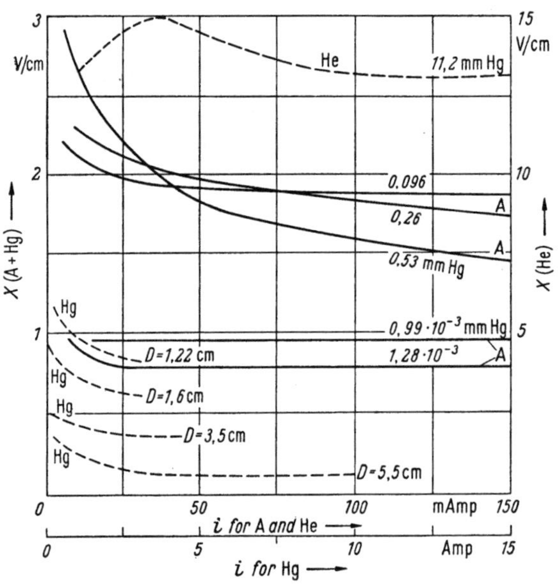

Fig. 36.

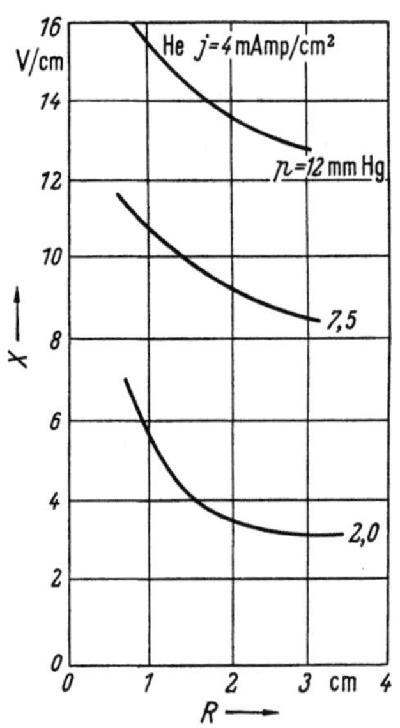

Fig. 37.

Fig. 36. Gradient of the positive column as a function of current in argon (GROOS) diameter 19.5 mm.; and helium (LOMPE and SEELIGER), diameter 30 mm. Field scale: A on left,, He on right. Also higher current discharges in Hg, in various size tubes (ELENBAAS).

Fig. 37. Variation of the gradient in the positive column of He, in tubes of different radii, R. (LOMPE and SEELIGER).

[1] H. SCHELLENMEIER: Z. Physik **122**, 269 (1944).

dependence of X on R (the radius of the tube) obeys the similarity laws, i.e.

$$\frac{X}{p} = f(p \cdot R) \tag{44.1}$$

but there are deviations at lower pressures. In very wide tubes the gradient becomes practically independently of the radius and GÜNTHERSCHULZE has used this property to define the so-called "normal gradient" [*152*], [*153*].

GÜNTHERSCHULZE also found, from measurements in cylindrical tubes that *over a restricted range* of pressure, current and radius ($p \simeq 5 \cdot 10^{-2}$ to 0.2 mms., varying somewhat with the gas, $i \simeq 10-100$ mA, $R \simeq 4-18$ mms.,) X is independent of the pressure, and follows the semi-empirical law

$$X = \frac{C}{R + a i} \tag{44.2}$$

where C is a constant depending on the ionisation potential and atomic weight of the gas. When X is in volts/cm., R in millimetres, i in mA, we get the following values:

Gas	C	a	Gas	C	a
He	45.4	$\simeq 0$	Kr	17.3	0.038
Ne	27.3	0.01	Xe	13.5	0.063
A[1]	39.1	0.134	Hg	6.3	$\simeq 0$?

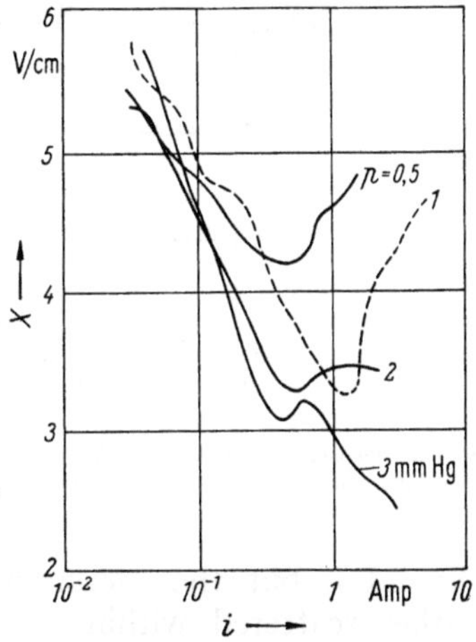

Fig. 38. Gradient as a function of the current in Neon. SCHELLENMEIER (tube diameter 10 mms.).

ELENBAAS[2], from measurements in the positive column of Hg vapour at slightly higher pressures (0.3 to 3 mm. Hg) finds another empirical equation for this restricted range of pressure, for currents of 0.5 to 8 amps and radii 6 to 27 mms., which with the units as before (R in mm., i in mA) we write:

$$X \text{ in volts/cm.} = \frac{0.28}{R^{0.69} \cdot i^{0.12}} . \tag{44.3}$$

There were also many earlier attempts to find similar empirical analytical expressions, valid over small ranges, notably by WILLS[3], PARTZSCH[4], and HOLM[5].

45. The electron temperature, T_e. This has been measured by a number of workers, with different discharge currents and gas pressures [*148*], [*150*], [*151*], [*155*], [*156*], [*158*] to [*161*]. The range of pressures which can be used is rather restricted; at $p > 1$ mm. collisions occur within the probe sheath, while at very low pressure the sheath extends into a considerable region of the discharge. Some results are shown in Figs. 39, 40, and 41. It is clear that T_e falls with increasing pressure, rises at first rapidly then slowly with increasing X/p, and changes very little with the current. The dependence of T_e on the tube diameter has not been specifically investigated; it could presumably be deduced from its variation with

[1] Argon apparently behaves differently from the other gases: the value of C does not fit into the pattern of the table, and there is much more dependence on i. The formula is not valid outside the ranges given. e.g. at 300 mA, X calculated does not agree with Fig. 35.

[2] W. ELENBAAS: Z. Physik **78**, 603 (1932).

[3] WILLS: Phys. Rev. **18**, 319; **19**, 65 (1904).

[4] PARTZSCH: Z. Physik **14**, 191 (1923); **15**, 287 (1923).

[5] R. HOLM: Wiss. Veröff. Siemens-Werk **3**, 159 (1923). See also ref. [*6*] 2nd edition, p. 327.

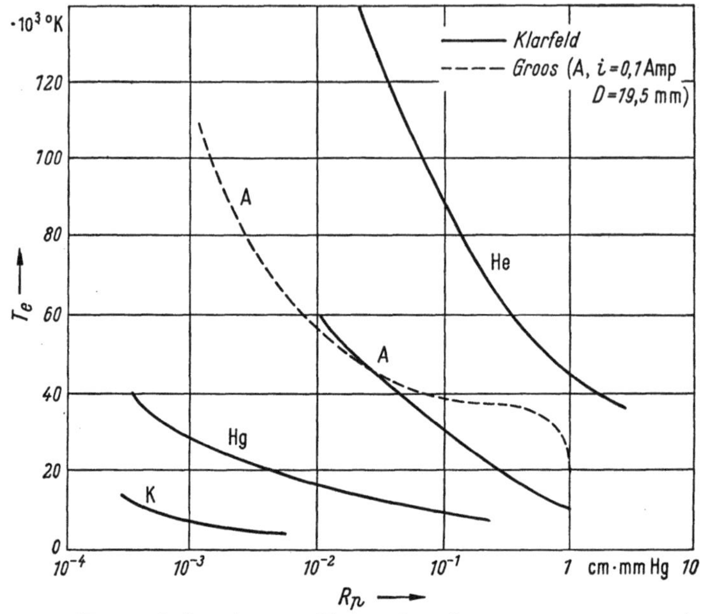

Fig. 39. Dependence of T_e upon Rp in various gases He Ar—column diameter 20 mm., current 0.3 A. K vapour—column diameter 32 mm., current 1 A: Hg vapour—column diameter 32 mm., current 3 A. —— KLARFELD, ---- GROOS (A, current 0.1 Amp., diameter 19.5 mm.).

X and i, but the measurements are rather scattered within the range of X, i, and R.

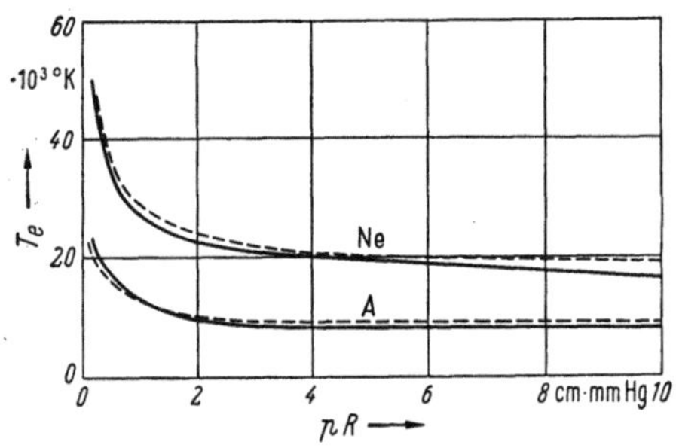

Fig. 40.

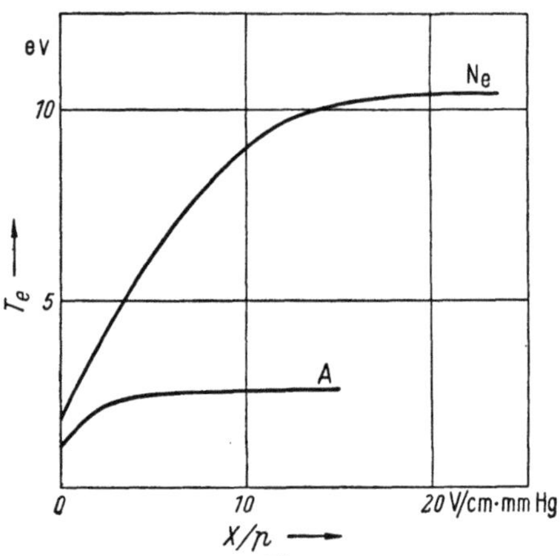

Fig. 41.

Fig. 40. Dependence of electron temperature on pR in Ne and A. —— measured; ----- calculated from eq. (51.6).

Fig. 41. Electron temperature as a function of X/p (SEELIGER and HIRCHERT). Diameter of tube = 4 cms., $i = 100$ mA.

c) The uniform column in molecular gases: main experimental results.

Molecular gases have not been studied so extensively as the rare gases, but certain results are known in O_2, N_2 H_2 air and H_2O[1,2] [*162*], [*163*].

46. The field. Fig. 42 shows the dependence of the gradient on pressure and tube radius. X depends little on the current between 5 and 100 mA; it increases with increasing pressure, and does not show the maxima and minima

[1] A. GÜNTHERSCHULZE: Z. Physik **41**, 718; **42**, 763 (1927) (N_2, H_2, H_2O, air, He, A, Kr, Xe, Hg).

[2] GEHLHOFF: Verh. phys. Ges. **21**, 349 (1919). For earlier measurements see refs. [*3*], [*6*]; these were nearly all made in narrow (capillary) tubes.

found at low pressures in rare gases; it decreases with increasing tube radius in common with other gases. The similarity laws apply well to molecular gases, i.e. $X/p=f(p\cdot R)$. A particular case of this is found in an empirical formula of GÜNTHERSCHULZE[1] (rather similar to that found in rare gases) which he writes as $X=C\frac{\lambda^{-\frac{2}{3}}}{R^{\frac{1}{3}}}$ where λ is the mean free path of gas molecules: it was thought

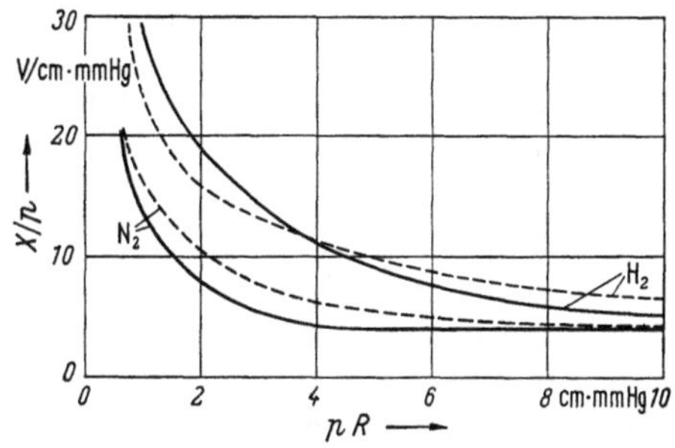

Fig. 42. X/p as a function of pR in N_2 and H_2. — measured, ----- calculated from equations (51.6) to (51.8).

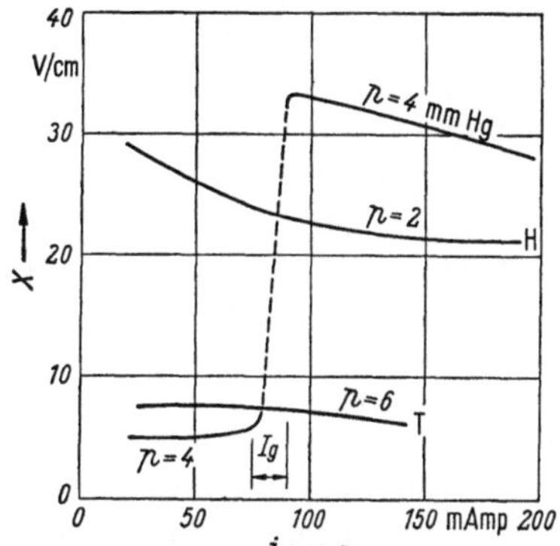

Fig. 43. Positive column gradient in oxygen as a function of the current [74]. Tube diam 3 cm.

that all gases would obey a law of the kind, namely $X=C\frac{\lambda^{n-1}}{R^{n}}$, with $n=1$ for rare gases, $n=\frac{1}{3}$ for molecular gases. HOLM later showed that this was an over-simplification, and gave the equation in the form:

$$X=X_0\frac{p^m}{R^{m-1}} \tag{46.1}$$

with the values:

H_2	$X_0=18.3$	$m=0.56$
N_2	32	0.65
Air	30	0.67

X is measured in volts/cm., p in mm. Hg, R in mm. Again it must be emphasised that this can be applied with certainty *only over a restricted range* of p, R and i ($p\sim 0.1$ mm. Hg, $R\sim 4$ to 20 or 40 mm. depending on the gas, i 10 to 100 mA).

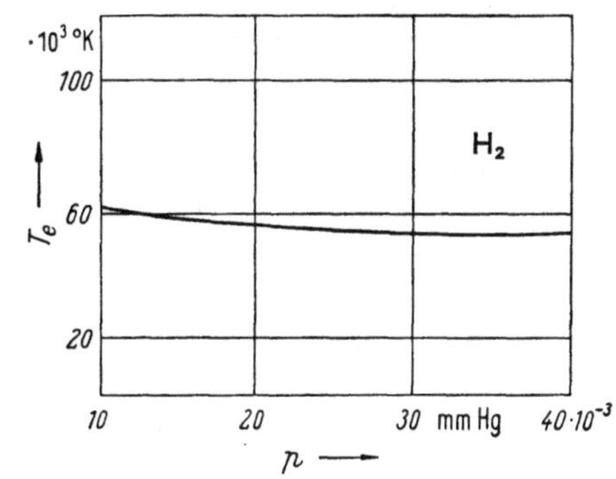

Fig. 44. Variation of electron temperature with pressure in H_2 (low pressure).

Two types of column can occur in oxygen; in one form there is a sudden change of the gradient as the current is increased, while the other shows a steady decrease of X as i increases, see Fig. 43. The form which shows the sharp increase of X also emits the spectrum of O_2 bands, whereas the other form gives the line spectrum of O. The dotted line and the interval I_g shows where the measurements are not reproducible, this range depending strongly on the cleanliness of the walls, being quite large ($\sim$200 mA) for very clean surfaces [74].

The two forms are also different in colour: one occurs only in narrow tubes, the other in wide. In medium diameter tubes the violet form can change to the grey-green form with a considerable increase in gradient. In H_2 a similar change occurs, but more slowly, the gradient increasing as the current increases: at low currents the column is dim and becomes brighter as the current increases.

[1] A. GÜNTHERSCHULZE: Z. Physik **41**, 718; **42**, 763 (1927) (N_2, H_2, H_2O, air. He, A, Kr, Xe, Hg).

47. The electron temperature. Few measurements of T_e have been made in molecular gases: some results by POTROVSKAYA[1] in the low pressure range are given in Fig. 44.

d) Gradient in very wide tubes.

48. The normal gradient X_n/p. As $R \to \infty$, $X \to$ a very small value. At one time, due to experiments of CLAUDE[2] and also as a result of SCHOTTKY's theory, it was thought that X would become zero in extremely wide vessels, but later work seems to show that X actually tends to a small constant value. When the tube is so wide that the walls no longer have any influence, the value of X/p measured at a gas pressure of 1 mm. Hg, and temperature 300 °K is defined as the normal gradient (X_n/p). It was at first believed[3] that in these circumstances the column behaves rather like a metallic conductor or electrolyte, and that X_n/p would be a constant having a fundamental significance somewhat analogous to specific resistance, but this was found not to be true. Measurements were made firstly in wide cylindrical tubes; later between two concentric spheres, the inner one being the cathode[4] (an unsatisfactory method because the results are obtained as the difference between the large quantities V_{total} — (anode fall + cathode fall). Finally[5] X_n/p was measured in a glow discharge between an anode and a movable cathode placed near the centre of a large spherical vessel, giving the results:

Table 7. *Normal gradient X_n/p measured at $p = 1$ mm., $T = 300°$ K.*

	Gas							
	He	Ne	A	O_2	N_2	H_2	H_2O	
X_n/p	< 0.12	$< 2.10^{-3}$	$< 2.10^{-3}$	13.1	3.25	2.23	50	volts/cm.·mm. Hg

It must be remembered that in very wide tubes the dark space is unusually long: the non uniform conditions at the head of the column may extend further than usual.

e) Theory.

49. Basic theory. The positive column is a region of constant low field strength along the axis and it follows from POISSONS' equation that $N^+ - N_e = 0$, along a filament parallel to the axis. Remembering the small variation of the radial field we write $N^+ - N_e \approx 0$. The concentrations of both ions and electrons are large, except at very low current densities, and we have what is sometimes called a quasi-neutral plasma, i.e. $N^+ \approx N_e \gg |N^+ - N_e|$. The relatively small drift current in the direction of the axis of the discharge is the difference between the components in opposite directions of two very much larger random currents. The electrons because of their greater speed carry nearly the whole discharge current. It might appear that since electrons and ions are produced at the same rate, more electrons than ions would leave each region of the column and a positive space charge build up. This does not happen because the column is being constantly fed by a supply of electrons from the FARADAY dark space, and by ions produced in the anode region and driven into the column by the field there. Equal numbers of electrons and ions flow out of the column into the anode and dark space respectively.

[1] A. S. POTROVSKAYA: J. techn. Phys. USSR. **21**, 617 (1951).
[2] CLAUDE: C. R. Acad. Sci. Paris **157**, 432 (1913); **158**, 479, 692 (1914).
[3] A. GÜNTHERSCHULZE: Z. Physik **22**, 70 (1924).
[4] A. GÜNTHERSCHULZE and F. KELLER: Z. Physik **77**, 703 (1932).
[5] A. GÜNTHERSCHULZE and H. BETZ: Z. Physik **81**, 283 (1933).

There is also a radial or transverse motion of charges, essential to the mechanism. When the column is first established electrons diffuse rapidly to the side walls charging them negatively, thus attracting the ions and repelling further electrons. Since the walls are insulating, equilibrium is established when ions and electrons flow in equal numbers to the wall, where they recombine. What is loosely termed the "wall current, i_ω" refers of course only to one, usually the ion, component of this flow: the total wall current is zero. Hence there exists a transverse field, potential gradient, and concentration gradient: the equipotential surfaces are not parallel to the electrodes but curved convex towards the cathode[1].

The theories of the positive column apply only to certain approximate ranges of pressure, radius etc., namely $p = 10^{-4}$ to 10 mm. Hg (in two steps, with an uncertain transition region) $R = 1$ to 10 cms., $i = 10^{-4}$ to 1 amp. At lower currents the rate of ionisation is too small to maintain the plasma properly neutral; at higher currents gas heating becomes appreciable. Even so there have been attempts to include both these effects in the theory.

All theories of the column have certain basic equations in common, although they appear in different guises not always easily recognisable. We shall list first the common assumptions, and then the equations in a formal way, referring to them later as they appear. The assumptions are:

Charge concentration[2] $N^+ = N_e = N$ (except in the theory of ECKER).
Electron energy distribution: MAXWELLian.
Electron temperature: Constant *across* the discharge i.e. $T_e \neq f(r)$.
Electron mobility: Given by LANGEVIN's equation.
Rate of ion production: $\propto N_e$ (except SPENKE).

The BOLTZMANN equation $\frac{N_r}{N_0} = e^{\frac{eV}{kT_e}}$ (derived by SCHOTTKY, assumed by TONKS and LANGMUIR).

The discharge current i and the tube radius R (in a cylindrical discharge, or the width d of a linear one) are regarded as given.

The unknown quantities are the electron temperature T_e, the longitudinal field X, and the concentration N. The production z of ion pairs per electron per cm.[3] is introduced mathematically as a further unknown. Four equations are required with their boundary conditions, and these fall into two pairs: in each pair one equation comes from an equilibrium condition of the discharge (Nos. I and III), and one from the assumptions listed above (II and IV). The equations are:

(I) Number Balance

No. of ions produced = No. flowing to the walls

$$z = f(p, R, T_e). \tag{49.1}$$

(II) Ion Production.

This equation simply gives the electron temperature necessary to maintain the above rate of ionisation (z) assuming the MAXWELLian distribution, and $z \propto N_e$ we get

$$z = F(p, T_e). \tag{49.2}$$

This pair of equations alone suffices to evaluate T_e and z.

[1] See R. HOLM: Phys. Z. **25**, 497 (1924).

[2] Since $N^+ = N_e$ we now omit the suffixes and consider simply the concentration N of charged particles of either sign. N_r is the concentration at a radius r.

(III) Energy Balance:

This may be written in two ways: for each cm. of the column

$$i X \cdot 1 = E_{\text{walls}} + E_{\text{radiation}} + E_{\text{gas heating}} \tag{49.3}$$

—equating the power input per cm. to the sum of the energies dissipated per sec. in the ways indicated, or:

$$i X = L_{\text{elastic}} + L_{\text{inelastic}} \tag{49.4}$$

—equating the energy gained by the electrons from the field to the energy lost in elastic and inelastic collisions[1]. The right hand side of the equation is usually a complicated function of N and T_e. Both forms lead to

$$i X = \Phi(N, T_e). \tag{49.5}$$

The second form of this equation becomes more difficult to apply when cumulative processes occur.

(IV) Current or mobility equation, neglecting the contribution of the ions to the discharge current, and calling N_{total} the *number* of electrons per cm. length,

$$i = N_{\text{total}}\, e \cdot v_{\text{drift}} = e \int_0^R N_r\, 2\pi\, dr \cdot b_e X = \Psi(N, T_e) \tag{49.6}$$

assuming LANGEVIN'S equation for the mobility, i.e.

(IVa)

$$b_e = \frac{3}{4} \frac{e \lambda_e}{m \bar{v}} \tag{49.7}$$

where $\bar{v}$ is the mean random velocity.

Theories based on the above assumptions lead to results which are independent of the current, and except for the theory of ECKER, and SPENKE'S extension of SCHOTTKY'S theory, the variation of X, T_e and N with i can be discussed only in qualitative terms. We consider now the various theories, and the approximate ranges they cover.

50. $\lambda_+, \lambda_e \ll R$: SCHOTTKY'S diffusion theory, medium pressure ($\sim 10^{-1}$ to 10 mm. Hg) [*164*]. Here the loss of ions and electrons is due almost entirely to ambipolar diffusion (coefficient D_a), volume recombination being neglected. Considering the elementary volume of unit length shown in Fig. 45, we get

$$\left.\begin{aligned} \text{Rate of loss} &= 2\pi (r + dr)\, D_a \left(\frac{dN}{dr}\right)_{r+dr} - 2\pi r D_a \left(\frac{dN}{dr}\right)_r \\ &= 2\pi D_a \left[r \frac{d^2 N}{dr^2} + \frac{dN}{dr}\right] \cdot dr. \end{aligned}\right\} \tag{50.1}$$

$$\text{Rate of production} = z \cdot N \cdot 2\pi r\, dr \tag{50.2}$$

where z is the number of ion pairs produced per electron per unit time. Equating these gives

$$\frac{d^2 N}{dr^2} + \frac{1}{r} \frac{dN}{dr} + \frac{z}{D_a} \cdot N = 0 \quad \text{(Equation I)}. \tag{50.3}$$

[1] This method has been applied by G. MIERDEL: Wiss. Veröff. Siemens-Werk **17**, 71 (1938). WASSERRAB has also given a similar treatment of the energy transfer by electrons in the positive column of Hg. T. WASSERRAB: Z. Physik **127**, 324 (1950).

The solution is the zero order BESSEL function

$$N = N_0 J_0\left(r\sqrt{\frac{z}{D_a}}\right) \tag{50.4}$$

where N_0 is the concentration at the axis ($r=0$).

By the properties of the BESSEL function, when $r\sqrt{\frac{z}{D_a}} = 2.405$, $J_0 = 0$, and for larger values of $r\sqrt{\frac{z}{D_a}}$ would be negative. This would have no physical meaning,

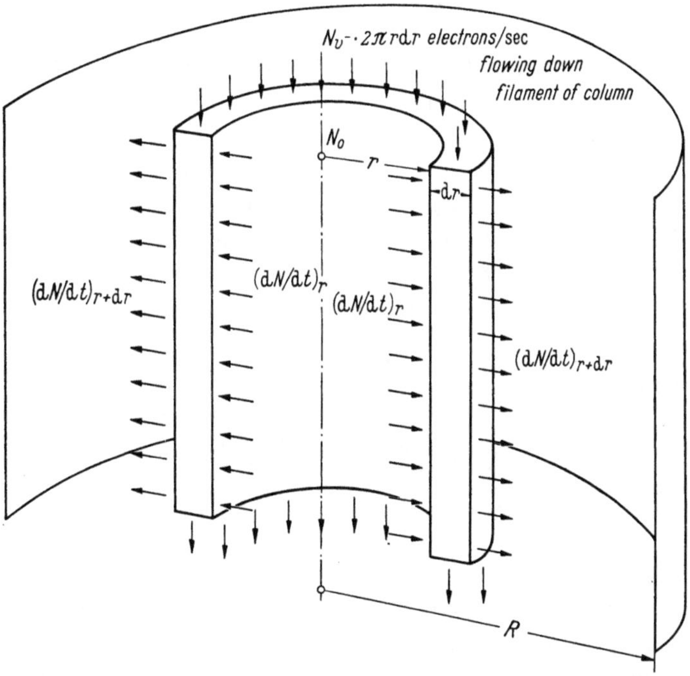

Fig. 45. Drift of charges down a filament of the positive column, and radial diffusion to the walls.

and so, assuming that the concentration is approximately zero at the walls[1], we use the largest value of $r\,(r=R)$ and impose the boundary condition

$$R\sqrt{\frac{z}{D_a}} = 2.405 \tag{50.5}$$

hence *the radial concentration*

$$N_r = N_0 J_0\left(2.405\,\frac{r}{R}\right). \tag{50.6}$$

The radial potential distribution can be found by considering the drift of electrons and ions to the walls; they move with equal drift velocities v due to the radial field X_r and their own separate diffusive motion, giving

$$v = -\frac{D^+}{N}\cdot\frac{dN}{dr} + b^+ X_r = -\frac{D_e}{N}\frac{dN}{dr} - b_e X_r \tag{50.7}$$

[1] More accurately, allowing for the wall charges

$$R\sqrt{\frac{z}{D_a}} = 2.405\Big/\left(1 + \frac{b^+\sqrt{T_e}}{71\cdot 10^9 R}\right).$$

FUNK and SEELIGER: Z. Physik **110**, 271 (1938).

(b^+, b_e ion and electron mobilities, D^+, D_e diffusion coefficients). Solving for X_r, neglecting D^+, b^+ compared with D_e, b_e and using the TOWNSEND equation

$$\frac{D_e}{b_e} = \frac{k T_e}{e} \tag{50.8}$$

which implicitly assumes that T_e is constant, independent of the radius, we get

$$\left.\begin{aligned} X_r &= -\frac{dV}{dr} = -\frac{k T_e}{e}\cdot\frac{1}{N}\frac{dN}{dr} \\ \therefore \frac{N_r}{N_0} &= e^{-\frac{e(V_0 - V_r)}{k T_e}} \end{aligned}\right\} \tag{50.9}$$

where V_r, V_0 are the potentials at r and at the axis respectively. *The wall current (of charges of one sign, e.g. ions)* per cm. of the tube is

$$i_w^+ = -2\pi R\cdot e D_a\left(\frac{dN}{dr}\right)_{r=R} = 2\pi R\cdot e\cdot b^+\frac{k T_e}{e}\left(\frac{dN}{dr}\right)_{r=R} = 7.83\, b^+ k T_e N_0. \tag{50.10}$$

The longitudinal field, X, was estimated by SCHOTTKY by equating the energy carried by the ion current to the wall per sec. per cm. of the tube with that lost by the total current i flowing along the tube:

$$i_w^+ V_i = \varkappa\cdot i X\cdot 1 \quad \text{(Equation III, in a restricted form).} \tag{50.11}$$

Each ion carries energy $e V_i$: $\varkappa$ is the fraction of energy lost by the main electron stream in making ionising collisions and is assumed to be a constant. He derives:

$$X \approx \frac{2\cdot 4}{R}\sqrt{\frac{V_i}{\varkappa}\cdot\frac{k T_e}{e}\cdot\frac{b^+}{b_e}}. \tag{50.12}$$

This is the "SCHOTTKY formula" mentioned in Sect. 48, indicating that $X \to 0$ as $R \to \infty$.

A different expression was later derived in another way by VON ENGEL and STEENBECK [*4*], after first deducing a relation for the electron temperature.

51. Extensions to the SCHOTTKY theory[1] [*4*], [*159*], [*165*]. *The electron temperature* is obtained by inserting expressions for D_a a nd z in equation (50.5).

Now

$$D_a \approx \frac{D_e}{b_e}\cdot b^+ = \frac{k T_e}{e}\cdot b^+ \tag{51.1}$$

z, the number of ion-pairs per electron per cm.³ is derived on two assumptions.

(I) That the ionisation function of the gas varies linearly with energy above the ionisation energy eV_i, and that ionisation occurs by direct collisions between electrons and gas molecules. This assumption introduces little error since the number of electrons having energy more than about 50 eV (where the function begins to depart from linearity) is usually negligible. The number of ion pairs produced per sec. by an electron having energy eV (velocity v) is then

$$F(V) = a'(V - V_i)\, v/\lambda_e = a p\,(V - V_i)\, v \tag{51.2}$$

where v/λ_e is the number of collisions per sec., a, a' constants.

(II) That the electrons have a MAXWELLian distribution of velocities (or energies

$$\frac{dN}{N} = \frac{4}{\sqrt{\pi}}\cdot\frac{v^2}{w^2}\, e^{-\frac{v^2}{w^2}}\, d\left(\frac{v}{w}\right) = \frac{2}{\sqrt{\pi}}\,\frac{\sqrt{eV}}{(kT_e)^{\frac{3}{2}}}\cdot e^{-\frac{eV}{kT_e}}\cdot d(eV) \tag{51.3}$$

[1] For a rather different approach see also J. S. TOWNSEND: C. R. Acad. Sci. Paris **186**, 55 (1928). — F. LL. JONES: Phil. Mag. **15**, 958 (1933).

where w is the most probable velocity and T_e is the electron temperature, defined by:

$$\left.\begin{aligned} \tfrac{1}{2} m w^2 &= k T_e, \\ \text{or, what is identical} \quad \tfrac{1}{2} m \overline{v^2} &= \tfrac{3}{2} k T_e. \end{aligned}\right\} \tag{51.4}$$

Hence the number of ion pairs formed by each electron per second,

$$z = \int\limits_{V_i}^{\infty} F(V) \frac{dN}{N} = a p \cdot \frac{m}{e} \frac{4}{\sqrt{\pi}} \left(\frac{2kT_e}{m}\right)^{\frac{3}{2}} e^{-\frac{eV_i}{kT_e}} \left[1 + \frac{1}{2} \frac{eV_i}{kT_e}\right] \quad \text{(Equation II)}. \tag{51.5}$$

Inserting these results gives finally a rather complicated equation, which reduces to

$$\frac{1}{\sqrt{\frac{eV_i}{kT_e}}} \cdot e^{-\frac{eV_i}{kT_e}} = 1.16 \cdot 10^7 c^2 p^2 R^2 \tag{51.6}$$

where $c^2 = \frac{a\sqrt{V_i}}{b^+ p}$, all constants of the gas, R in cm., p in mm. Hg, and other quantities in practical units.

Fig. 46 shows T_e/V_i plotted against $c \cdot p \cdot R$ giving a universal curve for all gases: it will be noted that these are similarity parameters. Table 8 gives the calculated values of c. Fig. 40 shows calculated curves for neon and argon.

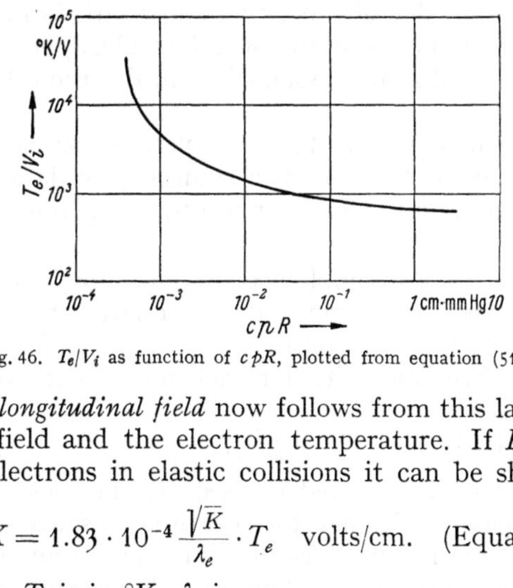

Fig. 46. T_e/V_i as function of cpR, plotted from equation (51.6).

Table 8.
Calculated values of c.

Gas	c
He . .	$3.9 \cdot 10^{-3}$
Ne . .	$5.9 \cdot 10^{-3}$
A . .	$5.3 \cdot 10^{-2}$
Hg . .	$1.1 \cdot 10^{-1}$
N_2 . .	$3.5 \cdot 10^{-2}$
H_2 . .	$1.05 \cdot 10^{-2}$
O_2 . .	$2.9 \cdot 10^{-2}$

The longitudinal field now follows from this last result, due to a relation between the field and the electron temperature. If $K(\frac{1}{2}mv^2)$ is the average energy lost by electrons in elastic collisions it can be shown that

$$X = 1.83 \cdot 10^{-4} \frac{\sqrt{K}}{\lambda_e} \cdot T_e \quad \text{volts/cm.} \quad \text{(Equation III and IVa combined)} \tag{51.7}$$

where T_e is in °K, λ_e in cm.

Since T_e is now known to be $f(pR)$ and since $\lambda_e \propto 1/p$ we get

$$\frac{X}{p} = F(p \cdot R), \tag{51.8}$$

the exact relation being again complicated, but obeying similarity. The values for N_2, H_2 are shown in Fig. 42. The treatment has also been extended to include inelastic collisions (K').

52. $\lambda_+ > R$: low pressure ($\sim 10^{-4}$ to $5 \cdot 10^{-3}$ mm. Hg). The theory of the low pressure column was derived by Tonks and Langmuir as part of a more general treatment which, by making different assumptions, can be split into two parts covering low and medium pressure ranges separately [*166*]. Their discussion of the medium pressure range leads to Schottky's solution. At low pressure there is no diffusion: a potential gradient exists between the axis and the wall and ions gain their energy at any radius r by free fall in this potential drop from the point (r_1) where they are formed.

Thus:

$$\tfrac{1}{2} m_+ v^2 = e(V_{r_1} - V_r). \tag{52.1}$$

They make no collisions with gas molecules, and are assumed to be distributed in any cross section according to the BOLTZMANN relation, which is derived in the diffusion theory. The treatment is complicated and we shall give here only the simpler results[1,2].

Equation (I) Sect. 49, gives

$$z = \text{const} \sqrt{\frac{T_e}{R}}. \tag{52.2}$$

(II) and (III) (production of ions and energy balance) are the same as in the SCHOTTKY theory.

(IV) is also the same as in the SCHOTTKY theory, but with a different numerical constant.

The ion current to the walls is

$$i_w^+ = 2.4\, N_0 e R \sqrt{\frac{k T_e}{m_+}}. \tag{52.3}$$

TONKS and LANGMUIR consider in detail the boundary conditions. The column does not extend to the wall but to the edge of a positive ion sheath formed on its surface. At the lowest pressure the thickness of this sheath can be appreciable, and the concentration N_w at its edge is not zero as assumed in the diffusion theory, but can be calculated.

Although the method of solution was given in 1929 complete numerical computation has been possible only fairly recently when certain atomic data have become accurately known. We shall refer to this later in discussing the results of KLARFELD and KARELINA.

53. The energy balance. Except in the simplest conditions (low pressure, small current, no cumulative processes) the energy balance equation has to be used in its most general form, and each of the terms E_{walls}, $E_{\text{radiation}}$, $E_{\text{gas heating}}$ needs to be calculated separately. We can write for each unit length of the column:

$$E_{\text{walls}} = i_{\text{wall}}^+ [V_i + (V_s - V_w) + \tfrac{4}{3}\overline{V}]. \tag{53.1}$$

The first two terms give the energy delivered per sec. by the ions. Each gains kinetic energy $e(V_s - V_w)$ in passing through the sheath (V_s = space potential at edge of sheath, V_w = wall potential), and loses this plus the recombination energy eV_i at the walls. At low pressure the ions falling freely have some energy when they reach the sheath: this must be added to the equation. $4/3\, e\overline{V}$ is the average kinetic energy lost by each electron hitting the wall.

The gas heating is mostly due to elastic collisions between electrons and gas molecules. An electron loses on the average a fraction $2m/M$ of its kinetic energy $\tfrac{1}{2} m v^2$ when it hits a gas molecule. The electrons are distributed in velocity [$dN_{v \to v+dv} = N \cdot f(v)\, dv$, the MAXWELLian distribution] and the heat generated *per cm.*³, per sec. at any given point r is

$$E_r = \int_{v=0}^{\infty} \frac{v}{\lambda_e} \cdot \frac{2m}{M} \cdot \frac{1}{2} m v^2 \cdot N f(v)\, dv \tag{53.2}$$

---

[1] Further comparison of the low pressure and diffusion theories can be found in [8] p 160.

[2] See E. H. LUDWIG: Z. angew. Phys. 5, 377, 421 (1953) for extensions of the theory.

v/λ_e gives the number of collisions per second. Since the electrons are also distributed across the radius $(N=F(r))$, and hence E_r is $\Phi(r)$ we get for the total heating in unit length over the whole cross section of the column:

$$E_{\text{gas heating}} = \int_{r=0}^{r=R} E_r \cdot 2\pi r\, dr. \tag{53.3}$$

The gas heating and the resultant temperature distribution across the column have recently been considered in some detail by WASSERRAB[1], who deals with the stationary and transient problems, assuming either constant gas density or gas pressure.

The energy radiated is in general most difficult to calculate, but in simple conditions, for example when cumulative processes are absent, we can assume that the number of quanta radiated is equal to the number of atoms excited by direct electron collision. If $p(v)$ is the probability of excitation to a particular level of energy $eV_{\text{exc.}}$ by electrons of velocity v, then the number of excitations per sec. per unit length of the column is:

$$\frac{dN^*}{dt} = \sum \int_{r=0}^{r=R} \int_{\frac{1}{2}mv^2=eV_{\text{exc.}}}^{\infty} p(v) \cdot N_r f(v)\, dv \cdot 2\pi r\, dr \tag{53.4}$$

the summation being over all possible levels of excitation. Usually excitation into a few levels (often the resonance levels) predominates. The energy radiated is given by multiplying by the requisite energy $h\nu$ of the line radiated from atoms in a particular level i.e.

$$E_{\text{radiation}} = \sum h\nu \int_{0}^{R} \int_{eV_{\text{exc.}}}^{\infty} p(v) \cdot N f(v)\, dv \cdot 2\pi r\, dr. \tag{53.5}$$

ROMPE and SCHÖN[2] have extended this treatment to calculate the energy radiated when other processes, e.g. cumulative ionisation, take place, and hence when the properties of the column vary with the current. Eq. (53.4) without the summation sign then gives the rate of production of a particular excited state. The rate of loss is given by $A \cdot N^*$, where A is the EINSTEIN constant of spontaneous emission (effectively corresponding to $1/\tau$ where τ is the lifetime of the excited state), and also by the number of collisions between electrons and excited states, which result in ionisation. We may write:

$$\left(\frac{dN^*}{dt}\right)_{\text{production}} = P_e \cdot p\, N_e \tag{53.6}$$

where P_e represents an integrated probability of excitation, N_e the concentration of electrons, p the gas pressure which is proportional to the density.

$$\left(\frac{dN^*}{dt}\right)_{\text{loss}} = \frac{N^*}{\tau} P_{e\to i} N^* \cdot N_e \tag{53.7}$$

where $P_{e\to i}$ represents the probability of an electron ionising an excited atom. The solution of these two equations gives N^*, whence (53.5) gives $E_{\text{radiation}}$. Inserting this in the energy balance equation, and solving simultaneously with the current equation leads to an equation giving the dependence of the gradient on the current.

[1] T. WASSERRAB: Z. Physik **129**, 190 (1951), where earlier literature is quoted.
[2] R. ROMPE and M. SCHÖN: Z. Physik **108**, 265; **111**, 345 (1938).

f) Extensions of the theory.

54. The linear column; and columns of general cross section. The SCHOTTKY treatment can be applied equally well when the positive column is confined between two plane parallel walls a distance h apart. SCHOTTKY gives the solution to this problem as a footnote in his original paper. The equation corresponding to (50.3) becomes

$$\frac{d^2 N}{d x^2} + \frac{z}{D_a} N = 0 \tag{54.1}$$

where x is the distance from the middle line of the column, measuring perpendicular to the walls. The solution, with the boundary conditions $N = N_0$ at $x = 0$ (the centre); $N = 0$ at $x = h/2$ and $x = -h/2$ (the walls) is

$$\frac{N_x}{N_0} = \cos \pi \frac{x}{h} \tag{54.2}$$

and the boundary condition gives

$$\frac{z}{D_a} = \frac{\pi^2}{h^2} \tag{54.3}$$

[compare eq. (50.4), (50.5), and see Fig. 78].

The rest of the theory is developed in analogous manner to that of the cylindrical column.

SPENKE and STEENBECK[1] have extended the treatment to columns of general cross section; they point out that the general equation

$$\nabla^2 N + \frac{z}{D_a} \cdot N = 0 \tag{54.4}$$

together with the boundary condition $N = 0$ at the walls is the same equation as for the vibrations of a plane membrane held fixed at its perimeter. $\sqrt{\frac{z}{D_a}}$ corresponds to ω/v where ω is the angular frequency of the vibration, v is the velocity of elastic waves in the membrane; N corresponds to the displacement. The equation has solutions only for certain discrete values of $\sqrt{\frac{z}{D_a}}$; the solutions for various shapes of cross section of the column (i.e. various shapes of membrane) can be obtained from standard works on sound. For example, for a column of rectangular section, sides a, b, we get

$$\frac{z}{D_a} = \frac{\pi^2}{a^2}\left(1 + \frac{a^2}{b^2}\right). \tag{54.5}$$

55. Diffusion theory including stepwise ionisation and volume recombination. SPENKE[2] has considered the theory of the column when ionisation is produced both by direct collisions between electrons and gas molecules, and also by two stage (stepwise) processes e.g. when an electron hits an already excited molecule. The rate of direct ionisation is zN, of two stage ionisation is $\propto N^2 = \alpha N^2$ (since the rate of production of excited molecules is $\propto N$, the concentration of electrons; and the rate at which these are ionised is again $\propto N$). The general equation (cf. 54.4) is then

$$D_a \nabla^2 N + zN + \alpha N^2 = 0. \tag{55.1}$$

[1] E. SPENKE and M. STEENBECK: Wiss. Veröff. Siemens-Werk **15**, 18 (1936).

[2] E. SPENKE: Z. Physik **127**, 221 (1950).

To solve the equation the parameter ν is introduced defined as:

$$\nu = \frac{\alpha N_0^2}{z N_0} = \frac{\alpha}{z} N_0 \tag{55.2}$$

which represents the relative proportions of two-stage and single-stage processes contributing to the total ionisation. ν can vary over the range 0 to ∞. For the plane column the equation becomes soluble in terms of elliptic functions for the ranges $\nu < \frac{1}{2}$ and $\nu > \frac{1}{2}$. For the cylindrical column the equations are more difficult and can be solved approximately only when ν is small or very large. In both cases the effect of introducing stepwise processes makes little difference the the variation of concentration across the column; the effect is slightly larger in the cylindrical column.

SEELIGER[1] has pointed out that inclusion of volume recombination into the theory would also give a term in N^2 leading to an equation formally the same as (55.1), and soluble in the same manner. He suggests using the parameter $\nu' = \frac{\alpha' N_0^2}{z N_0}$, the ratio of the charges removed by recombination to those produced by direct ionisation at the centre of the column: ν' has the range 0 to 1. By analogy with SPENKE's solutions it seems that recombination also would have little effect on the distribution, except when $\nu' \approx 1$: the boundary conditions are, however, probably different for the two types of processes and no complete solution has been attempted.

56. Diffusion theory in electronegative gases (negative ions present). HOLM[2] and SEELIGER[3] have considered the effect of electron attachment i.e. the formation of heavy negative ions, on the diffusion theory of the column. Here it is not permissible to use D_a the normal ambipolar diffusion coefficient, nor to take the net rate of production of electrons as zN, since they are also being removed by attachment. For the linear column SEELIGER writes the three basic equations of motion:

$$j_e = -e D_e \frac{dN_e}{dx} - X \cdot e N_e b_e \tag{56.1}$$

and similarly with suffixes $+$, $-$, for positive and negative ions respectively.

We have also for the charges flowing to the walls,

$$j_+ = j_e + j_- \tag{56.2}$$

and assume

$$b_+ = b_- \,(= b). \tag{56.3}$$

For the net rate of production per unit length of the column of both electrons, and negative ions we have the equations

$$\frac{dj_{e,-}}{dx} = Q_{e,-} - R_{e,-} \tag{56.4}$$

where Q, R are the rates of production and disappearance respectively. Electrons are formed by direct ionisation, and also by collisions between negative ions and gas molecules causing detachment. This gives

$$Q_e = z \cdot N_e + \varkappa N N_- \tag{56.5}$$

[1] R. SEELIGER: Z. Naturforsch. **8**a, 74 (1953).

[2] R. HOLM: Z. Physik **75**, 171 (1932).

[3] R. SEELIGER: Ann. Physik **6**, 93 (1949). We give here the more comprehensive treatment of SEELIGER. Also R. SEELIGER and A. KRUSCHKE: Phys. Z. **34**, 883 (1933).

where N is the concentration of neutral mols. $\varkappa$ is a constant. Putting

$$\lambda = \frac{N_-}{N_e}, \tag{56.6}$$

(this ratio assumed to be constant) gives

$$Q_e = z N_e + \varkappa N \lambda N_e, \quad Q_- = \beta N N_e, \tag{56.7}$$

$$R_e = \beta N N_e, \qquad R_- = \varkappa N \lambda N_e \tag{56.8}$$

where β also is a constant, representing the probability of electron attachment.

The solution of these equations gives

$$\frac{z}{D_a} = \frac{\pi^2}{h^2} + \frac{\pi^2}{h^2}\left(1 + \frac{T_+}{T_e}\right)\lambda \tag{56.9}$$

for a column of width h. The second term is zero if no negative ions are present. When the concentration of negative ions is small ($\lambda \to 0$) the solution is equivalent to that found by HOLM. By substituting for z and D_a one finds an expression for the electron temperature [analogous to eq. (51.6)]

$$\frac{1}{\sqrt{\frac{e V_i}{k T_e}}} \cdot e^{-\frac{e V_i}{k T_e}} \approx C \cdot \frac{\pi^2}{(p h)^2} (1 + \lambda). \tag{56.10}$$

For values of $p h \sim 1$ to 10 (mm. Hg. cm.) this has been found to give reasonable values of T_e.

57. Relativistic theory of a column carrying large currents. It is known that when large currents flow through the column (~ 100 A) the electrons and ions are more concentrated at the centre than is expected on the simple theories. This effect may be due to the selfmagnetic field of the current (see Sect. 90), or can also be described according to FETZ[1] by taking account of the relativity effect upon electrons which, at low pressure, can move at high speed between collisions. When this is done the theory predicts not a simple radial electric field, but an additional field depending on the co-ordinate system chosen; the extra terms can be inserted in the SCHOTTKY equation. Physically the result is that the force on the ions is purely electric, on the electrons electric and magnetic (i.e. a component appears depending on their velocity).

58. General theory of the low current (subnormal) column. Recently ECKER[2] has tried to develop a more general theory of the positive column, removing some of the assumptions common to the other theories, in particular the assumption that $N_e \approx N^+ \gg (N^+ - N_e)$ which is known not to be true at very low currents[3]. In this region N_e and N^+ are both relatively small, and the longitudinal field X rises with decreasing current. ECKER uses as his general equations

(I) MAXWELL'S, neglecting magnetic field, and time variation;
(II) Number balance for electrons and ions;
(III) BOLTZMANN'S transport equations.

These are in general not soluble, but become so if one makes the plausible assumption that the electrons and ions are likely to be distributed across the

[1] H. FETZ: Ann. Phys. **40**, 579 (1941). — Z. Naturforsch. **4**a, 627 (1949).
[2] G. ECKER: Proc. Phys. Soc. B **67**, 485 (1954), see also [*157*].
[3] G. MIERDEL and W. SCHMALENBERG: Wiss. Veröff. Siemens-Werk **15**, 60 (1936).

column in the same way, so that

$$\frac{N^+}{N_e} = \text{const}. \tag{58.1}$$

The results give the departure from quasineutrality and provides a means of defining formally the division between "normal" and "subnormal" columns. T_e is evaluated in the usual way, equating energy gained and lost in the field; X can be evaluated as a function of the current and the values are shown in Fig. 47 compared with MIERDEL and SCHMALENBERG's measured values in Hg at pressures about 1 to 10 mm. The theory also shows that for very small currents X tends to a limit X_0, obtained by writing D_e (electron diffusion) in place of D_a (ambipolar) in SCHOTTKY's equation.

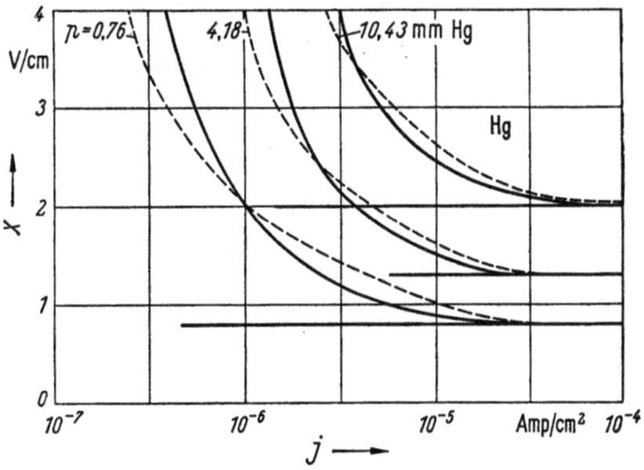

Fig. 47. Gradient of the positive column in Hg at very low current densities. ——— Measured. (MIERDEL and SCHMALENBERG.) ----- Calculated. (ECKER.)

g) Comparison of theory and experiment.

59. General. Most of the experiments designed to test the theories mentioned have been carried out over limited ranges of pressure, tube radius and current; a comprehensive comparison has been made only for the low pressure theory in Hg and Ne, which will be referred to later. However, the rather scattered evidence strongly suggests that the two basic theories are substantially correct within their respective ranges of pressure.

Many experiments test only one part of the theory e.g. current or mobility equation. The measured values of all the parameters except one, are inserted in the appropriate equation, the remaining one calculated and compared with its measured value.

Unless otherwise stated the measurements to be discussed have been made with LANGMUIR probes. T_e being derived from the slope of the $\log i - V$ curve, V_{space} from the kink in it, and N from the current flowing to the probe when it is held at space potential (see Sect. 9). When N is known z can be found by measuring the ion current to a probe arranged flush with the wall, and held at wall potential. From the characteristics of this probe $V_s - V_w$ (the term in E_{wall} in the energy balance equation) can be found.

60. Energy distribution, and LANGEVIN's mobility equation i.e. the assumptions. The constancy of T_e across the tube was confirmed to within 10% by KILLIAN (Hg, low pressure), DRUYVESTEYN (Ne, $p=1$ mm.), and SEELIGER and HIRCHERT (Ne, A, $p \sim 1$ to 10 mm.) [*151*], [*155*], [*159*]. KILLIAN also found that the observed rate of ionisation could be produced by a MAXWELLian distribution of electron energies except for $p < 10^{-3}$ mm. when more fast electrons must be present. HAIGH, SMITH, and MCCULLOCH[1, 2] have tried to measure the actual energy distribution by passing the electrons through a slit in a probe held at space potential and analysing their velocities with a magnetic field. They found, for

[1] C. HAIGH and T. SMITH: Phil. Mag. **41**, 557 (1950), A, air.
[2] C. HAIGH and R. K. MCCULLOCH: Proc. Leeds Phil. Soc. **6**, 82 (1952), A, air N_2.

example in A ($p=2.5\ 10^{-2}$ mm.) distributions which were markedly non-MAXWELLian, in fact approximating to DRUYVESTEYN'S, but with a mean energy considerably larger than that given by T_e. The probe itself, however, gave a linear plot indicating a MAXWELLian distribution. There was also a discrepancy of several volts between the space potential obtained from the probe characteristic, and that from extrapolating the observed distribution to zero energy. The reason for this is not known. BOYD[1] has observed a similar distribution in A at $p=4\cdot 10^{-3}$ mm., using a screened probe which separates the electron and ion components of the current to the probe; he finds fewer electrons of energy above 15 eV than in the MAXWELLian distribution.

It is believed that a MAXWELLian distribution is caused mainly by electron interaction and by appreciable cumulative ionisation; both these effects become small at low current densities and we should then expect deviations from this distribution. It is still not known whether these effects are numerically sufficient to explain the MAXWELLian distributions usually observed; it is possible that plasma oscillations may also play an important part. These would probably upset probe measurements because of the variation in space potential.

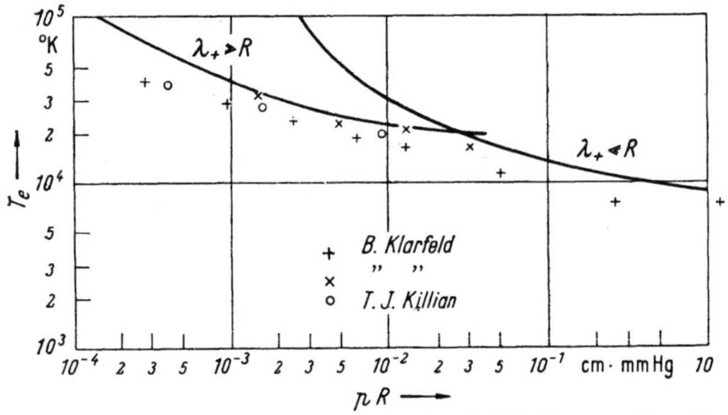

Fig. 48. Electron temperature as a function of pR in Hg; comparison of theory (full line) and experiment (points).

The LANGEVIN equation has been confirmed by KILLIAN and KLARFELD who calculated λ_e and found it within 30% of the values obtained from scattering experiments. BOECKNER and MOHLER [167] using a similar treatment in Cs found that the electrons are scattered more by the ions than by the atoms at high discharge currents. KLARFELD, using measured values of N and T_e calculated X as a function of the pressure in He, A, and found reasonable agreement with experiment.

In general the LANGEVIN formula seems to be confirmed, but must sometimes be modified to take account of the variation of λ_e with electron energy (RAMSAUER effect), the presence of ions[2], and at low pressure the effect of collisions with the walls[3].

61. Electron temperature. Fig. 40 shows measured and calculated values of T_e as a function of pR in Ne, A, for the SCHOTTKY theory; Fig. 48 compares them over an extended range of pressures for Hg: the calculations here are due to WASSERRAB[4]. It is seen that there is reasonably good agreement.

62. The energy balance, and the gradient [167] to [172], [150]. The major difficulties, both theoretical and experimental, lie in calculating or measuring the terms in the energy balance equation. This applies especially to the energy emitted in radiation because much of it is at wavelengths where measurements have been difficult or impossible, and further confusion arises because radiation

[1] R. L. F. BOYD: Nature, Lond. **165**, 228 (1950).
[2] S. GVOSDOVER: Phys. Z. Sowjet. **12**, 164 (1937); **13**, 133 (1938).
[3] W. G. POLLARD and R. D. PRESENT: Phys. Rev. **73**, 762 (1948).
[4] T. WASSERRAB: Z. Physik **128**, 312 (1950).

emitted by the discharge is often absorbed by the wall and thus appears in the measurements as a wall loss.

In a few relatively simple cases calculations have been possible. For example VON ENGEL and STEENBECK, by equating iX to the energy loss of electrons colliding with gas molecules found X/p as $f(pR)$ for N_2, H_2 in good agreement with experiment (Fig. 42). WASSERRAB has attempted a similar kind of calculation for Hg (Fig. 49). KLARFELD was able to calculate X when the only term in the energy balance equation was E_{wall} (low pressure). Experiments on the energy balance by SOMMERMEYER, MOHLER, DRUYVESTEYN and WARMOLTZ, KLARFELD and KARELINA lead to the general conclusion that at low pressure ($\sim 10^{-4}$ mm.) most of the energy is dissipated at the walls; at medium pressures ($\sim 10^{-3}$ to 1 mm.) radiation accounts for between 50 and 90%, and at pressures above about 1 cm. Hg (when the column contracts) gas heating becomes predominant. This is illustrated in Fig. 50: the pressure scale depends of course on the gas. The general features can be explained qualitatively [168]. At very low pressures inelastic collisions between electrons and gas molecules leading to ionisation predominate, and this energy, plus the appreciable kinetic energy of the electrons and ions is delivered to the walls when they recombine. Thus most of the energy lost goes into heating the walls. At higher pressures elastic impacts become relatively more frequent, T_e decreases and some fraction of the energy is used in heating the gas. At high pressures volume recombination and the formation of a relatively cold layer of gas near the walls reduce E_{walls} to a negligible amount. However, at considerably higher pressures when the discharge begins to contract $E_{gas\ heating}$ falls again, being replaced by radiated energy[1]. The decrease in T_e with increasing pressure results also in a larger proportion of exciting collisions and hence more energy emitted as radiation. The energy emitted in both resonance and non-resonance lines increases to maxima at slightly different

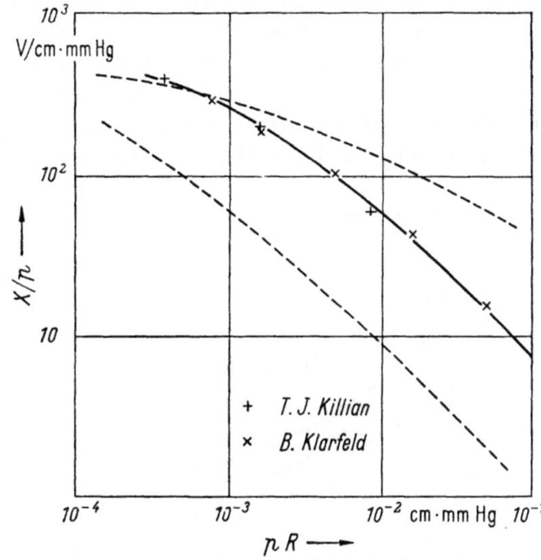

Fig. 49. X/p as $f(pR)$ in Hg. Comparison of theory ------- and experiment ———. (Footnote 4, p. 132.)

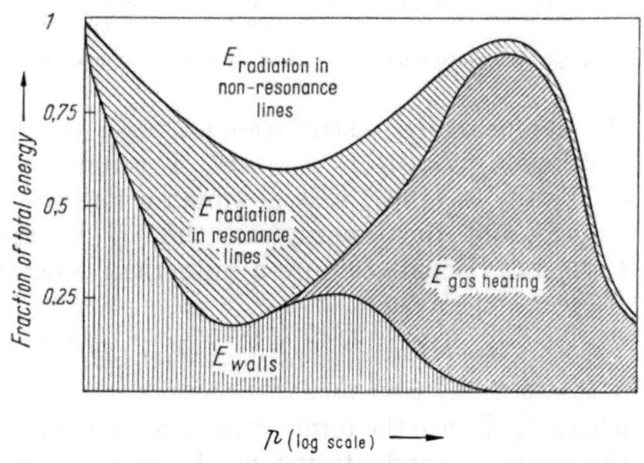

Fig. 50. Schematic representation of energy balance in the column at various pressures.

[1] W. ELENBAAS and W. DE GROOT: Physica, Haag 7, 807 (1935). — W. ELENBAAS: Physica, Haag 8, 859 (1936).

pressures[1, 2, 3] at low pressure resonance radiation predominates at high pressure the other.

MOHLER investigated the energy balance in Cs ($i \sim 0.1$ to 4 A, p 10^{-3} to 3 mm.): the radiation loss was measured by focussing the light from 1 cm. length of the column on to a calibrated thermopile. The transmission of the glass wall for Cs radiation was measured separately with a photoelectric cell. Nearly all the radiation was found to be in the resonance lines, and gas heating was found to be negligible for $i < 1$ A, $p > 0.03$ mm. SOMMERMEYER deduced the gas heating in He, Ne, A (i 50 to 200 mA, p 1 to 6 mm.) by measuring with a thermo-element

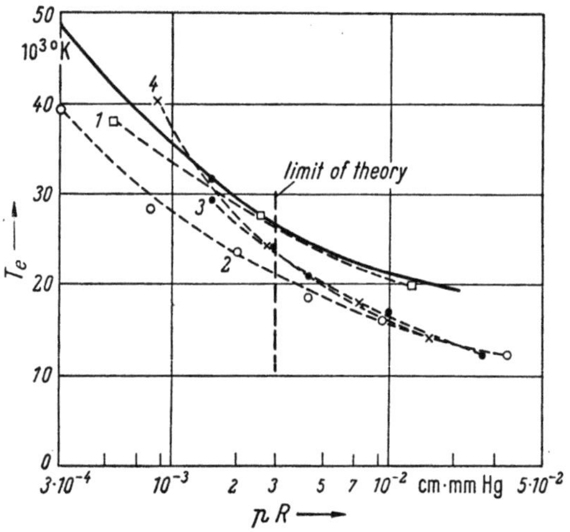

Fig. 51.

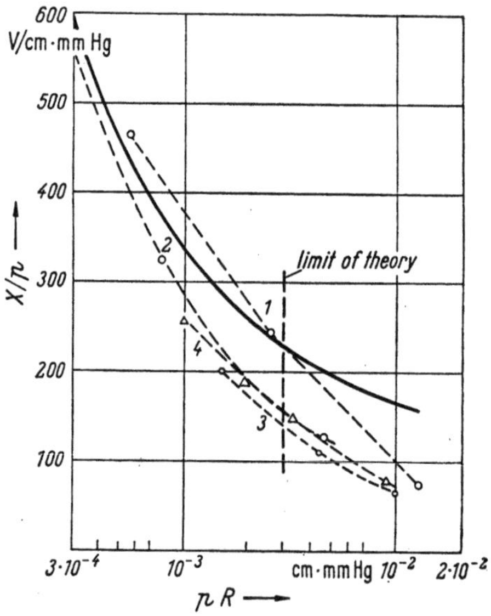

Fig. 52.

Fig. 51. Calculated curve and experimental data for the dependence of T_e upon the product pR in Hg, at low pressure. Theory: TONKS and LANGMUIR, KLARFELD applies to $pR < 3\cdot10^{-3}$ cm. · mm. Hg. Experiment: Curve 1: KILLIAN. tube diam. 6.2 cm., $i = 5$ A. Curve 2: GRANOVSKY diam. 6.5 cm. $i = 1$ A. Curve 3: KLARFELD diam. 3.2 cm. $i = 3$ A Curve: 4 KLARFELD diam. 3.2 cm. $i = 1$ A.

Fig. 52. Calculated curve and experimental data for dependence of X/p upon pR. Curves numbered as in Fig. 51.

the temperature distribution[4] across the discharge, which was found to be parabolic, giving $T_r = T_0 - a r^2$. This agrees with that expected from calculations of WARBURG (1895) who assumed a uniform generation of heat Q watts/sec./cm. length, and the assumption that Q is distributed in BESSEL-function form (SCHOTTKY) changes only a numerical constant. It follows that

$$a = \frac{T_0 - T_R}{R^2} = \frac{0.03\,Q}{k} \tag{62.1}$$

where T_0, T_R are the temperatures at the axis and the wall, k is a constant involving the thermal conductivity of the gas. SOMMERMEYER measured only the sum $(E_{\text{gas heating}} + E_{\text{walls}})$, deriving $E_{\text{radiation}}$ by subtraction. From continuity of heat flow from the column through the walls he derived that

$$E_{\text{gas heating}} + E_{\text{walls}} = f(T_R) \text{ only} \tag{62.2}$$

[1] H. KREFT and E. SEITZ: Phys. Z. **35**, 980 (1934).
[2] F. RÖSSLER and F. SCHÖNHERR: Phys. Z. **39**, 902 (1938).
[3] B. KLARFELD: Bull. Acad. Sci. URSS. **4**, 463 (1936).
[4] See also R. SEELIGER and H. STRÄHLER: Phys. Z. **27**, 732 (1926). — W. PUPP: Z. Physik **67**, 237 (1931).

and found the function by calibrating with a gas mixture (H_2 + air) which is known to emit hardly any radiation. His values of $E_{\text{gas heating}}$ were in good agreement with calculations from eqs. (53.2), (53.3) (corrected values, 1937) [*169*].

DRUYVESTEYN and WARMOLTZ used a calibrated thermopile to measure the energy radiated from discharges in Na, He, Ne, A and mixtures of Na with the rare gases, with pressures up to a few cm. Hg, $i \sim 0.2$ A. They found that gas heating becomes appreciable only at $p > 1$ cm. Hg, and is accounted for reasonably well by eq. (53.3), and also that the SCHOTTKY theory in general gives results for T_e, and X in quite good agreement with their observations.

63. General test of low pressure theory [*161*], [*166*], [*168*]. KLARFELD (Hg), KARELINA (Ne) have made extensive measurements and calculations in the range where the LANGMUIR-TONKS theory should apply. The calculations are detailed and rather complicated, the important new points being that an approximation more accurate than the linear rise with energy is used for the ionisation function; the change of λ_e with energy is taken into account in computing the mobility from LANGEVIN's equation, and the energy radiated is calculated by a method due to FABRIKANT[1] who uses an analytical approximation for the excitation functions of the main levels.

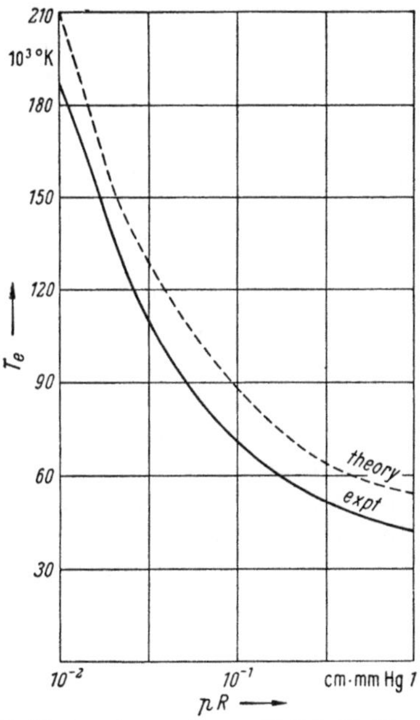

Fig. 53. Electron temperature as $f(pR)$ in He (KARELINA) $i = 0.3$ A, diam. = 32 mm. Low pressure positive column.

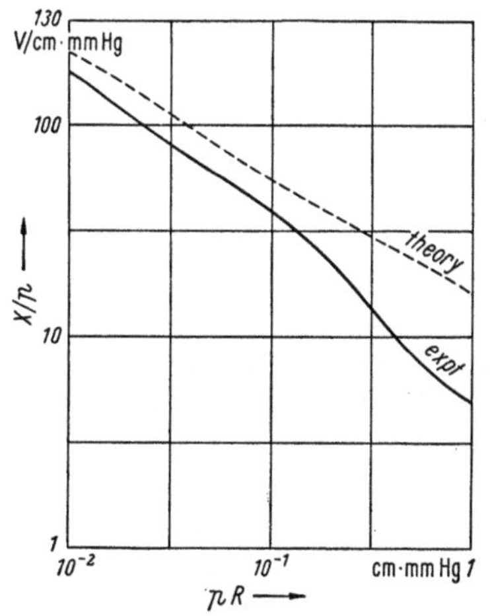

Fig. 54. Dependence of X/p upon pR in helium (KARELINA $i = 0.3$ A, diam. = 32 mm. Low pressure positive column.

Some of the results are shown in Figs. 51 to 54. POKROVSKAYA [*173*] has made more limited measurements in H_2 ($p \sim 9 \cdot 10^{-3}$ to $4 \cdot 10^{-2}$; $i \sim 0.3$ to 1 A) and similar calculations giving good agreement with experiment. He also shows that H_2^+ not H^+ ions are predominant in the discharge.

64. Radiation from the positive column; motion of ions. This is fully treated in Prof. R. G. FOWLER's article in this volume, and we consider here only those points which have a bearing on the mechanism of the discharge[2].

The spectrum emitted is usually the line spectrum of the gas atoms, the lines becoming rapidly fainter at higher energies, except at very low pressure when the electrons have a non-MAXWELLian distribution with a greater proportion of

[1] V. FABRIKANT: C. R. Acad. Sci. URSS. **15**, 451 (1937). — J. techn. Phys. USSR. **5**, 864 (1938).

[2] Many references will also be found in DRUYVESTEYN and PENNING, 1940, p. 163, 164.

fast ones. The recombination spectrum is absent except at high current densities[1]. In the molecular gases H_2 gives a continuous spectrum, and the N_2 band spectrum has been observed in discharges in narrow tubes when there is an appreciable increase both in X and in gas temperature (see Sect. 14).

At low current density the radiation is due only to direct excitation by electrons hitting gas atoms in the ground state, and the radiation in some lines is proportional to the current, and can be calculated with reasonable accuracy. For example at $p \sim 10^{-2}$ mm. the intensity of the line 1850 Å in Hg increases linearly with current, while 2537 Å shows a more complicated relation[2]. Calculations are outlined in Sect. 53: see also [174], [175] for 2537 Å line in Hg. At higher current densities collisions of the second kind, and cumulative excitation and ionisation processes occur, and the radiation emitted in any given line (e.g. resonance line) shows a variation also with the gas pressure[2,3,4] (see also Fig. 50). As the pressure increases from low values the radiation in the resonance line increases because fewer electrons are lost to the walls. At higher pressures (~ 1 mm.) a greater fraction of the power dissipated goes into gas heating, and other kinds of radiation via cumulative processes[5] (see Fig. 50).

At high current densities (1 to 50 A/sq. cm.) the spectrum of excited ions (spark lines) has also been observed in the rare gases[6]. Unlike the spark spectrum in the negative glow, excited by very fast electrons, this is almost certainly due to stepwise processes. Useful information about the motion of ions in the discharge can be obtained from this spectrum; the DOPPLER effect causes a broadening of the lines due to the random motion, and when viewed parallel to the discharge, a displacement of the lines due to the drift of the ions along the field. FRISCH and KAGAN [176], [177], [178] have found by this method that in A ($i = 350$ mA) the gas temperature $T \approx 1500°$ K, ion temperature $T_+ \approx 2400°$ K, $e\overline{V}_+ \approx 0.3$ eV and $v^+_{\text{drift}} \approx 3 \cdot 10^4$ cm./sec., i.e. the drift velocity is of the same order as thermal velocities. Similar results were found in neon. An approximate formula has been deduced for the ion velocity distribution[7] and also for the intensity, I, of the radiation in the ion lines[8]:

$$I = a\, N_e^2\, Q\,(V_m) \frac{q-1}{q^3} \left[2 + q \cdot \frac{e V_{ik}}{k\, T_e}\right] e^{-\frac{e V_{ik}}{k T_e}} \tag{64.1}$$

where

$$q = 1 - \frac{k\, T_e}{e\,(V_m - V_{ik})} \tag{64.2}$$

$a = \text{const}$, $Q\,(V_m)$ is the greatest value of the excitation function, $e\,V_m$ the energy at which it occurs, and $e\,V_{ik}$ the energy difference between the i-th and the k-th level.

The motion of ions in plasmas, their mobility and temperature and their contribution to wall currents has been discussed in mathematical detail by WENZL[9].

[1] E.g. F. L. MOHLER: Bur. Stand. J. Res. **10**, 771 (1933); **19**, 447, 559 (1937).
[2] F. A. BUTAEVA and V. A. FABRIKANT: Izv. Akad. Nauk. SSSR., Ser. Fiz. **13**, 271 (1949).
[3] Many references will also be found in DRUYVESTEYN and PENNING, 1940, p. 163, 164.
[4] F. RÖSSLER and F. SCHÖNHERR: Z. techn. Phys. **19**, 588 (1938).
[5] See e.g. FABRIKANT, BUTAJEWA and ZIRG: Phys. Z. Sowjet. **11**, 576 (1937).
[6] NUTTING and TUGMAN: Bull. Bur. Stand. **4**, 49 (1911), Argon. — See also HAMBURGER: Diss. Delft 1917.
[7] V. A. FOCK: J. exp. theor. Phys. USSR. **18**, 1049 (1948).
[8] V. A. FABRIKANT: C. R. Acad. Sci. URSS. **15**, 451 (1937).
[9] F. WENZL: Z. angew. Phys. **3**, 332 (1951).

h) The striated column.

65. Stationary striations. Striations in a great variety of patterns have been noticed in limited ranges of current and pressure in many gases. It seems likely that they were first observed by ABRIA; the literature contains a number of beautiful photographs of striated columns[1]. Many measurements have been made with the object of finding how the distance and the potential difference between striations, their radiation and associated space charge depends upon the nature and pressure of the gas, the current, and the diameter of the tube, and in order to classify the regions of current and pressure in which striations of particular types occur. It has been found that the properties of striated columns depend strongly on the purity of the gas, and it is not surprising that large differences exist between the measurements of earlier workers. For example striations in the purest hydrogen are usually broad and red, but with a slight trace of impurity present they change to narrow blue ones whose properties are quite different. Some workers believe that the blue striations occur only when impurities are present, but it is usually accepted that they do occur also in pure hydrogen. Here only those experiments which show most clearly how the properties of the striations depend on the parameters mentioned are presented. More complete references can be found in [3], [6], [8], [10], [165].

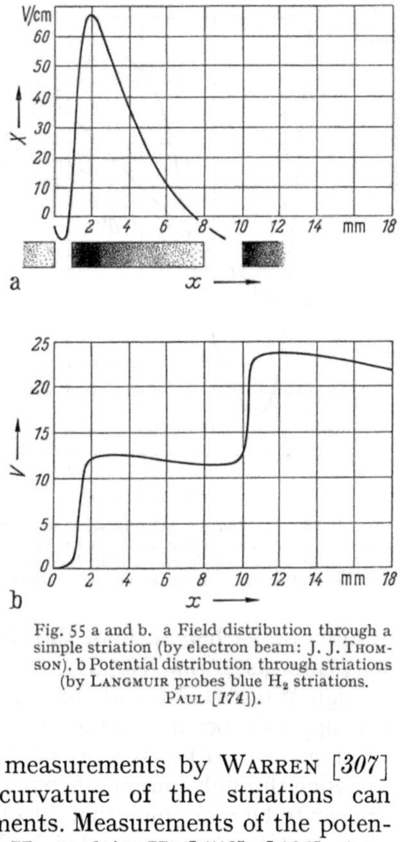

Fig. 55 a and b. a Field distribution through a simple striation (by electron beam: J. J. THOMSON). b Potential distribution through striations (by LANGMUIR probes blue H_2 striations. PAUL [174]).

Individual striations are curved convex towards the cathode, due to the effect of the negative wall charge. They are usually sharp and bright on their cathode side (front) and rather more diffuse and dimmer on the anode side. J. J. THOMSON[2] measured the electric field distribution through a striation by the deflection of an electron beam, finding the result shown in Fig. 55, with a small negative field just in front of the striation. Modern electron beam measurements by WARREN [307] have not shown this negative field. The curvature of the striations can sometimes confuse the results of such measurements. Measurements of the potential distribution with LANGMUIR probes[3,4] in Hg and in H_2 [179], [180] show a similar field distribution, but with a smaller or zero negative field (Fig. 55). The electron density N_e, has been found to vary in a similar way, being greatest in the bright regions[4,5]. H. A. WILSON deduced N_e from the current drawn to a pair of plates which impressed a small transverse electric field across the discharge. VAN DER POL[6] found a contrary result from measurements of conductivity using a LECHER wire oscillator, but these results appear unreliable

[1] See for example DE LA RUE and MÜLLER: Phil. Trans. Roy. Soc. Lond. **169**, 155 (1878), reproduced in J. J. and G. P. THOMSON; also R. HOLM: Phys. Z. **25**, 497 (1924).

[2] J. J. THOMSON: Phil. Mag. **28**, 441 (1909).

[3] W. H. MCCURDY: Phil. Mag. **48**, 898 (1924).

[4] COMPTON, TURNER and MCCURDY: Phys. Rev. **24**, 597 (1924).

[5] H. A. WILSON: Phil. Mag. **44**, 505 (1900).

[6] B. VAN DER POL: Phil. Mag. **38**, 360 (1919).

due to the large size of the wavelength compared with one striation. Later workers used LANGMUIR probes.

The first striation (nearest the cathode) is always slightly different from the others, being usually brighter than, and further than average distance from, its neighbour.

The potential difference, ΔV, between striations has been measured by moving the anode through them and noting the difference in the potential required to maintain constant current in the discharge (e.g. PENNING[1], in rare gases), and also from space potential measurements obtained by using LANGMUIR probes [*179*] to [*185*]. Earlier experiments of WEHNER[2], NEUBERT, and HOLM [*162*], [*181*], [*182*] show that as the current density increases, ΔV decreases to a constant value independent of the gas pressure and diameter of the tube. In rare gases, where only a few striations occur at the head of the column, PENNING found $\Delta V = 20$ v for He, 18.5 v for Ne (= excitation potential of the $2p$ state), and 11.9 volts for A (= excitation potential of the 1s state), while $\Delta V \approx 5$ v has been found for Hg. The broad red striations in hydrogen are, however, exceptional: for them ΔV increases almost linearly with pressure in the region of $p \simeq 1$ mm. (see Fig. 56).

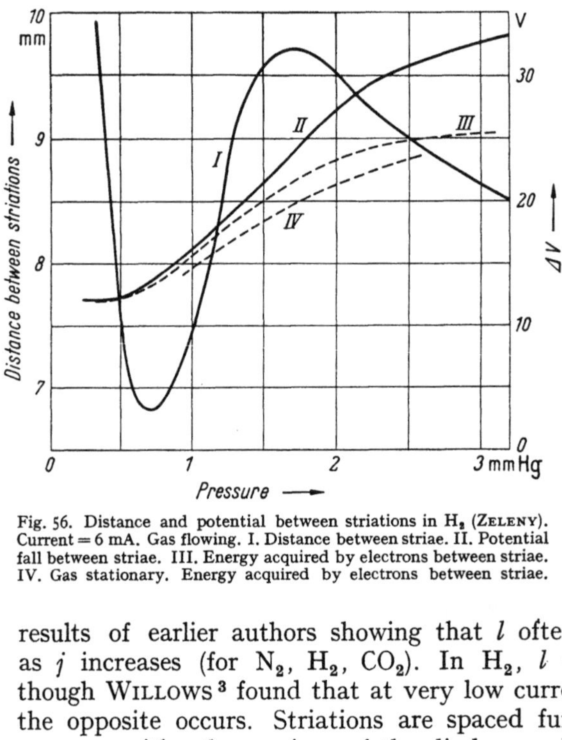

Fig. 56. Distance and potential between striations in H_2 (ZELENY). Current = 6 mA. Gas flowing. I. Distance between striae. II. Potential fall between striae. III. Energy acquired by electrons between striae. IV. Gas stationary. Energy acquired by electrons between striae.

The distance between striations l varies in a complicated way with current density, the results of earlier authors showing that l often approaches a constant value as j increases (for N_2, H_2, CO_2). In H_2, l decreases with increasing j (although WILLOWS[3] found that at very low currents l first increases), but in CO_2 the opposite occurs. Striations are spaced further apart at lower pressure in common with other regions of the discharge. Hydrogen is again an exception, l passing through a minimum when $p \sim 1$ mm (Fig. 56). At this point the striations appear to be more blue than at other pressures. A relation by GOLDSTEIN (1881) gives $l \propto 1/p^m$, later extended by WEHNER to include the radius R of the tube, in the form

$$\frac{l}{R} = C \cdot \frac{1}{(pR)^m} \tag{65.1}$$

m is a constant approximately equal to 0.5 (e.g. for H_2, $m = 0.53$), C depends on the current density. GOLDSTEIN's law conforms with similarity.

66. Conditions for the formation of striations are complicated, and different types of striation occur at different values of current and pressure. Holm has

[1] F. M. PENNING: Z. Physik **41**, 769 (1926), see also J. J. and G. P. THOMSON, Vol. 2, p. 386; [*3*].

[2] F. WEHNER: Ann. Phys. **32**, 49 (1910). — RIECKE: Ann. Phys. **26**, 282 (1905), see also references [*3*], [*181*], [*162*].

[3] WILLOWS: Proc. Cambridge Phil. Soc. **10**, 302 (1900).

classified the regions in which various kinds of striations are formed as closed areas on the $i - pR$ diagram (Fig. 57). GÜNTHERSCHULZE and MEINHARDT have studied the formation of striations in H_2 in some detail, and their results are summarised in Fig. 58. Wide red striations appear suddenly when conditions cross curve 1, and here $l \approx R$, $\Delta V \approx 50$ volts. As the value of pR is changed to approach curve 2, l and ΔV decrease gradually to $R/2$ and 20 volts respectively. Double striations occur suddenly at curve 2, the uneven ones (i.e. the front ones of each pair) being sharp, the others diffuse, and their distances apart alternating in the ratio 1:2. Going from curves 2 to 3 the striations become blue and equidistant and $\Delta V \approx 12$ v. They emit the continuous H_2 spectrum, of excitation potential 12.6 v, corresponding closely with ΔV measured here, and also that measured by PAUL. To these striations GOLDSTEIN's law applies. Curves 1 and 2 are given respectively by the equations

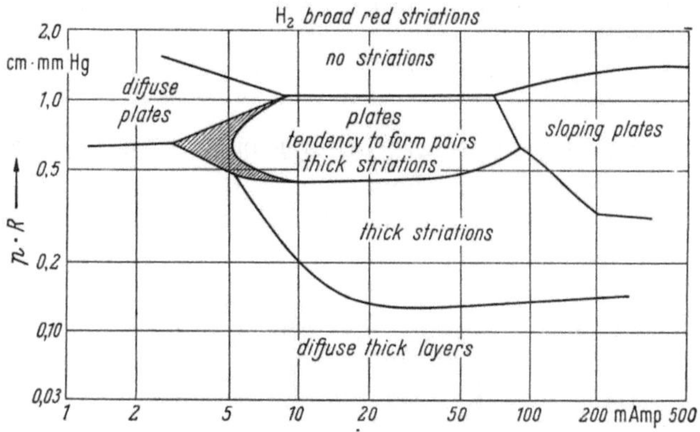

Fig. 57.
Conditions in which certain types of striations are formed in H_2 (HOLM).

$$\log i = A(pR) - B, \tag{66.1}$$

$$i = C(pR)^3, \tag{66.2}$$

where A, B, C are constants. Both conform with similarity. In general standing striations disappear at current densities $j >$ about 10^{-2} A/sq. cm.

The role played by the walls of the tube in maintaining striations is not really clear. LAU and REICHENHEIM think that in hydrogen recombination of H atoms to form H_2 at the walls is necessary for the appearance of striations [186]. More recently FOWLER[1] has noticed striations in a long tube having a side tube attached i.e. there was a gap in the wall of the main tube. No striations were ever found directly opposite this gap, and if they were made to move by slowly increasing the current they did so slowly and continuously except at the gap where they jumped abruptly across. He concludes that a complete solid boundary is vital for the striations to exist. GAMBLING[2], however, finds striations quite clear of the walls in H_2 at atmospheric pressure, and suggests that in FOWLER's experiment the product $p \cdot R$ at the gap (R very large here) was too great for striations to occur. It seems likely that these empirical laws for the existence of striations do not hold at higher pressures, where recombination probably takes place not at the walls but in the gas, as in the contracted uniform positive column.

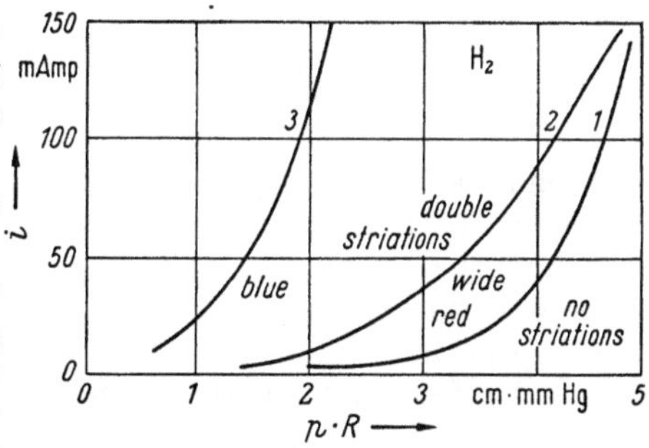

Fig. 58. Conditions for the appearance of striations in H_2 (GÜNTHERSCHULZE and MEINHARDT) [185].

[1] R. G. FOWLER: Phys. Rev. **84**, 145 (1951), H_2, low pressure.
[2] W. A. GAMBLING: Phys. Rev. **85**, 677 (1952).

67. Moving striations. Striations moving towards the cathode are found in molecular and rare gases, but have been studied chiefly in the latter. They occur only when the current and gas pressure are less than certain limiting values; moving striations disappear when $j >$ about 1 A/cm.², i.e. 100 times larger than the limit for stationary striations. PUPP found that in pure rare gases, the limiting current and pressure are almost independent of the tube radius and connected by

$$i = \frac{C}{p} \tag{67.1}$$

where C is a constant depending on the gas (see Fig. 59) [*187*].

Measurements on the velocity v, distance apart l, and frequency f, of moving striations have been made with rotating mirrors or in more modern experiments

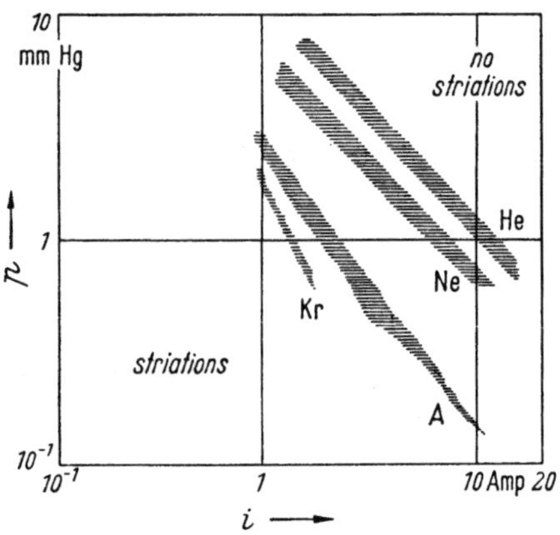

Fig. 59. Limiting condition for the formation of moving striations (tube diameters 1.5 – 5 cms.) (PUPP).

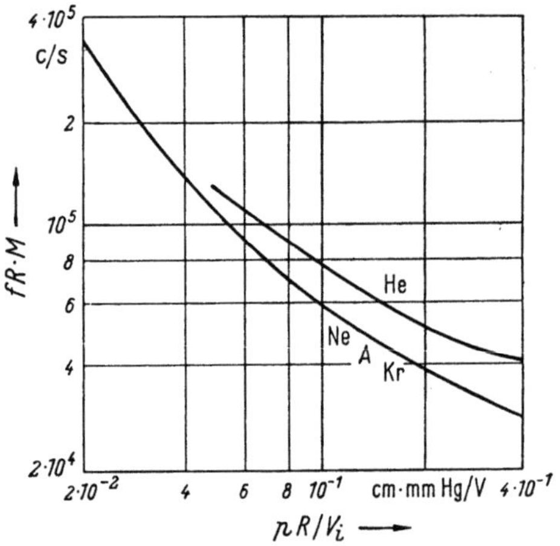

Fig. 60. Frequency f of moving striations as a function of pressure and tube radius (PUPP). M, V_i are the atomic mass and ionisation potential of the gas.

with photocells, and some workers have examined the associated fluctuations in the electrical properties of the discharge, using oscillographic techniques. For the usual regularly spaced striations $f = v/l$. Probe measurements show periodic variations of field, electron temperature and concentration along the tube. PUPP found that in long tubes having a large number of striations, so that end effects are small, the following relations hold:

$$\frac{l}{R} = F\left(pR, \frac{i}{R}\right), \tag{67.2}$$

$$fR = \varphi(pR), \quad \text{almost independent of } i. \tag{67.3}$$

The second formula conforms with similarity, but the first does not, owing to the factor i/R. When $f\,R \cdot M$ is plotted against pR/V_i all the rare gases lie on almost one curve (Fig. 60). M is the atomic mass, V_i the ionisation potential. The velocity of the striations towards the cathode is 10^3 to 10^5 cm./sec.

Recent extensive experiments on moving striations in all the rare gases, H_2, air and Hg by DONAHUE and DIEKE [*194*][1] have shown a more varied and complicated type of behaviour. Using a photomultiplier and oscilloscope they measured simultaneously the fluctuations in voltage, current and light intensity; the voltage fluctuation was used to trigger the oscilloscope so that the sweep

[1] See also DIEKE, LOH and CROSSWHITE [*195*], [*196*].

always began at the same instant of a cycle. In any given position the photomultiplier registers the variation of light intensity with time, giving the frequency of the striations. By moving it along the tube one observes the change in the trace, both in relative phase and in shape, from which one derives the growth or decay, and the velocity of the striations. The results, referring more particularly to argon, (but similar in the other gases) may be summarised thus:

(I) Fluctuations in voltage δV, and in current, are only a few percent of the steady value, but the light intensity varies from maximum to zero. Also δV seems to be a property of the discharge and not the circuit, and is usually equal to the lowest excitation potential of the gas.

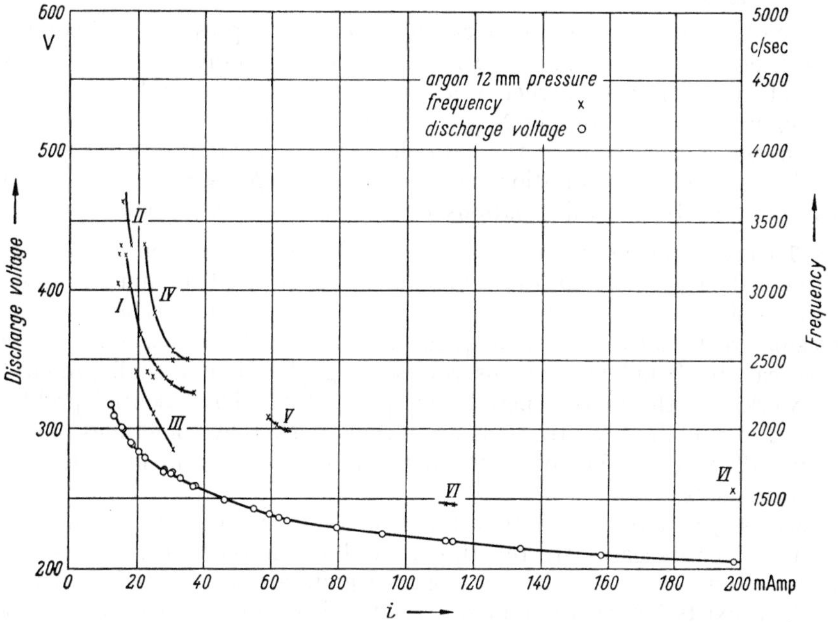

Fig. 61. Total discharge voltage ○ and frequency of oscillations × as functions of the current for an argon glow discharge at 12 mm. pressure. DONAHUE and DIEKE [194].

(II) A graph of frequency, f against i gives discontinuous curves, i.e. under different discharge conditions there are different modes of oscillation. Often striations will jump between several modes for a considerable time before settling into one; sometimes they never settle but continue in this unstable state. There are a very few values of current at which no striations occur: however, at large values of i, p no striations are found. (It is not said how these values compare with PUPP's.)

These results are shown in Fig. 61.

(III) The intensity pattern [I as $f(t)$, for fixed x] is quite different for various modes, and for any one mode it changes as the striation moves along the tube especially near the cathode. At low current some faint stationary striations can be seen near the cathode, their distances apart being related:

$$l_{\text{moving}} = 1 \text{ or } 2 \times l_{\text{stationary}}. \tag{67.4}$$

(IV) There are always two types of striations. Positive ones (+ space charge) go from anode to cathode with speeds from $3 \cdot 10^3$ to $3 \cdot 10^4$ cm./sec., usually about 10^4, and these are the striations often observed by eye. Their speed fluctuates considerably, even in apparently constant conditions, as was found also some years ago by ASTON and KIKUCHI [*142*]. At the same time negative striations with much smaller light intensity move much faster ($\sim 10^5$ cm./sec.) from cathode to anode. When the two kinds meet they stop for 20 to 100 μsecs., partially neutralising each other, and then proceed on their courses. If the meeting places are at fixed points in the tube the average light intensity there is large and the striations appear stationary[1]. The negative striations are believed to be bunches of electrons.

(V) Whenever a positive striation reaches the negative glow a negative one arrives: whenever a positive striation leaves the FARADAY dark space a negative one leaves also.

These observations have been clearest in argon at a pressure of about 12 mm. At lower pressures the results are more confusing. Many of these results are consistent with those of earlier workers, e.g. KIKUCHI found three types of (positive) moving striations in Ne at different pressures; at low pressures they moved faster as they reached the cathode. The application of a magnetic field affected only one type of striation and then only when $p \leq 5$ mm. Hg, making them appear fainter. In general striations travel faster at lower pressure.

68. Theory of the striated column. Although there have been many attempts to derive a theory to explain the behaviour of the striated column no convincing solution exists, and the matter can be discussed, except for certain calculations on plasma oscillations, only in a qualitative manner[2]. A discussion of earlier theories can be found in previous reviews, e.g. [*8*]; here *I* shall present only recent work and the more generally accepted ideas. The essential problem is, of course, to explain how striations are formed, why they sometimes move, and why an indefinite number of stationary striations are found in molecular but not in rare gases.

It seems generally agreed that the few standing striations at the head of the column in rare gases are repetitions of the FARADAY dark space—positive column pattern. Electrons in the negative glow are accelerated by the small field which exists between it and the anode. They first gain energy to excite, producing the bright front edge to the striation; further along the column they ionise. Having lost their energy by these inelastic collisions, and being retarded by the positive space charge they are no longer capable of exciting and have to travel some distance in the field (a dark region) until they gain enough energy to repeat the process. As they move along the tube the electrons become more spread out and no longer excite gas atoms in a thin layer; successive striations become more diffuse and soon merge into a uniform column. These striations are therefore analogous to the negative layers near the cathode (see Sect. 28).

In molecular gases and mixtures probe measurements show that there exists almost a field-free space between striations, in which electrons lose nearly all their energy. In Hg—A mixtures electrons appear to have MAXWELLian energy distributions in both dark and light regions, but with a higher mean energy in the latter[3]. But in H_2 recent measurements with a high frequency modulated

[1] They do not, however, have the appearance of the stationary striations usually found, for example in molecular gases, and it cannot be inferred that this is the only way in which such striations are produced.

[2] Some early attempts were made to give a mathematical treatment. For example J. J. THOMSON by considering the drift motion and ionising power of electrons along the tube, together with the effect of their space charge found a periodic solution for X and T_e below a critical pressure. No data were available for a numerical check. [Phil. Mag. **42**, 981 (1921).]

[3] N. L. OLESON and C. G. FOUND: J. Appl. Phys. **20**, 416 (1949).

probe indicate a group of fast electrons (energy $\sim\overline{V}+10$ e. volts) which are absent in the dark regions[1]. There exists a large field at the front of each striation in which the electrons are rapidly accelerated, producing the characteristic sharp bright front (see Fig. 55), and also giving rise to fine structure as levels of different energy are excited at different points along the tube[2]. It is not at all clear, however, how this state of affairs arises. It has been suggested that negative ions are required to produce this abrupt increase in field, by the formation of a thin double layer; although it is known that negative ions do have this effect (see Sect. 93) there is no evidence that their presence is necessary. The similarity principle applies to stationary striations and hence cumulative processes can be excluded.

Some authors have regarded the striated column as being in competition with the uniform one, and argue that whichever is the more economical in operation, i.e. produces a given ionisation for the least total potential, will actually exist[3]. This argument has a certain force, since it is known that ionisation in a non uniform field is a much more efficient process than in a uniform field; but no detailed comparison of the two different columns has been made, nor would such an argument disclose how a striated column is established. There are mainly three different views about the origin of striations; they are thought to arise:

(I) *At the anode or negative glow.* DONAHUE and DIEKE, from the results of their experiments, suggest that electrons from the negative glow may so accumulate that the field there becomes negative, preventing the diffusion of electrons into the positive column. They can then be released only be the arrival of a positive striation (number 1) moving from the anode. In the strong local field electrons are pulled out of the glow, and partially neutralise striation 1. Soon afterwards striation 2 moves along from the anode, pulls the electrons out of striation 1, and the resulting positive charge causes more electrons to escape from the glow. In this way bunches of electrons are handed on to successive positive striations. The obvious difficulty is to explain how the first positive striation leaves the anode. Even so this picture agrees with much experimental evidence: it explains the observed temporary neutralisation of positive and negative striations and the fact that negative striations are seen to originate from the rear of the negative glow. In addition one would expect that when electrons leave the glow the space charge there, becoming positive, would cause an increased field in the cathode dark space, and hence a burst of current at the cathode. This is in fact observed. There is also other evidence that the anode region is particularly sensitive for the production of fluctuations. Electrical oscillations in the discharge are sometimes found to be associated with a flickering light an the anode[4]; similar oscillations can be produced by applying a transverse magnetic field, but only near the anode[5]. It is well known that stationary striations can often be "grown" from the anode by moving it away from the cathode. Anode glows develop, become detached from the surface, and remain as striations in the gas[6].

The mechanism by which striations move (in Hg vapour only, not necessarily in other gases) is also suggested by the above mentioned authors (Fig. 62). It is thought that electrons first produce a layer of metastables and another layer of ions nearer the anode. Accelerated by this strong space charge succeeding electrons produce first another layer of metastables nearer the cathode, then ionise the existing

[1] R. L. F. BOYD and N. D. TWIDDY: Nature, Lond. **173**, 633 (1954).

[2] E. LAU and O. REICHENHEIM: Z. Physik **61**, 782 (1930).

[3] See e.g. J. J. and G. P. THOMSON [3], also DRUYVESTEYN and PENNING: Rev. Mod. Phys. **12**, 169 (1940).

[4] E. V. APPLETON and A. G. D. WEST: Phil. Mag. **45**, 879 (1923).

[5] J. L. STEINBERG: Nature, Lond. **160**, 833 (1947). — Rev. sci. Paris **85**, 607 (1947).

[6] See for example [3], vol. II, p. 386.

metastables after which they will have lost all their energy. The original ions will have dispersed somewhat by diffusion or recombination, and those remaining have their space charge partly neutralised by the slow electrons. The peak space charge density thus moves towards the cathode: the picture agrees with the observation that the peak intensity of the resonance line lies always nearer the cathode than for other lines.

Other evidence for the importance of the anode in forming striations comes from the works of PUPP, who found that they existed only when a positive anode fall was present; and also of TAKAMINE[1] who found that striations were generated by oscillations of the anode glow which grew until the glow detached itself. A theory by BAZAROV[2], based on plasma oscillations, predicts the existence of a few strata near the anode, under certain conditions, in agreement with observations of ZAITSEV[3].

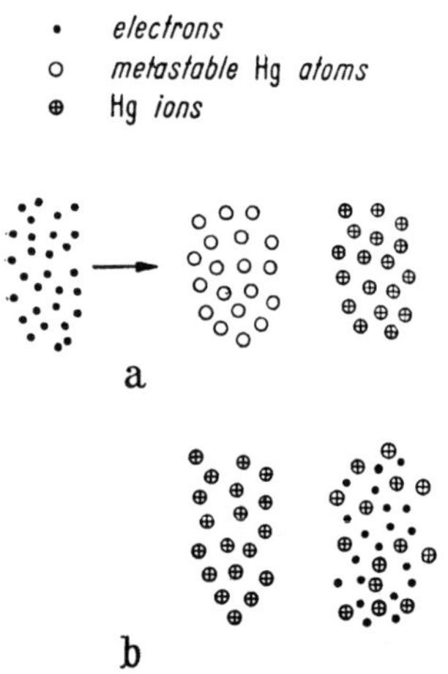

Fig. 62 a and b. Suggested mechanism of moving positive striations in Hg. The peak space charge density in (b) has moved to the left compared with (a).

Or (II) *in the plasma*. In the column, which is a nearly neutral plasma, groups of charges may oscillate due to their local electrostatic fields; these are called plasma oscillations and have been observed by many workers. These are discussed more fully in Sect. 95, but it has been shown theoretically that by this mechanism many kinds of spontaneous wave motions may be set up in a plasma. Their frequencies vary from about 10^9 c/s corresponding to electron motion, to 10^4 c/s or less due to oscillation of heavy ions. It has often been suggested[4] that striations occur at the nodes and antinodes of standing waves set up in a tube. Whether they are stationary or moving depends on the phase change when the wave is reflected at the ends of the column, and also whether the electron density there is constant or varying. Here there is less difficulty in describing how the striations start, since the plasma oscillation builds up from some favourable random fluctuation of charges in the gas. It is also argued that since striations occur in electrodeless high frequency discharges they cannot originate from electrodes, but this is less convincing since the walls of the vessel may have an effect similar to electrodes.

DRUYVESTEYN [*197*] considered the effect of superimposing on the steady values of N_e, X, T_e (corresponding to the uniform column) an oscillation whose amplitude varied with distance from the cathode. He found a solution was possible only if the striations (corresponding to the peaks of the oscillation) moved toward the cathode with a velocity $v \sim b^+ X$. Such striations are often found.

Calculations on the connection between strata and the propagation of electric oscillations in plasmas have been made by BAZAROV and more particularly by GORDEEV[5]. Strata are here considered to be wave group oscillations of potential, moving with the group velocity; the distance between standing striations corre-

[1] T. TAKAMINE, T. SUGA and A. YANAGIHARA: Inst. Phys. Chem. Res. Sci. Pap. **403** (1933).

[2] I. P. BAZAROV: J. exp. theor. Phys. USSR. **21**, 711 (1951).

[3] A. A. ZAITSEV: Moscow State Univ. Publ. **9** (1950).

[4] See e.g. G. D. MORGAN: Nature, Lond. **172**, 542 (1953).

[5] G. V. GORDEEV: Dokl. Akad. Nauk. SSSR. **79**, 771 (1951). See also A. A. ZAITSEV, Dokl. Akad. Nauk SSSR. **79**, 779 (1951), for a somewhat similar treatment for Ne, Ne A-molecular gas mixtures.

sponding to a monochromatic wavelength, between moving striations to the distance between maxima of wave groups. Reasonable agreement with experiment is found over a limited range for H_2, N_2, Hg. The connection between strata and electrical oscillations has recently been reviewed by ARMSTRONG, EMELEUS and NEILL [*198*].

Or (III) *at the cathode.* LOEB[1] regards the cathode as probably the seat of all instability and fluctuations in discharges, and suggests that striations may be due to fluctuations in the surface of the cathode, and to periodic movements of the space charge in front of the cathode. Electrons accelerated from the cathode to the negative glow produce a large positive space charge which as it moves towards the cathode increases the electric field in the dark space. This enhances the ionisation in subsequent avalances producing an unstable state with an excess of electrons in the negative glow. The observations on spontaneous electrical oscillations in the low pressure Hg arc[2] may be significant here: random and regular oscillations are found and it has been shown that the cathode is the source of the random oscillations (even when the spot is anchored), while the anode is the source of regular oscillations.

From results of his experiments KLARFELD[3] believes that any local concentration of charges, giving a steep rise in potential, whatever the cause, can generate striations. These may be continuous or damped. The process by which they are formed always occurs, however, on the cathode side of the striation. The strata are usually moving but it is found that their movement can be largely controlled by applying local fields through probes.

KLARFELD has observed striations in capillary tubes at high current densities and pressures ($j \sim 300$ A/sq. cm., $p > 760$ mm.). In such tubes, the column is however always continuous in the metal vapours Na, Cs, Cd, Hg, also in the rare gases with a small admixture of H_2, and in most gases at very low pressure.

VIII. The anode region.

69. General. The anode in the glow discharge acts simply as a collector, the current in the gas being transferred to the circuit by the electrons which strike its surface. Only in special circumstances do positive ions flow from the anode into the gas, that is when it is covered with a layer of alkali metal and heated. Then it acts as a KUNZMAN source, and performs a role analogous to thermionic emission from a heated cathode [*199*].

The properties of the anode region are different when the anode is in contact with a neutral plasma (the positive column) from those found when it is placed in a small negative space charge (the FARADAY dark space).

70. Anode in the positive column. When placed in the positive column the anode attracts electrons and repels positive ions so that a negative space charge builds up on its surface, forming a thin layer or sheath. The field here is greater then in the undisturbed column, and a potential drop exists across the sheath. This is called the *anode fall* in potential (V_a), the anode being appreciably more positive than the plasma. The thickness of the sheath, the potential drop V_a and the distribution of the field in this region are determined by a very simple condition. Electrons accelerated across this region must produce in unit time the same number of ions to flow into the positive column as flow out at the other

---

[1] L. B. LOEB: Phys. Rev. **76**, 255 (1949).

[2] B. GRANOVSKY and L. BYKHOVSKAYA: J. Phys. USSR. **10**, 351 (1946).

[3] B. KLARFELD: J. exp. theor. Phys. USSR. **22**, 66 (1952).

end into the FARADAY dark space. It is easy to see that if this condition were not fulfilled there would soon be too few ions at the anode end of the column, a larger negative space charge would grow in front of the anode, hence a larger anode fall and more ionisation would develop until enough ions were produced.

Electrons which have gained enough energy to ionise are also able to excite, and hence produce a layer of light very close to the anode surface—the *anode glow*. Since electrons start from the edge of the sheath with a small velocity (a few volts) we observe, starting from the plasma, a dark space in which the electrons are too slow to excite, next the anode glow and then sometimes an extremely thin dark space between the glow and the anode[1]. The sequence is similar to that forming the first cathode layer (Sect. 28).

The positive ion current through the column is very small, and so for the usual geometry (plane anode, cylindrical tube) we might expect V_a to be of the same order as V_i, the ionisation potential of the gas. This is on the assumption that electrons make very few collisions in the sheath, and that only single stage ionisation occurs: stepwise ionisation would naturally result in a smaller anode fall. It is in fact usually found in these circumstances that $V_a \approx V_i$.

Factors which affect the value of V_a are:

α) The size and shape of the anode. A large concave anode provides a large surface to collect electrons; more electrons are drawn in from the plasma, and the ion current to the column can still be provided even though each electron ionises less effectively than before. Losses by diffusion are also reduced, and V_a therefore decreases. Conversely a small anode requires a large anode fall, and a thick sheath. LANGMUIR and MOTT-SMITH [*26*], [*200*] deduce that the anode fall is positive, i.e. the anode is more positive than the plasma, when

$$(j_e)_r A < i \tag{70.1}$$

where $(j_e)_r$ is the random current density of the electrons, i the total discharge current flowing through the anode, and A its surface area. The thick sheath built up around a small (e.g. spherical) anode increases the effective area, and the outer surface of the sheath may be thought of as the collector. In their theory LANGMUIR and MOTT-SMITH suggest that when $(j_e)_r A > i$ a negative anode fall should occur, but this view has been criticised by von ISSENDORF [*201*].

A negative anode fall has been observed under artificial conditions (see δ); also when the anode is placed well into the negative glow its potential is less than plasma potential[2] (Sect. 39).

β) The current density. This factor is of course closely connected with the anode geometry. From similar reasoning we should expect the anode fall to increase as the current density rises. This has in fact been observed, and gives rise to the rather unfortunate nomenclature "normal" and "abnormal" anode fall by analogy with the cathode fall[3]. At high current densities the current and the anode glow often concentrate in small regions, regularly spaced about the anode surface, called *anode spots*.

γ) The presence of electronegative gases. These attach electrons, especially slow electrons forming heavy and slow negative ions. Both the electron current and the number of electrons available for ionising the gas are reduced, and a large

---

[1] First observed by F. W. ASTON: Nature, Lond. **89**, 218 (1912).

[2] When a small cathode is surrounded by a large concentric anode diffusion of ions may be an important factor; see [*213*], [*214*].

[3] This is confusing because V_a rises very slowly and steadily with increasing j, and the anode zone does not show those properties which distinguish the normal and abnormal cathode fall.

increase in the anode fall is necessary to compensate these effects. Very large values of V_a (up to 1000 volts) have been observed in the halogens. A particularly large anode fall ($\sim$2000 volts) has been observed when the anode contains a lithium halide, anode rays also being produced, cf. [3], Vol. 2, p. 405.

δ) The presence of an artificially produced space charge. When a large concentration of positive ions is artificially produced near the anode, a positive space charged is produced at its surface. The supply of ions is greater than is needed by the positive column and so there should exist a negative anode fall, drawing ions into the anode. This has been observed by PUPP[1].

ε) The proximity of the side walls. Electrons diffuse from the anode region to the side walls, and are thus no longer able to ionise in the gas. If there is also a flow of ions to the walls the electrons are neutralised; if not the electrons set up a radial field preventing further losses and concentrating the field at the anode. Wall effects may thus lead either to an increased or decreased anode fall. Little work has been done on this effect.

71. Effects analogous to the anode fall region; production of double layers LANGMUIR and MOTT-SMITH [26] observed that if the tube containing a low pressure discharge suddenly becomes narrower a luminous layer is formed at that point, and a voltage drop appears across it. The condition for the appearance of this layer is that $A\,(j_e)_r < i$ where A is the cross section of the narrower tube and i is the discharge current. Its production is favoured by high current densities, and the layer can act as an interrupting switch[2]. The mechanism of the effect has been studied in detail for Hg discharges by WASSERRAB [202].

72. Anode in the FARADAY dark space. In that part of the FARADAY dark space nearest to the negative glow there is an appreciable excess of electrons, which move outwards largely by diffusion. If sufficient electrons diffuse to the anode to carry the discharge current[3] no anode fall or anode glow develops, and the field right up to the anode surface remains almost as it would in the undisturbed dark space. It is not necessary for any ionisation to occur since there is no ion flow from the FARADAY dark space into the negative glow, except what would come from a positive column. The cathode dark space and negative glow maintain their own ion balance.

Clearly there must be a transition between these two types of anode region as the anode is moved further away from the cathode. It is found that an anode fall develops suddenly, at a critical distance D from the cathode, this point lying towards the far (anode) end of the FARADAY dark space. On the simple picture presented we would expect D to be large at low pressure (which aids diffusion) and when the cathode fall is large (which gives the primary electrons greater range, and lengthens the negative glow). This is found to be true [203].

73. Measurements of the properties of the anode region. The anode fall can be measured in the following ways:

(I) From the sudden rise in potential V across the electrodes as the anode is moved away from the cathode, with the current kept constant i.e. V as $f(x)$. (We could also measure V as $F(i)$ with p, x constant, or V as $\varphi(p)$ with x, i constant but the first method is preferable since the discharge conditions are there kept constant) [203], [204].

---

[1] W. PUPP: Phys. Z. **34**, 756 (1933).

[2] The $V-i$ curve of the discharge becomes vertical. See DÄLLENBACH: Z. Naturforsch. **10**a, 803 (1955). Cathode rays can sometimes emerge from narrow constrictions.

[3] Some authors think that fast primary electrons from the cathode carry the current.

Table 9. *The anode fall.*

Remarks.—Measurements in mercury vapour are uncertain [V_a between 13 and 3 V; MATTHIES, LANGMUIR [*10*], PARTZSCH[2]. When comparing V_a and V_i dissociation has to be considered. d calculated from equation (76.7).

Gas	p mm. Hg	j A/cm.²	V_a volts	V_i volts	d cm. Measured	d cm. Calculated	Reference
Hydrogen	0.2 1—16 1—5	3×10^{-3} $\approx10^{-4}$ $\approx10^{-4}$	$\approx$ 25 16.7 17—19	$\begin{cases}H_2: 15.5\\ H_1: 13.5\end{cases}$	— — 0.5—0.7	0.2 $\approx$ 1.1 $\approx$ 1.1	BRAMLEY [*180*] GÜNTHERSCHULZE [*203*] GÜNTHERSCHULZE [*208*]
Deuterium	1—5	$\approx10^{-4}$	16—18	Probably the same as V_i of hydrogen	0.6	1	GÜNTHERSCHULZE [*208*]
Nitrogen	0.8 0.5 0.01 to 0.05	4×10^{-4} — 10^{-4} to 10^{-3}	16.5 15.7	$\begin{cases}N_2: 15.8\\ N_1: 14.5\end{cases}$	— —	0.9 —	GÜNTHERSCHULZE [*203*] RUBENS [*211*]; anode spot CHAUNDY [*22*]
Oxygen	1 order 0.1 (?)	6×10^{-4} $\approx10^{-4}$	14.2 —	$\begin{cases}O_2: 12.5\\ O_1: 13.5\end{cases}$	— >0.12	0.4 1	GÜNTHERSCHULZE [*203*] ASTON[1]
Helium	10	1.6×10^{-3}	26	24.5	—	0.5	DRUYVESTEYN [*204*]
Neon	10 1.36	8×10^{-4} $\sim10^{-4}$	17 21.5	21.5 21.5	— —	0.5 —	DRUYVESTEYN [*204*] EMELEUS and BROWN [*116*]
Argon	20 0.5 0.1 to 0.5	1.6×10^{-3} 1.2×10^{-3} $\sim10^{-4}$	10 15.3 15.3	15.7	— — —	0.43 — —	DRUYVESTEYN [*204*] PARTZSCH[2] EMELEUS and HARRIS [*115*]

[1] F. W. ASTON: Nature, Lond. **89**, 218 (1917).
[2] A. PARTZSCH: Z. Physik **15**, 287 (1923).

(II) From space potential measurements using part of the anode surface as a probe [*205*] (see also [*115*], [*116*], [*180*]).

(III) From calorimetric measurements of the heat delivered to the anode by the electrons [*206*], [*207*].

(IV) From the integrated field measured by the deflection of an electron beam fired across the anode dark space [*22*].

The most reliable results are shown in Table 9. Fig. 63 shows results obtained by DRUYVESTEYN, using method (I) in rare gases: in neon and argon the anode fall first develops ($V_a \simeq V_i$) and then decreases to a value less than the excitation potential. The anode glow also changes (in Ne) from a yellow layer to a red sphere at the lower anode fall. It has been suggested that in these circumstances ionisation occurs by stepwise processes, but from the little evidence available the

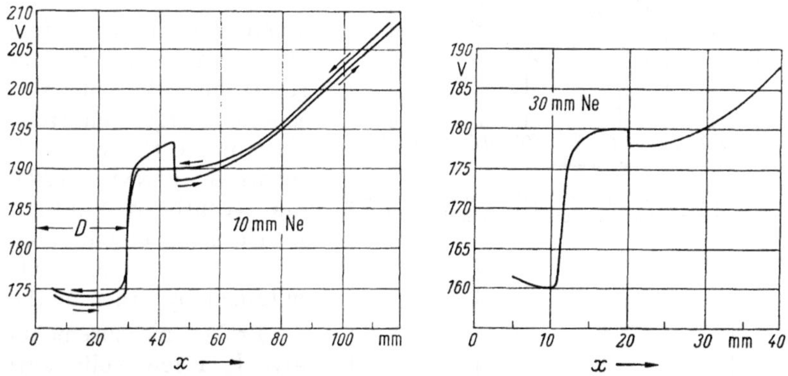

Fig. 63. Development of an anode fall in Ne as the anode is moved away from the cathode (distance x). $p = 10$ mm. Hg, $i = 10$ mA; $p = 30$ mm. Hg, $i = 30$ mA. Note the hysteresis effect at the lower pressure.

effect seems not to depend much upon the current. The effect may also be due to oscillations in the discharge, which would superimpose an appreciable fluctuating energy on to that gained by electrons in the anode dark space. The explanation is still not known[1].

GÜNTHERSCHULZE, who used all three variations of method (II) found that in N_2, O_2, H_2O

$$\frac{p\,D}{V^2} = \text{constant}. \tag{73.1}$$

and deduced that D corresponds to the range of primary electrons from the cathode[2], since both are $\propto V^2$. There are other interpretations, however, since an increased cathode fall would result both in a longer negative glow and an increased concentration of electrons, and hence a greater range for diffusing electrons. In rare gases also $D \gg l$, the length of the negative glow and it seems unlikely that the primary electrons could penetrate so far [*137*], [*8*]. For H_2, however, when $p < 1$ mm. Hg the relation is

$$\frac{p\,D}{V} = \text{constant} \tag{73.2}$$

See also Fig. 64.

GÜNTHERSCHULZE [*208*] found that the optically measured length of the anode glow d^+ obeyed the law $p\,d^+ = $ constant in H_2 and D_2 in the pressure range 0.6

[1] The small changes observed appear also to be in the wrong direction to be accounted for by cumulative processes. See also FUNK and SEELIGER: Z. Physik **113**, 203 (1939).

[2] R. WHIDDINGTON: Proc. Cambridge Phil. Soc. **16**, 321 (1911).

to 2 mm. Hg: at higher pressures [203] (up to 20 mm. Hg) in H_2, N_2, O_2, H_2O, d^+ was constant independent of p and i. The length d^+ was also measured electrically, as the distance through which the anode moves in building up the full anode fall, and this corresponds more closely to the actual thickness of the anode zone[1].

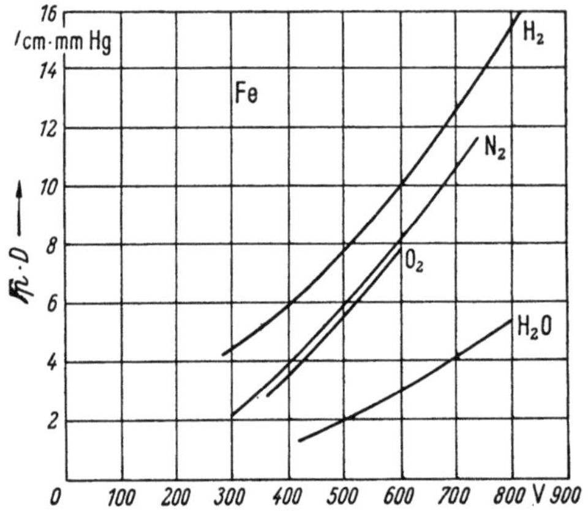

Fig. 64. The critical distance D from the cathode at which an anode fall appears, and its dependence upon the cathode fall V_c.

The power dissipated in heat at the anode is given by

$$\left.\begin{aligned} P &= i\left(V_a + \varphi + \frac{4}{3}\overline{V}\right) \\ &= i\left(V_a + \varphi + \frac{2kT_e}{e}\right) \end{aligned}\right\} \quad (73.3)$$

where φ is the work function of the surface, and the last term represents the mean random kinetic energy delivered by the electrons to the anode. Measurements by PENNING [206] in Ne, and recently by NERETINA [207] in Hg have given values of V_a in good agreement with those obtained by other methods.

Probe measurements [205] give V_a a few volts lower than those quoted for A, Ne, almost the same for H_2, H_2O, and some 10 volts larger for O_2. The reason for the discrepancy is not quite clear.

The electron beam method has given values of $V_a \approx V_i$, and lengths of the dark space in agreement with VON ENGEL's theory (see Sect. 76), and also shows a field distribution similar to that deduced from the results of GÜNTHERSCHULZE, BÄR and BETZ [209] (see Fig. 65).

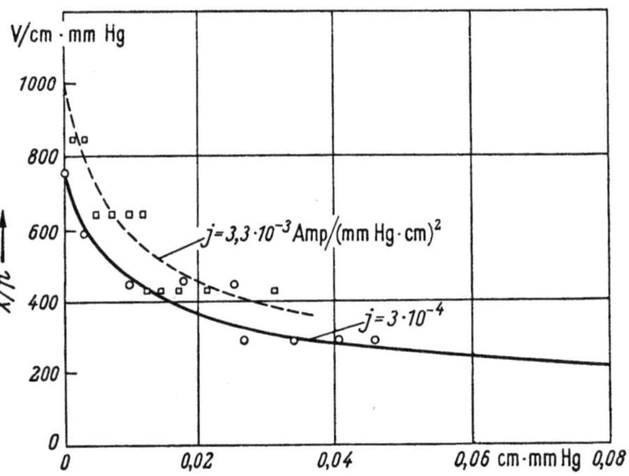

Fig. 65. The field strength as a function of the distance from the anode in nitrogen with aluminium electrodes [22].

For full line graph:

pressure	= 0.05 mm. of mercury
discharge voltage	= 600 V
j/p^2	= $3\cdot 10^{-4}$ A/(mm. Hg · cm.)^2
accelerating voltage on gun	= 8.9 kV

For dashed graph:

pressure	= 0.016 mm. of mercury
discharge voltage	= 950 V
j/p^2	= $3.3\cdot 10^{-3}$ A/(mm. Hg · cm.)^2
accelerating voltage on gun	= 4.1 kV

74. Electron energy and current distributions at the anode. SMITH, TAIT and WHIDDINGTON [210] have analysed by a magnetic method the energy distribution of electrons passing from the positive column through a small hole in the anode (low pressure 10^{-2} to 10^{-3} mm., He, H_2O, air). At lowest pressures they found a large group of primary electrons, having the full energy eV_{tube}, or having lost eV_i or eV_{exc} in one ionising or exciting collision. Two low energy groups were

[1] This is the length quoted in Table 9, for H_2 and D_2 and compared with theory.

found, one about 2 to 3 volts which they attributed to secondary emission from the anode, the other about 40 to 50 volts, corresponding to $\overline{V} + V_a$. Independent measurements of $\overline{V}$ by probes enable V_a to be found. The results give V_a independent of V_{tube} in the range 400 to 900 volts, and of current from 4 to 20 mA, and increasing slightly with increasing p. Values of up to 46 volts (air), and of 30 volts (He), 20 volts (H_2O) were found. These are rather larger than V_i, probably due to the effect of the side walls at these low pressures: V_a was found to depend strongly on R, the radius of the tube.

The distribution of current over the anode surface was measured by YARNOLD and HOLMES [*13*], who divided the anode into several concentric rings. They found a radial distribution of current given by the zero order BESSEL function, being maximum at the centre and almost zero at the walls. This corresponds to the electron density distribution in the positive column.

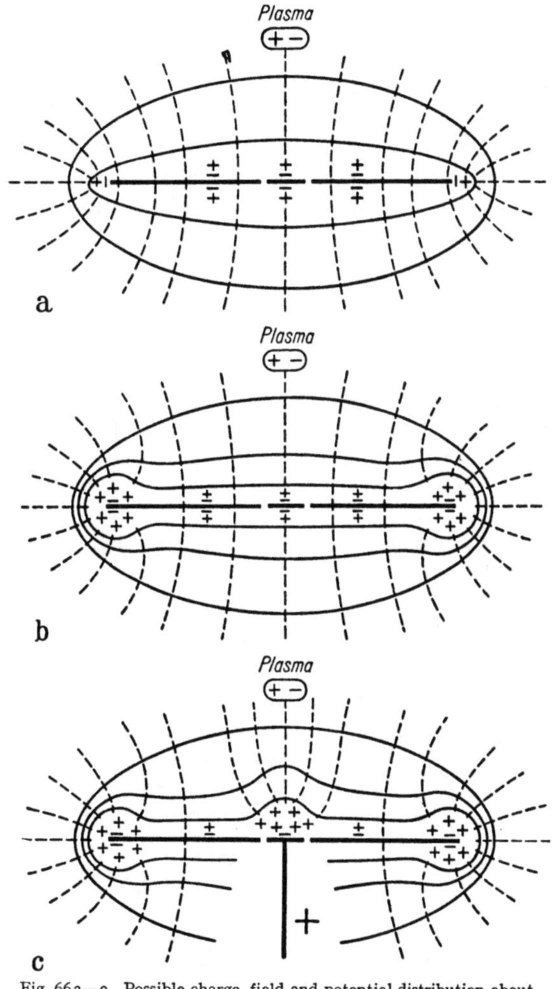

Fig. 66a—c. Possible charge, field and potential distribution about the disc anode. (a) When the current is small. The auxiliary anode is at the same potential as the anode proper. A thin electron sheath surrounded by a positive ion sheath exists near the anode. In the outer regions a plasma exists where the net space charge is nearly zero. (b) The same except that the current is large. (The field causes electrons to converge towards the edge producing intense ionization.) The increased positive space charge near the edge attracts still more electrons to this region. (c) The same as (b) except that the auxiliary anode is maintained positive with respect to the anode proper. Under these conditions a spot forms on this auxiliary anode for the same reason as at the edge.

75. Anode spots. When the current density to the anode is large the anode glow splits up into a series of bright rings or spots, arranged in a regular pattern often stationary but sometimes rapidly moving. The actual arrangement of spots depends upon the current density and the shape of the anode, and they are very sensitive to the presence of impurities in the gas and on the electrode surface. They were probably first observed by LEHMANN [1] in air; later by several other workers [2, 3, 4], and have since been studied in much detail. GÜNTHERSCHULZE, BÄR and BETZ [*209*] found both concentric rings and spots on a plane anode surface in a hydrogen discharge, and they explain their formation by the minimum principle i.e. that in this particular arrangement the necessary current is conveyed

[1] O. LEHMANN: Ann. Phys. **7**, 1 (1902).
[2] G. M. J. MACKAYE: Phys. Rev. **15**, 309 (1920).
[3] C. H. THOMAS and O. S. DUFFENDACK: Phys. Rev. **35**, 72 (1930).
[4] E. KIESSLING: Z. Physik **96**, 365 (1935).

to the anode by the least applied potential. The total anode surface covered by the spots is $\propto i/p$. They deduced from their explanation that spots should also occur in N_2 and found that they do. They also believe that anode spots and striations in H_2 have a common origin[1]. RUBENS and HENDERSON [211] found hemispherical spots on spherical and plane anodes in the nitrogen discharge, but only at pressures between 0.15 and 1.5 mm. Spots on a plane disc were distributed regularly around the edge. Probe measurements showed them to be regions of intense ionisation, containing a net positive charge, and the potential across them was approximately V_i. Using a small probe mounted flush with a plane anode they were able, by making it positive with respect to the anode, to grow a spot on its surface. When the probe was made negative the spot did not fade, but moved laterally away across the anode surface to some stable position. It is believed that the pattern of the spots is stable because parallel current beams flow to them: the consequent attractive force is balanced by electrostatic repulsion. Moving spots can be controlled by an external magnetic field. The mechanism by which these spots are formed on the edge of a plane disc has been explained in a way which accounts also for the action of the probe (or "auxiliary anode") already mentioned. This is shown in Fig. 66 (with the authors' description).

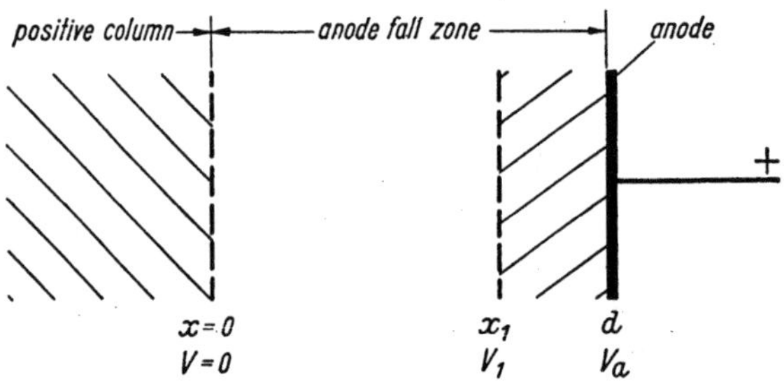

Fig. 67. Model of the anode fall region.

76. Theory of the anode region. A theory of the anode fall at a plane surface in a positive column has been developed by VON ENGEL [212], and is applicable when fairly low pressures ($p <$ about 0.1 mm.) and small current densities ($i < 10^{-3}$ A/sq.cm.) are used. All complicating secondary processes are ignored, and the simple model adopted of two parallel plates between which the field varies parabolically[2]. Then with the dimensions given in Fig. 67, the integration of such a field gives the potential at any point

$$V_x = 3V_a\left(\frac{x}{d}\right)^2\left(1 - \frac{2x}{3d}\right). \tag{76.1}$$

The number of ions dN produced in a distance dx by one electron is assumed to be [see also (51.2)].

$$dN = a\,p\,(V - V_i)\,dx = a\,p\,(\overline{V} + V_x - V_i)\,dx. \tag{76.2}$$

$\overline{V}$ being the mean kinetic energy of electrons crossing the boundary into the anode zone. From this is derived the total number N of ions produced in that small layer from x, where ionising collisions first become appreciable to the anode,

$$N = \frac{apd\,(\overline{V} + V_a - V_i)^{\frac{3}{2}}}{\sqrt{3V_a}}. \tag{76.3}$$

[1] See Sect. 68, and note also that striations can sometimes be produced by moving the anode away from the cathode. An anode glow grows, becomes detached and another grows. See J. J. and G. P. THOMSON, Vol. 2, p. 386.

[2] This variation was deduced from the results of GÜNTHERSCHULZE, BÄR and BETZ: Z. Physik **109**, 293 (1938), and has since been confirmed by direct measurement (Fig. 65).

If N is large the zone can be regarded as two parallel plates, one a copious source of ions, the other of electrons; then the ratio of the current densities is

$$\frac{j^+}{j_e} = \sqrt{\frac{m}{M}} \tag{76.4}$$

where M, m, are the masses of the positive ions and electrons. Also:

$$j_e = \frac{3.5}{16\pi}\sqrt{\frac{2e}{m}}\,\frac{V_a^{\frac{3}{2}}}{d^2}. \tag{76.5}$$

After some rearrangement and the approximation $V_a \sim V_i$ we get

$$\frac{j}{p^2} = \text{const}\,\sqrt{V_i\,(\overline{V} + V_a - V_i)^3}, \tag{76.6}$$

$$p\,d = \sqrt{\frac{3.5}{16\pi}\sqrt{\frac{2e}{m}}}\cdot V_i^{\frac{3}{4}}\sqrt{\frac{1}{(j/p)^2}}. \tag{76.7}$$

Here d is in cm., p in mm. Hg, e, j and V in electrostatic units. The equations naturally obey similarity, and values calculated from them are compared with experiment in Table 9[1]. Values of V_a agree reasonably well but too few measurements of d are available for proper comparison.

Druyvesteyn [*204*] has considered the diffusion of electrons to an anode placed in the Faraday dark space. He assumes as a first approximation a neutral plasma ($N_e = N^+ = N$) the concentration depending only on the distance x from the cathode, no ionisation and no effect from side walls. The electron and ion current densities j_e, j^+ to the anode are then:

$$j_e = -D_e\left(e\cdot\frac{dN}{dx}\right) + e N b_e X, \tag{76.8}$$

$$j^+ = eN\cdot b^+\cdot X. \tag{76.9}$$

From these eliminating X and integrating gives

$$N = N_0 - a\,x \tag{76.10}$$

applying only for $x \gg$ length of the cathode dark space. Hence at a critical distance D the electron and ion concentration become zero: at this point an anode fall develops. The critical distance is given by

$$D \sim \frac{1}{2}\frac{e N_0}{j_e} D_e = \frac{1}{2}\frac{e N_0}{j_e}\frac{1}{3}\lambda_e \bar{v} \tag{76.11}$$

where N_0, j_e are then electron concentrations and current densities at some arbitrary point (not defined, but probably near the edge of the negative glow), D_e the diffusion coefficient and $\bar{v}$ the mean random velocity of the electrons. The equation is not suitable for numerical calculation due to the indefinite nature of N_0 and j_e, but it shows that pD should be constant, as found by Güntherschulze for given V_c, provided p is large enough so that wall effects do not intrude. At low pressure these effects can no longer be neglected and then $D \to$ a constant $\approx 2R$, i.e. $\approx$ the diameter of the tube.

See also [*213*], [*214*], for the effect of the diffusion of positive ions in rather different geometry.

---

[1] Von Engel gives a similar table in his paper, and discusses the results in more detail.

IX. Cathode sputtering.

77. General description. It is a common experience that after some period of use the walls of a discharge tube become coated with a thin layer of the metal comprising the cathode. After a long time the layer becomes quite opaque being usually thickest near the cathode, and the cathode itself becomes noticeably smaller. The effect is called sputtering, and is due to the ejection of particles of the cathode substance by the impact of ions from the discharge. It was first observed in 1852 by GROVE and in 1858 by PLÜCKER, and is extensively used for the deposition of metallic layers [*215*] but the process is not fully understood. The effect has been observed on surfaces of quartz, rocksalt, glass and mica[1] as well as on most common metals. Metals combined in an alloy were thought to sputter as they would in the pure state, the cathode becoming richer in the component which is less easily sputtered[2]. Recent work has thrown doubt on this; in some Al alloys (e.g. Al—Cu—Mn—Mg) spectrographic analysis has shown that the deposited material has exactly the same composition as the cathode [*216*]. In these experiments, however, general evaporation as well as sputtering may have occurred, and so the interpretation of the result is not clear.

It has been found that most of the particles ejected from the cathode are uncharged atoms: the spectrum which they emit when excited is that of the neutral atom[3], and their paths through a gas at low pressure are not bent by a magnetic field[4]. Very fast ions are thought sometimes to eject larger particles [*217*]. There is evidence that sometimes part of the sputtered material may be in the form of negative ions[5]. The particles are ejected with a speed corresponding approximately to the melting temperature of the metal.

The rate of sputtering is defined as the number of atoms of the cathode material ejected per incident ion (written ϑ_0), or more precisely, per incident ion of a certain arbitrary energy, usually 500 eV (written ϑ_{500}). It is essentially measured by the change in weight of the cathode, or of part of the surface on which the metal is deposited, after a known charge has passed into the cathode. This change has been calculated by indirect methods when the cathode is a thin wire, e.g. by the change in its resistance, or in its thermionic emission [*218*], [*219*]. In an unusual arrangement GÜNTHERSCHULZE [*220*], [*228*] uses a wire anode surrounded by a cylindrical cathode. The sputtered material builds up an equilibrium pressure in the discharge, when the rate of sputtering per unit area from the cathode equals the rate of deposition per unit area. Assuming the pressure to be uniform this will equal also the rate of deposition per unit area everywhere else, in particular on the anode, whose increase in weight thus gives the sputtering.

Measurements of the sputtering rate have been confused by two factors. Firstly most experiments have been done in discharges with $p \sim 1$ mm., in which many of the ejected atoms diffuse back to the cathode. These measurements give only the difference between what is ejected and what settles back on the cathode (ϑ) and the results depend markedly on the gas pressure and the geometry of the tube[6]. Secondly the current measured at the cathode is

[1] STARK and WENDT: Ann. Phys. **81**, 921 (1912).
[2] ASADA and QUASEBARTH: Z. chem. Phys. A **143**, 435 (1929).
[3] A. v. HIPPEL: Ann. Phys. **81**, 1043 (1926).
[4] T. BAUM: Z. Physik **40**, 686 (1927), Ag cathode in H_2.
[5] K. SOMMERMEYER: Phys. Z. **34**, 582 (1933). See also Sect. 94.
[6] Experiments have been devised to illustrate chiefly the effect of back diffusion by placing obstacles near the cathode. H. FISCHER: Z. Physik **108**, 500 (1938).

not due to positive ions only but also to secondary electrons: this can sometimes be allowed for by the factor γ $(=j_e/j_+)$ if this is known. Thus we write ϑ_0 or ϑ_{500} for the true emission per ion; ϑ for the apparent emission per ion when back diffusion occurs, and $\frac{\vartheta}{1+\gamma}$ for the apparent emission when the current is also in error due to secondary emission.

Experiments have been devised to remove both these errors. The secondary emission has been measured by the application of retarding fields[1], or largely compensated by the application of a magnetic field which bends the electrons back to the cathode [*221*], and also gives the added advantage that a discharge can be maintained at a lower gas pressure. Back diffusion has been eliminated by using very low gas pressure ($\lambda \gg d$), or by using a very fine wire cathode in a cylindrical arrangement. Results obtained in arrangements of very different geometry are not really comparable; for example sputtering from a fine wire is 4 or 5 times that from a plane sheet of the same material, even in high vacuum when no back diffusion occurs[2]. The true sputtering rate can, of course, only be measured by bombarding clean degassed surfaces with ions of a single energy; in discharges the cathode has an adsorbed gas layer and the ions are distributed in energy, although in highly abnormal discharges most of them will have the full energy eV_c of the cathode fall.

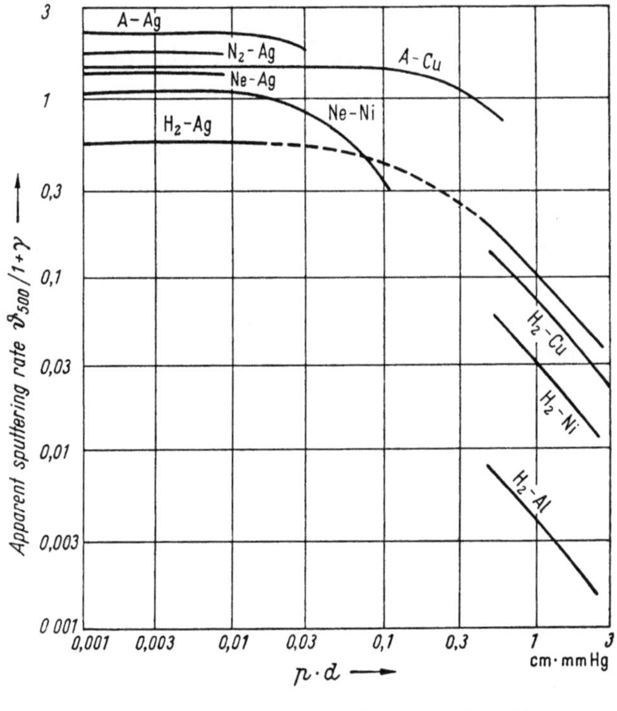

Fig. 68. Apparent sputtering rate as a function of the product of the pressure and cathode—anode separation, illustrating the effect of diffusion.

78. Experimental results. Fig. 68 shows values of sputtering rates measured by GÜNTHERSCHULZE and MEYER [*222*], [*223*], and PENNING and MOUBIS [*221*], compiled by the latter and in some instances extrapolated to an ion energy of 500 eV. The effects of both geometry and pressure can be combined in the parameter $p \cdot d$. The rate is constant at low pressure, but falls appreciably at higher pressure (when $\lambda \sim d$) due to back diffusion. It depends also on the following factors:

(I) *The gas.* The rate increases less than linearly with the mass of the bombarding ion. However, when chemical action occurs between the ion and the cathode material anomalous values are found. H_2, for example, gives unusually high sputtering rates. Such action usually results in oxides or hydrides (e.g.

[1] G. TIMOSHENKO: J. Appl. Phys. **12**, 69 (1941).
[2] H. FETZ: Z. Physik **119**, 590 (1942).

$Sb H_3$) being formed which are believed to diffuse into the gas and decompose some distance from the cathode [224]. In these circumstances very little metal is deposited on the walls near the cathode.

(II) *The material and surface of the cathode.* The sputtering rate varies considerably with the metal comprising the cathode, being greatest for silver and cadmium and least for tungsten. The state of the surface has an equally marked effect; Al, Zn and Mg were thought to give very little sputtering but it has been shown that this is due to very stable oxide layers formed on the surfaces of these metals. When these are removed the sputtering increases greatly[1]. Adsorbed layers of gas have the same effect, and it has been found that the sputtering from a fine Mo wire increases with time until a constant value is reached[2]; the surface layer has then been removed by ion bombardment. It has been suggested, however, that in these conditions the change in the cathode surface is due to sputtered material being redeposited in powder form[3], although there is little direct evidence for this. It appears that surface impurities always reduce the sputtering rate.

Some general results, taken from MASSEY and BURHOP [225] are given in table 10.

Table 10. *Sputtering by Ions of Energy 500 eV.*

Sputtering ion	Sputtered metal	Sputtering rate gm./amp. hr. for 500 eV ions	$\vartheta_{500}/(1+\gamma)$	ϑ_{500}	Heat of evaporation cals./gm.	References
He^+	Cu	0.97	0.40	—	50.8	7
Ne^+	Cu	4.06	1.67	—	50.8	7
Ne^+	Ag	5.5	1.34	—	45	4
A^+	Al	0.78	0.75	0.83	48	6
A^+	Cu	4.13	1.7	1.9	50.8	6
A^+	Cu	4.08	1.67	—	50.8	7
A^+	Ni	9.5	1.2	1.35	—	6
A^+	Ag	5.7	1.4	1.5	45	6
A^+	Ag	9.5	2.32	—	45	4
H_2^+	Cu	0.97	0.40	—	50.8	7
H_2^+	Ni	0.31	0.14	0.15	—	6
H_2^+	Ag	2.7	0.67	—	45	4
H_2^+	Ag	1.92	0.47	0.56	45	6
N_2^+	Ag	7.0	1.71	—	45	4
Hg^+	Be*	0.24	0.62	—	—	5
Hg^+	Al	0.55	0.53	—	48	5
Hg^+	Si	0.56	0.52	—	44	5
Hg^+	Cr*	1.75	0.89	—	—	5
Hg^+	Mn	1.13	0.54	—	—	5
Hg^+	Co*	1.72	0.77	—	—	5
Hg^+	Cu*	3.45	1.42	—	50.8	5
Hg^+	Zr	1.70	0.49	—	—	5
Hg^+	Mo	1.96	0.54	—	177	5
Hg^+	Pd*	5.21	1.28	—	45	5
Hg^+	Ag*	10.23	2.50	—	45	5
Hg^+	Ta	3.70	0.54	—	—	5
Hg^+	W	3.83	0.55	—	218	5
Hg^+	Pt*	7.88	1.07	—	127	5

1 KUECK and BREWER: Rev. Sci. Instrum. **3**, 427 (1932).
2 H. FETZ: Z. Physik **119**, 590 (1942).
3 A. GÜNTHERSCHULZE: Z. Physik **36**, 563 (1926).
4 GÜNTHERSCHULZE and MEYER: Z. Physik **62**, 607 (1930).
5 MEYER and GÜNTHERSCHULZE: Z. Physik **71**, 279 (1931).
6 PENNING and MOUBIS: Proc. Amst. Acad. **43**, 41 (1940).
7 A. GÜNTHERSCHULZE: Z. Physik **118**, 145 (1941).

(III) *The energy of the ions, eV.* Measurements made in discharges have given a general relation for the sputtering rate:

$$\vartheta = C(V - V_0) \tag{78.1}$$

where C, V_0 are constants for a given gas and cathode, although V_0 depends to a considerable extent on the state of the surface[1]. V_0 usually lies between 100 and 300 volts; earlier measurements gave $V_0 \sim 300$ to 400 volts, results which fostered the belief that sputtering does not take place in a normal discharge. In fact it can occur but to a small extent. Carefully cleaned surfaces in high vacuum are sputtered by ions of all energies, i.e. $V_0 = 0$ hence ϑ rises linearly with V up to about 1000 eV and afterwards more slowly.

Eq. (78.1) has also been expressed as

$$N = C'(V - V_0) \tag{78.2}$$

where N is the number of atoms liberated per coulomb of charge delivered *by the ions.* Some typical values are; for a copper cathode, with $j = 0.4$ A/cm.[2]

Gas	V_0	C'
He	303 volts	$1.5 \cdot 10^{16}$ atoms/volt. coloumb
Ne	305 volts	$6.2 \cdot 10^{16}$ atoms/volt. coulomb
A	305 volts	$6.2 \cdot 10^{16}$ atoms/volt. coulomb
H_2	303 volts	$1.5 \cdot 10^{16}$ atoms/volt. coulomb

GÜNTHERSCHULZE [*227*] has given an empirical relation for the mass M of material sputtered per second in a discharge, at relatively high pressure ($\lambda \ll d$), d here being the distance between parallel electrodes:

$$M p d = \text{const} \times V_c. \tag{78.3}$$

This can be derived approximately, since ϑ is proportional to energy of the ions, and here their energy is given more by the field ($X\lambda$, or X/p) than by the potential drop.

Thus

$$\vartheta \propto \frac{X}{p} \propto \frac{V_c}{pd}. \tag{78.4}$$

This applies strictly only to sputtering between infinite parallel plates: the effect of side walls has also been considered by GÜNTHERSCHULZE [*228*], and an analogous formula given for sputtering between concentric cylinders.

(IV) *The temperature of the cathode.* Adsorbed gas layers are removed by heating the cathode and the earlier results on the change of sputtering rate with temperature were thought to be due to this effect. However, FETZ has found that when a Mo wire is bombarded by Hg^+ ions the sputtering rate nearly doubled between 400 and 1000° C. The evidence on this point is conflicting.

The angular distribution of the sputtered material follows a cosine law when A^+ ions with energies from 5 to 10 keV hit Ag or gallium, independent of the angle of incidence[2]. It is not known whether this result is general. In some

[1] VON HIPPEL, however, found that ϑ passes through a maximum with increasing V. This apparent discrepancy has been discussed by GÜNTHERSCHULZE [*220*] and ECKER and EMELEUS [*226*].

[2] R. SEELIGER and K. SOMMERMEYER: Z. Physik **93**, 692 (1935).

instances it seems probable that the angle of incidence of the ions does affect the sputtering rate. The sputtering from a fine Mo wire by ions from a discharge at low pressure (10^{-3} mm.) is found to decrease as the diameter of the wire increases, an effect thought to be due to more ions striking almost normally. Here also the experimental evidence is scanty.

79. The effect of sputtering on the cathode surface. GÜNTHERSCHULZE and TOLLMIEN [*229*] have studied, by a method using the reflection of light, how the cathode surface changes during sputtering. They find that on many metals a regular pattern of upright equiangular sub-microscopic cones is formed, their angles being characteristic of the metal. The effect is marked in Mg, Cd, Zn, Al, Sn, Pb, Sb, Bi, less so in Ag, Cu, and does not occur at all in Au, Fe, Ni, Pt. The cones are most distinctly formed when the cathode is kept cold (room temperature) and when the cathode fall is low; the angles of the cones change a little with these parameters. The rate of sputtering drops as the cones are formed, and a simple theory is developed giving the relation between the rate of sputtering and the angles of the cones. The results agree well with experiment.

80. Sputtering efficiency. The sputtering efficiency is defined as the ratio of the energy required to extract one atom from the metal (energy of evaporation W), to the energy delivered by the $1/\vartheta_0$ ions necessary to eject this one atom. Since each ion delivers energy eV to the cathode, and since $V \propto \vartheta_0$ approximately, we get for the sputtering efficiency ε,

$$\varepsilon = \frac{W}{eV/\vartheta_0} \approx \text{const},$$

independent of V but differing of course for various gases and cathode surfaces. Most measurements agree in giving $\varepsilon \sim 1\,\%$.

81. Theories of sputtering. It was first suggested that sputtering was due to the heating of occluded gases in the metal, the pressure developed disrupting the surface. There is much evidence against this view: sputtering takes place in conditions when there is hardly any occluded gas, for example from liquid and very hot solid surfaces. KINGDON and LANGMUIR suggested a simple mechanical collision model, which they applied particularly to thorium surfaces. Ions hitting the metal surface cause depressions about the size of a few atoms; subsequent ions may be reflected in these depressions and retain enough energy to knock atoms out of the metal. The maximum energy transmitted from an incident ion of energy eV, mass m, to a metal atom, mass M is

$$4\,\text{eV}\,\frac{m\,M}{(m+M)^2};$$

at least two collisions are required to eject an atom, which has to acquire the energy of evaporation before it can escape. The theoretical calculations are in reasonable agreement with experiment for light ions only (except He^+). The theory predicts a minimum ion energy below which no sputtering should occur and there should also be an effect due to the angle of incidence of the ions [*219*]. The experimental evidence on these points is doubtful. Extensions of the theory, enabling the rate of sputtering to be calculated have been made by LAMAR and COMPTON [*230*].

A widely accepted theory, due originally to VON HIPPEL and BLECHSCHMIDT [*231*], [*232*] and developed later by TOWNES [*233*], regards sputtering as due to evaporation from small local regions where the impact of ions produces for a short time a very high temperature. An ion is thought to deliver its energy eV in a very short time t_0 ($\sim 10^{-16}$ sec.), and the rise in temperature $(T-T_0)$ of a point distant r from the point of impact, at time t is given by the standard equations of heat flow:

$$T-T_0=\frac{eV}{8J}\sqrt{\frac{s\varrho}{(\pi K t)^3}}\,e^{-\frac{s\varrho r^2}{4Kt}} \tag{81.1}$$

where ϱ, s, K are the density, specific heat and thermal conductivity of the cathode material. The vapour pressure can be calculated, and hence the number of atoms evaporated from a surface with the above distribution of temperature. The mean velocity of the evaporated atoms is assumed to be given by the temperature by the usual MAXWELLian relation, and this seems to agree with the measurements of BAUM who found Ag atoms with velocities of about $6\cdot 10^4$ cm./sec. corresponding to $T=1000°$ C. Other measurements have given different results[1], but their interpretation is not clear. It seems very doubtful whether it is legitimate to apply the macroscopic concepts of temperature and heat flow to regions containing only a few atoms; apart from that, application of the theory is difficult because the values of s and K are uncertain or unknown at temperatures near the melting point of metals. Even so the theory predicts correctly the dependence of ϑ on ion energy, and sometimes gives good absolute values of ϑ. The effect of thermionic electron emission at these temperatures might also influence the results[2,3].

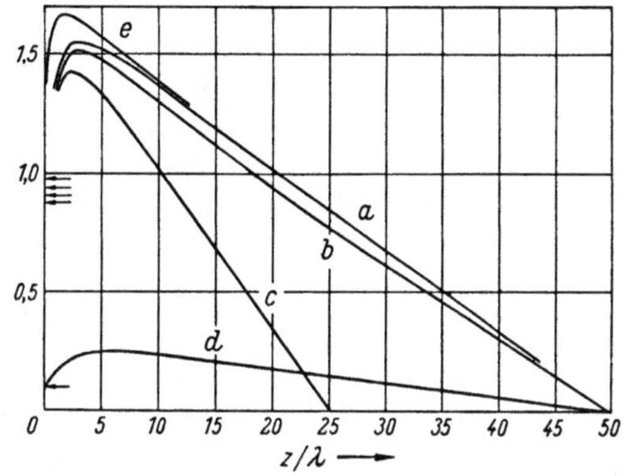

Fig. 69a. Theoretical axial distribution of sputtered material (Al cathode in Ne) under various conditions [226]. l = distance from anode to cathode, z = distance from cathode, R = tube radius, λ = mean free path of metal atoms in gas. The ordinate is a non-dimensional quantity proportional to the density of sputtered material on the cylindrical walls. The conditions are: Curve a: $2R=500\lambda$, $l=50\lambda$, $v_g/v_0=1$. Curve b: $2R=250\lambda$, $l=50\lambda$, $v_g/v_0=1$. Curve c: $2R=500\lambda$, $l=25\lambda$, $v_g/v_0=1$. Curve d: $2R=500\lambda$, $l=50\lambda$, $v_g/v_0=1/10$. v_0 is the mean velocity of emission of metal atoms. v_g is (presumably) the mean velocity of gas atoms. These curves should be compared in form with those of Fig. 69b.

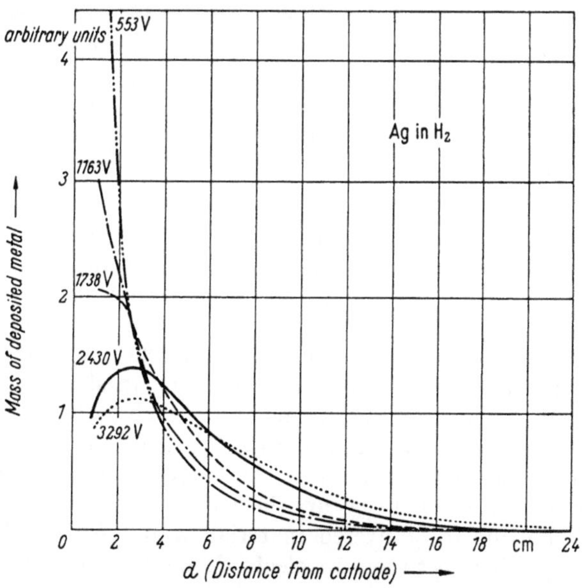

Fig. 69b. Experimentally measured axial density distribution of sputtered Ag in H_2 [228], with various applied potentials V.

Recently ECKER and EMELEUS [226] have pointed out that sputtering in a glow discharge can only be understood by considering three interdependent

---

1 BERKEY and MASON: Phys. Rev. **38**, 943 (1931).

2 COMPTON and LANGMUIR: Rev. Mod. Phys. **2**, 123 (1930).

3 OLIPHANT and MOON: Proc. Roy. Soc. Lond., Ser. A **127**, 388 (1930).

problems, namely how the atoms are released from the cathode surface, how they move through the gas, and how they are deposited on the walls. They believe that the inadequacy of previous theories arises from neglect of the difference in mass between metal and gas atoms, the high velocity with which sputtered atoms are ejected, and the radial variation of the discharge current. Taking account of these points they consider in detail the motion of metal atoms through the gas: the gas space is divided into two zones, a thin layer close to the cathode in which the majority of atoms make their first collision and in which their velocities are directed away from the cathode, and the rest of the space in which normal diffusion applies. The theory gives an expression for ϑ/ϑ_0, for the total cathode sputtering and its radial variation, and also gives the density distribution of metal atoms in the gas as a function of the distance from the cathode. This shows a maximum at a short distance in front of the cathode, its position depending on the velocity of ejection of the atoms (see Fig. 69). The shape of these curves corresponds to measured values of the density of wall deposits, obtained by GÜNTHERSCHULZE [*228*], who has also considered the motion of metal atoms through the gas on rather simpler assumptions and concluded that the thickness of the deposit should vary as $e^{-z/r}$ where z is the distance from the cathode, r the radius of the tube. SEELIGER [*234*] also discusses the motion of gas atoms in various kinds of experimental arrangement.

82. Other cathode effects; evaporation, bridging of gaps. HANAU has observed different forms of sputtering in abnormal discharges in He with Al alloy cathodes. The cathode was in the form of a hollow cylinder, and the sputtered material was deposited on the walls of the tube in concentric bands superimposed on a generally even distribution. The bands corresponded to atoms travelling in straight lines from the inside lower quarter of the cathode. He attributes these localised bands to true sputtering and the distributed deposit to general evaporation, but at the gas pressures used (1 to 10 mm. Hg) it seems equally likely that the entire effect is due to sputtering and the bands are formed by those atoms which leave the cathode with the greatest energy and have great persistence of velocity [*216*].

In glow discharges at relatively high pressure between electrodes placed close together ($\sim\frac{1}{20}$ mm.) the combined effects of sputtering and chemical action can cause a solid bridge to form across the gap. HAWORTH [*235*] has found the effect to be especially pronounced in a normal discharge in air ($p=1$ atmosphere) between Ag electrodes[1]. If the anode is placed in the FARADAY dark space a molten layer slowly changing in shape is observed to grow from it for several minutes. When this layer reaches the edge of the negative glow a solid bridge suddenly forms across the gap. A similar effect is observed between electrodes of Cu, Pd, Pt, Al but proceeding more slowly; with tungsten, although anode growths do occur, the gap is not bridged. The growths at first appear to be due to sputtered material from the cathode, but when the discharge is established between an electrode of Ag and one of W the bridge grows faster when the *anode* is made of silver, showing that sputtering is not the principal cause. X-ray analysis of the anode growths show them to be composed of Ag, $AgNO_2$ and Ag_2O in the relative volumes 5:10:1. Further experiments in N_2 and in N_2 with a trace of added O_2 show that the oxygen plays a dominant role, and that three processes occur namely cathode sputtering, and chemical action at both anode and cathode.

[1] See also H. FUPKE: Naturwiss. **39**, 426 (1952), N_2, Cu electrodes.

83. Absorption of gas. In a newly constructed discharge tube a considerable amount of gas is attached as surface layers to the walls and electrodes, and much of it is driven off when a discharge is established, due to bombardment of the surfaces by ions, electrons and excited atoms. After some period of use the pressure in the tube stops increasing and begins very slowly to decrease: gas is removed by a variety of agencies which usually result in some kind of chemical attachment to molecules of the walls or electrodes. The processes in the discharge e.g. the temporary dissociation of molecules into atoms, the formation of excited states etc. apparently favour types of chemical action at the surfaces which would not take place in an ordinary gas[1]. Conventional heating of the discharge tube (to a few hundred °C) does not release again all the gas absorbed, showing that some of it has become attached by quite strong forces.

Removal of the gas becomes much more rapid when cathode sputtering occurs. Early experiments showed that, with increasing potential applied to the discharge, no detectable decrease of the gas pressure occurred until a critical potential V_0 was reached, the same potential at which cathode sputtering began[2, 3, 4]. The effect is probably due to two main causes, chemical combination between the sputtered metal in atomic form, and the gas; and the trapping of gas molecules at the wall surface between layers of deposited metal. The detailed mechanism is not known and is probably very complex. Preferential absorption of certain gases often occurs and the effect is used in this way to purify small quantities of rare gas. The cathode surface is of course made very clean by sputtering while the deposited layer is contaminated by the gas it has absorbed. This explains how reproducible results for the cathode fall can be obtained using a cathode sputtered in this way [*53*], but not using the actual deposited metal as a cathode [*50*]. See Sect. 22.

The reduction in pressure is usually measured by conventional gauges, but when rapid determinations are needed MIERDEL[5] has proposed the elegant method of compensating the decrease by applying a magnetic field, previously calibrated against known changes in pressure.

X. Special types of discharge.

84. The obstructed discharge. If the anode is moved towards the cathode while the discharge current is kept constant, the potential necessary to maintain the discharge varies as shown in Fig. 6. In particular as the anode moves through the glow and just before it reaches the edge of the cathode dark space the voltage begins to rise rapidly, and the discharge is then obstructed. The effect is similar to the rise in the PASCHEN sparking potential curve when $pd < pd_{\min}$. The number of gas molecules between the plates decreases, electrons make fewer collisions and therefore need greater energy to produce the necessary number of ions. Over a small region, when the anode is still in the glow, the length of the dark space d is $\propto V$ (see Fig. 70), and the average field is constant: however, the ionisation may change due to a different distribution of the field [*236*].

In abnormal discharges the voltage-distance curve sometimes shows a pronounced minimum before obstruction begins (Fig. 6). GÜNTHERSCHULZE thought that adsorbed gas removed from the anode by fast primary electrons produces a dense

[1] See for example J. J. and G. P. THOMSON: Vol. 2, p. 466.
[2] VEGARD: Ann. Phys. **1**, 769 (1916).
[3] BRODETSKY and HODGSON: Phil. Mag. **31**, 478 (1916).
[4] GEC Research Staff, Phil. Mag. **40**, 585 (1920); **41**, 685; **42**, 227 (1921).
[5] G. MIERDEL: Z. Physik **122**, 614 (1944).

region in which more ionisation can occur. The minimum decreases when a grid anode is used. PENNING [237], however, showed that it still occurs when the anode is clean and degassed by heating, although not in all gases. The effect may be caused by changes in gas density due to heating of the anode [97], or to the disturbance of a favourable ion concentration gradient in the glow which, before the anode was inserted, caused ions to flow into the dark space [238]. This point has not been settled.

In helium the obstructed discharge shows an anomalous $V-d$ curve (Fig. 71) [239]. When $d<d_0$ the discharge cannot be maintained except by an abrupt and enormous increase in voltage. The reason is that at the potential corresponding to d_0 electrons have energy corresponding to the maximum of the ionisation

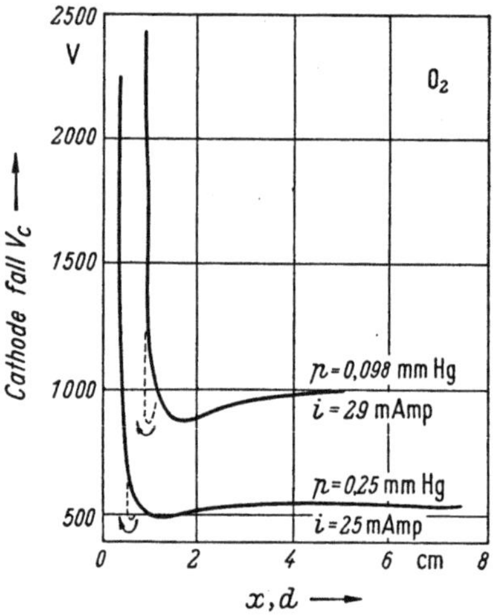

Fig. 70. The obstructed discharge in oxygen. The dependence of the cathode fall V_c and cathode dark space length d on the distance x of the anode from the cathode. ——— distance of anode (x); ------- length of dark space (d) [236].

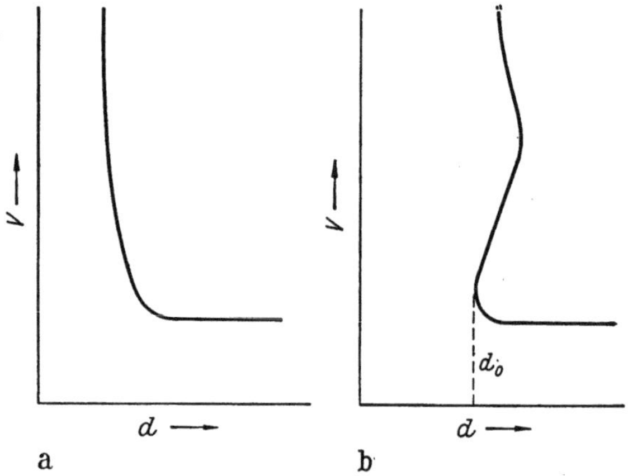

Fig. 71 a and b. Increase of the potential difference between the electrodes V as a function of the electrode separation d at constant current (obstructed glow discharge). (a) Most gases. (b) He (schematic).

function. Any increase in potential makes them less effective, and so cannot compensate for the fewer collisions which they make when $d<d_0$, not until it is so great that the positive ions have enough energy to ionise gas molecules (TOWNSEND's β effect), and also to give a much increased γ. We have then a canal ray discharge.

Measurements have also been made of the current in the discharge as the anode is moved towards the cathode, the voltage being kept constant, or nearly so[1]. The current drops linearly to zero as the anode moves from a point in the glow to the edge of the dark space. Sometimes, however, there is a slight maximum before the current begins to fall. The experiments show that at least part of the glow plays a vital part in maintaining the discharge, but whether due to photons hitting the cathode or to a supply of ions to the dark space cannot be decided.

At lower pressures ($\sim 10^{-2}$ mm. Hg) the discharge tends to concentrate towards the centre of the cathode: the field lines converge and radial fields become important.

85. Canal ray discharges. This, as has been mentioned, is a special case of a highly obstructed discharge, operated at very low pressure ($\sim 10^{-2}$ mm. or less),

[1] R. M. CHAUDHURI and S. H. ZUBERI: Indian J. Phys. **18**, 333 (1944). — M. K. HUSAIN: Indian J. Phys. **24**, 73 (1950).

and does not constitute a glow discharge at all. The applied voltage is usually 10 to 50 kV. We refer to earlier comprehensive reviews[1]. More recently KAMKE and DEUTSCHER[2] have made accurate measurements both on the radial distribution and the energy distribution of the ions arriving at the cathode. The calculated energy distribution, assuming a constant rate of ionisation by electrons, is considerably changed by the effects of charge transfer, an important process in these discharges. Quantitative theory[2] shows that the discharge must be maintained by ionisation in the gas both by electrons (α) and by positive ions (β), and a secondary process (γ) due to ions and fast neutral atoms hitting the cathode.

In the transition from an obstructed glow to a canal ray discharge the cathode glow extends outwards from the cathode and recedes from the side walls forming a thin pencil. There is a continuous change from the usual appearance to this state, and the sharpness of the pencil produced appears to be independent of the tube geometry. The luminous pencil passes without alteration through a hole in the cathode, and is due partly to excitation of the gas by positive ions (the canal rays), but mostly by neutral atoms, as can be shown by deflection experiments with a magnetic field[3].

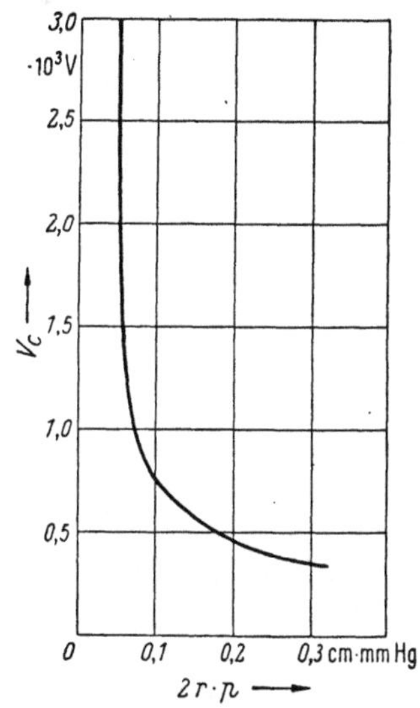

Fig. 72. Increase of cathode fall V_c in narrow tubes: the constricted discharge (air). r = tube radius.

86. The constricted discharge. The properties of the cathode dark space are usually independent of the side walls, but when the tubes are narrow, losses are large and the cathode fall begins to rise[4]. This occurs when the diameter of the tube $2r$ is roughly equal to the length of the dark space (see Fig. 72); the rise of voltage in the subnormal discharge occurs when a similar condition is satisfied, namely $2R \approx d$, where R is the radius of the discharge. The gradient in the positive column also increases in narrow tubes as expected from theory[5]. The effects of a sudden constriction in the tube are mentioned in Sect. 6, 71.

87. The spray discharge [*240*] to [*247*]. This was first observed when a semiconducting cathode was coated with a thin layer of highly insulating oxide grains, e.g. Al_2O_3 and the gas pressure was about 0.1 mm. The discharge showed no cathode fall or dark space, the negative glow beginning immediately at the cathode surface, and having the same spectral properties as a conventional glow discharge. The current density in the spray discharge is high ($j/p^2 > 1000\, j_n/p^2$), and is given by the equation for field emission $j = A V^2 e^{-B/V}$, where V is the total discharge potential. Most of the potential falls off across the oxide layer, and it is clear that positive ions charge up its outer surface and give average fields of order 10^6 volts/cm. The effect has also been observed when a relatively thick layer (2 to $10 \cdot 10^{-4}$ cms.) of chemically formed oxide is produced on an Al cathode: the layer is porous, however, and large space charges are believed to be present

[1] C. HAILER: Wiss. Veröff. Siemens-Werk **17**, 321 (1938).— WIEN: Handbuch der Experimentalphysik 1931.

[2] K. DEUTSCHER and D. KAMKE: Z. Physik **135**, 380 (1953).

[3] See for example J. D. CRAGGS: Proc. Phys. Soc. **54**, 245 (1942).

[4] v. ISSENDORF, SCHENKEL and SEELIGER: Wiss. Veröff. Siemens-Werk **9**, 73 (1930).

[5] See for example the measurements of GEHLHOFF: Verh. dtsch. phys. Ges. **21**, 349 (1919). MATTHIES and STRUCK: Verh. dtsch. phys. Ges. **14**, 83 (1912).

inside it. The average field seems rather low to produce field emission, which therefore probably takes place from local points. The electrons accelerated through the oxide layer lose about 20 to 50 eV energy, the amount increasing with applied potential.

The discharge cannot continue to burn for a long time as the oxide layer is destroyed, especially in the presence of oxygen when a continuous transition from the spray to the glow discharge occurs in about 10 minutes. If all traces of oxygen are continuously removed the layer may last for about 50 hours, in an atmosphere of H_2. A glow discharge in air, H_2 or Hg with a heated glass cathode causes a poorly conducting layer to form on the surface and a transition to a spray discharge occurs when its thickness is $\sim 5 \cdot 10^{-4}$ cms. The transition is favoured by decreasing cathode temperature, and increasing current density and layer thickness. The spray discharge shows a voltage-current variation corresponding roughly to subnormal, normal and abnormal regions; at low currents ion bombardment produces some of the current (γ process), at higher currents field emission predominates and the current is independent of gas pressure.

We refer to the work of LLEWELLYN JONES and colleagues on starting potentials, where they find that positive ions distributed over an oxide layer give a kind of field emission described by a modified FOWLER-NORDHEIM equation. The MALTER effect in secondary electron emission is an analogous process. NEWMAN has also found a similar phenomenon [*247*].

88. Discharges in magnetic fields. General. The effect of a magnetic field upon a glow discharge has been studied by many workers for almost a century. The early work, almost entirely qualitative, has been reviewed by STARK [*248*], and more modern work by SOMERVILLE [*249*]. Systematic quantitative experiments have been made only since about 1920, and even now the additional complication of a magnetic field makes adequate theoretical treatment difficult.

The basic effect of a magnetic field is to cause charged particles not already moving parallel to it, to describe helical paths around the lines of magnetic force. The radius of the helix decreases with increasing magnetic field. In most circumstances only the paths of electrons are altered, the ions being virtually unaffected. The electrons thus move a much longer total distance in the gas in order to move a given distance in the direction of the electric field. They hit gas molecules more often and thus have a greater chance of ionising. In this way the presence of a magnetic field acts like an increase in gas pressure. By considering the curvature of the free path of an ion of molecular weight M, and taking the chord between the extreme points as the effective free path it can be shown [*4*] that the apparent increase in pressure Δp (in mm. Hg) is:

$$\frac{\Delta p}{p} \approx 10^{-2} \frac{\lambda}{M T} \left(\frac{H}{p}\right)^2 \tag{88.1}$$

where λ is the mean free path in cm. at $p = 1$ mm., p the gas pressure in mm. Hg, T the ion temperature in °K, H the magnetic field in oersted. The numerical factor depends on the assumptions made in the derivation.

It must be remembered that this effective increase in pressure is apparent only in directions perpendicular to the magnetic field.

This effect has been used to measure the fall in pressure in a glow discharge due to gas being driven into the walls[1]. A magnetic field, previously calibrated in terms of pressure, is applied to restore the discharge current to its original value (see Sect. 83).

[1] G. MIERDEL: Z. Physik **122**, 614 (1944).

True differences in gas pressure have also been observed when a transverse magnetic field is applied, but these arise from other processes and are described in section 98 on electrophoresis.

The main effects of a magnetic field upon a glow discharge are:

(a) A decrease in the length of the cathode dark space when a transverse magnetic field is applied. This effect is very powerful because electrons have beam-like properties in this region.

(b) A change in the radial ion density in the positive column when subjected to a longitudinal magnetic field.

(c) Marked changes in the voltage-current characteristics of the discharge.

(d) A decrease in the starting potential. This does not refer to the fully developed discharge and will not be discussed in detail here[1].

There are other miscellaneous effects, for example the production of oscillations of high frequency including the gyro-frequency, and the main effects listed above can be combined with various electrode geometries to give discharges which rectify current or act as oscillators.

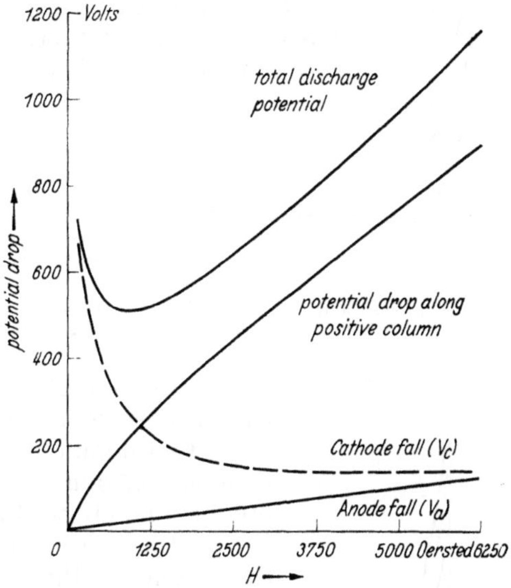

Fig. 73. Potential distribution along an unconfined glow discharge in the presence of a transverse magnetic field. Due to limitations of probe measurements the actual values may be in error by up to 50 volts. The true V_a is probably rather less, V_c rather more than shown in the graph.

Many of these general effects, together with photographs showing the appearance of the discharge, are illustrated in recent experiments of McBee and Dow on an unconfined glow discharge (i.e. without side walls) in air in the presence of a transverse magnetic field[2,3] ($p \sim 0.3$ to 10 mm. Hg, discharge current 0.05 to 2.5 Amp, $H = 0$ to 7000 Oe, the electrodes were copper cylinders with their axes perpendicular to the axis of the discharge). They found that with increasing field all the cathode regions are compressed, the negative glow becoming a thin layer of intense luminosity. Probe measurements showed that the anode fall and positive column gradient increase while the cathode fall first decreases and then becomes constant, the combination giving a minimum in the total discharge potential (see Fig. 73). The positive column and anode regions become more luminous, and anode spots more numerous; the sputtering rate increases greatly, random potential fluctuations increase in amplitude, their characteristic frequency first increasing and then decreasing as H increases. A transverse wind, directed according to the motor rule, blows from the positive column.

89. Influence of magnetic field on the cathode regions. *α) Cathode dark space.* In this region, most electrons move with relatively high speed normal to the cathode

[1] For a full account see [*249*], G. Valle: Nuovo Cim. **7**, 174 (1950) and J. M. Somerville: Proc. Phys. Soc. Lond. B **65**, 620 (1952).

[2] W. D. McBee and W. G. Dow: Commun. a. Electronics **72**, part 1, 229 (1953); also W. D. McBee: Ph. D. Thesis (Elec. Eng.) Univ. of Michigan 1951.

[3] See also W. H. Bergmann: Acta Phys. Austriaca **5**, 425 (1952).

surface. A longitudinal magnetic field, i.e. one parallel to the applied electric field, therefore has little effect upon the properties of the dark space, except to inhibit the radial motion of those electrons which are scattered by hitting gas molecules. Losses to the side walls are thus reduced.

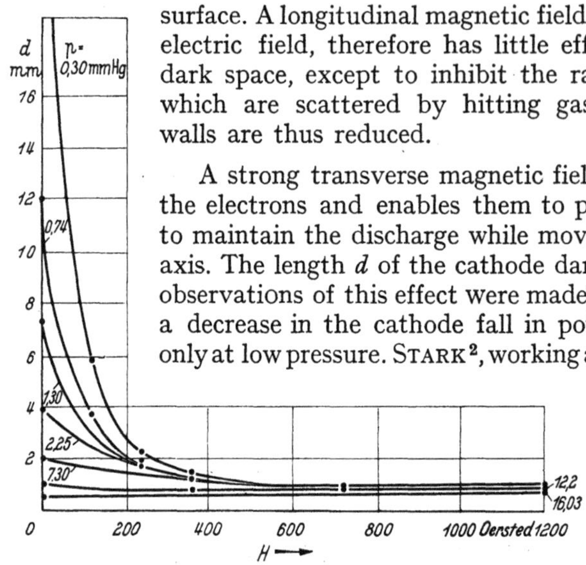

Fig. 74. Effect of a transverse magnetic field on the length of the cathode dark space d in a hydrogen discharge (Hg cathode).

A strong transverse magnetic field, however, bends the paths of the electrons and enables them to produce the necessary ionisation to maintain the discharge while moving a shorter distance along the axis. The length d of the cathode dark space is thus reduced. Early observations of this effect were made by WILLOWS[1], who also noticed a decrease in the cathode fall in potential, which occurs, however, only at low pressure. STARK[2], working at higher pressures observed only the change in the dark space length. Accurate measurements were made by GÜNTHERSCHULZE [*250*] on normal glow discharges in H_2, N_2, A, Hg, H_2O using a mercury cathode. Wall and temperature effects were excluded as far as possible. Working at gas pressures between 0.3 and 20 mm. Hg he found that the dark space length (measured visually) is decreased by the magnetic field, down to a minimum value d_{min} which for all the gases studied is about 0.7 mm. If $d < d_{min}$ initially then the magnetic field produces no effect. Typical results are shown in Figs. 74 and 75. In this range the normal cathode fall V_n is unchanged by the presence of the magnetic field. The current density increases with the magnetic field, and as far as the accuracy of the measurements allow appears to follow the law

$$j = \frac{C}{d^2} \qquad (89.1)$$

where C is a constant.

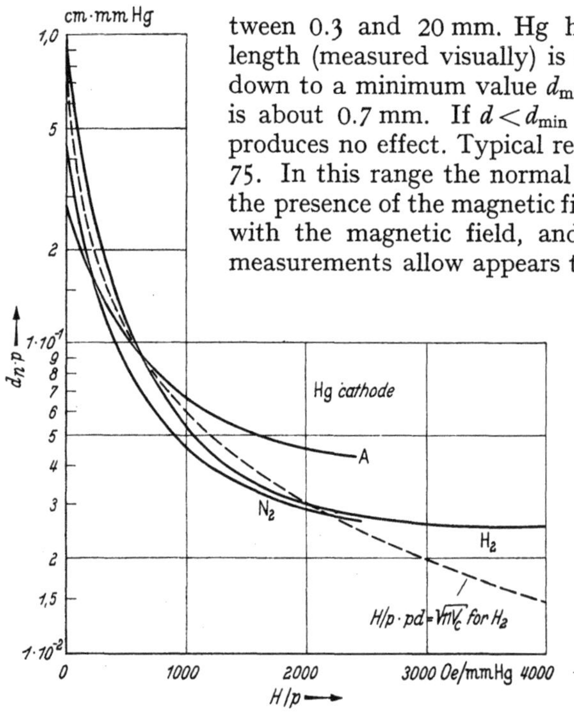

Fig. 75. Effect of a transverse magnetic field on the length of the cathode dark space d in various gases, plotted in similarity parameters.

The effect can be regarded as due to the apparent increase in pressure[3]. The processes obey the similarity laws since the value of the product pd is found to be a function of H/p, except at large values of H when gas heating occurs due to the large increase in current density. The current density j maintains itself pro-

[1] WILLOWS: Phil. Mag. **6**, 250 (1901).
[2] J. STARK: Ann. Phys. **12**, 31 (1903).
[3] See also T. TAKAMINE, T. SUGA and A. YANAGIHARA: Inst. Phys. Chem. Res. Tokyo Sci. papers 454, 69 (1933). They also noticed in a He discharge that a bright ring appears around the cathode, emitting the He_2 spectrum.

portional to $(p')^2$ where p' is the *apparent* gas pressure, for electrons, in the presence of the magnetic field.

The theory of the shrinking of the dark space has been treated rather more directly by THOMSON and THOMSON [3]. They begin with the equations of motion of an electron moving under the influence of an electric field X in the x direction (along the axis of the discharge) and a transverse magnetic field H in the y direction. These are:

$$m\frac{d^2x}{dt^2} = eX - eH\frac{dy}{dt}, \tag{89.2}$$

$$m\frac{d^2y}{dt^2} = eH\frac{dx}{dt}. \tag{89.3}$$

The electric field is assumed to decrease linearly with distance from the cathode, as it does when no magnetic field is present, i.e.

$$X = \frac{2V_c}{d}\left(1 - \frac{x}{d}\right). \tag{89.4}$$

There is some evidence for this in the early work of WILLOWS. The solution of these equations, with the boundary conditions that when $x=0$, $dy/dt = 0$, and $dx/dt = 0$, is

$$x = \frac{\frac{2V_c}{d}}{\frac{2V_c}{d^2} + \frac{eH^2}{m}}(1 - \cos\omega t) \tag{89.5}$$

where

$$\omega^2 = \frac{\frac{2eV_c}{d^2} + \frac{e^2H^2}{m}}{m}. \tag{89.6}$$

With these assumptions, which ignore collisions with gas molecules, an electron traverses a series of loops as shown in Fig. 76, reaching a maximum distance from the cathode

$$x_{\max} = \frac{\frac{4V_c}{d}}{\frac{2V_c}{d^2} + \frac{eH^2}{m}}. \tag{89.7}$$

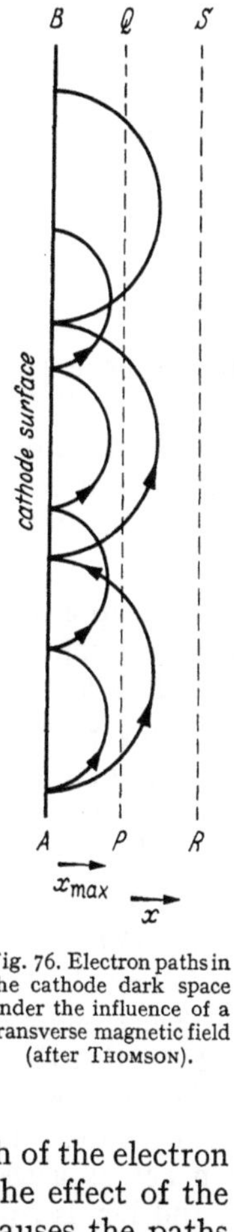

Fig. 76. Electron paths in the cathode dark space under the influence of a transverse magnetic field (after THOMSON).

If the length of the dark space is less than this distance the path of the electron in the dark space is only slightly increased. We should expect the effect of the magnetic field to be small. The application of a stronger field causes the paths to be bent more sharply, and at a critical value of H the electrons no longer pass out of the dark space, but perform a series of loops, their total path being enormously increased. The magnetic field then causes an appreciable reduction in dark space thickness, and the critical condition that the electron does not cross the dark space boundary is given by

$$x_{\max} < d \tag{89.8}$$

which gives

$$\frac{He^2}{m} > \frac{2V_c}{d} \tag{89.9}$$

or

$$Hd > \sqrt{11\,V_c} \tag{89.10}$$

where V_c is the cathode fall measured in volts, H is in oersted, d in cm.

We should therefore expect that a curve of d (as ordinate) plotted against H should show a short almost horizontal portion for small H, and then follow the hyperbola

$$H\,d = \sqrt{11\,V_c}. \tag{89.11}$$

The authors report having found a curve of this type in helium, giving good numerical agreement with the theory (taking $V_c = 153$ volts; the cathode material is not mentioned). GÜNTHERSCHULZE'S curves, however, do not show any flat regions, and they vary considerably with the pressure. THOMSON'S theory is, of course, an extreme simplification; the electrons are treated as though they move in vacuo, and they are considered to be simply reflected when the magnetic field returns them to the cathode surface.

We may note in passing that if Eq. (89.11) is written in the form

$$\left(\frac{H}{p}\right)\cdot p\,d = \sqrt{11\,V_c} \tag{89.12}$$

it fits the experimental results quite well for H_2 and N_2, except in strong fields, but less well for argon. The curve is plotted for H_2 as an example (Fig. 75). Eq. (89.12) must be regarded now *only as an empirical approximation:* This comparison with experiment is in no sense a check of THOMSON'S theory, which applies to motion in vacuo, ignoring the gas density, and therefore cannot be transformed in terms of similarity parameters.

The application of a transverse magnetic field can often reduce the potential across an initially obstructed discharge, by reducing the length of the dark space until it is less than the distance between the electrodes.

A magnetic field produces other effects in the cathode region. Due the normal electrodynamic force on a filament of current the discharge rotates over the cathode with a velocity which is a function of H/p. This property has been studied much more in arcs than in glow discharges.

β) The negative glow. Only qualitative experiments have been made in this region [*248*], [*3*]. In general the radial boundary of the negative glow tends to follow the lines of magnetic force, especially in strong magnetic fields where the electrons move in tight helices around these lines. A longitudinal magnetic field tends to lengthen the negative glow by preventing lateral motion of electrons, whilst a transverse field shortens it by bending the discharge against the walls of the tube. The wall, being negatively charged by the impinging electrons, can act as a secondary cathode and another discharge emerges normally from it and bends towards the anode. Experiments have also shown how a magnetic field alters the distribution of the glow over the surface of cathodes of various shapes[1,2].

90. Influence of a magnetic field on the positive column. A transverse magnetic field bends the positive column towards the wall of the tube, in a curve such as would be followed by a positively charged particle moving from the anode. The loss of charged carriers, and hence the potential required to maintain the column increases, and a sufficiently large magnetic field can put out the discharge. BECKMAN[3] has calculated in detail the relations between the gradient, electron temperature and magnetic field, and also the new electron density distribution when a cylindrical plasma is placed in a uniform transverse magnetic field. The

1 HITTORF: Pogg. Ann. **136**, 221 (1869).
2 SCHUSTER: Proc. Roy. Soc. Lond. **37**, 317 (1884).
3 L. BECKMAN: Proc. Phys. Soc. Lond. **61**, 515 (1948).

results agree fairly well with measurements in H_2, N_2, He, Ne. If the discharge hits the wall some distance from the anode the remaining space towards the anode is often filled with a number of striations, especially in relatively large fields ($H \sim 1000$ oersted or more), see Fig. 77. A magnetic field has the tendency to cause striations to form, or to increase the number present in an already striated column. The reasons for this effect have been discussed by THOMSON ([3], Vol. 2, p. 415).

Quite apart from this obvious effect of an electrodynamic force on the whole of the column, a magnetic field alters the inner structure of the column itself. Here the electrons are moving in all directions with random velocities very much larger than the drift velocity along the column. The magnetic field constrains all motion perpendicular to its direction; diffusion in particular is reduced, except along the lines of magnetic force, and there is also a redistribution of electron energies. In a longitudinal magnetic field these effects are not masked by any consequences of a large electrodynamic force on the column as a whole, e.g. the charging up of the side walls. Consequently losses by radial diffusion to the walls are reduced and the column can be sustained with a smaller gradient. The column also appears to be brighter in the centre and fainter near the walls.

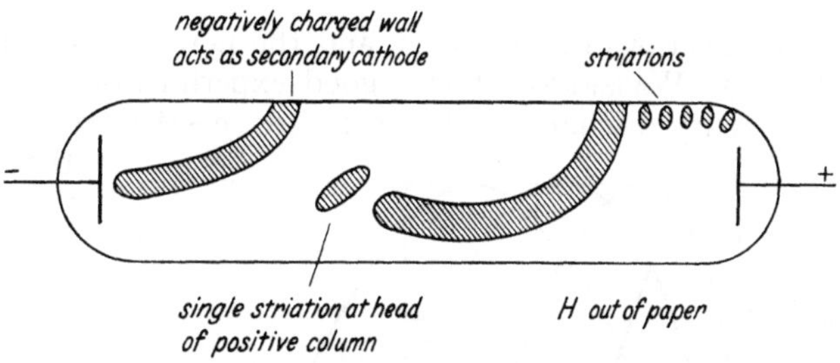

Fig. 77. Typical effect of a transverse magnetic field on a glow discharge (after THOMSON).

The motion of a random swarm of electrons moving in electric and magnetic fields, in the absence of space charge effects was worked out by TOWNSEND [251]. He later reaffirmed his results, after some controversy with TONKS [252], [253]. This treatment gives the coefficient of diffusion D_H, perpendicular to the magnetic field

$$D_H = \frac{D}{1 + \omega^2 \tau^2} \tag{90.1}$$

where D is the normal diffusion coefficient,

$$\omega = \frac{e H}{m c}, \text{ the gyro-frequency,} \tag{90.2}$$

$$\tau = \frac{\lambda_e}{v_{\text{random}}}, \tag{90.3}$$

where τ is the mean time of flight of an electron between collisions.

This equation has been verified experimentally by BAILEY who measured the lateral diffusion of a stream of electrons in parallel electric and magnetic fields [254]. He used photoelectric currents having negligible space charge.

TOWNSEND also showed that in parallel fields the drift velocity (i.e. mobility) in field direction is reduced by the same factor

$$b_H = \frac{b}{1 + \omega^2 \tau^2}. \tag{90.4}$$

These formulae have been used directly by WASSERRAB to calculate the ion and electron density in a plane positive column of width d (i.e. one contained between plane walls, see Sect. 54), subjected to a transverse magnetic field [*255*]. He uses a modified expression for the ambipolar diffusion coefficient, and following the usual extension of SCHOTTKY's treatment obtains instead of the normal equation[1]

$$N = N_{\max} \sin \frac{\pi x}{d} \tag{90.5}$$

the new ion density distribution

$$\frac{N}{N_{\max}} = \frac{\sin \frac{\pi x}{d}}{\sin\left(\operatorname{arc\,tan} \frac{2\pi}{a d}\right)} \exp\left\{\frac{a d}{2\pi}\left(\operatorname{arc\,tan} \frac{2\pi}{a d} - \frac{\pi x}{d}\right)\right\} \tag{90.6}$$

where

$$a d = \frac{H d b_{+} b_{-} X_y}{D_a}. \tag{90.7}$$

The directions of the fields, and the new density distribution can be seen from Fig. 78. WASSERRAB found good experimental agreement with these results in a mercury discharge at low pressure ($p \sim 10^{-3}$ mm. Hg), on the assumption that the light intensity, which he measured across the column, was proportional to the electron and ion concentration. At higher pressures agreement was not so good due to the appearance of striations and other factors.

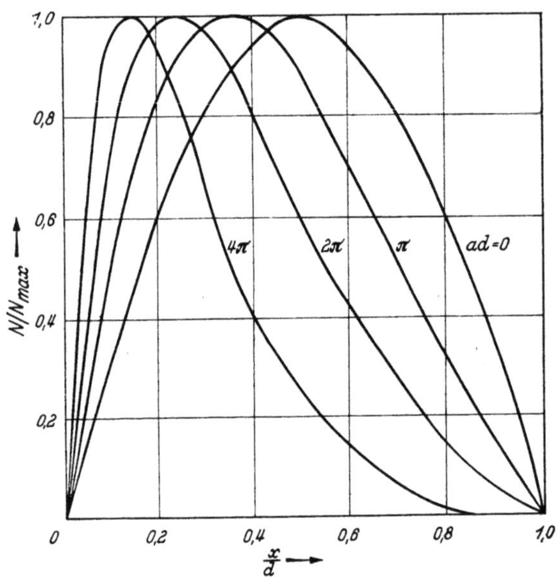

Fig. 78. Distribution of ions and electrons in Hg positive column with parallel walls, as a function of the transverse magnetic field in the z direction.

$$a d = \frac{H d b_{+} b_{-} X_y}{D_a}, \quad d = \text{distance between walls.}$$

x is measured perpendicular, y and z parallel to the walls.

Many other workers have also treated in detail the fundamental equations of motion of electrons in the presence of electric and magnetic fields and a concentration gradient [*256*] to [*261*]. Various authors obtain results differing slightly, sometimes by small numerical factors depending on assumptions made about electron energy distributions, and upon methods of averaging. The results of ALLIS, ALLEN, and TONKS are of most interest in applications to the glow discharge, and have been used to examine the behaviour of the positive column in a longitudinal magnetic field.

Not much work has been done on this problem and in a few points the results are inconsistent.

When a discharge tube is placed inside a solenoid producing a longitudinal magnetic field we must distinguish between end effects, which can be very prominent, and the real effect of a completely uniform longitudinal field. The end effects may be attributed either to the configuration of the magnetic field

[1] WASSERRAB measures x from one wall, not from the centre of the column: this gives a sine instead of a cosine function: compare Eq. (54.2).

or to the geometry of the cathode. If the discharge is placed in a short solenoid, electrons will tend to follow the converging lines of magnetic force at the ends, and the discharge will thus be concentrated at the centre of the tube. On the other hand, if the discharge is placed in a very long solenoid, any radial charge distribution fixed by the cathode will be propagated along the discharge, and hindered from dispersing perpendicular to the magnetic field. A combination of these effects can also occur.

In any experiments it is, therefore, necessary to ensure that the magnetic field is entirely free from radial components, and to examine only those regions of the discharge sufficiently far from the cathode. Assuming that these conditions have been satisfied in the positive column, it is clear that the magnetic field reduces the outward flow of electrons by diffusion. Their number can then be maintained by a lower rate of ionisation, i.e. a lower electron temperature T_e and hence a smaller gradient X. The outward flow of ions is made to balance that of the electrons by a readjustment of the radial electric field. Nearly all workers have found results consistent with this general picture, but there is disagreement on one important point, namely whether or not the magnetic field changes the radial distribution of ions and electrons from the normal BESSEL-function. CUMMINGS and TONKS *[262]*, *[263]*, in a detailed theory of the positive column in a longitudinal magnetic field, conclude that no change in the radial distribution is to be expected. The radial potential distribution should also be unaltered in shape but reduced in scale, although they calculate that in large magnetic fields the radial electric field may become zero or even reverse, i.e. the axis may become negative with respect to the walls. They believe that the apparent axial concentration of a discharge when placed in a uniform magnetic field is due entirely to a concentration being propagated from the cathode end. In the absence of a magnetic field this would have dispersed within a short distance. TONKS *[263]* has calculated approximately the dispersal effect along a cylindrical column. The solution for the radial electron and ion distribution is the sum of a series of zero order BESSEL functions. The first term, which is the normal distribution, is constant along the length of the column, while successive terms decrease with distance along the column at rates which are complicated functions of H and electron temperature.

CUMMINGS and TONKS tested their theory by experiments on the positive column of a mercury arc[1], at a pressure of $5 \cdot 10^{-3}$ mm. Hg, and a current of 4 amperes. The discharge tube, 60 cm. long, was placed entirely inside a solenoid, wound in sections through which suitable currents could be passed to produce a magnetic field of any desired configuration. The required parameters were measured with suitably placed probes. The tube had a cylindrical oxide coated cathode, divided into eight sections. It was found that in a true longitudinal field the current flowing from the cathode was collimated into a luminous beam which clearly showed the pattern of the cathode. The beam changed abruptly into a glow which filled the tube at a distance of about 30 cm. from the cathode. This glow was more intense at the axis than in the absence of a magnetic field. Probe measurements showed an increased electron density at the axis which was presumably propagated from the cathode because the distribution tended to the normal towards the anode. If, however, the magnetic field was made to diverge from the cathode the tube was completely filled by a glow apparently

[1] Although many of the experiments to be discussed here were performed on arc discharges they refer only to the positive column at not too large currents, and thus may be considered relevant.

indistinguishable from that obtained without magnetic field. Probe measurements in a limited range showed a normal electron density distribution but a modified radial potential distribution in reasonable agreement with the theoretical predictions,

$$N_e = N_0 \, e^{-\frac{eV}{k T_e \cdot \tau}} \tag{90.8}$$

where

$$\tau = \frac{\alpha D_e - \mu D^+}{\alpha D_e + \mu D^+ \cdot \frac{T_e}{T_+}} < 1. \tag{90.9}$$

Here μ is the ratio of the radial drift velocity of electrons and ions, and α a rather complicated function of H, T_e and p. The rate of decay of the initial density distribution with distance along the column also agreed within 30% with theory. This was found to depend on the gas pressure but very little on the magnetic field, a disturbance being propagated further at lower pressure. The probe characteristics were however badly distorted by the magnetic field at $H > 50$ oersted, and all measurements were taken at lower fields where the magnetic effects were small. The experiments demonstrate clearly the importance of end effects, but do not give conclusive results about the radial distribution of charges when these effects are absent.

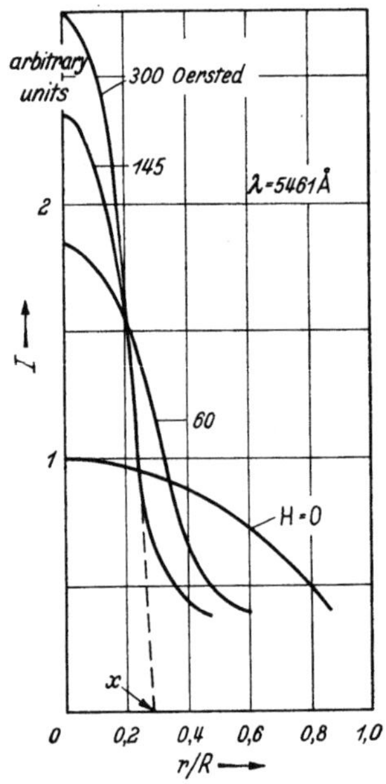

a

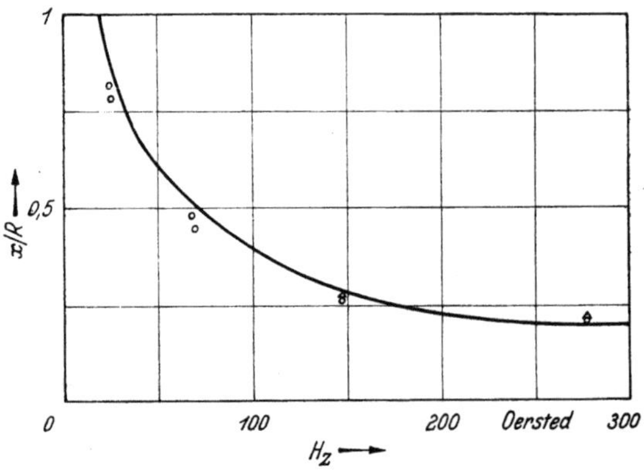

b

Fig. 79a and b. a) Intensity distribution of the mercury line ($\lambda = 5461$ Å) across the radius of the tube in the presence of a longitudinal magnetic field. b) Variation of the radius of the cord with longitudinal magnetic field (ROKHLIN). Curve: theoretical, after method of STÖRMER, △, ○ = experimental points.

Similar experiments were performed by ROKHLIN [*264*] who observed the effect of a magnetic field on the radiation from a mercury vapour discharge ($p \sim 10^{-3}$ mm. Hg, $i = 1.5$ to 4 Amp). This method avoids the disadvantages of probes, but indicates only the behaviour of fast electrons of energy greater than the excitation energy of the gas. He used two solenoids spaced a few cm. apart: the magnetic fields could be coincident, giving an almost uniform field between them, or opposite giving a distorted field having strong radial components. The image of a diameter section was observed in a spectroscope, and the intensities of the resonance lines (1850 and 2537 Å) measured by the brightness of a fluorescent probe placed in the tube.

With coincident fields the discharge is visibly constricted into a cord, at first rapidly and then more slowly with increasing H. (At higher pressures or currents the effect is less marked and finally ceases to be noticeable.) The cord follows the lines of magnetic force and can be moved about by displacing the solenoid coils or by the presence of a magnet. Fig. 79a shows a typical intensity distribution across the radius of the tube. The "radius" x, of the constriction (i.e. the cord) is derived arbitrarily by extrapolating the linear part of the curve, and its variation with the longitudinal field is compared in Fig. 79b with theoretical calculation following a method due to STÖRMER[1]. The intensity distribution of various spectral lines leads to the same values of this radius x, and also shows that the electron energy distribution is constant across a chord of the tube. Fig. 80 shows the change in relative intensity of several lines at the centre of the discharge, with increasing magnetic field. The fall in intensity is pronounced in lines from high excitation levels, indicating a decrease in the number of fast electrons. The maximum is due to two opposing effects, the increased concentration of electrons at the centre, and the decrease in their energy. ROKHLIN comes to the general conclusion that the constriction of the discharge is due to the radial components of the magnetic field on the cathode side of the plasma: the longitudinal part of the field, however, did not extend far enough for a proper assessment of its effect. Similar experiments were made by FABRIKANT and ROKHLIN [265], [266].

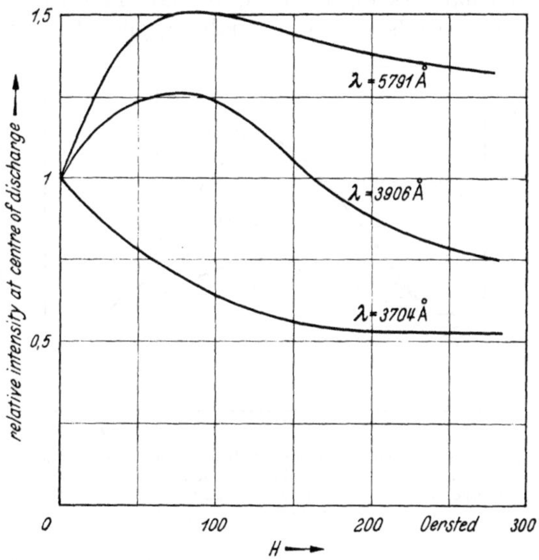

Fig. 80. Variation of intensity of radiation at centre of the discharge with longitudinal magnetic field.

REICHRUDEL and SPIWAK, however, find apparently different results [267]. They measure the radial variation of electron density and current density using plane probes and a concentrically divided anode, in discharges of Hg vapour or argon ($p \sim 10^{-3}$ to 1 mm. Hg). Particular attention is paid to the distortion of probe characteristics in a magnetic field[2]; amongst these is a curvature in the characteristics which can produce an error of up to a factor 4 in the estimate of electron density. Their results using short magnetic lenses with appreciable end effects are in agreement with those of ROKHLIN, but when the entire discharge tube is placed in a long solenoid producing a uniform longitudinal field a similar increase in current density towards the axis was found. Fig. 81 shows a typical result. The experiment was repeated in a normal glow discharge in argon at a considerably higher pressure (0.3 mm. Hg) with a plane cathode, to exclude the possibility that electron emission from a small mercury spot might be responsible for the concentration of electrons, and here also the results were similar. Although the tube was shorter than that used by CUMMINGS and TONKS (about 30 cm. compared with 60 cm.) the gas pressure was so much higher that

[1] See for example Handbuch der Experimentalphysik, vol. 25, part I, p. 418.

[2] See also C. FATALIEV: C. R. Acad. Sci. USSR. **23**, 891 (1939). — J. Exp. Theor. Phys. **9**, 813 (1939). — L. MIRLAS: J. Tech. Phys. **4**, 1522 (1934). — G. SPIWAK and E. REICHRUDEL: Phys. Z. Sowjet. **9**, 655 (1936). — J. Tech. Phys. USSR. **5**, 715 (1938). — Bull. Acad. Sci. URSS. **4** (1938). — C. R. Acad. Sci. USSR. **18**, 177 (1938).

it seems unlikely that the propagation of a cathode disturbance could account for these results. The fields used by REICHRUDEL and SPIWAK are considerably greater, however, and it may well be that the positive column behaves in different ways under different conditions of gas pressure, discharge current and magnetic field.

Very recently BICKERTON and VON ENGEL [*308*] have made measurements on the positive column in helium at low pressure, taking great care to eliminate end effects and using new methods of probe measurement in the presence of a magnetic field.

They find that the radial distribution of charges alters appreciably from the normal BESSEL function, and conclude that even in a low pressure column where LANGMUIR'S theory normally applies, the properties of the column in a magnetic field of sufficient strength are best described by SCHOTTKY'S theory of ambipolar diffusion. They discuss theoretically the radial potential distribution and the decrease in the gradient and electron temperature. However, when the gas becomes highly ionised the effect of the partial pressure of the electron gas may become important and both the gradient and the electron temperature should then be independent of the magnetic field. This had actually been observed earlier by DAVIES[1] who used a spectroscopic method to measure the electron temperature in the positive column of a low pressure caesium discharge. He found, contrary to expectations, a slight increase in the electron temperature when the magnetic field was applied. He used a high current density ($j = 5$ A/cm.²) in a narrow tube and it appears probable that the neutral gas near the axis may have been displaced by the electron gas so that the same or a larger gradient and electron temperature were necessary to maintain the ionisation. This view is supported by MOHLER'S observations[2] on a similar discharge without magnetic field: as the current density is increased at constant pressure both X and T_e increase.

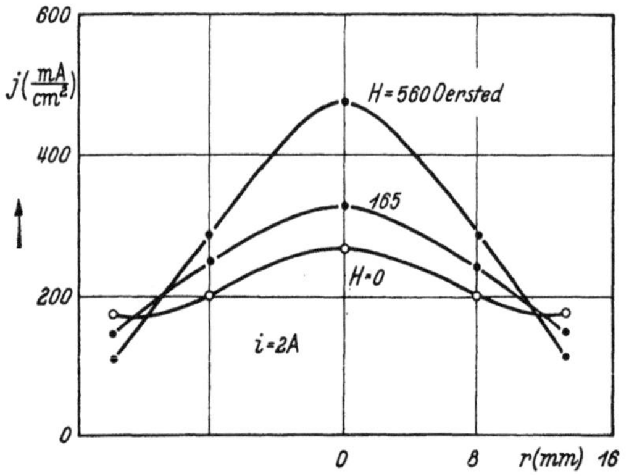

Fig. 81. Redistribution of current density across tube radius in the presence of a longitudinal magnetic field. The entire discharge is situated in the uniform part of a long solenoid. Gas Hg; $p = 0.037$ mm. Hg (?).

REICHRUDEL and his collaborators have also measured the starting potential, discharge current and potential distribution in glow discharges in a non-uniform electric field (spherical electrodes) and a superimposed axially symmetric magnetic field (Ne, A, air; $p \sim 10^{-4}$ to $3 \cdot 10^{-3}$ mm. Hg) [*269*]. The influence of the latter on the ionisation coefficient α is discussed. The potential distribution was measured by double probes, and oscillograms showed that both periodic and random oscillations are present in the plasma. The influence of the probes on the discharge was observed, being negligible when they are parallel to the axis of the tube, and large when they are perpendicular to the magnetic field. The effect of alternating fields was also observed. Wall currents in the positive column of a mercury arc subjected to an axial magnetic field were measured by KUGLER and ROGGENDORF[3].

---

1 L. W. DAVIES: Proc. Phys. Soc. Lond. B **66**, 33 (1953).
2 F. L. MOHLER: Bull. Res. Nat. Bur. Stand. **21**, 873 (1938).
3 G. A. KUGLER and A. ROGGENDORF: Phys. Z. **34**, 136 (1933).

It should also be mentioned that TONKS' theory has been used with some success to explain other magnetic properties of a plasma, in particular KILLIAN's observations on the effect of a transverse field on the positive column[1] (Hg discharge), and also STEENBECK's experiments and approximate theory of the diamagnetic susceptibility of a plasma[2]. Another effect is treated theoretically by TONKS, namely the contraction of the column away from the walls due to the self magnetic field of the discharge current. This is sometimes known as the "pinch effect", and occurs only at large values of current i.e. in arc discharges[3]. It is discussed more fully in the article on the arc, and has been studied both theoretically and experimentally by BENNETT[4], BLACKMAN[5], THONEMANN and COWHIG[6].

Experiments have been made on the effect of magnetic fields on the radiation from the column of a constricted discharge, e.g. in a capillary tube—the apparatus normally used to show the ZEEMAN effect. For example, KULKARNI[7], studying discharges in He, Ne, N_2 found that the intensity of a spectrum line reaches a maximum and then quickly decreases with increasing H. The value of H at the intensity maximum depends on the wavelength of the line and the presence of any foreign gas. For a given discharge voltage V there is a critical H above which the discharge goes out, and just below which it throbs (i.e. becomes intermittent). In these conditions the rare gases show the molecular spectrum (e.g. He_2) in regions near the electrodes. His experimental conditions were $V = 10$ to 15 kV, $H \leq 10000$ oersted, p not specified.

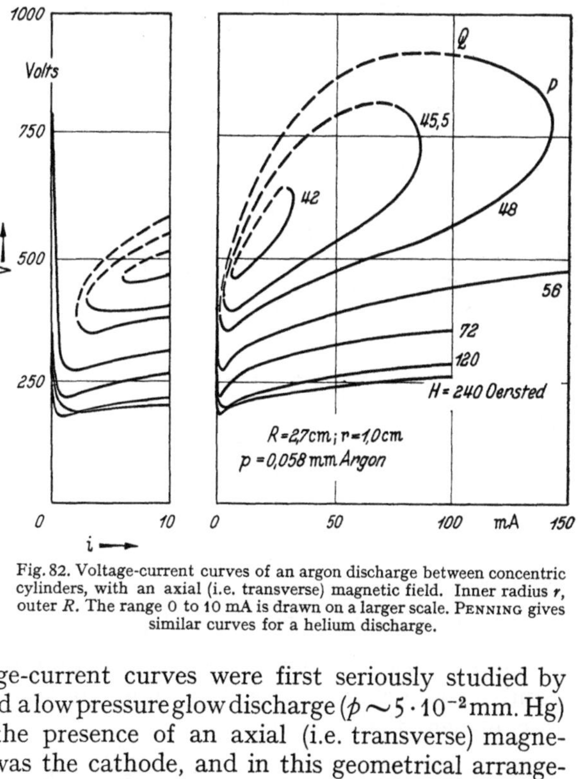

Fig. 82. Voltage-current curves of an argon discharge between concentric cylinders, with an axial (i.e. transverse) magnetic field. Inner radius r, outer R. The range 0 to 10 mA is drawn on a larger scale. PENNING gives similar curves for a helium discharge.

91. Influence of a magnetic field on current-voltage characteristics. The effect of a magnetic field upon the voltage-current curves were first seriously studied by PENNING [*270*], who investigated a low pressure glow discharge ($p \sim 5 \cdot 10^{-2}$ mm. Hg) between coaxial cylinders in the presence of an axial (i.e. transverse) magnetic field[8]. The outer cylinder was the cathode, and in this geometrical arrangement electrons leaving the cathode could describe very long cycloidal paths

[1] T. J. KILLIAN: Phys. Rev. **35**, 1238 (1930).
[2] M. STEENBECK: Wiss. Veröff. Siemens-Werk **15**, 1 (1936).
[3] See also L. TONKS: Trans. Electrochem. Soc. **72**, 167 (1937).
[4] W. H. BENNETT: Phys. Rev. **45**, 890 (1934), see also **91**, 1562 (1953).
[5] M. BLACKMAN: Proc. Phys. Soc. Lond. B **64**, 1039 (1951).
[6] P. C. THONEMANN and W. T. COWHIG: Nature, Lond. **166**, 903 (1950). — Proc. Phys. Soc. Lond. B **64**, 345, 618 (1951).
[7] S. B. KULKARNI: Current Sci. **13**, 254 (1944).
[8] A similar arrangement was used in earlier work, but only to measure the change in the starting potential of air in the region of the minimum starting potential. M. WEHRLI: Ann. Phys. **4**, 69, 285 (1922).

before reaching the anode. The magnetic field caused an apparent increase in pressure, up to a factor of 1000 and reduced the starting potential by a factor of up to 100. The starting potential was lower in non-uniform than in uniform magnetic fields, and also in cylinders with closed ends.

In the actual discharge, the current density can be increased by the magnetic field up to a factor of 10^4, due to the effective increase in pressure. The current-voltage curves have a peculiar shape, and in a certain range of p and H are composed of four sections having ordinary positive and negative characteristics (i.e. $dV/di \gtrless 0$), and also two other positive and negative regions called by PENNING "characteristics of the second kind" (respectively, the dotted regions, and e.g. PQ, in Fig. 82). Points on a positive section of the latter cannot be plotted since the discharge cannot be stabilised by a series resistance.

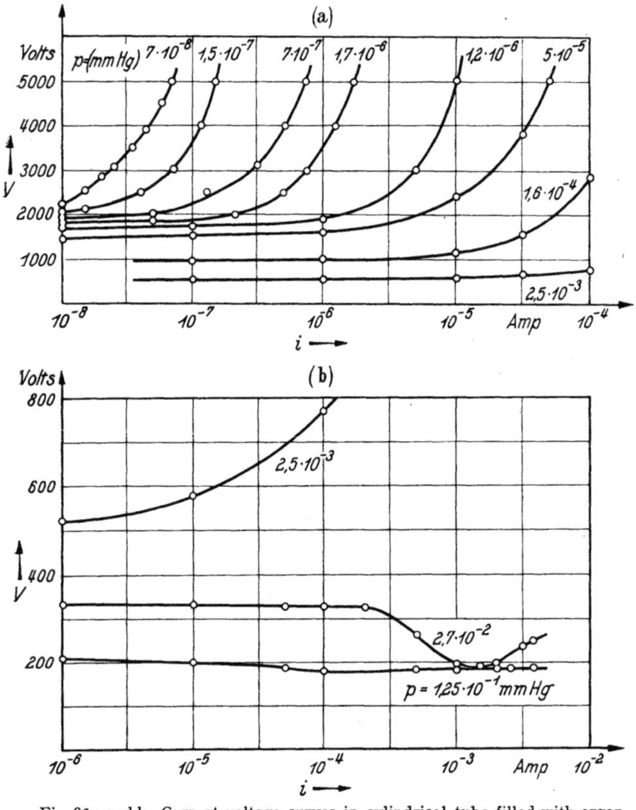

Fig. 83a and b. Current-voltage curves in cylindrical tube filled with argon in the presence of a transverse magnetic field $H = 1500$ oersted. a) Pressure range 10^{-3} to 10^{-7} mm. Hg, b) pressure range 10^{-1} to $2 \cdot 10^{-3}$ mm. Hg.

In recent years, the starting potential of discharges between cylinders in transverse magnetic fields has been studied comprehensively over a wide range of pressures (10 to 10^{-8} mm. Hg) by HAEFER [*271*]. Voltage-current curves have also been measured, and considered theoretically for argon, in the pressure range 10^{-1} to 10^{-7} mm. Hg (see Figs. 83a and b). At $p > 10^{-3}$ mm. Hg, the discharge behaves as a glow discharge with a falling characteristic, explicable by its positive space charge. At $p < 10^{-3}$ mm. Hg the discharge has a rising characteristic, obeying the relation

$$V - V_s = \text{const}\,\frac{i}{p} \tag{91.1}$$

where i is the discharge current, V_s the starting potential. The constant is a function of H. HAEFER extends the methods of treatment due to SCHADE[1], and VON ENGEL and STEENBECK[2].

Similar experiments have been done in dry air by SOMERVILLE and co-workers [*273*] at pressures 10^{-2} to 10^{-1} mm. Hg. They used inhomogeneous magnetic fields produced by a short electron lens, and found that the discharge tends to be bounded by the magnetic lines of force, and can be moved bodily along the

[1] R. SCHADE: Z. Physik **108**, 353 (1938).
[2] [*4*], Vol. 2, p. 51.

axis of tbe cylinder by moving the magnetic field. The effect of splitting the cylinders along a diameter plane and inserting a baffle between is different according as the inner cylinder is cathode or anode. When it is the cathode the discharge can be moved smoothly past the baffle; when it is the anode the discharge jumps abruptly. The arrangement can have a negative resistance and has been used as an oscillator.

92. Miscellaneous magnetic effects: oscillations, "magnetic rays". Many observers have noticed that the application of a magnetic field produces oscillations in the plasma regions of a discharge, particularly in the positive column and probably to some extent in the negative glow.

LAFFINEUR and PECKER have detected radiation of the gyro-frequency, i.e.

$$\nu = \frac{e H}{2\pi m c} \tag{92.1}$$

from the positive column[1], the intensity of the radiation being greatest at a pressure of 10^{-4} mm. Hg. Some years ago, TOWNSEND and GILL pointed out that a discharge in a magnetic field bore certain similarities to a high frequency discharge in which the electric field alternates at the gyro-frequency[2]. Many different frequencies can also be excited: THONEMANN and KING[3] detected radiations of about 1000 Mc/s from a mercury arc, COBINE and GALLAGHER[4] found others about 5 Mc/s. DENISSE and STEINBERG[5,6] have studied the generation of noise and oscillations in N_2 ($p = 1$ to $50 \cdot 10^{-3}$ mm. Hg) due to a localised transverse magnetic field, in the wavelength range about 1 to 3 m. They found that oscillations are produced only when the magnetic field is applied near the anode, increasing in intensity to a sharp maximum with increasing magnetic field. With constant H and gas pressure the oscillations increase with discharge current to a maximum and then die away rapidly to very small intensities. Many of these results contrast with the quite different behaviour of the lower frequencies studied by COBINE and GALLAGHER.

At one time it was though that many properties of discharges in magnetic fields, in particular the way in which a discharge follows the lines of magnetic force, could be explained only by the action of a new kind of ray. These were postulated by RIGHI[7], and called "magnetic rays": he believed that they were streams of particles (unit magnetic poles) each consisting of an electron circling around a positive ion, the two separating and behaving with their normal electrostatic properties in regions of no magnetic field. Later IVES[8] showed that the effects were explicable without this postulate, and also showed that many of the experimental results observed by RIGHI could be produced in suitable conditions without a magnetic field. Recently EHRENHAFT[9,10] and his collaborators have found what were thought to be new kinds of discharges in magnetic fields, explicable only in terms of some kind of magnetic ray, but this has been refuted by SIEBERTZ[11] who has given an approximate quantitative account of the results in terms of the normal behaviour of charged particles.

93. Discharges in electronegative gases. *α) General.* Electronegative gases are those which have a pronounced tendency to attach electrons forming heavy

[1] M. LAFFINEUR and C. PECKER: C. R. Acad. Sci. Paris **231**, 1446 (1950).
[2] J. S. TOWNSEND and E. W. B. GILL: Phil. Mag. **26**, 290 (1938).
[3] P. C. THONEMANN and R. B. KING: Nature, Lond. **158**, 414 (1946).
[4] J. D. COBINE and C. J. GALLAGHER: J. Appl. Phys. **18**, 110 (1947). See also Sect. 95.
[5] J. L. STEINBERG: Nature, Lond. **160**, 833 (1947). — Rev. Sci. Paris **85**, 607 (1947).
[6] J. DENISSE and J. L. STEINBERG: C. R. Acad. Sci. Paris **224**, 546 (1947).
[7] A. RIGHI: Phys. Z. **11**, 158 (1910); **15**, 529 (1914). See also PLÜCKER: Pogg. Ann. **103**. 88, 15 (1858).
[8] J. E. IVES: Phys. Rev. **9**, 349 (1917).
[9] F. EHRENHAFT: Phys. Rev. **75**, 1628 (1949).
[10] F. EHRENHAFT: Acta Phys. Austriaca **4**, 129 (1950).
[11] K. SIEBERTZ: Acta Phys. Austriaca **5**, 134 (1951).

negative ions. The halogens and some organic molecules are strongly electronegative, oxygen less so, and there is some evidence that many of the electronegative properties attributed to oxygen, e.g. the formation of O_2 depend upon the presence of small traces of water vapour. Earlier work has been summarised by SEELIGER [*6*], and rather later work by EMELEUS and SAYERS [*274*], who give a clear outline of the main effects of negative ions in a discharge (see also the monograph by MASSEY [*275*]). In general the addition of a relatively small quantity of an electronegative gas to a discharge in a rare gas results in the changes shown in Figs. 84a and b. All the negative zones contract towards the cathode, boundaries between dark and luminous regions become sharper and curved striations appear in the positive column, accompanied by an increase in the gradient. Often thin dark spaces appear in front of the anode and between the FARADAY dark space and positive column, and the surface of the anode may be covered with a thin luminous layer emitting light characteristic of the rare gas. Except for this anode luminosity a discharge in the pure electronegative gas looks very similar. In some gases (e.g. I_2 and organic vapours) the positive column becomes constricted, or appears like a mobile ribbon.

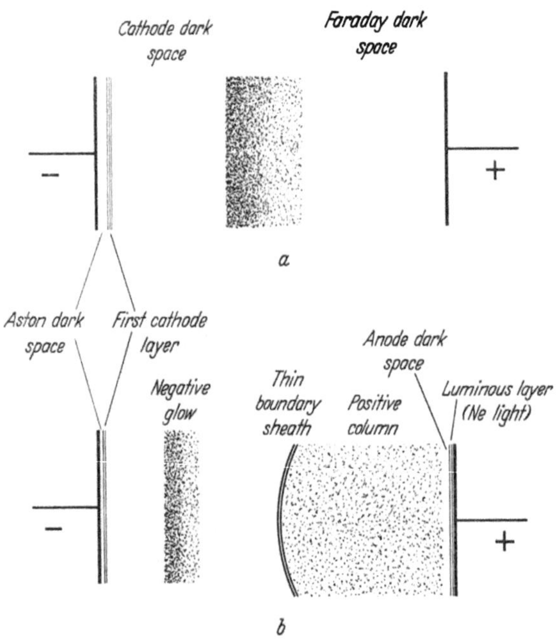

Fig. 84. Typical effect of adding an electronegative gas to a discharge in a pure rare gas. (a) Neon: $p = 2$ mm. Hg. (b) Neon plus a little chlorine. The boundary sleath and the anode regions are not always prominent in other mixtures of rare and electronegative gases.

Following EMELEUS and SAYERS we can trace the characteristic effects of electronegative gases as being due to the decrease in the average mobility of negative carriers, which now consist not only of electrons but also of some heavy negative ions. Three results follow, almost independent in their effects.

(a) An apparent increase in pressure: The drift velocity of electrons is reduced by their temporary attachment to gas molecules. This effect is equivalent to a more continuous retardation, such as would be caused by more frequent collisions with non-attaching gas molecules, i.e. by an increase in gas density. The contraction of the negative zones and of the spaces between striations, and the properties of the positive column are characteristic of a discharge containing only electrons and positive ions at higher gas pressure.

(b) Creation or concentration of regions of negative space charge: Since the negative ions are more massive than electrons they are less likely to diffuse away. Consequently they tend to produce regions of net negative space charge, and to make existing ones thinner. This accounts for the sharpening of boundaries between dark and luminous regions especially at the head of striations where a double layer (+ —) exists. The thin dark space at the head of the positive column is such a double layer, across which almost the whole potential drop of the FARADAY dark space appears. The dark space in front of the anode is believed to be due to a space charge of negative ions, which controls the current from the anode to the positive column. It was first noticed by ASTON[1] in an oxygen discharge, and was clearly visible when the anode was placed in the negative glow: it was faintly observed in air and nitrogen, but not at all in hydrogen. In O_2 the thickness d of the dark space appeared independent of gas pressure, and $\propto \frac{1}{\sqrt{i}}$.

(c) A change in the distribution of negative carriers in the plasma regions. Due to their very low mobility negative ions will tend to move along the lines of electric force: in the positive column this is towards the axis where, at not too low a pressure a concentration will thus be built up. (At very low pressure such concentration will be less marked because the ions will be carried by their own momentum across the centre to the opposite wall.) Destruction of negative ions in the gas phase is a very inefficient process[2] and thus large concentrations of negative ions can be built up even when little electronegative gas is present. An increased concentration of positive ions is then formed to balance the total negative space charge (negative ions plus electrons), and thus the whole column becomes concentrated at the axis, and may become unstable.

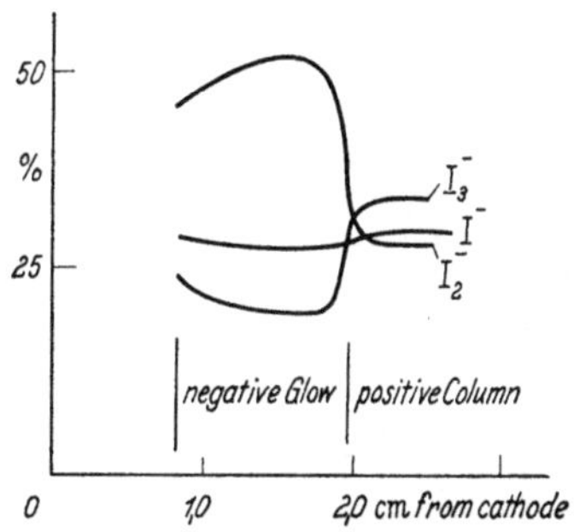

Fig. 85. Negative ion distribution in an iodine discharge [34].

β) Experimental results. The glow discharge in iodine has been studied in detail by SPENCER-SMITH [34], in the pressure range 0.1 to 0.5 mm. Hg. In this gas the FARADAY dark space is exceptionally short. Yellow striations form as the pressure is increased through the above range; at the highest pressure a red cathode glow appears, the cathode dark space becomes a deep purple and the positive column contracts to a ribbon along the axis of the tube. The gradient in the column is large and increases with pressure, e.g. $X \sim 30$ volt/cm. at $p = 0.12$ mm. Hg, $X \sim 100$ volt/cm. at $p = 0.5$ mm. Hg; $X/p \sim 200$.

Probe measurements are difficult to interpret in the presence of a large concentration of negative ions, but they show that in I_2 the positive ion concentration N^+ in the negative glow is about 100 times greater than the electron concentration N_e, while the electron temperature T_e is normal (~ 1 to 1.5 eV). In the positive column N_+/N_e is about the same, the absolute concentrations are lower, but T_e is considerably greater (~ 13 eV). SPENCER-SMITH also investigated the kind of ions produced, using a magnetic method for measuring e/M. He detected I^-, I_2^-, I_3^-, I^{++}, I^+, I_2^+, and I_3^+. I_2^- was present in larger concentration than had previously been found in ion beams[3]; I_2^+ was the most common positive ion, I^{++} being completely absent from the positive column. The distribution of negative ions along the discharge is shown in Fig. 85.

The changes caused by adding a halogen to a discharge in a rare gas are not however always due to the formation of negative ions, but often to the

[1] F. W. ASTON: Nature, Lond. **89**, 218 (1912).

[2] The destruction of negative ions has been considered theoretically by MASSEY and SMITH, Proc. Roy. Soc. Lond., Ser. A **155**, 472 (1936). See also [275], [276].

[3] HOGNESS and HARKNESS: Phys. Rev. **32**, 784 (1928).

ionisation of the halogen by metastable atoms of the rare gas. This occurs for example, in halogen-filled GEIGER counters: see also [*74*].

A similar analysis has been made in oxygen by LUNT and GREGG [*35*], at $p \sim 1$ mm. Hg. In the positive column $X/p \approx 80$, and the concentrations of negative ions and electrons are almost equal, only O^- ions being present. In the FARADAY dark space and negative glow, however, hardly any negative ions were detected: allowing for the sensitivity of the apparatus, $N^- < 1\%\ N_e$. It appears that in the positive column molecular ions are readily destroyed, elsewhere both molecular and atomic ions are destroyed. The magnetic method used to identify these ions is not capable of high resolution, and more accurate results have been obtained with a mass spectrometer probe using radio-frequency shutters, which has also the advantage that it can be moved across the section of the discharge[1]. Some results in O_2 are shown in Fig. 86.

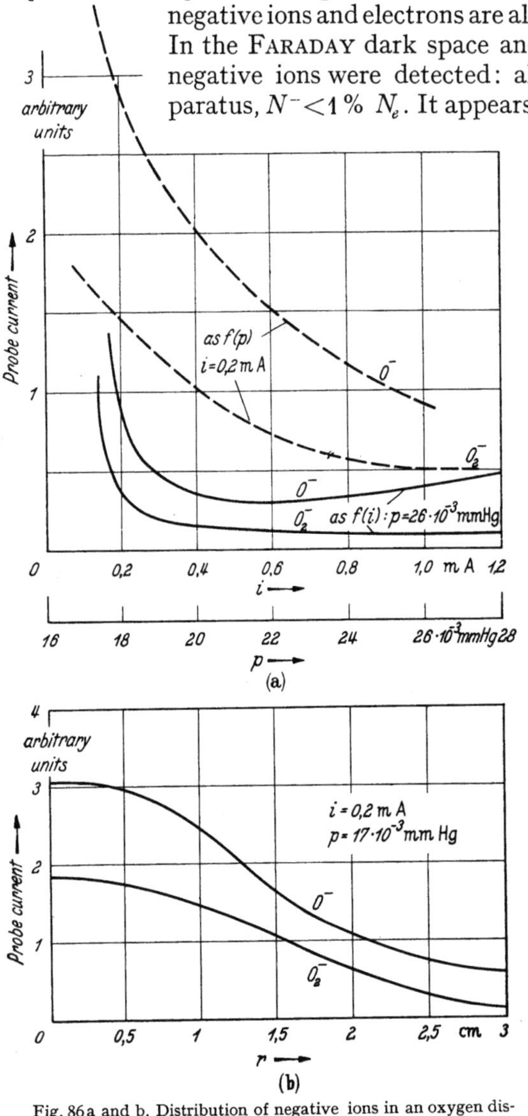

Fig. 86a and b. Distribution of negative ions in an oxygen discharge. a) At the centre, with different currents, i, and pressures, p. b) Variation across the radius, r.

Negative ions have also been detected in mercury discharges[2] (Hg^-), but it seems that they are formed only in the presence of some impurity. The presence of the smallest quantity of hydrogen gives rise to an ion which is thought to be $Hg\,H^-$. Negative ions are also known to be formed when certain metal surfaces are bombarded with positive ions. ARNOT and MILLIGAN observed Hg^- ions emitted with considerable velocity from a nickel surface when bombarded with Hg^+ ions, and they were thought to be formed by the extraction of two electrons from the metal, [*278*] to [*280*], but later work showed that the negative ion emitted need not be of the same element as the impinging positive ion. Adsorbed surface films, especially of oxygen are believed to play an important part in this process[3,4].

Earlier measurements of the properties of the positive column in oxygen have been mentioned in Sect. 46, and an attempt has been made to include the

[1] R. L. F. BOYD: Nature, Lond. **165**, 142 (1950).
[2] U. STILLE: Ann. Phys. **17**, 635 (1933).
[3] R. H. SLOANE and R. PRESS: Nature, Lond. **141**, 872 (1938). — Proc. Roy. Soc. Lond., Ser. A **168**, 284 (1938).
[4] F. L. ARNOT and C. BECKETT: Nature, Lond. **141**, 1011 (1938). — Proc. Roy. Soc. Lond., Ser. A **168**, 103 (1938).

presence of negative ions in a diffusion theory of the column, although this has not been extensively compared with experiment (Sect. 56, also [*309*]). Recent measurements show that the properties of the positive column in oxygen cannot be described by the simple SCHOTTKY theory [*309*].

One other type of discharge in electronegative gases, the *pipe discharge*, has been studied in some detail [*281*] to [*286*]. This has been observed in mixtures of organic vapours (e.g. naphthalene, alcohol, paraffin) or iodine with rare gases. The positive column contracts to a core, having a colour characteristic of the rare gas, surrounded by a different coloured mantle, often greenish-blue or white. A strong space charge of negative ions surrounding the inner core fulfils the function of the negatively charged walls usually found in low pressure positive columns. The theory has been considered by SEELIGER and SOMMERMEYER [*285*].

The presence of hydrocarbon vapour can sometimes cause the negative glow to jump periodically over the surface of the cathode. This is throught to be caused by the vapour condensing to form an insulating layer on the cathode[1].

94. Discharges in gas mixtures. The properties of discharges in mixtures of two or more gases have not been studied very extensively. Measurements have been made of the cathode fall, current density, length of cathode dark space, and of the gradient in the positive column, usually in mixtures of two gases.

α) Cathode region. Some of the earliest experiments were performed by WARBURG[2], who observed that the addition of traces of O_2 and H_2O to a discharge in N_2 caused a reduction of about 25% in the cathode fall. When water vapour is present in appreciable concentration, however, the cathode fall is increased. He also observed that complete removal of O_2 by means of a getter reduced the cathode fall in a nitrogen discharge by over 30% (Pt electrodes), and similarly in a hydrogen discharge. The extent of the reduction depends upon the electrode material.

More systematic measurements of the cathode fall have been made by GÜNTHERSCHULZE[3, 4], in binary mixtures of rare and molecular gases, the results of which are shown in Fig. 87. He also put forward a very simple theory to account for the effects of mixing gases in various proportions. If x is the fraction of gas 1 present, $(1-x)$ that of gas 2, λ_1 λ_2 their respective mean free paths and if an electron makes a total of N collisions per sec. the number N_1 which it makes with molecules of gas 1 is

$$N_1 = \frac{x\,\lambda_2}{\lambda_1+\lambda_2} N. \tag{94.1}$$

GÜNTHERSCHULZE multiplies this number by an average ionisation probability β_1 (β_2 for gas 2) to obtain the number of ionisations, noting that in each ionising impact the electron loses energy eV_{i1} where V_{i1} is the ionisation potential of gas 1. By considering the secondary process (γ) at the cathode, and the condition for sustaining the discharge he derives an expression for the cathode fall of the gas mixture:

$$(V_c)^2_{\text{mixture}} = \frac{\dfrac{\lambda_2}{\lambda_1} + \dfrac{1-x}{x}\,\dfrac{\beta_2}{\beta_1}\,\dfrac{V_{i2}}{V_{i1}}}{\dfrac{\lambda_2}{\lambda_1}\,\dfrac{1}{V_1^2} + \dfrac{1-x}{x}\,\dfrac{\beta_2}{\beta_1}\,\dfrac{V_{i2}}{V_{i1}}\,\dfrac{1}{V_2^2}} \tag{94.2}$$

[1] W. A. LEYSHON: Proc. Phys. Soc. Lond. **44**, 171, 188 (1932).
[2] E. WARBURG: Wied. Ann. **40**, 1 (1890).
[3] A. GÜNTHERSCHULZE: Z. Electrochem. **29**, 370 (1923).
[4] A. GÜNTHERSCHULZE: Z. Physik **21**, 50 (1924).

where V_1, V_2 are the cathode falls of the two pure gases[1]. The theory is very much oversimplified and as can be seen from the comparisons in Fig. 87 it does

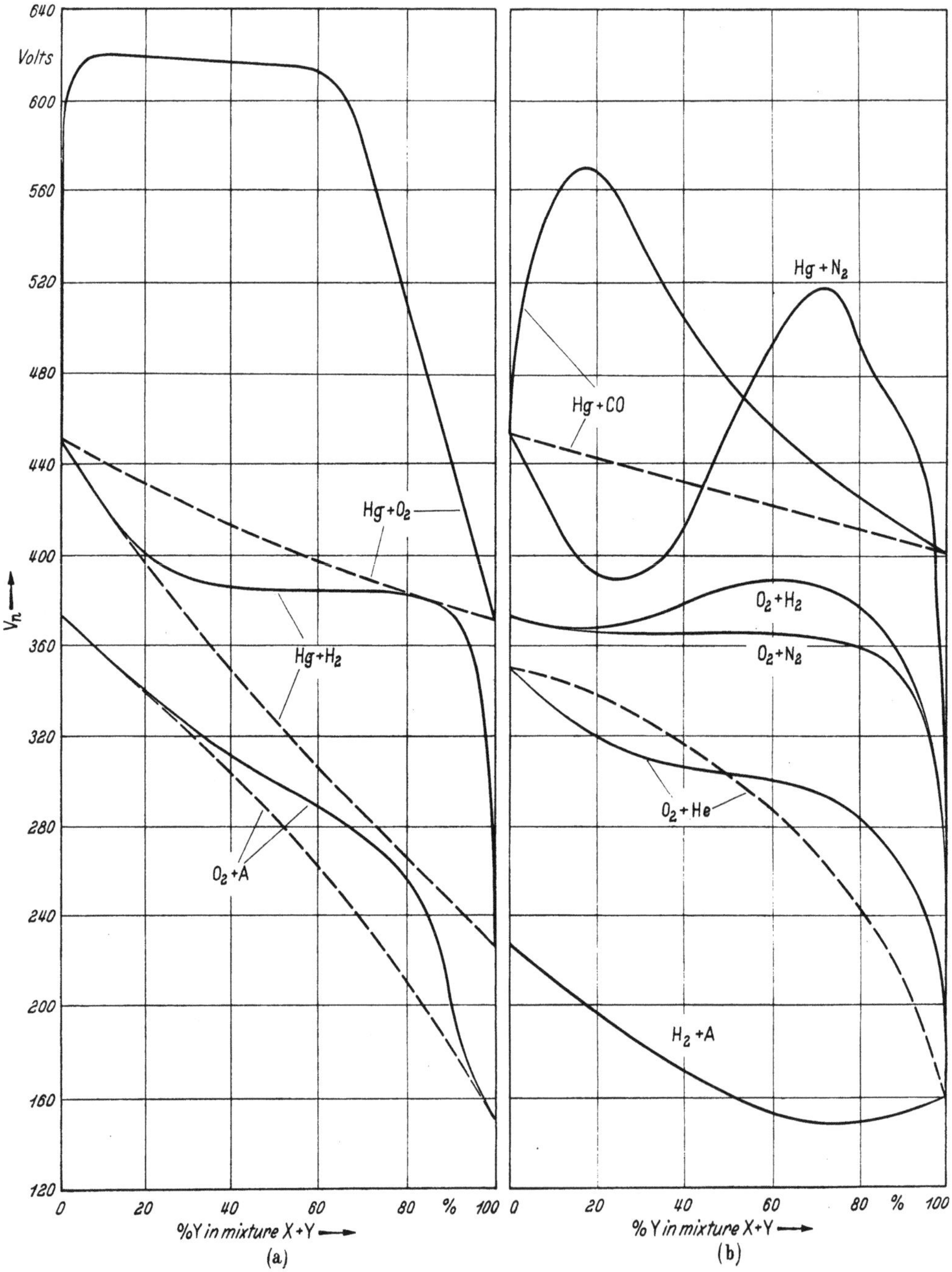

Fig. 87 a and b. The normal cathode fall V_n in various gas mixtures. Full lines measured; dotted lines calculated.

not fit the experimental facts very well, except for a few gases. One major reason for the discrepancy is the very marked change in the cathode surface

[1] The expression for $(V_c)_{\text{mixture}}$ given in the original paper contains a misprint.

caused by quite small concentrations of some gases, especially oxygen, and the associated change in γ. In the $Hg-O_2$ curve the sudden rise in V_c at small concentrations of O_2 is almost certainly a cathode effect: the effect of the properties of oxygen in the gaseous form becomes apparent only when it is present in large proportion (>60%).

Measurements of the normal cathode fall in Ne—A mixtures have been made more recently by PENNING and MOUBIS [53] and MEISSNER [61], whose results have already been discussed in Fig. 9. Similar measurements have also been made by CHRISTYAKOV[1] in mixtures of Ne—A and He—A.

The normal current density, length of cathode dark space and cathode fall in mixtures of N_2 and H_2 have also been measured by SEELIGER and REGER[2].

β) Positive column. The gradient in the positive column of gas mixtures has been measured by several workers, mostly in mixtures containing two rare gases, or a rare gas with mercury or an alkali metal vapour. HEADRICK and DUFFENDACK[3] studied the mixtures He—Ne, He—A, Ne—A, He—Hg, Ne—Hg, A—Hg, both electrically and spectroscopically, taking great care to maintain the purity of all the gases. (Pressure range 5 to 30 mm. Hg, currents 20 to 40 mA).

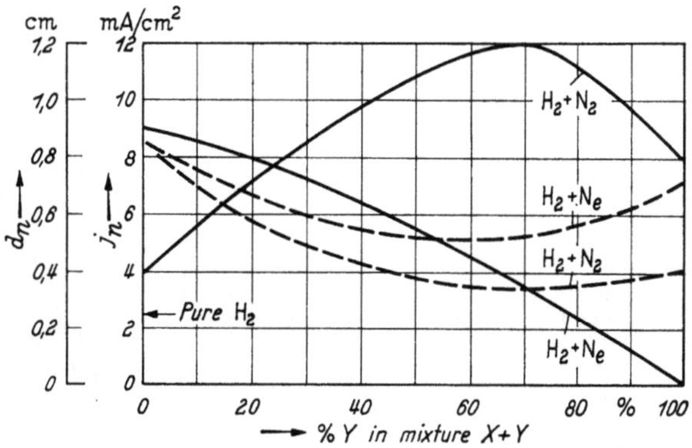

Fig. 87c. The normal current density j_n, and length of cathode dark space d_n in various gas mixtures. Full lines j_n, dotted lines d_n.

Beginning with a discharge in a pure gas, they found that the addition of a very small quantity of a different gas sometimes causes the gradient to change rapidly while the spectrum emitted becomes almost entirely that of the added gas. This occurs when the metastable states of the main gas have energies slightly greater than the ionisation or excitation levels of the second gas. Collisions of the second kind then cause the second gas to be ionised and excited. The effect can be seen for example in helium when very small amounts of neon or argon are added: here the field increases appreciably and the discharge emits predominantly the arc spectrum of neon or argon. Mercury, added in small quantities to neon causes a large decrease in the gradient, but when added to argon or helium produces a smaller increase. An increase in the gradient occurs when the metastable levels of the main gas are close to a number of excited levels of the added gas. Excitation then becomes very probable and energy which would have gone into ionisations of the pure main gas becomes diverted into radiation. A decrease in the gradient occurs when collisons of the second kind result primarily in ionisation of the added gas. Some results are shown in Fig. 88.

When, however, the metastable levels of the main gas lie below the excitation and ionisation potentials of the added gas then no pronounced effects occur when small quantities of the second gas are added. With increasing concentration the gradient of the column changes smoothly between those characteristic of

[1] P. N. CHRISTYAKOV: J. techn. Phys. USSR **19**, 1154 (1949).
[2] R. SEELIGER and M. REGER: Ann. Phys. **83**, 535 (1927); **86**, 1089 (1928).
[3] L. B. HEADRICK and O. S. DUFFENDACK: Phys. Rev. **37**, 736 (1931).

the two pure gases, in a manner roughly proportional to their relative concentrations. An example is the addition of helium to pure neon. (Fig. 88a, reading the abscissa from right to left.)

A trace of diatomic gas (e.g. N_2) added to a pure rare gas or a rare gas mixture can also have a pronounced effect: for example, a trace of nitrogen more than doubles the gradient in pure argon. Here, metastable argon atoms excite the N_2 band spectra by collisions of the second kind.

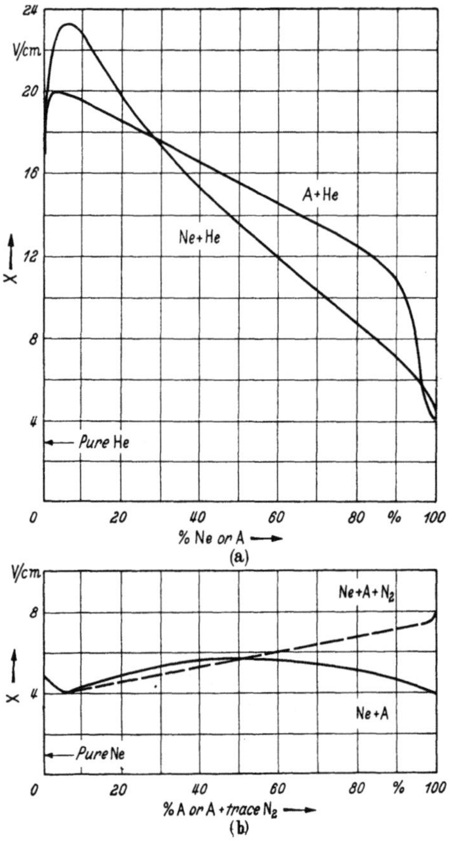

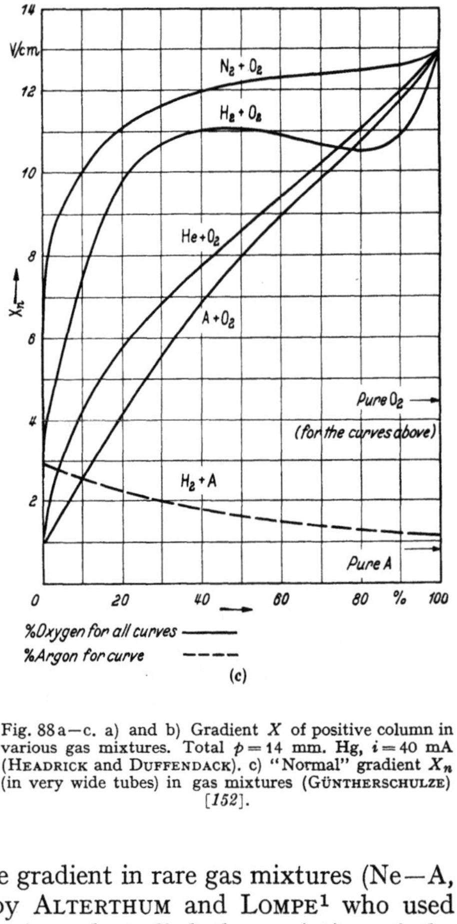

Fig. 88a—c. a) and b) Gradient X of positive column in various gas mixtures. Total $p = 14$ mm. Hg, $i = 40$ mA (HEADRICK and DUFFENDACK). c) "Normal" gradient X_n (in very wide tubes) in gas mixtures (GÜNTHERSCHULZE) [152].

Further extensive measurements of the gradient in rare gas mixtures (Ne—A, Ne—Na, Ne—Hg, A—Hg) were made by ALTERTHUM and LOMPE[1] who used tubes of various diameters (20 to 34 mm.), and studied the variation of the gradient with currents in the range 10 to 200 mA. Some typical results are given in Fig. 89. More recent measurements are due to ZAITSEV[2], for the mixtures Ne+A; Ne+Hg, A+Hg; see Fig. 90.

A theoretical treatment of the electron temperature in the positive column n mixtures of atomic gases has been given by DORGELO, ALTING and BOERS[3]. It is an extension of SCHOTTKY's theory, applying only to the pressure range where diffusion is a predominant process. They assume that the electrons have a MAXWELLian energy distribution, and that ions are produced only by direct electron collisions with gas molecules, and destroyed only by recombination at the walls.

[1] H. ALTERTHUM and A. LOMPE: Ann. Phys. **31**, 1 (1938).
[2] A. A. ZAITSEV: J. techn. Phys. USSR. **18**, 949 (1948).
[3] H. B. DORGELO, H. ALTING and C. J. BOERS: Physica, Haag **2**, 959 (1935).

If q_1, the number of ions of gas 1 formed per cm.³ per sec. is given by

$$q_1 = \alpha_1 N_e \tag{94.3}$$

where N_e is the electron concentration, they derive the equation, analogous to SCHOTTKY'S

$$\frac{k\,T_e}{e}\left(\frac{d^2 N_e}{d r^2} + \frac{1}{r}\,\frac{d N_e}{d r}\right) + \left(\frac{\alpha_1}{b_1} + \frac{\alpha_2}{b_2}\right)\cdot N_e = 0 \tag{94.4}$$

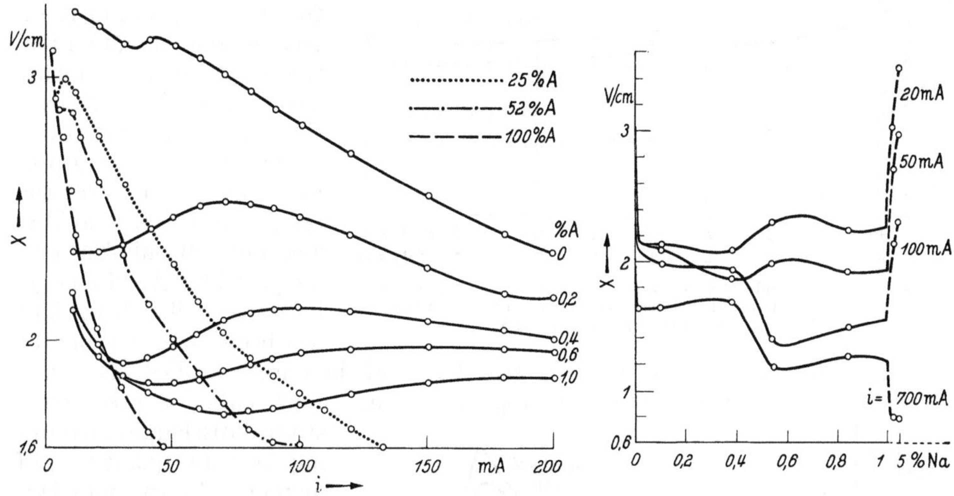

Fig. 89. Gradient of the positive column in mixtures of Ne—A, and Ne—Na (Ne at $p = 2$ mm. Hg) showing its dependence on the current and on the amount of added gas. Diameter 2 cm. (ALTERTHUM and LOMPE).

where b_1, b_2 are the ion mobilities. The solution is a zero order BESSEL function, which together with the boundary condition $N_e = 0$ at the wall $(r = R)$ leads to

$$\frac{\alpha_1}{b_1} + \frac{\alpha_2}{b_2} = \frac{k\,T_e}{e}\left(\frac{2\cdot 4}{R}\right) = \frac{2}{3}\,\overline{V}\cdot\left(\frac{2\cdot 4}{R}\right)^2 \tag{94.5}$$

where $e\overline{V}$ is the mean energy of the electrons. In general [see Eq. (51.5)]

$$\alpha = a\,p\,\frac{m}{e}\,\frac{4}{\sqrt{\pi}}\left(\frac{2k\,T_e}{m}\right)^{\frac{3}{2}}\left(1 + \frac{1}{2}\,\frac{e\,V_i}{k\,T_e}\right)e^{-\frac{e\,V_e}{k\,T_i}}$$

where V_i is the ionisation potential of the gas, and these equations suffice to determine $\overline{V}$ i.e. T_e. Qualitatively it is clear that the rate of ionisation depends strongly on V_i/T_e and hence the electron temperature will be low in gases having a low ionization potential. Fig. 91 shows the calculated values in reasonable agreement with their probe measurements of the mean electron energy.

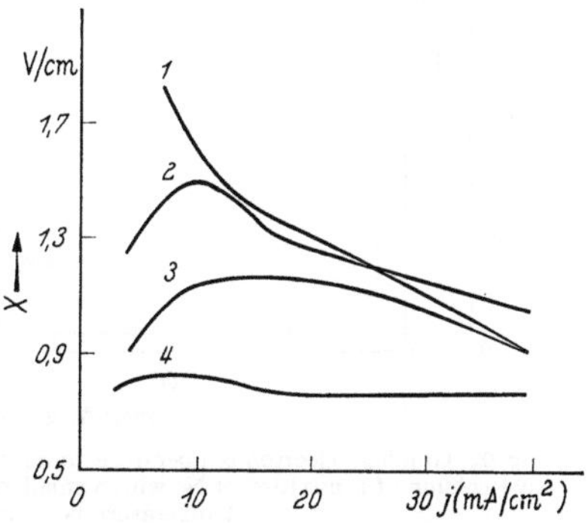

Fig. 90. Gradient of positive column in A—Hg mixtures (ZAITSEV). Diameter 2.8 cm.

Curve	p_{argon} (mm. Hg)	% Hg
1	7.5	$2.67\cdot 10^{-3}$
2	2.14	$9.4\cdot 10^{-3}$
3	0.8	$2.5\cdot 10^{-2}$
4	0.56	$3.55\cdot 10^{-2}$

Discharges in rare gases containing a little metal vapour show characteristic changes as the temperature of the wall, i.e. the pressure of the metal vapour, is increased. Perhaps the most striking example is the mixture Ne + Na, studied by UYTERHOEVEN and VERBURG[1], the behaviour of which is shown in Fig. 92. At low temperatures (200° C) hardly any sodium light is emitted, and only near the walls: the gradient and electron temperature are characteristic of pure neon. At 250° C light of both gases is emitted strongly, that of sodium still only near the wall, the gradient increases to a maximum and the electron temperature varies considerably over the cross section of the tube. At higher wall temperatures the gradient and electron temperature fall rapidly, and probably only Na^+ ions are present in the discharge. Exactly similar behaviour has been observed in the

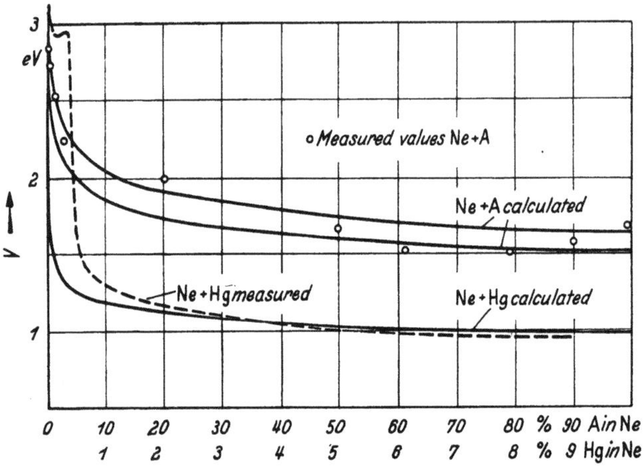

Fig. 91. Calculated and measured values of electron temperature (mean energy) in Ne—A and Ne—Hg mixtures (DORGELO, ALTING and BOERS). Neon pressure 5 mm. Hg. Current 1.5 A.

mixtures He—Na, A—Na, Ne—Hg, He—Cs[1], and the effects studied also when alternating potentials (50 c/s) are applied to the tube[2]. Other workers have studied discharges containing minute quantities of mercury ($\sim 10^{-2}$ mm. Hg) in a rare gas[3], mainly with regard to their spectroscopic properties. Such discharges can in certain conditions have an inner core emitting light of the rare gas (e.g. red neon) surrounded by a mantle of blue mercury light. SIEBERTZ[4] has given a comprehensive account of their properties considering probe and other electrical measurements as well as their light output. The metal vapour has an ionisation potential very much lower than that of the rare gas

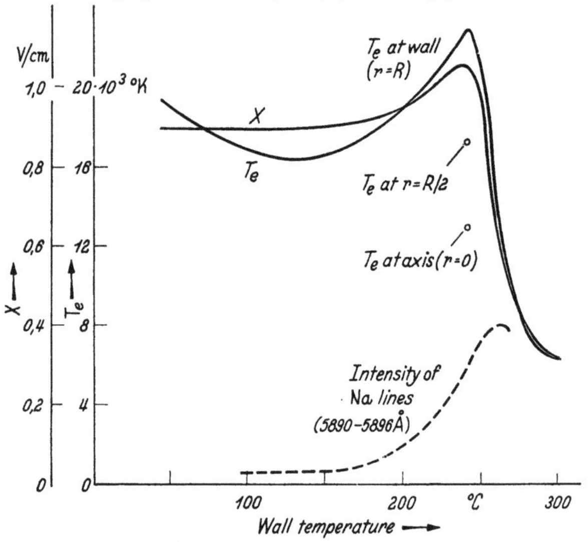

Fig. 92. Gradient, electron temperature, and intensity of Na lines in the positive column of a mixture of Ne with a small proportion of Na, as the wall temperature is varied.

(e.g. Na 5.1 volts, Hg 10.4 volts, Ne 21.5 volts) and is thus much more easily ionised. When the walls of the tube are relatively cool the metal atoms are

[1] W. UYTERHOEVEN and C. VERBURG: C. R. Acad. Sci. Paris **200**, 536 (1935); **202**, 1498 (1936). See also W. UYTERHOEVEN: Elektrische Gasentladungslampen (1938).

[2] W. UYTERHOEVEN and C. VERBURG: C. R. Acad. Sci. Paris **201**, 647 (1935); **207**, 1386 (1938); **208**, 503. 1939.

[3] W. GERLACH and K. SIEBERTZ: Z. Physik **91**, 37 (1934). — K. SIEBERTZ: Z. Physik **101**, 255 (1936). — F. M. PENNING: Physica, Haag **1**, 763 (1934). — H. SCHOBER: Z. techn. Phys. **16**, 67 (1935).

[4] K. SIEBERTZ: Ann. Phys. **27**, 421 (1936).

ionised as rapidly as they evaporate, and are driven back to the walls by the radial electric field. The spectrum of the metal vapour thus appears first near the walls. Even when the rate of evaporation is increased the outward radial flow of metal ions (electrophoresis, see Sect. 98) maintains a very large partial pressure of metal vapour near the walls.

95. Intermittent discharges. Oscillations. Oscillations in discharges can be divided broadly into two types: those whose frequency is determined by the external circuit in which the discharge is connected, and those whose frequency is, within wide ranges, independent of the circuit, in particular, independent of any condenser connected parallel to the discharge. Oscillations of the first kind are continuously variable in frequency with changes in external circuit elements (resistance, electromotive force, capacity); the discharge current becomes zero for an appreciable part of the cycle, i.e. the discharge is intermittent, and the behaviour can partly be understood by considering the characteristic curve [V as $f(i)$] of the discharge. Oscillations of the second kind are usually relatively small variations in the steady current, and arise from the internal structure of the discharge e.g. motion of space charges. Different modes of oscillation can occur in different ranges of current and voltage[1], so that even these oscillations are not entirely independent of the external circuit, but changes between the modes occur discontinuously with changing current, and bear no relation to the stability conditions of the discharge[2], as do oscillations of the first kind.

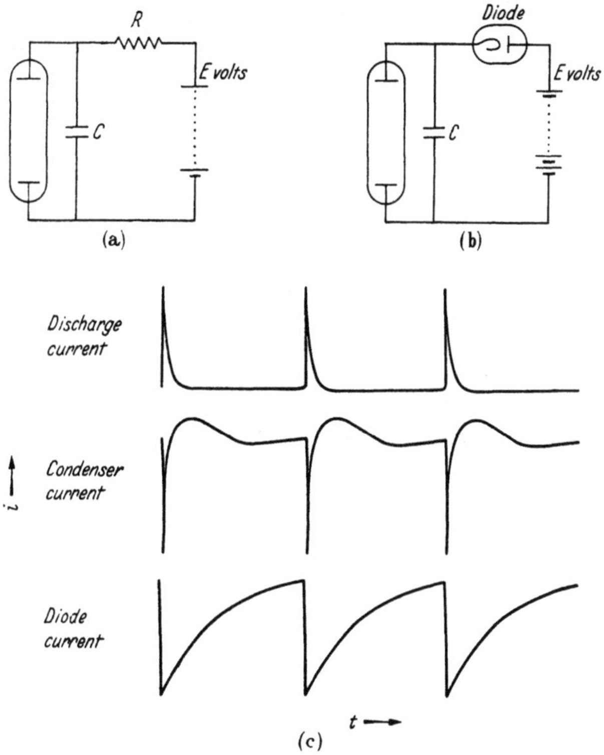

Fig. 93a—c. a) and b) Circuits producing intermittent (relaxation) oscillations. c) Currents flowing in various branches of circuit b).

When a discharge tube is connected in the circuit of Fig. 93a the discharge is often observed to be intermittent: in certain circumstances this can occur even when the condenser C is omitted. The effect was first noticed by GASSIOT in 1863, and studied later by HITTORF, RIGHI and many other workers. A full account of this work (before 1925) has been given by VALLE [*296*], and also in GEIGER-SCHEEL's "Handbuch der Physik" (1927).

The frequency of intermittence varies with the values of the resistance R and condenser C in the external circuit, and values have been measured from a few cycles per second to 150 kc/s. Audible frequencies have been measured by

[1] See e.g. DONAHUE and DIEKE [*194*], and Fig. 61.

[2] See Sect. 99, also the very full treatment of stability conditions by R. BÄR, GEIGER-SCHEEL's Handbuch der Physik, vol. 14, 1927, and the articles of R. SEELIGER and G. MIERDEL, Handbuch der Experimentalphysik, vol. 13, part 3, 1929.

connecting a telephone in the circuit, higher frequencies by a calibrated radio receiver or oscillator placed nearby. In more recent work not only the frequency, but also the wave form and the current or potential in various branches of the circuit have been measured by the use of the cathode ray oscilloscope.

The oscillations were explained simply by RIGHI[1] as being due to the repeated charging and discharging of the condenser C. The discharge tube is regarded essentially as a short-circuiting switch, which switches on when the condenser has been charged up to a potential V_s, its starting potential. The condenser is then rapidly discharged through the conducting gas until its potential has fallen to the extinction potential of the tube V_e, when the gas ceases to conduct and the condenser is charged again from the battery. RIGHI deduced that the period T, of these relaxation oscillations should be

$$T = CR\frac{E - V_e}{E - V_s} \tag{95.1}$$

where E is the electromotive force of the battery. This is based on the assumption that the time of development of the discharge is negligible compared with the interval of charging. The formula was verified experimentally by SCHALLREUTER[2]. Other workers have used the circuit of Fig. 93b in which the charging current of the condenser is the saturation current of the diode, and can be controlled by changing the temperature of the filament. Its value is usually assumed to be constant, leading to simpler circuit equations, but it is doubtful if this is legitimate. It is, however, always true that if $\bar{i}$ is the mean current flowing from the battery, then

$$\bar{i}T = C(V_s - V_e) \tag{95.2}$$

again assuming that the condenser is discharged almost instantaneously[3]. PENNING[4] verified this equation, measuring T, the period with a telephone, and V_s and V_e by a system of biased diodes. He found, however, that V_s and V_e were not the usual static values of sparking and extinction potential, and also that the transition from a continuous to an intermittent discharge did not obey KAUFMANN's criterion. It follows that the static V-i curves cannot be applied when the discharge is subjected to sudden changes. GILL [*297*] has studied intermittent discharges in air-filled tubes over a wide range of frequencies and pressures ($f = 0$ to $1.5 \cdot 10^5$ c/s., $p \sim 2$ to 85 mm. Hg), using the circuit of Fig. 93b. The period, determined mainly by the capacity and the circuit current, is given by the law

$$T = \frac{C + C'}{k\,i} \tag{95.3}$$

where k, C' are constants. C' represents the effective self capacity of the discharge; it depends slightly on the current and is larger than the geometrical capacity between the electrodes due to the space charges in the tube. This capacity is, of course, always present, and intermittent discharges can therefore occur even when the external condenser is removed ($C = 0$). The linear relation above also shows that the duration of the discharge is small compared with T (i.e. $\ll 10^{-5}$ sec.).

When the discharge tube becomes conducting the potential across it can fall until the current reaches a value i_1, say, after which the maintenance potential

[1] A. RIGHI: Rend. Accad. Sci. Bologna **6**, 188 (1902).
[2] SCHALLREUTER: Sammlung Vieweg **66** (1923).
[3] See for example H. GEFFKEN: Phys. Z. **26**, 241 (1925).
[4] F. M. PENNING: Phys. Z. **27**, 187 (1926).

is independent of the current. If the circuit current i, controlled by the diode, is less than i_1 the discharge goes out, but if $i > i_1$ the discharge falls into a continuous state. This explains the observation by GILL and others that the discharge is intermittent only when the circuit current is small. The critical current is larger at higher pressures; at $p \sim 2$ mm. Hg $i_{\text{crit}} \sim 10^{-4}$ A. These results have been confirmed by oscillographic studies by MITRA and SYAM[1] [298]; they show, however, that the current through the diode is not constant (see Fig. 93c). When a series resistance R is present they find that for relaxation oscillations to occur it must have a minimum value, given approximately by

$$\frac{R}{r} = \frac{E - V_e}{V_e} \tag{95.4}$$

where r is an "effective resistance" of the discharge, usually about a few thousand ohms. A similar result was noted by TAYLOR and CLARKSON[1]. Recently RITOW[2] studying an intermittent discharge in hydrogen has found the relation

$$\bar{i} \cdot T - C(V_s - V_e) = Q_0 + i_d \cdot T \tag{95.5}$$

where $\bar{i}$ is, as usual, the steady current delivered from the source, i_d is the "dark" current which flows between flashes of the discharge, and Q_0 is a constant, thought to be the net charge leaving the electrodes in each flash. i_d can sometimes be negative, especially at low pressure, rising to 5 to 20 times the charging current. This effect is believed to be caused by oscillations of a different kind in the discharge itself.

The experiments and theories already described do not explain what part the discharge actually plays in sustaining these oscillations, apart from the existence of starting and extinction potentials. Attempts have been made by other workers to explain the effect by considering the stability conditions of the discharge at the working point where the "resistance line"

$$V = E - R\,i \tag{95.6}$$

cuts the V-i curve of the discharge. The well known static curves do not apply, however, and the dynamic characteristics vary with the frequency of intermittence, i.e. with the values of the external circuit constants[3]. Oscillograms of voltage and current in the discharge branch, displayed simultaneously on perpendicular axes, have shown V-i curves with pronounced hysteresis, sometimes almost circular in shape. This is certainly due to the time of formation and decay of space and wall charges in the tube, but no simple theory can be given.

Dynamic characteristics have been measured also in non-intermittent circuits, for example by adding a relatively small sine-wave fluctuation to the static potential[4], or by causing abrupt changes in the discharge current by external

---

[1] For many other experiments and further references on intermittent discharges, studied by a variety of methods see: J. TAYLOR and W. CLARKSON: Proc. Phys. Soc. Lond. **36**, 269 (1924). — Phil. Mag. **49**, 336 (1925). — W. CLARKSON: Phil. Mag. **4**, 121, 849, 1002, 1341 (1927); **7**, 322 (1929). The effect of an inductance in series with C is discussed here and also by W. E. KOCK: Physica, Haag **4**, 359 (1933). — L. F. RICHARDSON: Proc. Roy. Soc. Lond. **162**, 293 (1937).

[2] H. RITOW: Phys. Rev. **86**, 130 (1952).

[3] H. GAWEHN: Ann. Phys. **20**, 601 (1934). — H. GAWEHN and G. VALLE: Ann. Phys. **23**, 381 (1935). — E. HAAK: Ann. Phys. **84**, 119 (1927).

[4] H. Y. LOH and G. H. DIEKE: J. Opt. Soc. Amer. **37**, 837 (1947).

electronic circuits[1]. The work on this very complicated subject has been reviewed by VALLE [*299*]. An extensive treatment with many references has more recently been given by MOUSSIEGT [*300*], [*301*]. He regards intermittence as due to the inaccessibility of the working point (P, Fig. 94): with increasing potential the operating point of the discharge jumps from A to A', with decreasing potential from B to B' so that the regime AB (in which P lies) is never traversed at all. MOUSSIEGT considers the different behaviour of the circuit for small and large condensers, and also derives expressions for the time of formation of the discharge.

Special kinds of intermittent discharges. The obstructed glow discharge is often intermittent, the period of the oscillations being particularly sensitive to the distance apart of the electrodes when this is nearly equal to the length of the cathode dark space. The oscillations become much slower as the anode is moved slightly nearer the cathode: the interval between flashes can be several seconds. This is called the *throbbing discharge* [*3*]. The effect is due partly to charges which accumulate on the side walls during the periods while the discharge is passing: they set up opposing fields which quench the discharge, after which they disperse and the discharge starts again. This also explains why the current in an obstructed discharge decreases steadily after it has been switched on: if the polarity is then reversed a large current will pass for a short time in the opposite direction, pass through a minimum and then increase slightly to its steady value. The discharge may pass through a throbbing stage as the current slowly changes, and in this state it is very sensitive to any external changes e.g. touching the tube may cause the throbbing to cease or may alter its rate. The process is analogous to polarisation of the electrodes when current passes through an electrolyte[2].

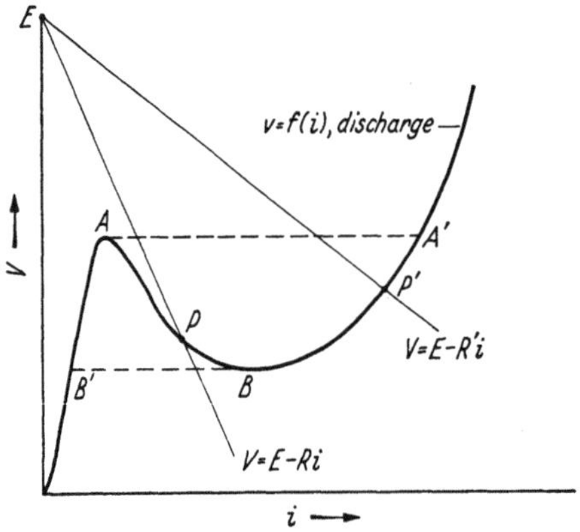

Fig. 94. Illustrating stability of a discharge. The working point is given by the intersection of the resistance line and the discharge curve. Circuit of Fig. 93a.

These oscillations may also be due partly to changes in the cathode surface, and hence in γ, while the discharge is passing. Removal of gas layers may reduce the efficiency of the secondary processes and cause the discharge to cease until they have formed again. Impurities in the gas, notably water vapour, also greatly affect the period of oscillation: this may be due to the conductivity produced by a film of water on the walls.

The accumulation and dispersal of wall charges were also shown to be responsible for the intermittent nature of a special kind of discharge in a magnetic field, first observed by RIGHI, and later explained by IVES. The electron stream

[1] B. T. BARNES: Phys. Rev. **86**, 351 (1952). — B. T. BARNES and S. EROS: J. Appl. Phys. **21**, 1275 (1950). — V. L. GRANOVSKY and V. A. OKHLOPKOV: C. R. Acad. Sci. URSS. **32**, 489 (1941). See also H. N. OLSEN and W. S. HUXFORD: Phys. Rev. **87**, 922 (1952), who measure fields in a rapidly changing Ne—A discharge by the STARK effect broadening of spectrum lines.

[2] Other examples of wall charges quenching a discharge, when the gas pressure is very low have been quoted by A. GÜNTHERSCHULZE: Z. Physik **31**, 606 (1925). E. BADAREU: Z. techn. Phys. **12**, 298 (1931).

from the cathode region was bent into a side tube, where a fluctuating luminosity was sometimes seen[1,2] (see also Sect. 88).

An unusual kind of interrupted discharge, called the *flashing discharge* has been observed in certain gas mixtures, notably Hg + A, with a little N_2 impurity, by KENTY[3]. (Hg at 27° C, A at 3 mm. Hg.) When the impurity is added the normal running striations disappear, and if the potential is considerably increased (e.g. by a factor 5) the discharge flashes showing both striations and large current fluctuations. KENTY believes that in these conditions the uniform column is not self sustaining and as ions diffuse away the average gradient rises and localises, forming pockets of potential in which electrons can gain 12 or 14 eV energy. These excite argon atoms, producing metastables and resonance radiation which ionise the mercury: intense light and a large current results, the gradient falls and the process is repeated. The oscillation is much less marked in the mixture Kr + Hg, and does not occur at all in X + Hg, because X* cannot ionise Hg.

Plasma oscillations. Oscillations which are essentially determined by displacements of space charge in the discharge itself and not by the external circuit were first observed, at about the same time by APPLETON and WEST[4] [striated discharge, in air (?)], CLAY[5] (H_2, He) and PENNING[6] (A), and later by PARDUE and WEBB[7] (air). Frequencies were observed varying from 10^3 to more than 10^6 c/s, depending upon the gas pressure, and also upon the presence of impurities in the gas.

The oscillations are thought to arise in their simplest form by the macroscopic displacement of a group of charges (ions or electrons) from their equilibrium position. The restoring force due to electrostatic attraction causes them to undergo simple harmonic motion. J. J. THOMSON [*302*] derived their frequency as

$$f = \sqrt{\frac{\frac{N_e \cdot e^2}{m} + \frac{N^+ e^2}{M}}{\pi}} \tag{95.7}$$

where N_e, m, N^+, M are the average concentrations and the masses of electrons and positive ions. TONKS and LANGMUIR [*303*] derive rather similar expressions, but discriminate between electronic oscillations which do not transmit energy, of natural frequency

$$f_e = \sqrt{\frac{N_e \cdot e^2}{\pi m}} \sim 10^9 \text{ c/s} \tag{95.8}$$

and positive ion oscillations which are much slower, and whose frequency varies with their wavelength λ, according to the relation

$$f_+^2 = \frac{f_0^2}{1 + f_0^2 \frac{\lambda^2 M}{k T_e}} \tag{95.9}$$

where

$$f_0 = \sqrt{\frac{N^+ e^2}{\pi M}}\,.$$

[1] A. RIGHI: Phys. Z. **11**, 158 (1910); **15**, 529 (1914).
[2] J. E. IVES: Phys. Rev. **9**, 349 (1917).
[3] C. KENTY: Phys. Rev. **80**, 96 (1950).
[4] E. V. APPLETON and A. G. D. WEST: Phil. Mag. **45**, 879 (1923).
[5] R. E. CLAY: Phil. Mag. **50**, 985 (1925).
[6] F. M. PENNING: Nature, Lond. **118**, 301 (1926). — Physica, Haag **6**, 241 (1926).
[7] L. A. PARDUE and J. S. WEBB: Phys. Rev. **32**, 946 (1928).

When the electron temperature T_e, is high, the frequency approaches f_0, analogous to electron oscillations. When the electron temperature is low, the electron concentration follows closely the periodic changes in ion concentration, and the waves are very like sound waves with a velocity

$$v^+ = \sqrt{\frac{k\,T_e}{M}} \sim 10^4 \text{ to } 10^5 \text{ cm./sec.}$$

in typical discharges. Thus for positive ion oscillations there is an upper limit of frequency, f_0, usually a few Mc/s, and a lower limit determined by v^+ and the dimensions of the tube. Similar expressions, with the addition of terms to include the random motion of the ions, have been deduced by J. J. and G. P. THOMSON[1]. LANGMUIR's equation for electron oscillations has also been deduced from a combination of the equations of diffusion and drift, with POISSON's equation, by SEELIGER[2]. These theoretical predictions have been largely verified, except that the upper frequency limit of electron oscillations is often much higher than that predicted[3].

Experiments have shown that a great variety of plasma oscillations occurs in discharges; they cover a wide range of frequencies, are sometimes progressive and sometimes standing waves, and are often though not always accompanied by striations[4]. Their frequencies, velocities and amplitudes are more varied than can be described by the simple theories just given. Such oscillations are found not only in the glow discharge, but in hot cathodes tubes, and arcs, probably in the atmosphere of the sun, indeed wherever a plasma with sufficiently high charge concentrations exists. No concise account of this widespread phenomenon can be given; the subject is in a state of rapid development both experimentally and theoretically, and we shall here mention only the main papers.

The theory of electron plasma oscillations has been extended by BORGNIS [*304*] who with LINDER[5], SEELIGER and STEENBECK[6] considered amongst other factors, the effect of electron pressure.

GABOR[7] has calculated the momentum and energy exchange of electrons with plasma waves. and together with ASH and DRACOTT[8] has measured by the deflection of an electron beam high frequency oscillations ($f \sim 10^8$ c/s) in the wall sheath of a low pressure discharge. Such an oscillating sheath exchanges energy very efficiently with electrons which penetrate it. This is believed to explain how electrons acquire a MAXWELLian energy distribution in very low pressure plasmas, where they hit the walls of the vessel more frequently than they hit gas molecules or other electrons.

A comprehensive treatment of plasma oscillations has been given by BOHM and GROSS [*305*], who removed many of the simplifying assumptions of earlier treatments. They consider not only the equations of steady oscillations but also their origin and growth, and the damping of oscillations by collisions between electrons and gas molecules. They discuss also the growth of oscillations in beams having a narrow velocity range[9]. A theory on similar lines although

[1] [*3*], Vol. 2, p. 353—358.

[2] R. SEELIGER: Z. Physik **118**, 618 (1942).

[3] See for example E. B. ARMSTRONG: Nature, Lond. **160**, 713 (1947).

[4] See Sect. 67, 68.

[5] E. G. LINDER: Phys. Rev. **49**, 645, 763 (1936).

[6] R. SEELIGER and M. STEENBECK: Z. Physik **113**, 526 (1939).

[7] D. GABOR: Proc. Roy. Soc. Lond., Ser. A **213**, 73 (1952). — Z. Physik **84**, 474 (1933).

[8] D. GABOR, E. A. ASH and D. DRACOTT: Nature Lond. **176**, 196 (1955).

[9] H. J. MERRIL and H. W. WEBB attribute the scattering of an electron beam traversing a plasma primarily to plasma oscillations deriving their energy from the fast electrons. Phys. Rev. **55**, 1191 (1939).

less extensive was developed by PIERCE, who also verified experimentally some of the results[1]. BAILEY has treated the spontaneous growth of waves in discharges[2] while SCHUMANN has analysed mathematically the formation of periodic distributions in regions containing freely moving ions and electrons[3], and the effect of applying sudden pulses to a plasma[4]. The effect of an applied external electric field on the oscillations has also been considered: HASIKUNI[5] has derived the dispersion relations in these conditions, and LANDAU[6] has considered the penetration of a periodic external field into a plasma, and also the frequency and damping of normal oscillations arising from an initial non-equilibrium state.

A complete review has been given by GABOR [*310*] and a review of the lower frequency oscillations due to the motion of positive ions by ARMSTRONG, EMELEUS and NEILL [*306*] who describe also experiments to produce plasma-ion waves by external stimuli. It was found, however, that small amplitude oscillations tended to be masked by electrical noise generated by the discharge[7].

For other references we refer to an earlier review[8], and also to the section on the striated positive column. Plasma oscillations can be generated also by the presence of a magnetic field (see Sect. 92), and are often connected with instability of anode spots.

96. The high pressure glow discharge. A glow discharge can exist at pressures up to and above atmospheric. The discharge is then contracted both axially and radially, and is independent of the walls of the tube. The properties of the cathode region appear at first sight no longer to obey similarity, since the current density is less than if j/p^2 were constant[9]. It has been shown that this is due to the heating of the gas and consequent change in gas density[10]. Since the current density is roughly proportional to p^2 and the length of the cathode dark space to $1/p$ the energy dissipated in a volume element close to the cathode increases as p^3. Assuming that the specific heat of the gas is proportional to the temperature, then the temperature of the gas in the cathode dark space can be calculated, and is found to be very high at atmospheric pressure ($\sim 10^3$ °K). It follows that j should be proportional to $p^{\frac{4}{3}}$, which is in good agreement with experiment. When the gas density is used the similarity laws hold up to 1 atmosphere.

GAMBLING and EDELS[11] have also measured the current density of a glow discharge in air at 1 atmosphere pressure, and discuss methods of measuring the cathode fall.

The contraction of the positive column at pressures above a few cm. Hg is a quite independent effect, the explanation of which is still not completely settled. The effect is not predicted by SCHOTTKY's theory, and it has been suggested that this is due to an over-simplification of the equations and boundary conditions. Attempts have been made to solve the problem by imposing boundary conditions

[1] J. R. PIERCE: J. Appl. Phys. **19**, 231 (1948). — Phys. Rev. **76**, 565 (1949).

[2] V. A. BAILEY: Nature, Lond. **161**, 599 (1948).

[3] W. O. SCHUMANN: Z. Physik **121**, 11, 629 (1943).

[4] W. O. SCHUMANN: Ann. Phys. **43**, 369 (1943).

[5] M. HASIKUNI: J. Sci. Hiroshima Univ. A **17**, 129 (1953).

[6] L. LANDAU: J. Phys. USSR **10**, 25 (1946). See also Phys. Z. Sowjet **10**, 154 (1936) for earlier work.

[7] See Sect. 13. This has been studied in some detail also by J. D. COBINE and C. J. GALLAHER: J. Appl. Phys. **18**, 110 (1947). — J. Franklin. Inst. **243**, 41 (1947).

[8] R. ROMPE and M. STEENBECK: Ergebn. exakt. Naturw. **18**, 257 (1939). A recent paper on plasma oscillations is R. A. BAILEY and K. G. EMELEUS: Proc. Roy. Irish Acad. A **57**, 53 (1955).

[9] H. THOMA and L. HEER: Z. techn. Phys. **13**, 404 (1932); **14**, 385 (1933).

[10] v. ENGEL, SEELIGER and STEENBECK: Z. Physik **85**, 144 (1933); also R. SEELIGER and K. BOCK: Phys. Z. **34**, 767 (1933). — O. BECKEN and R. SEELIGER: Ann. Phys. **24**, 609 (1935).

[11] GAMBLING and EDELS: Brit. J. Appl. Phys. **5**, 36 (1954), where most of the literature is given.

determined by the properties of wall sheaths[1], and by including volume recombination and two stage ionisation (see Sect. 55). Neither of these gives solutions appreciably different from the simpler theory. Other workers have suggested that heating of the gas near the axis[2], or the self magnetic field of the current causes the contraction[3]. Consideration of gas heating results in complex equations which are difficult to check experimentally, and although the self magnetic field is known to cause contraction in high current discharges, contracted columns can exist with very low currents.

Recently, FOWLER[4] has suggested a new reason, namely that inelastic collision between atoms and electrons are responsible not only for energy losses but also for a directional concentrating effect on the electrons, due to deceleration during such collisions. This can be expressed in the equations as a macroscopic force

$$F \propto \sum e\, V_m \sigma_m N,$$

where $e V_m$ is the mean energy loss in a collision of type m, σ_m the cross section, and N the density of neutral atoms. The resulting solution of the lateral electron and ion distribution N_x/N_0 (for the plane column) is a damped cosine function, which changes shape abruptly, over a certain small range of electron temperature, becoming more concentrated at the centre. There is reasonable agreement with experiment in helium.

POLETAEV[5] has calculated that there is an upper limit for the product pR (pressure $\times$ radius), above which the column becomes unstable and either extinguishes or possibly contracts. The results show a systematic disagreement with experiment in Hg and the rare gases.

97. Alternating glow discharges. Glow discharges to which an alternating potential is applied (usually 50 c/s) are widely used in technical applications, but little work has been done on the fundamental properties of the discharge. Dynamic characteristic curves have been measured, which are quite different from the static curves, showing marked hysteresis. The current lags behind the applied potential, due to the time necessary for ionisation to build up or decay. Consequently, the dynamic curves vary considerably with frequency[6]. Due to the residual ionisation left in the gas when the potential falls, re-ignition in the opposite direction is made easier, and when the cycles are repeated rapidly the working potential is reduced, especially when the current is large. The effect of residual ionisation was noted (at 50 c/s) in the high pressure glow discharge[7]. At high frequencies quite low fields suffice to maintain the discharge[8].

UYTERHOEVEN and VERBURG[9] measured the electron temperature in the positive column of 50 c/s discharges in Ne, and Ne—Na mixtures (Na vapour lamp). They used a LANGMUIR probe which was operated for a small interval at a given phase of the cycle. In pure neon the tube voltage and electron temperature varied approximately as a sine function, and corresponded closely to each other. In the mixtures a similar correspondence existed, but both V and

[1] W. P. ALLIS and D. J. ROSE: Phys. Rev. **93**, 84 (1954).

[2] W. ELENBAAS: The high pressure mercury vapour discharge. Amsterdam: N. Holland Publishing Co. 1951.

[3] P. C. THONEMANN and W. T. COWHIG: Proc. Phys. Soc. Lond. B **64**, 345 (1951).

[4] R. G. FOWLER: Proc. Phys. Soc. Lond. B **68**, 130 (1955).

[5] I. A. POLETAEV: J. techn. Phys. **21**, 1021 (1952).

[6] See for example H. J. REICH and W. A. DEPP: J. Appl. Phys. **9**, 421 (1938) and the literature quoted in Sect. 95.

[7] v. ENGEL, SEELIGER and STEENBECK: Z. Physik **85**, 144 (1933).

[8] F. KIRCHNER: Ann. Phys. **7**, 798 (1930).

[9] W. UYTERHOEVEN and C. VERBURG: C. R. Acad. Sci. Paris **207**, 1386, (1938); **208**, 269, 503 (1939).

T_e showed a much distorted time variation, T_e lagging behind the tube potential and falling sharply close to the wall.

BARNES[1] has measured dynamic characteristics of a low pressure glow discharge (rare gas + Hg) by superimposing square-wave changes upon the steady value of the current. He finds that after an instantaneous increase in current the discharge takes about 0.2 to 0.5×10^{-3} sec. to reach its steady state, while after a decrease in current it takes longer (1 to $3 \cdot 10^{-3}$ sec.) due to the slow diffusion of ions to the walls. The results appear to indicate that during the short period when the ion production is very large the electron density in the positive column does not obey the BESSEL function distribution. The change in current is accompanied by changes in gradient and radiation.

A theory of transient states of the low pressure discharge has been given by GRANOVSKY[2].

Two common forms of low pressure alternating discharge should be mentioned, the ring discharge [*3*] and the electrodeless discharge in a glass vessel placed between plane parallel external electrodes[3]. The latter shows the dark spaces, negative glows etc. characteristic of the glow discharge close to each end wall, and a central glow corresponding to the positive column.

XI. Miscellaneous effects.

98. Separation of gas mixtures. Pressure differences in a discharge (electrophoresis, cataphoresis). When a glow discharge burns for some time in a mixture of gases it is often found that the region around one electrode becomes enriched in one component, while that around the other electrode is correspondingly enriched in the other component. The effect is due to a drift of one gas towards the cathode, when it is termed cataphoresis, but such partial separation can also occur radially e.g. across the positive column, and the general term for the effect is electrophoresis. In a pure gas the passage of a discharge causes a pressure difference to be set up between anode and cathode. This effect appears very similar to the separation of the components of a gas mixture (which can be regarded as the establishment of differences in their partial pressures) but the explanation is not usually the same, and the two phenomena will therefore be treated separately.

α) Gas mixtures. The earliest literature on this subject has been summarised by LEHMANN[4]. Typical qualitative observations are those of J. J. THOMSON[5] who studied the glow discharge in a mixture of H_2 and Cl_2, and BALY[6], who used mixtures of hydrogen and several other gases. They noticed that the H_2 spectrum was most intense at the cathode, and that of the other gas most intense at the anode. THOMSON ascribed his observations to the migration of Cl_2 ions to the anode, but this explanation cannot apply to all the results of BALY. A systematic study of mixtures of rare gases was made by SKAUPY [*287*] who found the general rule that the gas with the lowest ionisation potential becomes more concentrated at the cathode, that with the highest ionisation potential at the anode. In mixtures of three rare gases the one having an intermediate ionisation

[1] B. T. BARNES: Phys. Rev. **86**, 351 (1952).
[2] B. GRANOVSKY: J. Phys. USSR **3**, 195 (1940).
[3] First noted and described by E. WIEDEMANN and EBERT: Wied. Ann. **1**, 1, 221 (1893). See also M. CHENOT: Ann. Phys., Paris **3**, 277 (1948) for a general description of these discharges in cylindrical vessels at high frequencies. High frequency discharges are discussed in another article in this Encyclopedia.
[4] O. LEHMANN: Elektrische Lichterscheinungen, p. 265. 1898.
[5] J. T. THOMSON: Proc. Roy. Soc. Lond. **58**, 246 (1895).
[6] E. C. C. BALY: Phil. Mag. **35**, 200 (1893).

potential has its greatest concentration somewhere near the middle of the tube. These were qualitative observations, the composition of the mixture being judged from the intensity of the spectra in various parts of the tube. Accurate quantitative measurements were made some years later by SKAUPY and BOBEK [*288*]. A known mixture of two rare gases was admitted into the middle of a long narrow tube connecting two spherical bulbs in which the electrodes were placed. When the discharge had been running long enough to establish a steady state, gas was drawn from one or other of the electrode regions by means of a TOEPLER pump and analysed. For some mixtures (e.g. A—He) the analysis could be performed quite accurately (to within $\frac{1}{2}$%) by measuring the refractive index in a gas interferometer[1], but for other mixtures spectroscopic analysis by comparison with known mixtures was necessary. The experiments gave the general result that separation of the components is favoured by low gas pressure and large current density[2] (at least in the range $p \sim 2$ to 16 mm. Hg and $i \sim 0.3$ to 1 A). Some typical results are given below, to illustrate the order of magnitude of the effect:

Table 11. *Helium—argon mixtures.*

Original mixture % argon	Total pressure mm Hg	Current A	Final mixture near anode % argon
8	5	0.5	0
12	4	0.8	0
12	4	0.3	4
25	10	0.5	17
32	10	0.5	29
52	16	0.5	48

The components can be completely separated in helium—neon mixtures, total pressure 2 mm. Hg, with a discharge current of 1.4 A, provided either that the original proportion of helium does not exceed 5%, or that the original proportion of neon does not exceed 2%. There are many other examples of such separation of components, notably when Hg or an alkali metal vapour is mixed with a rare gas, but no comprehensive numerical data.

The explanation of these phenomena is quite straightforward. The gas having the lowest ionisation potential is most strongly ionised: the ions are driven by the electric field to the cathode where they are discharged and remain in the gas phase as neutral molecules, thus enriching the mixture in that region. The composition reaches a steady state when the molecules of a particular kind diffuse back along the discharge as fast as they are transported to the cathode in the form of ions. The mixture becomes enriched in the other component at the anode in consequence, simply due to the removal of the first component, and not due to ions being transported to the anode[3]. It is clear that gases will be more completely separated when their ionisation potentials are widely different. Strictly, it is the ionisation probability rather than the ionisation potential which determines the relative number of ions of the two gases, and one would expect some separation (although perhaps small) amongst gases having almost equal ionisation potentials but different ionisation probabilities.

This effect of cataphoresis has been adapted as a powerful tool for removing minute traces of impurities from the rare gases, especially when the impurity

---

[1] See F. LÖWE: Phys. Z. **11**, 1047 (1910).

[2] More modern work, however, suggests that *high* gas pressure favours separation and this appears to be more probable.

[3] Except possibly in gases like the halogens where many negative ions are formed.

is itself another rare gas. The method, developed by REISZ and DIEKE [289], consists in running a glow discharge in a cylindrical tube connected to the bulb of the gas which is to be purified. The impurity, whatever its nature, becomes more concentrated in part of the discharge tube, from which it can be pumped away. The method is extremely effective, and in certain mixtures (e.g. He+ a trace of Ne) the neon pressure at the cathode is 100 times greater than in the main bulb of gas. This effective multiplication effect can also be used to analyse, by a previously calibrated spectroscopic method, the amount of impurity present in the gas. One part of Ne in $2\cdot 10^8$ parts of He can easily be detected. EMELEUS and SAYERS[1] have also used this method to purify gas samples.

REISZ and DIEKE find that the impurity is separated more effectively with increasing pressure and discharge current (at least up to $p=30$ mm. Hg, $i=30$ mA, the limiting values in their experiments). The magnitude of the cataphoresis effect decreases as the mass of the atoms increases. Some typical results are shown in Fig. 95.

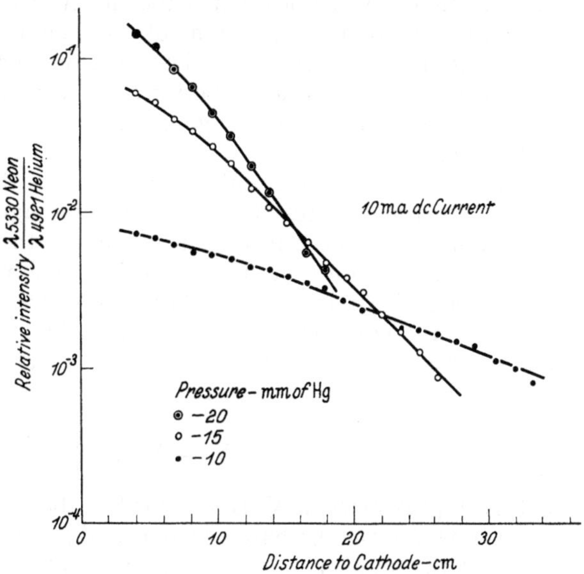

Fig. 95. Cataphoresis as a function of gas pressure at constant current for a commercial helium sample.

An approximate theoretical treatment of the separation of metal atoms from a rare gas has been given by DRUYVESTEYN [290]. Mathematical complexity made it necessary to consider separately the flow of metal atoms along the column to the cathode, and radially to the walls. Firstly, we assume that the concentration of metal vapour N atoms per cm.³ is constant over a section of the tube, of radius R. N^+, b^+ are the concentration and mobility of the metal ions, D the diffusion oefficient of metal atoms, and X the field. In the steady state the transport of metal vapour (as ions) to the cathode is balanced by an opposite flow due to diffusion:

$$-D\frac{dN}{dx}\cdot\pi R^2 = b^+ X N_0^+\int\limits_0^R J_0\left(\frac{2\cdot 4r}{R}\right)2\pi r\,dr = N_0^+ b^+ X\cdot 1.36 R^2 \qquad (98.1)$$

where x is the distance from the cathode, and N_0^+ is the ion concentration at the axis DRUYVESTEYN now distinguishes between two conditions: Either,

$$N^+ = N_e, \qquad (98.2)$$

N_e being the electron concentration. This applies near to the cathode, where the metal vapour concentration is relatively large and few ions of the rare gas are formed. Or,

$$N^+ \ll N_e \qquad (98.3)$$

which applies to regions further from the cathode, containing little metal vapour, and where rare gas ions are predominant. In the first case N^+ and X can be

[1] K. G. EMELEUS and J. SAYERS: Proc. Phys. Soc. Lond. A 65, 219 (1952).

eliminated by introducing the discharge current i. This gives a linear decrease in concentration of metal atoms with distance from the cathode

$$N = N_0 - 8.3 \cdot 10^{10} \frac{b^+ \sqrt{\overline{V}}}{\lambda_e D R^2} \cdot x. \tag{98.4}$$

Here $\overline{V}$ is the mean electron energy in volts, λ_e the mean free path of electrons in cm., and N_0 is the concentration at the cathode. The formula gives only the relative distribution of metal vapour, unless in certain circumstances the absolute concentration at some position is known. A further, more complicated formula is derived for the remaining regions of the tube further from the cathode, by introducing (after SOMMERMEYER[1]) the mean distance which a metal ion travels in the direction of the cathode, between its formation in the gas and its neutralisation on the wall. A knowledge of the ionisation function of the metal vapour is also necessary.

The linear fall in concentration predicted by eq. (98.4) has been checked against experimental results[2], with tolerable agreement.

A similar treatment has been applied to the radial separation of the metal vapour from the rare gas in the positive column neglecting the flow towards the cathode. Near the walls, where the concentration of metal atoms is large ($N^+ = N_e$) the equations are not easily soluble, but towards the centre of the column where most of the ions are those of the rare gas ($N^+ \ll N_e$) they result approximately in the BESSEL function distribution:

$$N_r = N_0 J_0\left(i \sqrt{\frac{\beta N_e}{D}} \cdot r\right). \tag{98.5}$$

In this approximation N_e is assumed to be constant across the inner regions of the tube, i is the discharge current and β is an average ionisation probability, such that the number of ionisations of metal atoms per cm.^{3} per sec. is given by $\beta N_e N$. N_0 is the concentration of metal atoms at the centre of the tube.

β) Pressure differences in pure gases. It was observed many years ago that the passage of a discharge in a single gas resulted in pressure differences between the electrode regions[3]. STARK[4] found, in air and nitrogen, that the pressure at the cathode was greater than that at the anode, a result confirmed later by WEHNELT and FRANCK[5] who used more refined techniques which avoided errors due to gas absorption at electrodes and walls. They used low pressures ($p < 0.1$ mm. Hg) and noticed that with decreasing pressure the pressure difference tended to a constant value proportional to the discharge current. This effect can easily be understood as due to the drift of gas ions to the cathode, where they are discharged and build up a gas pressure: the atoms then flow back at an equal rate by diffusion.

An extensive study of this phenomenon in several gases (A, Ne, He, N_2, H_2) was made first by HAMBURGER [*291*], [*292*] almost at the same time by SKAUPY [*293*], and afterwards by RÜTTENAUER [*294*], (rare gases only)[6]. Working at higher pressures ($p > 0.1$ mm. Hg) and currents ($j \sim 1$ A/cm.^{2}) they found a much larger effect in the opposite direction, namely an appreciably higher gas pressure at the anode than at the cathode: HAMBURGER found, in narrow capillary tubes,

[1] K. SOMMERMEYER: Phys. Z. **34**, 582 (1933).

[2] F. M. PENNING: Physica, Haag **1**, 763 (1934). Ne + 10% A + Hg; spectral measurements of the intensity of $\lambda = 5461$ Å near the cathode.

[3] For example E. WARBURG: Wied. Ann. **45**, 1 (1892).

[4] J. STARK: Ann. Phys. (BOLTZMANN-Festschrift) 399 (1904).

[5] A. WEHNELT and J. FRANCK: Verh. dtsch. phys. Ges. **12**, 444 (1910).

[6] HAMBURGER deduced the pressure difference from photographic measurements of the ultra-violet light emitted. SKAUPY and RÜTTENAUER used interconnected manometers.

that this difference could be as much as 30% of the total gas pressure, its value being approximately proportional to the current, and smaller in gases of low molecular weight. From his measurements in A, Ne, He, in a number of wide tubes of varying cross section S (0.45 to 2 cm.²) and length l (40 to 150 cm.) RÜTTENAUER deduced the empirical formula for the pressure difference Δp (in mm Hg) between anode and cathode:

$$\Delta p = f \cdot \frac{j X \sqrt{M}}{p} \cdot \frac{l}{S} = 2.4 \cdot 10^{-6} \frac{i X \sqrt{M}}{p R^4} \cdot l \tag{98.6}$$

where X is the gradient of the positive column, j the current density i the discharge current, M the molecular weight of the gas, and f is a constant[1].

The dependence of Δp on p given here is different from that originally found by HAMBURGER in capillary tubes: he observed that Δp was practically independent of the gas pressure, and inversely proportional to $\sqrt{S}$. HAMBURGER ascribes this to the different kind of gas flow in wide and narrow tubes, according as the mean free path of the gas atoms is much less than, or greater than the diameter of the tube. The back diffusion is then calculated by the formula of POISEUILLE or KNUDSEN respectively. Further HAMBURGER argues that the proper dependence of Δp on the gradient in the column should be expressed by some function $\varphi(X)$, which for certain ranges only is proportional to X, i.e. $\varphi(X) = kX$ where k is a different constant for each gas.

The reason for this effect was difficult to understand. HAMBURGER and SKAUPY separately suggested that gas was driven to the anode in the form of negative ions[2]. The total flow of gas molecules towards the anode is then given by

$$N^- v^- - N^+ v^+$$

where v^-, v^+ are the drift velocities of the ions. v^- is usually only slightly greater than v^+, and hence a very large concentration of negative ions would be necessary to produce the observed pressure difference. This is completely contrary to all the evidence, especially in the rare gases.

The explanation was first given by LANGMUIR [*295*]. In a uniformly ionised gas in which the charged particles move with constant drift velocity, all the momentum which the particles receive per unit time from the field is delivered to the gas atoms. If the numbers (and we assume, the total charge) of positively and negatively charged particles are equal, it follows that there is no net force on the gas, as long as the particles remain in the gas. The pressure difference must therefore be due to a wall effect. In the positive column the walls are negatively charged, repelling all electrons except the fastest ones, and attracting positive ions, so that the numbers of each arriving per unit time are equal. The momentum delivered to the wall by the ions is, however, very much greater than that delivered by the electrons, whose contribution is usually neglected[3]. Thus some of the momentum which the ions would have given to the gas is dissipated at the wall, whereas the electrons lose none in this way, and hence there is a net force on the gas towards the anode. LANGMUIR calculated the pressure gradient resulting from this effect, in rough agreement with RÜTTENAUER's results. A later treatment on the same lines is given by DRUYVESTEYN (*290*], for conditions

[1] It appears more correct to regard f as a constant having a different value characteristic of each gas. Here X is in V/cm, i in amps, l, R in cm., M in mass units.

[2] SKAUPY also put forward another theory, suggesting that electrons making elastic collisions deliver twice their momentum to the gas atoms, whereas the positive ions cannot deliver more than Mv. This reasoning was later shown to be false. [Z. Physik **2**, 213 (1920).]

[3] $M^+ v^+_{\text{drift}}$ is to be compared with $m_e (v_e)_{\text{drift}}$. Although $v_e \sim 10\, v^+$, $M^+ > 2000.\, m_e$, even for the lightest ions, and hence the contribution of the electrons is negligible.

in which the mean free path of electrons and ions (λ_e, λ_+) is very much less than the radius, R, of the tube. The unbalanced force due to the wall effect causes the gas near the wall to be forced towards the anode; nearer to the axis of the tube the gas moves in the opposite direction due to two effects (a) the concentration of ions N^+ is slightly greater than that of the electrons N_e, the difference giving rise also to the radial electric field. (b) the net transport of gas, as positive ions, towards the cathode.

By considering these factors, the viscous flow of the gas layer near the wall, and the viscous flow in the opposite direction due to back diffusion, an equation for Δp is derived:

$$\frac{\Delta p}{l} = 4.6 \cdot 10^{-3} \frac{i\,\lambda_+^3}{\lambda_e\, R^5}\left(1 - \frac{9\,\lambda_+}{2R}\right) - C \cdot 4 \cdot 10^9 \frac{X\,\overline{V}}{R^2} - 5 \cdot 10^8 \frac{i\sqrt{\overline{V}}\, b^+\,\eta}{N\,\lambda_e\, R^4} \quad (98.7)$$

where the first term represents the wall effect, and the next two represent (a) and (b) above.

Here i is the total discharge current (amps.), $\overline{V}$ the mean electron energy in volts, N the gas density (atoms per cm.^{3}), η the gas viscosity and C a numerical constant approximately equal to log $(R/2\lambda_+)$. The first term should be compared with eq. (98.6): the dependence on pressure and radius is different ($1/p^2R^5$ compared with $1/pR^4$) but by inserting reasonable values of the parameters DRUYVESTEYN's equation gives results in fair agreement with experiment. Too few measurements have been made for proper comparison, however, and the theory also relies on several assumptions. The gas flow across a section of the column is shown schematically in Fig. 96.

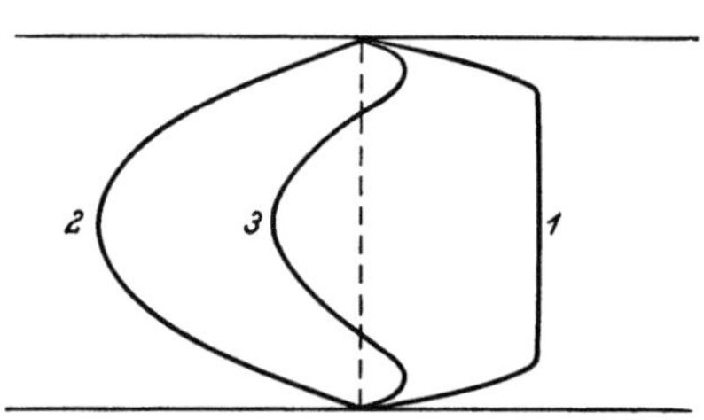

Fig. 96. Distribution of gas velocity across the positive column. (1) Due to wall effect, (2) compensating viscous flow, (3) total flow, whose integral across a section must be zero.

Two other observations of pressure differences in the glow discharge may be mentioned. DAUVILLIER[1] used the absorption of a beam of X-rays to measure density changes in discharges in He, Ne and A, and found a pressure difference within a layer about 0.5 mm. thick, on the cathode side of the negative glow.

SIRK[2] also found that when a magnetic field H is applied transversely across the positive column of a discharge, a pressure difference is set up perpendicular both to H and the current. The value is given by the simple relation

$$\Delta p = H \cdot j \cdot l \quad (98.9)$$

where j is the current density, and l the distance apart (perpendicular to H and j) of the points between which the pressure is measured. The effect is explained simply by the transport of ions in a curved path due to the normal electromagnetic force. The theory was given originally by SIRK, and recently in more concise form by DANNINGER and SIRK[3].

Pressure differences can also arise due to the electron pressure, a factor which STEENBECK has considered to calculate the pressure difference between the axis and the wall of a glow discharge. The total pressure, which is the sum of the partial pressure of electrons, ions and gas molecules ($p_e = N_e k T_e \ldots$ etc.), is constant at all points over a cross section[4]. The ion pressure is usually negligible, and since the electron pressure is much greater at the centre than at the wall the

[1] A. DAUVILLIER: C. R. Acad. Sci. Paris **182**, 575 (1926) (also near the positive column).
[2] H. SIRK: Sitzgsber. österr. Akad. Wiss. **122**, 417 (1913).
[3] R. DANNINGER and H. SIRK: Sitzgsber. österr. Akad. Wiss. (II A) **159**, 59 (1950).
[4] M. STEENBECK: Wiss. Veröff. Siemens-Werk **18**, 394 (1939).

neutral gas pressure is correspondingly less. STEENBECK used probes to measure N_e and T_e, and McLEOD gauges to measure the difference in neutral gas pressure, and found good agreement with calculations in an argon discharge at low pressure ($p = 4 \cdot 10^{-2}$ mm. Hg) and large current (up to $i = 10$ A). These are the conditions which would be expected to produce a large effect. In heavy current discharges the gas density at the axis decreases not only due to heating effects, but also due to the electron pressure. This effect also determines the greatest current which the discharge can carry: the pressure at any point cannot exceed the total gas pressure. This sets an upper limit on N_e (since $T_e \gg T_{\text{gas}}$, $N_e \ll N_{\text{gas}}$) and hence on the current which can flow.

99. Stability of discharges. When a discharge tube is connected to a d.c. battery in series with a resistance R (Fig. 93a) the potential V across the discharge is given by

$$V = E - Ri \tag{99.1}$$

where i is the current. The operating point is P (Fig. 94) where the "resistance line" (99.1) cuts the characteristic curve of the discharge. It is clear that, as the figure is drawn, any small fluctuations in current about P would tend to grow, whereas if a different resistance R' were used giving an operating point P' any small fluctuations in current would be compensated. Thus P' is a stable, P an unstable point, and the condition for stability may be written:

$$\frac{\partial V}{\partial i} > 0. \tag{99.2}$$

If a smaller resistance R and a lower battery potential E were used so that a less steep resistance line cut the curve in three points, it can easily be seen that the two outer operating points would be stable and the middle one unstable. We would then get the more general condition for stability

$$R + \frac{\partial V}{\partial i} > 0. \tag{99.3}$$

These are of course the simplest forms of stability which can be found: in other circuits, for example containing inductance, many more complex conditions can be found. This subject was exhaustively treated first by KAUFMANN in 1900 and later by DÄLLENBACH, and since it has been given in detail in previous handbooks[1] [*10*], we shall not repeat it here.

PENNING[2] observed certain discharges in helium which showed characteristic V–i curves having an almost closed loop (compare Fig. 82 showing similar curves obtained in a magnetic field). Such curves can be divided into four regions which, as we traverse them anti-clockwise, belong to alternating signs of dV/di: $-, +, -, +$. The first two would correspond to the regions AB, BA' of Fig. 94. and are called negative and positive characteristic of the first kind; the second pair are called similarly, but of the second kind. A discharge which has a positive characteristic of the second kind cannot be stabilised by a series resistance.

Attempts have been made to relate the circuit properties of the discharge, i.e. its effective resistance, capacity and inductance (which may be positive or negative) to the mechanism of the discharge itself. VAN GEEL has proposed and developed a theory which applies to small deviations from the stationary condition, which do not occur too quickly[3]. By using the quantities r $(= dV/di)$,

[1] Handbuch der Experimentalphysik, articles of SEELIGER and MIERDEL, vol. 13, part 3. 1929.

[2] F. M. PENNING: Proc. Roy. Acad. Amst. **34**, 1305 (1931).

[3] C. VAN GEEL: Physica, Haag **6**, 806 (1939); **8**, 361 (1941). — Ned. T. Natuurk. **8**, 437 (1941). — Self Induction and Inertia. Delft: Tech. Hogeschool 1955.

L, and T (a constant having the dimensions of time), all characteristic of the discharge, he is able to deduce its complex impedance and discuss various stability conditions. Many measurements, of these quantities, and an extension of the theory have been given by VERHAGEN[1]. The influence of after-effects, e.g. a changing gas temperature or the decay of metastable atoms, on the discharge characteristic have been measured and found to agree well with VAN GEEL's theory. The theory does not include wall currents, and so does not apply to discharges with a positive column.

100. Effect of external irradiation. It is well known that if the cathode of a dark discharge is strongly irradiated so as to give a relatively large photo-electric current the working potential of the discharge is reduced. In Fig. 1 the photo-electric current lines meet BC further to the right as the irradiation is measured, and may intersect in the region CD. One would expect that if a large enough initial current were produced the cathode fall necessary to sustain a glow discharge would be correspondingly reduced, and this has been found to be true when very intense light is shone on to photo-sensitive cathodes[2].

In mixtures (such as neon with a trace of argon) in which metastables of the impurity ionise the main gas, irradiation with light which destroys the metastables can cause changes in the discharge. For example the cathode fall increases (see Sect. 22) or, if the positive column is irradiated, its gradient increases or the current decreases. Very careful experiments by MEISSNER and MILLER[3] have shown that the spectral line which is most effective is that which raises the metastable atom into a nearby radiating state.

A curious effect has been observed when very low pressure and low current discharges are irradiated with X-rays[4] ($p \sim 10^{-4}$ mm. Hg, $i \sim 10^{-8}$A) the current at first instantaneously increasing, then decreasing, and falling to zero when the X-rays are removed. It is likely to be due to a photo-electric effect on the wall, and the subsequent accumulation of charge there.

For a full treatment of the effects of external radiation on ionised gases we refer to the article of A. VON ENGEL in vol. XXI of this Encyclopedia.

101. The glow discharge in moving gas. MAIER[5] has measured various properties of the glow discharge in argon and air with the gas flowing longitudinally or transversely with a velocity from 40 to 100 m./sec., that is with a drift velocity comparable with the thermal motion of the molecules. He found that the starting potential was increased by a few percent, but the time lag of starting was greatly reduced.

In a longitudinal gas stream the working voltage decreases appreciably when the gas flows towards the anode, and increases slightly when the gas flows towards the cathode. Typical values are $V = 540$ volts in still gas, 560 volts with gas flowing towards cathode, 430 volts with gas flowing towards anode (air, $p = 7.5$ mm. Hg, $i = 2.5$ mA, gas velocity 70 m./sec.). The gradient of the positive column becomes non uniform along its length: the column becomes shorter and the gradient everywhere less than the normal value in still gas when the gas stream flows towards the anode, and vice versa. The column changes its length always at the expense of the FARADAY dark space, the cathode regions being unaffected.

---

[1] C. J. D. M. VERHAGEN: Physica, Haag **8**, 361 (1941). — Thesis, Delft 1942.

[2] W. KLUGE: Z. angew. Phys. **5**, 173 (1953). — W. KLUGE and A. SCHULZ: Z. angew. Phys. **6**, 364 (1954). — Naturwiss. **42**, 40 (1955). — Z. Physik **142**, 83 (1955).

[3] K. W. MEISSNER and W. F. MILLER: Phys. Rev. **92**, 896 (1953).

[4] A. CANAC, G. MARET and L. GOLDZAHL: C. R. Acad. Sci. Paris **231**, 1054 (1950).

[5] E. MAIER: Z. Physik **93**, 65 (1934).

In a transverse gas stream the working potential of the discharge increases, the increase being proportional to the velocity of the gas, and is the larger the greater the pressure. The shape of the positive column is also deformed. MAIER discusses the effect of the motion of the gas upon the ions.

Similar work has been done by FUCKS[1] and his associates, who find changes in the discharge current between spherical electrodes when gas is blown across the gap. The effect has been used as a method of measuring rapid gas flow.

I wish to thank especially Dr. A. VON ENGEL for much helpful advice.

References.

General references.

[1] THOMSON, J. J.: Conduction of electricity, through gases, first edition. Cambridge Univ. Press 1906.
[2] TOWNSEND, J. S.: Electricity in Gases. Oxford: Univ. Press 1915.
[3] THOMSON, J. J. and G. P.: Conduction of electricity through gases, third edition. Cambridge Univ. Press 1933.
[4] ENGEL, A. v., and M. STEENBECK: Elektrische Gasentladungen. Berlin: Springer 1932, 1934.
[5] DARROW, K. K.: Electrical Phenomena in Gases. London: Baillière Tindall & Cox 1932.
[6] SEELIGER, R.: Physik der Gasentladungen. Leipzig: Johann Ambrosius Barth 1933.
[7] LOEB, L. B.: Fundamental processes of electric discharge in gases. New York: Wiley 1939.
[8] DRUYVESTEYN, M. J., and F. M. PENNING: Rev. Mod. Phys. **12**, 87 (1940).
[9] ENGEL, A. v.: Ionised Gases. Oxford: Univ. Press 1955.
[10] BÄR, R.: Handbuch der Physik, Bd. 14, S. 303. Berlin: Springer 1927.
[10a] MIERDEL, G.: Handbuch der Experimentalphysik, vol. 13, part 3, p. 313. Leipzig 1929.

Special references.

[11] CHILDS, E. C.: Phil. Mag. **9**, 529 (1930).
[12] CHIPLONKAR, V. T.: Proc. Indian Acad. Sci. A **12**, 440 (1940).
[13] YARNOLD, G. D., and S. HOLMES: Phil. Mag. **22**, 988 (1936). (Anode.)
[14] CHAUDHRI, R. M., and M. A. BAQUI: Proc. Phys. Soc. Lond., B **65**, 324 (1952). See also MOON and OLIPHANT: Proc. Cambridge Phil. Soc. **25**, 461 (1929).
[15] CRAGGS, J. D.: Proc. Phys. Soc. Lond. **54**, 245 (1942).
[16] CAMPAN, T. I.: Z. Physik **91**, 111 (1934).
[17] CHAUDHRI, R. M., and M. L. OLIPHANT: Proc. Roy. Soc. Lond. **137**, 662 (1932).
[17a] BREWER, A. K., and R. R. MILLER: Phys. Rev. **42**, 786 (1932).
The radial distribution of ion current is found in [14], [17a], older literature in [16], [17].
[18] ASTON, F. W.: Proc. Roy. Soc. Lond., Ser. A **84**, 526 (1911); **104**, 565 (1923).
[19] LITTLE, P. F., and A. v. ENGEL: Proc. Roy. Soc. Lond., Ser. A **224**, 209 (1954).
[20] STEIN, R. P.: Phys. Rev. **89**, 134 (1953).
[21] GEDDES, A. E.: Proc. Roy. Soc. Edinburgh **46**, 136 (1925).
[22] CHAUNDY, C. F. J.: Brit. J. Appl. Phys. **5**, 255 (1954).
[23] LITTLE, P. F., and A. v. ENGEL: Proc. Phys. Soc. Lond. B **65**, 459 (1952).
[24] BROSE, E.: Ann. Phys. **58**, 731 (1919).
[25] STENBING, W.: Ann. Phys. **10**, 296 (1931).
[26] LANGMUIR, I., and MOTT-SMITH: Gen. Electr. Rev. **26**, 731 (1923); **27**, 449, 583, 616, 726, 810 (1924).
[27] DRUYVESTEYN, M. J.: Z. Physik **64**, 793 (1930).
[28] GORCUM, A. H. VAN: Physica, Haag **3**, 207 (1936).
[29] BOYD, R. L. F., and N. D. TWIDDY: Nature, Lond. **173**, 633 (1954).
[30] WEIZEL, W., and W. OLMESDAHL: Z. Physik **114**, 719 (1939).
[31] EMELEUS, K. G., and SLOANE: Phil. Mag. **14**, 355 (1932).
[32] LANGMUIR, I.: J. Franklin Inst. **196**, 755 (1923).
[33] BOYD, R. L. F.: Nature, Lond. **165**, 228 (1950). — Proc. Roy. Soc. Lond., Ser. A **201**, 329 (1950).
[34] SPENCER-SMITH, J. L.: Phil. Mag. **19**, 806, 1016 (1935).
[35] LUNT and GREGG: Trans. Faraday Soc. **36**, 1062 (1940).

[1] W. FUCKS: Appl. Sci. Res. B **5**, 167 (1955) where other papers are quoted.

[36] HAIGH, C., and T. SMITH: Phil. Mag. **41**, 557 (1950).
[37] BOYD, R. L. F.: Nature, Lond. **165**, 142 (1950).
[38] IMAM, A., and S. R. KHASTGIR: Phil. Mag. **23**, 858 (1937).
[39] MAKINSON, R. E. B., P. C. THONEMANN, R. B. KING and J. V. RAMSAY: Proc. Phys. Soc. Lond. B **64**, 665 (1951).
[40] BIONDI, M. A.: Rev. Sci. Instrum. **22**, 500 (1951).
[41] GOLDSTEIN, LAMPERT and GEIGER: Electr. Commun. **29**, 243 (1952).
[42] CULLEN, A. L., and K. W. H. FOULDS: Nature, Lond. **169**, 237 (1952).
[43] HONERJÄGER, R., J. E. SCHULZE and DU BOIS: Naturwiss. **37**, 358 (1950).
[44] UDELSON, B. J., and J. E. CREEDON: Phys. Rev. **88**, 145 (1952).
[45] MILATZ, VREEDENBERG and BRAAK: Physica, Haag **10**, 433 (1943).
[46] GÜNTHERSCHULZE, A.: Z. Physik **49**, 358, 473 (1928), (He, Ne, A, H_2, N_2, O_2 on Fe cathode); also F. KÖPPEN: Z. Physik **75**, 654 (1932) (Hg on Al, Ni, Fe, W, Ag, Hg, Mo, Cu, C).
[47] GÜNTHERSCHULZE, A.: Z. Physik **59**, 433 (1930), (He, Ne, A, H_2, N_2, O_2 on Mg, Fe).
[48] GÜNTHERSCHULZE, A., and H. SCHNITGER: Z. Physik **103**, 491 (1936), (H_2, D_2 on Fe).
[49] GÜNTHERSCHULZE, A., and W. BÄR: Z. Physik **107**, 730 (1937), (He, Ne, A, Kr, Xe, H_2, N_2O_2, on Mg).
[50] PIKE, E. W.: Z. Physik **90**, 319 (1934).
[51] PENNING, F. M.: Physica, Haag **11**, 183 (1931).
[52] GROOT, W. DE: Naturwiss. **15**, 818 (1927).
[53] PENNING, MOUBIS and JURRIAANSE: Philips Res. Rep. **1**, 119, 225, 407 (1946).
[54] PENNING and MOUBIS: Physica, Haag **15**, 721 (1949).
[55] BOER, DE, and KRAAK: Rev. Trav. Chim. **56**, 1103 (1937).
[56] GRAAF, DE, and HAMAKER: Physica, Haag **9**, 297 (1942).
[57] CAMPBELL: Phil. Mag. **2**, 369 (1926).
[58] LANGMUIR, I.: Phys. Rev. **37**, 1139 (1915); **41**, 167 (1919).
[59] DRUYVESTEYN, M. J.: Z. Physik **73**, 727 (1932).
[60] GÜNTHERSCHULZE, A.: Z. Physik **24**, 52 (1924).
[61] MEISSNER: Z. Physik **117**, 325 (1941).
[62] ERNST, W. H.: Helv. phys. Acta **8**, 381 (1935). Older literature quoted.
[63] BROWN, W. L., and E. E. THOMPSON: Phil. Mag. **53**, 918 (1929).
[64] GÜNTHERSCHULZE, A., and F. KELLER: Z. Physik **71**, 246 (1931); **72**, 143 (1931); **75**. 105 (1932); **79**, 563 (1932).
[65] SPORN, H.: Z. Physik **112**, 278 (1939).
[66] MAYER, H.: Phil. Mag. **16**, 594 (1933).
[67] STARK and WENDT: Ann. Phys. **38**, 669 (1912).
[68] MURALT, A. v.: Ann. Phys. **85**, 1117 (1928) (air on Al cathode: j_n only).
[69] GÜNTHERSCHULZE, A., and H. SCHNITGER: Z. Physik **77**, 333 (1932).
[70] ROGOWSKI, W.: Arch. Elektrotechn. **25**, 551 (1931).
[71] FUCKS, W.: Z. Physik **87**, 139 (1933). See also M. STEENBECK on velocity of lateral spreading of the discharge. Arch. Elektrotechn. **13**, 732 (1932).
[72] ASTON, F. W.: Proc. Roy. Soc. Lond., Ser. A **79**, 80 (1907).
[73] ASTON and WATSON: Proc. Roy. Soc. Lond., Ser. A **86**, 168 (1912).
[74] SEELIGER, R.: Z. Naturforsch. **8**a, 74 (1953). — SEELIGER and WICHMANN: Ann. Phys. **9**, 235 (1951); also GÜNTHERSCHULZE: Z. Physik **42**, 763 (1927) (O_2).
[75] WILSON, H. A.: Phil. Mag. **4**, 608 (1902).
[76] CHAUDHRI, R. M.: Nature, Lond. **148**, 727 (1941).
[77] SEELIGER and LINDOW: Phys. Z. **25**, 337 (1924); **26**, 393 (1925). — Ann. Phys. **67**, 352 (1922).
[78] ASTON, F. W.: Proc. Roy. Soc. Lond., Ser. A **80**, 45 (1908).
[79] KOSSEL: Jb. Radioakt. Elektr. **18**, 326 (1921).
[80] GÜNTHERSCHULZE, A., and F. KELLER: Z. Physik **71**, 238 (1931).
[81] EMELEUS, BROWN and COWAN: Phil. Mag. **17**, 146 (1934), where many earlier references are given.
[82] HOLST and OOSTERHUIS: Physica, Haag **1**, 78 (1921). — Phil. Mag. **46**, 1117 (1923).
[83] PENNING and VEENEMANNS: Z. Physik **62**, 746 (1930).
[84] SEELIGER, R.: Naturwiss. **18**, 155 (1930).
[85] EMELEUS and HALL: Proc. Roy. Irish Acad. A **40**, 1 (1931).
[86] GÜNTHERSCHULZE, A.: Z. Physik **19**, 313 (1923); **30**, 175 (1924). — Z. techn. Phys. **11**, 49 (1930).
[87] LOMPE, SEELIGER and WOLTER: Ann. Phys. **36**, 9 (1939).
[88] LOMPE, A.: Z. Physik **109**, 310 (1938).
[89] PASCHEN, F.: Ann. Phys. **58**, 731 (1916).
[90] SCHULER, M.: Z. Physik **35**, 323 (1926).
[91] VAN VOORHIS, C. C., and A. G. SHENSTONE: Rev. Sci. Instrum. **12**, 257 (1941).

[92] CAMPBELL SWINTON: Proc. Roy. Soc. Lond. **61**, 79 (1897).
[93] ASTON, F. W.: Proc. Cambridge Phil. Soc. **19**, 317 (1919).
[94] BECK, J. W., and K. G. EMELEUS: Phil. Mag. **11**, 55 (1931).
[95] GÜNTHERSCHULZE, A.: Z. Physik **37**, 868 (1926).
[96] GÜNTHERSCHULZE, BÄR and WINTER: Z. Physik **107**, 642, 730 (1937); **108**, 780 (1938); **109**, 121 (1938); **111**, 208 (1938).
[97] FISCHER, H.: Z. Physik **110**, 197 (1938); **113**, 360 (1939).
[98] RUDOLPH, G.: Z. Physik **111**, 523 (1939).
[99] GÜNTHERSCHULZE, A.: Z. Physik **33**, 810 (1925).
[100] RYDE, J. W.: Phil. Mag. **45**, 1149 (1923).
[101] ROGOWSKI, W.: Arch. Elektrotechn. **26**, 643 (1932). — Z. Physik **114**, 1 (1939).
[102] SCHERZER, O.: Arch. Elektrotechn. **33**, 207 (1939).
[103] WEIZEL, ROMPE and SCHÖN: Z. Physik **112**, 339; **113**, 87, 730 (1939).
[104] COMPTON, K. T., and P. M. MORSE: Phys. Rev. **30**, 305 (1927).
[105] MORSE, P. M.: Phys. Rev. **31**, 1003 (1928).
[106] SEELIGER, R.: Z. Physik **115**, 111 (1940).
[107] MORTON, P. L.: Phys. Rev. **70**, 388 (1946).
[108] JOHNSON, G. W.: Phys. Rev. **73** (1948).
[109] SEELIGER, R., and J. SCHMEKEL: Ann. Phys. **73**, 249 (1924).
[110] COSTA, H.: Z. Physik **113**, 531 (1939).
[111] DEMBER and GEHLHOFF: Verh. phys. Ges. **8**, 264 (1906).
[112] WIEDEMANN, E.: Z. Elektrochem. **2**, 195 (1895).
[113] DAUVILLIER, A.: Phil. Mag. **2**, 1046 (1926).
[114] SEEMAN, H., and G. ORBAN: Ann. Phys. **23**, 137 (1935).
[115] EMELEUS and HARRIS: Phil. Mag. **4**, 49 (1927).
[116] EMELEUS and BROWN: Phil. Mag. **7**, 17 (1929).
[117] EMELEUS, GREEVES and MONTGOMERY: Proc. Roy. Irish Acad. **43**, 35 (1936).
[118] EMELEUS and BALLANTINE: Phys. Rev. **50**, 672 (1936).
[119] LANGMUIR, I.: Phys. Rev. **26**, 585 (1925).
[120] WEIZEL and FISCHER: Ann. Phys. **24**, 209 (1935).
[121] FISCHER, H.: Ann. Phys. **27**, 81 (1936).
[122] GEIGER, K.: Z. Physik **106**, 17 (1937).
[123] WEIZEL and FISCHER: Z. techn. Phys. **16**, 433 (1935).
[124] WEIZEL and FISCHER: Phys. Z. **36**, 826 (1935).
[125] DRUYVESTEYN, M. J.: Z. Physik **62**, 764 (1930).
[126] DRUYVESTEYN, M. J.: Physica, Haag **1**, 427 (1934).
[127] DRUYVESTEYN, M. J.: Z. Physik **57**, 292 (1929).
[128] DRUYVESTEYN, M. J., and N. WARMOLTZ: Z. Physik **68**, 378 (1931).
[129] DRUYVESTEYN, M. J.: Physica, Haag **11**, 129 (1931).
[130] EMELEUS, K. G., and O. S. DUFFENDACK: Phys. Rev. **47**, 460 (1935).
[131] DAUVILLIER, A.: J. de Phys. **7**, 369 (1926).
[132] EMELEUS, BROWN and COWAN: Nature, Lond. **127**, 593 (1931).
[133] EMELEUS, K. G.: Proc. Roy. Irish Acad. A **42**, 31 (1936), where many references are found.
[134] SEELIGER, R., and G. MIERDEL: Z. Physik **19**, 230 (1923).
[135] GROOT, W. DE: Naturwiss. **14**, 104 (1926). — Physica, Haag **6**, 53 (1926).
[136] KEYSTONE: Phil. Mag. **15**, 1162 (1933).
[137] BREWER, A. K., and J. W. WESTHAVER: J. Appl. Phys. **8**, 779 (1937).
[138] EMELEUS and KENNEDY: Phil. Mag. **18**, 874 (1934).
[139] LINDER, E. G.: Phys. Rev. **38**, 678 (1931).
[140] WÜLLNER: Pogg. Ann. (Jubelband) **1874**, 32.
[141] SPOTTISWOODE: Proc. Roy. Soc. Lond. **25**, 73 (1876).
[142] ASTON and KIKUCHI: Proc. Roy. Soc. Lond. **98**, 51 (1920).
[143] GÜNTHERSCHULZE, A.: Z. Physik **41**, 718 (1927); **42**, 763 (1927), (He, Ne, A, Kr, Xe, Hg), (O_2, H_2O, air, N_2, H_2).
[144] JOHNSON, P.: Phil. Mag. **10**, 921 (1930), (Ne).
[145] LLEWELLYN JONES, F.: Phil. Mag. **11**, 163 (1931), (He: a.c. measurement, see [*144*]).
[146] ELENBAAS, W.: Z. Physik **78**, 603 (1932), (Hg).
[147] LOMPE, A., and R. SEELIGER: Ann. Phys. **15**, 300 (1932), (He, Ne, A).
[148] GROSS, O.: Z. Physik **88**, 741 (1934), (A).
[149] ENDE, W.: Z. techn. Phys. **15**, 601 (1934), (Hg).
[150] KLARFELD, B.: J. techn. Phys. USSR. **4**, 44 (1937); **5**, 725 (1938), (He, Ne, A, Kr, K, Hg).
[151] SEELIGER, R., and R. HIRCHERT: Ann. Phys. **11**, 817 (1931), (Ne, A).
[152] GÜNTHERSCHULZE, A., and F. KELLER: Z. Physik **22**, 70 (1924); **77**, 703 (1932), "normal gradient".

[153] GÜNTHERSCHULZE, A., and H. BETZ: Z. Physik **81**, 283 (1933), "normal gradient".
[154] WILLS: Phys. Rev. **19**, 65 (1904).
[155] KILLIAN, T. J.: Phys. Rev. **35**, 1238 (1930), (Hg).
[156] GRANOVSKY, V.: C. R. Acad. Sci. URSS. **28**, 38 (1940) also Bull. Acad. Sci. URSS. **4**, 449 (1938).
[157] EMELEUS and BURNS: Brit. J. Appl. Phys. **5**, 277 (1954).
[158] SOMMERMEYER, K.: Z. Physik **90**, 232 (1934), (Ne).
[159] DRUYVESTEYN, M. J.: Z. Physik **81**, 571 (1933), (A).
[160] MIERDEL, G.: Wiss. Veröff. Siemens-Werk **17**, 71 (1938).
[161] KARELINA, N. A.: J. Phys. USSR. **6**, 218 (1942).
[162] HOLM, R.: Wiss. Veröff. Siemens-Werk **3**, 159 (1923).
[163] HOLM, R.: Z. Physik **75**, 171 (1932).
[164] SCHOTTKY, W.: Phys. Z. **25**, 342, 635 (1924).
[165] SEELIGER, R.: Phys. Z. **33**, 273, 313 (1932). — Naturwiss. **39**, 78 (1952) Reviews.
[166] TONKS, L., and I. LANGMUIR: Phys. Rev. **34**, 876 (1929).
[167] BOECKNER, C., and F. L. MOHLER: Bur. Stand. J. Res. **10**, 357 (1933).
[168] KLARFELD, B.: J. Phys. USSR. **5**, 155 (1941).
[169] SOMMERMEYER, K.: Ann. Phys. **13**, 315 (1932); **28**, 240 (1937), (He, Ne, A).
[170] MOHLER, F. L.: Bur. Stand. J. Res. **9**, 25 (1932); **21**, 697, 873 (1938), (Cs).
[171] DRUYVESTEYN, M. J., and N. WARMOLTZ: Phil. Mag. **17**, 1 (1934).
[172] DRUYVESTEYN, M. J.: Phys. Z. **33**, 823 (1932).
[173] POKROVSKAYA, A. S.: J. techn. Phys. USSR. **21**, 617 (1951).
[174] FOUND, C. G., and E. F. HENNELLY: Illum. Engr. N. Y. **36**, 400 (1941).
[175] KENTY, C.: J. Appl. Phys. **21**, 1309 (1950).
[176] FRISCH, S. E., and Y. M. KAGAN: J. exp. theor. Phys. USSR. **11**, 286 (1941); **17**, 577 (1947).
[177] FRISCH, S. E., and Y. M. KAGAN: J. Phys. USSR. **7**, 208 (1943).
[178] FRISCH, S. E., and Y. M. KAGAN: Izv. Akad. Nauk SSSR. **12**, 358 (1948).
[179] PAUL, H. H.: Z. Physik **97**, 330 (1935).
[180] BRAMLEY, A.: Phys. Rev. **26**, 794 (1925).
[181] HOLM, R.: Phys. Z. **25**, 497 (1924); **26**, 412 (1925).
[182] NEUBERT, P.: Ann. Phys. **42**, 1454 (1913). — Phys. Z. **15**, 432 (1914).
[183] MCCURDY, W. H., and P. DALTON: Phys. Rev. **27**, 163 (1926).
[184] ZELENY, J.: J. Franklin Inst. **209**, 625 (1950).
[185] GÜNTHERSCHULZE, A., and H. MEINHARDT: Z. Physik **110**, 95 (1938).
[186] LAU, E., and O. REICHENHEIM: Ann. Phys. **3**, 840 (1929); **5**, 296 (1930).
[187] PUPP, W.: Phys. Z. **33**, 844 (1932); **35**, 705 (1934); **36**, 61 (1935), Photocell measurements. Z. techn. Phys. **15**, 257 (1934).
[188] KIKUCHI, T.: Proc. Roy. Soc. Lond. **99**, 257 (1921).
[189] SAMSON, C.: Z. techn. Phys. **6**, 281 (1925).
[190] WHIDDINGTON, R.: Nature, Lond. **116**, 506 (1925); **126**, 467 (1929).
[191] ELENBAAS, W.: Nederl. Tijdschr. Natuurk. **1**, 16 (1934).
[192] MANEN, B. v.: Physica, Haag **1**, 968 (1934).
[193] GORCUM, A. H. v.: Physica, Haag **2**, 535 (1935).
[194] DONAHUE, T., and G. H. DIEKE: Phys. Rev. **81**, 248 (1951).
[195] LOH, H. Y., and G. H. DIEKE: J. Opt. Soc. Amer. **37**, 837 (1947).
[196] DIEKE, LOH and CROSSWHITE: J. Opt. Soc. Amer. **36**, 185 (1946).
[197] DRUYVESTEYN, M. J.: Physica, Haag **1**, 273, 1003 (1934).
[198] ARMSTRONG, EMELEUS and NEILL: Proc. Roy. Irish Acad. A **54**, 291 (1951).
[199] MAHLA, K.: Z. techn. Phys. **17**, 324 (1936).
[200] LANGMUIR, I.: Z. Physik **46**, 271 (1927).
[201] ISSENDORF, J. v.: Z. Physik **67**, 556 (1931).
[202] WASSERRAB, T.: Z. Physik **128**, 575 (1950).
[203] GÜNTHERSCHULZE, A.: Z. Physik **30**, 175 (1924); **40**, 414 (1927), (H_2, N_2, O_2, H_2O).
[204] DRUYVESTEYN, M. J.: Physica, Haag **4**, 669 (1937), (He, Ne, A).
[205] GÜNTHERSCHULZE, A., and F. KELLER: Z. Physik **81**, 799 (1933), (H_2, N_2, O_2).
[206] PENNING, F. M.: Physica, Haag **4**, 380 (1924); **5**, 217 (1925), (Ne).
[207] NERETINA, N. A.: J. techn. Phys. USSR. **16**, 353 (1946).
[208] GÜNTHERSCHULZE, A., and H. SCHNITGER: Z. Physik **104**, 395 (1936).
[209] GÜNTHERSCHULZE, BÄR and BETZ: Z. Physik **109**, 293 (1938).
[210] SMITH, TAIT and WHIDDINGTON: Proc. Leeds Phil. Soc. **5**, 207 (1949). See also L. P. SMITH and P. L. HARTMANN: Phys. Rev. **55**, 1145 (1939).
[211] RUBENS, S. M., and J. E. HENDERSON: Phys. Rev. **58**, 446 (1940).
[212] ENGEL, A. v.: Phil. Mag. **32**, 417 (1941).
[213] KELLER, F.: Z. Physik **97**, 8 (1935).
[214] ZIMMERMANN, G.: Z. Physik **91**, 767 (1934).

[215] HAIGH, C. P.: Electr. Engng. **15**, 61 (1942), an outline of technical methods.
[216] HANAU, R.: Phys. Rev. **76**, 153 (1949).
[217] KINGDON and LANGMUIR: Phys. Rev. **22**, 148, 357 (1923).
[218] G. E. C. STAFF: Phil. Mag. **45**, 98 (1923).
[219] KINGDON and LANGMUIR: Phys. Rev. **20**, 108 (1922); **21**, 210 (1923) and [217].
[220] GÜNTHERSCHULZE, A.: Z. Physik **118**, 145 (1941).
[221] PENNING, F. M., and J. H. A. MOUBIS: Proc. Roy. Acad. Amst. **43**, 41 (1940).
[222] GÜNTHERSCHULZE, A., and F. MEYER: Z. Physik **62**, 607 (1930).
[223] MEYER and GÜNTHERSCHULZE: Z. Physik **71**, 279 (1931).
[224] GÜNTHERSCHULZE: Z. Physik **38**, 575 (1926).
[225] MASSEY and BURHOP: Electronic and Ionic Impact Phenomena. Oxford Univ Press. 1952.
[226] ECKER, G., and K. G. EMELEUS: Proc. Phys. Soc.Lond. B **67**, 546 (1954). — Naturwiss. **41**, 185 (1954).
[227] GÜNTHERSCHULZE, A.: Z. Physik **36**, 563 (1926).
[228] GÜNTHERSCHULZE, A.: Z. Physik **119**, 79 (1942).
[229] GÜNTHERSCHULZE, A., and W. TOLLMIEN: Z. Physik **119**, 685 (1942).
[230] LAMAR and COMPTON: Science (Lancaster, Pa.) **80**, 541 (1934).
[231] HIPPEL, V., and BLESCHMIDT: Ann. Phys. **81**, 1043 (1926).
[232] HIPPEL, V., and BLESCHMIDT: Ann. Phys. **86**, 1006 (1928).
[233] TOWNES: Phys. Rev. **65**, 319 (1944).
[234] SEELIGER, R.: Z. Physik **119**, 482 (1942).
[235] HAWORTH, F. E.: J. Appl. Phys. **22**, 606 (1951).
[236] GÜNTHERSCHULZE, A.: Z. Physik **61**, 1, 581 (1930).
[237] PENNING, F. M.: Z. Physik **70**, 782.
[238] DRUYVESTEYN, M. J.: Physica, Haag **5**, 875 (1938).
[239] PENNING, F. M.: Phys. Z. **33**, 816 (1932). — Proc. Roy. Acad. Amst. **34**, 1305 (1931).
[240] GÜNTHERSCHULZE, A., and H. FRICKE: Z. Physik **86**, 451, 821 (1933).
[241] GÜNTHERSCHULZE, A.: Z. Physik **86**, 778 (1933).
[242] FRICKE, H.: Z. Physik **92**, 728 (1934).
[243] SCHNITGER, H.: Z. Physik **96**, 551 (1935); **102**, 163 (1936).
[244] GÜNTHERSCHULZE and BÄR: Z. Physik **106**, 662 (1930).
[245] WOLF, K.: Z. Physik **112**, 148 (1939).
[246] DASCOLA: Ill. Nuo. Cim. **4**, 63 (1947).
[247] NEWMAN, F. H.: Phil. Mag. **23**, 239 (1937).
[248] STARK, J.: Die Electrizität in Gasen. Leipzig: Johann Ambrosius Barth 1902.
[249] SOMERVILLE, J. M.: D. Sc. Thesis, University of Sydney 1952.
[250] GÜNTHERSCHULZE, A.: Z. Physik **24**, 140 (1924).
[251] TOWNSEND, J. S.: Proc. Roy. Soc. Lond., Ser. A 86, 571 (1912).
[252] TOWNSEND, J. S.: Phil. Mag. **25**, 259 (1938).
[253] TONKS, L.: Phys. Rev. **51**, 744 (1937).
[254] BAILEY, V. A.: Phil. Mag. **9**, 560, 625 (1930).
[255] WASSERRAB, T.: Z. Naturforsch. **2**a, 575 (1947).
[256] PIDDUCK, F. B.: Proc. Lond. Math. Soc. **15**, 89 (1915).
[257] PIDDUCK, F. B.: Phys. Rev. **53**, 197 (1938).
[258] GANS, R.: Ann. Phys. **20**, 293 (1906).
[259] TONKS, L., and W. P. ALLIS: Phys. Rev. **52**, 710 (1937).
[260] ALLIS, W. P., and H. W. ALLEN: Phys. Rev. **52**, 703 (1937).
[261] HUXLEY, L. G. H.: Phil. Mag. **23**, 210 (1937).
[262] TONKS, L.: Phys. Rev. **56**, 360 (1939).
[263] CUMMINGS, C. S. and L. TONKS: Phys. Rev. **59**, 514, 522 (1941).
[264] ROKHLIN, G. N.: J. Phys. USSR. **1**, 347 (1939).
[265] FABRIKANT, V., and G. N. ROKHLIN: C. R. Acad. Sci. USSR. **19**, 393 (1938); **20**, 437 (1938).
[266] FABRIKANT, V., and G. N. ROKHLIN: J. Exp. Theor. Phys. **9**, 804 (1939).
[267] REICHRUDEL, E., and G. SPIWAK: J. Phys. USSR. **4**, 211 (1941).
[268] REICHRUDEL, E.: J. Tech. Phys. **21**, 733 (1951).
[269] REICHRUDEL, E., A. V. CHERNETSKI, V. V. MIKHNEVICH and I. A. VASILEVA: J. Tech. Phys. **22**, 1945 (1952).
[270] PENNING, F. M.: Physica, Haag **3**, 873 (1936).
[271] HAEFER, R.: Acta Phys. Austriaca **7**, 52, 251 (1953).
[272] HAEFER, R.: Acta. Phys. Austriaca **8**, 213 (1954).
[273] SOMERVILLE, J. M., K. S. W. CHAMPION and E. K. BIGG: Australian J. Sci. Res. A **1**, 400 (1948).
[274] EMELEUS, K. G., and J. SAYERS: Proc. Roy. Irish Acad. A **44**, 87 (1938).
[275] MASSEY, H. S. W.: Negative Ions. Cambridge U. P. 1950.
[276] PRINGLE, D. H., and. W. E. J. FARVIS: Phys. Rev. **96**, 536 (1954).

[277] OLLENDORF F., and R. SEELIGER: Ann. Phys. **83**, 535 (1927); **86**, 1089 (1928).
[278] ARNOT, F. L.: Nature, Lond. **138**, 162 (1936); also Proc. Roy. Soc. Lond., Ser. A **158**, 137, 157 (1937).
[279] ARNOT, F. L., and J. C. MILLIGAN: Proc. Roy. Soc. Lond., Ser. A **156**, 538 (1936).
[280] MASSEY, H. S. W., and R. A. SMITH: Proc. Roy. Soc. Lond., Ser. A **155**, 472 (1936).
[281] SEELIGER, R.: Phys. Z. **33**, 316 (1932).
[282] WULFHEKEL, H., and R. SEELIGER: Phys. Z. **34**, 57 (1933).
[283] GÜNTHERSCHULZE, A.: Z. Physik **91**, 724 (1934).
[284] KELLER, F.: Z. Physik **97**, 8 (1935).
[285] SEELIGER, R., and K. SOMMERMEYER: Z. Physik **98**, 733 (1936).
[286] ZIMMERMANN, G.: Z. Physik **104**, 309 (1937).
[287] SKAUPY, F.: Verh. dtsch. phys. Ges. **18**, 230 (1916).
[288] SKAUPY, F., and F. BOBEK: Z. techn. Phys. **6**, 284 (1925).
[289] REISZ, R., and G. H. DIEKE: J. Appl. Phys. **25**, 196 (1954).
[290] DRUYVESTEYN, M. J.: Physica, Haag **2**, 255 (1935).
[291] HAMBURGER, L.: Diss. Delft 1917. — Z. wiss. Photogr. **18**, 1 (1919).
[292] HAMBURGER, L.: Proc. Roy. Soc. Amst. **20**, 1043 (1918); **23**, 379 (1921); **25**, 463 (1923).
[293] SKAUPY, F.: Verh. dtsch. phys. Ges. **19**, 264 (1917). — Z. Physik **2**, 213 (1920).
[294] RÜTTENAUER, A.: Z. Physik **10**, 269 (1922).
[295] LANGMUIR, I.: J. Franklin Inst. **196**, 751 (1923).
[296] VALLE, G.: Phys. Z. **27**, 473 (1926).
[297] GILL, E. W. B.: Phil. Mag. **8**, 955 (1929).
[298] MITRA, S. K., and P. SYAM: Phil. Mag. **14**, 616 (1932).
[299] VALLE, G.: Phys. Z. **43**, 473 (1942).
[300] MOUSSIEGT, J.: C. R. Acad. Sci. Paris **222**, 1280, 1479 (1946); **223**, 659 (1946); **224** 1490 (1947).
[301] MOUSSIEGT, J.: Ann. Phys., Paris **4**, 593 (1949) (main paper).
[302] THOMSON, J. J.: Phil. Mag. **11**, 697 (1931). See also J. PLACINTEANU: C. R. Acad. Sci., Paris **193**, 286 (1931).
[303] TONKS, L., and I. LANGMUIR: Phys. Rev. **33**, 195 (1929).
[304] BORGNIS, F.: Helv. phys. Acta **20**, 207 (1947).
[305] BOHM, D., and E. P. GROSS: Phys. Rev. **75**, 1851 (1949).
[306] ARMSTRONG, E. B., K. G. EMELEUS and T. R. NEILL: Proc. Roy. Irish. Acad. A **54**, 291 (1951).
[307] WARREN, R.: Phys. Rev. **98**, 1650, 1658 (1955); also Rev. Sci. Instrum. **26**, 765 (1955).
[308] BICKERTON, R. J., and A. von ENGEL: Proc. Phys. Soc. Lond. B **1956** (In Press).
[309] BURCH, D. S., G. H. DUNN, R. GEBALLE and V. M. MADSEN: Proc. 8th Gaseous Electronics Conference N.Y.
[310] GABOR, D.: Brit. J. Appl. Phys. **2**, 209 (1951).
[311] ENGEL, A. v.: Appl. Sci. Res. B **5**, 34 (1955).
[312] WAINFAN, N., WALKER and WEISSLER: J. Appl. Phys. **24**, 1318 (1953).

Very Recent Work

Cathode space charge. BRINI et al.: Nuovo Cim. **12**, 915 (1954), formative time. — *Probe theory etc.* KAGAN and PEREL: J. techn. Phys. USSR. **24**, 889 (1954), general. — SCHNEIDER: Acta. phys. Austriaca **9**, 107 (1955), several electron groups. — SCHULZE and BROWN: Phys. Rev. **98**, 1642 (1955), ion collection. — LOEB: Basic Processes of Gaseous Electronics, California U. P. 1955, full literature.

New method of measurement. INGRAM and TAPLEY: Research **7**, 63 (1954), magnetic resonance.

Negative glow. PRINGLE and FARVIS: Phys. Rev. **96**, 536 (1954). — Proc. Phys. Soc. Lond. B **68**, 836 (1955), screened probe in He.

Positive column. WILHELM: Ann. Phys. **15**, 148 (1955), theory for X. — WASSERRAB: Z. angew. Phys. **7**, 176 (1955), varying current column. — FARRIS: Proc. Phys. Soc. Lond. B **68**, 383 (1955). — J. Electronics **1**, 60 (1955), moving striations. — STEWART: J. Opt. Soc. Amer. **45**, 651 (1955), resonance with modulated V.

Magnetic field on plasma diffusion. SIMON: Phys. Rev. **98**, 317; **100**, 1557 (1955). — YASUOKA: J. Phys. Soc. Japan **10**, 1102 (1955).

Plasma oscillations. GELLER and LOW: Nature, Lond. **176**, 1021 (1955), growth in negative glow.

Pressure differences. ALLEN and THONEMANN: Proc. Phys. Soc. Lond. B **67**, 768 (1954), maximum current in column.

Stability. YEH: J. Appl. Phys. **27**, 98 (1956), ion transit time near cathode.

Radiation from Low Pressure Discharges.

By

R. G. Fowler.

With 17 figures.

Glossary of special usages.

Blockaded radiations: Radiations which are hindered in their passage through the gas by repeated reabsorption.

Concentration: The number of systems of a given type per unit volume.

Elatation: The consequence of energy transfer by inelastic collision (lat. elatio, lift or raise up).

Excitation: The production of discrete excited states.

Monochromatic radiant energy density: Radiant energy per unit volume per unit frequency interval.

Population: The concentration of systems having a given characteristic state.

Relaxation time: The mean radiation time of an oscillator.

Resonance radiation: Radiation from a state which has only a single choice of state below it, normally the ground state.

A. General aspects of gas radiance.

1. Introduction. The radiation emitted by gas molecules after stimulation in collision has been the paramount source of existing knowledge about the internal structure of the gas molecules, about the environment in which the gas molecules exist, and about the stimulating collisions themselves. Then, finally, by a synthesis if this knowledge, an ever improving understanding is being achieved of the means by which electrical conductivity is maintained in the manifold regimes which discharges display. At the present, the quantitative (and even in some cases the qualitative) internal consistency of the complete problem of the electrical and radiative behavior of discharges is far from satisfactory, although it is probable that the various processes and relations involved in the problem have now been completely identified in nearly every regime.

In addition to this fundamental physical knowledge about the intrinsic situation in gases, a great body of ancillary knowledge has been contributed to other fields of science and technology. To the chemist[1], radiation from gas discharges has supplied means of measuring atomic abundances, identities, and chemical reaction rates. To the astrophysicist[2] radiation has been the sole source of an imposing accumulation of facts about the cosmos. To the engineer[3] gas discharges are the basis of major sources of artificial illumination, whose design

[1] G. Glockler and S. C. Lind: The Electrochemistry of Gases etc. New York 1939. — G. R. Harrison, R. C. Lord and J. R. Loofbourow: Practical Spectroscopy. New York 1948.

[2] H. H. Plaskett: Mon. Nat. Roy. Astronom. Soc. **107**, 117 (1947). — L. H. Aller: Astrophysics. New York 1953. — A. Unsöld: Physik der Sternatmosphären. Berlin 1938.

[3] W. Uyterhoeven: Elektrische Gasentladungslampen. Berlin 1938. — M. Laporte: Les Lampes à Eclairs Lumière Blanche. Paris 1949.

and perfection have naturally profited largely by the growth of fundamental knowledge in this field.

2. Radiation theory. An understanding of the radiation from gas discharges rests on the elucidation of two separate stages in the process, (1) the release of energy as radiation by the gas molecules, and (2) the delivery of energy to the gas molecules. Both stages have been well explored by theoretical methods, and the problem of the radiation analyst is largely one of choosing from among a multiplicity of methods of energy delivery those particular ones which are salient in the case at hand. It will be convenient to discuss the release of radiation[1] first.

Although the general quantum theory of radiation is to some extent hampered by the problem of electron self-energy, the emission and absorption of electromagnetic radiation in the range of wavelengths which is important in gas discharges is not affected by this and leads to definite results. Each atomic or molecular species radiates a characteristic set of frequencies which are governed by the PLANCK relation

$$h\nu_{ji} = E_j - E_i. \tag{2.1}$$

The probability per unit time that such a transition will occur spontaneously in an atom excited to the jth state is given by

$$A_{ji} = \frac{64\pi^4 e^2 \nu_{ji}^3}{3hc^3} |x_{ji}|^2, \tag{2.2}$$

where $x_{ji} (= x_{ij})$ is the matrix element of the dipole moment for the transition from E_j to E_i. Most commonly occuring radiations are dipole radiations, but transitions which are forbidden in dipole radiation by the identical vanishing of x_{ji} (i.e., selection rules) can sometimes be observed weakly in quadripole or magnetic dipole radiation with a probability which is roughly in the ratio of $(a_j/\lambda)^2$ to that of an allowed transition. Here a_j is the orbital radius of the upper state. Values of A_{ji} for many radiators are given in LANDOLT-BÖRNSTEIN (Vol. 1, Part 1, 1950). Extraneous fields can permit forbidden transitions.

Counter to the spontaneous emission of radiation there is the reverse processes of absorption, governed by the probability, B_{ij}, that an absorption act will occur per unit time per unit density of monochromatic radiant energy in the vicinity of the absorber. It is found that

$$B_{ij} = \frac{8\pi^3 g_j e^2}{3g_i h^2} |x_{ij}|^2, \tag{2.3}$$

where g_i stands for the statistical weight of the state i.

Complementary to these two processes, EINSTEIN[2] introduced the process of enforced emission in the presence of radiation. For this process, the coefficient B_{ji} is defined as the probability that an enforced emission will occur per unit time per unit density of monochromatic radiant energy in the vicinity of the atom. As in the previous cases, theory shows that

$$B_{ji} = \frac{8\pi^3 e^2}{3h^2} |x_{ji}|^2. \tag{2.4}$$

In a state of thermal equilibrium between radiation and radiators which are not otherwise interacting, if $\varrho(\nu_{ji})$ is the monochromatic radiant energy density,

[1] W. HEITLER: The Quantum Theory of Radiation. Oxford 1952. — See also Vol. 31 of this Encyclopedia.

[2] A. EINSTEIN: Phys. Z. **18**, 121 (1917).

and N_i and N_j are the populations of the two states i and j, there is a detailed balancing of losses and gains between the states such that

$$(A_{ji} + B_{ji}\varrho)\, N_j = B_{ij}\varrho\, N_i. \tag{2.5}$$

In most of the practical situations in gas discharges at low pressure, however, the radiators are far from a state of thermal equilibrium with their radiations, and the number of acts of spontaneous emission greatly exceeds the number of absorptions, which in turn preponderates over the number of enforced emissions, so that these last may be completely neglected. Furthermore, except for transitions ending on the ground state, and in some instances on metastable states also, the absorption processes can be neglected as well. These exceptions become of increasing importance as particle densities increase.

It is frequently advantageous to compare the results of quantum radiation theory with those of classical theory. This is done by defining a quantity known as the *oscillator strength* f, the number of classical absorbers or radiators of unit amplitude which would be equivalent to one quantum absorber. This leads to the result

$$f_{ji} = \frac{m h \nu_{ji}}{\pi e^2} B_{ij} = \frac{g_j c^2 m A_{ji}}{g_j 8 \pi^2 \nu_{ji}^2 e^2}. \tag{2.6}$$

3. Broadening of spectral lines. The infinitely precise PLANCK frequencies envisioned in the foregoing theory of radiators, eq. (2.1), are never observed in reality, but instead narrow distributions of frequencies are found, centered about, or in the near vicinity of, the PLANCK frequency. These distributions are commonly referred to as "broadened spectral lines".

Classified causally, contributions to the broadening are found to fall into three species: those in which the FOURIER analysis of the emitted wave train contains other components than the PLANCK frequency because of phase discontinuities during the emission process; those in which the normal behavior of the radiators results in quanta which from the standpoint of the radiation field and the observer are not of standard PLANCK frequency; and those in which environmental perturbation of the radiator's energy levels leads to the emission of quanta which from the standpoint of an unperturbed atom are not of standard PLANCK frequency.

In the first two cases the average probability per radiator per unit time that a spontaneous emission will occur (as obtained by integrating the monochromatic probability over the distribution function of frequencies which the conditions permit) remains the same as for a PLANCK radiator, and is given by eq. (2.2). In the third case there is a slight dependence of the average probability upon the magnitude of the perturbation. If the distribution of perturbed energy levels can be approximated by a GAUSSian distribution, then because of the ν^4 dependence of the transition probability A_{ji}, the average probability will be increased in the second approximation by the factor $\left(1 + \frac{3}{4}\frac{\delta^2}{\nu^2}\right)$. Here δ is the half width, or frequency separation between those two points of the broadened line at which the intensity is half the maximum intensity.

α) Natural breadth. The wave trains emitted by an atomic or molecular radiator are of finite energy, and hence are never of the infinite duration or continuous phase that is required of truly sharp frequencies. The unavoidable beginning and end of such a finite train correspond under FOURIER transformation to a distribution of frequencies characterized by a half width δ_N, called the *natural* half width. This natural half width can be computed by a summation of all the transition probabilities linking the initial and terminal states of the transition to states lying below them, i.e.,

$$\delta_N = \frac{1}{2\pi}\left(\sum_{i=1}^{j-1} A_{ji} + \sum_{k=1}^{i-1} A_{ik}\right). \tag{3.1}$$

The distribution of frequencies itself can be expressed in terms of the probability $W(\nu)\,d\nu$ that radiation of frequency between ν and $\nu + d\nu$ will be observed from one radiator per unit time.

$$W(\nu)\,d\nu = \frac{A_{ji}}{2\pi}\,\frac{\delta_N\,d\nu}{(\nu-\nu_{ij})^2 + \delta_N^2/4}. \tag{3.2}$$

β) Impact broadening. Particles in the vicinity of a radiator can have an important influence on its behavior through their force fields. Two kinds of broadening may often be distinguished with advantage, even though they stem from this one common cause, and are amenable to a single all-embracing approach[1]. The permanent displacements of the energy levels of the radiators throughout their entire radiation period (relaxation time), by the fields of those neighbors whose motion is sufficiently small that they may be regarded as stationary, leads to a broadening termed *statistical*, since it depends on the statistical distribution of neighbors positionwise. The ephemeral displacements produced in the same manner, but by perturbers moving with such speed that their effects are felt for much less than the relaxation time of the radiator, serve only to introduce phase discontinuities in the emitted wave train which directly augment the natural breadth of the spectral lines by an amount symbolized by δ_c. This second effect is termed *impact* (or collision) broadening.

The theory of impact broadening has been formulated fully, but exact calculations are understandably difficult. An order of magnitude calculation which may be trusted within factors of order two is given by the WEISSKOPF[2] theory. If the frequency perturbation in the interaction is represented by the expression

$$\Delta\nu = \frac{C}{r^{\varkappa}}, \tag{3.3}$$

the half width δ_c is given by

$$\delta_c = \frac{2\sigma N}{\pi^2}\sqrt{\frac{2\pi k T}{\mu}}, \tag{3.4}$$

where σ (a function of C and $\varkappa$) is the impact cross section for the collision, N is the concentration of perturbers, and μ is the reduced mass of the colliding system. A tabulation of σ for the situations in which impact broadening figures importantly is given in Table 1. Experimental tests of the theory of impact broadening for optical levels confirm almost without exception the linear dependence of δ_c upon N.

Perturbations by an inverse square law interaction require special analysis. UNSÖLD[3] has shown that when the perturbers are electrons, the half width is

$$\delta_c = \frac{9h^2\alpha^2 N}{16 m^2 \bar{u}}\left(0.923 - \ln\eta_n + \frac{\eta_n^2}{24} + \cdots\right), \tag{3.5}$$

where

$$\eta_n = \frac{3h\alpha N^{\frac{1}{3}}}{4m\bar{u}}, \tag{3.6}$$

[1] L. SPITZER jr.: Phys. Rev. **58**, 348 (1940).
[2] V. WEISSKOPF: Z. Physik **75**, 287 (1932).
[3] A. UNSÖLD: Vjschr. Astr. Ges. **78**, 213 (1943).

Table 1. *Impact cross sections for various interactions.*

Radiator	Perturber	Interaction	$\varkappa$	σ
Neutral	Ion	STARK 1st order	2	See eq. (3.5)
Neutral	Identical neutral	Resonance	3	$\frac{4\pi^2 C}{\overline{u}}$
Neutral	Dipole	STARK 1st order	3	$\frac{4\pi^2 C'}{\overline{u}}$
Ion	Neutral	STARK 1st order	3	$\frac{4\pi^2 C''}{\overline{u}}$
Neutral	Ion	ZEEMAN $\left[\Delta\nu = \frac{C'''\,\overline{u}\,\sigma}{r^3}\right]$	3	$16\pi^3 (C''')^2$
Neutral	Quadripole	STARK 1st order	4	$\pi^3\left(\frac{C}{\overline{u}}\right)^{\frac{2}{3}}$
Neutral	Ion	STARK 2nd order	4	$\pi^3\left(\frac{C'}{\overline{u}}\right)^{\frac{2}{3}}$
Neutral	Unlike neutral	VAN DER WAALS (STARK 2nd order);	6	$\pi\left(\frac{6\pi^2 C}{8\,\overline{u}}\right)^{\frac{2}{5}}$

and α is the coefficient of STARK splitting for the radiator. GRIEM[1] has found it adequate to approximate α by $\frac{1}{2}n(n-1)$, n being the principal quantum number of the upper state.

The impact broadening produced by the collisions between the radiator and massive molecules is often termed pressure broadening. It is not significant below atmospheric densities, with ordinary dispersion.

Impact broadening, according to the foregoing discussion, leads to a probability $W(\nu)\,d\nu$ of observing radiation of frequency between ν and $\nu+d\nu$ per unit time per radiator which has the form:

$$W(\nu)\,d\nu = \frac{A_{ji}}{2\pi}\cdot\frac{(\delta_N+\delta_c)\,d\nu}{(\nu-\nu_{ji})^2+(\delta_N+\delta_c)^2/4}\,. \tag{3.7}$$

γ) Statistical broadening. Statistical broadening is analyzed by constructing the HOLTSMARK[2] probability function for the electric field which will be experienced at any given point in a randomly distributed swarm of static field centers. This function is found to depend only on the mean distance between perturbers and the character of their individual fields. This may be expressed by saying that the random field E is a function of the mean field existing at an action radius distance from one perturber. This average field was termed the *normal* field E_n by HOLTSMARK, in defining a dimensionless parameter $\beta = E/E_n$ in terms of which he plotted his statistical functions. For ionic perturbers he found

$$\left.\begin{aligned} w(\beta) &= \frac{4}{3\pi}\,\beta^2\left[1-0.4628\,\beta^2+0.1227\,\beta^4-0.0232\,\beta^6+\cdots\right], \quad \beta<1;\\ w(\beta) &= \frac{2.350}{\pi\beta^{\frac{5}{2}}}\left[1+\frac{5.106}{\beta^{\frac{3}{2}}}-\frac{7.4375}{\beta^3}+\frac{0.1}{\beta^{\frac{9}{2}}}-\cdots\right], \quad \beta>1. \end{aligned}\right\} \tag{3.8}$$

For ionic perturbers, HOLTSMARK found $E_n = 2.61\;eN^{\frac{2}{3}}$.

[1] H. GRIEM: Z. Physik **137**, 280 (1954).
[2] J. HOLTSMARK: Ann. Phys., Lpz. **58**, 577 (1919).

For dipole perturbers the function is a closed one;

$$w(\beta) = \frac{4\beta^2}{\pi(1+\beta^2)^2}. \tag{3.9}$$

Here $E_n = 4.54 \mathcal{M} N$, where $\mathcal{M}$ is the dipole moment of a perturber. HOLTSMARK also gives the function for quadripole perturbers. Plots of these functions are shown in Fig. 1.

In using the HOLTSMARK function we make use of the fact that the STARK shift $\Delta\nu_k$ of any component, say the kth, of a line whose undisturbed PLANCK frequency is ν_0 can be approximated by a series in E.

$$\Delta\nu_k = (\nu_k - \nu_0) = \alpha_k E + \gamma_k E^2 + \cdots. \tag{3.10}$$

The power to which E is raised is spoken of as the *order* of the STARK effect. Only radiators having nonpenetrating orbits exhibit first order effects. Hydrogen and ionized helium set the most striking examples, although the first order effect should also be observable in the fundamental series of alkali metals.

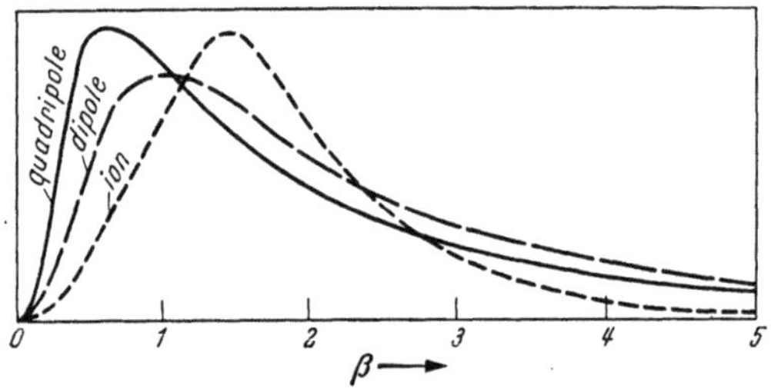

Fig. 1. The HOLTSMARK probability functions. [After HOLTSMARK: Ann. Phys., Lpz. 58, 38 (1919).]

If now the kth component of a spectral line normally has a profile $W'(\nu, \nu_0)$ centered about ν_0 according to some empirical expression or to theoretical forms such as eqs. (3.7) or (3.13), in the electric field it will be shifted by an amount $\Delta\nu_k$ to center about some new frequency ν_k. Then the probability $W_k(\nu)\,d\nu$ of radiation of frequency between ν and $\nu + d\nu$, per unit time per radiator will be given by

$$W_k(\nu)\,d\nu = \frac{d\nu}{E_n} \int_0^\infty W'(\nu, E)\, w(E)\, dE, \tag{3.11}$$

where ν_0 has been replaced first by ν_k, the shifted profile center, and then by E through the use of eq. (3.10).

The profile of the entire spectral line is lastly determinable by summing W_k over all STARK components, as has been shown by SCHMALJOHANN[1].

What the exact range of validity of the statistical theory really is, has not yet been determined with certainty. SPITZER[2] has shown that electrons almost always contribute to impact broadening rather than statistical broadening because of their high intrinsic speed. For heavy ions, the criterion of significance is more obscure. In terms of a parameter γ which must be infinite to achieve the HOLTSMARK distribution, for protons the ion temperature must fulfill the condition

$$T < \frac{3 \times 10^{-6} N^{\frac{2}{3}}}{\gamma^2}, \tag{3.12}$$

where N is the ion concentration. Assuming that the critical $\gamma \sim 10$ and $N \sim 10^{17}$, the theory would show failure at around 10^4 °K. There is some experimental evidence that this may be the case[3] in the anomalously long time buildup of ion concentration observed in low pressure sparks.

[1] P. SCHMALJOHANN: Diss. Kiel (1936), unpublished (see UNSÖLD, reference 3 of p. 212).
[2] L. SPITZER jr.: Phys. Rev. **53**, 39 (1939).
[3] R. G. FOWLER, W. R. ATKINSON and L. W. MARKS: Phys. Rev. **87**, 966 (1952).

GRIEM[1] has combined the joint electron and ion effects present when a plasma is under study by assuming that the electrons contribute via impact broadening to the profile $W'(\nu, E)$ of the STARK components, which profile he then distributes statistically according to eq. (3.11). This is reasonable as long as numerous electron impacts are experienced by the radiator during the relaxation period. His agreement with experiment is among the best achieved thus far.

Small first order STARK broadenings are symmetrical about the undisturbed PLANCK frequency but as E_n increases, the variation of transition probability with ν_0^4 produces an asymmetry. Simultaneously, and to the same approximation, the corrections upon the intensity distribution in consequence of second order STARK effect also produce an asymmetry[1].

Radiators which show no first order STARK effect present a problem in analysis since the impact electron broadening and second order statistical heavy ion broadenings will be comparable in magnitude. Strongly broadened asymmetrical lines, which suggest statistical second order broadening are observed in such dense regions of ionization as the cathode spot of a high pressure arc[2].

δ) Doppler broadening. Broadening is also observed in spectral lines because of the simple thermal motion of the radiators along the line of observation. By radiation reaction, a portion of the translational energy of the radiator is converted into radiation during the recoil. For a swarm of radiators this results in a statistical distribution of DOPPLER shifted quanta. If a radiator is in motion along the line of observation with a velocity component u, its PLANCK frequency ν_0 will be received by the observer as a frequency $\nu_0\left(1-\frac{u}{c}\right)$. If $\varphi(u)\,du$ is the fraction of radiators with a velocity between u and $u+du$ in the line of sight, then the probability of receiving a quantum ν will be obtained by averaging over the distribution $\varphi(u)$. If the radiators are in thermal equilibrium with respect to translation so that $\varphi(u)$ is the MAXWELL-BOLTZMANN distribution, and a natural broadening plus weak pressure broadening profile is combined with the DOPPLER shift, it follows that

$$W(\nu)\,d\nu = \frac{c A_{ji}}{\nu_{ji}} \sqrt{\frac{M}{2\pi k T}} \left[e^{-\omega^2} - \frac{(\delta_N+\delta_c)\,c}{\nu_{ji}} \sqrt{\frac{M}{2\pi k T}} \left(1 - 2\omega F(\omega)\right)\right] d\nu, \tag{3.13}$$

where $\omega = c\left(\frac{\nu-\nu_{ji}}{\nu_{ji}}\right)\sqrt{\frac{M}{2\pi k T}}$, and $F(\omega) = e^{-\omega^2}\int\limits_0^{\omega} e^{\omega^2}\,d\omega.$

The quantity $\frac{\nu_{ji}}{c}\sqrt{\frac{2kT}{M}}$ is a measure of the broadening caused by the DOPPLER effect alone. It ranges in magnitude from 0.01 Å to 0.1 Å over the gamut of normal discharge conditions.

A useful approximation to this expression for the monochromatic emission probability is

$$W(\nu)\,d\nu = \frac{c A_{ji}}{\nu_{ji}} \sqrt{\frac{M}{2\pi k T}} \left[e^{-\omega^2} + \frac{(\delta_N+\delta_c)\,c}{2\nu_{ji}} \sqrt{\frac{M}{2\pi k T}}\,\frac{1}{\omega^2}\right] d\nu. \tag{3.14}$$

The first term in this expression is significant for small values of ω, while for large values the second term dominates. The profile consists, therefore, of an error function core in the center of the usual profile for natural plus pressure broadening which decreases as $(\nu-\nu_{ji})^{-2}$ on the wings.

[1] H. GRIEM: Z. Physik **137**, 280 (1954).
[2] R. M. ST. JOHN: Phys. Rev. **98**, 1664 (1955).

DICKE[1] has demonstrated that during magnetic dipole transitions, the recoil momentum of the radiation may not be given to the radiator, but to a near neighbor, resulting in a narrower DOPPLER width than predicted above.

4. Statistics of radiating systems. If the probability per unit time of spontaneous emission of a frequency ν by a single radiator is known, it is in principle possible to compute the intensity of radiation of this frequency when an assemblage of radiators is under consideration. The problem of radiators in equilibrium with their own radiation was analyzed in eq. (2.5). In terrestrial discharges, however, it was pointed out that radiators are rarely in equilibrium with their radiation. They are much more likely to be in equilibrium with the free electron population, as LADENBURG[2] has shown for neon metastables. On the other hand they are usually not in equilibrium at all, but are merely in a steady state, with energy losses of one kind balanced agsinst energy gains of another kind.

Low pressure gas discharges, despite the real complexity of their energy transfer processes, are simple to analyze when compared with the vastly more complex situation in terrestrial high pressure discharges, or in stellar atmospheres whether at high or low pressure, these last being complicated by their vast extension over space. To treat this latter situation it is necessary to develop and solve the general *equation of transfer*[3] as a means of extending the calculation from the individual to the group, and in this equation mathematical difficulties of a high order are encountered.

A working criterion for a low pressure gas discharge, from the optical point of view, can be that it should not appreciably reabsorb its radiations, or in other words that spontaneous emission is its sole radiation process. Within any desired accuracy ε, the low pressure theory will in this sense apply to transitions ending above the ground state if

$$\frac{\lambda^2 \sigma i N_0 \ell}{A e} < \varepsilon, \tag{4.1}$$

where λ is the wavelength of the radiation, A is the transition probability, ℓ is the length of the radiation path through the gas, σ is the mean excitation cross-section, i is the exciting current density, and N_0 is the molecular concentration. For the conventional gas discharges, at an accuracy of 1%, the limiting concentration would lie near 10^{17} cm.$^{-3}$. For transitions ending on the ground state, the condition becomes

$$\lambda^2 N_0 \ell < \varepsilon, \tag{4.2}$$

and N_0 must in general be less than 10^9 cm.$^{-3}$ for absorption to be negligible. For transitions ending on metastable states, an estimate of $\sim 10^{12}$ cm.$^{-3}$ for the critical concentration would be obtained by eq. (4.2), but non-radiative loss processes may greatly increase this value.

It is the partial concentration of a given type of gas molecule which is important in reabsorption, so that low pressure theories with suitable modifications to include the secondary effects of the major components, can be applied to trace components in high pressure gas mixtures such as occur in the thermal arcs and sparks used in spectrochemical practice.

It is also apparent from the discussion of reabsorption that as a practical matter the definition of low pressure discharge must be relaxed from one in which all radiations escape to one in which upper terminal state radiations escape.

---

[1] R. H. DICKE: Phys. Rev. **89**, 472 (1953).

[2] R. LADENBURG: Rev. Mod. Phys. **5**, 243 (1933).

[3] S. CHANDRASEKHAR: Radiative transfer. Oxford 1950.

Ground state radiations will thus be blockaded in nearly every situation under discussion in this article, and if alternative terminal levels are attainable from the given excited state, the energy of the blockaded radiations will be gradually converted into transmissible form plus metastable energy, or resonance radiation, or both, depending on the structure of the radiator. If the radiator possesses a resonance level, then at concentrations above the escape value [eq. (4.2)], this level will be populated to nearly the same degree as a metastable level would be. An analysis of the propagation of blockaded radiations has been made by assuming that the problem was one of diffusion, with the possibility of defining a mean free path of radiation which could then be treated as equivalent to a mean free path for the translation of the excited state[1].

Recently[2] the possibility of defining a mean free path has been challenged, but it can be shown that the argument advanced against it is not inexorable, and that the propagation is adequately described by a diffusion process (see Sect. 7).

For those transitions which pass unhindered through the gas, the analysis of radiated intensity is straightforward. If N_j refers to the population of the jth excited state of the radiator, then the essential changes in this population will be brought about by radiative transition to lower states, radiative transitions from upper states, absorption of blockaded ground state radiations, and collision interactions with other particles, which last will not be analyzed further at present, but will be described by a *production function* P, which will be regarded as having both positive (gain) and negative (loss) aspects. Then a system of equations governs the energy level populations.

$$\frac{\partial N_j}{\partial t} = \left\{ D_j \nabla^2 N_j + B_{0j}\,\varrho\, N_0 - \sum_{i=0}^{j-1} A_{ji} N_j \right\} + \sum_{k=j+1}^{\infty} A_{kj} N_k + P_j \,. \tag{4.3}$$

D_j is the diffusion coefficient for state j radiators. Of the terms in the bracket (1) only the first occurs when j is a metastable state, (2) only the third occurs when j is a state which does not radiate to the ground state, and (3) both the second and third are needed when j radiates to the ground state, and a modification of eq. (4.3) is necessary when the radiation is blockaded.

When j is one of those states which radiate to the ground state, then assuming that the diffusion treatment of blockaded radiation is adequate, a second set of equations is needed, each being an extension of eq. (2.5).

$$\frac{\partial \varrho}{\partial t} = D_\nu \nabla^2 \varrho + h\nu_{j0} \left[A_{j0} N_j - B_{0j}\, \varrho\, N_0 \right]. \tag{4.4}$$

D_ν is a photon diffusion coefficient defined as $c/3k$, where c is the velocity of light and k is the photon absorption coefficient. Eliminating the absorption term between eqs. (4.3) and (4.4), and assuming that the gas density is large enough that the photons spend much less time in the free state than in the radiators, then $N_j > \varrho/h\nu$, and one set of equations can be used for the blockaded states to replace the two sets required above:

$$\frac{\partial N_j}{\partial t} = D_j \nabla^2 N_j - \sum_{i=1}^{j-1} A_{ji} N_j + \sum_{k=j+1}^{\infty} A_{kj} N_k + P_j \,. \tag{4.5}$$

[1] K. T. Compton: Phys. Rev. **20**, 283 (1922). — E. A. Milne: J. Lond. Math. Soc. **1**, 1 (1926). — C. Kenty: Phys. Rev. **42**, 823 (1932).

[2] T. Holstein: Phys. Rev. **72**, 1213 (1947); **83**, 1159 (1950).

When this complete set of equations [eq. (4.3)] as amended by eq. (4.5) is solved for the several N_j, the monochromatic power radiated in any one transition can be found by integrating over the active volume of the discharge if the discharge radiates isotropically as follows:

$$h\nu_{ji} A_{ji} \int_v N_j \, dv \,. \tag{4.6}$$

If steric factors must be considered, the power radiated per unit volume per unit solid angle is needed:

$$\frac{h\nu_{ji}}{4\pi} A_{ji} N_j \,. \tag{4.7}$$

The difficulties of the problem center around identification of the major processes contributing to those changes in N_j which have been collected under the heading *production*. These vary with the discharge studied, and it is this variety which leads to the individuality of discharges.

B. Excitation.

5. Introduction. The process of creating a population of radiators having a common discrete state is called *excitation*. This may be accomplished in many ways: in inelastic collisions between passing particles; through capture by ions of available electrons, either unattached ones, or those which are attached to

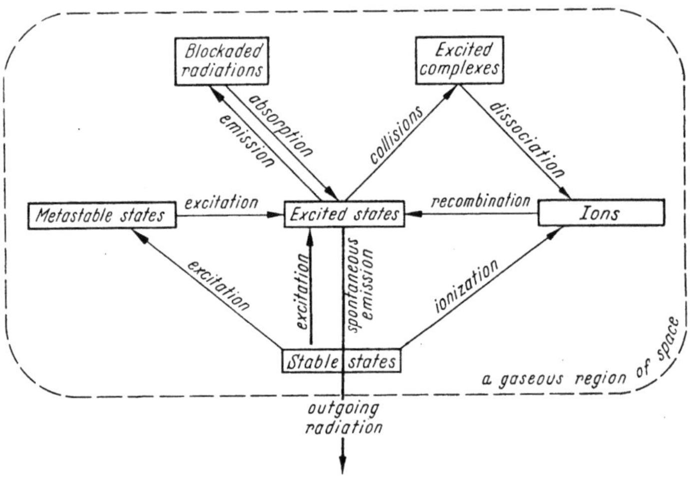

Fig. 2. A flow chart of radiation producing processes in a gas.

less possessive gas molecules; and finally by the interaction with radiation. Inelastic collision processes may be subdivided[1] into inelastic *collisions of the first kind*, in which the total kinetic energy of the colliding system is reduced by the collision; and inelastic *collisions of the second kind*, in which the internal potential energy of at least one member of the colliding system is reduced by the collision. These processes are depicted in Fig. 2 as they occur in any self maintained discharge.

None of these processes have excitation as their only consequence; ionization to one or more stages is also possible. The production of ionization and excitation by inelastic collisions of the first kind will be termed *elatation*.

[1] O. Klein and S. Rosseland: Z. Physik **4**, 46 (1921).

6. Excitation by particle impact. Elatation is the primordial source of all radiation from gas discharges. Radiators in any state may absorb the kinetic energy of impinging particles and go over into higher energy states with a minimum of restriction by selection rules. As a general thing theory shows that transitions between states for which the reverse radiative transition is allowed are somewhat more probable than to those which are forbidden in radiation.

Free electrons and presumably positrons are by far the most efficient impacting agents because relative to the atomic electrons they possess the most favorable mass ratio for energy transfer. Massive particles with a given velocity behave in the fashion that electrons of the *same velocity* would behave, since for particles beyond mass equality the velocity acquired by the struck particle in a collision increases at most by a factor of two. The ratio of the energies of massive particles and electrons which are capable of producing comparable excitations is thus proportional to their mass ratio. Because the least mass ratio (for hydrogen) is $\sim 2\times 10^3$, the relative efficiency of massive particles in elatation will be negligibly small except in discharge regimes involving very strong fields, such as in cathode fall spaces, or in gas filled x-ray tubes.

α) Electron impacts of the first kind. The study of the radiation yield of electron impacts began seriously with the discovery by Franck and Hertz[1] of the critical potentials in gases. A refined and beautiful version of these experiments was performed by Newman[2] who showed that the spectrum of Na can be excited one line at a time when pure metal is used, contrary to the findings of Foote, Meggers and Mohler[3] who had found that after the appearance of the D lines, it was necessary to exceed the ionization potential before any more lines appeared. Quantitative measurements of the efficiency with which electrons transfer energy to gas molecules was next undertaken. Collision processes are governed by a quantity called a collision cross section which is represented by σ and defined for binary collisions by the equation

$$\frac{dN}{dt} = \sigma N_1 N_2 \bar{u}\,, \tag{6.1}$$

where N_1 and N_2 are the concentrations of colliding particles of both species, $\bar{u}$ is their mean relative velocity, and dN/dt is the rate of production of the altered molecular state.

Experimental measurements of the cross sections which gas molecules offer to electrons for various exciting transitions have been carried out along two general lines. First, electrical measurements have been made[4] of the fraction of electrons which have lost discrete amounts of energy. Second, optical measurements have been made of the number of photons of a given species emerging from a gas through which a known charge has passed [*1*]. Both methods are difficult and the results have proved quantitatively discordant with each other and with theory whenever quantitative measurement has been possible. Additional experimental work is needed. Despite the admitted difficulties of the theory, in the present state of knowledge it is entirely possible that the theoretical cross sections are superior to the experimental ones.

In the electrical method, electrons which have lost large amounts of energy do not diffuse into a screened cage held at a variable retarding potential. The measurements are not meaningful beyond the ionization potential of the gas,

[1] J. Franck and G. Hertz: Verh. dtsch. phys. Ges. **16**, 512 (1914).
[2] F. Newman: Phil. Mag. **50**, 165 (1925).
[3] P. Foote, W. Meggers and F. Mohler: Astrophys. J. **55**, 145 (1922).
[4] H. Maier-Leibnitz: Z. Physik **95**, 499 (1935).

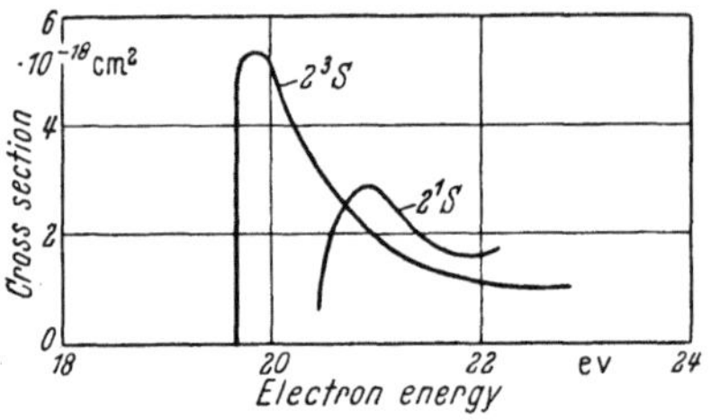

Fig. 3. Electrically determined excitation cross sections in helium. [After MAIER-LEIBNITZ: Z. Physik **95**, 499 (1935).]

and seem to display too sharp a maximum, when compared with theory. Only a few of the lowest states of certain favorable atoms can be investigated by this method. Fig. 3 gives two of the best curves obtained.

Optical measurements of the cross sections are a direct approach to the desired quantities. To make such measurements significant, an accurate understanding of the processes which interplay during the conversion of the electron energy to radiation, and an equally certain understanding of the electrical processes in the discharge employed are essential. In most of the experiments it has been assumed that grids introduced in the gas discharge would influence the discharge by the laws of vacuum electrostatics, and hence that the accelerating potentials were accurately known. This may be doubted, owing to the well known impotence of control grids in gas discharges because of sheath formation. It is probable that the entire discharge has been controlled by a LANGMUIR sheath about the cathode in all cases in which thermionic cathodes were used, rather than by the grids. Hence it is unlikely that the electron velocities were known with true precision.

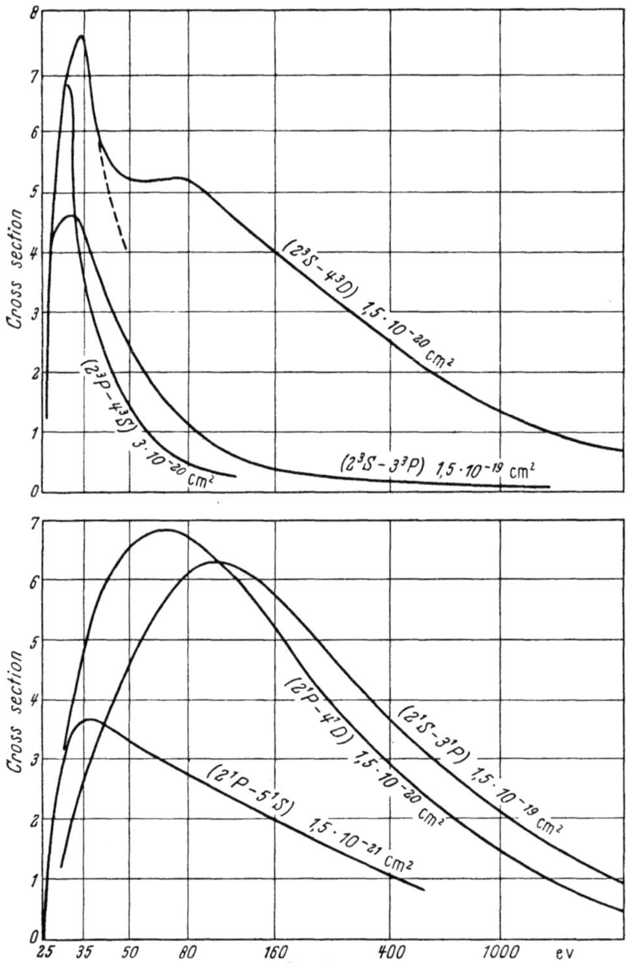

Fig. 4. Optically determined excitation cross sections in helium. The factor required to give the ordinates of each curve in cm.² is labelled on the curve. Curves for the ³*D* states show an enhancement believed to result from neutral impacts on ¹*P* states. The dotted line shows the probable course of the true function. [After LEES: Proc. Roy. Soc. Lond., Ser. A **137**, 173 (1932).]

In the matter of radiation processes it has been assumed that they were the simple electron excitations that it was desired to measure. It is hardly profitable to speculate on the importance of alternatives to this assumption in the absence of quantitative data. Possible alternatives are, however, (a) the conversion of imprisoned radiations; (b) the re-excitation of excited states to thermally proximate excited states by neutral particle impacts; (c) the formation of molecular complexes between the excited states and neutrals; (d) the population of the excited states by recombination in any of its various forms. The first two processes were suggested by LEES and SKINNER[1] in discussing their measurements.

The form of the experimental optical cross sections[1] is typified by Fig. 4. Agreement in form with theory cannot be accepted as a sufficient test, however, since whatever the process, the governing source of energy is fundamentally the primary electrons, and they will necessarily become increasingly efficient in elatation for small increases in energy above the threshold, followed by a decrease in efficiency because the time duration of the electron's perturbation decreases as its velocity increases. It is apparent, therefore, that all curves will show a rise and fall of the same general character.

To summarize the experimental measurements, more effort is needed to obtain absolute cross sections. In those cases in which absolute estimates have been possible, the electrical method predicts a sharp rise to a maximum and an equally sharp decline in cross section in a span of half an electron volt above the critical potential. The optical method while not agreeing with this behavior yields diverse results in the hands of different investigators. Despite the demonstration by NEWMAN[2] that sharp energy resolution is experimentally possible, no effort has been made to apply high resolution in the neighborhood of the critical potentials. There is no sign of the electrical peak in the helium curves of LEES or THIEME[1], even though it should be of the same order of magnitude as the observed optical maxima. Instead, two sorts of maxima are observed: broad 100 ev. wide maxima, with a slow decrease on the high energy side, for all helium singlet transitions, and 10 ev. wide maxima, with a rapid decrease, for all helium triplet transitions. The sharp maxima are consonant with theory qualitatively but not quantitatively. The broad maxima are only expected in the case of ^1P levels.

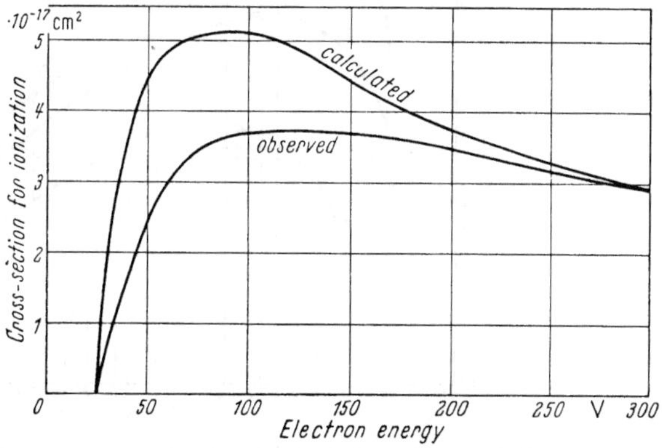

Fig. 5. Ionization by electron impact [2].

BATES and others[3] have reviewed the wave-mechanical theory of elatation by electrons recently and are unable to assess the relative merits of the various approximations needed in the solution of these collision problems because of the inadequacies and disagreements between experiments. A complete discussion of the theory will be found in volume 36 of this Encyclopedia. It is the general opinion that the BORN approximation, wherever its limitations are fulfilled, should give reliable results. It seems especially successful as a theory for the ionization process, where quantitative agreement with the experimental results[3,4] is excellent (Fig. 5).

Existing measurements of ionization cross section, being in good quantitative agreement with wave-mechanical theory, can in all probability be looked upon

[1] J. LEES and H. SKINNER: Proc. Roy. Soc. Lond., Ser. A **137**, 186 (1932). — J. LEES: Proc. Roy. Soc. Lond., Ser. A **137**, 173 (1932). — O. THIEME: Z. Physik **78**, 412 (1932).
[2] F. NEWMAN: Phil. Mag. **50**, 165 (1925).
[3] D. BATES, A. FUNDAMINSKY, J. LEECH and H. MASSEY: Philos. Trans. Roy. Soc. Lond., Ser. A **243**, 93 (1950)
[4] J. TATE and P. SMITH: Phys. Rev. **39**, 270 (1932).

as trustworthy. The measurement of charged events is more certain to be free of error than the measurement of radiation events, if only because a greater sensitivity of detection permits experimentation in a hundredfold lower gas density range. In addition to the well known work of MASSEY [1] and his co-workers, a classical calculation of the form of the ionization function based on the RUTHERFORD scattering formula has been proposed by DE LA RIPELLE[1] with striking success at fitting the experimental curves in a number of gases (Fig. 6).

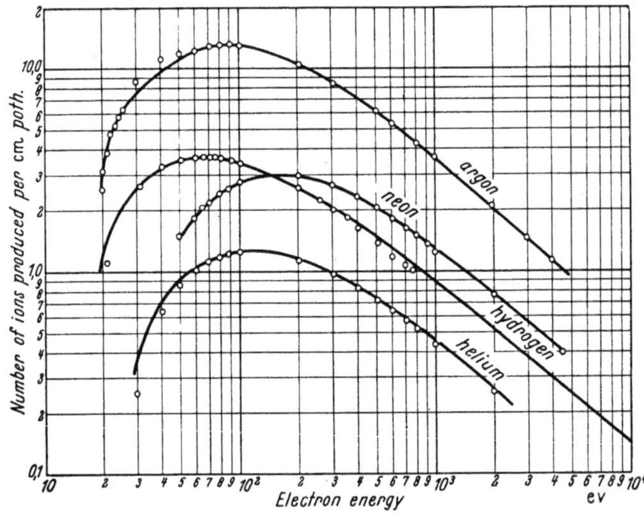

Fig. 6. Semiclassical theory of ionization by electron impact. The points are experimental data. [After DE LA RIPELLE: J. Phys. Radium 10, 319 (1949).]

MOTT and MASSEY [2] find that for optically allowed transitions, the cross-section is approximately given by

$$\sigma_{0j} \approx \frac{3}{4\pi} \cdot \frac{e^2 c^3}{h} \cdot \frac{A_{j0}}{u^2 \nu_{j0}^3} \ln \frac{2m u^2}{h \nu_{j0}}, \tag{6.2}$$

while for optically forbidden transitions which are allowed in quadripole radiation

$$\sigma_{0j} \approx \frac{3}{4\pi} \cdot \frac{e^2 c^3}{h} \cdot \frac{A_{j0}}{u^2 \nu_{j0}^3} \cdot \frac{m c^2}{2h \nu_{j0}}. \tag{6.3}$$

An expression having the same energy dependence as the cross-section for optically allowed transitions is found for the ionization cross-section by integrating over the continuum. DE LA RIPELLE, on the other hand, obtains a slightly different expression of the form

$$\sigma_{0+} = \frac{K}{V + \chi_0 V_+} \ln \frac{V}{V_+}, \tag{6.4}$$

where V_+ is the ionization energy of the radiator, and V is the electron energy, both in electron volts. Curve fitting gives the representative values of K and χ_0 set forth in Table 2.

Table 2. *Constants for eq. (6.4).*

Gas	K	χ_0
Argon	2.36×10^{-14}	2.9
Helium	3.55×10^{-15}	2.03

SKINNER and APPLEYARD[2] found that the process of electron excitation, being a time perturbation from the field of the passing electron, resembles in

[1] F. DE LA RIPELLE: J. Phys. Radium **10**, 319 (1949).
[2] H. SKINNER and E. APPLEYARD: Proc. Roy. Soc. Lond., Ser. A **117**, 224 (1928).

many ways the excitation found with the absorption of ground state photons. The emitted radiations were found to be partially polarized when the intensity of radiation passing an analyzer set with its plane parallel with the plane defined by the observation point and electron beam was compared with that analyzed at right angles to the beam. Defining a percentage polarization as

$$\Phi = 100 \frac{I_{\parallel} - I_{\perp}}{I_{\parallel} + I_{\perp}}, \tag{6.5}$$

it was found for mercury that both positive and negative values of Φ occur, as a function of electron velocity (Fig. 7). Theoretical analysis of the polarization was attempted by PENNEY[1], but use of the BORN-OPPENHEIMER approximation

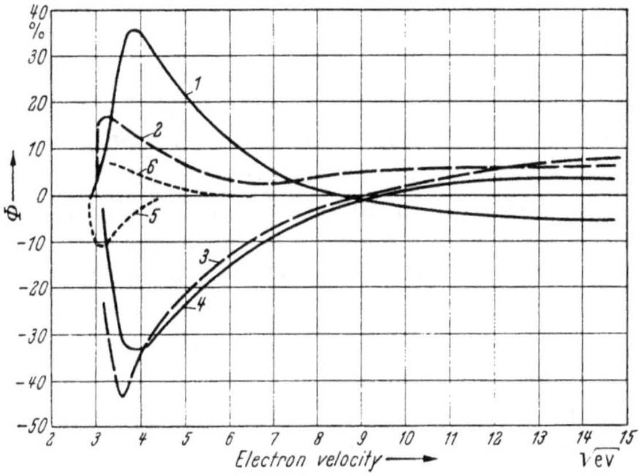

Fig. 7.

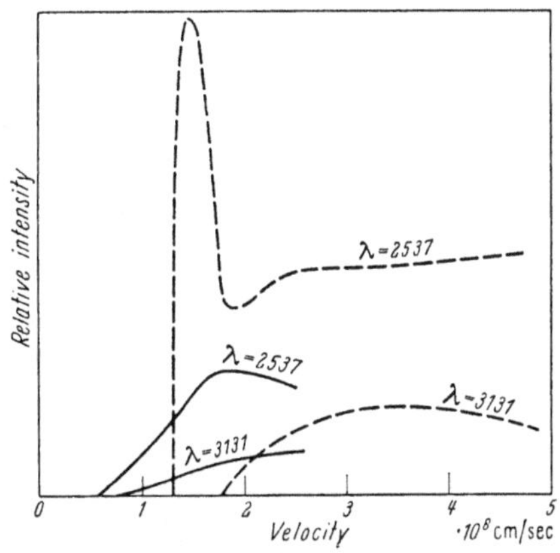

Fig. 8.

Fig. 7. Polarization of radiation excited by electron impact. [After APPLEYARD and SKINNER: Proc. Roy. Soc. A **117**, 224 (1928).]

Fig. 8. Relative cross sections for electron and hydrogen atom impacts (in mercury), compared on a velocity basis. [After GAILER: Ann. Phys., Lpz. **24**, 421 (1935).]

gave results in complete contradiction with the experiments. Further work on this phenomenon might assist in clarifying the excitation function paradox in general.

β) Excitation by massive particle impacts. Theory and experiment agree that the cross-sections for massive particle impacts compare with corresponding electron impacts on an equal velocity basis. The experimental measurements are very difficult. Radiation efficiencies are extremely low, and it is easy to set up subsidiary processes involving impacts of freed electrons which will mask the desired effects.

It seems probable that the excitation is velocity controlled partly because the inelastic interaction is between individual electrons of the two colliding systems. GAILER[2] has shown that hydrogen atoms are vastly more effective at velocity equality than naked protons. His curves for the excitation of mercury are typical of the general phenomenon observed. Electrons cannot produce elatation below critical potentials for lack of energy. Massive particles possess the necessary energy long before they reach velocity equality, and thus weak but increasing elatation is observed up to a peak of efficiency at velocity equality.

Massive particle impacts are of great importance in the deceleration of nuclear particles, but have less application in gas discharges. It is necessary to invoke

[1] W. PENNEY: Proc. Nat. Acad. Sci. **18**, 231 (1932).
[2] K. GAILER: Ann. Phys., Lpz. **24**, 421 (1935).

massive particle impacts in an explanation of meteoric light production. HERLOFSON[1] estimates that production of heat, light, and ionization are roughly in the energy proportions of $10^4:10^2:1$, and develops an expression for the stellar magnitude of meteorites of any given size. The luminosity observed in the region just behind a shock wave in the RAYLEIGH effect[2] and in standard shock tubes[3], seems certain to require molecular impacts as the agency in excitation, since the region from which the bulk of the radiation comes comprises gas molecules which had remained undisturbed until the instant at which they were swept up in the thermally violent advance of the shock wave. MEINEL's investigations[4] suggest that the polar aurora must be at least in part excited by up to 50 kV proton currents from the sun which impact on atmospheric gases.

Table 3. *Cross-sections for neutral impact excitation.*

Term	Cross-section in 10^{-15} cm.^{2}
$3\,^3P$	2.1
$3\,^3D$	11.9
$4\,^1D$	67.0
$4\,^3D$	15.0
$4\,^1S$	3.0
$4\,^3S$	1.5
$5\,^1D$	51.0
$5\,^3D$	27.0
$5\,^1S$	1.5
$5\,^3S$	0.9

WOLF and MAURER[5] in analyzing the conversion of blockaded ground state radiations in helium, found that not only the expected transitions which originate from the ^1P levels were observed, but practically the complete spectrum. This effect has also been observed by LEES and SKINNER, and others[6] in the diffuse glow about a collimated beam, and in the LANGMUIR dark space of a hot cathode discharge. Possibly it may also exist in the FARADAY dark space of a glow discharge[7], and the long exposure required to observe it may have been the reason that only one line was reported there. WOLF and MAURER proposed that the extra radiations are excited by impacts of neutral helium atoms on the ^1P population, transferring it (with the addition of kinetic energy) to the ^1D, ^3P, and ^3D states. The intensity of the extra lines is quite large. In every case the energy deficiency between the states is comparable with the mean thermal energy of the neutrals, and the probability of the process decreases with increasing energy discrepancy. WIGNER's rule is violated, but may not apply to this situation.

In their experiments WOLF and MAURER took every precaution against the diffusion of ions and electrons into the observation space. Analysis of the experimental results leaves little choice but to agree that the explanation of neutral impacts is the most logical possibility, despite the enormous cross-sections it implies. If this be accepted, it is probable that the same process must be important in active discharges in many cases to produce a profound alteration of the balance of population, and account in all probability for the peculiar behavior of the excitation functions for electron impact discussed in Sect. 6α. Cross-sections for the suggested process are given in Table 3. WOLF and MAURER suggest that reverse collisions of the second kind are also necessary to explain the abundance dependences of the extra transitions.

[1] N. HERLOFSON: Rep. Phys. Soc. Progr. Phys. **11**, 450 (1946/47).

[2] J. STRUTT: Proc. Roy. Soc. Lond., Ser. A **183**, 20 (1944). — R. FOWLER, W. ATKINSON, W. COMPTON and R. LEE: Phys. Rev. **88**, 137 (1952).

[3] E. RESLER, S. LIN and A. KANTROWITZ: J. Appl. Phys. **23**, 1390 (1952). — R. HOLLYER, A. HUNTING, O. LAPORTE, E. SCHWARCZ and E. TURNER: Phys. Rev. **87**, 911 (1952).

[4] A. MEINEL: Phys. Rev. **80**, 1096 (1950).

[5] R. WOLF and W. MAURER: Z. Physik **92**, 100 (1934); **115**, 410 (1940).

[6] J. LEES and H. SKINNER: Proc. Roy. Soc. Lond., Ser. A **137**, 186 (1932). — L. MAXWELL: J. Franklin Inst. **214**, 533 (1932). — R. FOWLER and O. DUFFENDACK: Phys. Rev. **76**, 81 (1949).

[7] K. EMELEUS and O. DUFFENDACK: Phys. Rev. **47**, 460 (1935).

7. Excitation by absorption of photons [*3*]. Discussion has already been given to the probability of photon absorption and the effect of self-absorption on the statistical balance in discharges (Sect. 4). As in the emission of radiation, the selection rules appear to govern its absorption rigorously. Apparent deviations have always resulted in the discovery of subsidiary processes involving other systems.

FÜCHTBAUER[1], irradiating mercury vapor with light from a powerful mercury arc, displayed the ultimate in excitation of a gas by radiation alone. WOOD[2] devised exacting tests of the correctness of FÜCHTBAUER's assumption that he was dealing with stepwise excitation by absorption. They found it possible to excite strongly the emission of those frequencies which required the absorption of two successive quanta. The success of this procedure is understandable since the intermediate levels in the two stage absorption are either metastable or resonance levels, and hence develop a rather large and relatively permanent population.

MAURER and WOLFE investigated the absorption of ground state radiations in helium as discussed in Sect. 6β, and found it to be an exciting process of order of 1% of the major beam processes even at a distance of >2 cm. from the exciting electron beam. It is certain that absorption in the beam itself it must be treated seriously at ordinary discharge pressures.

MOHLER[3] studied in caesium the ionization which can be accomplished by radiation. He found that there are probably three processes which lead to ionization. First, the direct absorption of quanta with energy exceeding the ionization energy results in the ejection of an electron from the atom. This is photoelectric action, and is the reverse process to recombination. Second, the absorption of discrete line radiations apparently results in ionization by collisions with neutrals while the excited state persists, according to the reaction

$$Cs^* + Cs \rightarrow Cs_2^+ + e.$$

Third, Cs_2 molecules apparently exist in the vapor and can be excited to high molecular states from which they spontaneously dissociate to molecular ions.

The behavior of blockaded resonance radiations which propagate through the gas by successive absorption and emission acts has been investigated chiefly in studies of the time required for a pulse of incident radiation to be released from the gas. ZEMANSKY[4] and recently ALPERT, MCCOUBREY and HOLSTEIN[5] have made experimental measurements of this decay time for mercury, which has one of the most conveniently accessible resonance lines occurring in nature. Fig. 9 is taken from the work of the latter authors combined with earlier work by WEBB and MESSENGER[6]. Gas purity is of great importance because the long effective duration of the excited states provides sufficient time for collisions of the second kind with neutrals to remove the energy by even rather improbable processes. At high densities of even the pure gas itself this quenching becomes important because of molecular formation. As the decay constant plotted in Fig. 9 falls below the linear behavior characteristic of a diffusive process, mercury band fluorescence sets in, indicative of losses of 6^3P_1 states by molecular formation,

[1] C. FÜCHTBAUER: Phys. Z. **21**, 635 (1920).
[2] R. WOOD: Proc. Roy. Soc. Lond., Ser. A **106**, 679 (1924).
[3] F. MOHLER and C. BOECKNER: U.S. Bur. Stds. Jour. Res. **5**, 51, 831 (1930).
[4] M. ZEMANSKI: Phys. Rev. **29**, 513 (1927).
[5] D. ALPERT, A. MCCOUBREY and T. HOLSTEIN: Phys. Rev. **76**, 1257 (1949).
[6] H. WEBB and H. MESSENGER: Phys. Rev. **33**, 319 (1929).

probably according to the reaction of MROZOWSKI[1],

$$\mathrm{Hg}\,(6\,{}^1S_0) + \mathrm{Hg}\,(6\,{}^3P_1) \rightarrow \mathrm{Hg}_2\,({}^3\Sigma_{1u}).$$

Much consideration has been given to theories of diffusion of imprisoned radiations which envision the process as a series of lightspeed jumps between atoms with more time spent in each radiation act than in transfer, so that the average speed of propagation is essentially l/τ, where l is the mean free path and

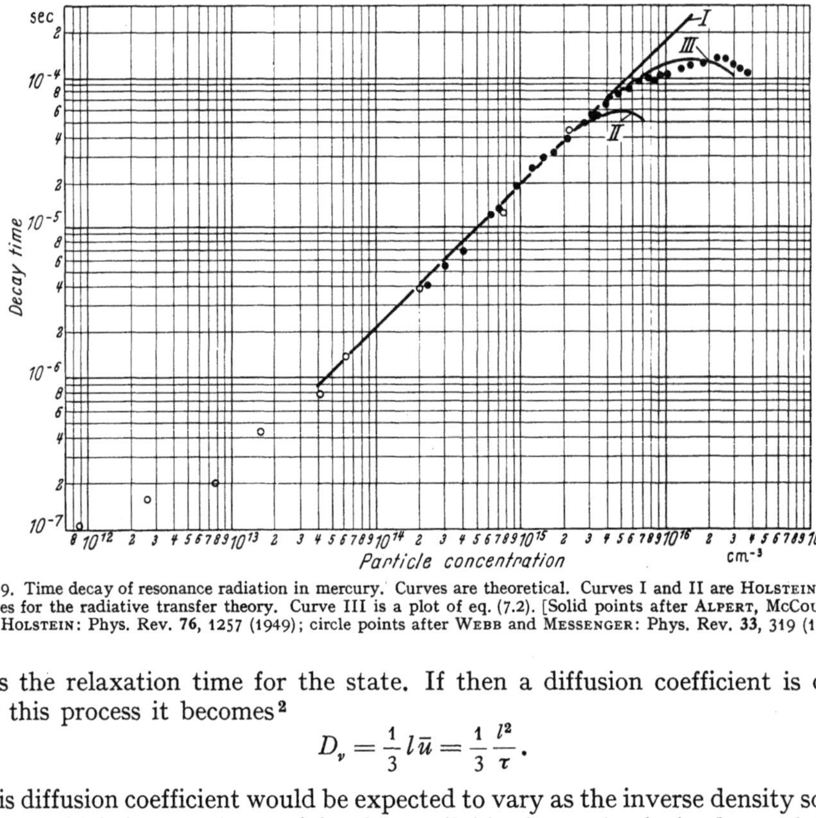

Fig. 9. Time decay of resonance radiation in mercury. Curves are theoretical. Curves I and II are HOLSTEIN's limiting curves for the radiative transfer theory. Curve III is a plot of eq. (7.2). [Solid points after ALPERT, McCOUBREY and HOLSTEIN: Phys. Rev. 76, 1257 (1949); circle points after WEBB and MESSENGER: Phys. Rev. 33, 319 (1929).]

τ is the relaxation time for the state. If then a diffusion coefficient is defined for this process it becomes[2]

$$D_\nu = \frac{1}{3}\, l\,\bar{u} = \frac{1}{3}\,\frac{l^2}{\tau}. \tag{7.1}$$

This diffusion coefficient would be expected to vary as the inverse density squared, a fact which is not observed in the available data. Analysis shows, however, that above 2×10^{13} molecules/cm.^{3}, mercury atoms will diffuse kinetically farther during their radiation time than the free path for resonance radiation of the central PLANCK frequency. Therefore, the data of Fig. 9 has been fitted to a curve combining simple diffusion of the excited states with loss by molecular formation at large densities. The equation deduced for this process is

$$\frac{1}{t} = \frac{\pi}{8}\left(\frac{2.405}{a}\right)^2 \frac{\bar{u}}{\sigma_1 N_0} + \sigma_2 N_0\,\bar{u}. \tag{7.2}$$

Here t is the decay time, σ_1 is the cross-section for diffusion of the excited states, σ_2 is the molecular formation cross-section, and a is the tube radius for the cylindrical geometry used experimentally. The cross-sections are found to be $\sigma_1 = 1.92\times10^{-15}$ cm.^{2} and $\sigma_2 = 1.34\times10^{-17}$ cm.^{2}. The kinetic cross-section is in reasonable agreement with the coefficient for self diffusion in mercury, which other methods give as about 5×10^{-15} cm.^{2}.

[1] S. MROZOWSKI: Z. Phys. **106**, 458 (1937). — Rev. Mod. Phys. **16**, 153 (1944).
[2] K. COMPTON: Phys. Rev. **20**, 28 (3122). — E. MILNE: J. Lond. Math. Soc. **1**, 40 (1926).

A complete analysis of the problem requires consideration of the several isotopic species of Hg atoms, and individual treatment of their diffusion and excitation transfers both radiative and non-radiative. HOLSTEIN, ALPERT, and MCCOUBREY[1] have extended the study of resonance radiation to an enriched isotope of Hg, and find that with suitable choice of constants the HOLSTEIN theory neglecting kinetic diffusion can be made to fit the new data over a limited range of densities. The diffusion theory described above is also capable of fitting the new data with proper choice of constants. It seems probable, therefore, that the HOLSTEIN theory should be regarded as a partial approach to the calculation of a diffusion coefficient for excited states which should be extended by the use of a full BOLTZMANN equation in place of the truncated form employed by HOLSTEIN[2].

8. Recombination into excited states [*4*]. The role of volume recombination in gas discharges has been the subject of much debate. While at one time it was favored as the paramount populating process, it has come generally to be regarded as a very minor effect both electrically and in the production of radiation at low pressures.

The major source of ion losses is now thought to be surface recombination. Surface recombination may result in luminescence, but little or nothing is known about this possibility. It is probable that the bulk of surface recombination results in mere heating of the surface.

Volume recombination takes place in five principal forms of interaction: Type I, an electron, positive ion, photon system; Type II, an electron, electron, positive ion system; Type III, an electron, positive ion, neutral atom system; Type IV, an electron, positive molecular ion system; and Type V, a positive ion, negative ion system. The major importance of the fourth process has only been recognized since the development if microwave methods for measuring recombination coefficients. Only the first, fourth, and fifth named processes need be looked upon as potentially leading to radiation.

Volume recombination is a random collision process. For binary systems the number of recombination acts per unit time is given by the expression $\overline{\sigma_r u}\, N_+ N_-$, where the average taken over cross-section and relative speed proceeds either over the distribution of the electrons if N_- is to represent an electron concentration, or over the MAXWELL-BOLTZMANN distribution for the neutrals if N_+ and N_- are both heavy ions. For ternary systems, the expression is $\overline{\eta\sigma_r u} N_+ N_- N_0$, where η is the action volume of the third party to the collision. It is common practice in either case to collect these factors into a single coefficient α, called the recombination coefficient, so that the recombination rate is expressed as $\alpha N_+ N_-$.

The unimportance of volume recombination processes hinges on the efficiency of diffusion processes for electron and ion removal. A rough criterion for the relative importance of recombination and excitation can be obtained by solving the statistical equation for ion concentration balancing by neglecting recombination, and then determining the ratio of the neglected term to the excitation rate by use of the solution. It is found that for recombination to be an important factor,

$$N_+^3 \gtrsim \frac{\bar{u}^2 \sigma_n e\, T_-^2}{\sigma_0^2 \sigma_+^2 \alpha_n i\, T^2 \Lambda^2}, \tag{8.1}$$

[1] T. HOLSTEIN, D. ALPERT and A. O. MCCOUBREY: Phys. Rev. **85**, 985 (1952).
[2] T. HOLSTEIN: Phys. Rev. **72**, 1212 (1947).

[σ_n = excitation cross-section, state n; σ_0 = elastic scattering cross-section for ions; σ_+ = total ionization cross-section; i = current density available for elatation; Λ = wall to tube center distance; T_- = electron temperature; $\bar{u}$ = mean molecular speed.] For hydrogen atoms this criterion indicates $N_+ \gtrsim 10^{17}/\text{cm.}^3$ for H_α at 10^{-3} amp/cm.^{2}, in ordinary discharge tubes where $a \sim 1$ cm. and $T_- \sim 10^4$ °K.

Volume recombination in any of its forms, while inconspicuous in active discharges, is the prominent and unique source of radiation from low pressure afterglows. After the cessation of the electric current, a rapid ($\sim 10\,\mu$sec.) decay in intensity sets in while the electron cloud is cooling and blockaded ground state radiations are being converted to transmissible radiations. Then, over a period of perhaps a hundred milliseconds duration, a low intensity afterglow is observable[1]. The statistical equation for the ion concentration reduces to three terms

$$\frac{\partial N_+}{\partial t} = D_a \nabla^2 N_+ - \alpha N_+^2. \tag{8.2}$$

Here the recombination term has been modified in accordance with the experimental fact that electrical discharges maintain a high degree of local neutrality ($N_+ = N_-$), which as far as radiation is concerned may be treated as absolute neutrality. D_a is the ambipolar diffusion coefficient defined by SCHOTTKY[2]. Explicit solutions of this equation are possible in terms of PAINLEVÉ[3] functions, but have never been utilized. Limiting solutions, however, are obtainable. For negligible recombination (at low pressures), the ion concentration is found to have an exponential decay in time, with a principal time constant of $\frac{\ell}{D_a R_1^2}$, where ℓ is the characteristic distance from the walls to the center of symmetry of the discharge, and R_1 is the first root of whatever function symmetry prescribes for the solution of $\nabla^2 f + f = 0$[4].

When diffusion is negligible, symmetry is irrelevant, and the ion concentration decays as t^{-1}, according to the equation

$$\frac{1}{(N_+)_0} - \frac{1}{N_+} = \alpha t. \tag{8.3}$$

This result applies only when the nature of the ions which are recombining does not change during the afterglow period. KUNKEL[5] has investigated the problem when progressive changes in the form of the recombination process are present, that is to say, changes from types 1, 2 or 3 to types 4 or 5. He finds that under these circumstances the $1/N_+$ dependence will not become linear with t as in eq. (8.3) until the conversion is complete.

The experimental investigations of α have been based on applying eq. (8.3) in the afterglow of a discharge. This information alone has not always been sufficient to identify the type of recombination which is active. Primitive techniques[6] using stroboscopic spectrography combined with LANGMUIR probe analysis resulted in some estimates of the recombination coefficient, but were far from precise. Unexplained continua, band spectra, and temperature and

---

[1] L. HAYNER: Z. Physik **35**, 365 (1925). — C. KENTY: Phys. Rev. **32**, 624 (1928).

[2] W. SCHOTTKY: Phys. Z. **25**, 342, 635 (1924).

[3] P. PAINLEVÉ: Bull. Soc. Math. France **28** 221 (1900). — C. R. Acad. Sci. Paris **135**, 411, 641, 857, 1020 (1902).

[4] W. ALLIS and D. ROSE: Phys. Rev. **93**, 84 (1954).

[5] W. KUNKEL: Phys. Rev. **84**, 218 (1951).

[6] L. HAYNER: Z. Physik **35**, 365 (1926). — C. KENTY: Phys. Rev. **32**, 624 (1928). — F. MOHLER: Rev. Mod. Phys. **1**, 216 (1929).

density dependences of the recombination coefficient indicated complex processes of recombination in nearly every case. Microwave techniques[1] have made possible a partial unravelling of this snarl, and a much more precise determination of the recombination coefficients then heretofore. The present indication is that free electron recombination is never observed, but that ambipolar diffusion sets in as the loss mechanism at much higher densities than those in which α for this process is measurable by electron loss methods. In the practical afterglow range, the process is probably of type four (molecular dissociation) at the lower densities, shifting to type three (THOMSON three body process) at high densities, although there may well be intermediate stages in which several different type four processes predominate successively.

The most disturbing factor about the existing situation is that a discrepancy of $\sim 10^3$ is always found between comparable measurements by the microwave-afterglow method and the probe-afterglow method, however carefully they have been carried out. No satisfactory explanation has been made.

One supposed characteristic of recombination spectra is the continuum. The correlation between continua and recombination processes is not a complete one, however. Continua are to be expected at the series limits of atomic spectra in the case of Type I and II recombination, but arise from other causes as well. Conversely, certain forms of recombination (Types III, IV and V) are devoid of continua in most cases. Among the latter processes lie the most important forms of recombination.

Continua originate in many ways[2]. In addition to atomic series limit continua, there are molecular dissociation continua, pressure continua from collision interference with line radiation, retardation radiation continua, attachment continua like the H^- spectrum observed astronomically and terrestrially[3]. Thus BALASSE and LEMMENS[4] identified four continua in the electrodeless mercury discharge between 4 and 156 mm. Hg: two continua being of the molecular dissociation type with different excitation levels, a third caused by the coexistence of two modes of recombination (alternatively the ionized molecule), and the fourth resulting from electron recombination directly with the singly ionized atom.

MOHLER[5], working with caesium, studied the series limit continua. From the total continuous emission he was able to deduce a recombination coefficient of 3×10^{-14} cm.^{3}/sec. for 2000° K electrons, into the 7^2P state, and 7×10^{-17} cm.^{3}/sec. into the 6^2S state. It is probable that these estimates of the probability of type I recombination are as good as any which have been made, and that electron radiative recombination is a very improbable process. It is striking that 3×10^{-14} cm.^{3}/sec. is almost exactly the value which CILLIE[6] finds for the recombination coefficient into the $n=6$ level in hydrogen-like atoms. It is in general very difficult to identify recombination by series limit continua, since most of the useful ones do not lie in regions accessible to quartz spectrographs. It would be extremely interesting, however, to have an experimental comparison between the ratio of population flux of a state arising as a series limit continuum, and the population flux in the particular transition which that continuum feeds, such as the BALMER continuum and LYMAN α.

---

[1] M. BIONDI and S. BROWN: Phys. Rev. **76**, 1697 (1949). — A. REDFIELD and R. HOLT: Phys. Rev. **82**, 874 (1951). — P. DANDURAND and R. HOLT: Phys. Rev. **82**, 868 (1951).

[2] For a complete discussion the reader is referred to this Encyclopedia, vol. 28, article by W. FINKELNBURG and TH. PETERS.

[3] W. LOCHTE-HOLTGREVEN: Naturwiss. Rdsch. **1**, 13 (1954).

[4] G. BALASSE and M. LEMMENS: Bull. Acad. roy. Belge, Cl. Sci. **37**, 56 (1950).

[5] F. L. MOHLER: Rev. Mod. Phys. **1**, 216 (1929).

[6] G. CILLIE: Mon. Not. Roy. Astronom. Soc. **92**, 820 (1931/32).

α) *Radiative electron recombination.* This process, in which a free electron and a positive ion collide and coalesce into an excited atom plus a photon possessing the surplus energy is called radiative recombination to acknowledge that the liberated photon is essential to the conservation of energy and momentum. On a statistical argument, Milne[1] finds that the probability of capture with radiation varies inversely as the electron kinetic energy for small speeds.

Table 4. $M(n, T_-)\times 10^8$, *for hydrogen, to final state n.*

n	T_-: 10^3	5×10^3	10^4	2×10^4	5×10^4
1	20.0	8.8	6.0	3.9	2.3
2	9.8	3.9	2.7	1.6	0.78
3	6.4	2.5	1.4	0.86	0.37
4	4.7	1.6	0.94	0.52	0.21
5	3.5	1.1	0.64	0.34	0.13
6	2.9	0.86	0.46	0.23	0.088
7	2.3	0.66	0.35	0.17	0.062
8	1.9	0.52	0.26	0.13	0.046
9	1.6	0.41	0.21	0.099	0.034
10	1.4	0.34	0.17	0.078	0.027
12	1.03	0.24	0.11	0.051	0.017
14	0.80	0.17	0.08	0.036	0.012
16	0.63	0.15	0.059	0.026	0.0085
18	0.52	0.10	0.050	0.022	0.0063
20	0.42	0.08	0.035	0.015	0.0038
30	0.19	0.03	0.013	0.0055	0.0015
100	0.0113	0.0015	—	—	—
200	0.0020	—	—	—	—

At one time it was erroneously supposed that the electron would have a higher probability of capture into the upper states of the atom than the lower, and hence that recombination would be recognizable by a striking enhancement of the upper transitions in each spectral series over what would be expected when the population was derived from electron impact excitation. Modern treatment of the problem of electron recombination has shown that the probability of capture into the lower states is increasingly more probable the lower the state. Were radiative electron recombination the only source of population, the variation of intensity with total quantum number would be approximately $(1/n^3)\log n$, while if electron impact excitation were the cause, the variation would be as $1/n^3$. The difference between these dependences is not sufficient to base an identification upon. Enhancement of population in upper states over impact excitation values is often observed, but it can in general be traced to collisions of the second kind or multiple body recombination processes which favor certain states from energy matching considerations[2].

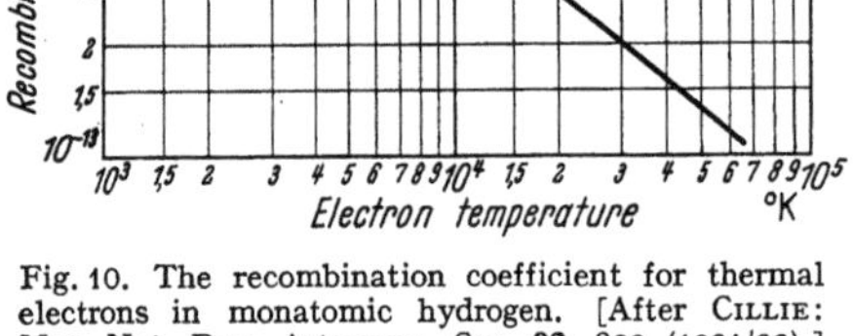

Fig. 10. The recombination coefficient for thermal electrons in monatomic hydrogen. [After Cillie: Mon. Not. Roy. Astronom. Soc. **92**, 820 (1931/32).]

Radiative electron recombination is an intrinsically improbable process because of the small likelihood that energy and momentum can be conserved simultaneously in the interacting system. Cillié[3], following Gaunt[4] and others[5] has calculated the recombination coefficient as a function of total quantum number

[1] E. Milne: Phil. Mag. **47**, 209 (1924).

[2] K. Emeleus and G. Smith: Proc. Phys. Soc. B **64**, 426 (1951).

[3] G. Cillié: Mon. Nat. Roy. Astronom. Soc. **92**, 820 (1931/32).

[4] J. Gaunt: Phil. Trans. Roy. Soc. Lond., Ser. A **229**, 200 (1929).

[5] H. Kramers: Phil. Mag. **46**, 836 (1923). — A. Maue: Ann. Phys., Lpz. **13**, 161 (1932). — D. Menzel and Ch. Perkeris: Mon. Nat. Roy. Astronom. Soc. **96**, 77 (1936).

and electron temperature in the case of hydrogen-like atoms. He found that

$$\alpha_n = 3.208 \times 10^{-6} M(n, T_-) Z^4, \tag{8.4}$$

where Z is the ionic valence,

$$M(n, T_-) = \frac{1}{n^3 T_-^{\frac{3}{2}}} \exp \frac{e V_+}{k T_-} \int\limits_{\frac{e V_+}{k T_-}}^{\infty} e^{-u} \frac{du}{u}, \tag{8.5}$$

and

$$3.208 \times 10^{-6} = \frac{2^9 \pi^5}{(6\pi)^{\frac{3}{2}}} \frac{e^{10}}{m^2 c^3 h^3} \left(\frac{m}{k}\right)^{\frac{3}{2}}. \tag{8.6}$$

The functions $M(n, T_-)$ are tabulated for hydrogen in Table 4, while the total $\alpha(T_-) = \sum_n \alpha_n(T_-)$ is given in Fig. 10.

CILLIÉ applied this theory to the intensity ratios of the BALMER series from planetary nebulae and found good agreement with observations, although, as previously noted, this result need not be conclusive.

UNSÖLD[1] has developed a similar approximate theory of the continuous absorption spectrum from which recombination coefficients can be calculated. He finds a slightly different expression than that of CILLIÉ.

BATES, BUCKINGHAM, MASSEY, and UNWIN[2] have evaluated this same constant for monoenergetic electrons by an exact calculation. Their data is given in Table 5 for hydrogen and Table 6 for oxygen.

Table 5. $\alpha_n \times 10^{13}$ *for hydrogen, to final state n* (cm³/sec).

n	Electron Energy in eV			
	0.28	0.13	0.069	0.034
1	2.54	3.54	5.12	7.30
2	1.41	1.90	2.76	3.98
3	0.89	1.27	1.87	2.71
4	0.59	0.91	1.38	2.03
5	0.43	0.68	1.07	1.62
6	0.31	0.53	0.86	1.34
7	0.23	0.43	0.70	1.12
8	—	0.35	0.59	0.96
9	—	0.28	0.50	0.82
10	—	0.22	0.43	0.71
12	—	—	0.31	0.55
14	—	—	0.23	0.44
16	—	—	—	0.36
18	—	—	—	0.29
20	—	—	—	0.23
$\sum_{n=1}^{\infty} \alpha_n$	7.2	11.4	18.6	29.7

β) Dielectronic recombination. This process is the inverse of autoionization. The excess energy of the cometary electron is expended against one of the bound electrons, exciting it to a stationary state so that the radiator possesses two excited electrons. It may then radiate in returning to equilibrium. The probability of the process has been investigated by BATES and MASSEY[3], and it is

Table 6. $\alpha \times 10^{12}$ *for oxygen* (cm³/sec).

Electron energy eV	0.7	0.36	0.175	0.10	0.044	0.020
α	0.224	0.379	0.683	1.17	1.98	3.34

[1] A. UNSÖLD: Ann. Phys., Lpz. **33**, 607 (1938).
[2] D. BATES, R. BUCKINGHAM, H. MASSEY and J. UNWIN: Proc. Roy. Soc. Lond., Ser. A **170**, 322 (1939).
[3] D. BATES and H. MASSEY: Phil. Trans. Roy. Soc. Lond., Ser. A **239**, 269 (1943).

found that the recombination coefficient for this process is at most no larger than that for single electron recombination (Type I).

γ) Three body electron recombination. This process was analyzed by J. J. THOMSON[1], who found that for negative and positive ions,

$$\alpha \approx \frac{64\,(2\pi)^{\frac{1}{2}}\,e^{6}\,m^{\frac{1}{2}}}{81\,(k\,T)^{\frac{5}{2}}\,M}, \tag{8.7}$$

over the range of densities which is under discussion here. Below 0.1 mm. Hg three body electron recombination should be even less significant than radiative electron recombination, being governed approximately by the expression $5\times 10^{-10}\,\frac{p}{M_w}$ cm.³/sec. where p is in mm. Hg, and M_w is the molecular weight of the gas, provided eq. (8.7) can be extended to electrons without modification.

δ) Dissociative recombination. The process of dissociative recombination has only recently been recognized for its true importance[2]. If a gas contains molecular ion aggregates formed by any one of several processes:

	$A^+ + 2A \rightarrow A_2^+ + A$	(three body),
	$A^+ + A \rightarrow A_2^+ + h\nu$	(binary, radiative),
or	$A^* + A \rightarrow A_2^+ + e$	(collision 2nd kind),

then the reaction

$$A_2^+ + e \rightarrow A^* + A$$

is of great probability if a dissociative curve crosses the energy curve of the molecular ion as shown in Fig. 11. BATES has estimated the coefficient to be expected of this process, and finds it $\sim 10^{-7}$ cm.³/sec. Thus this is vastly more important than any of the other processes of recombination, and, furthermore, is probably identifiable as a possible process in almost every gas. It is generally accepted as the explanation of the large coefficients measured by microwave methods, where data ranging from 10^{-6} to 10^{-9} cm.³/sec. have been obtained. While by any standards the coefficients for this process are more significant than those for the other forms of recombination, and hence explain the large coefficients observed in afterglows, there is no exact theoretical confirmation of the order of magnitude of the coefficients observed by either experimental method, so that neither side of the disagreement can be favored. A point made by REDFIELD and HOLT[3], that the measured values are strongly dependent upon the inverse of the power input, suggests that this fact may have some

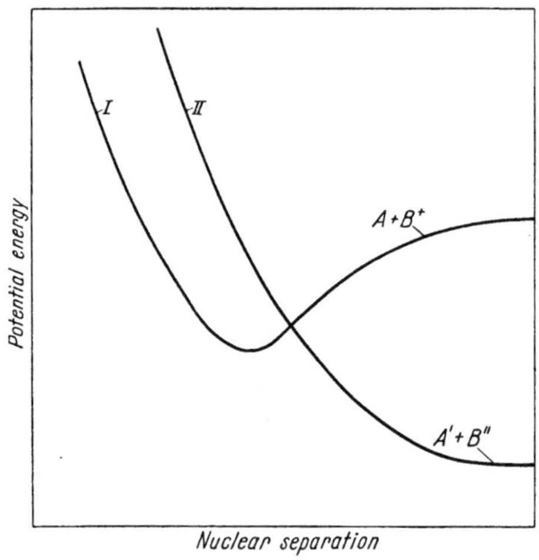

Fig. 11. Crossing of the potential energy curves for the dissociated excited states and for the molecular ion, near a normal state of the latter, favors dissociative recombination. [After MASSEY: Adv. Physics 1, 395 (1952).]

[1] J. J. THOMSON: Phil. Mag. **47**, 337 (1924).
[2] D. BATES: Phys. Rev. **78**, 492 (1950).
[3] A. REDFIELD and R. HOLT: Phys. Rev. **83**, 874 (1951).

bearing on the discrepancy, since the probe measurements are usually made in discharges having 0.01 to 0.1% ionization, while the microwave measurements are at a much lower level ranging from 10^{-4} to 10^{-3}% ionization. There seems to be an inverse variation between the values of α observed by these methods and the initial ion concentration, such as might be expected from a complex multistage recombination process.

BIONDI and HOLSTEIN[1] have shown that when a trace of argon is added to helium, the recombination decay chain is completely broken down, and charges are lost only by ambipolar diffusion over the same range of densities in which the large recombination coefficients are observed in pure helium. The explanation advanced is that helium ions undergo charge transfer to the lower ionization-potential argon ions which then have no loss mechanism because of the improbability of the electronic radiative process, and also of the three body $e + A^+ + He$ process; these being the only ways left open since the molecular ion AHe cannot exist.

ε) Mutual neutralization. Like dissociative recombination, both processes representing a preassociation of the particles of a three body collision, the process of mutual neutralization of positive and negative massive ions may itself have a higher probability than the electronic three body process. Nevertheless, in either case, it is necessary to create the partial assemblies (i.e. the molecular ions or the negative ions) in a first stage three body process which limits and controls the entire process by its intrinsic improbability. Experimental evidence and theory are both inadequate to discuss this final process, but it is clear that it will only be active in electronegative gases, so that the electron attachment coefficients measured by LOEB [*5*] are of primary importance in the analysis of the problem.

DANDURAND and HOLT[2] suggest that one phase of the complex recombination process in mercury at high densities may be attributed to Hg^- formation, with subsequent mutual neutralization. RICHARDSON and HOLT[3], investigating hydrogen, found principally continuous radiation from the discharge, and encountered another complex recombination process which may require recognition of H^- in the recombination role in some phases of the decay.

9. Cascading. Spontaneous transition from the upper states of a radiator to the lower states furnishes a significant portion, although not a major portion, of the population of each state. In complete generality, the problem of the flux of population of any one state must be dealt with by finding the fluxes for all states, i.e., by solving the complete set of equations for statistical balancing [eqs. (4.23), (4.25)]. In restricted cases, however, it is possible to define a cascading coefficient which makes the general solution unnecessary. These restricted cases are often important practical cases. To make this simplification possible it is necessary that the aboriginal population of each state be derived from a source which is independent of the individual excited state populations. This restriction is fulfilled by (1) electron excitation from the ground state, (2) recombination processes, and (3) collisions of the second kind between foreign radiators and ground state particles.

Assuming, as is usual, that the blockaded ground state radiations are wholly imprisoned, the first r equations (arranging levels by their energies) will be either

[1] M. BIONDI and T. HOLSTEIN: Phys. Rev. **82**, 962 (1951).
[2] P. DANDURAND and R. HOLT: Phys. Rev. **82**, 868 (1951).
[3] J. RICHARDSON and R. HOLT: Phys. Rev. **81**, 153 (1951).

for metastable or resonant levels, and of the form

$$\frac{\partial N_j}{\partial t} = D_j \nabla^2 N_j + \sum_{k=j+1}^{\infty'} A_{kj} N_k + P_j \,. \tag{9.1}$$

The levels above $j=r$ will be governed by the equations

$$\frac{\partial N_j}{\partial t} = \sum_{k=j+1}^{\infty'} A_{kj} N_k - \sum_{i=2}^{j-1} A_{ji} N_j + P_j \,. \tag{9.2}$$

In accordance with the restrictions, it will be assumed that P_j is of the form

$$P_j = \sigma_j \frac{i}{e} N_0 + \alpha_j N_+^2 + S_j N_j^* N_0 , \tag{9.3}$$

where N_j^* is the concentration of foreign metastables.

In obtaining a solution, it will also be assumed that diffusion loss can be approximated by a diffusion radius $\Lambda = \frac{\ell}{R_1}$ [see discussion of Eq. (8.2)]. Then $D_j \nabla^2 N_j$ is replaceable by $(D_j/\Lambda_j^2) N_j$. Solutions of the form

$$N_j = (N_j)_0 \sum_{l=j}^{\infty'} g_{lj} e^{-A_l t} + \varphi_j G_j \tag{9.4}$$

can now be found, where

$$A_l = \sum_{j=2}^{l-1} A_{lj} \,.$$

In these solutions, the functions φ_j are either

$$\varphi_j = \frac{P_j}{A_j} \quad \text{for} \quad j > r, \tag{9.5}$$

or

$$\varphi_j = \frac{P_j \Lambda_j^2}{D_j} \quad \text{for} \quad j \leq r \,. \tag{9.6}$$

The constants g_{lj} are characteristic radiator constants when $j > r$, formed by combination of the transition probabilities. They govern the transient response of the radiating system and will be called the *cascading constants*. They are

$$\left.\begin{aligned} g_{lj} = \frac{A_{lj}}{A_j - A_l} &+ \sum_{s=j+1}^{l-1} \frac{A_{ls} A_{sj}}{(A_1 - A_l)(A_s - A_l)} + \\ &+ \sum_{s=j+1}^{t=1} \sum_{t=j+1}^{l-1} \frac{A_{ls} A_{st} A_{tj}}{(A_j - A_l)(A_s - A_l)(A - {}_t A_l)} + \cdots , \end{aligned}\right\} \tag{9.7}$$

and

$$g_{jj} = 1 \,.$$

When $j > r$ the quantities G_j are characteristic radiator functions which comprise a union of the transition probabilities and the production cross-sections. They will be called the *tributary functions*. They are given by iterated summation from the state in question up to the highest stable state of the radiator. Since this is nominally infinity, but in actual fact is some large finite number specifying an orbit commensurate with the intermolecular action sphere, it will be denominated ∞'.

$$G_j = 1 + \sum_{s=j+1}^{\infty'} f_{sj} \left\{ p_{sj} + \sum_{t=s+1}^{\infty'} f_{ts} \left[p_{ts} + \sum_{u=t+1}^{\infty'} f_{ut} (p_{ut} + \cdots) \right] \right\}, \tag{9.8}$$

where

$$f_{kj} = \frac{A_{kj}}{A_k}, \quad \text{and} \quad p_{kj} = \frac{P_j}{P_k} = \frac{\sigma_k}{\sigma_j}.$$

The tributary functions and cascading constants are no longer characteristic of the *radiator only* when $j \leq r$, but can still be obtained by replacing the non-existent transition probabilities which eqs. (9.7) and (9.8) call for, by the corresponding diffusion parameters D_j/Λ_j^2.

Little experimental work has been done to measure the cascading functions and coefficients, undoubtedly because no entire family of transitions is ever to be found in a single suitable spectrographic region. It has been customary, therefore, either to ignore the problem of cascading entirely and assume that only the aboriginal populations contribute significantly to the radiation observed, or at most to consider one of the more important incursive transitions. Some countenance is given to this by the work of MOHLER[1] who examined the total populations in the upper states of caesium in an arc discharge. He found that the population fell so rapidly with n, that of the first three or four states it could be said that each state had a larger population than all those lying above it. Such a rapid decrease in population could easily render it unnecessary to consider cascading, but at the same time, such large decrements are not found in all discharges.

ORNSTEIN, BURGER, and KAPUSCINSKI[2] examined the approximate tributary function for the 2^3S, 2^3P, 3^3S, 3^3D states in helium, by computing the intensity ratio

$$\frac{I(2^3P - 3^3D) + I(2^3P - 3^3S)}{I(2^3S - 2^3P)}. \tag{9.9}$$

This ratio is a measure of the importance of the first order cascading terms in the tributary function for the 2^3P state. They found that it ranged in value from 0.22 to 0.05 as the helium density was changed over the range from 1 to 30 mm. Hg. Their conclusion that this density dependence and the small values of the ratio observed prove that the elatation process could not have been electronic recombination is not justified, but their measurements do show appreciable amounts of population derived from cascading. Too many processes contribute to population flux in the pressure range studied to state with certainty the cause of the pressure dependence observed, but it is possible that it is associated with the failure, as density increases, of the 3^3D state to form molecular ions, because of the large energy discrepancy between the 3^3D state and corresponding molecular ion state.

Use of the cascading theory in the steady state requires nothing but the tributary functions, which in turn vary with the populating process. In the transient state, of course, the cascading constants are also needed, Usually the transient[3] state follows a steady state in which the $(N_j)_0$ populations have been determined by the usual populating processes, so that

$$(N_j)_0 = \varphi_j G_j. \tag{9.10}$$

SLACK[4], and others[5], made use of the cascading transient state to estimate the total transition probabilities A_l. No corrections were made for cascading,

[1] F. MOHLER: U.S. Bur. Stds. J. Res. **16**, 227 (1936).

[2] L. ORNSTEIN, H. BURGER and W. KAPUSCINSKI: Z. Physik **51**, 38 (1928).

[3] It should be noted that this cascading transient state is not to be confused with the recombination afterglows, which are of such long duration that they may be regarded as *steady* states in this restricted sense. Rather, the rapid decay preceding the afterglow is at least partially a cascading transient state.

[4] F. SLACK: Phys. Rev. **28**, 1 (1926).

[5] D. SINCLAIR and H. WEBB: Phys. Rev. **50**, 440 (1936).

which could alter the probabilities reported by these and other authors sufficiently to account for the minor disagreements observed in the lifetimes measured by this and other methods.

10. Populating by collisions of the second kind [*1*], [*6*]. Much empirical and qualitative information exists about special collisions of the second kind, but little has been done from either a theoretical or a quantitative standpoint. Two rules govern collisions of the second kind in a very loose way: (1) the FRANCK resonance rule[1] that the most probable process is the one resulting in the least increase of total kinetic energy; (2) the WIGNER spin rule[2] that the most probable process is one with no change in the total spin momentum of the electron participants.

The characteristic feature of the populations resulting from collisions of the second kind is *enhancement*[3]: the selective population of one or more related levels of the radiator. Usually the energy donor is a metastable of a different species, since a long lifetime is conducive to favorable impacts. Metastables of the same species may interact, however, usually with the production of ionization, and radiative excited states also enter into collisions of the second kind when the density increases sufficiently that collisions are probable during their radiation period. Blockaded radiation states are also effective in collisions of the second. Owing to the miscellaneous nature of the information, a tabular summary of some of the observations has been made (Table 7).

From the standpoint of the radiation emitted by gas discharges, collisions of the second kind are a devious means of radiating the energy of non-radiating

Table 7. *Some populating collisions of the second kind.*

Donor	Radiator	Observation
Hg′	Tl	Selectively excited arc spectrum
He*, Ne*	N	Selectively excited atomic arc spectrum
He*, Ne*	CO^+	Comet tail bands, associated bands, first negative bands
He, A	C_2	SWAN bands
He	C	Arc lines
He*, Ne	N_2^+	Negative bands
He	O_2^+	Negative bands
He, Ne, A	O	Arc lines
H, He, Ne, A	Mg	Spark spectrum Mg II
He, Ne	Ag, Cn, Al, Au	Enhanced spectra in He, Ne enriched atmosphere
A*	H_2	LYMAN bands
Ne*	H_2	Upper BALMER series members
A*, Kr*	O	Auroral spectrum
N*	N_2^*	RAYLEIGH afterglows[4]
N	Hg	$4\,^3D$ lines excited but not $5\,^3D$
Hg′	Na	Selective excitation, $7\,S$ state
N_2^*	N_2'	Second positive bands
H	metals	Arc spectra
He″	He	Other states of very slightly lower energy radiated[5]

X* signifies a known metastable state; X′ signifies a known resonant state; X″ signifies a freely radiating state.

[1] J. FRANCK and P. JORDAN: Anregung von Quantensprüngen durch Stöße. Berlin 1926.
[2] E. WIGNER: Nachr. Ges. Wiss. Göttingen **1927**, 375.
[3] This term was introduced in a more restricted sense by DUFFENDACK and MANLEY [Phys. Rev. **47**, 56 (1935)], but it seems desirable to extend its meaning.
[4] O. OLDENBURG: Phys. Rev. **90**, 727 (1953).
[5] J. LEES and H. SKINNER: Proc. Roy. Soc. Lond., Ser. A **137**, 186 (1932). — R. WOLF and W. MAURER: Z. Physik **92**, 100 (1934); **115**, 410 (1940).

states. As such they must be considered in competition with diffusion in eq. (4.25). This means that rate of loss terms of the type

$$\overline{su}\, N_d\, N_r, \tag{10.1}$$

($d \equiv$ donor, $r \equiv$ receptor) must be introduced in these equations in addition to, and at large densities in supercedence of the diffusion term. In the cases that the cross-sections s are known they often exceed the kinetic theory elastic cross-sections by factors of 10 to 100.

In Table 8, a summary of the metastable and resonance states of the atomic gases is given. While these states represent the source of most energy donations, the blockaded states of the first principal series must not be ignored completely.

Table 8. *Atomic metastable states* [7].

Gas	Resonance states	Metastable states
H, Li, Na, K, Rb, Cs, Fr	$^2P_{\frac{3}{2}}$; $^2P_{\frac{1}{2}}$	none, except for $2^2S_{\frac{1}{2}}$ in hydrogen which is degenerately metastable, and in impact equilibrium with the other $n=2$ states
He	none	2^1S_0; 2^3S_0; 2^3P_2
Ne, A, Kr, Xe, Rn	$^3P_{1,0}$; $^1P_1$	$^3P_2$
Mg, Zn, Cd, Hg	$^1P_1$; $^3P_1$	$^3P_2$; $^3P_0$
N, P, As	none	$^2P_{\frac{3}{2}}$; $^2P_{\frac{1}{2}}$; $^2D_{\frac{5}{2}}$; $^2D_{\frac{3}{2}}$
O	$^5S_2$; $^3S_1$	$^1D_2$; $^1S_0$
S	$^5S_2$; $^3S_1$	$^1D_2$; $^1S_0$
F, Cl, Br, I (?)	$^4P_{\frac{5}{2},\frac{3}{2},\frac{1}{2}}$; $^2P_{\frac{3}{2},\frac{1}{2}}$	$^2P^0_{\frac{1}{2}}$

Molecular gases also possess metastable states, but the number and variety of molecular gases as well as the incompleteness of the state of knowledge about them requires their omission[1].

Two rate of loss analyses can be made which cast great light upon the importance of metastable and resonant states. In the first place, if a process exists for loss of metastables by collisions of the second kind, it may be an important method of exciting the receptor molecules even at small concentrations. Thus comparing roughly the rate of production of excited states by the action of metastables with the rate of production by electron excitation, the ratio varies from approximately the square of the minimum dimension of the gas container expressed in kinetic mean free paths at low densities, to N_0/N_r at high densities. In the second place, at low densities there is no loss process which can dispose as efficiently of metastables as spontaneous radiation disposes of radiative states. This can be expressed in terms of the relative lifetimes of radiative and metastable states, the latter being under attack by some foreign agency. Then the ratio of lifetimes is $\frac{\overline{Sv}\, N_r}{A} \sim 10^{-19} N_r$, so that the concentration of foreign receptors must be atmospheric or above before the rate of metastable loss is of radiative magnitude.

[1] A. Gaydon: Dissociation Energies. London 1953. — G. Herzberg: Spectra of Diatomic Molecules. New York 1950.

C. Depopulating processes.

11. General remarks. Every process for populating the energy states of a radiator has its inverse (Fig. 2). Although it is theoretically possible to have processes which only balance statistically in a cyclic fashion, DIRAC[1] asserts that such processes are not found among atomic radiators, but that simple inverses are the rule. In estimating the importance of many depopulating processes, especially collisions of the second kind, it is often useful to make use of the principle that to *make possible* detailed balancing when thermal equilibrium *does exist*, the probability of a forward process must be equal to that of its inverse even though equilibrium does not exist in the given situation. Thus the number of de-excitations per second of radiative states by slow electrons in a discharge is negligibly small even though the process is the inverse of the highly probable process of electron excitation, because in ordinary discharges the number of excited states is negligibly small ($\sim 10^{-10}$) with respect to the number of neutral atoms.

α) Spontaneous emission. The chief process for depopulating levels is almost invariably spontaneous emission. This has already been discussed in detail theoretically and also as a populating process in cascading (Sect. 9). The state of knowledge of the EINSTEIN A coefficients is in a reasonably good state of advancement both theoretically and experimentally[2], and sufficient agreement exists in the case of the simpler elements to warrant confidence. Nevertheless the knowledge is far from complete, and this represents a serious handicap both to astronomers[3] and discharge physicists. An estimate of the reliability of the experimental results can be formed by examining Table 9, in which the mean lifetime of Hg ($6^3P_1 - 6^1S_0$) is summarized from the results of sixteen different workers[4] using seven different methods of measurement. This transition is particularly adaptable to a variety of techniques, making such comparison possible.

Table 9. *Lifetimes of Hg* ($6^3P_1 - 6^1S_0$) $\times 10^7$ *sec.*

Depolarizing of resonance radiation by steady magnetic fields	0.88; 1.13; 0.98; 1.0
Depolarization by alternating magnetic fields	1.0
Absorption coefficient	1.03
Line profile	1.04; 1.0
Refractive index	1.14
Canal rays	0.98
Radiation decay at low density	1.08; 1.0

For most transitions only two methods are possible, either direct measurement of the transient decay of population, with its implied cascading analysis, or a measurement of natural line breadth, which determines a composite of the upper and lower state life times. Neither method gives the desired quantity without manipulation, so that it usually transpires that nothing definite can be reported until an entire atomic system has been analyzed. This in turn is almost impossible because of the variety of spectral regions in which the necessary data must be sought.

Forbidden transitions from metastable levels have assumed great interest in a number of astronomical low pressure "discharge" problems, in auroral and night

[1] P. DIRAC: Proc. Roy. Soc. Lond., Ser. A **106**, 581 (1924).
[2] LANDOLT-BÖRNSTEIN, 6. Aufl., Bd. I, Teil 1. 1950.
[3] H. PLASKETT: Mon. Nat. Roy. Astronom. Soc. **107**, 117 (1947).
[4] P. GARRETT: Phys. Rev. **40**, 779 (1932).

sky light, and in some special terrestrial devices. JACQUINOT[1] and PLUVINAGE[2] have shown that with interferometric methods it is possible to isolate weak radiations, corresponding to various forbidden intercombination lines of helium, in terrestrial gas discharges. They find that the forbidden transition probabilities are at most only 10^{-4} as large as the allowed ones. RUBINOWICZ[3] states that this low intensity is the only working tool available for identifying these multipole radiations in the cosmos. BARBIER[4] makes experimentally harmonious intensity calculations for various forbidden transitions based on assuming that the parent metastables are excited by electron impact. This result should be contrasted with the calculation of CILLIÉ[5], who obtained experimentally harmonious calculations of the allowed intensity ratios in hydrogen based on recombination as the source of population in nebulae, already discussed in Sect. 8α.

β) Collisions of the second kind. Collisions of the second kind are also a depopulating process as well as a populating process. Sometimes, as in the process in helium suggested by WOLF and MAURER (Sect. 5β), the result of the collision is an exchange of states in which both populating and depopulating processes figure. They found it necessary to assume that in addition to the exciting collisions of the first kind against neutrals which converted ^1P states into other adjacent levels, collisions of the second kind were needed to reverse this process and account for the complete density dependence observed. LOCHTE-HOLTGREVEN[6], following the work of WOOD[7] showed that various foreign gases could cause conversion of the population of either the $^2P_{\frac{3}{2}}$ or the $^2P_{\frac{1}{2}}$ in sodium vapor into the other state.

In other cases, the energy is carried away in a form which is not immediately available for radiation. Thus RÖSSLER and SCHÖNHERR[8] studied the radiation of Hg ($6^3P_1 - 6^1S_0$) as a function of pressure and current, and identified both pressure and current dependent losses which they attributed to collisions of the second kind with neutrals and electrons respectively.

The quenching of resonance radiations and metastable states is an extensively explored process. Exemplary of this is the work of DUFFENDACK and OWENS[9], in which they found that the quenching of Hg 6^3P_1 by hydrogen is probably a two step process involving the formation of HgH and HgH'; while the quenching process with nitrogen or carbon monoxide can be done in a single step in which the energy which must be absorbed to convert 6^3P_1 into 6^3P_0 is taken up as vibrational energy. The work of DARWENT[10] has provided additional information on a variety of quenching gases. KENTY[11] has devised a periodic, self-quenching glow discharge based on collisions of the second kind. In mixtures of argon (or krypton) and mercury, addition of a little N_2 or CO induced the periodic behavior. He proposes that at the outset an electrostatic field exists in which the noble gas is excited. Mercury ions are then formed at the expense of the excited gas, cancelling the field so that this ion production mechanism ceases.

---

[1] P. JACQUINOT: Proc. Phys. Soc. Lond. B **63**, 969 (1950).
[2] P. PLUVINAGE: J. Phys. Radium **13**, 405 (1952).
[3] A. RUBINOWICZ: Progr. Phys. **12**, 233 (1948/49).
[4] D. BARBIER: C. R. Acad. Sci. Paris **226**, 390 (1947).
[5] G. CILLIÉ: Mon. Nat. Roy. Astronom. Soc. **92**, 820 (1931/32).
[6] LOCHTE-HOLTGREVEN: Z. Physik **47**, 362 (1928).
[7] R. WOOD and L. DUNNOYER: Phil. Mag. **27**, 1018 (1914).
[8] F. RÖSSLER and F. SCHÖNHERR: Z. techn. Phys. **19**, 588 (1938).
[9] O. DUFFENDACK and J. OWENS: Phys. Rev. **46**, 417 (1934).
[10] B. DARWENT: J. Chem. Phys. **18**, 1532 (1950). — B. DARWENT and F. HURTIBISE: J. Chem. Phys. **20**, 1684 (1952).
[11] C. KENTY: Phys. Rev. **80**, 96 (1950).

The N_2 or CO impurity then suppresses all Hg resonance states so that the glow cannot be self maintained in Hg vapor and goes out, permitting a restoration of the original electrostatic field, and repetition of the cycle.

At large densities, many workers[1] have observed that a marked decrease sets in in the intensity of radiations from discharges. This is especially true of the inert gases. In helium this decrease sets in at approximately 2.5 mm. Hg. MEYEROTT[2] opened the way to an understanding of this process by the suggestion that the population of He_2 and He_2^+ and presumably other molecular ions might be much larger than previously estimated. BATES[3] suggested that the large microwave recombination coefficients could be understood as dissociative recombination with He_2^+. PHELPS and BROWN[4] isolated large quantities of He_2^+ from helium discharges at 5 mm. Hg pressure, but found little at 1 mm. Hg. HORNBECK and MOLNAR[5] suggested that the appearance potentials of molecular ions in noble gases could only by explained by the existence of a collision process of the second kind in which excited helium atoms formed molecular ions upon collision with neutral atoms. FOWLER and DUFFENDACK[6] had proposed an unidentified process of the second kind as one possible cause of the intensity decrease, but discounted the possibility because of the supposition that it would require degradation of the entire excitation energy into kinetic form, in defiance of the FRANCK-CONDON principle. There seems little doubt now that the intensity loss process is a collision of the second kind of the type suggested by HORNBECK and MOLNAR:

$$\mathrm{He}\,{}^1S_0 + \mathrm{He}' \rightarrow \mathrm{He}_2^{+\prime} + e\,.$$

One very convincing piece of evidence for this is the failure of the $3\,{}^3D$ level to form molecular ions with any great efficiency because the energy levels of the molecular ion in this state lie above those of the atomic state by more than thermal energies, a situation which is not found for any of the other levels with $n=3$ or above. This accounts for the pronounced shift of color in helium discharges from green to yellow with increasing density. Cross-sections for this process were estimated by FOWLER and DUFFENDACK as of the order of 10^{-16} cm.^{2}.

It seems certain that molecular processes of this kind must be invoked in many gases at pressures approaching 10 mm. Hg, as superceding spontaneous emission in importance.

D. Special discharges.

12. TOWNSEND discharges. Little or nothing has been reported about the TOWNSEND discharge as far as its radiation characteristics are concerned. Indirectly it is known[7] that high energy photons are present in the discharge and influence the characteristics of GEIGER counters (in which they are the active process) profoundly.

TOWNSEND discharges are low current discharges maintained from external sources by photons or particles of high energy which liberate electrons either in the gas or from surfaces immersed in the gas. Since the discharge is promoted by an electric field in the gas which is strong enough to accelerate the electrons

[1] W. HANLE: Ergebn. exakt. Naturw. **10**, 284 (1931).
[2] R. MEYEROTT: Phys. Rev. **70**, 671 (1944).
[3] D. BATES: Phys. Rev. **77**, 718 (1950).
[4] A. PHELPS and S. BROWN: Phys. Rev. **86**, 102 (1952).
[5] J. HORNBECK and J. MOLNAR: Phys. Rev. **84**, 621 (1951).
[6] R. FOWLER and O. DUFFENDACK: Phys. Rev. **76**, 81 (1949).
[7] J. CRAGGS and A. JOFFE: Nature, Lond. **159**, 369 (1947).

up to the ionization energy, but not so strong that the discharge becomes self-maintained, it is apparent that any electron may have sufficient energy to produce excitation also, and some radiation is to be expected. In fact the radiation should not differ greatly from that in any other discharge in the same gas, at the same pressure and in the same electric field. If α is the TOWNSEND ionization coefficient so that the number of electrons liberated in any single avalanche of length x is

$$N_- = e^{\alpha x}, \tag{12.1}$$

and if β_j is the number of excitations to state j per unit length of electron path, defined as

$$\beta_j = \frac{N_0}{U} \frac{\int\limits_0^\infty \sigma_j u^3 \varphi(u)\, du}{\int\limits_0^\infty u^2 \varphi(u)\, du}, \tag{12.2}$$

where U is the drift velocity for electrons in the electric field present; and $\varphi(u)$ is the electron distribution compatible with that field, probably the SMIT[1] distribution, then the energy of radiation in a transition ν_{ji} is given by

$$h\nu_{ji} A_{ji} \frac{\beta_j}{\alpha} (e^{\alpha x} - 1), \tag{12.3}$$

in energy per avalanche of length x. From this basic expression the power radiated can be calculated in the various eventualities which may arise, i.e., emission of the initial electrons from surfaces or in the volume, cascading, transfers among the states by neutral collisions, or possibly blockaded radiations. Lack of experimental data renders further discussion unprofitable.

13. The HOLST-OOSTERHUIS[2] discharge. When the TOWNSEND discharge passes over into an internally maintained discharge at small currents (10^{-7} amp) between plane parallel electrodes, and in noble gases, a beautiful laminar emission pattern is observed. Fig. 12, taken from DRUYVESTEYN[3] shows these layers in neon gas at about 25 mm. Hg. It is found that the number of lamina times the ionization potential of the gas is equal to the potential difference between the electrodes. No radiation analysis was made of this discharge, but GROTRIAN[4] examined a somewhat similar discharge in mercury spectrographically and found that the $6^3P_1 - 6^1S_0$ resonance line was the chief component of the radiation from the laminae except at high pressures where the green band fluorescence replaced it. In the case of the noble gases the theory of the lamina is that electrons, all of which originate from the cathode, acquire energy in the uniform field between the plates which they successively lose and gain in flights of one excitation mean free path. The current is so small that space charge does not distort the field, and the layers are regularly spaced. It is seen in Fig. 12 that as current increases, space charge distortion sets in and the layers first become compressed in front of the cathode, and finally coalesce into the cathode glow of the ordinary glow discharge. As the anode is approached the electrons begin to straggle in velocity with an accompanying loss of definition in the layers. The sharpness of the layers, and their potential increments of one ionization potential at a time lead to the conclusion that if they are indebted to electron

[1] J. SMIT: Physica, Haag **3**, 543 (1936).
[2] G. HOLST and E. OOSTERHUIS: Physica, Haag **1**, 78 (1921). — Phil. Mag. **46**, 1117 (1923).
[3] M. DRUYVESTEYN: Z. Physik **73**, 33 (1931/32).
[4] W. GROTRIAN: Z. Physik **5**, 148 (1921).

excitation for their radiations, then electron excitation must be a comparatively rare event when compared with ionization, since a 20% discrepancy in energy loss between loss processes of comparable importance would result in a complete disorganization of the electron beam in the course of four or five lamina, whereas as many as 20 have been observed. GROTRIAN's discharge was externally maintained by a heated cathode and the difference between laminae was equal to the first excitation potential rather than the ionization potential. DRUYVESTEYN remarks that as the current is increased and the transition to the negative glow sets in, the discharge becomes unstable and oscillatory. While the electrical explanation of the discharge seems satisfying, the experimental knowledge about the laminar discharge radiations is insufficient to justify such a simplified view

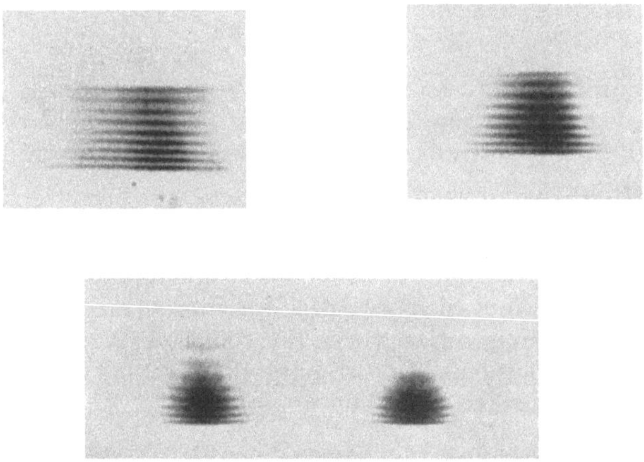

Fig. 12. The HOLST-OOSTERHUIS layer discharge (in neon), at increasing currents, showing the effect of space charge and the transition to the negative glow. Current increases from (a) to (d). The cathode is at the bottom of each picture. [After DRUYVESTEYN: Z. Physik 73, 33 (1931).]

of the radiation process, which may be quite complex. A second uncertainty is whether the cathode current is maintained by positive ion or photon impacts. Much evidence is accumulating for the latter process in the glow discharge, but the difference between the layer potentials found by GROTRIAN and DRUYVESTEYN, together with the transition behavior which the latter finds with increasing currents, suggests ionic cathode processes in the internally maintained laminar discharge, with a possible shift in process to photon impacts at the higher currents. It is also probable that any analysis of the radiation should regard the primary electrons as a thermal stream with its energy between the laminar zones of elatation controlled by elastic losses. Finally, at the pressures used by the above investigators, the dissociative recombination of molecular ions may be an important process in radiation production.

14. Monoenergetic electron discharges. In this category fall the cathode regions of glow discharges, low voltage arcs, and other thermionic discharges; beam discharges like those used for the measurement of optical cross-sections; and with some slight extension of the class probably polar aurorae also.

α) Polar aurorae. Polar aurorae are now understood as solar particle beam excitations of the atmospheric gases although some of the hydrogen radiation observed must certainly be radiated by the beam particles themselves. VEGARD[1]

[1] L. VEGARD: Nature, Lond. 170, 536 (1952).

proposes that the solar beam is a jet of neutral plasma. MEINEL[1] has found that there are DOPPLER shifted components to the observed H_α line which indicate emission by the incident hydrogen beam. PETRIE[2] suggests that electron impacts are also involved, possibly by the electron component of VEGARD'S plasma. Thus although details are lacking the picture is reasonably complete. It may also be worthwhile to note that 3000 megacycle radio waves have been observed emitted from the polar aurorae[3].

β) Electron Beams. Elatation by monoenergetic beams has been employed widely as a means of measuring atomic and molecular characteristics. In the discussion of some of these measurements it was brought out that a complete understanding of radiation processes is needed before the results become completely meaningful. In the experiments of LEES and SKINNER[4] this was carefully brought out. The main electron beam was found to be surrounded by a diffuse glow, which upon spectrographic analysis, was shown to contain most of the transitions of which the atom was capable, with the general exception of those originating on S states. Furthermore it was found that the total discharge intensity of all transitions is rigorously and accurately proportional to the primary electron current. While was not certainly established that the diffused radiation was independently linear with current, this last radiation represented a sufficient fraction of the total to have had a noticeable effect on the total discharge dependence if otherwise. Numerous explanations of the diffused glow are possible, including scattered electrons, but in view of the experiments of WOLF and MAURER[5], the most credible is that originally proposed by LEES and SKINNER: the blockaded radiations diffuse, are reabsorbed, and the resulting ^1P states are converted into other states by collisions with neutral atoms.

MAXWELL[6] investigated a number of atomic properties by means of electron beams which he controlled with strong magnetic fields parallel to the beam. Crossed electric fields of moderate strength then had little effect on the electrons which move under these forces in oscillatory paths whose maximum displacement from a straight line is $\frac{2mEc^2}{eB^2}$, being less than 0.05 mm. in his experiments. MAXWELL found that it was possible to detect the lifetimes of ionic excited states by the sidewise shift of their radiations in the applied electric cross-field. Absolute values of his measurements made in helium are in excellent agreement with theory, which, since ionized helium is isoelectronic with hydrogen, can be rigorously calculated. He finds evidence that $He^{+\prime}$ is excited and ionized in a single impact predominantly, and that ignored cascading plays a noticeable part in altering the apparent lifetime of a state, since longer lifetimes are observed with greater primary beam energies and consequent increases in the upper state populations.

γ) Thermionic cathode discharges. Cathode regimes are, in the case of thermionic cathodes *certainly*, and in the case of cold cathodes *very probably* governed by monoenergetic electron streams. DUFFENDACK and KOPPIUS[7] demonstrated by probe measurements that for a low voltage arc in mercury, the cathode dark-space (also termed a LANGMUIR sheath in this type of discharge) surrounding

[1] A. MEINEL: Phys. Rev. **80**, 1096 (1950).
[2] W. PETRIE: Canad. J. Res. **26**, 359 (1948).
[3] P. FORSYTHE, W. PETRIE and B. CURRIE: Nature, Lond. **164**, 453 (1949).
[4] J. LEES and H. SKINNER: Proc. Roy. Soc. Lond., Ser. A **137**, 186 (1932).
[5] R. WOLF and W. MAURER: Z. Physik **115**, 410 (1940).
[6] L. MAXWELL: Phys. Rev. **32**, 715, 721 (1928); **38**, 1664 (1931). — J. Franklin Inst. **210**, 427 (1930); **214**, 533 (1932).
[7] O. DUFFENDACK and O. KOPPIUS: Phys. Rev. **55**, 1199 (1939).

the cathode, embraced almost the entire fall in potential across the discharge, so that the negative glow was field free up to a point within as little as one or two millimeters of the cathode. At the electron energies employed only a single exciting impact was possible, and the negative glow was the only luminous regime in the discharge. The electron temperatures measured for the plasma electrons ranged from 4000 to 10000° K, indicating fields of somewhat less than one volt per cm. in the glow, agreeing with probe plots of the space potential distribution. Examining the radiations from this discharge they found that the intensities of the family of transitions ending on the 6^3P states increased according to an exponential saturation curve with mercury concentration, and increased linearly with tube current. At the densities studied no reversal whatsoever was observed. In admixtures of argon, in addition to the same saturation behavior, excitation was found to be apportioned between mercury and argon in the proportions of their relative abundances. Everything observed was in complete accord with the hypothesis that monoenergetic primary electrons in a fixed finite number were expended in single collisions to the extent to which the abundance of obstructing molecules permitted.

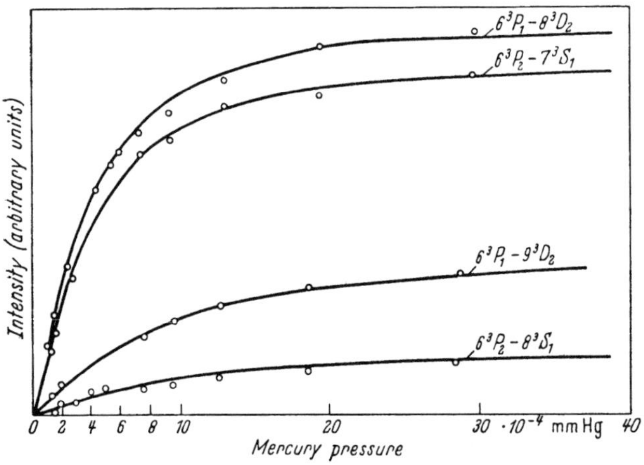

Fig. 13. Intensity vs. concentration curves for the mercury low voltage arc. [After DUFFENDACK and KOPPIUS: Phys. Rev. 55, 1199 (1939).]

For the theory outlined, it is possible to derive the expected intensity dependence. Assuming a monoenergetic stream of electrons having a particle current density of i/e, the number of losses from a unit area of the beam in a distance dx is

$$\frac{di}{e} = -\sigma \frac{i}{e} N_0 \, dx\,, \tag{14.1}$$

and is equal to the change of particle current density. Integrating,

$$i = i_0 e^{-\sigma N_0 x}, \tag{14.2}$$

is the equation of decrement of the primary electron stream. This result is generally applicable to monoenergetic electron currents which make single collisions. The cross-section σ is the cross-section for all significant energy losses, ionization plus excitation of all kinds. Eq. (14.2) can easily be generalized to allow more than one collision per electron.

The power radiated per unit volume in any transition excited by the electron stream is now given by

$$f_{ji}\, h\nu_{ji}\, \sigma_j \frac{i_0}{e} N_0 e^{-\sigma N_0 x}\,. \tag{14.3}$$

Integrated over the entire stream from cathode to anode, the power radiated is

$$f_{ji}\, h\nu_{ji} \frac{\sigma_j}{\sigma} \frac{I_0}{e} \left(1 - e^{-\sigma N_0 x}\right). \tag{14.4}$$

While the theoretical dependence conforms well with the experimental (Fig. 13), the total elatation cross-section σ required by DUFFENDACK and KOPPIUS to fit their curves is surprisingly large: 1.5×10^{-14} cm.^{2} for 30 ev electrons. Since this total cross-section will be preponderantly attributable to ionization, it would be expected to be of the same order of magnitude as the ionization cross-section, namely, 6×10^{-16} cm.^{2} as measured by BLEAKNEY[1]. Elastic cross-sections are the only ones approaching the large value found, but there is little or no reason to believe that elastic losses can contribute significantly to this phenomenon. In helium, similar measurements by FOWLER and DUFFENDACK[2] showed a total cross-section for elatation of the order of 8×10^{-17} cm.^{2}, with 70 ev primary electrons, while SMITH[3] found 3×10^{-17} cm.^{2} for the ionization cross-section alone. LAWSON[4] in a similar experiment in argon observed 3×10^{-16} cm.^{2} total elatation cross-section at 100 ev, while SMITH found 3×10^{-16} cm.^{2} for the ionization cross-section alone. It seems probable therefore that the density saturation method, when refined with a full understanding of the radiation processes active, can be expected to yield total elatation cross-sections which in combination with the usually reliable ionization cross-sections can give estimates of total excitation. In the case of mercury there are indications of a calibration error in the measurement of ionization cross-section, possibly associated with inaccurate estimates of the vapor density in the beam used.

In the radiations stimulated by monoenergetic electrons, general opinion favors direct electron excitation as the chief mechanism of population, rather than recombination. The debate[5] over the two processes has been an historic one, but seems to be resolving itself toward the former mechanism. The current dependence of radiation production has been tested over wide ranges ($\sim 10^5$), and except for transitions with particular causes for misbehavior, radiation is linear with the number of primary electrons. Furthermore, in studies of interrupted discharges, little indication is found of processes as intrinsically slow as recombination might be expected to be. Finally, recombination continua are rarely observed. None of these arguments can be said to be crucial, however, in view of the known process of dissociative recombination, which does not exhibit a continuum, has a high enough cross-section to be rapid at ion concentrations of the order found in these discharges, and finally, may be in concentrational equilibrium with the primary electron beam (just as LADENBURG[6] finds the metastable states of neon to be), so that their loss rate in recombination could therefore be dependent upon the rate of production of atomic ions, and hence linear with the primary current also.

In support of a recombination hypothesis there is the work of DEWEY[7], who showed that DOPPLER broadening of the radiation in the negative glow around a thermionic cathode in helium was considerably in excess of that observed in the positive column of the same discharge. She explained this by the argument that the speeds of the radiators were derived from their previous history as ions. She found also that the intensity of radiation varied roughly as the square of the electron concentrations measured by a probe.

Either theory of population will accord with the work of DUFFENDACK and KOPPIUS if the linear current requirement is fulfilled by the theory. Further work is therefore necessary to establish this point conclusively.

---

[1] W. BLEAKNEY and L. SMITH: Phys. Rev. **49**, 402 (1936).
[2] R. FOWLER and O. DUFFENDACK: Phys. Rev. **76**, 81 (1949).
[3] L. SMITH: Phys. Rev. **36**, 1293 (1930).
[4] M. LAWSON: Thesis, University of Oklahoma 1950.
[5] M. DRUYVESTEYN: Physica, Haag **11**, 129 (1931).
[6] R. LADENBURG: Rev. Mod. Phys. **5**, 243 (1933).
[7] J. DEWEY: Phys. Rev. **32**, 918 (1928).

δ) Cold cathode discharges. A complete analysis of the cathode radiation with cold cathodes must await a fuller understanding of the electrical processes present as well as of the radiation processes. Little and von Engel[1] have investigated the hollow cathode discharge, in which only cathode processes are observed, and propose that of the two competing theories of the electron maintaining process, photoelectric emission at the cathode is preferable to secondary emission by ion impacts.

Emeleus[2] has summarized the general phenomena observed near cold cathodes. A dark space next to the cathode is observed, known as the Aston or primary dark space, in which at most only transitions of the type and intensity observed

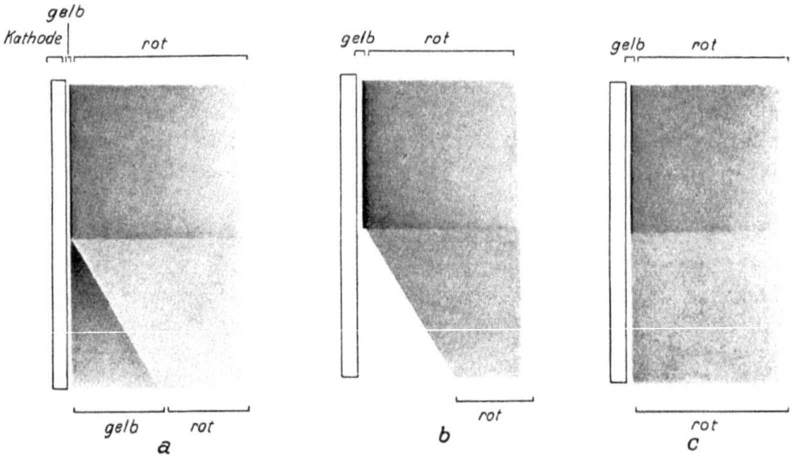

Fig. 14. Mirror photographs of the cathode regimes in a glow discharge after partial interruption by reducing tube potential below maintaining value. The cathode is at the left, and in figure (a) the tube potential was reduced by 20 volts; in (b) by 60 volts; and in (c) was reduced to zero. Note the flow phenomenon in the first two cases in which a speed of about 9 meters/sec. is observed. [After Druyvesteyn: Z. Physik 57, 292 (1929).]

by Wolf and Maurer[3] and presumably from the same cause (blockaded radiations) are observed. This dark space is attributed to lack of energy on the part of the newly liberated electrons. A short distance from the cathode is the *cathode glow* in which region presumably the electrons have acquired energies exceeding the resonance potentials and so produce excitation. Another dark space called the Crookes or cathode dark space is next observed, which is only relatively dark, and from which Dauvillier[4] has observed soft x-rays which show increasing penetrative powers as one approaches the edge of the negative glow. The negative glow lies on the opposite side of the cathode dark space from the cathode, and appears when viewed in the soft x-ray region of the spectrum to be essentially a gaseous anti-cathode. While the passage from the Crookes dark space to the negative glow often seems sharp and definite to the eye, a complete analysis of the radiation from both regions shows that no boundary can be established optically. The electrical characteristics, however, show a rather abrupt change at the edge of the negative glow, the rapid rise in potential of the dark spaces being replaced by a field free region in which most of the negative glow lies, much as was described with thermionic cathodes. It is now believed that the

[1] P. Little and A. v. Engel: Proc. Roy. Soc. Lond., Ser. A **224**, 209 (1954).
[2] K. Emeleus: Phil. Mag. **17**, 146 (1934).
[3] R. Wolf and R. Maurer: Z. Physik **115**, 410 (1940).
[4] A. Dauvillier: Phil. Mag. **2**, 1046 (1926).

CROOKES dark space arises because the primary electrons have acquired such velocity that their probability of exciting the optical levels of an atom has become vanishingly small. This is in accord with the known behavior of excitation cross-sections at high electron velocity. Excitation of the inner shell electrons with the emission of soft x-rays becomes more probable committantly, providing a source of the energetic photons which are required by LITTLE and VON ENGEL in their analysis of the cathode emission process.

As in the thermionically excited negative glows, the cold cathode negative glow is an excellent region to find recombination as a populative process if this is ever important. The weak electric fields present are conducive to large recombination coefficients.

DRUYVESTEYN[1] performed a series of interrupted discharge experiments on the negative glow in argon, neon and in mixtures of the two. He found that in both gases, all lines which decay immediately (10^{-4} sec.) to zero intensity belong to the spark spectrum. The arc lines $1s-2p$ in Ne and $1s-3p$ in A are weaker in the afterglow than in the active glow. Almost all other arc lines first become abruptly more intense than in the active glow, and then decay. Viewing the cathode region with a rotating mirror he observed remarkable behavior of the afterglow which depended on the tube potential that was maintained during the afterglow. These phenomena are shown in Fig. 14.

GÜNTHERSCHULZE and KELLER[2] observed that a luminous layer $\frac{1}{8}$ as thick as the primary darkspace is present on many cathodes (Mg, Zn, Cd, Tl, Na, Li, Cs) when the cathode fall is large, and that the spectrum of this layer is characteristic of the cathode material. This is probably related to sputtering of the cathode material, a problem which has recently been studied anew by ECKER and EMELEUS[3].

15. Thermal electron discharges. In this category fall the positive columns of glows, arcs, and sparks; high frequency discharges; and anode glows. The present discussion will be bounded somewhat artificially as dealing with the pressure range from atmospheric pressure downward. Over most of this pressure range the thermal electrons are describable by a temperature parameter called the electron temperature, but are not in thermal equilibrium with the gas. At the high pressure end of the range a fair degree of equilibrium between the electrons and the gas sets in, so that the distributions of both are described by a single temperature parameter. Discharges are never and can never be in true thermal equilibrium, however.

Analysis of the radiation from thermal discharges must be made on a basis of the electron concentration and its velocity distribution. To the highest precision it would be necessary to consider that this velocity distribution is space dependent but other difficulties usually mask this lacuna. Experiment shows and theory suggests that it is reasonably accurate to consider the electron temperature (which governs the velocity distribution) constant over large regions of the discharge. This is because the electron temperature is almost directly proportional to the electrostatic field in the gas, and the electrostatic field is tangentially constant, at least, from its conservative properties. Furthermore, in the absence of space charges, which are usually small in regions where thermal elatation predominates, there can be no change in the normal component either, thus establishing the conditions for constancy of electron temperature.

[1] M. DRUYVESTEYN: Z. Physik **57**, 292 (1929).
[2] A. GÜNTHERSCHULZE and F. KELLER: Z. Physik **75**, 105 (1932).
[3] G. ECKER and K. EMELEUS: Proc. Phys. Soc. Lond. B **67**, 546 (1954).

Therefore a fairly general theory of this type of regime can be based on an assumption of separability of the space and velocity dependences of the electron distribution to give the number of primary electrons dN_- at x, y, z having velocity u_x, u_y, u_z as

$$dN_- = N_-(x, y, z)\, dx\, dy\, dz\, \varphi(u_x, u_y, u_z)\, du_x\, du_y\, du_z. \tag{15.1}$$

The production function can now be written for elatations of a type governed by the cross-section $\sigma_j(u)$, per unit volume per unit time:

$$P_j = 4\pi N_- \int\limits_{u_{\min}}^{\infty} u^3 \sigma_j \varphi\, du, \tag{15.2}$$

where it has been assumed that the electron velocity greatly exceeds the molecular velocity, so that the relative velocity is entirely the former, and thereafter spherical coordinates can be introduced into velocity space, with $u = \sqrt{u_x^2 + u_y^2 + u_z^2}$. The integration then proceeds over all electrons which can make inelastic collisions, from the resonance potential up to the limit of the distribution. It is assumed here that the function φ has been normalized to unity.

Since the major portion of discharge current density is given by the expression

$$i = e N_- U \tag{15.3}$$

for the *electron current density*, at least in gases which do not attach electrons, the production is directly proportional to the current density, and if radiation is the chief energy loss mechanism, the radiation must be proportional to the current, just as in monoenergetic electron discharges.

Calculation of the integral in Eq. (15.2) is a straightforward operation when the particular theoretical expressions of Sect. 6 and the MAXWELL-BOLTZMANN distribution of velocities are used. In many cases this is certainly an excellent first approximation, but other distributions have been developed to meet special needs[1]. It is a peculiar fault of the MAXWELL-BOLTZMANN distribution that it makes no account of the energy losses which the inelastic collisions described in eq. (15.2) imply, and hence indicates the existence of more fast electrons than are ever actually observed in a gas. Recently, however, EASLEY[2] has shown that the deficiency of fast electrons may sometimes at least be an artifact of the probe technique. In mercury and mercury-noble gas positive columns she finds that the MAXWELL-BOLTZMANN distribution is preserved even by electrons at 10 ev energy.

Much qualitative[3] and some truly quantitative knowledge exists concerning the radiation from the glow discharge. A large part of this applies to the positive column which is by far the most spectacular regime of the discharge. ÅNGSTRÖM found that the radiation reaching a bolometer from the positive column was only a few percent of the energy supplied to the column electrically. This was probably partially because the tube walls failed to transmit the bulk of the radiations of the discharge. Modern measurements of the radiation efficiency of sodium discharges have ranged as high as 90%, but sodium gives a particularly favorable case. PENNING[4] has made an analysis of the energy losses by the thermal electron swarm in the spirit of eq. (15.2). He finds that it may lose energy elastically; in ionization, and secondary electron acceleration; and in excitation. His computations for argon, neon, nitrogen and hydrogen are given

[1] M. DRUYVESTEYN and F. PENNING: Rev. Mod. Phys. **12**, 87 (1940). — J. SMIT: Physica, Lond. **3**, 543 (1936).

[2] M EASLEY: J. Appl. Phys. **22**, 590 (1951).

[3] R. BÄR: Handbuch der Physik, Bd. 14, S. 171. 1927.

[4] F. PENNING: Physica, Haag **5**, 286 (1938).

in Fig. 15, and show that at high pressures where small E/p prevails, the energy supplied to the discharge goes mostly into gas heating, so that the luminous efficiency is low. At low pressures large proportions of energy go into elatation, but a considerable fraction is not radiated, so efficiencies of the order of 1 to 10% as reported by ÅNGSTRÖM are quite reasonable in many cases.

HODGES and MICHELS[1] examined the pressure dependence of radiations from the positive column of a helium discharge. They found that the emission of any given line increases and then decreases. Triplet transitions are favored over singlet transitions as pressure increases, and the higher levels of a triplet series are enhanced with pressure. The indications are that the two processes (1) dissociative recombination of molecular helium ions and (2) collisions of excited states with neutrals, suggested earlier in this chapter, are active here also. The enhancement of the upper triplet states is understandable if process (2) is active, since the energy deficiency between corresponding singlet and triplet levels is less for the higher levels.

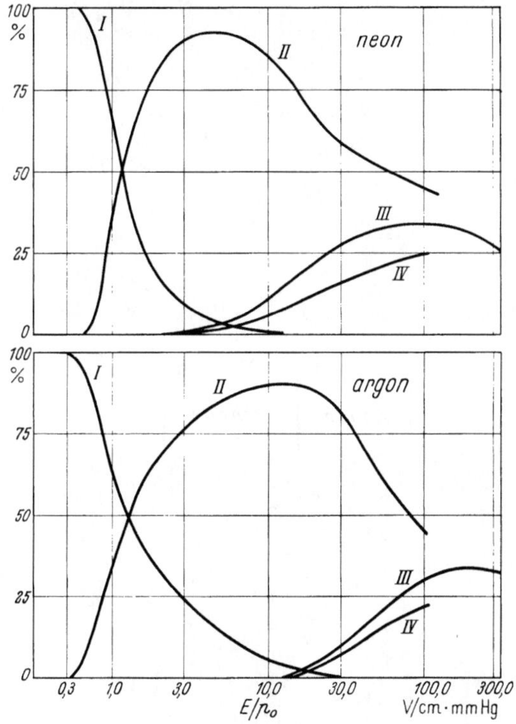

Fig. 15. Efficiency of various electron energy loss processes in neon and argon. [After PENNING: Physica, Haag 5, 286 (1938).]

PARKINSON[2] has observed interesting behavior in a 15 KC alternating current glow discharge in air and noble gases, and especially in helium, where once more the molecular ion seems to play an essential role. Observing the light in front of one electrode which was alternately a cathode and an anode, PARKINSON found that the neutral molecular band spectrum was strong during the anode period, absent during the cathode period. At the same time atomic radiations were observed which displayed the same afterglow-decay behavior as the molecular bands during the anode cycle, but followed the discharge current wave form during the cathode cycle. Low level singlet transitions showed a preponderance of cathode cycle current-governed response over anode cycle afterglow, but high level triplets had completely the reverse behavior. These facts are displayed in Figs. 16 and 17. PARKINSON found that all decays occured with the same time constant, 35 μsec. He found also that the intensity of the molecular and atomic afterglows decreased very rapidly with pressure, as would be expected if the three body process of molecular ion formation were the contributory cause.

LADENBURG[3] has investigated the population of metastable states, and the probabilities for transitions ending on these states, in neon, by a number of experimental methods. BUTAEVA and FABRIKANT[4] have studied the influence of current, and pressure on the mercury-in-argon glow discharge. They found

[1] C. HODGES and W. MICHELS: Phys. Rev. **32**, 913 (1928).
[2] W. PARKINSON: J. Opt. Soc. Amer. **41**, 619 (1951).
[3] R. LADENBURG: Rev. Mod. Phys. **5**, 243 (1933).
[4] F. BUTAEVA and V. FABRIKANT: J. techn. Phys. USSR. **18**, 1127 (1949).

that while the 1850 Å resonance line of the principal series of singlets had a linear dependence of intensity on current, the 2537 Å resonance intercombination line was not linear. KENTY[1] has compared a theoretical analysis with experiment in the production of Hg 2537 Å radiation by a positive column at low pressures, and finds that there is good agreement if one assumes that two-stage processes involving metastable and resonance states play a major role. FRISCH and KAGAN[2] have discussed the shift and broadening of ion lines in the positive column as a measure of the velocity and temperature of the ions in argon and krypton. TALLEY, LOWE, and SCANLON[3] have made one of the few investigations on record of the infra red emission spectra which a positive column emits. At between 20 and 700 mm. Hg pressure of CO_2, with a current of 20 ma, they found the ν_2 band of CO_2, lying at 2300 cm.$^{-1}$, emitted very strongly from a visually faint glow discharge. SAUFFERER[4] and HETTNER[5] have carried the search for long wave length radiations from glow discharges

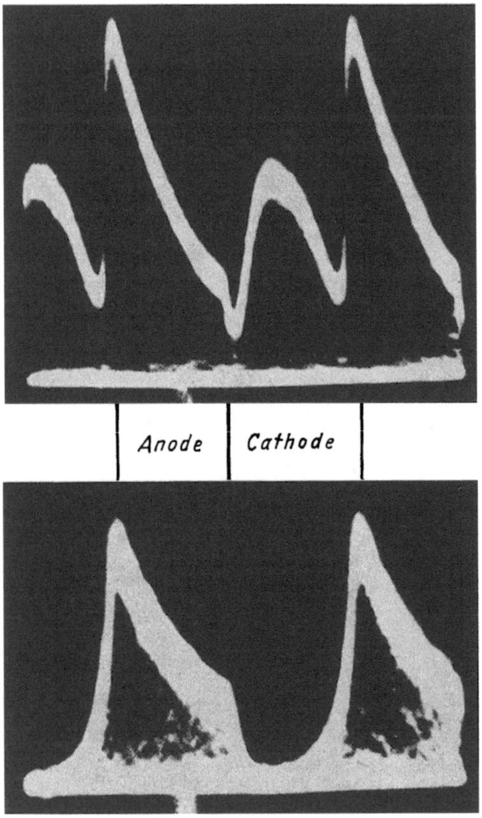

Fig. 16. Wave forms of a typical atomic line (above) and molecular band (below). Oscillogram covers about 120 μsec. [After PARKINSON: J. Opt. Soc. Amer. **41**, 619 (1951).]

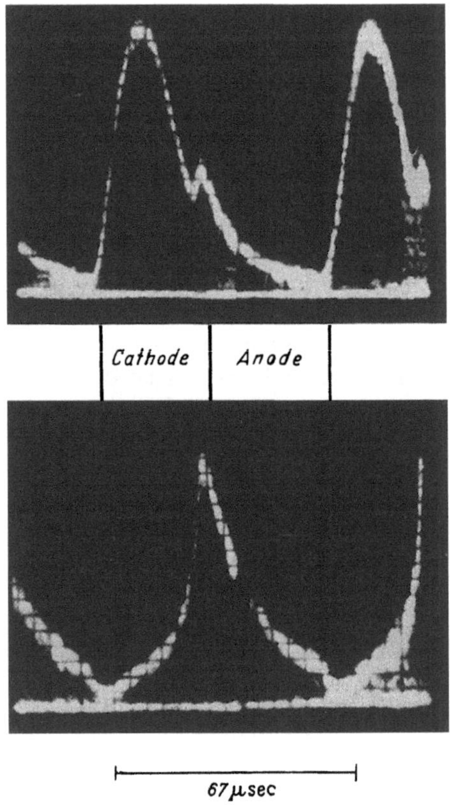

Fig. 17. Wave form of a low singlet line, 5016 Å (above) and of a fairly high triplet line, 4026 Å (below).

---

[1] C. KENTY: J. Appl. Phys **21**, 1309 (1950).

[2] S. FRISCH and Y. KAGAN: Izv. Akad. Nauk SSSR. **12**, 358 (1948). — J. exp. theor. Phys. USSR. **17**, 577 (1947).

[3] R. TALLEY, D. LOWE and W. SCANLON: J. Opt. Soc. Amer. **42**, 980 (1952).

[4] H. SAUFFERER: Z. Physik **131**, 376 (1951/52).

[5] G. HETTNER: Z. Physik **131**, 385 (1951/52).

to the near radio region, and found that about 10^{-4} watt/cm. of column of 0.1 mm. waves is observed from mercury arcs in the 10 to 100 mm. Hg pressure range, with 10 watts/cm. energy input.

Continuous spectra have been found in the positive column with various gases in the pressure range from 2 to 100 mm. Hg. HAHN and FINKELNBURG[1] found this continuum to be roughly independent of the nature of the gas, but dependent on the square of the discharge current, and they identified it as an electron retardation continuum. ROMARD and VODAR[2] have observed similar effects. RÖSSLER[3] observed a strong continuum with definite structure between 4300 to 4900 Å in mercury at 75 mm. Hg pressure. He assigns this to dissociation of the quasi-stable Hg_2 molecule. It is becoming apparent that the understanding of many radiation processes hinges on more knowledge about the reactions of such quasi-stable molecules as this in the discharge.

Complex stable molecules introduced into the discharge are almost invariably fragmented by electron and possibly ion impacts. Thus SCHÜLER and REINEBECK[4] found benzene chloride reduced to simple diatomic fragments in a glow discharge in noble gases. Simple stable molecules on the other hand often exhibit unusual radiation spectra in the glow discharge. HERMAN and HERMAN[5] found that the bands of neutral oxygen are strongly developed in excess of xenon, but normally unobserved in pure oxygen, and suppressed by helium or argon as diluent gases for the oxygen. This behavior is determined by the matching of energy levels in collisions of the second kind.

The arc positive column at low pressure and the glow positive column at similar pressures have many features in common, and are probably chiefly distinguishable by the fact that in the arc, cathode material is removed and supplied in vapor form to the column. KOFOID[6] has shown that the electrode material appears in the spectrum at a very definite pressure which increases as the arc length is reduced, and also as the current density is reduced.

Microwave and high frequency discharges have almost identical emission and temperature characteristics with steady glow discharges of the same power density. In a point by point comparison, BECK[7] found a steady glow discharge in mercury indistinguishable from a 100 megacycle discharge. MARGENAU[8], with HARTMANN, has shown that the theory of microwave discharges leads to this same conclusion of similarity, barring the space charge effects which are possible in steady discharges. CORLISS, BOZMAN, and WESTFALL[9] find that an electrodeless discharge at 3×10^9 cps. will excite the pure metal spectrum of involatile metals which have been introduced as pure halides, chiefly bromides and iodides. FRISCH and SCHREIDER[10] propose a similar discharge for quantitative spectrographic analysis of gas mixtures. The halide discharge method of Corliss, et al., might be extremely useful in the quantitative spectrographic analysis of rarely encountered alloys.

---

[1] O. HAHN and W. FINKELNBURG: Z. Physik **122**, 36 (1944).
[2] J. ROMARD and B. VODAR: C. R. Acad. Sci. Paris **224**, 1634 (1947).
[3] F. RÖSSLER: Z. Physik **133**, 80 (1952).
[4] H. SCHÜLER and L. REINEBECK: Naturwiss. **39**, 445 (1952).
[5] R. HERMAN and L. HERMAN: J. Phys. Radium **11**, 69 (1950).
[6] M. KOFOID: J. Appl. Phys. **19**, 399 (1948).
[7] H. BECK: Z. Physik **97**, 355 (1935).
[8] H. MARGENAU: Phys. Rev. **73**, 297 (1948). — H. MARGENAU and L. HARTMANN: Phys. Rev. **73**, 309 (1948). — L. HARTMANN: Phys. Rev. **73**, 316 (1948). — H. MARGENAU: Phys. Rev. **73**, 326 (1948).
[9] C. CORLISS, W. BOZMAN and F. WESTFALL: J. Opt. Soc. Amer. **43**, 398 (1953).
[10] S. FRISCH and E. SCHREIDER: Izv. Akad. Nauk SSSR **13**, 465 (1949).

16. Transient discharges. A number of discharges and discharge regimes exist which might have been treated in some one of the preceding sections, but which have some special characteristics because they are not steady discharges. These are the corona discharge, the striated positive column, spark discharges, and other related transient or afterglow effects such as the RAYLEIGH phenomenon in pulsed discharges.

The corona discharge has been studied extensively by LOEB[1] and his coworkers, both about positive electrodes and negative ones. The positive corona still presents unsolved problems, but the negative corona is now definitely understood to be a transient miniature glow discharge with all the appurtenances the term implies. ENGLISH[2] found the luminosity buildup of the negative corona to be simple and abrupt upon the application of potential to the electrode. The positive corona, on the other hand showed a slow, long enduring rise in luminosity, followed by an abrupt buildup. GAUNT and CRAGGS[3] found that the luminosity and current transients do not agree in phase exactly, but rather that the luminosity lags the current slightly, especially in hydrogen. This delay in buildup is reminiscent of the effects found by PARKINSON[4] in the glow discharge, who noted that above 100 KC the sharp distinction between anode and cathode behavior was washed out by a long continuing process. It is tempting to suppose that a slow onset process like recombination is needed to account for a large portion even of the current sensitive radiation, but further work on these transient discharges is necessary. Our final understanding of most discharge processes will come from the study of transient discharges of 10^{-6} to 10^{-8} second duration. It is quite possible, for example, that the reason that a microwave discharge radiates in every way like a steady glow discharge is merely that its energy gain occurs in such a short time interval (10^{-9} sec.) compared with its energy losses, that it is to all intents and purposes a steady discharge.

DONAHUE and DIEKE[5] have recently investigated the striations in the positive column of the glow discharge, a joint electrical and radiation phenomenon which has challenged many workers[6] yet remains incompletely understood. Both stationary and moving striations are observed. The stationary striations seem to admit of an explanation involving molecular reactions, although this is not always accepted. The moving striations, first reported by ABRIA[7], have been found by DONAHUE and DIEKE to consist of two kinds, the well known positive striations running from the anode to cathode, and a new negative striation moving in the contrary direction. The negative striations move faster than the positive ones. Positive and negative space charges accompany the positive and negative striations, hence the terminology. In mercury, spectroscopic observations were made on the striations. It was found that as the striations moved, the Hg 2537 radiation appeared well in advance of radiation from any of the higher energy levels. It was concluded that the atoms responsible for the 2537 radiation travelled ahead of those which yielded the other radiations, rather than that some radiation propagation effect existed.

Examining pulsed electrodeless discharges in hydrogen at 0.2 to 0.5 mm. Hg, STRUTT[8] observed that radiation which endured 10^{-5} sec. appeared in a side

[1] L. LOEB: Phys. Rev. **71**, 712 (1947).
[2] W. ENGLISH: Phys. Rev. **77**, 850 (1950).
[3] H. GAUNT and J. CRAGGS: Nature, Lond. **167**, 647 (1951).
[4] W. D. PARKINSON: J. Opt. Soc. Amer. **41**, 619 (1951).
[5] T. DONAHUE and G. DIEKE: Phys. Rev. **81**, 248 (1951).
[6] M. DRUYVESTEYN and F. PENNING: Rev. Mod. Phys. **12**, 87 (1940).
[7] J. THOMSON: Phil. Mag. **18**, 441 (1909).
[8] R. STRUTT (Lord RAYLEIGH): Proc. Roy. Soc. Lond., Ser. A **183**, 26 (1944).

tube as an advancing front. ZANSTRA[1] suggested a recombination afterglow, but LEE and FOWLER[2] showed that this luminosity must have been excited after the cessation of the main discharge. FOWLER, ATKINSON and MARKS[3] showed in connection with this belated luminosity (and CRAGGS and HOPWOOD[4] demonstrated the same to be true at atmospheric pressure), that the ion concentrations were so large that recombination was still the probable process even though the luminosity was certainly excited at the point where it was observed. Both groups used the STARK broadening of the BALMER lines as a tool in measuring ion concentrations. It was found that the radiated energy in H_β was quite accurately proportional to the square of the ion concentration so measured. FOWLER, ATKINSON, COMPTON, and LEE[5] showed that the low pressure effects observed by STRUTT (RAYLEIGH phenomenon) were caused by impact excitation of an unknown kind which accompanied the advance of a shock wave from the abruptly heated region in the main discharge outward along any side tubes open to the flow. Luminous shocks such as this can be produced also by purely mechanical means[6].

Low and intermediate pressure sparks similar to those described in the preceding paragraph can be made easily by the discharge of a capacitor through a gas. The resulting luminosity is very intense and of short duration. Many workers[7] have employed this source and investigated its properties. OLSEN and HUXFORD[8] used hydrogen in small quantities as a probe in argon and neon discharges to measure ion concentration. They concluded that recombination accounted for the bulk of the arc line radiation observed, but that the recombination coefficients needed therefor were exceedingly small: 2.5×10^{-13} cm.^{3}/sec. in argon, 5.3×10^{-14} cm.^{3}/sec. in neon. Strong continuous radiation was observed which they interpreted as primarily retardation radiation of the electron swarm. Spark line emission was coincident in time with the electron current, and was supposed to be by direct electron excitation. Much is expected of a further analysis of the low pressure spark discharge, which seems to be, radiationwise, an especially simple process in mechanism, the time of duration being too brief for collisions of the second kind and other secondary processes to be of any moment.

Major Reference Sources.

[1] MASSEY, H. S. W., and E. H. S. BURHOP: Electronic and Ionic Impact Phenomena. Oxford 1952.

[2] MOTT, N. F., and H. S. W. MASSEY: Theory of Atomic Collisions. Oxford 1950.

[3] PRINGSHEIM, P.: Fluorescence and Phosphorescence. New York 1949.

[4] MASSEY, H. S. W.: Adv. Physics **1**, 395 (1952).

[5] LOEB, L. B.: Fundamental Processes of Electrical Discharge in Gases. New York 1939. — Basic Processes of Gaseous Electronics. Berkeley 1955.

[6] WILLEY, E. J. B.: Collisions of the Second Kind. London 1937.

[7] BACHER, R. F., and S. A. GOUDSMIT: Atomic Energy States. New York 1932.

---

[1] H. ZANSTRA: Proc. Roy. Soc. Lond., Ser. A **186**, 236 (1946).

[2] R. LEE and R. FOWLER: Phys. Rev. **81**, 457 (1951).

[3] R. FOWLER, W. ATKINSON and L. MARKS: Phys. Rev. **87**, 966 (1952).

[4] J. CRAGGS and W. HOPWOOD: Proc. Phys. Soc. Lond. **59**, 755, 771 (1947).

[5] R. FOWLER, W. ATKINSON, W. COMPTON and R. LEE: Phys. Rev. **88**, 137 (1952).

[6] E. RESLER, S. LIN and A. KANTROWITZ: J. Appl. Phys. **23**, 1390 (1952). — R. HOLLYER, A. HUNTING, O. LAPORTE, E. SCHWARCZ and E. TURNER: Phys. Rev. **87**, 911 (1952). — H. PETSCHEK, P. ROSE, H. GLICK, A. KANE and A. KANTROWITZ: J. Appl. Phys. **26**, 83 (1955).

[7] H. EDGERTON: J. Opt. Soc. Amer. **36**, 390 (1936).

[8] H. OLSEN and W. HUXFORD: Phys. Rev. **87**, 922 (1952).

Elektrische Bögen und thermisches Plasma.

Von

W. Finkelnburg und H. Maecker[1].

Mit 104 Figuren.

1. Historisches. Um das Jahr 1808 scheinen zuerst Davy und Ritter einen zwischen horziontalen Holzkohlestäben brennenden elektrischen Lichtbogen erzeugt und beobachtet zu haben. Von der durch den thermischen Auftrieb nach oben gekrümmten Form dieser ersten Bogenentladung stammt der noch heute benutzte Name „Bogen". Bereits 1815 war die sehr hohe Temperatur des Lichtbogens bekannt und wurde zum Schmelzen schwerschmelzender Stoffe verwendet. 1821 beschreibt Davy die Wirkung eines Magnetfeldes auf den Bogen, nachdem ein Jahr vorher die erste Bogenlichtbeleuchtung vorgeführt worden war, die allerdings erst 24 Jahre später im großen technisch angewandt wurde. Um die Mitte des vorigen Jahrhunderts hatten Foucault und Fizeau bereits systematische Lichtmessungen am Bogen ausgeführt, waren Metallbögen, Vakuumbögen und die Möglichkeit der Zündung eines Bogens durch einen kondensierten Funken bekannt. Wenig später wurde auch der Wechselstrombogen zwischen Kohleelektroden beschrieben und gleichzeitig gefunden, daß ein solcher (bei den damals zur Verfügung stehenden geringen Mitteln) zwischen Metallelektroden nicht betrieben werden konnte. Um die gleiche Zeit, 1867, begann die systematische physikalische Untersuchung des Lichtbogens, und 20 Jahre später stellten Luggin und Lecher durch die ersten Sondenmessungen fest, daß die Bogenbrennspannung aus drei deutlich trennbaren Anteilen, dem Kathodenfall, dem Anodenfall und der Säulenspannung besteht. 1892 betrieb Arons den ersten Bogen im Vakuum zwischen Quecksilber-Elektroden und leitete damit die Untersuchung von Dampfbögen ein. 1898 entdeckte Simon den sprechenden und zwei Jahre später Duddel den selbstschwingenden Bogen. 1910 entdeckte Beck den nach ihm benannten Hochstromkohlebogen, der mit seiner überragenden Kraterleuchtdichte neue Möglichkeiten der Anwendung eröffnete. Schon wenige Jahre vorher hatten Küch und Retschinsky die erste Quecksilber-Hochdrucklampe gebaut, aus der in den 30er Jahren Elenbaas die Quecksilber-Höchstdruck-Kapillarlampe und Rompe und Thouret die Quecksilber-Höchstdruck-Kugellampe mit Wolframelektroden entwickelten. Um die gleiche Zeit veröffentlichten Elenbaas und Heller ihre Theorie der Lichtbogensäule, die allen neueren Säulentheorien zugrunde liegt. Ab 1940 schließlich entwickelte Schulz die ersten Edelgas-Hochdrucklampen.

In der neuesten Zeit finden wir einmal die Entwicklungstendenz, die für lichttechnische Zwecke statt freibrennender Bögen mit verdampfenden Elektroden die in Quarzkolben zwischen nicht verdampfenden Wolframelektroden brennenden Bögen bevorzugt. Wir finden weiter eine erhebliche Ausdehnung der Betriebsparameter, insbesondere in Richtung hoher Stromstärken und hoher Drucke, wobei sich herausstellt, daß die älteren Vorstellungen vom Mechanismus des Lichtbogens viel zu eng und einseitig waren. Wir finden zunehmend genauere Messungen der elektrischen und thermischen Daten, insbesondere auch der Temperatur in den verschiedenen Bogengebieten, und wir finden ein starkes Interesse der Astrophysik an der Lichtbogenforschung, die in zunehmendem Maße die Prüfung der Theorien des Hochtemperaturplasmas durch Messungen am und im Lichtbogen ermöglicht. Wir finden schließlich Versuche zu einer einheitlichen und systematischen theoretischen Beschreibung des Mechanismus der Vielzahl von Lichtbogentypen, obwohl gerade auf diesem Gebiet trotz der fast 150jährigen Geschichte des Lichtbogens unsere Kenntnis noch äußerst lückenhaft bleibt.

2. Schrifttum. Das Schrifttum über den Lichtbogen ist so ungeheuer umfangreich, daß nur wenige der neuesten Arbeiten hier erwähnt werden können und für die gesamte frühere Literatur auf ältere zusammenfassende Darstellungen verwiesen werden muß. Gute Darstellungen der älteren Arbeiten und der Literatur aus dem vorigen Jahrhundert und den

[1]) Die Kapitel A und B sind von W. Finkelnburg, die Kapitel C und D von H. Maecker in Zusammenarbeit mit Th. Peters, und die Kapitel E und F von W. Finkelnburg in Zusammenarbeit mit G. Busz-Peuckert bearbeitet worden.

ersten Jahren dieses Jahrhunderts finden sich in den Büchern von AYRTON (1902), STARK (1902), THOMSON (1903), MONASCH (1904), CHILD (1906) und RASCH (1910). Eine äußerst vollständige Darstellung der historischen Entwicklung der Kenntnis vom elektrischen Lichtbogen und besonders von den technischen Anwendungen bis zum Jahre 1906 gibt E. VON CZUDNOCHOWSKI in seinem umfangreichen Buch „Das elektrische Bogenlicht" [*3a*]. Eine Darstellung der Entwicklung bis 1923 mit aller wichtigen Literatur enthält die zweite Auflage des Beitrages von A. HAGENBACH „Der elektrische Lichtbogen" im Handbuch der Radiologie [*7a*], viel interessantes Material auch Bd. II des Werkes „Elektrische Gasentladungen" von v. ENGEL und STEENBECK [*5a*]. Aus neuester Zeit findet sich in dem Buch von W. WEIZEL und R. ROMPE „Theorie elektrischer Lichtbögen und Funken" [*18a*] eine kurze Darstellung der Lichtbogentheorie unter besonderer Berücksichtigung der Arbeiten beider Autoren und ihrer Mitarbeiter, wobei das Literaturverzeichnis aus zeitbedingten Gründen nicht ganz vollständig ist. Eine einigermaßen vollständige Zusammenfassung der neueren Veröffentlichungen über elektrische Lichtbögen wurde unter dem Titel "Bibliography on High-Pressure Electric Arcs" im Auftrage des Subcommittee on Fundamental Arc Research des American Institute of Electrical Engineers vom Battelle Memorial Institute in Columbus/Ohio herausgegeben. Außer diesen Veröffentlichungen über das Gesamtgebiet der elektrischen Lichtbögen gibt es noch eine Anzahl von Büchern mit Literaturangaben über Sonderformen von Lichtbögen. Zu ihnen gehören, außer einem Artikel über Metallbögen von GUYE und MONASCH [*73*] die Bücher „Der eingeschlossene Lichtbogen bei Gleichstrom" von K. STOCKHAUSEN [*16a*], „Untersuchungen über den elektrischen Lichtbogen, insbesondere über den unter Druck befindlichen" von W. MATHIESEN [*10a*], „Hochstromkohlebogen", von W. FINKELNBURG [*6a*], "L'arc électrique intensif" von J. PARISOT [*12a*] und "The High Pressure Mercury Vapour Discharge" von W. ELENBAAS [*4a*]. Alle diese Bücher enthalten jeweils neuere Literatur aus den behandelten Gebieten.

A. Allgemeine Lichtbogenphysik.

3. Definition des Lichtbogens. Als elektrischen Lichtbogen bezeichnet man im allgemeinen eine Entladung relativ großer Stromstärke (über 1 Amp) und relativ niedriger Brennspannung, insbesondere eine solche mit niedrigem Kathodenfall (kleiner als 20 V). Man unterscheidet damit den Lichtbogen von der Glimmentladung, die sich durch relativ geringe Stromstärke und hohe Brennspannung, insbesondere durch einen hohen Kathodenfall von mehreren hundert Volt auszeichnet. Der für jeden Lichtbogen charakteristische geringe Kathodenfall setzt eine Kathode voraus, die entweder selbst infolge hoher Temperatur ihrer Oberfläche, oder indirekt mittels einer ihr dicht vorgelagerten Gas- bzw. Dampfschicht hoher Temperatur eine Elektronenemission von (je nach dem Bogentyp) 10^3 bis 10^7 Amp/cm² zu unterhalten vermag. Diese hohe Temperatur der Bogenkathode bzw. der ihr vorgelagerten Plasmaschicht wird beim selbständigen Lichtbogen durch den Entladungsmechanismus selbst aufrechterhalten. Eine die Glühelektronenemission (allerdings wesentlich geringerer Emissionsdichte) ermöglichende Kathodentemperatur kann aber grundsätzlich auch durch Fremdheizung der Kathode erzeugt werden, und solche Glühkathodenentladungen werden wegen ihrer relativ niedrigen Brennspannung vielfach als Bogenentladungen bezeichnet. Eine einheitliche, allgemein angenommene Definition des Lichtbogens gibt es also noch nicht, wenn auch Übereinstimmung darüber besteht, daß *nur Entladungen mit geringem Kathodenfall als Bögen bezeichnet* werden sollten. In dem vorliegenden Bericht wollen wir uns im wesentlichen auf die Behandlung *selbständiger* Lichtbögen beschränken, Entladungen mit fremdgeheizter Kathode also ausschließen.

Unter den selbständigen Lichtbögen gibt es einige bei geringem Gas- bzw. Dampfdruck betriebene, bei denen im Gasraum der Bogensäule die Elektronentemperatur wesentlich höher ist als die Temperatur der Ionen und neutralen Gasteilchen. Solche Bögen kann man als nicht-thermische bezeichnen. Die große Mehrzahl aller selbständigen Lichtbögen dagegen zeichnet sich dadurch aus, daß in ihren Säulen Elektronentemperatur, Ionentemperatur und Gastemperatur praktisch gleich sind. Unsere Darstellung befaßt sich, vor allem in ihrem theoretischen Teil, fast ausschließlich mit solchen *thermischen* Lichtbögen, die in der Literatur vielfach auch kurz als *Hochdruckbögen* bezeichnet werden, weil bei einem Gas- bzw. Dampfdruck von einer Atmosphäre oder mehr wegen der großen Zahl der Stöße zwischen Elektronen, Atomen und Ionen die Bedingungen für thermisches Gleichgewicht besonders gut erfüllt sind.

4. Der Begriff des thermischen Gleichgewichtes. Unter Vorgriff auf die eingehendere theoretische Behandlung in Kapitel C seien noch einige Bemerkungen über das Lichtbogenplasma und die Frage des thermischen Gleichgewichtes gemacht. Da das Lichtbogengas sich je nach Typ und Stromstärke des betreffenden Bogens auf einer Temperatur zwischen etwa 5000 und 50000° K befindet, liegt sein Ionisierungsgrad zwischen etwa 1% und über 100% (unter Berücksichtigung der Mehrfachionisation!); die Lichtbogensäule besteht also aus einem stark strahlenden Gemisch von Elektronen, positiven Ionen und mehr oder weniger hoch angeregten neutralen Atomen, das man als *Plasma* bezeichnet. Spricht man speziell von einem *thermischen* Plasma, so ist damit gemeint, daß alle kinetischen und Reaktionsgleichgewichte (z.B. zwischen Molekülen und Atomen wie zwischen Ionen und Elektronen einerseits und neutralen Atomen andererseits einschließlich höherer Ionisationsstufen) ebenso wie die übrigen Plasmaeigenschaften (mittlere Teilchengeschwindigkeit, Strahlung, elektrische Leitfähigkeit, Wärmeleitfähigkeit usw.) eindeutige Funktionen der Temperatur sind, und zwar einer für sämtliche Plasmateilchen einheitlichen Temperatur. Solches thermisches Gleichgewicht liegt immer nur in einer gewissen Näherung vor, weil es durch Ausstrahlung, radiale Ladungsträgerdiffusion und Wärmeleitungsprozesse dauernd gestört wird. Die oben genannten Plasmaeigenschaften sind trotzdem in guter Näherung eindeutige Funktionen der Plasmatemperatur, und wir haben damit in dieser Näherung thermisches Gleichgewicht, wenn die Gleichgewichtsstörungen klein sind. Das ist der Fall, wenn der größte Teil der Ionisationsakte durch Rekombinationsprozesse rückgängig gemacht wird und nur ein geringer Prozentsatz der Ladungsträger durch Diffusion verloren geht; und wenn in gleicher Weise der überwiegende Teil der Atomanregungen durch Stöße zweiter Art rückgängig gemacht wird und nur wenige angeregte Atome ihre Energie irreversibel emittieren. Zu diesen Bedingungen für thermisches Gleichgewicht kommt hinzu, daß die Zahl der Stöße zwischen Elektronen und schweren Plasmateilchen so groß sein muß, daß die Elektronen trotz der geringen im Einzelstoß an Atome oder Ionen übertragbaren Energie ihre gesamte aus dem elektrischen Feld entnommene Überschußenergie an die schweren Teilchen übertragen können, so daß sich eine *einheitliche* Plasmatemperatur einstellt. In einem thermischen Lichtbogenplasma ist daher die ungeordnete thermische Geschwindigkeit der Plasmateilchen groß gegen die vom elektrischen Feld bewirkte Driftgeschwindigkeit[1].

Nach der gegenwärtigen theoretischen wie experimentellen Kenntnis liegt in den Säulen aller Hochdruckbögen in dem oben angedeuteten Sinne in guter Näherung thermisches Gleichgewicht vor. In den den Elektroden dicht vorgelagerten Gebieten hoher Feldstärke dagegen (Kathoden- und Anodenfallgebiet) erfolgt bei vielen Lichtbögen die Elektronenbeschleunigung so plötzlich, daß die genügende Stoßzahl zum thermischen Ausgleich fehlt. In diesen Fallgebieten erfolgt die Ionisierung dann nicht thermisch, sondern durch die überthermischen, im Feld beschleunigten Elektronen. Nur bei Lichtbögen sehr hoher Temperatur

[1] Für den Begriff Driftgeschwindigkeit und andere Begriffe aus der Theorie der Trägerbewegung vgl. auch den Beitrag von W. P. ALLIS in Bd. XXI dieses Handbuches.

(vgl. Ziff. 26) kann auch in den Fallgebieten die Feldstärke so klein werden, daß wir näherungsweise thermisches Gleichgewicht und damit auch thermische Ionisation haben.

5. Lichtbögen und verwandte Entladungsformen. Zwischen den thermischen Lichtbögen und den an anderer Stelle der Bde. XXI und XXII dieses Handbuches behandelten anderen Entladungsformen bestehen natürlich vielfache Beziehungen. So tritt (vgl. Ziff. 7) die Glimmentladung häufig als Vorstufe eines Lichtbogens auf, und zwischen beiden Entladungsformen gibt es als Übergangsfall den bei erniedrigtem Druck (etwa 80 Torr) stabilen *Glimmbogen*. Bei diesem entspricht der Kathodenansatz nach HAGENBACH [7a] dem des normalen Bogens, während der räumlich ausgedehnte Anodenansatz dem einer Glimmentladung gleicht und der Glimmbogen sich entsprechend durch eine gegenüber dem normalen Bogen erhöhte Brennspannung auszeichnet. Die vielfach auch als Bögen bezeichneten Niederdruckentladungen mit fremdgeheizter Kathode wurden oben bereits erwähnt; bei ihnen ist die Elektronentemperatur meist wesentlich höher als die Ionen- und Gastemperatur. Eine interessante Sonderform dieser Niederdruck-Glühkathodenentladungen sind die sog. *Niedervoltbögen*, deren an den Elektroden gemessene Brennspannung bis auf wenige Volt sinkt, ja in Sonderfällen sogar leicht negativ werden kann. In Ziff. 31 wird auf sie kurz eingegangen. Erwähnt sei schließlich die Beziehung zwischen den thermischen Lichtbögen und den durch die Aufladung von Kondensatoren in Gasen nicht zu geringen Druckes entstehenden Funken. Wenn man von den zur Ausbildung eines hochionisierten Funkenkanals führenden Durchschlagsvorgängen absieht, sich also auf die Entladung im schon ausgebildeten Funkenkanal beschränkt, so kann man den Hochdruckfunken als einen nichtstationären thermischen Lichtbogen auffassen, der nur deshalb nach relativ kurzer Brenndauer verlischt, weil die Spannungsquelle den Bogenstrom nicht über längere Zeit zu liefern imstande ist.

6. Die physikalischen Probleme des Lichtbogen-Mechanismus. In dem folgenden Kapitel B über die Lichtbogen-Phänomenologie wird eine Fülle empirischer Einzeltatsachen, Beobachtungen und Messungen über die Vielzahl verschiedener Lichtbogentypen und die Abhängigkeit der Lichtbogenerscheinungen von einer großen Zahl von Parametern aufgezählt werden müssen. Es scheint daher zweckmäßig, der späteren theoretischen Behandlung in den Kapiteln C, D, E und F vorgreifend, der Lichtbogenphänomenologie einen Überblick über die grundlegenden physikalischen Probleme des Lichtbogenmechanismus voranzustellen.

Jede elektrische Entladung läuft ja ab in einer elektrisch leitenden Gasstrecke mit beiderseitigem elektrischen Anschluß an den Stromkreis, d.h. an die Elektroden. Bei dem hier speziell zu behandelnden *thermischen Lichtbogen* beruht die elektrische Leitfähigkeit der Gasstrecke auf der hohen Temperatur des dann als Plasma bezeichneten Gases. Die große elektrische Leitfähigkeit der Bogenstrecke ist dann eine Folge der thermischen Ionisierung, und der elektrische Anschluß des Lichtbogens an die Elektroden ist darum ein besonderes Problem, weil dicht vor ihnen die Temperatur von dem hohen Wert im Säulenplasma auf den meist wesentlich niedrigeren der Elektrodenoberfläche abfallen muß. In diesen als Kathodenfall und Anodenfall bezeichneten Gebieten vor den Elektroden liegen daher völlig andere Verhältnisse vor als im ungestörten Plasma, woraus automatisch die Gliederung des Bogens in Säule, Kathodenfall und Anodenfall folgt. Die von einer Theorie des Lichtbogens zu beantwortenden Fragen sind also die folgenden:

a) Welche Ladungsträger leiten den Strom in den verschiedenen angeführten Bogengebieten?

b) Wie sieht die Ladungsträgerbilanz aus, und wie wird der durch axiale und radiale Abwanderung erfolgende Trägerverlust beim stationären Bogen ausgeglichen?

c) Wie entsteht ein stationärer Bogen, d.h. ein Bogen mit zeitunabhängigen Zustandsgrößen?

d) Wie hängt der elektrische Widerstand, den der Bogen darstellt, von den Materialeigenschaften und den äußeren Bedingungen, insbesondere von der eingestellten Stromstärke ab?

e) Welche Elementarprozesse finden im Gasraum und an den begrenzenden Elektrodenoberflächen statt und wie bestimmen sie die makroskopischen Entladungserscheinungen?

f) Wie sieht die Energiebilanz eines Lichtbogens aus, und wie hängt sie von den Parametern der Entladung ab?

a) Die wesentlichen Ladungsträger im Lichtbogen sind freie Elektronen und positive Atomionen; negative Ionen und Molekülionen spielen nur bei gewissen Sondertypen von Bögen sowie in den für den Ladungstransport weniger wichtigen Außenzonen eine Rolle. Die Ladungsträgerbewegung ist bei der hohen Temperatur und dem geringen Potentialabfall je freie Weglänge in thermischen Bögen eine Diffusion mit einer kleinen überlagerten feldbestimmten Driftgeschwindigkeit. Wirkt allein das angelegte elektrische Feld auf die Ladungsträger, vernachlässigt man also eventuell vorhandene Gasströmungen, so sind im Plasma, genügend weit von den Elektroden entfernt, Elektronen und positive Ionen im umgekehrten Verhältnis ihrer Beweglichkeiten an der Leitung beteiligt; der Strom wird also fast ausschließlich von den leichten Elektronen getragen. Das würde 100%ig der Fall sein, wenn die Ionen mit einer ihrer Feldgeschwindigkeit entgegengesetzt gleichen Geschwindigkeit von der Kathode fortgeblasen würden; die Ionen würden dann lediglich die negative Raumladung der Elektronen kompensieren. Der Anodenanschluß wäre dann kein Problem, da aus dem Plasma einfach Elektronen heraus- und in die Anode einströmen würden. Reichte dann die Raumladungskompensation bis zur Anodenoberfläche, so hätte man überhaupt keinen eigentlichen Anodenfall, sondern nur eine (vom Anodenfall allerdings nicht zu trennende) Feldstärkeerhöhung zur Überbrückung des Gebietes geringerer Temperatur und damit geringerer elektrischer Leitfähigkeit vor der Anode. Tatsächlich liegt meist vor der Anode eine nicht vollständig kompensierte negative Raumladung, zumal die Ionen meist langsam von der Anode wegströmen. Falls also die Anode nicht selbst positive Ionen emittiert, wird der Strom in der anodischen Grenzschicht allein von Elektronen getragen.

Vor der *Kathode* sind zwei Grenzfälle denkbar. Wenn die Kathode eine der Elektronenstromdichte in der Säule entsprechende Elektronenemission besitzt, würde auch in der kathodischen Grenzschicht der Strom im wesentlichen von Elektronen getragen. Zu erklären bleibt dann, woher die Kathode die zur Elektronenemission erforderliche Leistung erhält. Der zweite Grenzfall wäre der, daß die Kathode keine Elektronen emittiert, und der Strom in der Grenzschicht ausschließlich von den aus dem anliegenden Plasma in die Kathode einströmenden positiven Ionen getragen würde. Da diese in der Säule nur einen geringen Anteil des Ladungstransportes übernehmen, müßten sie aus der Plasmagrenze zur Erhaltung der eingestellten Stromstärke in entsprechend größerer Zahl austreten, müßten also im Plasma durch thermische Ionisierung erzeugt werden. Bei den meisten Bögen entspricht der Kathodenmechanismus keinem der beiden skizzierten reinen Grenzfälle: Man findet überwiegend eine positive Raumladung im Kathodenfall, daher eine Beschleunigung der aus dem Plasma kommenden

Ionen und eine Aufheizung der Kathode, durch die diese besonders unter Berücksichtigung des hohen Feldes zur sog. thermischen Feldemission von Elektronen befähigt wird. Da demnach im allgemeinen vor den Elektroden eines Lichtbogens unkompensierte Raumladungen liegen, zeigt die Potentialverteilung längs der Bogenachse den in Fig. 1 schematisch gezeichneten Verlauf.

Jeder Lichtbogen besitzt einen von den Entladungsparametern abhängigen elektrischen Widerstand R_B; bei eingestellter Stromstärke I liegt an ihm folglich ein Spannungsabfall, die Bogenbrennspannung:

$$U_B = I R_B. \tag{6.1}$$

Da R_B sich nicht gleichmäßig über die Bogenlänge verteilt, sondern vor der Anode und Kathode groß und in der Säule relativ klein ist, ist auch die elektrische Feldstärke in den verschiedenen Bogengebieten sehr verschieden groß. Infolge der elektrischen Feldstärke werden die Ladungsträger jeweils zwischen zwei Zusammenstößen in bzw. entgegen der Feldrichtung beschleunigt. Praktisch wichtig ist dabei die Beschleunigung der Elektronen, die ihrerseits die aus dem Feld entnommene kinetische Energie sehr schnell untereinander austauschen, etwas langsamer auch an die übrigen schwereren Plasmateilchen abgeben. Daher ist die Elektronentemperatur stets *etwas* größer als die Gastemperatur, doch bleibt die Differenz im allgemeinen klein, d.h. beträgt meist ein bis einige Prozent.

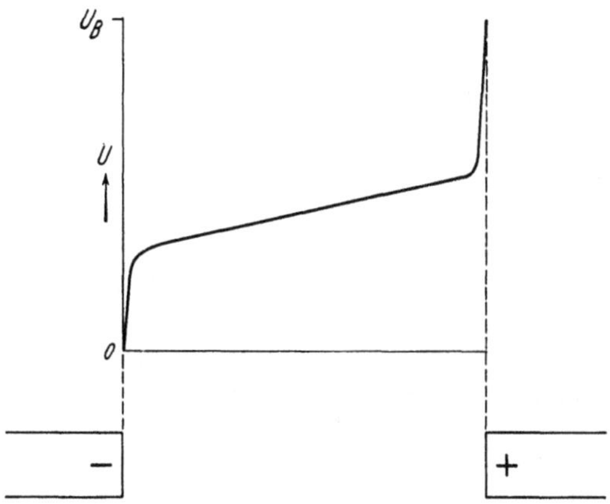

Fig. 1. Potentialverteilung zwischen den Elektroden eines Lichtbogens mit Kathoden- und Anodenfall (schematisch).

b) Jedes Längenelement der Bogensäule verliert Träger durch axiale Abwanderung, erhält aber wegen der überall konstanten Stromstärke die gleiche Zahl durch Zuwanderung von der jeweils entgegengesetzten Seite wieder. Die Säule verliert dagegen laufend Ladungsträger durch radiales Auswandern in kühlere Randzonen, wo Elektronen und Ionen rekombinieren. Dieser Ladungsträgerverlust muß durch Ionisierung von außen einwandernder neutraler Atome wettgemacht werden, was auch im thermischen Gleichgewicht vorwiegend durch Elektronenstoß geschieht. Im Gegensatz zur Säule verlieren die beiden Fallgebiete Ladungsträger vorwiegend durch einseitige Abwanderung zu den Elektroden, bzw. zur angrenzenden Bogensäule. Die Nachlieferung dieses sehr erheblichen Trägerverlustes erfolgt bei genügend großer Feldstärke in den Fallgebieten durch Stoßionisierung beschleunigter Elektronen oder, bei genügend hoher Temperatur, auf thermischem Wege. In den Fallgebieten ist daher im allgemeinen die Elektronentemperatur (bzw. die mittlere Elektronengeschwindigkeit) sehr viel größer als die Gastemperatur, falls diese nicht so hoch ist, daß auch in den Fallgebieten die Ionisierung rein thermisch erfolgen kann.

c) Da der spezifische Widerstand des Plasmas in den verschiedenen Bogengebieten, und mit ihm der gesamte Bogenwiderstand R_B, von der räumlichen Dichte und Beweglichkeit der Ladungsträger bestimmt wird, und da diese Größen von der Temperatur und der Feldstärke abhängen, ist R_B eine Funktion der Bogenparameter. Es ist daher keineswegs selbstverständlich, daß zu vorgegebenen

Daten des Stromkreises auch ein stationärer, sich zeitlich nicht ändernder Bogenzustand gehört. Stationarität ist vielmehr nur möglich, wenn zufällige Schwankungen der eingestellten Stromstärke sich durch Bogenvorgänge selbst oder durch Rückwirkungen auf den Stromkreis ausgleichen und nicht vergrößern. Die hieraus folgende Stationaritätsbedingung wird in Ziff. 8 im einzelnen behandelt.

d) Da der spezifische Widerstand des Plasmas der Ladungsträgerdichte und ihrer Beweglichkeit umgekehrt proportional ist, hängt der Bogenwiderstand R_B von der radialen und axialen Verteilung der Trägerdichte ab. In der Säule ist diese als Funktion der Temperatur berechenbar, nicht dagegen im allgemeinen in den für den gesamten Bogenwiderstand R_B entscheidenden Fallgebieten; hier sind die Verhältnisse von Bogen zu Bogen sehr verschieden. In der Säule hängt die radiale Verteilung der Trägerdichte, d.h. der effektive elektrische Bogendurchmesser, stark von der Kühlung ab, da durch Wegblasen der Ionen aus den Außengebieten der Säule, ebenso wie durch die Kühlwirkung fester oder flüssiger Wände, der Gradient der Trägerdichte dn/dr und mit ihm die radiale Träger-Abdiffusion vergrößert wird. Damit aber steigt dann R_B und bei konstanter Stromstärke die Bogenspannung U_B. Da die Trägerbeweglichkeit von der mittleren freien Weglänge der Elektronen abhängt, hängt R_B auch von der Gasart, vom Gasdruck und allen anderen die freie Weglänge beeinflussenden Parametern, besonders aber der Plasmatemperatur, ab.

e) Im stationär brennenden Bogen findet eine große Zahl von Elementarprozessen im Säulenplasma, in den Fallgebieten und an den Elektroden statt, deren Gesamtheit das Geschehen im Bogen bestimmt, und die in verschiedenen Bogentypen eine sehr unterschiedliche Rolle spielen können.

Hier ist zunächst die Anregung von Atomen (und Molekülen) besonders durch den Stoß thermischer Elektronen zu nennen. Diese Anregungsenergie wird entweder direkt in Form von Lichtquanten ausgestrahlt[1] oder vor der Emission noch in Stößen zweiter Art von Atom zu Atom weitergegeben. Emittierte Lichtquanten verlassen im allgemeinen das Bogenplasma nach außen und wirken infolge dieses Energieverlustes des Plasmas kühlend; lediglich die Energie scharfer Resonanzlinien oder der Mitten verbreiterter Resonanzlinien wird wiederholt absorbiert und reemittiert werden, bis sie meist in den Außenzonen des Bogens stecken bleibt. Diese Kühlung der Bogensäule durch Ausstrahlung ist um so wirksamer, je mehr Linien hoher Übergangswahrscheinlichkeit in dem Spektralgebiet liegen, in dem nach der PLANCKschen Kurve bei der Plasmatemperatur viel Energie ausgestrahlt werden kann. Im Plasma emittierte Photonen können natürlich auch von den Elektroden absorbiert werden und damit aufheizend wirken, an der Kathode auch Photoelektronen auslösen, doch scheint dieser letzte Effekt keine große Rolle zu spielen.

Der wohl wichtigste Elementarprozeß im Bogen ist die Ionisierung von Atomen und unter Umständen auch Ionen besonders durch thermische Elektronen. Im thermischen Gleichgewicht rekombiniert der überwiegende Teil der Elektronen und Ionen entweder im Dreierstoß, wodurch die Ionisierungsenergie wieder in das Plasma zurückgegeben wird, gelegentlich auch im Zweierstoß unter Emission der frei werdenden Energie. Ein Teil der Elektronen und Ionen aber diffundiert infolge des Konzentrationsgefälles auch radial ab und gibt seine Ionisierungsenergie in den kühleren Außenzonen des Bogens bzw. an der Wand

[1] Strahlungsfragen sind besonders in den Artikeln von G. L. WEISSLER in Bd. XXI, und von R. G. FOWLER in diesem Bande eingehend behandelt, ferner in Kap. C IX (Ziff. 65 bis 76) dieses Beitrages.

ab. Aus den anodennahen Plasmagebieten fließen ferner Elektronen zur Anode und heizen sie durch Abgabe ihrer kinetischen Energie und der beim Eintritt in die Anode frei werdenden Austrittsarbeit auf. Aus den kathodennahen Plasmagebieten fließen umgekehrt Ionen auf die Kathode und heizen diese durch Abgabe ihrer Neutralisierungsenergie und gegebenenfalls auch ihrer kinetischen Energie auf.

Es wurde schon erwähnt, daß die Elektronen, und in geringem Umfange auch die Ionen, aus dem elektrischen Feld kinetische Energie entnehmen und sie teilweise an die schwereren Plasmateilchen weitergeben. Zusammen mit diesen transportieren sie kinetische Energie aus der Säule radial in die Außenzonen des Bogens (klassische Wärmeleitung) und aus den elektrodennahen Plasmagebieten auch zu den Elektroden. Diese letzteren Prozesse sind je nach den Temperaturgradienten vor den Elektroden und der Größe der Ansatzfläche des Bogens von mehr oder weniger großer Bedeutung.

Die Kathodenoberfläche emittiert in dem aus der RICHARDSON-Gleichung für ihre Temperatur und Austrittsarbeit folgenden Umfang Elektronen, die zur Säule hin abfließen[1]. Bei hoher positiver Raumladung und damit Feldstärke vor der Kathode ist auch eine Feldemission von Elektronen[2] möglich; eine besondere Rolle scheint die kombinierte „thermische Feldemission" zu spielen.

Metallanoden emittieren im allgemeinen keine positiven Ionen, wohl dagegen Kohleanoden mit Metallsalzen, wenn die verdampfenden Metallatome eine geringe Ionisierungsenergie besitzen.

f) Aus dem Zusammenwirken aller dieser primären und sekundären Elementarprozesse folgt natürlich die Energiebilanz des gesamten Bogens.

Für die stationäre Bogensäule wird in Kapitel D gezeigt, wie die je Zentimeter Säulenlänge umgesetzte elektrische Leistung (Stromstärke mal elektrische Feldstärke) durch Ausstrahlung und radiale Wärmeleitung (einschließlich freier Konvektion und etwa vorhandener sonstiger Strömungen) abgeführt wird und sich daraus ein Gleichgewicht zwischen radialer Temperaturverteilung, radialer elektrischer Leitfähigkeit und axialer elektrischer Feldstärke einstellt. Je nach der Temperaturabhängigkeit der elektrischen und der Wärmeleitfähigkeit, den äußeren Kühlbedingungen und dem Anteil der Strahlungsemission erhält man die unten zu besprechenden verschiedenen Säulentypen.

Die Energiebilanz des Kathodenfallgebietes ist viel komplizierter als die der Säule. Die Kathode selbst wird aufgeheizt durch die auf sie auftreffenden positiven Ionen, die sich an ihr neutralisieren und dabei ihre Ionisierungsenergie als Neutralisierungsenergie abgeben. Ob auch ihre kinetische Energie an die Kathode abgegeben wird, hängt davon ab, ob die neutralisierten Ionen an der Kathodenoberfläche haften bleiben oder von ihr reflektiert werden; im letzteren Fall dient die kinetische Energie zur Aufheizung des der Kathode vorgelagerten Plasmas. Die Kathode absorbiert ferner einen Teil der auf sie auftreffenden Ionen und empfängt schließlich durch Wärmeleitung Energie aus den benachbarten Plasmaschichten, woraus wieder die Bedeutung der Temperaturverteilung vor der Kathode klar wird. Die der Kathode zugeführte Energie dient zur Deckung der Austrittsarbeit der emittierten Elektronen, zur Wärmeableitung in die kühleren Teile der Kathode, zur Wärmestrahlung (falls die Kathode genügend Temperatur hat) und, falls sie verdampft, zur Verdampfung von Elektrodenmaterial.

Entsprechend wird die Anode durch die in sie einströmenden Elektronen, die dort ihre kinetische Energie und ihre Austrittsarbeit abgeben, ferner durch

[1] Vgl. den Beitrag von W. B. NOTTINGHAM in Bd. XXI dieses Handbuches.

[2] Vgl. hierzu R. H. GOOD jr. u. E. W. MÜLLER in Bd. XXI dieses Handbuches.

Absorption von Säulenstrahlung sowie durch Wärmeleitung aus den angrenzenden Plasmaschichten geheizt. Die ihr zugeführte Leistung gibt die Anode durch Wärmeableitung, durch Wärmestrahlung und bei manchen Bogentypen auch durch Anodenverdampfung ab.

In den der Kathode bzw. der Anode vorgelagerten Plasmagebieten des Kathoden- und Anodenfalles (Fig. 1) findet wegen der hier meist viel höheren Werte der elektrischen Feldstärke ein viel größerer Energieumsatz je Volumeneinheit als in der Säule statt, besonders falls das Plasma vor der betreffenden Elektrode kontrahiert ist. In diesen Gebieten findet meist auch eine erhöhte Ionisierung (und oft auch Anregung) statt, da bei vielen Bogentypen hier die für den Mechanismus entscheidende Ladungsträgererzeugung vor sich geht. Ein Teil der oft großen in den Fallgebieten umgesetzten elektrischen Leistung wird von den in ihnen beschleunigten Ladungsträgern als kinetische und potentielle Energie den Elektroden zugeführt; der Rest geht wieder durch Ausstrahlung und radiale Wärmeleitung verschiedenster Art verloren. Bei manchen Bogentypen spielen schließlich auch Sondererscheinungen für die Energiebilanz des Kathoden- und Anodenfallgebietes eine Rolle, so bei starker Kontraktion die in Ziff. 96 behandelte Aufheizung von seitlich angesaugtem Gas bei der Ausbildung eines Plasmastrahles, beim anodisch verdampfenden Hochstromkohlebogen (Ziff. 121) auch die den überwiegenden Teil der Anodenfall-Leistung erfordernde Erzeugung und Aufheizung eines Anodenmaterial-Dampfstrahles.

7. Die Zündung von Lichtbögen. Es gibt grundsätzlich drei verschiedene Möglichkeiten, einen Lichtbogen zu zünden, nämlich

1. durch Berühren und anschließendes Auseinanderziehen der Elektroden,

2. durch Erzeugung eines den Elektrodenzwischenraum überbrückenden hoch ionisierten Dampfstrahles, meist mittels einer dritten Zündelektrode und

3. durch Anlegen einer zum Durchzünden ausreichenden hohen Spannung bzw. eines Hochspannungsstoßes.

Bei der Bogenzündung durch Berührung mit darauffolgendem Auseinanderziehen der beiden Elektroden werden die beiden Berührungspunkte so hoch erhitzt, daß der kathodische im allgemeinen zur Elektronenemission fähig wird und damit den eigentlichen Bogenmechanismus in Gang setzen kann. Gleichzeitig wird im allgemeinen ein wenig Elektrodenmaterial der letzten Verbindung zwischen den Elektroden durch den diese Brücke durchsetzenden Strom verdampft und so hoch erhitzt, daß damit das den Strom leitende Plasma für den in der Ausbildung begriffenen Bogen erzeugt ist. Erforderlich ist zur Zündung eine offenbar vom Elektrodenmaterial abhängige Mindeststromstärke, sowie ein Stromkreis, der auch beim Auseinanderziehen der Elektroden die Bogenstromstärke nicht unter einen gewissen Mindestwert sinken läßt. Statt der direkten Berührung der beiden Elektroden kann man diese selbstverständlich auch kurzzeitig durch eine dritte Elektrode überbrücken, die den Bogen sozusagen von der einen Elektrode zu anderen herüberzieht.

Kohlebögen von mehreren hundert Ampere Stromstärke kann man auch dadurch zünden, daß eine bei der Zündung anodisch geschaltete dritte Zündelektrode im abgeschalteten Zustand die Spitze der Kathode berührt und nach Anlegen der Spannung elektromagnetisch von dieser weggerissen wird. Bei Zündstromstärken von mehreren hundert Ampere wird an der Berührungsstelle Kathode/Zündelektrode nach Zeitlupenaufnahmen von GUILLERY [*6a*] dann eine Wolke hoch ionisierten Kohlenstoffdampfes erzeugt, die bei Erreichen der Anode den eigentlichen Bogen zündet.

Bei den beiden geschilderten Zündarten wurde die hohe Temperatur der Kathodenspitze gleichzeitig mit dem leitfähigen Bogenplasma erzeugt. Bei der Zündung eines Bogens durch dauerndes oder kurzzeitiges Anlegen einer genügend hohen Spannung an die Elektroden hat man zwischen dem die Zündung einer selbständigen Entladung einleitenden Stoßionisierungsmechanismus und dem nachfolgenden Übergang dieser Vorentladung in den Lichtbogen (unter Aufheizung der Kathode) zu unterscheiden.

Das Anlegen einer normalen Gleichspannung von etwa 220 V genügt höchstens bei gewissen Niederdruckentladungen wie Quecksilberlampen zur Zündung, bei denen sich erst während des Betriebes der für den thermischen Bogen erforderliche hohe Dampfdruck entwickelt. Bei diesen Lampen entstehen beim Anlegen der Zündspannung zunächst Elektronenlawinen, die sich durch Rücklaufen der positiven Ionen zur Kathode zu einer Art Glimmentladung mit noch kalter Kathode entwickeln. Durch die Ionen dieser Vorentladung wird dann der kathodische Bogenmechanismus eingeleitet, sei es der Aufbau einer positiven Raumladung vor der Kathode, deren elektrisches Feld Feldelektronenemission möglich macht, sei es die Aufheizung der Kathodenoberfläche, bis diese durch thermische Elektronenemission die für die Entladung benötigten Elektronen liefert. Da gleichzeitig mit zunehmender Stromdichte (durch Impuls- und Energieaustausch) das Entladungsplasma sich immer mehr aufheizt, tritt bei genügend hoher Stromstärke die thermische Ionisation im Gasraum an die Stelle der Ionisation durch die im Kathodenfall beschleunigten Elektronen. Für den Mechanismus dieser Entladung hoher Stromstärke ist also der anfänglich große Kathodenfall der Glimmentladung nicht mehr erforderlich, und wir finden mit zunehmender Stromstärke den Übergang von der feldbestimmten Glimmentladung zur Bogenentladung mit thermischer Elektronenemission der Kathode (oder einem der anderen in Kapitel E zu besprechenden Kathodenmechanismen) und thermischer Ionisierung im Gasraum, die sich daher auch durch einen kleinen Kathodenfall von der Größenordnung von 10 V auszeichnet. Fig. 2 zeigt diesen Übergang von der Glimmentladung in die Bogenentladung für den Fall einer 5 Atm-Xenon-Entladung, bei der die Glimmentladung durch entsprechende Vorschaltwiderstände stabilisiert worden war. Oberhalb 2 Amp war die Glimmentladung nicht mehr stabil.

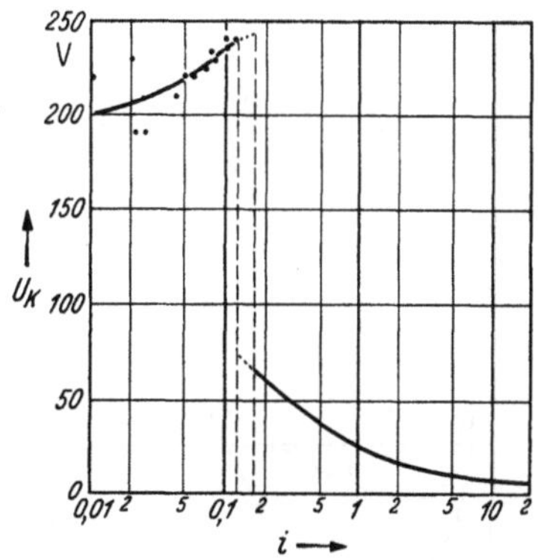

Fig. 2. Strom-Spannungs-Charakteristik einer 5 Atm-Xe-Entladung im Übergangsgebiet zwischen Glimmentladung und Bogenentladung, nach BAUER und SCHULZ [4].

Bei allen Hochdruckbögen reicht zur Zündung das Anlegen normaler Netzspannung nicht aus, sondern es ist ein Hochspannungsstoß erforderlich, der, wenn es sich um eine Kondensatorentladung handelt, durch Stoßionisierung und den sog. Streamer-Mechanismus[1] jedenfalls in vielen Fällen zur Ausbildung eines gut leitenden Funkenkanals[1] führt. Dieser entwickelt sich dann durch den weiteren Stromfluß in die stationäre Bogensäule, während gleichzeitig über die positiven Ionen der normale Bogenmechanismus eingeleitet wird.

Ein besonderer Zündmechanismus spielt nach neueren Arbeiten von GERMER und Mitarbeitern [*65*] bei der Zündung extrem kurzer Bögen eine Rolle, bei denen der Elektrodenabstand in der Größenordnung einer freien Elektronenweglänge

---

[1] Vgl. hierzu den Beitrag von L. B. LOEB in diesem Bande.

liegt. Bei diesen Bögen, deren Zündspannung bis 30 V herunter reicht, ist der übliche Entladungsaufbau über Elektronenlawinen nicht möglich, weil er eine große Zahl von Stößen zwischen den Elektroden voraussetzt. Man hat es vielmehr auch bei relativ hohen Gasdrucken mit einer Art Hochvakuumzündung zu tun. Diese wird durch Feldelektronenemission der Kathode (für welche deren Oberflächenrauhigkeit wichtig ist) eingeleitet, da die Feldstärke bei etwa 100 V über der Bogenstrecke von etwa 10^{-4} cm Länge dazu ausreicht. Die Elektronen durchfallen praktisch frei den Raum bis zur Anode, wo die hohe auffallende Leistung ($j^- U$) im wesentlichen zur Anodenmaterialverdampfung verbraucht wird, wodurch nach Rechnung vor der Anode Dampfdrucke bis zu 1000 Atm erzeugt werden sollen. Die weiteren von der Kathode kommenden Feldelektronen ionisieren einen Teil dieser Dampfatome, und die so erzeugten positiven Ionen laufen praktisch stoßfrei zur Kathode, wo sie durch ihre Raumladung die Feldstärke und damit die Feldelektronenemission verstärken, bis die Zündung einer richtigen stromstarken Bogenentladung erreicht ist. Die Trägervervielfachung erfolgt bei diesem Mikrofeldbogen also nach einem ganz anderen Mechanismus als bei Bögen großer Länge.

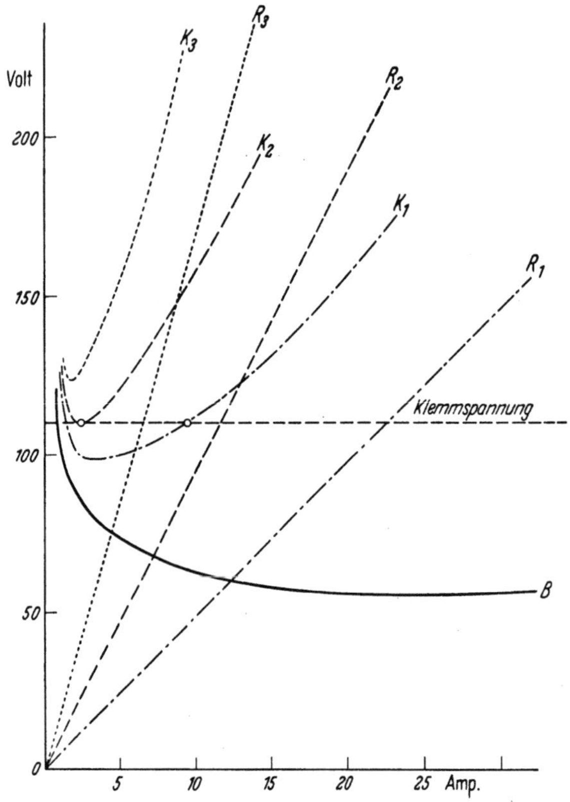

Fig. 3. Stabilisierung eines frei brennenden Lichtbogens mit fallender Strom-Spannungs-Charakteristik durch Vorschaltwiderstände verschiedener Größe.

8. Die Stabilitätsbedingung für stationäre Lichtbögen. Da die Brennspannung einer zündenden Bogenentladung mit wachsender Stromstärke sinkt und nur bei wenigen Bögen durch deren speziellen Mechanismus oberhalb einer gewissen Stromstärke wieder zu steigen beginnt, ist die Stabilisierung eines Lichtbogens auf eine stationäre Brennstromstärke im allgemeinen nur in Verbindung mit einem stabilisierenden Vorschaltwiderstand möglich. Denn bei fallender Charakteristik des Stromkreises würde ja jede zufällige Stromvergrößerung eine Spannungsabnahme und diese wiederum eine weitere Stromzunahme bewirken, so daß ein stationärer Brennzustand nicht möglich ist. Bei steigender Charakteristik des gesamten Bogenstromkreises aber, d.h. bei mit der Stromstärke zunehmender Spannung am Bogen selbst oder am Bogen plus dem vorhandenen Vorschaltwiderstand, werden zufällige Stromschwankungen nach beiden Seiten automatisch kompensiert.

Man übersieht diese Verhältnisse an Hand von Fig. 3. Hier stellt Kurve B die fallende Charakteristik eines frei brennenden Kohlebogens dar, während die Geraden R_1, R_2 und R_3 die Charakteristiken dreier in der angegebenen Reihenfolge zunehmender OHMscher Widerstände darstellen. Die Charakteristik des gesamten Kreises setzt sich additiv aus der des Bogens und seines Vorschaltwiderstandes zusammen, so daß wir für die drei Vorschaltwiderstände R_1, R_2, R_3 die zur gleichen Bogencharakteristik gehörenden Gesamtcharakteristiken K_1, K_2 und K_3 erhalten. Zeichnen wir nun als horizontale gestrichelte Gerade die Klemmenspannung U_0 der den Bogen speisenden Stromquelle ein, so muß im Gleichgewicht die Summe der Spannungsabfälle am Bogen (U_B) und am Vorschaltwiderstand (U_R) gleich der Klemmenspannung sein

$$U_B + U_R = U_0. \quad (8.1)$$

Der Bogen brennt also mit der Stromstärke, die dem Schnittpunkt der Gesamtcharakteristik K mit der Klemmenspannungsgeraden U_0 entspricht. Die Ordinaten von B und R bei dieser Stromstärke geben die Bogenbrennspannung und den Spannungsabfall am Vorschaltwiderstand für die betreffende Widerstandsstellung an. Aus Fig. 3 sieht man, daß der Vorschaltwiderstand mindestens so groß sein muß, daß eine in der Gegend des Arbeitspunktes steigende Gesamtcharakteristik K resultiert:

$$\frac{dU_B}{dI} + \frac{dU_R}{dI} > 0. \quad (8.2)$$

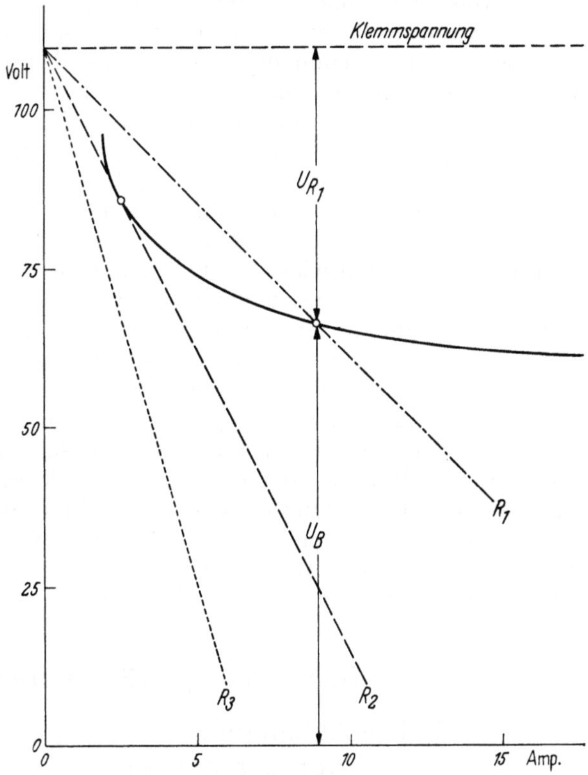

Fig. 4. Zur Stabilisierung eines Lichtbogens mit fallender Charakteristik durch Vorschaltwiderstände verschiedener Größe.

Dies ist die von KAUFMANN [*99*] gefundene Lichtbogenstabilitätsbedingung. Dabei werden zufällige Bogenschwankungen um so schneller ausgeglichen, je steiler die Gesamtcharakteristik K, je größer also der in Gl. (8.2) links stehende Ausdruck ist. Ein Lichtbogen brennt also um so stabiler (aber auch unwirtschaftlicher), mit je größerem Stabilisierungswiderstand er betrieben wird. Fig. 3 zeigt aber, daß es bei gegebener Klemmenspannung U_0 einen maximalen Widerstand (im Fall von Fig. 3 R_2) gibt, bei dem die Gesamtcharakteristik K eben noch die Klemmenspannungsgerade berührt. Die Abszisse dieses Punktes ist die kleinste Stromstärke, mit der der Bogen bei der gegebenen Klemmenspannung brennen kann. Um den Bogen bei noch kleineren Stromstärken betreiben zu können, oder bei größerer Stromstärke eine entsprechend größere Stabilität zu erreichen, muß man die Klemmspannung heraufsetzen.

Man stellt die gleichen Verhältnisse häufig auch durch Fig. 4 dar, in der die Widerstandsgeraden die Differenzen von Klemmenspannung und Spannungsabfall am Widerstand angegeben. Der Schnittpunkt einer Widerstandsgerade mit der Bogencharakteristik gibt den Arbeitspunkt des stationären Bogens bei diesem Widerstand, während der Berührungspunkt von R_2 mit der Bogencharakteristik den größtmöglichen Stabilisierungswiderstand und die Mindeststromstärke des Bogens bei der betreffenden Klemmenspannung angibt. In dieser Darstellung sieht man sehr schön, wie bei Veränderung der Klemmenspannung alle Widerstandsgeraden und ihre Schnittpunkte mit der Charakteristik sich verschieben, während umgekehrt aus Fig. 3 die Stabilisierungsbedingung deutlich wird, daß die Gesamtcharakteristik von Bogen plus Widerstand in der Gegend des Arbeitspunktes steigend sein muß.

Bei Kohlebögen und anderen Bögen mit fallender Charakteristik im Bereich geringer Stromstärke erhält man erfahrungsgemäß gute Stabilität, wenn man zum Betrieb eine Klemmenspannung verwendet, die ungefähr doppelt so groß ist wie die Bogenbrennspannung bei der gewünschten Stromstärke, wenn man also die Hälfte der gesamten umgesetzen Leistung im Vorschaltwiderstand verbraucht. Bögen mit positiver Charakteristik (Ziff. 17) wie etwa den BECK-Bogen kann man grundsätzlich natürlich ohne Vorschaltwiderstand betreiben, verwendet im allgemeinen aber gern eine Klemmenspannung, die um 10 bis 20% über der Bogenbrennspannung liegt und vernichtet diesen geringen Prozentsatz der umgesetzten Leistung im Vorschaltwiderstand, wodurch man die Stabilität des Bogens verbessert und eine Reguliermöglichkeit gewinnt.

Für die technische Anwendung von Bögen hoher Stromstärke mit fallender oder nur schwach steigender Charakteristik ist der Vorschaltwiderstand wegen seines Umfanges und Leistungsverlustes unerwünscht; solche Bögen, z. B. Schweißbögen oder Großscheinwerferbögen, speist man deshalb gern aus einem eigenen Gleichstromgenerator, dessen Stromspannungscharakteristik durch entsprechende Wicklung so gestaltet ist, daß die Gesamtcharakteristik von Bogen und Generator den gewünschten Steigungsgrad besitzt und die Stromstärke auch beim Zünden durch Berührung der beiden Elektroden einen vorgegebenen Maximalwert nicht übersteigt. Bei solchen Aggregaten reguliert man die Bogenstromstärke verlustlos durch Veränderung der Erregung des Generators.

B. Lichtbogen-Phänomenologie.

I. Die Abhängigkeit der Bogenerscheinungen von den äußeren Bedingungen.

9. Der Einfluß der Stromstärke. Es wurde in Ziff. 7 bereits erwähnt, daß der Übergang von der Glimmentladung in den thermischen Lichtbogen bei Vergrößerung der Stromstärke durch ein Überhandnehmen der thermischen Ionisation in der Bogensäule, und durch eine thermisch bestimmten Elektronenemission der Kathode bzw. der ihr vorgelagerten Plasmaschicht, gekennzeichnet ist. Entsprechend kann allgemein festgestellt werden, daß mit zunehmender Bogenstromstärke bzw. Stromdichte die Bedingungen für thermisches Gleichgewicht im gesamten Bogen immer besser erfüllt sind. Es ist noch nicht grundsätzlich klar, wie Stromstärke und Stromdichte in frei brennenden Bögen miteinander gekoppelt sind, und wie diese Kopplung durch Änderungen des Druckes verändert wird. Im Kohlebogen beispielsweise bleibt die Stromdichte mit zunehmender Stromstärke zunächst in erster Näherung konstant, um erst beim Auftreten des vielfach auch als kontrahierte Säule bezeichneten Bogenkernes (je nach Verhältnissen oberhalb 50 bis 80 Amp, vgl. Ziff. 25), mit zunehmender Stromstärke anzusteigen. Zunehmende Stromdichte der Bogensäule aber bedeutet grundsätzlich zunehmende Temperatur. So nimmt die Achsentemperatur des Kohlebogens mit der Stromstärke von etwas über 6000° beim 10 Amp-Bogen auf etwa 11000° beim 200 Amp-Bogen zu, und ähnlich liegt es bei anderen frei brennenden Bögen.

Zunehmende Temperatur aber bedeutet, wenn wir den Bereich oberhalb 15000° außer Acht lassen, auch zunehmende Leitfähigkeit der Bogensäule und damit abnehmende Säulenfeldstärke. Auf die eben erwähnte, mit zunehmender Stromstärke erfolgende Umbildung der Säule des Kohlebogens in die sog. Hochstromsäule, die durch den weißen Kern hoher Stromdichte und einen mit ihm gekoppelten, von der Kathode ausgehenden Gasstrahl ausgezeichnet ist, wird in Ziff. 25 näher eingegangen. Einen Extremfall einer durch Stromstärkesteigerung bewirkten Temperaturerhöhung und Brennspannungsverminderung haben Busz und Finkelnburg [*20*] in einem Argonbogen beobachtet, dessen minimale Brennspannung von über 30 V bei 6 Amp auf nur 7,5 V bei 150 Amp abnahm, während die Temperatur bis zu einem Maximalwert von 30000° K zunahm.

Über den Einfluß der Stromstärke auf die Bogensäule läßt sich auch theoretisch etwas aussagen. Während beim Niederstrombogen nämlich die radiale Temperatur- und Leitfähigkeitsverteilung hauptsächlich durch die Wärmeleitfähigkeit bestimmt ist, muß bei höheren Stromstärken der Einfluß des eigenen rotationssymmetrischen Magnetfeldes, das natürlich auch im Bogen selbst vorhanden ist, in Rechnung gestellt werden. Es bewirkt, da sich parallele Stromfäden je nach Stromdichte mehr oder weniger anziehen (pinch-Effekt), eine normalerweise nur geringe Druckerhöhung in der Bogensäule. Wenn dagegen Verengungen des Entladungsquerschnittes vorhanden sind, wie es fast immer an der Kathode, oft auch an der Anode der Fall ist, oder wenn durch äußere Maßnahmen eine künstliche Verengung der Säule vorgenommen wird, dann ist nach Maecker [*139*] an diesen Engstellen die Druckerhöhung durch das Eigenmagnetfeld wesentlich größer als in der übrigen Säule, so daß zur Entlastung dieses Überdruckes eine Plasmaströmung aus den Engstellen heraus entsteht. Speziell beim Hochstromkohlebogen geht eine solche heiße Plasmaströmung von rund 10^4 cm/sec Geschwindigkeit vom Kathodenbrennfleck aus in die Säule hinein. Da nun weiter nach dem Ohmschen Gesetz die Stromdichte bei gegebener Feldstärke dort am größten ist, wo die Leitfähigkeit am höchsten ist, verläuft der Hauptteil des Bogenstromes in dieser heißen und daher gut leitenden Strömung, die dadurch auf hoher Temperatur gehalten wird und sich als hell leuchtender Kern von den umgebenden Säulenteilen (Aureolen) abhebt.

Bei gewissen Bögen, z.B. den auch technisch wichtigen Hg- und Xe-Hochdruckbögen mit Wolframelektroden, findet bei zunehmender Stromstärke auch eine Änderung im Emissionsmechanismus der Kathode statt. Auf diesen zuerst von Weizel und Thouret [*226*], [*249*] festgestellten Wechsel vom Brennfleckbogen zum brennflecklosen Bogen wird in Ziff. 88 und 108 näher eingegangen.

Bei allen Lichtbögen, deren Elektroden nicht sorgfältig gekühlt werden, beginnen diese, und zwar vornehmlich die Anode, bei hinreichender Steigerung der Stromstärke zu verdampfen. Dadurch wird der Charakter des betreffenden Bogens grundsätzlich verändert, indem nunmehr der meist leichter ionisierbare Metall- oder Kohlenstoffdampf in zunehmendem Maße die Stromleitung ebenso wie die Lichtemission übernimmt. Der Bogen verwandelt sich also vom Gasbogen in einen Dampfbogen. Fördert man bewußt diese Entwicklung, indem man (vgl. Ziff. 25) etwa durch Verwendung relativ dünner Anoden die anodische Stromdichte erhöht, und damit die Anodenverdampfung verstärkt, so kann dies bei hinreichender Stromstärke zur Ausbildung eines die Anode verlassenden Dampfstrahles führen, der Geschwindigkeiten bis zu 10^4 cm/sec besitzen kann. Derartige Dampfstrahlen aber verändern, wie in Ziff. 121 gezeigt werden wird, rückwirkend den Mechanismus des Bogens, dessen Gesamtcharakteristik dadurch steigend werden kann. Schweißlichtbögen mit sich verbrauchender Kathode

(Ziff. 26) dürften ein weiteres Beispiel für frei brennende Lichtbögen sein, bei denen die Stromstärkesteigerung zu stürmischer Verdampfung und damit zu einer Änderung der Entladungsbedingungen führt.

Die Strahlung von Lichtbögen hängt primär von der Stromdichte ab, ändert sich aber mit der Stromstärke, weil deren Änderung im allgemeinen auch Änderungen der Stromdichte und Bogentemperatur bewirkt. Allgemein treten mit zunehmender Stromstärke Linien höherer Anregungs- und Ionisierungsstufen immer stärker hervor, ebenso ein vom Ionisierungsgrad abhängiges kontinuierliches Spektrum. Auf diese Plasmastrahlung wird in Kap. C IX näher eingegangen. Bei Höchstdruckbögen bis 1000 Atm schließlich [*165*] findet man bei genügend großer Stromstärke (100 bis 200 Amp) einen allmählichen Übergang von der überwiegend diskontinuierlichen Plasmastrahlung zur schwarzen Strahlung PLANCKscher Energieverteilung.

10. Der Einfluß des Elektrodenmaterials. Als Elektrodenmaterialien für Lichtbögen kommen in erster Linie Kohle in ihren verschiedenen Formen (Graphit, Koks und Ruß), gegebenenfalls mit Zusätzen besonders im Docht, sowie praktisch alle Metalle in Frage. Metallverbindungen, z.B. Metalloxyde, die bei höherer Temperatur genügende Leitfähigkeit besitzen, sind von RASCH [*13a*], [*171*] untersucht worden, haben aber mit einer in Ziff. 27 zu besprechenden Ausnahme keine Bedeutung erlangt. Das gleiche gilt für Bögen, deren eine Elektrode ein wässeriger oder geschmolzener Elektrolyt ist.

Die Art des Kathodenmaterials ist zunächst von Bedeutung für die Elektronenemission der Bogenkathode, wobei die Elektronenaustrittsarbeit, der Schmelzpunkt, der Siedepunkt und die Wärmeleitfähigkeit des Materials die entscheidenden Materialkonstanten sind. Nach der RICHARDSON-Gleichung hängt die thermische Elektronenemission exponentiell von der Austrittsarbeit des Kathodenmaterials ab. Da man nun für fast alle praktischen Zwecke feste Elektroden braucht, ist entscheidend die Elektronenemissionsdichte bei einer noch unter dem Schmelzpunkt liegenden Temperatur. Die Tatsache, daß Kohle bei den normal erreichbaren Drucken überhaupt keinen Schmelzpunkt besitzt, ist der Grund für die Beliebtheit von Kohlenstiften als Bogenelektroden. Ähnliches gilt für das Wolfram, dessen Schmelzpunkt mit etwa 3500° C genügend hoch liegt, um eine ausreichende Elektronenemission bei Temperaturen unter dem Schmelzpunkt zu gewährleisten. Metalle mit niedrigem Schmelzpunkt haben im allgemeinen einen relativ niedrigen Siedepunkt; solche Metalle können als Flüssigkeitskathoden (z.B. Hg) Verwendung finden. Auch die Wärmeleitfähigkeit des Kathodenmaterials beeinflußt das Verhalten eines Bogens entscheidend. Je geringer die Wärmeleitfähigkeit ist, desto größer ist die thermische Trägheit des Materials, und desto weniger schnell kühlt sich die erhitzte Kathode etwa in der Löschpause eines Wechselstrombogens ab.

Eine besondere, bei der Behandlung des Elektrodenmaterialeinflusses zu erwähnende Rolle spielen Oxydhäute oder Oxydflecken auf kathodischen Oberflächen. Da erhitzte Metalloxyde eine besonders hohe Elektronenemission besitzen, erleichtert ihre Anwesenheit den Bogenbetrieb, was das häufig beobachtete Wandern des Kathodenbrennflecks auf der Kathode und dessen Festsetzen an dickeren Oxydflecken verständlich macht. Lichtbögen in sehr reinen Edelgasen sollen nach SHENSTONE [*205*] (vgl. auch DOAN und THORNE [*30*]) zwischen oxydfreien, nicht bis zur Verdampfung gebrachten Metallelektroden überhaupt nicht brennen.

Allgemein ist der Siedepunkt des Elektrodenmaterials, genauer gesagt dessen Dampfdruckkurve, dadurch für die Bogenerscheinungen von Bedeutung, daß schon eine geringe Verdampfung einer der beiden Elektroden den Ionisierungsgrad, die elektrische Leitfähigkeit und die Emission des Bogens merklich beeinflussen können (Übergang vom Gasbogen zum Dampfbogen), während eine stärkere Verdampfung nach Ziff. 121 den Bogenmechanismus recht tiefgreifend beeinflussen kann. Eine solche Elektrodenverdampfung wird natürlich bei gleicher Energiezufuhr je Flächeneinheit um so schneller erfolgen, je schlechter die zugeführte Energie durch Wärmeleitung abgeführt werden kann, je geringer also die Wärmeleitfähigkeit des Elektrodenmaterials ist. Das früher so unverständlich erscheinende unterschiedliche Verhalten verschiedener Kohleelektroden bezüglich des Zischens (vgl. Ziff. 113) z.B. hat seine Erklärung in der sehr verschiedenen Wärmeleitfähigkeit von Graphit-, Koks- und Rußkohlen gefunden [*6a*].

11. Einfluß des Entladungsgases bzw. -dampfes. Bezüglich des Einflusses des Entladungsgases bzw. -dampfes auf das Bogenverhalten können wir zwischen der direkten Wirkung und einer über die Elektroden erfolgenden indirekten Wirkung unterscheiden. Eine solche indirekte Wirkung liegt besonders beim Sauerstoff vor, selbst wenn er nur spurenweise vorhanden ist, weil er auf der Kathodenoberfläche, wie Ziff. 10 erwähnt, Oxydflecke erzeugen und damit die Elektronenemission der Kathode beeinflussen kann. Auch das Abbrennen von Kohle- und anderen Elektroden ist vielfach, wenn auch keineswegs stets, auf einen chemischen Abbau seitens des Entladungsgases zurückzuführen. Beim Niederstromkohlebogen in Luft beispielsweise erfolgt der sog. Abbrand der Kohlen fast ausschließlich durch Oxydation und CN-Bildung, und die dabei entstehenden Moleküle beeinflussen ihrerseits den Ionisationsgrad des Bogenplasmas besonders in den Außenzonen.

Der direkte Einfluß der Materialkonstanten des Entladungsgases bzw. -dampfes wird bei der Besprechung der Theorie der Bogensäule deutlich werden, weshalb wir uns hier auf ein paar qualitative Bemerkungen beschränken. Bei Bögen in reinen Gasen oder Dämpfen ist die Ionisierungsspannung der Atome die wichtigste Größe; daneben spielt der Wirkungsquerschnitt der Atome gegenüber langsamen Elektronen, die Wärmeleitfähigkeit (besonderes Verhalten des Wasserstoffbogens [*160*]!), sowie für die Strahlung das Verhältnis der wesentlichen Anregungsspannungen zur Ionisierungsspannung eine Rolle. Der entscheidende Einfluß der Ionisierungsspannung auf den Bogen ist schon anschaulich leicht einzusehen. Um eine vorgegebene Ladungsmenge je Sekunde bei normaler Brennspannung durch den Bogen zu transportieren, benötigt man eine gewisse Ladungsträgerdichte, d.h. bei gegebenem Druck einen gewissen Ionisationsgrad. Im thermischenBogen ist der Ionisationsgrad aber nach Ziff. 33 durch die SAHA-Gleichung mit der Säulentemperatur verknüpft, so daß zur Einstellung des erforderlichen Ionisierungsgrades die Säulentemperatur um so höher sein muß, je größer die Ionisierungsspannung des Trägergases ist. Bei gleicher Stromstärke bzw. Stromdichte finden wir daher in Lichtbögen eine um so höhere Säulentemperatur, je höher die erste, bei Bögen hoher Stromstärke auch die zweite Ionisierungsspannung des Trägergases ist, wobei allerdings auch dessen Wärmeleitvermögen für die sich einstellende radiale Temperaturverteilung von Bedeutung ist. In der weitgehend durch die Ionisierungsspannung des Kohlenstoffes (11 V) bestimmten 200 Amp-Hochstromkohlebogensäule z.B. fand MAECKER [*138*] eine Achsentemperatur von 11000 bis 12000° K, während im Argonbogen (Ionisierungsspannung 15,7 V) gleicher Stromstärke die Achsentemperatur 25000° K erreichte [*20*]. Daß schon die Beimischung eines geringen Prozentsatzes von Gas oder Dampf niedriger Ionisierungsspannung die Bogentemperatur besonders bei niedrigen Stromstärken entscheidend verringern kann, ist nach dieser Überlegung klar. Erwähnt sei schließlich noch, daß es für das Verhalten eines Bogens wichtig sein kann, ob das Trägergas metastabile Zustände besitzt oder nicht. Solche metastabilen Atome können nämlich in die Fallgebiete diffundieren und dort wie die Beimischung eines Gases geringer Ionisierungsspannung wirken.

12. Der Einfluß des Druckes. Über den Einfluß des Gasdruckes auf die Eigenschaften frei brennender Lichtbögen liegen nur verstreute, wenn auch einigermaßen übereinstimmende Beobachtungen und Messungen vor, dagegen keine diese einheitlich und befriedigend beschreibende und erklärende Theorie, weil die Druckabhängigkeit etwa der Säulenfeldstärke von zu vielen selbst druckabhängigen Elementarvorgängen abhängt und daher je nach den Bedingungen theoretisch zwischen $p^{\frac{1}{2}}$ und $p^{\frac{1}{8}}$ liegen kann. Von allgemeiner Bedeutung ist, daß bei gegebenen Bogenabmessungen mit zunehmendem Druck die Zahl der freien Weglängen zwischen den Elektroden zunimmt und damit die Einstellung des thermischen Gleichgewichtes erleichtert wird, da die Feldstärke in allen Bögen weit weniger als proportional mit dem Druck wächst, der Gradient je freie Weglänge also abnimmt. Entsprechend findet man den Glimmbogen [7a] als Übergangsform zwischen Glimmentladung und thermischem Lichtbogen auf den Druckbereich unter 0,1 Atm beschränkt.

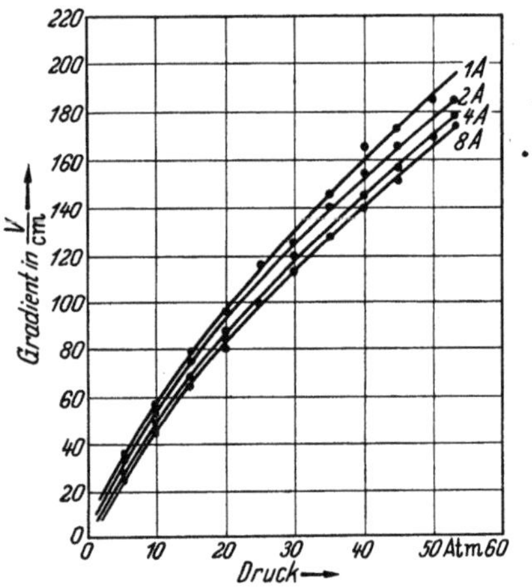

Fig. 5. Druckabhängigkeit der Säulenfeldstärke von Quecksilber-Höchstdrucklampen verschiedener Stromstärke, nach RÖMPE und THOURET [178].

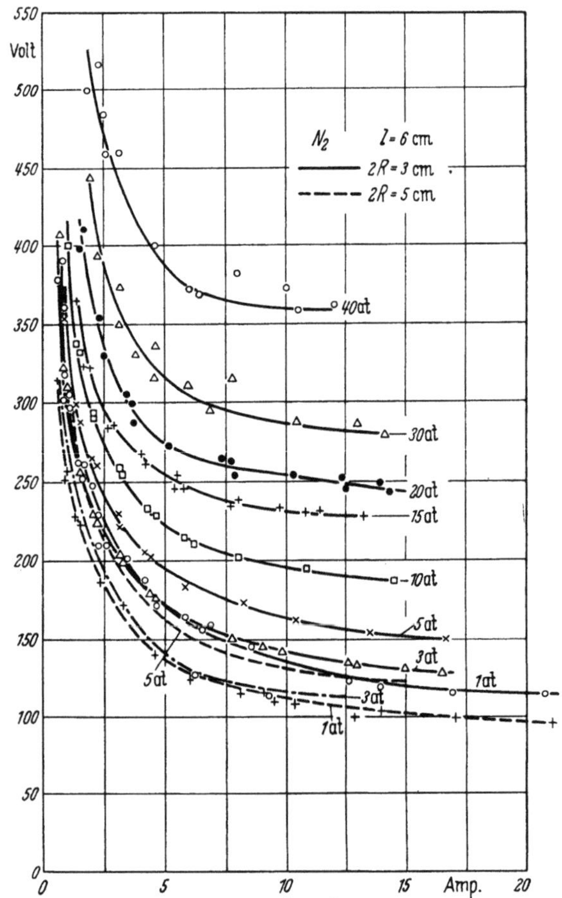

Fig. 6. Strom-Spannungs-Charakteristiken von Stickstoff-Wälzbögen verschiedenen Druckes, nach FOITZIK [54].

Aus zahlreichen, bis 1900 zurückreichenden Messungen (Literatur bei HAGENBACH [7a]) geht hervor, daß bei Kohle- wie Metallbögen von einigen Ampere Stromstärke in Luft wie in Stickstoff die gesamte Brennspannung, die Säulenfeldstärke und damit auch die Summe von Anoden- und Kathodenfall um etwa den Faktor 2 zunehmen, wenn man den Gasdruck von den niedrigst möglichen

Werten auf Atmosphärendruck erhöht. Der Druckbereich über einer Atmosphäre ist für den Kohlebogen in verschiedenen Gasen besonders von LUMMER [*132*] und später MATHIESEN [*10a*] eingehend untersucht worden. Nach letzterem nimmt z.B. die minimale Brennspannung eines 10 Amp-Homogenkohlebogens von 17 auf 34 V zu, wenn man den Druck von 1 Atm auf 10 Atm erhöht. Dabei sind die Angaben über das Verhalten von Anodenfall und Kathodenfall bei Drucksteigerung widerspruchsvoll und scheinen empfindlich von der speziellen Kohlenzusammensetzung abzuhängen. Die Temperatur in allen Bogengebieten und mit ihr auch die Strahlungsdichte nimmt mit steigendem Druck zu, entsprechend der allgemein zu machenden Beobachtung, daß alle Entladungsteile bei Druckzunahme sich zusammenziehen, der Leistungsumsatz je Volumeneinheit also erheblich zunimmt. Bei einem Reinkohle-Niederstrombogen stieg [*6a*] bei konstant gehaltener Bogenleistung die Kraterleuchtdichte bei Druckerhöhung auf 10 Atm von 14000 auf 48000 Stilb an, bei einem BECK-Bogen von 50000 auf 150000, in beiden Fällen also um rund den Faktor 3 bei Verzehnfachung des Druckes. An Quecksilber-Höchstdrucklampen haben ROMPE und THOURET [*178*], [*179*] eine Brennspannungszunahme von 40 auf 170 V bei Drucksteigerung von 4 auf 52 Atm bei 5 Amp Stromstärke gemessen und erwartungsgemäß eine sehr starke Zunahme der Strahlung (Leuchtdichte) mit dem Druck gefunden. Fig. 5 zeigt die von ihnen gemessene Zunahme der Säulenfeldstärke mit dem Druck.

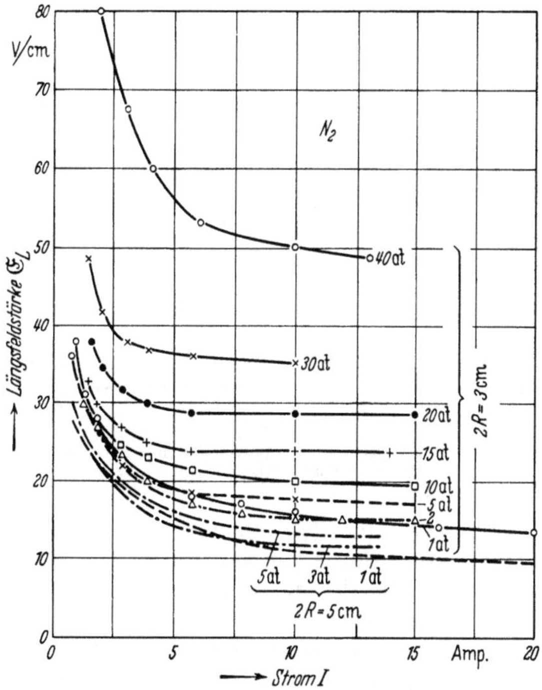

Fig. 7. Abhängigkeit der Säulenfeldstärke von der Stromstärke für Stickstoff-Wälzbögen verschiedenen Druckes, nach FOITZIK [*54*].

In einem Wälzbogen (vgl. Ziff. 28) hat FOITZIK [*54*] in dem Druckbereich von 1 bis 40 Atm festgestellt, daß die zur Stabilisierung des Bogens in der Achse erforderliche Mindestrotationsgeschwindigkeit des Glasrohres mit wachsendem Druck stark zunahm. Die Brennspannung des Wälzbogens in Stickstoff stieg nach Fig. 6 von 120 auf 360 V, wenn der Druck von 1 auf 40 Atm erhöht wurde. Dabei stieg die Säulenfeldstärke nach Fig. 7 von 14 auf 50 V/cm. In CO_2 konnte nur bis 10 Atm stabilisiert werden; bei Erhöhung auf diesen Druck stieg die Brennspannung von 150 auf 300 V, die Säulenfeldstärke von 20 auf 33 V/cm, wobei sich alle angegebenen Zahlen auf den Stromstärkebereich (rund 15 Amp) beziehen, in dem diese Werte von der Stromstärke unabhängig zu werden beginnen. An Bögen geringer Stromstärke (1 bis 7 Amp) und einer Bogenlänge von wenigen mm zwischen Kupferelektroden hat SUITS [*224*] in Stickstoff Messungen bis über 1000 Atm Druck ausgeführt. Fig. 8 gibt die Zunahme der Bogenbrennspannung mit dem Druck für vier verschiedene Stromstärken wieder und zeigt, daß hier die Brennspannung nur äußerst langsam mit dem Druck ansteigt.

Am Argonbogen hoher Stromstärke (Ziff. 26) fällt nach BUSZ-PEUCKERT und FINKELNBURG [*22*] die Brennspannung bei Drucksteigerung von 0,3 auf 0,8 Atm von 18 auf 15 V, um bei weiterer Steigerung bis 20 Atm auf etwa den dreifachen Ausgangswert, d.h. 45 V zuzunehmen. Der Durchmesser des frei brennenden Bogens verkleinert sich bei Drucksteigerung, während seine Achsentemperatur zunimmt, und die Intensität des kontinuierlichen Spektrums gegenüber der der Linien sehr stark wächst. Die von WEIZEL und ROMPE [*18a*] für Bogensäulen geringen Ionisierungsgrades und geringer Strahlung bei konstanter Temperatur theoretisch erwartete Feldstärkezunahme mit $p^{\frac{1}{2}}$ wurde von PETERS [*165*] an einem in Wasserdampf bei Drucken bis 1000 Atm brennenden 100 Amp-Bogen (an dem auch der Übergang von der teilweisen diskontinuierlichen Gasstrahlung zur rein kontinuierlichen Strahlung PLANCKscher Intensitätsverteilung gezeigt werden konnte) bestätigt, während nach obigem im allgemeinen mindestens die Bogenbrennspannung mit einer viel geringeren Potenz des Druckes als 0,5 wächst.

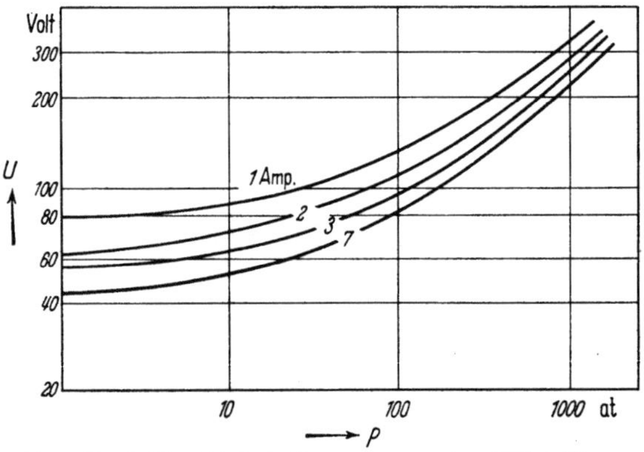

Fig. 8. Druckabhängigkeit der Brennspannung eines Stickstoffbogens verschiedener Stromstärke, nach SUITS [*224*].

13. Der Einfluß äußerer Temperaturänderungen. Bezüglich des Einflusses äußerer Kühlung oder Heizung auf einen Lichtbogen ist zwischen der Wirkung auf die Säule und der auf die Elektroden bzw. Fallgebiete zu unterscheiden. MATHIESEN [*10a*] hat einen mit 30 Amp und 50 V brennenden Niederstromkohlebogen dadurch von außen geheizt, daß er ihn in den Brennpunkt eines Hohlspiegels stellte, der die Strahlung einer anderen kräftigen Bogenlampe auf den Bogen konzentrierte. Er fand dabei Brennspannungsverkleinerungen bis zu 12%, wenn die positive Kohle geheizt wurde (deren pyrometrisch gemessene Temperatur dabei etwas abnahm), dagegen keinen merklichen Einfluß, wenn die Kathode durch Strahlung geheizt wurde. Versuche über eine alleinige Aufheizung einer Bogensäule scheinen nicht gemacht worden zu sein. Dagegen ist bekannt, daß die Kühlung einer Bogensäule durch äußeres Anblasen zu einer erheblichen Verkleinerung des Säulendurchmessers, einer Erhöhung der Säulentemperatur und einer Vergrößerung der Säulenfeldstärke führt. Dieser Effekt, von dem in manchen Schalterkonstruktionen Gebrauch gemacht wird, ist theoretisch verständlich. Durch die äußere Kühlung wird die elektrische Leitfähigkeit in den Außenzonen der Bogensäule herabgesetzt. Die durch die Stromstärke vorgegebene sekundliche Ladungsmenge kann daher nur transportiert werden, wenn in dem gekühlten Säulenteil entweder die Feldstärke größer oder die Leitfähigkeit (und damit die Temperatur) höher wird. Tatsächlich stellt sich ein Gleichgewicht ein, bei dem Feldstärke und Temperatur gerade zum Stromtransport ausreichen, und die je cm Säulenlänge zugeführte Leistung (Feldstärke mal Stromstärke) gerade den Energieverlust der Säule bei der betreffenden Temperatur zu decken vermag. Auch die extreme Säulentemperatur im GERDIEN-Bogen (Ziff. 28) hängt mit der intensiven Kühlung der Säule durch den sie begrenzenden Wasserwirbel zusammen.

Über die Wirkung einer Kühlung der Anoden von Lichtbögen ist außer der selbstverständlichen Verhinderung der Verdampfung nicht viel bekannt. Eine

Kühlung der Bogenkathode, z. B. durch Verwendung einer wassergekühlten Metallkathode, führt bei Bögen in Luft zu äußerst geringen Durchmessern des Kathodenbrennflecks, während umgekehrt die Aufheizung etwa einer großflächigen Kohlekathode durch einen Hilfsbogen nach MAECKER [*139*] zu einer erheblichen Vergrößerung des Kathodenbrennflecks und damit zur Verkleinerung der kathodischen Stromdichte (von der wiederum der in Ziff. 9 erwähnte kathodische Gasstrahl abhängt) Anlaß gibt.

14. Einfluß äußerer mechanischer oder magnetischer Kräfte auf den Lichtbogen. Da jeder Lichtbogen ein beweglicher Leiter ist, verändert er seine Länge und Form unter dem Einfluß äußerer Kräfte. Schon der Name Licht*bogen* rührt ja daher, daß der horizontal in freier Atmosphäre brennende Lichtbogen durch

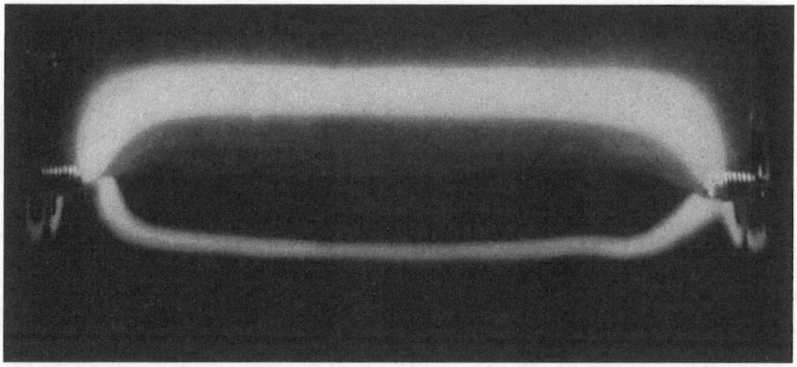

Fig. 9. Krümmung eines langen, in einem weiten Rohr brennenden Quecksilber-Bogens unter dem Einfluß des thermischen Auftriebs. Die Lichterscheinung in der unteren Hälfte des Bildes ist durch reflektiertes Licht bewirkt; nach WEIZEL [*18a*].

den Auftrieb der heißen Gase nach oben geblasen wird und daher im Gleichgewicht nicht in der Verbindungslinie der Bogenfußpunkte sondern nach oben gekrümmt brennt (Fig. 9). Auch beim senkrecht in freier Atmosphäre brennenden Lichtbogen hat der Auftrieb der heißen Gase eine wesentliche Wirkung, da er einen die Bogensäule umhüllenden und damit kühlenden aufwärtsgerichteten Luftstrom erzeugt. Wie stark Form und Brennspannung eines frei brennenden Bogens besonders bei niedriger Stromstärke durch Konvektion bestimmt sind, zeigt der bekannte Versuch von STEENBECK [*216*], der einen schwere- und damit konvektionsfreien Lichtbogen in einem frei fallenden Kasten untersuchte, und eine erhebliche Durchmesservergrößerung und Brennspannungserniedrigung gegenüber dem Normalfall mit Konvektion fand. Da nach Ziff. 22 bei fast allen Bögen auch eine gewisse Beweglichkeit der Bogenfußpunkte auf den Elektroden existiert, kann man durch äußeres Anblasen nicht nur Form und Durchmesser der Bogensäule verändern, sondern auch den gesamten Bogen zum Wandern bringen. Geeignete axial und tangential wirbelnde Gasströme werden nach Ziff. 28 auch zur Stabilisation langer Lichtbögen verwendet, und GRETENER [*70*] hat mit Erfolg axiale Luftströme benutzt, um die Anodenflamme eines Hochstromkohlebogens (vgl. Ziff. 25) in die Säule zu blasen und zu stabilisieren.

Zu den äußeren einen Lichtbogen beeinflussenden Kräften kann man auch die Zentrifugalkräfte rechnen, die in einem schnell rotierenden Bogen (Wälzbogen oder Wirbelbogen, Ziff. 28) wirksam sind und eine Stabilisierung solcher Bögen in der Achse bewirken (vgl. Fig. 10).

Äußere elektrische Felder sind auf den Bogen wie auch auf nichtstromführende Plasmen (z. B. Anodenflamme Ziff. 25) fast ohne Einfluß, da wir in allen thermischen Plasmen gleiche Dichte der Ladungsträger beider Vorzeichen und damit

Quasineutralität haben, und da auch kräftige elektrische Felder nur eine fast unmerkliche Ladungsträgertrennung bewirken.

Magnetische Felder dagegen üben auf Lichtbögen als bewegliche stromführende Leiter erhebliche Wirkungen aus, die am grundlegendsten wohl theoretisch von TONKS [228] untersucht worden sind. In einem axialen Magnetfeld krümmt sich ein langer Lichtbogen, da er fast stets geringe radiale Komponenten besitzt, spiralig und beginnt meist schon bei magnetischen Feldstärken oberhalb 10 Oe um seine Achse zu rotieren, wie schon 1841 von WALKER [238] beobachtet wurde. Dabei vergrößert sich nach ELENBAAS [4a] die Säulenfeldstärke beträchtlich, da die Rotation der Säule durch das relativ kühlere Gas der Umgebung einer Kühlung durch äußeres Anblasen gleichkommt. Eine solche magnetische

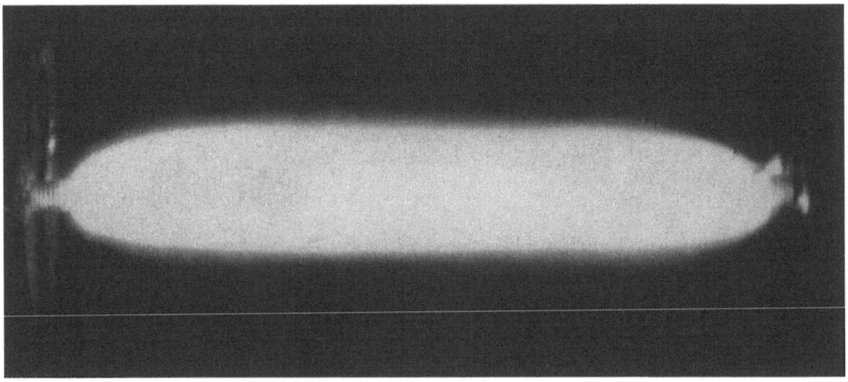

Fig. 10. Stabilisierung eines in einem weiten Glasrohr brennenden langen Quecksilber-Bogens (wie Fig. 9) durch schnelle Rotation des Rohres um seine Längsachse, nach WEIZEL [18a].

Rotation des Bogens ist auch technisch gelegentlich benutzt worden, um die mittlere Stromdichte und damit Leuchtdichte auf einer Kohleanode möglichst gleichförmig und zeitlich konstant zu halten. Gegenüber der durch ein axiales Magnetfeld bewirkten Bogenrotation spielt eine zweite Wirkung eines solchen Feldes offenbar eine geringe Rolle. Sie besteht darin, daß die Bahn jedes Elektrons schraubenförmig gekrümmt und dadurch seine mittlere freie Weglänge in Feldrichtung verkürzt wird.

Durch ein zur Säulenachse senkrechtes (transversales) Magnetfeld wird der Bogen wie jeder Leiter senkrecht zur Richtung des Stromes und des Magnetfeldes abgelenkt. Behalten die Bogenfußpunkte ihre Stellung bei, so kann man damit die Säulenform beeinflussen, z.B. die konvektionsbewirkte Bogenkrümmung kompensieren [6a]. Entsprechend verwendet man bei fast allen Kinobogenlampen und vielen sonstigen technischen Bogenanwendungen bewickelte oder durch den Bogenstrom selbst magnetisierte sog. „Blasmagnete", um die gewünschte Bogenform zu erzeugen und zu erhalten. Auf die große Vielzahl der hierbei möglichen Bogenerscheinungen kann nur hingewiesen werden; einen Überblick über die Wirkung äußerer Magnetfelder auf Gleich-, Wechsel- und Drehstrom-BECK-Bögen (Ziff. 25) findet man in der Monographie von FINKELNBURG [6a].

15. Der Einfluß von Bogenlänge, Elektrodenstellung und Elektrodenform auf Gestalt und Stabilisierung eines Lichtbogens. Die auf den ersten Blick verwirrende Fülle der verschiedenen Lichtbogenformen hängt außer von der Stromstärke in erster Linie von der Form und Stellung der Elektroden und deren Abstand, d.h. also der Bogenlänge ab. Allgemein brennt, von Ausnahmen abgesehen, ein Bogen um so ruhiger und stabiler, je kleiner seine Länge ist. Bei einem solchen kurzen Bogen ist die Bogenform auch unabhängig davon, ob der Bogen horizontal oder vertikal brennt. Bei längeren Bögen wird besonders das Mittelstück ihrer Säule immer beweglicher und verändert sich entsprechend leichter bei Einwirkung äußerer Kräfte, wie etwa der Konvektion der aufsteigenden heißen Gase. Die Stabilität von Bögen nimmt also im allgemeinen mit zunehmender Bogenlänge ab. Die meisten Bögen brennen ferner bei koaxialer Elektrodenstellung besser als bei Winkelstellung der Elektroden (die bei vielen Kohlebögen bevorzugt wird, um den positiven Krater ohne Behinderung durch die Kathode abbilden zu können), weil bei Winkelstellung die elektrodynamischen Kräfte des Stromkreises den Bogen nach außen krümmen und bei größeren Stromstärken sogar ausblasen können. Eine Ausnahme von der Regel großer Stabilität kurzer Bögen mit koaxialen Elektroden stellt der Hochstromkohlebogen mit stark verdampfender Anode (vgl. Ziff. 25) dar, weil bei ihm die Wechselwirkung des anodischen Dampfstrahles mit dem Ziff. 9 behandelten kathodischen Gasstrahl bei koaxialer Elektrodenstellung und besonders bei geringer Bogenlänge eine erhebliche Instabilität und Bogenunruhe bewirkt. Ein solcher Bogen brennt daher gemäß Fig. 11 nur bei sorgfältig erprobten Werten von Bogenlänge und Elektrodenwinkel stabil [6a].

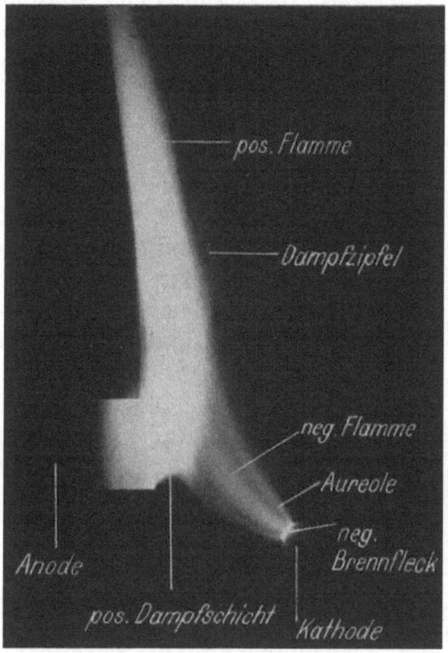

Fig. 11. Stabil brennender Hochstromkohlebogen mit stark verdampfender Anode (Beck-Bogen) [6a].

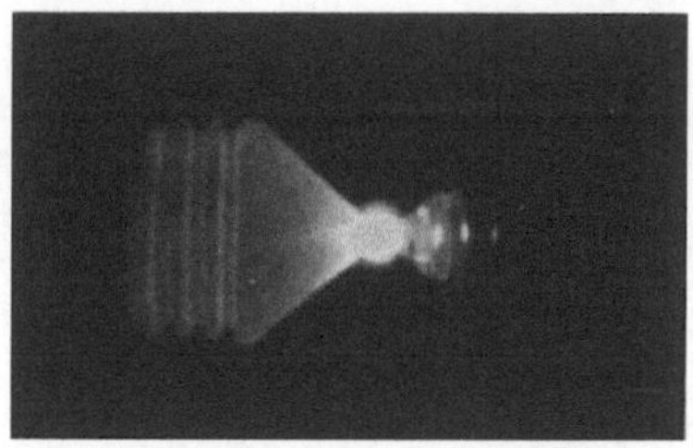

Fig. 12. Kurzer elektrodenstabilisierter Quecksilber-Höchstdruckbogen, nach Weizel und Rompe [18a].

Fast alle Lichtbögen brennen ferner um so stabiler, je dünner bzw. spitzer die Kathode ausgebildet ist, weil der Kathodenfußpunkt dann an dieser einen festen Sitz findet. Die Ausbildung der Anode ist für die Bogenruhe bei ausgedehntem Anodenansatz von geringerer Bedeutung, während bei konzentriertem Anodenansatz wiederum eine Anodenspitze vorteilhaft ist.

Nach der auf Elenbaas [37] zurückgehenden Theorie der Bogensäule ist für die Säulenstabilität die radiale Energieabfuhr von entscheidender Bedeutung. Ein langer Lichtbogen, der in einem Rohre brennend dessen Durchmesser gerade erfüllt, brennt daher wesentlich stabiler (sog. Wandstabilisierung nach Weizel [243]) als ein solcher, der ein Rohr größeren Durchmessers nicht völlig erfüllt und daher, Konvektionsströmungen folgend, im Inneren hin und her pendeln kann. Bei frei brennenden Lichtbögen fehlt die stabilisierende Wand völlig. Kurze frei stabil brennende Bögen (Beispiel Fig. 12) bezeichnet man nach Weizel als *elektrodenstabilisiert*, weil ihr Durchmesser noch von den Elektroden her bestimmt wird. Kurze Quecksilberhochdruckbögen (Ziff. 27) sind typische Vertreter dieser Bogengruppe. Lange frei brennende Bögen wie Fig. 9 bezeichnet WEIZEL demgegenüber als *konvektionsbestimmt*, weil die aufsteigenden Konvektionsströmungen die radial abfließende Wärme forttransportieren und damit die Rolle der Wand übernehmen. Wegen der Turbulenz der den Bogen dann begrenzenden Konvektionsströmungen kann man von einer eigentlichen Stabilisierung in diesem Fall nicht sprechen und benutzt zur Stabilisierung langer Bögen deshalb andere, in Ziff. 28 beschriebene Methoden. Bei nichtvertikal und besonders bei horizontal brennenden langen Bögen können die aufwärts gerichteten Konvektionsströme zu einer so starken Aufwärtskrümmung der Säule führen, daß diese periodisch in der Mitte abreißen (vgl. Fig. 13) und durch die dann automatisch ansteigende Elektrodenspannung nahe der Bogenachse wiederzünden kann. Derartige Bögen brennen äußerst unruhig und erzeugen ein knatterndes Geräusch; sie werden wegen ihrer äußeren Erscheinung als Flammenbögen bezeichnet. Bei Stromstärken oberhalb 50 bis 100 Amp beginnen, wie oben schon für den BECK-Bogen erwähnt, vom Bogen selbst ausgelöste Erscheinungen wie der Kathodengasstrahl und eventuell auch Dampfstrahlen für die Bogenstabilität eine immer größere Rolle zu spielen, während die Bedeutung der Konvektion immer mehr zurücktritt. Das zeigt sich beispielsweise durch die an Lichtbogenscheinwerfern gemachte Beobachtung, daß ein Hochstromkohlebogen seine Erscheinungsform unabhängig von seiner Stellung im Raum unverändert beibehält. Man kann hier also von einer „Eigenstabilisierung" oder auch „Eigenströmungsstabilisierung" der Hochstrombögen sprechen.

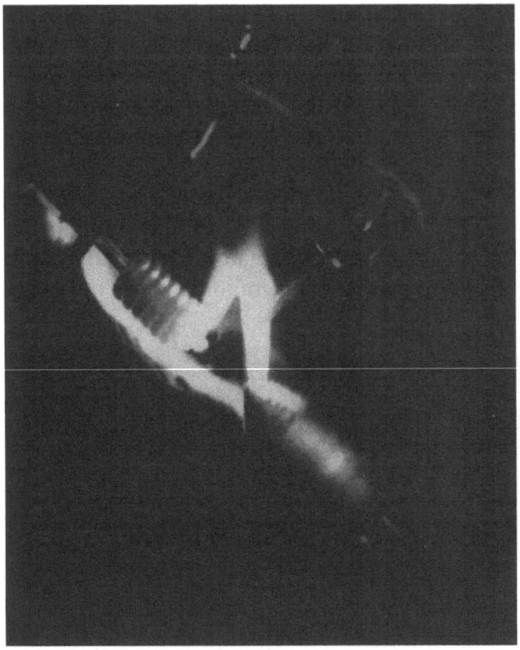

Fig. 13. Xenon-Hochdruck-Flammenbogen, nach WEIZEL (die im linken unteren Teil sichtbare Leuchterscheinung ist kein Bogen, sondern ein Reflex am Quarzkolben).

II. Allgemeine Eigenschaften stationärer und nichtstationärer Lichtbögen.

16. Strom-Spannungs-Charakteristiken. Als Charakteristik oder Kennlinie eines Lichtbogens bezeichnet man den Verlauf seiner Brennspannung mit der Stromstärke. Da die Brennspannung sich aus Kathodenfall, Säulenspannung und Anodenfall zusammensetzt, gibt die Charakteristik zwar keinen vollständigen Aufschluß über das elektrische Verhalten eines Bogens im einzelnen, beschreibt sein elektrisches Gesamtverhalten aber eindeutig und ist für die Auslegung der Stromversorgung wie die Stabilisierung jedes Lichtbogens von entscheidender Bedeutung.

Bei der Messung von Charakteristiken ist oft eine Anzahl von Vorsichtsmaßregeln übersehen worden, deren Beachtung unerläßlich ist, wenn man richtige und reproduzierbare Werte erhalten will. Dabei wird zunächst nur von der sog. statischen Charakteristik gesprochen, die sich aus den Brennspannungswerten zusammensetzt, die der betreffende Bogen im stationären Gleichgewicht bei verschiedenen Stromstärken besitzt. Da die Einstellung dieses Gleichgewichtes bei manchen Bögen beträchtliche Zeit (bis zu Minuten) erfordert, muß nach Einstellung einer bestimmten Stromstärke die Einstellung des Gleichgewichtes abgewartet werden, ehe der zugehörige Brennspannungswert abgelesen wird. Auf die bei schnellen Stromänderungen gemessene ,,dynamische Charakteristik" kommen wir in Ziff. 23 zurück. Bei der Messung von Charakteristiken ist ferner zu beachten, daß die wirkliche Bogenlänge während des Fortschreitens von kleinen zu großen Stromstärken konstant bleiben muß. Das ist bei frei brennenden Bögen dann nicht der Fall, wenn bei konstant gehaltenem Elektrodenabstand der Bogen nicht in der Verbindungslinie der beiden Fußpunkte brennt, sondern eine von der Stromstärke abhängige Krümmung zeigt. Bei Kohlebögen und besonders solchen mit Dochtkohle bildet sich ferner oft ein positiver Krater aus, dessen Tiefe von der Stromstärke abhängt und eine Änderung der effektiven Bogenlänge bewirkt, die um so mehr ins Gewicht fällt, je kleiner der Elektrodenabstand ist. Schließlich ist zu beachten, daß die an den beiden Klemmen einer Bogenlampe abgegriffene Spannung keineswegs mit der Bogenbrennspannung identisch ist, sondern der Spannungsabfall an den Elektroden berücksichtigt werden muß, der seinerseits wieder häufig von der Elektrodentemperatur und damit von der Stromstärke abhängt. Eine ohne Berücksichtigung des Spannungsabfalles in den Elektroden gemessene Charakteristik ist also häufig nicht nur in der absoluten Höhe, sondern auch in ihrem relativen Verlauf verfälscht. Bei der Beurteilung von Charakteristiken besonders aus älterer Zeit ist daher große Vorsicht geboten, besonders, wenn der Brennspannungsverlauf über einen großen Stromstärkebereich betrachtet wird. Zur Reduktion der an den Klemmen gemessenen Bogenspannung auf die wahre Brennspannung wird meist eine der beiden folgenden Methoden benutzt [*6a*]. Entweder schließt man den Bogen nach Messung der Brennspannung bei einer bestimmten Stromstärke durch schnelles Zusammenschieben der Elektroden kurz und mißt damit jeweils deren Spannungsabfall bei einer gegebenen Stromstärke, oder man mißt die wahre Brennspannung direkt, indem man die Klemmen des Voltmeters mit zwei Sonden verbindet, die zur Messung mit den Elektroden kurz hinter den beiden Bogenfußpunkten in Berührung gebracht werden.

Die Stromspannungs-Charakteristiken frei brennender Lichtbögen sind im allgemeinen, mindestens im unteren Stromstärkebereich, fallend, d.h. die Brennspannung nimmt mit zunehmender Stromstärke ab, wie Fig. 3 und 14 zeigen. Dies beruht darauf, daß mit zunehmender Stromstärke der Widerstand der Bogenstrecke bei frei brennendem Bogen im allgemeinen abnimmt, weil sowohl der Durchmesser der leitenden Bogensäule als auch (besonders bei vielen Hochstrombogensäulen) die Temperatur und damit die Leitfähigkeit der Säulen zunehmen.

Knickpunkte oder Sprünge in Bogencharakteristiken, wie sie z.B. von Ayrton [*1a*] nach Fig. 14 beim Einsatz des Zischens eines Niederstromkohlebogens, und von Finkelnburg [*6a*] beim Einsetzen des sog. Beck-Effektes beim Beck-Lichtbogen (Fig. 15) beobachtet worden sind, deuten auf Änderungen im Entladungsmechanismus bei gewissen Stromstärken hin und erlauben damit Rückschlüsse auf letzteren. Daß zahlreiche Lichtbögen besonders im Bereich größerer Stromstärke eine steigende, positive Charakteristik besitzen, kann durchaus verschiedene Ursachen haben, so daß der Schluß von einer steigenden Charakteristik auf einen bestimmten Bogenmechanismus nicht immer eindeutig ist. Brennt ein Bogen, wie z.B. die älteren Quecksilberbögen mit Hg-Elektroden oder Unterwasserbögen, in abgeschlossenen Gefäßen, so steigt mit zunehmender Stromstärke

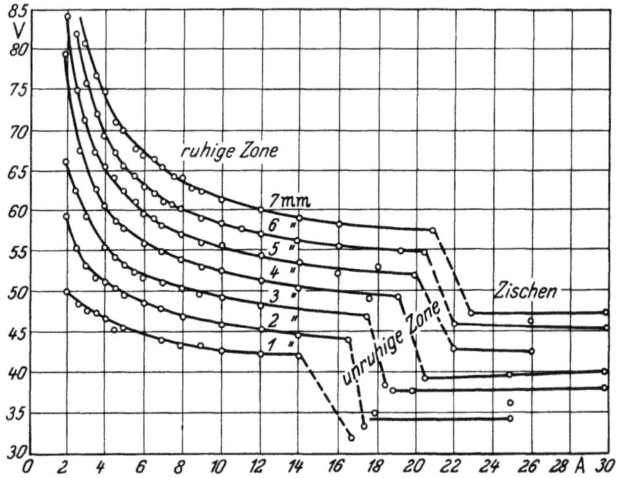

Fig. 14. Strom-Spannungs-Charakteristiken eines Niederstromkohlebogens verschiedener Länge mit Übergang vom ruhigen zum zischenden Bogen, nach AYRTON [1a].

der Gleichgewichtsdampfdruck und damit wegen der Druckabhängigkeit der Säulenfeldstärke auch die Bogenbrennspannung an, so daß für derartige eingeschlossene Bögen eine steigende Charakteristik resultiert. Auch der durch eine wasserumströmte Düse brennende GERDIEN-Bogen (Ziff. 28) besitzt nach Fig. 16 bei hinreichend kleinem Durchmesser einc steigende Charakteristik, die

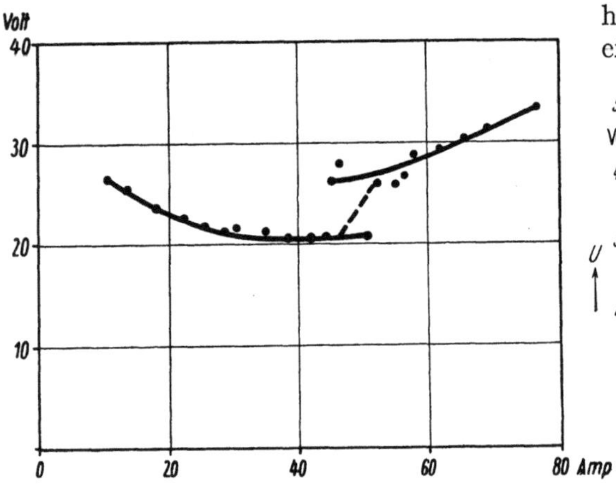

Fig. 15. Strom-Spannungs-Charakteristik eines BECK-Bogens mit Charakteristik beim Einsetzen des BECK-Effektes (Dampfausbruch aus der Anode), nach FINKELNBURG [6a].

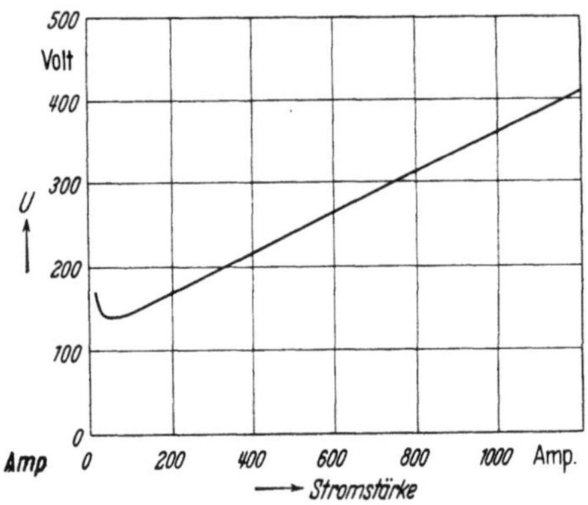

Fig. 16. Steigende Strom-Spannungs-Charakteristik eines GERDIEN-Wirbelbogens, nach BURHORN [18].

dadurch verursacht wird, daß mit zunehmender Stromstärke ein mehr als dieser proportionaler Anteil der Säulenleistung zur Erzeugung der aus den Düsenöffnungen austretenden Wasserdampf-Plasmastrahlen verwendet wird. In diesen beiden Fällen nimmt also (aus verschiedenen Gründen) die Säulenfeldstärke selbst mit der Stromstärke zu und bewirkt die steigende Charakteristik. Ganz anders ist es beim anodischen frei brennenden Hochstromkohlebogen. Bei ihm ist zwar die Säulenspannung im Bereich großer Stromstärke von dieser unabhängig, dagegen wächst [6a] mit zunehmender Stromstärke der Anodenfall wegen der gleichzeitig wachsenden Anodenverdampfung. Die in Fig. 15 gezeigte steigende Charakteristik des BECK-Bogens beruht, wie FINKELNBURG durch Sondenmessungen zeigen konnte, auf diesem mit der Stromstärke wachsendem Anodenfall.

17. Potentialverteilung, Kathodenfall, Anodenfall und Säulengradient. Wie in jeder Entladung bilden sich auch im Lichtbogen in der Säule und besonders vor den Elektroden Raumladungen aus, die eine sehr ungleichmäßige Potentialverteilung zwischen den Elektroden bewirken. Im allgemeinen liegt vor der Kathode ein Potentialsprung von etwa 10 V und vor der Anode ein solcher, der zwischen fast Null und über 30 V betragen kann, während der Rest des Potentialabfalles sich mehr oder weniger gleichmäßig auf die Säule verteilt. In den Gebieten weit außerhalb des Bogens dagegen herrscht ein ganz gleichmäßiger Potentialfall von der einen Elektrode zur anderen, so daß dicht an den Elektroden der Potentialgradient kleiner, im Mittelgebiet dagegen größer ist als im Bogen. Da die Äquipotentialflächen zusammenhängende Flächen sind, müssen in den Randgebieten des Bogens die in Fig. 17 schematisch dargestellten Äquipotentialflächen-Faltungen und damit radiale elektrische Felder auftreten, die durch Potentialsondenmessungen auch nachgewiesen worden sind [53].

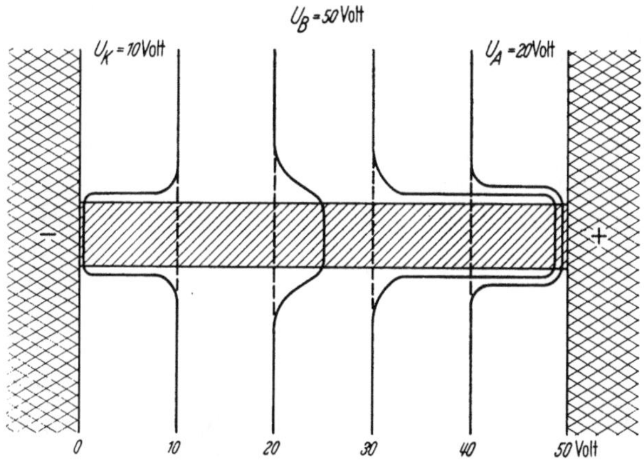

Fig. 17. Schematische Darstellung der Äquipotentialflächen-Faltungen in den Außenzonen eines Modellbogens von 50 V Brennspannung, unter der Annahme eines Kathodenfalles von 10 V und eines Anodenfalles von 20 V (vgl. FINKELNBURG und SEGAL [53]).

Die Potentialverteilung im Bogen kann überhaupt ganz allgemein mit Sonden abgetastet werden, die über ein Spannungsmeßgerät genügend hohen inneren Widerstandes (z.B. auch über einen Schleifen- oder Kathodenstrahl-Oszillographen) mit einer der beiden Elektroden verbunden sind. Solche Messungen sind zuerst 1887 von LECHER [122] und seitdem wiederholt mit immer verfeinerten Methoden durchgeführt worden. Für eine kritische Diskussion der Methoden und ihrer Leistungsfähigkeit vgl. [158]. Heute werden als Sonden meist dünne Drähte (0,3 mm) aus hochschmelzendem Metall verwendet, die mit solcher Geschwindigkeit durch den Bogen geführt werden, daß sie trotz der hohen Temperatur des durchstrichenen Plasmas nicht schmelzen, andererseits aber die oszillographische Aufzeichnung des Potentials an jedem Bogenort erlauben. Nach der für ein Niederdruckplasma entwickelten LANGMUIRschen Sondentheorie nimmt jede Potentialsonde (Ziff. 17) wegen der großen Elektronenbeweglichkeit im Plasma ein gegen ihre Umgebung negatives Potential an. Im Lichtbogen ist die Lage insofern noch komplizierter, als um die relativ zum Plasma kühle Sonde herum sich eine Zone niedrigerer Temperatur ausbildet, deren Wirkung auf das Sondenpotential theoretisch schwer zu übersehen ist. Neuere Erfahrungen [21] haben aber eindeutig ergeben, daß diese nicht genau bestimmbare Potentialdifferenz zwischen der Sonde und dem ungestörten Plasma innerhalb der Meßgenauigkeit von $\pm 0,5$ V nicht von der Temperatur abhängt und (2 ± 1) Volt betragen dürfte. Die älteren in der Literatur zu findenden Angaben sind demgemäß meist noch zu korrigieren, d.h. die Kathodenfallwerte um diesen Betrag zu erhöhen, die Anodenfallwerte zu erniedrigen. In neueren Arbeiten [*31*], [*56*], [*148*], [*161*] wurde die entsprechende Korrektur meist schon auf die verschiedensten Arten berücksichtigt. Trotz der in dieser neueren Literatur zu findenden weitgehenden Skepsis gegenüber Potentialmessungen in Lichtbögen sind wir auf Grund eigener Erfahrungen überzeugt, daß man Potentialverteilungen sehr zuverlässig messen kann und daß auch Absolutmessungen mit einem Fehler von ± 1 V oder maximal ± 2 V noch wertvolle Aufschlüsse bringen können.

Für den in Ziff. 26 erwähnten Argonbogen, dessen anodische Stromdichte mit zunehmender Bogenlänge abnimmt, fanden BUSZ-PEUCKERT und FINKELNBURG [*21*], daß der Anodenfall mit zunehmender Bogenlänge wächst und wegen der Kopplung der Bogenlänge mit der Plasmatemperatur vor der Anode nach Fig. 18 auch von dieser abhängt. Die vielfach angewendete Methode, die elektrische Feldstärke in der Bogensäule aus der Änderung der Brennspannung mit der Bogenlänge zu bestimmen, liefert nach diesem Ergebnis nur bei solchen Bögen richtige Werte, deren Säule annähernd zylindrisch und deren anodische Stromdichte von der Bogenlänge unabhängig ist. Aus dem genannten Ergebnis folgt weiter, daß man durch Extrapolation der Brennspannung auf die Bogenlänge Null zwar die Summe von Kathoden- und Anodenfall für einen extrem kurzen Bogen erhält, daß diese Summe wegen der Stromdichteabhängigkeit des Anodenfalles aber für längere Bögen nicht richtig zu sein braucht. Aus dem gleichen Grunde kann man auch die von BAUER und SCHULZ [*4*] nach einer neuen Methode aus der Energiebilanz der Elektroden bei Gleich- und Wechselstrombetrieb für kurze Bögen ermittelten Kathodenfall- und Anodenfallwerte nicht auf andere Bogenlängen übertragen. Von Einzelfällen abgesehen, können also heute Kathoden- und Anodenfall von Lichtbögen nur mit der oben angegebenen Unsicherheit aus Sondenmessungen ermittelt werden.

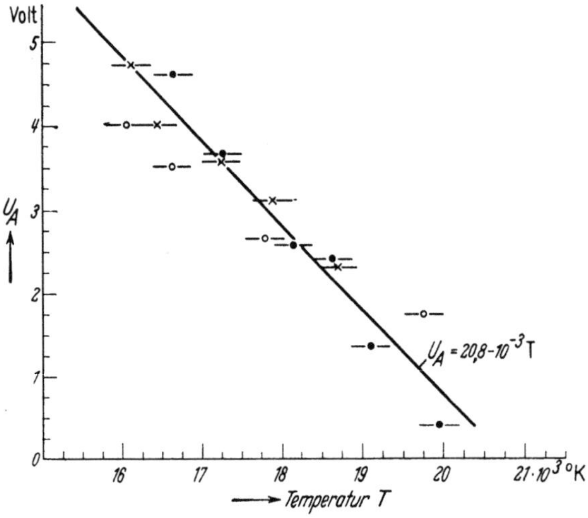

Fig. 18. Abhängigkeit des Anodenfalles U_A eines Argon-Hochtemperatur-Bogens von der Plasmatemperatur vor der Anode, nach BUSZ-PEUCKERT und FINKELNBURG [*21*].

Über die Abhängigkeit von Kathoden- und Anodenfallwerten von den physikalischen Bedingungen ist wenig Sicheres bekannt. Fest steht, daß im Gegensatz zu älteren Meinungen, Kathoden- und Anodenfall im Lichtbogen nicht direkt mit den Ionisierungs- oder Anregungsenergien der Ladungsträger verknüpft sind, insbesondere also wesentlich kleiner als diese sein können (vgl. Kapitel E und F). Für den Kathodenfall eines Xenonhochdruckbogens mit Wolframkathode haben BAUER und SCHULZ gefunden, daß er mit zunehmender Stromstärke abnimmt und einem Grenzwert von etwa 6,5 V zustrebt, der bei 20 Amp erreicht wurde. Für einen Hochstromkohlebogen haben MÜLLER und FINKELNBURG [*159*] durch Sondenmessungen gezeigt, daß der Kathodenfall mit zunehmender Stromstärke bis etwa 200 Amp auf 3 V abnimmt, um dann bei Steigerung auf 350 Amp konstant zu bleiben, während der Spannungsabfall im Kontraktionsbereich vor der Kathode noch leicht zunimmt.

18. Überblick über die Strahlungseigenschaften von Lichtbögen. Bezüglich ihrer Strahlungseigenschaften verhalten sich die verschiedenen Lichtbogentypen, und innerhalb jedes Bogens die verschiedenen Entladungsgebiete, oft außerordentlich verschieden. Im einzelnen wird die Strahlung eines thermischen Plasmas in Kap. C IX behandelt, so daß wir uns hier auf eine qualitative Übersicht beschränken. Intensität und Charakter der Strahlung (Banden, Linien, Kontinua) hängen von Gasart, Gasdruck und Temperatur (Ionisierungsgrad), und über die Temperatur auch von Stromdichte und Feldstärke in dem emittierenden Bogengebiet ab, während das Verhältnis von Wärmeleitvermögen zu Emissionsvermögen der Elektroden wie des Plasmas den strahlungslos abgeführten Anteil der dem Bogen zugeführten elektrischen Leistung bestimmen. Bei Bögen in Molekülgasen werden bei niedrigen Temperaturen (Niederstrombögen, sowie Außenzonen von Hochtemperaturbögen) Molekülbanden, in Bereichen höherer Temperatur die Linien der durch Dissoziation der Moleküle entstehenden Atome, bei Temperaturen über etwa 10000° K mit zunehmender Intensität auch Ionenlinien des Trägergases emittiert. Je höher der Ionisierungsgrad des Plasmas ist, um so mehr verbreitern im allgemeinen die emittierten Linien, um so größer wird gleichzeitig die Intensität des Elektronenbrems- und Rekombinationskontinuums und damit meist auch das mittlere Emissionsvermögen. Dabei ist für das gesamte Strahlungsverhalten eines Bogens stets entscheidend, in welchen Bereichen, je nach dem speziellen Bogenmechanismus, die größte Leistung umgesetzt wird.

Beim Niederstrombogen mit Rein- oder Homogenkohleanode z.B. werden wegen des großen Anodenfalls über 80% der gesamten Strahlung von der glühenden Anode ausgestrahlt, während der Kathodenbrennfleck wegen seiner geringeren Ausdehnung und Temperatur nur wenige Prozent zur Strahlung beiträgt und die Strahlung der Bogensäule fast vernachlässigbar ist. Letzteres liegt daran, daß in dem Wellenlängenbereich, in dem nach der Planckschen Formel bei der Temperatur der Bogensäule das Strahlungsmaximum liegt, nur wenige und meist scharfe Molekülbanden und Atomlinien zur Ausstrahlung zur Verfügung stehen. Durch „Färbung" der Säule, etwa durch Hinzufügung von Ca-, Ce- oder Fe-Salz zum Kohlematerial, kann man daher die Strahlung der Säule wesentlich vergrößern. Auch beim Übergang zu höherer Stromdichte (Hochstromsäule) vergrößert sich wegen des stärker werdenden kontinuierlichen Spektrums die Ausstrahlung der Säule, bleibt aber meist bei wenigen Prozent der der Säule zugeführten elektrischen Leistung. Beim Beck-Bogen mit stark verdampfender, geeignete Leuchtsalze enthaltender positiver Dochtkohle (Ziff. 25) übernimmt der im Anodenfall hoch erhitzte Anodendampf fast die gesamte Ausstrahlung, die hier bis zu 75% der gesamten, dem Bogen zugeführten Leistung erreichen kann. Die Lichtausbeute des Beck-Bogens ist daher mit maximal 90 Lumen die größte aller thermischen Lichtbögen.

Bei den zwischen Wolframelektroden brennenden Quecksilber- oder Edelgas-Hochdruckbögen spielt die Strahlung der Elektroden wegen deren guter Wärmeleitfähigkeit und auf etwa 3000° begrenzten Temperatur eine relativ geringe Rolle. Der wesentliche Teil der dem Bogen zugeführten Energie wird hier in der Säule umgesetzt (große Feldstärke bei hohem Druck!) und von dieser auch wieder ausgestrahlt. Das ist möglich, weil das Quecksilber- wie das Edelgasplasma bei den in den Lampen vorhandenen Temperatur- und Druckbedingungen in dem für die Strahlung nach der Planck-Formel in Frage kommenden Wellenlängenbereich intensive Linien und kontinuierliche Spektren besitzt. Im Xenon-Höchstdruckbogen hat Larché [*117*] entsprechend mit über 10^6 Stilb die höchste bisher erzeugte Leuchtdichte gemessen, allerdings nur in der sehr kleinen der Kathode vorgelagerten Plasmakugel. Bei dem Ziff. 28 behandelten, in einer wasserumströmten Düse brennenden GERDIEN-Bogen wurde umgekehrt in der eingeschnürten Säule bei 1500 Amp und 300 V/cm von BURHORN, MAECKER und PETERS [*18*] die enorme Leistungsdichte von 11000 kW/cm² erreicht, im wesentlichen aber strahlungslos zur Verdampfung und Aufheizung von Wasser auf die hohe Plasmatemperatur verbraucht. Nur ein äußerst kleiner Bruchteil der Leistung wurde ausgestrahlt, weil die geringe Zahl der bei 50000° K in dem Kanal von 2,4 mm Durchmesser vorhandenen emittierenden Zentren zur Abfuhr eines größeren Anteils der Leistung durch Strahlung nicht ausreicht.

Als Beispiel für einen Bogen, dessen Strahlung im wesentlichen von der Kathode emittiert wird, sei der in Ziff. 27 behandelte Bogen mit Zirkonoxyd-Kathode erwähnt, bei dem infolge eines im einzelnen noch nicht untersuchten Entladungsmechanismus eine die Kathode bedeckende dünne Zirkonschicht und das ihr vorgelagerte Dampfkissen fast ausschließlich für die Strahlung verantwortlich sind.

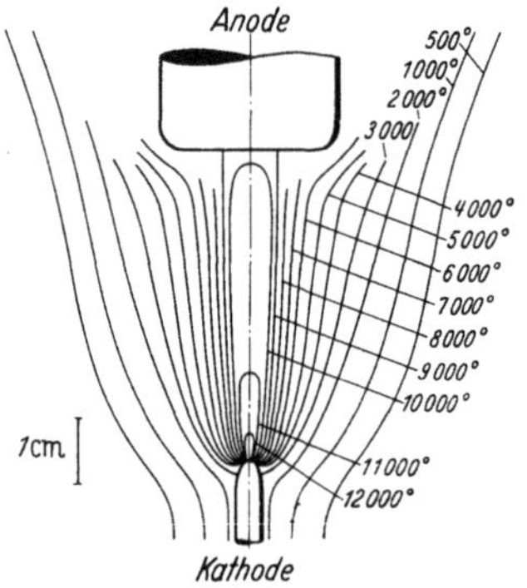

Fig. 19. Isothermenfeld der Säule eines in Luft brennenden 200 Amp-Kohlebogens, nach MAECKER [*138*].

Dieser kurze Überblick zeigt, wie entscheidend die Strahlungsemission eines Lichtbogens von dessen gesamten physikalischen Eigenschaften abhängt, und wie die Entwicklung elektrischer Lichtbögen zu wirtschaftlichen Strahlungsquellen darauf hinausläuft, jeweils in dem Bogengebiet den Leistungsumsatz und das Emissionsvermögen zu erhöhen, in dem bei dem betreffenden Bogentyp die Emission konzentriert ist.

19. Überblick über die Lichtbogentemperaturen. Das Problem der Lichtbogentemperaturen, auf das in Kapitel C und D für den Fall der Säule näher eingegangen wird, ist von allgemeiner Bedeutung, weil die Erzeugung von Temperaturen im Bereich von 4000 bis 50000° K für mehr als einige Mikrosekunden Dauer nur im elektrischen Lichtbogen möglich ist. Dabei ist wie bei der Strahlung festzustellen, daß die Temperaturen in verschiedenen Bogentypen wie in verschiedenen Gebieten eines Bogens in weiten Grenzen schwanken können. Allgemein ist die Temperatur eines Lichtbogenbereiches um so höher, je größer sein Leistungsumsatz jE an der betreffenden Stelle und je kleiner sein Emissionsvermögen und die Wärmeableitung aus dem betrachteten Gebiet sind. Unter sonst gleichen Bedingungen ist also die Temperatur eines Bogengebietes um so höher, je weniger es strahlt, weil Strahlung ja Leistungsabfuhr bedeutet. So besitzt die stark strahlende Anodenflamme des 200 Amp BECK-Bogens eine Temperatur von nur 6000 bis 7000° K, während in der infolge ihrer geringen Strahlung fast unauffälligen Bogensäule nach Fig. 19 Temperaturen zwischen 10 und 12000° herrschen. Entsprechend besitzt die wenig strahlende Säule eines Niederstromkohlebogens ([*5a*], [*146*]) eine Achsentemperatur von knapp 7000° K, während in den stärker strahlenden Säulen von Niederstrommetallbögen und -metallsalzbögen Achsentemperaturen von unter 5000° gemessen worden sind [*224*]. In den handelsüblichen Quecksilber- und Xenon-Hochdruckbögen liegen die Säulentemperaturen zwischen 7000 und 10000° K, während in einem wenig strahlenden Argon- und Stickstoffbogen bei Atmosphärendruck und einigen hundert Ampere Maximaltemperaturen bis 30000° K gemessen wurden [*20*]. Hier ist die Strahlung wegen der bei der hohen Temperatur und Atmosphärendruck geringen Plasmadichte sehr klein; ferner wirkt die hohe Ionisierungsspannung des Trägergases (im Vergleich etwa zu der niedrigeren mittleren Ionisierungsspannung im Luft-Kohlenstoffplama des Kohlebogens) temperaturerhöhend. Daß für diesen Bogen aber außerdem auch die Wärmeableitung klein sein muß, geht aus der Tatsache hervor, daß in dem oben erwähnten GERDIEN-Düsenbogen extremer Belastung trotz rund 50facher je cm Säulenlänge umgesetzter Leistung nach Fig. 20 „nur" die um 70% höhere Temperatur von 52000° K erreicht wird [*18*]. Die verschiedenen Methoden der Messung von Bogentemperaturen werden in Ziff. 81 behandelt.

20. Strömungs- und Materialtransportvorgänge in Lichtbögen. Mit dem Ladungstransport im Lichtbogen sind notwendig gewisse axiale wie radiale mikroskopische Bewegungen von Ionen und neutralen Gasteilchen verbunden. Bei einer Anzahl von Lichtbögen kommt zu diesem mikroskopischen Materialtransport ein makroskopischer hinzu, der durch Strömungsvorgänge bedingt wird, die ursächlich mit dem Bogenmechanismus zusammenhängen und je nach den Bedingungen von der Kathode oder von der Anode fort gerichtet sind. MANNKOPFF [*146*] hat zuerst nachgewiesen, daß die feldbedingte Wanderung der positiven Ionen zur Kathode hin eine Anreicherung der Elemente geringer Ionisierungsspannung dicht vor der Kathode bewirkt, weil diese Elemente besonders hochgradig ionisiert sind. Diese Anreicherung kann zum spektralanalytischen Nachweis solcher Elemente benutzt werden. Radiale Bewegungen, die aber zyklisch verlaufen und daher nicht zu einem eigentlichen Materialtransport führen, kommen dadurch zustande, daß positive Ionen und Elektronen aus der Säulenachse (wo ihre Konzentration groß ist) nach außen diffundieren und dort rekombinieren, während umgekehrt neutrale Atome nach innen diffundieren, wo sie ionisiert werden. In gleicher Weise diffundieren in Molekülgasen Moleküle radial in die Bogenachse, während aus ihr heraus die durch thermische Dissoziation der Moleküle entstandenen Atome diffundieren. Für die spezifische Wärme wie die Wärmeleitung des Plasmas der Säule sind diese Vorgänge (vgl. Kapitel C) von entscheidender Bedeutung.

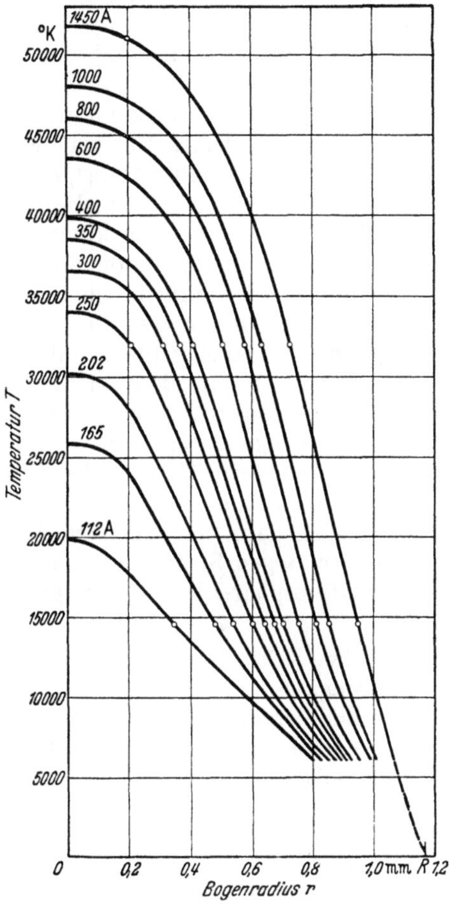

Fig. 20. Radiale Temperaturverteilung in der Säule eines GERDIEN-Wirbelbogens für verschiedene Stromstärken, nach BURHORN, MAECKER und PETERS [*18*].

Oberhalb einer je nach den äußeren Bedingungen zwischen 60 und meist 100 Amp liegenden kritischen Stromstärke bildet sich nach Ziff. 9 die Säule bei vielen Bögen in der Weise um, daß ein Kern hoher Temperatur entsteht, in dem eine makroskopische Plasmaströmung von der Kathode weg stattfindet, deren Geschwindigkeit mehrere 10^4 cm/sec erreichen kann [*139*], [*251*], [*252*]. Auf die Einzelheiten wird in Ziff. 94 eingegangen. Diese Plasmaströmung stellt also einen von der Kathode fort gerichteten Transport von Materie und Wärme dar, der der Ionendrift entgegen gerichtet ist und die Ionenwanderung zur Kathode unter Umständen völlig verhindern kann. Beim anodischen Hochstromkohlebogen beispielsweise gelangen Ionen des Anodenmaterialdampfes bei

Stromstärken oberhalb rund 100 Amp überhaupt nicht zur Kathode. Bei diesem Bogen, dessen Anode nach Ziff. 25 bei einer anodischen Stromdichte von einigen hundert Amp/cm² stürmisch verdampft, findet ein von der Anode fortgerichteter Transport von Anodenmaterial in Form eines Anodendampfstrahles statt, der ebenfalls Strömungsgeschwindigkeiten bis zu 10^4 cm/sec erreichen kann, durch die Gegenwirkung des Kathodenplasmastrahles aber im allgemeinen (Fig. 11) als Anodenflamme von der Kathode fortgelenkt wird. Verwendet man als Anode eine reine Kohleelektrode und stellt ihr in geringem Abstand eine nicht stromführende Elektrode gegenüber, so kann man auf dieser einen großen Teil des von der Anode abtransportierten Materials wieder auffangen (Pilzbildung, vgl. [*6a*]). Dabei findet der Materialtransport zu rund 95% in ungeladener Form statt.

Bei einem zwischen abschmelzenden Metallelektroden brennenden Bogen (z.B. Schweißbogen, vgl. Ziff. 26) kommt zum Materialtransport durch den Anodendampf noch der durch Tropfenübergang zwischen den Elektroden hinzu. Der Übergang solcher Metalltröpfchen von der meist negativ gepolten Schweißelektrode zum positiven Schweißstück ist für die Schweißtechnik von entscheidender Bedeutung und daher vielfach untersucht worden; er erfolgt (vgl. z.B. [*67*]) bei positiv gepolter Elektrode in kleinen, bei negativer Elektrode in großen schweren Tropfen. Bei dem früher viel verwendeten Schweißen mit einer Kohleelektrode verursacht übrigens bei positiver Polung dieser Elektrode der Anodendampftransport eine Vergrößerung des Kohlenstoffgehaltes des Schweißstückes, die, da oft unerwünscht, durch Umkehrung der Polarität beseitigt werden kann.

Allgemein kann man feststellen, daß nach neueren Untersuchungen in frei brennenden Bögen Strömungs- und Turbulenzerscheinungen (ähnlich wie in den Fixsternatmosphären!) eine viel größere Rolle spielen als man früher geglaubt hatte.

21. Chemische Vorgänge in Lichtbögen. Je nach der Art der Elektroden und des Bogengases bzw. -dampfes kann in Lichtbögen eine große Zahl verschiedener chemischer Vorgänge ablaufen, die teilweise den Bogenmechanismus beeinflussen, teilweise schließlich auch technisches Interesse gewonnen haben. Im Gegensatz zu älteren Anschauungen scheint es sich bei den chemischen Reaktionen im Lichtbogen ausschließlich um thermische Reaktionen zu handeln; maßgeblich ist also die Temperatur, während rein elektrische Vorgänge wie etwa die Dissoziation von Molekülen durch im elektrischen Feld beschleunigte Elektronen oder Ionen keine merkliche Rolle spielen. Man kann ohne viel Willkür chemische Vorgänge im Entladungsplasma und solche an den Elektroden unterscheiden. Zu den letzteren gehört besonders der chemische Abbau der Elektroden des in Luft oder anderen mit Kohle reagierenden Gasen brennenden Kohlebogens. Bei niedriger Stromstärke beruht dieser Kohleabbrand ausschließlich auf Reaktion mit dem Luftsauerstoff, nach MATHIESEN [*10a*] unter CN-Bildung auch mit Stickstoff. Nur bei Kohlebögen großer anodischer Stromdichte spielt die Kohleverdampfung neben der Verbrennung eine wesentliche Rolle, wobei der überwiegend in Form von C_2-Radikalen auftretende Dampf dann später mit dem Luftsauerstoff weiter reagiert. Besitzen die Kohlen metallsalzhaltige Dochte, so bilden sich bei der hohen Bogentemperatur oft Metallkarbide, die unter Umständen einen ruhigen Bogenbetrieb stören können. Bei Metallelektroden bilden sich beim Betrieb in Luft häufig Oxydoberflächen oder -flecken, die die Stabilität mancher Niederstrombögen günstig beeinflussen.

Beim Betrieb eines Lichtbogens in Luft ist die wichtigste im Gas ablaufende Reaktion die NO-Bildung, die technisch als erster Schritt der Salpetersäuresynthese wichtig ist. Bei der früher zur HNO_3-Fabrikation ausschließlich verwendeten Lichtbogenmethode bläst man Luft durch eine Bogenzone hoher

Temperatur in eine Zone geringer Temperatur, um die thermische Rückbildung des erzeugten NO zu verhindern. BIRKELAND und EYDE [*11*] bliesen einen zwischen wassergekühlten Kupferelektroden brennenden Drehstrombogen von mehreren 1000 kW Leistung magnetisch zu einem Lichtbogenrad von bis zu 3 m Durchmesser auseinander, durch das die Luft hindurchgeblasen wird. SCHÖNHERR [*197*] stabilisierte einen 6 m langen Lichtbogen (4000 V, 100 Amp) in einem senkrechten Rohr von 15 cm Durchmesser durch einen ihn umgebenden Luftwirbel, in dem die tangential schräg eingeblasene Luft im Bogen kurz erhitzt und dann wieder abgeschreckt wird. Bei modernen Ausführungen ist man bis zu Lichtbogenleistungen von 33000 kW gekommen. Bläst man statt Luft andere Gase in einen Lichtbogen ein, so kann man je nach dem vorliegenden Gleichgewicht die verschiedensten interessierenden Reaktionen erhalten. Von Bedeutung ist z.B. das Einblasen von Methangas CH_4, das dabei teilweise in das wertvolle Acethylen C_2H_2 umgewandelt wird.

Im anodischen Hochstromkohlebogen (Ziff. 25) kann man alle dem Docht der Positivkohle zugesetzten festen Substanzen verdampfen, auf Temperaturen von 5000 bis 8000° K erhitzen und dann mit den Bogengasen oder eingeblasenen Gasen reagieren lassen. Die große Zahl der in den verschiedenen Zonen des Bogens wie der Anodenflamme auftretenden Moleküle bzw. Radikale kann man dann spektroskopisch leicht nachweisen. Die Ausnutzung der hier ablaufenden Reaktionen in einer Hochstrombogenchemie dürfte mit zunehmender Beherrschung der Lichtbogenphysik immer erfolgversprechender werden [*48*].

Es sei noch darauf hingewiesen, daß auch in einem unter Flüssigkeiten brennenden Lichtbogen eine große Anzahl interessanter aber noch wenig untersuchter chemischer Prozesse abläuft. Praktisch ausgenutzt wird die Erzeugung von ziemlich hoch konzentriertem Azethylen bei Betrieb eines Lichtbogens unter irgendeinem Öl, wobei durch bewegliche Elektroden (z.B. zahlreiche Kohlekugeln zwischen zwei festen Kohlebarren) für einen schnellen Wechsel in Zündung und Löschung zahlreicher kurzer Lichtbögen gesorgt wird.

22. Bewegungen der Lichtbogen-Fußpunkte. Obwohl bei den meisten stationären Lichtbögen der kathodische wie der anodische Fußpunkt ihre Stellung auf den Elektroden beibehalten, beobachtet man bei manchen Bögen schon ohne äußeren Anlaß besonders aber unter dem Einfluß äußerer magnetischer oder mechanischer Kräfte eine mehr oder weniger schnelle Wanderung der Fußpunkte auf den Elektroden.

Allgemein hat sich gezeigt, daß der Anodenfußpunkt auf der Anodenoberfläche eine fast unbeschränkte Beweglichkeit besitzt, genauer gesagt, seine Bewegung unter dem Einfluß äußerer Kräfte nur durch die Reibung der in ihm endenden Bogensäule in der umgebenden Atmosphäre gehemmt wird. Diese Feststellung ist im Einklang mit der theoretischen Erwartung, daß die Anode ja im wesentlichen die einströmenden Elektronen aufnimmt, ohne selbst mit ihrer Oberfläche am Entladungsmechanismus entscheidend beteiligt zu sein.

Auf der Kathode liegen die Verhältnisse anders und sehr verschieden je nachdem, ob die Elektronenemission wesentlich von der Kathode selbst erfolgt oder für sie nach Ziff. 98 eine vor der Kathode liegende Plasmaschicht hoher Temperatur verantwortlich ist. Bei thermischer Elektronenemission der Kathode kann der kathodische Fußpunkt nur so schnell wandern, wie die von ihm überstrichenen Oberflächengebiete sich auf Glühtemperatur erhitzen können. Die maximale Wanderungsgeschwindigkeit des Kathodenansatzes auf einer Kohlekathode liegt daher bei etwa 10 cm/sec und dürfte auch bei brennflecklosen Bögen mit glühender Wolframkathode nur klein sein. Auf reinen Metallen wie Silber oder Quecksilber, die nicht in merklichem Umfange selbst thermisch

Elektronen emittieren, ist die maximale Wanderungsgeschwindigkeit des kathodischen Brennflecks entsprechend viel größer, und es sind (z.B. von Stolt [*221*]) hier Werte von 10^3 bis über 10^4 cm/sec gemessen worden. Auf verdampfenden Kathoden (Hg) findet man (Froome [*58*]) auch verschwindende und sich wieder neu bildende Kathodenbrennflecke in Verbindung mit deren schneller Bewegung. Sind die Metalloberflächen nicht ganz sauber oder finden die betreffenden Entladungen in einem oxydierenden Gas statt, so erfolgt die Wanderung des Kathodenfußpunktes, z.B. auf Kupfer, im allgemeinen nicht stetig, sondern sprunghaft von einem für die Elektronenemission günstigen Oxydfleck zum anderen. Auf solchen Kathoden findet man auch ohne Einwirkung äußerer Kräfte häufig ein sprunghaftes Hin- und Herlaufen eines, und in manchen Stromstärkebereichen auch mehrerer, Kathodenbrennflecke. Auch auf rauhen Metallkathoden wandert der Kathodenbrennfleck häufig, wenn er an mikroskopischen Spitzen günstigere Feldbedingungen vorfindet als an benachbarten Kathodenstellen. Bei Verwendung flüssiger Quecksilberkathoden kann man aber durch Anbringung kleiner Spitzen aus schwerschmelzendem Metall den sonst regellos umherlaufenden Kathodenfleck an solchen Spitzen richtig verankern. Wir kommen darauf in Ziff. 31 und 104 zurück.

Von der Wanderung der Elektrodenfußpunkte und mit ihnen des gesamten Bogens wird in zahlreichen Schaltern Gebrauch gemacht, in denen man (vgl. Ziff. 29) den oft sehr stromstarken Schaltbogen durch äußere Kräfte zum schnellen Wandern zwingt und schließlich löscht, bevor die Elektroden abschmelzen können, was bei stationären Bogenfußpunkten unvermeidlich wäre. Auch der bekannte Hörnerblitzableiter, bei dem ein Lichtbogen zwischen zwei nach außen gebogenen hörnerartigen Elektroden hochläuft, bis er erlischt, beruht auf dieser Wanderungsmöglichkeit der Bogenfußpunkte. Eine auch theoretisch besonders interessante Erscheinung, auf die wir in Ziff. 104 noch ausführlich zurückkommen, ist die der Ampereschen Regel widersprechende sog. rückläufige Bewegung des Kathodenfußpunktes auf der flüssigen Hg-Kathode eines Quecksilberbogens in einem zur Austrittsrichtung der Elektronen senkrechten magnetischen Feld. Als Beispiel für die Vielzahl der Wanderungserscheinungen und ihrer Ursachen sei noch die das Zischen des homogenen Kohlebogens verursachende unregelmäßige Wanderung des kontrahierten Anodenfußpunktes auf der Oberfläche einer verdampfenden Kohleanode erwähnt. Wir kommen auf diesen Effekt in Ziff. 113 zurück.

23. Dynamische Bogencharakteristiken und allgemeine Eigenschaften von Wechselstrombögen. Bisher war in unserer Darstellung der Lichtbogenphysik ausschließlich von gleichstrombetriebenen Bögen und entsprechend in Ziff. 16 von der sog. statischen Strom-Spannungscharakteristik die Rede. Tatsächlich ist der überwiegende Teil der wissenschaftlichen Untersuchungen an Lichtbögen mit Gleichstrom ausgeführt worden, weil dieser den entscheidenden Vorteil bietet, daß man es mit völlig stationären Verhältnissen zu tun hat und die verschiedenen Entladungsteile: Kathode, Kathodenfallgebiet, Säule, Anodenfallgebiet und Anode, zeitlich konstant und gut getrennt untersucht werden können. Für alle praktischen Anwendungen umgekehrt beginnen Wechselstrombögen sich mehr und mehr in den Vordergrund zu schieben. Deshalb sollen an dieser Stelle die Besonderheiten des Wechselstrombetriebes von Lichtbögen kurz besprochen werden. Wechselstrombetriebene Kohlebögen sind bereits seit 1892 untersucht worden; eine kurze Darstellung der älteren Ergebnisse findet sich bei Hagenbach [*7a*]. Speziell Simon [*14a*], [*207*] hat von 1905 an das besondere Verhalten von Wechselstrom-Kohlebögen eingehend untersucht. Nach ihm bezeichnet man als *dynamische Charakteristik* das Strom-Spannungsverhalten eines

Bogens bei so schneller Änderung der Stromstärke, daß der stationäre Gleichgewichtszustand sich nicht einstellen kann. Die dynamische Charakteristik hängt daher im Gegensatz zu der das Verhalten eines bestimmten Bogens eindeutig kennzeichnenden statischen Charakteristik stark von der Geschwindigkeit der Stromstärkeänderung, d.h. von dI/dt, beim Wechselstrombogen also von der Frequenz ab. Im Grenzfall $dI/dt \to \infty$ ändert sich der Lichtbogenzustand während der Stromstärkeänderung überhaupt nicht; der Bogen verhält sich wie ein OHMscher Leiter mit entsprechender positiver, hysteresefreier Charakteristik. Fig. 21 zeigt eine von SIMON aufgenommene dynamische Charakteristik, die den Zusammenhang der am Bogen abgegriffenen oszillographisch gemessenen Bogenspannung mit der Stromstärke während einer vollen Wechselstromperiode zeigt. Die aus Fig. 21 ersichtliche Hysterese erklärt sich zwanglos aus dem thermischen

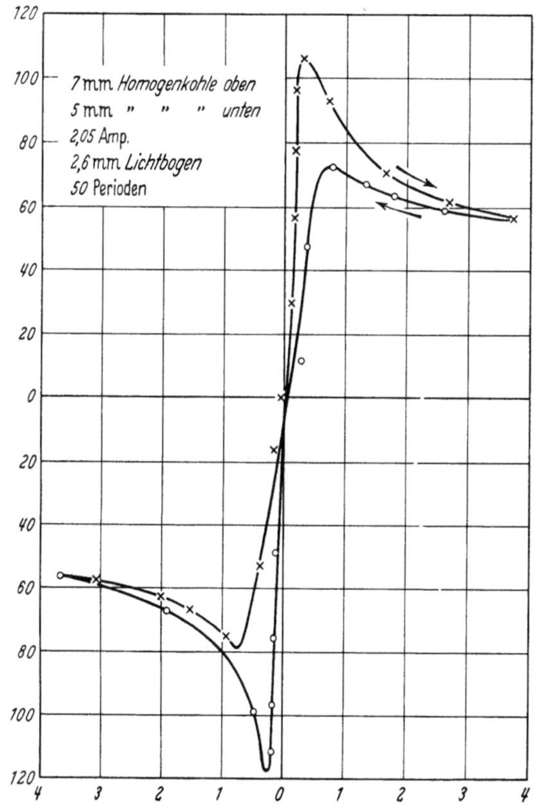

Fig. 21.

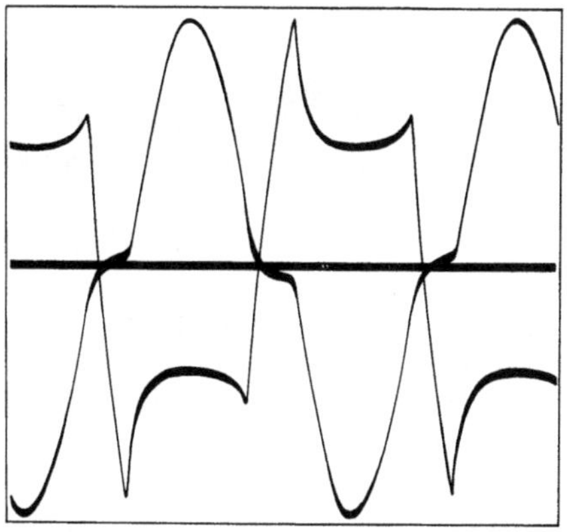

Fig. 22.

Fig. 21. Dynamische Charakteristik eines 50 Hz-Wechselstrombogens, nach SIMON [7a].

Fig. 22. Oszillographische Aufnahme des zeitlichen Verlaufs der am Bogen liegenden Spannung und der Bogenstromstärke für einen 50 Hz-Wechselstrombogen, nach SIMON (vgl. [7a]).

Verhalten des Bogens. Der in der Gegend des Stromdurchganges durch Null in jeder Periode verlöschende Bogen benötigt beim Wiederansteigen der Spannung eine Zündspannung (im Fall der Fig. 21 von 105 bis 115 V) und durchläuft nach der Zündung mit zunehmender Stromstärke die übliche fallende Charakteristik. Bei abnehmender Stromstärke werden dann, da Temperatur und Ionisierungsgrad der Bogenstrecke den vorher durchlaufenen höheren Stromstärkewerten entsprechen, Brennspannungswerte durchlaufen, die niedriger liegen als die während der Bogenaufheizung durchlaufenen. Sinkt die Bogenstromstärke unter einen Minimalwert (in Fig. 21 etwa 0,7 Amp), so genügt die momentane Spannung des Wechselstromes nicht mehr zur Aufrechterhaltung eines selbständigen Lichtbogens (vgl. Ziff. 8) und der Bogen erlischt, um in der nächsten Halbperiode des Wechselstromes den gleichen Zyklus mit umgekehrter Polarität wieder zu durchlaufen. Je größer die Bogenlänge ist, desto weniger wird die

Bogenstrecke von den erhitzten Elektroden her während der Löschpause nachgeheizt, und desto größer ist folglich auch die Zündspitze, desto ausgeprägter entsprechend die Hysterese. Fig. 22 zeigt eines der SIMONschen Oszillogramme, die das geschilderte Verhalten von Spannung und Stromstärke deutlich erkennen lassen. Bei genügend kurzen Kohlebögen hoher Stromstärke kann die thermische Trägheit des gesamten Bogens und die Nachheizung der Bogenstrecke von den Elektroden aus so groß werden, daß die Hysterese in der dynamischen Charakteristik und mit ihr die Wiederzündspitze praktisch vollkommen verschwinden. Dieses Verhalten hochbelasteter Wechselstrombögen zwischen stark glühenden Elektroden haben WEIZEL und Mitarbeiter [*226*], [*249*] auch als charakteristisch für die in Ziff. 106 noch zu besprechenden brennflecklosen kurzen Quecksilber- und Xenon-Hochdruckbögen zwischen hoch erhitzten Wolframelektroden gefunden. Fig. 23 zeigt ein Spannungsoszillogramm für einen solchen Bogen.

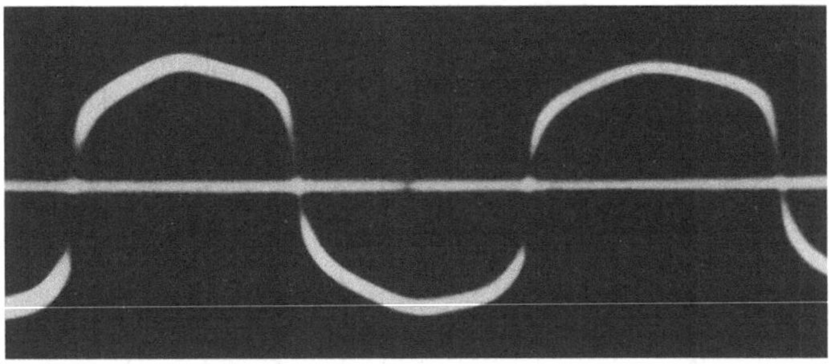

Fig. 23. Spannungsoszillogramm eines brennflecklosen kurzen Quecksilber-Hochdruckbogens ohne Zündspitzen, nach WEIZEL und Mitarbeitern [*226*].

Während in den besprochenen Fällen die thermische Trägheit der glühenden Elektroden so groß war, daß eine wesentliche Veränderung der Elektronen-Emissionsfähigkeit der Elektroden während der Löschpause jeder Periode nicht eintrat und das Verhalten des Wechselstrombogens daher weitgehend von der Bogensäule bestimmt wurde, findet man andere Verhältnisse für Wechselstrombögen niedriger Stromstärke zwischen Metallelektroden. Man hat sogar lange geglaubt, daß Wechselstrombögen zwischen Metallelektroden überhaupt nicht betrieben werden können, weil die schnelle Abkühlung der Elektroden in der Löschpause die Wiederzündung verhindern würde, wenn man nicht Betriebsspannungen von mehreren tausend Volt verwendet. Es gelang zwar v. ENGEL und STEENBECK [*5a*], [*41*] einen 10 Amp-Wechselstrombogen zwischen wassergekühlten Kupferelektroden von 3 mm Länge zu betreiben (220 V Netzspannung), aber nur, indem sie durch induktive Stabilisierung die stromlose Pause auf 5 bis 10% des Normalwertes herabdrückten. Selbst dann aber lag die Wiederzündspannung mit über 200 V noch bei dem achtfachen der normalen Brennspannung. Der Grund für dieses Verhalten liegt darin, daß es sich bei solchen Metallbögen um kathodische Brennfleckbögen (vgl. Ziff. 107) handelt, bei denen die Elektronenemission nur teilweise aus der Kathode selbst, zum anderen Teil aber aus einer dieser vorgelagerten Plasmaschicht hoher Temperatur stammt, die bei jeder Wiederzündung neu aufgebaut werden muß. Zwischen zwei gekühlten Metallelektroden kann also bei normaler Widerstandsstabilisierung ein Wechselstrombogen mit Brennspannungen unter 300 V überhaupt nicht betrieben werden. Zwischen einer glühenden Elektrode (Kohle oder Wolfram) und einer gekühlten Metallelektrode kann bei einer Bogenlänge von wenigen Millimetern ein Wechselstrombogen zwar betrieben werden, brennt aber nur in der Halbperiode, in der die glühende Elektrode Kathode ist. Ein solcher Bogen wirkt nach SIMON ([*14a*] vgl. auch BUSZ und FINKELNBURG [*20*]) folglich als Gleichrichter, der den Strom nur in der Richtung durchläßt, in der die glühende Elektrode Kathode ist (Fig. 24). Es wird in Ziff. 109 noch gezeigt werden, daß eine intensive Verdampfung einer metallischen Kathode die Elektronenerzeugung vor der Kathode stark begünstigt. Daher kann ein Wechselstrombogen, wie HEWITT [*83*] schon 1903 fand, zwischen einer Quecksilberelektrode und einer Eisenelektrode in Quecksilberdampf nur in der Halbperiode brennen, in der das leicht verdampfende Quecksilber Kathode ist. Diese Erscheinung liegt den bekannten Quecksilber-Gleichrichtern zugrunde.

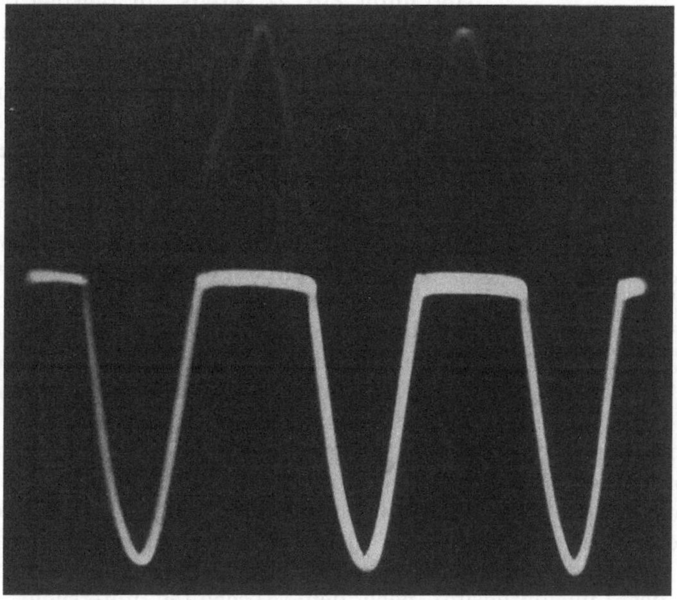

Fig. 24. Brennspannungs-Oszillogramm eines 50 Hz-200 Amp-Argonbogens mit Gleichrichterwirkung infolge glühender Wolfram-Kathode und kalter Cu-Anode, nach BUSZ-PEUKERT und FINKELNBURG [*20*].

Bei Wechselstromfrequenzen in der Gegend von 50 Hz unterscheidet sich die dynamische Charakteristik stromstarker Wechselstromlichtbögen also wegen der oben erwähnten verschwindenden Hysterese nicht grundsätzlich von derjenigen entsprechender Gleichstrombögen, wie man durch stroboskopische Beobachtung leicht verifizieren kann. Bei einem mit Wechselstrom betriebenen BECK-Bogen (Ziff. 25) beispielsweise kann man stroboskopisch den Wechsel der Polarität im einzelnen genau verfolgen und ein Augenblicksbild nicht von dem eines entsprechenden Gleichstrombogens unterscheiden. Je höher die Frequenz des benutzten Wechselstromes aber wird, um so mehr verändert sich das Bild, da die thermische Trägheit auch der Bogensäule eine immer größere Rolle zu spielen beginnt. Bei sehr hohen Frequenzen (deren Grenzwert von der Wärmeleitfähigkeit des Bogenplasmas abhängt), verändern sich schließlich Temperatur und Ionisierungsgrad, d.h. die elektrische Leitfähigkeit des Bogenplasmas, während einer Wechselstromperiode praktisch überhaupt nicht mehr. Dies beobachtet man am schönsten bei Überlagerung eines hochfrequenten Wechselstromes über einen Gleichstrombogen. Dann verhält sich nämlich der Bogen während einer Wechselstromperiode wie ein rein OHMscher Widerstand mit einem aus der statischen Charakteristik des Gleichstrombogens folgenden

Widerstandswert. Bei reinen Wechselstrombögen wird aus dem gleichen Grunde bei genügend hohen Frequenzen nach Fig. 25 die dynamische und als ihr Mittelwert auch die statische Charakteristik steigend [*5a*]. Die Trägheit des Bogenplasmas zeigt sich auch darin, daß die bei geringen Frequenzen mit der Stromstärke synchron sich ändernde Lichtstärke (Strahlungsemission) bei sehr hohen Frequenzen mehr und mehr der Stromstärke nachzuhinken beginnt, die Modulationsfähigkeit des Bogens also nachläßt [*106*], [*145*], [*206*], [*245*].

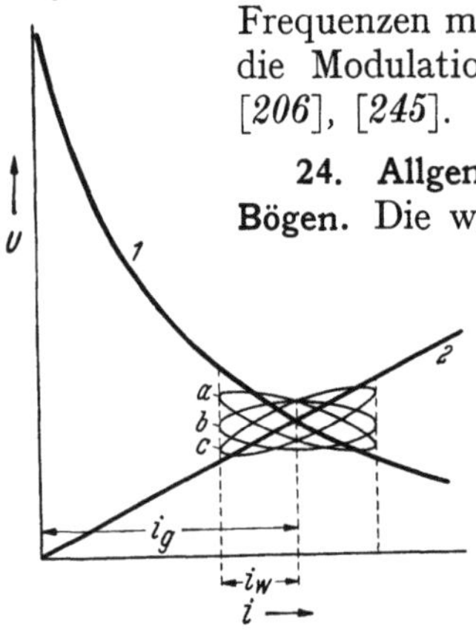

Fig. 25. Statische Gleichstrom-Charakteristik (*1*), OHMsche Charakteristik (*2*) und dynamische Charakteristiken zunehmender Frequenz ($a \rightarrow b \rightarrow c$) einer dem Gleichstrombogen überlagerten Wechselstromkomponente, nach V. ENGEL und STEENBECK [*5a*].

24. Allgemeine Eigenschaften unperiodischer nicht-stationärer Bögen. Die wichtigsten Eigenschaften nicht-stationärer Lichtbögen sind unter Ziff. 23 bereits am Beispiel des Wechselstrombogens besprochen worden. Zu den unperiodischen nicht-stationären Bögen zählen die verschiedenen Typen von Abreißbögen, ferner Funken nicht zu geringer Dauer, auf die wir unten noch zu sprechen kommen. Bei den *Abreißbögen*, wie sie in der Spektralanalyse und in den verschiedenen Formen von Schaltern (vgl. Ziff. 29) eine Rolle spielen, haben wir es meist mit Bögen zu tun, die während des größten Teils ihrer jeweiligen Brenndauer als stationäre Bögen anzusehen sind. Das entscheidend Neue bei ihnen sind die Vorgänge beim Verlöschen, das durch Abschalten, durch Bogenverlängerung bis zum Abreißen, durch Druckerhöhung, durch Ausblasen mittels Gasströmen oder magnetischen Feldern erfolgen kann, sowie das spätere Wiederzünden. Über die *Lichtbogenlöschung* ist außerordentlich viel gearbeitet worden, meist allerdings mehr mit dem Ziel, die experimentellen Bedingungen für eine möglichst schnelle Löschung zu ermitteln, als mit dem, die der Löschung zugrunde liegenden Vorgänge zu studieren. An einem Niederstromkohlebogen hat WITTE [*255*] das Abklingen der Linienintensität nach dem Abschalten untersucht und daraus

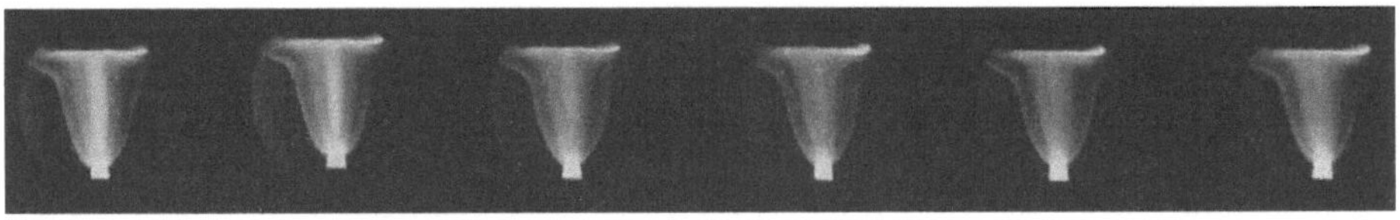

Fig. 26. Zeitliches Abklingen der Strahlung einer 200 Amp-Kohlebogensäule nach WIENECKE [*251*]. Die erste Aufnahme links wurde 25 μsec nach dem Abschalten gemacht, die folgenden Aufnahmen jeweils 22 μsec später.

auf die thermische Anregung im Bogen geschlossen. ROHLOFF [*174*] hat das Verhalten der Anodenflamme sowie des Säulenkerns eines Hochstromkohlebogens nach dessen Abschalten mit einer Zeitlupe untersucht, während neuerdings WIENECKE [*251*] den Abschaltvorgang verschiedener Hochstrombögen mit wesentlich verbesserter elektronischer Abschaltung und höher auflösender Zeitlupe untersucht hat (Fig. 26). Dabei wurden von ihm unter anderem Plasma-Abklingzeiten gemessen und die Wärmeleitfähigkeit von Plasmen als Funktion der Temperatur im Bereich der Lichtbogentemperaturen gemessen. Beim abgeschalteten Hochstromkohlebogen dauert es etwa $5 \cdot 10^{-3}$ sec, bis das leuchtende Plasma zwischen den Elektroden verschwunden ist. Daß hierbei eine Behinderung der Abkühlung durch die glühenden Elektroden eine Rolle spielt, wird durch die Beobachtung nahe gelegt, daß das Plasma eines zwischen wassergekühlten

Elektroden brennenden Bogens nach dem Abschalten schon in etwa $1 \cdot 10^{-3}$ sec verschwindet. Auch die Reaktion von Bögen auf kurzzeitiges Abschalten, auf Stromstöße kurzer Zeitdauer usw. konnten von WIENECKE genau verfolgt werden.

Über das Wiederzünden gelöschter Bögen sind sonst noch wenig Einzelheiten bekannt. Erfolgt die Wiederzündung genügend lange Zeit nach dem Abschalten, so unterscheidet sie sich nicht von einer Erstzündung, die in Ziff. 7 behandelt wurde. Beim Wiederanlegen der Spannung zu einem Zeitpunkt, zu dem noch ein Rest leuchtenden Plasmas zwischen den Elektroden vorhanden ist, erfolgt die Wiederzündung häufig nicht durch einfaches Wiederaufheizen des gesamten Bogens, sondern es bildet sich nach Zeitlupenaufnahmen von WIENECKE zunächst ein stark strahlender Kanal in dem Restplasma, der sich dann im Verlauf etwa einer Millisekunde wieder zu dem stationären Bogen entwickelt.

Zu den nicht-stationären Bogenerscheinungen gehört schließlich auch das unter gewissen Bedingungen reproduzierbar zu beobachtende Umschlagen einer Glimmentladung in den Bogen und umgekehrt (Fig. 27) sowie die erwähnte Entwicklung einer Bogensäule hoher Stromdichte als Endstadium eines Funkendurchschlages, das aber nicht direkt zum Thema dieses Beitrages gehört.

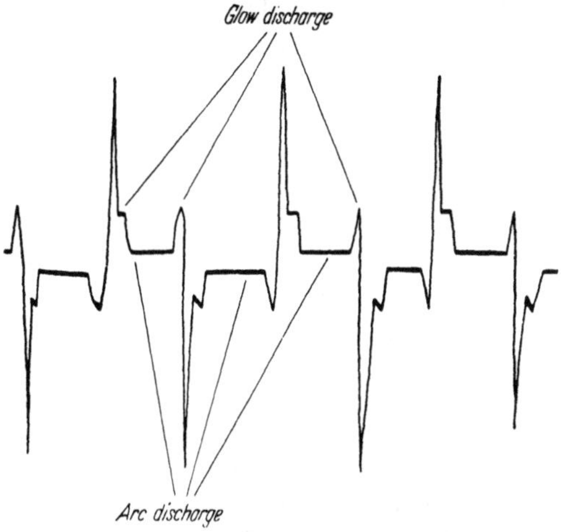

Fig. 27. Spannungsoszillogramm einer periodisch von der Glimmentladung in einen Bogen umschlagenden und wiederzündenden Entladung, nach MEEK und CRAGGS [*151b*].

III. Übersicht über die wichtigsten Lichtbogentypen.

25. Kohlelichtbögen. Der von alters her am meisten untersuchte Kohlelichtbogen ist ein zwischen homogenen Graphit- oder Koksstiften (anfänglich auch Holzkohlestiften) in freier Luft brennender Bogen. Die Säule dieses, bei Stromstärken bis zu 40 Amp und Spannungen zwischen 50 und 80 V brennenden *Homogenkohle-Niederstrombogens* besitzt eine Temperatur zwischen 6000 und 7000° K; ihre schwach violette Farbe stammt von den für die Strahlung fast allein verantwortlichen CN-Molekülen. Die Säulenstromdichte beträgt wenige hundert Amp/cm²; der Säulengradient etwa 15 V/cm. An der Kathode verengt sich die Entladung auf eine Stromdichte von etwa 500 Amp/cm². Der Kathodenbrennfleck der negativen Kohle besitzt eine Temperatur von 3200 bis 3600° K. Der anodische Ansatz des Bogens ist größer; die anodische Stromdichte beträgt meist etwa 40 Amp/cm² und kann nur bei sehr dünnen Positivkohlen höhere Werte annehmen, ohne daß das eigenartige Phänomen des Zischens einsetzt. Die Temperatur des Anodenbrennflecks der seiner Form wegen meist als positiver Krater bezeichnet wird, ist merklich höher als die des Kathodenbrennflecks und liegt meist bei 4000°, wobei der feste Kohlenstoff wegen seines mittleren Absorptionsvermögens von 76% eine Strahlung aussendet, die weitgehend der eines schwarzen Strahlers von 3820° K entspricht und in zunehmendem Umfange als Strahlungsnormal verwendet wird [*24*], [*45*]. Die Kohleelektroden eines solchen homogenen Niederstrombogens zeigen einen Abbrand, der zwischen 60 und 100 mm/Std beträgt und nach den Feststellungen von STEINLE [*218*] fast ausschließlich auf einen chemischen Abbau der Kohle beruht, in reinen Edelgasen (und vermutlich allen anderen mit Kohlenstoff nicht reagierenden Gasen) daher fehlt. Im positiven Krater beobachtet man, bei vergrößerter Abbildung, ein von LUMMER [*132*] als Fische und Waben bezeichnetes Bewegungsphänomen, das dieser als Schmelzerscheinung deutete. Nach STEINLE handelt es sich hier aber um beim Kohleabbau freigelegte kleine Graphitkristalle, die sich durch verändertes Reflexions- und Emissionsvermögen von der Umgebung abheben. Im Gegensatz zu einem reinen Graphitkrater, dessen Temperatur und Leuchtdichte in einem gewissen Bereich der Stromstärke von dieser unabhängig, d.h. konstant bleibt, hängen Temperatur und Leuchtdichte von Koks- und Rußkohlen in etwas unübersichtlicher Weise von der Strombelastung ab [*6a*].

Als *Reinkohlebogen* bezeichnet man einen Niederstromkohlebogen, dessen Anodendocht aus mit Teer verbackenem Kohlepulver und gewissen Zusätzen zur Bogenberuhigung besteht. Sein Verhalten ist dem des Homogenkohle-Niederstrombogens ähnlich.

Beim Überschreiten der oben erwähnten anodischen Normalstromdichte von etwa 40 Amp/cm² beginnt beim Homogen- und Reinkohlebögen das anodische Zischen, auf das in Ziff. 113 näher eingegangen wird.

Setzt man dem Docht der Positivkohle einen merklichen Prozentsatz von Metallsalzen (Oxyde, Fluoride usw.) zu, so gelangen die Metallatome aus den sich thermisch zersetzenden Salzen in die Bogensäule und verändern dort nach Ziff. 18/19 Temperatur und Strahlung grundlegend. Solche, früher gern bei großer Bogenlänge betriebene Bögen emittieren die Atom- und Molekülspektren des betreffenden Metalls und seiner Oxyde und werden *Effektkohlebögen* oder auch *Flammenbögen* genannt. Ihre Brennspannung liegt, da die Ionisierungsbedingungen auch vor der Kathode und Anode viel günstiger sind als im Homogenkohle- und Reinkohlebogen, weit unter der dieser beiden Bögen.

Bei Steigerung der Stromstärke eines Homogen- oder Reinkohlebogens vergrößert sich nach Ziff. 9 zunächst nur der Durchmesser der Bogensäule, ohne daß deren Temperatur und Strahlung sich wesentlich ändern. Oberhalb einer von den Bedingungen an der Kathode abhängigen Grenzstromstärke, die im allgemeinen zwischen 70 und 100 Amp liegt, nimmt die Bogensäule aber ein anderes Aussehen an. Von der Kathode aus schiebt sich dann in die bläulich-violette Säule eine weißliche Stichflamme mit wesentlich kontinuierlichem Spektrum und Atomlinien des C, N und O hinein, die mit weiter zunehmender Stromstärke sich schließlich bis zur Anode (oft mehrere Zentimeter Länge) ausbreitet und an Durchmesser zunimmt. Gleichzeitig beobachtet man [*118*], [*174*], daß diese als Säulenkern oder auch kontrahierte Säule bezeichnete „negative Stichflamme", in der die Stromdichte zwischen 500 und 1000 Amp/cm² liegt, einen Impuls auf ihr entgegenstehende Objeke und einen Rückstoß auf die Kathode ausübt, was zu der kürzlich von MAECKER [*139*] bestätigten Vermutung geführt hat, daß bei dieser Hochstrombogensäule ein Plasmastrahl von der Kathode ausgehend in der Säulenachse auf die Anode zuschießt. Mit dieser Umbildung der Niederstrom- in die Hochstrombogensäule ist eine erhebliche Temperaturerhöhung verbunden, indem in der Säulenachse nahe der Kathode bei einem 200 Amp-Bogen (Fig. 19) Werte von 12000° erreicht werden.

Gleichzeitig mit dieser Umbildung der Säule nimmt nach GUILLERY [*72*] die Stromdichte an der Kathode bis zu einem bei etwa 400 Amp erreichten Grenzwert von 5000 Amp/cm² zu. Bei Stromstärken über 1000 Amp scheint eine nochmalige Umbildung der Hochstromsäule stattzufinden, da dann nach GUILLERY und BECK (unveröffentlicht, vgl. [*6a*]) in der Achse ein gegenüber der

Umgebung sehr heller fadenartiger Kern sichtbar wird. Die Einzelheiten dieser Erscheinung werden zur Zeit von WIENECKE untersucht. Als *anodisch verdampfenden Hochstromkohlebogen* (englisch meist als high-intensity arc) bezeichnet man die besonders von FINKELNBURG in einer Monographie [*6a*] eingehend beschriebene Sondererscheinung eines Kohlelichtbogens hoher Stromstärke, die dadurch entsteht, daß man durch Verwendung relativ dünner Positivkohlen die anodische Stromdichte auf ein Mehrfaches der Normalstromdichte steigert. Die der Anode zugeführte Leistung kann dann nur noch zu einem kleinen Teil durch die Anode rückwärts abgeleitet werden, während ihr größerer Anteil zur Verdampfung der Anode dient. Von der Anodenstirnfläche strömt daher (vgl. Fig. 11) ein hoch erhitzter, hell leuchtender Dampfstrahl mit einer Strömungsgeschwindigkeit von 10^3 bis 10^4 cm/sec ab, dessen anfänglich mit der Anodenachse übereinstimmende Richtung durch die Wirkung des kathodischen Gasstrahles und meist auch des Eigenmagnetfeldes des Bogens in der aus Fig. 11 ersichtlichen Weise verändert wird. Bei Verwendung einer homogenen Kohle als Anode findet deren Verdampfung nicht gleichmäßig auf der gesamten Anodenstirnfläche statt, sondern der anodische Bogenansatz zieht sich zu einem kleinen Brennfleck hoher Stromdichte (bis 50000 Amp/cm², vgl. [*6a*], [*188*]) zusammen, der auf der Anodenstirnfläche hin und her läuft und das bekannte Zischen eines überlasteten Homogenkohlebogens bewirkt. Für alle Einzelheiten sei auf Ziff. 113 verwiesen.

Verwendet man als Anode des Hochstromkohlebogens eine Dochtkohle mit Leuchtzusätzen, so verdampft der Docht schneller als der Mantel, wodurch sich ein oft tief gehöhlter positiver Krater entwickelt, aus dem der Leuchtdampf abströmt. Dieser von leuchtendem Dampf erfüllte Krater hat nach der Erfindung von HEINRICH BECK 1913 [*5*] eine besonders hohe Leuchtdichte, wenn man dem Docht Oxyde und Fluoride des Cers zusetzt. Diese Sonderform des anodischen Hochstromkohlebogens bezeichnet man als BECK-Bogen. Er wird mit Anodendurchmessern von 6 bis 16 mm und Stromstärken zwischen 40 und 200 Amp als Kino-Projektionslichtquelle und bei Stromstärken bis über 1000 Amp und maximalen Leuchtdichten von über 200000 Stilb (Candela/cm² $\approx$ HK/cm²) als Lichtquelle für Großscheinwerfer verwendet. Im Gegensatz zur fallenden Charakteristik gewöhnlicher frei brennender Lichtbögen (Ziff. 16) ist die Charakteristik des anodisch verdampfenden Hochstromkohlebogens steigend, so daß er ohne Stabilisierungswiderstand betrieben werden kann. Nach Sondenmessungen [*6a*] ist die Ursache dieser mit der Stromdichte zunehmenden Brennspannung ein mit ersterer zunehmender Anodenfall, der bis zu 40 V betragen kann, und der mit der natürlich ebenfalls stromstärkeabhängigen Anodenverdampfung gekoppelt ist. Auf die Deutung aller dieser Erscheinungen des anodischen Hochstromkohlebogens, dessen Anoden-„Abbrand" also im Gegensatz zum wirklichen Abbrand der Niederstromanode auf Verdampfung beruht, wird im Zusammenhang mit der Theorie der anodischen Entladungsteile in Ziff. 117 näher eingegangen.

Für die technische Anwendung des BECK-Bogens verwendet man neuerdings Lampen, bei denen die sich schnell verbrauchende positive Kohle stückweise durch einen Magazinmechanismus aneinander gesetzt wird, während an Stelle der Negativkohle gelegentlich eine sich nur äußerst langsam verbrauchende Graphitkreisscheibe verwendet wird [*49*], [*50*]. Eine interessante Sonderentwicklung des BECK-Bogens ist der von GRETENER [*70*] stammende sog. Ventarc-Bogen, bei dem die Bogensäule als Lichtquelle dient und ihre Leuchtdichte dadurch erhöht wird, daß mittels einer äußeren Luftströmung die Anodenflamme in die Bogensäule geblasen wird.

26. Frei brennende Bögen zwischen Metallelektroden. Bei den unter Atmosphärendruck frei zwischen Metallelektroden brennenden Lichtbögen hat man zwischen zwei Gruppen zu unterscheiden. Bei der ersten Gruppe von Bögen findet eine mehr oder weniger starke Verdampfung der Elektroden statt, so daß die Stromleitung im wesentlichen im Elektrodenmaterialdampf erfolgt *(Metalldampfbögen)*. Bei der zweiten Gruppe wird die Verdampfung der Elektroden durch entsprechende Kühlung so weitgehend verhindert, daß der Stromtransport in dem die Bogenatmosphäre bildenden Gas erfolgt *(Gasbogen)*. Übergangsfälle zwischen beiden Gruppen stellen solche Bögen dar, bei denen nur eine der beiden Elektroden merklich verdampft und der Bogen, wenn die verdampfende Elektrode Anode ist, in seinem kathodischen Teil ein Gasbogen und in seinem anodischen überwiegend ein Dampfbogen ist. Zu dieser Übergangsgruppe gehört z.B. der praktisch wichtige Schutzgas-Schweißbogen, bei dem eine gemäß Fig. 28 von Argon umströmte praktisch nicht verdampfende Wolframkathode dem stark verdampfenden Schweißstück gegenüber steht. Die normalen Schweißbögen, bei denen eine abschmelzende Kathode aus einem dem Schweißgut gleichen oder ähnlichen Material verwendet wird, sind demgegenüber reine Metalldampfbögen.

Metalldampfbögen geringer Stromstärke sind zwar in der Anfangszeit der Lichtbogenphysik vielfach untersucht worden (vgl. HAGENBACH [*7a*]), haben aber keine besondere Bedeutung erlangt. Sie zeichnen sich meist gegenüber den Kohlebögen durch eine auffallende Instabilität und Unruhe aus. Diese scheint weitgehend darauf zu beruhen, daß solche Bögen im allgemeinen weder auf der Kathode noch auf der Anode feste Fußpunkte haben, sondern unruhig hin und her wandern bzw. (auf der Kathode) springen. In beiden Brennflecken ist die Stromdichte wesentlich größer als bei Kohlebögen gleicher Stromstärke und gleicher sonstiger Abmessungen. Eine Ausnahme macht wohl nur der für spektroskopische Zwecke zuerst von PFUND [*167*] entwickelte, später etwas abgewandelte Eisenbogen geringer Stromstärke (3 bis 15 Amp). Bei ihm gelingt es, einen sehr konstanten ruhigen Bogen zu erzielen, wenn man als Kathode eine in den ausgehöhlten Kathodenstift gelegte Eisenoxydperle verwendet, die sich im Betrieb laufend ergänzt, während als Anode ein sich von selbst mit einer Oxydhaut überziehender Eisenstift dient. Hier zeigt sich wieder die stabilisierende Rolle der Metalloxyde; ein zwischen reinen Eisenstiften in Luft oder anderen Gasen betriebener Bogen spritzt und brennt stets sehr unruhig. Der PFUND-Bogen dient immer noch weitgehend zur Erzeugung des als Vergleichsspektrum vielfach verwendeten Eisenspektrums. Seine Achsentemperatur bei 5 Amp Stromstärke wurde kürzlich von BURHORN [*16*] zu 6300° K bestimmt.

Bei höherer Stromstärke schmelzen und verdampfen sämtliche Metalle mit Ausnahme des Wolframs (das bei höherer Temperatur chemisch recht aktiv ist und daher nicht in Luft und H_2O-haltigen Gasen und Dämpfen verwendet werden kann) so stark, daß über das Verhalten von Metalldampfbögen hoher Stromstärke wenig Zuverlässiges bekannt ist.

Zu dieser Klasse von Lichtbögen gehören praktisch alle *Schweißbögen*, auf deren verschiedene Typen hier wenigstens kurz hingewiesen sei. Man unterscheidet grundsätzlich Schweißbögen, bei denen der elektrische Lichtbogen nur die hohe Temperatur erzeugt, die zum Schmelzen des in Drahtform seitlich zugeführten Schweißmaterials erforderlich ist, von solchen Schweißbögen, bei denen die eine Elektrode selbst abschmilzt und damit das die Naht bildende Schweißmaterial liefert. Zur ersten Gruppe gehört der Kohleschweißbogen, bei dem der Bogen zwischen einer meist negativ gepolten Kohle und dem Schweißstück brennt. Er wird heute mehr und mehr durch reine Metallbögen verdrängt, spielt aber mit Stromstärken von etwa 600 Amp in der automatischen Faßschweißerei, sowie mit Stromstärken über 1000 Amp. zum Zerschneiden dicker Platten etwa beim Abwracken von Schiffen, noch eine gewisse Rolle. In der letztgenannten Funktion wird der Kohleschweißbogen mehr und mehr durch das sog. „oxyarc-Verfahren" ersetzt, bei dem in den mit 200 Amp betriebenen Lichtbogen durch die bewegliche Elektrode Sauerstoff geblasen wird, der bei der hohen Temperatur vorwiegend chemisch das Abschneiden bewirkt. Zum eigentlichen Verschweißen von Eisenteilen dient in immer zunehmenderem Maße ein bei Stromstärken zwischen 100 und 500 Amp zwischen dem positiv gepolten Werkstück und einer negativen Eisenelektrode brennender Bogen relativ geringer Spannung (15 bis 25 V). Überwiegend verwendet man dabei Eisenelektroden, die mit gewissen anorganischen Pasten umhüllt sind, die bei der hohen Temperatur des Bogens verdampfen und einerseits den Bogen beruhigen und seine Brennspannung verkleinern, andererseits auch als Schutzgas bzw. -dampf den unerwünschten Zutritt des Luftsauerstoffs zur glühenden Schweißnaht und damit deren Verbrennen verhindern. In automatischen Schweißmaschinen betreibt man Bögen dieser Art bei Stromstärken bis zu 1000 Amp, in Sonderfällen (Eisenpulverschweißung) sogar bis 4000 Amp. In neuester Zeit läßt man den zwischen dem Schweißstück und einer blanken Eisendrahtelektrode brennenden Schweißbogen gern von einem Edelgas (Schweißargon, in USA auch Helium) umströmen. Zum Schweißen von Nichteisenmetallen und besonders von dem immer wichtiger werdenden Aluminium verwendet man die gleiche Schutzgasmethode, benutzt aber als Kathode einen sich praktisch nicht verbrauchenden Wolframstift und führt das zur Schweißnaht erforderliche Material dann wieder in Drahtform seitlich in den bei Stromstärken zwischen 100 und 400 Amp brennenden Bogen ein.

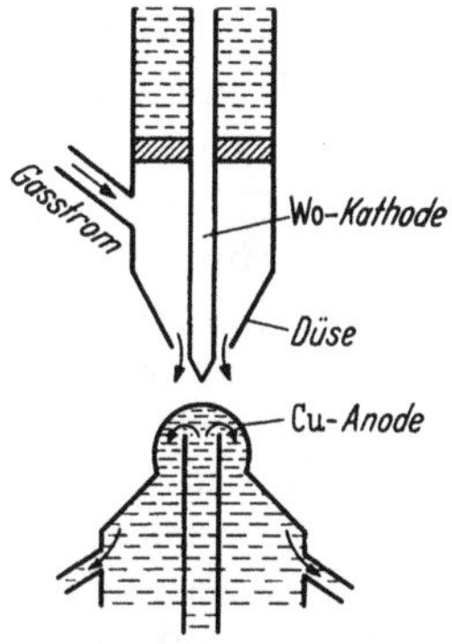

Fig. 28. Anordnung des Argon-Schutzgas-Bogens hoher Stromstärke, nach BUSZ-PEUKERT und FINKELNBURG [*20*].

Bei allen diesen Schweißbögen sind die physikalischen Bedingungen wegen der Unübersichtlichkeit der sich zeitlich ändernden Verhältnisse (Bogenlänge, Temperatur, Materialdampfanteil usw.) sehr wenig bekannt, ganz besonders natürlich bei Schweißbögen mit sich selbst verzehrender, d.h. abtropfender Elektrode. Auf die umfangreiche Literatur, die sich meist mit technischen Fragen, aber auch den physikalisch äußerst interessanten Problemen der Tropfenbildung, und des gesamten Materialdurchganges durch diese Bögen befaßt, kann hier nur hingewiesen werden [*154*], [*172*], [*257*]. In neuerer Zeit werden namentlich kombinierte kinematographische und oszillographische Methoden zum Studium der Schweißvorgänge verwendet.

Frei brennende Gasbögen zwischen gekühlten Metallelektroden bei Stromstärken bis zu 500 Amp sind kürzlich von BUSZ und FINKELNBURG [*20*] mit der in Fig. 28 dargestellten Anordnung untersucht worden. Während die Kupferanode möglichst gut gekühlt wurde, war die Wolframkathode so gekühlt, daß sie zwar nicht merklich verdampfte (keine Metallinien im Spektrum!), ihre Spitze sich aber auf Weißglut befand und damit zur Elektronenemission fähig war. Aus der die Kathode ringförmig umgebenden Düse strömte langsam das die Bogenatmosphäre bildende Argon oder Stickstoffgas aus. Spektroskopisch wurde in der Achse des Argonbogens bei 200 Amp eine Maximaltemperatur unmittelbar vor der Kathodenspitze von 25000° K, bei 500 Amp eine solche von 30000° K gemessen; während beim Stickstoffbogen die Temperaturen noch etwas höher liegen. Mit diesen für frei brennende Bögen ungewöhnlich hohen Temperaturen war eine sehr niedrige Brennspannung dieser Gasbögen verbunden. Die minimale Brennspannung des Argonbogens lag (bei 150 Amp) bei 7,5 V und selbst beim

10 mm langen Bogen erst in der Gegend von 17 V, während die minimale Spannung des Stickstoffbogens mit 14 V zwar merklich höher war, aber immer noch unter der Ionisierungsspannung des atomaren Stickstoffs liegt. In diesen Bögen wurde spektroskopisch auch die Anregung zwei- bzw. dreifach ionisierter Atome nachgewiesen, in Übereinstimmung mit der Theorie, nach der bei so hoher Temperatur z.B. im Stickstoff über 60% der Plasmateilchen aus freien Elektronen besteht, während der Rest sich zu gleichen Teilen aus N^+- und N^{++}-Ionen mit geringen Anteilen N und N^{+++} zusammensetzt.

Im Gegensatz zu dem auch an der Anode einigermaßen kontrahierten Stickstoffbogen setzt der Argonbogen an der Anode ohne Kontraktion diffus an. Seine anodische Stromdichte und die Temperatur vor der Anode nehmen daher mit zunehmender Bogenlänge merklich ab. Darauf beruht wohl der bei diesen Bögen von BUSZ-PEUKERT und FINKELNBURG [*21*] gefundene Effekt, daß der Anodenfall sich mit zunehmender Bogenlänge merklich vergrößert. Wir kommen in Ziff. 114 darauf zurück.

27. Eingeschlossene Lichtbögen. Während die bisher behandelten Bögen in freier Atmosphäre, wenn auch nicht durchwegs in Luft brannten, haben in den letzten Jahren die Lichtbögen zunehmend an Bedeutung gewonnen, die in Röhren oder Kugeln aus Glas oder Quarz unter völligem Abschluß von der Atmosphäre bei von den Betriebsbedingungen abhängigen Drucken brennen.

Schon im vorigen Jahrhundert hat man mit gutem praktischen Erfolg Kohlelichtbögen ebenso wie Metallbögen in geschlossene Glasglocken eingebaut und dadurch den für allgemeine Beleuchtungsanwendungen unbequem schnellen Abbrand besonders der Kohleelektroden auf wenige Prozent des Abbrandes in freier Luft vermindern können. In Ziff. 25 wurde bereits erwähnt, daß der Kohleabbrand bei dem zur Beleuchtung früher allein benutzten Niederstrombogen im wesentlichen ein wirklicher chemischer Abbrand ist, und dieser verschwindet nach Verbrennen des in der Glocke anfänglich vorhandenen Sauerstoffs weitgehend. Als „Dauerbrandlampen" oder unter ähnlichen Namen haben solche eingeschlossenen Kohlelichtbögen bis Anfang dieses Jahrhunderts eine beträchtliche Rolle in der Beleuchtungstechnik gespielt, haben aber heute fast jede Bedeutung verloren. Ihre Eigenschaften sind von STOCKHAUSEN in seiner 1907 erschienenen Monographie [*16a*] mit aller Literatur eingehend beschrieben.

Heute spielen eingeschlossene Lichtbögen fast ausschließlich in der Form von Quecksilber- und Edelgasbögen eine Rolle. Im Jahre 1906 haben KÜCH und RETSCHINSKY [*114*] zuerst einen in einem ausgepumpten Quarzrohr zwischen zwei flüssigen Quecksilberelektroden brennenden Lichtbogen beschrieben, der in den folgenden Jahrzehnten als „künstliche Höhensonne" besonders medizinisch viel verwendet worden ist. Bei etwa 1 cm Rohrdurchmesser, einigen Ampere Stromstärke und etwa 200 V Spannung erreichen diese Lampen einen Betriebsdruck von 1 bis 4 Atm. Ihre elektrische Charakteristik ist steigend, da mit zunehmender Stromstärke, d.h. Leistungszufuhr, die verdampfende Quecksilbermenge und damit der Dampfdruck zunehmen, und da allgemein mit zunehmendem Druck des Entladungsplasmas die Bogenbrennspannung wächst. Die Zündung dieser Lampen erfolgt durch Kippen, wobei das von der einen Elektrode zur anderen fließende Quecksilber kurzzeitig eine Brücke bildet.

In den heute gebräuchlichen technischen Formen verwendet man im allgemeinen keine flüssigen Quecksilberelektroden mehr, sondern Wolframelektroden, wobei der Kathode meist zur Erleichterung der thermischen Elektronenemission etwas Thorium zugesetzt wird (thorierte Kathode). Zur Erleichterung der Zündung gibt man in die gut evakuierte und ausgeheizte Röhre außer dem Quecksilber etwas Edelgas. Meist wird vor dem Abschmelzen in die Lampe eine Quecksilbermenge eingewogen, die bei völliger Verdampfung und der erwarteten Betriebstemperatur den gewünschten Dampfdruck ergibt. Die Hüllen solcher Lampen bestehen aus Quarz oder (nur bei geringer Beanspruchung) einem temperaturbeständigen Glas; moderne Quecksilberhochdrucklampen arbeiten bei Wandtemperaturen bis 1000° C.

Man kann drei verschiedene Typen solcher Lampen unterscheiden: 1. von der Entladung erfüllte Röhren mit einem Betriebsdruck in der Gegend von 1 Atm, 2. gekühlte Kapillarlampen mit Betriebsdrucken bis zu 100 Atm, und 3. Kugellampen mit Betriebsdrucken von etwa 25 Atm, bei denen im Gegensatz zu den vorgenannten Typen der Bogen die von ihm weit entfernte Kugelhülle nicht berührt. Eine besonders die Theorie berücksichtigende Darstellung dieser Quecksilber-Hochdruckbögen (mit starker Betonung der sog. wandstabilisierten Type 1) hat ELENBAAS in einer auch die Literatur bis 1950 sehr vollständig berücksichtigenden Monographie gegeben [*4a*].

Ein charakteristisches Beispiel für einen wandstabilisierten Quecksilberbogen vom Typ 1 ist der von KREFFT, RÖSSLER und RÜTTENAUER [*112*] stammende und von Osram fabrizierte „UV-Standard-Strahler", ein in einem Quarzrohr von etwa 100 mm Länge und 30 mm Durchmesser bei 2 Amp und 120 bis 130 V brennender Quecksilberbogen mit einem Dampfdruck von etwa 1 Atm. Bei Beachtung der vorgeschriebenen Betriebsbedingungen sendet dieser Bogen ein Linien- und kontinuierliches Spektrum relativ genau bekannter Intensität im Bereich 2100 bis 20000 Å aus, und stellt damit eine wertvolle Eichlichtquelle dar. Die Achsentemperatur des Bogens dürfte zwischen 5500 und 6000° K liegen.

Bei den besonders von ELENBAAS bei der Firma Philips entwickelten Kapillarlampen brennt der Bogen in einer Quarzkapillare von 2 mm lichter Weite und 10 bis 15 mm Länge zwischen Wolframelektroden bei Drucken bis zu 100 Atm. Bei einer Leistungsaufnahme bis zu 800 W/cm müssen diese Kapillaren mit Wasser oder komprimierter Luft gekühlt werden. Ihre Leuchtdichte ist hoch und soll Werte bis 180000 Stilb erreichen.

Die heute sich immer mehr als Lichtquellen durchsetzenden *Quecksilber-Höchstdruck-Kugellampen* sind zuerst bei Osram von ROMPE und THOURET [*178*], [*179*] entwickelt und von diesen gemeinsam mit SCHULZ, WEIZEL u. a. [*202*] experimentell und theoretisch untersucht worden. Es sind (vgl. Fig. 29) relativ kurze zwischen Wolframelektroden bei Quecksilber-Dampfdrucken von 30 bis 60 Atm brennende Bögen. Solche Kugellampen sind von Osram serienmäßig für Leistungen zwischen 150 und 500 W (im Versuch bis 15000 W) für Stromstärken zwischen 2 und 200 Amp bei Brennspannungen meist zwischen 60 und 90 V und Bogenlängen zwischen 0,5 und 10 mm hergestellt worden. Die Kolbendurchmesser dieser im Gegensatz zu den Kapillarlampen keine besondere Kühlung erfordernden Lampen betragen 30 mm für eine 500 W-Lampe und 75 mm für eine 5000 W-Lampe; die Lichtausbeute der Lampen liegt in der Gegend von 55 Lumen/W.

Physikalisch handelt es sich bei diesen kurzen Höchstdruckröhren im Gegensatz zu den wandstabilisierten Bögen der Typen 1 und 2 nach WEIZEL [*243*] um elektrodenstabilisierte Bögen. Ihre Achsentemperatur liegt nach SCHULZ und WEIZEL [*202*] zwischen 7000 und 8000° K. Je nach Dampfdruck und Bogenleistung unterscheidet man (ROMPE und SCHULZ [*176*]) zwei verschiedene Formen des Bogens, bei deren erster (geringe Leistung) der Säulendurchmesser mit der Wurzel aus der Leistung zunimmt, während bei höheren Drucken und Säulenleistungen (Form 2) der Säulendurchmesser konstant bleibt und nur seine Temperatur und Strahlungsdichte zunehmen. Im Gegensatz etwa zu den Kohlelichtbögen (Ziff. 25), bei denen die Strahlung der Säule gegenüber der der Elektroden fast vernachlässigbar ist und die Säulenleistung zum überwiegenden Teil durch Wärmeleitung abgeführt wird, strahlen alle Quecksilber-Hochdrucksäulen äußerst intensiv, und im Grenzfall sehr hoher Leistung je cm Säulenlänge wird die zugeführte Leistung ganz überwiegend durch Strahlung abgeführt.

Neben dem Verhalten bei Gleich- und Wechselstrombetrieb ist der Quecksilber-Hochdruckbogen auch auf seine Modulierbarkeit mittels überlagerten Wechselstromes im Bereich von 50 bis 10^7 Hz experimentell und theoretisch untersucht

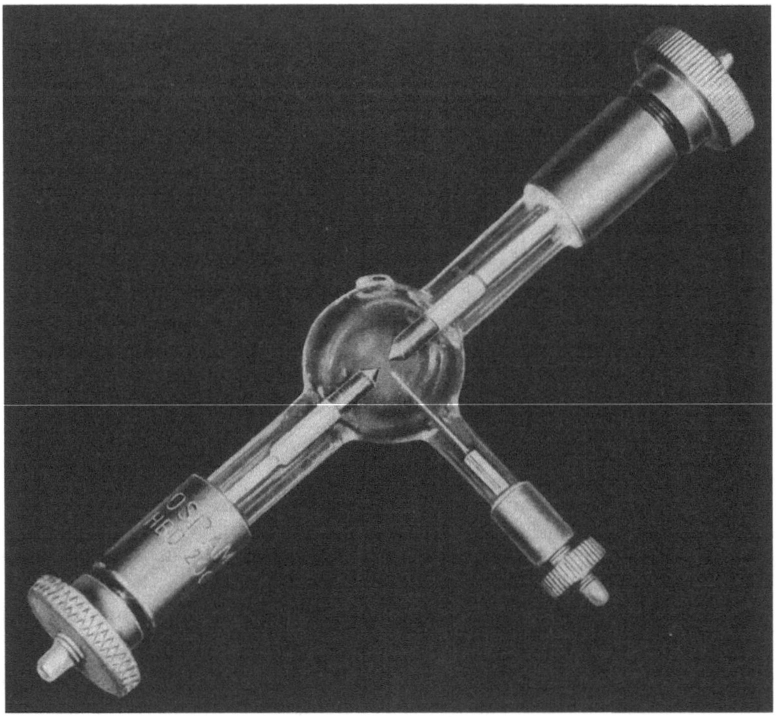

Fig. 29. Quecksilber-Höchstdruck-Kugellampe mit seitlicher Zündelektrode (Osram-Werkphoto).

worden [*245*], wobei die qualitativ erwartete Abnahme des Modulationsgrades mit zunehmender Frequenz interessante Rückschlüsse auf das physikalische Verhalten des Bogens erlaubte.

Die Entwicklung der wissenschaftlich besonders als Lichtquellen mit intensivem kontinuierlichem Spektrum bis ins ferne Ultraviolett sowie technisch als Kleinlichtquellen höchster Leuchtdichte mehr und mehr an Bedeutung gewinnenden *Edelgas-Hochdrucklampen* geht vor allem auf Arbeiten von SCHULZ [*200*] und LARCHÉ [*117*] zurück. Elektroden und Anordnung der Kugellampen entsprechen weitgehend der der Quecksilberkugellampe. Besonders der Xenon-Hochdruckbogen zeichnet sich durch eine intensive Strahlung fast rein weißer Farbe aus; die Leuchtdichte der Bogensäule in den serienmäßigen Osramlampen liegt zwischen 10000 und 65000 Stilb. Infolge der kleinen RAMSAUER-Querschnitte der Edelgasatome ist die Brennspannung des Xenon-Hochdruckbogens sehr niedrig; sie beträgt bei den serienmäßigen Lampen von 160 bis 2000 W nur 15 bis 28 V bei Stromstärken zwischen 8 und 70 Amp, einer Bogenlänge zwischen 1 und 4 mm und einem Betriebsdruck von 25 Atm. Der Bogen besitzt in diesen Lampen einen ziemlich kontrahierten Ansatz an der Kathode, vor deren Spitze ein von LARCHÉ als Plasmakugel bezeichnetes Gebiet relativ hoher Temperatur (nach den Angaben 10000° K) liegt, das bei ziemlich hohem Absorptionsvermögen (bis zu 86%) eine Leuchtdichte über einer Million Stilb besitzt. Nach der Anode zu fächert der Bogen aus, so daß dort die Leuchtdichte um mindestens den Faktor 20 kleiner ist.

Eine ganz neuartige Bogenentladung hat seit ihrer Veröffentlichung durch BUCKINGHAM [*15*] unter dem Namen "*Western Union Arc*" zunehmende Anwendung gefunden. Es handelt sich um einen Bogen, der in Argon von Atmosphärendruck bei einer Brennspannung von etwa 20 V und einer Stromstärke bis zu einigen Ampere zwischen einer Kathode aus Zirkon-Oxyd und einer beliebigen Anode brennt. Während die Bogensäule geringer Stromdichte wenig, die kühlgehaltene Anode gar nicht strahlt, entsteht auf der vom Bogen berührten Kathodenoberfläche eine sich laufend selbst erneuernde dünne Haut aus Zirkonmetall, vor der infolge langsamer Verdampfung eine hell leuchtende dünne Dampfschicht liegt. Die Strahlung dieser Kathode besteht aus einem Kontinuum, das angeblich von der auf 3000° K erhitzten Zr-Schicht herrühren soll, sowie Atomlinien des Argons und Zirkons. Die Leuchtdichte hängt von der kathodischen Stromdichte ab, und erreicht bei 900 Amp/cm^2 den Maximalwert von fast 10000 Stilb. Der gesamte Bogen ist in eine Glaskugel eingeschmolzen. Dieser neuartige Lampentyp wird für Leistungen zwischen 2 W und 12000 W hergestellt und soll Lebensdauern bis 1000 Std besitzen. Über die Physik des Bogens ist noch wenig bekannt.

28. Physikalische Sonderformen von Lichtbögen. Für die Untersuchung spezieller Eigenschaften von Lichtbögen und thermischen Plasmen, gelegentlich auch für besondere Anwendungen etwa der hohen Lichtbogentemperatur, ist von den Physikern eine Vielzahl von Sonderformen von Lichtbögen entwickelt worden, deren wichtigste hier erwähnt seien.

Zur Untersuchung der Frage, welchen Einfluß der Auftrieb der erhitzten Gase und die durch sie bewirkte Konvektion auf Form und Eigenschaften eines gewöhnlichen Kohlelichtbogens hat, hat STEENBECK [*216*] einen 10 Amp-Kohlelichtbogen der Wirkung des Schwerefeldes der Erde entzogen. Dazu hat er den Lichtbogen mit der ihn aufnehmenden Filmkamera in einem Kasten hochgeschleudert und frei wieder zurückfallen lassen. Er fand, daß der Bogen sich durch diese Verhinderung des thermischen Auftriebes der umgebenden Gase unter Absinken seiner Brennspannung stark ausdehnte.

Die Ausschaltung der Schwerkraftwirkung auf einen längeren Lichtbogen haben v. ENGEL und STEENBECK [*40*] und später FOITZIK [*54*] nach einer Anregung von HOLM dadurch erreicht, daß sie den Bogen als sog. *Wälzbogen* in einem schnell rotierenden Glas- oder Quarzrohr brennen ließen. Infolge der Zentrifugalwirkung stabilisiert sich die Achse der Bogensäule dann oberhalb einer von Gasart und Gasdruck abhängenden minimalen Drehgeschwindigkeit in der Achse des Rohres, und man erhält einen sehr gut zylinder-symmetrischen Bogen mit wohl definierter radialer Temperaturverteilung. WEIZEL [*243*] hat in ähnlicher Weise einen Quecksilber-Hochdruckbogen dadurch in der Verbindungslinie der Elektrodenspitzen stabilisiert, daß er die gesamte Kugellampe in schnelle Rotation um ihre Längsachse versetzte (Fig. 30), und PETERS [*165*] hat die gleiche Stabilisierung bei seinem gleich zu behandelnden Wasserdampf-Höchstdruckbogen verwendet.

HOLM, KIRSCHSTEIN und KOPPELMANN [*88*] haben einen in einem Glasrohr zwischen koaxialen Elektroden brennenden Lichtbogen dadurch zu stabilisieren versucht, daß sie das ganze Rohr axial von Luft durchströmen ließen. Sie erreichten damit eine befriedigende Stabilisierung, aber wegen der stärkeren

Säulenkühlung auch eine merkliche Feldstärkevergrößerung (von 10 auf 36 V je Zentimeter bei 50 Amp).

Eine viel benutzte Stabilisierung von Lichtbögen mit sehr langer Bogensäule (bis mehrere Meter Länge!) geht auf SCHÖNHERR [*197*] zurück und wird als *Wirbelstabilisierung* bezeichnet. Der Bogen brennt hierbei in einem mehr oder weniger weiten Rohr, in das nach Fig. 31 tangential Luft oder ein anderes zu untersuchendes Gas eingeblasen wird, wodurch die gesamte Gasfüllung des Rohres in schnelle Rotation versetzt wird und die Zentrifugalkräfte wiederum eine Stabilisierung des Bogens in der Rohrachse bewirken. Da die den Bogen umwirbelnden Gasmassen stets auch axiale Geschwindigkeitskomponenten besitzen, wird ständig kaltes Gas an den heißen Bogen hingeführt, der dadurch stark gekühlt wird.

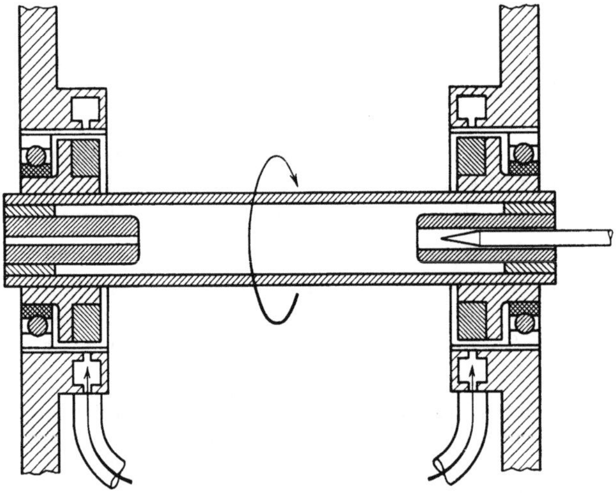

Fig. 30. Schematische Darstellung einer Wälzbogenanordnung, bei der der Bogen zwischen durchbohrten Elektroden (in der rechten der durchschiebbare Zündstift) in einem schnell rotierenden Rohr brennt.

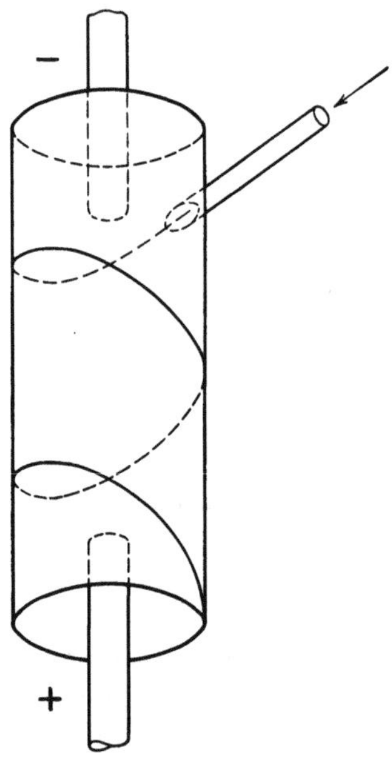

Fig. 31. Schematische Darstellung der Wirbelstabilisierung eines in einem langen Glasrohr brennenden Lichtbogens.

An Stelle des stabilisierenden Gaswirbels kann man auch einen stabilisierenden Wasserwirbel (oder allgemein Flüssigkeitswirbel) verwenden und erreicht damit zugleich eine äußerst wirkungsvolle Kühlung des begrenzenden Rohres. Dieser Wasserwirbelbogen geht auf GERDIEN und LOTZ [*62*] zurück, die schon 1922 einen Lichtbogen zwecks Erhöhung der Stromdichte und Temperatur durch eine enge Düse brannten und diese dadurch vor der Zerstörung durch die sehr hohe Temperatur der Bogensäule schützten, daß sie tangential Wasser in die Düse hineinspritzten, das eine isolierende sich stets erneuernde Schicht zwischen Düse und Bogensäule bildete und gleichzeitig durch die Rotationswirkung des Wirbels den Bogen in der Achse stabilisierte. Wegen der Verdampfung des Wassers brennt ein solcher GERDIEN-Bogen praktisch in Wasserdampf, und das hoch erhitzte Plasma schießt im Betrieb nach beiden Seiten aus der Düse heraus. In den letzten Jahren ist der GERDIEN-Bogen von LARENZ [*119*] sowie BURHORN, MAECKER und PETERS [*18*] in den verschiedensten Richtungen weiter entwickelt (Fig. 32) und zu Messungen verwendet worden. Insbesondere haben die letzteren Autoren in der Achse eines durch eine Düse von 2,5 mm Durchmesser brennenden 1500 Amp-Bogens die bisher höchste Lichtbogentemperatur von etwas über 50000° K gemessen. Wegen der intensiven Kühlung der Bogensäule durch den Wasserwirbel betrug die Säulenfeldstärke 300 V/cm; für andere

Bogenmessungen wurden auch Wasserwirbelbögen bis über 100 mm Länge und über 10 mm Durchmesser verwendet.

Zu den physikalischen Sonderformen von Lichtbögen kann man schließlich auch die Bögen rechnen, die zwischen einer festen Elektrode und der Oberfläche

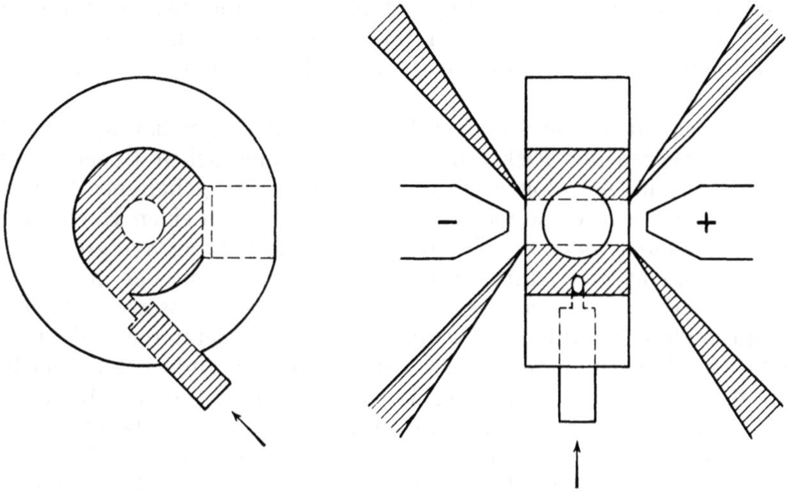

Fig. 32. Quer- und Längsschnitt des wasserwirbelstabilisierten GERDIEN-Bogens (schematisch, Wasser schraffiert).

einer genügend gut leitenden Flüssigkeit (Elektrolyt) brennen, sowie solche, die zwischen festen Elektroden ganz unter Flüssigkeiten betrieben werden. Dabei handelt es sich natürlich in Wirklichkeit um Dampfbögen, wobei lediglich der Bogen sich den Dampf selbst erzeugt. Ohne besondere Vorkehrungen werden die entstehenden Dampfblasen sich im allgemeinen aus dem Bogen fortbewegen, wodurch ein äußerst instabiles Brennen resultiert. Derartige Bögen in flüssigen Kohlenwasserstoffen werden aber zur Acetylenerzeugung aus Ölen technisch verwendet. Zu physikalischen Messungen hat PETERS [*165*] einen Wasserdampfbogen in einem mit hoher Drehzahl um die Längsachse rotierenden wassergefüllten Rohr betrieben. Dabei hielt die Zentrifugalkraft das Wasserdampfplasma in der mit der Rohrachse zusammenfallenden Bogenachse und bewirkte damit stabile Brennverhältnisse. PETERS [*166*] hat ferner das in einem Wasserbogen mit 50 atü Innendruck erzeugte hoch erhitzte Plasma aus einer durch die Anodenachse gebohrten LAVAL-Düse ausströmen lassen (Fig. 33) und damit einen Plasmastrahl von 8000° K erzeugt, dessen Strömungsgeschwindigkeit mit $6{,}5 \cdot 10^5$ cm/sec das 1,6fache der Schallgeschwindigkeit in diesem hoch erhitzten Plasma betrug.

Fig. 33. Anordnung von PETERS [*166*] zur Erzeugung eines unter hohem Druck brennenden Wasserdampf-Bogens, dessen Plasma durch die durchbohrte Anode nach oben ausströmt und dabei einen Plasmastrahl hoher Temperatur von Überschallgeschwindigkeit erzeugt.

29. Abreißbögen. Die bisher behandelten Lichtbogentypen zeichneten sich alle durch eine im wesentlichen konstante Bogenlänge und durch ein zeitunabhängiges Verhalten aus. Demgegenüber benutzt man gelegentlich bei der Spektralanalyse und in allen größeren Schaltern Lichtbögen, deren Länge absichtlich schnell verändert wird, häufig bis zum Abreißen des Bogens. In solchen Abreißbögen ändern sich also mit der Bogenlänge auch Brennspannung und Stromstärke mit der Zeit; im allgemeinen findet bei ihnen auch eine stromstärkeabhängige Elektrodenverdampfung statt, deren Betrag damit ebenfalls veränderlich ist.

Bei der Spektralanalyse von Metallen verwendet man häufig Abreißbögen, um das Abschmelzen der Elektroden beim kontinuierlichen Bogentyp zu verhindern. Dabei wird der Bogen entweder bei konstant gehaltener Länge durch einen rotierenden Schalter periodisch gelöscht und durch einen Hochspannungsstoß wieder gezündet, oder er wird periodisch durch Längenvergrößerung zum Verlöschen gebracht und durch Elektrodenberührung wieder gezündet. Im ersteren Fall verhält sich der Bogen während des Brennens wie ein mehr oder weniger stationärer Metallbogen, während wir es im letzteren Fall mit einem wirklichen Abreißbogen zu tun haben, der wie bei den einfachsten Gleichstromschaltern dann verlöscht, wenn bei Bogenverlängerung seine Mindest-Brennstromstärke (Ziff. 8) unterschritten wird. Dabei hat die Vergrößerung der Brennspannung eine doppelte Ursache. Einmal nämlich erhöht sich bei Bogenverlängerung der Spannungsabfall längs der Bogensäule, zum anderen aber bedingt die entsprechende Stromstärkeverminderung nach der statischen Charakteristik (Fig. 3) auch eine Vergrößerung der Säulenfeldstärke und vielleicht auch des Anodenfalles. Bei Schaltbögen erzwingt man zusätzlich vielfach eine Brennspannungserhöhung dadurch, daß man die Säule intensiv kühlt und damit ihre Feldstärke nach Ziff. 13 vergrößert oder auch durch Erzeugung hohen Druckes ein Ansteigen der Säulenfeldstärke erreicht.

In elektrischen Schaltern treten zwischen beweglichen Elektroden Lichtbögen variabler Länge von Stromstärken bis zu 10^5 Amp auf. Ihre Rolle in Gleichstrom- und Wechselstromschaltern ist aber eine äußerst verschiedene. Im Gleichstromschalter kann das Abschalten nur dadurch bewirkt werden, daß man zwangsweise die Bogenbrennspannung bis über den nach Fig. 4 (Ziff. 8) möglichen Maximalwert, d.h. bis zum Abreißen des Bogens erhöht. Dazu läßt man den Bogen z.B. zwischen zwei auseinander gebogenen Elektroden nach Art des Hörner-Blitzableiters durch die elektromagnetischen Kräfte des Stromkreises hochtreiben und sich damit bis zum Abreißen verlängern. Das Festbrennen der Bogenfußpunkte an Oxydflecken oder anderen Verunreinigungen muß dabei vermieden werden, weil es zur Zerstörung der Elektroden führen kann. Man kann den Bogen auch durch Gasströme oder magnetische Kräfte verlängern und gleichzeitig gegen kühlende Flächen blasen, wodurch die Säulenfeldstärke sich bis zum Verlöschen des Bogens vergrößert. Eine weitere der vielen Beeinflussungsmöglichkeiten ist, den Bogen gegen eine Anzahl senkrecht zu seiner Achse angeordnete isolierte Bleche zu blasen, wodurch der eine Bogen in eine ganze Anzahl hintereinander geschalteter Bögen aufgeteilt wird, deren gesamter Spannungsbedarf wegen der Vervielfachung des Kathoden- und Anodenfalles die Klemmenspannung übersteigt. Bei allen diesen Maßnahmen soll aber der Schaltvorgang nicht zu schnell verlaufen, da sonst wegen der im Kreis im allgemeinen vorhandenen Induktivität Überspannungen auftreten, die etwa die Isolation von Motoren ernstlich gefährden können.

Bei Wechselstromschaltern ist die Funktion des Bogens eine ganz andere als bei den behandelten Gleichstromschaltern. Da im allgemeinen der Beginn des Schaltvorganges nicht mit einem Null-Durchgang des Wechselstromes zusammenfällt, wird sich zwischen den Kontakten fast stets ein Lichtbogen bilden. Dessen Brennspannung braucht aber nicht bis zum Abreißen erhöht zu werden, da der Bogen beim folgenden Strom-Nulldurchgang automatisch stromlos wird. Es muß nur etwa durch Wegblasen des heißen Restplasmas verhindert werden, daß der Bogen beim Wiedererscheinen der Spannung erneut zündet. Das entscheidende Problem beim Gleichstromschalter ist also die Vergrößerung der Bogenspannung bis zum Abreißen, beim Wechselstromschalter dagegen das Verhindern der Wiederzündung nach dem Strom-Nulldurchgang durch vollständige und schnelle Entionisierung der Schaltstrecke in der Brennpause. Einzelheiten über Schaltbögen und Schalterkonstruktionen findet man etwa bei v. ENGEL und STEENBECK [*5a*] sowie bei KESSELRING und KOPPELMANN [*103*].

30. Lichtbögen für elektrische Öfen. Wegen seiner mit anderen Mitteln nicht erreichbaren Temperaturen (vgl. Ziff. 19) wird der elektrische Lichtbogen als Quelle hoher Temperatur für Schmelz- wie Reaktionsöfen in großem Umfang verwendet. Als erster hat wohl MOISSON [*157*] einen mit mehreren hundert Ampere Gleichstrom in einem Ofen betriebenen Kohlebogen zum Schmelzen und Verdampfen fester Stoffe benutzt. Auf die der Salpetersäureherstellung dienenden Lichtbogenreaktionsöfen nach BIRKELAND wurde in Ziff. 21 bereits hingewiesen, so daß wir uns hier auf die Schmelzöfen beschränken. In seiner einfachsten Form besteht ein solcher aus einem Kohletiegel oder einem mit Kohlematerial ausgekleidetem Tiegel, der bei Gleichstrombetrieb als Anode dient, und einer von oben in ihn hineinragenden zweiten, in vertikaler Richtung beweglichen, Kohleelektrode. Solche Öfen können für Experimentierzwecke mit lichten Weiten von 1 bis 3 cm und Stromstärken von 100 bis 200 Amp bei etwa 50 V Bogenspannung verwendet werden; für technische Zwecke werden riesige Tiegel mit Elektrodendurchmessern bis zu 1 m und Stromstärken um 50000 Amp bei 150 V Bogenspannung für die Herstellung von Calciumkarbid aus Kalk und Kohle verwendet. Auch die Phosphorfabrikation aus Calciumphosphat, Sand und Koks geschieht mittels ähnlicher Lichtbogenöfen von mehreren 10^4 kW Leistung. Bei modernen Ausführungen wird grundsätzlich mit Drehstrom gearbeitet, wobei dann drei Elektroden verwendet werden und der Tiegel nur das Schmelzgut enthält, gegen das der Bogen gelegentlich durch geeignete Magnetfelder geblasen wird. Über die in diesen Lichtbogenöfen ausgenutzten Bogentemperaturen scheint nichts Genaues bekannt zu sein. Da entweder Gase in den Bogen eingeblasen werden oder im Betrieb Dämpfe entstehen, die ihrerseits den Bogen zerblasen, dürfte sich trotz der extrem hohen Stromstärken keine stationäre kontrahierte Bogensäule mit der für sie charakteristischen Achsentemperatur von über 10000° K ausbilden. Wahrscheinlicher ist, daß die Temperatur in solchen zerblasenen Bögen je nach der Art des eingeblasenen Gases oder vom Schmelzgut ausgehenden Dampfes in der Gegend von 5000 bis 7000° K liegt. Auch zum Reinschmelzen gewisser sauerstoffempfindlicher Metalle wie Titan und Zirkonium werden neuerdings Lichtbogenöfen verwendet, in denen aber der Bogen entweder in einem Edelgas oder (fälschlich oft Vakuumbogen genannt) ausschließlich in dem betreffenden Metalldampf brennt. Die Öfen bestehen hierbei meist aus einem wassergekühlten vakuumdichten, anodisch gepolten Tiegel und einer von oben in ihn hineinragenden vertikal beweglichen, ebenfalls wassergekühlten Wolframkathode. Auch hierbei scheint es sich, wie bei den großen Kohlebogenöfen, um zerblasene Bögen mit wahrscheinlich relativ niedriger, für das Schmelzen der Metalle aber bei weitem ausreichender Temperatur zu handeln.

31. Unselbständige und nicht-thermische Lichtbögen. Abschließend sei noch eine Reihe von Bogenentladungen erwähnt, die nicht im Sinne unserer Definition von Ziff. 3 als selbständige thermische Lichtbögen anzusehen sind, weil entweder die Elektronen von einer fremdgeheizten Kathode emittiert werden oder man es in der Säule nicht mit einem thermischen Plasma zu tun hat, sondern die Elektronentemperatur wesentlich größer ist als die Gastemperatur. Da diese Entladungen wegen ihres geringen Kathodenfalles und ihrer entsprechend geringen Brennspannung aber vielfach als Bogenentladungen bezeichnet werden, seien sie hier kurz erwähnt. Es handelt sich um die Vakuum- und Niedervoltbögen, um Quecksilberleuchtröhren, Quecksilberdampf-Gleichrichter und schließlich die Wolfram-Punktlichtlampe.

Als Vakuumbögen bezeichnet man vielfach relativ stromstarke Entladungen mit fremdgeheizter Glühkathode, die in Metalldämpfen oder Gasen von geringem Druck (0,1 Torr oder weniger) brennen. In ihrer Säule stellt sich zwischen den Elektronen thermisches Gleichgewicht so hoher Temperatur ein, daß der Ersatz der durch Diffusion verloren gehenden Ladungsträger (Rekombination kann meist vernachlässigt werden) durch Stöße der energiereichsten Elektronen erfolgt, während wegen des geringen Druckes die Zahl der Stöße zwischen Elektronen und Gasteilchen zur Einstellung echten thermischen Gleichgewichtes nicht ausreicht. Vor der Kathode liegt, wie üblich, ein Gebiet überwiegend positiver Raumladung und damit ein Kathodenfall, der sich mindestens so hoch einstellt, daß die von der Glühkathode emittierten Elektronen in ihm auf die der Plasma-Elektronentemperatur entsprechende Energie beschleunigt werden. Kathode und kathodenseitige Plasmagrenze wirken also hier wie zwei ergiebige Elektroden für Elektronen bzw. Ionen, zwischen denen ein raumladungsbegrenzter Strom fließt, während die relativ weit entfernte Anode im wesentlichen nur die in sie einströmenden Elektronen aufnimmt. In weiten Gefäßen, d.h. bei seitlich nicht eingeschnürtem Säulenplasma, sind die Säulenverluste und damit auch der Säulengradient sehr klein, wachsen aber bei seitlicher Einschnürung, z.B. in den (natürlich nicht zu den Bögen im Sinne unserer Definition gehörenden) Leuchtröhren.

Wir bezeichnen solche fremdgeheizten Entladungen als unselbständig, weil ein wesentlicher Teil der Ladungsträger nicht von der Entladung selbst erzeugt, sondern durch eine äußere Energiequelle geliefert wird.

Weil der Kathodenfall solcher unselbständigen stationären Vakuumbögen weit unterhalb der Ionisierungsspannung des Trägergases liegen kann, nennt man sie gelegentlich auch *Niedervoltbögen*. Bei gewissen dieser Niederdruck-Niedervoltbögen hat man eine Potentialverteilung zwischen den Elektroden gefunden, bei der vor mindestens einer Elektrode ein solcher Potentialberg liegt, daß die maximale Potentialdifferenz im Plasma merklich größer ist als die zwischen den Elektroden. Da nach neueren Erkenntnissen (vgl. z.B. [*151*]) in solchen Entladungen auch Trägerschwingungen eine merkliche Rolle spielen können, ist ihr Mechanismus oft recht kompliziert.

Eine zunächst unselbständige nicht-thermische Bogenentladung kann nach der Zündung dadurch selbständig werden, daß die Kathode durch die aus dem Plasma auf sie einströmenden Ionen auf ihrer Emissionstemperatur gehalten wird. Dieser Fall liegt bei den meisten der heute zur Beleuchtung verwendeten Leuchtröhren vor, die demnach nach der Zündung zwar als selbständige, aber nicht-thermische Bogenentladungen angesprochen werden können. Da sie meist mit Wechselstrom betrieben werden, kommt allerdings hinzu, daß die Kathodenheizung zu einem wesentlichen Teil auch durch die einfallenden Elektronen in der Halbperiode erfolgt, in der die betreffende Elektrode gerade Anode ist.

Um eine nicht-thermische, aber nach der Zündung selbständige Bogenentladung handelt es sich auch bei der Wolfram-Punktlichtlampe. Sie besteht aus zwei Wolframklötzchen in Stickstoff von geringem Druck, die sich vor der Zündung leicht berühren, nach Erhitzung auf Glühtemperatur aber durch eine Bimetallhalterung so voneinander entfernt werden, daß sich zwischen ihnen eine Bogenentladung ausbildet. Auch in dieser Entladung dürfte die Elektronentemperatur wesentlich über der Gastemperatur liegen. Je nach Größe arbeiten diese Lampen bei Stromstärken von 5 bis 15 Amp; als Lichtquelle dient die kontinuierliche Strahlung der Wolframanode bzw. bei Wechselstrombetrieb eine der beiden dann gleichen Elektroden.

Fig. 34. Wolfram-Punktlichtlampe (Osram-Werkphoto).

Bei den in der Elektrotechnik eine sehr große Rolle spielenden Quecksilberdampf-Gleichrichtern handelt es sich ebenfalls um selbständige, aber nichtthermische Entladungen geringer Brennspannung (17 bis 20 V) und großer Stromstärke (bis über 1000 Amp) in Quecksilberdampf sehr geringen Druckes (Größenordnung 10^{-3} Torr). Als Kathode dient flüssiges Quecksilber, auf dessen Oberfläche der oder die Kathodenbrennflecke schnell hin- und herlaufen, wenn man sie nicht durch Anbringung kleiner Spitzen aus schwer schmelzendem Metall, die aus der Quecksilberoberfläche herausragen, „verankert". Statt solcher Spitzen kann man nach Lotz [*131*] auch Molybdänpulver geeigneter Korngröße mit dem Quecksilber mischen und dadurch zahllose Mikrospitzen der Entladung zur Verfügung stellen. Als Anode dienen meist große Graphitblöcke, gelegentlich auch Eisenelektroden. Bei Wechselstrombetrieb brennt die Entladung dann nur in der Halbperiode, in der das verdampfende Quecksilber Kathode ist.

C. Theorie des thermischen Plasmas.

I. Das thermische Gleichgewicht und Quasineutralität im Bogenplasma.

32. Gleichgewicht der Geschwindigkeitsverteilung. Das Gas, in dem ein Lichtbogen brennt, besteht aus Elektronen, Ionen, Atomen und Molekülen. In der Physik nennt man ein solches mit Ladungsträgern versehenes und daher leitfähiges Gas ein *Plasma*. Für das Plasma gelten die idealen Gasgesetze. Das

Lichtbogenplasma speziell hat darüber hinaus im Gegensatz zum Plasma einer Niederdruckentladung noch die Eigenschaft des lokalen thermischen Gleichgewichtes und der Quasineutralität.

Vollständiges thermisches Gleichgewicht liegt vor, wenn die Häufigkeit aller nur möglichen Energiezustände der BOLTZMANN-Verteilung gehorcht. Ein solcher Zustand herrscht z.B. in einem Hohlraum, dessen Wände und dessen Inneres bezüglich Strahlung und innerer Energie eine einheitliche Temperatur haben; oder in einer heißen Gasmasse, die so weit ausgedehnt ist, daß in diesem Bereich zahlreiche Energieaustauschprozesse, insbesondere Strahlungsabsorptions- und -emissionsvorgänge stattfinden.

Ungünstig für die Einstellung thermischen Gleichgewichtes sind solche Prozesse, die die Energie des Systems einseitig durch Energiezu- oder -abfuhr zu ändern suchen. Sind nun diese Vorgänge im System selten gegen die Energieaustauschprozesse, dann ist die Störung des Gleichgewichtes entsprechend gering. Wir wollen dies an den wichtigsten vorkommenden Fällen prüfen (vgl. hierzu: ORNSTEIN und BRINKMANN [*163*], [*164*], MANNKOPFF [*146*], ELENBAAS [*4a*], EDELS [*34*].

Die Elektronen (und Ionen) nehmen aus dem elektrischen Feld die Energie $e \cdot n_e \cdot \mathfrak{v}_e \mathfrak{E}$ je Zeit- und Volumeneinheit auf ($\mathfrak{v}_e$ = Driftgeschwindigkeit der Elektronen, n_e = Zahl der Elektronen je Volumeneinheit). Wären die Elektronen nur allein vorhanden, dann würde diese Energie nur als kinetische Energie des Schwerpunktes und nicht als Temperaturzunahme erscheinen. Erst die Existenz schwerer Teilchen, an denen die gerichtete Geschwindigkeit unregelmäßig, im Mittel rechtwinklig nach allen Seiten gestreut wird, ermöglicht die Umsetzung der kinetischen Energie in die unregelmäßige Temperaturbewegung. Die vollständige Unordnung, d.h. eine einheitliche Elektronentemperatur ist schon nach wenigen Stößen erreicht. Setzt man für die Stoßzeit der Elektronen gegen schwere Teilchen

$$\tau_{es} = \frac{1}{v_e(n_0 Q_{e0} + n_i Q_{ei})} \tag{32.1}$$

(n_0 bzw. n_i = Zahl der Ionen bzw. Neutralteilchen je Volumeneinheit, v_e = thermische Geschwindigkeit der Elektronen, Q_{e0} bzw. Q_{ei} = Querschnitt, den die Ionen bzw. die Neutralteilchen den Elektronen entgegenstellen) die in Lichtbögen vorkommenden Zahlen ein, dann ergibt sich, daß sich die Elektronentemperatur je nach den Verhältnissen in 10^{-12} bis 10^{-17} sec eingestellt hat (ROMPE und WEIZEL [*18a*]).

Zur Berechnung der Einstellzeit zwischen Elektronen- und Gastemperatur beachten wir, daß die Elektronen ihren im Felde $\mathfrak{E}$ erlangten Energieüberschuß (nicht ihre Gesamtenergie!) zum Bruchteil $2m_e/m_s$ (m_e = Elektronenmasse, m_s = Masse der schweren Teilchen) bei jedem Stoß abgeben. Es sind also $\frac{1}{2} m_s/m_e \approx 10^4$ Stöße oder 10^{-7} bis 10^{-5} sec zum Ausgleich erforderlich. Währenddessen nehmen aber die Elektronen laufend neue Energie aus dem Felde auf. Es stellt sich also eine über der Gastemperatur T_g liegende Elektronentemperatur T_e so ein, daß die von den Elektronen je Volumen- und Zeiteinheit aufgenommene Energie gleich der von ihnen auf Grund des Temperaturunterschiedes an das Gas durch Stöße abgegebenen ist. Es ist daher:

$$\frac{3}{2} k (T_e - T_g) \frac{2 m_e n_e}{m_s \tau_{es}} = e n_e b_e \mathfrak{E}^2, \tag{32.2}$$

worin $\mathfrak{v}_e = b_e \mathfrak{E}$ durch die Beweglichkeit $b_e = \frac{e}{m_e} \tau_{es}$ ausgedrückt ist. Rechnet man hieraus mit

$$v_e = \sqrt{\frac{8kT}{\pi m_e}} \quad \text{und} \quad \lambda_e = \frac{1}{\sum\limits_k n_k Q_{ek}} \tag{32.3}$$

den relativen Temperaturunterschied $(T_e - T_g)/T$ aus und verlangt, daß er klein gegen Eins sei, dann folgt

$$\frac{T_e - T_g}{T_e} = \frac{m_s}{4\,m_e} \cdot \frac{(\lambda_e\, e\, \mathfrak{E})^2}{(\frac{3}{2}\, k\, T_e)^2} \ll 1\,. \tag{32.4}$$

Dabei ist $(\lambda_e e \mathfrak{E})$ die von einem Elektron auf einer freien Weglänge aus dem Felde aufgenommene und $\frac{3}{2} k T_e$ seine thermische Energie. Es genügt also nicht, daß die von einem Elektron innerhalb einer freien Weglänge aus dem Felde aufgenommene Energie quadratisch klein gegen seine thermische ist, sondern es muß auch der in dem Verhältnis m_s/m_e ausgedrückte, sehr ungünstige Ausgleichsmechanismus berücksichtigt werden. Kleine Feldstärke, also hohe Stromstärke, und hoher Druck sind förderlich für die Ausbildung des thermischen Gleichgewichtes. Rechnen wir z.B. für den besonders günstigen Argon-Hochstrombogen (Busz und Finkelnburg [*14*]) mit $\mathfrak{E} \approx 13$ V/cm, $\lambda_e = 3 \cdot 10^{-4}$ cm, $m_A/m_e = 7 \cdot 10^4$ und $T = 30000°$ K die relative Temperaturdifferenz aus, dann folgt $\Delta T/T = 2\%$.

Die Einstellung des Gleichgewichtes wird allerdings günstiger, wenn man Anregungs- und Ionisationsprozesse mitberücksichtigt. Soweit dies durch die Umkehrprozesse (Stöße 2. Art, Dreierstoß-Rekombination) rückgängig gemacht werden, beschleunigen sie nur die Einstellgeschwindigkeit der *Elektronen*temperatur, die ohnehin schon sehr groß ist. Wenn dagegen Ausstrahlungen durch spontane Emission oder Rekombination unter Emission der frei werdenden Energie stattfinden, dann braucht diese verlorene Energie nicht mehr von den Elektronen auf die schweren Teilchen übertragen zu werden und muß daher von der im Feld gewonnenen Energie der Elektronen $\lambda_e e \mathfrak{E}$ abgezogen werden Bei Bögen unter Atmosphärendruck beträgt der Strahlungsanteil nur einige Prozent (Brinkmann [*2*], Finkelnburg [*6*]), kann also unberücksichtigt bleiben. Bei hohen Drucken kommt aber die abgestrahlte Leistung in die Größenordnung der zugeführten, so daß die nun noch auf die schweren Teilchen zu übertragende Energiedifferenz wesentlich verkleinert und daher T_e mehr an T_g herangerückt wird. Man könnte nun befürchten, daß die Maxwell-Verteilung der Elektronen dadurch wesentlich gestört wird, daß zur Anregung und besonders zur Ionisierung nur sehr schnelle Elektronen fähig sind. Der Überschuß an langsamen und der Mangel an schnellen Elektronen wird aber ziemlich schnell durch die zahlreichen Stöße der Elektronen unter sich ausgeglichen.

Eine tiefer gehende Behandlung des thermischen Gleichgewichtes müßte sich auf eine Bilanzbetrachtung im Phasenraum gründen, indem man fragt, wieviele Teilchen aus einem Volumenelement des Phasenraumes durch äußere Kräfte und durch Stöße aller Art herausgeworfen werden und wieviele durch entsprechende Vorgänge hineinkommen. Verlangt man für die Stationarität die Gleichheit beider Prozesse, dann ergeben sich daraus die Verteilungsfunktionen. Diese Überlegungen sind die Grundlagen für die Boltzmannsche Stoßgleichung, die sich daher als Ausgangspunkt für Gleichgewichtsbetrachtungen anbietet. Die ersten Schritte auf diesem Wege hat Kihara [*104*] gemacht, der in seiner Näherung zu den gleichen Ergebnissen kommt wie wir.

Experimentelle Bestimmungen der Elektronen- und Gastemperatur liegen in großer Zahl vor. Meistens wird die Elektronentemperatur als Anregungstemperatur aus der Intensität von Spektrallinien entnommen, während sich die Gastemperatur aus der Dichte bei bekanntem Druck mit den idealen Gasgesetzen ergibt. Die Dichte kann dabei aus der Schallgeschwindigkeit (Poritzky und Suits [*168*]), aus der Absorption von α- oder Röntgenstrahlen oder aus der Strahlungsintensität radioaktiver Isotope gefunden werden. Die Gastemperatur kann auch direkt aus der Intensitätsverteilung in Rotationsbanden oder aus der

Dopplerbreite geeigneter Linien ermittelt werden (s. dazu Abschnitt D, Ziff. 81). Auch das Abklingen des Leuchtens nach dem Abschalten des Bogens gibt ein sicheres Kriterium für das Temperaturgleichgewicht zwischen Elektronen und schweren Teilchen, denn die Elektronentemperatur und das von ihr erzeugte Leuchten bricht sehr schnell zusammen, während die Gastemperatur in rund 1 msec abklingt (MANNKOPFF [*146*], WITTE [*255*], WIENECKE [*251*]). All diese Messungen haben bisher zur Übereinstimmung zwischen Elektronen- und Gastemperatur im Rahmen der angegebenen Grenzen geführt (BURHORN [*16*], KOCH [*107*], JAGODZINSKI [*93*]).

33. Ionisationsgleichgewicht. Wenn wir nach dem Gesagten innerhalb einer gewissen Toleranz mit einer einheitlichen Temperatur $T_e \approx T_g$ für die kinetische Energie rechnen können, so liegen die Verhältnisse für die folgenden, noch zu definierenden Temperaturen ungünstiger.

Als Ionisationstemperatur T_I wollen wir diejenige Temperatur bezeichnen, die in die EGGERT-SAHAsche Gleichung [*35*], [*183*] zur Bestimmung der Konzentrationsverhältnisse

$$\frac{n_e n_i}{n_0} = \frac{2 Z_i}{Z_0} \frac{(2\pi m_e k T)^{\frac{3}{2}}}{h^3} e^{-\frac{E_I}{kT}} \tag{33.1}$$

(Z_i bzw. Z_0 = Zustandssumme des Ions bzw. Neutralteilchens, h = PLANCKsches Wirkungsquantum, E_I = Ionisierungsenergie) einzusetzen wäre. Im exakten Gleichgewicht mit vollständiger Hohlraumstrahlung ist natürlich $T_I = T$ zu setzen. Außerdem müßte dann die Zahl der Photoionisationen $n_0 q_I$ gleich der Anzahl der Photorekombinationen $n_i n_e q_R$ (horizontales Gleichheitszeichen im Schema), und die Anzahl der Ionisationen durch Elektronenstoß $n_0 n_e S_I$ gleich der Anzahl der inversen Prozesse, nämlich der Anzahl der durch Dreierstoß $n_i n_e^2 S_R$ erfolgenden Rekombinationen je Volumen- und Zeiteinheit sein (q_I, q_R bzw. S_I, S_R = Ionisations- und Rekombinationswahrscheinlichkeit unter Mitwirkung von Quanten bzw. durch Stöße).

Ionisation und Rekombination im Plasma

Photoionisation	Photorekombination
$n_0 q_I$	$n_i n_e q_R$
Ionisation durch Stöße	Dreierstoßrekombination
$n_0 n_e S_I$	$n_i n_e^2 S_R$

Betrachten wir nun für eine Abschätzung den Spezialfall, daß sich auch noch die Photoprozesse ebenso häufig ereignen wie die Stoßprozesse (vertikales Gleichheitszeichen im Schema) und entfernen wir nun den Hohlraum mit seiner Strahlung, dann entfallen zunächst die Photoionisationen, so daß die Rekombinationen ums Doppelte überwiegen und dadurch den Ionisationsgrad so lange abbauen, bis wieder die Zahl der Ionisationen (nur durch Stöße) durch die Zahl der Rekombinationen ausgeglichen ist. Das ist der Fall, wenn n_e und n_i um rund 20% und damit die Stoßionisationen um 10 bis 20%, die Photorekombinationen um etwa 40% und die Dreierstoßrekombinationen um über 60% heruntergegangen sind, wie man leicht aus den angegebenen Ausdrücken für die Häufigkeit abschätzt. Die SAHA-Gleichung würde also für diesen speziellen Fall n_e um 20% zu hoch liefern. Wenn wir diesen Fall als äußerste Grenze für die Gültigkeit der EGGERT-SAHAschen Gleichung ansehen, dann erhebt sich die Frage, bei welchen Verhältnissen etwa gleiche Häufigkeiten der Elementarprozesse realisiert sind. Nach ELWERT [*39*] hängt dieser Zustand praktisch nur von der Elektronenkonzentration ab und liegt bei knapp $n_e = 7 \cdot 10^{15}\ \mathrm{cm}^{-3}$, was bei Atmosphärendruck einem Ionisationsgrad von rund 10% entspricht. Bei Berücksichtigung der Ionisation aus angeregten Zuständen kann dieser Grenzwert noch merklich tiefer liegen. Liegt die Elektronendichte über dieser Grenze, sei es, daß man bei Atmosphärendruck einen höheren Ionisationsgrad (maximal $n_e = 2 \cdot 10^{17}\,\text{cm}^{-3}$) erreicht, oder, daß man höhere Drucke anwendet, dann ergibt die SAHA-Gleichung die richtige Elektronendichte. Ist die Elektronendichte jedoch kleiner als die angegebene Grenze, wie z. B. bei allen Niederstrombögen bei Atmosphärendruck, dann muß man die von ELWERT angegebene (etwas unsichere) Coronaformel verwenden, in der nur die Stoßionisationen und die Photorekombinationen erfaßt sind:

$$\frac{n_e}{n_0} = \frac{3^{\frac{3}{2}}}{16\alpha^3} \frac{\zeta_n}{n} \frac{E_{I,H}^2}{E_I^3} \cdot \frac{kT}{g} e^{-\frac{E_I}{kT}} \quad \text{für} \quad n_e < 7 \cdot 10^{16}\,\text{cm}^{-3} \tag{33.2}$$

(α = SOMMERFELDsche Feinstrukturkonstante, ζ_n = Zahl der Valenzelektronen, n = Quantenzahl der Valenzbahn, $E_{I,H}$ = Ionisierungsenergie des Wasserstoffes, g = Zahl zwischen 1,4 und 4, deren Größe der zitierten Arbeit zu entnehmen ist) Es sei schon hier darauf aufmerksam gemacht, daß zwar ganz allgemein Vorsicht angebracht ist, wenn man mit einer irgendwie bestimmten Temperatur die Elektronendichte berechnen will, daß dagegen die Ermittlung der Temperatur aus der irgendwie anders festgestellten Elektronendichte mit Hilfe der SAHA-Gleichung trotz aller Unsicherheiten weniger kritisch ist, weil die im Exponenten stehende Temperatur sehr unempfindlich gegen Fehler in der Elektronendichte ist.

34. Gleichgewicht der Anregungstemperatur. Wir fragen jetzt, wie weit die Anregungstemperatur mit der Elektronentemperatur übereinstimmt, d. h. ob die Zahl der im angeregten Zustand befindlichen Teilchen nach der BOLTZMANNschen Formel für die Energieverteilung

$$\frac{n_n^{\text{th}}}{n_0} = \frac{g_n}{g_0} e^{-\frac{E_n}{kT}} \tag{34.1}$$

(n_n^{th} = Zahl der im n-ten Zustand thermisch angeregten Teilchen, n_0 = hier Zahl der Teilchen im Grundzustand, $g = 2J + 1$ = statistisches Gewicht, J = Quantenzahl des gesamten Atomdrehimpulses, E_n = Anregungsenergie für den n-ten Zustand) mit T_e (oder T_g) berechnet werden darf. Das ist selbstverständlich erlaubt, wenn das Plasma von der Hohlraumstrahlung gleicher Temperatur durchsetzt würde. Dann sind Photoanregungs- und -emissionsprozesse unter sich gleich, und dasselbe gilt für die Stoßanregungen und Stöße 2. Art (horizontales Gleichheitszeichen im folgenden Schema).

Für diesen Fall ist die Zahl der spontanen und vom Strahlungsfeld u_ν erzwungenen Emissionen nach EINSTEIN $A\, n_n^{\text{th}}/[1 - \exp(-h\nu/kT)]$, worin A die Übergangswahrscheinlichkeit je Zeiteinheit ist. Für die gleich große Zahl der Absorptionsprozesse gilt $(n_0\, u_\nu A\, g_n/g_0) \cdot (c^3/8\pi h\nu^3)$, worin u_ν die Dichte des Strahlungsfeldes, im Falle der Hohlraumstrahlung $u_\nu = 4\pi B_\nu/c$ (B_ν = Intensität der Hohlraumstrahlung) ist.

Die Zahl der Anregungsstöße setzen wir mit der MAXWELLschen Geschwindigkeitsverteilung $f(v_e)$ zu $n_e n_0 \int v_e f(v_e)\, Q_{\text{Anr}}(v_e)\, d v_e$ an, bedenken aber, daß Q_{Anr} nur oberhalb der Anregungsenergie $\frac{1}{2} m_e v_e^2 > E_n$ von Null verschiedene Werte hat. Ziehen wir dort Q_{Anr} als konstant heraus, dann können wir von der Anregungsenergie bis ∞ integrieren und erhalten $n_e n_0 Q_{\text{Anr}} \frac{4}{\sqrt{2\pi m_e kT}} (E_n + kT) \exp(-E_n/kT)$. Dies setzen wir gleich der Zahl der Stöße 2. Art $n_e n_n S_2$ und erhalten daraus

$$S_2 = \frac{4}{\sqrt{2\pi m_e k T}} \frac{g_0}{g_n} Q_{\mathrm{Anr}} (E_n + k T).$$

Diese Ergebnisse sind im oberen Teil des folgenden Schemas zusammengestellt:

Thermisches Gleichgewicht mit Hohlraumstrahlung

Emission		Absorption
$n_n^{\mathrm{th}} \dfrac{A}{1 - e^{-h\nu/kT}}$	$=$	$\dfrac{n_0 B_\nu A g_n c^2}{2 g_0 h \nu^3}$
Stöße 2. Art		Anregungsstöße
$n_e n_n^{\mathrm{th}} \dfrac{4}{\sqrt{2\pi m_e k T}} \dfrac{g_0}{g_n} Q_{\mathrm{Anr}} (E_n + kT)$	$=$	$n_e n_0 Q_{\mathrm{Anr}} \dfrac{4}{\sqrt{2\pi m_e k T}} (E_A + kT)\, e^{-\frac{E_n}{kT}}$

je Raum- und Zeiteinheit.

Entfernt man die strahlenden Wände des Hohlraumes, dann sinkt die Zahl der Absorptionen mit der Abnahme von B_ν auf I_ν ab, wodurch die Zahl der angeregten Teilchen n_n so lange verarmt, bis wieder *die Summe* auf der linken Seite gleich der der rechten Seite geworden ist. Dann ist aber n_n prinzipiell nicht mehr durch die BOLTZMANN-Verteilung (34.1) bestimmt, sondern liegt darunter.

Elementarprozesse mit verdünntem Strahlungsfeld I_ν

$$\left.\begin{array}{c} \text{Emission} \\ n_n \dfrac{A}{1 - e^{-h\nu/kT}} \\ \text{Stöße 2. Art} \\ n_e n_n \dfrac{4}{\sqrt{2\pi m_e k T}} \dfrac{g_0}{g_n} Q_{\mathrm{Anr}} (E_n + kT) \end{array}\right\} = \left\{\begin{array}{c} \text{Absorptionen} \\ \dfrac{n_0 I_\nu A g_n c^3}{g_0 2 h \nu^3} \\ \text{Anregungsstöße} \\ n_e n_0 Q_{\mathrm{Anr}} \dfrac{4}{\sqrt{2\pi m_e k T}} (E_n + kT)\, e^{-\frac{E_n}{kT}} \end{array}\right.$$

Die Photoprozesse sind also nicht mehr untereinander gleich, ebensowenig wie die Stoßprozesse unter sich. Lediglich die Summe links ist der Summe rechts gleich.

Soll nun die Zahl n_n der angeregten Teilchen ohne umgebende Hohlraumstrahlung nicht viel kleiner als die entsprechende Zahl n_n^{th} mit Hohlraumstrahlung sein, so gibt es dazu zwei Möglichkeiten: Das Plasma schafft sich durch eine ausreichende Zahl von Emissions- und Absorptionsvorgängen die Hohlraumstrahlung selbst ($I_\nu \approx B_\nu$), wozu eine ausreichende „optische Tiefe" ($\varkappa_\nu$ = Absorptionskoeffizient mal Länge l) erforderlich ist. Bei Resonanzlinien genügt dafür eine Ausdehnung des Plasmas l von Bruchteilen von Millimetern, für etwas höher liegende Übergänge zum Grundzustand mag dafür der Durchmesser eines Bogens manchmal ausreichen, bei Übergängen zwischen hoch liegenden Termen ist im allgemeinen keine nennenswerte Strahlungsdichte mehr vorhanden, die Emission erfolgt aus optisch dünner Schicht, d.h. ohne Selbstabsorption. Hier kann aber die zweite Möglichkeit die Forderung $n_n \approx n_n^{\mathrm{th}}$ erfüllen: Wenn die Anregungsstöße sehr zahlreich sind und zum größten Teil durch Stöße 2. Art rückgängig gemacht werden, dann ist die Differenz beider Prozesse, nämlich die Anzahl der Emissionsprozesse klein gegenüber den Stoßprozessen und die Zahl der angeregten Teilchen ist nur wenig abgesunken. Dementsprechend ist zu fordern, daß

$$n_n \frac{A}{1 - e^{-h\nu/kT}} \ll n_n n_e \frac{4}{\sqrt{2\pi m_e kT}} \frac{g_0}{g_n} Q_{\mathrm{Anr}} (E_n + kT) \tag{34.2}$$

sei. Eine Abschätzung führt zu der Ungleichung

$$3 \cdot 10^7 \frac{n_e Q_{\mathrm{Anr}}}{A} (E_n + kT) \gg 1 \qquad [(E_n + kT) \text{ in eV}], \tag{34.3}$$

die uns zunächst zeigt, daß die Anregung um so mehr thermisch erfolgt, je größer die Elektronendichte und der Anregungsquerschnitt ist, und je kleiner die Übergangswahrscheinlichkeit der fraglichen Linie ist Setzt man als Beispiel Zahlenwerte ein, etwa $n_e = 10^{16}\,\mathrm{cm}^{-3}$, $Q_{\mathrm{Anr}} = 10^{-16}\,\mathrm{cm}^2$, $A = 10^7\,\mathrm{sec}^{-1}$, dann kommt bei einigen Elektrovolt Anregungsenergie rund 10 heraus. Unter Umständen können aber die Elektronendichte und vor allem der Anregungsquerschnitt tiefer liegen, und die Übergangswahrscheinlichkeit höher, dann wäre die thermische Anregung gefährdet. Ein wesentlicher Vorgang, der noch vernachlässigt ist, kann sich aber sehr günstig auswirken, nämlich die Anregung aus höheren Termen, vor allem dem Resonanzterm. Die Besetzungszahl ist hier zwar beträchtlich kleiner als im Grundzustand, dafür können aber schon verhältnismäßig langsame Elektronen eine höhere Anregung bewirken. Man kann den soeben für den Grundzustand durchgeführten Gedankengang nochmals für angeregte Terme verfolgen und kommt zu Ausdrücken, die mit denen in Gl. (34.3) übereinstimmen und dort additiv hinzuzufügen sind. In diesen Fällen wird zwar Q_{Anr} und E_n kleiner, aber auch A, so daß die weiteren Glieder von der gleichen Größenordnung wie das erste werden und so die Chancen zur Erfüllung von Ungleichung (34.3) mehr oder weniger verbessern. Für eine bestimmte Hg-Entladung hat ELENBAAS [*4a*] diese Verhältnisse durchgerechnet und gefunden, daß die Anregung vom Grundzustand aus durch Stöße nicht ausreichend für die thermische Besetzung der höheren Zustände ist; erst die Anregung vom Resonanzniveau aus schafft einen Überschuß von 2,5 im Sinne von Gl. (34.3). Eine nicht der BOLTZMANN-Verteilung gehorchende Anregung wäre am ehesten in den Randgebieten von Bögen zu erwarten, und zwar in dem Sinne, daß die Linienintensitäten zu schwach im Verhältnis zur Temperaturanregung erscheinen. Da dies aber noch nie beobachtet wurde, eher das Gegenteil, dürfen wir im allgemeinen mit Temperaturanregung rechnen.

35. Störung des Temperaturgleichgewichtes durch ein Temperaturgefälle. In allen Bögen herrscht ein Temperaturgefälle, das an sich immer eine Störung des Gleichgewichtes darstellt. Man kann aber schon aus der Anschauung heraus sagen, daß die Störung dann klein ist, wenn die Temperaturdifferenz über eine freie Weglänge klein gegen die Temperatur selbst ist:

$$\frac{\lambda_x \operatorname{grad} T_x}{T_x} \ll 1. \tag{35.1}$$

Allerdings ist hier für die freie Weglänge λ_x immer derjenige Wert einzusetzen, der für den fraglichen Ausgleichsprozeß maßgebend ist. Fragt man nach dem Ausgleich der Elektronentemperatur T_e oder der Gastemperatur T_g, dann ist entsprechend λ_e bzw. λ_g zu wählen. Wegen der Kleinheit dieser Größe bestehen im allgemeinen keine Bedenken. Für die Ionisations- und Anregungstemperatur T_I und T_A sind λ_I und λ_A schon größer, aber auch hier wird man bei den praktisch vorkommenden Temperaturgradienten mit einem lokal einwandfrei definierbaren Temperaturwert rechnen können. Nicht ganz so sicher liegen die Verhältnisse bei chemischen Reaktionen, wo die Reaktionspartner unter Umständen ziemlich große Weglängen zurücklegen können, bis sie rekombinieren; kann man doch ein heißes Plasma abschrecken und so den weiteren Ablauf der Rekombination unterbinden (Knallgaserzeugung im wasserstabilisierten Bogen, technische Zersetzung von CH_4 in C_2H_2 in wirbelstabilisierten Bögen). Das theoretische Studium dieser Relaxationserscheinungen (MEIXNER [*152*]) steckt aber noch so sehr in den Anfängen, daß wir uns hier mit dem Hinweis auf dieses Problem begnügen müssen.

36. Quasineutralität. Für den Lichtbogen als Ganzes muß man elektrische Neutralität annehmen, und zwar nicht nur, weil die Ladungsträger zu je gleichen Teilen aus neutralen Teilchen entstanden sind, sondern vor allem deswegen, weil bei den vorhandenen Ladungsträgerdichten viel zu große Kräfte notwendig wären, um auch nur einen geringen Prozentsatz von Ladungsträgern einer Sorte herauszuziehen (ROMPE und STEENBECK [*177*]). Denkt man z.B. an den Kern der Säule eines Hochstromkohlebogens von 200 Amp mit einem Kerndurchmesser von 7 mm und einer Elektronendichte von $5\cdot 10^{16}\,\mathrm{cm}^{-3}$, so verifiziert man leicht aus der POISSONschen Gleichung

$$\operatorname{div}\mathfrak{E} = 4\pi e\,(n_i - n_e) \tag{36.1}$$

durch einmalige Integration, daß bei seitlicher Verschiebung der Ladungsträger um die Strecke s eine Feldstärke von

$$\mathfrak{E} = 4\pi e\, n_e s \tag{36.2}$$

auftritt, was bei unserem Beispiel für $E = 10$ V/cm eine Verschiebung von $s = 3\cdot 10^{-10}$ cm ausmachen würde.

In den Randgebieten eines Bogens könnte allerdings die Ladungsträgerdichte so gering werden, daß schon schwache Kräfte relativ große Trennung der Ladungsträger verursachen könnten, so daß lokal keine Quasineutralität mehr vorhanden ist. Vor dieser Gefahr sind wir so lange sicher, wie der relative Unterschied der Dichte der beiden Ladungsträger klein gegen Eins ist. Dafür folgt aus der POISSONschen Gleichung

$$\frac{n_i - n_e}{n_L} = \frac{\operatorname{div}\mathfrak{E}}{4\pi e\, n_L}. \tag{36.3}$$

Als ladungstrennende Kraft kommt z.B. die schnelle Diffusion der Elektronen nach außen in Frage, wofür wir eine Abschätzung mit Näherungsformeln durchführen wollen. Wir werden später genauere Formeln kennen lernen. Im Gleichgewicht halten sich der Diffusionsstrom $+e\,D_e \operatorname{grad} n_e$ und der elektrische Strom im entstandenen Felde $-e n_e b_e \mathfrak{E}$ die Waage. Aus

$$-e\, n_e b_e \mathfrak{E} = e\, D_e \operatorname{grad} n_e \tag{36.4}$$

wird mit der TOWNSEND-Beziehung

$$\frac{D_e}{b_e} = \frac{k\,T}{e}, \tag{36.5}$$

$$\mathfrak{E} = -\frac{k\,T}{e\, n_e}\operatorname{grad} n_e = -\frac{k}{e}\,\frac{d\log n_e}{d\log T}\operatorname{grad} T. \tag{36.6}$$

Nun bedienen wir uns der Coronaformel (33.2) als Beziehung zwischen n_e und T

$$n_e \sim e^{-\frac{E_I}{kT}},$$

womit aus (36.6)

$$\mathfrak{E} = -\frac{E_I}{e} \operatorname{grad} \log T$$

wird. Dies in (36.3) eingesetzt, liefert

$$\frac{n_i - n_e}{n_L} = \frac{E_I}{4\pi e^2 n_L} \operatorname{div} \operatorname{grad} \log T = 3 \cdot 10^5 \frac{E_I[\mathrm{eV}]}{n_L} \operatorname{div} \operatorname{grad} \log T \ll 1. \tag{36.7}$$

Wählt man für div grad $\log T$ den schon sehr hohen Wert $100\ \mathrm{cm}^{-2}$ und für $E_I = 10$ eV, dann darf man noch bis zu einer Elektronendichte von $n_e \approx 10^9\ \mathrm{cm}^{-3}$ herab von Quasineutralität reden. Dies bedeutet für einen Bogen bei Atmosphärendruck und darüber eine sehr tiefe Grenze.

Genau so, wie die Diffusionskräfte auf die Elektronen die Ladungen zu trennen suchen, vermögen dies auch die magnetischen Kräfte des Eigenfeldes $\frac{1}{c}[\mathfrak{j}\,\mathfrak{H}]$, wenn sie auch nach innen gerichtet sind. Wir wiederholen daher die Frage, ob auch in diesem Falle eine Abweichung von der Quasineutralität herbeigeführt werden kann. Ähnlich wie vorher verlangen wir, daß

$$\frac{n_i - n_e}{n_L} = \frac{\operatorname{div} \mathfrak{E}_r}{4\pi e n_L} \ll 1 \tag{36.3a}$$

sein soll und setzen die radiale elektrische Feldkraft gleich der magnetischen je Volumeneinheit

$$e n_e \mathfrak{E}_r = \frac{1}{c}[\mathfrak{j}\,\mathfrak{H}]. \tag{36.8}$$

Daraus wird

$$\operatorname{div} \mathfrak{E}_r = \operatorname{div} \frac{[\mathfrak{j}\,\mathfrak{H}]}{e n_e c} = \operatorname{div} \frac{1}{c}[\mathfrak{v}_e \mathfrak{H}] = \frac{\mathfrak{v}_e}{c} \operatorname{rot} \mathfrak{H} = 4\pi e n_e \frac{\mathfrak{v}_e^2}{c^2}, \tag{36.9}$$

wenn man nacheinander

$$\mathfrak{j} \approx \mathfrak{j}_e = e n_e \mathfrak{v}_e \tag{36.10}$$

($\mathfrak{v}_e$ = Driftgeschwindigkeit der Elektronen in Stromrichtung),

$$\operatorname{div}\left[\frac{\mathfrak{v}_e}{c}\mathfrak{H}\right] = \frac{\mathfrak{v}_e}{c} \operatorname{rot} \mathfrak{H} - \mathfrak{H} \operatorname{rot} \frac{\mathfrak{v}_e}{c}, \tag{36.11}$$

$$\operatorname{rot} \mathfrak{v}_e \sim \operatorname{rot} \mathfrak{E}^a = 0 \tag{36.12}$$

und schließlich

$$\operatorname{rot} \mathfrak{H} = \frac{4\pi}{c} \mathfrak{j} \tag{36.13}$$

verwendet. Setzt man $\operatorname{div} \mathfrak{E}_r$ aus (36.9) in unsere Forderung (36.3a) ein, dann gilt

$$\frac{n_i - n_e}{n_L} = \frac{\mathfrak{v}_e^2}{c^2} \ll 1, \tag{36.14}$$

d.h. solange die Driftgeschwindigkeit der Elektronen wesentlich kleiner als die Lichtgeschwindigkeit ist, tritt keine nennenswerte Trennung der Ladungen auf. Dies ist bei allen Bogenentladungen erfüllt.

Damit ist die Quasineutralität für alle Bogenentladungen nachgewiesen.

37. Ideale Gasgesetze. Im Plasma sind die Teilchendichten wegen der hohen Temperaturen trotz höherer Drucke so klein, daß immer mit den idealen Gasgesetzen gerechnet werden kann. Trotzdem wollen wir die Grenzen abstecken, jenseits derer andere Gesetze in Kraft treten. Bei großen Dichten können trotz der hohen Bogentemperaturen die Ionen mit ihren weitreichenden COULOMB-Feldern so nahe zusammenkommen, daß die zwischen den Ionen herumfliegenden

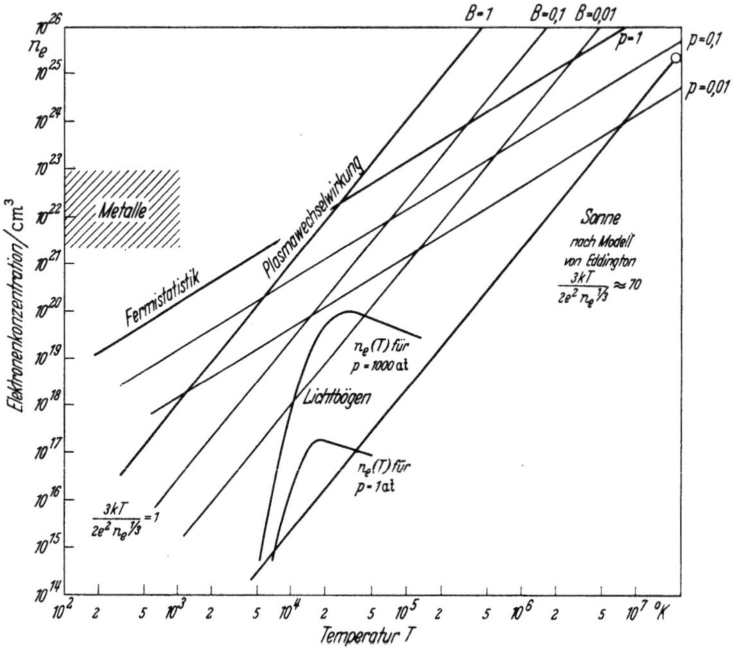

Fig. 35. Elektronendichte-Temperatur-Diagramm mit Existenzkurven für die Sonne, für Lichtbögen, für den Entartungsparameter der FERMI-Statistik $P = 1$; 0,1; 0,01 und der Plasmawechselwirkung $B = 1$; 0,1; 0,01. Außerdem der von den Metallen eingenommene Bereich.

Elektronen dauernd in nennenswertem Maße unter dem Einfluß der COULOMB-Kräfte stehen. Man kann trotz des stetigen Kraftgesetzes der Ionen diesen einen definierten Querschnitt gegenüber Elektronenstoß zuordnen (s. unten), nämlich

$$Q_{ei} = \frac{e^4}{2(k\,T)^2} \log\left[1 + \left(\frac{k\,T}{e^2 n_i^{\frac{1}{3}}}\right)^2\right] \tag{37.1}$$

und somit auch ein Volumen als die 3/2-Potenz von Q_{ei}. Ist nun das so definierte Volumen aller Ionen in der Volumeneinheit klein, also

$$n_i\, Q_{ei}^{\frac{3}{2}} = \frac{n_i e^6}{2^{\frac{3}{2}}(k\,T)^3} \log^{\frac{3}{2}}\left[1 + \left(\frac{k\,T}{e^2 n_i^{\frac{1}{3}}}\right)^2\right] \ll 1\,, \tag{37.2}$$

dann darf mit den idealen Gasgesetzen gerechnet werden. Gl. (37.2) hängt nur von dem Verhältnis der thermischen zur potentiellen Energie im COULOMB-Feld ab und zwar etwa mit der dritten Potenz. Wie Fig. 35 im $n_e - T$-Diagramm zeigt, nähert sich ein Bogenplasma erst bei extrem hohen Drucken der kritischen Grenze.

Es kann nun noch geschehen, daß die Elektronen bei sehr hohen Elektronendichten zu einem FERMI-Gas entarten wie bei den Metallen. Die Bedingung, daß dies nicht eintritt, ist, daß der Entartungsparameter P (Teilchenzahl/Zustandssumme der Elektronen) klein gegen 1 bleibt (siehe z.B. ZEISE [256])

$$P = \frac{n_e h^3}{2(2\pi m_e k T)^{\frac{3}{2}}} \ll 1. \tag{37.3}$$

Auch dieser Parameter ist in Fig. 35 für einige Werte eingetragen und bestätigt, daß wir bei Bogenentladungen keine Entartung zu befürchten haben.

Nachdem wir festgestellt haben, daß im Bogenplasma, von Grenzfällen abgesehen, thermisches Gleichgewicht und Quasineutralität herrschen, wollen wir weiterhin diese Voraussetzungen als erfüllt ansehen.

II. Einführung in die Thermodynamik irreversibler Prozesse.

38. Methodik. Die Aufgabe der Bogenphysik ist es nun, die Strömungsverhältnisse der Einzelkomponenten, wie Elektronen, Ionen, Atome und Moleküle, unter dem Einfluß der Kräfte, wie elektrische und magnetische Felder oder Druck- und Diffusionskräfte, zu ergründen und die Energieumwandlungen unter Berücksichtigung von Ionisation und Dissoziation und der Strahlung zu verfolgen. Aus diesen Gesetzmäßigkeiten lassen sich dann mit den gegebenen Randbedingungen im Prinzip Aussagen über Temperaturverteilung, Feldstärke und anderes mehr machen.

Da die Temperatur die entscheidende Variable zur Beschreibung des Bogenplasmas ist, muß die Wärmelehre zum Hauptträger unserer Untersuchungen gemacht werden. Der exakte Weg wäre der, daß man von der kinetischen Gastheorie ausgeht, damit von den mikroskopischen Elementarprozessen zur makroskopischen Thermodynamik mit Einschluß der irreversiblen Prozesse gelangt, und so das Verhalten des Bogenplasmas verstehen lernt. Die strenge gaskinetische Ableitung führt von den Einzelstößen zur BOLTZMANNschen Stoßgleichung, einer Integro-Differentialgleichung, die für Nichtgleichgewichtszustände (Zustandsvariable ändern sich mit Ort, Zeit und Geschwindigkeit) die Änderung der Verteilungsfunktion beschreibt. Mit der Verteilungsfunktion selbst als Lösung der BOLTZMANNschen Stoßgleichung kann man dann Diffusionsströme, auch unter dem Einfluß von Kräften, Wärmeströme, Reibung und anderes mehr berechnen. Leider ist aber diese Lösung bisher nur in einigen Fällen durchgeführt und erfordert einen erheblichen mathematischen Aufwand, weswegen wir uns hier die Diskussion dieser Methoden versagen müssen. Von Fall zu Fall werden wir auf die Ergebnisse solcher Rechnungen zurückgreifen und möchten hier nochmals betonen, daß für die strenge Durchrechnung eines speziellen Problems die Anwendung der exakten kinetischen Gastheorie unerläßlich ist (BOLTZMANN [*13*], ENSKOG [*42*], CHAPMAN und COWLING [*25*], WALDMANN [*236*] u.a.).

Hier wollen wir uns mehr der Thermodynamik auch der irreversiblen Prozesse bedienen (PRIGOGINE [*169*], DE GROOT [*71*], MEIXNER [*153*], HAASE [*76*], ONSAGER [*162*]) die zwar den Absolutwert der Koeffizienten offen läßt, dafür aber eine Übersicht über die vorkommenden Effekte bietet. Die Gültigkeit der Ansätze dieser makroskopischen Methodik ist durch Vergleich mit Ergebnissen gaskinetischer Rechnungen sichergestellt. Die Koeffizienten werden wir dann aus der Gaskinetik entnehmen[1] (s. a. [*140*]).

39. Definitions- und Bilanzgleichungen. Wir denken zunächst an eine Gasmischung, die aus k Komponenten besteht und in der auch chemische Reaktionen

[1] Die Grundlagen der kinetischen Gastheorie werden in den Beiträgen von H. GRAD und L. WALDMANN in Bd. XII, diejenigen der irreversiblen Thermodynamik von J. MEIXNER in Bd. III dieses Handbuches behandelt.

erfolgen können. Außerdem sind Teilchengradienten, Temperaturgradienten und Strömungsgeschwindigkeiten jeder Komponente zugelassen. Es ist nun in der Thermodynamik üblich und zweckmäßig, alle fraglichen Größen auf die Masseneinheit zu beziehen. Entsprechend haben wir eine Schwerpunktsgeschwindigkeit $\mathfrak{V}$, definiert durch die Gleichung

$$\varrho\,\mathfrak{V} = \sum_k \varrho_k\,\mathfrak{v}_k \tag{39.1}$$

($\varrho, \varrho_1 \ldots \varrho_k$ = Dichte der Mischung, der ersten, der k-ten Komponente, $\mathfrak{v}_1 \ldots \mathfrak{v}_k$ = Geschwindigkeit der ersten, der k-ten Komponente), eine Massenkonzentration

$$c_k = \frac{\varrho_k}{\varrho}, \quad \sum_k c_k = 1, \tag{39.2}$$

den mit der Schwerpunktsgeschwindigkeit verbundenen Massenstrom $\varrho\,\mathfrak{V}$ und die Massenströme der einzelnen Komponenten relativ zum Schwerpunkt

$$\mathfrak{J}_k = \varrho_k(\mathfrak{v}_k - \mathfrak{V}), \tag{39.3}$$

deren Summe definitionsgemäß verschwinden muß, wie man sich durch Ausrechnen leicht überzeugt

$$\sum_k \mathfrak{J}_k = 0. \tag{39.4}$$

Für dieses System können wir nun eine Reihe von Erhaltungs- und Bilanzgleichungen anschreiben:

1. Erhaltung der Masse: Die Dichte einer Komponente kann sich dadurch ändern, daß der absolute Massenstrom dieser Komponente eine Divergenz hat, oder daß Teilchen dieser Art durch chemische Reaktionen gebildet werden

$$\frac{\partial \varrho_k}{\partial t} = -\operatorname{div} \varrho_k\,\mathfrak{v}_k + \Gamma_k \tag{39.5}$$

(Γ_k ist die je Zeit- und Volumeneinheit durch die Reaktion umgesetzte Masse.) Die Summe über alle Komponenten muß zur Kontinuitätsgleichung

$$\frac{\partial \varrho}{\partial t} = -\operatorname{div} \varrho\,\mathfrak{V} \tag{39.6}$$

führen, worin die chemische Umwandlung nicht mehr erscheint, weil durch sie keine Masse neu geschaffen wird

$$\sum_k \Gamma_k = 0. \tag{39.7}$$

Mit Hilfe der substantiellen Differentiation

$$\frac{d}{dt} = \frac{\partial}{\partial t} + \mathfrak{V}\operatorname{grad}, \tag{39.8}$$

in der sich das Zeichen d/dt auf ein bewegtes Substanzelement, das Zeichen $\partial/\partial t$ dagegen auf den festgehaltenen Ort bezieht, lassen sich die Massenerhaltungssätze unter Verwendung der Identität $\operatorname{div} u\,\mathfrak{v} = u \operatorname{div} \mathfrak{v} + \mathfrak{v} \operatorname{grad} u$ auch schreiben

$$\frac{d\varrho_k}{dt} = -\varrho_k \operatorname{div} \mathfrak{V} - \operatorname{div} \mathfrak{J}_k + \Gamma_k \tag{39.9}$$

oder mit

$$\frac{d\varrho_k}{dt} = c_k \frac{d\varrho}{dt} + \varrho \frac{dc_k}{dt} \tag{39.10}$$

auch

$$\varrho \frac{d c_k}{d t} = -\operatorname{div} \mathfrak{J}_k + \Gamma_k; \tag{39.11}$$

schließlich ist

$$\frac{d \varrho}{d t} = -\varrho \operatorname{div} \mathfrak{V}. \tag{39.12}$$

2. *Kraftgleichung:* Der Schwerpunkt der Mischung kann nur durch äußere Kräfte $\mathfrak{F}_k$ (pro Gramm gerechnet!) wie Gravitation g, elektrische Kräfte $\frac{e_k}{m_k} \mathfrak{E}$, magnetische Kräfte $\frac{e_k}{m_k c} [\mathfrak{v}_k \cdot \mathfrak{H}]$ und andere mehr und durch einen Druckgradienten $-\operatorname{grad} p$ (eigentlich Drucktensor, wenn wir die Reibung räumlich benachbarter Schichten berücksichtigen) beschleunigt werden

$$\varrho \frac{d \mathfrak{V}}{d t} = -\operatorname{grad} p + \sum_k \mathfrak{F}_k \varrho_k. \tag{39.13}$$

Das ist die Grundgleichung der Hydrodynamik.

3. *Erhaltung der Energie:* Multiplizieren wir die vorigen Gleichung mit $\mathfrak{V}$, dann erhalten wir die Gleichung für die kinetische Energie

$$\frac{1}{2} \varrho \frac{d \mathfrak{V}^2}{d t} = -\mathfrak{V} \operatorname{grad} p + \mathfrak{V} \sum \mathfrak{F}_k \varrho_k. \tag{39.14}$$

Hierzu tritt noch die Wärmeenergie

$$\varrho \frac{d q}{d t} = \varrho \frac{d u}{d t} + \varrho p \frac{d v}{d t} = \varrho \frac{d u}{d t} + p \operatorname{div} \mathfrak{V} = -\operatorname{div} \mathfrak{W} + \sum_k \mathfrak{F}_k \mathfrak{J}_k, \tag{39.15}$$

wobei das Glied $p \operatorname{div} \mathfrak{V}$ aus der Erhaltungsgleichung (39.12) gewonnen wurde. Links steht die Änderung der inneren Energie und der Kompressionsarbeit ($v =$ Volumen pro Gramm), rechts die Divergenz des Wärmestromes $\mathfrak{W}$, der seinerseits sowohl die Wärmeleitung allein ohne Ströme als auch die von den einzelnen Massenströmen transportierte Energie enthält. Der letzte Term rechts ist die von den einzelnen Massenströmen in Reibungswärme verwandelte potentielle Energie. Die zeitliche Änderung der Gesamtenergie pro Volumeneinheit als Summe beider Gleichungen erhält dann den typischen Charakter einer Bilanzgleichung

$$\varrho \frac{d}{d t} \left(\frac{1}{2} \mathfrak{V}^2 + u \right) = -\operatorname{div} (p \mathfrak{V} + \mathfrak{W}) + \sum_k \mathfrak{F}_k \varrho_k \mathfrak{v}_k, \tag{39.16}$$

wonach sich die Gesamtenergie der Volumeneinheit durch die Divergenz eines Energiestromes und durch eine Energiequelle ändert.

4. *Entropiebilanz:* Wir führen jetzt noch die Entropie ein, mit deren Hilfe wir gewisse Aussagen über die Massen- und Wärmeströme gewinnen werden. Zu diesem Zweck gehen wir von Gl. (39.15) aus und ersetzen die linke Seite aus dem zweiten Hauptsatz in Form der GIBBSschen Gleichung bezogen auf die Masseneinheit:

$$\frac{d q}{d t} = \frac{d u}{d t} + p \frac{d v}{d t} = T \frac{d s}{d t} + \sum_k \mu_k \frac{d c_k}{d t}. \tag{39.17}$$

Hierin ist s die Entropie und μ_k das chemische Potential der k-ten Komponente pro Gramm. Dann wird mit Gl. (39.15)

$$\varrho T \frac{d s}{d t} = -\operatorname{div} \mathfrak{W} + \sum_k \mathfrak{F}_k \mathfrak{J}_k - \varrho \sum_k \mu_k \frac{d c_k}{d t}, \tag{39.18}$$

wobei der letzte Term mit Gl. (39.11) noch in

$$\varrho \sum_k \mu_k \frac{dc_k}{dt} = -\sum_k \mu_k \operatorname{div} \mathfrak{J}_k + \sum_k \mu_k \Gamma_k \tag{39.19}$$

aufgespalten werden kann. Nach einfachen Umformungen gelangen wir so zu der Entropiebilanz

$$\varrho \frac{ds}{dt} = -\operatorname{div} \frac{\mathfrak{W} - \sum_k \mu_k \mathfrak{J}_k}{T} + \frac{-\mathfrak{W}(\operatorname{grad} T)/T + \sum_k \mathfrak{J}_k \left(\mathfrak{F}_k - T \operatorname{grad} \frac{\mu_k}{T}\right) + \sum_k \mu_k \Gamma_k}{T}, \tag{39.20}$$

worin das erste Glied die Divergenz eines Entropiestromes $\mathfrak{J}_s$, das zweite eine Entropiequelle σ_s darstellt.

40. Anschauliche Erläuterungen. Um das Verständnis für diese Gleichungen und die noch folgenden phänomenologischen Ansätze zu erleichtern, wollen wir die Kraftgleichungen ähnlich wie bei SCHLÜTER [*189*] zunächst für nur zwei Einzelkomponenten hinschreiben und durchdiskutieren. Sie lauten:

$$\varrho_1 \frac{d\mathfrak{v}_1}{dt} + n_1 n_2 \varepsilon_{12} (\mathfrak{v}_1 - \mathfrak{v}_2) = \mathfrak{F}_1 \varrho_1 - \operatorname{grad} p_1 - \varrho_1 \varkappa_1 \frac{\operatorname{grad} T}{T}, \tag{40.1}$$

$$\varrho_2 \frac{d\mathfrak{v}_2}{dt} + n_1 n_2 \varepsilon_{21} (\mathfrak{v}_2 - \mathfrak{v}_1) = \mathfrak{F}_2 \varrho_2 - \operatorname{grad} p_2 - \varrho_2 \varkappa_2 \frac{\operatorname{grad} T}{T}. \tag{40.2}$$

(Zur Begründung des wesentlichen Teiles dieser Gleichungen aus der BOLTZMANNschen Stoßgleichung heraus vgl. M. H. JOHNSON [*94*].) Die Reibungskraft ist der Geschwindigkeitsdifferenz proportional angesetzt. Der Reibungsfaktor vor der Differenz wird als Wechselwirkung der beiden Teilchensorten ebenfalls mit dem Produkt der Teilchenzahlen $n_1 n_2$ anwachsen. Den Reibungskoeffizienten ε_{12} werden wir später aus der gaskinetischen Theorie entnehmen.

Auf der rechten Seite stehen die antreibenden Kräfte, also außer den äußeren Kräften $\mathfrak{F}_k \varrho_k$ pro Volumeneinheit der Partialdruckgradient und eine für die Thermodiffusion verantwortliche Thermokraft, deren Größe durch die später zu definierenden Koeffizienten $\varkappa_1$, $\varkappa_2$ gegeben ist. Bilden wir nun die Summe beider Gleichungen, vernachlässigen aber die Beschleunigung der Diffusionsströme relativ zum Schwerpunkt, so daß wir

$$\varrho_1 \frac{d\mathfrak{v}_1}{dt} + \varrho_2 \frac{d\mathfrak{v}_2}{dt} = \varrho \frac{d\mathfrak{B}}{dt} \tag{40.3}$$

schreiben dürfen, dann resultiert die Kraftgleichung (39.13) für den Schwerpunkt; denn die Reibungskräfte wirken nur *zwischen* den beiden Komponenten und können nicht den Schwerpunkt bewegen. Es muß daher

$$\varepsilon_{12} = \varepsilon_{21} \tag{40.4}$$

sein. Mit der gleichen Begründung für die Thermokräfte folgt

$$\varrho_1 \varkappa_1 = -\varrho_2 \varkappa_2. \tag{40.5}$$

Um zu den Massenströmen relativ zum Schwerpunkt $\mathfrak{J}_1 = \varrho_1 (\mathfrak{v}_1 - \mathfrak{B})$ zu kommen, setzen wir die Definitionsgleichung (39.1) für $\mathfrak{B}$ hier ein und beweisen damit, daß

$$\mathfrak{J}_1 = \varrho_1 (\mathfrak{v}_1 - \mathfrak{B}) = \frac{\varrho_1 \varrho_2}{\varrho} (\mathfrak{v}_1 - \mathfrak{v}_2) \tag{40.6}$$

ist. Den gleichen Ausdruck erhalten wir aus dem Reibungsglied der Einzelkraftgleichungen, wenn wir jede durch ihr ϱ_k dividieren, voneinander abziehen und nach $(\mathfrak{v}_1 - \mathfrak{v}_2)$ auflösen. Bringen wir noch das Beschleunigungsglied auf die rechte

Seite und fügen den Faktor $\varrho_1\varrho_2/\varrho$ an, dann folgt

$$\left.\begin{aligned}\mathfrak{J}_1 = \left(\frac{\varrho_1\varrho_2}{\varrho}\right)^2 \frac{1}{n_1 n_2 \varepsilon_{12}} \Big(\mathfrak{F}_1 - \mathfrak{F}_2 - \frac{d\mathfrak{v}_1}{dt} + \frac{d\mathfrak{v}_2}{dt} - \\ - \frac{1}{\varrho_1}\operatorname{grad} p_1 + \frac{1}{\varrho_2}\operatorname{grad} p_2 - x_1 \frac{\operatorname{grad} T}{T} + x_2 \frac{\operatorname{grad} T}{T}\Big).\end{aligned}\right\} \tag{40.7}$$

Zur Erläuterung der Gleichung für die innere Energie und Entropie greifen wir das Glied $\mathfrak{J}_k\mathfrak{F}_k/T$ in (39.20) heraus. Die Kraft pro Volumeneinheit $\mathfrak{F}_k\,\varrho_k$ multipliziert mit der Geschwindigkeit $\mathfrak{v}_k$ ergibt die von der Kraft in der Zeiteinheit geleistete Arbeit. Hiervon dient der Teil $\mathfrak{V}\,\mathfrak{F}_k\,\varrho_k$ zur Beschleunigung des Schwerpunktes und ist bereits in der Schwerpunktsgleichung erfaßt. Der Rest, nämlich

$$\mathfrak{F}_k\,\varrho_k(\mathfrak{v}_k - \mathfrak{V}) = \mathfrak{F}_k\,\mathfrak{J}_k \tag{40.8}$$

ist die durch Reibung in Wärme verwandelte Arbeit der Kraft am Massenstrom $\mathfrak{J}_k$. Wäre $\mathfrak{F}_k$ z.B. eine elektrische Feldkraft $\frac{-e}{m_e}\mathfrak{E}$ und $\mathfrak{J}_k = \frac{-m_e}{e}\mathfrak{j}_e$ der mit der Elektronenstromdichte $\mathfrak{j}_e$ verbundene Massenstrom, dann wäre $\mathfrak{F}_k\mathfrak{J}_k = \mathfrak{E}\,\mathfrak{j}_e$, also gleich der JOULEschen Wärme.

Auch der Ausdruck $-T\operatorname{grad}\frac{\mu_k}{T}$ ist eine mechanische Kraft, er enthält unter anderem den $\operatorname{grad} p_k$, so daß das Produkt $\mathfrak{J}_k T \operatorname{grad}\frac{\mu_k}{T}$ ebenfalls in Wärme verwandelte Reibungsarbeit darstellt. Bei dem später erscheinenden Glied $(\operatorname{grad} T)/T$ läßt sich die Trennung nicht so anschaulich durchführen, trotzdem bezeichnet man auch hier den Ausdruck $(\operatorname{grad} T)/T$ im übertragenen Sinne als „Kraft". Wichtig ist nur, daß alle so definierten „Kräfte", multipliziert mit den zugehörigen Strömen eine Wärmeproduktion pro Volumen- und Zeiteinheit bedeuten. Man bezeichnet diese Produkte sehr einleuchtend als Energiedissipation. Diese durch T dividiert ergeben dann die Entropieerzeugung oder -quelle pro Volumen- und Zeiteinheit. Sie muß nach dem zweiten Hauptsatz stets ≥ 0 sein.

Das Glied $\mu_k\Gamma_k$ in (39.20) enthält die chemische Affinität, die nur dann von Null verschieden ist, wenn sich das chemische Gleichgewicht nicht sofort auf die herrschende Temperatur einstellt. Mit diesem Glied kann man daher chemische Relaxationserscheinungen (MEIXNER [*152*]) behandeln. Wenn wir hier überall chemisches Gleichgewicht voraussetzen, fällt dieses Glied fort.

41. Phänomenologische Ansätze und ONSAGERS Reziprozitätssätze. Nach Gl. (39.20) läßt sich die Entropieproduktion σ_s als eine Summe von Produkten zwischen Strömen und „Kräften" schreiben. Als erste, aber in weiten Bereichen gültige Näherung für die Beziehungen zwischen Strömen und Kräften nimmt man lineare Zusammenhänge zwischen den Strömen und allen Kräften an. Damit schreiben sich die sog. phänomenologischen Ansätze:

$$\left.\begin{aligned}\mathfrak{J}_1 &= L_{11}\mathfrak{X}_1 + L_{12}\mathfrak{X}_2 + L_{1u}\mathfrak{X}_u,\\ \mathfrak{J}_2 &= L_{21}\mathfrak{X}_1 + L_{22}\mathfrak{X}_2 + L_{2u}\mathfrak{X}_u,\\ \mathfrak{W} &= L_{u1}\mathfrak{X}_1 + L_{u2}\mathfrak{X}_2 + L_{uu}\mathfrak{X}_u,\end{aligned}\right\} \tag{41.1}$$

worin $\mathfrak{X}$ eine Abkürzung für die „Kräfte"

$$\left.\begin{aligned}\mathfrak{X}_1 &= \mathfrak{F}_1 - T\operatorname{grad}\frac{\mu_1}{T},\\ \mathfrak{X}_2 &= \mathfrak{F}_2 - T\operatorname{grad}\frac{\mu_2}{T},\\ \mathfrak{X}_u &= -\frac{1}{T}\operatorname{grad} T\end{aligned}\right\} \tag{41.2}$$

und die L_{kl} Koeffizienten sind. Daß man sich mit diesen linearen Ansätzen begnügt, hat auf Grund gaskinetischer Rechnungen von ENSKOG die Bedingung zur Folge, daß sich innerhalb einer freien Weglänge die Zustandsgrößen nicht wesentlich ändern dürfen. Die gleichen Voraussetzungen mußten aber schon erfüllt sein, als die thermodynamischen Zustandsgrößen eingeführt wurden, weil diese sonst ihren Sinn verlieren. Für die meisten praktischen Aufgaben reicht diese Näherung aber vollständig aus, insbesondere, wie oben bereits gezeigt, für das Bogenplasma.

Über die Koeffizienten können wir nun einige Aussagen machen. Bilden wir in (41.1) die Summe der Massenströme $\sum_k \mathfrak{J}_k$, dann muß diese wegen Gl. (39.4) links und rechts verschwinden. Da aber die $\mathfrak{X}$-Kräfte völlig frei wählbar sind, kann das Verschwinden nur erreicht werden, indem man die Summe der übereinander stehenden, also zur gleichen $\mathfrak{X}$-Kraft gehörenden L-Koeffizienten Null setzt:

$$\sum_k L_{kl} = 0; \qquad \sum_k L_{ku} = 0. \tag{41.3}$$

Außerdem gelten die wichtigen, aus der kinetischen Statistik abgeleiteten Reziprozitätssätze von ONSAGER, nach denen die Koeffizientenmatrix symmetrisch ist, also

$$L_{kl} = L_{lk} \quad \text{und} \quad L_{ku} = L_{uk}, \tag{41.4}$$

vorausgesetzt, daß zusammengehörige Kräfte $\mathfrak{X}_k$ und Ströme $\mathfrak{J}_k$ so gewählt sind, daß ihr durch T geteiltes Produkt eine Entropiequelle darstellt. Das zu erreichen war aber gerade der Sinn der Einführung des zweiten Hauptsatzes mit der Entropie.

Durch Anwendung der ONSAGERschen Reziprozitätssätze erhalten wir für die horizontale Richtung

$$\sum_l L_{kl} = 0 \quad \text{und} \quad \sum_l L_{ul} = 0. \tag{41.5}$$

Verwenden wir diese Beziehungen, dann schreiben sich die phänomenologischen Ansätze

$$\left.\begin{aligned} \mathfrak{J}_k &= \sum_{l=1}^{n-1} L_{kl}(\mathfrak{X}_l - \mathfrak{X}_n) + L_{ku}\mathfrak{X}_u & \qquad & L_{kl} = L_{lk}, \\ \mathfrak{W} &= \sum_{l=1}^{n-1} L_{ul}(\mathfrak{X}_l - \mathfrak{X}_n) + L_{uu}\mathfrak{X}_u & \qquad & L_{ku} = L_{ul}. \end{aligned}\right\} \tag{41.6}$$

Auf eine Mischung mit zwei Komponenten angewandt, wird daraus

$$\left.\begin{aligned} \mathfrak{J}_1 = -\mathfrak{J}_2 &= L_{11}(\mathfrak{X}_1 - \mathfrak{X}_2) + L_{1u}\mathfrak{X}_u, \\ \mathfrak{W} &= L_{1u}(\mathfrak{X}_1 - \mathfrak{X}_2) + L_{uu}\mathfrak{X}_u. \end{aligned}\right\} \tag{41.7}$$

Mit diesem Zweiersystem wollen wir uns zunächst beschäftigen. Dazu schreiben wir die Kräfte ausführlich hin:

$$\mathfrak{X}_1 - \mathfrak{X}_2 = \mathfrak{F}_1 - \mathfrak{F}_2 - T\,\mathrm{grad}\,\frac{\mu_1 - \mu_2}{T}. \tag{41.8}$$

Darin ist μ_k (auch mit g_k bezeichnet) das chemische oder GIBBSsche Potential der k-ten Komponente pro Gramm:

$$\mu_k = u_k + p v_k - T s_k = h_k - T s_k. \tag{41.9}$$

u_k ist die innere Energie, s_k die Entropie der k-ten Komponente pro Gramm und v_k das Volumen, das ein Gramm der k-ten Komponente einnehmen würde, wenn es unter dem Gesamtdruck stünde. Also ist

$$v_k = \frac{p_k}{p} \cdot \frac{1}{\varrho_k} \quad \text{und} \quad p v_k = \frac{p_k}{\varrho_k} = \frac{kT}{m_k}. \tag{41.10}$$

Nun ist nach der statistischen Mechanik

$$s_k = \frac{u_k}{T} + \frac{k}{m_k} \log Z_k, \tag{41.11}$$

worin k die BOLTZMANNsche Konstante ist, und die Zustandssumme Z_k für einfache, unangeregte Teilchen

$$Z_k^{\mathrm{tr}} = \frac{e}{n_k} \frac{(2\pi m_k k T)^{\frac{3}{2}}}{h^3} \tag{41.12}$$

nur von den Translationsfreiheitsgraden gebildet wird. Der Faktor e/n_k muß wegen der Nichtunterscheidbarkeit der Teilchen angebracht werden.

In dem Ausdruck μ_k/T fällt die innere Energie heraus und es bleibt mit $n_k = p_k/kT$

$$\frac{\mu_k}{T} = \frac{k}{m_k}\left(-\frac{5}{2}\log T + \log p_k - \log K\right), \tag{41.13}$$

wovon der Gradient lautet:

$$\operatorname{grad}\frac{\mu_k}{T} = -\frac{5}{2}\frac{k}{m_k}\operatorname{grad}\log T + \frac{k}{m_k}\operatorname{grad}\log p_k \tag{41.14}$$

oder mit (41.10)

$$-T\operatorname{grad}\frac{\mu_k}{T} = h_k \frac{\operatorname{grad} T}{T} - \frac{1}{\varrho_k}\operatorname{grad} p_k. \tag{41.15}$$

Dabei bedeutet h_k die Enthalpie pro Gramm der Sorte k.

Eine andere Form von (41.15) erhält man, wenn man in (41.14) p_k mit p erweitert und den Molenbruch p_k/p zunächst mit γ_k bezeichnet. Der letzte Ausdruck in (41.14) mit T multipliziert wird dann

$$\left.\begin{aligned} \frac{kT}{m_k}\operatorname{grad}\log p_k &= \frac{kT}{m_k}(\operatorname{grad}\log p + \operatorname{grad}\log\gamma_k) = v_k \operatorname{grad} p + \frac{kT}{m_k \gamma_k}\operatorname{grad}\gamma_k \\ &= v_k \operatorname{grad} p + \frac{kT}{\overline{m}\, c_1}\operatorname{grad} c_1, \end{aligned}\right\} \tag{41.16}$$

wobei

$$\gamma_1 = \frac{c_1 m_2}{\overline{m}}, \quad d\gamma_1 = \frac{m_1 m_2}{\overline{m}^2} d c_1 \quad \text{und} \quad \overline{m} = c_1 m_2 + c_2 m_1 \tag{41.17}$$

nach Rückkehr zum Zweiersystem verwendet wurden. An Stelle von (41.15) kann man daher auch schreiben

$$-T\operatorname{grad}\frac{\mu_1}{T} = h_1 \frac{\operatorname{grad} T}{T} - v_1 \operatorname{grad} p - \frac{kT}{\overline{m}\, c_1}\operatorname{grad} c_1. \tag{41.18}$$

Setzen wir dies in die Gl. (41.7) ein, dann erhalten wir

$$\left.\begin{aligned} \mathfrak{J}_1 = -\mathfrak{J}_2 = L_{11}\Big[\mathfrak{F}_1 - \mathfrak{F}_2 - (v_1 - v_2)\operatorname{grad} p - \frac{kT}{\overline{m}\, c_1 c_2}\operatorname{grad} c_1\Big] + \\ + L_{11}(h_1 - h_2)\frac{\operatorname{grad} T}{T} - L_{1u}\frac{\operatorname{grad} T}{T} \end{aligned}\right\} \tag{41.19}$$

oder mit (41.15)

$$\left.\begin{aligned} \mathfrak{J}_1 = -\mathfrak{J}_2 = L_{11}\left[\mathfrak{F}_1 - \mathfrak{F}_2 - \frac{1}{\varrho_1}\operatorname{grad} p_1 + \frac{1}{\varrho_2}\operatorname{grad} p_2\right] + \\ + L_{11}(h_1 - h_2)\frac{\operatorname{grad} T}{T} - L_{1u}\frac{\operatorname{grad} T}{T} \end{aligned}\right\} \quad (41.20)$$

und

$$\mathfrak{W} = L_{1u}\left[\mathfrak{F}_1 - \mathfrak{F}_2 - \frac{1}{\varrho_1}\operatorname{grad} p_1 + \frac{1}{\varrho_2}\operatorname{grad} p_2\right] + L_{1u}(h_1 - h_2)\frac{\operatorname{grad} T}{T} - L_{uu}\frac{\operatorname{grad} T}{T}.$$

An der $\mathfrak{J}_1$-Gleichung sehen wir jetzt schon die Verwandtschaft mit der entsprechenden, aus den Einzelkraftgleichungen abgeleiteten $\mathfrak{J}_1$-Gleichung (40.7)

42. Überführungswärmen und Energieströme. Wenn kein Temperaturgradient, wohl aber irgend eine andere Kraft vorliegt, dann fließt nach Gl. (41.20) ein Massenstrom, aber auch ein Energiestrom $\mathfrak{W}$, da offenbar von jedem Massenstrom verschieden viel Energie transportiert wird. Den Überschuß ordnen wir $\mathfrak{J}_1$ zu und nennen die von einem Gramm des Massenstromes $\mathfrak{J}_1$ transportierte Überschußenergie die Überführungswärme Q^*, die positiv oder negativ sein kann. $\mathfrak{J}_2$ transportiert dann natürlich keine Energie mehr. Den Wert von Q^* erhalten wir aus dem Quotienten

$$Q^* = \frac{\mathfrak{W}}{\mathfrak{J}_1} = \frac{L_{1u}}{L_{11}}. \quad (42.1)$$

Diese Überführungswärme ist nicht nur ganz allgemein, auch wenn ein Temperaturgradient einen Thermodiffusionsstrom erzeugt, die vom Massenstrom geförderte Energie, sie macht als Faktor auch $-\operatorname{grad}\log T$ im letzten Glied von $\mathfrak{J}_1$ zu einer mechanischen Kraft, wie es bereits in dem Glied davor die Enthalpiedifferenz bewirkt hat. Da jeder Massenstrom mindestens seine eigenen Enthalpien mitnimmt, ist $Q^* - (h_1 - h_2)$ ein nur gaskinetisch zu verstehender Energiefaktor, der allein aus dem Temperaturgradienten eine mechanische Kraft macht, wie man an den beiden letzten Termen in (41.20) erkennt. Für die Massenströme erhalten wir nun zwei gleichwertige Gleichungen

$$\left.\begin{aligned} \mathfrak{J}_1 = -\mathfrak{J}_2 = L_{11}\Big[\mathfrak{F}_1 - \mathfrak{F}_2 - (v_1 - v_2)\operatorname{grad} p - \frac{kT}{\overline{m}\, c_1 c_2}\operatorname{grad} c_1 - \\ -(Q^* - h_1 + h_2)\frac{\operatorname{grad} T}{T}\Big], \end{aligned}\right\} \quad (42.2)$$

$$\mathfrak{J}_1 = -\mathfrak{J}_2 = L_{11}\left[\mathfrak{F}_1 - \mathfrak{F}_2 - \frac{1}{\varrho_1}\operatorname{grad} p_1 + \frac{1}{\varrho_2}\operatorname{grad} p_2 - (Q^* - h_1 + h_2)\frac{\operatorname{grad} T}{T}\right], \quad (42.3)$$

die wir wahlweise benutzen werden. Die letzte von beiden stimmt nun völlig mit der von uns aus der Mechanik abgeleiteten Gl. (40.7) überein, wenn wir noch

$$L_{11} = \left(\frac{\varrho_1 \varrho_2}{\varrho}\right)^2 \frac{1}{n_1 n_2 \varepsilon_{12}} \quad (42.4)$$

und

$$x_1 - x_2 = Q^* - h_1 + h_2 \quad (42.5)$$

setzen. Für den Wärmestrom findet man

$$\left.\begin{aligned} \mathfrak{W} = L_{1u}\left[\mathfrak{F}_1 - \mathfrak{F}_2 - (v_1 - v_2)\operatorname{grad} p - \frac{kT}{\overline{m}\, c_1 c_2}\operatorname{grad} c_1\right] - \\ - [L_{uu} - Q^* L_{11}(h_1 - h_2)]\cdot\frac{\operatorname{grad} T}{T}. \end{aligned}\right\} \quad (42.6)$$

Ersetzt man hier den ersten Ausdruck rechts aus Gl. (42.2), dann bekommt man

$$\mathfrak{W} = Q^* \mathfrak{J}_1 - (L_{uu} - L_{11} Q^{*2})\frac{\operatorname{grad} T}{T}, \quad (42.7)$$

d. h. es wird Energie vom Massenstrom und durch bloßen Wärmekontakt befördert. Legt man an die homogene Mischung zweier Gase nur einen Temperaturgradienten an, dann fließt nach Gl. (42.6) zunächst ein Wärmestrom (Diffusionsthermik nach CLUSIUS und WALDMANN [*234*], [*237*])

$$\mathfrak{W}_T = -\left[L_{uu} - Q^* L_{11}(h_1 - h_2)\right] \frac{\operatorname{grad} T}{T} = -\varkappa \operatorname{grad} T, \tag{42.8}$$

der nach Gl. (42.3) und (42.7) mit dem Anteil

$$\mathfrak{W}_{TD} = -Q^* L_{11}(Q^* - h_1 + h_2) \frac{\operatorname{grad} T}{T} \tag{42.9}$$

von dem Thermodiffusionsstrom

$$\mathfrak{J}_{TD} = -L_{11}(Q^* - h_1 + h_2) \frac{\operatorname{grad} T}{T} \tag{42.10}$$

getragen wird. Der Rest, also der Wärmestrom durch Kontakt ohne Massenströmung, ist natürlich mit dem letzten Glied von (42.7) identisch

$$\mathfrak{W}_{\mathfrak{J}=0} = -(L_{uu} - L_{11} Q^{*2}) \frac{\operatorname{grad} T}{T} = -\varkappa_{\mathfrak{J}=0} \operatorname{grad} T.$$

Dieser Wärmestrom stellt sich nach einiger Zeit dann ein, wenn der Thermodiffusionsstrom sich selbst durch den Aufbau eines Konzentrationsgradienten zum Verschwinden gebracht hat. Die Berechnungen liefern meistens $\varkappa$, während die Messungen im allgemeinen $\varkappa_{\mathfrak{J}=0}$ ergeben. Der Unterschied ist aber meistens klein außer bei leichten Teilchen, insbesondere Elektronen.

III. Diffusions- und Energieströme in reagierenden Gasen.

43. Vergleich mit den Resultaten der Gaskinetik. Um die Bedeutung der Koeffizienten zu verstehen, vergleichen wir unsere Gleichungen mit den aus der kinetischen Gastheorie für ein Gemisch zweier Gase ohne äußere Kräfte folgenden (ENSKOG [*42*], WALDMANN [*235*], [*236*])

$$\mathfrak{J}_1 = -\varrho D \operatorname{grad} c_1 - \alpha \varrho D c_1 c_2 \frac{\operatorname{grad} T}{T}, \tag{43.1}$$

$$\mathfrak{W} = -\varrho D \left(\frac{\alpha k T}{\overline{m}} + h_1 - h_2\right) \operatorname{grad} c_1 - \varkappa \operatorname{grad} T, \tag{43.2}$$

oder

$$\mathfrak{W} = \left(\frac{\alpha k T}{\overline{m}} + h_1 - h_2\right) \mathfrak{J}_1 - \varkappa_{\mathfrak{J}=0} \operatorname{grad} T. \tag{43.3}$$

Hierin ist D der Diffusionskoeffizient der beiden Komponenten ineinander, α der dimensionslose Thermodiffusionsfaktor (oft auch mit $D' T/D$ bezeichnet) und $\varkappa$ die Wärmeleitfähigkeit einschließlich der durch Thermodiffusion bedingten. Der Koeffizientenvergleich mit (42.2), (42.4), (42.5) und (42.6) ergibt

$$L_{11} \frac{k T}{\overline{m}\, c_1 c_2} = \varrho D, \tag{43.4}$$

$$L_{11} = \varrho D \overline{m}\, c_1 c_2 / k T, \tag{43.5}$$

$$\varepsilon_{12} = \frac{k T}{D(n_1 + n_2)}, \tag{43.6}$$

$$Q^* - h_1 + h_2 = x_1 - x_2 = \frac{\alpha k T}{\overline{m}}, \tag{43.7}$$

$$Q^* = \frac{\alpha k T}{\overline{m}} + h_1 - h_2, \tag{43.8}$$

$$\varkappa_{\mathfrak{J}=0} = \varkappa - Q^* \alpha \varrho D c_1 c_2 / T. \tag{43.9}$$

Damit erhalten wir endgültig folgende Gleichungen für die Massen- und Energieströme

$$\left.\begin{aligned}\mathfrak{J}_1 = -\mathfrak{J}_2 = \varrho\frac{D\overline{m}}{kT}c_1 c_2 \times \\ \times\left[\mathfrak{F}_1 - \mathfrak{F}_2 - (v_1 - v_2)\,\mathrm{grad}\,p - \frac{kT}{\overline{m}\,c_1 c_2}\mathrm{grad}\,c_1 - \frac{\alpha kT}{\overline{m}}\frac{\mathrm{grad}\,T}{T}\right],\end{aligned}\right\} \quad (43.10)$$

$$\left.\begin{aligned}\mathfrak{W} = \varrho D\frac{\overline{m}}{kT}c_1 c_2\left(\frac{\alpha kT}{\overline{m}} + h_1 - h_2\right) \times \\ \times\left[\mathfrak{F}_1 - \mathfrak{F}_2 - (v_1 - v_2)\,\mathrm{grad}\,p - \frac{kT}{\overline{m}\,c_1 c_2}\cdot\mathrm{grad}\,c_1\right] - \varkappa\,\mathrm{grad}\,T\end{aligned}\right\} \quad (43.11)$$

oder

$$\mathfrak{W} = \left(\frac{\alpha kT}{\overline{m}} + h_1 - h_2\right)\mathfrak{J}_1 - \varkappa_{\mathfrak{J}=0}\,\mathrm{grad}\,T \quad (43.12)$$

mit

$$(v_1 - v_2)\,\mathrm{grad}\,p + \frac{kT}{\overline{m}\,c_1 c_2}\mathrm{grad}\,c_1 = \frac{1}{\varrho_1}\mathrm{grad}\,p_1 - \frac{1}{\varrho_2}\mathrm{grad}\,p_2. \quad (43.13)$$

Bemerkenswert ist, daß mit der Bestimmung der Diffusionskonstante auch der Koeffizient L_{11} für *alle* anderen „Kräfte“ festgelegt ist. Der Thermodiffusionsfaktor α beherrscht sowohl die Thermodiffusion als auch die Diffusionsthermik. Der Kontaktwärmeleitkoeffizient bleibt noch zu bestimmen. Rein formal kann man aus dem Ausdruck $x_1 - x_2 = \alpha kT/\overline{m}$ und aus Gl. (40.5) $\varrho_1 x_1 = -\varrho_2 x_2$ die x-Werte einzeln ausrechnen. Es ergibt sich dann

$$x_1 = c_2\frac{\alpha kT}{\overline{m}} \quad \text{und} \quad x_2 = -c_1\frac{\alpha kT}{\overline{m}}, \quad (43.14)$$

wovon unten noch Gebrauch gemacht wird.

Wenn keine Schwerpunktsströmung existiert, dann muß im stationären Zustand bei nicht reagierenden Gasen nach der Erhaltungsgleichung für ϱ_k Gl. (39.9) auch div $\mathfrak{J}_1$ verschwinden, d.h. die Massenstrom*stärke* muß konstant sein. Ist darüber hinaus das System geschlossen, so ist auch $\mathfrak{J}_1 = 0$. Dann ist der Wärmestrom allein durch $\varkappa_{\mathfrak{J}=0}$ beherrscht, nachdem sich ein grad c aufgebaut hat. Eine geschlossene Konvektionsströmung kann allerdings der Entmischung entgegenwirken, so daß die Stationaritätsbedingung lautet

$$\mathrm{div}\,\varrho_1\mathfrak{V} = -\mathrm{div}\,\mathfrak{J}_1. \quad (43.15)$$

In diesem Falle kommt auch das von $\mathfrak{J}_1$ beherrschte Glied in der Wärmestromgleichung zum Tragen.

44. Massen- und Energieströme reagierender Gase. Wenn die Partner reagieren können, braucht div $\mathfrak{J}_1$ auch im stationären Zustand nicht zu verschwinden, sondern kann durch die chemische Erzeugung oder Vernichtung aufgenommen werden

$$\mathrm{div}\,\mathfrak{J}_1 = \Gamma_1; \qquad \mathrm{div}\,\mathfrak{J}_2 = \Gamma_2. \quad (44.1)$$

Unter diesen Umständen spielt sich folgendes ab (Beispiel $N_2 \rightleftharpoons 2N$): Wird in einem Molekülgas ein Temperaturgradient aufrecht erhalten, dann steigt mit zunehmender Temperatur auch der Dissoziationsgrad und die relative Massenkonzentration c_1 der leichteren und energiereicheren Komponente. Nach Gl. (43.10) bildet sich dann gleichzeitig ein (schwacher) Thermodiffusionsstrom und ein (starker) Diffusionsstrom aus, die beide völlig stationär erhalten bleiben, indem Atom*paare* nach kälteren Gebieten diffundieren und rekombinieren und ebensoviele Moleküle nach heißeren Stellen wandern und dissoziieren. Daher bleibt der Schwerpunkt erhalten, während der Begriff „Teilchenruhesystem" seinen Sinn verliert. Der Gewinn an Atomen $-\operatorname{div}\mathfrak{J}_a$ rekombiniert als Γ_a, die gleiche Masse erscheint bei den Molekülen als $+\Gamma_m$ und verläßt als $\operatorname{div}\mathfrak{J}_2$ das betrachtete Volumen in umgekehrter Richtung. Es ist also

$$\Gamma_a + \Gamma_m = 0 \quad \text{und} \quad -\operatorname{div}\mathfrak{J}_a = \operatorname{div}\mathfrak{J}_m, \quad \text{also} \quad \mathfrak{J}_a = -\mathfrak{J}_m. \tag{44.2}$$

Mit den Atomen wandert je Gramm die Enthalpie

$$h_a = \frac{5}{2}\frac{kT}{m_a} + u_a + \frac{1}{2}\frac{E_D}{m_a} \tag{44.3}$$

und mit den Molekülen

$$h_m = \frac{5}{2}\frac{kT}{m_m} + u_m, \tag{44.4}$$

wobei u_a die Elektronenanregung, u_m die Anregung von Rotation, Oszillation und Elektronensprung enthält und E_D die reine Dissoziationsenergie für ein Molekül bedeutet. Von $\mathfrak{J}_a$ wird daher je Gramm, abgesehen von $\alpha k T/\overline{m}$, der Energieüberschuß

$$h_a - h_m = \frac{1}{2m_a}\left(\frac{5}{2}kT + E_D\right) + u_a - u_m \tag{44.5}$$

übertragen, also gerade die zur Umsetzung der Masseneinheit erforderliche Gesamtenergie. Dem Strom $\mathfrak{J}_m$ darf dann natürlich keine Energie mehr zugeordnet werden. Die in Gl. (41.15) eingeführten Enthalpien h_k gelten nicht nur für die dort vorausgesetzten einfachen Teilchen, sondern allgemein für jede Art von Teilchen, wie eine ausführliche Rechnung zeigt. Da

$$\operatorname{grad} c_a = \left(\frac{\partial c_a}{\partial T}\right)_p \operatorname{grad} T \tag{44.6}$$

ist, erhalten wir für dieses Problem die Gleichungen

$$\mathfrak{J}_a = -\mathfrak{J}_m = -\left(\varrho D\frac{\partial c_a}{\partial T} + \alpha\varrho D c_a c_m/T\right)\operatorname{grad} T, \tag{44.7}$$

$$\mathfrak{W} = \left[\frac{1}{2m_a}\left(\frac{5}{2}kT + E_D\right) + u_a - u_m + \frac{\alpha kT}{\overline{m}}\right]\mathfrak{J}_a - \varkappa_{\mathfrak{J}=0}\operatorname{grad} T. \tag{44.8}$$

Führt man in die Wärmestromgleichung (44.8) den Diffusionsstrom $\mathfrak{J}_a$ aus (44.7) ein, dann ist der vor $\operatorname{grad} T$ stehende Faktor der totale Wärmeleitwert $\tilde{\varkappa}$ für ein reagierendes zweiatomiges Molekülgas:

$$\left.\begin{aligned} \mathfrak{W} &= -\left[\left(\frac{5kT}{2m_m} + \frac{E_D}{m_m} + u_a - u_m + \frac{\alpha kT}{\overline{m}}\right)\varrho D\left(\frac{\partial c_a}{\partial T} + \alpha c_a c_m/T\right) + \varkappa_{\mathfrak{J}=0}\right]\operatorname{grad} T \\ &= -\tilde{\varkappa}\operatorname{grad} T \qquad (\text{s. } \varkappa_D \text{ in Fig. 41}). \end{aligned}\right\} \tag{44.9}$$

Die Gln. (44.7) bis (44.9) spielen in den äußeren Bogenpartien für die Energiebilanz eine wesentliche Rolle, worauf Höcker und Finkelnburg [*86*] und in neuerer Zeit King [*105*] aufmerksam gemacht haben.

IV. Eigenschaften des reinen Plasmas.

45. Die Stromdichte. Ähnliche Überlegungen wollen wir nun auf ein reines Plasma, das nur aus Ionen der Masse m_i und der Ladung $+e$ und aus Elektronen der Masse m_e und der Ladung $-e$ besteht, anwenden. Rekombinationen sind also vorerst nicht zugelassen.

Die elektrische Stromdichte ist wegen $\mathfrak{J}_i = -\mathfrak{J}_e$

$$\left.\begin{aligned} \mathfrak{j} = \mathfrak{j}_i + \mathfrak{j}_e = e n_i \mathfrak{v}_i - e n_e \mathfrak{v}_e = \frac{e}{m_i}\mathfrak{J}_i - \frac{e}{m_e}\mathfrak{J}_e + e(n_i - n_e)\mathfrak{V} \\ = -e\left(\frac{1}{m_i} + \frac{1}{m_e}\right)\mathfrak{J}_e + \varrho_{el}\mathfrak{V}. \end{aligned}\right\} \tag{45.1}$$

Dazu gehört die Erhaltungsgleichung, die aus Gl. (39.9) zu berechnen ist,

$$\frac{\partial \varrho_{\text{el}}}{\partial t} = -\operatorname{div}\mathfrak{j}. \tag{45.2}$$

Sehen wir zunächst von einer Raumladung ab, dann folgt aus Gl. (42.3)

$$\mathfrak{j} = -e\left(\frac{1}{m_i} + \frac{1}{m_e}\right)L_{11}\left[\mathfrak{J}_e - \mathfrak{J}_i - \frac{1}{\varrho_e}\operatorname{grad} p_e + \frac{1}{\varrho_i}\operatorname{grad} p_i - (Q^* - h_e + h_i)\frac{\operatorname{grad} T}{T}\right]. \tag{45.3}$$

Legen wir jetzt nur eine elektrische Feldstärke an das reine Plasma an, so erhalten wir das OHMsche Gesetz

$$\mathfrak{j} = +e\left(\frac{1}{m_i} + \frac{1}{m_e}\right)L_{11}\, e\left(\frac{1}{m_i} + \frac{1}{m_e}\right)\mathfrak{E} = e^2\left(\frac{m_i + m_e}{m_i m_e}\right)^2 L_{11}\mathfrak{E} = \sigma\mathfrak{E}. \tag{45.4}$$

Mit einem beliebig gerichteten Magnetfeld erscheinen die beiden LORENTZ-Kräfte:

$$\mathfrak{j} = -e\left(\frac{1}{m_i} + \frac{1}{m_e}\right)L_{11}\left\{-e\left(\frac{1}{m_i} + \frac{1}{m_e}\right)\mathfrak{E} - \frac{e}{m_e}\left[\frac{\mathfrak{v}_e}{c}\mathfrak{H}\right] - \frac{e}{m_i}\left[\frac{\mathfrak{v}_i}{c}\mathfrak{H}\right]\right\}. \tag{45.5}$$

Hier ziehen wir die Leitfähigkeit

$$\sigma = e^2\left(\frac{1}{m_i} + \frac{1}{m_e}\right)^2 L_{11} \tag{45.6}$$

heraus und ersetzen $\mathfrak{v}_e$ und $\mathfrak{v}_i$ mit Gl. (45.1) und (39.1) durch $\mathfrak{V}$ und $\mathfrak{j}$ entsprechend den Gleichungen

$$\mathfrak{v}_e = \mathfrak{V} - \frac{m_i\mathfrak{j}}{e\varrho_L} \tag{45.7}$$

und

$$\mathfrak{v}_i = \mathfrak{V} + \frac{m_e\mathfrak{j}}{e\varrho_L}$$

mit

$$\varrho_L = \varrho_e + \varrho_i.$$

Faßt man die magnetischen Glieder zusammen, so folgt daraus

$$\mathfrak{j} = \sigma\left(\mathfrak{E} + \frac{1}{c}[\mathfrak{V}\mathfrak{H}] - \frac{1}{e n_L c}[\mathfrak{j}\mathfrak{H}]\right), \tag{45.8}$$

wobei $m_e \ll m_i$ und $n_L = n_e = n_i$ verwendet wurde. Durch das Magnetfeld kommt also zunächst die induzierte Feldstärke hinzu, sofern sich das Plasma als Ganzes bewegt. Eine solche Bewegung kann nun nach der Schwerpunktsgleichung (39.13) durch das Magnetfeld selbst hervorgerufen werden

$$\varrho\frac{d\mathfrak{V}}{dt} = \frac{1}{c}[\mathfrak{j}\mathfrak{H}] - \operatorname{grad} p. \tag{45.9}$$

Wenn nicht ein Hindernis die Ausbildung eines Druckgradienten bewirkt, dann wird also das Plasma durch die LORENTZ-Kraft beschleunigt, was aber nach Gl. (45.8) eine Gegeninduktion und damit ein Absinken der Stromdichte und der LORENTZ-Kraft zur Folge hat. Die Endgeschwindigkeit ist erreicht, wenn der

Strom auf Grund der Kompensation der angelegten Feldstärke durch die induzierte Feldstsärke verschwunden ist, wenn also

$$\mathfrak{E} = -\frac{1}{c}[\mathfrak{B}\,\mathfrak{H}] \tag{45.10}$$

geworden ist (Stromlosigkeit beim laufenden Motor ohne Belastung).

Ist aber die Bewegung gehemmt, so daß sich ein Druckgradient ausbilden kann, so ist dessen Einfluß aus Gl. (45.3) nach Gl. (45.8) zu übernehmen

$$\mathfrak{j} = \sigma\left\{\mathfrak{E} + \frac{1}{c}[\mathfrak{B}\,\mathfrak{H}] - \frac{1}{e\,n_L\,c}[\mathfrak{j}\,\mathfrak{H}] + \frac{m_e m_i}{e\,(m_e+m_i)}\left(\frac{1}{\varrho_e}\operatorname{grad} p_e - \frac{1}{\varrho_i}\operatorname{grad} p_i\right)\right\}. \tag{45.11}$$

Mit Hilfe der Schwerpunktsgleichung (45.9) eliminieren wir aus (45.11) $\frac{1}{c}[\mathfrak{j}\,\mathfrak{H}]$, verwenden

$$\operatorname{grad} p = \operatorname{grad} p_i + \operatorname{grad} p_e \tag{45.12}$$

und vernachlässigen m_e gegen m_i. Dann folgt

$$\mathfrak{j} = \sigma\left\{\mathfrak{E} + \frac{1}{c}[\mathfrak{j}\,\mathfrak{B}\,\mathfrak{H}] - \frac{1}{e\,n_L}\left(\frac{1}{2}\operatorname{grad} p + \varrho\,\frac{d\mathfrak{B}}{dt}\right)\right\}. \tag{45.13}$$

Wenn also z.B. das Plasma durch ein Magnetfeld so an eine Wand gedrückt wird, daß keine Bewegung mehr möglich ist, dann versucht zwar der zu $\mathfrak{j}$ senkrechte grad$p$ nach Gl (45.13) einen Strom in dieser Richtung zu erzeugen, bildet aber dadurch nach der Kontinuitätsgleichung (45.2) eine Raumladung und über die Poissonsche Gleichung

$$\operatorname{div}\mathfrak{E} = 4\pi\,\varrho_{el} \tag{45.14}$$

eine Feldstärke, die einen Stromfluß in dieser Richtung unterbindet und so jeglichen Einfluß des Magnetfeldes auf den Strom zunichte macht. Nur die Leitfähigkeit σ kann durch den Druck selbst etwas modifiziert sein, was praktisch aber keine Rolle spielt (voller Strom beim festgehaltenen Motor).

Schließlich kann auch der Temperaturgradient einen Strom erzeugen, dessen Größe nach Gl. (43.7) und (45.3)

$$\left.\begin{aligned} \mathfrak{i}_T = e\left(\frac{1}{m_i} + \frac{1}{m_e}\right) L_{11}\,(Q^* - h_e + h_i)\,\frac{\operatorname{grad} T}{T} &= \sigma\,\frac{m_i m_e}{e\,(m_i+m_e)}\,\frac{\alpha\,k\,T}{\overline{m}}\,\frac{\operatorname{grad} T}{T}\\ &= \frac{\sigma\,\alpha\,k}{2e}\operatorname{grad} T \end{aligned}\right\} \tag{45.15}$$

ist. Dieser Effekt ist nicht immer klein, denn bei einem grad$T$ von 10^4 Grad/cm beträgt die entsprechende, eingeprägte Feldstärke einige Volt je Zentimeter.

Als äußere Kräfte $\mathfrak{F}$ kann man außer der besprochenen Lorentz-Kraft auch Gravitationskräfte, Zentrifugalkräfte und andere mehr hinzuziehen. Genau wie oben setzt man diese Kräfte in die Strömungsgleichungen und in die Schwerpunktsgleichung ein, um die Stromdichte zu erhalten.

Man kann als äußere Kräfte auch die Trägheitskräfte je Gramm $-d\mathfrak{v}_e/dt$ und $-d\mathfrak{v}_i/dt$ wählen. Diese in Gl. (45.3) eingesetzt, liefern als Differenz

$$\frac{d}{dt}(\mathfrak{v}_i - \mathfrak{v}_e) = \frac{1}{e}\,\frac{d\,(\mathfrak{j}/n_L)}{dt} = 4\pi\,e\,\frac{m_e+m_i}{m_e m_i}\,\frac{d\,(\mathfrak{j}/\omega_L^2)}{dt} \tag{45.16}$$

unter Verwendung der Plasmafrequenz

$$\omega_L^2 = \frac{4\pi\,e^2\,\varrho_L}{m_e\cdot m_i} \approx \frac{4\pi\,e^2\,n_L}{m_e}. \tag{45.17}$$

Dieses Glied beschreibt die Plasmaschwingungen, auf die wir im optischen Teil noch kurz zurückkommen werden.

Zusammengefaßt lautet die Stromgleichung für das reine Plasma

$$\left.\begin{aligned} 4\pi\sigma\frac{d}{dt}\left(\frac{\mathfrak{j}}{\omega_L^2}\right)+\mathfrak{j}=\sigma\Big\{\frac{m_e}{e}(\mathfrak{F}_i-\mathfrak{F}_e)+\mathfrak{E}+\frac{1}{c}[\mathfrak{B}\,\mathfrak{H}]-\\ -\frac{1}{e\,n_L}\left(\frac{1}{2}\operatorname{grad}p+\varrho\frac{d\mathfrak{B}}{dt}\right)+\frac{\alpha k}{2e}\operatorname{grad}T\Big\},\end{aligned}\right\} \tag{45.18}$$

worin nun unter $\mathfrak{F}$ die äußeren Kräfte ohne elektrische und magnetische und ohne Trägheitskräfte gemeint sind.

Fassen wir die Ergebnisse noch einmal zusammen, so folgt, daß bei Existenz beliebiger Kraftfelder anfänglich von Ort zu Ort verschiedene Beschleunigungen $d\mathfrak{B}/dt$ und Stromdichten $\mathfrak{j}$ anlaufen, die aber wegen der Kontinuitätsgleichungen für Masse und Stromdichte zum Teil Gegendrucke und Gegenfelder erzeugen, bis bei zeitlich konstanten Kraftfeldern der Vorgang stationär, d.h. bis

$$\operatorname{div}\varrho\,\mathfrak{B}=0 \quad \text{und} \quad \operatorname{div}\mathfrak{j}=0 \tag{45.19}$$

geworden ist.

46. Energiestrom im reinen Plasma. Für den Energiestrom findet man aus Gl. (43.12) und (45.1)

$$\mathfrak{W}=-\left(\frac{\alpha kT}{\overline{m}}+h_e-h_i\right)\frac{m_e m_i}{e(m_e+m_i)}\,\mathfrak{j}-\varkappa_{\mathfrak{J}=0}\operatorname{grad}T. \tag{46.1}$$

Im stationären Zustand ist $\operatorname{div}\mathfrak{j}=0$, also

$$\operatorname{div}\mathfrak{W}=-\frac{m_e m_i}{e(m_e+m_i)}\frac{\partial}{\partial T}\left(\frac{\alpha kT}{\overline{m}}+h_e-h_i\right)\mathfrak{j}\operatorname{grad}T-\operatorname{div}(\varkappa_{\mathfrak{J}=0}\operatorname{grad}T). \tag{46.2}$$

Man sieht, daß auch mit dem elektrischen Strom ein Energiestrom verbunden ist, wenn in $\mathfrak{j}$-Richtung ein $\operatorname{grad}T$ existiert.

47. Bestimmung der Koeffizienten nach SPITZER. Wir benötigen nun noch den Wert für die Koeffizienten σ, α und $\varkappa_{\mathfrak{J}=0}$, die von SPITZER [*212*], [*213*] mit der BOLTZMANNschen Stoßgleichung für das reine Plasma ausgerechnet sind:

$$\left.\begin{aligned} \sigma&=\frac{2m_e v_e^3}{e^2 Z\log q}\left(\frac{2}{3\pi}\right)^{\frac{3}{2}}\gamma_E,\\ \alpha&=3\,\frac{\gamma_T}{\gamma_E},\\ \varkappa_{\mathfrak{J}=0}&=\frac{20m_e^2 k v_e^5}{3e^4 Z\log q}\left(\frac{2}{3\pi}\right)^{\frac{3}{2}}\delta_T\left(1-\frac{3}{5}\frac{\delta_E\gamma_T}{\delta_T\gamma_E}\right),\\ v_e^2&=\frac{3kT}{m_e};\qquad Z=\sum\frac{n_Z Z^2}{n_e};\qquad q=\frac{kT}{e^2 Z n_i^{\frac{1}{3}}}.\end{aligned}\right\} \tag{47.1}$$

Tabelle 1. *Werte für die Transportkoeffizienten.*

	$Z=1$	$Z=2$	$Z=4$	$Z=16$	$Z=\infty$
γ_E	0,5816	0,6833	0,7849	0,9225	1,000
γ_T	0,2727	0,4131	0,5714	0,8279	1,000
δ_E	0,4652	0,5787	0,7043	0,8870	1,000
δ_T	0,2252	0,3563	0,5133	0,7907	1,000

Z ist die mittlere Ionenladungszahl; von dem Faktor $\log q$ wird im nächsten Kapitel noch die Rede sein. Die Wärmeleitfähigkeit des reinen Plasmas ist wegen der großen Geschwindigkeit der Elektronen sehr groß und überragt, wie schon HÖCKER und SCHULZ [*87*], [*200*] vermutet hatten, die meisten anderen Wärmeleitprozesse (s.a. Fig. 41).

Außerdem ist das Wiedemann-Franzsche Gesetz gültig, man erhält nämlich durch Ausrechnung für $Z = 1$

$$\frac{\varkappa_{\mathfrak{J}=0}}{\sigma} = 2\,\frac{k^2}{e^2}\,T \tag{47.2}$$

in Übereinstimmung mit Lorentz. — Es mag noch darauf hingewiesen werden daß alle Koeffizienten von der Trägerdichte unabhängig sind.

V. Irreversible Prozesse im System mit drei Komponenten.

48. Phänomenologische Ansätze. Bisher haben wir nur ein Zweiersystem behandelt und die Theorie auf ein reagierendes Gas und ein reines Plasma angewendet. Da ein großer Teil der Gleichungen schon unter Verwendung des $\sum$-Zeichens geschrieben wurde, fällt es nun nicht schwer, sie formal auf ein Dreiersystem zu erweitern. Hierfür lauten die phänomenologischen Ansätze

$$\left.\begin{aligned}
\mathfrak{J}_1 &= L_{11}\mathfrak{X}_1 + L_{12}\mathfrak{X}_2 + L_{13}\mathfrak{X}_3 + L_{1u}\mathfrak{X}_u,\\
\mathfrak{J}_2 &= L_{21}\mathfrak{X}_1 + L_{22}\mathfrak{X}_2 + L_{23}\mathfrak{X}_2 + L_{2u}\mathfrak{X}_u,\\
\mathfrak{J}_3 &= L_{31}\mathfrak{X}_1 + L_{32}\mathfrak{X}_2 + L_{33}\mathfrak{X}_3 + L_{3u}\mathfrak{X}_u,\\
\mathfrak{W} &= L_{u1}\mathfrak{X}_1 + L_{u2}\mathfrak{X}_2 + L_{u3}\mathfrak{X}_3 + L_{uu}\mathfrak{X}_u,\\
\sum_k \mathfrak{J}_k &= 0; \quad \sum_k L_{kl} = 0; \quad L_{kl} = L_{lk}; \quad \sum_l L_{kl} = 0;\\
& \quad \sum_k L_{ku} = 0; \quad L_{ul} = L_{ku}; \quad \sum_l L_{lu} = 0.
\end{aligned}\right\} \tag{48.1}$$

Ersetzt man jeweils das dritte Glied mittels der Gleichungen

$$\left.\begin{aligned}
L_{k3}\mathfrak{X}_3 &= -L_{k1}\mathfrak{X}_3 - L_{k2}\mathfrak{X}_3\\
&\text{und}\\
L_{u3}\mathfrak{X}_3 &= -L_{u1}\mathfrak{X}_3 - L_{u2}\mathfrak{X}_2,
\end{aligned}\right\} \tag{48.2}$$

so folgt

$$\left.\begin{aligned}
\mathfrak{J}_1 &= L_{11}(\mathfrak{X}_1 - \mathfrak{X}_3) + L_{12}(\mathfrak{X}_2 - \mathfrak{X}_3) + L_{1u}\mathfrak{X}_u,\\
\mathfrak{J}_2 &= L_{21}(\mathfrak{X}_1 - \mathfrak{X}_3) + L_{22}(\mathfrak{X}_2 - \mathfrak{X}_3) + L_{2u}\mathfrak{X}_u,\\
\mathfrak{W} &= L_{u1}(\mathfrak{X}_1 - \mathfrak{X}_3) + L_{u2}(\mathfrak{X}_2 - \mathfrak{X}_3) + L_{uu}\mathfrak{X}_u,\\
\mathfrak{J}_3 &= -\mathfrak{J}_1 - \mathfrak{J}_2.
\end{aligned}\right\} \tag{48.3}$$

Denken wir uns nun in einem speziellen Versuch $\mathfrak{X}_u = 0$ gesetzt, dann fließen Diffusionsströme $\mathfrak{J}_k$, von denen jeder die Energie Q_k^* transportiert. Die Summe dieser Energieströme muß mit $\mathfrak{W}$ übereinstimmen

$$\left.\begin{aligned}
\mathfrak{J}_1 Q_1^* &= L_{11} Q_1^* (\mathfrak{X}_1 - \mathfrak{X}_3) + L_{12} Q_1^* (\mathfrak{X}_2 - \mathfrak{X}_3),\\
\mathfrak{J}_2 Q_2^* &= L_{21} Q_2^* (\mathfrak{X}_1 - \mathfrak{X}_3) + L_{22} Q_2^* (\mathfrak{X}_2 - \mathfrak{X}_3),\\
\mathfrak{J}_1 Q_1^* + \mathfrak{J}_2 Q_2^* &= (L_{11} Q_1^* + L_{21} Q_2^*)(\mathfrak{X}_1 - \mathfrak{X}_3) + (L_{12} Q_1^* + L_{22} Q_2^*)(\mathfrak{X}_2 - \mathfrak{X}_3),\\
\mathfrak{W} &= L_{u1}(\mathfrak{X}_1 - \mathfrak{X}_3) + L_{u2}(\mathfrak{X}_2 - \mathfrak{X}_3).
\end{aligned}\right\} \tag{48.4}$$

Da die $\mathfrak{X}$-Kräfte beliebig wählbar sind, müssen die beiden letzten Gleichungen gliedweise übereinstimmen und die entsprechenden Koeffizienten gleich sein

$$\left.\begin{aligned}
L_{u1} = L_{1u} &= L_{11} Q_1^* + L_{21} Q_2^*,\\
L_{u2} = L_{2u} &= L_{12} Q_1^* + L_{22} Q_2^*.
\end{aligned}\right\} \tag{48.5}$$

Verwenden wir dies in Gl. (48.3), dann folgt allgemein:

$$\left.\begin{aligned}\mathfrak{J}_1 &= L_{11}(\mathfrak{X}_1 - \mathfrak{X}_3 + Q_1^* \mathfrak{X}_u) + L_{12}(\mathfrak{X}_2 - \mathfrak{X}_3 + Q_2^* \mathfrak{X}_u),\\ \mathfrak{J}_2 &= L_{21}(\mathfrak{X}_1 - \mathfrak{X}_3 + Q_1^* \mathfrak{X}_u) + L_{22}(\mathfrak{X}_2 - \mathfrak{X}_3 + Q_2^* \mathfrak{X}_u),\\ \mathfrak{W} &= Q_1^* \mathfrak{J}_1 + Q_2^* \mathfrak{J}_2 + (L_{uu} - L_{1u} Q_1^* - L_{2u} Q_2^*)\,\mathfrak{X}_u.\end{aligned}\right\} \tag{48.6}$$

Das letzte Glied mit $\mathfrak{X}_u$ ist die Kontaktwärmeleitung, wenn also keine Ströme fließen. Q_1^* ist der unter Umständen auch negative Energieüberschuß, der von einem Gramm der Sorte „1" mehr transportiert wird als von einem Gramm der Sorte „3". Entsprechendes gilt für Q_2^*.

Auch an dieser Stelle wollen wir andeuten, daß man dieses Gleichungssystem für die Diffusionsströme aus den Einzelkraftgleichungen gewinnen kann wie es SCHLÜTER [*189*], allerdings ohne Thermokräfte, gemacht hat. Man hat sich nur für einen Augenblick jeweils eine Komponente des Dreiersystems wegzudenken und auf die Stoßprozesse, die zur Reibung und Thermokraft führen, die Gesetze des Zweiersystems anzuwenden. Dann kann man nach Verwendung von (43.14) schreiben

$$\left.\begin{aligned}&\varrho_1 \frac{d\mathfrak{v}_1}{dt} + n_1 n_2 \varepsilon_{12}(\mathfrak{v}_1 - \mathfrak{v}_2) + n_1 n_3 \varepsilon_{13}(\mathfrak{v}_1 - \mathfrak{v}_3)\\ &\qquad = \mathfrak{J}_1\varrho_1 - \operatorname{grad} p_1 - \varrho_1\left(\frac{\varrho_2\alpha_{12}}{m_1\varrho_2 + m_2\varrho_1} + \frac{\varrho_3\alpha_{13}}{m_1\varrho_3 + m_3\varrho_1}\right) k \operatorname{grad} T,\\ &\varrho_2 \frac{d\mathfrak{v}_2}{dt} + n_2 n_1 \varepsilon_{21}(\mathfrak{v}_2 - \mathfrak{v}_1) + n_2 n_3 \varepsilon_{23}(\mathfrak{v}_2 - \mathfrak{v}_3)\\ &\qquad = \mathfrak{J}_2\varrho_2 - \operatorname{grad} p_2 - \varrho_2\left(\frac{\varrho_1\alpha_{21}}{m_2\varrho_1 + m_1\varrho_2} + \frac{\varrho_3\alpha_{23}}{m_2\varrho_3 + m_3\varrho_2}\right) k \operatorname{grad} T,\\ &\varrho_3 \frac{d\mathfrak{v}_3}{dt} + n_3 n_1 \varepsilon_{31}(\mathfrak{v}_3 - \mathfrak{v}_1) + n_3 n_2 \varepsilon_{32}(\mathfrak{v}_3 - \mathfrak{v}_2)\\ &\qquad = \mathfrak{J}_3\varrho_3 - \operatorname{grad} p_3 - \varrho_3\left(\frac{\varrho_1\alpha_{31}}{m_3\varrho_1 + m_1\varrho_3} + \frac{\varrho_2\alpha_{32}}{m_3\varrho_2 + m_2\varrho_3}\right) k \operatorname{grad} T.\end{aligned}\right\} \tag{48.7}$$

Die Summe aller Gleichungen gibt wieder die bekannte Kraftgleichung für den Schwerpunkt. Die Summe der Reibungskräfte muß verschwinden, woraus zu schließen ist, daß $\varepsilon_{kl} = \varepsilon_{lk}$ ist. Aus dem gleichen Grunde muß aber für die Thermodiffusionsfaktoren $\alpha_{kl} = -\alpha_{lk}$ gelten. Auf der rechten Seite stehen, wenn man noch die Beschleunigungsglieder mit herübernimmt, die mit ϱ_k multiplizierten $\mathfrak{X}$-Kräfte. Nach etwas umständlichen Rechnungen kann man von den Reibungsgliedern des obigen Gleichungssystems zu den Gl. (48.6) für $\mathfrak{J}_1$ und $\mathfrak{J}_2$ kommen. Wir werden von dem System (48.7) später noch Gebrauch machen.

49. Stromdichte im Plasma mit Neutralgas. In Ziff. 48 haben wir im wesentlichen die Grundgleichungen für das Dreiersystem erhalten. Wenn wir nun den Index „1" auf die Elektronen, den Index „2" auf die Ionen und den Index „3" auf die Neutralteilchen eines Plasmas beziehen, ist es zweckmäßiger, an Stelle von $\mathfrak{J}_e$ und $\mathfrak{J}_i$ die elektrische Stromdichte

$$\mathfrak{i} = -e\left(\frac{\mathfrak{J}_e}{m_e} - \frac{\mathfrak{J}_i}{m_i}\right) \tag{49.1}$$

und den Massenstrom der Ladungsträger

$$\mathfrak{J}_L = \mathfrak{J}_e + \mathfrak{J}_i = -\mathfrak{J}_0 \tag{49.2}$$

einzuführen. Mit den Abkürzungen

$$\left.\begin{aligned} L_{e\mathfrak{j}} &= -e\left(\frac{L_{11}}{m_e}-\frac{L_{21}}{m_i}\right); \quad L_{i\mathfrak{j}} = -e\left(\frac{L_{12}}{m_e}-\frac{L_{22}}{m_i}\right),\\ L_{e\mathfrak{J}} &= L_{11}+L_{21}; \quad L_{i\mathfrak{J}} = L_{12}+L_{22}; \quad L_{12}=L_{21}\end{aligned}\right\} \tag{49.3 a}$$

ist dann:

$$\left.\begin{aligned} \mathfrak{j} &= L_{e\mathfrak{j}}(\mathfrak{X}_e-\mathfrak{X}_0+Q_e^*\mathfrak{X}_u)+L_{i\mathfrak{j}}(\mathfrak{X}_i-\mathfrak{X}_0+Q_i^*\mathfrak{X}_u),\\ \mathfrak{J}_L &= L_{e\mathfrak{J}}(\mathfrak{X}_e-\mathfrak{X}_0+Q_e^*\mathfrak{X}_u)+L_{i\mathfrak{J}}(\mathfrak{X}_i-\mathfrak{X}_0+Q_i^*\mathfrak{X}_u),\\ \mathfrak{W} &= (Q_i^*-Q_e^*)\frac{m_e}{e}\,\mathfrak{j}+(m_e Q_e^*+m_i Q_i^*)\frac{1}{m_0}\mathfrak{J}_L+\varkappa_{\mathfrak{J}=0}\,T\mathfrak{X}_u.\end{aligned}\right\} \tag{49.3 b}$$

Für die $\mathfrak{X}$-Kräfte schreiben wir wieder

$$\mathfrak{X}_k=\mathfrak{F}_k-T\operatorname{grad}\frac{\mu_k}{T}; \quad \mathfrak{X}_u=-\frac{1}{T}\operatorname{grad}T \tag{49.4}$$

und wählen als äußere Kräfte elektrische und magnetische Felder und die Gravitation:

$$\left.\begin{aligned} \mathfrak{F}_e &= -\frac{e}{m_e}\left(\mathfrak{E}+\left[\frac{\mathfrak{v}_e}{c}\mathfrak{H}\right]\right)+\mathfrak{g}-\frac{d\mathfrak{v}_e}{dt},\\ \mathfrak{F}_i &= \frac{e}{m_i}\left(\mathfrak{E}+\left[\frac{\mathfrak{v}_i}{c}\mathfrak{H}\right]\right)+\mathfrak{g}-\frac{d\mathfrak{v}_i}{dt},\\ \mathfrak{F}_0 &= \mathfrak{g}-\frac{d\mathfrak{v}_0}{dt}.\end{aligned}\right\} \tag{49.5}$$

Mit den $\mathfrak{X}$-Kräften erscheint in der $\mathfrak{j}$-Gleichung (49.3 b) die Summe

$$\left.\begin{aligned} &-L_{e\mathfrak{j}}\left(\frac{e}{m_e}\mathfrak{E}+\frac{e}{m_e}\left[\frac{\mathfrak{v}_e}{c}\mathfrak{H}\right]\right)+L_{i\mathfrak{j}}\left(\frac{e}{m_i}\mathfrak{E}+\frac{e}{m_i}\left[\frac{\mathfrak{v}_i}{c}\mathfrak{H}\right]\right)\\ &=e\left(\frac{L_{i\mathfrak{j}}}{m_i}-\frac{L_{e\mathfrak{j}}}{m_e}\right)\mathfrak{E}+\frac{e}{c}\left(\frac{L_{i\mathfrak{j}}}{m_i}[\mathfrak{v}_i\mathfrak{H}]-\frac{L_{e\mathfrak{j}}}{m_e}[\mathfrak{v}_e\mathfrak{H}]\right)\\ &=e\left(\frac{L_{i\mathfrak{j}}}{m_i}-\frac{L_{e\mathfrak{j}}}{m_e}\right)\left(\mathfrak{E}+\frac{1}{c}[\mathfrak{v}_L\mathfrak{H}]\right)+\frac{1}{c\varrho_L}\left(L_{i\mathfrak{j}}\frac{m_e}{m_i}+L_{e\mathfrak{j}}\frac{m_i}{m_e}\right)[\mathfrak{j}\mathfrak{H}],\end{aligned}\right\} \tag{49.6}$$

die mit den Definitionsgleichungen

$$\left.\begin{aligned} &\varrho_e+\varrho_i=\varrho_L; \quad \varrho_L\mathfrak{v}_L=\varrho_e\mathfrak{v}_e+\varrho_i\mathfrak{v}_i; \quad \mathfrak{j}=e\,n_L(\mathfrak{v}_i-\mathfrak{v}_e);\\ &n_e=n_i=n_L; \quad \mathfrak{v}_i=\mathfrak{v}_L+\frac{m_e}{e}\frac{\mathfrak{j}}{\varrho_L}; \quad \mathfrak{v}_e=\mathfrak{v}_L-\frac{m_i}{e}\frac{\mathfrak{j}}{\varrho_L}\end{aligned}\right\} \tag{49.7}$$

umgeformt wurde.

Die Gravitationskräfte $\mathfrak{g}$ heben sich in jeder Kraftdifferenz auf, werden aber später als Ursache für den Druckgradienten wiederkommen.

Die auftretende Summe der Trägheitsglieder

$$-L_{e\mathfrak{j}}\frac{d\mathfrak{v}_e}{dt}-L_{i\mathfrak{j}}\frac{d\mathfrak{v}_i}{dt}+(L_{e\mathfrak{j}}+L_{i\mathfrak{j}})\cdot\frac{d\mathfrak{v}_0}{dt} \tag{49.8}$$

geht unter Verwendung von Gl. (39.1) und (49.7) über in

$$(L_{e\mathfrak{j}}+L_{i\mathfrak{j}})\frac{\varrho}{\varrho_0}\frac{d}{dt}(\mathfrak{V}-\mathfrak{v}_L)-e\left(\frac{L_{i\mathfrak{j}}}{m_i}-\frac{L_{e\mathfrak{j}}}{m_e}\right)\frac{m_e m_i}{e^2}\frac{d(\mathfrak{j}/\varrho_L)}{dt}, \tag{49.9}$$

wobei die Dichten ϱ_k lokal konstant gehalten werden:

$$\varrho_L \frac{d\mathfrak{v}_L}{dt} \approx \varrho_i \frac{d\mathfrak{v}_i}{dt} + \varrho_e \frac{d\mathfrak{v}_e}{dt}; \quad \varrho \frac{d\mathfrak{V}}{dt} \approx \varrho_0 \frac{d\mathfrak{v}_0}{dt} + \varrho_L \frac{d\mathfrak{v}_L}{dt}. \tag{49.10}$$

Die Druckglieder

$$-L_{e\mathfrak{j}} \frac{1}{\varrho_e} \operatorname{grad} p_e - L_{i\mathfrak{j}} \frac{1}{\varrho_i} \operatorname{grad} p_i + (L_{e\mathfrak{j}} + L_{i\mathfrak{j}}) \frac{1}{\varrho_0} \operatorname{grad} p_0 \tag{49.11}$$

fassen wir mit Hilfe von

$$p_L = p_i + p_e \quad \text{und} \quad p = p_0 + p_L \tag{49.12}$$

zusammen zu

$$-\left(\frac{L_{e\mathfrak{j}}}{2\varrho_e} + \frac{L_{i\mathfrak{j}}}{2\varrho_i} + \frac{L_{e\mathfrak{j}} + L_{i\mathfrak{j}}}{\varrho_0}\right) \operatorname{grad} p_L + (L_{e\mathfrak{j}} + L_{i\mathfrak{j}}) \frac{1}{\varrho_0} \operatorname{grad} p. \tag{49.13}$$

Da der Druckgradient und die Schwerpunktsbeschleunigung eine Folge der äußeren Kräfte sind, können wir schreiben

$$(L_{e\mathfrak{j}} + L_{i\mathfrak{j}}) \frac{1}{\varrho_0} \left(\varrho \frac{d\mathfrak{V}}{dt} + \operatorname{grad} p\right) = (L_{e\mathfrak{j}} + L_{i\mathfrak{j}}) \frac{1}{\varrho_0} \left(\frac{1}{c} [\mathfrak{j}\,\mathfrak{H}] + \varrho\, \mathfrak{g}\right). \tag{49.14}$$

Alle Temperaturglieder ergeben schließlich

$$[L_{e\mathfrak{j}} (Q_e^* - h_e + h_0) + L_{i\mathfrak{j}} (Q_i^* - h_i + h_0)] \frac{\operatorname{grad} T}{T}. \tag{49.15}$$

Damit wird die Stromdichte

$$\left.\begin{aligned} \mathfrak{j} = e\left(\frac{L_{i\mathfrak{j}}}{m_i} - \frac{L_{e\mathfrak{j}}}{m_e}\right)\left[\mathfrak{E} + \frac{1}{c}[\mathfrak{v}_L \mathfrak{H}] - \frac{m_e m_i}{e^2} \frac{d}{dt}\left(\frac{\mathfrak{j}}{\varrho_L}\right)\right] + {}\\ + \left[\left(L_{i\mathfrak{j}} \frac{m_e}{m_i} + L_{e\mathfrak{j}} \frac{m_i}{m_e}\right) \frac{1}{\varrho_L} + (L_{i\mathfrak{j}} + L_{e\mathfrak{j}}) \frac{1}{\varrho_0}\right] \frac{1}{c} [\mathfrak{j}\,\mathfrak{H}] - {}\\ - (L_{i\mathfrak{j}} + L_{e\mathfrak{j}}) \frac{\varrho}{\varrho_0} \left(\frac{d\mathfrak{v}_L}{dt} - \mathfrak{g}\right) - \left(\frac{L_{i\mathfrak{j}}}{2\varrho_i} + \frac{L_{e\mathfrak{j}}}{2\varrho_e} + \frac{L_{i\mathfrak{j}} + L_{e\mathfrak{j}}}{\varrho_0}\right) \operatorname{grad} p_L - {}\\ - [L_{e\mathfrak{j}} (Q_e^* - h_e + h_0) + L_{i\mathfrak{j}} (Q_i^* - h_i + h_0)] \frac{\operatorname{grad} T}{T}. \end{aligned}\right\} \tag{49.16}$$

50. Der Strom der Ladungsträger als Ganzes. Die Gleichung für $\mathfrak{J}_L$ hat gemäß Gl. (49.3b) ganz dieselbe Form wie die $\mathfrak{j}$-Gleichung (49.16), nur ist der Index $\mathfrak{j}$ an den L-Koeffizienten durch den Index $\mathfrak{J}$ zu ersetzen. Sie wird aber handlicher, wenn man das erste Glied mit der Feldstärke usw. mit Hilfe der $\mathfrak{j}$-Gleichung eliminiert, so daß man dann leicht den $\mathfrak{J}_L$-Strom *in* Richtung von $\mathfrak{j}$, wo $\mathfrak{j}_{\parallel} = \mathfrak{j}$ ist, und *senkrecht* zu $\mathfrak{j}$, wo natürlich $\mathfrak{j}_{\perp} = 0$ ist, betrachten kann. Wir lösen also Gl. (49.16) nach $\left[\mathfrak{E} + \frac{1}{c}[\mathfrak{v}_L \mathfrak{H}] - \frac{m_e m_i}{e^2} \frac{d}{dt}\left(\frac{\mathfrak{j}}{\varrho_L}\right)\right]$ auf und setzen dies in die $\mathfrak{J}_L$-Gleichung ein. Nach einigen Rechnungen ergibt sich

$$\left.\begin{aligned} \mathfrak{J}_L = -\frac{\varrho_0 \varrho_L}{\varrho} b\,\mathfrak{j} + \frac{\varrho_0 \varrho_L}{\varrho} \cdot \frac{1}{a} \left\{\frac{1}{c}[\mathfrak{j}\,\mathfrak{H}] + \varrho_L \left(\mathfrak{g} - \frac{d\mathfrak{v}_L}{dt}\right) - \operatorname{grad} p_L - {}\right.\\ \left. - \frac{\varrho_L \varrho_0}{\varrho} \left[\frac{m_e}{m_0} (Q_e^* - h_e + h_0) + \frac{m_i}{m_0} (Q_i^* - h_i + h_0)\right] \frac{\operatorname{grad} T}{T}\right\}, \end{aligned}\right\} \tag{50.1}$$

wobei

$$\left.\begin{aligned} b = -\frac{\varrho}{\varrho_0 \varrho_L \sigma} (L_{i\mathfrak{j}} + L_{e\mathfrak{j}}); \quad \sigma = e\left(\frac{L_{i\mathfrak{j}}}{m_i} - \frac{L_{e\mathfrak{j}}}{m_e}\right); \\ \frac{1}{a} = \left(\frac{\varrho}{\varrho_0 \varrho_L}\right)^2 \frac{(m_e + m_i)^2}{m_e \cdot m_i} \left(L_{12} - \frac{L_{e\mathfrak{j}} L_{i\mathfrak{j}}}{\sigma}\right) \end{aligned}\right\} \tag{50.2}$$

ist. Mit denselben Abkürzungen schreibt sich

$$\left.\begin{aligned} \mathfrak{j} = \sigma\Big\{&\mathfrak{E} + \frac{1}{c}[\mathfrak{v}_L \mathfrak{H}] - \frac{m_e m_i}{e^2}\frac{d}{dt}\left(\frac{\mathfrak{j}}{\varrho_L}\right) - \left[b + \frac{m_i - m_e}{e\,\varrho_L}\right]\frac{1}{c}[\mathfrak{j}\,\mathfrak{H}] + \\ &+ b\,\varrho_L\left(\frac{d\mathfrak{v}_L}{dt} - \mathfrak{g}\right) + \left[b + \frac{m_i - m_e}{2e\,\varrho_L}\right]\operatorname{grad} p_L + \\ &+ \frac{1}{m_e + m_i}\Big[m_e\left(b\,\frac{\varrho_0\,\varrho_L}{\varrho} + \frac{m_i}{e}\right)(Q_e^* - h_e + h_0) + \\ &+ m_i\left(b\,\frac{\varrho_0\,\varrho_L}{\varrho} - \frac{m_e}{e}\right)(Q_i^* - h_i + h_0)\Big]\frac{\operatorname{grad} T}{T}\Big\}. \end{aligned}\right\} \tag{50.3}$$

Der Vollständigkeit halber schreiben wir noch die Energiestromgleichung hinzu:

$$\mathfrak{W} = (Q_i^* - Q_e^*)\,\frac{m_e}{e}\,\mathfrak{j} + (m_e Q_e^* + m_i Q_i^*)\,\frac{1}{m_0}\,\mathfrak{J}_L - \varkappa_{\mathfrak{J}=0}\cdot\operatorname{grad} T. \tag{50.4}$$

51. Bestimmung der Koeffizienten. Nun sind noch die fünf Koeffizienten σ, a, b, Q_e^* und Q_i^* zu bestimmen. Dazu gehen wir, dem Verfahren von SCHLÜTER [*189*] folgend, auf unsere Einzelkraftgleichungen (48.7) zurück, multiplizieren die erste mit ε_{23} und die zweite mit ε_{13} und subtrahieren die zweite von der ersten. Die Reibungsglieder lassen sich dann in die Stromdichte $\mathfrak{j} = e n_L(\mathfrak{v}_i - \mathfrak{v}_e)$ umwandeln, wobei rechts vor der als äußere Kraft eingeführten Feldstärke die elektrische Leitfähigkeit

$$\sigma = e^2\,\frac{n_L(\varepsilon_{i0} + \varepsilon_{e0})}{n_L(\varepsilon_{i0}\,\varepsilon_{ie} + \varepsilon_{e0}\,\varepsilon_{ei}) + n_0\,\varepsilon_{i0}\,\varepsilon_{e0}} \tag{51.1}$$

erscheint, die so auf die Reibungskoeffizienten zurückgeführt ist.

Nimmt man in dieser Rechnung noch die weiteren Kräfte außer der Thermokraft hinzu, dann ergibt die Ausrechnung

$$b = \frac{\varrho_e\,\varepsilon_{i0} - \varrho_i\,\varepsilon_{e0}}{e\,\varrho_L\,n_L(\varepsilon_{i0} + \varepsilon_{e0})} \tag{51.2}$$

in Übereinstimmung mit SCHLÜTER. In dessen $\mathfrak{j}$-Gleichung sind allerdings noch kleine Rechenfehler zu beheben.

Bei den Thermokräften verwenden wir für α_{ei} den von L. SPITZER [*212*], [*213*] berechneten Wert $\alpha_{ei} = 3\,\frac{\gamma_T}{\gamma_E}$, während für die übrigen α-Faktoren die theoretische Formel (siehe z.B. WALDMANN [*233*])

$$\alpha_{kl} = \frac{(m_k - m_l)(n_k + n_l)}{2(\varrho_k + \varrho_l)}, \tag{51.3}$$

die annäherungsweise für starre Kugeln gilt und für den Fall des LORENTZ-Gases von WALDMANN durch strenge Rechnungen bestätigt wurde, eingesetzt wird.

Die Ausrechnung liefert unter konsequenter Vernachlässigung von m_e gegen m_i und m_0

$$Q_e^* - h_e + h_0 = \frac{kT}{2m_e}(\alpha_{ei} - 1), \tag{51.4}$$

$$Q_i^* - h_i + h_0 = -\frac{kT}{2m_i}\left(\alpha_{ei} + \frac{\varrho_i}{\varrho_e + \varrho_0}\right). \tag{51.5}$$

Die weiteren Koeffizienten finden wir sehr einfach, wenn wir $\mathfrak{j} = 0$, d.h. $\mathfrak{v}_i = \mathfrak{v}_e$ setzen. Dann schrumpft das System der Einzelgleichungen zu einem Zweiersystem mit neuen Reibungsfaktoren und neuen Kraftkombinationen zusammen, wenn man die Summe der ersten beiden Gleichungen als neue Kraftgleichung für die Ladungsträger auffaßt. Es folgen dann die beiden Gleichungen

$$\left.\begin{aligned}\varrho_L\frac{d\mathfrak{v}_L}{dt}+n_L n_0\varepsilon_{L0}(\mathfrak{v}_L-\mathfrak{v}_0)&=\mathfrak{F}_e\varrho_e-\operatorname{grad}p_e-\varrho_e\left(\frac{\varrho_i\alpha_{ei}}{m_e\varrho_i+m_i\varrho_e}+\frac{\varrho_0\alpha_{e0}}{m_e\varrho_0+m_0\varrho_e}\right)k\operatorname{grad}T+\\&\quad+\mathfrak{F}_i\varrho_i-\operatorname{grad}p_i-\varrho_i\left(\frac{\varrho_e\alpha_{ie}}{m_i\varrho_e+m_e\varrho_i}+\frac{\varrho_0\alpha_{i0}}{m_i\varrho_0+m_0\varrho_i}\right)k\operatorname{grad}T,\\\varrho_0\frac{d\mathfrak{v}_0}{dt}+n_0 n_L\varepsilon_{0L}(\mathfrak{v}_0-\mathfrak{v}_L)&=\mathfrak{F}_0\varrho_0-\operatorname{grad}p_0-\varrho_0\left(\frac{\varrho_e\alpha_{0e}}{m_0\varrho_e+m_e\varrho_0}+\frac{\varrho_i\alpha_{0i}}{m_0\varrho_i+m_i\varrho_0}\right)k\operatorname{grad}T\end{aligned}\right\}\quad(51.6)$$

mit

$$\varepsilon_{0L}=\varepsilon_{L0}=\varepsilon_{e0}+\varepsilon_{i0}\qquad(51.7)$$

und

$$n_L=n_e=n_i.$$

An die Stelle von $\mathfrak{F}_1$ usw. im Zweiersystem (40.1) tritt also hier

$$\mathfrak{F}_L=\frac{\mathfrak{F}_e\varrho_e+\mathfrak{F}_i\varrho_i}{\varrho_e+\varrho_i}\quad\text{usw.}\qquad(51.8)$$

Dieselbe Umformung, die beim Zweiersystem von den Reibungsgliedern auf den Massenstrom $\mathfrak{J}_1$ führte, ergibt hier den gemeinsamen Strom der Ladungsträger gegenüber dem Neutralgas ohne $\mathfrak{j}$:

$$\left.\begin{aligned}\mathfrak{J}_L&=\varrho_L(\mathfrak{v}_L-\mathfrak{V})=\frac{\varrho_L\varrho_0}{\varrho}(\mathfrak{v}_L-\mathfrak{v}_0)\\&=\left(\frac{\varrho_L\varrho_0}{\varrho}\right)^2\frac{1}{n_L n_0\varepsilon_{L0}}\left(\frac{\mathfrak{F}_e\varrho_e+\mathfrak{F}_i\varrho_i}{\varrho_L}-\mathfrak{F}_0-\frac{d\mathfrak{v}_L}{dt}+\right.\\&\quad\left.+\frac{d\mathfrak{v}_0}{dt}-\frac{1}{\varrho_L}\operatorname{grad}p_L+\frac{1}{\varrho_0}\operatorname{grad}p_0+\frac{\varrho}{\varrho_0\varrho_e}\frac{k}{2m_0}\operatorname{grad}T\right)\end{aligned}\right\}\quad(51.9)$$

oder mit Gl. (43.13)

$$\left.\begin{aligned}\mathfrak{J}_L=\left(\frac{\varrho_L\varrho_0}{\varrho}\right)^2\frac{1}{n_L n_0\varepsilon_{L0}}&\left(\frac{\mathfrak{F}_e\varrho_e+\mathfrak{F}_i\varrho_i}{\varrho_L}-\mathfrak{F}_0-\frac{d\mathfrak{v}_L}{dt}+\frac{d\mathfrak{v}_0}{dt}-\right.\\&\left.-(v_L-v_0)\operatorname{grad}p-\frac{kT}{\overline{m}c_L c_0}\operatorname{grad}c_L+\frac{\varrho}{\varrho_0+\varrho_e}\frac{k}{2m_0}\operatorname{grad}T\right)\end{aligned}\right\}\quad(51.10)$$

mit

$$\overline{m}=m_L c_0+m_0 c_L;\quad m_L=\frac{1}{2}\cdot(m_e+m_i)=\frac{m_0}{2}.$$

Bei den Thermokräften muß α_{ei} herausfallen, weil erstere nicht den Schwerpunkt der Ladungsträger bewegen können. Gl. (51.10) stimmt mit Gl. (43.10) des Zweiersystems überein, wenn für den ambipolaren Diffusionskoeffizienten, mit dem die Ladungsträger und Neutralteilchen ineinander diffundieren,

$$D_{\text{amb}}=\frac{2kT}{n\,\varepsilon_{L0}}\qquad(51.11)$$

und für den entsprechenden Thermodiffusionsfaktor

$$\alpha_{\text{amb}}=-\frac{1}{4}\frac{\varrho+\varrho_L}{\varrho_0+\varrho_e}\qquad(51.12)$$

gesetzt wird.

Um nun von Gl. (51.9) auf Gl. (50.1) ohne das $\mathfrak{j}$-Glied zu kommen, ziehen wir noch einen Faktor $\varrho_L\varrho_0/\varrho$ in die Klammer von Gl. (51.9) hinein und verwenden die Kraftgleichung (45.9). Das Ergebnis liefert durch Koeffizientenvergleich

$$a=n_L n_0\varepsilon_{L0}\qquad(51.13)$$

Den Kontaktwärmeleitwert $\varkappa_{\mathfrak{J}=0}$ werden wir zusammen mit den Reibungskoeffizienten ε später mit Hilfe der kinetischen Gastheorie auf die Stoßquerschnitte zurückführen.

Mit den thermischen Koeffizienten und unter Vernachlässigung von m_e gegen m_i lauten die Gleichungen, wenn wir die zeitabhängigen Glieder noch auf die linke Seite nehmen und die Enthalpien

$$h_e = \frac{5}{2}\frac{kT}{m_e}, \tag{51.14}$$

$$h_i = \frac{5}{2}\frac{kT}{m_i} + \frac{E_I}{m_i}, \tag{51.15}$$

$$h_0 = \frac{5}{2}\frac{kT}{m_0}, \tag{51.16}$$

ohne Berücksichtigung der Anregungsenergie verwenden:

$$\left.\begin{aligned} &-\sigma b \varrho_L \frac{d\mathfrak{v}_L}{dt} + \sigma\frac{m_e m_i}{e^2}\frac{d}{dt}\left(\frac{\mathfrak{j}}{\varrho_L}\right) + \mathfrak{j} = \sigma\Big\{\mathfrak{E} + \frac{1}{c}[\mathfrak{v}_L\mathfrak{H}] - \left(b + \frac{m_i}{e\varrho_L}\right)\frac{1}{c}[\mathfrak{j}\mathfrak{H}] - \\ &-\varrho_L b\,\mathfrak{g} + \left(b + \frac{m_i}{2e\varrho_L}\right)\operatorname{grad} p_L + \left[\alpha_{ei} - \frac{e\,n_0\varrho_L}{\varrho_0+\varrho_e}\left(b + \frac{m_i}{e\varrho_L}\right)\right]\frac{k}{2e}\operatorname{grad} T\Big\}, \end{aligned}\right\} \tag{51.17}$$

$$\left.\begin{aligned} &\frac{1}{a}\frac{\varrho_0\varrho_L}{\varrho}\varrho_L\frac{d\mathfrak{v}_L}{dt} + \mathfrak{J}_L = -\frac{\varrho_0\varrho_L}{\varrho} b\,\mathfrak{j} + \\ &+\frac{\varrho_0\varrho_L}{\varrho}\frac{1}{a}\left\{\varrho_L\mathfrak{g} + \frac{1}{c}[\mathfrak{j}\mathfrak{H}] - \operatorname{grad} p_L + \frac{\varrho_L\varrho_0}{\varrho_0+\varrho_e}\frac{k}{2m_0}\operatorname{grad} T\right\} \end{aligned}\right\} \tag{51.18}$$

$$\mathfrak{W} = -\frac{kT}{2e}\left(5 + \alpha_{ei} + \frac{\varrho_0}{\varrho_0+\varrho_e}\right)\mathfrak{j} + \frac{kT}{2m_0}\left(5 + \frac{2E_I}{kT} - \frac{\varrho}{\varrho_0+\varrho_e}\right)\mathfrak{J}_L - \varkappa_{\mathfrak{J}=0}\operatorname{grad} T. \tag{51.19}$$

Mit Hilfe der Thermodynamik irreversibler Prozesse ist es also gelungen, über die Plasmadynamik von SCHLÜTER hinaus nicht nur die Thermodiffusion in der $\mathfrak{j}$- und $\mathfrak{J}_L$-Gleichung zu berücksichtigen, sondern vor allem die vollständige Energiestromgleichung hinzuzufügen, so daß nunmehr eine einheitliche Plasmadynamik und -thermodynamik vorliegt.

VI. Diskussion der Plasmagleichungen.

52. Elektrische Ströme. Zur Diskussion der Gleichungen beginnen wir mit beliebig verteilten Kraftfeldern und Bewegungen. In der $\mathfrak{j}$-Gleichung bewirken sie einen elektrischen Strom, der nicht nur von der lokalen Feldstärke $\mathfrak{E}$, sondern z. B. auch von der Induktionsspannung $\frac{1}{c}[\mathfrak{v}_L\mathfrak{H}]$ angetrieben wird. Die Kraft $\frac{1}{c}[\mathfrak{j}\mathfrak{H}]$ schiebt die Stromfäden grundsätzlich zur Seite, ebenso wie der grad p, wenn er durch das Magnetfeld erzeugt wird. Der Temperaturgradient bewegt die Ladungsträger durch Thermodiffusion gegeneinander, veranlaßt aber gleichzeitig über den mit grad p_L zusammenhängenden Gradienten des Ionisationsgrades

$$\operatorname{grad} x = \left(\frac{\partial x}{\partial T}\right)_p \operatorname{grad} T \tag{52.1}$$

eine verschieden starke Diffusion der Ladungsträger, was wiederum einem Strom entspricht. Der anfängliche Strom hat als instationärer Quellen und Senken, wodurch gemäß der Kontinuitätsgleichung

$$\operatorname{div} \mathfrak{j} = -\frac{\partial \varrho_{\mathrm{el}}}{\partial t} \tag{52.2}$$

Raumladungen auf- oder abgebaut werden, die dem ursprünglichen Strom durch ihr Feld

$$\operatorname{div} \mathfrak{E} = 4\pi \varrho_{\mathrm{el}} \tag{52.3}$$

gemäß der $\mathfrak{j}$-Gleichung entgegenwirken. Der Endzustand ist schließlich dann erreicht, wenn sich der Strom ein solches Raumladungsfeld aufgebaut hat, daß er stationär geworden ist, daß also $\operatorname{div} \mathfrak{j} = 0$ gilt. Die eingeprägten Kräfte tragen im stationären Zustand zum Teil wirklich zum Stromtransport bei, zum anderen Teil werden sie durch Raumladungsfelder kompensiert. Die Aufteilung geschieht eben gerade so, daß keine Stromstauungen mehr erfolgen. Bezeichnet man mit $\mathfrak{E}^a$ das von den Elektroden stammende angelegte Feld, dessen Divergenz verschwindet, ferner mit $\mathfrak{E}^r$ das durch Raumladungen gebildete Feld und faßt man die eingeprägten Kräfte mit $\mathfrak{E}^e$ zusammen, dann ist im stationären Zustand

$$\operatorname{div} \mathfrak{j} = \operatorname{div} \sigma (\mathfrak{E}^a + \mathfrak{E}^r + \mathfrak{E}^e) = \sigma \operatorname{div} (\mathfrak{E}^r + \mathfrak{E}^e) + (\mathfrak{E}^a + \mathfrak{E}^r + \mathfrak{E}^e) \operatorname{grad} \sigma = 0 \tag{52.4}$$

oder mit $\operatorname{div} \mathfrak{E}^r = 4\pi \varrho_{\mathrm{el}}$ und $\mathfrak{j} = \sigma (\mathfrak{E}^a + \mathfrak{E}^r + \mathfrak{E}^e)$

$$4\pi \varrho_{\mathrm{el}} = \mathfrak{j} \operatorname{grad} \frac{1}{\sigma} - \operatorname{div} \mathfrak{E}^e. \tag{52.5}$$

Da σ eine Temperaturfunktion ist, haben grad $1/\sigma$ und grad T entgegengesetzt gleiche Richtung. Herrscht also in $\mathfrak{j}$-Richtung ein Temperaturgradient, so ist das zwar im allgemeinen ein Anlaß zur Raumladungsbildung, wie z.B. in der Achse eines nicht zylindrischen Bogens, aber nur zur Glättung des Potentialgebirges, denn der Hauptteil dieses Gliedes kommt durch die Divergenz der eingeprägten Kräfte, in unserem Beispiel des grad T, nicht zur Auswirkung. In den Außenbezirken dagegen laufen die $\mathfrak{j}$-Linien etwa auf den Isothermen, das erste Glied rechts verschwindet praktisch, aber die eingeprägten Kräfte, etwa der Temperaturgradient, haben eine nennenswerte Divergenz. Sie wird durch gleichstarke Raumladungen kompensiert, so daß oder weil per definitionem kein Strom senkrecht zu $\mathfrak{j}$ fließt.

Aus den vielen Aussagen, die aus der Stromgleichung zu entnehmen sind, wollen wir zwei herausgreifen:

Im stationären Zustand ohne makroskopische Strömung kann erstens ein eigenes oder fremdes Magnetfeld, da es immer senkrecht auf den Strom wirkt, diesen weder stärken noch schwächen, auch nicht über den Druckgradienten, der ja ebenfalls senkrecht auf $\mathfrak{j}$ steht. Hierauf hat besonders SCHLÜTER [*189*] aufmerksam gemacht. Der Druckgradient wird durch eine entsprechende Raumladung kompensiert.

Zweitens, herrscht in Richtung von $\mathfrak{j}$ ein beachtlicher Temperaturgradient, dann kann er den Strom so stark antreiben, daß dort der Feldstärkebedarf sehr klein wird. Prinzipiell könnte sogar die Feldstärke ihre Richtung umkehren, so daß dieser Bereich als galvanisches Element wirkt. Dies kann in der Nähe von solchen Bogenanoden von Bedeutung sein, die keinen Brennfleck bilden. Auch in Kathodennähe vermuten GERTHSEN und SCHULZ [*66*] einen Antrieb der Elektronen durch einen temperaturbedingten Konzentrationsgradienten.

53. Massenströme. Wir haben bisher nur den Strom ins Auge gefaßt, müssen aber beachten, daß die eingeprägten Kräfte während des Einstellvorganges auch Massendiffusionsströme auslösen, wovon die $\mathfrak{J}_L - (= -\mathfrak{J}_0)$ Gl. (51.16) Zeugnis ablegt. Auch diese Ströme haben Quellen und Senken und können zur lokalen Anreicherung oder Verarmung einer Teilchenart nach der Bilanzgleichung für die Komponenten

$$\frac{\partial \varrho_k}{\partial t} - \Gamma_k = -\operatorname{div} \mathfrak{J}_k \tag{53.1}$$

Anlaß geben, die den erzeugenden Massenstrom über grad p_k verlangsamen. Hier ist der stationäre Zustand wieder bei div $\mathfrak{J}_k = 0$ erreicht, wenn nicht eine Reaktion Γ_k die Zu- oder Abfuhr von Materie umwandelt. Unter diesen Umständen lautet dann wie früher die Stationaritätsbedingung

$$\operatorname{div} \mathfrak{J}_k = \Gamma_k = -\Gamma_l = -\operatorname{div} \mathfrak{J}_l. \tag{53.2}$$

Im Gegensatz zum reinen Plasma, wo ein elektrischer Strom nur die Ladungsträger gegeneinander verschieben konnte ohne Bewegung des Schwerpunktes, erzeugt der Strom bei Anwesenheit des Neutralgases auch eine Bewegung der Ladungsträger als Ganzes relativ zum Neutralgas, was auf die unterschiedliche Reibung der beiden Ladungsträgerkomponenten am Neutralgas zurückzuführen ist. Daß die Leitfähigkeit durch die Anwesenheit des Neutralgases beeinträchtigt ist, versteht sich von selbst.

54. Energieströme. Ähnliche Überlegungen wie oben zur Ermittlung des stationären Zustandes müssen wir gleichzeitig für den Energietransport anstellen. Wenn in dem anfangs beliebigen Gebirge der Kräfte Energieströme fließen, dann führt dies nach der Energiegleichung zu Temperaturänderungen, bis die Divergenz des Energiestromes als Ganzes überall von der lokalen Energieerzeugung gespeist wird oder verschwindet

$$\varrho \frac{du}{dt} = -\operatorname{div} \mathfrak{W} + \sum_k \mathfrak{F}_k \mathfrak{J}_k = 0. \tag{54.1}$$

Wenn der Wärmestrom nur von grad T bewirkt wird und als äußere Kraft nur ein elektrisches Feld wirkt, dann entsteht hieraus die in der Bogenphysik bekannte ELENBAAS-HELLERsche Gleichung für den stationären Zustand

$$\operatorname{div} \tilde{\varkappa} \operatorname{grad} T + \mathfrak{E}\mathfrak{j} = 0. \tag{54.2}$$

Wenn eine Gesamtströmung $\mathfrak{V}$ existiert, dann läßt sich das zusätzliche Glied $-p$ div $\mathfrak{V}$ mit $\varrho\, du/dt$ vereinigen

$$\varrho \frac{du}{dt} + p \operatorname{div} \mathfrak{V} = \varrho \frac{du}{dt} - \frac{p}{\varrho}\frac{d\varrho}{dt} = \varrho \frac{dh}{dt} - \frac{dp}{dt} = -\operatorname{div} \mathfrak{W} + \sum_k \mathfrak{F}_k \mathfrak{J}_k. \tag{54.3}$$

Ist die Strömung stationär, dann kann man von den zeitabhängigen Gliedern einen entsprechenden Term mit Hilfe der substantiellen Differentiation abspalten

$$\varrho \frac{\partial h}{\partial t} - \frac{\partial p}{\partial t} = -\varrho \mathfrak{V} \operatorname{grad} h + \mathfrak{V} \operatorname{grad} p - \operatorname{div} \mathfrak{W} + \sum_k \mathfrak{F}_k \mathfrak{J}_k = 0. \tag{54.4}$$

Wegen der Abhängigkeit der Enthalpie von T und p läßt sich dafür auch schreiben

$$-\varrho c_p \mathfrak{V} \operatorname{grad} T + \left[1 - \varrho \left(\frac{\partial h}{\partial p}\right)_T\right] \mathfrak{V} \operatorname{grad} p - \operatorname{div} \mathfrak{W} + \sum_k \mathfrak{F}_k \mathfrak{J}_k = 0 \tag{54.5}$$

mit

$$c_p = \left(\frac{\partial h}{\partial T}\right)_p.$$

Da $\mathfrak{W}$ zum Teil von $\mathfrak{j}$ getragen wird, können wir auch hier noch die zugehörige Stationaritätsbedingung $\operatorname{div}\mathfrak{j}=0$ einführen. Fassen wir diesen Anteil in $\mathfrak{W}_\mathfrak{j}$ einschließlich des in $\mathfrak{J}_L$ steckenden $\mathfrak{j}$-Gliedes zusammen, dann ist

$$\mathfrak{W}_\mathfrak{j} = Q_\mathfrak{j}\,\mathfrak{j} = -\frac{kT}{2}\left[\frac{1}{e}\left(5+\alpha_{ei}-\frac{\varrho_0}{\varrho_0+\varrho_e}\right)-\frac{\varrho_0\varrho_L}{\varrho}\,\frac{b}{m_0}\left(5+\frac{2E_I}{kT}-\frac{\varrho}{\varrho_0+\varrho_e}\right)\right]\mathfrak{j} \tag{54.6}$$

und dessen Divergenz

$$\operatorname{div}\mathfrak{W}_\mathfrak{j} = Q_\mathfrak{j}\operatorname{div}\mathfrak{j}+\frac{\partial Q_j}{\partial T}\,\mathfrak{j}\operatorname{grad}T+\frac{\partial Q_j}{\partial p}\operatorname{grad}p\,. \tag{54.7}$$

Hier verschwindet das erste Glied wegen $\operatorname{div}\mathfrak{j}=0$. Bei Zylindersymmetrie verschwinden auch die beiden letzten Glieder, weil $\mathfrak{j}$ mit den Isothermen und Isobaren zusammenfällt, dagegen nicht mehr bei allgemeiner Rotationssymmetrie, wo auch entlang der Achse ein $\operatorname{grad}T$ und ein $\operatorname{grad}p$ existiert.

Der Rest des Energiestromes $\mathfrak{W}_{\mathfrak{j}=0}$, soweit er also nicht vom elektrischen Strom getragen wird, setzt sich mit

$$Q_{\mathfrak{J}_L}=\frac{kT}{2m_0}\left(5+\frac{2E_I}{kT}-\frac{\varrho}{\varrho_0+\varrho_e}\right) \tag{54.8}$$

zusammen aus

$$\left.\begin{aligned}\mathfrak{W}_{\mathfrak{j}=0} = Q_{\mathfrak{J}_L}\frac{\varrho_0\varrho_L}{\varrho}\cdot\frac{1}{a}\left(\varrho_L\mathfrak{g}+\frac{1}{c}[\mathfrak{j}\,\mathfrak{H}]-\operatorname{grad}p_L+\frac{\varrho_L\varrho_0}{\varrho_0+\varrho_e}\,\frac{k}{2m_0}\operatorname{grad}T\right)-\\ -\varkappa_{\mathfrak{J}=0}\operatorname{grad}T\,.\end{aligned}\right\} \tag{54.9}$$

Um eine bessere Vergleichsmöglichkeit mit den $\mathfrak{W}$-Gleichungen des Zweiersystems zu gewinnen, wählen wir an Stelle von Gl. (54.9) eine Schreibweise, die man am einfachsten auf Gl. (51.10) aufbaut:

$$\left.\begin{aligned}\mathfrak{W}_{\mathfrak{j}=0} &= \left(\frac{5kT}{2m_0}+\frac{E_I}{m_0}+\frac{\alpha_{\mathrm{amb}}kT}{\overline{m}}\right)\varrho D_{\mathrm{amb}}\left\{c_L c_0\frac{1+c_L}{p_L}\,\frac{1}{c}[\mathfrak{j}\,\mathfrak{H}]-\right.\\ &\quad -\left(c_L c_0\frac{1+c_L}{2p}+\frac{\partial c_L}{\partial p}\right)\operatorname{grad}p-\\ &\quad \left.-\left(c_L c_0\alpha_{\mathrm{amb}}/T+\frac{\partial c_L}{\partial T}\right)\operatorname{grad}T\right\}-\varkappa_{\mathfrak{J}=0}\operatorname{grad}T\\ \text{mit}\quad \overline{m} &= \frac{m_0}{2}(1+c_L)\,.\end{aligned}\right\} \tag{54.10}$$

Der Beitrag des Magnetfeldes und des Druckgradienten liegt in praktisch vorkommenden Fällen wesentlich unter den übrigen Gliedern, sind aber zum Teil den folgenden Gliedern entgegen gerichtet. Der Faktor vor $\operatorname{grad}T$ ist der totale Wärmeleitwert·

$$\varkappa_{\mathfrak{j}=0}=\left(\frac{5}{2}\,\frac{kT}{m_0}+\frac{E_I}{m_0}+\frac{\alpha_{\mathrm{amb}}kT}{\overline{m}}\right)\varrho D_{\mathrm{amb}}\left(c_L c_0\alpha_{\mathrm{amb}}/T+\frac{\partial c_L}{\partial T}\right)+\varkappa_{\mathfrak{J}=0}\,, \tag{54.11}$$

wenn kein elektrischer Strom in Richtung $\operatorname{grad}T$ fließt. Darin ist der Term mit $\partial c_L/\partial T$ charakteristisch für die ambipolare Diffusion, erzeugt durch den Konzentrationsabfall mit der Temperatur, und der Term $c_L c_0\alpha_{\mathrm{amb}}/T$ für die viel kleinere ambipolare Thermodiffusion. Die erste Klammer enthält die vom Massenstrom getragene Überführungsenergie, im wesentlichen die Ionisationsenergie. Beide Diffusionsvorgänge sind, wie wir noch sehen werden, im Gegensatz zu den Dissoziationsvorgängen *nicht* von Bedeutung (s. $\varkappa_I$ in Fig. 41), weil die Diffusionsgeschwindigkeit der Ladungsträger durch die großen Querschnitte der Neutralteilchen gegenüber Elektronen und auch Ionen stark gehemmt ist. Dagegen ist in $\varkappa_{\mathfrak{J}=0}$ die Wärmeleitfähigkeit der Elektronen enthalten, die bei größeren Ionisationsgraden sehr groß werden kann.

55. Magnetische Kräfte und Energiebilanz. Wir werfen nun noch einen Blick auf die Erhaltungsgleichung für die Energie als Ganzes mit Berücksichtigung der Bewegung in einem Magnetfeld. Dann wird die Änderung der inneren Energie des Plasmas beschrieben durch

$$\varrho \frac{du}{dt} = -p \operatorname{div} \mathfrak{V} - \operatorname{div} \mathfrak{W} + \mathfrak{j}\left(\mathfrak{E} + \frac{1}{c}\,[\mathfrak{V}\,\mathfrak{H}]\right). \tag{55.1}$$

Danach kann die innere Energie und somit die Temperatur erhöht werden durch eine Kompression $-p \operatorname{div} \mathfrak{V}$,durch Energiezufuhr seitens des Wärmestromes, und durch JOULEsche Wärme, in der sich die erzeugende Feldstärke aus der angelegten $\mathfrak{E}$ und der induzierten $\frac{1}{c}\,[\mathfrak{V}\mathfrak{H}]$ zusammensetzt.

Zur Änderung der inneren Energie kommt noch die der kinetischen:

$$\frac{\varrho}{2}\frac{d\mathfrak{V}^2}{dt} = -\mathfrak{V} \operatorname{grad} p + \mathfrak{V}\,\frac{1}{c}\,[\mathfrak{j}\,\mathfrak{H}]. \tag{55.2}$$

Diese Arbeit wird vom Druckgradienten und vom Magnetfeld geleistet. Die gesamte Energie des Plasma ist nun

$$\varrho \frac{d}{dt}\left(\frac{\mathfrak{V}^2}{2} + u\right) = -\operatorname{div}(p\,\mathfrak{V} + \mathfrak{W}) + \mathfrak{j}\,\mathfrak{E}. \tag{55.3}$$

Die Arbeit der magnetischen Kräfte ist herausgefallen, weil sie nicht in der Lage sind, Arbeit zu leisten, sondern nur kinetische Energie durch Bremsung in Wärme umzuwandeln (vgl. die Bremsung und Erwärmung eines Cu-Pendels im Magnetfeld) oder durch Ablenkung beider Ladungsträgerarten nach derselben Seite die potentielle Energie der Ladungsträger zum Teil in makroskopische kinetische Energie umzusteuern. Im einzelnen laufen die Vorgänge etwa folgendermaßen ab: Bewegt sich das Plasma im Magnetfeld, dann wird nach Gl. (51.15) die Spannung $\frac{1}{c}\,[\mathfrak{V}_L\,\mathfrak{H}]$ induziert, was zwei verschiedene Folgen haben kann. Ist nämlich die räumliche Ausdehnung des Magnetfeldes groß gegenüber der des Plasmas, dann wird an dessen Rändern durch den anlaufenden Strom eine Raumladung aufgebaut, deren Feld das indzuzierte kompensiert $\left(\mathfrak{E}^r = -\frac{1}{c}\,[\mathfrak{V}_L\,\mathfrak{H}];\ \mathfrak{j} = 0\right)$ und die Bewegung bleibt unbehindert. (Beispiel: volle Drehzahl eines unbelasteten Elektromotors.) In sämtlichen Energiegleichungen verschwinden die elektromagnetischen Glieder. Ist aber die Ausdehnung des Magnetfeldes klein gegenüber der des Plasmas, dann können sich Wirbelströme der Größe $\mathfrak{j} = \frac{\sigma}{c}\,[\mathfrak{V}_L\,\mathfrak{H}]$ ausbilden, die nach der Kraftgleichung eine Verzögerung bewirken. Der Verlust an kinetischer Energie erscheint zunächst scheinbar als eine Art Gewinn an potentieller Energie des Magnetfeldes $\mathfrak{V}/c \cdot [\mathfrak{j}\,\mathfrak{H}]$. Derselbe Betrag mit dem umgekehrten Vorzeichen $(\mathfrak{A}\,[\mathfrak{B}\,\mathfrak{C}] = -\mathfrak{B}\,[\mathfrak{A}\,\mathfrak{C}])$ geht aber in die Gleichung für die Produktion innerer Energie ein. In diesem Fall hat also das Magnetfeld die Umwandlung kinetischer Energie in innere veranlaßt, ohne die Gesamtleistung verändert zu haben.

In einem anderen Fall möge ein vom Strom $\mathfrak{j} = \sigma\,\mathfrak{E}$ durchflossenes Plasma von einem Magnetfeld nach der Kraftgleichung beschleunigt werden. Dann wird diese Arbeit nach der kinetischen Energiegleichung vom Magnetfeld geleistet. Die Stromdichte sinkt dabei wegen der Gegeninduktion ab: $\mathfrak{j} = \sigma\left(\mathfrak{E} - \frac{1}{c}\,[\mathfrak{V}\,\mathfrak{H}]\right)$. Die Erhöhung der inneren Energie durch die JOULEsche Wärme $\mathfrak{j}\,\mathfrak{E}$ wird nun einmal dadurch verringert, daß $\mathfrak{j}$ kleiner geworden ist (für unser Problem eine

Nebenerscheinung), zum anderen dadurch, daß eine magnetische Gegenenergie $-\mathfrak{j}\,[\mathfrak{B}\mathfrak{H}]/c$ die JOULEsche Wärme herabsetzt. Das Magnetfeld war also der Mittler dafür, daß nicht mehr die ganze elektrische Energie in Wärme verwandelt, sondern ein Teil davon in kinetische umgesetzt wurde. Die Gesamtleistung wird durch das Magnetfeld wiederum nicht verändert. (Festgehaltener Motor wird losgelassen.)

56. Stationäre Wirbelströme. Es sei hier noch vermerkt, daß die stromerzeugenden Kräfte nicht immer ein Potential haben müssen, d.h. durch ein Gradientenfeld zu beschreiben sein müssen, sondern daß sie sehr wohl auch eine Rotation haben können, wie z.B. die magnetischen Kräfte. Zur Erreichung eines stationären Zustandes können daher nur die Potentialkräfte durch eine Raumladung kompensiert werden, deren Felder immer rotationsfrei sind: rot $(\mathfrak{E}^a + \mathfrak{E}^r) = 0$. Kräfte ohne Potential müssen durch stationäre Ströme ausgeglichen werden, die eine Rotation haben, deren Divergenz aber verschwindet. Dazu reicht außer beim zylindersymmetrischen Fall im allgemeinen nicht der große Hauptstromkreis aus, dessen Stromlinien durch ein abgegrenztes Gebiet divergenzfrei hindurchtreten, sondern man muß bei jeder nichtzylindersymmetrischen Entladung mit lokalen Wirbelströmen rechnen. Genau so können auch, ursächlich damit verknüpft oder nicht, Wirbel der Strömungen und der Energieströme stationär vorhanden sein.

VII. Berechnung der thermodynamischen Zustandsgrößen.

57. Zustandssummen. Zur Vollständigkeit der Plasmagleichungen gehört noch eine Anweisung für die Berechnung der thermodynamischen Größen, soweit das noch nicht geschehen ist, insbesondere der chemischen Gleichgewichte für Dissoziation und Ionisation. Dazu benötigen wir als wichtigste Größe die totale Zustandssumme für das ganze System Z_s, die sich aus der Zustandssumme Z für das Einzelteilchen gemäß

$$Z_s = \frac{Z^N}{N!} \tag{57.1}$$

zusammensetzt. Die Statistik liefert

$$Z = \sum_n g_n\, e^{-E_n/kT}, \tag{57.2}$$

worin g_n das statistische Gewicht und E_n die Energie des n-ten Teilchenzustandes ist. Wenn wir nun näherungsweise die Energie in die Summe von Translations-, Rotations-, Oszillations-, Elektronenanregungs- und Trennungsenergie aufteilen können, dann kann auch die Zustandssumme aus den entsprechenden Faktoren zum Produkt zusammengesetzt werden

$$Z = Z_{\mathrm{tr}}\, Z_{\mathrm{rot}}\, Z_{\mathrm{osz}}\, Z_e\, e^{-E_i/kT}. \tag{57.3}$$

Hierin ist E_i allgemein die Trennungsarbeit, soweit sie dem betrachteten Teilchen zuzuschreiben ist. Die Zustandssumme über die ungequantelte Translationsenergie kann für ein ideales Gas zu

$$Z_{\mathrm{tr}} = \frac{V}{h^3}\,(2\pi\, m\, k\, T)^{\frac{3}{2}} \tag{57.4}$$

berechnet werden, so daß

$$Z_s = \frac{e}{n\,h^3}\,(2\pi\, m\, k\, T)^{\frac{3}{2}}\, Z_{\mathrm{rot}}\, Z_{\mathrm{osz}}\, Z_e\, e^{-E_i/kT} \tag{57.5}$$

wird. Faßt man das Molekül näherungsweise als starren Rotator auf, dann ist die Rotationsenergie gegeben durch

$$E_{\mathrm{rot}} = \frac{h^2}{8\pi^2 \Theta} J(J+1), \tag{57.6}$$

worin Θ das Trägheitsmoment und J die Rotationsquantenzahl ist. Damit wird

$$Z_{\mathrm{rot}} = \frac{1}{2} \sum_J (2J+1)\, e^{-\frac{h^2 J(J+1)}{8\pi^2 \Theta k T}} \approx \frac{8\pi^2 \Theta k T}{2h^2}, \tag{57.7}$$

worin noch die Symmetriezahl 2 im Nenner für gleichatomige Moleküle angegliedert wurde. Will man der Tatsache Rechnung tragen, daß bei höheren Rotationen und Oszillationen der Atomabstand und damit das Trägheitsmoment vergrößert wird, dann muß man eine höhere Näherung anstreben (vgl. J. O. FAST [*46*]).

Die Energiezustände eines harmonischen Oszillators werden durch die Gleichung

$$E_{\mathrm{osz}} = h\,\nu\,(v + \tfrac{1}{2}) \tag{57.8}$$

gegeben, worin ν die Eigenfrequenz des Oszillators und v die Schwingungsquantenzahl ist. Hieraus ergibt sich die Zustandssumme zu

$$Z_{\mathrm{osz}} = \sum_v e^{-\frac{h\nu(v+\frac{1}{2})}{kT}} = \frac{1}{2\,\mathrm{Sin}\,(h\nu/2kT)}. \tag{57.9}$$

Auch hierin kann man die Änderung der Eigenfrequenz durch höhere Rotationen und Oszillationen und die Anharmonizität durch bessere Näherungen berücksichtigen.

Die Berechnung der Elektronenanregung geht für Moleküle und Atome in gleicher Weise nach der Formel

$$Z_e = \sum_n g_n\, e^{-E_n/kT}. \tag{57.10}$$

Diese Summe muß gliedweise aufaddiert werden, wobei die Energiewerte aus dem Termschema des fraglichen Teilchens zu entnehmen sind und das statistische Gewicht nach der Gleichung

$$g = 2J + 1 \tag{57.11}$$

aus dem Gesamtdrehimpuls J zu berechnen ist. Da die Summe sehr schnell konvergiert, genügt es oft, nur das statistische Gewicht des Grundzustandes oder einige weitere Glieder, je nach Höhe der unteren Terme, mitzunehmen. Höchstens jedoch darf die Summierung bis zu dem Term vorgenommen werden, der noch nicht durch die Wechselwirkung mit den Nachbarteilchen zerstört ist. Für Wasserstoff ist die Quantenzahl n^* des Grenzterms von D. E. INGLIS und E. TELLER [*92*] zu

$$7{,}5 \operatorname{Log} n^* = 23{,}26 - \operatorname{Log} 2 n_e \tag{57.12}$$

berechnet worden.

58. Die thermodynamischen Größen und das Massenwirkungsgesetz. Aus der Gesamtzustandssumme hatten wir bereits die innere Energie

$$U = k T^2 \frac{\partial}{\partial T} (\log Z_s), \tag{58.1}$$

die Entropie

$$S = k \log Z_s + \frac{U}{T}, \tag{58.2}$$

die Enthalpie

$$H = U + p\frac{V}{N} \tag{58.3}$$

und das GIBBSsche Potential

$$G = H - TS \tag{58.4}$$

für ein Teilchen im Mittel angegeben. Alle diese Größen beziehen wir nun auf die Masseneinheit der Sorte k durch Division mit m_k, führen die ideale Gasgleichung

$$p_k = \frac{\varrho_k \cdot kT}{m_k} \tag{58.5}$$

ein und teilen die Zustandssumme auf in

$$Z_s = \frac{e}{n_k} Z_{\mathrm{tr}} Z_k^{(i)} e^{-E_i/kT} = \frac{e}{n_k h^3} (2\pi m_k k T)^{\frac{3}{2}} Z_k^{(i)} e^{-E_i/kT}. \tag{58.6}$$

Dann ist

$$u_k = \frac{kT}{m_k}\left(\frac{3}{2} + \frac{E_i}{kT} + T\frac{\partial \log Z_k^{(i)}}{\partial T}\right), \tag{58.7}$$

$$s_k = \frac{k}{m_k}\left\{\log\left[(2\pi m_k k T)^{\frac{3}{2}} \frac{kT}{h^3 p_k} Z_k^{(i)}\right] + T\frac{\partial \log Z_k^{(i)}}{\partial T}\right\}, \tag{58.8}$$

$$h_k = \frac{kT}{m_k}\left(\frac{5}{2} + \frac{E_i}{kT} + T\frac{\partial \log Z_k^{(i)}}{\partial T}\right), \tag{58.9}$$

$$\mu_k = g_k = \frac{kT}{m_k}\left(\frac{E_i}{kT} - \log\left[(2\pi m_k k T)^{\frac{3}{2}} \frac{kT}{h^3 p_k}\right]\right). \tag{58.10}$$

Um nun zum Massenwirkungsgesetz für die Reaktion

$$\text{„1“} + \text{„2“} = \text{„3“} \tag{58.11}$$

zu kommen, ist zu fordern, daß sich für gegebene Werte von T und p ein solcher Trennungsgrad

$$x = \frac{n_1}{n_2 + n_3} \tag{58.12}$$

einstellt, daß das chemische Potential μ der Mischung ein Minimum, also

$$\left(\frac{\partial \mu}{\partial x}\right)_{p,T} = 0 \tag{58.13}$$

wird. Das Gesamtpotential μ setzen wir aus den Einzelpotentialen im Verhältnis ihrer Massenkonzentrationen zusammen

$$\mu = \mu_1 c_1 + \mu_2 c_2 + \mu_3 c_3. \tag{58.14}$$

Für die Ausrechnung ist es zweckmäßig, den Trennungsgrad in μ_k einzuführen, mit den Beziehungen

$$\left.\begin{aligned} c_1 &= \frac{m_1}{m_3} x; \quad c_2 = \frac{m_2}{m_3} x; \quad c_3 = 1 - x; \\ p_1 &= p_2 = \frac{x}{1+x} p; \quad p_3 = \frac{1-x}{1+x} p. \end{aligned}\right\} \tag{58.15}$$

Damit wird, wenn die Trennungsarbeit nur einem der getrennten Partner zugeschrieben wird,

$$\left.\begin{aligned}\mu = \frac{kT}{m_3}\Big\{x\frac{E_i}{kT} &- x\log\left[Z_1^{(i)}(2\pi m_1 kT)^{\frac{3}{2}}\frac{kT}{h^3 p}\frac{1+x}{x}\right] - \\ &- x\log\left[Z_2^{(i)}(2\pi m_2 kT)^{\frac{3}{2}}\frac{kT}{h^3 p}\frac{1+x}{x}\right] - \\ &- (1-x)\log\left[Z_3^{(i)}(2\pi m_3 kT)^{\frac{3}{2}}\frac{kT}{h^3 p}\frac{1+x}{1-x}\right]\Big\},\end{aligned}\right\} \qquad (58.16)$$

wovon die partielle Abteilung nach x verschwindet bei

$$\frac{x^2}{1-x^2} = \frac{Z_1^{(i)} Z_2^{(i)}}{Z_3^{(i)}}\left(2\pi\frac{m_1 m_2}{m_3}kT\right)^{\frac{3}{2}}\frac{kT}{h^3 p}e^{-E_i/kT}. \qquad (58.17)$$

Im Falle der Dissoziation ist dies das Massenwirkungsgesetz, wenn man $E_i = E_D$ setzt.

Setzt man dagegen „1“ = „e“; „2“ = „i“; „3“ = „0“, ferner $Z_e^{(i)} = 2$ wegen des Elektronenspins und $m_i \approx m_0$, dann entsteht die EGGERT-SAHAsche Gleichung für das Ionisationsgleichgewicht:

$$\frac{x^2}{1-x^2}n = \frac{2Z_i^{(i)}}{Z_0^{(i)}}\frac{(2\pi m_e kT)^{\frac{3}{2}}}{h^3}e^{-\frac{E_I}{kT}}. \qquad (58.18)$$

Die Ionisierungsenergie ist um den kleinen Betrag

$$\Delta E_i = 7\cdot 10^{-7}\sqrt[3]{n_e}\,\mathrm{cm\,eV} \qquad (58.19)$$

herabzusetzen, weil die höchsten Terme infolge der Wechselwirkung mit den Nachbarionen nicht existenzfähig sind und so hoch angeregte Elektronen bereits frei beweglich sind.

59. Berechnung der inneren Energie und der Enthalpie. Zur Berechnung des totalen Wärmeleitwertes benötigen wir noch die innere Energie pro Gramm der Mischung. Sie setzt sich aus den Einzelenergien im Verhältnis der Massenkonzentrationen zusammen wie

$$\left.\begin{aligned}u &= u_1 c_1 + u_2 c_2 + u_3 c_3 = \frac{m_1}{m_3}x u_1 + \frac{m_2}{m_3}x u_2 + (1-x)u_3 \\ &= \frac{kT}{m_3}\left\{\frac{3}{2} + \frac{3}{2}x + \frac{E_i}{kT}x + xT\frac{\partial}{\partial T}\log\left(\frac{Z_1^{(i)} Z_2^{(i)}}{Z_3^{(i)}}\right) + T\frac{\partial}{\partial T}\log Z_3^{(i)}\right\}.\end{aligned}\right\} \qquad (59.1)$$

Diese Formel kann in gleicher Weise für die Dissoziation wie für die Ionisation benutzt werden. Während aber bei der Dissoziation in den auftretenden Zustandssummen neben der Elektronenanregung auch die wesentlichen Rotationen und Oszillationen der Moleküle enthalten sind, treten im Falle der Ionisation nur die unwesentlichen Anregungszustände der Atome und Ionen auf.

Für die Umrechnung der Änderung der inneren Energie mit der Zeit braucht man noch das totale Differential der inneren Energie als Funktion von Temperatur und Druck

$$du = \frac{\partial u}{\partial T}dT + \frac{\partial u}{\partial p}dp. \qquad (59.2)$$

Es lohnt aber nicht, die Differentiation analytisch durchzuführen, weil die Gleichungen sehr lang werden. Man rechnet besser graphisch oder numerisch. Dabei kann man im allgemeinen auf den geringen Druckeinfluß verzichten.

An Stelle von u kann man nach den Gln. (58.7) und (58.9) auch die Enthalpie h verwenden, deren totales Differential

$$dh = c_p\,dT + \left(\frac{\partial h}{\partial p}\right)_T dp \tag{59.3}$$

aus der Enthalpie der Mischung

$$h = h_1 c_1 + h_2 c_2 + h_3 c_3 \tag{59.4}$$

wieder am besten durch graphische oder numerische Differentiation gewonnen wird. Die Druckabhängigkeit kann man auch hier meist vernachlässigen.

60. Die Schallgeschwindigkeit des Plasmas. Für spätere Untersuchungen fügen wir hier die Berechnung der Schallgeschwindigkeit ein. Sie ist durch die Gleichung

$$c = \sqrt{\left(\frac{\partial p}{\partial \varrho}\right)_s} \quad \text{oder} \quad \frac{1}{c^2} = \left(\frac{\partial \varrho}{\partial p}\right)_s = \left(\frac{\partial \varrho}{\partial p}\right)_T + \left(\frac{\partial \varrho}{\partial T}\right)_p \left(\frac{\partial T}{\partial p}\right)_s \tag{60.1}$$

bestimmt. Mit der Zustandsgleichung eines reagierenden Gases

$$\varrho = \frac{p}{1+x}\,\frac{m_3}{k\,T} \tag{60.2}$$

und dem Massenwirkungsgesetz $x = f(T, p)$ wird für ein Plasma bei Vernachlässigung der inneren Anregungen

$$\left(\frac{\partial \varrho}{\partial p}\right)_T = \frac{\varrho}{p}\left[1 + \frac{1}{2}x(1-x)\right], \tag{60.3}$$

$$\left(\frac{\partial \varrho}{\partial T}\right)_p = \frac{-\varrho}{T}\left[1 + \frac{1}{2}x(1-x)\left(\frac{5}{2} + \frac{E_i}{k\,T}\right)\right], \tag{60.4}$$

$$\left(\frac{\partial T}{\partial p}\right)_s = \frac{T}{p}\,\frac{2 + x(1-x)\left(\frac{5}{2} + \frac{E_i}{k\,T}\right)}{5 + x(1-x)\left(\frac{5}{2} + \frac{E_i}{k\,T}\right)^2}, \tag{60.5}$$

so daß sich für die Schallgeschwindigkeit

$$c^2 = \frac{5 + x(1-x)\left(\frac{5}{2} + \frac{E_i}{k\,T}\right)^2}{3 + x(1-x)\left[\left(\frac{3}{2} + \frac{E_i}{k\,T}\right)^2 + \frac{3}{2}\right]}\,\frac{p}{\varrho} = \gamma_s \frac{p}{\varrho} \tag{60.6}$$

ergibt. Der Isentropenexponent γ_s ist um den Faktor

$$\frac{\gamma_s}{\gamma} = 1 + \frac{1}{2}x(1-x) \tag{60.7}$$

größer als der Adiabatenexponent $\gamma = c_p/c_v$, was maximal gut 10% ausmachen kann. Für nicht regierende Gase bleibt natürlich

$$c^2 = \gamma\,\frac{p}{\varrho}. \tag{60.8}$$

VIII. Rückführung der Koeffizienten auf die Stoßquerschnitte.

61. Diffusionsgrößen. Um nun wirklich mit den Gleichungen der Plasmatheorie zahlenmäßig rechnen zu können, müssen noch die Koeffizienten auf die Stoßquerschnitte zurückgeführt werden. Wir bedienen uns dazu der einfachen kinetischen Gastheorie, die aber soweit korrigiert ist, daß die Abweichungen von den strengen Resultaten der BOLTZMANNschen Stoßgleichung nur einige Prozent betragen. Als wichtigste Größe für alle Diffusionsvorgänge benötigen wir die Diffusionskonstante, die die Diffusion zweier Gase gegeneinander beherrscht. Den Begriff der Diffusionskonstante kann man aus der Teilchendiffusionsgleichung im Teilchenruhesystem

$$\nu_1 = -D \operatorname{grad} n_1; \quad \nu_2 = -D \operatorname{grad} n_2; \quad \operatorname{grad} n_1 + \operatorname{grad} n_2 = 0 \tag{61.1}$$

entnehmen. ν_k ist die pro Zeit- und Flächeneinheit durch eine Kontrollfläche, durch die in beiden Richtungen absolut gleich viele Teilchen strömen, hindurchtretende Teilchenzahl der Sorte k. Voraussetzung für diese Diffusionsgleichung sind konstante Werte für Druck und Temperatur. Man kann aber auch die Gleichung für Massendiffusionsströme $\mathfrak{J}_k$ im Schwerpunktssystem wählen:

$$\mathfrak{J}_1 = -\varrho D \operatorname{grad} c_1; \quad \mathfrak{J}_2 = -\varrho D \operatorname{grad} c_2; \quad \operatorname{grad} c_1 + \operatorname{grad} c_2 = 0. \tag{61.2}$$

Beide Diffusionskonstanten sind identisch. Ihr Wert wird von der kinetischen Gastheorie mit der Formel

$$D = \frac{3}{8}\sqrt{\frac{\pi}{2}}\,\frac{\sqrt{k T\left(\frac{1}{m_1} + \frac{1}{m_2}\right)}}{Q_{12}(n_1 + n_2)} \tag{61.3}$$

auf den Stoßquerschnitt zurückgeführt [*84*]. Hieraus gewinnen wir nach Gl. (43.6) den in den Einzelkraftgleichungen auftretenden Reibungskoeffizienten

$$\varepsilon_{12} = \frac{8}{3}\sqrt{\frac{2}{\pi}\,k T\,\frac{m_1 m_2}{m_1 + m_2}}\,Q_{12}, \tag{61.4}$$

der eine für den Einzelstoß charakteristische Größe ist und nicht mehr von der Konzentration abhängt. Damit ist dann auch die für alle $\mathfrak{X}$-Kräfte gemeinsame Größe

$$L_{11} = \left(\frac{\varrho_1 \varrho_2}{\varrho}\right)^2 \frac{1}{n_1 n_2 \varepsilon_{12}} \tag{61.5}$$

festgelegt. Der entsprechende Reibungsfaktor $n_1 n_2 \varepsilon_{12}$ hat die anschauliche, gaskinetische Bedeutung der Zahl aller Stöße pro Volumen- und Zeiteinheit zwischen verschiedenartigen Teilchen der Sorte „1" und „2". Der Reibungskoeffizient behält auch dann noch seinen Wert, wenn weitere Gaskomponenten hinzutreten, weil in ε_{12} nur der Impulsaustausch zwischen „1" und „2" durch Stöße enthalten ist.

Wir haben oben gesehen, daß man den Strom der Ladungsträger zusammen gegenüber dem Neutralgas $\mathfrak{J}_L$ ebenfalls als dem Zweiersystem gehorchend ansehen kann. An die Stelle des Reibungskoeffizienten ε_{12} im Zweiersystem trat

$$n_L n_0 \varepsilon_{L0} = n_L n_0 (\varepsilon_{e0} + \varepsilon_{i0}) = a. \tag{61.6}$$

Setzen wir nun in Gl. (51.16) $\operatorname{grad} p_L = 2kT \operatorname{grad} n_L$ unter Nichtbeachtung von Thermoeffekten und setzen den gesamten Koeffizienten vor $\operatorname{grad} n_L$ gleich D_A, dann wird

$$\frac{\mathfrak{J}_L}{m_L} = -\frac{\varrho_0 \varrho_L}{m_L \varrho}\,\frac{1}{a}\,2kT \operatorname{grad} n_L \approx -\frac{3}{4}\,\frac{\sqrt{\pi k T/m_0}}{(n_L + n_0)\,Q_{i0}} \operatorname{grad} n_L = -D_A \operatorname{grad} n_L, \tag{61.7}$$

wenn man m_i gegen m_e und noch ε_{e0} gegen ε_{i0} vernachlässigt. Weiter überzeugt man sich leicht an Hand von Gl. (61.3), daß

$$D_A \approx 2 D_{i0}, \tag{61.8}$$

womit diese bekannte Beziehung bestätigt ist [*5*].

Nun gilt weiterhin die allgemein gültige TOWNSEND-Beziehung

$$\frac{D_e}{b_e} = \frac{kT}{e}, \tag{61.9}$$

worin b_e die Beweglichkeit der Elektronen, also ihre Geschwindigkeit bei der Feldstärke Eins, relativ zum Schwerpunkt des Plasmas ist. Wir verifizieren diese Gleichung, indem wir die Diffusionskonstante D_e aus der Gl. (51.15) mit (45.1)

$$\mathfrak{j} = -e\left(\frac{1}{m_e} + \frac{1}{m_i}\right)\mathfrak{J}_e = \sigma\left(b + \frac{m_i}{2e\varrho_L}\right)\operatorname{grad} p_L \tag{61.10}$$

herausholen. Mit der gleichen Umformung wie in Gl. (61.7) wird

$$\frac{\mathfrak{J}_e}{m_e} = \nu_e = -\frac{2\sigma}{e}\left(\frac{m_i}{2e\varrho_L} + b\right) k\,T \operatorname{grad} n_L = -D_e \operatorname{grad} n_L. \tag{61.11}$$

Bildet man nun den Quotienten von D_e und $b_e = -\sigma/e n_L$, dann fällt σ heraus, und es bleibt

$$\frac{D_e}{b_e} = \frac{k\,T}{e}\left(1 + 2\frac{m_e\varepsilon_{i0} - m_i\varepsilon_{e0}}{m_0(\varepsilon_{i0} + \varepsilon_{e0})}\right). \tag{61.12}$$

Der zweite Summand ist klein gegen Eins, so daß damit der gewünschte Beweis erbracht ist.

Macht man nun auch in der Formel für die Leitfähigkeit Gl. (51.5) gleichartige Vernachlässigungen wie oben, dann folgt zunächst

$$\sigma = \frac{e^2 n_L}{n_0\varepsilon_{e0} + n_L\varepsilon_{ei}}. \tag{61.13}$$

Setzt man hier den Ausdruck von Gl. (61.4) für die ε ein, dann resultiert

$$\sigma = \frac{e^2 n_L}{n_0 Q_{e0} + n_L Q_{ei}}\,\frac{3}{8}\sqrt{\frac{\pi}{2kT m_e}} = e\,n_L\frac{3}{4}\frac{e\lambda_e}{m_e v_e}, \tag{61.14}$$

worin

$$b_e = \frac{3}{4}\frac{e\lambda_e}{m_e v_e} \tag{61.15}$$

der von LANGEVIN berechnete Ausdruck für die Beweglichkeit der Elektronen ist. Genauere Berechnungen erhöhen den Zahlenfaktor etwas [*149*], so daß es berechtigt erscheint, wenn viele Autoren [*18a*], [*144*] zur Messung von Querschnitten den Zahlenfaktor $\frac{3}{4}$ weglassen.

62. Der Wärmeleitwert ohne Ströme. Es ist nun noch die Kontaktwärmeleitung $\varkappa_{\mathfrak{J}=0}$ zu berechnen, was zunächst nur für schwere Teilchen geschehen soll. Auch hier wollen wir die aus der einfachen kinetischen Theorie stammenden Formeln, aber mit Korrekturen nach der strengen Theorie verwenden. Für die Gaskomponente k gilt die einfache Formel

$$\varkappa^k_{\mathfrak{J}=0} = \tfrac{1}{3} n_k \lambda_k v_k \left(\tfrac{5}{2} c^k_{\mathrm{tr}} + c^k_i\right), \tag{62.1}$$

in der c^k_{tr} die spezifische Wärme der Translationsfreiheitsgrade und c^k_i die der inneren Freiheitsgrade je Teilchen ist. Es ist

$$c^k_{\mathrm{tr}} = \tfrac{3}{2} k, \tag{62.2}$$

während c^k_i mit Gl. (58.1) aus der Zustandssumme je Teilchen berechnet werden muß. Der Faktor $\frac{5}{2}$ vor c^k_{tr} ist die sog. EUCKEN-Korrektur [*44*], die der Tatsache Rechnung trägt, daß die schnelleren durch die Kontrollflächen fliegenden

Teilchen auch mehr kinetische Energie transportieren. n_k bzw. v_k sind die Teilchenzahlen je Volumeneinheit bzw. die mittlere thermische Geschwindigkeit der Sorte k.

In der freien Weglänge λ_k sind die Querschnitte aller gestoßenen Partikel einschließlich der eigenen mitzunehmen. Außerdem ist der Faktor $\sqrt{\frac{1}{m_k}+\frac{1}{m_l}}$, der die Relativgeschwindigkeit der stoßenden Partner berücksichtigt, an den Querschnitten anzubringen. Schließlich muß auch noch die Persistenz, also die Tendenz der Teilchen, auch nach dem Stoß eine Geschwindigkeitskomponente in der Ursprungsrichtung beizubehalten, in Rechnung gestellt werden. Alles zusammen führt dann zu der freien Weglänge

$$\frac{1}{\lambda_k}=\left(1-\frac{1}{2}P_{kk}\right)n_k Q_{kk}\sqrt{\frac{1}{m_k}+\frac{1}{m_k}}+\left(1-\frac{1}{2}P_{kl}\right)n_l Q_{kl}\sqrt{\frac{1}{m_k}+\frac{1}{m_l}}\ldots, \tag{62.3}$$

worin die Persistenz nach der Formel

$$P_{kl}=\frac{m_k-0{,}2\,m_l}{m_k+m_l} \tag{62.4}$$

zu berechnen ist. In die Formel für die Zähigkeit η eines Gases geht genau die gleiche freie Weglänge ein, weswegen man auch schreiben kann

$$\varkappa^k_{\mathfrak{J}=0}=\frac{\eta}{m_k}\left(\frac{5}{2}c^k_{\mathrm{tr}}+c^k_i\right). \tag{62.5}$$

Schließlich hat man noch über alle Einzelwerte zu summieren

$$\varkappa_{\mathfrak{J}=0}=\sum_k \varkappa^k_{\mathfrak{J}=0}. \tag{62.6}$$

Mit diesen Korrekturen erhält man aus Gl. (62.1) sehr brauchbare Werte für die Kontaktwärmeleitung schwerer Teilchen (s. $\varkappa_m$, $\varkappa_a$ und $\varkappa_i$ in Fig. 41).

Während bei Mischungen *schwerer* Teilchen die Thermodiffusion ohne wesentlichen Einfluß auf die Wärmeleitung ist, muß der Thermodiffusionsfaktor der Elektronen auch in die Kontaktwärmeleitfähigkeit des Plasmas Eingang finden. Für ein reines Plasma haben SPITZER und Mitarbeiter (l. c.) diesen Schritt durchgeführt; ihre Ergebnisse haben wir bereits behandelt (Ziff. 47). Für ein Plasma mit kleinem Ionisationsgrad stützen wir uns auf L. WALDMANN [*236*], der unter der Voraussetzung weniger, leichter Teilchen in einem schweren Grundgas (LORENTZ-Gas) den Thermodiffusionsfaktor und seinen Einfluß auf den Kontakt-Wärmeleitwert aus der BOLTZMANNschen Stoßgleichung berechnet hat. Er findet

$$\alpha=\left(\frac{\partial\log D}{\partial\log T}\right)_p-2. \tag{62.7}$$

Für das zwischenmolekulare Kraftgesetz

$$K\sim r^{-\nu} \tag{62.8}$$

wird

$$\alpha=\frac{5-\nu}{2(\nu-1)} \tag{62.9}$$

und

$$\varkappa^e_{\mathfrak{J}=0}=\frac{1}{3}n_e\lambda_e v_e k\left(\frac{5}{2}+\alpha_e\right)=\frac{1}{3}n_e\lambda_e v_e k\,\frac{2\nu}{\nu-1}. \tag{62.10}$$

Für starr elastische Kugeln berechnet sich α_e zu

$$\alpha_e=-\tfrac{1}{2}\quad\text{(starre Kugeln)} \tag{62.11}$$

und für geladene Teilchen mit $\nu = 2$ zu

$$\alpha_e = \tfrac{3}{2} \quad \text{(geladene Teilchen).} \tag{62.12}$$

Der Vergleich mit dem α-Wert für das reine Plasma $\alpha = 3\frac{\gamma_T}{\gamma_E} = 1{,}4$ fällt erstaunlich gut aus, obwohl die WALDMANNsche Voraussetzung $n_e \ll n_s$ hier gar nicht mehr erfüllt ist. Wir setzen nun für ein sehr schwach ionisiertes Gas als Kraftgesetz zwischen Elektronen und den hauptsächlich gestoßenen Neutralteilchen starre Kugeln an, tragen aber der mit steigendem Ionisationsgrad zunehmenden Ionenzahl durch einen Faktor $(1 + x)$ Rechnung, der wenigstens den Ausdruck

$$\left(\frac{5}{2} + \alpha\right) = \frac{2\nu}{\nu - 1} \tag{62.13}$$

zum etwa richtigen Endwert des reinen Plasmas führt. Für den so korrigierten Wärmeleitwert kann man nicht das gleiche verlangen, weil hier die gemachten Voraussetzungen doch einschneidender sind. Es ist deswegen zweckmäßig, die WALDMANNsche Formel

$$\varkappa^e_{\mathfrak{J}=0} = \tfrac{1}{3} n_e \lambda_e v_e k \, 2 (1 + x) \tag{62.14}$$

für kleine Ionisationsgrade, die SPITZERsche für das reine Plasma zu verwenden, und das Zwischengebiet durch graphische Interpolation zu überbrücken (s. $\varkappa_e$ WALDMANN in Fig. 41).

63. Die Stoßquerschnitte. Nennen wir vorübergehend r_1 und r_2 die Radien zweier sich stoßender Teilchen, dann können sich die beiden Mittelpunkte nur auf den Abstand $\sigma_{12} = r_1 + r_2$ nähern. Der Stoßvorgang verläuft also genau so, als ob das eine Teilchen punktförmig sei und das andere den Querschnitt

$$\pi \sigma_{12}^2 = Q_{12} \tag{63.1}$$

hätte.

Beim Stoß neutraler Teilchen gegeneinander ist der Querschnitt Q_{12} noch mit dem SUTHERLAND-Faktor $1 + C/T$, der die größere Eindringtiefe der Stoßpartner bei höherer Geschwindigkeit und damit höherer Temperatur berücksichtigt, zu versehen. Die Größenordnung von C liegt bei $100°$, während die der Querschnitte selbst einige 10^{-16} cm^2 beträgt.

Stoßen Elektronen auf Neutralteilchen, dann darf nicht der gaskinetische Querschnitt genommen werden, wie das früher manchmal geschah, sondern es müssen die oft über zwei Zehnerpotenzen größeren RAMSAUER-Querschnitte eingesetzt werden. Das gleiche gilt für den Stoß von Ionen gegen Neutralteilchen, was bisher nicht genügend beachtet wurde. Denn auch für diesen Stoßprozeß liegen die Querschnitte zwei Zehnerpotenzen über den gaskinetischen. Ist für einen Stoßvorgang außer dem Wirkungsquerschnitt Q_0 auch die Winkelstreuung bekannt, d.h. kennt man die Wahrscheinlichkeit $p(\vartheta) \sin\vartheta \, d\vartheta \, d\varphi$ dafür, daß ein Elektron in das Raumwinkelelement $\sin\vartheta \, d\vartheta \, d\varphi$ abgelenkt wird, dann ist

$$Q_0 \, p(\vartheta) \sin\vartheta \, d\vartheta \, d\varphi = I_0(\vartheta) \tag{63.2}$$

der differentielle Wirkungsquerschnitt für die Streuung in das betrachtete Raumwinkelelement. Es ist daher

$$Q_0 = \int_0^{\pi} \int_0^{2\pi} I_0(\vartheta) \sin\vartheta \, d\vartheta \, d\varphi \tag{63.3}$$

der Wirkungsquerschnitt, der angibt, innerhalb welchen Kreises um das zu stoßende Teilchen das Zentrum des stoßenden ankommen muß, damit überhaupt eine Berührung stattfindet. Dagegen kommt es für die Diffusionsvorgänge auf die „auslöschenden“ Stöße und den entsprechenden Transportquerschnitt an, der so definiert ist, daß nach dem Stoß der Impuls in der Ursprungsrichtung im Mittel verschwunden ist. Das führt zu der Definition

$$Q_{12} = \int_0^\pi \int_0^{2\pi} I_0(\vartheta)\,(1-\cos\vartheta)\sin\vartheta\, d\vartheta\, d\varphi. \tag{63.4}$$

Der Querschnitt der Ionen gegenüber Elektronenstoß ist wegen des COULOMB-Feldes besonders groß. Viele Autoren haben sich mit diesem Problem beschäftigt. Das Ergebnis der jüngsten und genauesten Berechnung von L. SPITZER und Mitarbeitern haben wir bereits zitiert. Danach ist die elektrische Leitfähigkeit

$$\sigma = \frac{4\,(k\,T)^{\frac{3}{2}}\,\gamma_E}{\pi\sqrt{\pi/2}\,m_e\,e^2\,Z\log q}, \tag{63.5}$$

woraus man mit der allgemeinen Leitfähigkeitsformel

$$\sigma = \frac{e^2\,n_L}{m_e\,v_e}\,\frac{1}{n_L\,Q_{ie}} \tag{63.6}$$

den Ionenquerschnitt zu

$$Q_{ie} = \frac{0{,}567\,Z^2\,e^4}{\gamma_E\,(k\,T)^2}\log q \tag{63.7}$$

berechnen kann. Der Parameter q enthält die Grenze, bei der man für die Integration das COULOMB-Feld abschneiden muß, weil das Integral nicht konvergiert. Hier hat nun SPITZER einen Wert von CHAPMAN und COWLING [*25*] übernommen, der dem Abschneiden beim nächsten Ion entspricht. Indessen ist dieser Bereich sicher zu groß, vielmehr sollte man weniger als die Hälfte dieses Abstandes erwarten. Dies wird durch Messungen von MAECKER und PETERS [*142*], [*144*] am wasserstabilisierten Hochleistungsbogen, in dem bei einer Temperatur von 50000° K ein reines teilweise mehrfach ionisiertes Plasma vorliegt, bestätigt. Um mit den Messungen in Übereinstimmung zu kommen, muß das COULOMB-Feld bei $^1/_4$ des Abstandes zweier benachbarter Ionen abgeschnitten werden. Daraus ergibt sich

$$q = \frac{k\,T}{Z\,e^2\,n_i^{\frac{1}{3}}} \tag{63.8}$$

und

$$Q_{ie} = \frac{0{,}567\,Z^2\,e^4}{\gamma_E\,(k\,T)^2}\log\frac{k\,T}{Z_e^2\,n_i^{\frac{1}{3}}}. \tag{63.9}$$

Z ist hier die effektive Ionenzahl Gl. (47.1); γ_E ist der Tabelle 1 in Ziff. 47 zu entnehmen.

Für einfache Ionen gilt in guter Näherung

$$Q_{ie} = \frac{e^4}{(k\,T)^2}\log\frac{k\,T}{e^2\,n_i^{\frac{1}{3}}}. \tag{63.10}$$

64. Materialfunktionen von Stickstoff und Xenon. Als Beispiel für die behandelten Materialfunktionen mögen einige davon für ein *Stickstoffplasma* in graphischer Darstellung bei einem Druck von 1 Atm gezeigt werden.

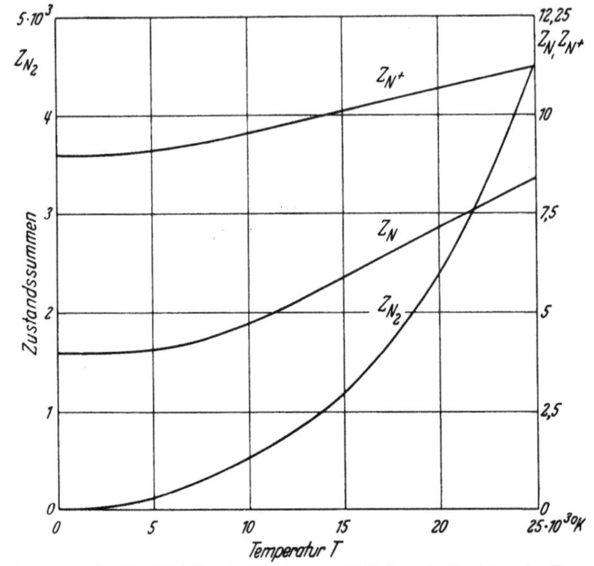

Fig. 36. Innere Zustandssummen des N_2-Moleküls, des N-Atoms und N^+-Ions als Funktion der Temperatur, nach FAST [46]

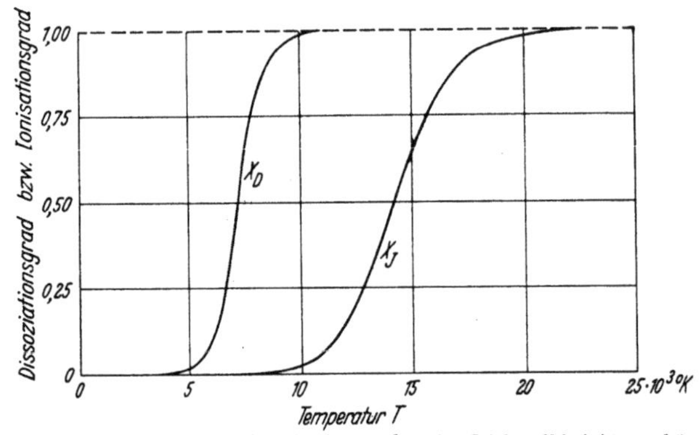

Fig. 37. Dissoziationsgrad x_D und Ionisationsgrad x_I für Stickstoff bei Atmosphärendruck.

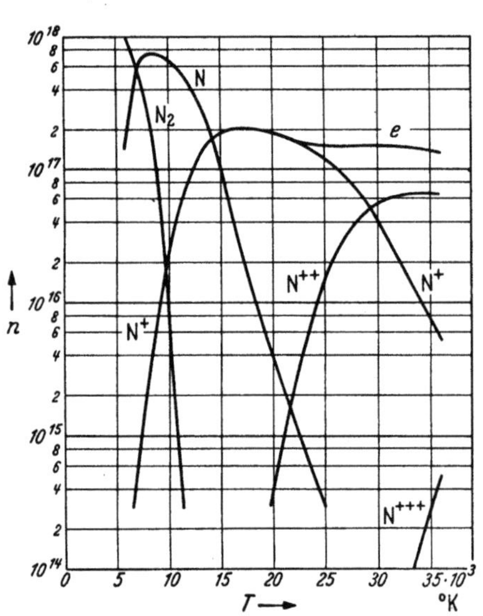

Fig. 38. Teilchenzahlen/cm³, für Stickstoff bei Atmosphärendruck.

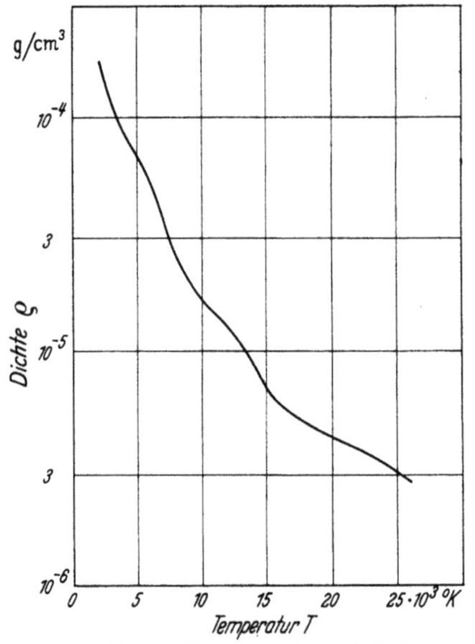

Fig. 39. Dichte von Stickstoff bei Atmosphärendruck.

Fig. 36 zeigt den Verlauf der inneren Zustandssummen mit der Temperatur nach Rechnungen von Fast [46].

Mit dem Massenwirkungsgesetz kann daraus der Dissoziations- und Ionisationsgrad berechnet werden (Fig. 37). Dabei ist die Dissoziationsenergie

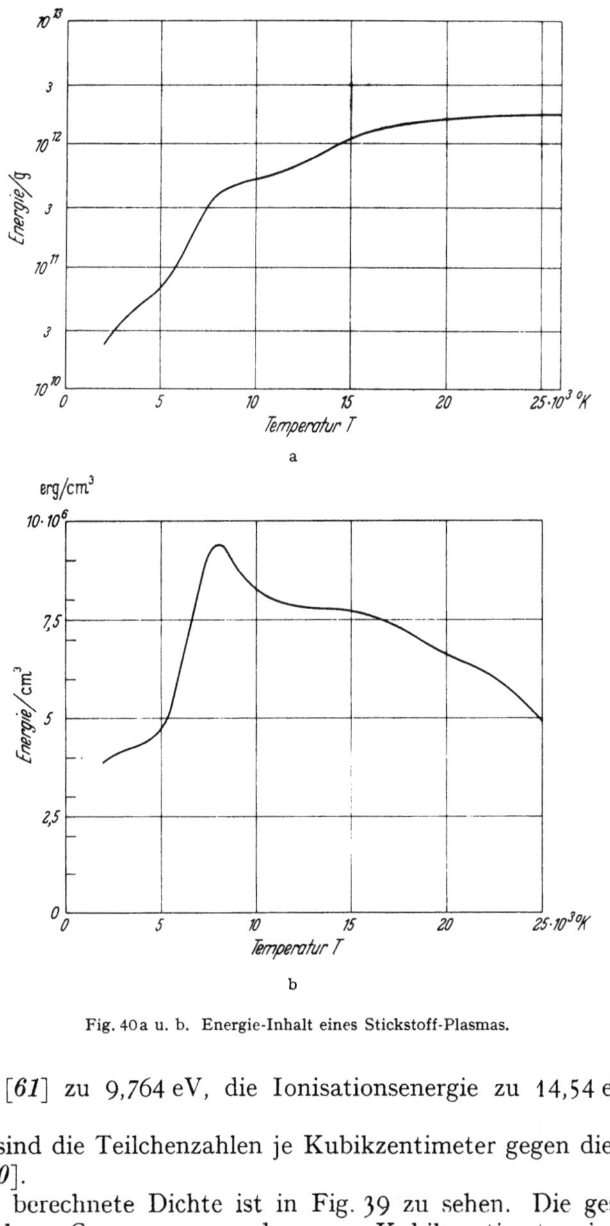

Fig. 40a u. b. Energie-Inhalt eines Stickstoff-Plasmas.

nach Gaydon [61] zu 9,764 eV, die Ionisationsenergie zu 14,54 eV eingesetzt worden.

In Fig. 38 sind die Teilchenzahlen je Kubikzentimeter gegen die Temperatur aufgetragen [20].

Die daraus berechnete Dichte ist in Fig. 39 zu sehen. Die gesamte innere Energie, einmal pro Gramm, zum anderen pro Kubikzentimeter sind in Fig. 40a und b wiedergegeben.

Die Wärmeleitung setzt sich aus einer Reihe von Summanden zusammen, und zwar aus dem Ruhewärmeleitwert $\varkappa_{\mathfrak{J}=0}$ der Moleküle, der Atome, der Ionen und Elektronen und aus dem Diffusionsbeitrag der Dissoziations- und Ionisationsenergie, die alle in Fig. 41 zu sehen sind.

Die elektrische Leitfähigkeit steigt zwar demnächst rasch mit der Temperatur an, wird aber bei wachsendem Ionisationsgrad durch die Ionen stark beeinträchtigt (Fig. 42).

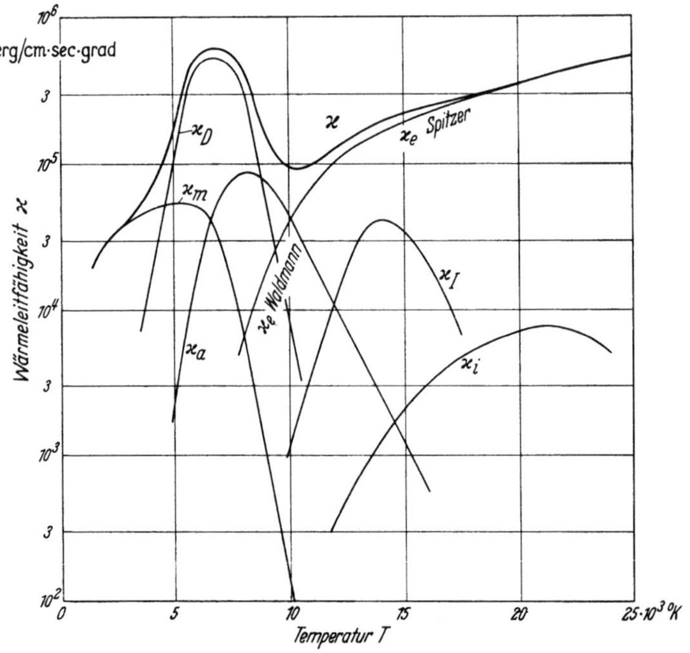

Fig. 41. Die Kontaktwärmeleitfähigkeit der Moleküle $\varkappa_m$, der Atome $\varkappa_a$, der Ionen $\varkappa_i$ und Elektronen $\varkappa_e$, ferner die Wärmeleitung durch Diffusion der Dissoziationsenergie $\varkappa_D$ und der Ionisationsenergie $\varkappa_I$. Die Summe aller ist der totale Wärmeleitfähigkeitskoeffizient $\varkappa$.

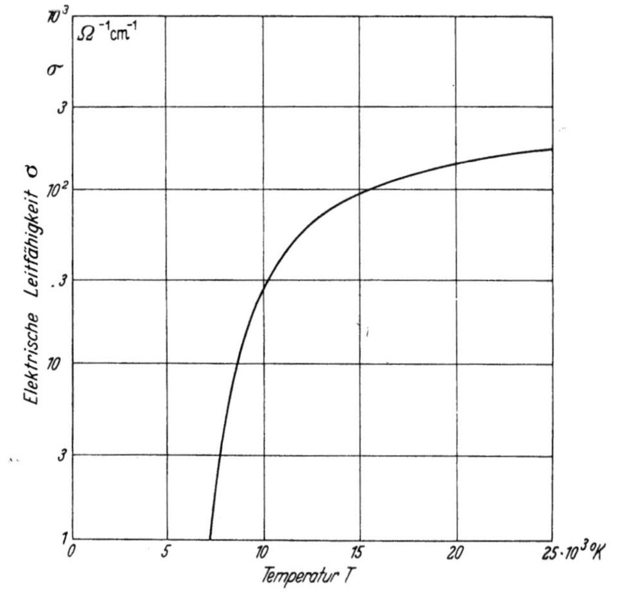

Fig. 42. Elektrische Leitfähigkeit eines Stickstoffplasmas bei Atmosphärendruck.

Als Prototyp eines einatomigen Gases wählen wir Xenon, dessen Eigenschaften von SCHIRMER [*187*] durch exakte Lösung der BOLTZMANNschen Stoßgleichung berechnet worden sind. Gegenüber den Molekülgasen hat Xenon natürlich keinen Dissoziationsberg in der Wärmeleitfähigkeit (Fig. 43). Die elektrische Leitfähigkeit ist in Fig. 44 für 25 atü dargestellt.

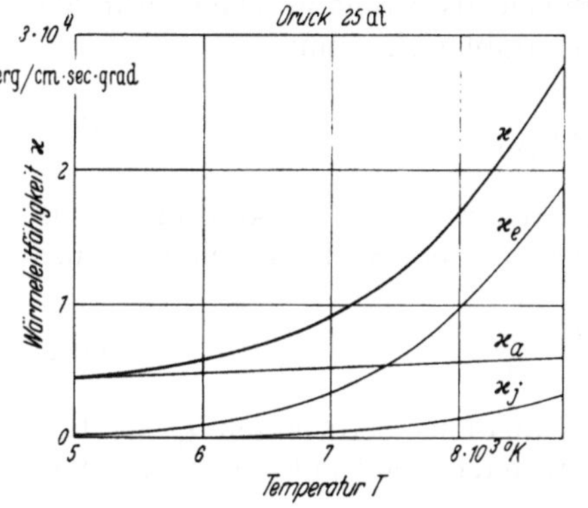

Fig. 43. Wärmeleitfähigkeit von Xenon bei 25 at Druck.

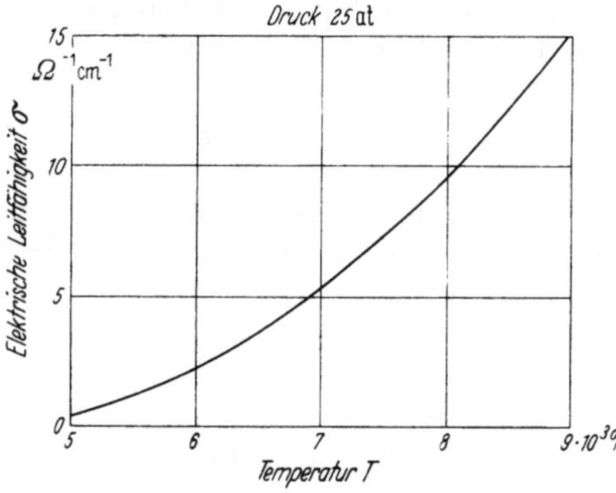

Fig. 44. Elektrische Leitfähigkeit bei 25 at Druck.

IX. Strahlung des Plasmas.

65. Linienintensitäten. Die im Laboratorium herstellbaren Plasmen sind mit vereinzelten Ausnahmen (Plasmakugel im *X*-Hochdruckbogen von LARCHÉ [*117*] oder wasserstabilisierter Hochdruckbogen von PETERS [*165*]) Linienstrahler mit kontinuierlichem Untergrund. Von vollständigem thermischem Gleichgewicht hinsichtlich der Strahlung kann also wegen des Fehlens der Hohlraumstrahlung keine Rede sein. Wohl aber kann innerhalb einiger zum Grundzustand führender oder anderer intensiver Linien die Strahlungsdichte mit der Hohlraumstrahlung übereinstimmen. Diese „Emission aus optisch dicker Schicht" werden wir später behandeln. Die meisten Linien werden aus optisch dünner Schicht emittiert. Die Gesetze für die Intensitäten solcher Linien wollen wir jetzt zusammenstellen (vgl. UNSÖLD [*231*]).

Die Zahl der Lichtquanten $h\nu$ [erg] einer Spektrallinie, die die Volumeneinheit eines Plasmas in der Zeiteinheit nach allen Seiten hin spontan emittiert, ist der Übergangswahrscheinlichkeit A_n^m [sec^{-1}] und der Zahl der im angeregten Zustand m befindlichen Atome proportional. In die Raumwinkeleinheit wird davon nur der 4π-te Teil ausgestrahlt. Dies ist der *Emissionskoeffizient* i_L dieser Linie

$$i_L = \frac{1}{4\pi} A_n^m n_m h\nu \left[\frac{\text{erg}}{\text{sec cm}^3 \text{ sterad}}\right]. \tag{65.1}$$

Stehen l cm³ gleicher Art hintereinander, so tritt aus dieser Säule in Längsrichtung die l-fache *Intensität*

$$I_L = \frac{1}{4\pi} A_n^m n_m h\nu l \left[\frac{\text{erg}}{\text{sec cm}^2 \text{ sterad}}\right]. \tag{65.2}$$

Die Zahl der angeregten Atome je Volumeneinheit ist im vollständigen thermischen Gleichgewicht nach der BOLTZMANNschen Energieverteilung

$$n_m = n_0 \frac{g_m}{Z_0^{(i)}} e^{-E_m/kT}, \tag{65.3}$$

worin n_0 die Teilchendichte ein- und desselben Ionisationszustandes, aber aller Anregungszustände und $Z_0^{(i)}$ die Zustandssumme über die Anregungen nach (57.10) ist.

Bei Fehlen der Hohlraumstrahlung darf diese Formel nur dann angewendet werden, wie wir in Ziff. 34 gesehen hatten, wenn die Zahl der anregenden Stöße groß gegen die Zahl der Emissionsakte ist. Dann sprechen wir von *thermischer Anregung*. Diese Bedingung wollen wir im Bogen als erfüllt ansehen. Für die Intensität einer Linie aus homogener Schicht ergibt sich somit

$$I_L = \frac{1}{4\pi} A_n^m n_0 \frac{g_m}{Z_0^{(i)}} e^{-E_m/kT} h\nu l \left[\frac{\text{erg}}{\text{sec cm}^2\,\text{sterad}}\right]. \tag{65.4}$$

66. Kontinuumintensität. Um auch einen Ausdruck für die kontinuierliche Strahlungsintensität zu gewinnen, schneiden wir aus dem Spektrum ein schmales Stück der Breite

$$d\nu = -c\frac{d\lambda}{\lambda^2} \tag{66.1}$$

heraus und definieren den kontinuierlichen Emissionskoeffizienten ε_ν als Leistung, die von der Volumeneinheit in die Raumwinkeleinheit je Frequenzintervall ausgestrahlt wird

$$\varepsilon_\nu = \frac{dE}{dt\,d\tau\,d\nu\,d\omega} \left[\frac{\text{erg sec}}{\text{sec cm}^3\,\text{sterad}}\right]. \tag{66.2}$$

Bei einer Schichtdicke l vergrößert sich die Intensität proportional, solange keine Selbstabsorption merklich wird:

$$I_\nu = \varepsilon_\nu l \left[\frac{\text{erg}}{\text{cm}^2\,\text{sterad}}\right]. \tag{66.3}$$

Diese Intensität ist nach Art und Dimension der eines Flächenstrahlers gleich, beide sind daher miteinander vergleichbar (Strahlungsnormal). Den kontinuierlichen Emissionskoeffizienten selbst hat KRAMERS [*111*] für nackte ein- und mehrwertige Ionen im Hinblick auf die Röntgenstrahlen berechnet. Er bediente sich dabei zunächst der Aussage aus der klassischen Physik, daß ein Elektron, welches von einem Ion abgelenkt wird, eine Leistung dR/dt abstrahlt, die dem Quadrat der Beschleunigung b proportional ist:

$$\frac{dR}{dt} = \frac{2}{3}\frac{e^2}{c^3} b^2. \tag{66.4}$$

Durch FOURIER-Analyse des bei einem Vorbeigang ausgesandten Strahlungsimpulses konnte er den spektralen Verlauf des frei-frei Kontinuums angeben. Um auch den Fall des Elektroneneinfanges, also die Seriengrenzkontinua zu erfassen, führte er eine halbquantentheoretische Betrachtung mit dem Korrespondenzprinzip durch, deren Ergebnisse durch eine Reihe späterer quantenmechanischer Rechnungen im wesentlichen bestätigt wurden. UNSÖLD [*230*] hat diese Theorie auf das kontinuierliche Spektrum wasserstoffähnlicher Elemente erweitert und, nachdem FINKELNBURG [*47*], [*79*] das kontinuierliche Spektrum intensiver Gasentladungen als Folge der Wechselwirkung zwischen Elektronen und Ionen gedeutet hatte, auf diese angewendet. Die Ergebnisse lassen sich etwa folgendermaßen zusammenfassen: Das Gesamtkontinuum, frei-frei-Strahlung und Grenzkontinua, haben bei kleinen Frequenzen eine konstante, von der Frequenz unabhängige Intensität. Nach größeren Frequenzen hin, wo die Grenzkontinua tiefer liegender Terme am Ende der Linienserien neu hinzutreten, bleibt zwar der Mittelwert immer noch konstant, aber der Verlauf löst sich mehr und mehr in einzelne Sägezähne auf, deren Abstände und Amplituden immer größer werden, bis schließlich das Spektrum mit dem Hauptserien-Grenzkontinuum als letztem Sägezahn ausläuft. Die einzelnen Grenzkontinua setzen schon vor der theoretischen Grenze ein, weil die letzten Linien durch Verbreiterung zu einem vorgezogenen Kontinuum zusammenfließen. Aber auch dort, wo noch Linien zu erkennen sind, überlagern sich doch, wie GÖING, MEIER und MEINEN [69] gezeigt haben, die Linienflügel, so daß in Wirklichkeit die Kontinuumsintensität in der Umgebung einer Seriengrenze nicht sprunghaft hochschnellt, sondern eine stetige Änderung zeigt, was oft ihre Erkennbarkeit beeinträchtigt.

Im photographischen Bereich kann der Verlauf des Kontinuums über der Frequenzskala je nach Art des Termschemas noch konstant sein, oder mit zunehmender Frequenz stark abfallen, je nachdem, ob die fraglichen Terme bzw. die Sägezähne noch sehr dicht liegen (z.B. Fe) oder schon weit auseinander fallen (z.B. H). Hierauf haben MAECKER und PETERS [143] besonders hingewiesen. Formelmäßig ergeben die Theorien folgende Zusammenhänge:

Bezeichnen wir die Energiedifferenz im Termschema von der Ionisierungsgrenze bis herab zu dem Term, der den Abschluß einer hinreichend dichten Termfolge bildet, als Grenzfrequenz $h\nu_g$, dann gilt für den entsprechenden, einigermaßen glatten Kontinuumsverlauf im Spektrum von kleinen Frequenzen bis ν_g

$$\varepsilon_\nu = \frac{32\pi^2}{3\sqrt{3}} \frac{e^6 \overline{(Z+s)^2}}{c^3 (2\pi m)^{\frac{3}{2}}} \frac{n_e n_i}{\sqrt{kT}} = 6{,}36 \cdot 10^{-47}\, \overline{(Z+s)^2}\, \frac{n_e n_i}{\sqrt{kT}} \quad \text{für} \quad \nu \leqq \nu_g. \tag{66.5}$$

Die Intensität dieses Kontinuums ist also unabhängig von der Frequenz. Zwischen den dann folgenden Maxima fällt die Intensität ab mit

$$\varepsilon_\nu \sim \frac{n_e n_i}{\sqrt{kT}}\, e^{-h\nu/kT} \quad \text{für} \quad \nu > \nu_g. \tag{66.6}$$

Der absolute Betrag von ε_ν für die einzelnen Kanten hinter der Grenzfrequenz hängt von individuellen Konstanten des betreffenden Atoms ab.

Für atomaren Wasserstoff gilt

$$\varepsilon_{\nu,H} = 6{,}36 \cdot 10^{-47}\, \frac{n_i n_e}{\sqrt{kT}}\, e^{-h\nu/kT} \left\{ e^{h\nu_g/kT} + 2\, \frac{E_{iH}}{kT} \sum \frac{1}{n^3}\, e^{E_n'/kT} \right\} \tag{66.7}$$

(n und E_n' = Quantenzahl und Energie für den unteren Zustand des betreffenden Grenzkontinuums, von der Ionisierungsgrenze gerechnet). Die Summe ist über die Grenzkontinua, die jenseits von ν_g an der Stelle ν beitragen, zu nehmen.

Der Zahlenwert für $\overline{(Z+s)^2}$ darf nicht aus den Termdefekten gegen Wasserstoff entnommen werden, weil es dabei für die Energieberechnung auf die Mittelung über $1/r$ (r = Radiusvektor der Elektronenbahn) ankommt, während zur Bestimmung der Übergangswahrscheinlichkeiten über r^2 zu mitteln ist. Hier bekommen also die kernfernen Gebiete relativ großes Gewicht und $\overline{(Z+s)^2}$ wird kleiner. Man wird also auf jeden Fall sagen können, daß $\overline{(Z+s)^2}$ für das Kontinuum zwischen dem aus dem Termdefekt nach der Formel

$$\overline{(Z+s)^2} = n^2\, \frac{E_i - E_n}{E_{iH}} \tag{66.8}$$

bestimmten Wert und Z^2 liegen sollte. Mit $\overline{(Z+s)^2} \approx Z^2$ trifft man die bisherigen Meßergebnisse am besten.

Wie man sieht, bietet der glatte Teil des Kontinuums eine gute Möglichkeit, die Ladungsträgerdichte durch eine Absolutmessung der Kontinuumsintensität zu bestimmen, woraus über die SAHA-Gleichung die Temperatur mit großer Genauigkeit bestimmt werden kann. Der unbedeutende Nenner $\sqrt{kT}$ kann durch Schätzung oder einmalige Iteration berücksichtigt werden. Auch der durch Gl. (66.6) beschriebene Intensitätsabfall ist von der Temperatur abhängig, jedoch sehr viel weniger empfindlich. Alle angegebenen Formeln können natürlich mit der SAHA-Gleichung auf reine Temperatur- und Druckabhängigkeit umgerechnet werden.

67. Dispersionstheorie des Plasmas. Da der kontinuierliche Emissionskoeffizient für lange Wellen nach Gl. (66.5) konstant ist, steigt der Absorptionskoeffizient nach dem KIRCHHOFFschen Satz mit der reziproken KIRCHHOFF-PLANCK-Funktion B_ν, also etwa mit $1/\nu^2$ an:

$$\varkappa_\nu = \frac{\varepsilon_\nu}{B_\nu} = 6{,}36 \cdot 10^{-47} \frac{n_e n_i}{(kT)^{\frac{1}{2}}} \frac{c^2}{2\nu^2 kT} \tag{67.1}$$

mit

$$B_\nu \approx \frac{2\nu^2 kT}{c^2} \quad \text{für lange Wellen.} \tag{67.2}$$

Er hat unter Bogenverhältnissen in der Gegend von $\nu = 10^{11} \div 10^{12}\,\text{sec}^{-1}$ den Wert 1 erreicht, d.h. für kleinere als diese Frequenzen kommt bei einer Ausdehnung der Lichtquelle von der Größenordnung 1 cm keine Strahlung mehr aus dem Bogeninneren heraus, nur eine gewisse Oberfläche strahlt schwarz ab. Es könnte uns also gleichgültig sein, ob bei langen Wellen der Absorptionskoeffizient noch irgendwie modifiziert wird oder nicht. Trotzdem wollen wir einen kleinen Seitenblick auf den Gang des Absorptionskoeffizienten im Radiofrequenzgebiet werfen, weil wir dadurch einen Einblick in die Plasmaschwingungen erhalten.

Wir setzen mit SCHLÜTER [*190*] die Stromdichtegleichung (45.18)

$$4\pi\sigma \frac{d}{dt}\left(\frac{\mathfrak{j}}{\omega_L^2}\right) + \mathfrak{j} = \sigma\,\mathfrak{E}$$

an. Die Leitfähigkeit σ drücken wir nach Gl. (51.1) durch den Reibungskoeffizienten

$$\varepsilon_{ie} = \frac{e^2}{\sigma} \tag{67.3}$$

aus und führen diesen wieder auf die mittlere Zahl „auslöschender" Stöße

$$\gamma = \varepsilon_{ie} \frac{\varrho_i + \varrho_e}{m_i m_e} \tag{67.4}$$

zurück, so daß

$$\frac{\partial}{\partial t}\left(\frac{\mathfrak{j}}{\omega_L^2}\right) + \gamma \frac{\mathfrak{j}}{\omega_L^2} = \frac{1}{4\pi}\,\mathfrak{E} \tag{67.5}$$

entsteht. Außerdem geben die MAXWELLschen Gleichungen für ein reines Plasma, in dem keine Polarisation vorhanden sein kann ($\varepsilon = 1$) (DARWIN [*29*], BURKHARDT [*19*]):

$$\Delta\,\mathfrak{E} - \operatorname{grad}\operatorname{div}\mathfrak{E} = \frac{4\pi}{c^2}\frac{\partial\mathfrak{j}}{\partial t} + \frac{\partial^2\mathfrak{E}}{c^2\partial t^2}\,. \tag{67.6}$$

Die Entstehung von Raumladungen ($\operatorname{div}\mathfrak{E} \neq 0$) und von Magnetfeldern ($\operatorname{rot}\mathfrak{E} \neq 0$) sind also noch zugelassen. Wir nehmen jetzt für $\mathfrak{j}$ und $\mathfrak{E}$ räumliche Wellen an, die sich in Richtung des Vektors $\mathfrak{n}$ mit dem Betrag

$$|\mathfrak{n}| = n(1 - ik) \tag{67.7}$$

und

$$k = \frac{c\varkappa}{2\omega n} \tag{67.8}$$

ausbreiten:

$$\mathfrak{j} = \mathfrak{j}_1 \cdot e^{i\omega\left(t - \frac{\mathfrak{n}\mathfrak{r}}{c}\right)}\,; \qquad \mathfrak{E} = \mathfrak{E}_1^{i\omega\left(t - \frac{\mathfrak{n}\mathfrak{r}}{c}\right)}\,, \tag{67.9}$$

wobei die Amplituden $\mathfrak{j}_1$ und $\mathfrak{E}_1$ komplex sein können. Führt man dies in die beiden oberen Gleichungen ein, so ergibt die Ausrechnung:

$$4\pi(i\omega+\gamma)\,\mathfrak{j}_1=\omega_L^2\,\mathfrak{E}_1, \tag{67.10}$$

$$\omega(1-\mathfrak{n}^2)\,\mathfrak{E}_1+\omega\,\mathfrak{n}(\mathfrak{n}\,\mathfrak{E}_1)=4\pi i\,\mathfrak{j}_1. \tag{67.11}$$

Nach der ersten Gleichung müssen $\mathfrak{j}_1$ und $\mathfrak{E}_1$ parallel sein. Daraus folgt, für die zweite, daß auch $\mathfrak{E}_1$ in Richtung der Ausbreitung liegen muß (Longitudinalwellen), oder daß $\mathfrak{E}_1$ senkrecht auf $\mathfrak{n}$ steht, damit $(\mathfrak{n}\,\mathfrak{E}_1)$ verschwindet (Transversalwellen).

Für die Longitudinalwellen (Langmuir) ist rot $\mathfrak{E}_1=0$, es bleibt also von Gl. (67.11) nur die rechte Seite, was bedeutet, daß die Ladungsverschiebungen in Fortpflanzungsrichtung $\mathfrak{j}$ gerade durch den Maxwellschen Verschiebungsstrom $-\frac{1}{4\pi}\frac{\partial\mathfrak{E}}{\partial t}$ kompensiert werden, so daß kein Magnetfeld und auch keine Abstrahlung entsteht. Die verschobenen Ladungen werden wieder zu ihrem Ausgangspunkt zurückgezogen. Faßt man für $\mathfrak{E}_1\parallel\mathfrak{n}$ Gl. (67.10) und (67.11) zusammen, dann fällt der Brechungsindex heraus und es wird

$$\omega(\omega-i\gamma)=\omega_L^2. \tag{67.12}$$

ω_L bekommt also die Bedeutung einer echten Eigenfrequenz für longitudinale Plasmaschwingungen auf Grund der auf die verschobenen Ladungen wirkenden rücktreibenden Kräfte.

Bei transversalen elektromagnetischen Wellen mit $\mathfrak{E}_1\perp\mathfrak{n}$ ist div $\mathfrak{E}=0$, es bilden sich also geschlossene Ströme mit ihren Magnetfeldern ohne Erzeugung von Raumladungen aus, rücktreibende Kräfte treten nicht auf, nur die Massenträgheit der Elektronen und die Reibungskraft. Aus Gl. (67.10) und (67.11) folgt jetzt die bekannte Ecclesche Beziehung

$$1-\mathfrak{n}^2=\frac{\omega_L^2}{(\omega-i\gamma)}, \tag{67.13}$$

in der nicht der für die Dispersionstheorie sonst typische Resonanznenner, etwa von der Form $\frac{1}{\omega_L^2-\omega^2}$ steht. Rechnet man den Absorptionskoeffizienten aus dem Brechungsindex nach Gl. (67.7) und (67.8) aus, so kommt dennoch so etwas wie eine Plasmalinie heraus, deren Maximum je nach Dämpfung von ω_L in Richtung kleinerer Frequenzen liegt und deren „blauer" Dämpfungsflügel in die oben angegebenen Kontinuumsformeln von Kramers einmündet.

Für die Leitfähigkeit findet man (Burkhardt [*19*]) aus Gl. (67.10)

$$\sigma=\frac{\omega_L^2}{4\pi\cdot(i\omega+\gamma)}\approx\frac{n_L e^2}{m_e}\,\frac{1}{(i\omega+\gamma)}. \tag{67.14}$$

Der spezifische Widerstand des Plasmas als Kehrwert von σ setzt sich also aus dem Ohmschen Anteil $R=\gamma\frac{m}{n_L e^2}$ und dem induktiven Anteil $\omega L=\omega\frac{m}{n_L e^2}$ zusammen. Die Dielektrizitätskonstante ist $\varepsilon=1$.

68. Andere Kontinua. Außer dem normalen Elektronenkontinuum können noch weitere Kontinua in Erscheinung treten.

Das sind einmal die Kontinua, die durch die Elektronenaffinität neutraler Atome bedingt sind. Bei manchen Atomen wird nämlich das Coulomb-Feld des Kernes nicht voll durch die Elektronenhülle abgeschirmt, sondern greift über diese hinaus in den Außenraum. So, wie nun bei Ionen ein Grenzkontinuum oder ein Frei-frei-Kontinuum dadurch erzeugt wird, daß ein herannahendes Elektron eingefangen oder abgelenkt wird, so können auch bei Neutralatomen mit einem nennenswert herausgreifenden Kernfeld Anlagerungen oder Ablenkungen von Elektronen Anlaß zu einem Affinitätskontinuum geben. Für Wasserstoff hat zuerst WILDT [*263*] ein solches gefordert, um das Kontinuum im Sonnenspektrum erklären zu können. CHANDRASEKAR [*23*] hat dann für dieses Problem den Absorptionskoeffizienten $\varkappa_\nu$ berechnet, aus dem erwartungsgemäß hervorgeht, daß die Intensität dieses Kontinuums proportional mit $n_H n_e$ geht, also nur bei mäßigen Ionisationsgraden auftreten kann, und daß sich das Frei-frei-Kontinuum vom Ultrarot her bis ins Sichtbare mit abfallender Intensität erstreckt, dem sich eine Art Grenzkontinuum mit dem Maximum bei 8000 Å überlagert. Die Existenz dieses H^--Kontinuums ist von FUCKS [*59*], LOCHTE-HOLTGREVEN und NISSEN [*128*], NISSEN [*160*] und PETERS [*165*] an verschiedenen Entladungen in Wasserstoff und Wasserdampf überzeugend nachgewiesen.

Für andere Elemente, insbesondere Sauerstoff und die Halogene, sind ähnliche Vorgänge zu erwarten, bisher aber noch nicht gefunden.

Zum anderen entstehen bei der Wechselwirkung reaktionsfähiger Neutralteilchen Rekombinationskontinua, die in den Bogenaureolen eine Rolle spielen, wie MAECKER [*138*] für den Hochstrombogen und BURHORN [*16*] für das FeO-Molekül im Eisenbogen wahrscheinlich gemacht haben.

Quantitative Angaben liegen hierüber noch nicht vor, es mag aber in diesem Zusammenhang darauf hingewiesen werden, daß man sich bei einem Bogenspektrum mit der Umrechnung von der gemessenen Querverteilung $I(x)$ auf die radiale Strahlungsdichte $i(r)$ durch das Molekülkontinuum in den Aureolen zum Elektronenkontinuum in Achsennähe hindurchrechnen kann, so daß das Molekülkontinuum zu eliminieren ist.

69. Selbstabsorption. Die bisherigen Zusammenhänge gelten ohne Selbstabsorption, d.h. bei Emission aus optisch dünner Schicht. Vor Gebrauch der Formeln ist also bei jeder Lichtquelle nachzuweisen, daß diese Voraussetzung erfüllt ist. Dazu müssen wir noch die Absorptionsvorgänge betrachten. Sie werden beherrscht durch den Absorptionskoeffizienten $\varkappa_\nu$, der angibt, wie groß an jeder Stelle l die relative Abnahme der Intensität mit dem Weg ist:

$$\varkappa_\nu = -\frac{d I_\nu}{I_\nu \, d l}. \tag{69.1}$$

Bei homogener Schicht der Länge l ist also von der eingestrahlten Intensität $I_{\nu,0}$ nur noch der Bruchteil

$$\frac{I_\nu}{I_{\nu,0}} = e^{-\varkappa_\nu l} \tag{69.2}$$

vorhanden. Ist nun die optische Tiefe $\varkappa_\nu l \ll 1$, dann ist die absorbierte Strahlungsleistung vernachlässigbar klein. Ist dagegen $\varkappa_\nu l \gg 1$, dann ist praktisch die ganze einfallende Intensität absorbiert. Der Absorptionskoeffizient ist eine Materialkonstante, die aber noch von Druck, Temperatur, Strahlungsdichte und vor allem von der Frequenz abhängt. Im vollständigen thermischen Gleichgewicht gilt der KIRCHHOFFsche Satz:

$$\varepsilon_\nu = \varkappa_\nu B_\nu, \tag{69.3}$$

worin B_ν die Intensität der Hohlraumstrahlung ist:

$$B_\nu = \frac{2 h \nu^3}{c^2} \frac{1}{e^{h\nu/kT} - 1}. \tag{69.4}$$

Bei Fehlen der Hohlraumstrahlung ist der KIRCHHOFFsche Satz auch dann noch anwendbar, wenn die Zahl der absorptionsfähigen Atome erhalten geblieben ist, d.h. wenn nach wie vor die BOLTZMANNsche Energieverteilung für die Anregung gilt. Unter ε_ν ist hier der vollständige Emissionskoeffizient, der sich aus dem für spontane Emission ε'_ν und dem durch die Strahlungsdichte erzwungenen Emissionsanteil zusammensetzt, gemeint. Nach EINSTEIN [36] gilt

$$\varepsilon_\nu = \frac{\varepsilon'_\nu}{1 - e^{-h\nu/kT}} \tag{69.5}$$

und damit für den KIRCHHOFFschen Satz

$$\varepsilon'_\nu = \varkappa_\nu \left(1 - e^{-h\nu/kT}\right) B_\nu = \varkappa'_\nu B_\nu . \tag{69.6}$$

Betrachten wir nun in einem homogenen endlichen Plasmavolumen einen cm³, dann nimmt in diesem die Intensität um ε_ν zu, wird aber gleichzeitig durch Absorption um den Betrag $-I_\nu \varkappa_\nu$ geschwächt, so daß die Intensitätsänderung an jeder Stelle

$$\frac{dI_\nu}{dl} = \varepsilon_\nu - I_\nu \varkappa_\nu = \varkappa_\nu (B_\nu - I_\nu) \tag{69.7}$$

ist. Die Integration vom Anfang des Volumens entlang irgend einer Richtung bis l ergibt zunächst

$$I_\nu = B_\nu - C\, e^{-\varkappa_\nu l}. \tag{69.8}$$

Da für $l = 0$ auch noch $I_\nu = 0$ ist, wird $C = B_\nu$ und

$$I_\nu = B_\nu \left(1 - e^{-\varkappa_\nu l}\right). \tag{69.9}$$

Für kleine $\varkappa_\nu l$ führt die Reihenentwicklung zu

$$I_\nu = \varkappa_\nu l B_\nu \quad \text{für} \quad \varkappa_\nu l \ll 1 , \tag{69.10}$$

oder wenn wir noch die erzwungene Emission in Abzug bringen,

$$I_\nu = \varkappa'_\nu l B_\nu . \tag{69.11}$$

Für kleine optische Tiefen wächst die Intensität also proportional mit der Schichtdicke. Für große optische Tiefen ist dagegen die Intensität

$$I_\nu = B_\nu \qquad (\varkappa_\nu l \gg 1), \tag{69.12}$$

d.h. ein Plasma kann höchstens die Hohlraumstrahlung seiner Temperatur aussenden.

70. Linienemission und -absorption. Solange wir diese Betrachtungen auf ein spektrales Kontinuum beziehen, ist nichts hinzuzufügen. Aber auch auf Linien, die immer mehr oder weniger verbreitert sind, treffen die angegebenen Formeln zu, wenn man nur das betrachtete Frequenzintervall genügend klein macht. Man könnte dann an jeder Stelle der Linie den Vergleich mit einem Flächenstrahler durchführen und durch Integration über ν die Linienintensitäten erhalten. Leider kann man aber die Kontur der Linien nur in den wenigsten Fällen auflösen, weswegen wir gleich auf das Integral über das Linienprofil

$$\int\limits_L \varepsilon_\nu \, d\nu = i_L \tag{70.1}$$

übergehen, das im allgemeinen unabhängig von der Linienform ist. Den Zusammenhang zwischen i_L und A_n^m hatten wir in Gl. (65.1) schon angegeben. Wir wollen hier noch den Begriff der Gesamtabsorption einführen, der häufig für Intensitätsberechnungen von Linien verwendet wird.

Es ist der relative Verlust einer einfallenden Strahlung $I_{\nu,0}$ bis zur Stelle l bei optisch dünner Schicht

$$A_\nu = \frac{I_{\nu,0} - I_\nu}{I_{\nu,0}} = \varkappa_\nu l. \tag{70.2}$$

Die Gesamtabsorption der Linie wird also

$$A_L = l \int_L \varkappa_\nu \, d\nu. \tag{70.3}$$

Nun liefert die Dispersionstheorie von H. A. LORENTZ (s. unten) für die Gesamtabsorption einer Linie die von der Verbreiterung und Frequenz unabhängige Beziehung

$$A_L = \frac{\pi e^2}{m c} \mathfrak{N} \cdot l = \frac{\pi e^2}{m c} n_n f_L l, \tag{70.4}$$

in der $\mathfrak{N}$ die Zahl der Ersatzoszillatoren je Volumeneinheit und f_L die für jede Linie charakteristische Oszillationsstärke ist. f_L und A_n^m können bei verbotenen Linien veränderlich werden, wenn nämlich äußere Störungen oder Felder die Kopplung zwischen Bahndrehimpuls und dem Elektronenspin im Atom schwächen.

Der Ableitung gemäß kann man auch auf A_L den KIRCHHOFFschen Satz anwenden und kommt so wieder zur Linienintensität:

$$I_L = A_L B_\nu (1 - e^{-h\nu/kT}) = \frac{\pi e^2}{m c} n_0 \frac{g_n}{Z_0^{(i)}} \cdot e^{-E_n/kT} f_L l \frac{2 h \nu^3}{c^2} \frac{1 - e^{-h\nu/kT}}{e^{h\nu/kT} - 1}. \tag{70.5}$$

Mit der EINSTEINschen Gleichung

$$E_m - E_n = h\nu \tag{70.6}$$

und durch Erweiterung des letzten Bruchs mit $\exp(-h\nu/kT)$ wird

$$I_L = \frac{2\pi e^2 h \nu^3}{m c^3} n_0 f_L \frac{g_n}{Z_0^{(i)}} e^{-E_m/kT} l. \tag{70.7}$$

Durch Vergleich mit der anderen Intensitätsformel Gl. (65.4) ergibt sich

$$f_L = \frac{g_m}{g_n} \frac{m c^3}{8\pi^2 e^2 \nu^2} A_n^m. \tag{70.8}$$

Die Größen f_L und A_n^m müssen berechnet oder gemessen werden und können für eine Anzahl von Linien Tabellenwerken entnommen werden.

71. Dämpfungs- und Stoßverbreiterung[1]. Die spektrale Verteilung der Intensität einer Linie ist bei optisch dünner Schicht dem spektralen Verlauf des Absorptionskoeffizienten $\varkappa_\nu$ proportional. Die häufigste Ursache für die Linienverbreiterung ist die Dämpfung. Dem Grundgedanken der Disperionstheorie von H. A. LORENTZ folgend, geht man von der Kraftgleichung eines schwingenden Elektrons

$$m_e \frac{d^2 x}{d t^2} + m_e \omega_0^2 x + m_e \gamma \frac{d x}{d t} = -e \mathfrak{E} \tag{71.1}$$

aus. Darin ist d^2x/dt^2 die Beschleunigung, $m_e \omega_0^2 x$ die rücktreibende Kraft, ausgedrückt durch die Eigenfrequenz, $m_e \gamma \, dx/dt$ die Reibungskraft mit der Dämpfungskonstante γ und $-e\mathfrak{E}$ die elektrische Kraft des Strahlungsfeldes. Die Lösung dieser Gleichung mit der einfallenden Welle $\mathfrak{E} = \mathfrak{E}_0 \exp(i\omega t)$ und dem

[1] Die Grundlagen für die hier und in den folgenden Ziffern kurz behandelten Fragen sind eingehend in dem Beitrag von R. G. BREENE in Bd. XXVII dieses Handbuches dargestellt.

Ansatz $x = x_0 \exp(i\omega t)$ führt über die Polarisierbarkeit eines brechenden und vor allem absorbierenden Mediums zu dem Verlauf des Absorptionskoeffizienten

$$\varkappa_\nu = \frac{\pi e^2}{m c} \Re \frac{2\gamma}{(\varDelta\omega)^2 + (\gamma/2)^2} \quad \text{mit} \quad \varDelta\omega = 2\pi\varDelta\nu, \tag{71.2}$$

wo $\varDelta\nu$ der Frequenzabstand von der Linienmitte und der zweite Bruch eine Gewichtsfunktion mit dem Integral Eins ist. In der Linienmitte $(\varDelta\omega)^2 \ll (\gamma/2)^2$ ist $\varkappa_\nu$ erwartungsgemäß konstant, fällt aber mit wachsendem Abstand von der Mitte rasch ab und läuft für $(\varDelta\omega)^2 \gg (\gamma/2)^2$ mit $1/(\varDelta\omega)^2 \sim 1/(\varDelta\lambda)^2$ aus, ein Charakteristikum für die Dämpfungsflügel.

Aus Gl. (71.2) berechnet sich die (ganze) Halbwertsbreite der Linie zu

$$2(\varDelta\omega)_{\frac{1}{2}} = 2\pi(2\varDelta\nu)_{\frac{1}{2}} = \frac{2\pi c}{\lambda_0^2}(2\varDelta\lambda)_{\frac{1}{2}} = \gamma. \tag{71.3}$$

Die Halbwertsbreite in Kreisfrequenzeinheiten ist also gleich der Dämpfungskonstanten. Außerdem hat γ noch die Bedeutung der reziproken Abklingzeit.

Für ungestörte Atome kommt als Dämpfung nur der Strahlungsverlust des beschleunigten Elektrons gemäß Gl. (66.3) in Frage. In diesem Fall ist

$$\gamma = \frac{2e^2}{3m_e c^3}\omega_0^2 = \frac{8\pi e^2 \nu_0^2}{3m_e c^3} = \frac{8\pi e^2}{m c \lambda_0^2} \tag{71.4}$$

und die (natürliche) Halbwertsbreite in Wellenlängeneinheiten

$$2(\varDelta\lambda)_{\frac{1}{2}} = \frac{4\pi e^2}{3 m c^3} = 1{,}18 \cdot 10^{-4}\,\text{Å} \tag{71.5}$$

wird unabhängig von der Wellenlänge und ist so klein, daß sie in einem Plasma durch andere Verbreiterungen überdeckt ist.

Dagegen ist die Stoßdämpfung (H. A. LORENTZ [*130*]) bei den im Bogenplasma vorhandenen Drucken wichtig. Die auf ein schwingendes Elektron kommenden Stöße anderer Teilchen wirken im Mittel ebenfalls wie eine Reibung, und γ erhält nun die Bedeutung der Stoßzahl je Zeiteinheit. Nach der kinetischen Gastheorie ist

$$\gamma \approx v_k n_k Q_{ak} = 2(\varDelta\omega)_{\frac{1}{2}}, \tag{71.6}$$

wo v_k bzw. n_k die thermische Geschwindigkeit bzw. die Zahl je Volumeneinheit der stoßenden Teilchen und Q_{ak} der für die Stoßdämpfung maßgebliche Querschnitt des leuchtenden Atoms ist. Über Q_{ak} ist wenig bekannt. Um einmal ganz grob die Verbreiterung abzuschätzen, wählen wir als stoßende Teilchen Elektronen mit $v_e \approx 10^8$ cm/sec, $n_e \approx 10^{16}\,\text{cm}^{-3}$ und $Q \approx 100 \cdot 10^{-16}\,\text{cm}^2$. Dann ist $\gamma \approx 10^{10}\,\text{sec}^{-1}$, also 100mal größer als die Strahlungsdämpfungskonstante mit rund $10^8\,\text{sec}^{-1}$. Genau so verhalten sich die Linienbreiten.

Die Wirkung der Dämpfung auf die Linienbreite kann man sich auch so vorstellen, daß die Frequenz des strahlenden Elektrons durch das Abklingen der Amplitude oder durch Phasensprünge infolge von Stößen verändert wird. Die FOURIER-Analyse gibt dann die Verteilung der Frequenzen.

Die Quantentheorie dagegen sagt nichts über den Vorgang der Ausstrahlung aus. Vielmehr wird dort die Verbreiterung dem oberen und unteren Term entsprechend den zugehörigen, ungestörten Lebensdauern der Terme zugeschrieben. Demnach ist γ durch $\gamma_m + \gamma_n$ zu ersetzen.

72. DOPPLER-Verbreiterung. Als weitere Verbreiterungsursache fügen wir den thermischen DOPPLER-Effekt ein, der darauf beruht, daß sich die leuchtenden Atome thermisch bewegen und dadurch die Frequenz je nach Größe und Richtung der Geschwindigkeit relativ zum Beobachter durch den DOPPLER-Effekt verändert wird. Ist v_y die Geschwindigkeitskomponente des Atoms auf den Beobachter zu, dann ist der Bruchteil der Teilchen, deren Geschwindigkeit zwischen v_y und $v_y + d v_y$ liegt, gegeben durch

$$\frac{dn}{n} = \frac{1}{\sqrt{\pi}} e^{-\frac{v_y^2}{v_{y,0}^2}} \frac{d v_y}{v_{y,0}}, \tag{72.1}$$

wenn $v_{y,0} = \sqrt{2kT/m_a}$ ist.

Der DOPPLER-Effekt wird beschrieben durch

$$\frac{\Delta\nu}{\nu} = \frac{\Delta\lambda}{\lambda} = \frac{v_y}{c} \tag{72.2}$$

und für $v_{y,0}$ durch

$$\frac{\Delta\nu_D}{\nu} = \frac{\Delta\lambda_D}{\lambda} = \frac{v_{y,0}}{c}, \tag{72.3}$$

woraus entsteht

$$\frac{\Delta\nu}{\Delta\nu_D} = \frac{\Delta\lambda}{\Delta\lambda_D} = \frac{v_y}{v_{y,0}}. \tag{72.4}$$

Da die Linienintensität mit der Teilchenzahl proportional wächst, gilt

$$\frac{dn}{n} = \frac{dI_L}{I_L} \tag{72.5}$$

und

$$\frac{dI_L}{d(\Delta\nu)} = \frac{dI_L}{d\nu} = I_\nu. \tag{72.6}$$

Die Verteilung der Intensität bei reiner DOPPLER-Verbreiterung ist danach

$$I_\nu = I_L \frac{1}{\sqrt{\pi}(\Delta\nu)_D} \cdot e^{-(\Delta\nu/\Delta\nu_D)^2}, \tag{72.7}$$

sie hat also die Form der GAUSSschen Fehlerverteilung. Die Intensität ist auf die Hälfte gesunken, wenn

$$\tfrac{1}{2} = e^{-(\Delta\nu/\Delta\nu_D)^2}, \quad \text{also} \quad 2(\Delta\nu)_{\frac{1}{2}} = 2\sqrt{\log 2}\,\Delta\nu_D = 1{,}665\,\Delta\nu_D \tag{72.8}$$

ist. Die DOPPLER-Breite hat bei Bogenverhältnissen die Größenordnung 0,1 Å, sie ist daher immer groß gegen die natürliche Linienbreite, wird aber vergleichbar mit der Stoßverbreiterung, meistens sogar von dieser übertroffen. Eine genaue Diskussion der Überlagerung von Stoß- und DOPPLER-Verbreiterung zeigt, daß in der Mitte meist der DOPPLER-Kern dominiert, weil der Abfall mit der e-Funktion viel schneller erfolgt als bei der Dämpfung, die ihrerseits hauptsächlich den Linienflügeln ihr Abfallgesetzt mit $1/(\Delta\lambda)^2$ aufprägt.

73. Druckverbreiterung. Bei der Stoßdämpfung handelt es sich um kurzzeitige Störungen, die im ausgesandten Wellenzug Phasensprünge von der Größe Eins und darüber hervorrufen müssen, um wesentlich zu sein.

Ist dagegen der Störvorgang von längerer Dauer (LENZ [*124*]), so daß sich eine quasistationäre Frequenzverstimmung und Linienverschiebung einstellen kann, dann spricht man von Druckverbreiterung, für die die Bedingung gilt

$$t_s \Delta\omega \gg 1, \tag{73.1}$$

wobei t_s die Störungszeit ($t_s \approx$ kleinster Abstand der Störungspartner/Geschwindigkeit) ist.

Für diesen Effekt kommen in erster Linie Ionen und auch Elektronen mit ihrer Wirkung auf STARK-Effekt-empfindliche Linien, vor allem Wasserstoff in Frage. Für H_β wollen wir die STARK-Effekt-Verbreiterung angeben.

Der mittlere Abstand der Ionen in einem Plasma ist etwa $n_i^{-\frac{1}{3}}$ und die COULOMB-Feldstärke ungefähr $F \approx e n_i^{\frac{2}{3}}$.

Um von der Ionendichte unabhängig zu werden, führen wir die Normalfeldstärke

$$F_0 = 2{,}61\, e n_i^{\frac{2}{3}} \tag{73.2}$$

ein und beziehen alle statistisch vorkommenden Feldstärken F auf diese. Die reduzierte Feldstärke β ist also

$$\beta = \frac{F}{F_0}. \tag{73.3}$$

Die relative Häufigkeit $W(\beta)\, d\beta$, mit der β vorkommt, ist zuerst von HOLTSMARK [89] für diese Zwecke berechnet worden. Sehr schwache und sehr starke Felder sind sehr selten, die Verteilung muß also bei mittleren Feldern ein Maximum haben. Die Verschiebung der einzelnen STARK-Effekt-Komponenten ist bei Wasserstoff symmetrisch und der jeweiligen Feldstärke proportional. Man verbreitert nun jede einzelne Komponente mit der Feldverteilungsfunktion $W(\beta)\, d\beta$ und addiert diese verwaschenen Einzellinien, gewichtet mit ihrem Intensitätsanteil, zusammen auf. Dadurch erhält man das in Fig. 45 für H_β dargestellte Linienprofil $S(\Delta\lambda/F_0)$, dessen Gesamtfläche einschließlich der linken Hälfte auf Eins normiert ist (JÜRGENS [95]). Als Abszisse ist $\Delta\lambda/F_0$ aufgetragen, die Kurve zeigt also das Linienprofil bei der Normalfeldstärke $F_0 = 1$.

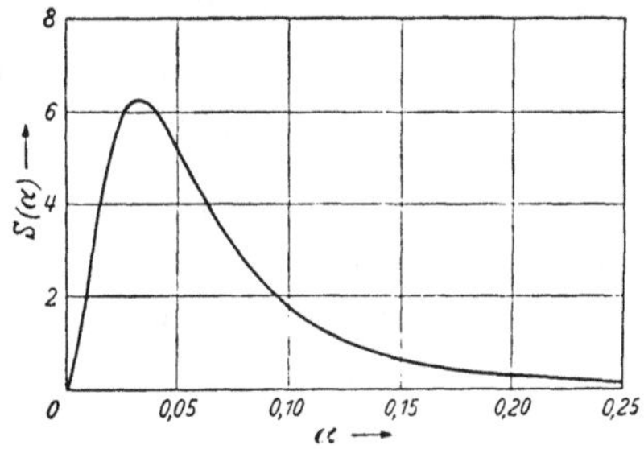

Fig. 45. Kontur der Wasserstoff-Linie H_β 4861 Å unter dem Einfluß der STARK-Effekt-Verbreiterung. Abszisse: Abstand von der Linienmitte $\Delta\lambda$/Normalfeldstärke F_0. Fläche beider Linienhälften auf Eins normiert.

Bei größeren Normalfeldstärken (Ionendichten) muß der Abszissenmaßstab proportional gereckt und der Ordinatenmaßstab in gleichem Maße gestaucht werden, um die Normierungsbedingung einzuhalten. Anwendungsbeispiele folgen später. In der Linienmitte sollte nach Fig. 45 die Intensität verschwinden, weil H_β keine unverschobene Komponente hat. In der Tat beobachtet man dort bei H_β eine deutliche Einsenkung. Bei den anderen Wasserstofflinien wird die leere Mitte durch die dämpfungsverbreiterten Nullkomponenten, die im elektrischen Feld nicht verschoben werden, ausgefüllt.

Die Linien aller anderen Elemente haben einen mehr oder weniger großen *quadratischen* STARK-Effekt. Da aber weder die Größe der Verschiebung noch die Intensitäten der einzelnen Komponenten im allgemeinen bekannt sind, kann man nicht solche normierten Linienprofile wie bei den Wasserstofflinien zeichnen. Man muß sich daher auf Relativmessungen beschränken, indem man für eine anderweitig bekannte Normalfeldstärke F_0 das gemessene Profil, auf Fläche Eins normiert, über $\Delta\lambda$ aufträgt und das zweite Profil dieser Linie, aus dem das entsprechende F_0 bestimmt werden soll, im gleichen Diagramm darstellt. Da nun für quadratischen STARK-Effekt

$$\left(\frac{\Delta\lambda}{F_0^2}\right)_1 = \left(\frac{\Delta\lambda}{F_0^2}\right)_2 \quad \text{oder} \quad \frac{\Delta\lambda_1}{\Delta\lambda_2} = \frac{{}_1F_0^2}{{}_2F_0^2} \tag{73.4}$$

ist, gibt der Faktor, mit dem das zweite Profil unter Einhaltung der Normierungsbedingung zu stauchen ist, um mit dem ersten zur Deckung zu kommen, das Verhältnis der F_0^2 an. Eine andere Möglichkeit ist, in ein und demselben Plasma die Linienform irgend eines Elementes mit der Linienkontur von zugesetztem Wasserstoff zu vergleichen und zu eichen.

Ist in einzelnen Fällen die STARK-Effekt-Konstante, definiert durch die Gleichung

$$\Delta \nu = \frac{C}{e^2} F^2 \tag{73.5}$$

bekannt, dann setzt sich nach UNSÖLD [*229*] und LINDHOLM [*125*] die ganze Halbwertsbreite aus dem Ionenanteil

$$\delta_i = 50.3 \cdot 10^{-10}\, T^{\frac{1}{6}} C^{\frac{2}{3}} \left(\frac{1}{\mu_a} + \frac{1}{\mu_i}\right) n_i, \ \mathrm{cm}^{-1} \tag{73.6}$$

(μ = Atomgewicht) und dem entsprechenden Elektronenanteil zusammen. Anwendungen auf den Hg-Bogen findet man bei ROMPE und SCHULZ [*201*], [*175*] und KOCH und DUNSTÄDTER [*108*]. In ähnlicher Weise wird auch eine Linienverschiebung durch den quadratischen STARK-Effekt verursacht (SCHULZ [*198*]).

74. Verbreiterung und Absorptionskoeffizient. Den absoluten Absorptionskoeffizienten $\varkappa_\nu$ an irgend einer Stelle der Linie finden wir durch Differentiation der Gesamtabsorption pro cm A_L/l nach ν oder, was auf dasselbe hinauskommt, indem wir die auf das Integral Eins reduzierte Verbreiterungsfunktion von ν an die Gesamtabsorption pro cm als Faktor anhängen. So ergibt sich für die Dämpfungsverbreiterung wieder

$$\varkappa_\nu = \frac{\pi e^2}{m c} n_n f_L \frac{\gamma}{(2\pi \Delta\nu)^2 + (\gamma/2)^2} \tag{74.1}$$

mit dem Bruch als Gewichtsfunktion. Für den DOPPLER-Effekt wird entsprechend

$$\varkappa_\nu = \frac{\pi e^2}{m c} n_n f_L \frac{e^{-\left(\frac{\Delta\nu}{\Delta\nu_D}\right)}}{\sqrt{\pi}\,\Delta\nu_D}, \tag{74.2}$$

wo ebenfalls der letzte Bruch die Gewichtsfunktion darstellt. Bei der STARK-Effekt-Verbreiterung ist das auf Eins normierte Linienprofil über $\Delta\lambda/F_0 = \alpha$ aufgetragen. Es ist also

$$\int_{-\infty}^{+\infty} S(\alpha)\, d\alpha = \int_{-\infty}^{+\infty} S(\alpha) \frac{d\alpha}{d\lambda} \frac{d\lambda}{d\nu} d\nu = \int_{-\infty}^{+\infty} S(\alpha) \frac{1}{F_0} \frac{\lambda_0^2}{c} d\nu = 1. \tag{74.3}$$

Mit dem Integranden als Gewichtsfunktion ist dann

$$\varkappa_\nu = \frac{\pi e^2}{m_e c} n_n f_L \frac{\lambda_0^2}{F_0 c} S(\alpha) = \frac{\pi e^2}{m_e \nu_0^2 F_0} n_n f_L S(\alpha). \tag{74.4}$$

Die Emissionskoeffizienten ε'_ν gewinnt man aus den Absorptionskoeffizienten wieder nach dem KIRCHHOFFschen Satz (69.6).

75. Wachstumskurve. Aus allem geht hervor, daß $\varkappa_\nu$ wesentlich von dem Maß der Verbreiterung abhängt, weil die Gesamtintensität der Linie im allgemeinen nicht durch die Verbreiterung beeinflußt wird. Daher hängt die wichtige Frage, ob eine Linie schon durch Selbstabsorption vor allem in der Mitte geschwächt ist, unmittelbar mit ihrer Breite zusammen. Solange die Schicht noch optisch dünn ist, wächst die Intensität an jedem Punkt der Linie mit der Schichtdicke proportional an. Bei zunehmender optischer Tiefe macht sich zuerst in der Linienmitte die Absorption bemerkbar, bis schließlich bei größeren Tiefen die Linienmitte wie ein schwarzer Körper dieser Temperatur strahlt, während die Linienflügel noch ungeschwächt sind. Steigert man die optische Tiefe noch mehr, dann haben schon größere Bereiche in der Umgebung der Linienmitte die KIRCHHOFF-PLANCK-Funktion erreicht und die Linie ist abgeflacht. Die Gesamtintensität der Linie I_L wächst daher, wie V. D. HELD [82] gezeigt hat, zunächst proportional mit der Schicht an, bleibt dann, wenn das Gros der Linie „angestoßen" ist, etwa konstant und wächst schließlich früher oder später, je nach Größe der Verbreiterung, mit der Wurzel aus der Schichtdicke weiter (Wachstumskurve). Da man meistens die Linienkontur nicht auflösen kann, gibt der Gang der Intensität einer Linie mit der Schichtdicke wichtige Aufschlüsse über die Selbstabsorption.

76. Strahlung aus inhomogenen Schichten. Bisher haben wir nur homogene Schichten betrachtet. Liegt ein Temperaturfeld vor, so gilt auch jetzt noch

$$\frac{d I_\nu}{d l} = \varepsilon_\nu - \varkappa_\nu I_\nu; \qquad (69.7)$$

die Integration kann aber nicht mehr, da nun $\varkappa_\nu$ und B_ν von $T(l)$ abhängen, geschlossen durchgeführt werden. Betrachten wir zunächst den Fall optisch dünner Schicht, dann ist

$$I_\nu = \int \varepsilon_\nu(T_l)\, d l. \qquad (76.1)$$

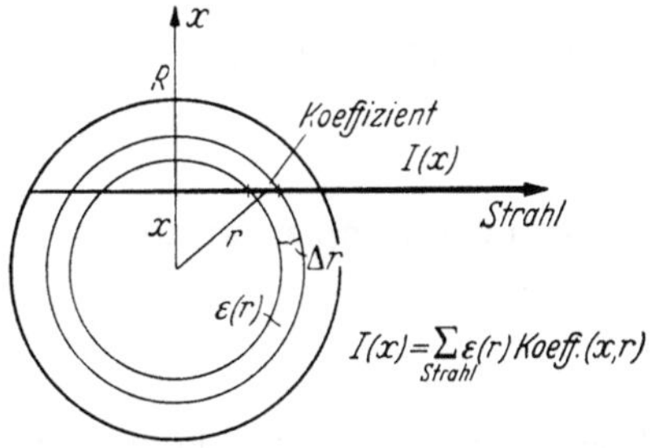

Fig. 46. Die beobachtete Intensität $I(x)$ setzt sich bei rotationssymmetrischen Lichtquellen aus den Beträgen der einzelnen Kreisringe verschiedener Emissionskoeffizienten zusammen.

Ist die Lichtquelle rotationssymmetrisch, so hängt ε_ν nur von r ab, und es wird, wenn man die Emissionskoeffizienten der ganzen Linie $i_L = \int_L \varepsilon_\nu\, d\nu$ ins Auge faßt,

$$I_L(x) = 2 \int_{r=x}^{\infty} i_L(r) \frac{r\, d r}{\sqrt{r^2 - x^2}}, \qquad (76.2)$$

wie man aus Fig. 46 ablesen kann.

Meistens wird die Querverteilung $I_L(x)$ einer Linie beobachtet und nach $i_L(r)$ gefragt. Dann muß diese ABELsche Integralgleichung — alle Umformungen bringen keine Vereinfachung — numerisch gelöst werden, indem man innerhalb kleiner Kreisringe $i_L(r)$ als konstant vor das Teilintegral zieht.

Wenn in einer inhomogenen Lichtquelle auch noch teilweise Selbstabsorption auftritt, dann müßte man die ganze Theorie des Strahlungsgleichgewichtes hinzuziehen, wovon wir hier Abstand nehmen wollen.

In einigen Fällen wird aber der Zustand interessant, in dem so starke Absorption in der Lichtquelle stattfindet, daß man lokal mit der Strahlungsdichte des Hohlraumes rechnen kann. Dann „diffundiert" die Strahlung durch dauernde Umwandlung von Lichtquanten in Anregungsenergie und umgekehrt nach Gebieten kleinerer Strahlungsdichte und somit kleinerer Temperatur. Nach außen dringt dann nur die schwarze Strahlung einer gewissen Mantelfläche, von der aus der optische Weg $\varkappa_\nu l$ nach außen von der Größenordnung Eins ist. Diesem Bild entsprechend machen wir einen Analogieschluß von der Teilchendiffusion zur Strahlungsdiffusion.

Von einem Strahl mit der Anfangsintensität

$\mathfrak{j}_0$	$I_{\nu,0}$
Teilchen/cm² sec	Strahlungsenergie/cm² sec

ist an der Stelle l nur noch der Bruchteil

$$\frac{\mathfrak{j}}{\mathfrak{j}_0} = e^{-nQl}, \qquad \frac{I_\nu}{I_{\nu,0}} = e^{-\varkappa_\nu l} \tag{76.3}$$

vorhanden. Demnach ist die freie Weglänge

$$\lambda_{\mathfrak{j}} = \frac{1}{nQ}, \qquad \lambda_\nu = \frac{1}{\varkappa_\nu}. \tag{76.4}$$

Beim Vorgang der Diffusion gilt nach dem FICKschen Gesetz

$$\mathfrak{j} = -\frac{\lambda v}{3}\,\mathrm{grad}\,n, \qquad I_\nu = -\frac{c}{3\varkappa_\nu}\,\mathrm{grad}\,u_\nu. \tag{76.5}$$

Da die Strahlungsdichte

$$u_\nu = \frac{8\pi h\nu^3}{c^3}\,\frac{1}{e^{h\nu/kT}-1} \tag{76.6}$$

nur von der Temperatur abhängt, kann man auch schreiben

$$I_\nu = -\frac{c}{3\varkappa_\nu}\,\frac{d u_\nu}{d T}\,\mathrm{grad}\,T. \tag{76.7}$$

Gegen diese Ableitung könnte man den Einwand machen, daß materielle Teilchen sofort nach dem Stoß in eine beliebige Richtung weiterfliegen, während Lichtquanten nach der Absorption eine Zeit lang in Form der Anregungsenergie im absorbierenden Teilchen verbleiben und auch danach durch stufenweise Emission als Lichtquanten anderer Frequenzen erscheinen oder durch Stöße zweiter Art in kinetische Energie umgewandelt werden können. Demgegenüber muß aber an unsere Voraussetzung lokalen thermischen Gleichgewichtes einschließlich der Hohlraumstrahlung erinnert werden, derzufolge für jedes verschwindende Quant ein neues in beliebiger Richtung ausgesandt wird.

Die Strahlungsdiffusion leistet also einen Beitrag zur Wärmeleitung. Bei großem Absorptionskoeffizient ist dieser aber wegen der kleinen freien Weglänge λ_ν gering (bei großer freier Weglänge, d. h. optisch dünner Schicht natürlich auch), während bei mittelgroßen Absorptionskoeffizienten, etwa in den Flügeln starker Linien oder in einem verhältnismäßig starken Kontinuum dieser Beitrag mit dem Wärmeleitwert von Atomen vergleichbar wird, wie Untersuchungen von ROMPE und SCHULZ [*176*] und MANNKOPFF [*147*] gezeigt haben.

D. Die Bogensäule.

I. Allgemeines.

77. Definition. Unter der Bogensäule möge der Teil der Bogenentladung verstanden werden, der zwischen den Elektroden-Fallgebieten liegt. Die Grenzen sollen so weit von den Elektroden entfernt bleiben, daß mit Sicherheit keine Beeinträchtigung des thermischen Gleichgewichtes durch die Elektrodenvorgänge zu befürchten ist. Nach den Überlegungen im Anfang von Abschnitt C können wir dann mit einer lokal einheitlichen Temperatur rechnen und die unter dieser Voraussetzung gültigen Gesetzmäßigkeiten anwenden.

In der Säule stellen sich von selbst solche Zustände, insbesondere eine solche Temperatur ein, daß der durch den äußeren Stromkreis bestimmte Strom fließen kann. Daß dies möglich ist, d.h. daß überhaupt ein Bogen existiert, ist auf die Tatsache zurückzuführen, daß ein neutrales Gas durch Energiezufuhr ionisiert, also leitfähig gemacht werden kann. In diesem Zusammenhang wird die Energiebilanz von Interesse, denn von ihr hängt in hohem Maße die Temperatur und damit die Leitfähigkeit ab.

Meistens herrscht im ganzen Bogen ein einheitlicher Druck, denn eine kleine Druckschwankung würde sich mit etwa Schallgeschwindigkeit des Plasmas (einige 10^5 cm/sec) ausgleichen. Lokale Druckänderungen können von äußeren Kräften bewirkt werden, von denen nur die magnetischen LORENTZ-Kräfte praktische Bedeutung für die Bogensäule haben. Ein elektrisches Feld kann keinen Druck in der Bogensäule erzeugen, da sie nahezu quasineutral ist.

78. Bestimmung der optischen Schichtdicke. Viele Eigenschaften der Säule werden aus dem Spektrum entnommen. Die zu diesem Zweck verwendeten Gesetze gelten aber meistens entweder für Emission aus optisch dünner oder seltener aus optisch dicker Schicht. Es ist daher wichtig, in jedem Fall vor der Messung zu prüfen, ob die jeweilige Voraussetzung erfüllt ist.

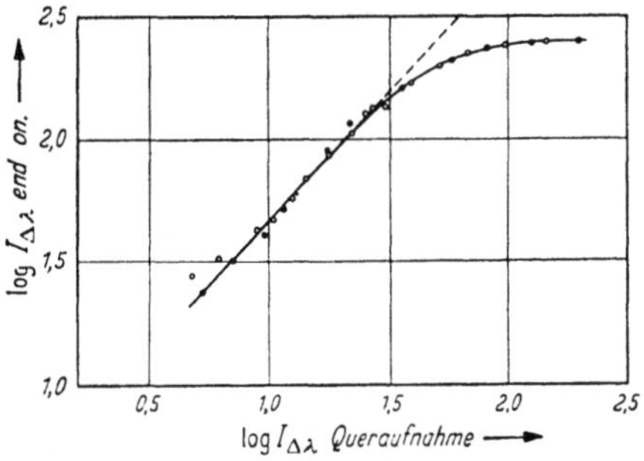

Fig. 47. Intensität in Achsenrichtung als Funktion der Querintensität beim wasserstabilisierten Rohrbogen für verschiedene Wellenlängen innerhalb der Linie H_α. Solange die optische Tiefe $\varkappa_\nu l \ll 1$ ist, wachsen beide Intensitäten proportional mit $\varkappa_\nu$ (45°-Neigung). Für $\varkappa_\nu l \gg 1$ erreichen beide Intensitäten wegen Selbstabsorption die schwarze Strahlung (Endpunkt der Kurve).

Zum Nachweis der Emission aus optisch dünner Schicht kann dies einmal dadurch geschehen, daß die geometrische Schichtdicke l der homogenen Schicht variiert wird und gleichzeitig festgestellt wird, ob sich die Intensität gemäß Gl. (69.11) proportional mitändert. Dabei ist zur Erzielung ausreichender Genauigkeit eine möglichst große Variation der Schichtdicke anzustreben. Zu diesem Zweck kann z.B. die emittierte Strahlung mit einem geeigneten Hohlspiegel nochmals durch die Lichtquelle geschickt werden, wodurch die Intensität verdoppelt werden müßte, wenn keine Selbstabsorption erfolgt. Besser ist es, die Längsausdehnung der Lichtquelle um einen noch größeren Faktor unter Konstanthaltung aller anderen Parameter zu verändern und das proportionale Anwachsen der Intensität nachzuweisen. Hat die Lichtquelle in verschiedenen Richtungen verschiedene Dicken, so muß sich die in verschiedenen Richtungen gemessene Intensität wie die entsprechenden geometrischen Schichtdicken verhalten. Ein schönes Beispiel hierfür bietet die von JÜRGENS [*95*] durchgeführte Messung des Intensitätsverhältnisses zwischen end-on- und side-on-Beobachtung an dem von MAECKER [*136*] entwickelten, wasserstabilisierten Rohrbogen. Trägt man nämlich für eine Reihe von Frequenzen innerhalb der stark verbreiterten Wasserstofflinie H_α die side-on- gegen die end-on-Intensität im logarithmischen System auf (Fig. 47), dann wachsen beide für die kleineren Intensitäten in den Linienflügeln proportional, die Kurve ist eine Gerade mit einer Neigung von 45° gegen die Achsen. Setzt aber für größere Intensitäten, also mehr zur Linienmitte hin, in der end-on-Richtung Selbstabsorption ein, dann verflacht sich die Kurve und verläuft schließlich horizontal, weil dann trotz zunehmender side-on-Intensität die end-on-Intensität auf dem konstanten Wert der schwarzen Strahlung gemäß Gl. (69.12) bleibt. Die end-on-Strahlung erfolgt also nur solange aus optisch dünner Schicht, wie die beiden Intensitäten zueinander proportional sind.

Eine weitere Möglichkeit zur Prüfung der Emission aus optisch dünner Schicht ist die Durchstrahlung der zu untersuchenden Lichtquelle mit Fremdlicht und Bestimmung des Absorptionsgrades, wie es LARCHÉ [*118*] an der *X*-Hochdrucklampe gemacht hat. Ist der Absorptionsgrad klein gegen Eins, dann ist die Lichtquelle transparent und umgekehrt.

Schließlich ist noch folgendes Prüfungsverfahren von Bedeutung: Ist die Absolutintensität an jeder Stelle einer Linie, insbesondere in der Linienmitte klein gegen die Intensität der schwarzen Strahlung gleicher Temperatur, dann ist die Lichtquelle nach Gl. (69.9) optisch dünn. Eine entsprechende Messung kann man bei stark verbreiterten Linien ohne weiteres durchführen. Bei optisch dicker Schicht erreicht die Linienmitte die Intensität der schwarzen Strahlung nach Gl. (69.12) und ist daher abgeflacht, was man an den Aufnahmen von SCHNAUTZ [*196*] am Cu-Bogen und von JÜRGENS am wasserstabilisierten Rohrbogen (Fig. 48) erkennen kann.

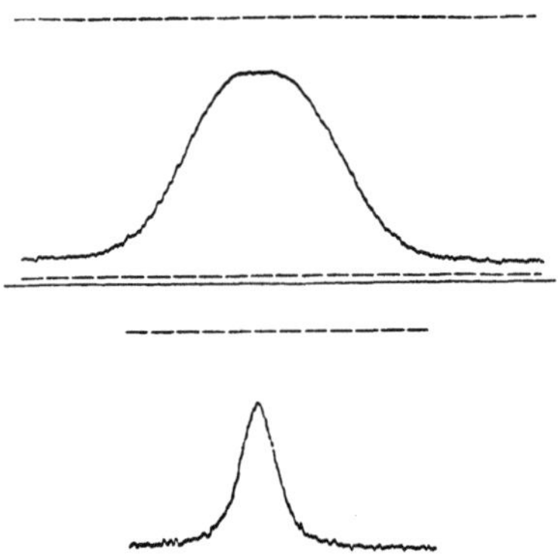

Fig. 48. Linienkontur von H_α bei optisch dünner Schicht (unten) und optisch dicker Schicht (oben). Hier ist die Linie wegen Erreichung der schwarzen Strahlung abgeflacht.

Kann dagegen das Linienprofil nicht aufgelöst werden, dann setzt man als ungünstigsten Fall die DOPPLER-Verteilung (72.7) an und berechnet das Verhältnis der Intensität in der Linienmitte zur Intensität der schwarzen Strahlung gleicher Temperatur.

Die gleichen Überlegungen gelten natürlich auch für ein Kontinuum.

79. Inhomogene Schichten. Alle Bogensäulen haben eine inhomogene Temperaturverteilung. In den günstigsten Fällen liegt Zylindersymmetrie vor (wandstabilisierter Bogen, Wälzbogen, durch Gas- oder Flüssigkeitswirbel stabilisierter Bogen). Bei diesen Bögen herrscht wenigstens in Achsenrichtung für einen kleinen Raumwinkel vollständige Homogenität, was für exakte Messungen sehr wertvoll ist. Bei Beobachtung senkrecht zur Achse kann man näherungsweise mit dem Kanalmodell arbeiten, d.h. den wahren radialen Intensitätsabfall durch eine Rechtecksverteilung ersetzen, deren Höhe gleich dem Emissionskoeffizienten in der Achse ist, und deren Breite so zu bemessen ist, daß aus dem Produkt beider die beobachtete Intensität resultiert.

Für genaue Messungen muß die Umrechnung der beobachteten Querverteilung der Intensität auf den radialen Verlauf des Emissionskoeffizienten mit Hilfe der ABELschen Integralgleichung durchgeführt werden, die in Ziff. 76 beschrieben ist. Die Umrechnung liefert definitionsgemäß i_L (je Längeneinheit). Mit der Umrechnung sieht man gewissermaßen durch die umhüllenden Schichten hindurch auf die gewählte Schicht oder auf die Achse.

80. Absolutmessungen. Bei manchen Untersuchungen müssen Absolutmessungen vorgenommen werden. Zu diesem Zweck nimmt man neben dem zu untersuchenden Spektrum noch das eines Strahlungsnormals über den gleichen Lichtweg mit gleicher Belichtungszeit auf. Als Strahlungsnormal ist außer einer geeichten Wolframlampe [*232*] oder dem UV-Normal [*112*], [*181*] besonders der Kohlebogen zwischen reinen feinkörnigen Graphitelektroden nach MCPHERSON [*134*] geeignet, dessen Anodenkrater nach Messungen von EULER [*45*] eine schwarze Temperatur von 3995° K hat, wenn er mit einer kurz unterhalb der Zischgrenze liegenden Stromstärke betrieben wird. Sollte der Intensitätsunterschied zwischen dem zu untersuchenden Spektrum und dem des Normals zu groß sein, so müssen die Intensitäten meßbar angeglichen werden, was durch genügend schnell rotierende Sektoren zur Vermeidung des Intermittenzeffektes, durch Graufilter bekannter Durchlässigkeit oder durch Abblenden an solchen Stellen des Strahlenganges, bei denen nachweislich eine gleichmäßige Ausleuchtung vorhanden ist (z. B. im parallelen Strahlengang des Spektralapparates), geschehen kann.

Da die Normale im allgemeinen Kontinuumsstrahler sind, können damit zunächst auch nur Kontinua gemessen werden. In diesem Sinne gehören dazu auch verbreiterte, vom Spektralapparat aufgelöste Linien. Nichtaufgelöste Linien, die durch die Apparatefunktion verbreitert erscheinen, sind ebenfalls wie ein kleines Kontinuum zu behandeln, es darf aber nur von dem Integralwert der Intensität über die ganze Linie Gebrauch gemacht werden.

II. Messung der Zustandsgrößen in der Säule.

81. Direkte Temperaturmessung mit spektroskopischen Methoden. Die Temperatur ist die wichtigste Zustandsgröße in der Bogensäule. Es gibt viele Methoden, um sie zu messen. Ein großer Teil davon macht sich das Eigenleuchten des Bogens zunutze, weil dieses Verfahren noch am einfachsten ist und die Entladung nicht stört.

In diesem Falle bauen die Methoden meistens auf der folgenden Formel für die Intensität einer Spektrallinie bei Emission aus optisch dünner Schicht auf:

$$I_L = \frac{1}{4\pi} A_n^m n_0 \frac{g_m}{Z_0^{(i)}} \cdot e^{-E_m/kT} \cdot h\nu l. \tag{65.4}$$

Die naheliegendste, wenn auch nicht einfachste Methode zur Temperaturmessung nach dieser Gleichung ist die Messung der Absolutintensität einer Linie. Dazu muß I_L als Funktion der Temperatur, die außer im Exponenten der e-Funktion noch in der Teilchenzahl n_0 und ganz schwach in der Zustandssumme $Z_0^{(i)}$ steckt, beschrieben werden können, d.h. es müssen alle anderen Faktoren bekannt sein. Für ν, E_m und g_m kann man die Zahlenwerte aus dem Termschema entnehmen. Die Länge l kann bei homogener Schicht (Achsenrichtung bei Zylindersymmetrie) direkt gemessen werden, oder wird bei Rotationssymmetrie durch die Umrechnung mit der ABELschen Integralgleichung aus Ziff. 76 auf Eins reduziert. Die Teilchenzahl als Funktion der Temperatur muß mit dem Massenwirkungsgesetz unter Benutzung des Gesamtdruckes berechnet werden.

Das größte Hindernis für diese Methode ist die Tatsache, daß die Übergangswahrscheinlichkeiten A_n^m nur für eine begrenzte Zahl von Linien bekannt sind. Besonders günstig in dieser Beziehung ist der Wasserstoff, für den alle wichtigen Übergangswahrscheinlichkeiten streng berechenbar sind. Auch für einige andere Elemente liegen Werte vor, wenn auch nur zum Teil Näherungswerte.

Beispiele für die Temperaturbestimmung durch Messung der Absolutintensität sind zu finden bei ELENBAAS [*4a*] für den Hg-Bogen und bei JÜRGENS für den wasserstabilisierten Rohrbogen [*95*]. Umgekehrt kann man natürlich bei bekannter Temperatur Übergangswahrscheinlichkeiten aus der Absolutintensität einer Linie ermitteln (MAECKER [*137*], JÜRGENS [*96*]).

Die weiteren Temperaturmeßmethoden, die sich der Gl. (65.4) bedienen, erfordern nur Relativmessungen.

Hat man im Spektrum verschiedene Linien eines Elementes, von denen mindestens die relativen Übergangswahrscheinlichkeiten bekannt sind, dann kann man den für alle Linien gemeinsamen Faktor $\frac{1}{4\pi} n_0/Z_0^{(i)} h l = K$ herausziehen und erhält nach Logarithmieren und Umformen

$$\log \frac{I_L}{A_n^m g_m \nu} = -\frac{E_m}{k} \frac{1}{T} + \log K, \tag{81.1}$$

wofür wir kürzer schreiben

$$Y = -\frac{1}{T} X + K'. \tag{81.2}$$

Trägt man also für möglichst viele, mindestens aber zwei Linien Y gegen X auf, so resultiert eine Gerade, deren Neigung durch $-1/T$ gegeben ist. Auch für diese Messung muß das Spektrum eines Strahlungsnormals aufgenommen werden, um die Intensität von Linien in verschiedenen Spektralgebieten miteinander vergleichen zu können. Man kann auch Linien verschiedener Elemente heranziehen, wenn nur das Mischungsverhältnis und natürlich die Atomkonstanten bekannt sind. Sehr genau ist diese ganze Methode aber nicht, weil die Neigung nur umgekehrt proportional mit der Temperatur geht. Diese Methode ist z.B. von MOHLER [*156*] auf die Cs-Entladung und von JÜRGENS [*95*] auf den wasserstabilisierten Rohrbogen angewendet worden (Fig. 49).

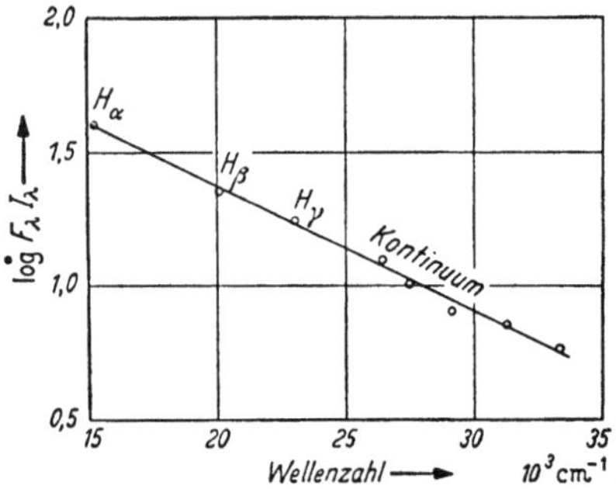

Fig. 49. Logarithmus der reduzierten Linienintensität der BALMER-Linie als Funktion der Anregungsenergie. Die Gerade ist proportional $1/T$.

Auch für die Linien einer Rotationsbande ist Gl. (65.4) anwendbar. Da nämlich die Anregungsenergie des Rotationszustandes mit der Quantenzahl J bis auf eine additive Konstante durch Gl. (57.6) zu

$$E_m = \frac{J(J+1)\,h^2}{8\pi^2\Theta} \tag{57.6}$$

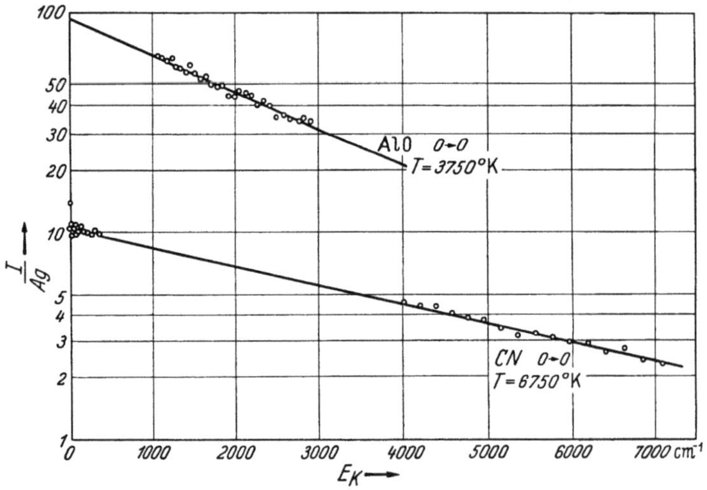

Fig. 50. Logarithmus der reduzierten Intensität von Linien einer Rotationsbande gegen die Rotationsenergie aufgetragen. Die Neigung der Geraden ist wieder $1/T$ proportional. (Nach BRINKMAN [*2a*].)

und das statistische Gewicht durch $g_m = 2J+1$ gegeben ist, sind die Linienintensitäten näherungsweise proportional zu

$$I_L \sim (2J+1)\, e^{-J(J+1)\,h^2/8\pi^2\Theta kT}. \tag{81.3}$$

Durch geeignete logarithmische Auftragung erhält man wie vorher eine Gerade, deren Neigung $-1/T$ ist (Fig. 50). Dieses Verfahren wurde unter anderem von ORNSTEIN [*164*] und Mitarbeitern und von LOCHTE-HOLTGREVEN und MAECKER

[*127*] auf die Bogensäule angewendet. Für die Genauigkeit gilt das oben Gesagte. Außerdem versagt das Verfahren für höhere Temperaturen, weil dann die Moleküle dissoziiert sind.

Weiterhin eignet sich Gl. (65.4) auch dazu, den relativen Verlauf der Temperatur aus dem relativen Intensitätsverlauf ein und derselben Spektrallinie zu finden, wenn nur in irgendeinem Punkt die Temperatur festliegt. Bildet man nämlich das Verhältnis zwischen den Intensitäten der Linie an einer Stelle mit der Temperatur T und der Stelle mit der bekannten Temperatur T_0, dann ist das meßbare Intensitätsverhältnis

$$V = \frac{I(T)}{I(T_0)} = \frac{n(T) Z^{(i)}(T_0)}{n(T_0) Z^{(i)}(T)} e^{-\frac{E_m}{k}\left(\frac{1}{T} - \frac{1}{T_0}\right)}, \tag{81.4}$$

woraus an jeder Stelle T mit V am besten durch graphische Auftragung ermittelt werden kann (Beispiele: HÖRMANN [*9o*], ELENBAAS [*4a*], MAECKER [*138*]) (Fig. 51, vgl. auch Fig. 54). Man sieht an Gl. (81.4) ferner, daß die Intensität einer Linie um so steiler abnimmt, je höher die Anregungsenergie E_m des Ausgangsterms ist.

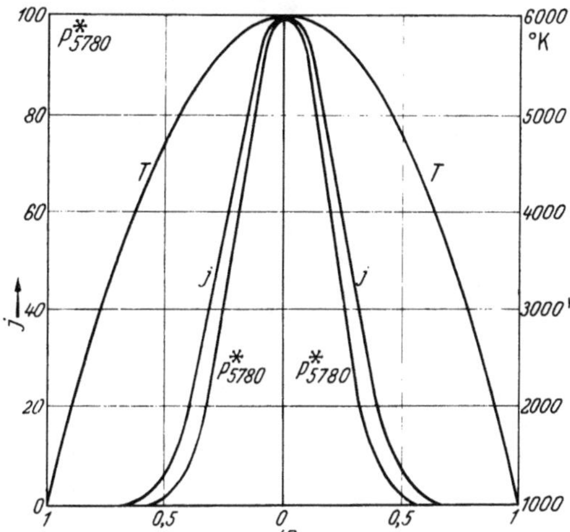

Fig. 51. Radialer Intensitätsabfall der Linie 5780 Å im wandstabilisierten Hg-Bogen; daraus radialer Temperatur- und Stromdichteverlauf.

Diese Relativmethode haben LARENZ und BARTELS zu einer bedeutungsvollen Absolutmethode für hohe Temperaturen ausgebaut [*121*], [*119*]. Beachtet man nämlich nur die Temperaturabhängigkeit der Intensität einer Linie

$$I_L = K \frac{n(T)}{Z^{(i)}(T)} e^{-\frac{E_m}{kT}}, \tag{81.5}$$

dann nimmt die Intensität einerseits mit dem BOLTZMANN-Faktor, der asymptotisch an den Wert Eins herangeht, zu. Dem wirkt aber andererseits die mit steigender Temperatur abnehmende Teilchenzahl entgegen; denn die emittierenden Teilchen werden durch Dissoziation oder Ionisation mehr und mehr zerstört. Als Folge dieser beiden gegeneinander arbeitenden Funktionen durchläuft die Intensität einer Linie mit steigender Temperatur ein Maximum, für das die zugehörige Temperatur exakt berechnet werden kann. Wird also in einer Bogensäule diese Maximumstemperatur überschritten, dann kann man in der Tat Intensitätsmaxima außerhalb der Bogenachse beobachten und diesem Ort die richtige absolute Temperatur zuordnen. Von diesem Punkt aus erschließt dann eine Relativmessung nach der vorangehenden Methode den ganzen Temperaturverlauf nach innen und außen in der Umgebung des Maximums. Mit diesem Verfahren hat zuerst LARENZ [*119*] im GERDIEN-Bogen bei 500 A und 4,5 mm Kanaldurchmesser aus den Maxima der Linien neutraler und einfach ionisierter Teilchen die Temperatur zu 35000° K bestimmt (Fig. 52). Später konnten BURHORN, MAECKER und PETERS [*18*] durch weitere Einengung des Bogens auf 2,3 mm Durchmesser und Erhöhung der Stromstärke auf 1500 A die Temperatur auf 50000° K steigern, was sie mit dem LARENZschen Verfahren am Maximum einer Linie des zweifach ionisierten Sauerstoffs nachgewiesen haben (Fig. 20).

Gleichzeitig ist durch dieses Verfahren klargestellt, wie das Zustandekommen der bekannten Bogenaureolen zu verstehen ist (Fig. 53 und 55): Mit von außen her ansteigender Temperatur folgen nacheinander die Intensitätsmaxima der

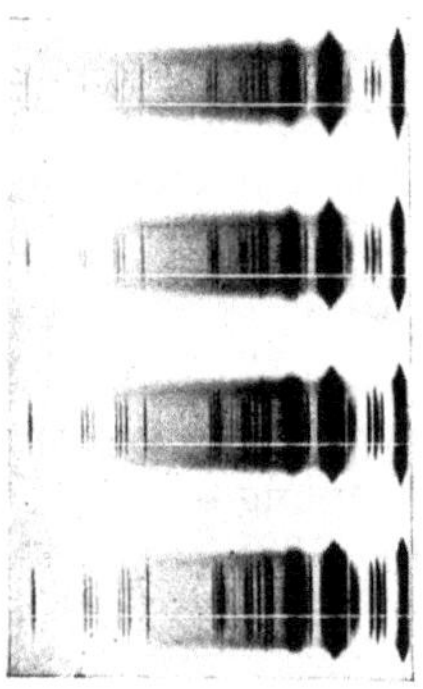

Fig. 52.

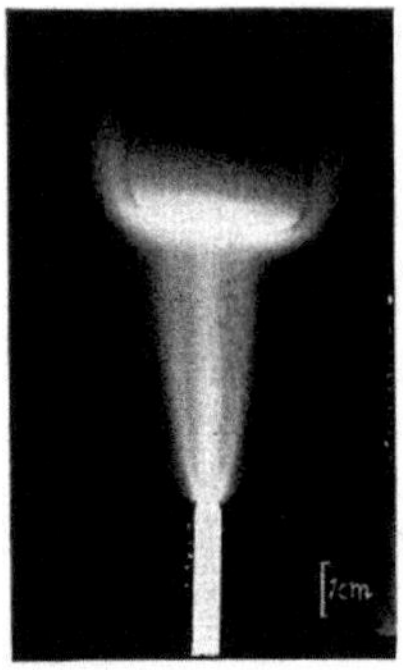

Fig. 53.

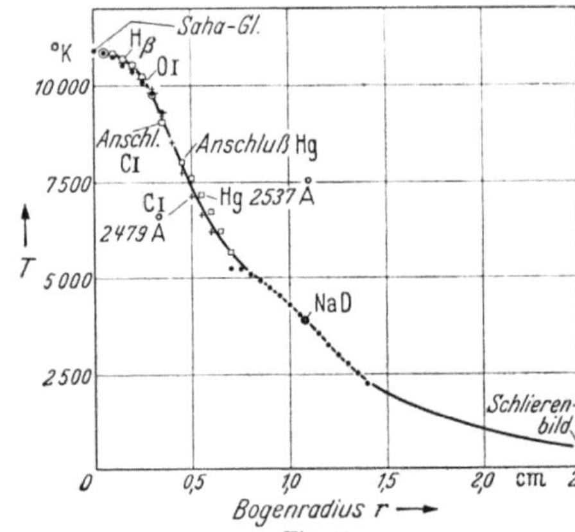

Fig. 54.

Fig. 52. Spektren quer zum H_2O-GERDIEN-Bogen. Die Maximalintensität der BALMER-Linien und des Grenzkontinuums liegt außen bei etwa 15000° K. Die Intensität der O II-Linien rückt mit (nach unten) steigender Stromstärke ebenfalls weiter nach außen. Auch O III-Linien zu erkennen.

Fig. 53. Vertikaler Hochstromkohlebogen 200 A. 5 mm Durchmesser-Kathode unten, 35 mm Durchmesser-Anode oben. Außen gelbe CaO-Aureole, innen CN-Anode, Säulenkern in der Achse.

Fig. 54. Temperaturverteilung quer über die Säule des Hochstromkohlebogens 1,42 cm über der Kathode. SAHA-Gleichung: Absolute Temperatur aus n_e mit C I- und C II-Linien. H_β, O I, C I und Hg: Relativer Temperaturverlauf aus Intensitätsabfall dieser Linien. NaD: Absolute Temperatur nach Methode von LARENZ. Schlierenbild: Absolute Temperatur aus Fig. 63 nach MAECKER [128].

verschiedenen Teilchenarten mit ihren typischen Farben, angefangen mit den schwach gebundenen Molekülen bis hinauf zu mehrfach geladenen Ionen (Fig 55). Die Aureolen sind in guter Näherung Isothermen (Fig. 56).

Eine weitere allerdings sehr beschränkt anwendbare Methode, die ebenfalls Emission aus optisch dünner Schicht voraussetzt, ist die Messung der DOPPLER-Verteilung einer Linie gemäß Gl. (72.7). Man erhält aus dieser Messung direkt die

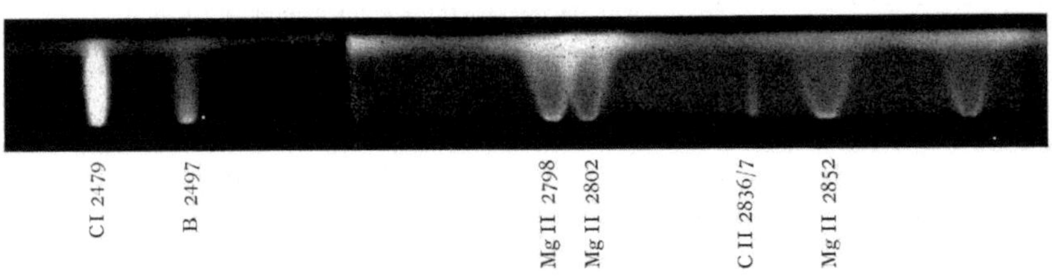

Fig. 55. Weit mehr Aureolen als in Fig. 53 erschließt man durch „Bildspektrogramme" ohne Spektrographenspalt. Man erkennt die breite Mg I-2852 Å-Aureole, die schmaleren Mg II-2798/2802 Å- und Al II-Aureolen, das in der Mitte nicht mehr eingesenkte Bogenbild im Licht von B 2497/8 Å, den breiten Bogenkern bei C I 2479 Å und schließlich das nur in Kathodennähe vorhandene Leuchten von C II 2836/7 Å.

Temperatur der schweren Teilchen, die bei thermischem Gleichgewicht natürlich mit der Plasmatemperatur übereinstimmt. Da aber fast alle Bogenlinien durch andere Effekte verbreitert sind, eignen sich nur Übergänge innerhalb der inneren Schale von Übergangselementen für diese Methode, weil die inneren Elektronen durch die äußeren vor Störungen geschützt sind. Der geringen DOPPLER-Breite wegen muß natürlich ein PEROT-FABRY-Interferometer benutzt werden Dieses Verfahren hat BURHORN [16] am Fe-Bogen zum Nachweis thermischen Gleichgewichtes angewendet (Fig. 57).

Erfolgt bei einer homogenen Schicht, also bei Beobachtung in Achsenrichtung an einem zylindersymmetrischen Bogen, die Emission aus optisch dicker Schicht, dann ist nach Gl. (69.12) die Intensität gleich der der KIRCHHOFF-PLANCK-Funktion Gl. (69.4). Eine Absolutmessung der Intensität liefert also direkt die Temperatur (SCHNAUTZ [*196*], JÜRGENS [*95*], LARCHÉ [*117*]), (Fig. 48).

Ist die Schicht zwar optisch dick, aber inhomogen wie bei side-on-Beobachtung an einem zylindersymmetrischen Bogen, so ist die Linie in der Mitte durch Selbstumkehr eingesenkt. Die zu beiden Seiten der Mitte aufragenden Intensitätsmaxima („Kuppen") geben nach BARTELS [*1*] ebenfalls Auskunft über die Temperatur, die Ableitungen der entsprechenden Beziehungen erfordern aber die Anwendung der Theorie des Strahlungsgleichgewichtes. Eine nach diesem Verfahren durchgeführte Messung findet man bei GÖING [*68*], (Fig. 58).

Fig. 57.

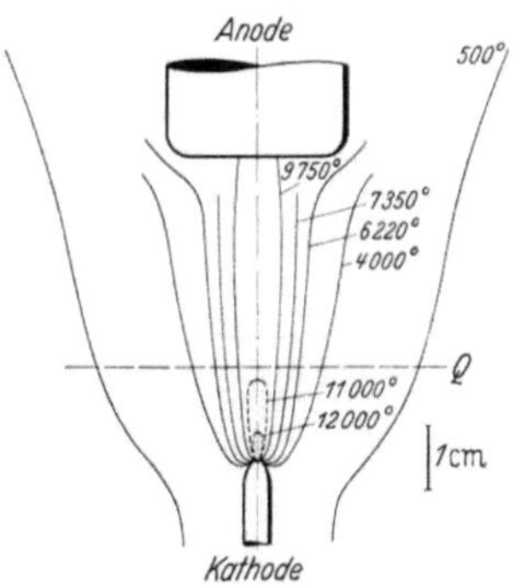

Fig. 56.

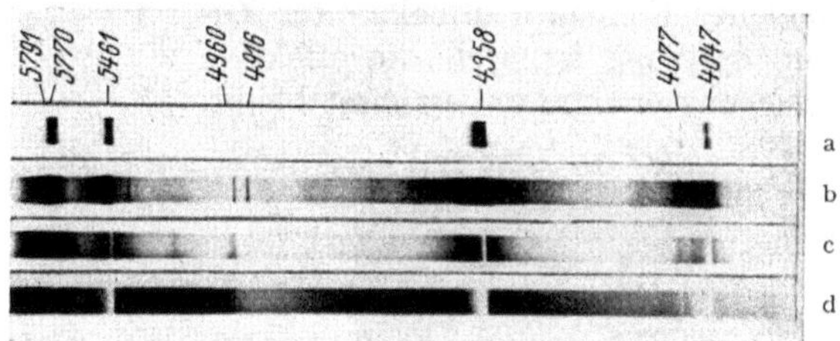

Fig. 58.

Fig. 56. Isothermenfeld für den 200 A-Kohlebogen, entnommen aus den Fig. 53 und 63 und bei Q angeschlossen an Fig. 54.

Fig. 57. Interferenzspektrogramm quer zum Fe-PFUND-Bogen. Man beachte, daß einige Linien scharfe Interferenzen, andere, auch schwache dagegen wegen Stoßverbreiterung keine solche zeigen.

Fig. 58a–d. Spektrum des Hg-Höchstdruckbogens (Osram HBO 200). a Schwach belichtet, b stark belichtet ($T = 7700°$ K), c bei schwacher Stoßbelastung ($T = 11\,700°$ K), d bei starker Stoßbelastung ($T = 16000°$ K). Man erkennt die Selbstumkehr der Linien 5461, 4358, 4077 und 4047 Å bei Stoßbetrieb.

Schließlich könnte man die Tatsache, daß die Intensität des Elektronenkontinuums zwischen zwei Seriengrenzen nach Gl. (66.5) mit $\exp(-h\nu/kT)$ abfällt, zur Temperaturmessung heranziehen, jedoch ist dieser Weg noch nicht eingeschlagen worden und ist auch nicht sehr genau.

Neuerdings hat LARENZ [*120*] das jedem Widerstand eigene Rauschen zur Temperaturmessung in Bogensäulen herangezogen. Bekanntlich liefert jeder Widerstand R unabhängig von seiner Art ein mittleres Rauschspannungsquadrat der Größe

$$u^2 = 4kT\,R\,\Delta\nu, \tag{81.6}$$

worin $\Delta\nu$ die Breite des gewählten Frequenzbandes ist. Bezieht man diese Formel auf ein Volumenelement und integriert über die ganze Bogenlänge, wobei natürlich nur die leitfähigen Bereiche zum Tragen kommen, dann kann man die an den Elektroden abnehmbare Rauschleistung in Abhängigkeit von der mittleren Bogentemperatur berechnen und so die Temperatur aus der gemessenen Rauschleistung angeben. Voraussetzung für die Anwendbarkeit dieser Methode ist die Existenz thermischen Gleichgewichtes überall, andernfalls müssen wenigstens die thermischen Bereiche groß gegen die nicht thermischen, vor allem die elektrodennahen Bereiche sein. Deswegen eignet sich dieses Verfahren besonders für ausgedehnte Bogensäulen, für die es eine mittlere achsennahe Temperatur liefert. Für das UV-Standard findet LARENZ auf diese Weise 6500° K und für die Osram-Lampe HQA 500 7000° K, beides sehr plausible Werte.

Damit sind die direkten Temperaturmeßmethoden an sich erschöpft, wir werden aber noch einige indirekte, die primär die Elektronendichte und erst über die SAHA-Gleichung die Temperatur liefern, kennen lernen.

82. Messung der Elektronendichte. Obwohl bei bekannter Temperatur die Elektronendichte mit Hilfe der EGGERT-SAHAschen Theorie berechnet werden kann, ist es doch aus verschiedenen Gründen wünschenswert, die Elektronendichte als einen wesentlichen Faktor für die elektrische Leitfähigkeit direkt zu messen. Die Elektronendichte hängt nämlich so empfindlich von der Temperatur ab, daß sich der Fehler vervielfacht. Außerdem reicht die SAHA-Gleichung allein nicht für den Fall aus, daß das Bogengas eine Mischung verschiedener Elemente unbekannter Zusammensetzung ist, weil die effektive Ionisierungsenergie unbestimmt bleibt.

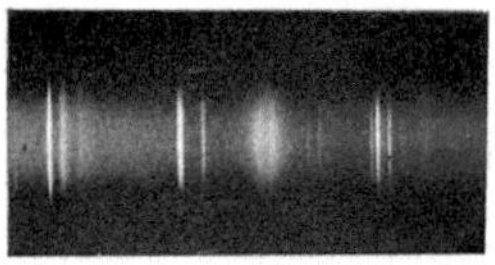

Fig. 59. Spektrum quer über den Kern des Hochstromkohlebogens. Durch intermolekularen STARK-Effekt verbreiterte Wasserstofflinie H_β mit Einsenkung (nicht Selbstumkehr). Neben einigen scharfen C- und N-Linien das Elektronenkontinuum als Untergrund.

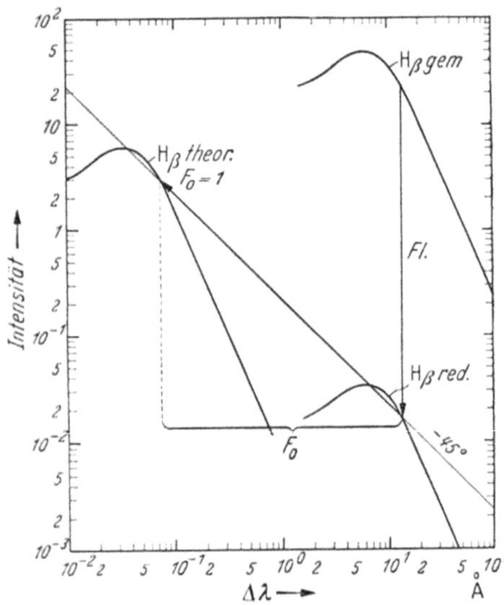

Fig. 60. Um aus der STARK-Effektverbreiterung von H_β die Ionendichte zu finden, braucht man nur das relativ gemessene Profil im logarithmischen Koordinatensystem auf das theoretische ($F_0 = 1$) zu schieben. Der Abszissenwert der Verschiebung ist dann $F_0 = 2{,}61\, e\, n_i^{\frac{2}{3}}$.

Die direkteste Methode zur Messung der Elektronendichte ist die Ausnutzung der in Ziff. 73 behandelten STARK-Effektverbreiterung geeigneter Spektrallinien, die sogar noch erlaubt ist, wenn kein thermisches Gleichgewicht vorliegt. Ist Wasserstoff das Bogengas selbst oder als Zusatz vorhanden, dann ist die Linie H_β (Fig. 59) besonders einfach und genau für die Auswertung zu handhaben. Sie hat sich beispielsweise bei der Messung der Elektronendichte im wasserstabilisierten Rohrbogen (JÜRGENS [*95*]) und im Hochstromkohlebogen (MAECKER [*138*]), bewährt. Hinsichtlich der Genauigkeit lehrt die Abhängigkeit der Linienbreite von der Elektronendichte mit der Potenz $\frac{2}{3}$ gemäß Gl. (73.2), daß sich ein Fehler in der Messung der Linienbreite mit dem 1,5fachen Wert in der Elektronendichte auswirkt. Der großen Bedeutung wegen möge das Verfahren für die Linie H_β an Hand der Fig. 60 erläutert werden. Man zeichnet im doppelt logarithmischen Maßstab das theoretische Profil von $H_\beta (F_0 = 1)$ und in willkürlichen Intensitätseinheiten das gemessene Profil ein. Durch Verschieben des letzteren werden beide Profile zur Deckung gebracht. Die Komponente der Verschiebung in Abszissenrichtung ist das gesuchte F_0, aus dem nach Gl. (73.2) die Ionendichte zu berechnen ist. Man kann sich nämlich die Verschiebung in zwei Schritten vorgenommen denken, und zwar zunächst in Ordinatenrichtung um den Faktor $1/Fl$ (Fl = Gesamtfläche unter der gemessenen Kurve), wodurch das Integral über die ganze Linie auf Eins normiert wird, und dann in Richtung der $-45°$-Geraden bis zur theoretischen Kurve, wodurch unter Einhaltung der Normierungsbedingung der gesuchte Faktor F_0 angebracht wird.

Außer den Linien des Wasserstoffs mit ihrem linearen STARK-Effekt eignen sich auch Übergänge zwischen hoch liegenden Termen mit quadratischem STARK-Effekt, deren Breite durch Gl. (73.6) mit der Zahl der Ladungsträger verknüpft ist. Man muß allerdings sicher stellen, daß die Eigendruckverbreiterung keine Rolle spielt. Deswegen haben SCHULZ [*201*] und ROMPE und SCHULZ [*175*] für den Quecksilberbogen aus dem Übergang $2^1P_1 - 2^1S_0$ 10140 Å zunächst die Eigendruckverbreiterung des Terms $2\,^1P_1$ gemessen (der 2^1S_0 ist praktisch nicht verbreitert) und dann aus den Breiten der beiden gelben Linien $2^1P_1 - 3^1D_2$ 5770 Å, bzw. $2\,^1P_1 - 3\,^3D_2$ 5791 Å nach Abzug der Eigendruckverbreiterung des Terms $2\,^1P_1$ die Elektronendichte gemessen (Fig. 61).

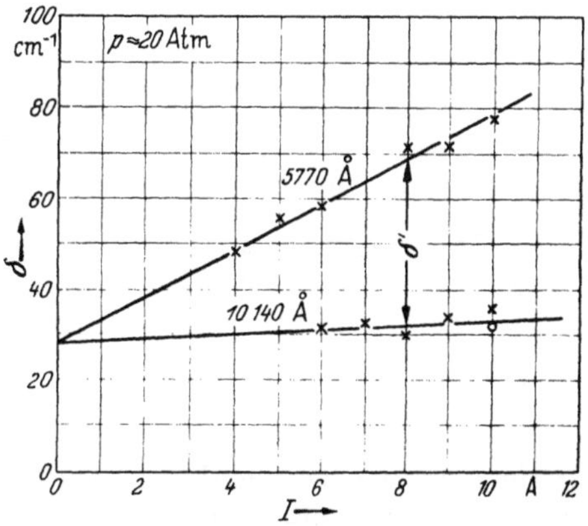

Fig. 61. Stromstärkeabhängigkeit der Linienbreiten von Hg 5770 Å und 10140 Å im Hg-Hochstrombogen. Erstere Linie wird hauptsächlich durch quadratischen STARK-Effekt, letztere im wesentlichen durch Stöße mit gleichartigen Teilchen verbreitert.

Ähnliche Messungen, aber unter Ausnutzung der Linienverschiebung, hat ebenfalls SCHULZ [*198*] durchgeführt.

Eine weitere Methode zur Messung der Elektronendichte bietet eine Absolutmessung des Elektronenkontinuums in einer Spektralgegend, in der mit Sicherheit die Kanten der Grenzkontinua noch so dicht liegen, daß Gl. (66.5)

$$\varepsilon_\nu = 6{,}36 \cdot 10^{-47}\, \overline{(Z+s)^2}\, \frac{n_e n_i}{\sqrt{kT}} \qquad (66.5)$$

noch gilt (vgl. das Kontinuum in Fig. 59). Für s kann man in den meisten Fällen in guter Näherung Eins setzen, während man für den nicht sehr wichtigen Faktor $1/\sqrt{kT}$ einen Schätzwert wählt, den man durch Iteration mit der SAHA-Gleichung verbessern kann. Ein Fehler in der Kontinuumsmessung wirkt sich nur mit dem halben Gewicht auf die Elektronendichte aus. Diesem Vorteil steht die gewisse Unsicherheit in der Abschirmungszahl s gegenüber. Anwendungsbeispiele findet man bei MAECKER [*138*], BURHORN [*16*] oder GÖING, MEIER und MEINEN [*69*].

Ist nun die Elektronendichte mit einer der beschriebenen Methoden bestimmt, dann läßt sich daraus in Verbindung mit der SAHA-Gleichung und der idealen Gasgleichung die Temperatur berechnen, und zwar sehr genau, weil sich die Elektronendichte empfindlich mit der Temperatur ändert. Allerdings ist dies Verfahren nur auf solche Plasmen anwendbar, deren Zusammensetzung im Ausgangszustand bekannt ist, das sind insbesondere einatomige Gase, dann Molekülgase, aber auch Gasmischungen mit bekanntem Verhältnis (F. BURHORN [*16*]).

Ist dieses Verhältnis nicht gegeben, dann muß zusätzlich noch das Intensitätsverhältnis einer Ionenlinie zu einer Atomlinie desselben Elementes gemessen werden. Mit Hilfe der aus Gl. (65.4) folgenden Beziehung

$$\frac{I_L^+}{I_L^0} = \frac{{}^+A_n^m\; n_+\; g_m^+\; Z_0^{(i)}\; \nu_+}{{}^0A_n^m\; n_0\; g_m^0\; Z_+^{(i)}\; \nu_0}\, e^{-\frac{E_m^+ - E_m^0}{kT}} \qquad (82.1)$$

gewinnt man zunächst n_+/n_0 für das gewählte Element als Funktion der Temperatur. Multipliziert man dieses Verhältnis mit dem vorher gemessenen n_e, dann erhält man die linke Seite der SAHA-Gleichung für dieses Element X:

$$\frac{n_+^X}{n_0^X} n_e = \frac{2Z_+^{(i)}}{Z_0^{(i)}} \frac{(2\pi m_e kT)^{\frac{3}{2}}}{h^3} e^{-E_I^X/kT}. \tag{82.2}$$

Man kann nämlich die SAHA-Gleichung auch auf eine Komponente X einer Mischung anwenden, dann ist aber $n_+^X \neq n_e$. Die Vereinigung der beiden Gln. (82.1) und (82.2) liefert

$$\frac{I_L^+}{I_L^0} n_e = 2 \frac{{}^+A_n^m}{{}^0A_n^m} \frac{g_m^+}{g_m^0} \frac{\nu_+}{\nu_0} \frac{(2\pi m_e kT)^{\frac{3}{2}}}{h^3} \cdot e^{-\frac{E_I^X + E_m^+ - E_m^0}{kT}}, \tag{82.3}$$

woraus unabhängig von der Mischung und den Zustandssummen die Temperatur sehr genau abgelesen werden kann, wenn die rechte Seite als Funktion der Temperatur dargestellt ist. Mit dieser Methode hat MAECKER [*138*] die Temperatur im Hochstromkohlebogen zu 11000° K gemessen (Fig. 54).

83. Dichtemessungen. Eine Gruppe weiterer Methoden zur Messung der Temperatur nimmt den Weg über die Dichte. Denn diese ist bei bekanntem Druck eine eindeutige Funktion der Temperatur:

$$\varrho = n_1 m_1 + n_2 m_2 \ldots. \tag{83.1}$$

Hierin sind die Teilchenkonzentrationen nach Maßgabe der Massenwirkungsgesetze Funktionen der Temperatur und des Druckes (siehe z.B. Fig. 38).

Mit Hilfe der Absorption von α-Strahlen oder Röntgenstrahlen, die unabhängig von der chemischen Bindung ist, sich aber proportional mit der Dichte, also umgekehrt mit der Temperatur ändert, läßt sich die Temperatur im Vergleich zur Außentemperatur messen. Messungen dieser Art haben v. ENGEL und STEENBECK [*41*], KOCH [*107*], KENTY und KARASCH [*101*] durchgeführt. RAMSAUER [*170*] hat ähnliche Untersuchungen mit 100000 eV-Elektronen angestellt.

Während bei den Absorptionsmessungen mit α-, Röntgen- oder schneller Elektronenstrahlung die chemische Bindung keine Rolle spielt, ist sie bei der Bestimmung der Dichte aus dem Brechungsindex durchaus von Einfluß, weil die Polarisierbarkeit der Einzelteilchen die Molekularrefraktion bestimmt. Daher sind Methoden, die die Dichte aus dem Brechungsindex ermitteln, nur bis zu solchen Temperaturen anwendbar, bei denen noch keine nennenswerte Zerstörung der Teilchen stattgefunden hat. Diese Grenze liegt bei Molekülgasen naturgemäß tiefer als bei einatomigen Gasen. Man kann allerdings mit der Regel über Addition von Atomrefraktionen noch höher hinaufrechnen. Den Brechungsindex kann man entweder aus Interferenzaufnahmen, die den durch die Bogensäule bedingten Gangunterschied in den Lichtwegen als Streifenverschiebung sichtbar machen, oder aus Schlierenaufnahmen finden, die zunächst den Gradienten des Brechungsindex und erst durch Integration den Brechungsindex selbst und schließlich mit der Molekularrefraktion

$$R = (n-1)\frac{2}{3}\frac{\mu}{\varrho} = \text{const} \qquad (\mu = \text{Molekulargewicht}) \tag{83.2}$$

die Dichte liefern (SCHARDIN [*185*]). In jedem Falle muß natürlich Rotationssymmetrie vorliegen und die Umrechnung mit der ABELschen Integralgleichung

(Ziff. 76) durchgeführt werden. Bei hohen Temperaturen werden diese Verfahren immer ungenauer, weil die Dichte im Vergleich zur Umgebung sehr klein wird. Das Interferenzverfahren wurde zuerst von SCHMITZ [*192*] auf den Niederstromkohlebogen (Fig. 62) und das Schlierenverfahren von SPERLING [*211*] auf den gleichen Bogen, später von MAECKER [*138*] auf den Hochstromkohlebogen angewendet (Fig. 63).

Eine weitere Möglichkeit, die Dichte zu finden, hat ECKER [*32*] speziell für den Hg-Bogen angegeben. Danach hat man das zur Füllung benützte Hg mit dem radioaktiven Isotop Hg^{203} anzureichern, so daß man quer zu einem Bogenquerschnitt

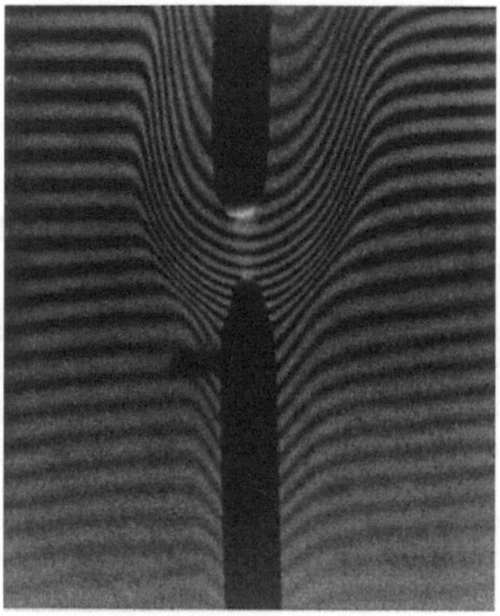

Fig. 62. Interferenzaufnahme vom 2,9 A-Kohlebogen. Die Streifenverschiebung führt über den Gangunterschied zum Brechungsindex an jeder Stelle.

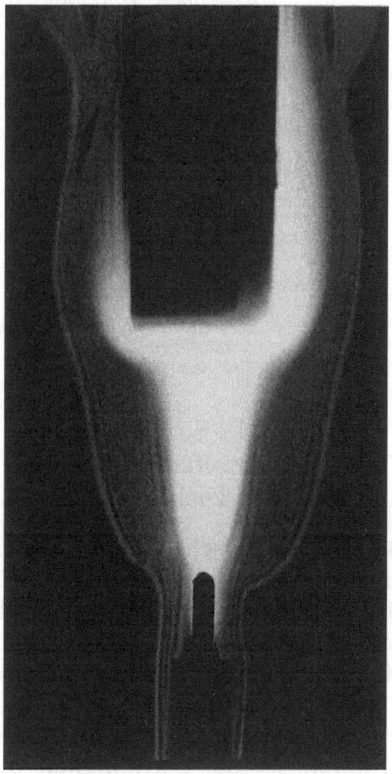

Fig. 63. Schlierenaufnahme vom 200 A-Kohlebogen. Innen sieht man das Eigenleuchten des Bogens, während die grauen Flächen die erfolgte Ablenkung des durchgeschickten Fremdlichtes anzeigen.

an jeder Stelle aus der radioaktiven Strahlung die Dichte als Funktion des Bogenradius nach Umrechnung mit der ABELschen Integralgleichung ermitteln kann.

Schließlich kann man nach SUITS [*168*], [*222*] auch aus der gemessenen Schallgeschwindigkeit nach Gl. (60.6) die Dichte errechnen. Die Genauigkeit dieses Verfahrens wird zwar durch die geringe Abhängigkeit der Schallgeschwindigkeit von der Dichte beeinträchtigt, dafür läßt sich aber die Schallgeschwindigkeit recht genau messen. Zu diesem Zweck löst man durch einen Funken in der axialen Bohrung einer Elektrode eine Knallwelle aus und registriert deren Ankunftszeit in der Bohrung der anderen Elektrode oszillographisch. Für den Niederstrombogen in Stickstoff findet SUITS eine mittlere Achsentemperatur von 6100° K.

Da die Knallwelle als Verdichtungsstoß die Helligkeit hinter ihrer Front stark erhöht, läßt sie sich auch direkt auf rotierenden Film aufnehmen, so daß die Zeit-Weg-Kurve der Knallwelle registriert wird. Der Differentialquotient ds/dt an jeder Stelle ist dann die Geschwindigkeit [*223*].

Eine Übersicht über die vorkommenden Temperaturen und Elektronendichten vermittelt die folgende Darstellung.

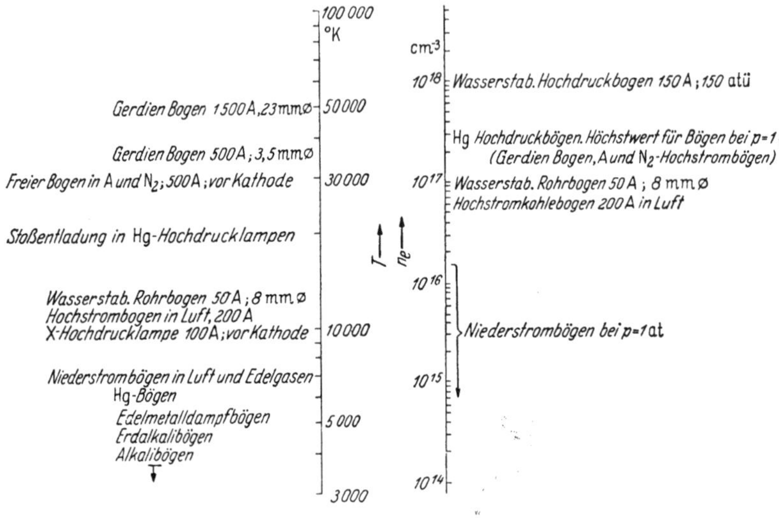

Damit ist die Reihe der direkt meßbaren Zustandparameter, nämlich der Temperatur, der Dichte und der Elektronendichte, erschöpft.

84. Die elektrische Leitfähigkeit. Im Abschnitt C war gezeigt worden, daß in einem Stromkreis, zu dem auch die Bogensäule gehört, elektromotorische Kräfte wirken, also anliegende und eingeprägte Feldstärken, die die Ladungsträger in Bewegung zu setzen versuchen. Ihnen entgegen wirken die Reibungskräfte untereinander und gegenüber den Neutralpartnern, so daß sich als Resultat eine gewisse Stromdichte einstellt. Der Koeffizient, mit dem die Resultante aller elektromotorischen Kräfte multipliziert werden muß, um die Stromdichte zu erhalten, ist die Leitfähigkeit σ gemäß dem OHMschen Gesetz

$$\boldsymbol{j} = \sigma(\mathfrak{E}^a + \mathfrak{E}^r + \mathfrak{E}^e)\,, \tag{84.1}$$

worin $\mathfrak{E}^a$ das divergenzfreie, angelegte Feld, $\mathfrak{E}^r$ das durch Raumladungen erzeugte Feld und $\mathfrak{E}^e$ ein Symbol für alle eingeprägten Kräfte ist. Im folgenden wollen wir von dem Beitrag der Ionen zur Stromdichte absehen, da dieser wegen des großen Massenunterschiedes zwischen Ionen und Elektronen nur in der Größenordnung von 1% liegt. Nach Gl. (61.14) ist dann

$$\sigma = e\, n_e \frac{e\,\lambda_e}{m_e v_e} = \frac{e^2 n_e}{\sqrt{8 m_e k T/\pi}} \cdot \frac{1}{\sum\limits_k n_k\, Q_{e k}} \tag{84.2}$$

eine Funktion der Temperatur und des Druckes. Die Teilchenzahlen kann man in bekannter Weise berechnen. Das schwierigste Problem bilden die Querschnitte der von den Elektronen gestoßenen Partner Q_{ek}. Soweit es sich um gestoßene Partikel handelt, die auch bei normalen Temperaturen bis zu einigen hundert Grad einer Messung zugänglich sind, sind sie als RAMSAUER-Querschnitte bekannt. In der Literatur findet man entweder die Wirkungsquerschnitte als Funktion der Elektronengeschwindigkeit, die nach Gl. (63.4) auf Transportquerschnitte umzurechnen sind, oder die bei kleinen Drucken gemessenen Driftgeschwindigkeiten als Funktion von E/p, aus denen man den Querschnitt berechnen kann. Um diese Werte auch bei hohen Temperaturen verwenden zu können, müssen die von der Geschwindigkeit der Elektronen abhängigen Querschnitte über die MAXWELLsche Geschwindigkeitsverteilung gemittelt werden, wie das für einige Gase v. ENGEL und STEENBECK [*5a*] getan haben. Bei weniger großen Genauigkeitsansprüchen genügt die Benutzung des Querschnittes für die mittlere thermische Geschwindigkeit. Zahlenwerte kann man den Tabellenwerken von LANDOLT-BÖRNSTEIN, BORN und ALLIS [*14*], MASSEY und BURHOP [*149*] und anderen entnehmen.

Zur Messung des Querschnittes von Partikeln, die nur bei hohen Temperaturen existieren, z.B. Atome von Molekülgasen oder Ionen, läßt sich die Stromtransportgleichung heranziehen: Der in einer etwa zylindrischen Bogensäule fließende Strom J ist nach dem über dem Querschnitt integrierten OHMschen Gesetz

$$J = E\,2\pi \int \sigma\, r\, d r = E \frac{2\pi e^2}{\sqrt{8 m_e/\pi}} \int \frac{n_e}{\sqrt{k T}} \frac{1}{\sum\limits_k n_k Q_{e k}} r\, d r. \tag{84.3}$$

Hat man nun die radiale Verteilung der Elektronendichte und der Temperatur gemessen und ist unter den $Q_{e\,k}$ nur ein unbekannter Querschnitt vorhanden, dann stellt die Stromtransportgleichung eine Bestimmungsgleichung für den Mittelwert des unbekannten Querschnittes dar. Zur Erreichung einer hinreichenden Genauigkeit ist es notwendig, daß die Teilchen mit dem unbekannten Querschnitt den Hauptanteil an der Bremsung der Elektronen tragen. Mit dieser Methode konnte die Gültigkeit einiger Querschnitte, deren Werte aus anderen Messungen bekannt waren, für hohe Temperaturen gesichert werden (z. B. ELENBAAS [*4a*] für Hg und SCHULZ [*199*] für A). Aber auch an Atomen von Molekülgasen, die sonst der Messung unzugänglich sind, können Querschnittsmessungen durchgeführt werden.

Zur Messung des Querschnittes der Wasserstoffatome hat PETERS [*165*] den wasserstabilisierten Hochdruckbogen bis zu Drucken von mehreren hundert Atmosphären verwendet. Dieser Bogen brennt in einem zylindrischen, von Wasser umgebenen Kanal, der in einem druckfesten, wassergefüllten Gefäß durch Rotation gebildet wird (Fig. 33). Durch den Bogen verdampft das Wasser und erzeugt einen hohen Druck von mehreren hundert Atmosphären. Bei einer Stromstärke bis zu 200 A im 5 mm Durchmesser-Kanal steigt die Elektronenkonzentration auf 10^{18} cm^{-3}, obwohl weder die Temperatur ($\approx 12000°$ K), noch der Ionisationsgrad (≈ 1%) sehr hoch sind. Unter diesen Verhältnissen werden die Elektronen fast nur von den Neutralatomen gebremst. Nachdem die Elektronendichte und die Temperatur als Funktion des Radius aus der Linienform der Linie H_β und des Kontinuums gemessen ist, kann man diese Funktionen in die Gl. (84.3) einsetzen, und den Querschnitt der bremsenden Atome so wählen, daß die gemessene Stromstärke resultiert. Es ergibt sich

$$Q_{2H,0} = 90 \cdot 10^{-16}\,\text{cm}^2. \tag{84.4}$$

Benutzt man aber den wasserstabilisierten Rohrbogen bei Atmosphärendruck mit 50 A, dann steigt nach Messungen von JÜRGENS [*95*] der Ionisationsgrad auf 11%, wodurch neben den Atomen auch die Ionen mit ihrem weitreichenden COULOMB-Feld für die Bremsung der Elektronen verantwortlich werden. Setzt man nun den obigen Wert für den Querschnitt der Neutralatome und den im theoretischen Teil Gl. (63.9), (63.10) angegebenen Querschnitt für die Ionen in die Stromtransportgleichung ein, so führt die Integration zur gemessenen Stromstärke.

Steigert man nun durch Verengung des Rohrdurchmessers auf 2,3 mm und Erhöhung der Stromstärke bis auf 1500 A die Stromdichte bis 50000 A/cm²

(Gerdien-Bogen), dann ist die Leitfähigkeit nur noch durch die ein- oder mehrfach geladenen Ionen gegeben, und die Stromtransportgleichung liefert nach Einsetzen der Temperaturverteilung die Bestätigung für die in Gl. (63.9) wiedergegebene Formel (Maecker, Peters und Schenk [*144*]).

Der Absolutbetrag des Ionenquerschnittes lag zur Zeit der Messungen (Maecker und Peters [*142*]) um den Faktor 0,55 tiefer als die damaligen Theorien (z.B. Gvosdover [*75*]) ihn erwarten ließen. Durch die genaueren Rechnungen von

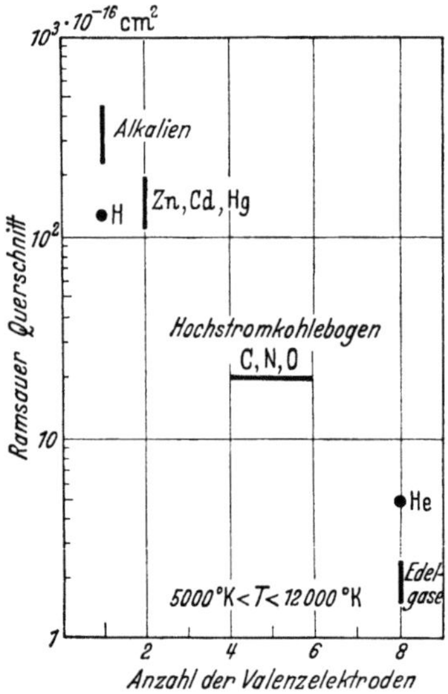

Fig. 64. Übersicht über die Wirkungsquerschnitte einiger Elemente, nach der Valenzelektronenzahl geordnet.

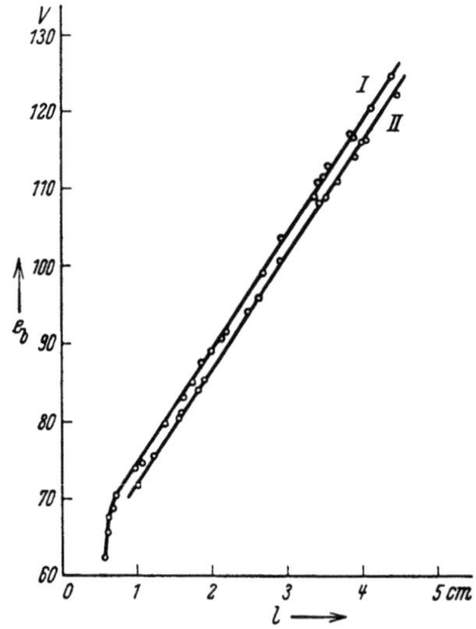

Fig. 65. Bogenspannung als Funktion der Bogenlänge für den 10 A-Kohlebogen. *I* Gereinigte Kohlen, *II* ungereinigte Kohlen. Die Steigung der Geraden ist die Feldstärke.

Spitzer und Mitarbeitern [*212*], [*213*] wurde dann der Einklang zwischen Experiment und Theorie wieder hergestellt. Zu ähnlichen Meßergebnissen kommen auch Schulz [*199*] am Argonbogen, Elenbaas [*4a*] und Hayaux am Hg-Bogen [*91*].

Durch ganz andersartige Untersuchungen von Kantrowitz [*98*] an Stoßwellen in Argon bei Temperaturen bis zu 16000° K sind die bisherigen Messungen und Rechnungen für die Ionenquerschnitte ebenfalls bestätigt worden.

Weitere Leitfähigkeitsmessungen sind am Hochstromkohlebogen mit der Temperaturverteilung von Fig. 54 für eine Mischung der Elemente N, O und C [*144*] und am Eisenbogen [*144*], [*16*] vorgenommen worden. Eine überschlägige Zusammenstellung der meisten bekannten Werte findet man in Fig. 64, in der eine Ordnung nach Valenzelektronenzahlen vorgenommen wurde. Dabei scheint sich eine gewisse Systematik der Querschnitte als Funktion der Valenzelektronenzahl anzudeuten.

Diese Beispiele zeigen, daß das Problem der Plasmaleitfähigkeit im Prinzip geklärt ist, nur müssen natürlich die Wirkungsquerschnitte der von den Elektronen gestoßenen Atome und Moleküle bekannt sein. Ist diese Voraussetzung aber erfüllt, dann kann man außer der elektrischen Leitfähigkeit sogar die Temperatur aus der Stromtransportgleichung berechnen (Schirmer [*187*]).

85. Die elektrische Feldstärke. Für die Benutzung der Stromtransportgleichung und der später folgenden Energiebilanz muß die Feldstärke gemessen sein. Für den Fall der Zylindersymmetrie kann in erster Näherung die Bogenspannung nach Abzug eines geschätzten Betrages für die Elektrodenfälle durch die Bogenlänge dividiert werden, was um so richtigere Resultate liefert, je länger die Säule ist.

Unter der Voraussetzung der Zylindersymmetrie ist das exakte Meßverfahren für die Feldstärke die Bestimmung der Brennspannungsänderung mit der Längenänderung. Ein Beispiel hierfür gibt Fig. 65.

Hat der Bogen lediglich Rotationssymmetrie, dann ist das Verfahren nur zulässig, wenn man annehmen darf, daß die Bogenteile, die vor der Längenänderung bereits existieren, insbesondere also die kathoden- und anodennahen Gebiete, durch die Längenänderung nicht beeinflußt werden. Nur unter dieser Voraussetzung gilt, wie BUSZ-PEUCKERT und FINKELNBURG [*21*] gezeigt haben, daß die Feldstärke für den neu geschaffenen Säulenteil durch den Differentialquotienten der Bogenspannung nach der Länge gegeben ist.

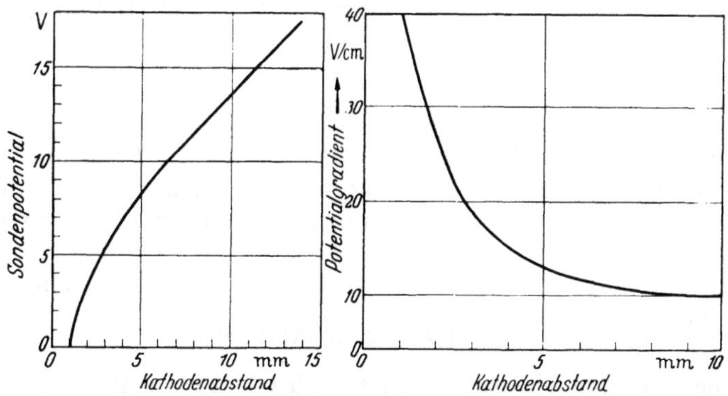

Fig. 66. Sondenpotentiale, gemessen entlang der Achse eines 200 A-Kohlebogens und bezogen auf das Potential 1 mm vor der Kathode. Daraus Feldstärke als Funktion des Abstandes von der Kathode.

Ein anderes bewährtes Mittel zur Ermittlung der Feldstärke ist die Messung des Potentialverlaufes in der Säule mit Hilfe von Sonden. Zwar bleibt der Absolutwert des Potentials unbestimmt, weil es bis jetzt nicht gelungen ist, das Kontaktpotential zwischen der kalten Sonde und dem Plasma zu bestimmen — offenbar spielen hier Nichtgleichgewichtsvorgänge die entscheidende Rolle; wenn man aber dieses Kontaktpotential als konstant, unabhängig von dem Plasmazustand annimmt, dann liefert die Differentiation der Potentialkurve doch recht zuverlässige Werte für den Feldstärkeverlauf. Als Anwendungsbeispiel für diese Methode möge in Fig. 66 der Feldstärkeverlauf in der Säule des Hochstromkohlebogens gezeigt werden (FINKELNBURG und SEGAL [*52*]).

Man kann schließlich auch, wenn in einem Bogen das Temperaturfeld und die Wirkungsquerschnitte zuverlässig bekannt sind, die Feldstärke aus der Stromtransportgleichung berechnen oder wenigstens ihren relativen Verlauf als Funktion der Bogenhöhe (BUSZ-PEUCKERT und FINKELNBURG [*21*]) finden.

86. Bestimmung der Wärmeleitfähigkeit aus der Energiebilanz. Die Energiegleichung (54.5) schrumpft im stationären Zustand ohne Strömung und ohne äußere Kräfte außer den elektrischen unter Vernachlässigung der Strahlung zu der Form

$$\operatorname{div} \varkappa \operatorname{grad} T + \mathfrak{j}\,\mathfrak{E} = 0 \tag{86.1}$$

zusammen, d.h. die einem Volumenelement pro Volumen- und Zeiteinheit zugeführte elektrische Energie wird durch Wärmeleitung abgeführt. Bei Rotationssymmetrie wird hieraus

$$\frac{\partial(\varkappa \operatorname{grad}_z T)}{\partial z} + \frac{1}{r}\,\frac{\partial(r\varkappa \operatorname{grad}_r T)}{\partial r} + \mathfrak{j}\,\mathfrak{E} = 0 \tag{86.2}$$

und bei annähernder Zylindersymmetrie

$$\frac{1}{r}\,\frac{\partial(r\varkappa\,\mathrm{grad}_r T)}{\partial r}+\mathfrak{j}\,\mathfrak{E}=0. \tag{86.3}$$

Einmalige Integration ergibt

$$-2\pi r\varkappa\frac{\partial T}{\partial r}=J_r E \tag{86.4}$$

mit

$$J_r=2\pi\cdot\int_0^r j\,r\,dr=2\pi E\int_0^r \sigma\,r\,dr \tag{86.5}$$

wobei J_r die von dem Kreis mit dem Radius r eingeschlossene Stromstärke bezeichnen soll. Ist nun die radiale Temperaturverteilung und die elektrische Leitfähigkeit gemessen, dann kann der Wärmeleitwert als Funktion des Radius und also auch der Temperatur ermittelt werden:

$$\varkappa=-\frac{E\,J_r}{2\pi r\,\partial T/\partial r}. \tag{86.6}$$

Für den Bereich außerhalb der Leitfähigkeitszone wird $E\,J_r$ konstant gleich der gesamten elektrisch zugeführten Leistung je Längeneinheit der Säule, so daß hier der Wärmeleitkoeffizient dem Temperaturabfall im $T-\log r$-Diagramm umgekehrt proportional wird.

Mit diesem Verfahren hat beispielsweise ELENBAAS [*4a*] den Wärmeleitwert von Hg zu

$$\varkappa=8{,}4\;T^{\frac{3}{4}}\;\mathrm{erg/cm\;grad\;sec} \tag{86.7}$$

bestimmt. Die primitive Theorie verlangt nach Gl. (62.1) einen Gang von $\varkappa$ mit $T^{\frac{1}{2}}$. Da aber die noch anzubringende SUTHERLAND-Konstante den Exponenten auf jeden Fall vergrößert, ist der Wert $\frac{3}{4}$ sehr plausibel.

87. Die Wärmeleitfähigkeit aus dem Abklingen abgeschalteter Bögen. Eine andere Methode zur Messung der Wärmeleitfähigkeit hat WIENECKE [*251*] angewendet.

Wenn man einen Lichtbogen plötzlich durch geeigneten Kurzschluß stromlos macht, dann sinkt das Temperaturgebirge durch Wärmeleitung zusammen. Die für diesen instationären Vorgang maßgebliche Energiegleichung heißt für konstanten Druck

$$\varrho\frac{dh}{dt}=\varrho c_p\left(\frac{\partial T}{\partial t}+\mathfrak{V}\,\mathrm{grad}\,T\right)=\mathrm{div}\,\varkappa\,\mathrm{grad}\,T. \tag{54.4}$$

Im Vergleich zu Gl. (86.6) ist zwar die elektrische Leistung $\mathfrak{j}\,\mathfrak{E}$ fortgefallen, dafür aber das zeitabhängige Glied $\varrho\,dh/dt$ hinzugekommen. Die Geschwindigkeit $\mathfrak{V}$ ergibt sich aus der Kontinuitätsgleichung

$$\frac{\partial\varrho}{\partial t}=-\mathrm{div}(\varrho\,\mathfrak{V}) \tag{39.12}$$

durch Integration von der Bogenachse her mit $\mathfrak{V}=0$ nach außen zu

$$\mathfrak{V}(r)=-\frac{1}{r\,\varrho(T)}\int_0^r\frac{r\,\partial\varrho}{\partial t}\,dr, \tag{87.1}$$

wenn man sich auf Zylindersymmetrie beschränkt. Man kann dann die Energiegleichung (54.4) integrieren und erhält für die Wärmeleitfähigkeit

$$\varkappa(r)=\frac{1}{r\,\partial T/\partial r}\int_0^r r\,\varrho\,c_p\left(\frac{\partial T}{\partial t}+\mathfrak{V}\frac{\partial T}{\partial r}\right)dr. \tag{87.2}$$

Zur Ermittlung von $\varkappa$ nach dieser Gleichung muß das Temperaturfeld als Funktion der Zeit nach dem Kruzschluß bekannt sein. Zu diesem Zweck wird eine von BARTELS und EISELT [2] beschriebene Zeitlupenanordnung verwendet, wie sie in Fig. 67 dargestellt ist. Der Lichtbogen wird durch das Objektiv L_1 auf den schnell rotierenden Planspiegel Dr. Sp. abgebildet. Von hier wird das Licht auf neun nebeneinanderstehende Planspiegel Sp. G. geworfen, die so justiert sind, daß das von ihnen reflektierte Licht in das Kameraobjektiv L_2 trifft. Dieses bildet das auf dem Drehspiegel vorhandene Bogenbild über die Spiegelgalerie auf die Photoplatte Pl. ab. Dort erscheinen also nebeneinander neun zeitlich aufeinanderfolgende Bilder des Lichtbogens. Bevor der vom Drehspiegel reflektierte Lichtstrahl die Spiegelgalerie erreicht, erzeugt er in einem Photomultiplier SEV einen Impuls, der ein zum Bogen parallel geschaltetes Thyratron zündet, wodurch der Bogenkurzschluß innerhalb 10 μsec herbeigeführt wird (s. auch Fig. 26).

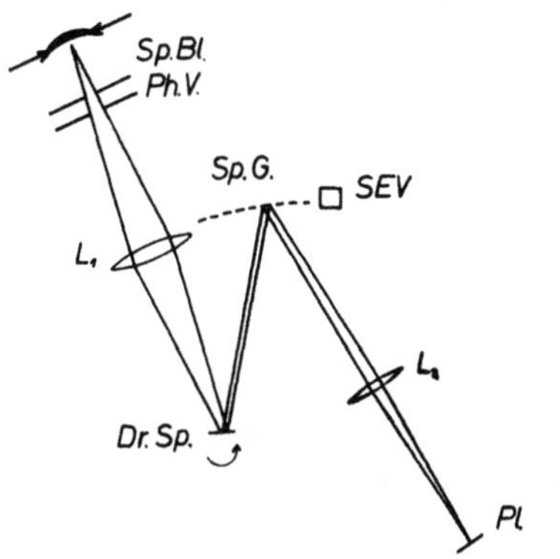

Fig. 67. Zeitlupe zur Aufnahme eines abgeschalteten Lichtbogens. Der Bogen wird auf dem Drehspiegel abgebildet, von wo das Licht nacheinander von den einzelnen Planspiegeln in die Kamera geworfen wird.

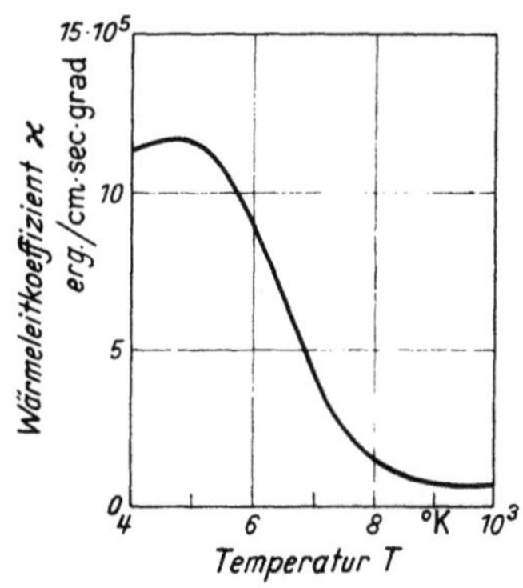

Fig. 68. Wärmeleitfähigkeit der Luft als Funktion der Temperatur, gemessen aus der Abklinggeschwindigkeit eines abgeschalteten Bogens.

Mit dieser Apparatur hat WIENECKE das zeitliche Abklingen des Leuchtens in der Säule des Hochstromkohlebogens verfolgt und aus der jeweiligen Lage der Aureolen, die gleichzeitig Isothermen bekannter Temperatur (s. Fig. 53 und 56) sind, das Temperaturfeld als Funktion der Zeit bis zu 1 msec nach Kurzschließen bestimmt. Mit den Gln. (59.3), (59.4), (58.9) und (39.2) können c_p und ϱ als Funktion der Temperatur bei bekannter Gaszusammensetzung berechnet werden, so daß aus dem gemessenen zeitlichen Verlauf des Temperaturfeldes mit Gl. (87.1) zunächst die Geschwindigkeit und ferner mit Gl. (87.2) die Wärmeleitfähigkeit ermittelt werden kann. Das Ergebnis für die Säule des Hochstromkohlebogens, dessen Zusammensetzung MAECKER [138] zu N:O:C = 56:14:30 gemessen hat, in Fig. 68 zeigt wieder den typischen Dissoziationsberg im Bereich von 4 bis $6 \cdot 10^3$ °K, hinter dem die Wärmeleitfähigkeit wieder auf etwa den klassischen Wert absinkt.

III. Die Bogensäule als Ganzes.

88. Kanalmodell und STEENBECKsches Minimumprinzip. Nach Untersuchungen von ONSAGER [162], PRIGOGINE [169], DE GROOT [71] und HAASE [76], [77] ist der stationäre Nicht-Gleichgewichtszustand eines thermodynamischen Systems ausgezeichnet durch einen positiven Minimalwert der Entropieproduktion. Demnach ist nicht das Prinzip der Entropievermehrung das entscheidende Gesetz, welches den Ablauf eines thermodynamischen Prozesses regelt — die Entropie eines offenen Systems kann z. B. durchaus abnehmen — sondern die Tendenz, eine minimale Entropieerzeugung anzustreben, bestimmt das Geschehen. WEGELAND und MAZUR haben dieses Extremalprinzip bei allgemeinen kontinuierlichen Systemen behandelt und gezeigt, daß sich als zugehörige EULER-LAGRANGEsche Gleichungen dieses Variationsproblems die Erhaltungssätze für Masse, Ladung und Energie ergeben. Auf den thermischen Lichtbogen angewandt bedeutet dieses, daß die ELENBAAS-HELLERsche Energiegleichung und der Ausdruck für die stationäre Stromverteilung $\operatorname{div} \mathfrak{j} = 0$ als EULERsche Gleichungen des Prinzips der minimalen Entropieproduktion angesehen werden können.

Betrachten wir nun mit PETERS[1] die Entropiebilanz im Gesamtraum eines stationären Systems, so gilt nach (39.20) unter Verwendung des GAUSSschen Satzes:

$$\Theta = \int \sigma_S \, d\tau = \int \operatorname{div} \mathfrak{J}_S \, d\tau = \oint \mathfrak{J}_S \, d\mathfrak{f} = \oint \frac{1}{T}\Big(\mathfrak{W} - \sum_k \mu_k \mathfrak{J}_k\Big) d\mathfrak{f}. \qquad (88.1)$$

Das heißt: die im Inneren des gesamten Systems produzierte Entropie ist gleich dem Integral der Entropiestromdichte über die Begrenzungsfläche. Als solche wählen wir eine Isothermenfläche, auf der die Temperatur gleich der Umgebungstemperaur T_R ist. Dann wird

$$\Theta = \frac{1}{T_R} \oint \Big(\mathfrak{W} - \sum_k \mu_k \mathfrak{J}_k\Big) d\mathfrak{f} = \frac{1}{T_R} \int \operatorname{div}\Big(\mathfrak{W} - \sum_k \mu_k \mathfrak{J}_k\Big) d\tau. \qquad (88.2)$$

Unter Berücksichtigung der Energiebilanz (54.1) und der Massenerhaltung (39.11) für den stationären Fall geht dieser Ausdruck (bei Vernachlässigung einer Entropieerzeugung durch chemische Reaktionen) über in

$$T_R \Theta = \int \sum_k (\mathfrak{J}_k, \mathfrak{F}_k - \operatorname{grad} \mu_k) \, d\tau. \qquad (88.3)$$

Bei der Anwendung dieser Gleichung auf den elektrischen Lichtbogen ist zu beachten, daß als äußere Kraft praktisch nur die angelegte elektrische Feldstärke in Frage kommt, also $\mathfrak{F}_k = \frac{e_k}{m_k} \mathfrak{E}$ zu setzen ist und daß demgegenüber bei annähernder Zylindersymmetrie der Gradient des chemischen Potentials, der implizit den Einfluß eines Druck-, Konzentrations- und Temperaturgradienten auf den Stromtransport enthält, in guter Näherung zu vernachlässigen ist, weil dann $\mathfrak{j}$ und $\operatorname{grad} \mu_k$ etwa senkrecht zueinander stehen. Dann wird aus Gl. (88.3) mittels der Beziehung $\mathfrak{j} = \sum_k \frac{e_k}{m_k} \mathfrak{J}_k$ für die elektrische Stromdichte

$$T_R \Theta = \int (\mathfrak{j}, \mathfrak{E}) \, d\tau. \qquad (88.4)$$

Die Gln. (88.3) und (88.4) lauten in Worten:

Umgebungstemperatur × Entropieerzeugung = Verlustleistung

(Satz von GOUY und STODOLA [*220*]).

Die Entropieerzeugung im elektrischen Lichtbogen ergibt sich also zu

$$\Theta = \frac{IU}{T_R} \quad \text{bzw.} \quad \Theta = \frac{IE}{T_R} \qquad (88.5)$$

pro Zentimeter Säulenlänge eines zylindrischen Lichtbogens. Da nun die Entropieproduktion im stationären Zustand einen Minimalwert annimmt, ist $\delta\Theta = 0$ und

[1] Th. PETERS: Physik. Verh. **6** (9), 173 (1955).

mit vorgegebener Stromstärke I und fester Randtemperatur T_R auch

$$\delta U = 0 \quad \text{bzw.} \quad \delta E = 0. \tag{88.6}$$

Das heißt die Brennspannung bzw. die elektrische Feldstärke nimmt bei vorgegebenem Strom und gegebenen Randbedingungen einen Minimalwert an. Diese Aussage ist in der Gasentladungsphysik seit langem bekannt als das STEENBECKsche Minimumprinzip [*215*], welches sich somit als Spezialfall eines echten thermodynamischen Extremalprinzips erweist. Praktische Verwendung findet dieses Minimumprinzip hauptsächlich bei dem Kanalmodell des elektrischen Lichtbogens. Hierbei wird die Säule gewöhnlich aufgeteilt in eine elektrische Leitfähigkeitszone einheitlicher Temperatur T vom Durchmesser $2r$ und in eine Energieableitungszone, in der die Temperatur bis zur Randtemperatur T_R am Bogenrand R absinkt. Die Stromtransportgleichung

$$I = \sigma(T)\, E\, \pi\, r^2 \tag{88.7}$$

und die Energiegleichung, die sich zu der Form

$$I\, E = g(r)\, h(T) \tag{88.8}$$

vereinfachen läßt (STEENBECK [*217*]), stellen bei konstantem I und vorgegebenen Randbedingungen ($T = T_R$ für $r = R$) dann zwei Beziehungen dar zwischen den drei Größen T, r, E. Eliminiert man T oder r, so bleibt noch eine unendliche Mannigfaltigkeit von Bogenzuständen mit der Abhängigkeit $E(r)$ bzw. $E(T)$ übrig. An dieser Stelle greift das Prinzip der minimalen Feldstärke ein, indem gemäß Gl. (88.6)

$$\frac{dE}{dr} = 0 \quad \text{bzw.} \quad \frac{dE}{dT} = 0$$

gefordert wird, womit die Temperatur und die Breite des Kanals eindeutig festgelegt sind.

Diese Anwendung des STEENBECKschen Prinzips hat seinerzeit zahlreiche Diskussionen ausgelöst. ELENBAAS [*38*] stellte sich auf den Standpunkt, daß Feldstärke, Stromstärke und Temperaturverteilung bei Integration der exakten Energie- und Stromtransportgleichung völlig bestimmt und daher die Anwendung einer dritten Gleichung, etwa in Form eines Minimalprinzips überflüssig sei. Die Verfechter des Kanalmodells (KESSELRING und KOPPELMANN [*103*]) hielten hingegen die Einführung des STEENBECKschen Prinzips für unerläßlich, um die Bogendaten eindeutig zu bestimmen. ROMPE und WEIZEL [*180*] kamen in ihrer Darstellung dieses Problems zu dem Schluß, daß der ELENBAASsche Standpunkt seine Berechtigung hat, denn die Forderung nach einer dritten Bestimmungsgleichung ist ja nur im Näherungscharakter des Kanalmodells begründet. Sie diskutierten dann weiterhin die Möglichkeit, daß es sich beim STEENBECKschen Prinzip um ein echtes thermodynamisches Extremalprinzip handeln könnte, wobei dann die $E-H$-Gleichung als EULERsche Gleichung dieses Variationsproblems erscheinen müßte. Mit den von ihnen durchgeführten Ansätzen kamen ROMPE und WEIZEL aber nicht zum Ziel, und sie glaubten daher, daß man das Minimumprinzip selbst mit Rücksicht auf seine unbestreitbaren Erfolge lediglich als eine Art praktischer Faustregel ansehen könnte. Indessen haben, wie oben erwähnt, WERGELAND [*250*] und MAZUR [*150*] neuerdings gezeigt, daß man bei der Variation der Entropieproduktion zu dem gewünschten Ergebnis kommt.

Das STEENBECKsche Prinzip ist in zahlreichen Fällen (z.B. von FOITZIK [*54*]) an Bogenentladungen geprüft worden und hat sich dabei überraschend gut

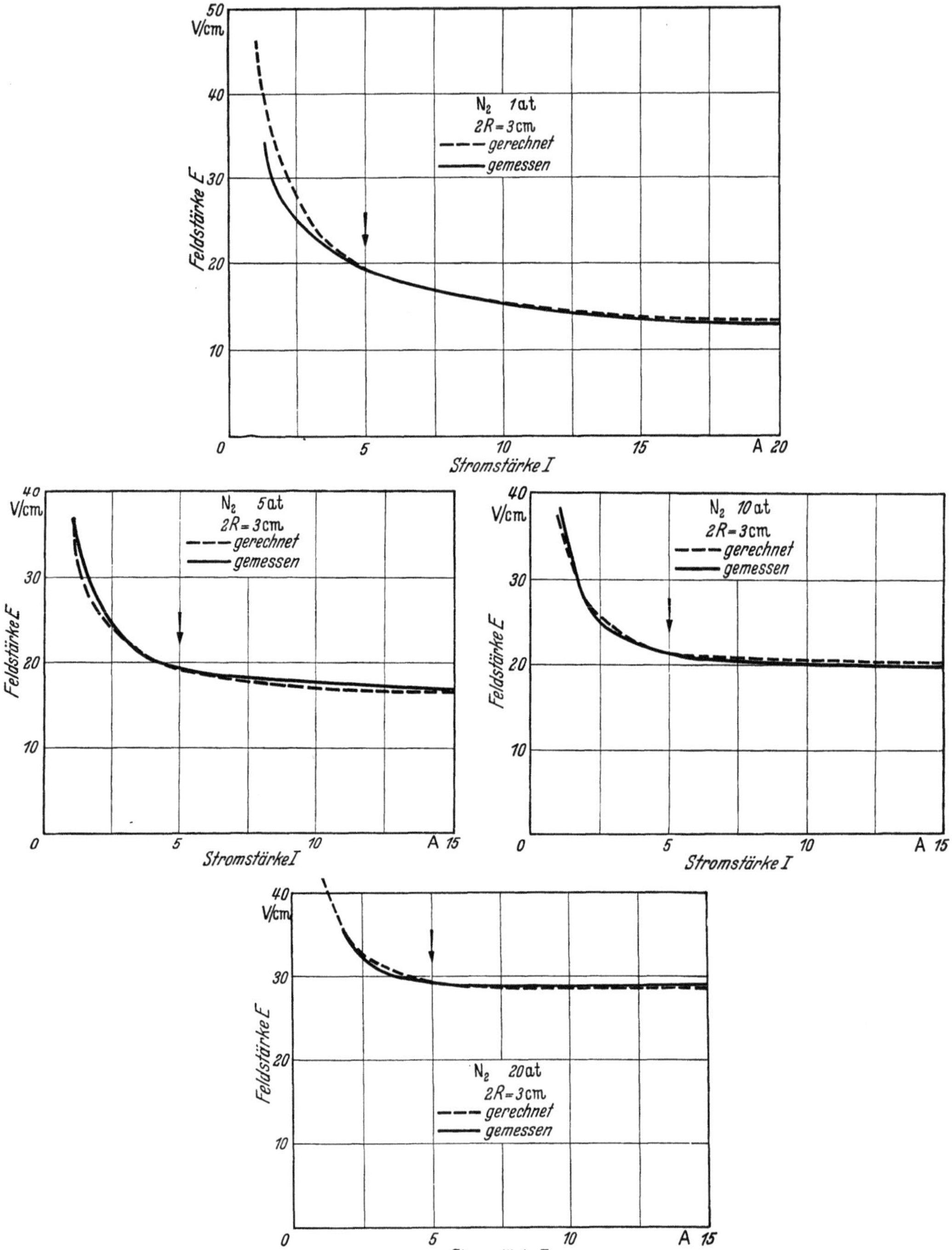

Fig. 69. Charakteristiken des Wälzbogens in N_2, gemessen von FOITZIG [54] (ausgezogen) und nach dem Minimumprinzip berechnet von STEENBECK [202] (gestrichelt).

bewährt, wie Fig. 69 zeigt. Wir dürfen diese Bestätigung daher auch als experimentellen Beweis für die Gültigkeit des Prinzips der minimalen Entropieerzeugung ansehen.

89. Temperaturverlauf außerhalb der Leitfähigkeitszone. Um einen Überblick über den radialen Temperaturverlauf von annähernd zylindersymmetrischen Bögen ohne wesentliche Strahlung zu erhalten, formen wir Gl. (86.4) nach SCHMITZ [*194*] mit den Beziehungen

$$\varkappa\, d\,T = d\left(\int \varkappa\, d\,T\right) \tag{89.1}$$

und

$$\frac{1}{r}\, d\,r = d\,(\log r) = d \log \frac{r}{r_0} \tag{89.2}$$

um in

$$-\frac{d\left(\int \varkappa\, d\,T\right)}{d \log \frac{r}{r_0}} = \frac{1}{2\pi}\, J_r\, E\,. \tag{89.3}$$

Die erste Beziehung ist gerechtfertigt, da $\varkappa$ nur von der Temperatur abhängt. Außerhalb des Leitfähigkeitskanals ist $J_r = J$. Für dieses Gebiet besagt Gl. (88.3),

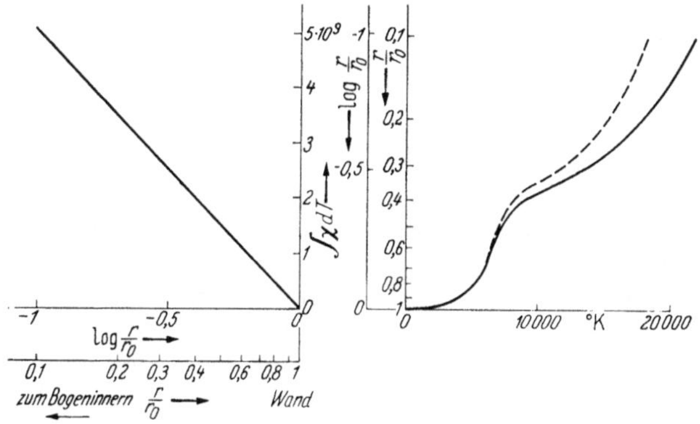

Fig. 70. Außerhalb der Leitfähigkeit eines zylindrischen Bogens ist die Temperaturfunktion dem negativen Logarithmus des relativen Radius proportional mit der Leistung/2π als Steigung (links). Daher kann man in einem $\int \varkappa\, d\,T$, T-Diagramm (rechts) an die Ordinate den $\log r/r_0$ antragen und erhält so den radialen Temperaturverlauf außerhalb der Leitfähigkeit.

daß das Integral $\int \varkappa\, d\,T$ über $\log r$ aufgetragen eine Gerade mit der Neigung $-\frac{1}{2\pi}\, J\,E$ liefert. Dies wird durch eine weitere Integration bestätigt, die die Gleichung

$$\int_{T_W}^{T} \varkappa\, d\,T = -\frac{1}{2\pi}\, J\,E \log \frac{r}{r_0} \tag{89.4}$$

liefert, welche wir durch

$$Y = -\frac{1}{2\pi}\, J\,E\,X \tag{89.5}$$

abkürzen wollen (Fig. 70 linke Seite). r_0 ist der Radius des Rohres und $\int \varkappa\, d\,T$ soll an der Wand verschwinden.

Wenn nun $Y(T)$ aus theoretischen Berechnungen bekannt ist, dann kann man diese Funktion unter Beibehaltung der Ordinate von Gl. (89.5) über der neuen Abszisse T auftragen (Fig. 70 rechts). Wegen des linearen Zusammenhangs zwischen Y und X nach Gl. (89.5) kann man weiter die Ordinate Y mit einer linearen X-Skala gemäß der Gleichung

$$-\frac{2\pi}{J\,E}\, Y = X \tag{89.6}$$

versehen, wobei der der Wand entsprechende X-Wert auf der Abszisse zu liegen kommt und ein Fortschreiten nach oben eine Annäherung an das Bogeninnere bedeutet (negative X-Werte). Man sieht also, daß die über T aufgetragenen $\int \varkappa\, dT$-Kurven bereits die radiale Temperaturverteilung außerhalb der Leitfähigkeit wiedergeben, wenn man sie nur um 90° dreht und sich anstelle von $\int \varkappa\, dT$ den $\log r/r_0$ aufgetragen denkt, und zwar mit dem Wandwert am Ursprung beginnend.

Hieraus kann man sich schon einige Vorstellungen vom radialen Temperaturverlauf der achsenfernen Gebiete machen.

Wäre der Wärmeleitwert $\varkappa$ konstant, dann würde $\int \varkappa\, dT$ eine Gerade sein, d.h. die Temperaturen würden logarithmisch vom Rande her ansteigen.

Befolgt der Wärmeleitwert ein Potenzgesetz

$$\varkappa = \varkappa_0 T^n, \tag{89.7}$$

wie man das für den Hg-Bogen (ELENBAAS [*4a*], KOCH [*109*]) und näherungsweise für alle einatomigen Gase annimmt, dann ist

$$Y = \int \varkappa\, dT = a\, T^{n+1} \tag{89.8}$$

oder mit Gl. (89.4)

$$T = a' \left(\log \frac{r}{r_0}\right)^{1/(n+1)} \tag{89.9}$$

Die Temperatur wird also von der Wand her zunächst schnell ansteigen, dann aber um so flacher nach innen laufen, je höher der Exponent n ist.

Ist der Verlauf des Wärmeleitwertes mit der Temperatur nicht monoton, sondern wellig, wie bei allen Molekülgasen (siehe z.B. N_2 in Fig. 41 und Fig. 68), dann wird auch der Temperaturverlauf Wendepunkte aufweisen und zwar in dem Sinne, daß bei kleinem $\varkappa$ der $-\operatorname{grad} T$ groß ist und umgekehrt.

Für qualitative Betrachtungen darf man noch ein Stück in die Leitfähigkeitszonen hineingehen, hat aber zu beachten, daß die „Konstante" in Gl. (89.5) $-\frac{1}{2\pi} \cdot JE$ nicht mehr konstant bleibt, sondern mit dem jeweils noch eingeschlossenen Strom abnimmt. Dies hat ein Abbiegen der Temperaturkurve in die zur Ordinate Y parallele Richtung zur Folge, wie es sein muß, da sich die Temperatur in Achsennähe nicht mehr mit dem Radius ändert (Fig. 70 gestrichelt).

Die Überlegungen dieser Ziffer führen also zu der Aussage, daß Bögen in einatomigen Gasen einen mehr oder weniger steilen, monotonen Temperaturanstieg vom Rande her haben, entsprechen dem gleichmäßigen Verlauf des Wärmeleitwertes mit der Temperatur. Dagegen bewirkt der Dissoziationsberg in der Wärmeleitwertkurve der Molekülgase in diesem Bereich einen besonders flachen Temperaturverlauf, der erst weiter innen, bedingt durch das Wärmeleitwertminimum zwischen Dissoziationsberg und Plasmawärmeleitfähigkeit wieder steil ansteigt. So ist es zu verstehen, worauf R. A. KING [*105*] besonders aufmerksam gemacht hat, daß Bögen in Molekülgasen, im Gegensatz zu solchen in monoatomaren Gasen einen ausgesprochenen Kern von großer Helligkeit zeigen, vorausgesetzt, daß die Achsentemperatur überhaupt einen Wert oberhalb des Dissoziationsberges erreicht. Dies hängt in erster Linie von der Stromstärke, aber auch von gewissen Strömungsverhältnissen ab, von denen unten noch die Rede sein wird. Bei kleinen Stromstärken ist die Voraussetzung hinreichend hoher Achsentemperatur jedenfalls nicht erfüllt, weswegen man hier auch keinen Kern beobachtet. Überschreitet jedoch die Stromstärke einen gewissen Wert, der beim Kohlebogen in Luft oder N_2 je nach den äußeren Umständen zwischen 50 und 100 Amp und mehr liegt, dann erscheint fast plötzlich der Kern, der den schnellen Anstieg der Achsentemperatur anzeigt. Nimmt man zur Deutung dieser schnellen Kernentstehung in erster Näherung an, daß die Charakteristik in diesem Stromstärkebereich horizontal verläuft, die Leistung JE also proportional mit der Stromstärke steigt, dann nimmt nach Gl. (89.5) auch die Neigung der Geraden $\int \varkappa\, dT$ gegen $\log r/r_0$ gleichmäßig mit der Stromstärke zu und ebenso der $\int \varkappa\, dT$-Wert im Inneren des Bogens. Die Temperatur selbst dagegen wächst im Bereich des Dissoziationsberges nur langsam (Niederstrombogen), im Bereich des Wärmeleitungsminimums aber sehr schnell (Kern im Hochstrombogen).

90. Temperaturverlauf zylindrischer Bögen ohne Strahlungsleistung. Will man exakte Aussagen über den gesamten Temperaturverlauf von Bögen haben, deren ausgestrahlte Leistung gegenüber der elektrischen Leistung zu vernachlässigen ist, was wohl bei allen unter Atmosphärendruck brennenden erlaubt ist, dann muß man die Energiegleichung (86.4) mit einem geeigneten Näherungsverfahren lösen. Randbedingungen dafür sind für $r=0$, $dT/dr=0$ und für $r=r_0$, $T=T_W$.

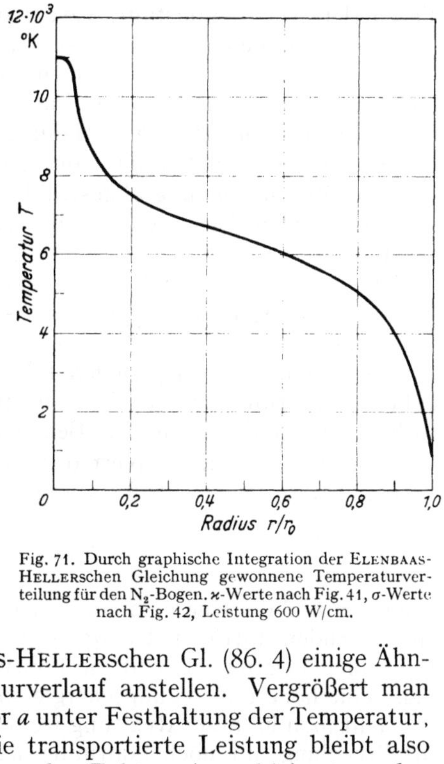

Fig. 71. Durch graphische Integration der ELENBAAS-HELLERschen Gleichung gewonnene Temperaturverteilung für den N_2-Bogen. $\varkappa$-Werte nach Fig. 41, σ-Werte nach Fig. 42, Leistung 600 W/cm.

Fig. 71 zeigt einen berechneten Temperaturverlauf für Stickstoff der erwartungsgemäß am Rande sehr steil abfällt wegen des kleinen Wärmeleitkoeffizienten bei tiefen Temperaturen, auf dem Wege zur Mitte aber stark abflacht, bedingt durch den Dissoziationsberg im Wärmeleitkoeffizienten und sich schließlich in Achsennähe zum gut leitenden Bogenkern hoher Temperatur erhebt.

In monoatomaren Gasen zeigen die Bögen einen glatten, etwa parabelförmigen Temperaturabfall entsprechend den Überlegungen der vorigen Ziffer (Fig. 51).

Weiterhin lassen sich mit der ELENBAAS-HELLERschen Gl. (86. 4) einige Ähnlichkeitsbetrachtungen über den Temperaturverlauf anstellen. Vergrößert man nämlich die Variable r überall um den Faktor a unter Festhaltung der Temperatur, dann ändert sich die linke Seite nicht, die transportierte Leistung bleibt also die gleiche, während rechts die Feldstärke um den Faktor $1/a$ verkleinert werden muß:

$$-2\pi a r \frac{dT}{a\,dr} = 2\pi \left(\frac{E}{a}\right)^2 \int_0^{ar} \sigma \cdot a r \cdot a\,dr. \tag{90.1}$$

Da die Leistung $J_r E$ erhalten bleibt, muß die Stromstärke um a vergrößert werden. Es gilt daher: Zylindrische Bögen ohne wesentliche Strahlung haben für verschiedene Durchmesser bei gleicher Leistung die gleiche relative Temperaturverteilung. Die Stromstärke steigt, die Feldstärke fällt proportional mit dem Durchmesser bei gleicher Leistung.

91. Strahlungsleistung. Wegen der inhomogenen Struktur der Bogensäule erfolgt auch die Strahlung von Schichten verschiedener Temperatur. Solange keine Reabsorption stattfindet, tritt die gesamte abgestrahlte Leistung je Volumeneinheit $4\pi \int \varepsilon_\nu\, d\nu$ als zusätzliches Glied in der Energiegleichung auf. Bei starker Absorption dagegen diffundiert die Strahlung nach außen und erhöht auf diese Weise nach Ziff. 76 den totalen Wärmeleitkoeffizienten. Dieser Prozeß ist z.B. bei allen Resonanzlinien, in Einzelfällen auch bei höheren Übergängen wirksam, ferner im gesamten langwelligen Gebiet oberhalb einer Wellenlänge von $^1/_{100}$ bis $^1/_{10}$ mm und schließlich in der von LARCHÉ [*117*] beschriebenen Plasmakugel vor der Kathode des Xenon-Hochdruckbogens mit seiner extrem hohen Leuchtdichte. Auch der von PETERS [*165*] untersuchte, wasserstabilisierte Hochdruckbogen mit seiner fast schwarzen Strahlung von 12000° K bei 1000 Atm ist hier zu nennen. Bei Bögen unter Atmosphärendruck macht die Strahlung nur wenige Prozent des Energieumsatzes aus und kann daher meistens vernachlässigt werden. Selbst bei sehr hohen Temperaturen nimmt die Dichte des Plasmas so stark ab, daß der Strahlungsanteil klein bleibt. Hält man aber die mittlere Dichte konstant, wie das bei allen Bögen in abgeschlossenen Gefäßen mit fester Füllmenge der Fall ist, dann gewinnt die Strahlung mehr und mehr an Bedeutung und kann bei hohen Drucken den wesentlichen Teil der zugeführten Leistung abstrahlen. So finden KERN [*102*] und ELENBAAS [*4a*], daß in der wandstabilisierten Hg-Entladung 10 Watt durch Wärmeleitung abgeführt werden und zwar unabhängig von den Entladungsparametern, und daß der Rest der zugeführten Leistung abgestrahlt wird. Bei 100 W/cm zugeführter Leistung, was einem hohen Druck entspricht, wird also 90 W durch Strahlung nach außen geleitet.

Um die abgestrahlte Leistung auch theoretisch zu erfassen, müßte man im Falle der Transparenz den Emissionskoeffizienten ε_ν über alle Linien und über das kontinuierliche Spektrum integrieren, im allgemeinen ein aussichtsloses Unternehmen. Beim Hg-Bogen hilft man sich daher so, daß man alle Terme außer dem Grundterm zu einem einzigen zusammenschiebt und dessen Energie so bestimmt, daß er mit den Beobachtungen übereinstimmt. Dieses Verfahren führt zu dem plausiblen Wert der mittleren Anregungsenergie von $\bar{E}_m \approx 7{,}8$ eV. So kann die ausgestrahlte Leistung $S(T)$ zu

$$S(T) = K_2 \frac{1}{T} e^{-E_m/kT} \tag{91.1}$$

angesetzt und die Energiebilanz eingeführt werden, wobei K_2 dem Druck und der zu messenden mittleren Übergangswahrscheinlichkeit proportional zu setzen ist. Bei hohen Leistungen wird die kontinuierliche Ausstrahlung wirksamer, so daß dann eine höhere Anregungsenergie zu wählen wäre.

Geht man in dieser Richtung weiter, dann erreicht man einen Zustand, in dem die Strahlung nur noch kontinuierlich ist, obwohl sie noch aus optisch dünner Schicht emittiert wird, und in dem die Linienstrahlung relativ klein wird. Dieser Fall ist im Xenon-Hochdruckbogen (LARCHÉ [*117*]) realisiert, wenn man die Plasmakugel vor der Kathode ausschließt. Um auch in diesem Fall den Beitrag der Strahlung zur Energiebilanz zu erfassen, wenden wir Gl. (66.4) an und multiplizieren ε_ν, anstatt zu integrieren, mit $4\pi\nu_g$, um die gesamte Ausstrahlung je Volumeneinheit zu erhalten:

$$S(T) = 4\pi \int \varepsilon_\nu \, d\nu = 4\pi \, \varepsilon_\nu \, \nu_g. \tag{91.2}$$

Dabei haben wir die Absorption im ganz langwelligen Gebiet vernachlässigt, was gar keine Rolle spielt. Will man noch den kleinen Anteil des abfallenden Kontinuums jenseits der Grenzfrequenz ν_g mitnehmen, dann kann man die Grenze noch etwas hinausschieben.

Sofern noch ein Elektronenaffinitätskontinuum entstehen kann, muß dessen Ausstrahlung noch gesondert hinzugerechnet werden. So kann z.B. das H^--Kontinuum bei ausreichend hohen Drucken ein Vielfaches des normalen Elektronenkontinuums ausmachen, wie PETERS [*165*] am wasserstabilisierten Hochdruckbogen gezeigt hat. Außer dem H^--Kontinuum als einzigem bisher nachgewiesenen Affinitätskontinuum sind auch für viele andere Elemente mit großer Elektronenaffinität, also besonders für die Halogene und für Sauerstoff derartige Kontinua zu erwarten.

Mit der abgestrahlten Leistung $S(T)$ nimmt die Energiegleichung für Zylindersymmetrie die Form

$$\sigma\,\mathfrak{E}^2 - S(T) + \frac{1}{r}\,\frac{\partial}{\partial r}\left(r\,\varkappa\,\frac{\partial T}{\partial r}\right) = 0 \tag{91.3}$$

an. Die erforderliche numerische Integration haben zuerst WEIZEL und SCHMITZ [*248*], [*195*], später FRANCIS [*57*], im Hinblick auf die Hg-Entladung mit der Annahme eines konstanten Wärmeleitkoeffizienten und kleinen Ionisationsgrades durchgeführt. Neuerdings haben dann KOCH [*109*], LESEMANN und WALTHER [*109*] die Rechnungen mit einer automatischen Rechenmaschine wiederholt und dabei $\varkappa$ dem Ansatz (88.7) entsprechend von der Temperatur abhängen lassen. Aus der Fülle der Ergebnisse greifen wir das Charakteristikendiagramm Fig. 72 heraus. Hier ist $E R^{\frac{3}{2}}$ gegen $I R^{-\frac{3}{2}}$ mit pR^2 als Parameter aufgetragen. An den Kurven sind die reduzierten Temperaturen

$$\vartheta = k\,T/E_I \tag{91.4}$$

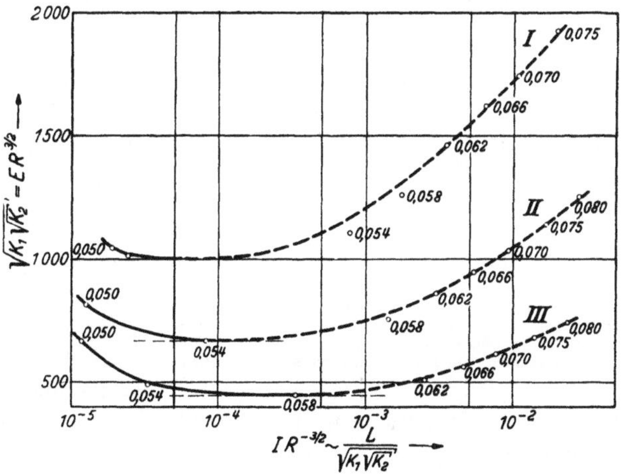

Fig. 72. Für einen wandstabilisierten Hg-Bogen aus der ELENBAAS-HELLERschen Gleichung berechnete Charakteristiken, mit Werten für die reduzierte Temperatur. *I*: $K = 4$; *II*: $K = 2$; *III*: $K = 1$ ($K \sim pR^2$).

als Beiwerte angeschrieben. Man erkennt bei kleinen Stromstärken den für fast alle Gasentladungen typischen Abfall der Charakteristiken, der auf die Zunahme der Leitfähigkeit infolge stärkerer Ionisierung bei wachsender Stromstärke und somit steigender Temperatur, vor allem aber auf die Zunahme des Bogenquerschnittes zurückzuführen ist. Wenn dagegen bei hohen Stromstärken der Bogen das Rohr im wesentlichen ausfüllt und die Ionisierung einen hohen Grad erreicht hat, dann nähert sich der Bogen in seinen elektrischen Eigenschaften einem Draht mit steigender Charakteristik und konstantem spezifischen Widerstand. Wachsender Druck treibt den Spannungsbedarf hinauf, aber nur mit einer schwachen Potenz, etwa mit der Wurzel aus dem Druck. Außerdem setzt der Druck bei festgehaltener Stromstärke die Temperatur herab. Diese und noch weitere Resultate der Rechnungen werden von den Beobachtungen teils qualitativ, teils quantitativ bestätigt.

Mit der ELENBAAS-HELLERschen Energiegleichung werden alle wandstabilisierten Bögen erfaßt, solange die Konvektion keine Rolle spielt.

Läßt man die Beschränkung auf eine Koordinate fallen und schreibt die ELENBAAS-HELLER-Gleichung in der Form

$$\sigma\,\mathfrak{E}^2 - S(T) + \operatorname{div}\varkappa\operatorname{grad}T = 0, \tag{91.5}$$

dann lassen sich damit im Prinzip auch die elektroden-stabilisierten Bögen behandeln. Bei diesen Typen trägt man der Tatsache Rechnung, daß sich an der Kathode, oft auch an der Anode ein Brennfleck oder wenigstens eine Kontraktion ausbildet, wodurch die Form und die Lage des Bogens festgelegt wird. Diese Art der Stabilisierung findet sich bei sehr kurzen Bögen und bei solchen, in denen der Einfluß der Strömung durch Rotation oder freien Fall (STEENBECK [*216*]) ausgeschaltet ist. Es bildet sich dann eine Form von ellipsoidähnlicher Gestalt aus, vor allem deswegen, weil die Feldlinien zwischen den Elektroden diesen Verlauf haben. Die Integration der Energiegleichung für diesen wesentlich schwierigeren Fall hat ebenfalls SCHMITZ [*195*] durch Einführung elliptischer Koordinaten ausgeführt mit dem Ergebnis, daß der elektrodenstabilisierte Bogen mit seiner endlichen Länge im Gegensatz zum unendlich lang angenommenen, wandstabilisierten Bogen keine Begrenzung des Volumens durch eine Wand braucht, um sich auf einen stationären Zustand, insbesondere auf ein stationäres Temperaturfeld einzustellen, weil nun die Wärmeleitung die dem Bogen zugeführte elektrische Energie etwa kugelsymmetrisch abführen kann. Für den stromführenden Kanal gilt näherungsweise wieder die ELENBAAS-HELLERsche Gleichung für die zylindrische Form, woraus z.B. folgt, daß auch die elektrodenstabilisierten Bögen, obwohl sie durch keine Wand eingeengt sind, bei hinreichender Stromstärke eine steigende Charakteristik haben (Fig. 73) (siehe z.B. GERTHSEN und SCHULZ [*66*]). Hier übernehmen nämlich die Elektroden die Aufgabe der Wand, den Bogen zusammenzuhalten.

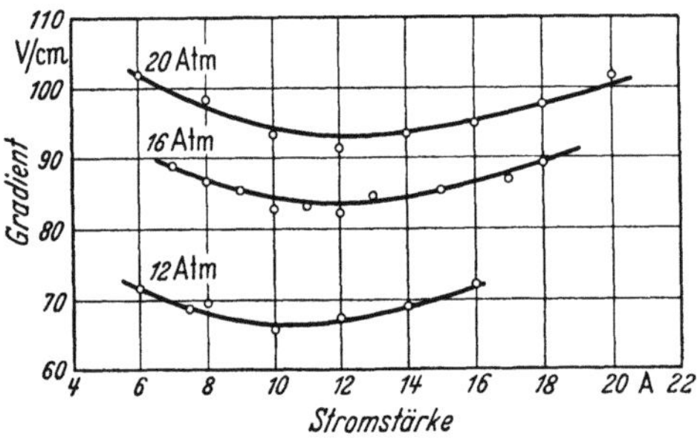

Fig. 73. Charakteristiken eines kurzen, elektrodenstabilisierten Hg-Bogens von ellipsoid-ähnlicher Gestalt. Die Elektrodenbrennflecke halten den ganzen Bogen zusammen, so daß auch ohne Rohr die Kurven steigen können.

92. Energiebilanz und Auftriebsströmung an vertikalen Bögen. Sobald nun Strömungen irgendwelcher Art ins Spiel kommen, spricht man von konvektionsbestimmten Bögen. In der Energiegleichung kommt dann das stationäre Strömungsglied $\varrho\, c_p\, \mathfrak{V}\,\mathrm{grad}\, T$ hinzu, wenn wir von kleinen Druckabweichungen absehen. Es gilt also

$$\sigma\, \mathfrak{E}^2 - S(T) - \varrho\, c_p\, \mathfrak{V}\,\mathrm{grad}\, T + \mathrm{div}\, \varkappa\, \mathrm{grad}\, T = 0. \qquad (92.1)$$

Ehe wir mit dieser Gleichung spezielle Probleme behandeln, wollen wir noch einige allgemeine Betrachtungen anstellen.

Das Strömungsglied verschwindet nicht nur, wenn $\mathfrak{V}$ verschwindet, sondern auch dann, wenn $\mathfrak{V}$ in Richtung der Isothermen läuft.

Sind $\mathfrak{V}$ und grad T wenigstens mit einer Komponente gleichgerichtet, dann wird das betrachtete Volumenelement gekühlt. Sind beide aber zum Teil gegeneinander gerichtet, dann bedeutet das eine Erwärmung durch die Strömung.

Bei frei brennenden Bögen im Gravitationsfeld erzeugt sich der Bogen selbst eine Auftriebsströmung. Für die ungestörte Atmosphäre ist nämlich nach der Kraftgleichung

$$\varrho_L \frac{d\mathfrak{V}}{dt} = -\,\mathrm{grad}\, p + \varrho_L\, \mathfrak{g} = 0, \qquad (92.2)$$

also

$$\mathrm{grad}\, p = \varrho_L\, \mathfrak{g}. \qquad (92.3)$$

Durch die elektrische Energiezufuhr im Bogen wird das Bogengas aufgeheizt und die Dichte auf ϱ_B verringert. Im Bogen gilt dann näherungsweise

$$\varrho_B \frac{d\mathfrak{V}}{dt} = -\,\varrho_L\, \mathfrak{g} + \varrho_B\, \mathfrak{g} \qquad (92.4)$$

oder

$$\frac{d\mathfrak{V}}{dt} = -\frac{\varrho_L - \varrho_B}{\varrho_B}\,\mathfrak{g}, \tag{92.5}$$

d.h. das Bogengas wird nach oben beschleunigt. Da aber der Bogen durch die feststehenden Elektroden an seinem Ort gehalten wird, können wir jedenfalls für qualitative Überlegungen so tun, als ob der Bogen durch eine vertikale Strömung angeblasen würde.

Wir wollen nun den Übergang einmal vom wandstabilisierten, zum anderen vom elektrodenstabilisierten Lichtbogen zum freibrennenden, durch den Auftrieb bestimmten, machen. Dazu denken wir uns zunächst einen vertikal durch ein Rohr begrenzten Bogen, dessen Temperaturverteilung der ELENBAAS-HELLERschen Gleichung gehorcht, d.h. daß auch an der Wand ein der Leistung entsprechender Temperaturgradient vorhanden ist. Wird nun das Rohr entfernt, dann stellt der Knick der Temperaturverteilung dort, wo das Rohr war, eine Divergenz des radialen Wärmestromes dar, die an dieser Stelle zur Aufheizung und damit zur Verflachung der ganzen Temperaturverteilung führt. Lassen wir nun eine vertikale Konvektion durch Auftrieb zu, dann würde sie nichts an dem Vorgang ändern, weil die Strömung mit den Isothermen parallel läuft. Deswegen stellen sich die Isothermen schief nach außen, bis sich ein solcher Winkel zwischen Strömung und Isothermen eingestellt hat, daß die elektrisch zugeführte Energie von dem Glied $\varrho c_p \cdot \mathfrak{V} \cdot \operatorname{grad} T$, über den Raum integriert, abgeführt werden kann. Aus diesem Grunde muß sich ein vertikaler, konvektionsbestimmter Bogen immer aufweiten (s. hierzu auch Fig. 85). Dabei übernimmt die Konvektion die Aufgabe der Wand, die Energie abzuführen. Da zur Bogenmitte hin der Winkel zwischen den Isothermen und den Vertikalen immer kleiner wird, verliert das Konvektionsglied mehr und mehr an Bedeutung, so daß man innerhalb der Leitfähigkeitszone nach wie vor mit der ELENBAAS-HELLERschen Gleichung rechnen darf.

Gerade das Umgekehrte vollzieht sich beim vertikalen, elektrodenstabilisierten Bogen. Ohne Konvektion spreizen die Isothermen und Stromlinien von der unteren Elektrode her stark auf. Bei Einsetzen einer vertikalen Konvektion kühlt diese die aufspreizenden Bogenteile in Elektrodennähe besonders stark, weil Strömung und Temperaturgradient etwa parallel gerichtet sind. Dadurch wird der Bogen in Elektrodennähe zusammengedrückt.

Beide Wirkungen der Konvektion zusammengenommen haben eine kegelartige Form des vertikalen Bogens zur Folge (s. Fig. 62, 63).

Zur quantitativen Behandlung dieser Vorgänge müßte man die Kraftgleichung (39.13) mit der Gravitation als äußerer Kraft unter Beachtung der Kontinuitätsgleichung und wegen der variablen Dichte auch der Energiegleichung lösen, um das Strömungsfeld zu erhalten, was wegen der mathematischen Schwierigkeiten unmöglich ist.

Für experimentelle Untersuchungen muß man daher neben dem Temperaturfeld auch das Geschwindigkeitsfeld ausmessen. Dies hat HAGENAH [*78*] nach dem Vorbild von KENTY [*101*] am Niederstromkohlebogen durch Einführen sehr kleiner Kohleteilchen und kinematographischer Beobachtung im Dunkelfeld durchgeführt (Fig. 74). Die maximale Geschwindigkeit betrug 2 m je Sekunde. Zur Auswertung der Energiebilanz konnte er sich der Temperaturfeldmessung von HÖRMANN [*90*], SCHMITZ [*192*] und SPERLING [*211*] bedienen. Das Ergebnis war, daß die Strömung in der Leitfähigkeitszone keine Rolle spielt, vielmehr die Wärmeleitung allein die elektrische Leistung nach außen führt. Hier kann dann die Strömung wegen der nun geneigten Isothermen die Energie übernehmen und nach oben abführen (Fig. 75).

Um trotz aller theoretischen Schwierigkeiten auch für diese Probleme etwas vorhersagen zu können, wollen wir die Energiegleichung (92.1) über eine (etwas gekrümmte)Scheibe der Höhe Eins integrieren, die parallel zum Temperaturgradienten begrenzt ist und so weit nach außen reicht, daß die Außentemperatur

Fig. 74. Dunkelfeld-Filmaufnahmen (von rechts und links) von einem 10 A-Kohlebogen, in dem kleine Kohleteilchen von zwei Schüttelrinnen eingeführt werden.

erreicht ist und kein Temperaturgradient mehr existiert. Vor der Integration verwandeln wir das Strömungsglied in einen Divergenzausdruck:

$$\varrho\, c_p\, \mathfrak{V} \operatorname{grad} T = \varrho\, \mathfrak{V} \operatorname{grad} h = \varrho\, \mathfrak{V} \operatorname{grad} h + h \operatorname{div} \varrho\, \mathfrak{V} = \operatorname{div} (h\, \varrho\, \mathfrak{V}) \qquad (92.6)$$

wegen $\operatorname{div} \varrho\, \mathfrak{V} = 0$.

$h \varrho \mathfrak{V}$ ist der von der Konvektion getragene Energiestrom. Das Raumintegral über die Scheibe

$$\int_V \sigma\, \mathfrak{E}^2\, d\tau - \int_V S(T)\, d\tau + \int_V \operatorname{div} \varkappa \operatorname{grad} T\, d\tau + \int_V \operatorname{div} h\, \varrho\, \mathfrak{V}\, d\tau = 0 \qquad (92.7)$$

enthält zunächst die zugeführte elektrische Leistung JE und die abgestrahlte Leistung. Die beiden anderen Integralausdrücke verwandeln wir mit dem GAUSSschen Satz in Oberflächenintegrale und erhalten:

$$\left.\begin{aligned} &JE - \int_V S(T)\, d\tau + \\ &+ \oint \varkappa \operatorname{grad} T\, d\mathfrak{f} + \oint h\, \varrho\, \mathfrak{V}\, d\mathfrak{f} = 0. \end{aligned}\right\} \qquad (92.8)$$

Hierin verschwindet das Wärmeleitglied, weil $\operatorname{grad} T\, d\mathfrak{f}$ überall an der Oberfläche verschwindet, so daß also die zugeführte elektrische Energie, soweit sie nicht ausgestrahlt wird, letzten Endes nur von der Konvektion abtransportiert werden muß.

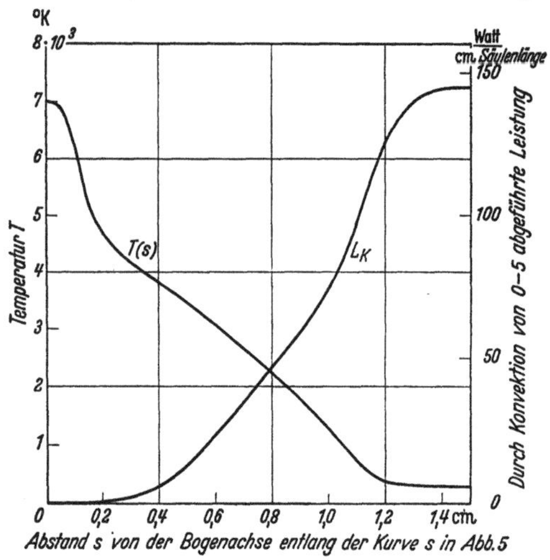

Fig. 75. Temperaturverteilung im 10 A-Kohlebogen $T(s)$. Die von der Konvektion bis zum Abszissenwert s abgeführte Leistung L_K.

Diese Tatsache benutzend, haben SUITS und PORITZKY [168] die Erfahrungen der Wärmetechnik zu Rate gezogen. Danach gilt für die Wärmeabgabe von einem starren Zylinder mit dem Durchmesser $2r_0$ in einem Gas mit der Dichte ϱ, dem Ausdehnungskoeffizienten $\beta = 1/T$, der Zähigkeit η, der spezifischen Wärme c_p und der Wärmeleitfähigkeit $\varkappa$ in einem Gravitationsfeld

$$Nu = f\,(Gr, Pr) \qquad (92.9)$$

oder

$$\frac{2r_0 h}{\varkappa} = f\left(\frac{(2r_0)^3 \beta\, \varrho^2 g\, \Delta T}{\eta^2}, \frac{c_p \eta}{\varkappa}\right), \qquad (92.10)$$

worin h die Wärmeübergangszahl, also die pro Flächeneinheit und Einheit der Temperaturdifferenz abgegebene Leistung ist. Letztere kann man mit der Bogenleistung JE durch

$$h = \frac{JE}{2r_0 \pi \Delta T} \tag{92.11}$$

(wenn man von der Strahlung absieht) ersetzen und noch die Stromtransportgleichung in der Form

$$J = j\, r_0^2 \pi \tag{92.12}$$

hinzuziehen.

Die Funktion f ist eine über viele Zehnerpotenzen gemessene Funktion und kann in gewissen Bereichen durch

$$Nu = \text{const} \cdot (Gr \cdot Pr)^{\alpha} \tag{92.13}$$

angenähert werden, worin Pr nach der kinetischen Gastheorie konstant ist und in die vordere Konstante hineingezogen werden darf. Dann entsteht

$$J\,E = \varkappa\, \Delta T \,\text{const} \cdot \left(\frac{(2r_0)^3 \varrho^2 \mathfrak{g}\, \Delta T}{T\, \eta^2}\right)^{\alpha}. \tag{92.14}$$

Für die temperaturabhängigen Größen verwendet man eine Mitteltemperatur. Beschränkt man sich auf Niederstrombögen, so kann man die Temperatur etwa konstant setzen außer in der Stromdichte, wo die SAHA-Gleichung eingeht.

Bei gegebener Gasart usw. kann man aus der allgemeinen Funktion f den Exponenten α für den interessierenden Bereich entnehmen (0,1 bis 0,2), und findet so einige Beziehungen für das Verhalten des konvektionsbestimmten Bogens als Ganzes. Für die Charakteristik gilt z.B.

$$E = \text{const}\, J^{-(2-3\alpha)/(2+3\alpha)} \tag{92.15}$$

speziell für Luft

$$E = \text{const}\, J^{-0,6}. \tag{92.16}$$

Weiterhin können Beziehungen zwischen Stromstärke, Feldstärke, Druck, Durchmesser, Gravitation und Anströmung aufgestellt werden, die, soweit sie nachprüfbar sind, ganz gut mit der Erfahrung übereinstimmen.

Auch ELENBAAS beschäftigt sich eingehend mit der Strömung im Hg-Bogen und dem Einfluß auf die Energiebilanz. Das Geschwindigkeitsfeld erhält er einmal durch Integration der Kraftgleichung mit Berücksichtigung der (äußeren) Reibung

$$\frac{\partial \mathfrak{V}}{\partial t} + \mathfrak{V} \operatorname{grad} \mathfrak{V} = \mathfrak{g} - \frac{1}{\varrho} \operatorname{grad} p - \frac{\eta}{\varrho} \operatorname{rot} \operatorname{rot} \mathfrak{V}. \tag{92.17}$$

Zum andern verfolgt er experimentell kleine Teilchen in der Entladung und kommt auf beiden Wegen zu Geschwindigkeiten von etwa 50 cm/sec. Der Energietransport wird teils direkt mit dem Strömungsglied in der Energiegleichung, teils mit den Methoden der Wärmetechnik wie bei SUITS erfaßt.

93. Auftriebsströmung an horizontalen Bögen. Wenn ein zylindrischer Bogen quer zu seiner Achse angeblasen wird, was durch einen künstlichen, seitlichen Gasstrom geschehen kann oder beim horizontalen Bogen schon spontan durch den Auftrieb erfolgt, dann wird er nach einer anfänglichen Verschiebung von den Elektroden in einer neuen, bogenförmigen Bahn festgehalten. Im Vergleich zu einem nicht angeblasenen Bogen, bei dem die Energiegleichung

$$\sigma\, \mathfrak{E}^2 + \operatorname{div} \varkappa \operatorname{grad} T = 0 \tag{93.1}$$

gilt, hat der angeblasene Bogen auf der konkaven Luvseite gemäß der Energiegleichung

$$\sigma \mathfrak{E}^2 + \operatorname{div} \varkappa \operatorname{grad} T = \varrho c_p \mathfrak{B} \operatorname{grad} T \tag{93.2}$$

einen stärkeren Durchhang der Temperaturkurve. Da auf dieser Seite nämlich $\mathfrak{B}$ und grad T parallel gerichtet sind, muß die dadurch bewirkte Abkühlung wieder durch Senken von Wärmestromlinien ausgeglichen werden. Auf der konvexen Leeseite dagegen sind $\mathfrak{B}$ und grad T entgegengesetzt gerichtet; es findet daher eine Aufheizung durch die Strömung statt, die eine Aufblähung der Temperaturkurve zur Folge hat. Es entsteht also eine positive Divergenz des Wärmestromes, die die zugeführte Energie stationär ableiten kann. Diese typischen Merkmale eines quer angeblasenen Bogens erkennt man deutlich an den Fig. 76 und 77. Die Temperatur fällt vom Achsenwert nach der konkaven Seite steil, nach der konvexen Seite dagegen besonders langsam ab [*9a*].

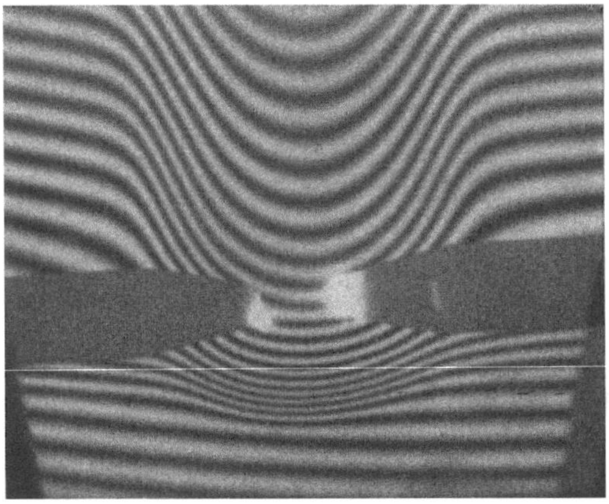

Fig. 76. Interferenzaufnahme eines horizontalen Niederstromkohlebogens. Wegen der Anströmung von unten durch den Auftrieb fällt die Temperatur nach unten sehr steil, nach oben dagegen sehr flach ab.

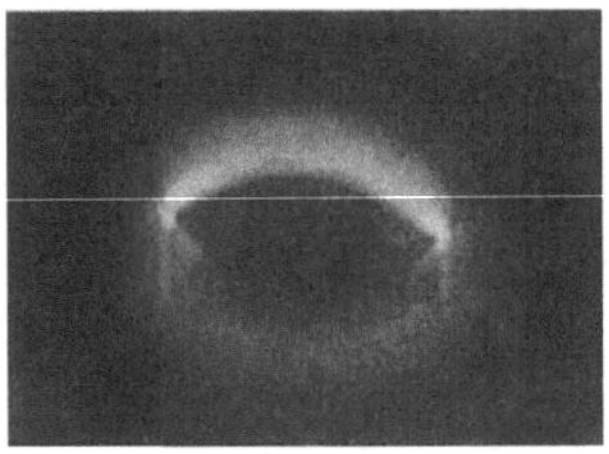

Fig. 77. Horizontaler Hg-Hochdruckbogen wird durch den Auftrieb nach oben gebogen und zeigt nach unten einen steilen, nach oben einen flachen Temperaturabfall.

94. Plasmaströmung infolge eigenmagnetischer Kompression. So wie sich ein freier Bogen im Gravitationsfeld durch den Auftrieb des erhitzten Plasmas eine Strömung selbst schafft, so kann er das gleiche auch durch sein eigenes Magnetfeld bewirken, wie MAECKER [*139*] gezeigt hat; denn nach der Kraftgleichung mit dem magnetischen Glied

$$\varrho \frac{d\mathfrak{B}}{dt} = -\operatorname{grad} p + \frac{1}{c}[\mathfrak{j}\,\mathfrak{H}] \tag{94.1}$$

übt das Magnetfeld auf ein stromführendes Volumen eine Kraft aus, die entweder den Druck erhöht oder das Plasma in Bewegung setzt oder beides. Über die Aufteilung in Druckgradient und Beschleunigung entscheidet die Kontinuitätsgleichung.

Da die magnetischen LORENTZ-Kräfte sowohl senkrecht auf $\mathfrak{j}$ als auch auf $\mathfrak{H}$ stehen, sind sie beim zylindrischen Bogen immer radial zur Achse gerichtet (parallele Stromfäden ziehen sich an). In dieser Richtung kann keine stationäre Strömung fließen, weswegen sich ein Druckfeld aufbaut, dessen Gradient die LORENTZ-Kräfte kompensiert:

$$\operatorname{grad} p = \frac{1}{c}[\mathfrak{j}\,\mathfrak{H}]\,. \tag{94.2}$$

Mit

$$\operatorname{rot} \mathfrak{H} = \frac{4\pi}{c}\,\mathfrak{j} \tag{94.3}$$

oder bei Zylindersymmetrie

$$\mathfrak{H} = \frac{4\pi}{c} \frac{1}{r} \int \mathfrak{j}\, r\, dr \tag{94.4}$$

wird der Überdruck

$$p = \frac{4\pi}{c^2} \int_r^{r_a} \frac{\mathfrak{j}}{r} \int_0^r \mathfrak{j} \cdot r\, dr. \tag{94.5}$$

Für einen homogenen Kanal mit der Stromstärke J, dem Radius r_a und der Stromdichte $j = J/r_a^2 \pi$ ist der Überdruck

$$p(r) = \frac{J j}{c^2} \left(1 - \frac{r^2}{r_a^2}\right) \tag{94.6}$$

und speziell in der Achse $J j/c^2$. Diesen Druck kann man messen, indem man die Elektrode mit einer kleinen axialen Bohrung versieht und ein Manometer anschließt. Die entstehenden Drucke (Fig. 78) sind bei Bögen unter Atmosphärendruck sehr klein und spielen als solche keine Rolle im Gegensatz zu Niederdruckentladungen hoher Stromstärke, wo die *relative* Drucksteigerung durch das Eigenmagnetfeld erhebliche Werte annehmen kann und dadurch die Entladung kontrahiert (pinch-Effekt). Hierüber hat vor allem TONKS [227] ausführliche Arbeiten veröffentlicht.

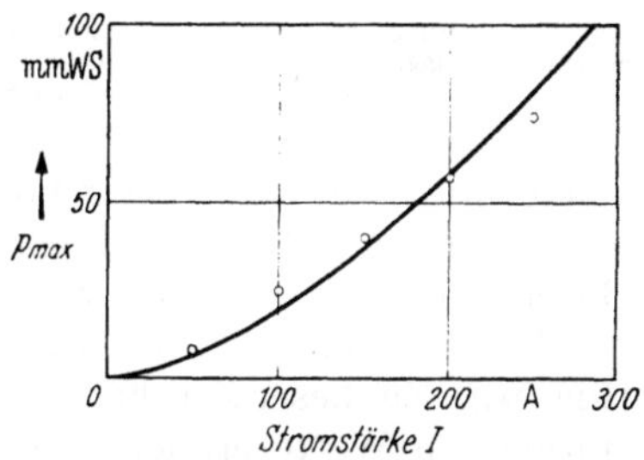

Fig. 78. Überdruck in einer axialen Kathodenbohrung als Funktion der Stromstärke. Die Punkte sind gemessen, die Kurve ist berechnet unter Berücksichtigung der Störung durch die Bohrung.

Zeigt aber der Bogen eine Aufweitung, wie das bei fast allen Bögen vom Kathodenbrennfleck auf die Säule zu der Fall ist, dann nimmt die magnetische Feldstärke, die Stromdichte und folglich auch der Druck in Achsenrichtung ab, es entsteht also ein negativer Druckgradient entlang der Achse vom Brennfleck her. Da das Magnetfeld in der Achse verschwindet, verursacht der Druckgradient eine stationäre Strömung zur Säule hin:

$$\tfrac{1}{2}\varrho \operatorname{grad} \mathfrak{V}^2 = -\operatorname{grad} p. \tag{94.7}$$

Nehmen wir an, daß die Aufweitung zur Säule hin ausreicht, um den Überdruck praktisch ganz verschwinden zu lassen, und integrieren längs der Achse, dann ist

$$\frac{1}{2} \bar{\varrho} \mathfrak{V}_{\max}^2 = p_{\max} = \frac{J j}{c^2}. \tag{94.8}$$

Für den 200 Amp-Hochstromkohlebogen mit $\bar{\varrho} \approx 1{,}5 \cdot 10^{-5}$ g/cm³ ist $\mathfrak{V}_{\max} \approx 3{,}5 \cdot 10^4$ cm/sec, also trotz des kleinen Überdruckes vor der Kathode eine ansehnliche Geschwindigkeit. Man versteht nun auch, warum die Hochstrombögen eine so hohe Richtungsstabilität aufweisen.

Die Plasmaströmung kann man durch zahlreiche Experimente nachweisen, z. B. durch Einführen kleiner Kohleteilchen, die unabhängig von der Stellung des Bogens immer von der Kathode fortgeschleudert werden. Durch kinematographische Aufnahmen kann man auch das Geschwindigkeitsfeld jedenfalls in den äußeren Bogenpartien ausmessen. In Achsennähe sind dagegen die Beschleunigungen zu groß, als daß die Teilchen folgen könnten.

Deswegen hat WIENECKE [252] mit der bereits in Ziff. 87 beschriebenen Apparatur die Geschwindigkeit in der Bogenmitte auf folgende Weise gemessen (Fig. 79).

Parallel zum Bogen wird über eine Feinsicherung ein Thyratron geschaltet. Wird durch den Impuls vom Multiplier das Thyratron gezündet, so wird die Bogenstrecke spannungs- und damit stromlos. Eine definierte Zeit später brennt die Feinsicherung durch, der Parallelschluß wird unterbrochen und der Bogen

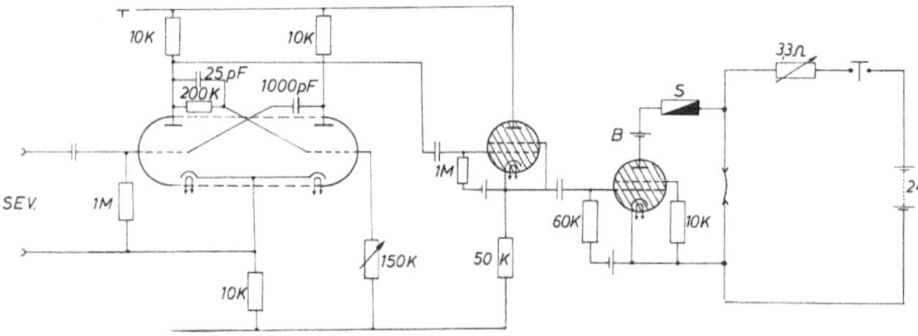

Fig. 79. Schaltung zur Erzeugung einer kurzen Stromunterbrechung am Bogen. Wenn das parallel zum Bogen (rechter Stromkreis) liegende Thyratron zündet, übernimmt es den Bogenstrom. Kurz danach brennt die Feinsicherung S durch und der Strom geht wieder durch den Bogen.

zündet wieder durch. Dieser Vorgang am Bogen wird mit der Zeitlupenanordnung aufgenommen. Die Unterbrechungszeit muß so lang gewählt werden, daß der Bogenkern vorübergehend verschwunden ist. Bei der Wiederzündung bildet sich der Plasmastrahl von der Kathode her neu aus und schießt zur Anode durch (Fig. 80). Zu dieser Zeit haben Strom und Spannung am Bogen schon beinahe die stationären Werte erreicht. Aus dem Vorwachsen der Kernfront kann die Geschwindigkeit im Kern bestimmt werden. Wenn auch die so vorgenommene

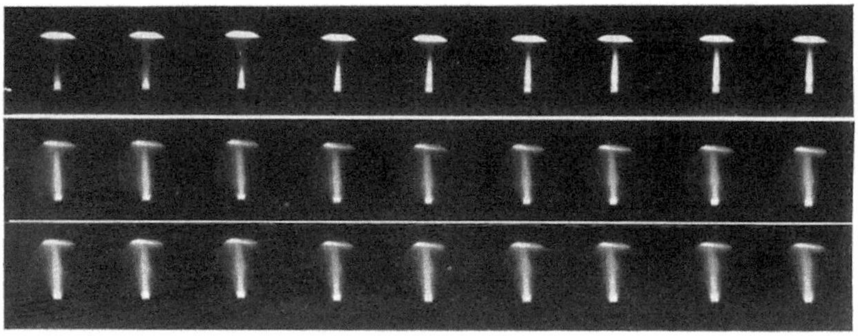

Fig. 80. Zeitlupenaufnahme bei Wiederzündung des Bogens (obere Reihe: $\Delta t = 19\,\mu$sec, mittlere und untere Reihe: $\Delta t = 26\,\mu$sec).

Störung etwas grob erscheint, so hat sie sich doch als beste der Möglichkeiten zur Kennzeichnung eines Bogenelementes erwiesen. Außerdem rechtfertigen die Ergebnisse die Anwendbarkeit der Methode. Das gemessene Geschwindigkeitsfeld ist in Fig. 81a und b dargestellt. Die höchste Geschwindigkeit dicht vor der Kathode liegt über 300 m/sec in guter Übereinstimmung mit dem näherungsweise nach Gl. (94.8) berechneten Wert. Zur Seite hin fällt die Geschwindigkeit am Rande des Kernes sehr steil ab, besonders in Kathodennähe. Zur Kontrolle der Meßergebnisse wurde noch die Massenbilanz

$$\operatorname{div}(\varrho\,\mathfrak{V}) = 0 \tag{94.9}$$

im ganzen Bogengebiet geprüft und innerhalb einer Streuung von maximal $\pm 10\%$ als erfüllt gefunden.

Durch die Erzeugung der Strömung wird auf die Kathode ein Rückstoß ausgeübt, den schon ROHLOFF [*174*] an einer schwimmenden Kathode beobachtet und gemessen hat. Zur theoretischen Berechnung hat man die Kraftgleichung (94.1) über das ganze Bogenvolumen zu integrieren; denn das Produkt Dichte mal Beschleunigung integriert über das ganze Bogenvolumen, ist gerade die Trägheitskraft der Bogenmasse, und nach dem NEWTONschen Prinzip von actio gleich reactio wirkt auf die Kathode die gleiche Kraft. Die Ergebnisse der Messungen und Rechnungen finden sich in Fig. 82.

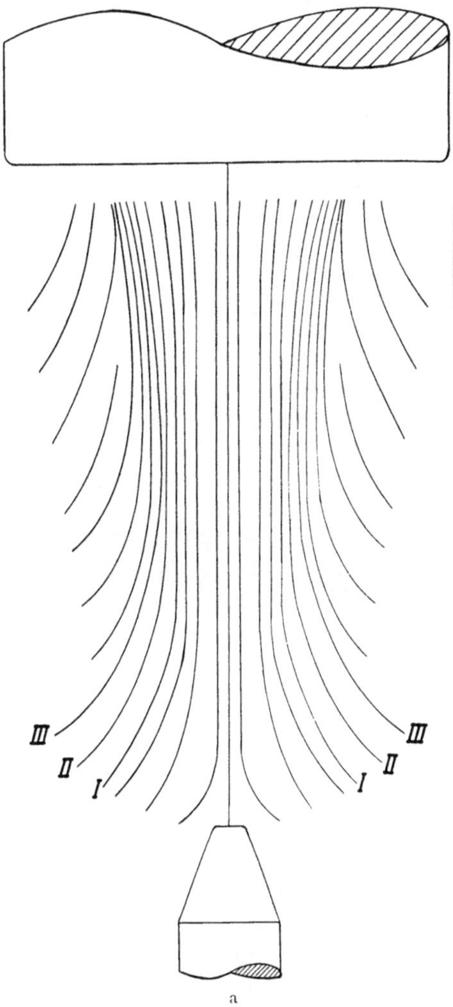

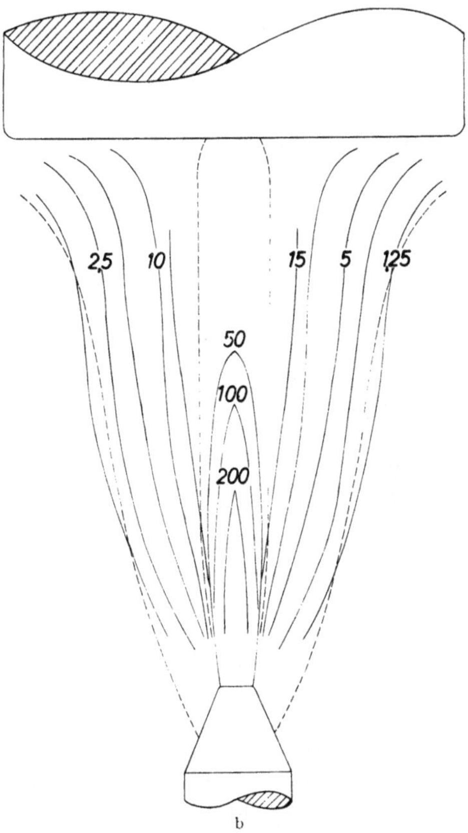

Fig. 81 a u. b. a Stromlinienbild im 200 A-Hochstromkohlebogen. b Kurven gleicher Geschwindigkeit (Zahlen in m/sec).

Ferner gilt, daß der gesamte in der Umgebung der Kathode erzeugte Impuls durch jeden genügend fernen Querschnitt des Bogens als Impulsstrom $\int \varrho \mathfrak{V} \cdot \mathfrak{V}\, d\mathfrak{f}$ hindurch gehen und daß dieser wieder gleich der Kraft auf die Kathode sein muß. Führt man diese Integration an dem von WIENECKE gemessenen Geschwindigkeitsfeld für den 200 Amp-Hochstromkohlebogen durch, dann ergibt sich ein Impulsstrom und damit eine Kraft auf die Kathode von 0,44 g-Gewicht in Übereinstimmung mit den gemessenen und nach der Kraftgleichung berechneten Werten in Fig. 82.

Wenn nun die Erklärung der von der Kathode ausgehenden Plasmaströmung mit der eigenmagnetischen Kompression am Kathodenbrennfleck durch die

Versuchsergebnisse gesichert erscheint, dann muß auch an jeder anderen Engstelle des Bogens, wie auch immer sie entstanden ist, eine Strömung von der Engstelle fort entstehen Dies kann man sehr schön zeigen, wenn man die Bogensäule durch ein kleines Loch in einer gekühlten Cu-Platte einengt. Durch die magnetische Kompression in der Engstelle wird das Plasma nach beiden Achsenrichtungen hinausgeschleudert. Dabei trifft der zur Kathode gerichtete Plasmastrahl mit dem von dort kommenden zusammen und beide bilden einen Plasmateller (Fig. 83) [*141*]. Durch zusätzliches Heizen der Kathode oder durch Verwendung von Dochtkohlen kann man den Kathodenbrennfleck vergrößern und so die kathodische Strömung unterbinden, so daß auch der Plasmateller verschwindet.

Wenn sich an der Anode ein Brennfleck bildet, ist dieser ebenfalls Ursache für eine Strömung, die mit der kathodischen zusammenprallt.

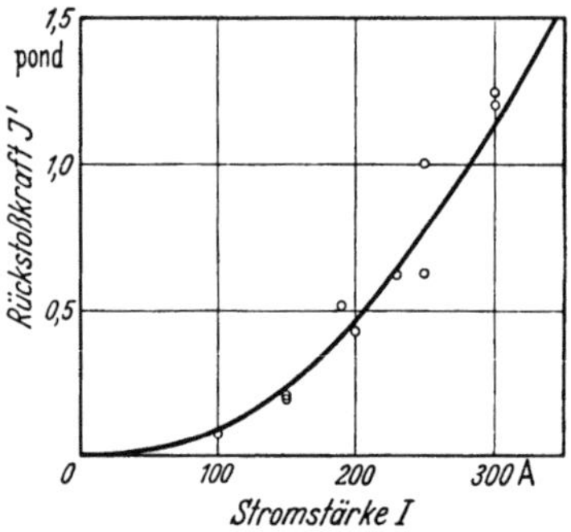

Fig. 82.

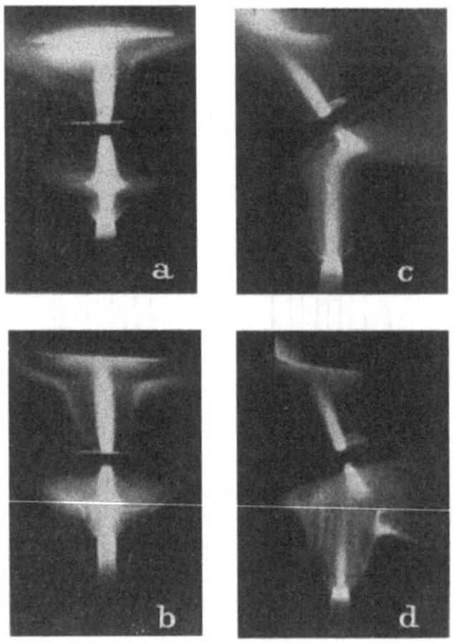

Fig. 83.

Fig. 82. Rückstoß auf die Kathode durch die Strömungserzeugung. Punkte gemessen, Kurve berechnet.

Fig. 83a—d. Bogensäule eines 100 A-Bogens, durch eine gekühlte Cu-Platte mit 2,5 mm Durchmesser-Loch in der Mitte eingeengt. a Wegen des in der Engstelle magnetisch erzeugten Druckes geht nach oben und unten eine Plasmaströmung, wovon die letztere mit der von der Kathode aus der gleichen Ursache kommenden Strömung zusammenprallt und einen Plasmateller bildet. b Kleine Kohleteilchen zeigen den Weg der Strömung. c Bei schiefer Cu-Platte bildet sich eine resultierende Strömung aus, was wieder durch kleine Teilchen nachweisbar ist (d).

Auch bei allen anderen, stromstarken Entladungen beobachtet man von der Kathode und von der Anode, wenn dort ein kleiner Bogenansatz existiert, derartige schnelle Strömungen. So hat LARCHÉ [*118*], der als erster auf diese Phänomene aufmerksam gemacht hat, einen Plasmastrahl aus der kathodischen Plasmakugel im X-Hochdruckbogen nachgewiesen. In einem kurzzeitigen Hg-Bogen mit über 100 Amp hat HAYNES [*80*] eine Geschwindigkeit von nahezu 10^5 cm/sec von der Kathode und unter Umständen auch von der Anode gemessen. Schließlich zeigen Aufnahmen von NISSEN [*160*] im H-Hochdruckbogen sehr deutlich das Zusammenstoßen einer kathodischen mit einer anodischen Strömung.

95. Plasmaströmung und Phänomenologie des Bogens. Da die beschriebenen Plasmaströmungen ihrem magnetischen Ursprung entsprechend immer von Engstellen ausgehen, müssen sie wegen der hohen Stromdichte dort notwendigerweise sehr heiß sein. Sie heben daher auf ihrem Wege zunächst die Achsentemperatur des Bogens an, im Vergleich zu einem Bogen ohne diese Strömung. Demgegenüber ist die Wärmeleitung bestrebt, den Plasmastrahl abzukühlen, was unter Umständen zur schnellen Temperaturerniedrigung führt, wie z.B. im A-Hochstrombogen [*20*]. Ist nämlich die elektrische Leitfähigkeit im Strahl trotz der hohen Temperatur nicht wesentlich größer als in der Umgebung, dann wird das Bild der elektrischen Stromlinien durch die Strömung kaum verändert. Wenn aber die Leitfähigkeit im Strahl merklich größer als in der Umgebung ist, dann wird hier auch die Erzeugung JOULEscher Wärme stärker sein und die Temperatur des Plasmastrahls hochhalten. In Molekülgasen erreicht auf diese Weise die Achsentemperatur bei genügender Stromdichte vor der Kathode solche Achsentemperaturen, daß sich der typische Kern ausbilden kann. Wenn man den Brennfleckdurchmesser durch Heizen der Kathode vergrößert, dann verschwindet die heiße Strömung und damit der Kern, wie man an Fig. 84 sieht. Auch wenn man dem Kern die JOULEsche Wärme entzieht, etwa dadurch, daß man den Bogen durch eine mit einem großen Loch versehene Platte brennen läßt und diese als Anode schaltet, verliert er seine Existenzmöglichkeit wie Fig. 85 zeigt. Gleichzeitig sieht man an dieser Aufnahme, daß der obere Teil des Bogens von der Kegelstumpf-Form in die zylindrische übergeht, wenn in ihm kein Strom fließt und die Konvektion außen keine Leistung abzuführen braucht.

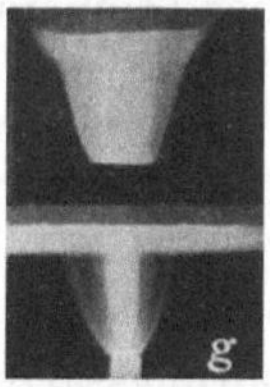

Fig. 84. Kathodenströmung und Kern entstehen nur, wenn ein *kleiner* Kathodenbrennfleck existiert. Heizt man aber die Plattenkathode des oberen Bogens durch den unteren Bogen auf, dann verbreitert sich der Bogenansatz an der oberen Plattenkathode so sehr, daß weder Strömung noch Kern trotz 200 A Stromstärke erzeugt werden.

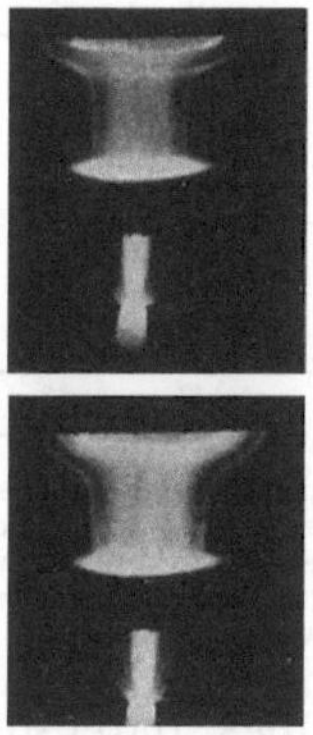

Fig. 85. Der Bogen brennt durch eine große Öffnung in einer Kohleplatte hindurch (unteres Bild). Schaltet man die Anodenleitung auf diese Platte, dann verschwindet der Kern im oberen Bogenteil trotz bleibender Strömung, weil die JOULEsche Wärme zur Aufrechterhaltung der hohen Temperatur gegenüber der Wärmeableitung fehlt.

96. Plasmaströmung und Energiebilanz. Die Energiebilanz unterscheidet sich formelmäßig nicht von der für konvektionsbestimmte Bögen. Bei der Anwendung z. B. auf einen Querschnitt der voll ausgebildeten Säule des Hochstromkohlebogens zeigt sich jedoch, daß die Energiezufuhr durch den Plasmastrahl im Kern gering ist, weil in dem Glied $\varrho\, c_p \cdot \mathfrak{V} \cdot \operatorname{grad} T$ der Faktor $\operatorname{grad} T$ in Achsenrichtung sehr klein ist. Wie man sich aber überlegen kann, ist diese Art der Auswertung gar nicht in der Lage, im Falle einer Strömung die Herkunft der hohen Temperatur zu beschreiben. Man muß vielmehr ein Massenelement auf seinem Wege von außen zur Achse hin verfolgen und feststellen, in welchem Maße es von den einzelnen Energieströmen Energie zugeführt bekommt. Dann wird man finden, daß ein im Kern befindliches Massenelement zunächst in kaltem Zustand an der Kathode hochgezogen ist, dann in dem Gebiet höchster Stromdichte beschleunigt und hoch aufgeheizt wurde, und daß es auf seinem Wege zum betrachteten Ort im Kern einerseits viel Energie durch Wärmeleitung verloren, andererseits aber auch durch die JOULEsche Wärme gewonnen hat, so daß es im Endeffekt immer noch eine verhältnismäßig hohe Temperatur behalten hat.

Als weitere Wirkung der Strömung auf die Erscheinung des Hochstrombogens kommt noch hinzu, daß durch die Zähigkeit auch das Gas außerhalb des Kerns mitgenommen und dadurch eine kalte Anströmung des Bogens verursacht wird, die den Bogen viel mehr einschränkt, als es bei reiner Auftriebsströmung der Fall wäre. Das geht deutlich aus Fig. 86 hervor.

Das alles zeigt, welche Bedeutung das magnetische Eigenfeld als Ursache der Plasmaströmung für Hochstrombögen hat.

97. Beziehungen der Bogenphysik zu anderen Zweigen der Physik. Durch die Feststellung, daß die magnetisch angetriebene Plasmaströmung das äußere Bild und den Energiehaushalt der Hochstrombögen wesentlich beeinflußt, rückt die eigentliche Dynamik des Plasmas mehr in den Vordergrund. Während der Lichtbogen bisher mit Ausnahme des konvektionsstabilisierten Bogens im makroskopischen Sinne als statisches Gebilde angesehen wurde, weisen die neueren Untersuchungen darauf hin, daß es sich hier ganz allgemein um ein Problem der Gasdynamik hoher Temperaturen handelt. Dieses Gebiet ist nicht nur für den Lichtbogenphysiker, sondern seit Jahren auch für den Astrophysiker und den Aerodynamiker bzw. den Strahlantriebsphysiker von wachsendem Interesse.

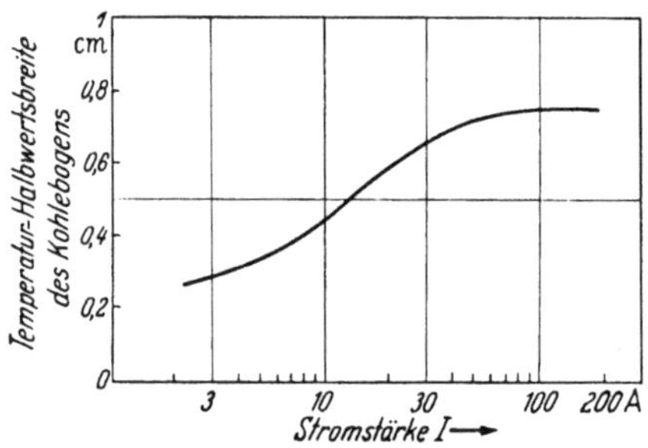

Fig. 86. Während der Durchmesser des Niederstromkohlebogens mit der Stromstärke ansteigt, bleibt der des Hochstrombogens etwa konstant, weil die eigenmagnetisch bedingte Anströmung von außen eine Aufweitung verhindert.

Die eingangs dargestellte Theorie mit den Erhaltungssätzen für Masse, Ladung, Impuls und Energie, den MAXWELLschen Gleichungen und den phänomenologischen Ansätzen für die Massenströme, die elektrischen Ströme und den Wärmestrom ist eine formale Erweiterung der normalen Gasdynamik (Aerodynamik) in das Gebiet hoher Temperaturen, in dem der Bewegungszustand und die energetischen Eigenschaften der nunmehr elektrisch leitenden Gase weitgehend durch elektrische und magnetische Felder beeinflußt werden. Der wesentliche Unterschied gegenüber der Dynamik normaler Gase besteht darin, daß man wegen des Einflusses elektromagnetischer Felder und der ständigen Absorption und Emission von Strahlung nur in Ausnahmefällen mit adiabatischen Zustandsänderungen in der Gasströmung rechnen kann. Die irreversiblen Prozesse sind auch nicht näherungsweise mehr zu vernachlässigen[1].

Alle aus dieser dynamischen Theorie sich ergebenden Phänomene sind nun offensichtlich in viel allgemeinerem Ausmaß als in dem elektrischen Lichtbogen in der Sonnenatmosphäre zu beobachten. Typische Beispiele sind die wirbelnden Bewegungen in den Sonnenflecken im Zusammenhang mit starken Magnetfeldern (bis zu 3000 Gauß), die vertikalen Strömungen in der Granulation, die Bewegungsmechanismen der Fackeln und Protuberanzen, die Aufheizung der Corona usw.[2]. Die Bewegung der solaren Materie ist dabei eher turbulent als laminar, ferner bestehen günstige Bedingungen für die Entstehung von Stoßwellen.

Auch in der interstellaren Materie spielen nach den Erkenntnissen der letzten Jahre Stoßwellen und Turbulenzeffekte im Zusammenwirken mit Magnetfeldern eine erhebliche Rolle. Auf diesem Gebiet herrscht bereits eine enge Zusammenarbeit zwischen Astrophysikern und Aerodynamikern [*98*]. Experimentelle Untersuchungen liegen zur Zeit auf diesem Gebiet kaum vor. Eine von BIERMANN und SCHLÜTER [*10*] stammende Theorie über die Entstehung von Magnet-

[1] Vgl. auch die Behandlung der Gasdynamik in Bd. IX dieses Handbuches.
[2] Vgl. die Behandlung der Sonnenphysik in Bd. LII dieses Handbuches.

feldern in turbulent strömenden Plasmen ist von LOCHTE-HOLTGREVEN und SCHILLING [*186*] an turbulenten, elektrisch leitenden Flammen geprüft und im Rahmen der Vergleichsmöglichkeit bestätigt worden.

Eingehende experimentelle Untersuchungen über instationäre Stoßwellen haben, wie bereits erwähnt, KANTROWITZ und Mitarbeiter in Argon durchgeführt [*97*].

Zur Erzeugung stationärer Stoßwellen in Plasmen und zur Untersuchung weiterer plasmadynamischer Effekte ist eine Plasmaströmung mit Überschallgeschwindigkeit erforderlich. Diese kann nach PETERS [*166*] durch Expansion eines Hochdruckplasmas, das durch einen Bogen in einem weitgehend mit Wasser gefüllten Druckkessel (Fig. 33) erzeugt wird, durch eine Lavaldüse hindurch realisiert werden. Ausgehend von einem Kesselzustand von 12500° K und 50 Atm erhält man hinter der Düse bei Atmosphärendruck einen Plasmastrahl von 8000° K und 6500 m/sec, was einer MACH-Zahl $\mathfrak{v}/c = 1{,}6$ entspricht. Durch Störung des Strahles am Rande entsteht eine stationäre Stoßwelle (Fig. 87). Derart hohe Gasgeschwindigkeiten sind auch für die Strahlantriebe von Interesse, deren wichtigste Kenngröße die Austrittsgeschwindigkeit ist. Da die bei adiabatischer Expansion erreichbare Geschwindigkeit $v = \sqrt{2\Delta h}$ von der Differenz der Enthalpien/Gramm abhängt, kann eine wesentliche Steigerung über die bislang mit chemischen Reaktionen im Brennraum erzielte Geschwindigkeit von etwa 3000 m/sec hinaus nur durch Erhöhung der Enthalpie im Brennraum, also durch Aufheizung bis zum Plasmazustand erreicht werden.

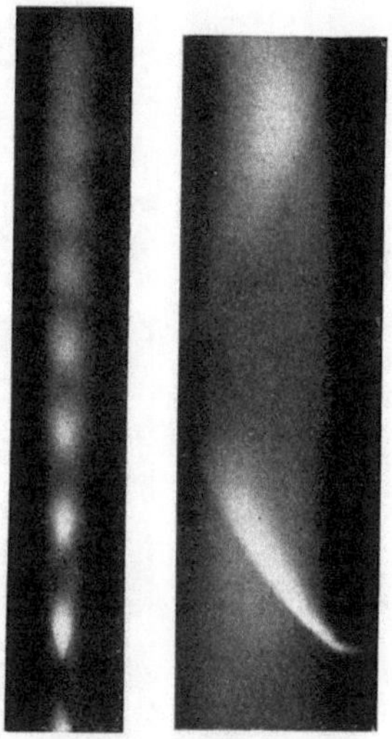

Fig. 87. Plasmastrahl mit Überschallgeschwindigkeit: $\mathfrak{v}/c = 1{,}6$; $\mathfrak{v} = 6500$ m/sec; $T = 8000°$ K. Stationärer Verdichtungsstoß im Plasmastrahl, ausgelöst durch eine Cu-Schneide am Strahlrand.

Hierüber hat SÄNGER-BREDT [*184*] ausführliche Rechnungen angestellt.

E. Kathodische Erscheinungen und Kathodenmechanismus von Lichtbögen.

I. Kathodische Erscheinungen und empirische Daten.

98. Empirisches über Lichtbögen mit verschiedenem Kathodenmechanismus. Die Beobachtung des Kathodenansatzes und seines Verhaltens bei verschiedenen Lichtbögen läßt bereits phänomenologisch die Existenz dreier verschiedener Kathodenmechanismen erwarten. Zur ersten Gruppe, die aus unten verständlich werdenden Gründen *brennflecklose Bögen* genannt wird, gehören Bögen mit Kohlekathode oder Wolframkathode sehr hoher Temperatur. Zur zweiten Gruppe von Bögen, die meist einfach *Brennfleckbögen* genannt werden, und sich durch einen an der Kathode ziemlich stark kontrahierten, aber zeitlich und örtlich stationären Kathodenbrennfleck auszeichnen, gehören Quecksilber- und Edelgas-Hochdruckbögen mit Wolframkathode bei nicht zu hoher Strombelastung. Die dritte Gruppe von Bögen, die man als „*nichtstationäre Brennfleckbögen*" bezeichnen könnte, zeichnet sich durch einen äußerst stark kontrahierten, auf der Kathodenoberfläche in schneller und meist unruhiger Bewegung befindlichen Brennfleck aus, der nach neueren Zeitlupen-Untersuchungen sogar nur eine sehr begrenzte Lebensdauer besitzt und bei seinem Verlöschen durch einen neu entstehenden ähnlichen Brennfleck (oder deren mehrere) abgelöst wird. Die Existenzbedingungen und das Verhalten der beiden erstgenannten kathodischen Bogentypen wurden von WEIZEL, THOURET und GÜNTHER [226] an Quecksilber- und Xenon-Höchstdruckbögen (20 bis 35 Atm, Stromstärke der Größenordnung 10 Amp) mit Wolframelektroden genauer untersucht. Bei der im niedrigeren Stromstärkebereich stabilen Form des Brennfleckbogens beobachtet man nach diesen später von BAUER [3] bestätigten Beobachtungen einen sehr kleinen Kathodenbrennfleck hoher Leuchtdichte, bei dem offenbar das kontrahierte Plasma bis dicht an die Kathodenoberfläche heranreicht; auch bei stark vergrößerter Abbildung ist keine Andeutung eines Dunkelraumes zwischen Kathodenoberfläche und Brennfleck erkennbar. Bei Steigerung der Stromstärke schlägt der Bogen plötzlich, aber ohne wesentliche Brennspannungsänderung und daher wohl ohne größere Änderung der Kathodenfallspannung, in die sog. brennflecklose Form um. In dieser ist der kathodische Ansatz der Bogensäule viel ausgedehnter und die Stromdichte folglich wesentlich kleiner; die maximale gemessene Leuchtdichte liegt um etwa den Faktor 2 unter der des Brennfleckbogens. Das Säulenplasma reicht ferner viel weniger nahe an die Kathodenoberfläche heran; bei vergrößerter Abbildung beobachtet man einen Dunkelraum von etwa 0,1 mm Ausdehnung zwischen Kathodenoberfläche und Plasma. Schließlich ist bei der brennflecklosen Bogenform die Temperatur der Wolframkathode selbst höher als bei der Form mit kontrahiertem Kathodenbrennfleck. Der Umschlag des Brennfleckbogens in den brennflecklosen Bogen bei Stromstärkeerhöhung erfolgt stets plötzlich, aber mit einer gewissen Hysterese; bei abnehmender Stromstärke liegt der Umschlagpunkt tiefer als bei zunehmender. Offenbar hängt der Umschlag mit der Oberflächentemperatur der Kathode zusammen, da erstens bei obenstehender Kathode die Umschlagsstromstärke niedriger liegt als bei untenstehender, durch die aufsteigenden Gase gekühlter Kathode, und da zweitens die Umschlagsstromstärke beim Wechselstrombogen wesentlich höher liegt als beim Gleichstrombogen. WEIZEL und Mitarbeiter deuten den brennflecklosen Bogen als einen Bogen mit wesentlich thermischer Elektronenemission der Kathode, den Brennfleckbogen dagegen als einen solchen, bei dem das vor der Kathode liegende kontrahierte Plasma hoher Temperatur für den Kathodenmechanismus entscheidend ist, weil es zur Kathode hin positive Ionen und zur Säule hin Elektronen abgibt. In Übereinstimmung

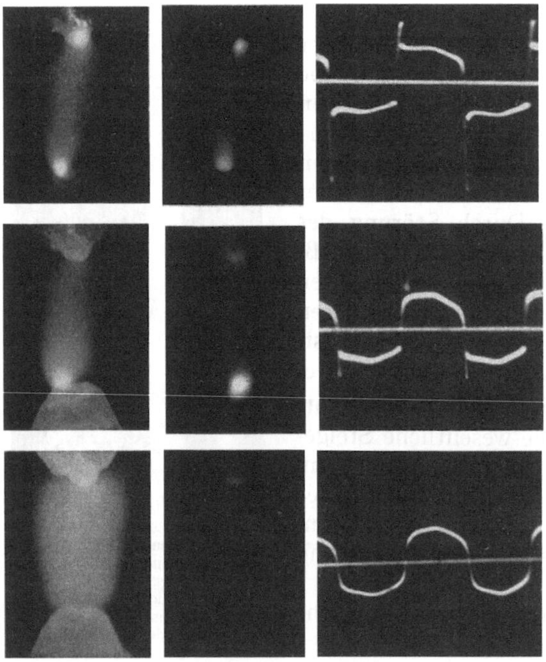

Fig. 88. Photographische Aufnahmen und Spannungsoszillogramme von Quecksilber-Wechselstrombögen mit kathodischem Brennfleck an beiden Elektroden (oben), mit Brennfleck nur an der unteren Elektrode (Mitte) und von einem brennflecklosen Bogen (unten), nach THOURET, WEIZEL und GÜNTHER [226].

mit dieser Deutung steht die Beobachtung (Fig. 88), daß bei einem brennflecklosen Wechselstrombogen das Wiederzünden ohne Zündspitze erfolgt, während beim Brennfleckbogen stets eine hohe Zündspitze beobachtet wird. Beim brennflecklosen Bogen ist eben die thermische Trägheit der hoch erhitzten Kathode so groß, daß beim Wiedererscheinen der Spannung der Stromfluß wieder beginnen kann. Beim Brennfleckbogen dagegen führt die geringe thermische Trägheit des Plasmas zu einem so schnellen Abbau der für den Mechanismus entscheidenden kathodischen Plasmaschicht, daß sie beim Wiedererscheinen der Spannung nach dem Nulldurchgang neu aufgebaut werden muß und die hierfür erforderliche Energie sich in der „Zündspitze" äußert.

Die Beobachtungen scheinen also tatsächlich eine Unterscheidung zwischen dem thermischen Kathodenmechanismus und einem solchen mit Plasma-Brennfleck zuzulassen. Erwartungsgemäß ist der kathodische Mechanismus von Kohlebögen wegen deren hoher Kathodentemperatur ebenso wie der Kathodenmechanismus der in Ziff. 26 behandelten Hochtemperaturbögen mit Wolframkathode ein thermischer, während der stationäre Kathodenbrennfleckbogen offenbar nur bei solchen Hochdruckbögen mit Metallkathode vorkommt, bei denen sich ohne Kathodenverdampfung eine Plasmaschicht genügend hoher Dichte und Temperatur dicht vor der Kathode ausbilden kann.

Der Typ des kathodisch nicht stationären Bogens mit regellos wanderndem Brennfleck ist besonders von Güntherschulze [*74*], Seeliger und Mitarbeitern [*204*], Wehrli [*240*], Becken und Sommermeyer [*7*] sowie neuerdings Froome [*58*] genauer untersucht worden. In diesen schnell wandernden, häufig auch mehrfachen, oft verschwindenden und wieder entstehenden nichtstationären Brennflecken hat Froome Stromdichten bis über 10^6 Amp/cm^2 gefunden. Dieser Brennflecktyp findet sich bei allen Niederdruckbögen mit Metallkathode von niedrigem Siedepunkt, besonders also auch bei dem als Quecksilberdampf-Gleichrichter viel benutzten, Ziff. 31 behandelten Niederdruckbogen mit lebhaft verdampfender Hg-Kathode. Auch Bögen mit Wolfram- und sogar Kohlekathode sollen aber gleich nach der Zündung bei noch kalter Kathode, oder bei Gasdrucken kurz oberhalb des Umschlagpunktes zur Glimmentladung, gelegentlich diesem Typ angehören. Nach Seeliger und Mitarbeitern läßt sich bei Drucken zwischen 10 und 100 Torr der Umschlag vom nicht-stationären Brennfleckbogen in den brennflecklosen Bogen und umgekehrt gut verfolgen, wobei der nicht-stationäre Brennfleckbogen stets eine je nach dem Gasdruck um 5 bis 50 V niedrigere Brennspannung besitzt als der brennflecklose thermische Bogen. Gasbeladung der Kathode, besonders mit Sauerstoff, begünstigt die Ausbildung des Brennfleckbogens. Es wird in den Ziff. 105 bis 109 gezeigt werden, daß für die geschilderten drei Bogentypen auch drei verschiedene Kathodenmechanismen verantwortlich sind. Während beim brennflecklosen Bogen der Strom vor der Kathode im wesentlichen von Glühelektronen und beim nichtstationären Brennfleckbogen mit Kathodenverdampfung anscheinend überwiegend von kathodischen Feldelektronen getragen wird, sind beim stationären Brennfleckbogen die aus dem Plasma kommenden positiven Ionen für den Strom vor der Kathode entscheidend.

99. Kathodenfallwerte. Im Gegensatz zur Bogensäule fehlt es für den Kathodenfall und den Kathodenmechanismus des Lichtbogens noch an einer geschlossenen quantitativen Theorie. Das liegt nicht nur daran, daß es für verschiedene Lichtbogentypen mit Sicherheit grundsätzlich verschiedene Kathodenmechanismen gibt, sondern auch daran, daß ein zuverlässiges empirisches Zahlenmaterial als Grundlage quantitativer Theorien noch fehlt. Das gilt für die seit vielen Jahrzehnten ausgeführten Kathodenfallmessungen ebenso wie für die gleich zu besprechenden kathodischen Stromdichten und die Temperaturen an und vor der Kathodenoberfläche.

Aus den meist mit Sonden (vgl. Ziff. 17) ausgeführten Kathodenfallmessungen geht ziemlich eindeutig hervor, daß im Stromstärkebereich von etwa 1 bis 40 Amp der Kathodenfall bei Stromstärkesteigerung abnimmt. Über die Absolutwerte dagegen herrscht noch keine Übereinstimmung, und es scheint sehr schwierig, Ursachen für die zum Teil erheblichen Abweichungen festzustellen, obwohl eine der Ursachen in der von den äußeren Bedingungen abhängigen Zusammensetzung, Dichte und Temperatur der der Kathode direkt vorgelagerten Plasmaschicht zu suchen sein dürfte. So sind von verschiedenen Autoren, z.B. [*161*], [*31*], [*56*], [*148*], für den Niederstromkohlebogen mit Homogenkohlekathode Kathodenfallwerte zwischen 5 und 13 V angegeben worden, obwohl die Mehrzahl der Messungen auf einen Kathodenfallwert in der Gegend von 9 V für einen Bogen von einigen Ampere Stromstärke hinweist und mit zunehmender Stromstärke ein Abfall um 2 bis 3 V wahrscheinlich ist. Ersatz der Homogenkohlekathode durch eine Dochtkathode scheint innerhalb der großen Streuung der angegebenen Werte den Kathodenfall nicht zu ändern. Bei Metallbögen werden für den gleichen Stromstärkebereich Kathodenfallwerte zwischen 5 und 20 V angegeben, wobei in zahlreichen Fällen der Vergleich von Werten durch Fehlen der Angaben über die Entladungsdaten erschwert wird. So gibt für den Quecksilberbogen eine Reihe von Autoren Kathodenfallwerte von 5 bis 6 V, zahlreiche andere Autoren solche von 8 bis 11 V an, ohne daß eine Ursache für diese Diskrepanz festzustellen wäre. Allerdings scheint sich eine Abnahme des Kathodenfalles mit zunehmendem Quecksilber-Dampfdruck anzudeuten, z.B. [*8*] eine solche von 11 auf 8 V bei Drucksteigerung von 0,6 auf $100 \cdot 10^{-3}$ Torr. Für Kathoden aus Natrium, Tantal und Cadmium werden für den Stromstärkebereich von einigen Ampere Kathodenfallwerte zwischen 5 und 9 V angegeben, während für Kathoden aus Kupfer, Eisen, Zink und Wolfram überwiegend Werte von 15 bis 16 V veröffentlicht wurden, z.B. [*26*], [*31*], [*161*]. Genaue Messungen der Abhängigkeit des Kathodenfalles von der Stromstärke haben kürzlich BAUER und SCHULZ [*4*] für eine Xenon-Höchstdruckentladung veröffentlicht; sie finden bei 30 Atm Gasdruck und Wolframelektroden eine Abnahme des Kathodenfalles von 13 auf etwas unter 10 V bei Stromstärkezunahme von 3 auf 10 Amp.

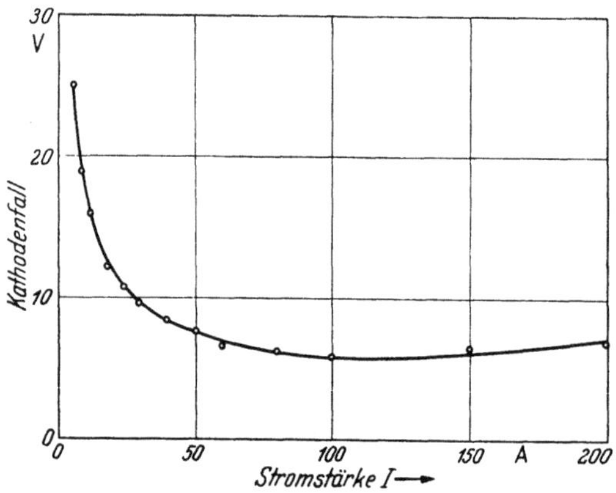

Fig. 89. Stromstärke-Abhängigkeit des Kathodenfalles eines Argon-Bogens, nach BUSZ-PEUCKERT und FINKELNBURG [*22*].

Für einen 200 Amp-Kohlebogen folgt aus neuen Sondenmessungen von MÜLLER und FINKELNBURG [*159*] der sehr niedrige Kathodenfallwert von (3 ± 1) Volt. An dem in Ziff. 26 erwähnten, bei Atmosphärendruck brennenden Argon-Bogen mit Wolframkathode zeigt nach Sondenmessungen von BUSZ und FINKELNBURG [*22*] der Kathodenfall den in Fig. 89 wiedergegebenen Verlauf, fällt also von recht hohen Werten bei kleiner Stromstärke zu den für Hochtemperaturbögen offenbar charakteristischen Werten weit unterhalb selbst der ersten Anregungsspannung des Trägergases. In Übereinstimmung damit liegt auch der Kathodenfall des Stickstoff-Hochtemperaturbogens gleicher Art im Stromstärkebereich zwischen 50 und 200 Amp bei (7 ± 2) Volt.

100. Kathodische Stromdichtewerte. Zuverlässige Daten der kathodischen Stromdichte bei verschiedenen Bogentypen, die zusammen mit zuverlässigen Kathodenfallwerten die empirische Grundlage einer quantitativen Kathodenfalltheorie bilden könnten, fehlen noch weitgehend; immerhin lassen sich wenigstens größenordnungsmäßige Angaben machen. Die Schwierigkeiten der Stromdichtebestimmung liegen darin, daß bisher direkte Messungen, etwa mit Sonden, noch nicht ausgeführt worden sind und mit der erforderlichen Genauigkeit auch sehr schwierig sein dürften. Die vorliegenden Bestimmungen der kathodischen Stromdichte beruhen vielmehr auf indirekten Methoden. Dabei wird entweder die Größe des sichtbaren Kathodenansatzes, z.B. auf Zeitlupenaufnahmen, vermessen und damit der optische Kathodenansatz dem elektrischen gleichgesetzt, oder es wird aus der Größe kathodischer Schmelzspuren bei magnetisch bewegtem Kathodenansatz auf die kathodische Stromdichte geschlossen. Der Schluß vom optischen auf den elektrischen Durchmesser des Bogenansatzes bzw. Brennfleckes enthält erhebliche Fehlerquellen, deren Größe von der Temperatur des Plasmas vor der Kathode abhängt, weil Strahlung und elektrische Leitfähigkeit in verschiedener Weise von der Temperatur abhängen. Der Schluß von der Größe der kathodischen Schmelzspuren auf die kathodische Stromdichte aber enthält die Unsicherheit, daß bei gleicher Stromdichte die Schmelzspurgröße nicht nur von der Wärmeleitfähigkeit des Kathodenmaterials, sondern auch von der Dauer der Einwirkung abhängen muß. Schließlich wird man auch im kathodischen Ansatz nicht von einer einheitlichen Stromdichte sprechen können, da diese von der Säulenachse nach außen abnehmen muß.

Bei Kohlelichtbögen nimmt die kathodische Stromdichte nach GUILLERY [*72*] mit wachsender Stromstärke von einigen hundert Ampere je Quadratzentimeter bis zu einem bei etwa 400 Amp Bogenstromstärke erreichten Grenzwert von 5000 Amp/cm² zu, um bei größerer Stromstärke konstant zu bleiben. Bei Lichtbögen mit Wolframkathode erhält man nach Ziff. 106 bei genügender Kathodentemperatur den ziemlich ausgedehnten kathodischen Ansatz des sog. brennflecklosen Bogens, dessen kathodische Stromdichte nach eigenen Messungen und solchen von BAUER [*3*] zwischen 1000 und 3000 Amp/cm² liegt. Bei etwas geringerer Temperatur der Wolframkathode und bei praktisch allen Bögen mit niedrig schmelzenden Metallelektroden dagegen findet man den sog. kathodischen Brennfleckbogen mit kathodischen Stromdichten zwischen 10^4 und 10^7 Amp/cm². Während für den Wolframbogen mit stationärem Brennfleck BAUER [*3*] kathodische Stromdichtewerte zwischen 3 und $7 \cdot 10^4$ Amp/cm² ermittelt hat, fanden COBINE und GALLAGHER [*27*] und besonders FROOME [*58*] für die Stromdichte schnell bewegter Brennflecke nicht-stationärer Bogenentladungen besonders mit Quecksilberkathode Werte bis 10^7 Amp/cm². Dabei ergaben Beobachtungen von SCHMIDT [*191*], FROOME u.a., daß zum mindesten bei flüssigen Bogenkathoden aus Quecksilber oder Alkalimetallen stationäre, zeitlich konstante Brennflecke überhaupt nicht existieren. So beobachtet man nach SCHMIDT an einer Hg-Kathode bei Stromstärken merklich über 15 Amp stets mehrere laufend entstehende und wieder verschwindende kathodische Brennflecke, deren Stromstärken zwar in weiten Grenzen schwanken können, deren jeder im Mittel jedoch nur eine Stromstärke von 5 bis 6 Amp besitzt, wobei aber nach FROOME die Stromdichte unabhängig von der zufälligen Fleckgröße konstant 2 bis $10 \cdot 10^6$ Amp/cm² sein soll.

101. Temperaturen an und vor der Bogenkathode. Zuverlässige Angaben über die Temperatur im kathodischen Bogenansatz lassen sich nur für brennflecklose Bögen (Ziff. 106) mit vorwiegend thermischer Elektronenemission machen, während im Kathodenbrennfleck von Bögen mit kontrahiertem Kathodenansatz hoher Stromdichte Messungen nicht gemacht wurden und zudem nach Ziff. 102 die Bedeutung von Temperaturangaben hier recht zweifelhaft ist.

Beim brennflecklosen Kohlebogen nimmt die Kathodentemperatur von rund 3200° K bei wenigen Ampere [*7a*] bei Stromsteigerung auf 400 Amp nach Guillery [*72*] bis zum Siedepunkt des Kohlenstoffs bei 4000° K zu. Beim brennflecklosen Xenonbogen mit Wolframkathode findet Bauer [*3*] Maximaltemperaturen um 3000° K, Thouret, Weizel und Günther [*226*] etwas niedrigere Werte. Für den kontrahierten Brennfleck eines Xenonbogens mit 40000 Amp/cm² Stromdichte gelangt Bauer aus Messungen in Verbindung mit Wärmeleitungsrechnungen, und in Übereinstimmung mit der Beobachtung feiner Schmelzspuren, zu Werten hart bei der Schmelztemperatur des Wolframs, während die Angaben über die Temperatur im nicht-stationären Brennfleck von Hg-Bögen mit flüssiger Kathode zwischen einigen hundert Grad (z.B. A. Lutz [*133*]) und rund 2300° K (Literatur bei Wasserrab [*239*]) schwanken.

Über die Temperatur und Temperaturverteilung im Plasma *vor* der Kathode ist aus Beobachtungen lediglich bekannt, daß beim Brennfleckbogen dicht vor der Kathode ein Gebiet besonders hoher Strahlungsdichte und damit wohl auch Temperatur liegt, während die Beobachtung eines deutlichen Dunkelraumes vor der Kathode brennfleckloser Bögen zeigt, daß bei diesen umgekehrt die Temperatur relativ langsam von dem niedrigen Wert der Kathodenoberfläche auf den hohen der Bogensäule ansteigt.

102. Das Problem des „Zustands" der Kathodenoberfläche. Bei allen theoretischen Überlegungen und Rechnungen über Kathodenfallvorgänge wird Gebrauch gemacht von einer großen Zahl makroskopischer Eigenschaften der Kathodenoberfläche wie z.B. Elektronenaustrittsarbeit, Temperatur, Wärmeleitfähigkeit, Verdampfungswärme und Akkomodationskoeffizient. In gleicher Weise rechnet man für das der Kathodenoberfläche direkt vorgelagerte Plasma mit makroskopischen Begriffen wie Temperatur und Druck, wobei im allgemeinen stillschweigend angenommen wird, daß man zwischen einer klar definierten Festkörperoberfläche und dem ihr vorgelagerten Plasma scharf unterscheiden kann. Loeb [*8a*] hat zuerst darauf hingewiesen, daß angesichts der außerordentlichen Leistungskonzentration im Kathodenbrennfleck von Bögen hoher kathodischer Stromdichte die Anwendbarkeit der genannten makroskopischen Begriffe und Daten äußerst fraglich ist. Bei einem Bogen mit der kathodischen Stromdichte von 10^6 Amp/cm² z.B., bei dem 10% des Kathodenstromes von Ionen getragen werden möge, treffen sekundlich $6\cdot 10^{23}$ Ionen die Flächeneinheit der Kathode. Jedes von ihnen besitzt eine kinetische Energie von 5 bis 10 eV und eine potentielle Energie gleicher Größenordnung, die bei der Neutralisation an der Kathode frei wird. Jedes der $6\cdot 10^{23}$ Ionen transportiert also eine Energie zur Kathodenoberfläche, die ein Mehrfaches der Bindungsenergie der Kathodengitterbausteine beträgt. Da 1 cm² der Kathodenoberfläche aber nur einige 10^{15} Oberflächenatome besitzt, wird jedes dieser Oberflächenatome 10^8mal je Sekunde von einem seiner eigenen Bindungsenergie gegenüber äußerst energiereichen Ion getroffen. Unter diesen Umständen ist die Erhaltung einer geordneten Gitterstruktur der Kathodenoberfläche im Bereich des Kathodenbrennfleckes ersichtlich unmöglich. Die Kathodenoberfläche befindet sich vielmehr im Bereich des Brennfleckes, soweit sich überhaupt Aussagen machen lassen, auf einer Tiefe von mehreren Atomlagen in einem Zustand so extremer Auflockerung, daß es zweifelhaft erscheint, ob man vom Zustand eines stark gestörten Festkörpers oder dem eines höchst komprimierten Plasmas reden soll. Die Leistungsdichte in diesen

Oberflächenschichten des Kathodenbrennfleckes liegt ja in dem gewählten Beispiel bei 1000 kW je Quadratzentimeter, und da diese große Leistung nicht durch Wärmeleitung genügend schnell abgeführt werden kann, muß eine beträchtliche Kathodenmaterial-Verdampfung stattfinden. Dabei ist aber wiederum sehr zweifelhaft, ob in den fraglichen Grenzschichten zwischen Kathode und Plasma überhaupt thermisches Gleichgewicht herrscht und man daher von wirklicher Verdampfung sprechen darf, oder ob nicht vielmehr die Grenze zwischen Festkörper und Plasma fließend ist und der Materialabbau direkt durch die erwähnten zahlreichen Stöße der auftreffenden Ionen erfolgt, deren Energie weit über kT liegt. Ist letzteres der Fall, hat man es also wirklich mit einem Übergangsfall zwischen einem hocherhitzten Festkörper und einem Plasma hoher Partikeldichte und gleichzeitig extremer Energiedichte zu tun, dann verlieren für dieses Übergangsgebiet die normalen Begriffe der Austrittsarbeit und mit ihnen die für den Kathodenmechanismus so wichtigen Prozesse der thermischen Elektronenemission bzw. der Feldemission weitgehend ihre Bedeutung. Die von manchen Seiten (z.B. Smith [*209*], Rothstein [*182*]) hieraus für den Kathodenmechanismus gezogenen Folgerungen scheinen noch extrem unsicher. Es dürfte aber kein Zweifel darüber bestehen, daß der geschilderte besondere „Zustand" der Kathode besonders in den nicht-stationären Brennflecken bei jeder vollständigen Kathodenfalltheorie berücksichtigt werden müßte und dementsprechend näherer Untersuchung bedarf.

103. Die Wirkung des kathodischen Plasmastrahles auf die kathodischen Entladungsteile. In Ziff. 94 wurde bereits ausführlich theoretisch begründet, daß an Entladungsstellen starker Kontraktion und damit hoher Stromdichte radial nach innen und teilweise auch von der Engstelle weg gerichtete magnetische Kräfte des Bogenstromes wirksam sind. Durch sie wird nach Maecker [*139*] Gas aus der Umgebung der Engstelle angesaugt, im Bogen aufgeheizt und als Plasmastrahl mit Geschwindigkeiten von 10^3 bis 10^5 cm/sec axial in die Richtungen geringerer Säulenkontraktion abgeschoben. Eine solche Engstelle ist stets der Kathodenansatz des Bogens, wo die Stromdichte wesentlich größer ist als in der anschließenden Bogensäule. Bei genügend hoher Stromstärke und kathodischer Stromdichte geht also von der Kathode der z.B. von Larché [*118*] beobachtete Plasmastrahl aus. Daß dieser nur oberhalb einer gewissen Stromstärke beobachtet wird, liegt daran, daß der ihn erzeugende magnetische Druck dem Produkt aus Stromstärke und Stromdichte proportional ist. Die Wirkung dieses Plasmastrahles auf die Säule ist in Ziff. 95/96 behandelt worden. Seine Wirkung auf die kathodischen Entladungsteile besteht darin, daß er die Stromdichte im Plasma vor der Kathode, wie im Kathodenansatz selbst, dadurch vergrößert, daß die seitlich angesaugten relativ kalten Gase die Kathodenoberfläche wie das kathodische Plasma intensiv kühlen. Ob, bzw. in welchem Umfange der Kathodenfall selbst beeinflußt wird, ist nicht bekannt. Dagegen erhöht das Ansaugen kalten Gases aus der Umgebung die Feldstärke in dem der Kathode vorgelagerten Plasma, weil bei vorgegebener Stromstärke die zur Aufheizung dieses kalten Plasmas erforderliche Leistung nur durch Feldstärkeerhöhung aufgebracht werden kann. Mißt man nun, wie das nicht selten geschehen ist, als Kathodenfall den über den letzten Millimetern vor der Kathode liegenden Potentialabfall abzüglich des normalen Säulengradienten, so wird die vom Plasmastrahl bewirkte Spannungsvergrößerung vor der Kathode als Teil des Kathodenfalles mitgemessen. Müller und Finkelnburg [*159*] haben an der Kathode eines Hochstromkohlebogens im Stromstärkebereich von 50 bis 350 Amp durch Sondenmessungen den eigentlichen Kathodenfall und die plasmastrahlbedingte Zusatzspannung getrennt und für letztere Werte bis 2 V gefunden.

104. Die rückläufige Bewegung des Kathodenbrennflecks in einem magnetischen Querfeld. Zu den noch nicht befriedigend erklärten kathodischen Bogenerscheinungen, deren Deutung aus einer noch fehlenden geschlossenen Kathodenfalltheorie folgen müßte, gehört die rückläufige Bewegung ("retrograde motion") des Kathodenbrennflecks in einem transversalen Magnetfeld. Der Effekt wurde zuerst 1928 von MINORSKY [*155*] beschrieben und später vor allem von C. G. SMITH [*209*] sowie ST. JOHN und WINANS [*219*] genauer untersucht: seine Deutungsmöglichkeiten gehören zu den meist diskutierten Problemen der Bogenphysik.

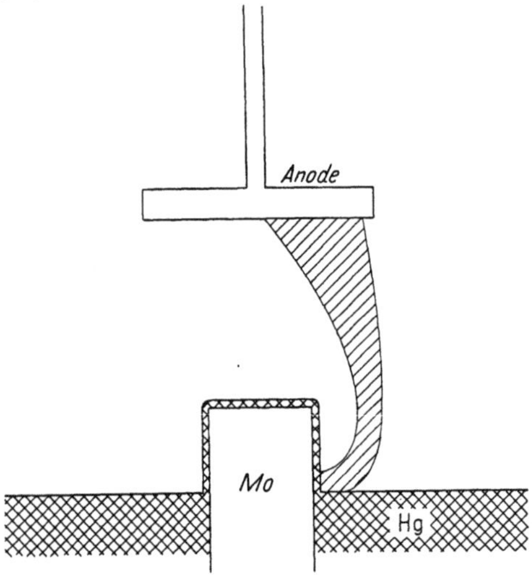

Fig. 90. Schematische Anordnung zum Studium der rückläufigen Bewegung eines Quecksilber-Bogens mit Verankerung des Kathodenbrennflecks an einer in Quecksilber stehenden Mo-Kathode; schematisch nach ST. JOHN und WINANS [*219*].

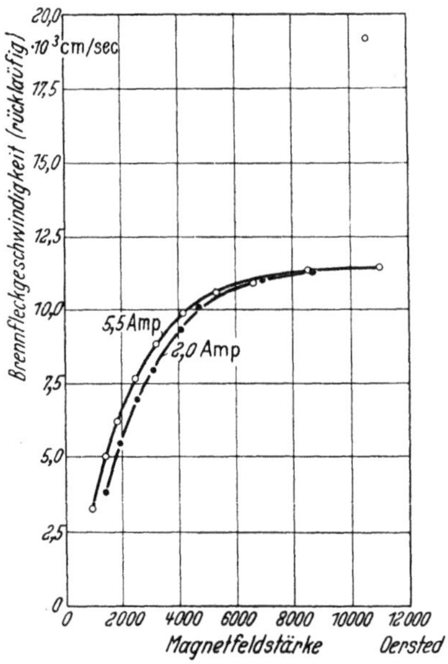

Fig. 91. Abhängigkeit der (rückläufigen) Brennfleck-Geschwindigkeit von der Magnetfeldstärke für zwei verschiedene Bogenstromstärken, nach ST. JOHN und WINANS [*219*].

Die Erscheinung ist zuerst an einer Quecksilber-Niederdruckentladung mit „Kathodenfleck-Anker" beobachtet und an ihr auch weiter untersucht worden, wurde aber von GALLAGHER und COBINE [*60*] auch an Hg-freien Bögen mit verschiedenen festen Kathoden in einer ganzen Anzahl von Gasen nachgewiesen. Verwendet man als Kathode einer Niederdruck-Bogenentladung in Quecksilberdampf einen Mo- oder Ta-Zylinder, der gemäß Fig. 90 etwas aus der Hg-Oberfläche hervorragt, so ist dessen gesamte Oberfläche vom Quecksilber benetzt, und der Kathodenbrennfleck setzt bevorzugt an der kreisförmigen „Hohlkehle" zwischen Metallzylinder und normaler Quecksilberoberfläche an. Die Richtung des kathodischen Säulenansatzes steht dann zunächst senkrecht auf dem mit der Elektrodenachse zusammenfallenden Magnetfeld, das bei den Versuchen zwischen 1000 und 12000 Oe betrug. Nach der AMPEREschen Regel sollte der Kathodenansatz dann in einer bestimmten Richtung um den Metallzylinder herumlaufen. Die entscheidende Beobachtung ist nun, daß zwar bei einem brennflecklosen Bogen (Ziff. 106) der Bogenansatz richtig mit einigen hundert Zentimeter je Sekunde in der theoretisch erwarteten Richtung um den dann weißglühenden Metallzylinder umläuft, bei einem Brennfleckbogen aber der Umlauf des Brennfleckes mit Geschwindigkeiten der Größenordnung 10^4 cm/sec in der *umgekehrten* Richtung um den bei dieser Bogenform kühleren Metallzylinder erfolgt.

In einem gasfreien Hg-Niederdruckbogen (Ziff. 31) steigt bei konstanter Stromstärke diese rückläufige Umlaufsgeschwindigkeit mit zunehmender Magnetfeldstärke gemäß Fig. 9 bis etwas über $1 \cdot 10^4$ cm/sec, um dann bei einer stromstärkeabhängigen kritischen Feldstärke zwischen 9000 und 11000 Oe plötzlich auf etwa den doppelten Wert zu springen. Unterhalb des Sprungpunktes steigt die rückläufige Geschwindigkeit bei niedriger Magnetfeldstärke langsam mit zunehmender Bogenstromstärke, um bei hohen Magnetfeldstärken fast unabhängig von der Stromstärke zu werden. Bei Zusatz von Edelgas zu dem meist bei 0,1 Torr liegenden Quecksilberdampfdruck sinkt die rückläufige Geschwindigkeit des Kathodenbrennfleckes gemäß Fig. 92, um bei einem von der Magnetfeldstärke abhängigen Gasdruck Null zu werden und bei noch höherem Gasdruck geringe Werte in der normalen, der AMPEREschen Regel entsprechenden Umlaufsrichtung anzunehmen. Fig. 93 zeigt, wie dieser Punkt des Wechsels vom rückläufigen zum normalen Umlauf von Stromstärke, Fremdgasdruck und Magnetfeldstärke abhängt. Interessanterweise ist nach GALLAGHER und COBINE der Druck, bei dem der Umschlag stattfindet, von der Ionisierungsspannung des Entladungsgases bzw. -dampfes abhängig. Erwähnt sei schließlich, daß spektroskopische Aufnahmen des Säulenansatzes an der Kathode bei geringer Feldstärke das Spektrum des Hg-Atoms, bei hoher Feldstärke zusätzlich Linien des Hg^+ ergaben, während der Kathodenbrennfleck selbst ein intensives kontinuierliches Spektrum emittiert. Diese letzte Beobachtung kann im Zusammenhang mit Ziff. 102 von Interesse sein.

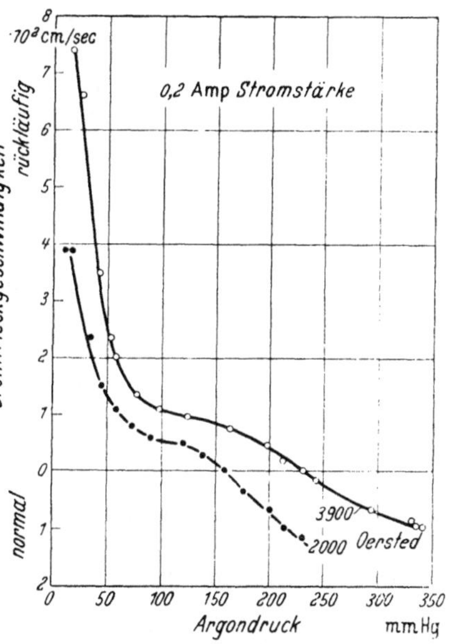

Fig. 92. Abhängigkeit der Brennfleckgeschwindigkeit vom Argon-Gasdruck für zwei verschiedene Magnetfeldstärken nach ST. JOHN und WINANS [219].

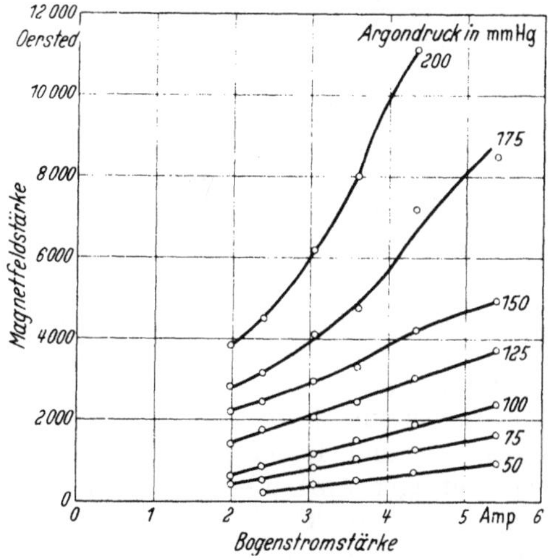

Fig. 93. Abhängigkeit der Magnetfeldstärke, bei der der Umschlag zwischen rückläufiger und normaler Brennfleckbewegung erfolgt, von der Bogenstromstärke für verschiedene Werte des Argon-Gasdrucks, nach ST. JOHN und WINANS [219].

Daß der Effekt der rückläufigen Bewegung mit dem Kathodenmechanismus zusammenhängt und damit auch für dessen Theorie von Interesse ist, geht schon aus der Tatsache hervor, daß die rückläufige Bewegung nur beim Brennfleckbogen, nicht aber beim brennflecklosen Bogen auftritt. Die noch fehlende geschlossene Theorie des kathodischen Brennfleckbogens muß also automatisch auch die Erscheinungen der rückläufigen Bewegungen im Quermagnetfeld erklären. Die bisher vorliegenden Deutungsversuche (TANBERG, LONGINI, ROTHSTEIN, ST. JOHN und WINANS, ROBSON und v. ENGEL) gehen von ganz verschiedenen Voraussetzungen aus und keiner von ihnen vermag sämtliche oben erwähnten experimentellen Ergebnisse zwanglos zu erklären.

II. Kathodenfalltheorien.

105. Allgemeines über Kathodenfalltheorien. Das Grundproblem jeder Kathodenfalltheorie ist das in Ziff. 6 bereits angedeutete, nämlich der mit den physikalischen Gesetzen verträgliche elektrische Anschluß des kathodenseitigen Säulenendes an die mit dem negativen Pol der Spannungsquelle verbundene metallische Kathode. Von einer vollständigen Kathodenfalltheorie wird man aber darüber hinaus verlangen müssen, daß sie alle experimentell sichergestellten kathodischen Phänomene zwanglos zu erklären vermag. Für keinen der oben besprochenen verschiedenen Bogentypen liegt bisher eine solche quantitativ befriedigende Theorie vor. Im Gegenteil deutet z.B. die Tatsache, daß die in Ziff. 104 behandelte rückläufige Bewegung des Kathodenbrennfleckes in einem magnetischen Querfeld von keiner gegenwärtigen Kathodenfalltheorie erklärt werden kann, auf deren Unzulänglichkeit hin. Diese einleitende Bemerkung soll den Wert der unten zu behandelnden theoretischen Versuche in keiner Weise schmälern, sondern nur den gegenwärtigen Stand der Kenntnis andeuten.

Die zur elektrischen Überbrückung des Raumes zwischen Säulenende und Kathodenoberfläche erforderlichen Ladungsträger können grundsätzlich auf drei verschiedene Weisen erzeugt werden. Bei genügend hoher Temperatur der Kathodenoberfläche könnte der Strom vor der Kathode ausschließlich von Elektronen getragen werden, die von der Kathodenoberfläche thermisch emittiert werden. Bei genügend hoher elektrischer Feldstärke vor der Kathodenoberfläche könnten diese Elektronen auch durch Feldemission von der Kathodenoberfläche geliefert werden. Es besteht drittens die Möglichkeit, daß der Strom zwischen Säulenende und Kathode ausschließlich oder im wesentlichen von positiven Ionen getragen wird, wenn das kathodenseitige Ende der Säule sich auf so hoher Temperatur befindet, daß es die erforderliche Ionenstromdichte abzugeben vermag, und wenn die Raumladungen gerade diesen Strom zulassen. Es wird unten gezeigt, daß bei wirklichen Bögen keine der drei angedeuteten Möglichkeiten allein für den Strom vor der Kathode verantwortlich ist. Bei den brennflecklosen Bögen hat man es vielmehr vorwiegend mit einer Kombination von thermischer und Feldemission zu tun, zu der aber eine gewisse Ionenstromkomponente von der Plasmagrenze her notwendig hinzukommt. Bei ausgesprochenen Brennfleckbögen umgekehrt bestimmt zwar der von der Plasmagrenze, d.h. dem kathodenseitigen Säulenende herkommende Ionenstrom die Verhältnisse, doch sind auch hier Feld- und thermische Emission fast stets am Strom vor der Kathode beteiligt.

Vor dem näheren Eingehen auf die einzelnen Theorien deuten wir kurz die wichtigsten physikalischen Probleme bei den drei oben skizzierten Mechanismen an. Bei der thermischen Elektronenemission der Kathode ist zu untersuchen, woher die Kathode die Leistung bekommt, um sie trotz Ausstrahlung, Wärmeableitung und Kühlung durch die von ihr unter Leistung der Austrittsarbeit emittierten Elektronen auf ihrer zur thermischen Emission erforderlichen Temperatur zu halten. Bei der Feldemission der Kathode ist das Problem, welche Raumladungsverteilung vor der Kathode das erforderliche hohe elektrische Feld erzeugt. Bei der Ionenemission von der Plasmagrenze zur Kathode hin schließlich ist zu untersuchen, wie die Plasmaschicht auf der erforderlichen hohen Temperatur gehalten wird, und wie die Raumladungsverhältnisse beschaffen sind, die den Stromfluß zwischen Säulengrenze und Kathode bestimmen.

Historisch haben zuerst STARK [*15*], [*214*] und genauer K. T. COMPTON [*28*] die thermische Kathodentheorie, vom Beispiel des Kohlebogens ausgehend, entwickelt. Da die thermische Theorie auf Bögen mit Quecksilber- und Kupfer-kathode wegen deren niedriger Siedepunkte und hoher kathodischer Stromdichten nicht zutreffen konnte, haben LANGMUIR [*115*] und MACKEOWN [*135*] die Feldemissionstheorie vorgeschlagen, wobei sie annahmen, daß der Kathodenfall von größenordnungsmäßig 10 V über einer freien Weglänge vor der Kathode läge. Da die Rechnungen trotzdem zu kleine Elektronenstromdichten ergaben, hat zuerst SLEPIAN [*208*] die Ionenemission aus dem kathodenseitigen Ende der Säule untersucht und die erste kathodische Ionenstromtheorie aufgestellt, die später von WEIZEL, ROMPE und SCHÖN [*244*] sowie ECKER [*33*] weiter ausgebaut wurde. Letzterer und BAUER [*3*] haben dann die Kombination von thermischer und Feldemission der Kathode untersucht und zusammen mit der Ionenstromtheorie zur Deutung der verschiedenen empirischen Kathodenmechanismen herangezogen. Nicht näher ausgeführt dagegen ist ein nur teilweise veröffentlichter Vorschlag von ROTHSTEIN [*182*], der im Sinne der Diskussion von Ziff. 102 die Plasmaschicht, die zur Kathode hin positive Ionen, zur Säule hin Elektronen emittiert, in die Kathodenoberfläche selbst verlegen will, und den aus verschiedenen Gründen nicht recht befriedigenden Versuch gemacht hat, auf dieser Grundlage auch die in Ziff. 104 behandelten Erscheinungen zu klären.

Grundsätzlich lassen sich natürlich die elektrischen Verhältnisse im Kathodenfallgebiet (wie im Anodenfallgebiet) durch das für alle stationären Entladungen geltende Gleichungssystem von Ziff. 45 beschreiben und bei genügender Kenntnis aller Größen und Funktionen auch quantitativ berechnen. Es ist aber für die Fallgebiete besonders charakteristisch, daß in ihnen der Strom über eine Schicht relativ geringer Trägerdichte durch eine hohe Spannung transportiert werden muß. Hier stehen sich etwa eine Elektronen emittierende Kathode und die Ionen emittierende kathodenseitige Säulengrenze in geringem Abstand gegenüber, und zwischen beiden Schichten liegt die Kathodenfallspannung U_k. Nach LANGMUIR [*116*] ist nun auch bei für die Träger beliebig ergiebigen Flächen die Stromdichte in der Schicht durch die Raumladung des Stromes selbst begrenzt. Wir betrachten zunächst nur den Elektronenstrom und nehmen an, daß die Elektronen ohne Anfangsgeschwindigkeit aus der Kathode kommen. Zählen wir dann x von der Kathode zur Säulengrenze hin, so gibt die POISSON-Gleichung

$$d^2 U/dx^2 = -4\pi \varrho \tag{105.1}$$

den Zusammenhang zwischen dem Potential U und der Raumladungsdichte ϱ im Fallgebiet. Nun sind aber Stromdichte j und Geschwindigkeit v der Elektronen mit deren Raumladungsdichte ϱ durch

$$\varrho = j/v \tag{105.2}$$

und Elektronengeschwindigkeit v und durchfallenes Potential U durch

$$\frac{m}{2} v^2 = e\,U \tag{105.3}$$

verknüpft. Durch Einsetzen von (105.2) und (105.3) in (105.1) und Integration folgt, wenn E_0 die Feldstärke an der Kathode ist

$$\left(\frac{dU}{dx}\right)^2 = E^2 = E_0^2 + \frac{16\pi j}{\sqrt{2e/m}} \sqrt{U}. \tag{105.4}$$

Da E_0 durch die Raumladung der aus der Kathode austretenden Elektronen verkleinert wird, erhält man die maximale Sättigungsstromdichte j_s für $E_0 = 0$ und damit nach nochmaliger Integration

$$j_s = \frac{1}{9\pi} \sqrt{\frac{2e}{m}}\, \frac{U^{\frac{3}{2}}}{d^2}, \tag{105.5}$$

wenn d die Dicke des Raumladungsgebietes ist. Das ist die bekannte LANGMUIRsche Raumladungsformel.

Für einen ambipolaren, sich kreuzenden Strom von Elektronen und Ionen folgt aus einer entsprechenden Rechnung nach MACKEOWN [135]

$$E^2 = \frac{16\pi}{\sqrt{2e}} \left(j^+ \sqrt{M} - j^- \sqrt{m}\right) \sqrt{U}, \tag{105.6}$$

Eine gewisse Stromdichte j im Kathoden- bzw. Anodenfallgebiet erfordert also erstens eine hinreichende Trägerergiebigkeit der betreffenden Elektrode bzw. der ihr gegenüberstehenden Plasmaschicht, und zweitens nach (105.5) bzw. (105.6) eine hinreichend große Saugspannung bzw. Feldstärke im Fallgebiet.

106. Die thermische Feldemissionstheorie für den brennflecklosen Bogen[1]. Der brennflecklose Bogen besitzt nach Ziff. 101 eine Kathode sehr hoher Temperatur, vor dieser einen Dunkelraum von vielen freien Weglängen Ausdehnung und an diesen anschließend die Bogensäule. Die Temperaturverteilung im Übergangsgebiet von der Säule zur Kathodenoberfläche ist nicht bekannt; der Kathodenansatz ist relativ ausgedehnt, die kathodische Stromdichte hat die Größenordnung 10^3 Amp/cm^2. Aus der Säule strömen Ionen auf die Kathode zu, werden im Kathodenfall beschleunigt und erhitzen die Kathode durch Abgabe eines gleich noch zu behandelnden Anteils ihrer Energie. Entsprechend ihrer hohen Temperatur T_K ist die Kathodenoberfläche im Stande, einen Elektronensättigungsstrom der Größe

$$j^- = -\frac{4\pi e m k^2}{h^3} T^2 \cdot e^{-\frac{e\Phi - \sqrt{e^3 E}}{kT}} \tag{106.1}$$

zu emittieren. Gl. (106.1) ist die bekannte RICHARDSON-Gleichung, im Exponenten korrigiert für die Erniedrigung der Austrittsarbeit $e\Phi$ durch das vor der Kathode liegende elektrische Feld E. Die auf die Kathode zuströmenden positiven Ionen bilden nämlich wegen ihrer geringen Beweglichkeit eine den Elektronenstrom von der Kathode her überkompensierende positive Raumladung, die sich in einem entsprechenden elektrischen Feld im Kathodenfallraum äußert, dessen Berechnung nach Gl. (106.1) nur bei genauer Kenntnis der Raumladungsverteilung möglich ist. Ist dieses Feld hoch genug, d.h. größer als 10^6 V/cm, so kommt zu dem durch Gl. (106.1) gegebenen Glühelektronenstrom nach FOWLER und NORDHEIM [55] noch ein Elektronen-Feldemissionsstrom der Sättigungsdichte

$$j^- = 1{,}55 \cdot 10^{-6} \frac{E^2}{e\Phi} \cdot e^{-\frac{6{,}9 \cdot 10^7 (e\Phi)^{\frac{3}{2}}}{E}} \tag{106.2}$$

hinzu. Dabei gilt Gl. (106.1) nur für hohe Temperaturen und nicht zu hohe Feldstärken, da es sich um eine auf klassischer Basis abgeleitete Formel handelt, Gl. (106.2) dagegen streng nur für den absoluten Nullpunkt der Temperatur, d.h. für die Emission unangeregter FERMI-Elektronen mittels Tunnel-Effekts. Im Zwischengebiet höherer Temperaturen *und* höherer Felder, d.h. beim Zusammenwirken beider Emissionsmechanismen, spricht man heute von „thermischer Feldemission". Fig. 94 zeigt, wie für diese die Sättigungsstromdichte j_s von der Kathodentemperatur und der Feldstärke vor der emittierenden Fläche abhängt. Beim brennflecklosen Bogen wird allerdings die Feldstärke vor der Kathode im allgemeinen so klein sein, daß der Feldemissionsanteil vernachlässigbar ist. Bei Übergangsfällen zum Brennfleckbogen spielen aber beide Anteile eine wichtige Rolle.

---

[1] Vgl. hierzu auch die Beiträge von NOTTINGHAM und von MÜLLER und GOOD in Bd. XXI dieses Handbuches.

Das Verhältnis von Elektronen- zu Ionenstrom im Kathodenfallgebiet des brennflecklosen Bogens läßt sich grundsätzlich aus der Energiebilanz an der Kathode berechnen, wenn alle Anteile der Bilanz genügend genau erfaßbar sind. In älteren Darstellungen wird meist angenommen, daß die aus der Säule kommenden Ionen nach Durchfallen des Kathodenfalles ihre gesamte kinetische Energie und die bei ihrer Neutralisation frei werdende Energie (Differenz der Ionisierungsspannung V_i und der Austrittsarbeit Φ des zur Neutralisation benötigten Elektrons) an die Kathode abgeben und daß diese der Kathode zugeführte Energie zur Bestreitung der Austrittsarbeit der thermisch emittierten Elektronen dient. Aus dieser Bilanz ergibt sich ersichtlich das Verhältnis von Elektronenstrom zu Ionenstrom vor der Kathode zu

$$\frac{j^-}{j^+} = \frac{V_K + V_i - \Phi}{\Phi}. \quad (106.3)$$

Mit den üblichen Werten für V_K, V_i und Φ folgt hieraus ein Ionenstromanteil zwischen 20 und 50%. Bei dieser Bilanz ist angenommen, daß der Energieverlust der Kathode durch Ausstrahlung und Wärmeableitung durch die anderen nicht berücksichtigten Effekte wie Wärmeleitung vom Plasma zur Kathode, Absorption von Plasmastrahlung durch die Kathode und andere gerade gedeckt wird. Das ist aber sicher nicht allgemein der Fall. Der Gl. (106.3) liegt ferner die Annahme zugrunde, daß die auf die Kathode aufprallenden Ionen dort nicht nur ihre Neutralisationsenergie, sondern auch ihre kinetische Energie abgeben, d.h. an ihr zunächst gleichsam haften bleiben (Akkomodationskoeffizient 1). Bei der hohen Temperatur der Kathodenoberfläche ist der Akkomodationskoeffizient aber im allgemeinen wesentlich kleiner als 1, und entsprechend wird ein mehr oder weniger großer Anteil, der die Kathode treffenden Ionen dort nur seine Neutralisierungsenergie abgeben, dann aber von der Oberfläche reflektiert werden und als neutrales Atom mit der vorher als Ion im Kathodenfall gewonnenen kinetischen Energie wieder in Richtung Säule zurückfliegen. Die im Mittel von einem Ion an die Kathode abgegebene Energie ist daher kleiner als die in Gl. (106.3) angegebene. Eine zuverlässige Energiebilanz an der Kathode und eine auf ihr basierende Bestimmung des Ionenstromanteiles vor der Kathode scheitert vorläufig an unserer mangelhaften Kenntnis solcher Daten wie des Akkomodationskoeffizienten als Funktion der Temperatur, der Temperaturverteilung vor der Kathode, der Wärmeleitfähigkeit des Plasmas vor der Kathode usw.

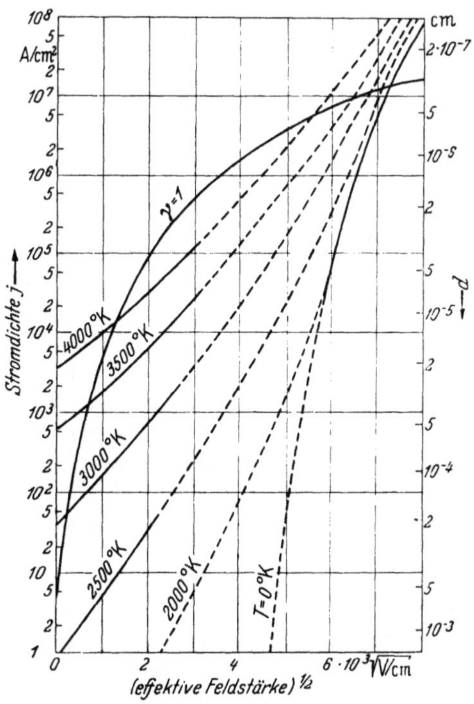

Fig. 94. Abhängigkeit der Elektronen-Emissionsstromdichte von der Wurzel aus der effektiven Feldstärke vor der Kathode für Wolframkathoden verschiedener Temperatur und (voll ausgezogene Kurve) raumladungsbegrenzte Sättigungsstromdichte für Oberflächenrauhigkeit Null, nach BAUER [3].

Für den Fall eines brennflecklosen Hg-Hochdruckbogens mit Wolframelektrode kommen WEIZEL und THOURET [249] auf Grund theoretischer Überlegungen zu dem Ergebnis, daß der Elektronenstromanteil vor der Kathode sogar 97,5%

beträgt. Nach den älteren Vorstellungen, Gl. (106.3), wäre ein so geringer Ionenstromanteil wie 2,5% unmöglich gewesen. Aus Energiebilanzgründen müssen WEIZEL und THOURET deshalb annehmen, daß ein wesentlicher Teil der Kathodenenergie durch Wärmeleitung von der Säule auf die Kathodenoberfläche übertragen wird.

107. Die Plasmakontraktionstheorie für den stationären Brennfleckbogen. Beim Brennfleckbogen mit nicht oder nicht merklich verdampfender Kathode und ruhendem Brennfleck, der besonders am Beispiel eines 6 Amp- 35 Atm-Hg-Bogens mit Wolframelektroden untersucht worden ist, beobachtet man vor der Kathode eine starke Kontraktion der normal etwa 2 mm Durchmesser besitzenden Säule. Ein Dunkelraum zwischen der Kathodenoberfläche und der ihr vorgelagerten Plasmagrenzschicht (oft auch Plasmakugel genannt) ist nicht mit Sicherheit zu beobachten.

Im Anschluß an einen ersten Versuch von SLEPIAN [*208*] nehmen WEIZEL, ROMPE und SCHÖN [*244*] sowie besonders ECKER [*33*] an, daß eine wesentliche Elektronenemission der Kathode bei diesem Bogentyp zwar möglich, für den Kathodenmechanismus aber nicht entscheidend ist. Entscheidend soll vielmehr sein, daß die der Kathodenoberfläche in einer Entfernung von einer oder einigen wenigen freien Weglängen gegenüberstehenden Plasmaschicht hoher Temperatur Ionen zur Kathode hin und Elektronen zur Säule hin abgibt. Die Plasmaschicht wird aufgeheizt durch die von der Kathode herkommenden, im Kathodenfall beschleunigten Elektronen, sowie durch die als neutrale Atome von der Kathode mit ihrer vollen kinetischen Energie reflektierten neutralisierten positiven Ionen. Trotz dieser Energiezufuhr kann die der Kathode vorgelagerte Plasmaschicht die zu ihrer hohen Stromergiebigkeit erforderliche hohe Temperatur aber nur halten, wenn sie wenig Energie durch Leitung an die Kathode abgibt. In dieser Tatsache liegt der physikalische Grund für die starke Kontraktion des Brennfleckbogens: nur wenn die Grenzfläche zwischen hoch erhitztem Plasma und Kathodenoberfläche sehr klein ist, kann die Plasmagrenzschicht trotz der Nähe der relativ kalten Kathode ihre hohe Temperatur halten. Diese auf ECKER zurückgehende physikalische Begründung der Kontraktion, daß nämlich ein solcher Bogen überhaupt nur mit starker Kontraktion vor der Kathode existieren kann, ergänzt das ältere theoretische Ergebnis von WEIZEL, ROMPE und SCHÖN, daß ein so kleiner Kathodenfall wie der beobachtete der Größenordnung 10 V nur bei Kontraktion vor der Kathode möglich sei. WEIZEL, ROMPE und SCHÖN hatten nämlich bei dem Versuch, die beobachteten Tatsachen hoher Stromstärke und niedrigen Kathodenfalles beim Brennfleckbogen unter der Voraussetzung ausschließlichen oder überwiegenden Ionenstromes vor der Kathode zu erklären, einen um den Faktor 10 zu großen Kathodenfall errechnet und geschlossen, daß dieser sich nur durch Säulenkontraktion vor der Kathode ermäßigen könne.

ECKER hat deshalb den Versuch gemacht, die Verhältnisse exakt durchzurechnen. Er nimmt an, daß vor der Kathode ein Raumladungsgebiet von etwa einer freien Weglänge Ausdehnung liegt, in dem ein durch die LANGMUIRsche Raumladungsgleichung [*116*] begrenzter Ionenstrom fließen soll, der größenordnungsmäßig dem Gesamtstrom entspricht. Das etwa vorhandene Defizit soll als Elektronenstrom durch thermische Feldemission von der Kathode gedeckt werden. Zwischen dem kontrahierten Raumladungsgebiet einerseits und der Säule normalen Durchmessers andererseits soll ein sog. Kontraktionsgebiet liegen. Es wird als quasineutral und damit als Teil der Säule angesehen, unterscheidet sich aber von ihr im Potential- und Temperaturverlauf, weil der Entladungsquerschnitt hier von dem der Säule auf den sehr geringen des Raumladungsgebietes abnimmt. ECKER nimmt als erste Näherung an, daß wegen der kleinen Grenzfläche zur Kathode axiale Leistungsverluste des kathodenseitigen Endes des Kontraktionsgebietes vernachlässigbar sind, und daß als Energieverlust im wesentlichen die Strahlung zu berücksichtigen ist, die (vgl. auch LARCHÉ [*118*]) bei sehr hoher Kontraktion näherungsweise als schwarz anzusehen sein soll, in Richtung auf die Säule zu mit abnehmendem Ionisierungsgrad aber rasch „optisch dünn" wird. Am säulenseitigen Ende des Kontraktionsgebietes umgekehrt soll die Strahlung unwesentlich sein und die Energiebilanz wie in der Säule wesentlich durch Wärmeleitungsverluste, und zwar in dem gerechneten Modell vorzugsweise zur Kathode hin, bestimmt werden. Der Rechnung liegt ferner die Annahme zugrunde, daß im Kontraktionsgebiet kein radiales Temperatur- und Trägerdichtegefälle herrscht (Kanalmodell). Die schon erwähnte letzte Annahme, daß der Strom vor der Kathode ein reiner Ionenstrom sei, ist praktisch nicht folgenschwer, weil gezeigt werden kann, daß selbst ein erheblicher Elektronenstromanteil wegen der großen Beweglichkeit und damit Geschwindigkeit der im Kathodenfall beschleunigten Elektronen die durch die Ionen bestimmte Raumladung nicht wesentlich verändert.

Die auf dieser Grundlage durchgeführte Rechnung ergibt für das Kontraktionsgebiet des erwähnten Hg-Bogens bei Annäherung an die Kathode eine Zunahme der Temperatur auf das 1,5fache der Säulentemperatur, während die Leuchtdichte in Übereinstimmung mit der Beobachtung auf das 10- bis 15fache zunimmt. Auch der aus dem Feldstärkeverlauf berechnete Potentialabfall über dem Kontraktionsgebiet setzt sich mit dem über der Raumladungszone liegenden Potentialfall zu einem Kathodenfall der richtigen Größenordnung (≈ 15 V) zusammen. Dabei wurde zunächst der Betrag der maximalen Kontraktion vor der Kathode aus dem STEENBECKschen Minimumprinzip bestimmt, nach dem ein Bogen sich stets so einstellt, daß er mit der kleinstmöglichen Brennspannung brennt.

Eine nicht unwesentliche Verbesserung erzielte ECKER noch durch Berücksichtigung der Tatsache, daß aus dem kathodenseitigen Ende der kontrahierten Säule thermische Elektronen gegen das Feld anlaufend in das Raumladungsgebiet eindringen und die Raumladung der positiven Ionen teilweise kompensieren. Dadurch kann sich je nach den Bedingungen der Potentialabfall über dem Raumladungsgebiet auf ein Drittel bis ein Sechstel des ohne Berücksichtigung dieser thermischen Elektronen berechneten erniedrigen. Daß im Gegensatz zu den von der Kathode emittierten Elektronen die von der Säulengrenze kommenden einen so großen Einfluß auf die Raumladung ausüben, liegt an ihrer geringeren Geschwindigkeit und größeren Zahl.

Daß die Übereinstimmung zwischen Theorie und Erfahrung zunächst immer noch nicht befriedigend war, zeigte sich aus der Tatsache, daß die Stromstärke- und Druckabhängigkeit des Brennfleckverhaltens von der Theorie noch nicht richtig wiedergegeben wurde. Insbesondere konnte die Theorie den Umschlag des stabilen Brennflecks in einen auf der Kathodenoberfläche herumlaufenden Brennfleck bei Druckerniedrigung unter einen bestimmten Wert nicht erklären.

Die Behebung dieser Unstimmigkeiten gelingt ECKER [*33*] durch eine erweiterte und exaktere Berechnung der Endkontraktion vor der Kathode ohne Benutzung des ohnehin nicht bewiesenen Minimumprinzips allein aus den allgemeinen Entladungsgesetzen. Nach LANGMUIR [*116*] läßt sich nämlich einem Plasma der Ionendichte n_i und der mittleren thermischen Ionengeschwindigkeit v_i maximal die Sättigungsstromdichte

$$j_s = \frac{e\, n_i v_i}{4} \tag{107.1}$$

entziehen. Der Sättigungsstrom J_s läßt sich daher als Funktion des Gasdruckes p, der Temperatur T, des Ionisierungsgrades $\alpha(T, p)$ und des Atomgewichtes M der Ionen

$$J_s = \sqrt{\frac{3\pi^2 e^2}{16 k M}} \frac{p R_e^2 \alpha(T_R, p)}{\sqrt{T_R}} \tag{107.2}$$

schreiben. Da hier die Temperatur T_R am kathodenseitigen Ende der kontrahierten Säule als Funktion der Endkontraktion R_e ebenso wie der Ionisierungsgrad α als Funktion von T_R und p berechenbar sind, läßt sich auch J_s als Funktion der Endkontraktion R_e berechnen. Das ist in Fig. 95 für drei verschiedene Drucke geschehen. Es zeigt sich, daß der der Plasmagrenze maximal zu entziehende Ionensättigungsstrom oberhalb 0,3 mm Säulenradius wegen des bei dieser geringen Kontraktion sehr geringen Ionisierungsgrades äußerst klein ist, bei einem Radius von etwas weniger als 0,1 mm durch ein Maximum geht und bei noch stärkerer Kontraktion nach Gl. (107.2) wegen des abnehmenden Radius und der zunehmenden Temperatur bei sich wenig änderndem Ionisierungsgrad wieder abnimmt.

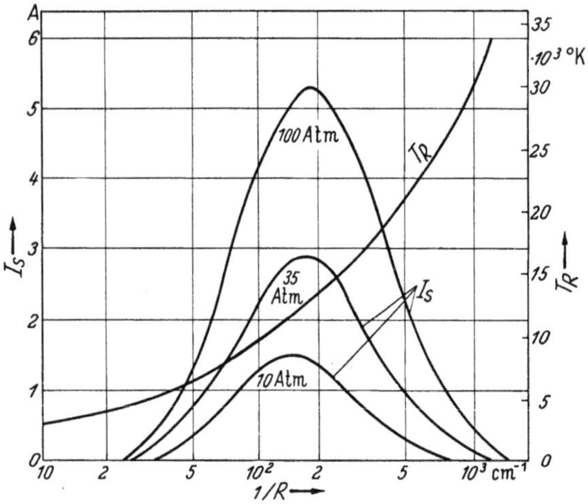

Fig. 95. Verlauf der Plasma-Temperatur vor der Kathode als Funktion der Endkontraktion (T_R) sowie Verlauf des Sättigungsstromes als Funktion der Endkontraktion für drei verschiedene Dampfdrucke eines Hg-Bogens, nach ECKER [33].

Aus diesen Kurven und ähnlichen für verschiedene Bogenstromstärken berechneten zieht ECKER eine Reihe wichtiger Schlüsse. Obwohl er darauf hinweist, daß die erwähnten der Rechnung zugrunde liegenden Vernachlässigungen einen quantitativen Vergleich von Theorie und Experiment unmöglich machen, folgt aus Fig. 95, daß ein Brennfleckbogen nur in dem Kontraktionsbereich brennen kann, in dem der Ionensättigungsstrom J_s des Plasmas einen wesentlichen Teil des Gesamtstromes (von im durchgerechneten Fall 6 Amp) zu decken vermag. Ein Brennfleckbogen muß also im Bereich des Maximums der J_s-Kurve brennen. Bei zu geringem Gas- bzw. Dampfdruck p bleibt der maximale Ionenstrom auch im Maximum so klein, daß der geschilderte Mechanismus nicht bzw. nur mit wesentlicher Beteiligung von kathodischer Elektronenemission zu funktionieren vermag. Wichtig ist ferner, daß auch beim brennflecklosen Bogen (Ziff. 106) in Übereinstimmung mit der Beobachtung eine gewisse Säulenkontraktion vor der Kathode erforderlich ist, wenn der zur Aufheizung der Kathode benötigte Ionenstromanteil aus dem thermischen Plasma entnommen werden soll.

Hier zeigt sich eine gewisse Einseitigkeit der ECKERschen Theorie, da es nach WASSERRAB [239] durchaus möglich ist, daß bei merklicher Emission kathodischer Elektronen diese nach ihrer Beschleunigung im Kathodenfall durch Stoßionisierung in der kathodenseitigen Plasmagrenzschicht nicht-thermisch Ionen erzeugen und die Ionenergiebigkeit dieser Schicht daher in Wirklichkeit größer ist als sie nach der ECKERschen rein thermischen Theorie erscheint. Als Vorzug der verbesserten Form seiner Theorie führt ECKER noch an, daß sie nun auch die Vorgänge vor der *Anode* richtig beschreibt, wo ja die Elektronenergiebigkeit des

Plasmas für den Strom zur Anode aufkommen muß. Wegen $\sqrt{M}$ im Nenner von Gl. (107.2) ist vor der Anode aber in Übereinstimmung mit der Beobachtung eine viel geringere Kontraktion erforderlich, um die Bedingung Elektronen-Sättigungsstrom gleich Bogenstromstärke zu erfüllen.

Bisher war ausschließlich von dem räumlich stationären Brennfleck des Hochdruckbogens die Rede. Nach ECKER beginnt dieser aber bei Unterschreitung einer gewissen Mindeststromstärke (bei seinem Bogen 4 Amp) auf der Wolframkathode zu wandern. Ferner werden im Übergangsgebiet zwischen stationärem Brennfleckbogen und brennflecklosem Bogen auf der gleichen Kathode häufig mehrfache Kathodenbrennflecke gefunden. ECKER deutet beide Erscheinungen als beginnende kathodische Instabilität. Er zeigt insbesondere, daß bei zu geringer Bogenstromstärke der Kontraktionsmechanismus den notwendigen Ionenstrom nicht mehr liefern kann, für die Existenz des Bogens also nunmehr eine Erhöhung der kathodischen Elektronenemission notwendig ist. Diese erzeugt der Bogenansatz durch Wanderung zu Mikrospitzen der Kathode mit erhöhter Feldelektronenemission.

Im Gegensatz zu ECKER faßt BAUER [*3*] das Problem des Brennfleckbogens von der Seite der kathodischen Elektronenemission her an. Unabhängig von ECKER berechnet er fast gleichzeitig die Elektronenergiebigkeit einer Wolframkathode verschiedener Temperatur als Funktion der effektiven Feldstärke vor der Kathode (Fig. 94). Wegen der im allgemeinen vorhandenen Rauhigkeit der Oberfläche ist die effektive Feldstärke größer als die aus der Kathodenfallspannung und der zu einer freien Weglänge angenommenen Kathodenfalldicke berechnete. Aus der Potentialtheorie berechnet er weiter für einen ambipolaren Trägerstrom die dem LANGMUIRschen $U^{\frac{3}{2}}$-Gesetz entsprechende Elektronen-Sättigungsstromdichte, wobei er allerdings Annahmen über den Ionenstromanteil vor der Kathode und den Rauhigkeitsgrad der Kathodenoberfläche machen muß. Da ein stationärer Bogen nur mit solchen Werten von kathodischer Feldstärke und Elektronenstromdichte existieren kann, die der Emissionsbedingung bei der Kathodentemperatur und der Raumladungsbedingung genügen, trägt BAUER den Emissionsstrom und den raumladungsbegrenzten Sättigungsstrom gegen die kathodische Feldstärke auf (Fig. 94 für Rauhigkeit Null) und findet zwei Schnittpunkte. Der untere Punkt geringer Emissionsstromdichte und Kathodenfeldstärke entspricht dem brennflecklosen Bogentyp, der obere Schnittpunkt hoher Stromdichte und Kathodenfeldstärke entspricht dem Brennfleckbogen. BAUER weist darauf hin, daß der untere Schnittpunkt nur realisierbar ist, wenn der zugehörige Ionenstromwert nach der ECKERschen Bedingung (Fig. 95) von der Säule geliefert werden kann. Ist das nicht der Fall, so existiert nur ein Brennfleckbogen. In einem solchen Fall gelangt man aber gemäß Fig. 94 durch Temperaturerhöhung der Kathode zu einer höher liegenden Emissionsstromkurve, bei der nunmehr auch der untere, dem brennflecklosen Bogen entsprechende Schnittpunkt physikalisch realisierbar ist. BAUERs Theorie liefert also für den Bogen zwei Existenzbereiche, aber keine Bedingung dafür, wann ein Umschlag erfolgen muß.

108. Zur Theorie des Überganges zwischen Brennfleckbogen und brennflecklosem Bogen. Schon vor den Arbeiten von ECKER und BAUER haben WEIZEL und THOURET [*249*] den Versuch gemacht, die in Ziff. 98 besprochenen experimentellen Befunde über den Umschlag des Brennfleckbogens in den brennflecklosen Bogen theoretisch verständlich zu machen. Sie stellten die Frage, wie der brennflecklose Bogen ohne starke Kontraktion vor der Kathode mit einem ebenso geringen Kathodenfall auskommt wie der Brennfleckbogen, dessen niedriger

Kathodenfall nur durch die starke Kontraktion möglich ist. Als Ergebnis der notwendigerweise noch zahlreiche Vernachlässigungen enthaltenden Rechnung finden sie, daß dies nur möglich ist, und der Umschlag in den brennfleckiosen Bogen folglich erst erfolgen kann, wenn bei zunehmender Kathodenerhitzung das Verhältnis von Elektronen- zu Ionenstrom vor der Kathode dem Verhältnis der Trägerbeweglichkeiten in der Raumladungsschicht hoher Feldstärke gleichkommt, d.h. für

$$\frac{j^-}{j^+} \geq \frac{b^-}{b^+} = \sqrt{\frac{\lambda_e}{\lambda_i}}\sqrt{\frac{M}{m}}, \qquad (108.1)$$

wo M und m wie üblich die Massen der Ionen und der Elektronen sind. Während also beim Brennfleckbogen die Verhältnisse im Fallgebiet selbst bei einem Elektronenstromanteil von 90% noch durch die Raumladung der Ionen wegen deren geringer Beweglichkeit bestimmt bleibt, werden oberhalb des durch Gl. (108.1) gegebenen Elektronenstromanteils die Elektronen raumladungsbestimmend. Da das Verhältnis (j^-/j^+) für Quecksilber bei 40 liegt, muß der Ionenstromanteil vor der Kathode nach WEIZEL und THOURET also unter 2,5% sinken, bevor der Umschlag in den brennfleckiosen Bogen erfolgt, der dann keine starke Kontraktion mehr benötigt. Ein so hoher Elektronenstromanteil wurde (vgl. Ziff. 106) früher für unmöglich gehalten. WEIZEL und THOURET vertreten aber die Ansicht, daß wegen der großen räumlichen Ausdehnung des Kathodenfallgebietes beim brennflecklosen Bogen die gesamte von den Elektronen im Kathodenfall gewonnene Energie mit Ausnahme der in Strahlung umgesetzten durch mancherlei Prozesse schließlich im Gas dissipiert und durch Wärmeleitung zur Kathode transportiert wird. Im Gegensatz zu der früher allgemein vertretenen Auffassung (vgl. z.B. ENGEL-STEENBECK [*41*]) soll also die Energiebilanz der Kathode nicht mehr maßgeblich durch die auf sie auftreffenden Ionen, sondern durch den Rückstrom der von den Elektronen im Kathodenfall aufgenommenen Energie bestimmt sein.

Mit Hilfe dieser im einzelnen mangels experimenteller Kenntnis vieler zugrunde liegender Prozesse und Daten quantitativ schwer nachprüfbaren Theorie gelingt es WEIZEL und THOURET nun, die in Ziff. 98 angeführten experimentellen Ergebnisse über den Umschlag zu deuten. Insbesondere erklärt sich, wie übrigens auch aus BAUERs späterer Theorie, die Hysterese des Umschlages leicht aus der thermischen Trägheit der Kathode in Verbindung mit deren Energiebilanz, während der Unterschied der Umschlagstromstärken bei Quecksilber und Xenon darauf zurückgeführt wird, daß bei ersterem nach dem Energieniveauschema der Energieverlust durch Strahlung viel größer ist als bei letzterem.

109. Zur Kathodenfalltheorie der Bögen mit nicht-stationärem Brennfleck auf verdampfender Kathode. In Ziff. 98 wurde gezeigt, daß die äußere Erscheinung wie das Verhalten des Kathodenbrennfleckes im Fall der von der ECKERschen Theorie beschriebenen Hochdruckbögen mit stationärem Brennfleck und der Niederdruckbögen mit schnell beweglichem Brennfleck und meist verdampfender Kathode äußerst verschieden ist. Die Brennfleckstromdichten liegen im letzteren Fall um eine Größenordnung über denen der stationären Brennflecke. Brennfleck-Wanderungsgeschwindigkeiten von 10^4 cm/sec, Vielfachheit der Brennflecke mit jeweils nur kleiner Lebensdauer, rückläufige Brennfleckbewegung in einem magnetischen Querfeld (Ziff. 104) und intensives kontinuierliches Spektrum sind weitere charakteristische Eigenschaften des nicht-stationären Brennfleckes, die von der Theorie verständlich gemacht werden müssen, bisher aber einer einheitlichen Deutung noch harren. Es ist auch noch nicht sicher, ob die in Ziff. 98 erwähnten schnell wandernden Kathodenbrennflecke von Niederdruckbögen mit Eisen-, Kupfer- und Kohlekathode in jeder Beziehung mit den auf flüssigen

Kathoden (Quecksilber und Alkalimetallen) beobachteten identisch sind oder nicht. Insbesondere ist der Zusammenhang der Erscheinung der schnell beweglichen nicht-stationären Brennflecke mit der Verdampfung zwar für flüssige Kathoden, nicht aber allgemein, nachgewiesen.

Der Mechanismus des nicht-stationären Brennfleckbogens scheint auf den ersten Blick von dem in Ziff. 107 behandelten des stationären Brennfleckbogens grundsätzlich verschieden zu sein, weil die Existenz des stationären Brennfleckes nach ECKER an die Bedingung eines Gas- bzw. Dampfdruckes von mehreren Atmosphären geknüpft ist, während nicht-stationäre Brennflecke auf der Kathode auch in sog. Vakuumbögen bei Hg-Dampfdrucken der Größenordnung 10^{-3} Torr beobachtet werden. Dieser Unterschied könnte sich aber als ein nur scheinbarer erweisen. Infolge der intensiven Kathodenverdampfung im Brennfleck herrscht vor diesem eine so hohe Dampfdichte, daß ein Mechanismus ähnlich dem eines stationären Brennfleckbogens nicht ausgeschlossen erscheint. Es muß aber betont werden, daß für einen solchen nicht-stationären Brennfleck auf niedrig siedender Kathode das in Ziff. 102 Gesagte über den schlecht definierten und physikalisch noch kaum bekannten „Zustand" der Grenzschicht zwischen Flüssigkeit bzw. Festkörper und Plasma in viel stärkerem Maße gilt als für jeden anderen Kathodenfleck. Es scheint daher höchst zweifelhaft, ob bzw. in welcher Näherung man von einer normalen Elektronenaustrittsarbeit, von thermischem Gleichgewicht, normaler Verdampfung und Sättigungsdampfdrucken vor der Kathodenoberfläche überhaupt sprechen darf. Man bedenke nur, daß nach den für einen nicht-stationären Quecksilber-Kathodenfleck bekannten Daten von dessen Oberfläche sekundlich 10^9 Atomlagen abgebaut werden und im wesentlichen in Form energiereicher Ionen sie wieder treffen. Das bedeutet, daß eine ganze Atomlage in einer Zeit abgebaut wird, die im allgemeinen zur Einstellung von thermischem Gleichgewicht längst nicht ausreicht, während in der gleichen Zeit eine entsprechende Zahl schneller Ionen die im Abbau begriffene Oberfläche treffen. Bei dieser Lage kann man kaum erwarten, daß eine auf konventioneller Grundlage aufgebaute Theorie die Erscheinungen richtig beschreibt und wird erst einer Theorie Vertrauen zu schenken geneigt sein, die außer der hohen Stromdichte, dem geringen Kathodenfall und dessen schneller Wanderung auch eine so auffallende Eigenschaft des wandernden Brennfleckes wie die seiner rückläufigen Bewegung im magnetischen Querfeld (Ziff. 104) zu deuten vermag.

Es ist schon von LANGMUIR [*115*] an überwiegend die Meinung vertreten worden, daß der Strom vor der Kathode nicht-stationärer Brennfleckbögen wesentlich von Feldemissionselektronen getragen würde und die in Ziff. 98 erwähnten Beobachtungen, nach denen eine Herabsetzung der Austrittsarbeit der Kathode einen Umschlag in diese Bogenform begünstigt, sprechen für diese Deutung. Es wurde aber schon erwähnt, daß wegen der hohen Dampfdichte vor der Kathode auch für den nicht-stationären Brennfleckbogen der in Ziff. 107 erwähnte Mechanismus in Frage kommt, bei dem die Raumladungsverhältnisse vor der Kathode von den aus dem Plasma auf die Kathode einfallenden Ionen bestimmt werden.

WASSERRAB [*239*] hat kürzlich darauf hingewiesen, daß Kathodenfleck-Stromdichten der von FROOME gemessenen Größe von über 10^6 Amp/cm² nach der Raumladungsgleichung [*116*] eine kathodische Feldstärke von 10^7 bis 10^8 V je Zentimeter erfordern, und daß bei dieser Feldstärke notwendig eine erhebliche Feldemission der Kathode stattfindet. Am Strom vor der Kathode müßten also Elektronen und Ionen maßgeblich beteiligt sein. Dabei erzeugen die zur Kathode strömenden Ionen vor dieser das hohe Feld, das die Elektronenemission möglich macht, und infolge der Energiezufuhr durch die Ionen steigt die Temperatur des Kathodenflecks, bis die durch Wärmeableitung und Verdampfung abgeführte Energie der Energiezufuhr gleich wird. Die aus der Kathode austretenden und im Kathodenfall beschleunigten Elektronen ionisieren nach ein bis drei freien Weglängen die aus der Kathodenfläche verdampften Hg-Atome, die den Daten nach fast 100%ig als Ionen zur Kathode zurückwandern sollen. WASSERRAB versucht dann, für einen Quecksilber-Kathodenbrennfleck, ohne Berücksichtigung der oben erwähnten Schwierigkeiten bezüglich des „Zustands" der Kathode, die Kathodenverhältnisse quantitativ zu erfassen. Zur Bestimmung der drei Unbekannten, nämlich der kathodischen Feldstärke E_K, der kathodischen Stromdichte j_K und des Verhältnisses von Ionenstrom zu Elektronenstrom (j^+/j^-) benutzt er drei Gleichungen, nämlich die Feldemissionsgleichung (106.2), die Raumladungsgleichung für einen ambipolaren Trägerstrom (105.6) und die Energiebilanz der Kathodenoberfläche. Seine Arbeit enthält einige Vernachlässigungen, durch die die zahlenmäßigen Ergebnisse etwas verfälscht werden, weshalb wir hier die in Übereinstimmung mit dem Verfasser korrigierten Werte zitieren. So ergibt sich aus den korrigierten WASSERRABschen Formeln ein Verhältnis von Ionen- zu Elektronenstrom $j^+/j^- = 1$ und mit diesem Wert die außerordentlich hohe kathodische Feldstärke von $1{,}3 \cdot 10^8$ V/cm sowie eine kathodische Stromdichte von $2 \cdot 10^7$ Amp/cm². Aus der Wärmebilanz der Kathode schließt WASSERRAB auf eine Kathodenflecktemperatur von etwa 2100° K und weiter, immer unter der zweifelhaften Annahme wirklichen thermischen Gleichgewichtes, auf einen Quecksilberdampfdruck vor der Kathode von 2000 Atm, entsprechend einer Atomdichte von etwa 40% der des flüssigen Quecksilbers! Wegen der oben erwähnten Unsicherheiten über den Zustand der Hg-Oberfläche kommt diesen Ergebnissen wohl kaum mehr als größenordnungsmäßige Bedeutung zu; letztere aber liegt in der zu erwartenden Größenordnung.

Die aus den Untersuchungen von FROOME [*58*] hervorgehende Tatsache, daß der scheinbare Kathodenbrennfleck in Wirklichkeit bei höherer Stromstärke aus einer Front zahlreicher laufend entstehender und wieder vergehender kleiner Brennflecke von näherungsweise konstanter Stromdichte besteht, ist schon von GÜNTHERSCHULZE [*74*] mit der gegenseitigen Störung des jeden Brennfleck verlassenden Dampfstrahles und der gegenläufigen Ionenbewegung in Verbindung gebracht worden, und auch die schnelle Bewegung der Flecke mit einer Maximalgeschwindigkeit von rund 10^4 cm/sec dürfte mit diesem Wechselspiel zusammenhängen, harrt aber noch der wirklichen Erklärung.

F. Anodische Erscheinungen, anodisch bestimmte Sonderformen von Lichtbögen und Anodenfallmechanismen.

I. Anodische Erscheinungen und empirische Daten.

110. Anodenfallmessungen. Die Tatsache, daß unser Verständnis der anodischen Erscheinungen und des Anodenmechanismus von Lichtbögen noch so lückenhaft ist, beruht zu einem wesentlichen Teil darauf, daß zuverlässige Messungen des Anodenfalles von Lichtbögen in Abhängigkeit von den Entladungsbedingungen immer noch fehlen. Das ist um so überraschender, als zahlreiche Autoren, von LECHER angefangen, in den letzten 70 Jahren nach den verschiedensten Methoden aus Sondenmessungen Anodenfallwerte zu bestimmen versucht haben, und zum Teil für ihre Ergebnisse sehr gute Genauigkeiten in Anspruch nehmen. Die älteren Messungen sind vor allem bei HAGENBACH [*7a*] besprochen; für neuere Daten vergleiche [*81*], [*159b*]. Wie Ziff. 17 ausführt, sind Potentialsondenmessungen in thermischen Lichtbögen von Atmosphärendruck oder höherem Druck zwar noch mit gewissen grundsätzlichen Ungenauigkeiten behaftet, geben aber doch Ergebnisse, die nicht nur den Gang etwa des Anodenfalles bei Veränderung gewisser Parameter recht genau wiedergeben, sondern auch im Absolutwert auf besser als ± 2 V richtig sind. Die sehr viel größeren Diskrepanzen zwischen den Meßergebnissen verschiedener Autoren in älterer wie in jüngster Zeit dürften daher nicht methodisch bedingt sein, sondern reale physikalische Ursachen haben. Wir wissen aber noch nicht, von welchen Parametern der mit Potentialsonden gemessene Anodenfall abhängt.

Beim Niederstromkohlebogen zeigen alle vorliegenden Anodenfallmessungen Häufungspunkte bei 36 und bei 20 V und ausgedehnte unveröffentlichte Messungen von FINKELNBURG haben auch immer wieder auf Werte von entweder (20 ± 1) oder (36 ± 3) Volt geführt. Bei manchen Messungen mit der in [*53*] beschriebenen Methodik schien es, als setze sich der Anodenfall des Homogenkohle-Niederstrombogens aus zwei räumlich etwas getrennten Anteilen von rund 20 und 16 V zusammen, und dieser Befund ergab sich sehr deutlich bei Messungen des Anodenfalles des in Ziff. 116 zu besprechenden BECK-Bogens. Bei diesem setzt sich der Anodenfall offenbar aus einem der Anode ganz dicht vorgelagerten stromstärkeunabhängigen Fall von etwa 20 V und einem nach der Säule zu anschließenden räumlich etwas ausgedehnteren, mit der Stromstärke stark zunehmendem zweiten Potentialabfall zusammen. Auf Einzelheiten kommen wir in Ziff. 116 zurück. Beim Niederstromkohlebogen mit Metallsalze enthaltender Anode findet man nach MATHIESEN u.a. [*7a*] sehr viel kleinere und mit zunehmender Stromstärke (gemessen von 10 bis etwa 40 Amp) von 16 auf etwa 10 V abnehmende Anodenfälle, was mit der Tatsache zusammenhängen dürfte, daß solche Anoden selbst positive Ionen zu emittieren vermögen, die dann nicht im Anodenfallgebiet erzeugt zu werden brauchen. Messungen über die räumliche Ausdehnung des Anodenfallgebietes scheinen bisher nur für den Niederstromkohlebogen mit Sonden ausgeführt worden zu sein, wo BLOCK und FINKELNBURG [*12*] eine axiale Ausdehnung von nur etwa einer Elektronenweglänge gefunden haben.

Bei Niederstrombögen mit Metallanode oder Metalloxydanode liegen die meist in Luft bzw. den entsprechenden Metalldämpfen gemessenen Anodenfallwerte im allgemeinen zwischen 3 und 12 V. Dabei ist von besonderem Interesse eine alte Untersuchung von SCHULZE [*203*], in der die Anodenfallwerte einer großen Anzahl von Metallen bzw. Metalloxyden in einem 4 Amp-Bogen mit Kohlekathode gemessen wurden. SCHULZE fand einen ausgeprägten Zusammenhang zwischen den gemessenen Anodenfallwerten und den Ionisierungsspannungen der betreffenden Metalle, und zwar in dem Sinne, daß zwar die Anodenfallwerte um 20 bis 100% über den Ionisierungsspannungen lagen, innerhalb jeder Gruppe des periodischen Systems aber der Gang der Anodenfallwerte dem der Ionisierungsspannungen entsprach. Dieses Ergebnis stimmt mit der in Ziff. 119 behandelten Theorie schön überein.

Beim Übergang zu höheren Stromstärken und anscheinend auch höheren Drucken findet man im allgemeinen eine Abnahme der Anodenfälle, die in Ziff. 120 mit dem Übergang von der Feldionisierung zur thermischen Ionisierung im Anodenfallgebiet erklärt werden wird. Ausnahmen bei höheren Drucken können auf ausschließlich thermische Ionisierung zurückgeführt werden. Bei den sonst so gut untersuchten Quecksilber-Höchstdrucklampen scheinen Anodenfallwerte nicht gemessen zu sein, doch liegt nach ELENBAAS [*37*] die Summe von Kathoden- und Anodenfall bei 15 V, woraus man wohl auf einen Anodenfall von (6 ± 2) Volt schließen kann. Überraschend ist demgegenüber, daß BAUER und SCHULZ [4] an einem 30 Atm-Xenonbogen im Stromstärkebereich von 3 bis 9 Amp nach einer Energiebilanzmethode einen fast konstanten Anodenfall von nur ungefähr 1 V gemessen haben. Ähnlich kleine Anodenfallwerte sind von den in Ziff. 31 behandelten nicht-thermischen Niederdruck-Quecksilberbögen und Niedervoltbögen her bekannt.

Im Bereich hoher Stromstärken haben BUSZ-PEUCKERT und FINKELNBURG [22] an dem in Ziff. 26 erwähnten Stickstoffbogen einen im Stromstärkebereich von 50 bis 200 Amp linear von 17 auf 2 V abfallenden Anodenfall gemessen. Bei dem entsprechenden Argon-Hochtemperaturbogen hängt der Anodenfall zusätzlich von der Bogenlänge ab, und zwar nimmt er bei Vergrößerung der Bogenlänge, und d.h. hier auch Verminderung der anodischen Stromdichte und Plasmatemperatur, zu. Gegen die Plasmatemperatur vor der Anode aufgetragen nimmt der Anodenfall gemäß Fig. 18 von 5 V bei 16000° auf etwa 1 V bei 20000° ab. Als Beispiel für die unter Umständen recht komplizierte Abhängigkeit des Anodenfalles zeigt Fig. 96 den Verlauf des Anodenfalles des Argonbogens von 10 mm Länge im gesamten Stromstärkebereich von 6 bis 200 Amp. Der nicht-monotone Verlauf hängt hier mit der Tatsache zusammen, daß der Argonbogen unterhalb 30 Amp an der Anode kontrahiert ansetzt, bei höheren Stromstärken dagegen nicht kontrahiert. Beim 200 Amp-Hochstromkohlebogen ohne Anodenverdampfung schließlich liegt der Anodenfall im Gegensatz zu älteren Messungen nach MÜLLER und FINKELNBURG [159] bei (1 ± 2) Volt.

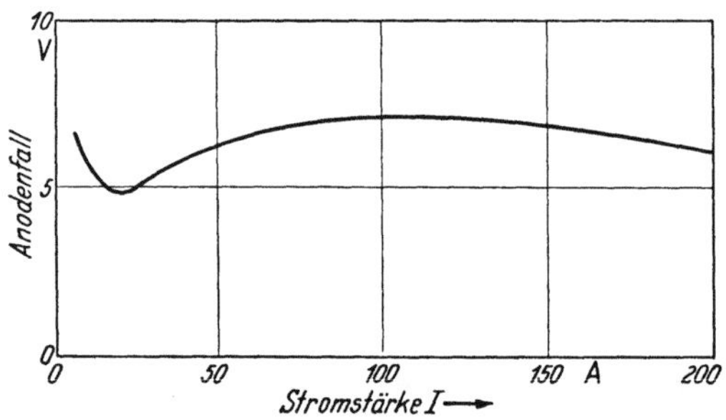

Fig. 96. Abhängigkeit des Anodenfalles eines atmosphärischen Argon-Bogens von der Stromstärke; nach BUSZ-PEUCKERT und FINKELNBURG [22].

111. Anoden- und Anodenplasmatemperatur. Soweit man bisher übersehen kann, hängt der Anodenfall eines Lichtbogens einerseits vom Material und wohl auch der Temperatur der Anode, und andererseits von der Zusammensetzung und Temperatur des der Anode vorgelagerten Plasmas ab. Wir erwarten also einen verschiedenen Anodenfall für eine Cu-Anode in Luft und eine solche in Kupferdampf. Dabei dürfte besonders bei höheren Stromstärken, d.h. im Bereich der in Ziff. 120 noch zu besprechenden thermischen Ionisierung im Anodenfall, die Temperaturverteilung im Plasma unmittelbar vor der Anode die entscheidende Rolle spielen. Leider liegen zuverlässige Messungen noch nicht vor.

Beim Niederstromkohlebogen mit Homogen- oder Reinkohleanode ist nach EULER [45] die Anodentemperatur (4000 ± 15)° K, und dieser dicht unterhalb des Siedepunktes des Kohlenstoffs bei Atmosphärendruck liegende Wert kann auch bei anderen Kohlebögen nicht wesentlich überschritten werden. Lediglich bei hohen Drucken kann die Anodentemperatur wegen der Zunahme des Sublimationspunktes mit dem Druck merklich höher liegen. Ältere Ergebnisse von LUMMER [132], dessen angebliche Kratertemperaturen bis 7000° K 1914 Aufsehen erregten, und MATHIESEN [10a] scheinen aus verschiedenen Gründen wenig Vertrauen zu verdienen, während aus neuerer Zeit nur eine vorläufige Messung von FINKELNBURG [6a] vorliegt, der für die Anodentemperatur eines 10 Atm-Niederstromkohlebogens in Luft 5000° K angibt. Über die Anodentemperatur von Kohlebögen mit Metallsalzen im Docht ist nichts Sicheres bekannt, doch dürfte sie niedriger liegen als bei ungesalzenen Kohleanoden. Bei Metallbögen ist die Anodentemperatur natürlich durch den Schmelzpunkt der Metalle begrenzt und damit einigermaßen festgelegt.

Die Plasmatemperatur vor der Anode ist wesentlich durch Leistungsumsatz, Wärmeleitung und Ausstrahlung des Plasmas bestimmt und hängt deshalb stark davon ab, ob die Säule vor der Anode kontrahiert oder nicht. Zuverlässige Messungen fehlen noch vollkommen.

112. Die normale anodische Stromdichte beim Niederstrom-Kohlebogen. Beim gewöhnlichen Niederstrom-Kohlebogen mit Homogen- oder Reinkohleanode setzt der Bogen, solange er noch nicht zischt (vgl. Ziff. 113), an der Anode mit einer konstanten Stromdichte an, die bei den gebräuchlichen Kohlesorten und Kohledurchmessern bei 40 Amp/cm² liegt. Bei geringer Stromstärke beobachtet man daher einen kleinen, meist auf der Anodenstirnfläche umherwandernden Anodenbrennfleck, dessen Durchmesser sich bei Stromstärkesteigerung allmählich vergrößert, bis er bei immer noch konstanter Stromdichte (die wir deshalb die normale anodische Stromdichte nennen) die ganze Anodenstirnfläche bedeckt. Bei weiterer Stromsteigerung kann der Anodenbrennfleck nun entweder unter Erhaltung der normalen Stromdichte auf die Seitenflächen der Kohleanode übergreifen, oder der Bogen kontrahiert infolge der Stromsteigerung an der Anodenstirnfläche, was zu der gleich zu besprechenden Erscheinung des Zischens Anlaß gibt. Zur Erklärung der normalen anodischen Stromdichte liegt nur ein erster Versuch von SCHLUGE [*6a*] vor, während ein Deutungsversuch für den Umschlag in den kontrahierten Zischbogen erst kürzlich von HÖCKER und BEZ [*85*] gemacht worden ist. Bei Kohlebögen mit einer Kohleanode, die Metallsalzzusätze enthält, kontrahiert der Bogen an der Anode überhaupt nicht; Steigerung der anodischen Strombelastung führt hier zu einer kontinuierlichen Zunahme der anodischen Stromdichte mit allmählichem Beginn einer gleichmäßigen Anodenverdampfung. Auf die mit dieser zusammenhängenden Erscheinungen wird in Ziff. 115/116 eingegangen.

113. Die anodische Kontraktion beim zischenden Kohlebogen und der Mechanismus des Zischens. Wird bei einem Homogenkohlebogen oder Reinkohlebogen (Positivkohle mit einem im wesentlichen aus reinem Kohlenstoff bestehenden Docht) bei Stromstärkeerhöhung die normale anodische Stromdichte von meist 40 Amp/cm² merklich überschritten, so beginnt der Bogen zu zischen. Unter Absinken seiner Brennspannung um 8 bis 10 V (Fig. 14) tritt ein mit zunehmender Belastung stärker werdendes zischendes Geräusch, verbunden mit Rundfunkstörungen auf, während die Anodenstirnfläche bei unruhiger Lichtverteilung sich mit einer grünlich-bläulichen Kohlenstoffdampfschicht überzieht, die sich bei weiterer Steigerung der Strombelastung allmählich zur Anodenflamme des zischenden Homogenkohle-Hochstrombogens entwickelt. Nach WEIZEL und FASSBENDER [*246*] sowie SCHLUGE und FINKELNBURG [*188*] ist der Einsatz des Zischens mit der Ausbildung eines sich mit großer Geschwindigkeit auf der Anodenstirnfläche umherbewegenden Anodenbrennflecks verbunden. Diese anodische Kontraktion bleibt aus, wenn das Kohlematerial Metallsalz enthält, hängt also wohl mit der Tatsache zusammen, daß Metallsalz-Kohleanoden positive Ionen zu emittieren vermögen, reine Kohleanoden dagegen nicht.

Die Zischerscheinungen hängen von der Art des Kohlematerials ab, und zwar tritt das Zischen am gleichmäßigsten bei Homogenkohlen von hohem spezifischen Widerstand und kleiner Wärmeleitfähigkeit auf. Beim Zischen handelt es sich also, im Gegensatz zu älteren Anschauungen, eindeutig um anodische, mit der Kontraktion des Anodenfußpunktes zusammenhängende

Erscheinungen und nicht um Oxydationsvorgänge. Von diesem Zischen zu unterscheiden ist eine Anzahl von Vorgängen, die mit mehr knatterndem als zischendem Geräusch verbunden sind und besonders bei größerer Bogenlänge und plötzlicher Überlastung auftreten. Bei diesen Vorgängen handelt es sich nicht um anodische, sondern um Säulen- bzw. Flammenvorgänge, insbesondere um das periodische Zusammenschlagen der zu zwei flammenartigen Gebilden zerreißenden Hälften der Säule eines sog. Flammenbogens (vgl. Ziff. 15).

Je nach der verwendeten Kohlensorte der Anode liegt die hörbare Zischfrequenz zwischen 1000 und 2000 Hz; sie ist bei gleicher Stromstärke um so größer, je größer die Wärmeleitfähigkeit des Kohlematerials ist, und steigt allgemein mit der Stromstärke. Das Zischen ist verbunden mit Spannungsschwankungen, die je nach

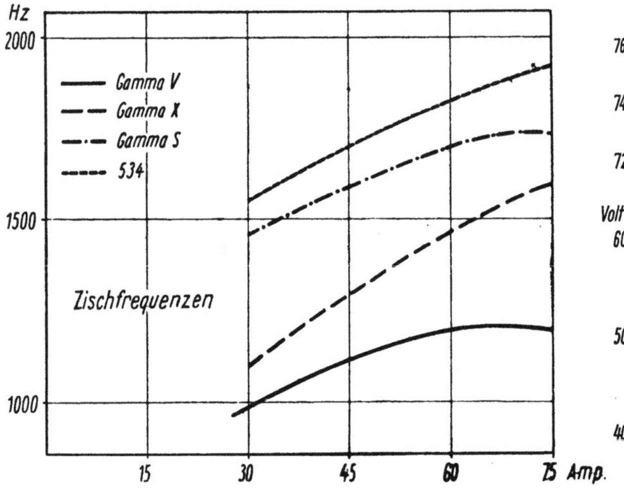

Fig. 97.

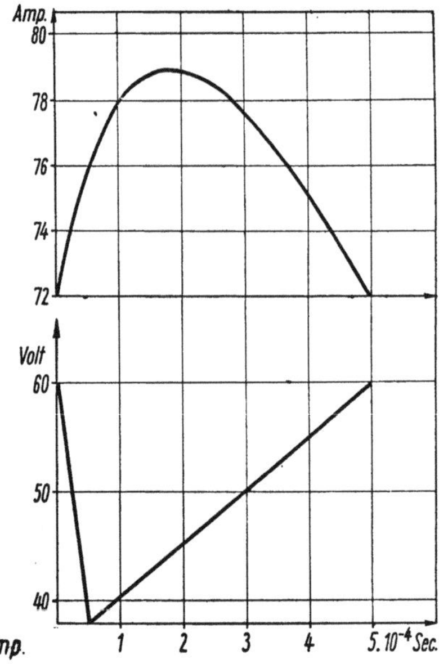

Fig. 98.

Fig. 97. Abhängigkeit der Zischfrequenz eines Kohlebogens mit 7 mm-Positivkohle von der Stromstärke für verschiedene Kohlematerialien, nach SCHLUGE und FINKELNBURG [*6a*], [*188*].

Fig. 98. Zeitlicher Verlauf der Stromstärke (oben) und der Brennspannung (unten) während einer Zischperiode des stark zischenden Homogen-Kohlebogens, nach SCHLUGE und FINKELNBURG [*6a*], [*188*].

Kohlesorte und Stromstärke zwischen 5 und 30 V liegen und mit Stromstärkeschwankungen zwischen 3 und 13 Amp. Fig. 98 zeigt den Zusammenhang von Spannung und Stromstärke während einer Periode für den Fall hoher Stromstärke. Nach SCHLUGE [*188*] ist mit jedem Spannungszusammenbruch ein Maximum der Schallstärke verbunden. SCHLUGE fand ferner, daß dem hörbaren Zischen noch höherfrequente elektrische Schwankungen mit Amplituden bis zu 3 V und 0,5 Amp bei Frequenzen von 50 bis 80 kHz überlagert sind, und daß diese mit der Ausbildung eines im scheinbaren Anodenbrennfleck umherlaufenden Mikrobrennflecks verbunden sind. In diesem Mikrobrennfleck fand er Stromdichten bis zu 50000 Amp/cm^2.

Während die Bewegung des mit Geschwindigkeiten bis zu $3 \cdot 10^4$ cm/sec sich bewegenden Mikrobrennfleckes anscheinend mit den Dampf-Eruptionen der Anode als Folge der hohen anodischen Stromdichte zusammenhängt, ist das mit zunehmender Belastung immer regelmäßiger werdende niederfrequente Zischen mit Bewegungen des Anodenfußpunktes über die ganze Anodenoberfläche verbunden. Besonders beim Bogen in Winkelstellung konnte auf Zeitlupenaufnahmen verfolgt werden, wie der Bogen (vgl. Fig. 99) im Rhythmus des Zischens von Stellung Nr. I in Stellung II wandert, worauf er an der spannungsgünstigeren Stellung I wieder zündet usw. Alle Einzelheiten dieses Vorganges konnten durch die gleichzeitige Untersuchung der elektrischen, akustischen und optischen Zischerscheinungen geklärt werden. Zu dem Mechanismus paßt auch der Befund, daß ein die Wanderungsgeschwindigkeit von I nach II vergrößerndes äußeres Magnetfeld zu einer Frequenzerhöhung des Zischens führte. LIST und JONES [*126*] haben in Übereinstimmung mit einer älteren Beobachtung von WEIZEL gefunden, daß kurz vor Einsatz des eigentlichen Zischens periodische Schwankungen von Stromstärke und Spannung mit Frequenzen von 50 bis 400 Hz auftreten, die mit der Ausbildung eines kontrahierten auf der Anodenstirnfläche kreisenden Anodenfußpunktes verbunden sind.

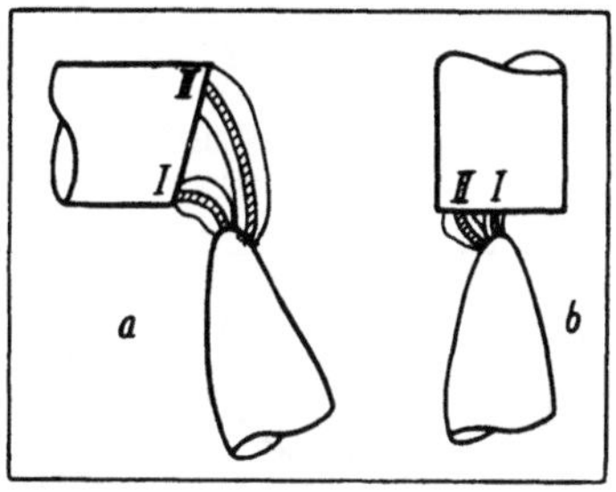

Fig. 99. Schematische Darstellung der Bewegung des anodischen Säulenfußpunktes auf der Anodenoberfläche beim zischenden Kohlebogen, die zu den periodischen Brennspannungs- und Stromstärkeänderungen Anlaß gibt; nach SCHLUGE und FINKELNBURG [*6a*], [*188*].

114. Anodische Stromdichte und anodische Kontraktion an Metallanoden. Über die physikalischen Verhältnisse an der Anode von Metallbögen ist wenig bekannt. Der Erscheinungsform nach kann man Bögen mit und ohne Kontraktion deutlich unterscheiden, ohne daß bisher genauer untersucht wäre, unter welchen Bedingungen Kontraktion auftritt bzw. nicht auftritt. Im allgemeinen scheinen in Luft oder anderen oxydierenden Gasen brennende Metallbögen an der Anode mit einem ziemlich stark kontrahierten Brennfleck anzusetzen, der häufig mit großer Geschwindigkeit auf der Anodenoberfläche umherläuft und nicht selten infolge der relativ hohen Stromdichte (Größenordnung 10^3 Amp/cm²) zu Dampferuptionen aus der Anode Anlaß gibt. Messungen der anodischen Stromdichte solcher Metallbögen scheinen nicht ausgeführt worden zu sein. An einem in Luft zwischen einer Kohlekathode und einer schnell bewegten Metallbandanode bei Stromstärken zwischen 40 und 400 Amp brennenden Bogen haben JONES und KOUWENHOVEN [*110*] kürzlich aus der Größe der anodischen Brennspuren auf die anodische Stromdichte geschlossen und je nach Material, Stromstärke und Geschwindigkeit der Anodenbewegung Werte zwischen 2000 und 13000 Amp/cm² gefunden. Im Gegensatz zu diesen Bögen zeigen Metallbögen in Edelgasen und reinem Stickstoff nur bei kleinen Stromstärken ($I < 30$ Amp) eine anodische Kontraktion mit einer Stromdichte von mehreren hundert Ampere je Quadratzentimeter, bei höheren Stromstärken dagegen keine anodische Kontraktion, sondern einen mehr oder weniger gleichförmigen diffusen Ansatz an der Anode [*21*]. Bei einem Xenon-Hochdruckbogen geringer Stromstärke folgen aus Daten von BAUER und SCHULZ [*4*] anodische Stromdichtewerte, die je nach Bogenlänge und Stromstärke zwischen 40 und 400 Amp/cm² liegen. Im Argon-Hochtemperaturbogen wurden je nach Bogenlänge und Stromstärke (100 bis 500 Amp) anodische Stromdichten zwischen 200 und 1000 Amp/cm² gemessen, in Stickstoffhochtemperaturbögen solche von 500 bis 600 Amp je Quadratzentimeter [*20*].

II. Anodisch bestimmte Lichtbögen und ihre Eigenschaften.

115. Metallsalz-Kohlebögen ohne anodische Kontraktion und anodische Hochstromkohlebögen. Steigert man die mittlere Stromdichte der Anodenstirnfläche eines Homogenkohlebogens über die normale Stromdichte (Ziff. 112) hinaus, so beobachtet man den in Ziff. 113 besprochenen Übergang zu der mit anodischer Kontraktion verbundenen zischenden Form des Kohlebogens. Im Gegensatz dazu läßt sich nach BECK [5] die anodische Stromdichte eines Kohlebogens, dessen Anode mit Metallsalz (meist in Dochtform) versetzt ist, ohne Kontraktion und Zischen fast beliebig steigern. Bei positiven Dochtkohlen bildet sich dabei wegen der leichteren Verdampfbarkeit des Dochtmaterials ein tiefer positiver Krater aus, während nur beim Entladungsansatz am Homogenkohlemantel gelegentlich Zischen auftreten kann. Nach Ziff. 16 zeigt die Strom-Spannungscharakteristik eines solchen Bogens zunächst den üblichen Abfall, um nach Überschreiten der normalen Stromdichte positiv zu werden. Das elektrische und optische Verhalten solcher Bögen ist besonders für den Fall der Oxyde und Fluoride des Cers und anderer Seltener Erden im positiven Docht enthaltenden BECK-Bögen von FINKELNBURG und Mitarbeitern [6a] untersucht worden. Im steigenden Teil der Charakteristik wurde bei einer anodischen Stromdichte, die je nach dem Anodendurchmesser zwischen 80 und 120 Amp/cm² liegt, ein Knick (vgl. Fig. 15) gefunden. Gleichzeitig schob sich aus dem positiven Krater eine mit zunehmender Anodenbelastung immer länger werdende hell leuchtende „Anodenflamme" hervor, die sich spektroskopisch als Anodenmaterialdampf mit intensiver Strahlung der Atome und teilweise auch der Ionen des Cers und der Seltenen Erden erwies. Fig. 15 zeigt, daß es sich bei dem erwähnten „Knick"

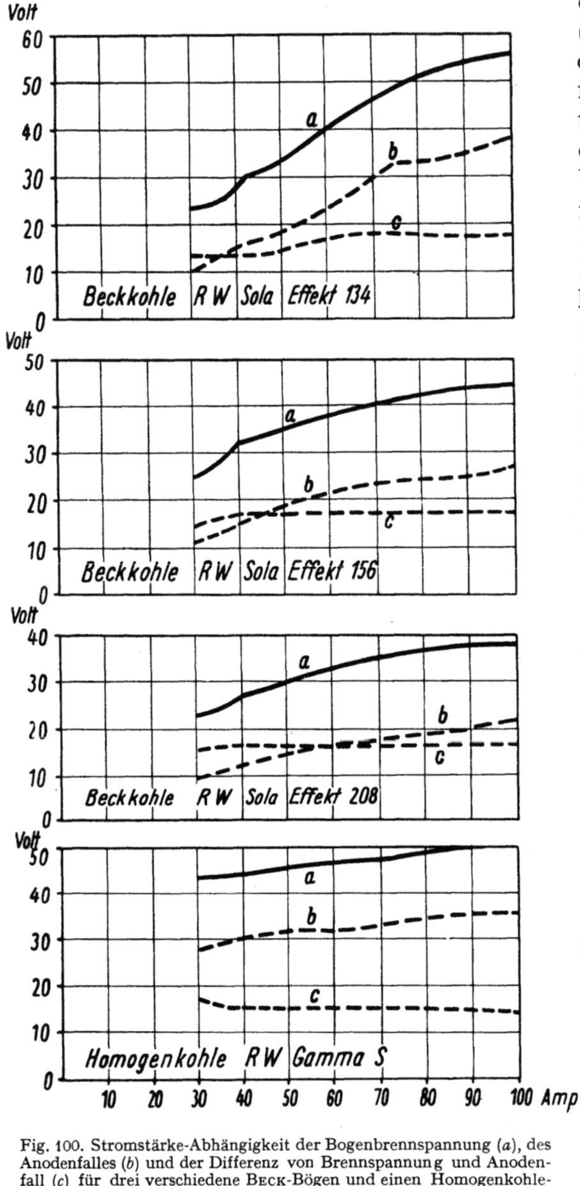

Fig. 100. Stromstärke-Abhängigkeit der Bogenbrennspannung (*a*), des Anodenfalles (*b*) und der Differenz von Brennspannung und Anodenfall (*c*) für drei verschiedene BECK-Bögen und einen Homogenkohle-Hochstrombogen, nach FINKELNBURG und SCHLUGE [6a], [51].

der Charakteristik in Wirklichkeit um einen statistisch erfolgenden Übergang zwischen zwei verschiedenen Brennformen des Bogens handelt, und die Beobachtung bestätigte, daß stets zur oberen Kurve die weiße Flamme (BECK-Effekt), zur unteren aber der gewöhnliche Niederstrombogen ohne Dampfemission gehört. Messungen an Bögen verschiedener Länge sowie an solchen mit verschiedenen Kathoden und Anoden ergaben, daß die Brennspannungszunahme mit der Stromstärke vor der Anode lokalisiert ist, und Potentialsondenmessungen bestätigten, daß die für einen frei brennenden Kohlebogen ungewöhnliche steigende Charakteristik des anodischen Hochstromkohlebogens ihre Ursache in einem mit der Stromstärke zunehmenden anomalen Anodenfall hat. Wie in Ziff. 17 schon erwähnt, ergaben unveröffentlichte Sondenmessungen mit verfeinerter Methodik, daß dieser anomale Anodenfall des voll entwickelten BECK-Bogens anscheinend aus einem konstanten Fall von etwa 20 V über der letzten freien Weglänge vor der Anode und einem ihr vorgelagerten stromstärkeabhängigen über ein etwas ausgedehnteres Gebiet sich erstreckenden Potentialabfall besteht. Fig. 100 zeigt Messungen der Brennspannung und des gesamten Anodenfalles als Funktion der Stromstärke für verschiedene BECK-Bögen. Gleichzeitige Messungen des Abbrandes der entsprechenden Positivkohlen sowie der von den Kraterdämpfen emittierten Strahlung ergeben einen so engen Zusammenhang von Anodenfall, Anodenverdampfung und Kraterstrahlung, daß man das gesamte vom gewöhnlichen Kohlebogen so abweichende Verhalten dieser überlasteten anodisch nicht kontrahierten Metallsalzbögen als durch den Anodenmechanismus bestimmt ansehen muß. Während man speziell als BECK-Bogen die besondere lichttechnisch interessierende Form dieses Bogens mit Cersalzen im Docht bezeichnet, möchten wir die Gesamtheit dieser nicht kontrahierenden sich sehr ähnlich verhaltenden Kohlebögen mit Anodenflamme und steigender Charakteristik deshalb als ***anodische Hochstromkohlebögen*** bezeichnen. Als Grenzfall gehört zu ihnen auch der stark überlastete zischende Kohlebogen.

116. Allgemeine Eigenschaften des BECK-Bogens. Fig. 11 zeigt einen voll entwickelten 200 Amp-BECK-Bogen mit seinen verschiedenen Entladungsteilen. Es handelt sich also um einen teilweise in Luft und teilweise im Anodenmaterialdampf brennenden Lichtbogen. Daß trotz der Drift im elektrischen Feld die Cer-Ionen aus dem dampferfüllten positiven Krater nicht in die eigentliche kontrahierte Luftsäule gelangen, dürfte auf dem in Ziff. 94 behandelten MAECKERschen Plasmastrahl beruhen, der die feldbestimmte Driftgeschwindigkeit überkompensiert. Wegen des außerordentlichen Linienreichtums der Cer-Ionen und -Atome im sichtbaren (und ultravioletten) Spektralgebiet strahlt der den positiven Krater erfüllende Anodendampf ebenso wie die aus ihm abströmende Anodenflamme sehr viel intensiver als der in Luft verlaufende Teil der Bogensäule. Da bei nicht zu langen Bögen der größte Teil der Bogenleistung im Cerdampf umgesetzt wird, ist die Strahlungsausbeute des BECK-Bogens mit 73% der gesamten Bogenleistung sehr hoch [*6a*]. Die meisten eingehenderen Strahlungsmessungen beziehen sich auf den von vorn beobachteten, von Cerdampf erfüllten positiven Krater, weil dieser als Lichtquelle eine erhebliche technische Bedeutung besitzt. Seine Gesamtstrahlungsdichte kann nach Fig. 101 bis nahezu 6000 W je cm^2 scheinbarer Krateroberfläche gesteigert werden und damit die Gesamtstrahlungsdichte der Sonnenoberfläche erreichen. Entsprechend mißt man mit maximal 200000 Stilb auch Kraterleuchtdichten, die die der Sonne noch übersteigen und berechnet aus der Gesamtstrahlungsdichte nach dem STEFAN-BOLTZMANNschen Gesetz schwarze Temperaturen des BECK-Kraters von 4500 bis 5500° K. Fig. 102 zeigt die mit einem Doppelmonochromator geringer Auflösung gemessene spektrale Energieverteilung der Kraterstrahlung eines mäßig und eines relativ hoch belasteten BECK-Bogens und zum Vergleich die die Kraterstrahlung eines Niederstrom-Reinkohlebogens recht gut wiedergebende PLANCK-Kurve für die schwarze Temperatur von 3700° K. Aus der Tatsache, daß die BECK-Bogenstrahlung ein steiles Maximum in der Gegend der höchsten Augenempfindlichkeit zeigt, folgt der gute lichttechnische Wirkungsgrad dieses Bogens, der über 80 Lumen/W erreicht. Fig. 102 zeigt ferner, daß das WIENsche Verschiebungsgesetz zum mindesten qualitativ erfüllt ist. In Ermangelung einer besseren Meßmethode wurde deshalb die Annahme gemacht, daß die aus der Wellenlänge des Intensitätsmaximums folgende „WIENsche Temperatur" der wahren mittleren Temperatur der Kraterdämpfe gleichgesetzt werden kann. Diese Annahme ist theoretisch sinnvoll für ein Spektrum wie das des Cers, das im Bereich des Intensitätsmaximums eine gleichförmige Liniendichte besitzt, wenn

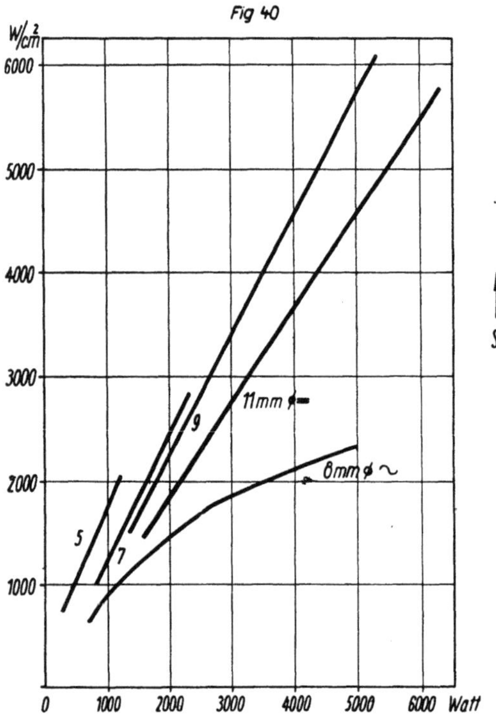

Fig. 101.

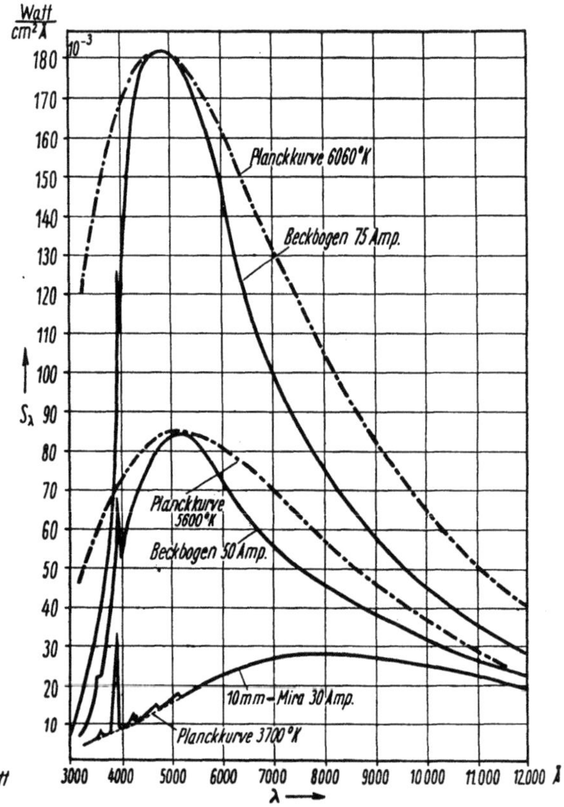

Fig. 102.

Fig. 101. Abhängigkeit der Gesamtstrahlungsdichte des positiven Kraters einiger Gleichstrom-BECK-Bögen und eines Wechselstrom-BECK-Bogens von der Bogenleistung; nach FINKELNBURG und SCHLUGE [6a], [51].

Fig. 102. Spektrale Energie-Verteilung der Kraterstrahlung eines 7 mm-BECK-Bogens für zwei verschiedene Belastungen, sowie eines 10 mm-Niederstromkohlebogens. Zum Vergleich sind die PLANCK-Kurven der schwarzen Strahler eingezeichnet. deren Strahlungsmaximum mit dem der Bögen übereinstimmt, nach FINKELNBURG und SCHLUGE [6a], [51].

die Linienintensitäten nach der PLANCKschen Formel durch die Temperatur bestimmt sind. Dafür aber liegen beim BECK-Bogen empirische Hinweise vor. Aus den nach Messungen wie in Fig. 102 bestimmten Intensitätsmaxima folgen mittlere Kraterdampf-Temperaturen zwischen 5500 und 7000° K. Da die wahre Dampftemperatur T_W mit der oben erwähnten aus der Gesamtstrahlungsdichte bestimmten schwarzen Kraterdampf-Temperatur T_S und mit dem mittleren Absorptionsvermögen der Kraterdämpfe a durch die Beziehung

$$T_S^4 = a \cdot T_W^4 \qquad (116.1)$$

zusammenhängt, folgen aus den genannten Werten für T_W und T_S Werte für das Absorptionsvermögen der 10 bis 15 mm dicken Kraterdampfschicht, die mit 0,3 bis 0,4 für ein so liniendichtes und den Messungen nach in den intensiveren

Linien schwarz strahlendes Plasma vernünftig erscheinen. Auf völlig anderer Grundlage, nämlich aus dem Verlauf der von ihm gemessenen Strömungsgeschwindigkeit längs der Anodenflamme, hat ROHLOFF [*174*] unter gewissen vereinfachenden Annahmen die Plasmatemperatur im Krater berechnet und mit Werten, die je nach der Strombelastung zwischen 6000 und 8500° K liegen, eine befriedigende Übereinstimmung mit den obigen Daten erzielt. Daß auch im stromführenden Kraterplasma die Temperaturen niedriger liegen als in der nach der Kathode zu sich anschließenden Hochstromsäule in Luft, liegt an der in Ziff. 13 schon behandelten Kühlung durch die Ausstrahlung.

Während beim Niederstromkohlebogen der „Abbrand" der Kohleelektroden stets im wesentlichen ein echter chemischer Abbau ist, erfolgt beim anodischen

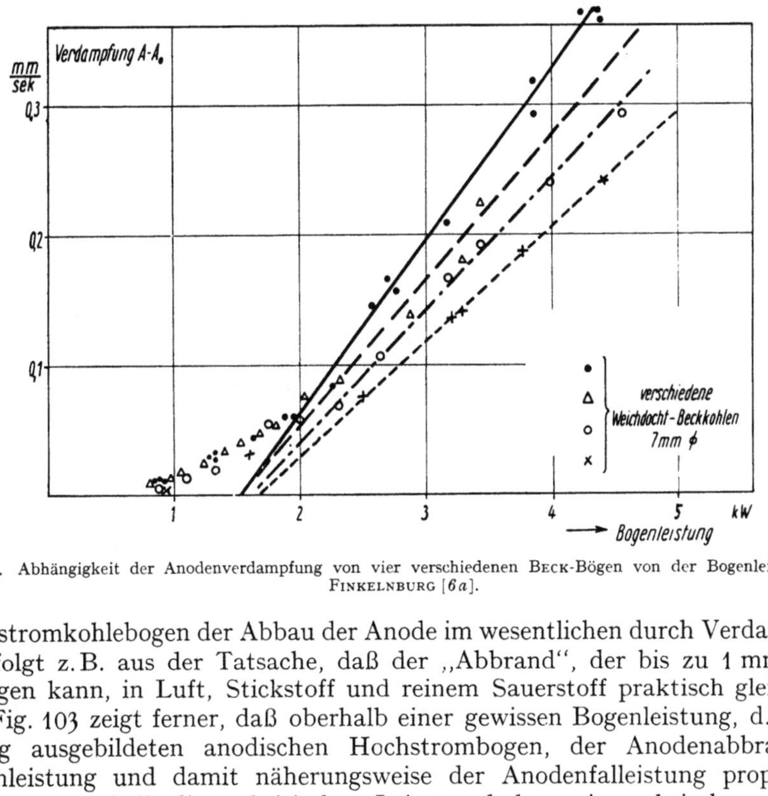

Fig. 103. Abhängigkeit der Anodenverdampfung von vier verschiedenen BECK-Bögen von der Bogenleistung, nach FINKELNBURG [*6a*].

Hochstromkohlebogen der Abbau der Anode im wesentlichen durch Verdampfung. Das folgt z.B. aus der Tatsache, daß der „Abbrand", der bis zu 1 mm je sec betragen kann, in Luft, Stickstoff und reinem Sauerstoff praktisch gleich groß ist. Fig. 103 zeigt ferner, daß oberhalb einer gewissen Bogenleistung, d.h. beim richtig ausgebildeten anodischen Hochstrombogen, der Anodenabbrand der Bogenleistung und damit näherungsweise der Anodenfalleistung proportional wächst. Unterhalb dieser kritischen Leistung haben wir praktisch noch einen anodischen Niederstromkohlebogen und entsprechend eine nur geringe Beteiligung der Verdampfung am Anodenabbrand. Damit stimmt überein, daß die Extrapolation der Meßpunkte zu sehr kleiner Bogenleistung hin auf den für Niederstromkohlebögen charakteristischen Abbrandwert von etwa 0,02 mm/sec führt. Weiter gibt die durch die extrapolierte Gerade auf der Abszisse abgeschnittene Bogenleistung den im wesentlichen konstanten Bogenleistungsbetrag an, der nicht für Anodenverdampfung und Aufheizung der Anodenflamme aufgewendet wird. Rechnet man den Anodenabbrand von mm/sec auf g/sec um und macht die nach den Messungen vernünftige Annahme, daß die mit der Stromstärke wachsende Anodenfalleistung im wesentlichen zur Verdampfung des Anodenmaterials und Aufheizung des als Anodenflamme abströmenden Dampfes auf die Plasmatemperatur von 6000 bis 8000° K dient, so folgt aus der Steigung der Kurven in Fig. 103, daß zur Verdampfung und Aufheizung auf die Kraterdampf-Temperatur je Gramm Anodenmaterial der Energiebetrag von $1{,}3 \cdot 10^5$ Wsec aufgewandt wird.

Es sei noch erwähnt, daß der BECK-Bogen auch mit Wechsel- und Drehstrom betrieben werden kann, da alle anodischen Vorgänge so schnell verlaufen, daß der Bogen sich in jedem Augenblick so verhält, wie es der jeweiligen Stromstärke entspricht. Für Einzelheiten vgl. [*6a*].

117. Die Deutung der Eigenschaften des anodischen Hochstromkohlebogens. Aus Ziff. 115 und 116 geht hervor, daß der wesentliche Unterschied des anodischen Hochstromkohlebogens gegenüber dem Niederstromkohlebogen mit nicht oder nicht merklich verdampfender Anode auf dem Anodenmechanismus beruht, während Kathodenmechanismus und Säule sich nicht von denen des gewöhnlichen Kohlebogens gleicher Stromstärke unterscheiden. Nach der in [*6a*] im einzelnen begründeten Vorstellung beginnt sich der anodische Niederstromkohlebogen in den anodischen Hochstromkohlebogen umzubilden, wenn bei gesteigerter Anodenbelastung die anodische Stromdichte den in Ziff. 112 besprochenen Normalwert überschreitet. Dann kann die der Anode zugeführte Leistung

$$L_A = J\,(U_A + U_{au}) + \text{Beträge von Strahlung und Wärmeleitung} \qquad (117.1)$$

nicht mehr vollständig durch Wärmeleitung bzw. Strahlung des glühenden Anodenendes abgeführt werden. Der mit der Stromstärke zunehmende, nicht durch Leitung und Strahlung abgeführte Leistungsanteil wird dann zur Verdampfung von Anodenmaterial und zur Aufheizung des durch den Anodenfall und die anschließenden Säulenteile hindurchströmenden Dampfes auf die Krater-Plasmatemperatur von 6000 bis 8000° K verwandt. Da die Verdampfung und Aufheizung des Anodenmaterials einer Volumenvergrößerung um den Faktor 10^5 entspricht, folgt aus einer Anodenverdampfung von 0,5 mm/sec ein den positiven Krater mit einer Geschwindigkeit von rund $5 \cdot 10^3$ cm/sec verlassender Anodendampfstrahl. Dies ist die Deutung der Anodenflamme, die folglich dadurch zustande kommt, daß die laufend neu entstehenden Dampfschichten die vorher entstandenen vor sich her von der Anodenstirnfläche fortschieben. Die relativ scharfe seitliche Begrenzung der Anodenflamme wird aus dieser Deutung noch nicht verständlich; alle übrigen Befunde dagegen passen sich ihr gut an. So nehmen Strahlungsdichte, Temperatur und nach ROHLOFF [*174*] auch Strömungsgeschwindigkeit der Anodenflamme mit zunehmendem Abstand vom Krater in der für einen sich abkühlenden Dampfstrahl zu erwartenden Weise ab. Mit zunehmender Anodenbelastung nimmt ferner die Anodenflammenlänge und mit ihr wiederum die von ROHLOFF mit einer Zeitlupenanordnung gemessenen Strömungsgeschwindigkeit zu. Diese am Kraterrand gemessenen Strömungsgeschwindigkeiten liegen mit 2 bis $7 \cdot 10^3$ cm/sec in der von FINKELNBURG vor ihrer Messung aus der obigen Deutung vorhergesagten Größe.

Es wird bei der Besprechung der Anodenfalltheorien in Ziff. 121 gezeigt, daß die mit der Stromstärke wachsende Strömungsgeschwindigkeit der Anodendämpfe durch Fortblasen der positiven Ionen von der Anode den Ionenbedarf vor dieser vergrößert und damit den gemessenen, mit der anodischen Strombelastung zunehmenden Anodenfall bewirkt. Die Kopplung zwischen anomalem Anodenfall und Anodenverdampfung scheint damit qualitativ verständlich. Da mit wachsendem Anodenfall und gesteigerter Stromdichte der Leistungsumsatz im Anodenfallgebiet stärker wächst als die Strömungsgeschwindigkeit der Anodendämpfe, werden diese mit zunehmender Stromstärke auf immer höhere Temperatur aufgeheizt, wodurch der empirische Befund seine Deutung findet, daß mit steigender Stromstärke auch Temperatur und Strahlungsdichte der Anodendämpfe zunehmen. Empirisch wurde (vgl. Fig. 104) zwischen der Kraterstrahlungsdichte B, der Bogenleistung W, und dem auf Verdampfung beruhenden Anodenabbrand $A - A_0$, eine für eine große Zahl verschiedenartig zusammengesetzter und

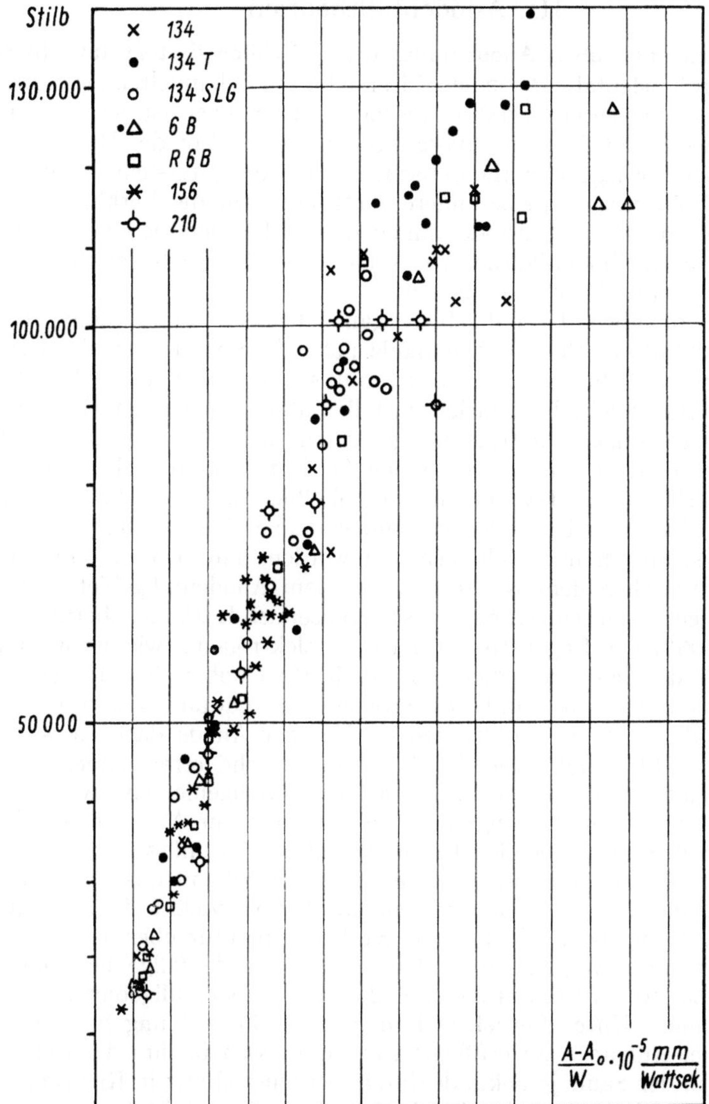

Fig. 104. Zusammenhang von Kraterleuchtdichte und Abbrand pro Watt für sieben verschiedene 7 mm-BECK-Bögen, nach FINKELNBURG [6a].

verschiedenen Abbrand zeigenden Positivkohlen gültige Beziehung gefunden, deren erster in einem weiten Bereich linearer Verlauf sich durch die Beziehung

$$B\,(\text{Stilb}) = 1{,}4 \cdot 10^9 \, \frac{A - A_0}{W} \, (\text{mm/Wsec}) \qquad (117.2)$$

darstellen läßt. In diesem Bereich ist die Kraterleuchtdichte also proportional der je Wattsekunde erzeugten Leuchtdampfmenge. Diese empirische Beziehung kann vorläufig theoretisch noch nicht abgeleitet werden, dürfte aber als empirische Grundlage einer späteren Theorie von Interesse sein. Für die nähere Diskussion vgl. [*6a*].

III. Anodenfalltheorien.

118. Allgemeines über Anodenfalltheorien, Feldionisierung und thermische Ionisierung im Anodenfall. Die in den Ziff. 110 bis 117 behandelten experimentellen Ergebnisse sind so vielseitig, daß es bisher nicht gelungen ist, sie mit einer einheitlichen Theorie der Anodenfallvorgänge zu klären. Vor der Besprechung der wenigen bisher vorliegenden theoretischen Versuche untersuchen wir ganz allgemein die Problemlage. Insbesondere vergleichen wir die Verhältnisse vor der Anode mit denen vor der Kathode, um zu zeigen, daß sich die Überlegungen und Ergebnisse von Kathodenfalltheorien keineswegs ohne Schwierigkeit auch auf das Anodenfallgebiet übertragen lassen.

Nach Ziff. 6 haben die Fallgebiete ganz allgemein die Aufgabe, den elektrischen Anschluß der thermischen Säule hoher Temperatur an die Elektroden geringer Temperatur herzustellen. Dazu muß sich in ihnen ein solcher Potentialverlauf bzw. eine solche Feldstärke einstellen, daß in jeder Schicht der axialen Dicke dx die Energiezufuhr $\mathfrak{E}dx$ den gesamten Energieverlusten dieser Schicht gleich ist, wobei in den elektrodennahen Schichten häufig, aber nicht immer, die Energieverluste zur Elektrode hin einschließlich der in der Schicht erzeugten und aus ihr abströmenden Ionen gegenüber dem radialen Energieverlust überwiegen. Es ist aber häufig nicht beachtet worden, daß doch ein entscheidender Unterschied zwischen dem Kathoden- und dem Anodenfallgebiet besteht. Im Kathodenfallgebiet einschließlich der es begrenzenden Kathodenoberfläche müssen die zu rund 99% den Bogenstrom tragenden Elektronen sowie der weitaus überwiegende Teil der den Strom dicht vor der Kathodenoberfläche tragenden Ionen erzeugt werden, und so stand das Problem der Ladungsträgerproduktion im Mittelpunkt der Kathodenfalldiskussion. Vor der Anode dagegen ist die Lage ganz anders. Rund 99% der die Anodenoberfläche erreichenden Elektronen strömen ja einfach aus der Säule in das Anodenfallgebiet ein, wo nur größenordnungsmäßig 1% der Ladungsträger in Form der zur Anode fließenden Elektronen und zur Säule fließenden Ionen erzeugt werden müssen.

Im folgenden sei wie üblich als Säule jedes Gebiet mit quasi-neutralem thermischen Plasma bezeichnet, dessen Temperatur über viele freie Weglängen konstant bleibt. Steigt oder fällt also die Säulentemperatur über den Bereich einer oder weniger freier Weglängen in dem an das Anodenfallgebiet angrenzenden Plasmabereich, so wird dieses zweckmäßig zum Anodenfallgebiet und nicht zur Säule gerechnet. Unter Berücksichtigung dieser Festsetzung hat das Anodenfallgebiet drei Aufgaben zu erfüllen: Erstens müssen in ihm die Ionen erzeugt werden, die in der Säule je Sekunde durch den Querschnitt in Richtung Kathode diffundieren, zweitens muß im Anodenfallgebiet die gerichtete Geschwindigkeit der im Anodenfall beschleunigten Ionen in thermische Geschwindigkeitsverteilung verwandelt werden, so daß die Ionen mit einer der Säulen- und Plasmatemperatur entsprechenden Geschwindigkeitsverteilung ohne merklich gerichtete Komponente in die Säule eintreten. Drittens schließlich muß im Anodenfallgebiet der Übergang von der relativ niedrigen Temperatur der Anode zu der meist hohen Plasmatemperatur der Säule erfolgen [*9*].

Anschaulich ist die Entstehung des Anodenfalles damit leicht einzusehen. Da, von wenigen Ausnahmen abgesehen, die Bogenanoden nicht in merklichem Umfange positive Ionen emittieren, ist der Strom unmittelbar vor der Anodenoberfläche ein reiner Elektronenstrom. Hier herrscht also eine unkompensierte (und in den anschließenden Gebieten auf die Säule zu durch die hier erzeugten Ionen nur teilweise kompensierte) negative Raumladung, die den Potentialabfall vor der Anode, den Anodenfall bewirkt. In diesem Fallgebiet werden die es durchlaufenden Elektronen zur Anode hin und die in ihm durch Feldionisierung oder thermische Ionisierung erzeugten Ionen zur Säule hin beschleunigt, wobei die von ihnen dem Anodenfall entnommene kinetische Energie zur Deckung der Verluste dieses Gebietes über die normalen radialen Verluste der Säule hinaus führt. Da wegen des Beweglichkeitsunterschiedes von Elektronen und Ionen von größenordnungsmäßig 10^3 die Anode erreichenden Elektronen nur *ein* Ion im Anodenfall erzeugt werden muß, spielt der Ionisierungsaufwand in der Energiebilanz des Anodenfallgebietes eine geringe Rolle gegenüber der von den Elektronen (und in geringem Maße auch durch Wärmeleitung von den anderen Plasmateilchen) auf die Anode übertragenen Energie.

Betrachten wir die Ionisierung von der Säule ausgehend etwas genauer: In der Säule halten der thermischen Ionisierung die Rekombination und die radiale Ionenabwanderung das Gleichgewicht. Dabei ist im allgemeinen der radiale Ionenverlust relativ gering gegenüber der Zahl der rekombinierenden Ionen, da die Zahl der Ionisationen in erster Näherung gleich der der Rekombinationen ist. Nur in diesem Fall ist man ja berechtigt, von einem isothermen Plasma bestimmter Temperatur zu sprechen, dessen Ionisierungsgrad durch die SAHA-Gleichung und die Plasmakonstanten eindeutig bestimmt ist. Genau genommen muß aber, da die Zahl der Ionisationen die der Rekombinationen überwiegt, auch in der Säule die Elektronentemperatur nach Ziff. 6 etwas über der Neutralgastemperatur liegen. Im Anodenfall tritt nun zur radialen Ionenabwanderung noch die in axialer Richtung hinzu. Je stärker diese axiale Ionenabwanderung die Ionenbilanz beeinflußt, um so mehr muß die Übertemperatur der Elektronen über die Plasmatemperatur ansteigen. Die zur T_e-Erhöhung erforderliche Energie kann aber nur dem Feld entnommen werden, und so gelangt man wieder zu dem Ergebnis einer mit Annäherung an die Anodenoberfläche zunehmenden Feldstärke.

Bezüglich des Ionisierungsmechanismus im Anodenfallgebiet kann man zwei Grenzfälle unterscheiden, je nachdem, ob der wesentliche Teil des Potentialabfalles vor der Anode über einer oder mehreren freien Elektronenweglängen liegt. Ist (erster Fall) die Anodenfalldicke von der Größenordnung von nur einer Elektronenweglänge, so werden die Elektronen während ihrer Beschleunigung praktisch keine Stöße erleiden; ihre Beweglichkeit ist dann durch die Verhältnisse des freien Falles zu beschreiben. Stoßionisierung ist nur möglich, wenn der die Elektronen beschleunigende Anodenfall von der Größenordnung der Ionisierungsspannung des Trägergases, oder bei Stufenionisierung mindestens von der Größenordnung der ersten Anregungsspannung des Trägergases ist. In diesem Fall spricht man von *Feldionisierung*. Erstreckt sich dagegen (zweiter Fall) der Anodenfall über mehrere freie Elektronenweglängen, so werden die im Anodenfall beschleunigten Elektronen im allgemeinen in Stößen mit schweren Plasmateilchen ihre gerichtete Bewegung in eine ungeordnete verwandeln, wozu wegen der geringen Elektronenmasse schon wenige Stöße ausreichen. Man kann dann in erster Näherung sagen, daß die Elektronen im Anodenfallgebiet auf eine unter Umständen sehr hohe Elektronentemperatur aufgeheizt werden, und die Ionisierung vor der Anode nunmehr durch die energiereichsten Elektronen des „MAXWELL-Schwanzes" erfolgt. Man spricht dann von *thermischer Ionisierung*. Wie hoch die Elektronentemperatur und der Anodenfall bei einem bestimmten Bogen mit thermischer Anodenfallionisierung sich einstellen, und wie hoch die Elektronentemperatur über der Plasmatemperatur des säulenseitig angrenzenden Gebietes liegt, hängt vom Einzelfall ab und läßt sich schwer allgemein sagen.

Eine geschlossene Anodenfalltheorie ist bisher nur für den Niederstromkohlebogen mit anodischer Feldionisierung von BEZ und HÖCKER [9] vorgelegt und später [85] auch auf andere Bögen übertragen worden. Im übrigen liegen nur qualitative Ansätze vor, auf die wir unten zu sprechen kommen.

119. Die Theorie des normalen Anodenfalles mit Feldionisierung. Nach Ziff. 118 ist Feldionisierung im Anodenfall eines Lichtbogens nur möglich, wenn erstens die Dicke des Anodenfallgebietes nur von der Größenordnung einer freien Elektronenweglänge ist und wenn zweitens die Transportweglänge der positiven Ionen, die sich im wesentlichen aus den Stößen mit Neutralteilchen berechnet, merklich kleiner als die der Elektronen ist, da andernfalls die Ionen nicht mit der thermischen Geschwindigkeitsverteilung der Neutralgasteilchen am Säulenrand ankommen können. Diese beiden Bedingungen grenzen nach der Anodenfalltheorie von BEZ und HÖCKER [9] den Bereich der Plasmatemperatur vor der Anode ein, in dem Feldionisierung möglich ist.

In ihrer Theorie teilen BEZ und HÖCKER das gesamte Anodenfallgebiet in vier Bereiche ein: das Säulengrenzgebiet, das Übergangsgebiet, das Beschleunigungsgebiet und das Ionisationsgebiet. Die vier Gebiete unterscheiden sich in verschiedener Weise bezüglich des Verhaltens der Elektronen und der Ionen in ihnen. Im Säulengrenzgebiet beginnt die Feldstärke von ihrem sehr kleinen Wert in der Säule langsam anzusteigen; gleichzeitig beginnt die für die Säule charakteristische Elektronendrift sich in einen freien Fall zu verwandeln. Übergangs- und Beschleunigungsgebiet sind vom Standpunkt der Elektronen aus nicht zu unterscheiden und werden unter Energiegewinn frei durchfallen, bis am anodenseitigen Ende des Beschleunigungsgebietes die Elektronen die zur Ionisierung erforderliche Energie erreicht haben. Im Ionisationsgebiet findet dann durch Feldionisierung die Erzeugung der erforderlichen Ionen statt. Vom Standpunkt der im Ionisationsgebiet erzeugten Ionen aus ist die Lage anders, weil deren freie Weglänge um einen Faktor kleiner ist als die der Elektronen. Die Ionen durchfallen das Beschleunigungsgebiet noch frei, stoßen aber jenseits der Grenze zum Übergangsgebiet bereits mit Neutralteilchen und wandeln dabei ihre gerichtete Geschwindigkeit schrittweise in eine ungeordnete thermische Geschwindigkeit um, die sich im Säulengrenzgebiet der Plasmatemperatur vollständig angleicht.

BEZ und HÖCKER berechnen den zur Erzeugung des Ionenstromanteils erforderlichen Anodenfall unter der Annahme, daß die Ionisierungswahrscheinlichkeit der Größe $(U(x) - U_i)$ proportional ist, wo $U(x)$ die vom säulenseitigen Ende des Anodenfalles aus gerechnete Beschleunigungsspannung der Elektronen, und U_i die Ionisierungsspannung des Trägergases ist. Aus den empirisch bekannten Werten der anodischen Stromdichte j und des Ionenstromanteils j^+/j^- in der Säule berechnen die Verfasser unter der Annahme des freien Falles der Elektronen den Verlauf der negativen Raumladung $\varrho^-(x)$ und in etwas komplizierterer Weise mittels der Theorie der Ladungsträgerbewegung im inhomogenen elektrischen Feld den Verlauf von $\varrho^+(x)$. Die Differenz beider Kurven ergibt dann den Verlauf der negativen Überschußraumladung $\varrho(x)$ und aus ihr durch Integration den Verlauf der Feldstärke $E(x)$ und des Potentials $U(x)$ im gesamten Anodenfallgebiet. Die Dicke des Anodenfallgebietes und der Gesamtwert der Anodenfallspannung U_A ergeben sich dann aus der Bedingung, daß gerade die aus Gesamtstromdichte und Beweglichkeitsverhältnis von Ionen zu Elektronen folgende Zahl von Ionen je Sekunde im Anodenfallgebiet erzeugt werden muß. Für den Niederstromkohlebogen (Ziff. 110) berechnen BEZ und HÖCKER daraus einen Anodenfallwert von 19 V und eine Ausdehnung des Anodenfalles von $3{,}4 \cdot 10^{-3}$ cm, in bester Übereinstimmung mit den in Ziff. 110 angegebenen experimentellen Daten von FINKELNBURG, SEGAL und BLOCK [*12*], [*53*].

Die hier dargestellte Theorie basiert auf der Annahme einer keine positiven Ionen emittierenden Anode. Bei Kohlebögen, deren Anode mit leicht ionisierbaren Metallsalzen versetzt ist, ist bei hoher Temperatur aber eine schwache Ionenemission zu erwarten; vor der Anode herrscht in diesem Fall ferner stets ein gewisser Dampfdruck leicht ionisierbarer Metallatome. Der Spannungsbedarf für die Ionenerzeugung ist bei solchen Metallsalzbögen also aus zwei Gründen erniedrigt, und die in Ziff. 110 erwähnten niedrigen Anodenfälle sind damit theoretisch verständlich.

Mit zunehmender anodischer Stromdichte, und d.h. bei thermischen Bogensäulen auch mit zunehmender Plasmatemperatur vor der Anode, muß die Anodenfallspannung wachsen, weil die größere Elektronenstromdichte nach der LANGMUIR-Gleichung [*116*] eine höhere Saugspannung zur Anode hin benötigt. HÖCKER und BEZ versuchen in einer zweiten Arbeit [*85*] besonders gewisse Eigenschaften unterbelasteter BECK-Bögen auf dieser Grundlage zu verstehen.

120. Der Übergang von der Feldionisierung zur thermischen Ionisierung im Anodenfall. Nach Ziff. 118 ist Feldionisierung nur möglich, wenn erstens der wesentliche Teil der Elektronenbeschleunigung über nur einer freien Elektronenweglänge erfolgt, und wenn zweitens letztere ein Mehrfaches der mittleren freien Ionenweglänge beträgt. Bei zunehmender anodischer Stromdichte und anodischer Plasmatemperatur wird aber wegen des zunehmenden Ionisierungsgrades und der großen Wirkungsquerschnitte von Ionen gegenüber Elektronen deren freie Weglänge immer kleiner, während gleichzeitig die Transportweglänge der Ionen (das ist die freie Weglänge für Stöße mit Neutralteilchen) immer größer wird. Es gibt deshalb eine kritische Temperatur, oberhalb welcher Feldionisierung im Anodenfall nicht mehr möglich ist, und diese kritische Temperatur hängt wegen der verschieden großen Wirkungsquerschnitte und Ionisierungsausbeuten wesentlich von der Gasart und vom Druck des Entladungsplasmas ab. Für den Kohlebogen berechnen HÖCKER und BEZ [*85*], daß Feldionisierung im Anodenfall nur im Temperaturbereich von 6400 bis 9500° K möglich ist, während für Argon bei Atmosphärendruck die obere Grenze sogar bei 7500° K liegen soll, falls es hier überhaupt Feldionisierung gibt, was nach dem experimentellen Befund nicht der Fall zu sein scheint.

Oberhalb dieser kritischen Temperatur erleiden die Elektronen im Beschleunigungsraum Stöße, durch die die Ausbildung einer gerichteten Geschwindigkeit verhindert wird, während gleichzeitig die Ionen nach Durchfallen des gesamten Anodenfalles die anodenseitige Säulengrenzschicht mehr und mehr aufzuheizen beginnen. Die Verhältnisse werden damit in zunehmendem Maße thermisch, und da nun die Ionisierung vor der Anode durch Elektronen erfolgt, die in der MAXWELL-Verteilung bei der erhöhten Elektronentemperatur eine Energie oberhalb der Ionisierungsenergie besitzen, kann die mittlere Elektronenenergie und damit der für diese verantwortliche Anodenfall weit unter die Ionisierungsspannung sinken. Zahlreiche Beispiele sprunghafter Bogen-Brennspannungserniedrigung um mehrere Volt bei stetiger Stromstärkeerhöhung werden von HÖCKER und BEZ auf diesen Umschlag von Feldionisierung zur thermischen Ionisierung im Anodenfall zurückgeführt, während beobachtete Beispiele geringer Anodenfälle als Folge eines thermischen Anodenfall-Mechanismus gedeutet werden. Letzterer soll besonders für die geringen Anodenfälle von Hochdruckbögen auch relativ geringer Stromstärke, z.B. des von BAUER und SCHULZ [*4*] genau untersuchten Xenon-Hochdruckbogens verantwortlich sein, da oberhalb

15 Atm Druck Feldionisierung im Anodenfall überhaupt nicht mehr möglich ist. In diesem Zusammenhang sei besonders darauf hingewiesen, daß ein großer Teil der ECKERschen Überlegungen [33] zum Kathodenfall (Ziff. 107) sich auf die Anodenfallverhältnisse übertragen läßt und manche beobachteten Erscheinungen sehr schön zu erklären vermag.

Es wurde oben erwähnt, daß im thermischen Gleichgewicht die Elektronentemperatur die Plasmatemperatur um so mehr übersteigt, je größer der Ionenentzug eines Plasmagebietes verglichen mit der durch Rekombination verschwindenden Ionenzahl ist. In der Nähe des Umschlages von der Feldionisierung zur thermischen Ionisierung ist bei den angegebenen relativ niedrigen kritischen Temperaturen die Ladungsträgerergiebigkeit des Anodenfallgebietes noch so klein, daß beträchtliche Übertemperaturen der Elektronen zu erwarten sind, wobei die Aufheizenergie aus dem hier noch relativ hohen und wenig ausgedehnten Anodenfall stammt. Je höher aber die Plasmatemperatur im anodennahen Gebiet wird, desto größer ist die Zahl der sekundlich gebildeten und wieder rekombinierenden Ionenpaare, desto geringer ist folglich prozentual der Ionenverlust infolge axialen Abströmens. Mit zunehmender Plasmatemperatur vor der Anode erwarten wir daher eine Abnahme der Elektronenübertemperatur und damit des diese verursachenden Anodenfalles. Gleichzeitig dürfte sich das Anodenfallgebiet in axialer Richtung immer mehr ausbreiten, so daß im Grenzfall der Unterschied zwischen der Säule mit ihrer normalen Feldstärke und dem Anodenfall verschwinden würde. Wie weit die bei anodischen Plasmatemperaturen von 10000 bis 20000° K von BUSZ-PEUCKERT, FINKELNBURG und MÜLLER [22], [159] gemessenen, bis Null abnehmenden Anodenfallwerte auf dieser theoretisch zu erwartenden Abnahme des Anodenfalles mit zunehmender Plasmatemperatur beruhen, ist schwer zu entscheiden, da hier noch ein gleich in Ziff. 121 zu behandelnder Effekt ins Spiel kommt. Es sei aber des grundsätzlichen Interesses wegen erwähnt, daß bei thermischen Lichtbögen auch leicht negative Anodenfallwerte nicht unmöglich erscheinen. Würde nämlich der infolge statistischer Fluktuationen im Hochtemperaturplasma die Anodenoberfläche treffende Elektronenstrom die eingestellte Bogenstromstärke übersteigen, so müßte die Anode sich wie eine im Plasma stehende Sonde so hoch negativ aufladen, daß der Elektronendiffusionsstrom zu ihr gerade gleich dem Bogenstrom wird.

121. Zur Theorie des anomalen Anodenfalles von Bögen mit axialem Gas- oder Dampfstrahl. Nach Ziff. 118 hängt der Anodenfall von den Raumladungsverteilungen $\varrho^+(x)$ und $\varrho^-(x)$ und diese wieder bei gegebenen Stromdichten j^+ und j^- von den Trägergeschwindigkeiten $v^+(x)$ und $v^-(x)$ im Anodenfallgebiet ab. Werden also die Trägergeschwindigkeiten (besonders im Übergangs- und Säulengrenzgebiet) durch makroskopische axiale Plasmaströmungen beeinflußt, so muß sich das in Änderungen des Anodenfalles äußern. Sind diese Plasmaströmungen stromstärkeabhängig, so erhält man anomale, d.h. stromstärkeabhängige Anodenfälle.

Das erste und bisher einzige genauer untersuchte Beispiel für solche Bögen mit anomalem Anodenfall ist der Hochstromkohlebogen mit schnell verdampfender Anode [6a]. Nach Ziff. 116 wächst bei diesem Bogen die Anodenverdampfung schneller als proportional mit der Stromstärke, und das verdampfte Material strömt als Anodendampfstrahl mit Geschwindigkeiten von 1 bis $7\cdot 10^3$ cm/sec von der Anodenstirnfläche ab. Mit dieser Geschwindigkeit werden also beim Hochstromkohlebogen Ionen und Elektronen von der Anode fortgeblasen. Während diese Blasgeschwindigkeit aber die Wanderungsgeschwindigkeit der schnellen Elektronen und deren Raumladung $\varrho^-(x)$ kaum verändert, wird die zur Kathode hin gerichtete Geschwindigkeit der schweren Ionen, deren feldbestimmte Driftgeschwindigkeit im Säulengrenzgebiet nur von der Größenordnung 10^2 cm/sec ist, durch den Anodendampfstrahl stark vergrößert. Dadurch wird in dem gesamten vom Anodendampfstrahl bestimmten Säulenteil der Ionenstromanteil stark erhöht. Es müssen folglich im Anodenfallgebiet auch mehr Ionen erzeugt werden, und dazu ist ein mit der Blasgeschwindigkeit zunehmender Anodenfall erforderlich. Dieser wird vom Anodendampfstrahl dadurch erzeugt, daß er vor der Anode die positive Raumladung $\varrho^+(x)$ verkleinert und entsprechend die den Anodenfall ergebende negative Überschußraumladung $\varrho(x)$ vergrößert. Die experimentell bekannte Zunahme des Anodenfalles mit der Stromstärke und dem auf Verdampfung beruhenden Positivkohlen-„Abbrand" scheint damit qualitativ klar. Eine quantitative Durchführung dieser auf STEENBECK, HEINZMANN und FINKELNBURG [*6a*] zurückgehenden Vorstellungen steht noch aus.

Ein von der Anode axial abströmender Plasmastrahl bildet sich als Folge der kontrahierenden Wirkung des Eigenmagnetfeldes nach Ziff. 94 auch bei jedem Bogen mit kontrahiertem Anodenbrennfleck hoher Stromdichte aus [*139*]. Ob dadurch der auf wenige freie Weglängen begrenzte Anodenfall im Sinne der obigen Ausführungen verändert wird, scheint aber zweifelhaft, da das als Plasmastrahl von der Anode abströmende Gas ja im Kontraktionsbereich vor der Elektrode radial angesaugt wird und das eigentliche Anodenfallgebiet dadurch wahrscheinlich wenig beeinflußt wird.

Eine starke Wirkung auf den Anodenfall hat dagegen der vom Kathodenbrennfleck axial abströmende Plasmastrahl. Wenn nämlich seine Geschwindigkeit vor der Anode gegenüber der Driftgeschwindigkeit v^+ der Ionen überwiegt, verhindert er das Abfließen der Ionen in die Säule und beeinflußt gleichzeitig durch Heranführen hoch erhitzten Plasmas an die Anodenoberfläche die Temperaturverteilung vor dieser in solcher Weise, daß die Säule gleichsam bis zur Anode reicht. Ein Anodenfall ist dann, da keine abströmenden Ionen mehr durch Ionisierung ersetzt werden müssen, nicht mehr erforderlich, es sei denn zur Überbrückung der in diesem Fall sehr dünnen Schicht geringerer Temperatur dicht vor der Anodenoberfläche. Diese Überlegungen MAECKERs [*139*] finden eine Bestätigung in Sondenmessungen von MÜLLER und FINKELNBURG [*159*], die an einem 200 Amp-Kohlebogen mit großflächiger nicht-verdampfender Anode, aber mit gut ausgebildetem kathodischen Plasmastrahl, einen Anodenfall von nur (1 ± 2) Volt fanden. Auch die von BUSZ-PEUCKERT und FINKELNBURG [*21*] am Argon-Hochtemperaturbogen gefundene Abnahme des Anodenfalles mit zunehmender Stromstärke bis zu einem nahe Null liegenden Wert dürfte mit dem kathodischen Plasmastrahl zusammenhängen, doch liegen die Verhältnisse bei diesem Bogen wegen der von der Länge und Stromstärke abhängenden anodischen Stromdichte sehr unübersichtlich.

Bibliographie.

Bücher über elektrische Lichtbögen.

[*1a*] AYRTON, H.: The Electric Arc. London 1902.
[*2a*] BRINKMANN, H.: Optische Studie van de Elektrische Lichtboog. Dissertation. Utrecht Amsterdam 1937.
[*3a*] CZUDNOCHOWSKI, W. v.: Das elektrische Bogenlicht, seine Entwicklung und seine physikalischen Grundlagen. Leipzig 1906.
[*4a*] ELENBAAS, W.: The High Pressure Mercury Vapour Discharge. Amsterdam 1951.
[*5a*] ENGEL, A. v., u. M. STEENBECK: Elektrische Gasentladungen. Berlin 1934.
[*6a*] FINKELNBURG, W.: Hochstromkohlebogen. Berlin-Göttingen-Heidelberg 1948.
[*7a*] HAGENBACH, A.: Der elektrische Lichtbogen. Leipzig 1924.

[8a] LOEB, L. B.: Fundamental Processes in Gas Discharges. New York 1947.
[9a] MAECKER, H.: Der elektrische Lichtbogen. Ergebn. exakt. Naturw. **25**, 293 (1951).
[10a] MATHIESEN, W.: Untersuchungen über den elektrischen Lichtbogen, insbesondere den unter Druck befindlichen. Leipzig 1921.
[11a] MONASCH, B.: Der elektrische Lichtbogen bei Gleichstrom und Wechselstrom und seine Anwendungen. Berlin 1904.
[12a] PARISOT, J.: L'Arc Electrique Intensiv. Paris 1950.
[13a] RASCH, E.: Das elektrische Bogenlicht. Physikalisch-technische Grundlagen der Lichterzeugung durch elektrische Entladungsvorgänge. Braunschweig 1910.
[14a] SIMON, TH.: Der elektrische Lichtbogen. Leipzig 1911.
[15a] STARK, J.: Die Elektrizität in Gasen. Leipzig 1902.
[16a] STOCKHAUSEN, K.: Der eingeschlossene Lichtbogen. Leipzig 1907.
[17a] THOMSON, J. J.: Elektrizitätsdurchgang in Gasen. Leipzig 1906.
[18a] WEIZEL, W., u. R. ROMPE: Theorie elektrischer Bögen und Funken. Leipzig 1949.

Spezielle Literatur.

[1] BARTELS, H.: Z. Physik **127**, 243 (1950); **128**, 546 (1950).
[2] BARTELS, H., u. B. EISELT: Optik **6**, 56 (1950).
[3] BAUER, A.: Z. Physik **138**, 35 (1954).
[4] BAUER, A., u. P. SCHULZ: Z. Physik **139**, 197 (1954).
[5] BECK, H.: ETZ **42**, 993 (1921).
[6] BECKEN, O., u. R. SEELIGER: Ann. Phys. **24**, 609 (1935).
[7] BECKEN, O., u. K. SOMMERMEYER: Z. Physik **102**, 551 (1936).
[8] BERTELE, H. v.: Brit. J. Appl. Phys. **3**, 127 (1952).
[9] BEZ, W., u. K. H. HÖCKER: Z. Naturforsch. **9**a, 72 (1954).
[10] BIERMANN, L., u. A. SCHLÜTER: Z. Naturforsch. **5**a, 65, 237 (1950).
[11] BIRKELAND and EYDE: Electrotechn. Industry **2**, 399 (1904).
[12] BLOCK, M. J., u. W. FINKELNBURG: Z. Naturforsch. **8**a, 758 (1953).
[13] BOLTZMANN, L.: Vorlesungen über Gastheorie. Leipzig 1896.
[14] BROWN, S. C., and W. P. ALLIS: Basic Data of Electrical Discharges. Technical Report Nr. 283, 1954. Massachusetts Inst. of Technology.
[15] BUCKINGHAM, W. O., and C. R. DEIFERT: J. Opt. Soc. Amer. **36**, 245 (1946).
[16] BURHORN, F.: Z. Physik **140**, 440 (1955).
[17] BURHORN, F., u. H. MAECKER: Z. Physik **129**, 369 (1951).
[18] BURHORN, F., H. MAECKER u. TH. PETERS: Z. Physik **131**, 28 (1951).
[19] BURKHARDT, G.: Ann. Phys. **5**, 373 (1950).
[20] BUSZ, G., u. W. FINKELNBURG: Z. Physik **139**, 212 (1954).
[21] BUSZ-PEUCKERT, G., u. W. FINKELNBURG: Z. Physik **140**, 540 (1955).
[22] BUSZ-PEUCKERT, G., u. W. FINKELNBURG: Z. Physik **144**, 244 (1956).
[23] CHANDRASEKHAR, S.: Astrophys. J. **102**, 401 (1945).
[24] CHANEY, HAMISTER and GLASS: Trans. Electrochem. Soc. **67**, 201 (1935).
[25] CHAPMAN, S., and T. G. COWLING: The Mathematical Theory of Non Uniform Gases. Cambridge 1939.
[26] CHILD, C. D.: Jb. Elektronik **3**, 189 (1906).
[27] COBINE, J. D., and C. J. GALLAGHER: Phys. Rev. **74**, 1524 (1948).
[28] COMPTON, K. T.: Phys. Rev. **21**, 266 (1923).
[29] DARWIN, C. G.: Proc. Roy. Soc. Lond. **146**, 17 (1934); **182**, 152 (1944).
[30] DOAN, G. E., and A. M. THORNE: Phys. Rev. **46**, 49 (1934).
[31] DÜRRWANG, J.: Helv. phys. Acta **8**, 333 (1935).
[32] ECKER, G.: Z. Physik **130**, 585 (1951).
[33] ECKER, G.: Z. Physik **132**, 248 (1952); **135**, 105 (1953); **136**, 1, 556 (1954).
[34] EDELS, H.: British Electrical and Allied Research. Association Report *L/T* 230, 1951. Proc. Phys. Soc. Lond. **64**, 354 (1951).
[35] EGGERT, J.: Phys. Z. **20**, 570 (1919).
[36] EINSTEIN, A.: Phys. Z. **18**, 121 (1917).
[37] ELENBAAS, W.: Physica **3**, 947 (1936).
[38] ELENBAAS, W., F. KESSELRING u. F. KOPPELMANN: ETZ **57**, 1497 (1936).
[39] ELWERT, G.: Z. Naturforsch. **7**a, 432, 703 (1952).
[40] ENGEL, A. v., u. M. STEENBECK: Wiss. Veröff. Siemens-Konz. **10**, 155 (1931).
[41] ENGEL, A. v., u. M. STEENBECK: Wiss. Veröff. Siemens-Konz. **12**, 74, 89 (1933).
[42] ENSKOG, D.: Kinetische Theorie der Vorgänge in mäßig verdünnten Gasen. Uppsala 1917.
[43] ERDMANN, G., u. F. JESNITZER: Werkstoffe und Schweißung. Berlin 1951.

[44] EUCKEN, A.: Lehrbuch der chemischen Physik, Bd. II, 1. Allgemeine Grundlagen für Gase. Leipzig 1950.
[45] EULER, J.: Ann. Phys. **11**, 203 (1953).
[46] FAST, J. D.: Philips Res. Rep. **2**, 382 (1947).
[47] FINKELNBURG, W.: Z. Physik **88**, 297 (1934).
[48] FINKELNBURG, W.: Chem. Techn. **15**, 141 (1942).
[49] FINKELNBURG, W.: Z. angew. Phys. **6**, 239 (1954).
[50] FINKELNBURG, W., and J. P. LATIL: J. Opt. Soc. Amer. **44**, 1 (1954).
[51] FINKELNBURG, W., u. H. SCHLUGE: Z. Physik **112**, 305; **113**, 562 (1939); **114**, 734 (1940); **116**, 214 (1941); **117**, 344 (1942); **119**, 206 (1943); **122**, 714 (1944).
[52] FINKELNBURG, W., and S. M. SEGAL: Phys. Rev. **80**, 258 (1950).
[53] FINKELNBURG, W., and S. M. SEGAL: Phys. Rev. **83**, 582 (1951).
[54] FOITZIK, R.: Wiss. Veröff. Siemens-Konz. **19**, 28 (1940).
[55] FOWLER, R. H., and L. W. NORDHEIM: Proc. Roy. Soc. Lond., Ser. A **119**, 173 (1928).
[56] FRAGSTEIN, C. v., u. M. ARNDT: Ann. Phys. **33**, 532 (1938).
[57] FRANCIS, V. J.: Phil. Mag. **37**, 433, 653 (1946); **40**, 435 (1949).
[58] FROOME, K. D.: Proc. Phys. Soc. Lond. **60**, 424 (1948); **62**, 805 (1949); **63**, 377 (1950).
[59] FUCKS, R.: Z. Physik **130**, 69 (1951).
[60] GALLAGHER, C. J., and J. D. COBINE: Phys. Rev. **71**, 481 (1947). — J. Appl. Phys. **21**, 768 (1950).
[61] GAYDON, A. G., and W. G. PENNEY: Proc. Roy. Soc. Lond., Ser. A **183**, 374 (1945).
[62] GERDIEN, H., u. A. LOTZ: Wiss. Veröff. Siemens-Konz. **2**, 489 (1922). — Z. techn. Phys. **4**, 157 (1923).
[63] GERDIEN, H., u. A. LOTZ: Z. techn. Phys. **5**, 515 (1924).
[64] GERDIEN, H., u. A. LOTZ: Wiss. Veröff. Siemens-Konz. **14**, 25 (1935).
[65] GERMER, L. H., and F. E. HAWORTH: J. Appl. Phys. **20**, 1085 (1949). — HAWORTH, F. E.: Phys. Rev. **80**, 223 (1950). — GERMER, L. H.: J. Appl. Phys. **22**, 955, 1133 (1951). — GERMER, L. H., and J. L. SMITH: J. Appl. Phys. **23**, 553 (1952). — KISLIUK, P.: J. Appl. Phys. **25**, 897 (1954).
[66] GERTHSEN, P., u. P. SCHULZ: Z. Physik **140**, 510 (1955).
[67] GILETTE, R. N., and R. T. BREYMEIER: Welding J., Res. Suppl. **1** März (1951).
[68] GÖING, W.: Naturwiss. **37**, 558 (1950). — Z. Physik **131**, 603 (1952).
[69] GÖING, W., H. MEIER u. H. MEINEN: Z. Physik **140**, 376 (1955).
[70] GRETENER, E.: Z. techn. Phys. **18**, 95 (1937). — J. S. Mot. Piet. Eng. **55**, 391 (1950)
[71] GROOT, R. S. DE: Thermodynamics of Irreversible Processes. Amsterdam 1951.
[72] GUILLERY, P.: Z. Naturforsch. **10**a, 248 (1955).
[73] GUYE, CH. E., et B. MONASCH: Eclair. Electrique **34**, 305 (1903).
[74] GÜNTHERSCHULZE, A.: Z. Physik **11**, 74 (1922).
[75] GVOSDOVER, S. D.: Phys. Z. Sowjet. **12**, 164 (1937).
[76] HAASE, R.: Ergebn. exakt. Naturw. **26**, 56 (1952).
[77] HAASE, R.: Z. Naturforsch. **8**a, 729 (1953).
[78] HAGENAH, W.: Z. Physik **128**, 279 (1950).
[79] HAHN, O. T., u. W. FINKELNBURG: Z. Physik **122**, 36 (1944).
[80] HAYNES, J. R.: Phys. Rev. **73**, 891 (1948).
[81] HEATLY, A. H., and R. S. SOANES: Trans. Electrochem. Soc. **72**, 281 (1938).
[82] HELD, E. F. M. VAN DER: Z. Physik **70**, 508 (1931).
[83] HEWITT, P. C.: ETZ **24**, 87, 187 (1903).
[84] HIRSCHFELDER, J. O., C. F. CURTISS and R. B. BIRD: Molecular Theory of Gases and Liquids. New York u. London 1954.
[85] HÖCKER, K. H., u. W. BEZ: Z. Naturforsch. **10**a, 706, 714 (1955).
[86] HÖCKER, K. H., u. W. FINKELNBURG: Z. Naturforsch. **1**, 305 (1946).
[87] HÖCKER, K. H., u. P. SCHULZ: Z. Naturforsch. **4**a, 265 (1949).
[88] HOLM, R., B. KIRSCHSTEIN u. F. KOPPELMANN: Wiss. Veröff. Siemens-Konz. **13**, 63 (1934).
[89] HOLTZMARK, J.: Ann. Phys. **58**, 577 (1919). — Phys. Z. **20**, 162 (1919); **25**, 73 (1924).
[90] HÖRMANN, H.: Z. Physik **97**, 539 (1935).
[91] HOYAUX, M.: Appl. Sci Res. **5** (1955).
[92] INGLIS, D. E., and R. TELLER: Appl. J. **90**, 439 (1939).
[93] JAGODZINSKI, H.: Z. Physik **120**, 318 (1943).
[94] JOHNSON, M. H.: Phys. Rev. **84**, 566 (1951).
[95] JÜRGENS, G.: Z. Physik **134**, 21 (1952).
[96] JÜRGENS, G.: Z. Physik **138**, 613 (1954).
[97] KANTROWITZ, A.: Gas Dynamics of Cosmic Clouds, Chapt. 16. Amsterdam 1955.
[98] KANTROWITZ, A., S. C. LIN and E. L. RESLER: J. Appl. Phys. 26, 95 (1955).
[99] KAUFMANN: Ann. Phys. **2**, 158 (1900).

[*100*] Kenty, C.: J. Appl. Phys. **9**, 53 (1938).
[*101*] Kenty, C., and W. J. Karash: Phys. Rev. **60**, 66 (1941); **78**, 625 (1950).
[*102*] Kern, J.: Z. techn. Phys. **18**, 249 (1938).
[*103*] Kesselring, F., u. F. Koppelmann: Arch. Elektrotechn. **30**, 71 (1936).
[*104*] Kihara, T.: Rev. Mcd. Phys. **24**, 1 (1952).
[*105*] King, L. A.: Nature, Lond. **174**, 1008 (1954).
[*106*] Koch, B.: Ann. Phys. **43**, 91 (1943).
[*107*] Koch, O.: Z. Physik **126**, 507 (1949).
[*108*] Koch, O., u. H. Dunstädter: Z. Physik **129**, 140 (1951).
[*109*] Koch, O., K. J. Lesemann u. A. Walther: Z. Physik **127**, 153 (1950).
[*110*] Kouwenhoven, W. B., and T. B. Jones: Electr. Engng. **68**, 834 (1949).
[*111*] Kramers, H. A.: Phil. Mag. **46**, 836 (1923).
[*112*] Krefft, H., F. Rössler u. A. Rüttenauer: Z. techn. Phys. **18**, 20 (1937).
[*113*] Kruskal, M., and M. Schwarzschild: Phys. Rev. **89**, 348 (1953).
[*114*] Küch, R., u. T. Retschinsky: Ann. Phys. **20**, 563 (1906).
[*115*] Langmuir, I.: Gen. Electr. Rev. **26**, 731 (1923).
[*116*] Langmuir, I.: Gen. Electr. Rev. **27**, 449 (1924).
[*117*] Larché, K.: Z. Physik **132**, 544 (1952). — Elektrotechn. Z. **72**, 427 (1951).
[*118*] Larché, K.: Lichttechn. Z. **2**, 41 (1951). — Z. Physik **136**, 74 (1953).
[*119*] Larenz, R. W.: Z. Physik **129**, 327, 343 (1951).
[*120*] Larenz, R. W.: Physik. Verh. **6** (9), 176 (1955).
[*121*] Larenz, R. W., u. H. Bartels: Naturwiss. **37**, 164 (1950).
[*122*] Lecher, E.: Wien. Ber. (2) **95**, 992 (1887).
[*123*] Lehmann, O.: Wied. Ann. **55**, 361 (1895).
[*124*] Lenz, W.: Z. Physik **83**, 139 (1933).
[*125*] Lindholm, E.: Diss. Uppsala 1942.
[*126*] List, B. H., und T. B. Jones: Electr. Engng. **1953**.
[*127*] Lochte-Holtgreven, W., u. H. Maecker: Z. Physik **105**, 1 (1937).
[*128*] Lochte-Holtgreven, W., u. W. Nissen: Z. Physik **133**, 124 (1952).
[*129*] Longini, L.: Phys. Rev. **71**, 642 (1947).
[*130*] Lorentz, H. A.: Versl. Amsterd. Akad. **14**, 518, 577 (1905).
[*131*] Lotz, A.: Unveröffentlicht.
[*132*] Lummer, O.: Die Verflüssigung der Kohle. Braunschweig 1914.
[*133*] Lutz, A.: C. R. Acad. Sci. Paris **228**, 912 (1949). — Rev. Gén. Electrique **34**, 153 (1950).
[*134*] McPherson, J.: J. Opt. Soc. Amer. **30**, 89 (1940).
[*135*] MacKeown, S. S.: Phys. Rev. **34**, 611 (1929).
[*136*] Maecker, H.: Z. Physik **129**, 108 (1951).
[*137*] Maecker, H.: Z. Physik **135**, 13 (1953).
[*138*] Maecker, H.: Z. Physik **136**, 119 (1953).
[*139*] Maecker, H.: Z. Physik **141**, 198 (1955).
[*140*] Maecker, H.: Z. Physik **144**, 586 (1956).
[*141*] Maecker, H.: Z. Naturforsch. **11**a, 32 (1955).
[*142*] Maecker, H., u. Th. Peters: Z. phys. Chem. **198**, 318 (1951).
[*143*] Maecker, H., u. Th. Peters: Z. Physik **139**, 448 (1954).
[*144*] Maecker, H., Th. Peters u. H. Schenk: Z. Physik **140**, 119 (1955).
[*145*] Mangold, H.: Elektr. Nachr.-Techn. **17**, 57 (1940).
[*146*] Mannkopff, R.: Z. Physik **76**, 396 (1932); **86**, 161 (1933).
[*147*] Mannkopff, R.: Z. Physik **120**, 301 (1943).
[*148*] Mason, R. C.: Phys. Rev. **51**, 28 (1937).
[*149*] Massey, H. S. W., and E. S. H. Burhop: Electronic and Ionic Impact Phenomena. Oxford 1952.
[*150*] Mazur, P.: Bull. Acad. Roy. Belgique, Cl. Sci. (5) **38**, 182 (1952).
[*151*] Medicus, G., and G. Wehner: J. Appl. Phys. **22**, 1389 (1951).
[*151b*] Meek, J. M., and J. D. Craggs: Electrical Breakdown of Gases. Oxford 1953.
[*152*] Meixner, J.: Z. Naturforsch. **4**a, 594 (1949); **7**a, 553 (1952).
[*153*] Meixner, J.: Ann. Phys. **39**, 333 (1941); **40**, 165 (1941); **41**, 409 (1942); **43**, 244 (1943). Z. phys. Chem. **53**, 235 (1943). — Termodynamic irreversibler Prozesse. Aachen 1954.
[*154*] Meller, K.: Elektrische Lichtbogenschweißung. Leipzig 1932.
[*155*] Minorsky, N.: J. Phys. Radium **9**, 127 (1928).
[*156*] Mohler, F. L.: Bur. Stand. J. Res. Wash. **17**, 45, 849 (1936/37).
[*157*] Moisson, H.: Le Four Electrique. Paris 1897.
[*158*] Müller, G., u. W. Finkelnburg: Z. Naturforsch. **1956** (im Druck).
[*159*] Müller, G., u. W. Finkelnburg: Naturwiss. **10**, 294 (1955).
[*159b*] Myer, J. L.: Z. Physik **87**, 1 (1933/34).
[*160*] Nissen, W.: Z. Physik **139**, 638 (1954).

[161] NOTTINGHAM, B. W.: J. Franklin Inst. **206**, 43 (1928); **207**, 299 (1929).
[162] ONSAGER, L.: Phys. Rev. **37**, 405 (1931); **38**, 2265 (1931).
[163] ORNSTEIN, L. S., u. H. BRINKMAN: Physica, Haag **1**, 797 (1934).
[164] ORNSTEIN, L. S., u. H. BRINKMAN: Proc. Amst. **34**, 498 (1930).
[165] PETERS, TH.: Z. Physik **135**, 573 (1953).
[166] PETERS, TH.: Naturwiss. **24**, 571 (1954).
[167] PFUND, A. H.: Astrophys. J. **27**, 297 (1908).
[168] PORITZKY, H., and C. G. SUITS: Physics **6**, 190 (1935).
[169] PRIGOGINE, J.: Etude Thermodynamique des Phenomènes Irréversibles. Paris u. Lüttich 1947.
[170] RAMSAUER, C.: Elektrotechn. u. Masch.-Bau **51**, 189 (1933).
[171] RASCH, E.: ETZ **22**, 155, 305 (1901).
[172] RIETZ, D. DU u. H. KOCH: Praktisches Handbuch der Lichtbogenschweißung. Braunschweig 1947.
[173] ROBSON, A. E., and A. v. ENGEL: Phys. Rev. **93**, 1121 (1954).
[174] ROHLOFF, E.: Z. Physik **126**, 175, 224 (1949).
[175] ROMPE, R., u. P. SCHULZ: Z. Physik **110**, 223 (1938).
[176] ROMPE, R., u. P. SCHULZ: Z. Physik **112**, 691 (1939); **113**, 10 (1939).
[177] ROMPE, R., u. M. STEENBECK: Ergebn. exakt. Naturw. **18**, 257 (1939).
[178] ROMPE, R., u. W. THOURET: Z. techn. Phys. **17**, 377 (1936).
[179] ROMPE, R., u. W. THOURET: Z. techn. Phys. **19**, 352 (1938).
[180] ROMPE, R., u. W. WEIZEL: Z. Physik **120**, 31 (1943).
[181] RÖSSLER, F.: Ann. Phys. **34**, 1 (1939).
[182] ROTHSTEIN, J.: Phys. Rev. **73**, 1214 (1948); **78**, 331 (1950).
[183] SAHA, M. N.: Phil. Mag. **40**, 472 (1920). — Z. Physik **6**, 40 (1921).
[184] SÄNGER-BREDT, J.: Z. Naturforsch. **6**a, 103 (1951); **8**a, 796 (1953).
[185] SCHARDIN, H.: Ergebn. exakt. Naturw. **20**, 303 (1942).
[186] SCHILLING, P. O., u. W. LOCHTE-HOLTGREVEN: Z. Naturforsch. **9**a, 520 (1954).
[187] SCHIRMER, H.: Z. Physik **142**, 1, 116 (1955).
[188] SCHLUGE, H., u. W. FINKELNBURG: Z. Physik **122**, 714 (1944).
[189] SCHLÜTER, A.: Z. Naturforsch. **5**a, 72 (1950); **6**a, 73 (1951).
[190] SCHLÜTER, A.: Ann. Phys. **10**, 418 (1952).
[191] SCHMIDT, E.: Ann. Phys. **4**, 246 (1949).
[192] SCHMITZ, G.: Z. Physik **126**, 1 (1949).
[193] SCHMITZ, G.: Z. Physik **127**, 209 (1950).
[194] SCHMITZ, G.: Z. Naturforsch **5**a, 571 (1950).
[195] SCHMITZ, G.: Z. Physik **132**, 23 (1952).
[196] SCHNAUTZ, H.: Spectrochim. Acta **1**, 173 (1939).
[197] SCHÖNHERR, O.: EZT **30**, 365 (1909).
[198] SCHULZ, P.: Z. techn. Phys. **19**, 585 (1938).
[199] SCHULZ, P.: Ann. Phys. **1**, 318 (1947).
[200] SCHULZ, P.: Ann. Phys. **1**, 95, 107 (1947). — Z. Naturforsch. **2**a, 583 (1947).
[201] SCHULZ, P.: Ann. Phys. **3**, 280 (1948).
[202] SCHULZ, P., u. W. WEIZEL: Z. Physik **122**, 697 (1944).
[203] SCHULZE, G.: Ann. Phys. **12**, 828 (1903).
[204] SEELIGER, R., u. H. SCHMICK: Phys. Z. **28**, 605 (1927).
[205] SHENSTONE, A. G.: Phys. Rev. **46**, 49 (1934).
[206] SICHLING, G.: Ann. Phys. **40**, 330 (1941).
[207] SIMON, TH.: Phys. **6**, 297 (1905). — Jb. drahtl. Telegr. **1**, 16 (1907).
[208] SLEPIAN, J.: Phys. Rev. **27**, 407 (1926).
[209] SMITH, C. G.: Phys. Rev. **62**, 48 (1942); **69**, 96 (1946); **84**, 1075 (1951).
[210] SOMMERMEYER, K.: Z. Physik **90**, 232 (1934).
[211] SPERLING, J.: Z. Physik **128**, 269 (1950).
[212] SPITZER, L., R. S. COHEN and C. R. ROUTLY: Phys. Rev. **80**, 230 (1950).
[213] SPITZER, L., and R. HÄRM: Phys. Rev. **89**, 977 (1953).
[214] STARK, J.: Ann. Phys. **12**, 673 (1903).
[215] STEENBECK, M.: Phys. Z. **33**, 809 (1932).
[216] STEENBECK, M.: Phys. Z. **38**, 1099 (1937).
[217] STEENBECK, M.: Wiss. Veröff. Siemens-Konz. **19**, 59 (1940).
[218] STEINLE, H.: Z. angew. Mineral. **2**, 28 (1939).
[219] ST. JOHN, R. M., and I. G. WINANS: Phys. Rev. **94**, 1097 (1954).
[220] STODOLA, A.: Dampf- und Gasturbinen. Berlin 1924.
[221] STOLT, H.: Z. Physik **31**, 240 (1925).
[222] SUITS, C. G.: Physics **6**, 315 (1935).
[223] SUITS, C. G.: Physics **6**, 190 (1935).

[224] SUITS, C. G.: J. Appl. Phys. **10**, 203, 728 (1939). — Phys. Rev. **55**, 561 (1939).
[225] TANBERG, R.: Nature, Lond. **124**, 371 (1929).
[226] THOURET, W., W. WEIZEL u. P. GÜNTHER: Z. Physik **130**, 621 (1951).
[227] TONKS, L.: Theory and Phenomena of High Current Densities in Low Pressure Arcs. The Electrochemical Society. Reprint 72—13. St. Luis Meeting, 1937.
[228] TONKS, L.: Phys. Rev. **56**, 360 (1939).
[229] UNSÖLD, A.: Z. Astrophys. **12**, 56 (1936).
[230] UNSÖLD, A.: Ann. Phys. **33**, 607 (1938).
[231] UNSÖLD, A.: Physik der Sternatmosphären. Berlin-Göttingen-Heidelberg 1955.
[232] VOS, J. C. DE: The Emissivity of Tungston Ribbon. Diss. Amsterdam 1953.
[233] WALDMANN, L.: Z. Naturforsch. **1**, 12 (1946).
[234] WALDMANN, L.: Z. Naturforsch. **1**, 59 (1946).
[235] WALDMANN, L.: Z. Naturforsch. **4**a, 105 (1949).
[236] WALDMANN, L.: Z. Naturforsch. **5**a, 322 (1950).
[237] WALDMANN, L., u. K. CLUSIUS: Naturwiss. **30**, 711 (1942).
[238] WALKER, G. W.: Ann. Phys. **54**, 514 (1841).
[239] WASSERRAB, T.: Z. Physik **130**, 311 (1951).
[240] WEHRLI, M.: Helv. phys. Acta **1**, 333 (1928).
[241] WEHRLI, M.: Helv. phys. Acta **3**, 180 (1930).
[242] WEHRLI, M., u. P. BÄCHTIGER: Helv. phys. Acta **4**, 290 (1931).
[243] WEIZEL, W.: Z. techn. Phys. **24**, 90 (1943).
[244] WEIZEL, W., R. ROMPE u. M. SCHÖN: Z. Physik **115**, 179 (1940).
[245] WEIZEL, W., R. ROMPE u. P. SCHULZ: Z. Physik **117**, 545 (1941); **119**, 237 (1942).
[246] WEIZEL, W., u. J. FASSBENDER: Z. techn. Phys. **21**, 391 (1940). — Z. Physik **120**, 252 (1943).
[247] WEIZEL, W., u. G. ECKER: Z. Physik **127**, 495 (1950).
[248] WEIZEL, W. u. G. SCHMITZ: Phys. Z. **44**, 383 (1943).
[249] WEIZEL, W., u. W. THOURET: Z. Physik **131**, 170 (1952).
[250] WERGELAND, H., u. K. NORSKE: Vidensk. Selsk. Forhandl. **23**, 110 (1950).
[251] WIENECKE, R.: Z. Physik **143**, 118 (1955).
[252] WIENECKE, R.: Z. Physik **143**, 128 (1955).
[253] WILDT, R.: Astrophys. J. **89**, 295 (1939).
[254] WINTER, A.: Z. Physik **126**, 281 (1949).
[255] WITTE, H.: Z. Physik **88**, 415 (1934).
[256] ZEISE, H.: Thermodynamik. Leipzig 1944.
[257] ZEYEN, K. L., u. W. LOHMANN: Das Schweißen der Eisenwerkstoffe. Düsseldorf 1943.

Electrical Breakdown of Gases with Steady or Direct Current Impulse Potentials.

By

Leonard B. Loeb.

With 9 Figures.

Foreword.

The compression of the vast material under this title into a compact article of some one hundred printed pages in a readable form, places an unbelievably difficult task on any writer. It precludes any attempt at historical summary and requires omission of much worthy work of earlier date which is superseded by the vast array of new information that post World War II techniques have yielded. In this article, there are presented the essential definitions and broad generalizations together with the basic relations, in as clear a fashion as is possible without derivation. There is no room for a discussion of the basic processes underlying breakdown which have undergone tremendous advance in understanding during the last ten years. This article presents enough of what the writer considers the essential features of breakdown processes, with critique, so that interested readers can use the extensive list of references in an effective fashion. Enough of the observations on breakdown in various systems where actions are understood, are given to illustrate the principles earlier propounded.

In thus writing a very condensed summary, the author suggests that the interested reader will find three recent books to which text references are generally inappropriate of considerable value. These are *Electronic and Ionic Impact Phenomena* by H. S. W. Massey and E. H. S. Burhop, Clarendon Press, Oxford, 1952, dealing with the underlying theory of many fundamental processes by two leading theoretical workers in the field. *Basic Processes of Gaseous Electronics*, L. B. Loeb, University of California Press, Berkeley, 1955, treating of the most recent experimental advances in the study of the processes, in many cases, in more detail than is shown in published articles where publication space is limited. By all odds, the most complete recent compilation on the *Electrical Breakdown of Gases* is to be found in an excellent book of that title by J. M. Meek and J. D. Craggs, Clarendon Press, Oxford, 1953. This represents the most extensive and comprehensive treatment of the subject by two active workers in the field since World War II.

I. Introduction. Basic definitions.

1. Primary and secondary mechanisms. Under normal conditions gases show little conductivity. What conductivity there is is produced by external agencies such as cosmic rays, γ-radiations and radioactive traces from container walls. These create electrons and some 20 ions/cm.^{3} per sec. in air leading to currents of $\sim 10^{-15}$ ampere from 100 cm.^{3} of volume. Use of any one of the many agencies which cause liberation of electrons and/or ions from surfaces or liberate electrons from atoms and molecules of the gas can augment these ionizing events to such an extent that with adequate sweep fields applied across the cathode and anode of an electrode pair ionization currents of some magnitude will traverse the gas, e.g., 10^{-11} ampere for X-ray ionization from 100 cm.^{3} volume. These conduction currents are adequately discussed in the writers recent book [*1*] where further references may be found. Usually with such sources the currents increase gradually as applied fields, or better the ratio of applied field in volts/cm., X, to the pressure, p, in mm. of Hg, X/p, increases. It is unlikely that the ratio X/p across any electrode system can be raised sufficiently to draw out all the carriers of each sign produced before other processes occur. These other processes consist of a *primary* action in which each electron creates an *avalanche* of new electrons and ions of magnitude $\exp \int_x^a \alpha dx$ in the field direction in traversing the distance from x to the anode, $x = a$, and a *secondary* action to be later defined. The primary process is called *ionization by collision* and the *first* TOWNSEND *coefficient*, α, represents the number of new ion pairs created per electron per cm. in the field direction. The integral yields the total ionization and represents the number of electrons that arrive at the anode. It is called the TOWNSEND *integral*. Generally speaking α/p is a unique function of X/p, which can experimentally be evaluated and for which convenient functional form approximations can be made to enable α/p to be evaluated for any point x in the field leading to an evaluation of the TOWNSEND integral [2]. The temporal duration of the avalanche as regards direct impact ionization is the order of the electron transit time, $\tau_e = \int_x^a \frac{dx}{v(x)}$, with $v(x)$ the *drift velocity* of the electrons. If, however, the avalanche electrons result not only by direct impact ionization, but possibly in equal measure through secondary interactions of excited states such as by the HORNBECK-MOLNAR, the PENNING effect, the HALE or MOHLER processes [3], there will be diffusion of the avalanche in space and time which in effective instances may be several times the avalanche transit time.

The effect of the primary process then is to increase the initial externally induced ionization current i_0 to that of a *field intensified current* i such that;

$$i = i_0 \exp \int_x^a \alpha\, dx = i_0 (M + 1) \tag{1.1}$$

where M is called the *amplification factor*. The value of M is

$$M = \exp \int_x^a \alpha\, dx - 1$$

and when $\exp \int_x^a \alpha\, dx$ is large it is essentially $\exp \int_x^a \alpha\, dx$.

If in the process of creating the avalanche of $i_0 M$ electrons and ions indicated above there are created *secondary agencies proportional* in number to $i_0 M$ which through action at the cathode or in the gas can *regeneratively* produce $\gamma\, i_0 M$ new electrons at strategically placed locations to start new avalanches the quantity i will rise above i_0 to a *field intensified current*, i, of larger magnitude. This quantity γ is a fractional multiplier the value of which may increase as X/p and hence as M increases. It is called the *second* TOWNSEND *coefficient*. The field intensified current is under many instances given by an expression of the form

$$i = \frac{i_0 \exp \int_x^a \alpha\, dx}{1 - \gamma M}. \tag{1.2}$$

As long as γM is less than unity i represents a field intensified current of increased magnitude over that caused by the primary $i_0 (M + 1)$ process alone.

In essence γ *is a probability* representing the *average value* of the chance that some event in an avalanche will produce a new propitiously placed electron that can yield a successor to the initial avalanche. While in general the expression γ has been reserved for actions at the cathode, because of their greater effectiveness, it is for the present being used in a somewhat broader sense. It must be emphasized that the *second coefficient derives from some agency* of the *primary process,* is *proportional* to it and in general *is a relatively small fractional multiplier* acting, as will be seen, in a *regenerative capacity.* Thus for a true secondary process $\exp \int_x^a \alpha \, dx = 1/\gamma$ with $1 \gg \gamma > 0$.

It may also be indicated that the ionizing events in an avalanche M, are subject to statistical fluctuation as well as events connected with γ. This follows since α is only the average of a large number of ionizing collisions governed by the electrons in an energy distribution in a field that have the energy to ionize and the chance that they do so in a given sequence of collisions with molecules in the field. Thus the quantity γM is subject to fluctuations in any given case. WIJSMAN [*4*] has shown that ignoring any disturbing factors, such as space charge distortion, if above the point where the *average* $\overline{\gamma M} = 1$, the probability that it is unity is given by

$$P_s = 1 - \frac{1}{\gamma M}. \tag{1.3}$$

It is essential that the statistical character of the values of γ and M in individual cases be recognized though in general average values will be used.

2. Secondary processes listed. It is next necessary briefly to indicate the nature of some of the secondary processes active to yield values of γ [*5*]. These are:

α) Active at the cathode. 1. Emission of electrons by the positive avalanche ions at the cathode, coefficient γ_{ci}. Usually nearly all ions reach the cathode. If the ionization potential E_i of the ion is greater than the work function φ_e of the surface, emission in vacuum from outgassed clean metal surfaces, termed *potential liberation,* yields a value of $\gamma_{ci} \sim 0.1$ and not exceeding 0.25 for singly charged ions. *All* ions irrespective of E_i appear to yield values of γ_{ci} if they have some tens of volts of kinetic energy *and* the *surfaces are somewhat gassy.* Yields increase to near unity on such surfaces, even if ionization potentials are lower than work functions, reaching values of γ_{ci} in vacuum of the order of unity at 2000 volts. In the presence of gas, γ_i, the observed number of electrons per positive ion is given by $\gamma_i = \nu \gamma_{ci}$, where γ_{ci} is the vacuum value at the cathode. The quantity ν is a fractional loss by back diffusion which can now roughly be calculated as a function of X/p and average energy of electron emission [*6*]. If the cathode is in the presence of gas and is being bombarded by ions the equilibrium state of the surface changes and γ_i, as observed, will undergo changes of increase, or decrease, in time and with current density and positive ion energy of unpredictable nature. In all γ_i action there is a delay in the emission, of the order of an ion transit time of ions from the anode ($x = a$) where produced to the cathode ($x = b$), $\tau_i = \int_a^b \frac{dx}{v_i(x)}$, following an avalanche which *sets* the *time scale of events.*

2. Excited states of gases emitting photons, whose energy $h\nu$ is greater than the work function φ_e of the surface, will after attenuation by absorption in the gas according to $e^{-\mu x}$, when created near the anode and by geometrical factors, g, (i.e., half the photons in many cases going towards the anode), liberate a fraction γ_{cp} of photo-electrons of which the fraction ν escape back diffusion loss. These excited states are created to the number of ω per cm. of path of the electron avalanche in the field direction while at the same time α ions and electrons are

created. Thus the quantity γ_i for positive ion action at the cathode is for photon action, γ_p, determined by $f=\omega/\alpha$ diminished by the loss factors, so that $\gamma_p = f e^{-\mu x} g\, \nu\, \gamma_{cp}$ now replaces γ_i. At lower values of X/p, f can be as great as 10 or perhaps more. The other factors vary but, for photons of 10 ev energy and more, γ_{cp} can approach 0.1 and $e^{-\mu x}$ may not be too large. Usually, though not always, the photo electric action at the cathode is about an order of magnitude less than γ_i. In the first place γ_{cp} and γ_{ci} vary differently with gas coating and changes in cathode condition. Thus a trace of O_2 may reduce γ_{ci} by a factor of 10 to 100 while it may even slightly enhance γ_{cp}. In many cases photons are emitted within 10^{-7} sec. of the avalanche creation so that a new avalanche of $\gamma_p M$ electrons reaches the anode virtually after an *electron* transit time τ_e. However, in many cases, the photons that are active having $h\nu > \varphi_e$ will not have a μ that is negligible. Thus attenuation of resonance lines in inert gases is very serious [*7*]. However, active photons may be created through pressure dependent secondary actions in which there is a time delay $> \tau_e$ [*8*].

3. Metastable atoms act at the cathode as if they were positive ions in liberating secondary electrons as long as the energy E_m exceeds the work function, φ_e. These are created as a fraction $f_m = \omega_m/\alpha$ in relation to ions. There is with metastables a geometrical loss factor g as for photons since diffusion is in three dimensions. There is also attenuation by destructive collisions in the gas in reaching the cathode. The chance of emission γ_{cm} at the cathode is of the same order as for ions of the same potential energy. The value of γ_m is then given by $\gamma_m = f_m g\, e^{-\mu_m x}\, \nu\, \gamma_{cm}$ in close parallelism with the photon action. In general f_m is larger than unity, and μ_m is much less than it is for photons in pure gases. If gases are impure μ_m can be large and the cathode γ_m does not figure, the action of the metastables being to produce secondary electrons in situ in the gas equivalent to an increase in α. There is one difference between photons, ions and metastables and this is that the *metastables must diffuse to the cathode.* The diffusion velocity per unit pressure gradient, and thus diffusion time, bears a fixed ratio to the mobility. Therefore since diffusion partial pressure gradients are low compared to the pressure of electrical fields on ions, diffusive movement of metastables is slow. Thus in general *the transit time for metastables from anode to cathode is of the order of 10 to 100 times the ion crossing time.* In consequence γ_m produces the slowest and most diffuse component of the secondary cathode liberation. Thus only for gases yielding metastable atoms or molecules and then only under exceptional conditions is the contribution of γ_m an important secondary factor in gaseous breakdown.

The remaining proper secondary factors operate *in the gas* [*5*], they are:

β) Active in the gas: 4. Ionization by positive ions on impact with molecules in the gas. The chance of this is of the order of 0.01 in relation to ionic gas kinetic impacts and usually requires energies of ions, or neutralized ions, of the order of 10^2 volts. It is not often likely as a secondary mechanism in gaseous breakdowns where X/p has values so low that the positive ions only rarely get tens of volts. Before this action can occur one of the other mechanisms is already much more effective.

5. Ionization by metastable atoms in the gas. Where this occurs it augments the first coefficient by the PENNING *effect* [*3*]. Under special conditions photons can be produced by interactions of metastables in pure gases leading to a γ_p action at the cathode [*8*]. Otherwise the chance that metastable action *in the gas* will do more than increase α is very remote and no evidence for such action has been observed.

6. *The action of photoionization in the gas*, for which photon creation is proportional to the ionization in the avalanche following the relation $f=\omega/\alpha$ in *uniform fields*, does lead to some field intensification. However, the effect is small except under most unlikely combinations of conditions so that it need not be considered except as with metastables in augmenting α. In *asymmetrical fields* whether caused by electrode geometry, or by ion space charge distortion, *the photoionization in the gas can lead to self sustaining processes causing a breakdown of the gas, or even a spark.* In such situations distances for photoionization from the point of maximum ionization are short, thus while μ is high x is mall and photon absorption leads to ionization that is strategically, or effectively, placed. Here again production is proportional to avalanche ionization, $f=\omega/\alpha$. There is a geometrical loss factor, back diffusion does not enter but f_1, the chance that the absorbed photon produces a new photo electron appears. In geometrically asymmetrical fields, as with the coaxial cylindrical positive wire GEIGER counter, threshold relations are in principle not different from those for γ_p [*9*]. However, where streamer formation occurs additional conditions are imposed by augmentation of the asymmetrical fields by avalanche created positive ion space charge. Here again the relations between primary and secondary processes fulfill all criteria characterizing photoionization in the gas as a proper secondary mechanism [*9*].

γ) Designation of chief processes and elimination of further processes. The forgoing section on secondary mechanisms may be properly summed up by saying that so far most *breakdown* observed in gases is caused primarily by the *two, cathode mechanisms* γ_i and γ_p and by *photoionization in the gas.* These mechanisms always play a *truly secondary rôle* in that they are small relative to the primary factor M, are proportional to it, serving merely to augment the effect and in their most important function to act in a regenerative capacity to sustain primary action. Under some breakdown conditions especially involving suitable time scales γ_p and γ_i act in concert, and over longer time intervals γ_i, γ_p and γ_m act together. Under varying conditions of cathode state, gaseous species, pressure, fields etc., either γ_i or γ_p will predominate. Where conditions are suitable in asymmetrical fields the photoionization in the gaseous phase will occur under conditions of such efficiency as to supersede the cathode processes so that anode breakdown will predominate. An excellent example results from the introduction of 4% O_2 into pure N_2 which suppresses γ_i, slightly augments a low γ_p and encourages photoelectric ionization in the gas [*10*].

There has been an inclination to add new secondary processes to the list, a favored one being field emission [*11*] and perhaps at other times thermionic emission. Properly speaking this usage is incorrect. Both processes are conditioned by external agencies, or, at most, are augmented by high current densities in already broken down gaps. If caused by external agencies in a gap *before breakdown*, they, like the photoemission used to insure a uniform gap current in all sorts of studies, furnish an i_0 created by an external agency, i.e., an ultraviolet source, a heating current in a filament, or a very high localized field yielding a FOWLER-NORDHEIM emission current of electrons [*11*]. Each of these is a *triggering agency* in this capacity [*11*]. Now it appears that for very short gaps at very high pressures, or densities, or for vacuum gaps of short length with very high fields the mechanism of breakdown has recently been delineated [*81*]. Under these conditions the field emission current from points becomes quite high. For the high pressure gap the electrons *occasionally* ionize a molecule in transit. This gives M small but finite values, e.g., of the order of 10^{-2}. The positive ions created under these conditions, few though they are, augment the localized fields at the sub microscopic points from X to $X' = X + \Delta X$. Placing the new X' into the exponent of the FOWLER-NORDHEIM theory increases the field emission electron flow. A simplified solution of the problem then yields an expression for breakdown *where the dominant field emission current term Γ multiplied by M yields a product ΓM which when it exceeds unity represents the instability leading to breakdown.* It has been suggested [*81*] that this condition, akin to the relation $\gamma M = 1$ for secondary-primary processes in ordinary breakdown, makes the multiplier Γ for field emission represent a *sort* of *secondary mechanism.* The analogy is far fetched indeed, for here *M has the true secondary rôle* with a value less than the multiplier Γ acting regeneratively on X, while the term Γ for field emission is NOT directly in proportion to ionizing events in M. This field emission breakdown with increase of field by M is dominantly a *breakdown by field emission,* which is augmented by the TOWNSEND integral as the secondary factor. The field emission increases until a power arc of metallized vapor builds up. For the vacuum breakdown the field emission current heats and vaporizes anode material which finally furnishes an ionized plasma yielding a low M and eventually leading to an arc. It is best therefore not to confuse an already complicated set of processes by associating them to factors in a breakdown mechanism of a different type through drawing parallels between artificially similar terms.

One more remark must be made in this connection and that is that *secondary processes leading to the break down of a gas are not to be confused with those sustaining discharge in brokendown gaps.* While the formally designated secondary mechanisms for breakdown may sustain glows and arcs the maintainance of these may include other factors, thermionic emission certainly being one and possibly field emission the other. The distinction of primary and secondary processes thus differs between breakdown and maintenance conditions. In what follows thermionic and field emission will *not* be considered *proper secondary breakdown* mechanisms.

3. Concepts of thresholds for breakdown and spark. *α) Breakdown threshold.* Returning now to the expression for the field intensified current with primary and secondary mechanisms active it is seen that as long as $\gamma M < 1$ the field intensified current has a definite value i greater than i_0 and greater than what it would have with primary coefficient α alone acting. If i_0 is cut off there is no current, that is, the whole current is *induced by an external* agency, is *field intensified* but is *not self sustaining.* Disregarding the equation for i at this point it is necessary to note that when

$$\gamma M = 1\,, \tag{3.1}$$

a condition is produced in which on the average for each avalanche of M electrons there is produced a new electron at the cathode capable of yielding a new successor avalanche. Ignoring statistical fluctuation it is then seen that; $\gamma M = 1$ represents the *threshold condition* of a *self sustaining discharge.* For if $\gamma M = 1$ an avalanche started by a single electron becomes self perpetuating and independently of i_0 the discharge maintains itself. If $\gamma M > 1$ the situation is one in which the succession of avalanches creates more carriers than needed to perpetuate the sequence and the current will grow in magnitude.

It is clear then that when the *regenerative* action of the secondary mechanism transcends some value such that γM equals unity the *field intensified* current has been replaced by a *self sustaining* current or discharge. The gap previously nonconducting without external influence has undergone an *electrical breakdown* to a *conducting state* and is *capable of discharging* the energy stored in a condenser, or an electrical circuit. The point characterizing the transition $\gamma M = 1$ is called the *breakdown threshold*, or *discharge threshold* and for a given gap form is characterized at a given pressure and temperature by characteristic values of X/p across the gap and an over all characteristic applied potential V_B called the *breakdown potential*.

Now it is clear that these transitions from a non conducting to a conducting state can take on a wide variety of characteristic forms. For example it could be the transition from a current of 10^{-14} ampere from a positive point to a negative plane to the breakdown current of 10^{-7} ampere with the blue glow at the point in air at 760 mm. characteristic of the burst pulse corona. On the other hand it could take the form of a uniform glow springing up from a current of 10^{-12} amp between parallel plates 10 cm. in diameter and 1 cm. apart in argon at 20 mm. pressure, to a current of the order of milliamperes with certain light and dark zones near the electrodes. Finally it could take the form of a brilliant fine filamentary flash between two electrodes at atmospheric pressure with its crackling noise characteristic of the common spark. Thus to the term breakdown may be coupled descriptive words indicating its nature, or the underlying mechanism active. Thus one would speak of a burst pulse corona breakdown, a TOWNSEND breakdown in argon or a spark breakdown and designate the thresholds by such terms. It is essential to point out, however, that the *term breakdown* and/or, as we shall later see, *spark breakdown* refer to the *transition mechanism*, whereas the term *discharge* applies more specifically to the *character of the conduction current mechanism to which the breakdown leads*. Thus it is more proper to speak of the corona discharge, the glow, or TOWNSEND discharge, and the arc discharge as the *end result* and reserve the word breakdown or spark breakdown for the actual transition mechanism.

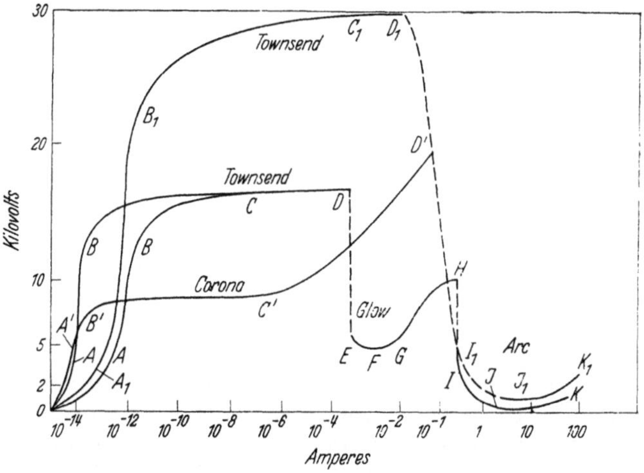

Fig. 1. Diagram of currents as abscissae and potential as ordinates for glow discharge tube geometry, letters A B C D etc., Point to plane corona gap geometry, letters A' B' C' D' and for uniform field gap geometry, letters A_1 B_1 C_1 D_1 as they might appear for N_2 at 1 mm., air at 760 mm., and N_2 at 760 mm., 1 cm. gap respectively.

β) The breakdown diagram, and definition of a spark. Before turning to the differentiation between breakdown and spark breakdown it is expedient to review by graphical representation the potential current relationships for various electrode, and/or, pressure conditions since in these certain very interesting principles are brought to light. In Fig. 1 there are portrayed a set of curves with potential plotted as ordinates and current plotted as abscissae to a mainly logarithmic scale, somewhat distorted in the range of low currents, to cover the wide range of currents. These curves apply to a number of gaps, letters A, B, C, etc. applying to a gap between electrodes with considerable separation at a pressure of about 1 mm. of N_2, the interelectrode distance having been reduced somewhat in the 1 to 100 ampere range, again in order to include all the phenomena in one curve. The letters A', B', C', etc. apply to a highly asymmetrical gap, e.g., a point to plane gap, of several cm. length at 760 mm. and letters A_1, B_1, C_1, D_1 apply to a uniform field gap in N_2 at 760 mm. pressure about 1 cm. long. The two curves $ABCD$ starting at the origin and merging at higher potentials represent two different currents, the lower current applying to an i_0 as with natural ionization of 10^{-14} ampere and the other to an i_0 of 10^{-12} ampere induced by ultraviolet light.

Consider first the lower pressure curve $ABCD$. If the initial current i_0 is 10^{-14} or 10^{-12} ampere the curve rises from zero at zero potential and reaches a quasi saturation value of 10^{-14} or 10^{-12} ampere at A. As the potential reaches some 7000 volts the field intensified currents begin at B. Currents increase rapidly to the low order self sustaining discharge at C where the two curves merge again. Further increase in potential increases the current to some point D. At this point space charge accumulation owing to the self sustaining discharge at $\gamma M \gg 1$, caused by an increase in cathode potential fall, leads through the dotted section to a more efficient discharge known as the GEISSLER, or glow discharge. The current remains high at the reduced potential at E following the decline. The region EF is still one in which current increases with falling potential. In fact the region DEF is an *unstable* region, with $dV/di = R$ negative. This negative *resistance* characteristic cannot *directly* be observed, but it means that the current will grow with decrease in potential except for the control of current by external resistance. The existance of such negative resistance is noted in the current behaviour as the external resistance is varied and the potential V_d across the discharge is constant given by $V_a = V_d + iR_e$ where V_a is the applied potential, i is the current and R_e is the external resistance. The horizontal region F is the region of the normal cathode fall in a glow discharge. Here current increases *at constant potential* through spread of the discharge over the cathode surface. At G the abnormal region of the glow discharge is reached. Here further increase in current can only be achieved by raising the potential. This regime continues to a point H. Here local heating of the cathode leads to a new type of discharge dependent on thermionic emission, or its equivalent in the low pressure arc. This new mechanism can operate with greater economy at high current and still lower potential. Here again the region HIJ is a region of instability and negative resistance with current up to J controlled only by external resistance. Increasing the current beyond J requires again increase in voltage to achieve the high current density arc with its great radiative losses.

The curve $A_1 B_1 C_1 D_1$ represents the same situation for the 760 mm. pressure uniform field gap. Here field intensification begins at B_1. At C_1 there is a low order self sustaining TOWNSEND discharge which is stable over a very small range of voltage [12]. Immediately above this point space charge accumulation causes a field distortion which leads to the photoelectric streamer propagation in mid gap ending directly in a power arc via a filamentary spark. The higher pressure metal vapor arc resulting $I_1 J_1 K_1$ requires higher potentials and goes over to the high current density arc with positive resistance characteristic in the region $I_1 K_1$.

The curve $A'B'C'D'$ for the positive point to plane corona in air with electrons from the plane by means of i_0 begins at about 9000 volts with a self sustaining current of a microampere at C'. The current increases as indicated up to D' at which point breakdown streamers appear leading to a power arc as with plane parallel electrodes.

What is to be noted is that all these gap forms at some point *begin with a field intensified current,* which with relatively little further increase in potential, except for the corona with its extensive low field region at the plane, go over to current forms that by space charge formation and disposition, or increase in current density, lead to decrease in potential for the same current. The lower pressure gap shows this best with two distinct discharge mechanisms operating at 5000 volts and milliamperes and some hundred volts and amperes. In all these transitions in which a more effective current mechanism is sustained at lower potential there is the region of dotted lines with negative $dV/di = R$, indicating a *negative resistance* and *instability* which leads to a *unidirectional irreversible current change*. This *change* is *characteristic of the process* which we will term the *spark breakdown*.

It is thus possible to define a *spark breakdown*, or *spark*, for short, as a class of *transient occurrences* in which a given existing conduction current in a gas *suddenly* and *irreversibly* changes to a current of a higher magnitude, operating more efficiently by a different mechanism under the imposed conditions which rendered the lower current unstable. Thus all the transitions at D_1, D', D and H are sparks. These sparks materialize as a result of increase in potential and charge generation or current in the gap. They could occur equally well, and do occur by changes in secondary emission with ion bombardment of the cathode at constant potential, or though pressure changes. DRUYVESEYN [*13*] has succeeded in producing the transition $H-I$ by means of an *externally* heated cathode in a *reversible* fashion; this would *not* constitute a spark, as it is not self induced and is not unidirectional and irreversible. Under simplifying assumptions v. ENGEL and STEENBECK [*13*] using thermionic emission and heating by ion bombardment of the cathode, calculate the transition $V_d - i$ relation without external heating but with current controlled by external resistance in agreement with observation. This is also *not* strictly a spark since it operates reversibly by external resistance current.

It is now clear that there have been defined the *breakdown* and the *spark breakdown*, or *spark*. These, it is clear, are distinct and separate phenomena. It is noted that in Fig. 1 the breakdown always precedes the more dramatic and striking spark breakdown. It has only been in the last four years that the precedence of a spark by the less notable breakdown phase has been established for uniform fields [*12*], [*14*]. In the spark breakdown of uniform field gaps the two thresholds in some cases are so near together in potential that the initial breakdown was at first not realized. In very pure argon gas, in fact just above threshold for the TOWNSEND breakdown, the current due to this is so large that the spark transition with current limited by external resistance is far less conspicuous than the initial breakdown [*15*].

A few more remarks must be made in relation to the definition of the spark just given.

First it must be noted that the conditions for a spark in a given gap may be realized and the spark breakdown may occur without it being possible, for the final stable discharge form of current such as at F, I or J_1 to be thoroughly established owing to circuit limitations. Thus it is perfectly possible for a condenser of low capacity to be charged up to the point D_1 and proceed to the point I_1 without there being enough power to maintain the arc at J that would otherwise occur. This in fact is the case with the frequently encountered static spark of every day life. In other words the ultimate end of the spark breakdown transition may be any sort of a conduction current condition, irrespective of the instability and the current transition defining the spark, depending on external circuitry. Thus care must be taken in defining the spark breakdown and studying its mechanism to disassociate its mechanism completely from the ultimate nature of the current to which the condition could lead.

This at once raises the question as to the exact points in a sequence of events that mark, or delimit, the breakdown, or spark. In other words, at what point in the sequence of events of the process is the breakdown, or spark, said to be completed? While there is some element of aesthetic choice the basic phenomenology of the breakdown should enable limits to be defined for common usage without difficulty. Thus the corona breakdown process, or transition, begins with the application of the breakdown potential to the initial current i_0 of 10^{-14} ampere and extends to the point where the corona current of 10^{-6} ampere has been established. In other cases the breakdown may have arbitrarily to be defined relative to a given application or stage in a process. Thus for instance spark breakdown has been defined as extending from the application of the breakdown potential, (including or excluding statistical delays), to the point where the brilliant light appears, or to the point where the potential across the gap has fallen to half the final, or even near the final values. In describing the detailed mechanism of the streamer spark the writer for example considers breakdown to extend from the start of the avalanche of the initiatory electron through streamer formation, the bridging of the gap by the streamer and the *completion* of the *brilliant return stroke* which shorts the gap with a filament having some 10^{17} or more electrons/cm.^{3}. The reason for terming this effective breakdown is that this sequence produces the change in conductivity required to permit the arc and is all that involves the processes and factors influencing the initial breakdown. This example should clarify the question sufficiently so that no great difficulty is encountered in other applications.

In what has gone before, it is clear that the definition of breakdown and spark breakdown are in no way associated with the temporal sequence of events, nor in fact with the time taken to complete the transition. Since each breakdown mechanism involves sequences of primary and secondary ionizing actions, each element of which involves characteristic time intervals such as τ_i, τ_e, etc., it is clear that each breakdown will always be associated with a *time* which is characteristic of the process. In fact the same gap in an inert gas may have three characteristic spark breakdown thresholds, (X/p values, or values of V at constant p) depending on the allowed time of potential application permitting γ_m, γ_i and γ_p, or γ_i and γ_p, or just γ_p to act. Time intervals may reach into the seconds, as with a low pressure TOWNSEND discharge at its ultimate threshold potential, or it may go down to 10^{-9} sec. as with well overvolted spark gap and an adequate i_0 to reduce statistical lags. Even in a given situation actual times may fluctuate about a mean value and statistical initiatory fluctuations may also occur further to diffuse temporal data. It is seen, however, that these temporal characteristics do not alter the definition and usage of the term breakdown although they are obviously closely associated with the particular processes active.

γ) Statistical fluctuations and effect of space charge. Referring again to Fig. 1 it is seen that there are various threshold regions for onset of self sustaining discharge such as C, C' and C_1 and for sparks such as D, D', D_1 and H. Now it will be recalled that in the threshold condition for breakdown $\gamma M = 1$ it was pointed out that despite an average value $\overline{\gamma M}$ the product γM is open in individually initiated avalanche sequences to relatively large statistical fluctuations. Regarding therefore the fact that a single avalanche sequence at the average threshold is capable of leading to breakdown, and by regarding Fig. 1, noting that this is in general independent of i_0, it is seen that viewed from the aspect of single electron avalanche initiation the achievement of a spark by the avalanche is open to considerable doubt. Thus even if i_0 has a finite value, as to possible initiating electrons viewed on a microsecond time scale, there may be quite a statistical time lag or *jitter* in the arrival of initiating electrons, and near threshold, of those electrons that appear only relatively few will lead to a breakdown sequence. Even with the appreciable currents flowing in the regions D_1, D', D and H, there will, because of statistical fluctuations in the building up sequences of the streamers, etc., be a *tendency to failure* of *initiating electrons*, or event sequences, in consummating the spark.

It is clear that the higher the overvoltage, that is the potential above the threshold value, and thus the larger the average γM above unity, the more certain is the event to occur. Hence at the *thresholds* there should be a tendency for *interruption of the discharge sequence.* In fact the WIJSMAN theory indicates zero chance for a continuous sequence at $\overline{\gamma M}=1$. In consequence where the breakdown does not lead to at change in mechanism, the threshold should be marked by interrupted or intermittent discharges, and spark transition by a jitter in time between arrival at the transition voltage and consummation of the transition. Such tendencies are in fact observed in an exaggerated form since in addition to the natural statistical fluctuations these are exaggerated, or modified, by space charge effects. As will later be indicated space charges accumulate when-ever $\gamma M > 1$ and when the value of X/p over the gap is not adequate to remove the surplus charge as fast as it accumulates. It is favored further by the great difference in drift velocities of electrons and positive ions. Action of *space charges is* at *times inhibitory* to the *breakdown process* by lowering fields at crucial points. Under other circumstances *space charge aids breakdown,* by advantages in field distribution, insuring against interruption by statistical fluctuations in long avalanche sequences. The following effects should be noted.

a) In *uniform fields,* as long as the first TOWNSEND coefficient α/p increases more rapidly than proportional to X/p, positive space charge accumulation favors avalanche sequences above $\overline{\gamma M}=1$, insures against interruption by statistical factors and lowers the threshold [*16*]. Where α/p varies more slowly than linearly with X/p positive space charge formation interferes with the avalanches and raises the sparking threshold.

b) In *asymmetrical fields* the space charge formation acts in more complicated fashion.

1. Outward of a highly stressed positive electrode the positive space charge acts to decrease the anode field and increase the very low cathode field. Under these conditions it is very effective in emphasizing statistical fluctuations above threshold by causing extinction or intermittence of discharge, and its action is the most effective near threshold. It further is effective in two regions of such gaps.

Near the anode the charge from the first set of avalanche sequences interrupts these until it is removed rapidly from the high field region permitting resumption of the discharge after short time intervals called *dead times.* After a sequence of such interrupted discharges the *accumulated* space charge in the *low field regions* reduces the effective field at the anode so that discharge cease until virtually all the space charges clear from the gap in the *clearing time* interval [*17*].

2. For the highly stressed negative electrode the positive ion space charge created near the electrode by avalanches enhances the cathode field and ensures continuity of discharge despite statistical fluctuation [*18*]. In fact it can act as a means of introducing an instability leading to negative resistance and a spark breakdown in gases whose electrons do not make negative ions and are rapidly removed in low fields [*18*]. If the electrons attach beyond the positive space charge but still near the cathode as in air then again an intermittent current is produced with the reduction of field by the slow ions [*19*].

3. Where in consequence of convergent geometry near a highly stressed anode, or in an asymmetrical field, avalanches are so intense that when the electron cloud is drawn away the positive ion space charge density *augments* the anode field, discharge is favored in the event that photoelectric ionization occurs in the gas [*20*]. This causes streamer formation, which unless it leads to breakdown, will leave behind in the low field regions space charge inhibiting further streamers [*20*]. If the streamers cross intermittence will not occur and a spark will result from space charge stimulated action [*20*].

It is thus to be seen that there are to be added to the consideration of breakdown and spark thresholds the effects of statistical fluctuations as augmented, or counteracted, by space charge action as well as the influence of space charge action itself. These actions, however, while modifying thresholds and complicating sparking sequences do not modify the basic definitions developed.

4. The accurate formulation of breakdown threshold. As indicated in the introduction, quite generally the threshold of the breakdown and/or the spark is set when the product of the avalanche size M and the probability of a regenerative secondary electron emitted as a result of the actions in creating the avalanche is unity. This follows since if γM is less than unity the current terminates as soon as the external initiating or triggering current i_0 ceases. If γM is greater than unity the charges accumulate in the gap leading to some change. This definition of the threshold was first introduced by HOLST and OOSTERHUIS in 1923, who working for the first time with relatively pure inert gases and clean electrodes also indicated that γ_i was the active secondary mechanism and varied with cathode material. The same condition $\gamma M = 1$ had in 1903 been indicated by TOWNSEND from his expression for i as a function i_0 using his concept of ionization by positive ions in the gas as the secondary mechanism. In this event for uniform fields the equation reads

$$\frac{i}{i_0} = \frac{e^{(\alpha-\beta)x}}{1 - \frac{\beta}{\alpha}\left[e^{(\alpha-\beta)x} - 1\right]}. \tag{4.1}$$

If β/α be replaced by the more likely γ_i action, an equation deduced independently both by J. J. THOMSON and TOWNSEND for uniform fields reading

$$\frac{i}{i_0} = \frac{e^{\alpha x}}{1 - \gamma\left(e^{\alpha x} - 1\right)} = \frac{(M+1)}{1 - \gamma M} \tag{4.2}$$

gives the specific form for i/i_0 in terms of γ and M.

Here TOWNSEND noted that when the denominator of equation (4.1) and/or (4.2) becomes indefinitely small, i/i_0 becomes indefinitely large and *he set this as the criterion for the spark threshold* [*22*]. With his experimental values of α/p as a function $f(X/p)$ and of β/p as another function $f_1(X/p)$ near breakdown he indicated that in fact the condition

$$1 - \frac{\beta}{\alpha}\left[e^{(\alpha-\beta)x} - 1\right] = 0 \tag{4.3}$$

was satisfied within limits of experimental error. From equation (4.2) the condition becomes

$$\gamma_i\left(e^{\alpha x} - 1\right) = \gamma_i M = 1. \tag{4.4}$$

There is an important fundamental difference in the derivation of the condition (3.1) $\gamma_i M = 1$ by HOLST and OOSTERHUIS and of its equivalent (4.3) or (4.4) by TOWNSEND. Equations (4.1) and/or (4.2) are ones for a steady state current i induced by i_0 and i is proportional to i_0. Setting the condition that a spark occurs when i/i_0 becomes *infinite* makes the threshold depend on i_0 and makes it indeterminate. HOLST and OOSTERHUIS regard the threshold from the viewpoint of *one* necessary initiating electron and in so far as average values, aside from statistical fluctuations, go their threshold (3.1) is determinate and finite. As will later be seen the more complete theories involving temporal growth of the discharge indicate that in fact i_0 on basic principles does *not* enter into the threshold condition as set by processes involved, its function being to influence the time of growth. Thus given adequate time, the condition of HOLST and OOSTERHUIS sets the threshold. Very recently a generalized mathematical theory of the process has again confirmed this definition [*23*]. Observationally at very high values of i_0 the sparking threshold in uniform fields has been shown to be influenced by the value of i_0. However, in this instance the effect has been shown by several investigators to be ascribed to a factor not included in the basic theory of breakdown so far considered, that is the positive ion space charge distortion of the uniform field in the gap, which as noted earlier together with statistical fluctuations can alter the threshold [*16*].

II. The growth of the TOWNSEND discharge.

5. STEENBECK-BARTOLOMEYCZYK analysis; without space charge. The analysis of the temporal growth of the TOWNSEND discharge, in uniform fields, i.e., the breakdown using TOWNSEND's basic concepts without the intervention of space charge was first undertaken by STEENBECK [*24*] and is to be found in some detail in VON ENGEL and STEENBECK's excellent monograph, Vol. II. It was extended later by BARTOLOMEYCZYK [*24*] to include both γ_i and γ_p instead of just γ_i. Some extension of this derivation to include metastable action was carried out by ROGOWSKI [*24*]. SCHADE [*24*] in an experimental study of the temporal growth of the breakdown at lower pressures independently arrived at an approximation to STEENBECK'S solution. Later more extensive studies of the earlier stages of the breakdown process for action of fast (γ_i and γ_p) secondary components at the cathode and slow (γ_m) components were made by R. R. NEWTON [*25*], by J. A. HORNBECK [*25*], by R. N. VARNEY [*25*] and by J. P. MOLNAR [*25*] for use in the evaluation of the second TOWNSEND coefficients. All of these apply to uniform field geometry and omit the action of space charges. Probably the solution of R. N. VARNEY [*25*] for the study of currents for evaluation of γ_i is the most extensive. However, of all the solutions leading to an evaluation of the course of a TOWNSEND breakdown in this geometry, in the absence of space charge fields the most cogent is that of STEENBECK and BARTOLOMEYCZYK. Space precludes its detailed development.

Let n_+ and n_- represent number densities of positive and negative ions while j_+ and j_- represent the corresponding current densities. The electron and ion drift velocities are $v_+ = X k_+$ and $v_- = X k_-$. Assume initially a single cathode process γ_i. Then the differential equations may be written

$$\left.\begin{aligned} \frac{\partial n_-}{\partial t} &= -\frac{\partial(v_- n_-)}{\partial x} + \alpha v_- n_-, \\ \frac{\partial n_+}{\partial t} &= +\frac{\partial(v_+ n_+)}{\partial x} + \alpha v_+ n_+, \end{aligned}\right\} \tag{5.1}$$

where x is distance in the direction of the field X and α is the first TOWNSEND coefficient. These lead to the current density relations

$$\left.\begin{aligned} \frac{\partial}{\partial t}\left(\frac{j_-}{v_-}\right) &= \frac{\partial j_-}{\partial x} + \alpha j_-, \\ \frac{\partial}{\partial t}\left(\frac{j_+}{v_+}\right) &= \frac{\partial j_+}{\partial x} + \alpha j_+. \end{aligned}\right\} \tag{5.2}$$

Integration is facilitated by a further assumption as to the temporal growth of current. Toward this end, STEENBECK intuitatively set the growth in t as an exponential one about an average characteristic build up time τ dependent on X/p, x, α, drift velocities of carriers and γ. This is justified since ionization increases exponentially with x across the gap and if the velocities are constant in the uniform field the growth of carriers with time will also be exponential. In a recent article [23], AUER has shown on rigorous generalized mathematical analysis that this assumption is justified. Thus both STEENBECK and BARTOLOMEYCZYK set the current density as a function of x and t as a new current density of x alone, in the form;

$$j_{1-}(x,t) = e^{t/\tau} j_-(x); \quad j_{1+}(x,t) = e^{t/\tau} j_+(x). \tag{5.3}$$

Setting $\frac{1}{v} = \left(\frac{1}{v_+} + \frac{1}{v_-}\right)$, integration yields

$$\left.\begin{aligned} j_-(x) &= C\, e^{\int\limits_0^x \left(\alpha - \frac{1}{\tau v_-}\right) dx}, \\ j_+(x) &= e^{\int\limits_0^x \frac{dx}{\tau v_+}} \left[C' - C \int\limits_0^x \alpha\, e^{\int\limits_0^x \left(\alpha - \frac{1}{v\tau}\right) dx} dx\right]. \end{aligned}\right\} \tag{5.4}$$

If j_0 is the externally induced photo-current at the cathode, boundary conditions set $j_-(0) = j_0 + \gamma_i j_+(0)$, $j_+(d) = 0$ at cathode and anode respectively. At $x = 0$ the negative electron contribution neglecting external ionization is $\gamma_i j_+(0)$ and at the anode $x = d$, $j_+(d) = 0$. The generalized sparking threshold then reads

$$1 = \frac{j_0}{j_-(0)} + \gamma_i \int\limits_0^d \alpha\, e^{\int\limits_0^x \left(\alpha - \frac{1}{v\tau}\right) dx} dx. \tag{5.5}$$

If photoelectric liberation is included, H. L. VON GUGELBERG set the boundary condition as

$$1 = \frac{j_0}{j_-(0)} + \gamma_i j_{1+}(0) + g\,\gamma_p \int\limits_0^d j_{1-}(x) \int\limits_0^x x\, e^{-\mu x} dx, \tag{5.6}$$

yielding the complete solutions

$$1 = \frac{j_0}{j_-(0)} + \gamma_i \int\limits_0^d \alpha\, e^{\int\limits_0^x \left(\alpha - \frac{1}{\tau v}\right) dx} dx + \frac{\nu \gamma_{cp}}{g} \int\limits_0^d \omega\, e^{\int\limits_0^x \left(\alpha - \frac{1}{\tau v_-}\right) dx - \mu x} dx. \tag{5.7}$$

Near breakdown j_0/j_{-0} should vanish compared to unity. That is, since it takes only a few electrons to trigger the breakdown, j_0 is small and even when j_0 is 10^{-12} ampere, since at breakdown $j_{-0} \sim 10^{-2}$ ampere, the ratio can be neglected relative to unity. Since v is essentially v_i in these relations, integration for a gap length d leads to the conditions

$$1 = \gamma_i \alpha \frac{\exp\left(\alpha - \frac{1}{\tau v_i}\right) d - 1}{\alpha - \frac{1}{\tau v_i}}, \tag{5.8}$$

$$1 = \frac{\alpha \gamma_i}{\alpha - \frac{1}{\tau v_i}} \left[e^{\left(\alpha - \frac{1}{\tau v_i}\right) d} - 1\right] + \frac{g\,\omega\,\gamma_p}{\alpha - \mu - \frac{1}{\tau v_i}} \left[e^{\left(\alpha - \mu - \frac{1}{\tau v_-}\right) d} - 1\right]. \tag{5.9}$$

If τ is large so that $1/\tau v_i$ is small relative to the values of α which range from 1 up, i.e., if $\tau v_i \sim 10^{-3}$ cm. or less, ($v \sim 10^6$ cm./sec., $\tau > 10^{-2}$ sec.), equation (5.8) reduces to equation (4.4) which is the same in value as the threshold condition (3.1) $\gamma_i M = 1$ defined by HOLST and OOSTERHUIS. It does not contain j_0, for as indicated above $j_0/j_{-0} \ll 1$ and so leads to a term negligible compared to unity. Actually here neglect of i_0 is based on the empirical observation that $i_0 \sim 1$ electron in $\Delta\tau$ and thus far less than $i_-(0)$.

It is now seen that when the temporal sequence of events is considered the breakdown condition is modified by the inclusion of a quantity $1/\tau v_i$ or $1/\tau v_-$ which subtracts from α. The quantity τv represents a characteristic length depending on carrier velocity and a time imposed by the TOWNSEND model of the process as well as the values of quantities characteristic of the process, i.e., α, v, γ and d. It is thus essential to evaluate and discuss the quantity τv and especially τ. Regarding equation (5.8) it is seen that since in general $1/\tau v_i$ must be less than α the relation makes

$$\tau_i = \frac{1}{v_i\left[\alpha - \frac{1}{d}\left(\log\frac{1}{\gamma_i} + 1\right)\right]} \tag{5.10}$$

for the simple case with just γ_i. In the case of equation (5.9) for both γ_i and γ_p, there are two expressions for τ, one depending on ion action alone and the other on photon action alone. These expressions solved for τ become, for ions and photons

$$\tau_{i,p} = \frac{1}{v_i\left[\alpha - \frac{1}{d_i}\log\left(\frac{1}{\gamma_i}\right)\left\{1 - \frac{g\,\gamma_{cp}\,\nu\,\omega}{\alpha-\mu}\,e^{(\alpha-\mu)\,d_i}\right\}\right]} \tag{5.11}$$

as $1/\tau_i v_-$ is very much less than α, and for photons alone;

$$\tau_p = \frac{1}{v_-\left\{\alpha - \mu - \frac{1}{d_p}\log\frac{\alpha-\mu}{g\,\gamma_{cp}\,\nu\,\omega}\right\}}\,. \tag{5.12}$$

It is seen that if absorption enters into the expression for photon action the value of d_p in equation (5.12) which without absorption can be set as

$$d_p = \frac{1}{\alpha}\log\left(\frac{\alpha}{g\,\gamma_{cp}\,\nu\,\omega} + 1\right) \tag{5.13}$$

becomes

$$d_p' = \frac{1}{\alpha-\mu}\log\left(\frac{\alpha-\mu}{g\,\gamma_{cp}\,\nu\,\omega} + 1\right). \tag{5.14}$$

If d becomes very large then the relations (5.10) and (5.12) become approximately

$$\tau_{i\infty} = \frac{1}{v_i(1+\gamma_i)\,\alpha} \tag{5.15}$$

and

$$\tau_{p\infty} = \frac{1}{v_-(\alpha - \mu + g\,\gamma_{cp}\,\nu\,\omega)}\,. \tag{5.16}$$

Regarding first the very simple relation (5.10), the characteristic time constant is the time for the slowest moving carrier to cover a slightly modified ionizing free path, i.e., $\tau_i \sim 1/\alpha v_i$. Owing to the influence of gap length and the value of γ_i, it is seen that, the smaller d and the smaller γ_i, the more the *effective* α *will be less than the true* α, and thus the longer will be τ, If d is very large, then the expressions for τ_p are essentially $1/v_-\alpha$ for photon action and $\tau_i = 1/v_i\alpha$ for ion action with the time slightly diminished by any increase in γ_i or $g\,\gamma_{cp}\,\gamma\,\omega$, and diminished by photon absorption.

It is thus to be noted that the breakdown process has a characteristic build-up time that is typical of the primary mechanism active, of the smallest carrier velocity and is modified by secondary processes present and by the gap length. The time is, however, principally set by the carrier drift velocities which are $\sim 10^5$ cm./sec. for ions and $\sim 10^7$ cm./sec. for electrons and which generally in sparking fields increase as $\sqrt{X/p}$. This time also critically depends on $1/\alpha$ with α increasing in some regions nearly exponentially and at other times at least parabolically with X/p. Thus as X/p increases, τ in all cases will rapidly decrease.

It must next be noted that, if the time Δt over which the potential is applied in a given gap is less than τ for the processes at work, then the spark will not materialize by that mechanism. Thus, by applying to a gap of constant d a pulsed potential of varying length, beginning with values of Δt great enough so that the actions of γ_i and γ_p both are brought into play, the threshold V_s for the spark will be lowered as a indicated by equation (5.9) relative to (5.8) in which γ_i alone is active, and the same will hold for $X s/p$. If Δt is reduced so that γ_i action cannot occur, then breakdown will only occur by the γ_p mechanism and the threshold will be higher. That is, X/p and thus the potential must increase. Similar relations to those for γ_i and γ_p could be deduced for γ_m but would be more involved. Thus it is possible that merely by applying the potential for Δt to a gap of constant d in such a way first that γ_m, γ_i and γ_p are active, breakdown will occur at a *low* X/p and thus at a given characteristic low V at a given d. If now Δt decreases below τ_m, then a higher threshold potential will be observed, and if finally Δt is shorter than τ_i, a breakdown at a potential corresponding to photon action alone will be observed that is very high.

H. L. VON GUGELBERG [*24*] demonstrated these relations by applying a constant X/p to a gap. Using a variable gap length he increased the length to a value d_{mip} at which for the inert gas the combined action of γ_m, γ_i and γ_p caused a breakdown. This breakdown corresponded to one with a very long time constant τ_m. Increasing d to d_{ip} a spark occurred at a higher potential corresponding to the combined action shown by equation (5.9) with a much shorter τ_i. Finally, in a still longer gap d_p, at a much higher potential he got breakdown by γ_p alone with a short τ_p corresponding to equation (5.12).

6. The effect of space charge. It was probably STEENBECK [*26*] who first indicated the influence of positive ion space charge distortion on the breakdown potential of a uniform field gap. Slightly later, independently, from an entirely different line of attack, VARNEY, WHITE, LOEB and POSIN [*16*] arrived at exactly the same conclusion in observing the upcurving of the log i/i_0, x-curves in a TOWNSEND evaluation of α caused by a high j_0 with large gap lengths. Later study of this matter with greater precision was carried out by CROWE, BRAGG and THOMAS [*16*]. The conclusion is, as indicated in chapter I, a lowering of the threshold, if $\alpha/p = f(X/p)$ rises more rapidly than proportional to X/p. The effect will not be worked out in detail but enough will be given to indicate the reasoning. Call X_0 the undistorted field, and let the field at any point x become $X(x) = X_0 + \Delta(x)$ as a result of the space charge. This alters the potential by an amount $\Delta V = \int_0^d \Delta(x)\,dx$. To simplify discussion, d will be given a *form similar* to that *usually observed*, by a function which can be fitted to it at *some point*, but *never accurately fits over the whole range*. This is

$$\alpha = A\,p\,e^{-Bp/X}. \tag{6.1}$$

The Δx from here on will symbollically be represented by Δ. As α varies with $X(x)$, and thus with x, let us write

$$\alpha_x = A p \exp\left[-\frac{B p}{X_0} \frac{1}{1+\Delta/X_0}\right].$$

As $X_0 \gg \Delta$, we expand in a power series

$$\left.\begin{aligned} \alpha_x &= A p \exp - \frac{B p}{X_0}\left[1 - \frac{\Delta}{X_0} + \frac{\Delta_2}{X_0^2}\right] \\ &= \alpha_0\left[1 - \frac{B p}{X_0^2}\Delta + \frac{B p}{X_0^4}\left(\frac{B p}{2} - X_0\right)\Delta^2\right], \end{aligned}\right\} \tag{6.2}$$

yielding

$$\int_0^d \alpha_x\, dx = \alpha_0 d + \frac{\alpha_0 B p}{X_0^2}\Delta V + \frac{\alpha_0 B p}{X_0^4}\left(\frac{B p}{2} - X_0\right)\int_0^d \Delta^2 dx. \tag{6.3}$$

Now Δ^2 is always positive and thus the whole righthand term is positive if $\frac{B p}{2} > X_0$, and negative if it is less. Thus

$$\int_0^d \alpha_x\, dx - \alpha_0 d \gtrless 0, \quad \text{if} \quad \frac{X_0}{p} \lessgtr \frac{1}{2} B. \tag{6.4}$$

It is desired to get the potential ΔV which will restore the avalanche and multiplication to the value in the undistorted field so that $\int_0^d \alpha_x\, dx = \alpha_0 d$. This yields

$$\Delta V = \left(X_0 - \frac{B p}{2}\right)\frac{1}{X_0^2}\int_0^d \Delta^2 dx.$$

Thus

$$\Delta V \lessgtr 0 \quad \text{if} \quad \frac{X_0}{p} \lessgtr \frac{1}{2} B.$$

Now $\frac{1}{2} B$ marks the point of inflection in the α/p, X/p-relation chosen, below which $\frac{d^2\alpha}{dX^2} > 0$ and for which the value of α/p increases faster than X_0/p. Above this point, α/p increases more slowly than X_0/p.

Thus ΔV is negative when α/p increases more rapidly than X_0/p. That is, the threshold of the spark lies below V_0 corresponding to X_{0d} by ΔV. Thus positive space charge accumulation in the gap lowers the breakdown threshold and reduces the statistical jitter caused by fluctuations in γM. It is shown that in fact on the present simplified theory

$$V = V_0 - a j^2 \quad \text{with} \quad a = \frac{4}{3}\pi^2 \frac{p}{X_0^4}\frac{d^3}{b^2}\left(\frac{B}{2} - \frac{X_0}{p}\right). \tag{6.5}$$

The change in V is not very large but ranges into the hundreds of volts with a lowering of about 1%. Actual lowering of V_s by a j_0 has been observed for very high j_0 values, where reduction in spark threshold may have values of 10%, by WHITE [*27*] and independently by ROGOWSKI and WALRAFF [*27*]. Calculations along this line were made by WHITE [*27*] and also by MEEK [*27*] using streamer threshold. The calculation has recently been carried out in connection with experimental data for spark breakdown in air by H. W. BANDEL [*12*]. The general problem has also recently been treated by CROWE, BRAGG and THOMAS [*16*].

As early as 1933, VON ENGEL and STEENBECK [*26*] considered the effect of space charge distortion and lowering of the threshold on the formative time lag t. Employing the theory above cited and considering the effect of the space charge again as a minor perturbation, calculation yielded the result that, owing to the time necessary for the space charge to build up, the formative lag at the lowered potential would amount to only 1.19 τ where τ is the formative time lag. Thus the space charge build up will, where occurring, increase the formative time lag by a relatively small factor, of the order of perhaps two

In relatively recent years a study of time lags has been made with carefully stabilized potential sources, beginning with the studies of FISHER and his students [*14*]. As will be indicated under the heading of time lags, the actual spark breakdown threshold in air and other gases is hard to establish within 0.2% with any precision over a series of days and hours owing to changes in electrodes, etc. However, in air, at a potential of some 2% above threshold, breakdown occurs in under 0.1 μsec. As the threshold potential is approached, the formative time lag goes up to some 100 μsec. In argon it is nearer 1000 μsec. at threshold and does not go down to less than a microsecond until the potential is raised to twice the threshold value. It was postulated by FISHER and the writer that the long time lags were required for a low order TOWNSEND breakdown to build up a space charge that would cause an ultimate streamer spark [*14*]. Subsequent oscilloscope, studies of BANDEL [*12*] in air and of MENES [*15*] in very pure argon have proven this beyond a shadow of doubt. The theory for the build up of the TOWNSEND discharge with space change distortion has been carried out by BANDEL [*12*] and MENES [*15*], thus extending the elementary treatment of STEENBECK considerably further. This material will be treated at another point.

III. Properties of the threshold equation.

7. PASCHEN'S law. It is now of interest to draw some simple conclusions from the elementary theory of TOWNSEND without space charge complications. The condition for the threshold of the spark, $\gamma M = 1$, will be treated using the effective value of M as for a uniform field gap and a second coefficient of nature unknown, except that it varies with X/p according to $\gamma = \Phi(X/p)$. In turn, let $M = e^{\alpha d}$ with d the gap length and with α expressed as $\alpha/p = f(X/p)$, where $f(X/p)$ is a functional form unspecified. The sparking threshold is then set as

$$\gamma_s e^{\alpha_s d} = 1, \tag{7.1}$$

where the subscript s refers to the values at threshold. It then follows that

$$e^{p f\left(\frac{X_s}{p}\right) d} = 1/\Phi\left(\frac{X_s}{p}\right), \tag{7.2}$$

$$\log\left(1/\Phi\left(\frac{X_s d}{p d}\right)\right) = p d f\left(\frac{X_s d}{d p}\right). \tag{7.3}$$

Since for uniform field gap $X_s d$ equals V_s, the sparking threshold potential, we find

$$f\left(\frac{V_s}{p d}\right) = \frac{1}{p d} \log\left(1/\Phi\left(\frac{V_s}{p d}\right)\right). \tag{7.4}$$

This states that

$$V_s = F(p d) \tag{7.4}$$

is a unique function of pd which in principle is proportional to the *total number molecules* in an *avalanche path across the gap*. The law follows, irrespective of whether the field is uniform, and has been deduced on quite general terms by

J. S. TOWNSEND. It was first enunciated in 1889 by PASCHEN [28] and is known as PASCHEN'S law. It is a direct consequence of the notorious principle of similitude [29] which in ignorance of discharge mechanisms was early invoked to predict the behaviour of gaseous discharge systems. Experimentally the law has been observed to hold for thresholds of breakdown and even for most sparking thresholds over a wide range of conditions to within the limits of experimental accuracy. These limits are actually not very accurate for, as studies of BANDEL [12] and of FISHER [14] show, sparking potentials are not too reproducible, especially over considerable ranges of p and d. It is probable that the law is amenable to much closer study with really outgassed electrodes in very pure inert gases such as those obtained by PENNING and his associates at the Philips Laboratory in Holland and as are current at Bell Telephone Research Laboratories and at Westhinghouse Research [30]. There are circumstances where it will not hold. If a breakdown threshold depends on *carrier density* and not on the number of ionizable molecules per cm. in the field direction, the law will fail, and it will fail where the threshold is altered by space charge distortion, as indicated by VARNEY, WHITE, LOEB and POSIN [16] or by LOEB and MEEK [31] in connection with sparks the thresholds, for which are set by streamer formation. In both of these cases, the deviations will be small and may not be detected. Again, in any cases in which α or γ is not solely a function of X/p but a function of p as well the law will fail. It now appears likely over considerable ranges of X/p that about half the value of α is not caused by direct impact ionization, but by secondary actions in which highly excited atoms of inert gas in single impacts with neutral atoms make molecular ions. Below pressures of some 3 mm. Hg or slightly more of argon gas, the formation of A_2^+ becomes pressure dependent so that in extreme cases α/p could to 50% vary not only with X/p but with p as well [3]. Again, above the maximum of the (α/X), (X/p)-, or η, (X/p)-curves to be discussed, α is a function of X/p and of the distance x covered in the gas. Finally, while γ_i will always vary with X/p, there are quantities like γ_m and γ_p where factors of attenuation, such as destruction and absorption coefficients, vary with pressure while ν will vary as X/p and γ_p in the COLLI-FACCHINI [8] effect is pressure dependent. Thus there is no insurance that PASCHEN'S law will hold, especially at lower pressures, though most measurements verifying it have over limited ranges been carried out under conditions where it does hold. For very low pressures, very short gaps, and very high gas densities, the law definitely fails [32]. Such breakdowns, however, do not follow the TOWNSEND model and the law, therefore, would be expected to fail.

8. The STOLETOW, or PHILIPS, efficiency function. It is instructive to consider the STOLETOW [33], or PHILIPS efficiency [33] form of the (α/p)-, (X/p)-relation. If the first TOWNSEND coefficient α/p is divided by X/p, then the quantity $\eta = \frac{\alpha/p}{X/p} = \frac{\alpha}{X}$ represents some useful properties. Since X represents volts per cm. along the field direction, and α represents ions per cm. along the same direction, α/X represents ions created per volt of energy available to the electron from the field. The quantity $1/\eta = X/\alpha$ represents the volts to create an ion pair. If $\eta = \alpha/X$ is plotted against X/p, it takes on a form that rises rather steeply from low X/p values, where owing to many elastic or other impacts much energy is lost, to a fairly pronounced peak and then declines somewhat less rapidly as shown in Fig. 2. The decline occurs where the electrons gain so much energy that their ionizing cross sections begin to decline with further increase in energy. Thus where α/X shows a peak the inverse curve of volts per ion, X/α, shows a minimum. Compared to the maximum ionization efficiency of monoenergetic electrons, the value of volts per ion pair, even at the minimum, is about twice that for electrons freely accerlerated and interacting with the gas. Beyond the maximum of the α/X curves the electrons *gain* energy faster than they can dissipate it in impacts. As they progress across the gap and pass the distance x_0 from the cathode at which they exceed the most efficient energy for ionization, the further along x they progress the more energy they gain, and the less efficient they become, so that α/X continues to decrease as the distance x covered increases. α/p is then no longer a unique function of X/p but varies with x as well.

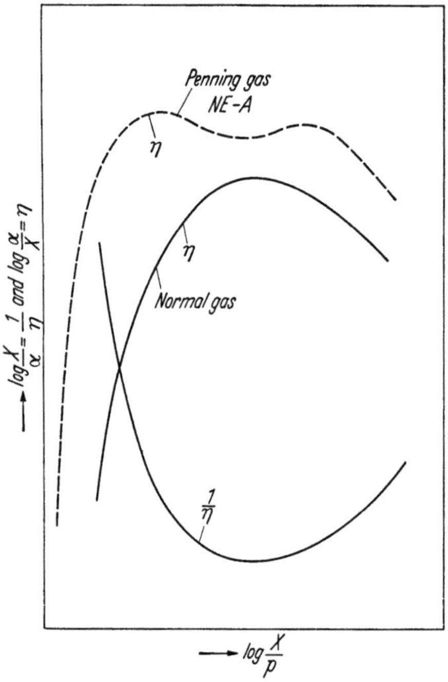

Fig. 2. Plot of the STOLETOW function $\alpha/X = \eta$ and $1/\eta$ as a function of X/p for a simple gas and for a PENNING two component gas.

It is perhaps of interest to add that if the simple (α/p)-, (X/p)-curve with its S shape is plotted, since at constant p the curve plotted is α against X, the quantity $\eta = \alpha/X$ represents a series of straight lines through the origin as seen in Fig. 3. It will be seen that in general this straight line cuts the α, X-curve over a limited range of slopes at *two points*. These correspond to the intercepts of the constant η lines with the η, (X/p)-curves. There are, however, values of η at which the curves are not intersected. Of interest is the limiting, or maximum value, of η at which the α, X-curve is intercepted, i.e., the *point of tangency* of the α/X line with the α, X-curve. This is the *maximum* value of η and corresponds to the peak of the η, (X/p)-curves.

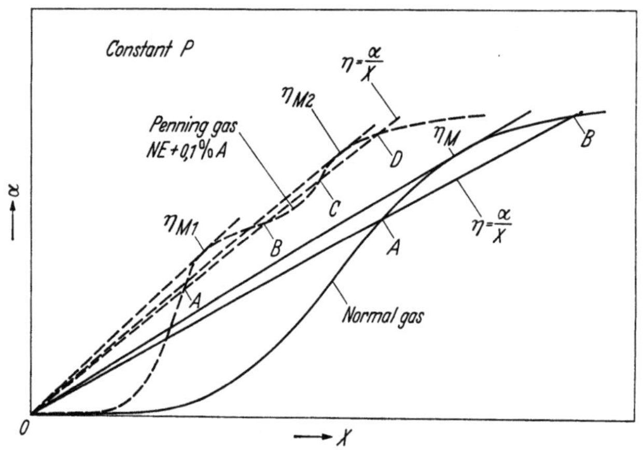

Fig. 3. Plot of α against X at constant p for two gases, a simple one and a PENNING two component gas. Note that the plots for $\alpha/X = \eta$ are straight lines from the origin cutting the curve at two or four points and tangent at the minimum sparking potential points.

9. The minimum sparking potential. Since $\alpha = \eta X$, so that the threshold condition, $\gamma e^{\alpha d} = 1$, becomes

$$\gamma\, e^{\eta X d} = \gamma\, e^{V_s \eta} = 1\,, \quad (9.1)$$

it is seen that the STOLETOW function is of interest in connection with sparking threshold. From this, it is possible to write

$$V_s = \frac{1}{\eta} \log \frac{1}{\gamma}\,.$$

It is seen that if γ is constant, or sensibly so, since $\log 1/\gamma$ changes slowly with X/p, V_s will be a *minimum* when $\alpha/X = \eta$ is a maximum or when $1/\eta$ is a minimum. This is seen in Fig. 2. It is clear that below the minimum value of X/α there can be *no smaller value* and since this is the dominant term yielding V_s, there is a *minimum sparking potential* V_{sm} corresponding to the maximum value of η_m at some value of $(X/p)_m$. It is also clear that for a conventional (α/p), (X/p)-curve below η_m there will be two X/p values for each η value, or correspondingly two intersections of the α, X-curve for each value as seen in Fig. 3. Thus, since for constant p there are two X/p values for a constant gap d, there will be two values of the sparking potential V_s corresponding to the two values of X, one below and one above the minimum.

Thus, if there are two values of V_s, and V_{s1} and V_{s2}, for a spark and since $\frac{V_{s1}}{p\,d}=\frac{X_{s1}\,d}{p\,d}$ and $\frac{V_{s2}}{p\,d}=\frac{X_{s2}\,d}{p\,d}$ and V_s is a function of pd, if a *constant* V_s is chosen *above* V_{sm}, there will be two values of pd, $(pd)_1$ and $(pd)_2$, having the same V_s. Then if $V_{s1}>V_{s2}$, if these are translated into pd corresponding to a constant V_s, $(pd)_1<(pd)_2$. Thus the V_s, pd-curve will be transposed a long the axis of ordinates relative to the $1/\eta$, X/p-curve from which it is derived. The curves for V_s as a function of pd are shown in Fig. 4 to be contrasted with those of V_s or $1/\eta$ as a function of X/p in Fig. 2. PENNING and DRUYVESTEYN [33] give a very nice geometrical construction through which the $V_s=\left(\frac{1}{\eta}\right)\log\frac{1}{\gamma}$ plots taken from η and γ can be transformed to V_s, pd-plots, showing the minimum sparking potential at one value of $pd=(pd)_m$.

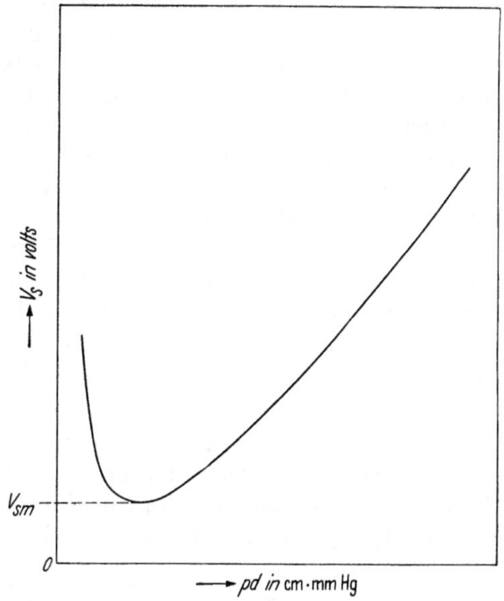

Fig. 4. A curve of the character of $1/\eta$ against X at constant γ as shown in Figure (2) translated to one for V_s against (pd).

E. E. DODD has shown by a simple calculation that the peak of the η, (X/p)-curve at $(X/p)_m$ yields $(X/p)_m$ for the minimum sparking potential.

If

$$\eta=\frac{\alpha}{X}=\frac{p}{X}\,f\left(\frac{X}{p}\right),\qquad \frac{1}{\eta}=\frac{X}{p}\,\frac{1}{f(X/p)}$$

then

$$\frac{d\,(1/\eta)}{d\,(X/p)}=\frac{1}{f(X/p)}-\frac{X}{p}\,\frac{f'(X/p)}{f^2(X/p)}=0$$

sets the condition for a minimum of $1/\eta$ as

$$\frac{X}{p}\,\frac{f'(X/p)}{f(X/p)}=1\,.$$

If the simple expression (6.1) be assumed for $f(X/p)$,

$$f'\,\frac{X}{p}=\frac{B}{(X/p)^2}\,f\left(\frac{X}{p}\right)$$

from which it follows that $1/\eta$ is a minimum, or η is a maximum, at $X/p=B$.

If α/p is a more involved function $f(X/p)$ than the simple appearing S shaped curve exemplified by equation (6.1), such as arises from the PENNING or HALE effect with two or more maxima to the α, (X/p)-curve, the V_s, pd-curve will also have *two or more minima* and more values of pd for the same value of V_s in limited regions above V_{sm}. This is nicely illustrated in Fig. 2 and 3 for the PENNING gas.

Physically, the significance of the minimum is quite clear in terms of the expression $\gamma e^{\alpha d}=1$. Since γ changes relatively slowly, the dominating term is the αd in the exponent. This gives the number of ionizing events in an electron avalanche transit across the gap, and with fixed γ this requires a given value of αd for the threshold. At large pd there are many molecules in the length d. Thus α/p can be low and yet yield enough ions. As pd decreases there are less molecules, therefore α/p must increase to compensate in αd for the decreased number of molecules encountered. As long as α/p increases rapidly with X/p, with increasing efficiency, V_s can decrease with decrease in pd. When, however, the potential drop across a section of the gap ceases to be effectively used in ionization, since with increase in X/p the probability of ionization decreases, and too few molecules are present, the energy is delivered to the anode and not to ionization. Then decrease in pd requires a more rapid increase in potential to provide the ions needed. Thus at the point where the ionization efficiency is a maximum and begins to decline there is a minimum of the sparking potential, V_{sm}. It is located at some value at pd_m with $(X/p)_m$ fixed by the minimum in $1/\eta$ or the tangent to the (α/p)-, (X/p)-curve from the origin. This will of course be slightly shifted with changes in γ.

10. The reduced sparking equation. V. ENGEL and STEENBECK by use of equation (6.1) show that if each gas differs only in the constants A, B and γ the *sparking equation* has a *universal form* expressed by a *reduced* sparking potential $V_R=V_s/V_{sm}$ and a *reduced pd* value $(pd)_R=pd/(pd)_m$. Then

$$V_R=\frac{(pd)_R}{1+\log(pd)_R}. \tag{10.1}$$

It is true that the V_s, pd-curve including the minimum applies quite generally to all types of TOWNSEND-like breakdown even including the microwave studies. It is also true that the curves are all of a generally similar shape, except of course those where α/p is a more complicated function of X/p, as in the PENNING effect [34]. However, since in general there is *no single functional form* that like the simplified expression (6.1) describes α/p as a function of X/p, since the electron energy distribution changes with X/p, the simplifications, of a reduced sparking potential relation and other idealized consequences seen in the use of equations like (10.1), are all illusory. Accurate study of observed breakdowns allows of no such simplification though there are general shapes of curves that give a superficial resemblance to such situations.

11. Agreement between observation and theory. The minimum sparking potentials have been measured in the past for many gas-electrode combinations. In view of what is known today concerning gaseous purity control and cathode cleanliness, none of the past data have more than an orienting significance. Reference should be made to older texts for such data. It is perhaps of value to comment briefly on the experimental verification of the threshold equation. Very early TOWNSEND [22], using data on the coefficients α and β experimentally determined, calculated V_s and found agreement with observation satisfactory to perhaps two percent. Numerous tests of the theory have since been made in various ways. Perhaps a good example can be shown in Fig. 5 which represents studies of HALE [35] and EHRENKRANTZ [35] using gases and electrodes of comparable purity. It is noted that sparking potentials differ even with comparable techniques by about 5% and that HALE's observed and calculated values show some differences even for the same gas. The difficulties are most pronounced near the minimum where both slight variations in γ as well as in α are of importance. That is, as α gets less efficient variations in γ become important. It can be indicated that, as first pointed out by HOLST and OOSTERHUIS [*21*] and the writer [*36*], at values of pd well above the minimum the value of the threshold is dominated by the α term and at most even large changes in γ will produce but small changes in V_s, while near the minimum γ can produce changes in V_s by as much as 50%. It is likely that using modern purification, outgassing and vacuum techniques better agreement between theory and observation could be obtained

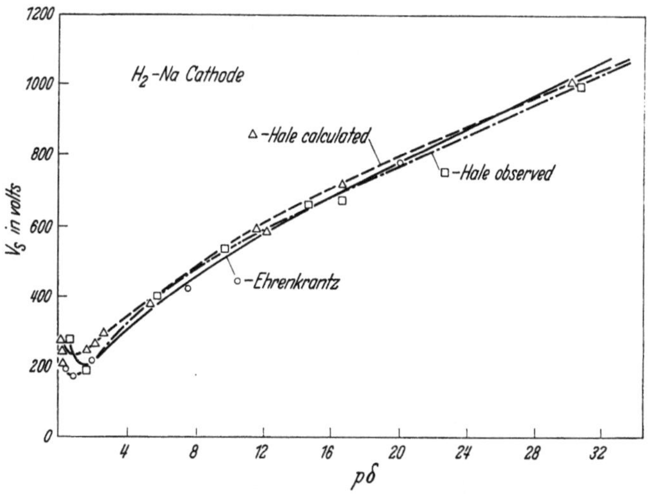

Fig. 5. Test of TOWNSEND threshold equation for $\gamma M = 1$ from data of EHRENKRANTZ and HALE on gases of comparable purity. HALE observed d, γ and V_s while EHRENKRANTZ observed V_s under similar conditions. The reproducibility of values of V_s are well illustrated as well as the reliability of carefully observed γ and α.

for the TOWNSEND breakdown thresholds where space charge distortion does not intervene. Spark thresholds involving space charge distortion will still be subject to uncertainty.

From what has gone before, it seems clearly indicated that the common practice of evaluating γ or secondary coefficients by using V_s and values of α/p as a function of X/p, even if determined in the same gas, is of questionable accuracy. This is true, since any errors in α occur in an exponent which causes changes in M of much greater value than major changes in γ, and since α in itself is inferred from values of V_s in which the inherent error is at least 2%.

It may be remarked in general that agreement between any calculated values of V_s, using observed values of γ and α/p as a function of X/p by insertion into a given theory following some model for the threshold process, are of *little significance as a verification of the model. All models* must *use* the *quantity* $M = e^{\alpha d}$ in which α is inferred for an observed value of V_s. The equivalent of the γ term coming from the theory based on the model will vary slowly compared to the variation of M, with the errors inherent in the observation of V_s and the *many conflicting erroneous values quoted in the literature*. Thus it is not difficult to get "*agreement*" with a *given theory*, using some model, by *searching the literature for earlier values of* V_s which give the desired quantities in agreement with the value of γ deduced. The adequacy of any model for V_s can only be determined by *the overall consistency* between the *predicted* and observed *breakdown pattern* with all the different phenomenological aspects of breakdown.

12. Breakdown thresholds with electron attachment; form of x_s, pd-curve. The discovery of the very large cross sections for dissociative electron attachment in air and certain spark suppressing gases by R. GEBALLE [37] and his associates has necessitated a study of the influence of electron attachement on the threshold for spark breakdown. Analysis for the multiplication factor in a TOWNSEND avalanche in a uniform field, when electrons attach to form negative ions in the field, leads to the expression

$$\frac{i}{i_0} = \frac{1}{\alpha - \vartheta}\left[\alpha\, e^{(\alpha-\vartheta)d} - \vartheta\right] \tag{12.1}$$

for uniform field geometry in place of equation (1.1) of TOWNSEND in which ϑ represents the number of dissociative attachments per cm. advance of an electron in the field direction. The relation always makes $i > i_0$, but the increase is small until X/p reaches values such that $\alpha \gtrsim \vartheta$. If experimental data cover enough range in d, then ϑ and α may be evaluated from equation (12.1). When there is a cathode γ active, GEBALLE and LIN evaluate i/i_0 to be

$$\frac{i}{i_0} = \frac{\left[\left(\frac{\alpha}{\alpha-\vartheta}\right) e^{(\alpha-\vartheta)d} - \frac{\vartheta}{\alpha-\vartheta}\right]}{1 - \frac{\gamma\alpha}{\alpha-\vartheta}\left[e^{(\alpha-\vartheta)d} - 1\right]}. \tag{12.2}$$

The value of i/i_0 becomes indefinitely great when the denominator vanishes that is when

$$\frac{\gamma\alpha}{\alpha-\vartheta}\left[e^{(\alpha-\vartheta)d} - 1\right] = 1. \tag{12.3}$$

This must be compared to equation (3.1) or (4.4) in the absence of attachment. To ascertain whether equation (12.3) actually satisfies the threshold condition, it was also deduced from a dynamic time analysis analogous to that of STEENBECK. To derive the consequences of this change, it is best to introduce α/p and ϑ/p into equation (12.3), such that

$$\frac{\gamma\alpha}{\alpha-\vartheta}\left[e^{\left(\frac{\alpha}{p}-\frac{\vartheta}{p}\right)pd} - 1\right] = 1. \tag{12.4}$$

When $\alpha/p \geq \vartheta/p$ breakdown is possible for a sufficiently large value of pd irrespective of values of α/p, ϑ/p and γ so that $V_s = F(pd)$ following PASCHEN's law. If $\alpha/p < \vartheta/p$ with increasing pd the relation approaches an asymptotic form which is independent of pd, which reads

$$\frac{\alpha}{p} = \frac{\vartheta}{p}(1+\gamma). \tag{12.5}$$

This relation depending on X/p alone fixes a limit on this quantity *below which no spark can occur* regardless of pd. The value of X/p at which this occurs can be set as $(X/p)_{\text{lim}}$.

The data of GEBALLE [37] for α/p and ϑ/p as a function of X/p can then be applied to equation (12.4) to calculate V_s for the gases CCl_2F_2 and O_2. The agreement obtained for electron attaching gases is as satisfactory as that for the non attaching gases. A study of equation (12.5) indicates that since $\gamma \sim 10^{-3}$ or less the limiting condition becomes $\alpha/p = \vartheta/p$. This at once enables $(X/p)_{\text{lim}}$ to be evaluated from tables of these quantities which was done for a number of attaching gases by GEBALLE and LIN [37]. There are no adequate data for large pd with which to compare the calculated values. It was noted that the order of the values of $(X_s/p)_{\lim}$ for the gases parallels the so called dielectric strength as given in the literature from engineering studies.

It is of interest in this connection to point out a general property of the ratio X_s/p relative to pd for all cases of TOWNSEND breakdown. Since $e^{\alpha d} = e^{\left(\frac{\alpha}{p}\right)pd}$ and since $\alpha/p = f(X_s/p) = f(V_s/pd)$, equation (3.1) for the threshold reads

$$\gamma\, e^{f\left(\frac{X_s}{p}\right)pd} = 1$$

or

$$f\left(\frac{X_s}{p}\right) = \frac{1}{pd}\log\frac{1}{\gamma} \tag{12.6}$$

which relates X_s/p, the sparking field strength to pressure ratio, to pd. This indicates that depending on the analytical form of $f(X_s/p)$, X_s/p will slowly or rapidly decrease along some sort of an hyperbola as pd increases, provided one remains on the high pd side of the minimum sparking potential, i.e., as long as α/p increases faster than linearly with X/p. At lower pd, X_s/p will be higher than at larger values. But at low values of X_s/p, $f(X/p)$ varies as $A\,e^{BX/p}$ i.e., rapidly and, as X_s/p increases, $f(X/p)$ is proportional to $(X/p)^2$, later becoming proportional to X/p, and finally varying as $\sqrt{X/p}$ at the highest values of X/p above the minimum sparking potential. In consequence the decline of the (X_s/p), (pd)-curve will be hyperbolic and steepest at low pd when X/p is high, varying as roughly $1/(pd)^2$, and as pd increases it will decrease successively as $1/pd$, $1/\sqrt{pd}$ and beyond this as $1/\log pd$. Thus X_s/p curves will in general decline sharply at pd values beyond the minimum value and for most scales of plotting *appear to level off to a nearly constant value as pd becomes large.* Under these conditions it becomes very hard to distinguish the approach of the (X_s/p), (pd)-curve to the limiting value $(X_s/p)_{\lim}$ in air or other attaching gases from the general trend of all (X_s/p), (pd)-curves to level off at large pd.

IV. Statistical fluctuations of TOWNSEND or other breakdowns and the various time lags.

13. Statistical fluctuations and basic concepts of time lags, theories of LAUE and WIJSMAN. Among the characteristics of any breakdown model which help delineate the process active for a given gap is the formative time lag τ as shown by the calculations of STEENBECK, [equations (5.10) and (5.11)] in the development of the TOWNSEND theory. It is now of interest to consider this matter in more detail.

As indicated in discussion of the simple TOWNSEND condition for a breakdown threshold, which in form at least represents a prototype for all breakdown threshold mechanisms in which gaseous ionization plays the leading rôle, the quantity γM is subject to considerable statistical fluctuation for individual avalanche sequences. Since, on the definition of breakdown threshold as $\gamma M = 1$ here used, it was implied that a single electron liberated from the cathode was capable of initiating a breakdown when conditions set the average $\overline{\gamma M} = 1$, it is necessary to regard a series of electron avalanches at the threshold and above and to count to see what fraction of them cause a breakdown to materialize. Thus, if for a succession of avalanches n_T the number n produce breakdown, the *breakdown probability* is defined as $P_s = n/n_T$. Since as indicated, in the absence of space charge, P_s starts at zero at the threshold $\overline{\gamma M} = 1$ and with increasing values of γM greater than unity increases according to WIJSMAN [4], it will be expected that P_s will rise as $\Delta V = V - V_s$ increases along some sort of a curve. If space charge accumulation aids the breakdown process, then the distribution will be altered in some measure. In any event it is clear that P_s is some function of *over-voltage*, generally expressed in per cent as $\frac{\Delta V}{V_s} \times 100$.

The concept of the *breakdown probability* of a spark was first suggested by LAUE and ZUBER [*38*] in 1925 as a result of their study of the appearance of sparks at the sparking potential. The problem as it involves fluctuations of γ was first studied by BRAUNBEK [*38*] in 1926 and later investigated in relation to SCHADE's [*24*] study of the formative time lag of TOWNSEND discharges by BRAUNBEK [*38*] and by HERTZ [*38*]. Neither of these workers recognized the importance of the fluctuations of $e^{\alpha d}$. The complete problem was investigated anew on the

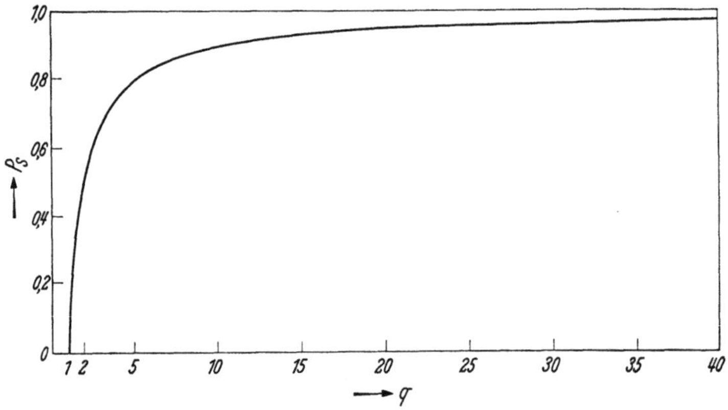

Fig. 6. The WIJSMAN theory of probability P_s of a breakdown for any electron avalanche due to statistical fluctuations of $q = \gamma M$ about $\overline{\gamma M}$ above $q = 1$. Note that even for $q = 10$ P_s is only 0.9.

insistence of the writer by WIJSMAN [*4*], [*38*] with rather interesting results. This investigation confirmed the relations of BRAUNBEK and HERTZ as far as they went. Assuming a TOWNSEND discharge at low pressure, *with no space charge distortion*, WIJSMAN found that the breakdown probability determined by the combined fluctuation of γ and $e^{\alpha d}$, with the condition $\gamma(e^{\alpha d} - 1) = 1$, was

$$P_s = 1 - \frac{1}{q}. \tag{13.1}$$

Here q represents the value

$$q = \gamma(e^{\alpha d} - 1),$$

that is, the value of the threshold product.

It is seen from the curve of P_s plotted against q as shown in Fig. 6 that, when q is less than unity, $P_s = 0$ and when q is greater than unity, (i.e., above V_s), P_s rises from 0 at V_s, or $q = 1$ to values approaching 0.9 at $q = 10$, that is at considerable *overvoltages*, $\Delta V/V_s$. It is at first sight quite surprising that the fluctuations should give P_s rising as slowly with q, in fact so slowly that to cause 99% of the electrons to yield sparks, $q = 100$. Even with α/p changing very rapidly with X/p in some regions, the increase in V above V_s to give this effect might be expected to be rather high. The second feature that seems at first surprising is that *there are no sparks to be expected below* V_s, as it might be suspected that fortunate sequences of events should lead to a spark below $q = 1$. The result of WIJSMAN, however is to be expected on the basis of the *simplifying assumptions* which were made to enable computation. This involved a consideration *of an infinite succession of pure* Townsend *avalanches with no space charge accumulations.* Accordingly, if no space charges are introduced, then the sequence of avalanches, any of one which might interrupt the sparking, continue indefinitely. Consequently, there can be no sequences below V_s with $q=1$ which will not ultimately break off the discharges.

In actual practice the sequences do build up a space charge which *insures the realization* of a spark and thus *only a finite number of avalanches are needed.* Accordingly, if the time of build up is large compared to the time for ions to cross the gap (many avalanche sequences), then the law deduced is approximately correct and $P_s=0$ at $q=1$. With a *short sequence of avalanches before the space charges begin to insure the spark*, it is possible that the values of P_s might begin below V_s and continue at less than unity above V_s as suggested earlier. The effect of space charges and relatively short ionizing sequences would be to displace the P_s, V-curve toward values below V_s and to shorten the region above V_s where P_s has low values. However, the growth of space charge would have to be rapid indeed to achieve this and it is *not observed.* Thus it must be noted that the effect of statistical fluctuation in the basic criteria for the threshold has two consequences. First, it leads to some uncertainty in picking the true value of V_s from experiment, since V_s is observationally not sharply defined within a range of values possibly below and certainly above V_s. Secondly, it leads to *fluctuations of appearance of sparks in time.* Hence if n_0 electrons are liberated per second on the average, only $P_s n_0$ sparks will occur and thus the average time between sparks will be lengthened.

The fact that not every electron starting from the cathode in a Townsend type discharge will lead to a spark at and above $V=V_s$ connotes that actual observations of sparking will be accompanied by certain time delays. Thus, assuming a certain value of V applied in a region where a spark can form and that the external source of ionization liberates a current i_0 of n_0 electrons per second from the cathode, it would be expected that on the average some time will elapse between application of the potential and the observed time of the appearance of the spark. Since electrons are liberated purely at random in time, the time of appearance of sparks will also be random in time on the basis of this action. This lapse of time was early observed, especially where natural ionization of the gas in the gap gave a small n_0 and a long delay. The average value of this time will be called the *observational time lag*, designated by T_0. In individual applications of potential the observed lag will rarely be just exactly T_0, for the emission of the n_0 electrons per second is quite a random affair in time. Thus the actual lags will fluctuate yielding an average value T_0 which will depend on V, on the nature of the gap, on n_0, and on the point at which the n_0 electrons are liberated in the gap, for in the more general case the electrons may be liberated in the volume of the gas instead of at the cathode.

Experimentally the problem was first investigated by Zuber [*38*] in 1925 with relatively crude techniques compared to later work, but data were sufficiently good so that Laue [*38*] could analyze the problem. It was there pointed out that the observational time lag T_0 consisted of two portions, first the *formative time lag*, T_f, and secondly the *statistical time lag*, T_s. Thus it is essential to write $T_0=T_f+T_s$.

The quantity T_f is the time taken for the spark to materialize, or for the gap to break down. It is related to τ, the characteristic time constant of current growth deduced by Steenbeck for the Townsend discharge.

The characteristic time τ as defined by equations (5.10) and (5.11) represents the characteristic *unit time constant* of the process for the mechanisms active.

The actual time for the current change representing the growth is obviously some multiple of this characteristic function, even without intervention of space charge accumulation. Thus, for example, at the GEIGER counter threshold it represents the time from the first avalanche started by one electron to the full pulse development by spread down the wire as seen in Sect. 18β. In the TRICHEL pulse corona in air 760 mm. it represents a sequence of some five or ten unit processes to achieve maximum current as seen in Sect. 34. In the TOWNSEND discharge at low pressure it represents the time for the processes to build up the discharge spreading across the whole gap.

If avalanches start simultaneously over the whole electrode area as for γ_i and γ_p with perhaps slight overvolting, especially with the effective γ_i, VON GUGELBERG [*24*] has shown that T_f is close to the τ predicted by equations (5.10) and (5.11). On the other hand, very near threshold, SCHADE [*24*] at low pressures in fairly clean Ne and H_2 and BANDEL [*12*] at 760 mm. in air with uniform fields and known triggering circuits have found τ_f very long compared to τ as defined by STEENBECK and VON GUGELBERG. In BANDEL'S case, there was no space charge but a γ_p of low value, observation agreeing with a more complete theory than that of VON GUGELBERG. Thus it is seen that the actual τ_f characteristic of a breakdown process while closely related to τ is not necessarily equal to it and is sometimes larger than τ by many fold. If the breakdown requires space charge accumulation, then, as VON ENGEL and STEENBECK indicate theoretically [*26*] and as BANDEL shows by observation and theory for spark breakdown in air τ_f will exceed τ several fold.

Finally, it must be recognized that between the *observed* T_f and the true formative lag τ_f for the breakdown process there is an essential difference which generally is quite indefinite. T_f is *the observed time between application of the potential and the initiating photoelectron liberation by* i_0 *from the cathode to the "observation" of a breakdown.* Thus T_f depends on how far the transition must proceed to be registered by the recording devices. This in turn may involve the sharp initial fall of potential, the complete collapse of the field, the achievement of a certain registering current, the detection of a luminosity through a KERR cell shutter or photomultiplier, or the bright flash of the spark. Thus, in relating T_f to τ_f and τ, care must be taken in recognizing the temporal significance of these observational techniques relative to the definition of the *true* breakdown time. Thus T_f will always be longer than τ, but may be longer or shorter by appreciable intervals than τ_f, the true formative lag of the process.

Just as fluctuations in avalanches produce fluctuations of V for sparks about V_s and cause a P_s, so also at a given V there should be fluctuations of τ_f for individual breakdowns about $\bar{\tau}_f$. The importance of this factor is not known and the corresponding fluctuations of individual values of T_f about $\overline{T}_f$ are not known. This is especially true, since limitations of experimental observing techniques in the short time intervals make segregation of the overlying T_s fluctuations and T_f fluctuations difficult, and since jitter of instrumental origin will mask fluctuations in T_f for the shorter values usually encountered, as noted by FISHER [*14*].

The average statistical time lags will involve the average time interval between liberation of electrons that can trigger the discharge. Ideally, if electrons all came from the cathode, each of them could yield $e^{\alpha d}$ electrons in the avalanche. If they are liberated in the volume of the gas then the avalanches will be shorter and the chance of extinction greater. While LAUE suggested that the number and location of the electrons as well as P_s might influence the statistical time lag, WIJSMAN'S analysis makes it possible to include in his calculation of P_s the location of the triggering electrons in the gap if this is required. In practice triggering electrons purposefully are usually made to come from the cathode. Then quite simply if there are n_0 electrons liberated from the cathode per seconds, P_s will be as calculated by WIJSMAN, or as determined by space charges. In consequence

$$T_s = \frac{1}{n_0 P_s}. \tag{13.2}$$

Thus the observational time lag T_0 will be composed of a statistical time lag T_s superposed on a formative time lag. It is accordingly possible to write,

$$T_0 = T_f + \frac{1}{n_0 P_s} = a\tau + \frac{1}{n_0 P_s}. \tag{13.3}$$

Here a is a factor converting computed build up times τ to the actual formative time lag τ_f. This lag with space charge action will be relatively sharply defined.

14. Problems in the experimental study of time lags. So far, for simplicity, the quantities T_0, T_f and $1/(n_0 P_s)$ have been treated as *average values* easily deriveable directly from measurements of the time of the occurrences when single electrons are liberated. Actually, the fluctuations in n_0 and due to P_s are purely chance phenomena as may also be the fluctuations in T_f. Such fluctuations yield values of individual times of all magnitudes. Accordingly, except in orders of magnitude, it is *not easy or accurate to pick out the values of the average quantities from the observed data.* It is thus essential to consider the exact analysis of data in terms of the theory. To appreciate this it is necessary to note what experimental data are available for such analysis.

Experimentally, measurements may be made by observing the time between application of the potential, or triggering of the gap, and the appearance of a spark. Then the observed data may be broken up into finite time intervals of duration t_1, t_2, t_3, etc., starting from $t=0$ with $t_1<t_2<t_3$ and the number of sparks out of the total n_T occuring after *these times* can be counted as $n_{t_1 a}$, $n_{t_2 a}$, $n_{t_3 a}$, etc. Thus sparks *after* a time t_{1a} should represent nearly all the sparks if t_{1a} is short. The ratio of the number of sparks n_{t_a} to the total sparks n_T observed *after* any given time t should according to LAUE be given by the well known *survival equation, provided the occurrences are perfectly random in time,* i.e., governed by chance with no regular trend. If, therefore, only statistical time lags occur,

$$\frac{n_{t_a}}{n_T} = e^{-n_0 P_s t} = e^{-t/\tau_s}. \tag{14.1}$$

It should be noted that such random occurrences give n_{t_a}/n_T as a monotonic declining exponential curve in t in which T_s is the average, or scale factor and corresponds to the value of t at which $n_{t_a}/n_T = 1/e = 0.367$. It is sometimes more convenient to apply a potential for a given time t and count n_t, the number of sparks liberated in the interval *between* $t=0$ and $t=t$, as t increases, out of the total number of applications of the potential n_T. In this event, the longer t the nearer n_t/n_T approaches unity so that

$$\frac{n_t}{n_T} = 1 - e^{-n_0 P_s t} = 1 - e^{-t/\tau_s}. \tag{14.2}$$

15. Studies of STRIGEL. Further definitions. Equation (14.1) was initially tested experimentally by ZUBER in 1925 for lags ranging *down* to $t=10^{-3}$ sec. With improvement in techniques for the measurement of very short time intervals, it became possible to push studies to still shorter times. Thus STRIGEL in 1932 [*39*] and TILLES [*39*] under the writer's direction 1934, studied lags down to $t = 10^{-4}$ sec. For longer intervals both observers found that log n_{t_a}/n_T plotted against t gave linear plots starting at the origin, the slope of which evaluated T_s, or $1/(n_0 P_s)$. As the time intervals approached 10^{-4} sec., both Strigel and Tilles noted that the decline of the log n_t/n_T plots began at a time $t = T_f$, later than zero, remaining at unity from $t = 0$ to $t = T_f$. They both correctly indicated that the T_f thus observed must be the formative time lag for their gaps.

In consequence when Strigel [39] resumed his studies with better equipment in 1936 he adopted a modification of the equation which he attributes to Laue. and which takes account of the formative time lag. An exactly similar relation was deduced independently by Wijsman [4] in his statistical analysis. Thus equations (14.2) and (14.3) become

$$\frac{n_{ta}}{n_T} = e^{-n_0 P_s (t - T_f)} = e^{-(t - T_f)/T_s} \tag{15.1}$$

and

$$\frac{n_t}{n_T} = 1 - e^{n_0 P_s (t - T_f)} = 1 - e^{(t - T_f) T_s} \tag{15.2}$$

in which $t \geqq T_f$. Strigel interprets T_f as "the shortest possible breakdown time below which no discharge can occur". On the other hand, he ascribes to T_s "the properties of a probability having the dimensions of a time, which embraces the chance factors and with them the associated lags in the build up of the discharge". This nearly literal translation of Strigel's definition would make it appear that T_f in equation (15.1) is a constant representing minimum time of spark development. Thus the *fluctuations in the formative lag* τ_f and thus T_f are *lumped* with the *statistical fluctuations* included in T_s. Strigel further sets $T_s = \frac{\Sigma t_0}{n_T} - T_f$, where Σt_0 is the sum of all observed lags and $\Sigma t_0/n_T = T_0$ is the average observational time lag T_0 referred to in equation (13.3). This relates the quantities defined earlier roughly as $T_0 = T_f + T_s$ in agreement with equation (13.3). However, this more rigorous definition requires a new interpretation of T_f and T_s. Thus T_f is now *a constant minimum observed time characteristic* of the process showing no fluctuation. Therefore T_s must be extended to include in $n_0 P_s$ any time flucutuations involved in the *formative lag*. In consequence the value of $P_s = n_0/T_s$ inferred by multiplying the reciprocal of the observed T_s by n_0 may not represent what originally was intended as the *probability of the spark at* V, but includes as well the chance that the spark takes the minimum observed formative lag T_f. The relations may therefore be written as

$$\frac{n_{ta}}{n_T} = e^{-\frac{t - T_f}{T_0 - T_f}} = e^{-\frac{t - T_f}{T_s}} = e^{-n_0 P_s (t - T_f)} \tag{15.3}$$

with modified interpretations of P_s and T_f.

It is seen that the form of equation (15.3) is such that the log (n_{t_a}/n_T), t-plots show just the behaviour noted by Tilles and Strigel, and thus that basically the theory of Laue is correct.

This clearly indicates how the study of time lags should be made. Observations must be made of n_t/n_T against t, the quantities T_0, T_f and T_s are evaluated if the time resolution of the method is sufficient to indicate T_f. If T_f is to be observed, T_s must be decreased by increasing n_0 and perhaps P_s so that T_f shows up in the measurement. Further analysis of $T_s = n_0 P_s$ can be made by measuring i_0 and deriving $n_0 = i_0/\varepsilon$ where ε is the electron charge. Studies of T_f and T_s relative to overvoltage, $\frac{\Delta V}{V_s} = \frac{V - V_s}{V_s}$, can then be carried out. Care must be used in the interpretation of P_s as noted. It must also be recognized that T_f will not be measurable unless T_s is commensurate with it, that is, n_0 must be increased until T_s is of the same order as T_f, so that T_f may clearly be resolved. P_s is very sensitive to $\Delta V/V_s$ so that by increasing $\Delta V/V_s$ as well as n_0, T_s can be reduced materially. To study P_s as depending on $\Delta V/V_s$ it should be possible to decrease n_0 to a point where T_s is much greater than T_f. Then variation in T_s with $\Delta V/V_s$ will depend largely on factors effecting statistical variations of the composite P_s. On the contrary, by reducing T_s by increasing n_0, the influence of T_f becomes large and fluctuations in T_s with $\Delta V/V_s$ will reflect fluctuations of P_s depending both on statistical factors in formative time lag as well as in the probability of a spark. Such an analysis of P_s has not to date been attempted and *little is known of the statistical fluctuations of the formative time lag.* All of this procedure seems straight forward and simple. Experimentally the proper study is not only extremely tedious but one presenting great technical difficulties.

16. Experimental investigations. *α) Introduction.* Many studies have been carried out on the time lags involved in sparking. These are hard to summarize briefly owing to the diversity of techniques employed as well as the specialized and limited objectives of many studies. Thus many studies have been made of the formative time lags as time resolving techniques improved, while the studies neglected to control such variables as overvoltages, initiating electron liberation and statistical lags [*40*]. Relatively few studies were carried out in such a fashion as to control and analyze the statistical fluctuations, or so as to enable application of equations (15.1) and (15.2). Most earlier studies and many later studies were limited to investigation of spark breakdown, relatively few dealt with the TOWNSEND breakdown. Finally, the techniques of observation varied widely, breakdown in some cases being indicated by rapid increase in current or decline in potential, in other cases by the appearance of luminosity in the gap. The triggering of the breakdown signal came from application of the full breakdown potential across the gap, the application of a step potential on top of a steady approach potential near breakdown value, or the sudden liberation of triggering electrons from the cathode of an overstressed auxiliary spark gap. In some cases time lags were measured by the interval between application and break down as indicated by oscilloscopes, or else through the arrival of signals controlled by delay lines.

It is impossible to report much of this work in the scope of this article. However, an extensive analysis of this has recently been published by the writer [*41*]. As indicated the initial studies were made in 1925 by LAUE in collaboration with ZUBER who furnished the basic data. Disregarding theory, a large number of studies beginning with those of PEDERSEN in 1923 using LICHTENBERG figures and extending with many varied techniques beyond 1928 with fast oscilloscopes in the laboratories of ROGOWSKI in Aachen and elsewhere revealed the very short formative time lags of the order of 10^{-7} sec. and less for spark breakdown using relatively high values of $\Delta V/V_s$ [*40*]. These observations indicated the apparent inadequancy of the normal TOWNSEND breakdown model with time lags dependent on ion motion. They resulted in numerous theoretical analyses of the breakdown process by many workers which led ultimately to the streamer mechanism of RAETHER [*42*] and of the writer [*20*] and MEEK [*43*].

The first more complete studies were the independent ones of STRIGEL in 1932 with further work published in 1936 [*39*] and work begun in 1932 by TILLES under the writer's direction.

β) STRIGEL's *Observations.* STRIGEL used the fast cathode ray oscillograph at the Siemens laboratories. He studied impulse potentials as well as steady breakdown. He verified the LAUE-ZUBER distribution of statistical lags in the microsecond region using large overvoltages and the formative time lags as predicted by LAUE's theory. He observed an *exponential* decrease in T_s with overvoltage. He showed that P_s depended on $\Delta V/V_s$ without clearly separating n_0 and P_s in the definition of T_s. His work in 1936 following publication of TILLES' work revealed the distribution law for the statistical lags and allowed him to evaluate the formative lag as a function of $\Delta V/V_s$ using high overvoltages and gaps of from 1 to 6 cm. length with spheres from 5 to 15 cm. diameter. His curves for T_f against $\Delta V/V_s$ in air were bracketed by later observations of WHITE [*40*]. The values of T_f ranged from 2.8×10^{-7} to 3×10^{-8} sec. for $\Delta V/V_s$ ranging from 0.05 to 0.065. T_f seemed to be practically independent of gap length. WHITE found that it took higher values of $\Delta V/V_s$ to cause breakdown of the much shorter gaps in the same time. T_f was observed to be independent of i_0 in the range of values used in contrast to observations by WHITE who found a decrease in T_f if i_0 was very high.

γ) TILLES' observations. TILLES observed T_0 for gaps 0.0683 to 0.17 cm. in air from the ZUBER range to 10^{-5} sec. using an impulse wave of short duration, or an approach potential of 0.96 V_s and an added potential step ΔV which gave V from what was *assumed* to be 1.01 V_s to 1.05 V_s. Linear plots of log n_{ta}/n_T against t were observed, and T_f was much less than T_s. Varying $\Delta V/V_s$ the character and magnitude of P_s could be observed. Varying i_0 at constant $\Delta V/V_s$, i_0/ε could be observed and thus T_s was delineated, although i_0 was not directly measured in the gap. The current density $j_0 = i_0/A$ (A = area) was varied over a range of 10^5. For the gap indicated in air, TILLES inferred that $T_s = 1/(P_s n_0)$ $= 0.37 \times 10^{-3}\, e^{-39\,\Delta V/V_s}\, i_0$. It is seen that P_s therefore increases roughly exponentially with $\Delta V/V_s$, although actually V_s and thus ΔV may not have been accurately established relatively to later observations of FISHER [*14*] and BANDEL [*12*]. The variation of n_0 as $i_0^{-\frac{3}{4}}$ must not be interpreted as indicating a failure of the theory. Under the probable average emission frequencies of TILLES' initiating electrons even the lowest $1/n_0$ was comparable with the values of T_s studied. Thus, owing to the random nature of individual electron liberation in time as current increased, increasing numbers of initiating electrons within i_0 will have failed to act as effectively counted triggers, because they were emitted within time intervals below the resolving power of the measurements. That is, doubling i_0 would not have reduced T_s in proportion since some of the $2n_0$ electrons would have been emitted within less than the 10^{-4} sec. resolving time and would thus not have added to the probability of the spark as measured. With impulse potential breakdown at lower illumination, T_s was proportional to i_0. Generally speaking no accurate correlation between $1/n_0$ and T_s is possible. Photoelectric emission is difficult to control, especially in air, owing to changes in the cathode surface with breakdown occurring. TILLES' formative lags T_s were very long even with a γ_i mechanism active. With gaps from 0.6 to 1.7 mm. long, even ion crossing times should be in the microsecond range. No doubt TILLES was working much closer to the real V_s than most others and was thus observing the long formative time lags caused by slow space charge build up in the short gaps as indicated by later work of FISHER [*14*] and his associates.

δ) Studies on TOWNSEND *thresholds of* SCHADE and VON GUGELBERG. The work of SCHADE in 1937 [*24*] and VON GUGELBERG [*24*] in 1947 represent the only studies of formative time lags approaching the conditions for a true TOWNSEND breakdown with γ_i and γ_p at lower pressures. Here statistical variations were avoided. SCHADE showed that a simplified version of equation (5.10) of STEENBECK by expansion in series applied. Near V_s the lags were quite long and increase in overvoltage decreased the lags down to the anticipated 10^{-5} sec. in slightly soiled Ne and H_2 gases. The study of VON GUGELBERG [*24*] was concerned largely with verifying the basic equations (5.10) and (5.11) for τ_i and τ_p. It yielded the correct order of magnitude for the values of the formative time lags as modified by space charge distortion and enabled rough estimates to be made of γ_i and γ_p.

ε) KERR cell studies of DUNNINGTON, LAWRENCE and WHITE. Of considerable interest relative to certain aspects of spark breakdown were the remarkably beautiful studies of the growth of the spark in time as revealed by KERR cell used by LAWRENCE and DUNNINGTON, and of DUNNINGTON [*40*]. A KERR cell between crossed Nicol's was placed in the path of the light from the spark. The light could come directly from the spark at a distance, L_e, or from the image of the spark in a mirror, a measured distance $L\acute{e}$ behind the Spark. When the potential was applied to the sphere gap with a current i_0 from the cathode, the pulse travelling over a delay line of length L_w reached the shutter and opened it. The light from the developing spark then traveled through the shutter until the decline of the potential from the condenser across the gap caused the shutter to close again after traversing the delay line. By slightly tilting the mirror, this yielded two images of the spark channel side by side after intervals of $t'_p = \frac{L_w - L'_e}{c}$ and $t_p = \frac{L_w - L_e}{c}$ (c = velocity of light]. Pressures in air varied from 200 to 760 mm. with gap lengths ranging from 1 to 10 mm., values of t_p ranged over times from 7×10^{-9} to 3.15×10^{-8} sec. Time value ranges were greatest for low pressures and long gaps and least for 1 mm. gaps at atmospheric pressure. At low pd values the luminosity started from the cathode and moved across the gap to the anode, to be joined by a bright spot that moved from the anode before complete breakdown. Above a value of $pd = 200$ cm. $\times$ mm. Hg the breakdown luminosity proceeded from mid gap usually accompanied by a slight anode spot. The mid gap streak moved towards cathode more rapidly than towards the anode. Interpretation is not certain. Only the brighter phases of the spark breakdown process could be seen. It appears as if in these overvolted gaps the large pd value breakdown proceeds via a midgap streamer as later observed by RAETHER in cloud track pictures. If so, this observation of LAWRENCE and DUNNINGTON is the first observation of mid gap streamer breakdown. It is to be noted that these luminosities are *part* of the spark breakdown mechanism only and thus constitute a fraction of T_f or τ_f only. It is also probable that errors inherent in the techniques rendered the time scales of events as inferred from t_p inaccurate so that actual time scales of events may have been considerably longer.

The work of DUNNINGTON was extended by WHITE [*40*] who improved techniques of time measurement, improved the KERR cells as well as extended studies to other gases. In H_2 all the way from 3 to 10 mm. gaps and atmospheric pressure, the luminosity progressed from the cathode to be joined by an anode streamer at $\frac{3}{4}$ gap length, the luminosity was faint. With N_2 as in air, both types of breakdown were observed with transition to midgap streamers at pd values well above those considered by DUNNINGTON. In He the filamentary discharge was preceded by a glowlike TOWNSEND discharge filling the gap, that is, breakdown of the TOWNSEND type preceded streamer formation. Similar results were obtained recently in very pure argon by MENES [*15*]. Formative time lags in A and He were several times longer than in the molecular gases. The velocities of the streamers in H_2 and N_2 were 3.2×10^7 cm./sec. in H_2 and 1.6×10^7 cm./sec. in N_2 at 760 mm. For an 8 mm. gap, the times for the luminosity to cross the gap were 1.8×10^{-8} sec. in H_2 and 2.7×10^{-8} sec. in N_2. There appeared to be branching and multiple filament effects in WHITE's work. These were most prominent where cathode and mid gap streamers met in N_2, O_2 and mixtures. CO_2 showed very tortuous paths with branching towards the cathode. WHITE used materially higher overvoltages than DUNNINGTON.

WHITE studied the influence of the photo-current i_0 on the time lags for high values of i_0 and higher overvoltages. Here the gap was raised to the firing overvoltage with a long T_s owing to absence of triggering electrons. By illuminating the cathode with the light from an impulse spark gap at a given instant, the time elapsed between the triggering photo-electric current i_0 in the gap from the auxiliary spark to appearance of the light in the main gap was measured. Thus T_0 could be directly observed. An approach voltage very near but below V_s was applied, then an overvoltage ΔV was applied, followed in some 10^{-2} sec. by the triggering spark in the auxiliary gap. The illuminating flash rose from zero intensity to a peak in some 10^{-7} sec. and declined. This flash liberated 10^7 electrons in some 10^{-7} sec. It is probable that not enough electrons were liberated in the first 2×10^{-8} sec. to trigger the gap, thus, as overvoltages increased WHITE's observed time lag, measurements could not well have extended below values of T_f of some 3×10^{-8} sec. For gaps from 1 to 5 mm. long in air, He and CO_2 at 760 mm. the values of T_f as a function of $\Delta V/V_s$ and of i_0 controlled by absorbing screens were measured. Time lags decreased rapidly from around 1.4×10^{-7} sec. with 10% overvoltage to 4×10^{-8} sec. or less at 50% overvoltage. Decreasing i_0 increased the value of $\Delta V/V_s$ to give a specific time lag. Decreasing i_0 from 1 to 0.002 at 10% overvoltage increased the time lag from 6.8×10^{-8} to 1.3×10^{-7} sec. These observations are in good agreement with predictions from the TOWNSEND theory although the sparks were undoubtedly streamer sparks. In He, higher i_0 and larger $\Delta V/V_s$ values were required to produce time lags comparable to those for air and CO_2, e.g., at full i_0 $\Delta V/V_s = 7\%$ for air to give $T_f = 1\times10^{-7}$ sec., while in He it took $\Delta V/V_s = 40\%$ as later discovered by FISHER [14]. The most interesting observation was perhaps that at really high i_0 the static sparking potential V_s could be reduced some 10% [27]. This was nearly simultaneously discovered by ROGOWSKI and WALRAFF [27]. This lowering at the heavy field intensified i_0 of WHITE is in conformity with the theory of STEENBECK as indicated by equation (5.7). Rough calculations by WHITE [27] and by MEEK [27] using his streamer theory rendered such results reasonable. Unfortunately, ROGOWSKI [27] and some of his co-workers ascribed the phenomenon to a much more complicated mechanism and developed a rather elaborate theory of breakdown in this connection which appears not to have gained acceptance.

ζ) WILSON's and NEWMAN's studies. The work of WHITE was carried further by WILSON [40] who attempted to circumvent the inconvenience of the rise time of WHITE's triggering spark. He illuminated the cathode of the sphere gap with constant ultraviolet light from an Hg arc. A constant approach voltage of $0.99\,V_s$ was applied to the gap. By means of a vacuum switch, the appropriate overvoltage could be placed across the gap in series with the approach voltage. The potential across the vacuum switch kept the KERR cell open. As soon as the vacuum switch closed, the potential fell to zero. This reduction in field travelled along a variable transmission line to the KERR cell, so that shutter closed at various intervals following application of the potential. The observed time lag could then be established by increasing the path length until the light could be seen. Owing to the fact that the illumination i_0 was now less than in WHITE's case, statistical fluctuations T_f came in. To facilitate observations a definite time delay was set on the delay line and the overvoltage was gradually raised until a spark was observed. With the statistical factors involved however, WILSON raised overvoltage in steps. At each step the overvoltage was applied 100 times and n_t, the number of sparks observed, over the total number n_T of applications was noted as the experimental probability of the spark *within* an interval t at a given $\Delta V/V_s$. This is a convenient, but not an orthodox, method of observing. From these data, he was able to plot n_t/n_T against $\Delta V/V_s$ expressed in per cent for several different values of t. What is really required, are data of n_t/n_T plotted against t for several values of overvoltage which should read

$$\frac{n_t}{n_T} = 1 - e^{-P_s n_0 (t - T_f)} = 1 - e^{-\frac{t - T_f}{T_s}}. \tag{16.1}$$

To get such data and evaluate T_f the relatively few points for single values of $\Delta V/V_s$, each one averaged for 100 readings apiece, were plotted as indicated. Cross plots were then made choosing $\Delta V/V_s$ fixed and measuring the intercepts on the (n_t/n_T), $(\Delta V/V_s)$-curves for different values of t. To get accurate cross plots many such curves for many values of t would have been required leading to a prohibitively lengthy task, for already the taking of 100 readings for each point was more extensive than any previous investigators had done. WILSON thus *interpolated intermediate curves by eye*. It is unfortunate that he did so, for the curves of (n_t/n_T) vs. $(\Delta V/V_s)$, with few points, especially near the low n_t/n_T-values, were too inaccurate for this procedure. They appeared to rise assymptotically to yield an S-shaped curve about $n_t/n_T = 0.5$. In consequence, WILSON'S *derived* (n_t/n_T), t-curves were not only *inaccurate* but *erroneous as to form* in that n_t/n_T-values appeared to rise asymptotically from zero to a saturation value of 1 in an S-shaped curve as a function of t. This indicated that the n_t/n_T relationship for sparking *no longer followed the* LAUE-WIJSMAN *law* but that $\frac{d(n_t/n_T)}{dt}$ yielded a GAUSSian distribution of time lags about a mean value at $n_t/n_T = 0.5$ as indicated by the writer in 1948 [*40*], [*61*]. This would mean that not only contrary to theory, perhaps because of space charge effects, WILSON'S sparks followed laws completely different from those indicated by all other observers. When the writer and WIJSMAN in 1951 discovered this contradiction between WILSON'S data and all other work, his curves were subjected to careful scrutiny. It was shown that while under reasonable assumption the (n_t/n_T), $(\Delta V/V_s)$-curves might have an S-shape, the inference of S-shaped (n_t/n_T), t-curves at constant $\Delta V/V_s$ was *not* a necessary consequence, *if the cross plots* had been made from *accurately drawn curves* [*41*]. However, careful study of WILSON'S data as reported indicated that there is really no certainty that any of his (n_t/n_T), $(\Delta V/V_s)$-curves even had an asymptotic rise near the origin. Thus the cross plots derived by WILSON were erroneous and misleading and most conclusions based on this work measuring T_f from the $n_t/n_T = 0.5$ level as for a GAUSSian distribution must be considered irrelevant. WILSON did, however, observe that decreasing the intensity of illumination i_0 on the (n_t/n_T), $(\Delta V/V_s)$-curves for a fixed t did not shift the origin of $\Delta V/V_s$-values. All it did, was to spread out the curves along the $\Delta V/V_s$ axis as the number of triggering electrons decreased. Thus, at a given lag t, the threshold of the spark sets in at a given overvoltage independent of his relatively low i_0, but the scatter of statistical lags requires higher overvoltage to observe n_t/n_T near unity. This means that P_s and T_f are *not sensitive to illumination at this over voltage*, but, since n_0 is decreased, $\Delta V/V_s$ must be increased to increase P_s so that $P_s n_0$ shall remain the same. It would be legitimate to plot a set of values using the feet of WILSON'S (n_t/n_T), $(\Delta V/V_s)$-curves as giving $t = T_f$ as a function of overvoltage. This would cause T_f to decrease fairly rapidly along some sort of a hyperbolic curve as $\Delta V/V_s$ increases *in general agreement with* STRIGEL'S *work*, but unlike WHITE'S data involving no limitation towards shorter time lags as overvoltage increases.

NEWMAN [*40*] nearly contemporaneously with WILSON used a 10000 volt sphere gap in air at 760 mm. with a quartz Hg lamp close to the spheres. At a given instant a negative pulse of high amplitude was started from one end of a transmission line. Simultaneously, an impulse signal of an overvoltage type was placed across the sphere gap with an approach voltage of from 50 to 98% superposed at the opposite end of the transmission line. If the gap broke down at once the positive and negative pulses met in midgap. The delay in breakdown T_0 caused the positive pulse to be met by the negative pulse closer to the sphere. Detecting elements for the positive pulse moved along the line, then gave the distance of advance of the positive pulse when it met the negative pulse. NEWMAN went to higher overvoltages than did WILSON, and his curves showed a steeper rise so that $\Delta V/V_s$ appeared to rise exponentially as T_0 decreased. The longer time lags of NEMWAN at lower overvoltages lay close to WHITE'S values with low illumination and the *overvoltages were one half or less than those of* WILSON with his normal illumination. This resulted from WILSON choosing an n_t/n_T value of 0.5 to yield his values of $\Delta V/V_s$ assuming a GAUSSian distribution about a formative time lag T_f.

Further progress in the study of time lags in uniform field or other type breakdown came in the post World War II period. Since this progress is intimately connected to the streamer theory of spark breakdown it is desirable to postpone further discussion until this important mechanism has been presented.

V. Breakdown in non-uniform fields and near highly stressed anodes. Photoionization in gases. Streamers and burst pulses.

17. Deficiencies of the TOWNSEND model in spark breakdown. In the preceding discussion of the TOWNSEND breakdown in nearly uniform fields, there are remarkable agreements with theory derived from the model as regards many properties of the phenomena observed including even the time lag studies. However, any successful model must satisfy *all* essential observed elements in order to be acceptable. There are in fact certain very important observations which cannot in any case be understood on the TOWNSEND model. These are:

1. The observation that in spark breakdown under most circumstances where gas densities are sufficiently high, so that thermal diffusion of electrons and carriers cannot spread the discharge over the whole electrically stressed volume, the first manifestations of the breakdown are in the narrow filamentary and often crooked spark channels. Although with adequate currents and current densities regions of electrical discharges may suffer constrictive actions, it is inconceivable that the diffused TOWNSEND breakdown in a uniform field gap should contract to the narrow spark and often crooked spark channels, observed within time scales active. Recently, BANDEL [*12*] has achieved *semi-quantitative proof of the necessary breadth of the* TOWNSEND *discharge in uniform fields* and studies in coaxial cylinders in pure [*45*], N_2 [*10*] and A [*44*], [*46*] at higher pressures confirm this conclusion.

2. As noted with improving techniques, the formative time lags appear to decrease indefinitely from times of the order of 10^{-7} sec. to several times 10^{-9} sec. in molecular gases as overvoltage increases from 2 to 60%, and similarly for overvoltages of the order of 100% in inert gases. Even for a uniform gap of 1 cm. length in air at normal density, it is very difficult to reconcile an observed T_f of 10^{-7} sec. at 2% overvoltage with any known γ_p process since electron transit

times are not less than 10^{-8} sec. and a sequence of 10 cumulative events would hardly cause breakdown. It was in fact this circumstance, first noted in the early nineteen hundred and twenties, that initiated the search which ultimately led to the streamer model. The still shorter time lags noted above and their indefinite decrease with overvoltage indicate the presence of a new and faster mechanism.

3. TOWNSEND early recognized that the corona type breakdowns from highly stressed anodes such as at wires and points, with very distant cathodes, could *not* be accounted for by any secondary cathode action. It was for this reason, among others, that he adhered so tenaciously to the β mechanism of ionization by positive ions *in the gas*. In principle TOWNSEND'S uncanny intuition was correct in insisting that the secondary mechanism in these situations resided *in the gas*. He could not until many years later, have known that *photoelectric ionization in the gas* was the active agency since the magnitude and circumstance of occurrence of this mechanism was completely ususpected.

18. Photoionization in gases, pulses and streamers. It is not strange, therefore, that, beginning with the observation of the intense photoionization *in air* by corona and spark discharge by CRAVATH [*47*] and by DESCHÈNE [*47*] in 1934, interest should have begun to focus on this secondary mechanism. By the same token it was to be expected that more careful investigations of the intermittent discharges from positive point-to-plane coronas in air and in the coaxial cylindrical geometry of positive wire GEIGER counter systems should have led to the discovery of the anode surface discharges propagating by photoelectric ionization in the gas and extinguishing by local space charge accumulation [*17*], [*48*]. Finally, the beautiful visualization of electron avalanches and the sequence of events in overvolted uniform field gaps with potential pulses of controlled duration in a cloud chamber by RAETHER [*42*] in 1935 led FLEGLER and RAETHER [*42*] to propose a midgap streamer breakdown mechanism operating by photo-ionization in the gas. Quite independently, observations of the discharge from a needle point corona gap by the writer and LEIGH [*49*], reported at the Physical Society meeting of December 1936, indicated the presence of streamers. The later studies of burst pulse corona and the preonset streamers by TRICHEL [*49*] and by KIP [*49*] enabled the writer to further develop the concepts of anode breakdown and photoelectric action as well as establishing many properties of the streamer propagation. This information enabled the writer and KIP roughly to calculate the field increase by positive space charge creation for an anode streamer and to indicate that the streamer mechanism was probably one of the modes of spark breakdown [*49*]. In 1940 MEEK [*43*] in the writer's laboratory, starting from the calculations of the writer and KIP, set up his semi-empirical relation for anode streamer propagation which he then applied to various forms of spark breakdown in air as indicated in the "Mechanism of the Electric Spark" by LOEB and MEEK, published in 1941 [*43*]. Quite independently, RAETHER [*42*] in 1941 formulated an equivalent semi-empirical relation for the uniform field gap. There was one difference between MEEK'S and RAETHERS'S relations. Working in a laboratory where the cathode directed positive corona streamer was observed and studied, MEEK developed his relations for that streamer. RAETHER, with his somewhat overvolted gaps, had observed both cathode and anode directed positive and negative streamers with midgap breakdown. Even though the positive streamer was faster and appeared to develop earlier in RAETHER'S two most revealing published photographs, RAETHER developed his theory for the anode directed negative streamer. The calculations in principle were independently based on the same crude type of analysis used by LOEB and KIP. Actually, as can be seen in the writer's long report, calculations of FLETCHER [*50*] and KUNKEL [*41*] as well as calculations of correction for image and dipole fields by DODD [*41*] indicate that the methods of MEEK and RAETHER are probably as satisfactory as the essentially simplified assumptions warrant. There is one more point to be made. Since MEEK's positive streamer crosses the whole gap from anode to cathode its propagation condition sets a threshold for the spark. In contrast to this, RAETHER's mid gap streamer relation gives a series of values of the avalanche length x_c needed to start a streamer for any value of V above V_s applied. If x_c were set equal to the gap length d, then the sparking threshold would be obtained. However, formally this criterion cannot apply since then the electron cloud streamer lies in the anode. Pushed to this limit, RAETHER's threshold will be slightly lower than that of MEEK. However, RAETHER did not use his condition in that fashion and for a streamer starting from the anode MEEK's criterion is the proper one.

Neither of these theories explicitly included the absorption coefficient for photoionization though it was stated that such ionization had to be adequate. In 1948, the writer, using the absorption coefficient, first deduced the relation for the threshold of the burst pulse corona [*51*], and then with WIJSMAN [*51*] deduced the threshold for a cathode directed positive streamer in uniform fields including the photo-ionization in the gas. The latter theories are no longer semi-empirical for they are based on the self sustaining, or perpetuating, discharge criterion in the absence of gross space charge distortion common to all breakdown threshold theory. While the relations for streamer threshold of MEEK, RAETHER and the writer were derived under the simplifying assumptions of uniform field geometry, nothing prevents their being applied to non uniform fields except the added mathematical complexities. It is next essential to remark that, in general, conditions deriving from photoelectric ionization in the gas are quantitatively not usually as efficient as those for secondary process at the cathode in uniform fields. It is thus to be anticipated, as was later discovered by FISHER [*14*] and his associates and others, that the threshold for uniform field sparks at other than suitable overvoltages with impulse potentials, will be set by an antecedent TOWNSEND discharge which creates streamer breakdown conditions in the space charge distorted gap. This, however, does not preclude spark and other breakdown thresholds in non-uniform field geometry, especially with highly stressed anode from being set by the streamer mechanism. It is thus of value to derive the following three *threshold relations applicable to photoionization in the gas* leading to (α) the burst pulse corona threshold, (β) the velocity of spread of burst pulse or GEIGER counter pulse discharge and (γ) the breakdown streamer threshold as derived for uniform field geometry.

α) The burst pulse corona threshold [*51*]. Consider a highly stressed anode in some geometrical configuration such that the field X_r at some distance r from a focal point, or the center of a hemispherically capped cylindrical point, or the axis of a wire of radius a, can be computed from the applied potential V. The surrounding gas is at a pressure p at a fixed temperature, usually room temperature. Then, since the TOWNSEND function α/p is known for the gas as a function of X_r/p it is possible to evaluate the size of the TOWNSEND avalanche $M = \exp \int_a^{r_0+a} \alpha\, d r$.

Here r_0 is the distance from the anode surface at which the fields are adequate so that effective ionization can begin and is usually far less than the distance to the cathode. Arrived at the anode, the M electrons diffuse to it and are removed, leaving a relatively small space charge of positive ions outward in the gap with greatest density at the anode. Accompanying the M electrons, there are created fM photons of such a wave length that they can photo-ionize the gas. At lower X/p, f will be low and increase with X/p so that it is sometimes materially greater than unity, eventually again decreasing as higher values of X/p are reached. When the applied potential reaches a value V_g such that the photons produced, when absorbed in the gas beyond a distance r_0 normal to the anode, can on the average create a new avalanche producing electrons for each preceding avalanche, then the threshold for a self sustaining discharge *burning at the anode* is set. Owing to the inhibiting influence of the positive ion space charge as it slowly moves outward from the gap and to the statistical character of fM and the chance of photoionization, the discharge sequence will after some time be interrupted leading to a succession of pulses. Such pulses were first observed by TRICHEL [*49*] for positive points, and called *burst pulses* in point-to-plane corona. They also constitute the GEIGER counter pulses of the type observed by STEVER [*17*] in coaxial cylindrical geometry. The threshold will be set by the condition that

$$\beta f M = 1, \tag{18.1}$$

where β is the chance that a photoelectron capable of producing an avalanche of size M is created beyond r_0 normal to the anode surface in the field. It is possible roughly to set

$$\beta = 0.5 f_1 e^{-\bar{r}_0 \mu_1 p_1/p}, \tag{18.2}$$

where the factor 0.5 comes from the fact that half of the photons created will be directed towards the anode, or into regions where they will not produce avalanches, $\bar{r}_0$ is an average value of r_0 needed to insure that the electron created traveses a radial distance r_0 from $r+a$ to a in the field, μ is the absorption coefficient of the gas for photoionizing photons at 1 mm. pressure, p_1/p is the partial pressure of the photoionizable gas present, and f_1 is the chance that a photon absorbed will photoionize the atom or molecule. For simplicity it is assumed that there is primarily one photon species of coefficient $\mu = \mu_1 p_1/p$ principally active. Rough calculation [*51*] sets $\bar{r}_0 = 1.86\ r_0$ and more refined claculation by WIJSMAN indicates that for a normal range of values of parameters the constant is not far different. Thus the condition for the first type of discharge, a localized one at the anode surface called the burst pulse, is given by

$$\beta f M = 0.5 f_1 f e^{-1.86 \mu_1 r_0 p_1/p} e^{\int_a^{a+r_0} \alpha\, dr} = 1. \tag{18.3}$$

In principle, this threshold does not differ from the TOWNSEND threshold for γ_p action except that $g f_1 f e^{-\mu d}$ is now replaced by β. This discharge will continue to burn until statistically caused fluctuation in the $\beta f M$ sequence, or lowering of the effective V_g by positive ion space charge in the gap near the anode terminates it. It will re-ignite when the next external stray electron starts a new avalanche beyond $r_0 + a$, after space charge sufficiently clears.

β) *The spread of the* GEIGER *counter pulse.* It was noted in GEIGER counter studies that the discharge spread along the anode wire in coaxial cylindrical geometry. This spread was first indicated to be photoelectric in the gas by GREINER [*48*] and later proven by STEVER [*17*]. The theory for the spread was developed by ALDER, BALDINGER, HUBER and METZGER [*52*] and leads to an equation which is of interest. In the original derivation they used a value for M which is not quite satisfactory. In the notation used here, their expression with M

properly expressed reads [53]

$$0.5 f f_a \frac{p_a}{p} e^{\int_a^{r_{0g}+a} \alpha\, dr} \left[e^{-\mu_1 \vartheta v p_1/p} + \left(\mu_1 \vartheta v \frac{p_1}{p}\right) E\,i\left(-\mu_1 \vartheta v \frac{p_1}{p}\right)\right] = 1. \tag{18.4}$$

Here all terms are the same as those in equation (18.3) except that f_1 is replaced by $f_a \frac{p_a}{p}$, where f_a is the value of f_1 for 1 mm. pressure and p_a/p is the partial pressure of the active gas. The symbol $E\,i$ stands for the exponential integral and the term comes from the effect of distances measured along the wire instead of radially. In principle, for the spread along the wire, it is analogous to the factor 1.86 used in (18.3). The term $E\,i$ furnishes a relatively unimportant correction. The only terms that are strange are two quantities ϑ and v. The quantitity ϑ is the longest time required for the emission of photoionizing photons. It is given by

$$\vartheta = \int_a^{r_{0g}+a} \frac{dr}{v(r)} + \tau_p. \tag{18.5}$$

Here $v(r)$ is the drift velocity of the avalanche electrons which varies with r, and τ_p is the time of emission of active photons which is usually short compared to the creation of the avalanche. The quantity v is the *velocity* of *propagation of the discharge along the wire.* The quantity $\vartheta v = x_0$ defines a fictive distance of spread for each successive step. Actually, ϑ is set by physical characteristics of the fields and nature of the gas. The quantity v is an observable which varies through relatively small limits. The *maximum* velocity v_m is set by $\vartheta = r_{0g}/v_m$ since the time must be in excess of ϑ for the avalanche to be completed and photons radiated. The *minimum* velocity of spread is set by the condition that the discharge cannot continue in one place for more than n successive events without building up an annihilating space charge. Near threshold, n is not very large. Then v_l, the lower limit to the velocity, is set by $r_{0g}/n\vartheta = v_l$, or $v_l = r_{0g}/n\vartheta$. Thus, $r_{0g}/\vartheta > v > r_{0g}/n\vartheta$. The quantity $v\vartheta = x_0$, the characteristic length, enters into the negative exponential term as the important product $x_0 \mu_1 p_1/p$. If the threshold is otherwise to remain constant as $\mu_1 p_1/p$ decreases, x_0 and thus v must increase. That is, the longer the absorbing free path $1/\mu$, the faster will be the pulse spread. This has been observed in coaxial geometry for O_2—N_2 mixtures with O_2 the active gas at constant potential [*10*], [*53*].

Thus there is not only a *threshold* for a *localized burst pulse*, given by equation (18.3), but a threshold for a GEIGER pulse spreading over the anode surface before it extinguishes.

γ) The thresholds for streamer propagation of MEEK, RAETHER, WIJSMAN *and* LOEB [*20*]. It has been observed by KIP [*49*] that, as the potential in positive point-to-plane corona increases above burst pulse threshold, especially with air contaminated with water vapor and/or CO_2, and depending on geometry, but below the potential where space charges clear so fast that the onset of steady corona is reached, there appear a series of much heavier pulses. These have some 10^{10} ions/cm. length of luminosity per pulse compared to the 10^7 odd in single burst pulses. In these the discharge is seen to propagate out into the gap along the field axis instead of spreading over the point surface as in the burst pulses. These are the *preonset streamers.* They have now been oscillographically studied in much detail yielding the ions per cm. of path length and their velocity of propagation which ranges from 1 to 6×10^7 cm./sec. These streamers project some cm. into the gap, depending on geometry, until they are slowed and deflected by encountering the space charge from antecedent streamers.

The appearance of streamers in coaxial geometry has been demonstrated first by MILLER [45] and the writer in pure O_2 and more completely and accurately verified in O_2—N_2 mixtures of some 60 or more % O_2 at 760 mm. and down to 200 mm. in pure O_2 by HUBER [10]. When the potential is raised considerably above the onset value for steady corona, such that the space charge is pretty well cleared from the low field region of the gap, nearly as fast as it accumulates, streamers reappear. If the gap is more uniform and short, this occurs at about twice or three times the onset potential. Then, the first streamer to reappear crosses the gap to give a spark. If the gap is long with high asymmetry, such as with a needle point of 10^{-3} cm. radius and a 5 cm. gap, spark potentials are 5 or more times the onset potential and streamers can be seen again for quite a range of potentials before one succeeds in crossing the gap and yielding a spark. These streamers have been studied in considerable detail by AMIN and by HUDSON [54], and the sequence of events in the quasi space charge fouled gaps is quite involved. A sequence of streamers of different velocities, ranging from 2×10^6 cm./sec. up to 2×10^8 cm./sec. depending on conditions, were observed. Since streamers have been observed by high speed photography in long impulse sparks by ALLIBONE and MEEK [55] and studied more in detail by MEEK and SAXE [55] in 10 to 50 cm. gaps with photomultiplier, since they have been observed directly in the cloud track pictures of RAETHER and perhaps even earlier in the KERR cell shutter studies of DUNNINGTON and LAWRENCE and WHITE as well as in positive point corona studies, there can be little doubt on two points. The first is that the streamers represent a real phenomenon in many gases, and the second is that they are closely associated with spark breakdown in most, if not all, systems at higher pressures when they cross the gap so as to make a conducting channel which, by means of the return stroke potential wave, yields a channel with close to 100% ionization at the existing density. Under these conditions it is essential to set up a criterion for streamer advance. As indicated above, this will be done for the uniform field case [51].

In this deduction as in all breakdown studies, the condition for threshold will be set when the sequence of events yielding favorable space charge formation and photoionization in the gas, initiated by a single avalanche electron travelling towards the anode, creates enough new electrons, so auspiciously placed as to sustain the sequence and in this case advance the discharge away from the anode out into the gap instead of spreading it over the anode. Before proceeding to the analysis, it must be noted that, as indicated under β), the discharge will spread over the anode, provided the absorbing mean free path for photoionizing radiation is adequate to permit spread with a velocity v. If this cannot occur and if by convergent geometry the avalanches build up a localized space charge field at the anode that augments the existing applied field, and if the photoionizing free path is small enough to feed electrons into this confined region, the discharge which cannot spread *over* the *surface* will spread *radially out into the gap*. Under usual conditions in air or N_2—O_2 mixtures the values of μ for burst pulse spread range around 10 cm.$^{-1}$ with free path $1/\mu$ of 1 mm., while for streamer propagation $\mu\sim 100$ cm.$^{-1}$ and $1/\mu=0.1$ mm.

The initiating electron, moving from cathode to anode d cm. distant in a uniform field, creates an avalanche of $e^{\alpha d}$ electrons and ions. The electrons arrived at d diffuse into the anode leaving behind the positive ions created, which are relatively immobile. The density of positive ions is greatest at the anode. They are distributed in some statistical fashion extending out to an average distance ϱ, radially from the avalanche axis along the field direction and increasing in linear density with x, such that in dx at a distance x the ions are $\alpha e^{\alpha x}dx$ in number.

The radial expansion of the avalanche is caused by the random diffusion of the ionizing avalanche electrons owing to their average energy in the field. The value of ϱ should be taken [56] as

$$\varrho = \sqrt{6Dt} = \sqrt{6 \cdot \frac{2}{3} u_t \frac{x}{X}} = \sqrt{\frac{4}{3} \frac{\lambda_0^2}{p^2} \frac{\varepsilon X x}{v^2 m}}. \tag{18.6}$$

Here D is the diffusion coefficient and t is the time to reach a point x at which ϱ is required. The quantity u_t is the terminal energy in volts of the avalanche electrons in the field X and λ_0 is the mean free path of the electrons at u_t at 1 mm. pressure in the gas, (the RAMSAUER free path), which can readily be found for known gases if u_t is roughly known. The electron drift velocity v in X is usually known and ε/m is the specific electronic charge. This average value of ϱ corresponds to the point at which the electron density has fallen to $1/e$ of its value, i.e., some 67% of the electrons created lie within ϱ.

Now at the anode this accumulation of positive ions creates a space charge field X_1 which must be added to the imposed field X. To get this field, it is not seriously incorrect to assume that most of the ions lie in a sphere of radius ϱ_c next to the cathode. This may be computed from the density of the positive ions N in the avalanche at some distance x from the cathode given by

$$N = \frac{\alpha e^{\alpha x} dx}{\pi \varrho^2 dx} = \frac{\alpha}{\pi \varrho^2} e^{x\alpha} \tag{18.7}$$

which when multiplied into the volume $\frac{4}{3}\pi\varrho^3$ yields a number n of ions

$$n = \tfrac{4}{3} \alpha \varrho e^{\alpha x}, \tag{18.8}$$

of charge ε, each producing a field X' at a distance r from the charge center, (and if $x = d$ with ϱ_c from the center of the last ionizing free path $1/\alpha$ from the anode), of value

$$X' = 4\pi \cdot \tfrac{4}{3} \varepsilon \alpha \varrho_c e^{\alpha d}/(4\pi r^2) = \tfrac{4}{3} \varepsilon \varrho_c \alpha e^{\alpha d}/r^2 = E/r^2. \tag{18.9}$$

Thus the total field on the cathode side of the space charge, along the axis of the avalanche end of the field X, is

$$X_1 = X + X' = X + \frac{E}{r^2}. \tag{18.10}$$

This, with minor corrections, is the calculation made by the writer [49] and KIP in 1939 and underlies the basic theories of the streamer advance of MEEK and RAETHER. MEEK, then assuming photoionization adequate, set the condition for the cathode directed positive streamer as being given by $X' = KX = KE/r^2$ with $r = \varrho_c$, and K an arbitrary constant which he first set equal to unity. Later, from preonset streamer thresholds in air, it appears that K should perhaps have been as low as 0.1. Omitting this correction, MEEK's condition at once led to an expression for streamer advance in the form of

$$X_s = \frac{4\varepsilon\alpha e^{\alpha d}}{3K\sqrt{6\frac{D}{k_e}\left(\frac{d}{X_s}\right)}} = \frac{4}{3}\frac{\varepsilon}{K}\alpha e^{\alpha d}\frac{1}{\sqrt{\frac{4}{3}\frac{\varepsilon}{m}X_s\frac{\lambda_0^2 d}{p^2 v^2}}}, \tag{18.11}$$

using the more correct $\sqrt{6Dt}$ for ϱ_c in place of the value $\sqrt{2Dt}$ assumed in the earlier work of the writer [49] and MEEK. Using the original constants of MEEK, the equation with $\sqrt{2Dt}$ in place of $\sqrt{6Dt}$ in logarithmic form used by MEEK becomes;

$$\frac{\alpha}{p}(pd) + \log\frac{\alpha}{p} = 14.85 + \log\frac{X_s}{p} - \frac{1}{2}\log(pd) + \log d. \tag{18.12}$$

Since RAETHER considered the electron cloud advancing towards the anode from midgap x_c, he had $e^{\alpha x_c}$ electrons in a spherical volume of radius ϱ_R in which ϱ_R is $\sqrt{3Dt}$, as used by RAETHER. This gave RAETHER a field at a distance r from the streamer head when the streamer formed at x_c from the cathode as

$$X'_{Rr} = \frac{4\pi\varepsilon e^{\alpha x_c}}{4\pi r^2} = \frac{\varepsilon}{r^2} e^{\alpha x_c}. \tag{18.13}$$

This leads to a field at $r = \varrho_R$ of

$$X_{1R} = \frac{\varepsilon e^{\alpha x_c}}{\varrho_R^2} = \frac{\varepsilon e^{\alpha x_c}}{3Dt} = \frac{\varepsilon e^{\alpha x_c}}{u_t x_c / X_{Rs}} = \frac{m e^{\alpha x_c}}{\varepsilon \frac{\lambda_0^2}{p^2}\frac{x_c}{v^2}\, X_{Rs}}. \tag{18.14}$$

With $X_{1Rs} = K X_{Rs}$ the condition for streamer advance of RAETHER becomes

$$e^{\alpha x_c} = \frac{K x_c}{3{,}2 \times 10^{-8}} \frac{\varepsilon}{m} \frac{\lambda_0^2 X_{Rs}^2}{2v^2 p^2} \tag{18.15}$$

and in logarithmic form, correcting a slight numerical inconsistency in RAETHER'S deduction, of $u_t\, x/X$ from $3Dt$:

$$\alpha x_c = 16.4 + \log u_t x_c + \log K. \tag{18.16}$$

Both MEEK'S and RAETHER'S equations are logically and dimensionally correct. For positive ion accumulation, MEEK'S expression depends on the rate of ionization with α in his numerator and X_1 depends on $\alpha\,\varrho_c$ in equation (18.9) which is dimensionally correct as α is a reciprocal length. RAETHER'S expression depends on $1/\varrho_R^2$ assuming ionization uniform over ϱ_R. The question of validity really hinge on relative values of $1/\alpha$ and ϱ. If $1/\alpha$ is larger than ϱ, MEEK'S procedure is legitimate and necessary for positive streamers, RAETHER'S calculation being also correct for negative streamers, for the electron cloud comprises $e^{\alpha x}$ electrons at x and these are largely distributed by diffusion radially and along the direction of movement. *Space charge fields* will thus *really differ in the two cases*. Where $\varrho = 1/\alpha$, MEEK'S and RAETHER'S relations become the same. With ϱ larger than $1/\alpha$, neither procedure is accurate. In air, at V_s, at 760 mm., $\varrho \sim \frac{1}{2\alpha}$.

Calculation of the fields for other than spherical volumes were carried out by WEISSLER [*57*] and produced only small variations in the constant terms of much less significance than other errors. DODD [*41*] calculated errors due to dipole and to image force fields in the anode. Since $1/\alpha$ ranges from $2\varrho_c$ upwards, the correction to X_1 will be a reduction of 11% downwards only. The dipole field reduces the calculated field by 30% at $1/\alpha = \varrho_c$ and rapidly diminishes at $1/\alpha > \varrho_c$.

The more serious questions as to whether, when photoionization and fields are adequate for positive and negative streamer advance in midgap breakdown, the streamers are capable of advancing owing to dipole effects has been analyzed semigraphically by RAETHER and in more complete mathematical form by KUNKEL [*41*] and by FLETCHER [*50*]. These analyses leave no doubt that the conditions for streamer advance being satisfied will permit both, cathode directed positive streamers from the anode and positive and negative streamers to proceed from midgap. These more exact calculations indicate that the order of magnitude approximations used above are quite as satisfactory as more esoteric relations the solutions of which present formidable difficulties.

Returning now to the space charge distorted anode field of equation (18.10), this can be applied to derive the relation for a self propagating discharge. If X_1 has reached a suitable value, then it is essential to consider what happens to photo-electrons created along the axis of the avalanche by avalanche photons in the enhanced field X_1. Accompanying the formation of the $n = \frac{4}{3}\alpha\,\varrho_c e^{\alpha x}$ electrons in the volume of radius ϱ_c, there will be $n f$ photons which may be assumed relatively constant for the small range of changes in X/p occurring. Of those photons arriving in a sensitive region R directed towards the cathode, some will succeed in liberating photoelectrons from the gas. Those created within R, now in the enhanced field X_1 will create avalanches that feed into the tip of the space charge. This region R may properly be chosen as a cylinder of radius ϱ_c about the axis of the initial avalanche in the field direction. The electrical field at any point is $\vec{X}_1 = \vec{X} + \vec{X}'$, the vector sum of imposed and space charge fields. For most points of importance within R, X' is so close to X that X_1 can be set as in equation (18.10).

Let μ be the absorption coefficient, if necessary written as $\mu = \mu_1 p_1/p$ if there is a partial pressure of photo-ionizable gas. Then, in a region between r and $r + dr$ lying entirely in R, the number of absorbed photons is approximately

$$n f \left(\frac{\varrho_c^2}{4r^2}\right) e^{-\mu r} \mu\, dr. \tag{18.17}$$

If a fraction f_1 of these lead to photo ionization, then

$$f_1 f n \left(\frac{\varrho_c^2}{4r^2}\right) e^{-\mu r} \mu\, dr. \tag{18.18}$$

new electrons will be created. Each of these causes an avalanche of which the tip at $r = \varrho_c$ contains a number of ions equal to

$$\tfrac{4}{3}\bar{\alpha}'\,\varrho' \exp \int_{\varrho}^{r} \alpha'\, dr' \tag{18.19}$$

in analogy to the initial avalanche, with, however, α' the first TOWNSEND coefficient in the field X_1. The quantity $\bar{\alpha}'$ is the value of α' in X_1 when $r = \varrho_c$. The quantity ϱ' represents a *new lowered diffusion radius*, since the electron run in the enhanced field may well be less than d. If the drift velocity is taken as relatively constant for small field changes (it varies as $\sqrt{X/p}$), ϱ' is roughly given as $\varrho' = \varrho_c \left(\frac{r}{d}\right)^{\frac{1}{2}}$. Combining the relations (18.8), (18.18) and (18.19) and integrating from ϱ_c to d gives the total number of ions created by all avalanches feeding into the tip from R as

$$\int_{\varrho}^{d} \frac{4}{3} f_1 f \alpha\,\varrho_c e^{\alpha d} \left(\frac{\varrho_c^2}{4r^2}\right) e^{-\mu r} \mu\, dr \cdot \frac{4}{3}\alpha'\,\varrho_c \left(\frac{r}{d}\right)^{\frac{1}{2}} e^{\int_{\varrho_c}^{r} \alpha'\, dr'}. \tag{18.20}$$

This approximation is invalid if $1/\mu$ is not appreciably greater than ϱ_c. The condition for streamer propagation is now set if the quantity in equation (18.20) is equal to, or exceeds, the initial number of electrons n inside of the sphere of radius ϱ_c or R. Thus, streamers will propagate if the field and other conditions are such that

$$\tfrac{1}{3} f_1 f \{\alpha^1 \varrho_c^3 \mu\, d^{-\frac{1}{2}}\} \int_{\varrho_c}^{d} r^{-\frac{3}{2}} e^{-\mu r} e^{\int_{\varrho_c}^{r} \alpha'\, dr'} dr = 1. \tag{18.21}$$

This is now again a typical threshold relation leading to a streamer discharge, or to a spark breakdown by streamer mechanism.

Solution of this transcendental equation follows the pattern of the solution of all such breakdown equations including TOWNSEND's. For a given gap geometry, gas type, and pressure, there are available values of α/p as a function of X/p, values of the electron drift velocities v, their RAMSAUER free paths λ_0 and either calculable or observed relations between u_t the energy and X/p. It must be assumed that the values of μ for the photoionization are known as well as those of f and f_1. For a solution of the TOWNSEND relations the quantities above and γ_i or γ_p etc. should be known. The relation for α/p as a function of X/p in question, around the value of V_s and X_s/p for sparking to be anticipated, can be approximated by a suitable algebraic relation of convenient form for integration over the limited range needed in performing integrations of $e^{\int \alpha dr}$ using the variation of X and hence α with r resulting from the geometry of the gap and pressure p. A value of X is chosen by guess. This enables the values of α/p, ϱ_c, λ_0, v, f and γ to be determined. X' is then calculated and yields $\bar{\alpha}'$. The data are then at hand for integration of the terms. It is then possible to note whether the integral on the left hand side in the streamer or TOWNSEND's relation is greater or less than 1. From the result, a new value of V or X is chosen and the process is repreated. Plotting the values of the left hand side against V, or X a curve results, leading to the choice of values for which the left hand term reduces to unity.

δ) Properties of streamers. As indicated, streamers have been observed in the corona preonset region and in the corona breakdown to a spark with highly stressed anode. The properties have been fairly well studied in the preonset region oscilloscopically and otherwise [*20*] and are being studied in the corona breakdown region [*54*]. They have been studied in the cloud track investigations for an overvolted uniform field gap and the velocities of the tips have been measured by RAETHER who observed values lying around 2 to 4×10^7 cm./sec. for the negative anode directed streamers, and from 4×10^7 to 2×10^8 cm./sec. for the positive cathode directed streamers, velocities appearing to increase as the tips crossed the gap [*42*]. The breakdown streamers have been studied with moving film camera by ALLIBONE and MEEK [*55*] for overvolted impulse breakdown from positive and negative long (1 m. on down) point-to-plane gaps, and velocities have been measured. These ranged from 4×10^7 cm./sec. downward in the long gaps. MEEK and SAXE [*55*] have measured currents and velocities of streamers in positive point-to-plane gaps oscillographically with photomultiplier cell from 55 cm. down to 10 cm. with rising potential at the point. Currents ranged from 1 to 11 amperes and speeds were low from 2×10^6 cm./sec. to 2×10^7 cm./sec. Photomultiplier showed broad luminosity, while photographs of the spark channel following in the return stroke indicated diameters under a few mm. In lightning discharges[1] SCHONLAND [*58*] and his associates have observed negative streamer-like processes in the broad stepped leader strokes of lightning discharge. Here near junction of steps with the ground positive streamers have been seen to meet the negative steps. The resulting spark channels where heavy ionization persisted were $\sim$20 cm.

In all long sparks and in the spark breakdown from corona gaps, ALLIBONE and MEEK, MEEK and SAXE and HUDSON in the writer's laboratory in analogy to lightning strokes have *observed the crossing of the gap by the streamer to be followed by the rapid intensely ionizing return stroke,* first observed in lightning discharge by SCHONLAND [*58*] and his group. RAETHER's technique is not adequate to

[1] Cf. the article of B. F. J. SCHONLAND in this volume.

resolving this process with its speed in excess of 3×10^8 cm./sec. It is the return stroke which changes ion densities from 10^{13} or more ions./cm.^{3} in the center of the streamer channel to the 10^{17} ions/cm.^{3} characteristic of the brilliant spark.

Some properties of breakdown streamers have recently been delineated by the writer [*59*]. These streamers represent a dynamic system with tip advancing at a high velocity through field distortion at the tip and photoelectric ionization in advance. Thus in the streamer tip the short ionizing electron avalanches, with velocities high, long range high fields and photo-ionization in advance covers longer distances in shorter times than is the case for normal avalanche tip process. There is a current flowing along the streamer channel to the anode of sufficient magnitude to require a marked potential gradient to maintain it. This gradient is rather high and requires that X/p be of the order of 10 to 25 volts/cm. per mm. Hg to keep the flow going and ionization active to make up for losses. While streamer tip velocities v_t are high, electron velocities *in* the *channel* v_c are low. For the current $i = n\varepsilon v_t = n_c \varepsilon v_c$ such that the n number of electrons/cm.^{3} in the tip are less than n_c, that in the streamer channel. Once a streamer gets underway, the number of positive ions created per cm. advance of the streamer is constant and characteristic of the streamer. Thus the net charge per cm. length of streamer is 8×10^9 for preonset corona streamers, 5.5×10^{12} for breakdown impulse streamers of MEEK and SAXE, and perhaps 5×10^{13} in the negative stepped lightning streamer. The space charge of positive ions left behind in the streamer channels is thus such that radial fields about the axis of the streamer channel cause ionizing avalanches to feed electrons in radially behind the tip, augmenting the flow up the streamer. This causes a *radial expansion of the streamer channel* behind the tip. Evidence for this comes from the breadth of the stepped leader strokes in lightning and equally from MEEK's and SAXE's photomultiplier studies. Indications of this for preonset streamers were observed by AMIN in the shoulders of his oscillograms. AMIN and more so HUDSON have noted this in the study of breakdown streamers. Here, where the tip velocities are very high and streamer travel is long, the tip luminosity separates from the luminosity caused by the radial fields and current flow after some few mm. Thus the heavy ionization and longer duration luminosity and ionization progress as a secondary broad streamer at from one half to a third the primary tip velocity. It is only when the secondary tip with its high conductivity reaches the cathode that the return stroke follows, see Section 37 for details.

VI. The rôle of the streamer in uniform field breakdown and the long time lags of sparking.

a) Application of MEEK's theory to sporks in air.

19. With the development of the RAETHER and MEEK theories for the threshold of the streamer spark, MEEK [*60*] put the theory to test by calculating thresholds for breakdown in air for a wide range of conditions in uniform and non-uniform field sphere gaps. In view of the dominance of the exponential term in the amplification factor the theory gave remarkably good values for the breakdown threshold with properly chosen data for air. However, more than this, by its means for the first time it was possible by a *single semi-empirical sparking relation* to account for anomalies with observed non-uniform fields including the HUETER [*60*] anomaly and the TOEPLER [*60*] discontinuity. In fact, it can be stated that for various geometrical gap forms *in air*, where fields are calculable, that the MEEK *theory represents* the *best all around semi-empirical relations for calculation* *of sparking thresholds.* It thus serves a useful purpose and may continue to be used in engineering applications for air at atmospheric and not too low pressures, nor too large gap lengths. As the fields not only become non-uniform and asymmetrical, with the anode highly stressed, the theory may be considered still more useful. However, if the *real* threshold in *air* for *uniform fields* is considered, the *threshold calculated by its means is probably 2.5% too high.* Since in industrial and engineering applications sparking threshold conditions are rarely such as to yield the *true threshold* value, his calculations have validity.

It is fortunate that the theory was not applied to calculate the threshold for really pure inert gases under clean electrode conditions, as it would have yielded values of the order of twice the observed threshold and prematurely thrown a proper mechanism into disrepute. The reason for this statement will presently become clear.

b) Search for transition from streamer to TOWNSEND spark.

With the introduction of the streamer theory both RAETHER [*42*] and MEEK and the writer [*43*], attempted to set the limit of pd for which the streamer mechanism should give way to the TOWNSEND mechanism [*61*]. For it was recognized that as soon as avalanches got to be so small, $e^{\alpha d}$ small, or ϱ_c so large, owing to electron diffusion, that the condition $KX_s = X'$ could not be fulfilled before $\gamma e^{\alpha d} = 1$ is reached, the threshold would be set by a TOWNSEND discharge. It was further predicted that at a given pressure p for longer gaps, since X_s/p tends towards a constant value as pd increases in consequence of equation (12.6) there was an upper limit to the gap length d beyond which the sparking potential should cease to follow MEEK's theory and would suffer an upward inflection owing to the inadequate density of ionization. Predictions as to the regions of transition to the TOWNSEND threshold at low pd were also possible from the disappearance of DUNNINGTON and LAWRENCE's mid-gap streamers in observations with KERR cells [*40*]. In consequence, MEEK and the writer [*43*] set the value for the transition of the streamer determined threshold to the TOWNSEND determined threshold at $pd \sim 200$. RAETHER set it at $pd \sim 1000$. FISHER first attempted to observe the upward inflection of the curves predicted at longer pd, but was unsuccessful owing in part to instrumental limitations and in part to circumstances to be indicated shortly [*61*].

20. The investigation of GÄNGER. The next investigation was that of GÄNGER [*62*] in 1949 in Karlsruhe who used fast oscilloscopes to study time lags of breakdown in order to extend STRIGEL's study and to use the formative timelag to check RAETHER's predicted transition to TOWNSEND discharge at $pd \sim 1000$. His potential pulse was nearly square with sharp rise but a 1% ripple on top. His circuits were elaborate and he used a peculiar anode with an auxiliary spark gap back of a hole for illuminating his cathode with ultraviolet light. He, howeλer, alternately used a quartz lamp at grazing incidence. His overvoltage control factor was not better than $\pm 4\%$. He thus worked at higher overvoltages. He was careful of his statistics and used the LAUE law for evaluating all data. He studied air at pressures from 25 to 760 mm. with potentials ranging from a few to 50 KV. He used large sphere electrodes with gap lengths from 0.3 to 9.0 cm. long giving nearly uniform fields for shorter gaps. He could follow his lags to less than 10^{-8} sec. He found the LAUE law to hold throughout and even at $pd = 80$ cm. × mm. Hg he *observed no increase* in T_f and no transformation to a TOWNSEND type discharge. All his sparks were sharp filamentary type sparks. In fact, FISHER in his later study indicated that it was not until pressures of 20 mm. were reached with $pd = 20$ cm. × mm. Hg that the spark channel began to have blurred outlines and expand as in a TOWNSEND discharge. GÄNGER found that *formative time lags increased rapidly as overvoltages decreased irrespective of the value of pd* and his trend indicated that formative time lags would become large indeed if over voltages were small enough. This was true *at all pd values* and not a result of the expected transition at $pd \sim 1000$. He suspected that, as the overvoltage $\Delta V/V_s$ decreased, it was possible that a streamer spark occurred, but that the threshold was set by an antecedent TOWNSEND discharge which lowered the threshold and resulted in a streamer spark. Contrary to STRIGEL and WHITE, he observed the time lag to increase with increasing gap length for lower overvoltages but not quite in proportion. The value of T_f was not influenced by pressure down to 25 mm. Data at 760 mm. and a 3 mm. gap agreed with the lower limit of NEWMAN'S values. The increases of T_s with decrease in overvoltage over the range observed might have been exponential as with TILLES' but appeared to be more nearly hyperbolic.

21. Discovery by FISHER of long time lags near threshold. In view of FISHER'S [*57*], [*61*] preoccupation with the limitations of MEEK'S sparking equation, it is not surprising that he should, independently and nearly contemporaneously with GÄNGER but with a different technique, have used oscilloscopic analysis to attempt to locate the transition from streamer to TOWNSEND breakdowns by means of the formative time lag. Together with BEDERSON, FISHER [*14*] used a very carefully stabilized source of constant potential on which a square wave pulse yielding an overvoltage could be suddenly impressed. The time from the application of the step potential to the spark was observed on an oscilloscope. The ionization chamber was a large metal one previously used by SANDERS for evaluating α in air. Uniform field gaps were used with clean dry mercury-free air at pressures ranging from 100 to 760 mm. and gap lengths from 0.6 to 1.4 cm. The center of the cathode was continuously illuminated with ultraviolet light from a stabilized quartz arc which gave from 1.3 to 6 electrons per microsecond from the cathode. The power supply permitted an approach voltage of 2 to 4 KV below V_s to be used. The additional potential V_p from batteries was applied to raise the potential V above V_s. With his control, overvoltages from 5%, but usually only 2%, on down to 0.015% could be supplied as a relatively square wave. The oscilloscope used had a time resolution from 0.5 microsecond and out to 100 microseconds. Time lags could be read directly. The advantage of this system was that static breakdown voltages could be established to within 0.015% of the threshold. The *sparking* usually *altered* the *threshold slightly,* but with the *schedule of measurement used* [*12*], after the gap was stabilized for the run, it was relatively constant. The threshold was recorded at the start of a series and several times during a run as well as at the end. This information was used to correct the overvoltage. From day to day *the threshold in air varied by 0.5% or more* depending on the state of the cathode. This came from alterations in γ_p, as BANDEL [*12*] later observed his i_0 to varyas a result of sparking. Correct values of the potentials were not easy to obtain because of the drop of potentials across capacities and the thyraton. Thus the sum of approach and pulse voltages was adjusted until failure to spark was obtained consistently. This was assumed to be the threshold. Time lag data showed considerable spread, but studies indicated that these spreads were instrumental and relatively little statistical. The spread was greater at low overvoltages where T_f was longer and P_s was less. At high $\Delta V/V$, T_f was of the order of a few microseconds and 5 readings sufficed to fix T_f. Later studies of N_2 and A carried out by FISHER and KACHICKAS used log (n_{ta}/n_T) plots. They took the LAUE-WIJSMAN form when T_s was large. Most of their data taken with adequate n_0 had T_s less than T_f.

In air at about 2.5% overvoltage the formative time lags were $\sim 10^{-7}$ sec. *characteristic of a streamer breakdown* for a 1 cm. gap, but as the overvoltage decreased to near threshold the time lags increased progressively so that at 0.02% they ranged in the order of 100 microseconds. This decrease followed a curve analogous to the decline observed by GÄNGER, but transcended it in that it extended to much longer intervals. The curves were independent of pressure as were GÄNGER'S down to 100 mm. The formative time lag in the longer intervals and low overvoltages was proportional to gap length. When the number of triggering electrons increased from 1 to 20 per microsecond the value of T_f was not altered, nor was T_f altered by changes in the approach voltage down to 4000 in some 30 KV.

In air the threshold appeared to be quite clearly defined, despite WIJSMAN'S theory, owing unquestionably to the favorable action of positive ion space charge in this range of X/p. Clean inert gases, such as argon, differed from air and N_2 in that the formative time lags in argon at threshold approached 1000 μsec. so that FISHER'S equipment was incapable of evaluating the lags and by the fact that it took values of $\Delta V/V_s \sim 1$ to reduce formative lags in inert gases to the order of 1 μsec. or less. Relatively pure N_2 which consisted of 99.89% N_2, 0.08% CO_2 and 0.03% O_2, by mass spectrographic analysis, acted much like argon gas with long lags at reasonably high overvoltages. Below a pd of 400 cm. × mm. Hg the gas did not directly breakdown to a spark, but a partial breakdown with a glow at the anode first appeared which led to a streamer spark only with an increase in potential. This was not observed with a pulsed step applied for as long as 100 μsec. Slightly impure N_2 acted like air.

22. Interpretation of the long time lags. The results of FISCHER were independently interpreted by FISHER and by writer [*14*] in 1951. It appears that in all gases considered with uniform field geometry, the threshold for a low order TOWNSEND discharge with a cathode γ_p action occurs at values of X/p below that sufficing for streamer action. In air, owing to the efficient photo-ionization of O_2, the threshold for a streamer spark with $T_f \sim 10^{-7}$ sec. lies only at 2.5% overvoltage above the TOWNSEND threshold. This circumstance explains the relative success of MEEK'S semi-empirical theory, since V_s is never reproducible or clearly defined in air to better than ± 0.5% and usually is still more uncertain without FISHER'S techniques. For A and for pure N_2 the threshold of the TOWNSEND discharge lies *far below that for the streamer.* Impure N_2 [*10*], air [*10*] and H_2 [*63*] will give an appreciable γ_p and lead to a TOWNSEND discharge that in 100 μsec. or so builds up a distorted field leading to a mid gap streamer breakdown. In relatively pure N_2 [*10*] there is no very effective γ_p on a time scale equivalent to an electron crossing time, but there is a γ_i leading to longer formative time lags than with γ_p.

In argon the situation is even more confused. There is a delayed pressure dependent γ_p from excited argon molecules created by metastable atoms which is effective down to some 100 or more mm. [*8*]. Below this, γ_i begins to be effective on a clean cathode [*8*], [*62*]. Thus argon will require a longer build-up time as the excited A_2 molecules must be created. Furthermore, measurable values of α appear at very low X/p in argon but M, the amplification factor, is so low that threshold for the TOWNSEND discharge sets in very early with its long formative lags in building up the space charge at values of X/p about one half those for a streamer spark. It is also probably that much current must flow in A and N_2 to make an efficiently photo-ionizable gas for the streamer spark, while in air this is not the case and may not be so much so in H_2.

On the basis of this concept, FISHER and KACHICKAS [*14*] calculated the value of T_f assuming a γ_p in soiled N_2 gas and assuming that a single initiating electron must create some N electrons to achieve the space charge for a streamer spark. With N chosen as 10^5, calculation yielded fairly good agreement with half their experimental data for $\Delta V/V_s$ from 0.25 to 2.0 with lags increasing from 0.8 μsec to 14 μsec. Obviously, this theory is rather weak, but the results were encouraging. Results for air and argon did not agree nearly as well.

23. The study of FLETCHER at high overvoltage. Before proceeding further, it is necessary to mention another study of formative time lags in heavily overvolted gaps. In 1949, FLETCHER [*50*] applied potential pulses with rise time in the milli-microsecond range to a sphere gap with electrode diameters of 2.34 cm. placed in a 5.75 cm. diameter transmission line, the cathode being illuminated by light from a triggering spark gap through quartz optics at grazing incidence. Gap lengths were 0.1 to 0.3 cm and the illuminating spark was fired 6×10^{-8} sec. before the impulse potential was applied. Clean room air at 22° C and 40% relative humidity was used. The threshold breakdown potentials were consistent with SCHUMANN'S [*64*] published values and $\Delta V/V_s$ ranged from 0.66 to 1.00. Formative time lags with heavy i_0 could directly be measured. On reducing i_0 statistical lags obeying the LAUE law were found at times greater than T_f. From his observations at high $\Delta V/V_s$, P_s was assumed equal to unity, allowing him to calibrate his i_0 as ranging from 0.63×10^{-8} to 9.6×10^{-12} amp/cm.^{2} with absorbing screens. T_f lay between 0.5 and 50×10^{-9} as indicated from the time of the application of potential to the sharp fall in potential which FLETCHER chose as the spark. Actually the potential never really fell to zero and there could be some doubt about what was observed. FLETCHER calculated the space charge densities of positive and negative ions in the avalanche. His first approach was to calculate the avalanche uninfluenced by space charge in a fashion that paralleled that of KUNKEL [*41*], though his approximation had an added term to permit the breadth of the avalanche to be larger than $1/\alpha$. Making approximations as to how u_T and v vary with field, the values of the densities of ions N_+ and electrons N_- for the avalanches were computed. With $1/\alpha$ less than ϱ, the positive and negative space charge centers nearly coincided. As $1/\alpha$ exceeds ϱ, the negative space charge exceeds the positive and the centers are materially displaced as earlier indicated in the elementary theories of MEEK and RAETHER. Under simplifying assumptions the space charge densities were computed when the sparking field X_s was created and streamers started to form. The number of electrons, the space charge density and the radius of the avalanche were computed at a time $t=t_t$ when the drift velocity of the space charge field becomes equal to the electron diffusion velocity. A second relation between space charge density and number of electrons was computed from the modified solution of the relations of KUNKEL and FLETCHER. After further reductions and simplification, the relations obtained yielded the avalanche size n, and the critical space charge field X_{1s}. This, in turn, led to an evaluation of the critical avalanche size n_c reducing the internal field of the avalanche to zero. The formative lag can then be calculated as

$$T_f=\frac{1}{\alpha v}\log n_c. \tag{23.1}$$

With this relation a good agreement was obtained between the calculated formative time lag as a function of the applied field and FLETCHER'S observations.

The calculation indicated that T_f should be a function of X alone independent of gap width. This was observed to occur for higher, but not lower, fields. For a given gap width, or applied voltage, below a certain critical field the time lags become larger for the longer gap width. This field FLETCHER believed to be the field for which the initial avalanche takes the entire gap width to reach its critical size. Thus for a smaller field a single avalanche cannot create a heavy enough field to cause breakdown. The critical fields were computed. The computed critical fields for 3 voltages agreed within expected accuracy. The agreement is not too good as the data used, such as RAETHER's values of v, are not accurate. FLETCHER states that the agreement does not serve to distinguish between the radial space charge field of the writer and MEEK, and the axial field used by FLETCHER. The influence of the value of the criterical avalanche size n_c as it affects T_f is minor as it enters as $\log n_c$, and a change in this by a factor of 10 changes the calculated lag only 12%. The value of the critical avalanche size was 10^8 electrons which is again surprisingly close to the values arrived at by the writer and MEEK, and by RAETHER.

FLETCHER's observed points fall severely below those of NEWMAN taken at constant gap length. It is possible that NEWMAN's data were taken from the bottom of the potential rise while FLETCHER's were from the top, with a rise time of NEWMAN's pulses of 10^{-8} sec., a matter not impossible before World War II techniques. WHITE says that his measurements agreeing with NEWMAN's could have been complicated by oscillations. The photo-electric current used by FLETCHER was too low by a factor of 100 to have influenced T_f.

Probably, the most important contribution of this study in extending formative lags in air to 100% overvoltage is the fact that for such high fields a more rigorous calculation of the streamer forming conditions leads to no serious contradictions and in general confirms the basically simpler derivations cited. It must be noted further that while the calculation of formative time lag was that for the *creation of the streamer forming avalanche*, it does *not* mean, as DICKEY [*65*] later implies, that this time does not account for breakdown. For, in the short gaps and with the streamer tip velocities and those of the return stroke, the interval from streamer formation to spark will not add a great increment of time to the curves to achieve full breakdown.

24. BANDEL's measurement of current growth during long lags. In the spring of 1954, BANDEL in the writer's laboratory had overcome the formidable technical switching difficulties in order to record the growth of the spark breakdown current beginning at 10^{-7} amp and rising up to 10^{-2} amp in a FISHER long time lag breakdown of from 20 to 80 microseconds in clean, mercury free, dry air. The measurement requires switching out the charging current produced by the step voltage of 1000 volts applied above the stabilized approach voltage of 50000 volts, which goes to charge the electrodes of the uniform field gap used, and of measuring the current growth oscillographically with cut off to protect the instruments from the heavy sparking current that follows. A combination of a rifle bullet switch with electronic switching solved the problem and yielded satisfactory results. The triggering electron current i_0 came from a monitored quartz arc shining on a contoured brass cathode. Sparking as noted above changed the photo-emission from the cathode and raised the value of V_s as observed by FISHER and BEDERSON. With the elaborate switching arrangements of BANDEL, the stability possible through the rapid sparking repetition technique of FISHER could not be achieved. However, using a regular temporal schedule of sparking, a reproducible set of data was obtained. Fig. 7 shows the curves for current in amperes as a function of time with an i_0 of 60 electrons per microsecond for a 1 cm. gap at 724 mm. of air at 22° C indicating various time lags corresponding to different overvoltages. The upper ends of the curves terminate within 1 microsecond of the filamentary spark. That is, the curves represent to a semilogarithmic scale the current changes during the building up of the space charge, but do not register the *growth of the* *streamer* spark during the last microsecond, or less. The charging current, lasting 2 μsec., is switched out. Fig. 8 shows the same curves replotted as *fractions* of the time to spark breakdown, i.e. of T_f, indicating in a sense the uniformity of

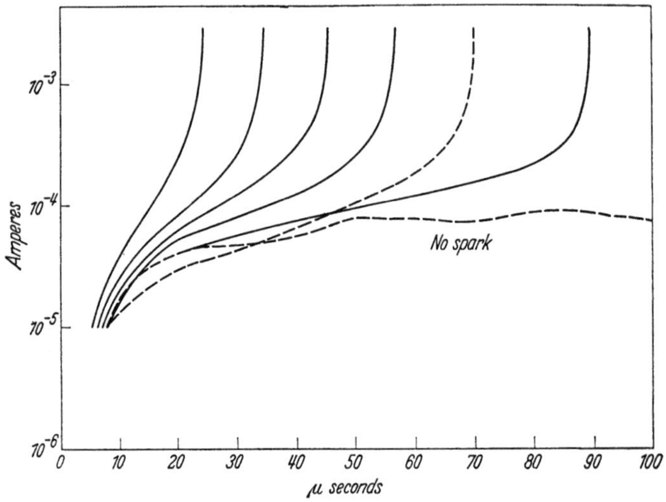

Fig. 7. BANDEL's curves for current in amperes against time in microseconds, with $i_0 = 60$ electrons/μsec., for a 1 cm. gap at 724 mm. air pressure at 22° C with different time lags corresponding to different overvoltages. The initial 3 μsec. of time are not indicated. The steep final rise is followed by a spark in 1 μsec. The growth of a field intensified current just below breakdown is shown in the flat curve.

the trend. The dashed curve in Fig. 7 at 70 μsec. shows by its wobble the maximum scatter of data at shorter times. The flat dashed curve is that for growth of self sustaining field intensified current just *below* the sparking threshold.

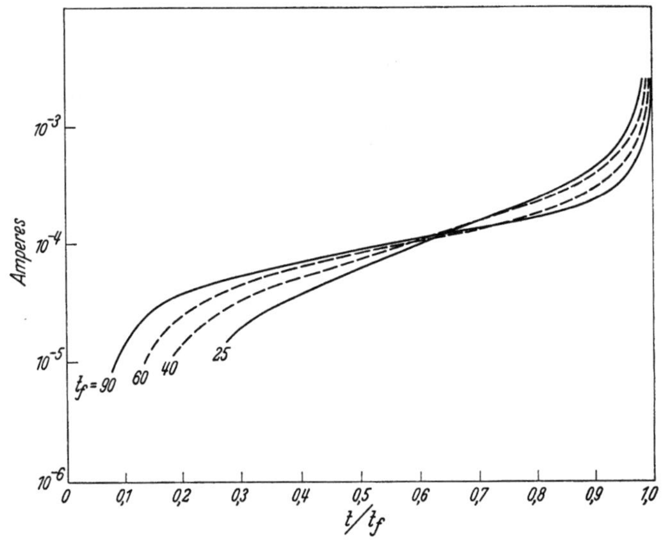

Fig. 8. Shows the curves of Fig. 7 replotted as current against fractions of the whole lag instead of times. This mode of plotting reveals the similarity of the curves for different lags.

Decreasing the current i_0 from 60 electrons/μsec. as in Fig. 7 to 6 electrons per microsecond increased the scatter of individual curves, but still a *mean curve* could be drawn. This indicates that with $i_0 = 6$ the curve for $T_f = 40$ μsec. passes 10^{-6} amp at the same time that the curve with $i_0 = 60$ passes 10^{-5} amp.

That is, an *increase in i_0 increases the rate at which the current builds up*, as is to be expected. These changes produced by i_0 apply only to the *early part of the curves* for the build up of the TOWNSEND current.

There can be no question that if V_s were constant, changes in i_0 would move the entire curves up or down, at least up to the point where space charge distortion starts the final up-curv ng, and the time lag for a given overvoltage would be changed. FISHER and BEDERSON did not observe this, because their variations in i_0 were small and the resultant changes in time lag *vs* overvoltage were less than the experimental accuracy.

A rough qualitative picture of the general shape of these curves is as follows. The current measured by the external circuit is due to the motion of all of the ions and electrons in the gap at any particular time. Because the electron velocity is so much higher, electrons make a much larger contribution to the current for the short time they are in the gap, but the positive ions remain in the gap for a correspondingly longer time. In a steady state, the contribution of the positive ions and that of the electrons are in the ratio of the average distances they each travel in the field, which is approximately $(\alpha d-1)/1$. For the values of αd at onset this would be a contribution by the positive ions of about ten times as great as that of the electrons. When any change occurs in the rate of ion, and electron, production, the electron current follows the change quickly, while the ion current requires an ion crossing time to adjust to the new value. Thus, whenever the measured current is increasing, a larger fraction of it is due to electrons than would be the case for steady state; and the faster it is increasing, the larger is the fraction of it due to electrons. This makes direct interpretation of the measured current difficult.

In the absence of any secondary mechanism, the measured current would increase rapidly at first as the constant rate of production increased the number of positive ions in the gap. It would level off gradually and become constant at one ion crossing time when the rate of loss became equal to the rate of creation. With secondary mechanisms acting, this initial build-up is of course greater, and the current cannot become constant by one ion crossing time. The more or less *exponential increase* in the *middle* of the *time lag* is due *to secondary mechanisms*, and the final up-curving which leads to breakdown is due to space-charge distortion of the field, as will be shown later.

Since the work of FISHER and KACHICKAS, there have been several attempts to account for the dependence of time lag on overvoltage. The first was that of DUTTON, HAYDON and JONES [*66*] who indicated no recognition of the streamer as the final mechanism, or of the consequent necessity of expressing a true criterion in terms of field distortion by positive ions. They used the approximate solution for a TOWNSEND discharge due to BARTOLOMEYCZYK and to DAVIDSON [*66*] and took a critiral current at the cathode as a criterion for breakdown which obviously is an improper criterion. By varying overvoltage, that is α, they obtained curves of roughly the same shape for T_f as a function of $\Delta V/V_s$ as FISHER and BEDERSON, and their solution is akin to that of FISHER and KACHICKAS only some what more elaborate.

RAETHER [*42*] in 1953 was the only one to attempt a criterion based on space charge distortion. His basic model of the physical situation was good, (the spreading of the discharge was verified in BANDEL's study) but he used crude approximations to carry the work through.

25. DICKEY's calculation of TOWNSEND breakdown in highly overvolted gaps. One more attempt has recently been made by DICKEY [*65*] to calculate the value of T_f, or better the oscilloscope currents observed in this instance in the *highly* overvolted gaps used by FLETCHER. As indicated earlier, FLETCHER had attempted to account for these on a streamer mechanism using a more complete theory, analogous to that set up by KUNKEL. In ignorance of the full mechanism of streamer breakdown and time scale involved, DICKEY criticised FLETCHER'S analysis. Assuming that the field in the gap remains uniform in space, but not in time, being altered by the avalanche without secondary actions expressed by the equation $n=n_0\cdot\exp\int\limits_0^t \alpha v\,dt$, which continues up to and perhaps beyond the point where potential across the gap begins to fall, DICKEY set the *circuit* equation for current in the gap between electrodes as

$$C\frac{dV}{dt} = -\frac{\varepsilon v n_0}{d} e^{\int\limits_0^t \alpha v\,dt} + \frac{V_0 - V}{2R_c}. \tag{25.1}$$

Here C is the capacity of the gap, v is the velocity of the electrons, n_0 is their number, V_0 is the initial potential applied and R_c is the characteristic resistance of the line. Using FLETCHER'S rough expression for v as a function of X/p and SANDER'S values for α, and considering α and v essentially constant over the narrow range of values in the avalanche, DICKEY'S, solution yielded the temporal variation of potential $(V)\,t$ as

$$V(t) = V_0 - \frac{\frac{\varepsilon v N_0}{C d} e^{\alpha v t}}{\alpha v + \frac{1}{2R_c C}}. \tag{25.2}$$

At the high overvoltages, "reasonable" agreement, (within a factor of 2), between the curves observed by FLETCHER and those calculated by equation (25.2) were obtained. At lower overvoltages, agreement was not satisfactory. It should be noticed that DICKEY'S analysis is not based on any model except cumulative ionization. It merely shows that *charge separation and ionization in the gap at such high fields*—irrespective of any detailed mechanism—*will account for the partial collapse of potential observed* by FLETCHER. That is, it appears that the very short intervals between high overvoltage application and partial collapse of the potential on the electrodes *of small capacity* can be accounted for by ionization and polarization, or displacement currents, in the gap, in an avalanche only, irrespective of the creation of a conducting channel and the ultimate discharge of the capacity. Actually, agreement between FLETCHER'S theory of the time to form a streamer is more accurate than DICKEY'S calculations which usually yield a longer time. The calculations do not prove anything except that, in somewhat the same or slightly longer times than observed, a fall of potential such as observed could be achieved by the avalanche alone. There is nothing inconsistent with the ionization as calculated by FLETCHER in the calculation by DICKEY. If DICKEY'S times were uniformly equal to FLETCHER'S observed times it might indicate that FLETCHER'S special avalanches and subsequent rapid shorting of the gap are not needed to account for the observations. In this case, however, as in FLETCHER'S work, both approaches indicate difficulties of the same sort when overvoltages decrease beyond a certain point, for then ionization in a single avalanche of length d is not adequate for short gaps, which brings the conditions right back to the problems of field distortion. However, irrespective of all of DICKEY'S argument, observations of DUNNINGTON, WHITE and WILSON indicate the actual streamer nature of the breakdown.

26. BANDEL's calculation of current growth and space charge action. Disregarding these earlier theoretical attempts, BANDEL [*12*] set up the complete differential equations for the growth of a TOWNSEND discharge using γ_p and γ_i. With the aid of the approximations made in the solution of the resulting equations, using only a γ_p, BANDEL calculated the solid curves of Fig. 9 which are to be compared with the dashed experimental curves taken for an i_0 of 12 electrons per μsec. The theory is as follows.

Consider the case of a uniform field, taking $x=0$ at the cathode and $x=d$ at the anode, the balance equations are

$$\left.\begin{aligned} \frac{1}{v_-}\frac{\partial i_-}{\partial t} &= -\frac{\partial i_-}{\partial x} + \alpha i_-, \\ \frac{1}{v_+}\frac{\partial i_+}{\partial t} &= \frac{\partial i_+}{\partial x} + \alpha i_-, \end{aligned}\right\} \tag{26.1}$$

where $i_-(x,t)$ and $i_+(x,t)$ are the electron and positive ion currents, and v_- and v_+ the electron and ion velocities, assumed constant for the range of overvoltages involved. Considering only α, γ_p, and γ_i mechanisms, the boundary conditions are

$$\left.\begin{aligned} i_-(0,t) &= i_0 + \gamma_i \alpha i_+(0,t) + \gamma_p \alpha \int_0^d i_-(x,t)\,dx, \\ i_+(d,t) &= 0. \end{aligned}\right\} \tag{26.2}$$

γ_i is the average number of electrons released from the cathode per impinging positive ion, and γ_p is the number of secondary electrons leaving the cathode through photoelectric action per positive ion.

BARTHOLOMEYCZYK solved these equations omitting i_0 from the boundary conditions. DAVIDSON [*66*] modified BARTHOLOMEYCZYK's solution to include i_0. He also pointed out that it does not satisfy the initial conditions and that it would be necessary to obtain an accurate solution in order to known how good or bad that one was. He then proceeded to get an exact solution; unfortunately this last is too complex to be of any practical help[1].

It has proven worth while to attack this problem from a different direction. If it is possible to get a solution for $i_-(0,t)$, the electron current at the cathode, then

$$i_-(x,t) = i_-\left(0, t - \frac{x}{v_-}\right) e^{\alpha x}. \tag{26.3}$$

The positive ion current created in dx' at x' and t is

$$d i_+ = \alpha i_-(x',t)\,dx' = \alpha i_-\left(0, t - \frac{x'}{v_-}\right) e^{\alpha x'}\,dx'. \tag{26.4}$$

This reaches x at a time $(x'-x)/v_+$ later, so the current at x and t coming from dx' is

$$d i_+(x,t) = \alpha i_-\left(0, t - \frac{x'-x}{v_+} - \frac{x'}{v_-}\right) e^{\alpha x'} = \alpha i_-\left(0, t + \frac{x}{v_+} - \frac{x'}{v}\right) e^{\alpha x'} \tag{26.5}$$

where $1/v = 1/v_+ + 1/v_-$ or $v = v_+ v_-/(v_+ + v_-)$. Then

$$i_+(x,t) = \int_x^{v\left(t+\frac{x}{v_+}\right) \text{ or } d} \alpha i_-\left(0, t + \frac{x}{v_+} - \frac{x'}{v}\right) e^{\alpha x'}\,dx. \tag{26.6}$$

[1] In a recent article DAVIDSON, Phys. Rev. **99**, 1072 (1955), has rearranged his equations in more tractable form.

The upper limit on this integral assumes that no electrons leave the cathode until $t = 0$, (time of application of the voltage), and then takes into account the fact that x' makes no contribution to the positive ion current at x, until there has been time for an electron to go up to x' and a positive ion to go back to x. So, the upper limit is $v(t - x/v_+)$ until this quantity equals d, and then remains d thereafter. Similarly there can be no contribution by γ_p from any x until electrons have had time to reach x from the cathode. With this, (26.3) and (26.6) may be used in (26.2) to obtain the integral equation

$$i_-(0,t) = i_0 + \gamma_i \alpha \int\limits_0^{v\,t \text{ or } d} i_-\left(0, t - \frac{x'}{v}\right) e^{\alpha x'}\, dx' + \gamma_p \alpha \int\limits_0^{v_- t \text{ or } d} i_-\left(0, t - \frac{x}{v_-}\right) e^{\alpha x}\, dx. \tag{26.7}$$

Now, trying

$$i_-(0,t) = i_0 (A - B\, e^{t/\tau}) \tag{26.8}$$

in (26.7) gives

$$\left.\begin{aligned} i_-(0,t) = i_0 + \gamma_i \alpha\, i_0 \Bigg[& \frac{A}{\alpha} \{\exp[\alpha (v\,t \text{ or } d)] - 1\} - \\ & - \frac{B\, e^{t/\tau}}{\alpha - \frac{1}{(v\,\tau)}} \left\{\exp\left[\left(\alpha - \frac{1}{v\,\tau}\right)(v\,t \text{ or } d)\right] - 1\right\}\Bigg] + \\ + \gamma_p \alpha\, i_0 \Bigg[& \frac{A}{\alpha} \{\exp[\alpha\,(v_- t \text{ or } d)] - 1\} - \\ & - \frac{B\, e^{t/\tau}}{\alpha - \frac{1}{v_- \tau}} \left\{\exp\left[\left(\alpha - \frac{1}{v_- \tau}\right)(v_- t \text{ or } d)\right] - 1\right\}\Bigg]. \end{aligned}\right\} \tag{26.9}$$

For times shorter than $v\,t = d$, this equation cannot be satisfied for any constant τ or A, and for longer times it is just the equation obtained by DAVIDSON. However, we now know that the solution is exact for longer times and can see when it begins to fail. Unfortunately, since $v \approx v_+$, it does not become good until one ion crossing time, so the whole solution is not of much help. On the other hand, in case $\gamma_i = 0$ the remaining equation can be satisfied for times greater than one-electron crossing time. In this limiting case, A and τ are given by

$$A = \frac{1}{1 - \gamma_p (e^{\alpha d} - 1)} \tag{26.10}$$

and

$$\gamma_p\, e^{\alpha d}\, e^{-\frac{d}{v_- \tau}} = 1 - \frac{1}{\alpha\, v_- \tau} + \gamma_p. \tag{26.11}$$

KUNKEL has suggested another solution of the integral equation (26.7). If $(d i_-/dt) \cdot (d/v_-) \ll i_-$, then $i_-(0,t)$ can be used in the γ_p term instead of $i_-(0, t - x/v_-)$, and in this case he was able, by means of a LAPLACE transform, to obtain the solution, for $d/v_- \leq t \leq d/v$,

$$\left.\begin{aligned} i_-(0,t) = & \frac{i_0}{1 + \gamma_i - \gamma_p (e^{\alpha d} - 1)} \\ & \left[1 + \frac{\gamma_i}{1 - \gamma_p (e^{\alpha d} - 1)} \times \exp\left(\frac{1 + \gamma_i - \gamma_p (e^{\alpha d} - 1)}{1 - \gamma_p (e^{\alpha d} - 1)}\, \alpha\, v\, t\right)\right] \end{aligned}\right\} \tag{26.12}$$

and for $t \geq d/v$,

$$i_-(0,t) = \frac{i_0}{1 - (\gamma_i + \gamma_p)(e^{\alpha d} - 1)} - i_0\, B\, e^{t/\tau} \tag{26.13}$$

where τ is the root of

$$\gamma_i e^{\alpha d} e^{-\frac{d}{v\tau}} = \frac{-1}{\alpha v \tau}\left[1 - \gamma_p (e^{\alpha d} - 1)\right] + 1 + \gamma_i - \gamma_p (e^{\alpha d} - 1).$$

The condition $(d i_-/dt)\cdot(d/v_-) \ll i_-$ means that the current change during one electron transit time is much less than the current itself. If we consider the case of $i_0 = 1$ electron per electron transit time (i.e., ~ 10 electrons per μsec.), this is approximately the same as saying $\gamma_p(e^{\alpha d} - 1) \ll 1$. This does not hold at the sparking potential, so again the whole solution is not good for the present case, but this solution is exact for the limiting case of $\gamma_p = 0$. In this case the constant B, evaluated-by equating (26.12) and (26.13) at $t = d/v$, can be written in a simple form by neglecting γ_i compared to 1, and 1 compared to $e^{\alpha d}$:

$$B \approx \frac{\gamma_i e^{\alpha d}\left(1 - \frac{1}{\alpha v \tau}\right)}{1 - \gamma_i e^{\alpha d}}. \tag{26.14}$$

Moreover, this can be used to evaluate the constant B in the case of $\gamma_i = 0$ above. From (26.7) it is seen that the integral equation for the two limiting cases is the same with only the change of subscripts on γ and v. Therefore, letting $\gamma_p = 0$ in (26.12), and then changing γ_i to γ_p and v to v_- gives the solution for $\gamma_i = 0$ and $t \le d/v_-$. It follows that, with this change of subscripts, (26.14) gives the constant B in the case of $\gamma_i = 0$. Further, this solution for $t \le d/v_-$ is good even if γ_i is not zero, because the ions cannot contribute in so short a time.

At this point, solutions have been obtained for the case when both γ_i and γ_p are important for $t \le d/v_-$ and for $t \ge d/v$, though for the latter B has not been evaluated, and for the limiting cases $\gamma_i = 0$ or $\gamma_p = 0$ for all times.

For $d/v_- < t < d/v$ an approximate solution for both γ's might be obtained by taking the one for $\gamma_i = 0$ in the early times before many ions can reach the cathode and somehow joining it to the one for both γ's which becomes valid at $t = d/v$. If no mathematical expression could be obtained, the two could still both be plotted into the intermediate region where they both lose validity and then possibly be joined graphically. However, this would have to be studied carefully, because there seems to be no a priori reason why the first derivative need be continuous at one ion crossing time. With both γ's quite unknown, both velocities a bit indefinite, and even α and the required overvoltage of doubtful applicability, as will be seen, the use of this method to try to fit the experimental data would require more trial and error calculation than seems worth while at present.

The data, taken for the cases when sparks did not occur, indicate that steady state, (if it may be called that in spite of the fluctuations), was achieved in 50μsec. or less. To account for the steady-state value of the current with the known voltage and SANDER's data for α requires that secondary mechanisms increase the current by one or two orders of magnitude. If the major part of this were due to γ_i, it would take thousands of microseconds to reach steady state, therefore calculations have been made here only for $\gamma_i = 0$.

The above solutions are equally good for $\gamma e^{\alpha d} \lessgtr 1$ because τ changes sign with the denominator.

Having once obtained an expression for $i_-(0, t)$, then (26.3) and (26.6) give $i_-(x, t)$ and $i_+(x, t)$. The current measured by the external circuit is due to the motion of all the charges in the gap, and for infinite parallel planes it is

$$i_t = \frac{1}{d}\int_0^d \{i_-(x, t) + i_+(x, t)\}\, d x. \tag{26.15}$$

The assumption of infinite electrodes is to assure not only uniformity of field, but also that all lines of force from a charge in the gap end on either the electrodes or other charges in the gap, not on chamber walls, but it is a reasonable approximation for this experiment.

For the case of $\gamma_i = 0$, when (26.10) and (26.14), (with change of subscripts), are used in (26.8) and this is used with (26.3) and (26.6) in (26.15), neglecting 1 compared to $e^{\alpha d}$ and taking $v = v_+$, the final result for the measured current is, for $t \leq d/v_+$,

$$i(t) = \frac{i_0}{\alpha d(1-\gamma_p e^{\alpha d})} \left[\left\{1 - \gamma_p e^{\alpha d} \frac{\Phi}{\Psi}(1 - \alpha v_+ \tau)\right\} e^{\alpha d} + \alpha v_+ e^{\alpha d} t - \right. \\ \left. - \left\{1 - \gamma_p e^{\alpha d} \frac{\Phi}{\Psi}\right\} e^{\alpha v_+} - \gamma_p e^{\alpha d} \frac{\Phi}{\Psi} e^{\Phi d} \left\{\frac{\alpha (v_+ \tau)^2 \Phi - 1}{v_+ \tau \Phi}\right\} e^{t/\tau} \right] \tag{26.16}$$

and for $t \geq d/v_+$,

$$i(t) = \frac{i_0}{\alpha d(1-\gamma_p e^{\alpha d})} \left[\alpha d e^{\alpha d} - \gamma_p e^{\alpha d} \frac{\Phi}{\Psi} \left\{ e^{\Phi d} \left(\frac{\alpha (v_+ \tau)^2 \Phi - 1}{v_+ \tau \Phi} \right) - \alpha v_+ \tau e^{\Psi d} \right\} e^{t/\tau} \right] \tag{26.17}$$

where $\Phi = \left(\alpha - \frac{1}{v_- \tau}\right)$, $\Psi = \left(\alpha - \frac{1}{v_+ \tau}\right)$, and τ is given by (26.11). In order to use these expressions to plot curves for comparison with the experimental data, values are needed for α, v_-, v_+, and γ_p. The expression,

$$\alpha = 1.048 \times 10^{-4} \left(\frac{X}{p} - 27.38\right)^2 p \tag{26.18}$$

was adjusted to fit SANDERS' data at $X/p = 34$ and 40, and is quite close at $X/p = 50$. X_s is not definitely known because of the way the sparking potential varied, but was nearly 28.8 kv/cm. for the data to be compared with it. For breakdown X/p is about 40 volts/cm. $\times$ mm. Hg; if the work of BRADBURY and NIELSEN for electron velocity in air is extrapolated to this X/p, it gives $v \approx 1.6 \times 10^7$ cm./sec. From VARNEY's work on drift velocities in N_2 and O_2, $v_+ = 6 \times 10^4$ cm./sec. was taken as a reasonable value for air at $X/p \approx 40$.

Using these values, and $p = 722$ mm. Hg and $i_0 = 12$ electrons/μsec., some curves were calculated for comparison with the data of Fig. 9 as follows. From X_s and p, (26.18) gives α. If $\gamma_p e^{\alpha d} < 1$, then for long times the solution (26.17) goes over into the familiar steady-state expression, $i = i_0 e^{\alpha d}/(1 - \gamma_p e^{\alpha d})$. Taking the steady-state value of the current for the "no spark" curve in Fig. 9 as 2×10^{-5} amp, this equation gives γ_p. Then τ can be obtained from (26.11), γ_p is then assumed constant for the small range of overvoltages studied, and α and τ are calculated for successively higher voltages. The curves calculated using these values were too widely separated vertically at the first shoulder, for a given change of final slope. A better fit was obtained by taking $X_s = 28.5$ kv/cm. and $v_+ = 8 \times 10^4$ cm./sec. (These gave $\gamma_p = 1.5 \times 10^{-5}$.) This change is greater than the experimental uncertainty in X_s, but α is really the quantity being varied; and it is more uncertain. The curves in heavy lines in Fig. 9 were calculated using these values.

The calculated curves have a similar shape and change with overvoltage in a similar manner, but there are two main differences which cannot be reconciled by adjustment of the parameters. The first is a difference in time for the initial current rise; the fit of the early part of the curves would be quite good if the experimental curves could be translated 5 μsec. to the left. The second is that the final slope of the calculated curves increases much faster with overvoltage than that of the experimental curves if FISHER and BEDERSON's data for overvoltage are applied to the time lags observed here using SANDER's data for α. The curve labeled $\tau = 4.9\,\mu$sec. was calculated for an overvoltage of only 0.03%. From the data of FISHER and BEDERSON, this overvoltage would give a time lag of 50 μsec. However, it is somewhat doubtful that their data apply because they probably had a lower γ_p due to "conditioning" of the electrodes by repeated sparking, as discussed above. In fact, it was noted in the course of this work that the overvoltages necessary seemed less than theirs, but with sparks 5 min. apart and V_s as capricious as it was, it was nearly impossible to get data on this. Further, although it was argued above that positive ions cannot contribute the major portion of the secondary electrons because of the apparent leveling off of the "no spark" curves, these curves fluctuate far too much to be able to rule out the positive ion mechanism entirely. If γ_i does contribute appreciably in the longer formative times, then the overvoltage must be correspondingly larger for the short time lags since then the ions cannot contribute significantly. All in all, this point is not too serious, because both theory and experiment are still too inexact and incomplete.

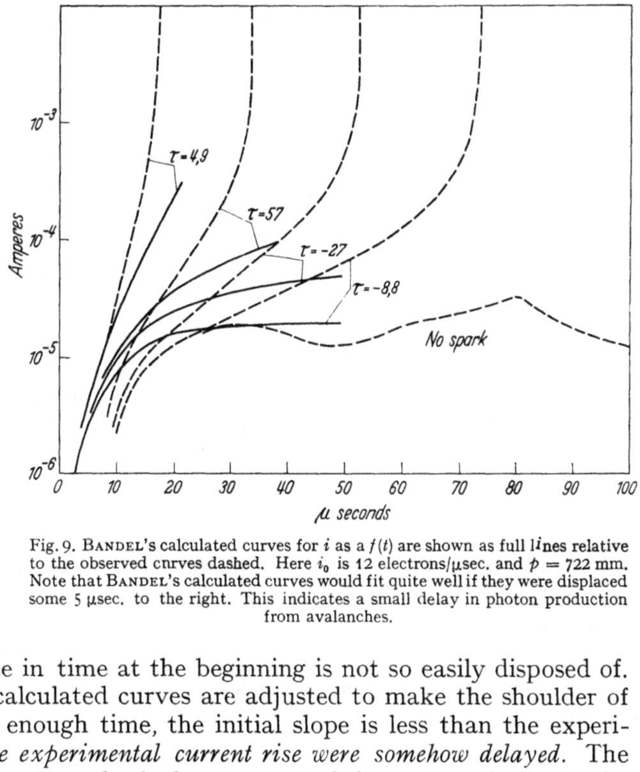

Fig. 9. BANDEL's calculated curves for i as a $f(t)$ are shown as full lines relative to the observed cnrves dashed. Here i_0 is 12 electrons/μsec. and $p = 722$ mm. Note that BANDEL's calculated curves would fit quite well if they were displaced some 5 μsec. to the right. This indicates a small delay in photon production from avalanches.

However the difference in time at the beginning is not so easily disposed of. If the parameters of the calculated curves are adjusted to make the shoulder of the curve come at a late enough time, the initial slope is less than the experimental; *it is as though the experimental current rise were somehow delayed.* The operation of the apparatus was rechecked as to start of the sweep relative to the pulse, linearity of the sweep, and rise time of the amplifiers; 1 μsec. seems an outside limit for experimental error.

The assumption in the derivation was that photon emission is in temporal synchronism with avalanche ionization. However, evidence is rapidly accumulating that much ionization and photon emission is caused by secondary processes with considerable delay time. If on the average enough photons are delayed the equivalent of an avalanche crossing time, the discrepancy could be accounted for. Evidence by HUBER [*10*] on the secondary processes in O_2 and N_2 mixtures as well as recent work by COLLI and FACCHINI [*8*] on argon indicate that delays of this order may be anticipated since the γ_p process may depend on a very selective group of photons.

As earlier stated, it was proposed that space charge distortion of the field is required to make streamer propagation possible; no mechanism considered in the foregoing analysis could account for the hyper-exponential increase in current observed just before breakdown. It is thus essential to see whether the observed increase in the rate of rise of the current can be shown quantitatively to be due to space charge.

In order to get an expression for the field due to space charge in the gap, consider infinite parallel plane geometry with charge distributed uniformly in the y and z directions in a slab of thickness dx' at x' and both planes at zero potential. If the volume charge density is given by $q_v(x)$, then the charge per unit area of this slab is $q_v(x')\,dx'$. If a GAUSSian surface is taken in the form of a cylinder parallel to the field, the use of the facts that with no variation in the y and z directions the field lines must be parallel, and that the integral of $X_{sc}\,dx$ across the gap (X_{sc} means: space charge field) must be zero, yields, in mks. units, $X_{sc} = -q_v(x')(d-x')\,dx'/k_0 d$, and $X_{sc} = q_v(x')\,x'\,dx'/k_0 d$ for the regions $x \leq x'$ and $x \geq x'$, respectively. Assuming the discharge to be spread uniformly over the electrodes and neglecting the few electrons present, $q_v(x) = i_+(x)/v_+ A$, where A is the area of the electrodes; so, finally,

$$X_{sc}(x) = \int_0^x \frac{i_+(x')\,x'}{v_+ A\,k_0\,d}\,dx' - \int_x^d \frac{i_+(x')\,(d-x')}{v_+ A\,k_0\,d}. \tag{26.19}$$

This will be in volts/cm. if we use amp, cm., sec., and $k_0 = 8.85 \times 10^{-14}$ farad/cm.

Now, to show whether or not space charge distortion of the field is enough to appreciably affect $\exp\left(\int_0^d \alpha\,dx\right)$, some expression for the current must be used in the above expression for X_{sc} and the integrations performed; then the resulting $X_{sc}(x)$ plus the applied field are used in (26.18) for α where the whole quantity is squared and again integrated across the gap. To do this with the rather complicated expression for $i_+(x,t)$ used above, involves a prohibitive amount of labor. However, in the expression for the final steady-state current when $\gamma e^{\alpha d} < 1$, i.e., $i = i_0 e^{\alpha d}/(1-\gamma e^{\alpha d})$, the effect of the action is the same as an increased i_0, and for the longest time lags studied when the current is increasing slowly the distribution of charge in the gap should not be so very different from what it would be for a steady current of that value. Thus the expression $i_+(x) = i_0(e^{\alpha d} - e^{\alpha x})$ should give a reasonable approximation for that case. When the integrations are carried through using this expression for i_+, the final result for the increment in the exponent is

$$\Delta \int_0^d \alpha\,dx \approx \frac{1.05 \times 10^{-4}}{p} \left(\frac{i}{v_+ k_0 A}\right)^2 \left(-\frac{1}{\alpha^4 d^4} + \frac{2.5}{\alpha^3 d^3} - \frac{1}{\alpha^2 d^2} + \frac{1}{12}\right) d^3 \tag{26.20}$$

where $i = i_0 e^{\alpha d}$ is the measured current with the assumed expression for $i_+(x)$.

For the long time lags, values of α near threshold are of interest; the change in α between the lower two calculated curves in Fig. 9 was ≈ 0.015. Certainly a smaller change than this would give an appreciable change in the current. However, since the effect of a change in α on the measured current appears gradually over one ion crossing time, a change of say one-tenth this much could hardly give a visible change of slope. Something around 0.005 seems a reasonable figure. Equating the above expression to this figure and solving for i, the current at which field distortion should cause a noticeable upcurving, the extent of variation in i for v_+ from (6 to 8) $\times 10^4$ cm./sec. and α from 11.1 to 11.8 (i.e., V_s from 28.5 to 28.8 kv) is from 2.8 to 3.7×10^{-4} amp. Experimentally, the semilogarithmic plots of current for the long time lags begin to curve upward around $(2 \text{ to } 4) \times 10^{-4}$ amp.

This agreement not only supports the proposed space charge mechanism but also shows *that the discharge is spread more or less uniformly over the electrodes as assumed in the calculations.* If the discharge were localized to an electrode area of a few cm.², the upcurving would have had to appear for currents more than an order of magnitude smaller. As seen above, the dependence on v_+ and α are not critical, and since i is proportional to the square root of the value chosen as a sufficient change in α, this is not overcritical either; there cannot be any order of magnitude error. This is also in agreement with the experimental observation that when the light was focused on the center of the cathode, sparks appeared to occur randomly over the whole electrode, independent of the illuminated area.

27. MENES' study in pure argon. The work of BANDEL was paralleled at a somewhat later date by MENES [*15*] in the Westinghouse research laboratories. MENES succeeded in outgassing an all metal sparking chamber so that really pure argon gas could be studied. Using a stabilized voltage supply and with the conveniently long formative time lags of an argon breakdown at low overvoltages he oscilloscopically measured the growth of current from inception to the point of inflection representative of beginning breakdown by space charge accumulation observed by BANDEL. The switching problems were simplified by the slow rise. Pressures from several hundred mm. down to 25 mm. were studied. The observed current growth was exponential except for the final stage. The data could be calculated assuming a delayed γ_p action based on the formation of excited molecular states of argon in triple impact with argon atoms, following COLLI and FACCHINI [*8*]. At the lower pressures, the rate of growth changed presumably because the photon creation declined while either resonance radiation reached the cathode or γ_i became active.

The measurements were not pursued to the point where a spark occurred. Current was limited and in his published abstract MENES stated that, while space charge field distortion was present near the end, *no* streamer mechanism such as described for air was seen in argon. In discussion, it transpired that with the long buildup time in argon the discharge developed into a general visible glow filling the gap. Current growth was definitely limited and, just before the end of the pulse, MENES admitted seeing one or more bright filamentary streamers cross the gap despite the glow. These however, were *not indicated by any sharp rise in current* as noted by BANDEL in air. Since sparks in argon are again filamentary streamer sparks, it is suspected that with the high current carrying capacity of the argon glow, the current limitation and perhaps the low overvoltage, the *increases in current* in the *streamer channels were too small* to be perceived against the heavy background of glow current. With more overvoltage and perhaps different current cut-off, it is certain that the streamer breakdown would have been observed as well in argon as in air.

It may thus be concluded that uniform field spark breakdown in all gases, *above* the few mm. pressure where diffusion governs, will have its threshold set by a TOWNSEND type of predischarge based on some secondary cathode action, presumably usually a γ_p. This builds up a space charge of such nature as to increase avalanche size, create a two component photoionizable gas and ends up as an anode, streamer initiated, filamentary spark. At appropriately low pressures, where diffusion dominates, the TOWNSEND threshold leads to a glow discharge which at sufficiently high current densities may go over to a power arc again via the phenomenon designated as a spark, but neccessarily of streamer origin.

VII. Breakdown in asymmetrical fields at steady potential: The coronas, glows and sparks.

28. Introduction, and general remarks. The study of the breakdowns in very asymmetrical fields much simplifies the discussion of symmetrical but non-uniform fields. Some general remarks may be made concerning this problen. The use of highly asymmetrical fields renders either the cathode or anode much more highly stressed, thus leading to a breakdown at that electrode with a minimum influence of the other electrode. It also creates a large range of potentials between the initial breakdown at the highly stressed electrode and the general spark breakdown of the gap owing to the extensive low field region at the one electrode. Again, such fields encourage the accumulation of space charges in the low field regions which will like external resistance act to control or limit current flow. This throws into high relief a difference in highly stressed positive and negative electrodes. In all cases the partial breakdown at the anode yields positive ions created most densely at the anode and moving very slowly in the high field regions relative. to electrons. Thus, in all gases, the positive space charge in the low field regions acts as a controlling resistance stabilizing the discharge and even interrupting it near threshold. For highly stressed cathodes, on the other hand, it is only in electron attaching gases that space charges in the lower field regions play an important rôle, so that free electron gases act differently from ion forming gases. Again, while the positive ions are created in high concentration at the anode where they can under appropriate conditions lead either to streamers or under adverse conditions act in the gap to control discharge, at highly stressed cathodes the heavy positive ion cloud is formed fairly near to the cathode *augmenting the fields*. This facilitates a γ_i process and cleans the cathode to augment γ_i and γ_p by bombardment of the surface. With highly stressed anodes, any cathode emission occurs in low fields, thus γ_i and γ_p are reduced by back diffusion loss and clean up is slow. Thus especially in *free electron gases highly stressed cathodes tend to lead to very unstable discharges, with a negative resistance characteristic*, while highly stressed anodes favor local breakdowns through photoionization in the gas. Instability may be complicated since, while ion bombardment cleans, it can with some gases disrupt the surface so badly as to adversely affect the discharge. Electrodes of this type are usually pitted by chemical sputtering. One more point should be made. With highly stressed anodes, there are always triggering electrons in adequate number entering the high field region from the large cathode or the volume of the gas. With the small highly stressed cathode, *very few triggering electrons occur in the high field regions* unless ultraviolet light is focussed on the cathode. Thus the field intensified currents below threshold are orders of magnitude lower for the highly stressed cathode than for a corresponding anode.

29. Geometrical gap forms. In considering such gaps, the geometrical forms available should briefly be mentioned. Most convenient for observation and most extreme are the hemispherically capped *cylindrical point-to-plane gaps*, with plane normal to the cylinder axis. Easy to fabricate and adjust and localizing discharges for most convenient visual observation and analysis, they have the drawback that only for a single ratio of point radius to gap length has the field along the axis been computed [*67*], and off axis fields are unknown. *Confocal paraboloid systems* and *hyperbolic point-to-plane* geometry are capable of accurate computation, but are limited in convenient ranges of focal lengths, hard to manufacture and very hard to adjust [*67*]. They also generally do not yield the extreme asymmetries of the point-to-plane. The most convenient geometry for calculation and also for quantitative study are the fine wire, large outer cylinder, *coaxial* *cylindical systems*, such as used in GEIGER counters. Here, the conservations of photons ect., and other cathode actions for positive wire bring the secondary cathode actions into much greater prominence. They have the inconvenience of extensive spread of discharge and multiple spots that make analysis difficult. In the discussions to follow, the coaxial cylinders and point-to-plane will principally be considered because of the quantitative data derivable from the former and the isolation of the stressed electrode in the latter.

Much work has been done with the convenient point-to-plane geometry, much also with coaxial cylinders, though largely in the GEIGER counter field of study. Only one study with hyperboloids by BENNETT and a few studies with confocal paraboloids largely hinging about avalanche size and sparking potentials have been carried out to date. Some discussion of this aspect has been given by the writer [*67*]

30. Highly stressed cathodes in free-electron gases. *α) General remarks.* Quantitative data are hard to acquire in this field of study even with coaxial geometry. It is difficult to control an i_0 at the cathode from outside sources. i_0 is very small and it has so far been impossible to produce a pulsed emission of electrons from the cathode in 10^{-8} sec. of sufficient magnitude to reveal the oscilloscopic details yielded by the α particle track triggering of COLLI and FACCHINI [*8*], and LAUER [*63*]. Free electron gases, pure N_2, H_2 and argon have been studied in the modern fashion somewhat by WEISSLER with point-to-plane geometry [*68*], by MILLER and the writer [*45*] with coaxial cylinders, and by LAUER [*63*]. The secondary actions in lower fields at the nickel or brass cathodes have been delineated by LAUER, COLLI and FACCHINI [*8*], and HUBER [*10*] with positive wire in coaxial cylinders.

Although the points or wires used were largely Pt, the nature of events is not changed much, since work functions of the Ni and Pt are akin and high and since the action of gases like O_2 is analogous on most surfaces. Point diameters ranged from 0.25 cm. down to 0.1 cm. and gap lengths ranged about 3 to 4 cm. Wire diameters ranged about 0.012 cm. and outer cylinders around 3 cm. Temperatures were usually around 20 to 22°. Pressures studied ranged from 12 to 760 mm. One difficulty in coordination and interpretation of results lies in establishing gaseous purity and the initial cleanliness of electrodes. The gases of LAUER [*63*], COLLI and FACCHINI [*8*], in the writer's laboratory, and of HUBER [*10*] were clean, and electrodes outgassed at 900° were initially clean. In MILLER'S work the gases were fairly but not uniformly clean. In WEISSLER'S work it is believed, but not certain, that they were clean. The electrodes in MILLER'S and WEISSLER'S work were relatively clean but probably not adequately outgassed so as to be initially clean. Thus conditioning discharges of low threshold were required.

β1) Pure nitrogen, coaxial cylinders. The studies of HUBER [*10*] showed that Ni and presumably Pt shows only a γ_i of the order of 10^{-2} corrected for back diffusion with no γ_p. With slight contamination, a low order γ_p appears. MILLER observed that in pure N_2 at all pressures it was impossible to observe a corona *without external resistance to control currents*. At about 4100 volts a heavy discharge occurred that melted the wire or blew a 10 ma fuse. The discharge rose rapidly from a current of 10^{-13} ampere to breakdown. If a series resistor was used and a discharge run for 20 min. at 800 μa, the gas was contaminated sufficiently to yield a negative low field space charge resistance and the events could be studied without external resistor. Use of a limiting resistor with clean gas indicated the discharge to start at a potential appropriate to the pressure as a diffuse glow which in time contracted to one or more spots with little increase in current. Contraction occurred the faster the higher the potential. Below 50 mm. the glow remained and no spot formed. Within limits of error the sustaining potential across the gap could be determined by deducting the iR drop in the resistor from the applied potential and remained constant with increasing current. Usually, the *potential* at which the *discharge appeared* was *well above* the maintenance threshold, so that once the glow had contracted to the spots, potential could be lowered until one spot after the other extinguished and the discharge ceased as the last spot went out. The *overshoot* in *potential* with these cathodes was often several hundred volts. It is not impossible that with pure gas the overvolting could have been the agency that led to the instability and complete breakdown. If the potential is raised a minute or two after extinction, overshoot is less and the discharge starts at one or several of the previous spots instead of as a glow. The structure of the glows as miniature glow discharges can not clearly be discerned with wires, and coaxial cylinder.

β2) Pure N_2 point-to-plane geometry. As with coaxial cylinders, the discharge started as a uniform glow and contracted to a spot discharge at the tip at 90 μa and 3700 volts [*68*]. With this geometry, owing to some series resistance and either slight contamination, or because of the long low field region, the instability observed by MILLER was prevented from going straight to a spark at threshold. The discharge remained fixed at one spot. The current-potential curve followed the form characteristic of a space charge controlled current of the TOWNSEND form to be given later, the space charge being probably largely electronic in long low field regions. The spot in this geometry showed a typical glow discharge structure with CROOKES dark space, negative glow, FARADAY dark space, and faint positive column. The current increased until it broke into a filamentary spark at 14000 volts. Conditioning of limited areas and overshoot were not as noticeable in this geometry as with coaxial cylinders.

It must be concluded that with slightly soiled surfaces potentials must be raised to a high value to start a discharge of the γ_i type all over the surface. Minor gas contamination may contribute some γ_p action. In any event, with dirty surface a low order discharge sets in which in due time cleans the surface in spots with high γ_i and the discharge contracts to one or several spots depending on the overshoot in potential. Reduction in potential causes current decrease and the spots to extinguish one by one. With MILLER's more favorable coaxial geometry, the cathode instability was such that the discharge increased in current and space charge distribution in the gap so as to lead to a streamer-like spark. Since γ_i action is localized in space, it is not impossible that the field distortion led to a localized narrow discharge path that gradually passed into an arc without intervention of a spark with coaxial cylinder. Speculation here is futile.

γ1) Pure H_2, coaxial cylinders. In H_2 the action on the nickel cathode at low fields appears to be a pure γ_p according to LAUER [*61*]. γ_i has been observed by HALE at high X/p as well as by PARKER for H^+ and H_2^+ ions with 2 volts of energy on Pt and Ta. It was not observed in H_2 by GUGELBERG [*25*]. For the negative wire currents rose from 10^{-13} to 10^{-11} ampere yielding a weak self sustaining discharge. Above 50 mm., as potential increased, the current suddenly jumped to 100 μa and with localized spots on the wire was limited only by external resistance. Here again, the potential could be lowered after the jump and the spots disappeared one by one. The threshold for negative corona was slightly below that for positive corona, but with overshoot such comparison is meaningless. No limit for spark-over was noted since this occurred if external resistance was removed.

γ2) Pure H_2, point-to-plane geometry. Threshold occurred at 2600 volts with a self sustaining discharge spread over the surface. In time, it contracted to a spot discharge with a thousand-fold increase in current. At 2800 volts the glow quickly contracted to a spot with characteristic glow discharge structure and no oscillations at 140 μa. Unlike that in N_2, the discharge spot wandered continually over the surface and would then suddenly cease as a spot to spread over the surface as a glow, again later to contract with no changes in current. The cathode surface after the discharge was found covered with small pits, in contrast to N_2 where a single cathode spot crater was observed. WEISSLER did not succeed in getting his H_2 to yield a spark up even above 24000 volts.

Notable in H_2, as contrasted with N_2, are first its reluctance to spark breakdown with a γ_p and less favorable point-to-plane geometry compared to coaxial cylinders, and secondly the *deconditioning effect of the heavy discharge.* It is clear that, while the diffused discharge cleaned the surface, the intense current densities of ions at perhaps 10 or more volts energy disrupt the Pt surface and cause the discharge to wander. Even 0.1% H_2 in an argon discharge has the same effect. This is related to chemical effects of H_2 and is termed by some authors *chemical sputtering.*

δ1) Pure argon, coaxial cylinders. In pure argon, above 150 mm., the self sustaining discharge occurs at low field nickel or brass cathodes through a delayed γ_p action by 1250 Å photons from slightly metastable A_2 molecules, created from $^3P_2$ metastable A atoms in triple impact [*8*]. At lower pressures on a clean surface of Ni the action is a γ_i action [*63*], [*8*].

In argon, above 200 mm. pressure, no current above 10^{-13} amp was observed by LAUER until at some critical potential a spark occurred, though the current rise to the spark could well have been in the millisecond time range or longer. At lower pressures, a corona started as a blue glow spread uniformly over the cathode wire. This glow either with slight rise in potential, or spontaneously in time, went over to a spark. No localized glows were observed.

The potential increase to change the glows to sparks in argon at lower pressures was generally slight and ranged from 10% to 1%. The threshold for negative corona and/or spark with stressed cathodes was about one half that for corona from stressed anode and the streamer spark threshold with stressed anode lay about 30% above the corona threshold.

δ2) Pure argon, point-to-plane geometry. In pure argon WEISSLER observed no corona. Independently of gap length, at 3500 volts the discharge appeared as an arc across the gap with steady current and 1000 μa, apparently controlled by external resistance. After the arc was achieved, the potential could be lowered appreciably before the arc went out. This again indicates some surface contamination and an overshoot in threshold. This is consistent with the γ_p action followed by a more effective γ_i at the highly stressed cathode.

ε) The sputtering phenomena at highly stressed cathodes. It had been observed by HUDSON [*61*] in air that the point cathode after running a corona had a crater-like pit which was pressumably caused by *sputtering* of the cathode material and redeposition around the discharge spot area by back diffusion at high gas pressures. WEISSLER investigated this phenomenon for cathodes of W, Pt, Cu, Al and Pb, using both N_2 and H_2 [*68*]. All points except Al after 12 hours at 1250 μa and about 1 amp/cm.² of crater area current density showed the characteristic pit. For Al the whole surface was roughened by fine holes. Possibly the irregular oxide layer on the Al led to this behaviour.

Sputtering has been well known in connection with the high CROOKES dark space fields in low pressure glow discharges, but its observation in connection with high pressure corona points came as a surprise. Recent studies of sputtering by positive ions of controlled energy in vacuum by BRADLEY [70] and of WEHNER [70] by probes in higher pressure discharges indicates sputtering to be associated in most cases with momentum transfer from rapidly moving ions to the lattice atoms. Correlation between sputtering and elastic constants of the lattice (velocity of sound) for momentum transfer between ion and fixed atom seems indicated. Preferential directions for sputtering on the lattice have been observed. In some cases, possibly for H_2 and Pt, solubility of the gas and/or chemical reactions interfere. SCHINDLER and WEISSLER [71] measured the ion impact energy by the heating at the cathode in a corona discharge and established the energy of positive ions to be of the order 10 ev or more. This is adequate to account for the phenomenon.

Obviously, sputtering of the surface atoms including adsorbed layers when ions strike the surface with high energy is intimately connected with "conditioning", or "clean up", altering γ_i and γ_p, as well as modifying the discharge and it must be reckoned with in considering all highly stressed cathodes.

31. Effect of small amounts of O_2. *α) Coaxial cylinders, N with less than 1% O_2.* The threshold was at once raised by 100% even with 1% O_2. Below 200 mm. pressure, the discharges still posessed the negative resistance characteristic, since space charge was then not adequate to furnish the internal current limiting action. In air it is possible to go to 27 mm. without limiting resistor. Overshoot of threshold was present perhaps in more exaggerated form as surface deactivation by O_2 was insured. Threshold was abrupt from 10^{-10} ampere to 100 μa. After the discharge had run for some time, potential could be reduced, and the current of 20 μa was periodically interrupted at from 500 to 9400 times a second by space charge fouling in the low field region of the gap. Frequency varied with potential and current increase rising from 10 to 90 μa. Above 100 μa, and the corresponding potential, space charges caused no more interruptions. Interpretation is not difficult. The O_2 reduces γ_i by a factor of 5 or more, yields a γ_p which can initiate a clean-up discharge and, after suitable cathode bombardment, yields a discharge with γ_i. The presence and nature of the negative ion space charge is indicated by its effect on internal resistance, by its inadequacy in this action below 200 mm. and by its throttling pulse producing action in low field regions at lower potentials.

β) Point-to-plane in N_2, H_2 and A with less than 1% O_2. In H_2 and N_2 addition if 0.1% O_2 led to increased thresholds by some 400 volts in 2700, currents for the same potentials were reduced to one half [68]. Between 0.3 and 0.4% O_2 led to further increases in threshold and the coronas were interrupted by space charge formation leading to pulses like the TRICHEL pulses in air.

Addition of H_2 or N_2 to the extent of 0.1% in argon *reduced* the threshold for the spark in pure argon by as much as 1000 volts to 2500 yielding glow thresholds which evidenced a considerable potential overshoot. Potentials could be reduced from threshold value to 1300 volts before they went out. Increase of potential led to a current increase and a continuous transition to an arc at around 4100 volts. Addition of 0.1% O_2 to argon gave a threshold about the same as argon with N_2 with 1 μa current, but the discharge showed TRICHEL pulses. With 0.4% O_2 threshold was about the same, the current 0.1 μa showing pulses ultimately leading to a spark breakdown above 10000 volts.

In general, it is clear that O_2 alters γ_i, if it is present in adequate quantities, it causes negative ion space charge intermittence of the current and thereby and perhaps otherwise, reduces current. It also raises breakdown potential probably by altering the breakdown mechanism, the precise nature of which appears at present unknown.

32. Highly stressed cathodes in air and O_2: Coaxial cylinders in air. The increased O_2 content such as in air complicated phenomena still more [*45*] than traces, conditions varying from day to day depending on the past history of the cathode. Currents rose abruptly from 10^{-12} ampere to 0.1 to 1 μa. At threshold the rise required some time after application of the potential. The discharge started as a glow over the whole wire. After some time the glow contracted to one or more spots, perhaps with increase in current. With such spots at threshold, with little overshoot, pulses appeared. If interrupted for a short time, the discharge started at the same spot. The TRICHEL discharge spots showed the crater-like pits and local sputtering observed on negative points. The spread of the glow along the wire in these mixtures is associated with ready photoionization of the gas with absorption coefficients of 10 cm.$^{-1}$ in air and less in leaner O_2 mixtures. Glows and spots appeared in air simultaneously. Spots repel other spots, and glows are separated from spots by dark spaces. This results from the nature of the potential distribution between the regions of discharge and no discharge as indicated by the writer [*72*] and FINKELNBURG [*72*].

The behaviour of pure O_2 is still more remarkable [*45*]. It is readily photoionized and the characteristic absorption coefficient at 760 mm. is of the order of 100 cm.$^{-1}$. Thus, photoionization in the gas in the high field region can become effective and can in a sense, with such geometry, supplant γ_p. This is possible until pressures below 200 mm. are reached at which point O_2 should resemble air.

Negative ion formation by dissociative attachment in O_2 should be more rapid and occur closer to the wire than in air. Thus, choking of the discharge could well take place as readily as with point-to-plane corona in air. Dissipation of space charge sufficient for a new discharge should be fairly rapid, even in the less favorable geometry of coaxial cylinders, so that dead times rather than clearing times will yield the frequency of the pulses. Thus, while the duration of TRICHEL pulses in point-to-plane corona in air at 760 mm. is of the order of 10^{-8} sec., or less, with MILLER's wires in pure O_2, the duration was 3 to 5 microseconds. The threshold for the negative wire corona in pure O_2 was higher by perhaps 200 volts than for the positive corona and obviously higher than in air. This circumstance is indicative that γ action is reduced and the process is augmented by a photoionization in the gas in the high field region. There was some γ_p action *since the past history of the cathode did alter observations.* After running the cathode in O_2 for some time to clean out N_2, etc., the discharge began as a low order self-sustaining discharge setting in at 10^{-10} amp and *smoothly going up towards a breakdown to higher currents.* There were no pulses in this region, and the threshold of the more important discharge with currents of the order of a microampere consisted of a minor jump in current, by a factor of two or so, and the appearance of pulses. At lower pressures, transitions were more abrupt. Visibility of discharges in O_2 are low, so that no light is seen, except at very high currents. There appeared to be a fairly uniform conditioning discharge all along the wire even with TRICHEL pulses. TRICHEL pulses appeared at a spot, but in pure O_2 this went out and then later a spot reappeared at some other point. At higher current, spots appeared to anchor and maintain themselves by clean up at one point. Except at 27 mm., no overshoot of potential was observed with the high pre-threshold current. At lower pressures and currents above 10 ma the pulses followed each other with hardly a break. Space charge choking ceased and the discharge went to a power arc. At higher pressures raising potential increased current by multiple spot formation. Nothing is known of the spark breakdown mechanism.

33. The highly stressed anode. α) *Coaxial cylinders, free electron gases.* In these discharges, rather complete study of actions taking place are possible for cathode action by *triggering single discharge sequences by single well collimated α-ray tracks* parallel to the cathode cylinder and close to it. These release a cloud of electrons that ionize in a thin sheet of avalanches, lying in the plane of anode wire and α-ray track. Time resolution is very high because of the short high field region and low dispersion in the α-ray track. The technique was devised by LAUER [*63*] from use of uncollimated α-particles by COLLI and FACCHINI in a similar study. It has since been used by HUBER [*10*], and by COLLI and FACCHINI [*8*]. Gases studied have been A, H_2, N_2 and O_2 and O_2-N_2 mixtures. Purity control was excellent. The method permits determination of the cathode agency at threshold, identifies ions by their drift velocity, indicates electron attachment within measure, checks on electron transit times. It cannot resolve very slow time scale phenomena, i.e., such as diffused metastable action or a γ_p with 10^{-4} sec. imprisonment of radiation [*7*]. Some studies can be made with anode processes. It does not lend itself to any study of the *spark* breakdown process.

LAUER [*43*] studied pure H_2 with a low field nickel cathode with pressures from 25 to 600 mm. The positive ions had a drift velocity of 13.4 cm²/volt sec. and were presumably H_3^+. They gave no γ_i greater than 5×10^{-6}. There was a γ_p of the order of 4.9×10^{-5} to 69×10^{-5} which, corrected for back diffusion γ_{cp} by THEOBALD, ranged from 0.014 to 0.0015 with X/p decreasing from 1.43 volts per cm. per mm. Hg to 0.66 at 650 mm. The γ_p gave a self-sustaining corona which near threshold showed the fluctuations created by space charge accumulations. The discharge was distributed all along and about the wire by diffusion of photons which, if not absorbed in the gas (which probably occurs to some extent), are conserved by multiple reflection by the cylindrical cathode. The discharge became visible at self-sustaining corona around 10 μa of current as a tightly adhering glow on the wire surface. The potential could not be raised high enough to yield spark breakdown. From observations of WEISSLER [*68*] on point-to-plane corona, it is known that ultimately streamers appear and breakdown occurs with a streamer starting from the anode. Currents are up in the milliampere range and higher, and much H as well as H_2 is present when streamers form.

In pure argon the situation is complex. At higher pressures, COLLI and FACCHINI [*8*] have shown that the corona starts with a γ_p from photons of ~1250 Å created by formation of A_2^* from excited atoms by triple impact of $^3P_2$-metastable atoms with A atoms. This effect is delayed some 3.4 μsec. owing to slight metastability of the A_2^* at higher pressures and has a characteristic time which increases as pressure decreases below 150 mm. and is pressure dependent owing to formation of A_2^*. Above 150 mm. it yields a self-sustaining corona. The value of γ_p ranges from 1 to 5×10^{-3} between 150 and 1000 mm[1]. A. If the gas is very pure and the Ni cathode well outgassed above 900° C for hours, a γ_i begins to appear at values of pressure below 200 mm. For the ions LAUER measured the drift velocity and identified the ion as A_2^+. Both he and COLLI and FACCHINI, at 50 mm. and below, observed some pulses possibly indicating A^+ ions as well. The values of γ_i ranged from 7.8×10^{-4} at 25 mm. to 0.2×10^{-4} at 400 mm. As corrected by THEOBALD for back diffusion, γ_{ci} ranged from 0.035 to 0.050 from 25 to 100 mm. and around 0.006 at 200 and 400 mm. It is not unlikely that the lower values at high p came from minor contamination of the cathode by traces of impurity in the high pressure argon. At around 100 mm. the secondary action at threshold is probably equally split between γ_i and γ_p and below 50 mm.

[1] Uncorrected for back diffusion.

the self-sustaining discharge is largely owing to γ_i. If even slightly contaminated argon was used, with less than 10^{-1}% of impurity, γ_i did *not* appear and the cathode required prolonged outgassing at 900° C to restore it in the presence of pure gas.

Breakdown in pure argon followed very readily by streamer spark by increasing the potential some 20 to 30% above threshold for self sustaining corona. In argon the current-potential curves rose fairly smoothly from 10^{-13} amp to around 10^{-11} amp before the steep rise leading to threshold. No overshoot was observed as conditioning of the cathode at low fields and current densities did not appear.

Successions of regular γ_p or γ_i pulses after triggering did not last indefinitely at onset because the diffusion of events in space and time with the *very many repetitions* spreads the discharge uniformly all around the wire. Simliarconsiderations apply to pure H_2 and N_2 as well.

β) Space charge limited currents under steady discharge. TOWNSEND'S *law.* The current was fairly steady with minor fluctuations. The theory of this current limited by space charge for positive ions was worked out in 1914 by TOWNSEND [*73*]. It has also been extended by SVEN WERNER [*73*] in GEIGER counter study and by others [*46*]. The relation of TOWNSEND reads

$$V - V_g = \frac{p\,i\,A^2}{2KV} \log\frac{A}{a} \tag{33.1}$$

or

$$i = \frac{2K}{p\,A^2 \log A/a}\, V(V - V_g) = BV(V - V_g)\,. \tag{33.2}$$

Here V_g is the threshold potential, V the potential applied, p the gas pressure, A and a the radii of cathode and anode cylinders, K the reduced mobility of the ions of the space charge, and i the current. It is noted that near threshold V_{g+} for small changes in V, i increases linearly with $V - V_g$. This region is called the OHM'S law regime. Over larger ranges, with $V \gg V_g$, the current increases parabolically as $V(V - V_g)$ as indicated by so many curves. It must also be noted that this applies to all ion space charge limited discharges of this type, irrespective of sign. The algebraic form of variation of current with V will be altered for different geometries, but equations will follow this type.

How the ion production and current increase as the potential is raised, while the amplification factor in the ionizing zone from r_{0g} to a remains constant as implied by the equation, was indicated by the writer and later confirmed by exact calculation by COLLI and FACCHINI and their associates [*46*]. This theory also accounts for current fluctuations observed [*46*]. The theory is equally applicable to negative wire corona as for positive corona where space charge choking does not occur.

Note, with the relatively low currents in H_2 and A at threshold, space charge choking occurs over so narrow a range of potential that counter action and pulses have been ignored in this discussion.

As pressure decreases and/or increasing potential begins to widen the zone of active ionization, r_0, much beyond r_{0g}, then deviations from TOWNSEND'S equation occur. When r_0 begins to approach A in value, then space charge control of current ceases and the negative resistance characteristics appear. At higher pressures and with creation of a two component gas, spark breakdown usually sets in by streamer formation at the anode before r_0 reaches A. At lower pressures, the extension of the ionizing zone r_0 to near the radius A enhances cathode fields and action and the instability mentioned results in negative resistance and arc formation, probably without streamers.

γ) Coaxial cylinders, pure N_2. The study of pure N_2 by HUBER is also revealing [*10*]. Here no evidence of a γ_p process could be observed, despite the inferences of the writer from MILLER's data [*45*]. The ion involved was N_4^+ identified by its drift velocity. γ_i on outgassed Ni ranged from 0.6×10^{-3} at 600 mm. to 3.4×10^{-3} at 50 mm., leading to a nearly constant value of γ_{ci} of about 2.5×10^{-2} corrected for back diffusion. This value is of the same order of magnitude as observed by PARKER [*74*] on N_2 covered surfaces of Ta and Pt. Self-sustaining corona threshold began with a regular sequence of γ_i pulses after triggering, which after many repetitions, gave a steady slightly fluctuating corona as seen in argon with glow all around the wire. HUBER did not study steady corona in N_2 and no current potential curves were taken by HUBER, which is unfortunate as MILLER's data are of doubtful validity. MILLER probably did not outgas his cathodes at 900° C and there are indications that his N_2 may have been slightly soiled as he flashed *W* filaments as "getters", and HUBER found that instead of "gettering" they contaminated the gas to varying degrees. MILLER's discharge showed a rather heavy predischarge current, rising from 10^{-12} ampere nearly exponentially to from 10^{-10} to 10^{-8} ampere, and then yielding a current of 10^{-5} ampere with multiple bright spots on the anode. Reducing the potential, caused the spots to disappear one by one ending with 2.5 μa at one spot. Further reduction in potential extinguished the spot and reduced the current by a factor of 5 or more. This action definitely fixes the discharge as a *cathode* determined discharge dependent on clean-up by ion bombardment and resulting in uneven clean-up as spots on the cathode. The localization in spots on the anode in MILLER's study, which may have also occurred above threshold under HUBER's conditions but was not looked for, definitely points to a γ_i action emanating from spots on the cathode. HUBER [*10*] with her very clean cathodes got a general γ_i from the surface, aided no doubt by her triggering all along the cathode. With multiple spots in MILLER's overshot discharge the oscilloscope showed a fluctuating pattern. Single spots yielded a relatively smooth current.

As HUBER showed with slight gas contamination there was a photon γ_p which did not lead to a self-sustaining discharge. Since MILLER's curves show a conditioning predischarge current, it appears as if his N_2 had not been too clean.

δ) Coaxial cylinders, pure gases; effect of small quantities of O_2. It was universally reported, first by WEISSLER [*68*] using point-to-plane geometry, and later by MILLER, LAUER and HUBER that addition of as much as 0.1 to 1% of O_2 altered the phenomena.

In H_2 LAUER reported that traces of O_2 caused some smearing and a more rapid attenuation of the γ_p pulse series just below threshold. The ion pulses, caused by the crossing of the positive ion sheath created at the anode from the initial α-particle-created electrons, instead of falling smoothly to low values, as the ions crossed the gap, and then sharply to zero, as the ions reached the cathode, had a number of random spikes superposed on them at 0.1% O_2. With 1% the irregularities increased in magnitude so that there was even a marked increase just before the ion sheath reached the anode. LAUER ascribed these irregularities to groups of electrons that would, in pure gas, have increased the initial current spike, but here were trapped by O_2 to make negative ions in midgap and arrived at the anode in time to create new avalanches and pulses while the ions were in transit. Rough statistical analysis by LAUER yielded the correct order of magnitude of the attachment coefficient values for O_2^- ions in air on this basis. The O_2^- ion detaches its electron in high fields so that the electrons can yield somewhat decreased avalanches at the anode. However, the nature of LAUER's curves with 1% O_2 in conjunction with HUBER's studies in N_2 indicate that some of these irregularities may have been due to incipient burst, or GEIGER counter pulses. It is to be regretted that LAUER did not evaluate γ_p with the more rapidly attenuating pulses with 1% O_2, but it seems likely that γ_p may have been decreased perhaps by 50% or so by the O_2. There may also have been some delayed active photons produced by secondary reactions in the gas. Unfortunately, LAUER did not report on the thresholds with O_2, but WEISSLER found that the threshold was 3500 volts for pure H_2, 5000 volts at 0.1% O_2 and 4000 volts with small burst pulses at 1% O_2 in point-to-plane corona.

No data in argon with 0.1% of O_2 were taken, but the effect of traces has been indicated by COLLI and FACCHINI in wiping out γ_i.

In N_2 HUBER's careful study has revealed the very radical changes produced by addition of different amounts of O_2. These changes run the whole gamut from a self-sustaining corona with a γ_i, through a mixed corona, through a burst pulse GEIGER counter discharge, to a straight anode streamer corona with 80% or more O_2. Addition of the order of 1% O_2 to the N_2 produced the following changes:

1. The γ_i pulses below threshold, were smaller and fewer in number at a given potential and pressure, than in pure N_2.

2. The secondary γ_i pulses began to lose their smooth shape as potential increased and became ragged.

3. Threshold for self-sustaining corona was 10% higher than in pure N_2.

4. With faster sweep, a new secondary pulse was observed spaced at intervals of about 1 microsecond suggesting a γ_p which did not, as with LAUER in H_2, become continuous at threshold.

5. On slow sweep, secondary pulses were superposed on the pulses due to ion crossing, as in LAUER's work in H_2, effectively coming from the initial peak and appearing later. The lowering effect of O_2 on γ_i did not result at once on admission of O_2 to the N_2 but required time, or a few discharges. The lowering could be rapidly achieved by warming the cathode somewhat in the soiled gas. This is in keeping with MILLER's findings relative to the disappearance of activated or clean spots on Ni in contaminated gas with low concentration of O_2.

The O_2 produced a lowering of γ_i by a factor of 5 in conformity with PARKER's [*73*] observations of γ_i for N^+ and N_2^+ ions on clean Ta covered with N_2 when exposed to O_2. The photosensitivity of the surface increased, leading in part to the appearance of a γ_p. This had been noted in the action of O_2 by THEOBALD [*6*] and by GEBALLE [*75*] and his associates. It is also likely that the presence of O_2 in N_2 led to the creation of new photons active on the Ni which were not absorbed by the N_2. With pure N_2 active photons from the anode region apparently never reached the cathode in appreciable numbers and/or without inordinate delay. That the short microsecond peaks with fast sweep were caused by a photoelectric γ_p at the cathode was indicated by the measurement of electron transit times which agreed with those computed from BRADBURY and NIELSON's values of drift velocities in N_2 within 2%. Increase in threshold would be caused by a reduction in γ_i even with the appearance of a γ_p. While in pure N_2 at a threshold of 2660 volts at 200 mm. γ_i was of the order of 10^{-3}, with 5% O_2 secondary action at the threshold of 2960 volts in the mixture gave a γ_i and γ_p together of perhaps 10^{-4}. However, there was in addition a new mechanism active in setting the threshold through photo-ionization in the gas. The relatively small increase in threshold with fivefold decrease in γ_i is perhaps surprising, except for the very rapid increase of M with small increase in V. The probability of attachment and a smearing of the ion crossing pulses cannot be overlooked. This was more evident at 5% O_2 where the statistical scatter in the midgap crossing period was no longer present but currents for ions in midgap were several times those for clean N_2 while the initial spike for electrons was reduced. The irregulatities around the γ_i peaks were *not caused by attachment.* With increasing O_2 up to 5%, these fluctuations became more and more certainly the appearances of discharges spreading over small regions of the anode by photo-ionization in the gas. With 5% O_2, at threshold, while the *initial* γ_i peaks did appear there were many typical GEIGER counter like discharges or burst pulses present.

δ) Coaxial cylinders, air and pure O_2. In air, at 20% O_2, the α-particle triggering produced irregular bursts of ionization that resemble burst pulses but break off shortly. These were discharges propagating along the wire by photo-ionization and quenching by space charge accumulation. At threshold, the sequences continued as the steady but slightly fluctuating burst pulse corona current. *No secondary cathode action could be detected.* Threshold at 200 mm. went *down to 2145 volts* indicating a new more efficient mechanism to be active. The quenching intervals in the corona before onset of the steady process was about a quarter of an ion crossing time. This is the time required for a positive ion to move across half the gap from the high field region.

Air			O_2-N_2 mixtures	
			$V = 2720$ volts, $p = 300$ mm	
p mm	V volts	v cm./sec.	% O_2	v cm./sec.
100	1580	3×10^7	0	no spread
200	2190	3×10^7	5	9×10^7
300	2720	4×10^7	10	6×10^7
400	3240	6×10^7	21	4×10^7
			40	no α-pulses

The existence of the GEIGER counter discharge propagating down the wire in mixtures from 5 to 40% O_2 in N_2 was established by HUBER [*10*]. Pure N_2 showed *only proportional counting* for pulses of α-particles fired *parallel* to the wire, or *transverse* to the wire at one end. With 5% and up to 40% O_2 the pulses were of the *same amplitude for the disparate parallel and transverse tracks and* the velocity of propagation of pulses down the wire were measured and yielded results as follows.

In pure O_2 the duration of single α pulse ionization well below threshold gave the mobility of the O_2 ions at 20° C and 760 mm. as (2.2 ± 0.1) cm.²/volt sec. which agrees with VARNEY's estimate of (2.25 ± 0.1) for the assumed O_2^+ ion. At a somewhat higher potential, the ion pulse showed irregularities on it that could only be interpreted as very localized discharges at several points of the wire that did not propagate down the wire but extinguished themselves locally. This sort of action is indicated in Sect. 18, α and β. Below 200 mm. this type of pulse made the discharge self-sustaining as a simple burst pulse, or GEIGER counter pulse, corona. Above 200 mm., as potential increased still more, very remarkable pulses appeared. These consisted in a series of regular pulses of the order of 200 or more times as large as the α-particle triggered pulses. They occurred whether α-particles were present or not. They were spaced at uniform intervals and were approximately constant in amplitude at a given potential and pressure. Their duration was less than 10^{-7} sec. Currents in these pulses were as high as 10^{-4} amp, and each pulse contained or created 10^8 electrons in contrast to the 10^7 electrons or less in a burst pulse. These are characteristic properties of short pre-onset streamers that do not cross the gap. They were first noted in this geometry by MILLER [*45*]. They had been discovered and studied with point-to-plane geometry in air by KIP [*49*]. The streamer spacing was close to the ion transit time, in fact to $\frac{2}{3}$ of that time. This means that the last streamer must have traveled 82% of the distance from wire to cylinder before the next one could propagate. The length of the streamers out into the gap was some 2 mm., so that the ions from the tip of the antecedent streamer reached the cathode before the next streamer. The streamers appeared to be localized at some one point along the wire for a sequence. Eventually, this sequence interrupted itself and a new sequence occurred at another point, in agreement with MILLER's observations. They caused a glow to surround the wire perhaps of 3 mm. radius instead of adhering closely to the wire as with burst pulses.

Below 200 mm. pressure the pure O_2 acted at first like air with mixed streamers and burst pulses at an appropriate potential and at still lower pressures only burst pulses occurred with no streamers, in conformity to discussion in Sect. 18, β and γ.

No spark breakdown studies were made with these gases. However, the experiences with point-to-plane coronas in these same gases indicate that, at appropriately high potentials and currents, when space charge dissipation is rapid, streamers again appear and cross the gap, leading to a power arc via a spark.

ζ) Coaxial cylinders, variation of threshold potential from pure N_2 to pure O_2. It is perhaps important to note, that as the gas changes from N_2 to O_2 and mechanisms change, the threshold potential starts fairly low for pure N_2 and *rises* 10% or more with 1% O_2 [*10*]. Above 5% O_2 the potential *falls* to a value 20% *below* pure N_2 [*10*]. Somewhere around 30% O_2, or slightly more, the threshold again rises to around the value for pure N_2, or higher, in pure. O_2 when streamers form [*10*]. That is, with clean cathode the γ_i discharge takes a fairly high potential. It is increased even with the appearance of a γ_p by the decrease in γ_i, by traces of O_2 on the Ni. It begins to decrease when photoelectric action *in the gas* leads to GEIGER counter action. It begins to increase again when streamers are the only discharge possible, since streamer action requires larger avalanches than burst pulses. Thus it is seen that great caution must be used in drawing conclusions relative to mechanisms on the basis of changes in breakdown potential alone.

η) Point-to-plane, free electron gases [*68*]. In H_2, threshold occurred at 3500 volts and 4μa with a feeble spot of light just at the tip of the point, adhering closely to it and 0.005 cm. diameter [*68*]. There were no regular pulses. As potential increased, the discharge current increased along a curve analogous to the TOWNSEND current curve of equation (33.2). At 10000 volts pre-breakdown streamers occurred followed by a streamer spark at 18000 volts. Note the ready breakdown to a spark by anode streamer at 18000 volts from a positive point, while *no breakdown appeared* for the same negative point in pure H_2 below 24000 volts and more.

Pure N_2 yielded a threshold at 4800 volts for a 4.6 cm. gap in contrast to 3700 volts for a 3.1 cm. gap with negative point. The discharge was localized at the surface in a spot. The current was steady. Increase in potential resulted in a brighter corona which extended further towards the cathode in a luminous channel. At 11000 volts visually faint streamers appeared with the accompanying oscilloscopic pattern. These feeble streamers appeared to be able to cross the gap for a considerable range in potentials before a spark resulted. At 20000 volts the streamers became strong enough to yield a spark. It is interesting to note that the γ_i conditioned discharge from the negative point led to an arc at very much lower potentials than the streamer spark in N_2, while in H_2 the γ_p discharge from the highly stressed point did not yield nearly the instability and low field spark breakdown that the γ_i in N_2 caused in this geometry.

In argon with positive point, a streamer spark crossed a gap of 4.6 cm. at 3500 volts without any corona. This is the same potential at which the much overvolted negative point yielded an arc-like breakdown. However, the negative point did show a discharge with lowered offset potential once the arc formed. It appears that, irrespective of point polarity, when the potential reaches 3500 volts at 760 mm. a discharge appears, the temporal growth of which has not yet been studied. The photon γ_p is high and at a highly stressed cathode γ_i should appear. A predischarge starts, builds up metastable atom concentrations and in some fashion ends in a streamer-like or filamentary breakdown to an arc. Once the arc operates, the metastables created will carry the current at reduced potential.

ϑ) Point-to-plane, free electron gases with traces of impurity [*68*]. The addition of O_2 to H_2 from 0.1% up at once yields pre-onset corona streamers. Here the streamers increased in length with potential and current, ending in a spark at 8000 volts, i.e., below that for pure H_2. Threshold occurred at 5000 volts instead of 3500 volts. With above 1% of O_2, the threshold was lowered to 4000 volts and 1 μa, and the burst pulse corona replaced the streamers. Breakdown streamers occurred at 15000 volts and a spark at 20000 volts. It seems that streamers increased in intensity as the O_2 content was raised to 1%, but the space charge after the first streamer choked off further streamers and the lateral spread of the burst pulse set in. This indicates that in H_2 the 1% O_2, or else the H_2O formed, produces exceedingly effective photons with high absorption coefficient of the same order as for pure O_2. Reduced currents in H_2 contaminated by O_2 probably arise from lowered Townsend coefficients, interruption of the discharge, as well as reduced drift velocities of ions.

Addition of 0.2% O_2 to N_2 resulted in phenomena closely paralleling those in H_2. A streamer corona at 4500 volts and 1 μa replaced the glow discharge of pure N_2 and lowered threshold 300 volts. Pre-breakdown streamers appeared at 9300 volts. With 1% O_2, burst pulses appeared at 4800 volts and 1 μa with pre-breakdown streamers at 12500 volts, and streamer spark at 20000 volts.

The addition of 0.1% of O_2 to argon resulted in pre-onset streamers at 3800 volts, with no burst pulses. Breakdown streamers occurred at 4100 volts with a spark at 4100 volts. The addition of 0.4% O_2 raised the threshold of a streamer and small burst pulse corona to 4500 volts at 0.6 μa with breakdown streamers and a spark at 7000 volts. With 1% O_2 the burst pulse corona set in at 4000 volts at 0.6 μa and no streamers. Breakdown streamers led to a spark at 15000 volts.

It can be concluded that depending on the initial mechanism in the pure gas the addition of 0.1% O_2 leads apparently to streamers with the geometry used, at either a lowered or a slightly elevated potential. In general 1% O_2 leads to burst pulse corona, some times at decreased potentials. In general the burst pulse corona threshold is some two hundred volts below the pre-onset streamer potential. Thus it would be expected that at the optimum percentage of O_2 either for streamers, or for the beginning of burst pulses, the potential would be lower for the burst pulses. Unfortunately, Weissler's data were not taken at the optimum concentrations to bear this out.

Spark breakdown *in all cases was caused by positive breakdown streamers* starting from the anode and occurred at markedly lower values for positive point than for negative point in the relatively short gaps when photon γ_p was the secondary mechanism, but for γ_i from a highly stressed cathode the inherent cathode instability leads to a lower threshold from negative point—*provided* build—up time is adequate. For argon with its multiple mechanisms and its rapid accumulation of metastables as well as its very low threshold values of X/p for excitation and ionization, it is clear that positive and negative point lead to arcs at the same threshold with mechanism undefined.

34. The negative point-to-plane in air, Trichel pulse corona. The study of these discharges in room air and in carefully dried air is quite extensive. Beginning with their discovery by O'Day and their first investigation by Trichel [*76*] and later by HUDSON, the writer and KIP [*69*], further studies have been carried out by ENGLISH [*76*] using various cathodes including water points and by ENGLISH and MOORE [*76*] on impulse corona. Later studies by BANDEL in 1951 on standard point-to-plane gaps over extensive ranges in dry and moist air also succeeded in clarifying the anomalies in photoelectric triggering observed by earlier workers [*76*]. Negative corona from ice points was discovered by BANDEL [*76*], although a water point does not give one. The final clarification came through the time analysis of AMIN [*19*] with fast oscilloscope. The variation of thresholds with gap geometry and pressure are not of too much significance and are discussed in various places, especially by BANDEL [*76*]. The mechanism is primarily of interest. The effect of the O_2 in air on the values of γ, the nature of the initial discharge mechanism and clean up processes make the negative point thresholds peculiarly intensitive to cathode material. Ultraviolet triggering is helpful in initiating discharge, but care must be used to keep i_0 *adequately low*, otherwise the field intensified heavy i_0 alters thresholds by space charge fouling of the gap as shown by BANDEL in 1951 [*76*]. H_2O has no γ, so that water points show no negative corona. Ice which at 1000 Å yields a photoelectric current of the same magnitude as CuO does yield a negative corona of as high a current as the feeble conductivity permits. ENGLISH even suspects negative corona can occur from highly stressed glass points.

As potential is raised on a point of some 0.25 cm. diameter or less and gaps of 4 cm. length, currents remain around 10^{-13} to 10^{-11} ampere and rise abruptly to from about 10^{-9} to 10^{-7} ampere at threshold. At this point, pulses quite random in time appear at irregular intervals. As potential increases and current rises towards 10^{-7} ampere the pulses become so remarkably regular in time that they resemble relaxation oscillator pulses. The frequency will increase up to 16^6 per second before pulsations cease, and the discharge alters to the condition near spark breakdown by a streamer process, first indicated by KIP [*76*] and confirmed by BANDEL and by GREENWOOD [*77*]. As indicated for pure gases, the discharge has the characteristic visual features of a glow discharge which must go on and off whith each pulse. The dimensions of the various features, such as CROOKE's dark space, negative glow, FARADAY dark space and column have, been measured by ENGLISH and have been analyzed as to field distribution, as well as the mechanism of the compression of the glow and flaring of the column by the writer [*72*] and confirmed in part by WEISSLER and SCHINDLER [*71*]. Further confirmation of these properties cam from studies of FINKELNBURG [*72*] on arcs. Calculations of the values of the avalanche size and development have also been made by the writer [*72*]. ENGLISH, with microscillograph, in 1950 established that the duration of the pulse at 760 mm. near threshold was of the order of 4×10^{-8} sec. or less. How a current can rise from a few electrons per microsecond to a density of 100 ampere/cm.^{2} in a spot some 0.02 cm. in diameter and be choked by negative ion space charge, remained an enigma until the cross sections for dissociative attachment to O_2 were established by GEBALLE and HARRISON in 1952 [*37*]. With these data, the writer in 1952 was able to account for what occurred [*76*]. The discharge at a 0.19 cm. diameter point initiates with an avalanche of 6×10^4 electrons and a photon γ_p of the order of 5×10^{-5} and builds up to the maximum current in 10 ionizing sequences of 2×10^{-8} sec. to the 3.6×10^9 electrons in the pulse. Some 60% of the electrons attach dissociatively with 3 volts energy beyond the negative glow at a distance between 0.06 and 0.12 cm. from the point. The field distortion created by these slow ions is sufficient to reduce the cathode field so that ionization ceases. The discharge is over, until the negative space charge moves far enough away so that one of the last positive ions left in the space charge near the cathode reaches the cathode and liberates an electron and starts a new pulse. Clearing time will be long at threshold, but as potential increases it will be short and pulses will follow in regular sequence with assured ion triggering.

By observing the pulse with the electrical signal and with photomultiplier, AMIN [*19*] was able to separate the electrical and luminous events. The electrical and luminous pulses were quite parallel, except that electrical pulses indicated a current due to movement of the ions in the high field region, superposed on ionizing events. The photon pulse for a 0.2 cm. diameter point and 4 cm. gap at 760 mm. pressure had a duration of about 10^{-8} sec. for *its bright* phase. This of course omits the build-up time which is not luminous and is not inconsistent with the rough calculations above made. Lowering the pressure, or raising the potential, which spreads out the negative ion space charge in length and increases the distances to be traversed more than it does electron drift velocities, increase the build-up time as it requires more ions to choke it off. The oscillograms show that, after the initial rapid build-up by γ_p, the positive ions reach the cathode and produce a continuation of electron liberation by the γ_i process until choking occurs. The positive ions, however, produce a longer, narrower negative ion space charge because the diffusion is less than for γ_p. Decrease in pressure thus produces a reduction in the initial photoelectric peak and an extension of the following plateau due to γ_i as more ions are needed to throttle the discharge from greater distances. Clearing time between pulses will decrease below those at higher pressures as negative drift velocities have increased. Repeat rates are increased for equivalent potentials. Increasing potential increases the frequency of pulses in proportion to the increase in current indicating a constant number of electrons per pulse. As the potential increases much further, the current grows more slowly than the repeat rate and a gradual decreases in charge per pulse appears at higher potential. Here again, as in positive wire corona, in the region where TOWNSEND'S equation (33.2) holds, increase in potential does not alter the ionizing sheath. How this occurs is indicated by the writer and COLLI and FACCHINI for the positive corona [*46*]. The increased potential just clears the space charge more rapidly. This leads to the notorious OHM'S law regime near corona onset. Further increase in potential begins to alter the active region, extending the space charge further into the gap. Fields near the cathode increase, positive ions reach the cathode and extend the discharge still further. The photon spike decreases, and ion plateau and pulse duration increases. The narrow confinement of the discharge aids quenching. and pulses cease with less total ions. Increase of frequency with increase in potential is expected since it is the time to remove the ions from near the cathode and not quenching time that affects frequency. Increase in point radius extends the strong field further into the gap. Thus, more ions are needed to choke the discharge, there are more ions per pulse and a decrease in frequency for constant potential and pressure.

The most interesting observation had to do with the "conditioning" of the point. ENGLISH, BANDEL [*76*] and MILLER [*45*] had observed that as the currents were run for same time at higher potentials, or extending the potentials near regions of pulseless discharge, the character of the pulses changed. The "conditioned" point showed a curve requiring *higher potential for the same frequency* of pulses and, what is worse, a much *lower current for the same frequency*. The change was in the point surface which, however, was not altered in physical dimensions. It recovered on standing 24 hours or more. The gas was not involved. Since all previous conditioning studies indicated *cleaning of the surface* and increased γ_i and/or γ_p this action of the "*conditioned*" point giving *lower charge per pulse* represented a paradox. Oscilloscopic analysis at once clarified the situation.

The "conditioned" point gave pulses with an *initial rise* 50% *greater* and a much *higher* γ_p *emission* leading to a *very heavy initial* current, which produced a compact and efficient negative space charge throttled by fewer ions in less time. Thus *increase* of γ_p by its efficiency, creates a pulse wfth *less* ions per pulse. However, the efficiently choked smaller pulse is placed nearer the point and thus requires more time to clear. This indicates why higher potentials are required to give the same frequency by 16%. It also shows how urgent it is that careful study be made of changes such as indicated, for while in fact the "conditioning" is due to *increased* γ_p action, owing to the clean up, it *resulted* in *lowered currents* and *apparently less yield*, when it was *anticipated* that such action *would increase emission and current.*

Comparison of the transit of the ions in burst pulses and TRICHEL pulses [*19*] leads to the conclusion that most of the negative ions are O^- ions formed near the point, but that some O_2^- ions formed in the low field regions add to the TRICHEL pulse and cause diffusion.

For the breakdown of the negative TRICHEL pulse corona to sparks, the investigations of BANDEL may be regarded [*76*]. Here it was observed that as potential, TRICHEL pulse frequency and current increased, the curves following the characteristic shape of the TOWNSEND equation began to rise more steeply. In this region often the spot began to wander over the surface. With larger points, multiple glows, or spots, appeared as with the coaxial cylinders. That is, one spot can carry just so much current and further current increase will occur by new spots if there is room enough. The wandering indicates instability that precedes multiple spots. Larger electrodes and higher pressures favor multiple spots. Above the region of instability, or possible at higher currents, the "conditioning" discussed occurred *causing a drop in current for constant potential* and then a new increase in current along the same sorts of curve only at higher potentials. In this region of current densities the complete interruption of the current between pulses ceased, and the *space charge was no longer able completely to choke off the discharge,* but caused it to oscillate about a mean current. Visual observation showed that, when this occurred, the prominent fan-like positive column began to fade. Within a few volts more, the oscillations ceased and the discharge showed a CROOKES dark space, negative glow and a faintly luminous column over the whole gap. Generally after, but sometimes before the fan disappeared, the streak looking like an incipient streamer reported by KIP [*76*], and ENGLISH and MOORE [*76*] with pulsed negative point corona, and by GREENWOOD [*77*], appeared in BANDEL'S work. This spike has been considered an incipient anode directed streamer occurring in the highly stressed region. As potentials still further increased, the streak began to move back to the cathode and near this potential the filamentary spark appeared. Whether in fact the streak is an anode directed negative streamer that leads to a spark cannot be stated at this time. In any event, the sparks observed always partook the characteristics of streamer sparks.

35. The positive point-to-plane burst pulse corona in air. As the positive potential is raised on the 0.25 mm. diameter point, the field intensified current of some 10^{-13} ampere from ionization in the gap rises fairly abruptly at some potential to in the neighborhood of 10^{-11} ampere at 7.30 kv [*17*]. At this point randomly spaced single pulses at a few milliseconds interval are observed on the oscilloscope. With fast sweep individual pulses having a duration of less than about 1.2×10^{-7} sec. can be observed [*51*], both with oscilloscope and electrically, as a current pulse having the same shape and duration as the luminous pulse, except for the effect of ion movement in the high field region. These *primary* pulses have from 10^7 electrons and ions in them at threshold to about 2×10^7 near onset of steady corona. The appearance of these pulses is associated with the uniform burst pulse glow, spread over the point surface as a tightly adhering film. Increasing the potential slightly increased the frequency of these pulses and the current correspondingly. At 7.32 kv the initial pulse was followed by a *secondary pulse* with about 10^6 electrons and ions. The secondary pulse was of shorter duration and followed some 10 microseconds after the first pulse. As potentials still further increased the intervals between primary pulses became shorter and more secondaries followed the primary before the pulses ceased. Just before the onset of streamers at 7.7 kv the secondaries follow primaries as close as 2, though averaging 4, microseconds apart and ranged up to 100 or more in a sequence. Single burst pulses composed of such sequences had a total duration of the order of 500 to 1000 μsec and comprised 10^8 ions. With intermediate sweep rate the collection of resolved pulses on fast sweep constituting single bursts took on the form of STEVER's [*17*] primary and secondary pulses in fast GEIGER counters, and with still slower sweep they were the burst pulses noted by TRICHEL and KIP [*49*]. Raising potential higher, the intervals between these bursts became shorter and bursts longer, streamers disappeared and eventually one burst continued indefinitely with occasional interruptions and mostly fluctuations at the *onset* of *steady corona.*

The burst pulses are typical GEIGER counter pulses in air on a small point surface. At the threshold indicated by equation (18.3) they begin and propagate over the surface according to equation (18.4). Once they have burned for some time, the 10^7 ions create a positive ion space charge which interrupts the discharge. After this clears slowly a new primary burst pulse starts. At higher potential, the primary does not quench so rapidly and, by the time it has quenched at its starting point, it is still burning at some other point. Thus, with higher sweeping field near the point, space charge of the first region quenched dissipates from that region shortly after, the last of the primary burst over the rest of the point ceases. After the "*dead time*" of 10 μsec.thus the secondary burst pulse appears in the cleared region, but is prevented from spreading far because space charge has not cleared from the whole point. Increase in potential thus increases sweep rates both near the point, (dead times shorter), leading to more secondaries with shorter intervals and also decreases clearing times for the space charge of the whole burst sequence from the gap. The length of the clearing times are of the order of the time for ions across 80% or more of the gap. Until clearing is effective the pre-onset streamers do not appear. They require potentials above the burst pulse threshold. As indicated by AMIN [*20*], these streamers initiate from a *primary* burst pulse. They may also trigger or be followed by a burst pulse. But once a burst pulse has developed, streamers are not able to develop. Either by initiating avalanches as soon as streamers clear, (KIP [*20*]), by using impulse potentials of such short duration that space charge fouling cannot occur, (ENGLISH and MOORE [*76*]), or using pure O_2 where burst pulses cannot spread [*10*], can regular pre-onset streamer corona be observed.

36. Breakdown in spark suppressing gases, the halogens, freon 1–2 air mixtures. Very little study has been made of these very reactive gases. There has been some work on the first coefficient and the effect of electron attachment on the threshold in uniform fields shown in Sect. 12. Some work has been done in inert gases with admixtures of the halogens in coaxial geometry in connection with GEIGER counters which is of little help [*79*]. Perhaps the most interresting study was made by WEISSLER and MOHR [*78*] with point-to-plane geometry in freon 1–2—air mixtures.

The very strong chemical reactivity of the halogens makes these gases difficult to work with through their action on most metal electrodes. Tin and Pt are rather resistant and have been used. Recently CLARK [79] has made SnCl coatings with resistances of the order of 100 Ω and more on glass tube walls of cylindrical geometry GEIGER counters. These have indefinite life and apparently a very low γ although the matter is being currently investigated in the writer's laboratory. Obviously, formation of halides on the electrodes will reduce materially and may alter γ. In addition fluorine and HF will attack glass walls forming SiF_4 and BF_3 that are volatile which can hardly be removed short of dismantling and proper chemical cleansing.

Aside from this action, the halogen molecular gases dissociatively attach electrons at thermal energies to form very stable negative atomic ions of heat of formation between 3 and 4 volts. Cross sections are large even at considerable dilution. These will not liberate their electrons in fields below an X/p ratio of 300 or more. These gases thus effectively remove electrons. The complex halogen-carbon and sulfur compounds such as freon 1–2, CCl_2F_5, CF_3SF_2, CSF_8 etc. will at relatively low energies of electron impact dissociatively attach electrons to create F^- and Cl^- ions and perhaps form $CCl_2F_2^-$ ions and will also dissociate, to yield atoms and molecules of Cl and F. Electron attachment reduces electron avalanches relative to non-attachment and in addition creates slowly moving atomic ions which can either produce negative space charges in the anode regions or act to neutralize or reduce positive ion space charges.

WEISSLER and MOHR [78] observed with Pt point cathodes in point-to-plane geometry that with from 0.1% freon in air to 100% freon there was no marked effect of the freon on the *initial* thresholds. A TRICHEL pulse corona appeared as in dry air with improper triggering in bursts at about 6000 volts and gave way to a continuous self-sustaining corona after passing through an intermitent regime whose threshold of 12500 volts in air was the higher the more freon there was present up to 15000 volts. The currents for the same potential were reduced 20% for 0.1% freon and to 0.14 by pure freon relative to air. In this, no doubt, the decreasing drift velocities of the ions as well as reduced γ played a rôle. Most important was the fact that after the discharge had run for some time, so that accumulations of halogen dissociation products appeared, the breakdown thresholds were *increased* materially, e.g., from 6000 to 9000 volts. If the gas stood 30 min. the lower thresholds reappeared. This appears to be a gas effect and not primarily a contamination of the cathode, and the Pt has appeared rather immune to attack by the halogens.

For positive point corona the thresholds were not altered initially below 1% of freon. Above this, the thresholds were increased even near the start. They were much further increased if the discharge continued for a while, but recovered on standing. This action is clearly not a cathode change, but a change in the first coefficient by halogen gases since with positive point the discharge is in the gas, as will be noted. More important was the discovery that even with traces of freon and/or the contamination products from a previous filling of the chamber, the corona threshold was *not* one of burst pulses but of *streamers* and that, when freon was present, the streamers persisted right into the region of steady corona. The optimum concentration of freon for streamers was between 1 and 3%. Increasing the amount of freon beyond this decreased the length of streamers and increased burst pulse corona.

The action of the freon here was no doubt twofold. It introduced a readily photoionizable gas, e.g., the halogens etc., if not freon itself. It further produced stable negative ions throughout the gap such that the positive space charge which inhibits streamers, as ENGLISH and MOORE and AMIN found, is in a large measure neutralized. It should also be noted that, when the negative ion space charge cloud segregates from the positive ion cloud and approaches the cathode, it acts to increase the field and enhance the discharge. When, however, too much freon is added, negative ion forming molecules increase negative space charges to a point where they cease to be useful and mobilities and first coefficients decrease. The discharge degenerates and sparking threshold is raised.

37. Positive point-to-plane, breakdown streamers in air. The properties of streamers have been outlined in Sect. 18, δ. However, at this writing, the studies of AMIN have been extended, at first by AMIN and HUDSON and later by HUDSON [*54*], to the breakdown streamers from positive point corona in air. Initially, streamers in the breakdown region but somewhat below spark breakdown were studied as they crossed the gap by photomultiplier slit placed at various distances from the point. Triggering came from the electrical pulse. The streamers crossed most of the gap. What appeared of interest was that, as the tip advanced across the gap, it separated into two streamer processes. The tip itself with high speed advanced and left behind a much slower luminous process which had considerable duration, or temporal length, high intensity and much slower velocity than the tip. It was postulated by the writer that with long traverse the tip advanced rapidly across the gap, say at 4×10^7 cm./sec. and that the slower radial expansion of the channel and its ionization could not advance as rapidly as the tip, owing to relatively lower radial fields and thus, in distances of 0,5 cm. or so, lagged behind the fast tip at about 10^7 cm./sec.

In considering what occurs, it must be remembered that first the point is not riding on a rising potential wave, as is the case with the streamers in MEEK and SAXE's impulse breakdown. On the contrary, the tip advances progressively into a low field region saturated with space charge from previous streamer channels, thus reducing radial fields causing the expansion of the channel the further it advances.

The secondary expansion will thus attenuate before crossing. For this reason, the prebreakdown streamers appear over some range of potential and even in the breakdown region several abortive streamers may occur before one succeeds in giving a spark.

HUDSON's studies in room air, at 760 mm. pressure at 20° C (controls showed humidity unimportant), extend from needle points or hemispherically capped cylinders of diameter ranging from 0.02 to 1 cm. and from there on to spheres of 3 to 10 cm. diameter opposite large plane cathodes. Gap lengths ranged from 0.8 to 3 or more cm. With smaller points sparking was preceded and accompanied by corona, while for 1.0 cm. diameter points upward, low order TOWNSEND discharges probably preceded breakdown. To reduce electrical noise, the anode tip was scanned by a photomultiplier cell through a slit while a second photomultiplier with slit recorded the arrival of the luminous streamer at various points on a 517 Tektronix synchroscope, the sweep of which was triggered by the first photomultiplier. The slits varying from 0.3 to 0.1 mm. high and from 2 to 0.1 cm. wide scanned the gap at various anode distances as well as horizontally. With longer gaps and finer points, preonset streamers could be observed and on one occasion, such streamers were observed with the 1.2 cm. diameter sphere and a longer gap.

In *all* sparks observed, which for short gaps were axial and straight, but for longer gaps were straight for some 5 mm. or more from the anode and were later crooked, the spark was initiated by a streamer sequence proceeding *from the anode tip* and crossing to the cathode. When the streamer channel had become sufficiently conducting and arrived at the cathode, there followed the brilliant return stroke characterizing the spark. For longer space charge fouled gaps from smaller points, the breakdown was preceded by several abortive streamer sequences culminating in a successful sequence leading to the spark. These abortive sequences were separated by some 100 μ secs. corresponding roughly to the space charge clearing time. In extreme cases, there were perhaps ten such sequences. As gaps approached to 1 cm. length, only *one* sequence led to the spark. These sequences began as a single luminous pulse near the anode, which after 3 to 5 mm. separated into a bright fast streamer tip of short length and duration called the *primary* and a slower pulse of high luminosity, longer duration and lower speed, the form of which altered and attenuated as it crossed the gap, called the *secondary*. In longer gaps, the secondary attenuated so that it could not be observed much beyond 1.5 cm. In the shorter gaps, it could be followed to the cathode, on its arrival being followed instantaneously by the rise of luminosity of the bright return stroke. In the longer gaps, between the time of disappearance of the luminosity of the secondary in midgap until the rise of the return stroke, there was a delay of from 2 μ sec. down, corresponding to the time for the secondary to reach the cathode at its initially measured speed. Speeds of the primaries ranged from 4×10^7 to 2×10^8 cm./sec., and for the secondary from 10^7 to 10^6 cm./sec. The speed of the return stroke was too high to measure and its direction of motion could not certainly be established, though in some cases, it appeared to be from cathode to anode. Secondary and primary streamers were less than 0.5 mm. in diameter. In longer gaps and for smaller points, the preonset streamers indicated that after the first 0.8 cm. of advance along the field axis, the primaries and secondaries branched and spread considerably. The primary tips could visually and photographically be followed across the gap producing flashes on impact at the cathode. Secondaries could be followed along the branches for some 0.5 cm. before their lower and spatially diffused visibility rendered them undetectable. With more uniform fields and shorter gaps, time resolution was inadequate to observe events in detail, but observations left little doubt about the presence of primary and secondary sequences *from the anode* followed by the return stroke except for the shortest gaps.

The branching with longer gaps reconcile these streamer studies with recent observations of ALLIBONE and those of MEEK and SAXE with broad radial luminosities and low tip velocities if it be assumed that with their lower time resolution and relative detector insensitivity, they were observing the secondaries of the branched streamers crossing the gap which with impulse breakdown should be visible across the gap.

VIII. Breakdown in non-uniform but symmetrical fields at steady potentials.

38. Enough has been learned about breakdown from highly stressed anodes and cathodes to permit general conclusions as to the cause of breakdown in symmetrical fields. Such fields may be made by two opposing points, by two opposing planes of limited area, or in symmetrical sphere gaps where the distance is larger than the sphere diameter. With such sphere gaps, until the distance becomes materially greater than the electrode diameter, the breakdown will partake more of the nature of a uniform field breakdown.

What will happen first is that, depending on the gas and the field geometry, one of the two electrodes, the cathode, or the anode, will initiate a localized discharge, e.g., a negative pointlike discharge from the cathode for pure free electron gases, or a burst pulse or streamer corona from the highly stressed anode as in O_2 or air. If the midgap field region is very low and distances are great, it is very likely that *discharges* will *start at different thresholds from both electrodes.* Under such circumstances, and otherwise quite generally with long gaps, the potentials of surrounding objects and the fields to the enclosing walls must be carefully regarded. Omitting any such disturbing influence, it is clear that the independent and isolated electrodes will undergo local breakdowns at probably different potentials, each *projecting the appropriate space charges into the surrounding gas.* General spark breakdown will then occur through the media of the localized discharges, (e.g., anode streamers), the space charge fields, or a midgap spark akin to a lightning stroke between oppositely charged ion space charge clouds.

This is in fact precisely what is observed. It is for this reason that the point-to-point gap is such a useless means for control or measurement of breakdown conditions, for it is affected by so many more extraneous variables owing to ill defined fields. Actually, it is impossible a priori to predict what will happen although, from what has gone before, suitable conjectures can be made in certain cases where gas type, field distortion by the walls and gas and geometry are clearly defined.

For the case sphere gaps with distances between electrodes not too great in air, some general conclusions may be drawn from MEEK's equation as earlier indicated [*60*]. This will not apply for very pure, or inert gases.

IX. Impulse breakdown in non-uniform fields.

39. Under these circumstances, a good deal is kniwn from the many engineering studies, but especially from the studies of ALLIBONE and MEEK [*55*] and of MEEK and SAXE [*55*] in air using moving film cameras or photomultiplier and oscilloscopes. Impulse potentials are assumed to rise so rapidly, and to lie so far above the steady state breakdown potentials, that the long formative time lag phenomena, such as pre-threshold TOWNSEND discharges and low order corona thresholds, cannot remain active very long. It appears also that for understandable results impulse breakdown should occur with a rising potential characteristic. It is then clear that conditions particularly favor the highly stressed anode. Thus, as ALLIBONE and MEEK, and MEEK and SAXE have so beautifully demonstrated, with highly stressed anode, *only* the breakdown streamer forms rapidly, advances across the low field region with speeds of the order of 2×10^6 to 5×10^7 cm./sec. On reaching the cathode, an ionizing wave of potential readjustment sweeps up the streamer channel from cathode to anode at speeds in excess of 3×10^8 cm./sec. and leaves a conducting channel behind.

In the case of the highly stressed cathode, even though the anode be a large conducting plane, the breakdown potentials are nearly twice those for the similar configuration with highly stressed anode. In this event, some discharges start from the overstressed cathode appearing as negative streamers, but usually poorly directed relative to field axis [*55*]. Each successive breakdown seems to reactivate and prolong the initial path. Eventually, the cathode streamer advances far enough so as to create a sufficiently intense image force field at the anode plane to initiate a *positive streamer.* Once the positive streamer starts, it progresses more rapidly than the negative streamer. The spark is consummated by junction of the two streamers within $\frac{1}{3}$ the gap length or less from the cathode, the highly ionizing potential waves sweeping toward anode and cathode from the junction point at high speed [*55*].

The fields created at the anode plane with impulse breakdown from highly stressed cathodes have been calculated both by MEEK [*80*] and more recently by LIAO [*80*]. They are certainly adequate to cause the positive streamers to advance.

With point to point impulse fields, with impulse fields at reduced pressures and perhaps in some relatively impure inert gases the observations are incomplete and diffusion and other effects enter certainly giving wierd appearances which need further study. In fact, there remains much to be done in the study of impulse breakdown. The study so far has been largely carried out in engineering laboratories where the high potentials and other facilities are available, but never under such circumstances and with such understanding of the basic physical processes at work that proper variables can be controlled and some understanding of the processes can be gained. Excepted in this statement are obviously the excellent studies of ALLIBONE and MEEK, and of MEEK and SAXE.

X. Sparks in very short gaps, at high pressures and in vacuum.

40. It is somewhat too early to attempt to explain either vacuum sparks, sparks for very short gaps and at high densities or in vacuum. Recently BOYLE, KISLIUK and GERMER [*81*] at Bell laboratories have made real advance in understanding the short sparks at high pressures and in vacuum. In the cases studied, the *primary* and *principal process* consists of high density field emission currents in fields of 10^7 volts/cm. or more from points. If gas is present, the electrons from these currents ionize the gas molecules, field amplification M in avalanches being *very small*, e.g., of the order of 1.01 because of the low value of pd. It happens that the few positive ions created by the high density currents suffice to augment the field emission from the fine points and an equation between field emission increase, times the amplification factor causing it, lead to a product ΓM which sets the breakdown threshold with $\Gamma M = 1$. While a *formal* analogy to the conventional breakdown condition $\gamma M = 1$ can be devised by appropriate approximations, it must be noted that the *primary process is field emission* involved in the complex term Γ, while the *secondary process* is the field intensification M produced by the avalanche. For short gaps in vacuum, the primary agent is again field emission. Here, the dense electron current heats and vaporizes the anode material in a local spot which then furnishes the vapor to be ionized as in the case of the dense gas. Both such breakdown processes lead to power arcs via sparks but acting by different mechanisms, quite foreign to any gaseous electronic system of the conventional sort.

The true vacuum spark for longer gaps is not completely understood. The work of DYKE [*82*] and associates indicates that it starts as a field emission current from a fine point which reaches such intensity that it heats and vaporizes the small point. The intense field emission current can ionize this vapor from the cathode and in some fashion build this into a spark bridging the gap. How this occurs is today not known.

It may be remarked that the spark, which materializes through concentration of some glow discharge at the cathode to the point, where thermionic emission from the cathode of a refractory material furnishes the electron flow to maintain an arc, does qualify as a legitimate spark transition. However, here again, the *primary mechanism furnishing most of the electrons* in transition is the thermionic emission of the cathode in which concentration of the glow discharge into a spot by increase in current density and possibly magnetic effects is a factor. In this sense it does not fall within the purview of the subject of this article.

References.

[1] LOEB, L. B.: Basic Processes in Gaseous Electronics, chap. VII. Berkeley: University of California Press 1955.
[2] See any text on Electrical Conduction in Gases. Most recent complete data, L. B. LOEB, Basic Processes, loc. cit. chap. VIII.
[3] LOEB, L. B.: Basic Processes, loc. cit. chapt. VIII, Sect. 6, p. 823ff.
[4] WIJSMAN, R. A.: Phys. Rev. **75**, 833 (1949).
[5] For latest summary L. B. LOEB: Basic Processes, loc. cit., chap. IX, also Brit. J. Appl. Phys. **3**, 341 (1952).
[6] THEOBALD, J. K.: J. Appl. Phys. **24**, 123 (1953). — R. N. VARNEY to be published 1956 probably in Phys. Rev.
[7] HOLSTEIN, T., D. ALPERT and A. O. MCCOUBREY: Phys. Rev. **76**, 1257 (1949); **85**, 985 (1952). — HOLSTEIN, T.: Phys. Rev. **72**, 1212 (1947); **83**, 1159 (1951).
[8] COLLI, L., and U. FACCHINI: Phys. Rev. **96**, 1 (1954). — COLLI, L.: Phys. Rev. **95**, 892 (1954).
[9] LOEB, L. B.: Phys. Rev. **73**, 798 (1948); **97**, 275 (1955). — LOEB, L. B., and R. A. WIJSMAN: J. Appl. Phys. **19**, 797 (1948).
[10] HUBER, E.: Phys. Rev. **97**, 267 (1955).
[11] LOEB, L. B.: Basic Processes, loc. cit., chap. IX.
[12] BANDEL, H. W.: Phys. Rev. **95**, 1177 (1954), Figs. 2 and 5. — DAVIDSON, P. M.: Phys. Rev. **99**, 1072 (1955).
[13] DRUYVESTEYN, H. J.: Z. Physik **73**, 727 (1932). — ENGEL, A. v., and M. STEENBECK: Elektrische Gasentladungen, Bd. 2, S. 134ff. Berlin: Springer 1934.
[14] LOEB, L. B.: Phys. Rev. **76**, 250 (1949). — FISHER, L. H., and B. BEDERSON: Phys. Rev. **81**, 109 (1951). — KACHICKAS, G. A., and L. H. FISHER: Phys. Rev. **88**, 878 (1952); **91**, 775 (1953). — LOEB, L. B.: Phys. Rev. **81**, 287 (1951). — MENES, M., and L. H. FISHER: Phys. Rev. **86**, 134 (1952); **94**, 1 (1954).
[15] MENES, M.: 7 Conference on Gaseous Electronics, New York, October 14—16, 1954, paper.
[16] VARNEY, R. N., H. J. WHITE, L. B. LOEB and D. Q. POSIN: Phys. Rev. **48**, 818 (1935). — ENGEL, A. v., u. M. STEENBECK: Elektrische Gasentladungen, Bd. II, Ziff. 21, S. 48ff. Berlin: Springer 1934. — CROWE, R. R., J. K. BRAGG and V. G. THOMAS: Phys. Rev. **96**, 10 (1954).
[17] STEVER, H. G.: Phys. Rev. **61**, 38 (1942). — AMIN, M. R.: J. Appl. Phys. **25**, 210 (1954).
[18] LOEB, L. B.: J. Appl. Phys. **19**, 882 (1948). — Phys. Rev. **76**, 250 (1949). — MILLER, C. G., and L. B. LOEB: J. Appl. Phys. **22**, 614 (1951).
[19] LOEB, L. B.: Phys. Rev. **86**, 256 (1952). — AMIN, M. R.: J. Appl. Phys. **25**, 627 (1954).
[20] LOEB, L. B., and A. F. KIP: J. Appl. Phys. **10**, 142 (1939). — LOEB, L. B.: Fundamental Processes Electrical Discharge in Gases, loc. cit., p. 426ff. — MEEK, J. M.: Phys. Rev. **57**, 722 (1940). — AMIN, M. R.: J. Appl. Phys. **25**, 358 (1954).
[21] HOLST, G., and E. OOSTERHUIS: Phil. Mag. **46**, 1117 (1923).
[22] TOWNSEND, J. S.: Electrician, April 3, 1903. See also TOWNSEND, J. S.: Electricity in Gases, chaps. VIII, IX, X. Oxford: Clarendon Press 1915.
[23] AUER. P. L.: Phys. Rev. **98**, 320 (1955. More complete analysis see Y. MIYOSHI to appear Phys. Rev. **1956**.
[24] STEENBECK, M.: Wiss. Veröff. Siemens-Konz. **8**, 84 (1930); **9**, 42 (1930). — Arch. Elektrotechn. **26**, 306 (1932). — ENGEL, A. v., u. M. STEENBECK: Elektrische Gasentladung, loc. cit., Bd. II, Ziff. 71, S. 178ff. — BARTHOLOMEYCYZK, W.: Z. Physik **116**, 235 (1940). — SCHADE, R.: Z. Physik **104**, 487 (1937); **111**, 437 (1938). — GUGELBERG, H. L. v.: Helv. Phys. Acta **20**, 307 (1947). — ROGOWSKI, W.: Z. Physik **115**, 257 (1940).
[25] NEWTON, R. R.: Phys. Rev. **73**, 570 (1948). — MOLNAR, J. P.: Phys. Rev. **83**, 933, 940 (1951). — HORNBECK, J. A.: Phys. Rev. **80**, 297, (1950); **83**, 374 (1951); **84**, 615 (1951). — VARNEY, R. N.: Phys. Rev. **93**, 1156 (1954). — For more detail see LOEB, L. B.: Basic Processes, loc. cit., chap. IX. — DAVIDSON, P. M.: Phys. Rev. **99**, 1072 (1955).
[26] ENGEL, A. v., u. M. STEENBECK: Elektrische Gasentladungen, loc. cit., Bd. II, Ziff. 21, S. 51ff. und 184.
[27] WHITE, H. J.: Phys. Rev. **48**, 113 (1935). — ROGOWSKI, W., u. A. WALLRAFF: Z. Physik **97**, 755 (1935). — FUCKS, W., u. W. SEITZ: Z. Physik **103**, 1 (1936). — ROGOWSKI, W., u. W. FUCKS: Arch. Elektrotechn. **29**, 362 (1935). — FUCKS, W.: Z. Physik **98**, 66 (1936). — MEEK, J. M.: Proc. Phys. Soc. **52**, 547, 822 (1940).
[28] PASCHEN, F.: Wied. Ann. **37**, 69 (1889).
[29] TOWNSEND, J. S.: Electricity in Gases, loc. cit., p. 322. — SEELIGER, R.: Einführung in die Physik der Gasentladungen, S. 88—91. Leipzig: Johann Ambrosius Barth 1923. — THOMSON, J. J.: Cond. Electricity Through Gases, Vol. II, pp. 531, 598. Cambridge:

Cambridge Univ. Press 1933. — ENGEL V. A., u. M. STEENBECK: Elektrische Gasentladungen, Bd. II, S. 95—102. Berlin: Springer 1939. — DRUYVESTEYN, M. J., and F. M. PENNING: Rev. Mod. Phys. **12**, 128 (1940). — HOLM, R.: Wiss. Veröff. Siemens-Konz. **3**, 159 (1923). — It was analyzed anew and distributed as an O. N. R. Report by E. E. DODD in October 1950.

[30] ALPERT, D., and R. T. BAYARD: Rev. Sci. Instrum. **21**, 571 (1950). — ALPERT, D.: J. Appl. Phys. **24**, 810 (1953). — ALPERT, D., and R. S. BURITZ: J. Appl. Phys. **25**, 202 (1954).

[31] LOEB, L. B., and J. M. MEEK: Mechanism of the Electric Spark, pp. 8, 47, 81. Stanford: Univ. Press 1941.

[32] SKILLING, H. H., and W. C. BRENNER: Trans. Amer. Inst. Electr. Engrs. **60**, 112 (1941).— HOWELL, H. H.: Trans. Amer. Inst. Electr. Engrs. **58**, 193 (1939). — TRUMP, J. G., F. J. STAFFORD and R. W. CLOUD: Trans. Amer. Inst. Electr. Engrs. **60**, 132 (1941). — YOUNG, D. R.: J. Appl. Phys. **21**, 222 (1950).

[33] DRUYVESTEYN, M. J., and F. M. PENNING: Rev. Mod. Phys. **12**, 97 (1940). — ENGEL, A. v., u. M. STEENBECK: Elektrische Gasentladungen, loc. cit., Bd. I, S. 93 u. 100ff. — STOLETOW, J.: Phil. Mag. **1**, 198 (1901). — THOMSON, J. J.: Conduct. of Elect. Through Gases, loc. cit., 2nd Ed. 1906, p. 484ff.

[34] KRUITHOFF, A. A., u. F. M. PENNING: Physica, Haag **3**, 515 (1936); **4**, 430 (1937). See especially LOEB, L. B.: Basic Processes, loc. cit., chap. VIII, Sect. 6.

[35] HALE, D. H.: Phys. Rev. **54**, 241 (1938); **56**, 815 (1939). — EHRENKRANTZ, F.: Phys. Rev. **55**, 219 (1939).

[36] LOEB, L. B.: Rev. Mod. Phys. **8**, 276 (1936).

[37] GEBALLE, R., and M. A. HARRISON: Phys. Rev. **85**, 372 (1952); **91**, 1 (1953). — GEBALLE, R., and M. A. REEVES: Phys. Rev. **92**, 867 (1953).

[38] LAUE, M. v.: Ann. Phys. **76**, 261 (1925). — ZUBER, K.: Ann. Phys. **76**, 231 (1925). — BRAUNBEK, W.: Z. Physik **39**, 6 (1926); **107**, 180 (1937). — HERTZ, G.: Z. Physik **106**, 102 (1937). — WIJSMAN, R. J.: Phys. Rev. **75**, 833 (1949), also FURRY, W. H.: Phys. Rev. **52**, 569 (1937).

[39] STRIGEL, R.: Wiss. Veröff. Siemens-Konz. **11**, 52 (1932); **15**, 1 (1936). — TILLES, A.: Phys. Rev. **46**, 1051 (1934).

[40] PEDERSON, P. O.: Ann. Phys. **71**, 371 (1923). — BEAMS, J. W.: J. Franklin Inst. **206**, 809 (1928). — TOROK, J. J.: Trans. Amer. Inst. Electr. Engrs. **47**, 177 (1928); **48**, 46 (1930). — BUROWAY, O.: Arch. Elektrotechn. **16**, 40 (1926). — TAMM, R.: Arch. Elektrotechn. **20**, 235 (1928). — ROGOWSKI, W.: Arch Elektrotechn. **20**, 99 (1928). — LAWRENCE, E. O., and F. G. DUNNINGTON: Phys. Rev. **35**, 396 (1930). — DUNNINGTON, F. G.: Phys. Rev. **38**, 1535 (1931). — STREET, J. C., and J. W. BEAMS: Phys. Rev. **38**, 416 (1931). — HAMOS, L. v.: Ann. Phys. **7**, 857 (1930). — STRIGEL, R.: Wiss. Veröff. Siemens-Konz. **11**, 52 (1931); **12**, 53 (1932). — WHITE, H. J.: Phys. Rev. **46**, 99 (1934); **49**, 507 (1936). — WILSON, R. R.: Phys. Rev. **50**, 1082 (1936). — NEWMAN, M.: Phys. Rev. **52**, 657 (1937). — LOEB, L. B.: Rev. Mod. Phys. **20**, 151 (1948).

[41] LOEB, L. B.: Spark Breakdwon in Uniform Fields, Office of Naval Research Report, Naval Research Laboratory Press. Washington 1954.

[42] RAETHER, H.: Z. Physik **94**, 576 (1935). — Z. techn. Phys. **16**, 435 (1935). — RAETHER, H., u. E. FLEGLER: Z. Physik **99**, 635 (1936); **103**, 315 (1936), proposal of streamer theory. — RAETHER, H.: Phys. Z. **37**, 560 (1936). — Z. Physik **107**, 91 (1937), avalanche velocities measured. — Z. techn. Phys. **18**, 564 (1937). — RAETHER, H.: Z. Physik **112**, 464 (1939). streamers shown. — Arch. Elektrotechn. **34**, 49 (1940), theoretical analysis. — Z. Physik **117**, 375, 524 (1941) quantitative treatment of streamer advance. — Ergebn. exakt. Naturw. **1949**, 73—119, summary of theory. — Z. angew. Phys. **5**, 211 (1953). — Naturwiss. **17**, 400 (1954).

[43] LOEB, L. B., and J. M. MEEK: Mechanism of the Electric Spark. Stanford Press 1941. — J. Appl. Phys. **11**, 438, 459 (1940). — MEEK, J. M.: Phys. Rev. **57**, 722 (1940).

[44] LAUER, E. J.: J. Appl. Phys. **23**, 300 (1952).

[45] MILLER, C. G., and L. B. LOEB: J. Appl. Phys. **20**, 494; **22**, 615 (1951).

[46] LOEB, L. B.: Phys. Rev. **90**, 144 (1953). — COLLI, L., U. FACCHINI, E. GATTI and A. PERSANO: J. Appl. Phys. **25**, 429 (1954).

[47] CRAVATH, A. M.: Phys. Rev. **47**, 254 (1935). — DECHÈNE, C.: J. Phys. Radium **7**, 533 (1936).

[48] GREINER, E.: Z. Physik. **81**, 543 (1933). — TRICHEL, G. W.: Phys. Rev. **55**, 382 (1939). — KIP, A. F.: Phys. Rev. **54**, 139 (1938). — RAETHER, H.: Z. Physik **110**, 611 (1938). — COSTA, H.: Z. Physik **113**, 531 (1939). — PAETOW, H.: Z. Physik **111**, 770 (1939).

[49] LEIGH, W., and L. B. LOEB: Phys. Rev. **51**, 149A (1937). — TRICHEL, G. W.: Phys. Rev. **55**, 382 (1939). — KIP, A. F.: Phys. Rev. **54**, 139 (1938); **55**, 549 (1939). — LOEB, L. B., and A. F. KIP: J. Appl. Phys. **10**, 142 (1939).

[50] FLETCHER, R. C.: Phys. Rev. **76**, 1501 (1949).
[51] LOEB, L. B.: Phys. Rev. **73**, 798 (1948); **74**, 2 (1948). — WIJSMAN, R. A., and L. B. LOEB: J. Appl. Phys. **19**, 797 (1948).
[52] ALDER, F., E. BALDINGER, P. HUBER and F. METZGER: Helv. phys. Acta **20**, 73 (1947).
[53] LOEB, L. B.: Phys. Rev. **97**, 275 (1955).
[54] Work in the writer's laboratory started by M. R. AMIN and G. G. HUDSON, now in progress by HUDSON: 8th Gaseous Electronics Conf. Schenedady N. Y. Act. 20—22, 1955, paper A 5.
[55] ALLIBONE, T. E., and J. M. MEEK: Proc. Roy. Soc. Lond., Ser. A **166**, 97 (1938); **169**, 246 (1938). — MEEK, J. M., and R. F. SAXE: Nature, Lond. **162**, 263 (1948). Report. Brit. Allied Industries, L/T, 183, Fall 1948.
[56] Basic Processes of Gaseous Electronics, loc. cit., chap III, p. 271ff. for u_T and Sect. 6A, p. 292ff. for further discussion of ϱ. See also Report of L. B. LOEB on Spark Breakdown in Uniform Fields, loc. cit. pp. 74ff., 78ff.
[57] FISHER, L. H., and G. L. WEISSLER: Phys. Rev. **66**, 95 (1944). Calculations were carried out in connection with this work but details not published.
[58] SCHONLAND, B. F. J., and H. COLLENS: Proc. Roy. Soc. Lond., Ser. A **143**, 654 (1934). — MALAN, D. J.: Proc. Roy. Soc. Lond., Ser. A **152**, 595 (1935); **162**, 175 (1937); **164**, 132 (1938). — HODGES, D. B., and H. COLLENS: Proc. Roy. Soc. Lond., Ser. A **166**, 56 (1938). — MALAN, D. J., and SCHONLAND: Proc. Roy. Soc. Lond., Ser. A **191**, 485 (1947); **206**, 145 (1951); **209**, 158 (1951).
[59] LOEB, L. B.: Phys. Rev. **94**, 227 (1954).
[60] LOEB, L. B., and J. M. MEEK: Mechanism of the Electric Spark, loc. cit., chap. III, p. 107ff.
[61] LOEB, L. B.: Proc. Phys. Soc. Lond. **60**, 561 (1948). — MILLER, C. G., and L. B. LOEB: Phys. Rev. **73**, 84 (1948). — FISHER, L. H.: Phys. Rev. **65**, 153 (1944); **69**, 530 (1946); **72**, 423 (1947). — FISHER, L. H., and B. BEDERSON: Phys. Rev. **75**, 1615 (1949).
[62] GÄNGER, B.: Arch. Elektrotechn. **39**, 508 (1949).
[63] LAUER, E. J.: J. Appl. Phys. **23**, 300 (1952).
[64] SCHUMANN, W. O.: Durchbruchsfeldstärke von Gasen. Berlin: Springer 1923.
[65] DICKEY jr., F. R.: J. Appl. Phys. **23**, 1336 (1952).
[66] DUTTON, HAYDON and F. LLEWELLYN-JONES: Brit. J. Appl. Phys. **4**, 170 (1953).
[67] LOEB, L. B., J. H. PARKER, E. E. DODD and W. N. ENGLISH: Rev. Sci. Instrum. **21**, 42 (1950). — BENNETT, W. H.: Phys. Rev. **60**, 714 (1941). — KIP, A. F.: Phys. Rev. **54**, 139,(1938). — FITZSIMMONS, K. E.: Phys. Rev. **61**, 175 (1942).
[68] WEISSLER, G. L.: Phys. Rev. **63**, 96 (1943).
[69] LOEB, L. B., A. F. KIP, G. G. HUDSON and W. H. BENNETT: Phys. Rev. **60**, 714 (1941).
[70] BRADLEY, R. C.: Phys. Rev. **93**, 719 (1954). — WEHNER, G. K.: Phys. Rev. **93**, 633 (1954). — J. Appl. Phys. **25**, 270 (1954). — WEHNER, G. K., and G. MEDICUS: J. Appl. Phys. **23**, 1035 (1952); **25**, 698 (1954). — PENNING, F. M., and J. H. MOUBIS: Proc. Amst. Acad. **43**, 41 (1940). — TIMOSHENKO, G.: J. Appl. Phys. **12**, 69 (1941). — KEYWELL, F.: Phys. Rev. **87**, 160 (1952).
[71] WEISSLER, G. L., and M. SCHINDLER: J. Appl. Phys. **23**, 844 (1952).
[72] LOEB, L. B.: J. Appl. Phys. **19**, 882 (1948). — FINKELNBURG, W., and S. M. SEGAL: Phys. Rev. **83**, 582 (1951).
[73] TOWNSEND, J. S.: Phil. Mag. **28**, 83 (1914). — SVEN WERNER: Z. Physik **90**, 384 (1934); **92**, 705 (1934).
[74] PARKER jr., J. H.: Phys. Rev. **93**, 1148 (1954).
[75] BURCH, D. S. C., R. C. IRICK and R. GEBALLE: Phys. Rev. **93**, 650A (1954).
[76] TRICHEL, G. W.: Phys. Rev. **54**, 1078 (1938). — ENGLISH, W. N.: Phys. Rev. **74**, 170 (1948); **74**, 179 (1948). — LOEB, L. B.: J. Appl. Phys. **19**, 882 (1948). — ENGLISH, W. N., and D. B. MOORE: J. Appl. Phys. **20**, 370 (1949). — ENGLISH, W. N.: Phys. Rev. **77**, 850 (1950). — BANDEL, H. W.: Phys. Rev. **84**, 92 (1951). — J. Appl. Phys. **22**, 984 (1951). — LOEB, L. B.: Phys. Rev. **86**, 256 (1952).
[77] GREENWOOD, A.: Phys. Rev. **88**, 91 (1952).
[78] WEISSLER, G. L., and E. I. MOHR: Phys. Rev. **72**, 289, 294 (1947).
[79] PENNING, F. M.: Z. Physik **46**, 335 (1928). — LIEBSON, S. H., and H. FRIEDMAN: Rev. Sci. Instrum. **19**, 303 (1948). — LIEBSON, S. H.: Phys. Rev. **72**, 181 (1947). — Rev. Sci. Instrum. **20**, 483 (1949). — CLARK sen., L. B.: Rev. Sci. Instrum. **24**, 641 (1953).
[80] LOEB, L. B., and J. M. MEEK: Mechanism of the Electric Spark, loc. cit. chap. II, p. 92ff. — LIAO, T. W.: Phys. Rev. **93**, 650A (1954).
[81] BOYLE, W. S., P. KISLIUK and L. H. GERMER: Phys. Rev. **97**, 255 (1955).
[82] DYKE, W. P., J. K. TROLAN, E. E. MARTIN and J. P. BARBOUR: Phys. Rev. **91**, 1043, 1054 (1953).

Breakdown in Gases: Alternating and High-Frequency Fields[1].

By

Sanborn C. Brown.

With 33 Figures.

Table of symbols.

α first Townsend ionization coefficient (cm.$^{-1}$)
B magnetic field (gauss)
β attachment frequency
C collision source term
c velocity of light
D diffusion coefficient
E rms electric field
e $+1.6\times 10^{-19}$ coulomb
F distribution function
f energy dependent part of F
Γ electron flow density
η ionization coefficient (volt^{-1})
J electron current density
L parallel-plate separation
l mean free path
Λ characteristic diffusion length
λ wavelength
M mass of gas atom or molecule
m mass of electron
μ mobility
N total number of collisions during an electron lifetime
n electron density
ν collision frequency
ω radian frequency
P power and probability of collision
p pressure (mm. Hg)
q appearance frequency
R radius
S source function
u energy
v velocity
ζ a.c. ionization coefficient

Subscripts.

b cyclotron
c collision
e effective
i ionization
m momentum transfer
p peak
x excitation.

A. Diffusion controlled breakdown.

I. Behavior of the average electron.

1. General considerations. In an ultra-high-frequency gas discharge breakdown, the primary ionization from the electron motion is the only production phenomenon that controls the breakdown, and for this reason it is the simplest type to consider. If one calculates the maximum kinetic energy in the oscillatory motion of an electron at the minimum field intensities for breakdown experimentally determined, one finds that this energy corresponds to about 10^{-3} volt. It is therefore obvious that the energy of oscillation is insufficient to account for breakdown.

It is well known that a free electron in a vacuum under the action of an alternating field oscillates with its velocity 90° out of phase with the field in the steady state, and thus takes no power, on the average, from the applied field.

[1] This work was supported in part by the Signal Corps, the Office of Scientific Research (Air Research and Development Command), and the Office of Naval Research, of the USA.

The electron can gain energy from the field only by suffering collisions with the gas atoms, and it does so by having its ordered oscillatory motion changed to random motion on collision. The electron gains random energy, on the average, on each collision until it is able to make an inelastic collision with a gas atom. The electron continues to gain energy in the field, on the average, despite the fact that it may either move with or against the field. The energy absorbed is proportional to the square of the electric field and hence independent of its sign. The rate of gain of energy of the electron from the electric field E is $P = (-eE_p v_p)_{\text{real}}/2$. We may express the average drift velocity v in terms of the a.c. mobility in the following way. Writing NEWTON's second law for the motion of the electrons in the form

$$m\,(dv/dt) + (m\nu_m)\,v = -\,eE_p \exp(j\omega t), \tag{1.1}$$

the velocity may be determined as

$$v = -[e/(j\omega m + m\nu_m)]\,E_p \exp(j\omega t) \tag{1.2}$$

and ν_m is the collision frequency for momentum transfer. Thus the a.c. mobility takes the form $\mu = -e/(j\omega m + m\nu_m)$. We may write for the power absorbed by the electrons from the a.c. electric field:

$$P = ne\mu E^2_{\text{real}} = \frac{ne^2E^2}{m\nu_m}\left(\frac{\nu_m^2}{\nu_m^2+\omega^2}\right), \qquad E^2 = \frac{E_p^2}{2}. \tag{1.3}$$

This equation may be written in terms of an effective field E_e:

$$P = \frac{ne^2E_e^2}{m\nu_m}, \tag{1.4}$$

where E_e is the effective field that would produce the same energy transfer as a steady field and is given by

$$E_e^2 = E^2\,\frac{\nu_m^2}{\nu_m^2+\omega^2}. \tag{1.5}$$

The gas discharge breakdown occurs when the gain in electron density resulting from ionization of the gas becomes equal to the loss of electrons by diffusion, recombination, or attachment. When the loss is by diffusion, the problem becomes simple. We shall discuss this case first.

Diffusion occurs whenever a particle concentration gradient exists. The total flow of particles out of a volume of high concentration may be written from ordinary kinetic theory considerations as

$$\boldsymbol{\Gamma} = -\,D\nabla n, \tag{1.6}$$

where D is the diffusion coefficient for electrons, n is the electron density, and $\boldsymbol{\Gamma}$ is the electron flow density in electrons per second per unit area.

We shall develop the breakdown conditions for a region bounded by walls that absorb electrons. A radioactive source near the discharge tube provides a small amount of ionization S in the tube. A detailed study of the build-up of the discharge is obtained from considering the continuity equation for electrons

$$\partial n/\partial n = \nu_i n + S - \nabla\cdot\boldsymbol{\Gamma} \tag{1.7}$$

or

$$\partial n/\partial t = D\nabla^2 n + \nu_i n + S. \tag{1.8}$$

In Eq. (1.8) the term $D\nabla^2 n$ describes the loss of electrons by diffusion. The term $\nu_i n$ is the rate of gain of electrons by ionization; S is the rate at which electrons are produced by an external source. For the case of inifinite parallel plates

$$\partial n/\partial t = D(\partial^2 n/\partial x^2) + \nu_i n + S. \tag{1.9}$$

Assuming that the approach to breakdown is so slow in time that $\partial n/\partial t$ may be neglected,

$$-S = D(\partial^2 n/\partial x^2) + \nu_i n. \tag{1.10}$$

This is a characteristic value problem which may be solved under the conditions that S, D, and ν_i are uniform throughout the cavity and that the boundary condition on the electron density is zero at the walls. Rigorous boundary conditions require the concentration to be small at a boundary and to extrapolate to zero outside the boundary at a distance of the order of a mean free path. In the range of pressures to be considered, the mean free path is very small compared to cavity dimensions, and the condition of zero concentration on the cylinder walls is imposed. The solution of Eq. (1.10) tells us that the electron density just before breakdown, at any distance x from the median plane between parallel plates of separation L, may be written

$$n = (4S/\pi)\cos(\pi x/L)/[D(\pi/L)^2 - \nu_i]. \tag{1.11}$$

Breakdown can be defined by the condition that the electron density goes to infinity, which occurs when $\nu_i = D(\pi/L)^2$.

If we write $\nu_i/D = (\pi/L)^2 = 1/\Lambda^2$, for parallel plates, the quantity Λ is known as the characteristic diffusion length and is very useful on describing the shape of the gas container when discussing diffusion problems. One other example of a very useful boundary condition is the case of a cylinder of height L and radius R, with flat ends. This geometry leads to the relation that $1/\Lambda^2 = (\pi/L)^2 + (2.4/R)^2$, where the diffusion to the end plates is given by the first term on the right; the second term describes the diffusion to the cylindrical walls.

2. Ionization coefficients. Gas-discharge phenomena under the action of d.c. fields are often described in terms of TOWNSEND ionization coefficients. If one considers that electrons in a d.c. field create α new electrons in a path one centimeter long in the field direction, the increase of electrons, dn, produced by n electrons in a distance dx will be $dn = \alpha n\, dx$, and $n = n_0 e^{\alpha x}$, where n_0 is the initial electron concentration. The quantity α is called the first TOWNSEND coefficient. This first TOWNSEND coefficient may also be written as the ionization produced by an electron falling through a potential difference of one volt rather than traveling one centimeter. This coefficient is given the symbol η and is related to α by $\eta = \alpha/E$.

These TOWNSEND first ionization coefficients may be given in terms of an "ionization" collision frequency. Since α is the number of electrons produced by the collisions of the primary electrons traveling one centimeter, one can write $\alpha = \nu_i/v$, where ν_i is the frequency of ionization, and v is the average drift velocity of the electrons in the field. The average drift velocity $v = |\mu E|$, and one may write $\alpha = \nu_i/\mu E$ or

$$\eta = \nu_i/\mu E^2. \tag{2.1}$$

By analogy with the first TOWNSEND coefficient for d.c. ionization, where the electron loss is controlled by mobility, we may define a coefficient for high-frequency discharges, where the loss is by diffusion, as

$$\zeta = \nu_i/D E_e^2. \tag{2.2}$$

From our previous discussion of diffusion we saw that at breakdown $\nu_i/D = 1/\Lambda^2$. Thus we may measure the a.c. ionization coefficient by measuring the breakdown field in tubes of known size, since $\zeta = 1/\Lambda^2 E_e^2$.

There is a very close physical relation between the a.c. and the d.c. ionization coefficients. If one divides Eq. (2.1) by Eq. (2.2), there results the relation $\eta/\zeta = D/\mu$. TOWNSEND showed that the ratio of D/μ was a measure of the average energy of the electrons and determined this average energy as a function of E/p experimentally. Thus, in principle, one can determine η from ζ, or vice versa, from these TOWNSEND-like measurements. There is difficulty in carrying this out exactly, however, because the actual values depend on the manner in which the averaging of the energy is carried out. Since the electron energy-distribution functions are different for the a.c. and the d.c. cases, one might expect mathematical complications to arise. However, in the two cases[1,2] in which distribution function calculations have connected the d.c. and a.c. ionization coefficient, very satisfactory results have been obtained.

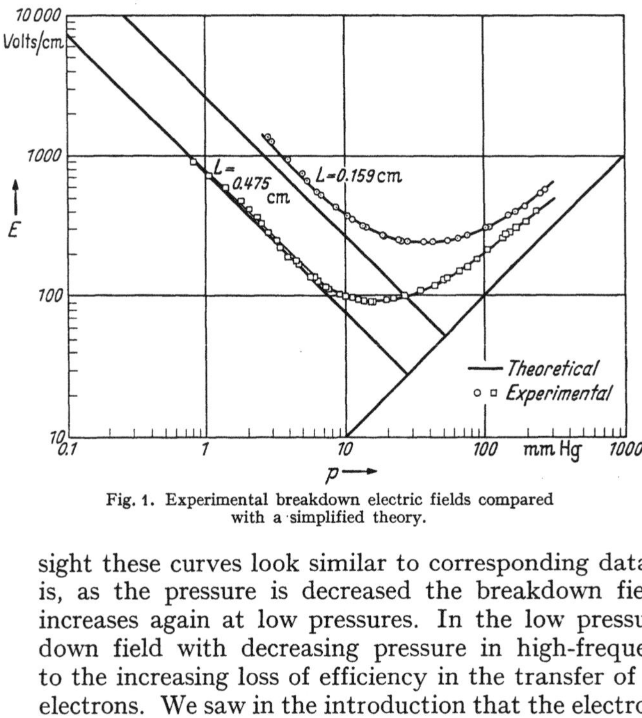

Fig. 1. Experimental breakdown electric fields compared with a simplified theory.

3. Breakdown fields. Typical of the behavior of the breakdown field at high frequency with changes in gas pressure are the curves shown in Fig. 1. At first sight these curves look similar to corresponding data taken with d.c. fields, that is, as the pressure is decreased the breakdown field first decreases and then increases again at low pressures. In the low pressure region, the rising breakdown field with decreasing pressure in high-frequency discharges corresponds to the increasing loss of efficiency in the transfer of energy from the field to the electrons. We saw in the introduction that the electron only gained energy insofar as it made collisions with the gas atoms and that between collisions it oscillated out of phase with the field and hence gained no energy. Thus, as the pressure is reduced, one must increase the field to make up for the loss of efficiency by just the factor of the effective field given in Eq. (1.5). In the high-pressure region, the reason for the rising breakdown field with increasing pressure in high-frequency discharges is the same as in the d.c. case. As the pressure increases, the electron mean free path decreases, and the energy gained per mean free path decreases as the square of the mean free path—as can be seen from Eq. (1.4). Since at these high pressures, most of the energy losses result from recoil losses in collision with the gas molecules, and the average electron energy is practically independent of the pressure, the energy gain per mean free path is proportional to the mean free path at constant E field. The field must increase inversely proportionally to the mean free path, or directly proportionally to the pressure. The minimum corresponds essentially to the point at which the frequency of

[1] S. KRASIK, D. ALPERT and A. O. MCCOUBREY: Phys. Rev. **76**, 722 (1949).
[2] L. J. VARNERIN and S. C. BROWN: Phys. Rev. **79**, 946 (1950).

collision between electrons and gas atoms is equal to the frequency of the applied *rf* field. Thus at low pressure, where the electron makes many oscillations per collision, its behavoir is governed by strictly a.c. considerations. At high pressure, where the electrons make many collisions per oscillation, their behavior is the same as in a d.c. discharge.

The remarkable feature of the breakdown curves, for those used to d.c. phenomena, is the fact that the greater the electrode spacing, the easier it becomes to cause a breakdown. This, of course, is a necessary result of the breakdown condition of the balance between energy gained from the field and electron loss by diffusion. As the electrode spacing becomes less, the diffusion loss becomes greater; therefore, the field must increase to make up for the increased loss.

Curves of gas-discharge breakdown as a function of pressure are often plotted in d.c. work as PASCHEN curves in which, for a particular gas, the breakdown voltage V is found to be a function of pd independent of the magnitude of the electrode spacing d. The same type of quantities may be introduced in the high-frequency case, where for the breakdown voltage we write $E\Lambda$, the product of the field and the diffusion length, and for pd we use $p\Lambda$. In the case of high-frequency phenomena we have one more variable than for the d.c. case, namely, the frequency, and this may be introduced as the variable $p\lambda$, where λ is the wavelength of the applied field. However, if we express the electric field in terms of its effective value E_e, we can take care of the frequency variations. The effective field is strictly correct only when ν_m is constant, but a similar equation would be valid for different variations of ν_m with energy. With $E_e\Lambda$ (the breakdown voltage) plotted as a function of $p\Lambda$, experimental results for ultra-high-frequency breakdown are shown in Fig. 2.

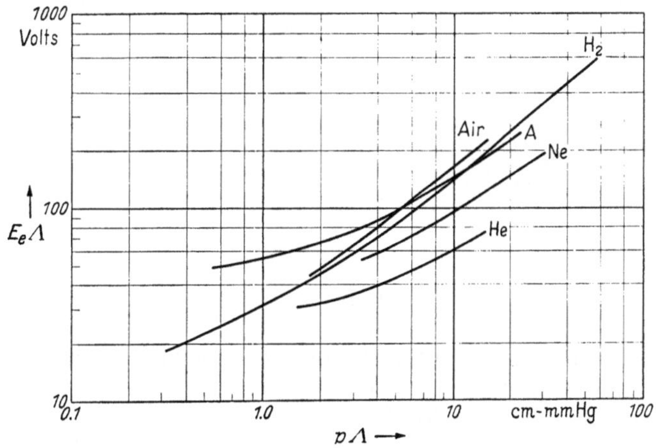

Fig. 2. PASCHEN curves for high-frequency breakdown.

The simplest gas to discuss theoretically is helium containing small admixtures of mercury vapor. This mixture, which for convenience we call Heg gas, has the advantage of acting as a gas of atoms without excitation levels. Helium has a metastable level at 19.8 volts, and transitions from this level to the ground state by radiation are forbidden. Since the metastable states have mean lives of the order of thousands of microseconds, practically every helium atom which reaches an energy of 19.8 volts will collide with a mercury atom and lose its energy by ionizing the mercury. Therefore, each inelastic collision will produce an ionization, and the effective ionization potential will be $u_i = 19.8$ volts. Furthermore, for Heg the collision frequency may be considered constant, having a value $\nu_m = 2.37 \times 10^9\, p$.

Although an accurate description of the breakdown measurements can be given theoretically only by taking into account the electron energy distribution, a physical picture of the mechanisms involved can be seen qualitatively in the following way. Let us consider a gas at high pressure. The power that goes into the electron from the electric field is dissipated in elastic collisions between the electrons and the gas molecules. For this case

$$\text{energy/collision} = \frac{e^2 E^2}{m \nu_m^2} = e \frac{2m}{M} \bar{u} \tag{3.1}$$

where $\bar{u}$ is the average energy. From this we may calculate the electric field to be

$$E = \left(\frac{m}{e} \frac{2m}{M} \bar{u}\right)^{\frac{1}{2}} \cdot \nu_m = 0.94\, p\, (\bar{u})^{\frac{1}{2}} \tag{3.2}$$

for the case of Heg gas in which nearly all the loss goes into non-ionizing collisions. Taking the average energy equal to one-third of the ionization potential, $E = 2.4p$. It can be seen that the high-pressure breakdown measurements of Fig. 1 tend to approach this line at high pressures.

In the low-pressure region, the electrons make many oscillations per collision and the breakdown field may be determined by equating the number of collisions to ionize to the number of collisions to diffuse out of the tube. Since we are discussing Heg gas, where all of the inelastic collisions are ionizing ones, let us consider that all of the input gpower goes into ionization, and write

$$\nu_i = \frac{P}{e u_i} = \frac{e E^2}{m u_i \nu_m} \left(\frac{\nu_m^2}{\nu_m^2 + \omega^2}\right). \tag{3.3}$$

The breakdown condition is $\nu_i = D/\Lambda^2$, and substituting the value of the diffusion coefficient $D = l\bar{v}/3$ we find $\nu_i = l\bar{v}/3\Lambda^2$. If we multiply the numerator and the denominator of this expression by the velocity, we obtain an expression in terms of the average energy which we may set equal to the previous expression for ν_i, in which $\omega^2 \gg \nu_m^2$, to obtain

$$\nu_i = \frac{\bar{v}^2}{3\Lambda^2 \nu_m} = \frac{e E^2 \nu_m}{u_i m \omega^2}. \tag{3.4}$$

If we solve this expression for the electric field, writing ω in terms of λ and $2\bar{u} = m\bar{v}^2/e$, we get

$$E = \frac{2\pi c}{\Lambda \lambda \nu_m} \left(\frac{2}{3} \bar{u} u_i\right)^{\frac{1}{2}}. \tag{3.5}$$

The ionization potential is 19.8 volts, and we assume a very low pressure, where all of the input power goes into ionization, and the average energy is of the order of the ionization potential. Calculating the electric field under these approximations leads to a relation $E = 1284/p\Lambda\,\lambda$, which agrees fairly well with the low-pressure measurements shown in Fig. 1.

4. Proper variables. If a gas contained in a vessel is placed in an alternating electric field, for a certain value of the electric field the gas will break down into an electrical discharge. This breakdown field may be expressed as

$$E_b = E(u_i, \Lambda, \lambda, p). \tag{4.1}$$

The term Λ has the units of length, and its appearance in explicit calculations also involves various known dimensionless ratios describing the shape of the vessel. It is customary to measure pressure in millimeters of mercury; the mean free path, which is inversely propertional to pressure, is measured in centimeters. A relation between pressure and mean free path is obtained by introducing the quantity, P_m, which is the number of collisions per unit length at a pressure of 1 mm. Hg. Thus, P_m may be regarded as having the units of reciprocal length, even though this is not its true dimension.

Treating the breakdown problem dimensionally, there are five variables with but two fundamental dimensions: volts and centimeters. This leads to three independent dimensionless variables between which there is a functional relation[1]. It is often convenient in physical problems to introduce variables that are not dimensionless but are, nevertheless, proper variables for dimensional analysis because the completely dimensionless variables contain one or more physically invariant quantities that need not be carried along in a practical discussion. There are a number of sets of such proper variables in a gas discharge problem that may be transformed into one another, and their relative convenience depends on the purpose for which they are to be used.

One very useful set of proper variables is

$$E\Lambda, \quad p\Lambda, \quad p\lambda. \tag{4.2}$$

The advantage of these variables lies in the fact that p, Λ, and λ are the experimentally independent parameters that determine the dependent variable E, the observed breakdown field.

Another set of proper variables that we shall use is obtained by dividing the first variable by the second and obtaining

$$E\Lambda, \quad E/p, \quad p\lambda. \tag{4.3}$$

This set has the particular advantage, in a discussion of breakdown phenomena, that we may define the a.c. ionization coefficient, $\zeta = 1/E^2\Lambda^2$, which is a function of E/p and $p\lambda$.

For the cases of a d.c. field, or when the pressure is high, so that in an a.c. field the electrons make many collisions per oscillation, the wavelength variation does not enter as a significant variable. Breakdown can then be described by the other two variables such as $E\Lambda$ and $p\Lambda$, as illustrated in Fig. 2.

5. Limits of diffusion theory. Certain basic assumptions are made in the calculations of breakdown as a balance between the ionization rate and the loss of electrons by diffusion. We examine here the limits which the assumptions place on the application of the theory to various experimental conditions[2]. These limits can be discussed in terms of the proper independent variables, $p\Lambda$ and $p\lambda$. One can plot on the $p\Lambda - p\lambda$ plane the conditions for all breakdown data for a single gas and derive limits in these variables which will define the applicability of the diffusion theory.

At low frequencies, the experimental measurements of breakdown are always taken in vessels whose dimensions are small compared to a wavelength of the exciting power. For this case, the uniform field assumption may be very good. At very high frequencies, there exists a limit to the size of the discharge tube consistent with the uniform field assumption of the diffusion theory. This limit can be written in terms of the size of vessel allowable to sustain a single loop of a standing wave of the electric field. The relation between the parallel plate separation, the wavelength, and the diffusion length may be written as

$$\frac{\lambda}{2} = L = \pi\Lambda. \tag{5.1}$$

Thus one arrives at a limit that can be written in terms of $p\lambda$ and $p\Lambda$ as

$$p\lambda = 2\pi(p\Lambda). \tag{5.2}$$

This limiting line is plotted in Fig. 3 and designated the *Uniform Field Limit*.

[1] P. W. Bridgeman: Dimensional Analysis, Chap. 4. New Haven 1922.

[2] S. C. Brown and A. D. MacDonald: Phys. Rev. **76**, 1629 (1949).

The diffusion theory will not apply of the electron mean free path becomes comparable to the tube size. In the limiting case, this can be expressed as the mean free path, l, being equal to Λ. The probability of collision, P_m, is equal to $1/pl$. To plot this condition in Fig. 3 we may write

$$p\Lambda = 1/P_m. \tag{5.3}$$

The value of P_m is not a constant, but depends upon the electron energy. If we assume that the average electron has an energy equal to one-third of the ionization potential, the average electron energy would be 5 volts for hydrogen. Using measured values for the probability of collision in hydrogen for the average electron, we get $P_m = 49$ (cm.-mm. Hg)$^{-1}$. With this value, we obtain the horizontal line in Fig. 3 marked *Mean Free Path Limit.*

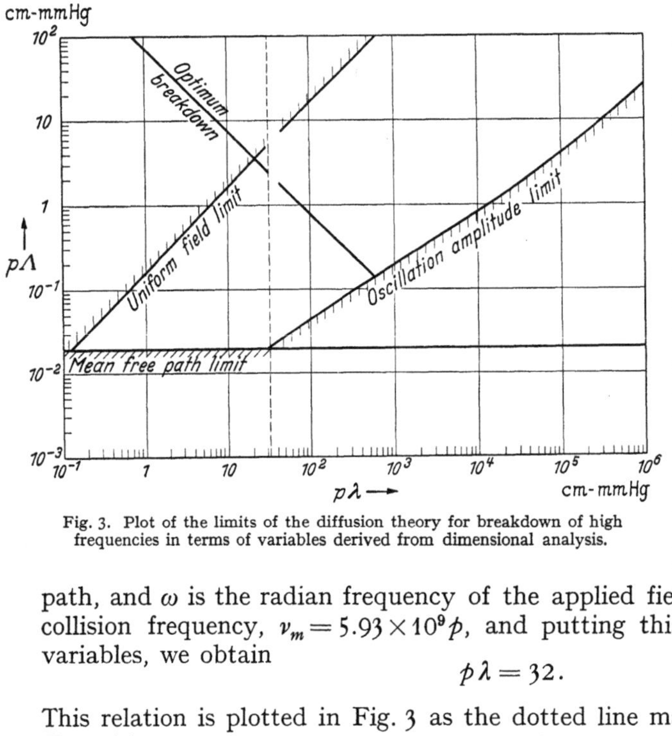

Fig. 3. Plot of the limits of the diffusion theory for breakdown of high frequencies in terms of variables derived from dimensional analysis.

Within the limits of experimental conditions in which diffusion theory adequately explains the breakdown behavior, several different phenomena may occur. One of the phenomenological changes that is important is the transition from many collisions per oscillation of the electron to many oscillations per collision. This can be written as the condition that $\nu_m = \omega$, where ν_m, the collision frequency, is the ratio of the average velocity v to the mean free path, and ω is the radian frequency of the applied field. Using a value of the collision frequency, $\nu_m = 5.93 \times 10^9 p$, and putting this in terms of the proper variables, we obtain

$$p\lambda = 32. \tag{5.4}$$

This relation is plotted in Fig. 3 as the dotted line marked *Collision Frequency Transition.*

We can calculate a line on the $p\Lambda - p\lambda$ plane corresponding to the minimum breakdown field for any given container size. At low pressure we have seen that the breakdown field approaches the condition that $E = 2\pi c (2\bar{u} u_i/3)^{\frac{1}{2}}/\Lambda \lambda \nu_m$. For hydrogen, $\nu_m = 5.93 \times 10^9 p$, $u_i = 15.4$ volts, and we consider $\bar{u} = u_i$. This leads to a value of $E = 400/\Lambda \lambda p$. Using Eq. (3.2) for hydrogen leads to $E = 5.3 p$. Eliminating the field between these two equations will allow us to calculate the pressure at which breakdown will occur most easily. In terms of the variables of Fig. 3 this leads to the equation

$$p\lambda = \frac{75}{p\Lambda}. \tag{5.5}$$

This relation is plotted in Fig. 3 as the line marked *Optimum Breakdown.*

When the amplitude of the electron oscillation in the electric field is sufficiently high, the electrons can travel completely across the tube and collide with the walls on every half-cycle.

Integrating Eq. (1.2), we obtain the displacement of the electron oscillation:

$$x = \frac{e E_p \exp(j\omega t)}{j\omega(j\omega + \nu_m) m}. \tag{5.6}$$

The amplitude of the oscillation is

$$\frac{e E_p}{m\omega\sqrt{\omega^2 + \nu_m^2}} = \frac{\sqrt{2}\, e E_e}{m\omega\nu_m}.$$

The limiting case on the diffusion mechanism in which all of the electrons will hit the walls would be calculated by setting the oscillation amplitude equal to one-half of the electrode separation. Thus, the oscillation amplitude limit becomes equal to

$$\frac{\sqrt{2}\, e E_e}{m\omega\nu_m} = \frac{L}{2}. \tag{5.7}$$

Substituting λ in terms of ω, v/l in place of ν_m, and $1/p P_m$ for l, we obtain

$$p\lambda = \left(\frac{\pi m c}{e} v P_m\right) \frac{pL}{\sqrt{2}(E_e/p)}. \tag{5.8}$$

Putting in numerical values and using the parallel plate relation that $L = \pi\Lambda$, one has

$$p\lambda = 10^6 \left(\frac{p\Lambda}{E_e/p}\right). \tag{5.9}$$

This equation can be solved numerically if the experimental values of the breakdown field are available. Experiments of this sort have been carried out for parallel-plate geometry with hydrogen gas, and numerical values can therefore be determined. This calculation yields the *Oscillation Amplitude Limit* of Fig. 3. On crossing the amplitude of oscillation limit new mechanisms occur; they will be discussed in Chapter C.

II. Theories based on the BOLTZMANN transport equation.

a) High-frequency electric fields.

Up to this point we have discussed the phenomena associated with breakdown from the point of view of the behavior of the average electron. This presents a simple physical picture of breakdown, but does not give us a reliable mathematical basis for theoretical prediction of the behavior of the gas. We therefore turn to a more exact description of breakdown, restricting our discussion to those gases in which diffusion is the sole loss mechanism for electrons. First let us work out a fairly simple solution[1] applicable to any gas and a wide pressure range, and then discuss the specific problems that arise for individual gases.

6. BOLTZMANN equation. When a high-frequency electric field, $E = E_p \exp(j\omega t)$, is applied to a gas, the velocity distribution $F(\overline{\boldsymbol{v}}, \boldsymbol{r})$ of the free electrons is determined by the BOLTZMANN equation:

$$(\partial F/\partial t) = C - \nabla_r \cdot \boldsymbol{v} F + \nabla_v \cdot e\boldsymbol{E} F/m \tag{6.1}$$

where C represents the effect of collision, and ∇_r and ∇_v are the gradient operators in configuration and velocity spaces. This equation is solved by expanding the

[1] W. P. ALLIS and S. C. BROWN: Phys. Rev. **87**, 419 (1952).

distribution function in spherical harmonics in velocity space and in FOURIER series in time[1]:

$$\left.\begin{aligned} F &= \sum_l \sum_k F_k^l P_l(\cos\vartheta) \exp(j k \omega t) \\ &= F_0^0 + \boldsymbol{v} \cdot \left[\boldsymbol{F}_0^1 + \boldsymbol{F}_1^1 \exp(j \omega t)\right]/v. \end{aligned}\right\} \tag{6.2}$$

All terms except the three indicated may be dropped[2] when the geometry, pressure, and frequency fall within the limits discussed in Sect. 5. These limits require that the mean free path be less than any dimension of the cavity, that the frequency be sufficiently high so that the electrons do not lose appreciable energy between cycles, and that the average motion of the electrons resulting from the action of the field and of collisions be sufficiently small so that the field does not clear the electrons out of part of the tube in each half-cycle.

In evaluating the collision term it is convenient to replace the customary mean free paths by the collision frequencies for momentum transfer, for excitation, and for ionization defined by

$$\nu_m = \iint v p P_m(\vartheta)(1 - \cos\vartheta) \sin\vartheta \, d\vartheta \, d\varphi, \tag{6.3}$$

$$\nu_x = v p P_x, \quad \nu_i = v p P_i, \tag{6.4}$$

where P_x and P_i are the experimentally defined probabilities of excitation of all levels and of ionization, and $P_m(\vartheta)$ is the probability of scattering into a unit solid angle inclined at ϑ to the original direction[3].

If the temperature of the gas is negligible compared to that of the electrons, the latter will lose a fraction $2m/M$ of their energy, per momentum transfer collision, to recoil of the molecule with which they collide. The mass of the molecule is M and that of the electron m. This fraction will be increased if there is appreciable transfer of energy to rotation or vibration but here it is assumed to be negligible.

A collision producing electronic excitation differs from that producing recoil, rotation, and vibration in that the colliding electron loses practically all of its energy instead of a very small fraction of it. These processes are treated mathematically as though fast electrons disappeared at the rate $(\nu_x + \nu_i) F_0^0$, and slow electrons appeared at the rate $b F_0^0$. Making use of these ideas in the evaluation of the collision term, introducing Eq. (6.2) into Eq. (6.1), and equating coefficients of similar time and angle functions yields one scalar and two vector equations:

$$(\nu_x + \nu_i - q) F_0^0 = -\frac{v}{3} \nabla_r \cdot \boldsymbol{F}_0^1 + \frac{1}{v^2} \frac{\partial}{\partial v} \left[\frac{e v^2}{6m} (\boldsymbol{E}_p \cdot \boldsymbol{F}_1^1)_{\text{real}} + \frac{m}{M} \nu_m v^3 F_0^0\right], \tag{6.5}$$

$$\nu_m \boldsymbol{F}_0^1 = -v \nabla_r F_0^0, \tag{6.6}$$

$$(\nu_m + j\omega) \boldsymbol{F}_1^1 = (e \boldsymbol{E}_p/m) \, \partial F_0^0 / \partial v. \tag{6.7}$$

7. Distribution in space. The direct and alternating currents densities,

$$\boldsymbol{J}_0 = \int_0^\infty -e \boldsymbol{F}_0^1 \frac{4\pi v^3}{3} dv = e \nabla_r D n, \tag{7.1}$$

$$\boldsymbol{J}_1 = \int_0^\infty -e \boldsymbol{F}_1^1 \frac{4\pi v^3}{3} dv = e \mu \boldsymbol{E} n, \tag{7.2}$$

---

[1] For the mathematical details refer to W. P. ALLIS: Motion of Ions and Electrons, vol. XXI of this Encyclopedia, hereafter referred to as ALLIS.

[2] ALLIS: Eqs. (19.2), (23.1). For the justification for dropping F_1^0 see ALLIS, paragraph preceding Eq. (24.12).

[3] In this article, ν_m is the same as ALLIS' ν_{cl}, later simplified there to ν_c. See ALLIS: Eqs. (27.5) to (27.8) and associated discussion.

are fully determined by the components $\boldsymbol{F}_0^1$ and $\boldsymbol{F}_1^1$ of the distribution function. These are, in turn, found to be derivatives of F_0^0. Substitution of Eqs. (6.6) and (6.7) into Eqs. (7.1) and (7.2) serves to determine the diffusion and a.c. mobility coefficients in terms of $\boldsymbol{F}_0^0$:

$$n D = \int_0^\infty \frac{v^2}{3\nu_m} F_0^0 \, 4\pi v^2 \, d v, \tag{7.3}$$

$$n \mu = \int_0^\infty \frac{4\pi}{3} \frac{e}{m} F_0^0 \frac{d}{d v}\left(\frac{v^3}{\nu_m + j\omega}\right) d v. \tag{7.4}$$

The components $\boldsymbol{F}_0^1$ and $\boldsymbol{F}_1^1$ can be eliminated from Eq. (6.5) by substitution from Eqs. (6.6) and (6.7) ,and this yields the differential equation for F_0^0:

$$(\nu_x + \nu_i - q)\, F_0^0 = \frac{v^2}{3\nu_m} \nabla_r^2 F_0^0 + \frac{1}{v^2} \frac{\partial}{\partial v}\left[\frac{e u_c}{3m} \nu_m v^2 \frac{\partial F_0^0}{\partial v} + \frac{m}{M} \nu_m v^3 F_0^0\right]. \tag{7.5}$$

The energy, in electron volts, $u_c = e E_p^2 / 2m(\nu_m^2 + \omega^2)$, which is introduced here, turns out to be the average energy transferred from the field to an electron between collisions, and $\nu_m u_c$ is the power transfer. It is, in general, a function of the electron's energy through the collision frequency ν_m, and it is also a function of the external parameters E, p, and $\lambda = 2\pi c/\omega$. The power transfer has a maximum for $\nu_m = \omega$, and this corresponds to the pressure for optimum breadkown. At pressures above this value, u_c varies as $(E_p/p)^2$.

The total excitation and ionization rates may be defined by

$$n \bar{\nu}_x = \int_0^\infty \nu_x F_0^0 \, 4\pi v^2 \, d v, \tag{7.6}$$

$$n \bar{\nu}_i = \int_0^\infty \nu_i F_0^0 \, 4\pi v^2 \, d v, \tag{7.7}$$

and, since every exciting collision yields one, and every ionizing collision two, slow electrons, we have

$$\int_0^\infty q F_0^0 \, 4\pi v^2 \, d v = n(\bar{\nu}_x + 2\bar{\nu}_i). \tag{7.8}$$

Use is made of these relations in multiplying Eq. (7.5) by $4\pi v^2 d v$ and integrating over all velocities. The term in brackets vanishes at both limits, and one obtains

$$n \bar{\nu}_i + \nabla_r^2 D n = 0. \tag{7.9}$$

This is a diffusion equation and expresses the fact that at breakdown the ionization rate equals the diffusion rate. For a uniform electric field (which will be the assumption from here on), it has a solution which is everywhere positive only if $\bar{\nu}_i = D/\Lambda^2$, where Λ is a diffusion length. This may be called the breakdown condition.

8. Distribution in energy. If the function F_0^0 is assumed to be the product of a function $n(x)$ of space and a function $f(u)$ of the energy u, defined by $u = m v^2/2e$, we can make use of Eq. (7.9) to replace $\nabla_r^2 F_0^0$ by $-n f/\Lambda^2$ and obtain the following equation for $f(u)$:

$$\left(\nu_x + \nu_i - q + \frac{2 e u}{3 m \nu_m \Lambda^2}\right) f = \frac{2}{3\sqrt{u}} \frac{d}{d u}\left[\nu_m u^{\frac{3}{2}}\left(u_c \frac{d f}{d u} + \frac{3m}{M} f\right)\right]. \tag{8.1}$$

The terms on the left side represent the electrons leaving unit volume of phase space through excitation, ionization, and diffusion, and their reappearance at low energy at the rate qf. On the right are the terms attributable to energy gained from the field and lost to recoil.

The excitation frequency ν_x sets in discontinuously at a potential u_x so that it is always necessary to divide the energy range into two parts and solve Eq. (8.1) for two functions, f_e and f_{xi}, appropriate to the elastic and inelastic ranges and join them at u_x. On the other hand, since $(\nu_x + \nu_i)$ may generally be approximated by a continuous function, it is not necessary to join functions at u_i. The method of solution appropriate to the two ranges is quite different, so that they must be discussed separately.

9. Inelastic range. When inelastic collisions are possible, they dominate all other collision processes because of the large energy losses involved. Accordingly, the recoil and diffusion terms may be left out of Eq. (8.1). We may also neglect q in this range. The equation to be solved is then

$$2\frac{d}{du}\left(\nu_m u_c u^{\frac{3}{2}} \frac{d}{du} f_{xi}\right) = 3(\nu_x + \nu_i)\, u^{\frac{1}{2}} f_{xi}. \tag{9.1}$$

The conditions imposed on the solution are somewhat contradictory; we must choose that solution that vanishes at infinity, but we want greatest accuracy just above u_x, where most of the excitations take place. The conventional asymptotic expansion does not satisfy the second requirement without an unreasonable number of terms, and the *WKB* method diverges at u_x, but we can use a somewhat similar approximation[1]. Setting

$$u^{\frac{1}{2}} f_{xi} = e^{-s}, \tag{9.2}$$

the equation for s is

$$s'^2 - s'' - \frac{s'}{2u} - \left(s' + \frac{1}{2u}\right)\frac{d}{du}\ln(\nu_m u_c) = \frac{3(\nu_x + \nu_i)}{2\nu_m u_c u}. \tag{9.3}$$

An analytic approximation to the experimental data must now be substituted for $\nu_m u_c$ and $\nu_x + \nu_i$, and a power series in $1/u$ is substituted for s', the last term in the series being reserved to obtain exact agreement at u_x.

Knowing s, the average ionization frequency per electron is given by

$$\bar{\nu}_i = 2\pi(2e/m)^{\frac{3}{2}} \int_{u_i}^{\infty} \nu_i u^{\frac{1}{2}} f_{xi}\, du, \tag{9.4}$$

and the total inelastic frequency is given [see Eq. (10.1)] by

$$(\bar{\nu}_x + \bar{\nu}_i) = -(4\pi/3)(2e/m)^{\frac{3}{2}} (\nu_m u_c u^{\frac{3}{2}} d f_{xi}/du)_x. \tag{9.5}$$

The subscript indicates that the quantity in the parenthesis is to be taken at u_x. Both of these expressions contain an unknown normalization constant and so cannot be evaluated as they stand, but their ratio,

$$N_{xi} = 1 + \frac{\bar{\nu}_x}{\bar{\nu}_i} = -\frac{2}{3}\left(\nu_m u_c u^{\frac{3}{2}} \frac{d f_{xi}}{du}\right)_x \Big/ \int_{u_i}^{\infty} \nu_i u^{\frac{1}{2}} f_{xi}\, du, \tag{9.6}$$

can be evaluated and has a physical meaning. Because one electron must leave the tube for every one produced, $1/\bar{\nu}_i$ is the average lifetime of a free electron from its liberation at an ionization to its absorption at the walls; $\bar{\nu}_x/\bar{\nu}_i$ is the average number of excitations produced by an electron during its lifetime; and the number N_{xi} represents the total inelastic collisions during an electron's free lifetime.

[1] For an equivalent approximation see ALLIS: Eq. (33.5).

Because of the exponential nature of f_{xi}, the number N_{xi} depends primarily on $\exp(s_x - s_i)$. From Eq. (9.3) it is seen that the main part of s' is given by $[3(\nu_x+\nu_i)/2\nu_m u_c u]^{\frac{1}{2}}$ which, when $\nu_m > \omega$, is proportional to p/E_p. Accordingly, the variation of N_{xi} with p/E_p is given very nearly by

$$N_{xi} = \alpha \exp(\beta p/E_p), \tag{9.7}$$

where α and β are constants obtainable from Eq. (9.3).

10. Elastic range. Below u_x, the excitation and ionization frequencies are zero, but we must discuss the appearance rate qF_0^0. In order to calculate it as a function of energy it is necessary to take the product of the distribution function and the excitation function of each level and shift the product down the energy scale by the energy of the particular level. Since the excitation functions of allowed transitions have a sharp maximum just above the excitation potential, the scattered electrons have very little energy. The excitation functions of forbidden transitions have a maximum far above the excitation potential, but there are a negligible numberof electrons with sufficient energy to excite these. Accordingly, most inelastically scattered electrons have very little energy, and no appreciable error is made in assuming that q is a delta-function at zero energy.

Multiplying Eq. (8.1) by $4\pi v^2 dv$ gives the net rate of loss of electrons from the spherical shell of thickness dv. Integrating and making use of Eq. (7.8) gives

$$\bar{\nu}_x + 2\bar{\nu}_i - \frac{4\pi}{3\Lambda^2}\int\limits_0^v \frac{f_e}{\nu_m} v^4\, dv = -\frac{4\pi}{3}\nu_m v^3\left(u_c \frac{df_e}{du} + \frac{3m}{M} f_e\right), \tag{10.1}$$

where $\bar{\nu}_x + 2\bar{\nu}_i$ is the rate of appearance of electrons at small velocities. The integral represents losses by diffusion of electrons of speeds between zero and v, and its value at infinity would, by Eq. (7.9), equal $\bar{\nu}_i$. The difference represents the rate at which electrons pass the energy u in the upward direction in order to supply the inelastic processes occurring at higher energies. Eq. (10.1) was derived by SMIT[1] directly from this principle of balance between electrons going up in energy and the rate of inelastic collisions; however, SMIT includes the thermal energy of the gas but does not include diffusion. In glow discharges the diffusion term is much larger than the thermal one. At very low pressures the diffusion term is quite large, and one must solve the second order Eq. (8.1) for f_e; but in most cases the diffusion term is small and there is then a great advantage in replacing the integral by an approximation such as

$$\frac{4\pi}{3\Lambda^2}\int\limits_0^v \frac{f_e v^4}{\nu_m}\, dv = \frac{\bar{\nu}_i \nu_m v^3}{(\nu_m v^3)_0}. \tag{10.2}$$

This expression gives the full diffusion loss $\bar{\nu}_i$ at the velocity v_0, which corresponds to the energy u_0 (to be defined shortly), and the third power of the velocity was actually found to be the best in the case for which extensive numerical calculations were made. With this substitution, Eq. (10.1) becomes a first-order inhomogeneous equation. If we define an energy variable w by $dw = 3m\,du/M u_c$, the solution of the homogeneous part[2] is

$$f_M = A e^{-w}. \tag{10.3}$$

[1] J. A. SMIT: Physica, Haag 3, 543 (1936).

[2] M. J. DRUYVESTEYN and F. M. PENNING: Rev. Mod. Phys. 12, 87 (1940). — H. MARGENAU: Phys. Rev. 69, 508 (1946).

The solution of the complete equation for f_e is

$$f_e = \frac{3}{4\pi}\left\{(\bar{\nu}_x + 2\bar{\nu}_i)\, e^{-w}\int_u^{u_0} e^w \frac{du}{\nu_m u_c v^3} - \bar{\nu}_i\left(\frac{M}{3m}\right)\left(\frac{e^{w_0 - w} - 1}{(\nu_m v^3)_0}\right)\right\}, \tag{10.4}$$

which, at higher pressures, can be replaced by the simpler function

$$f_e = \frac{3}{4\pi}(\bar{\nu}_x + \bar{\nu}_i)\, e^{-w}\int_u^{u_0} e^w \frac{du}{\nu_m u_c v^3}. \tag{10.5}$$

The constant of integration in both of these expressions appears as an energy u_0 at which the function f_e crosses the axis when extended beyond u_x. The meaning of this energy is seen by noting that f_e would be unchanged if the actual excitation and ionization functions ν_x and ν_i were replaced at u_0 by delta-functions with the proper relative magnitudes so that all inelastic collisions would take place at exactly that energy. Thus u_0 is the equivalent single excitation potential. By this equivalence the diffusion should also vanish above u_0 and hence the integral (10.2) must equal $\bar{\nu}_i$ at $v = v_0$.

The potential u_0 is determined by equating the logarithmic derivatives of f_e and f_{xi} at u_x. In general, the extrapolation $u_0 - u_x$ is small, and when this is so a linear extrapolation formula may be used. The first-order derivative may be eliminated from Eq. (8.1) by the standard transformation

$$g = (\nu_m u_c)^{\frac{1}{2}}\, u^{\frac{3}{4}}\, e^{w/2} f. \tag{10.6}$$

Then $g'' = 0$ when $g = 0$. The function $g(u)$ has a point of inflection at u_0 and may be extrapolated linearly back to u_x, giving

$$-\frac{1}{u_0 - u_x} = \left(\frac{g'}{g}\right)_x = \frac{1}{2}\frac{d\ln(\nu_m u_c)}{du} + \frac{1}{4u_x} + \frac{3m}{2Mu_c} - s'_x, \tag{10.7}$$

the whole right-hand side being taken at u_x. When this extrapolation is valid, the effective excitation potential u_0 may be calculated from the inelastic function f_{xi} alone.

11. Breakdown equation. The diffusion coefficient D may be calculated by substituting f_e and f_{xi} in Eq. (7.3). A negligible error is made by integrating f_e from 0 to u_0 and not using f_{xi}, the difference being readily computed and shown to be small. Using Eq. (10.4) we get

$$D = \frac{2\pi}{3}\left(\frac{2e}{m}\right)^{\frac{5}{2}}\int_0^{u_0} f_e\, u^{\frac{3}{2}}\frac{du}{\nu_m} = \frac{u_0^2}{E^2}\left[(\bar{\nu}_x + \bar{\nu}_i)\,\mathfrak{D} + \bar{\nu}_i\,\delta\right], \tag{11.1}$$

where $\mathfrak{D}$ and δ are two dimensionless functions of v_0, and E is the rms value of the electric field:

$$\mathfrak{D} = \frac{4}{v_0^4}\int_0^{v_0}\frac{v^4 e^{-w}}{\nu_m}\,dv\int_v^{v_0}\frac{\nu_m^2 + \omega^2}{\nu_m v^2}\, e^w\, dv, \tag{11.2}$$

$$\delta = \mathfrak{D} - \frac{ME^2}{3m\,u_0^2(\nu_m v^3)_0}\int_0^{v_0}\frac{v^4}{\nu_m}\left[e^{w_0 - w} - 1\right] dv. \tag{11.3}$$

The breakdown condition is then

$$\Lambda^2 E^2/u_0^2 = N_{xi}\,\mathfrak{D} + \delta, \tag{11.4}$$

where Eq. (9.6) or Eq. (9.7) can be substituted for the inelastic collision number.

This general outline for a theory of a.c. breakdown has been made specific by a number of workers. For any particular gas, specific assumptions are made to render the problem mathematically tractable. These assumptions will now be discussed briefly.

12. Hydrogen. For hydrogen gas[1] the collision frequency ν_m is nearly independent of energy and is given by $\nu_m = 5.9 \times 10^9 p$ at all energies above 4 volts. The effective field E_e, defined by Eq. (1.5), is a constant, and in terms of this the average energy gain per collision is $u_c = 5 \times 10^{-5} (E_e/p)^2$ electron-volts.

When ν_m is constant the variable $2w/3$ is the ratio of recoil loss to energy gain per collision, the loss exceeding the gain if w is greater than $\frac{3}{2}$. At the higher pressures breakdown is observed for w_i approximately 4, so that in these cases the electrons are losing more energy to recoil, in the average, than they gain from the field, over most of the energy range. There is a sufficient number of statistically lucky electrons, however, to overcome this handicap and reach the ionization potential, producing breakdown.

The integrals in Eqs. (11.2) and (11.3) can be evaluated in terms of incomplete gamma-functions or, more conveniently, by the series

$$\left.\begin{aligned} \mathcal{D} &= \frac{\nu_m^2 + \omega^2}{\nu_m^2} w_0^{-2} \int\limits_0^{w_0} w^{\frac{3}{2}} e^{-w} \int\limits_w^{w_0} t^{-\frac{3}{2}} e^t \, dt \, dw \\ &= \frac{\nu_m^2 + \omega^2}{\nu_m^2} 4! \sum_0^\infty \frac{(k+1)!}{(2k+5)!} (4 w_0)^k, \\ \delta &= 3 \frac{\nu_m^2 + \omega^2}{\nu_m^2} 4! \sum_0^\infty \frac{(k+1)!}{(2k+5)!} \frac{(4 w_0)^k}{2k+7}. \end{aligned}\right\} \tag{12.1}$$

Ramien[2] has measured the excitation and ionization probabilities in hydrogen, and his data can be represented by the functions

$$(\nu_x + \nu_i)/\nu_m = h_0 u - h_1 - h_2/u, \tag{12.2}$$

$$\nu_i/\nu_m = h_i (u - u_i), \tag{12.3}$$

with the constant $u_i = 16.2$ volts, $h_i = 9.2 \times 10^{-3}$ volt^{-1}, $u_x = 8.9$ volts. The values $h_0 = 8.7 \times 10^{-3}$ volt^{-1} and $h_1 = 76 \times 10^{-3}$ are in agreement with his data. Agreement with the breakdown data could not, however, be obtained if the losses observed by Ramien below 8.9 volts and ascribed to the excitation of vibrations were included in the theory.

With the above inelastic collision functions we set

$$s = a u - b \ln u + c/u. \tag{12.4}$$

The coefficients a and b are determined in the usual way for series near infinity,

$$a^2 = 3 h_0 / 2 u_c, \quad b = 3 h_1 / 4 a u_c - \tfrac{1}{4}. \tag{12.5}$$

The coefficient c is used to obtain exact agreement at u_x. This gives

$$c/u_x = a u_x - b + \tfrac{3}{4} - (2 a u_x - b + \tfrac{9}{16})^{\frac{1}{2}}. \tag{12.6}$$

The approximation is then tested by substituting Eq. (12.4) with these constants in Eq. (9.1) and solving for $(\nu_i + \nu_x)$. This gives the theoretical excitation frequency

---

[1] W. P. Allis and S. C. Brown: Phys. Rev. **87**, 419 (1952).

[2] H. Ramien: Z. Physik **70**, 353 (1931).

for which Eqs. (9.2) and (12.4) are the exact solution, and it must agree closely with the experimental data for P_x and P_i, particularly between u_x and u_i. Substitution of Eqs. (9.2) and (12.4) in Eq. (9.6) gives the number of inelastic collisions per electron:

$$N_{xi} = \frac{h_0}{h_i} \frac{a u_x - b + \frac{1}{2} - c/u_x}{\mathcal{P}(u_i)} \left(\frac{u_x}{u_i}\right)^b \times \exp\left[(a - c/u_i u_x)(u_i - u_x)\right], \qquad (12.7)$$

where

$$\mathcal{P}(u_i) = \sum_{k=0}^{\infty} \sum_{l=0}^{\infty} \frac{(k+l+1)!}{k!\,(a u_i)^k} \frac{(b-l)!}{(b-k-l)!} \frac{(c/a u_i^2)}{l!}.$$

This function agrees very well with the approximation (9.7) with $\alpha = 2$, $\beta = 71.7$ volts/cm. mm. Hg, for almost the whole range of the measurements. The limit $N_{xi} = 2$ as p/E approaches zero comes from the near equality of h_i and h_0, so that at high energies there are about equal numbers of excitations and ionizations. Introducing Eq. (12.4) into Eq. (10.7) we find the effective excitation potential from

$$1/(u_0 - u_x) = a - 3m/2M u_c - (b + \tfrac{1}{4})/u_x - c/u_x^2. \qquad (12.8)$$

Expressions (12.1), (12.7), and (12.8) may then be introduced in Eq. (11.4) to obtain a direct comparison with the quantities measured at breakdown. The results are shown in Fig. 4. The agreement is good over a wide range of pressure for several different values of Λ and at two different frequencies.

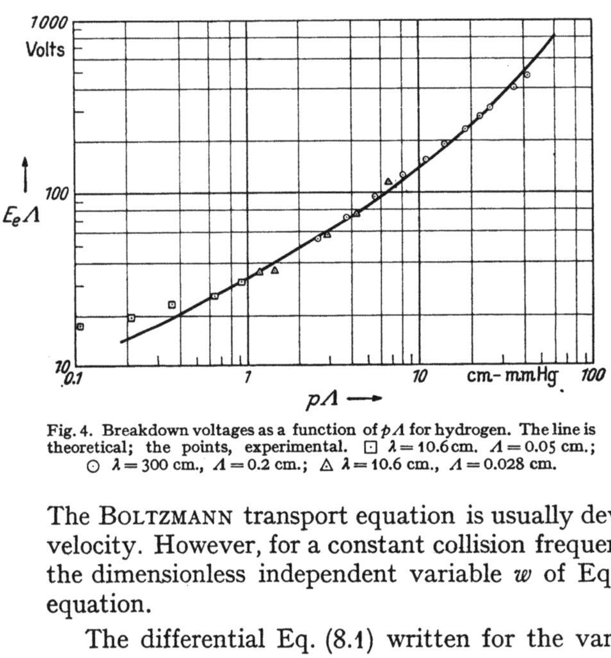

Fig. 4. Breakdown voltages as a function of $p\Lambda$ for hydrogen. The line is theoretical; the points, experimental. ⊡ $\lambda = 10.6$ cm. $\Lambda = 0.05$ cm.; ⊙ $\lambda = 300$ cm., $\Lambda = 0.2$ cm.; △ $\lambda = 10.6$ cm., $\Lambda = 0.028$ cm.

The disagreement at low pressures results from neglecting the diffusion term in f_{xi} and to the approximations made in the formula for the effective excitation potential u_0; at these pressures the more exact confluent hypergeometric functions of Sect. 13 should be used.

13. Rigorous Theory for Constant Collision Frequency.

The BOLTZMANN transport equation is usually developed in terms of the electron velocity. However, for a constant collision frequency it is convenient to introduce the dimensionless independent variable w of Eq. (10.3) in the final differential equation.

The differential Eq. (8.1) written for the variable $w = 3mu/Mu_c$ is

$$\left.\begin{aligned} f'' + f'\left[1 + (3/2w)\right] + f\left[(A/w) - (u_E/\Lambda E_e)^2\right] = 0, \\ A = \tfrac{3}{2} - (M/2m)(\nu_x + \nu_i - q)/\nu_m \end{aligned}\right\} \qquad (13.1)$$

where $u_E = u_c(M/3m)$. This equation is of second order. It has a pole of first order at the origin, $w = 0$, and an essential singularity at infinity, $w = \infty$. Since Eq. (12.2) is not valid for $u < u_x$, we have to divide the energy scale at $w = w_x$ and to discuss separately the integration of Eq. (13.1) in the two regions.

In the elastic collision range, below w_x, Eq. (13.1) takes the form

$$f_e'' + f_e'\left[1 + (3/2w)\right] + f_e\left[(3/2w) - (u_E/\Lambda E_e)^2\right] = 0. \qquad (13.2)$$

The solution in terms of confluent hypergeometric functions is[1]

$$f_e(w) = C_{eN} \exp\left[-(g+1)\frac{w}{2}\right]\left[M_1(w) + C_e w^{-\frac{1}{2}} M_2(w)\right], \tag{13.3}$$

where

$$M_1(w) = F(\alpha, \tfrac{3}{2}; gw); \qquad M_2(w) = F(\alpha - \tfrac{1}{2}, \tfrac{1}{2}; gw),$$
$$\alpha = \tfrac{3}{4}(g-1)/g; \qquad g^2 = 1 + 4(u_E/\Lambda E_e)^2$$

and C_e and C_{eN} are integration constants.

In the range of inelastic collisions between the electrons and the gas molecules, when the electron energy lies above u_x, Eq. (13.1) may be written as

$$\left.\begin{aligned} &f''_{xi} + f'_{xi}[1 + (3/2w)] - f_{xi}\,G(w) = 0,\\ &G(w) = \tfrac{1}{2}(g^2-1)^{\frac{1}{2}} + h_0 u_c (M^2/6m^2) - [\tfrac{3}{2} + M h_1/2m](1/w) - (3h_2/2u_c)(1/w)^2. \end{aligned}\right\} \tag{13.4}$$

There exists only one asymptotic solution for Eq. (13.4) which fulfills the boundary condition that the distribution function shall vanish at infinity. It is

$$\left.\begin{aligned} &f_{xi}(w) = C_{eN}\,C_i\, w^{(K-\frac{3}{4})}[\exp(-a_1 w)] \times \left[1 + \sum_{n=1}^{\infty}\prod_{r=1}^{n} p^{(r)}/n!\,(a_0 w)^n\right],\\ &\text{where}\\ &a_0 = [g^2 + 6h_0(u_E^2/u_c)]^{\frac{1}{2}}; \quad a_1 = \tfrac{1}{2}(a_0+1);\\ &K = (M/2m)\,h_1/a_0, \qquad p^{(r)} = \tfrac{1}{16} - 3h_2/2u_c - [K + \tfrac{1}{2} - r^2];\\ &\prod_{r=1}^{n} p^{(r)} = p^{(1)} p^{(2)} \dots p^{(n)}. \end{aligned}\right\} \tag{13.5}$$

The integration constants C_e and C_i can be calculated by joning f_e and f_{xi} in value and slope at $w = w_x$. As a result of these processes we find

$$\left.\begin{aligned} &C_e = -(w_x)^{\frac{1}{2}}(M_{1x}/M_{2x})(T_1/T_0),\\ &C_i = (1/2T_0 M_{2x} w_x^{K+\frac{1}{4}}) \exp[(a_0+g)\,w_x/2],\\ &T_0 = \Sigma'_x + \Sigma_x[(K-\tfrac{1}{4})/w_x - (a_0-g)/2 - g M'_{2x}/M_{2x}],\\ &T_i = T_0 + \Sigma_x[g(M'_{2x}/M_{2x} - M'_{1x}/M_{1x}) - 1/2w_x],\\ &\Sigma_x = \sum_{r=0}^{\infty} P_r; \quad P_0 = 1; \quad P_r = P_{r-1}(p^{(r)}/r a_0 w_x),\\ &\Sigma'_x = -(1/w_x)\sum_{r=0}^{\infty} r P_r; \quad M_{1x} = M_1(w_x); \quad M_{2x} = M_2(w_x),\\ &g M'_1 = \partial M_1(w)/\partial w; \quad g M'_2 = \partial M_2(w)/\partial w. \end{aligned}\right\} \tag{13.6}$$

The breakdown condition can be written, $D = \Lambda^2 \bar{\nu}_i$, where both the diffusion coefficient and the ionization frequency terms depend on the distribution function. They are calculated from

$$\left.\begin{aligned} D &= \int_0^{\infty} (2e/3m\nu_m)\,u f^*(u)\,du,\\ \bar{\nu}_i &= h_i \int_{u_1}^{\infty} \nu_m (u-u_i)\, f^*(u)\,du \end{aligned}\right\} \tag{13.7}$$

[1] Cf. vol. I of this Encyclopedia, article of J. Meixner, Sect. 5.

where $f^*(u)\,du = 4\pi v^2 f(w)\,dv$, this function being the fraction of electrons within the energy limits $u \pm \frac{1}{2}du$. In the elastic range, the diffusion integral can be represented by confluent hypergeometric functions. In the inelastic range, we approximate the functions by TAYLOR series around w_i and w_x.

Inserting D and $\bar{\nu}_i$ in the breakdown equation yields a transcendental equation for $E_e\Lambda$ and E_e/p which may be evaluated in the form

$$\frac{(M/m)\,(h_i/u_i)\,(C_i/C_e)\,(u_E/a_1)^2\, w_1^{k+\frac{3}{4}}\,\Sigma_i \exp(-a_1 w_1)}{1 + \left(2w_x^{\frac{3}{2}}/C_e\,C_{eN}\right)(f_{ex} + \partial f_{ex}/\partial w)} \times (1-\psi) = 1,$$

$$\psi = (2m/M)\,(u_i/h_i)\left[a_i/(\Lambda E_e)^2\right](u_x/u_i)^{k+\frac{3}{4}} \times (\Sigma_D/\Sigma_i)\exp\left[a_1(w_i - w_x)\right],$$

$$\left.\begin{aligned}
&\Sigma_i = \sum_{n,r=0}^{\infty} B_n^{(r)};\quad B_n^{(0)} = (u_x/u_i)^n P_n;\\
&B_n^{(r)} = B_n^{(r-1)}\left[(r+1)/r\right]\left[(k+\tfrac{3}{4}-n-r)/(a_1 w_i)\right],\\
&\Sigma_D = \sum_{n,r=0}^{\infty} A_n^{(r)};\quad A_n^{(0)} = P_n;\\
&A_n^{(r)} = A_n^{(r-1)}\left[(k+(7/4)-n-r)/(a_1 w_x)\right].
\end{aligned}\right\} \qquad (13.8)$$

The quantity ψ may be neglected for $E_e/p < 10$ v/cm. mm. Hg. The first equation of (13.8) represents the ratio of the net number of electrons produced in the

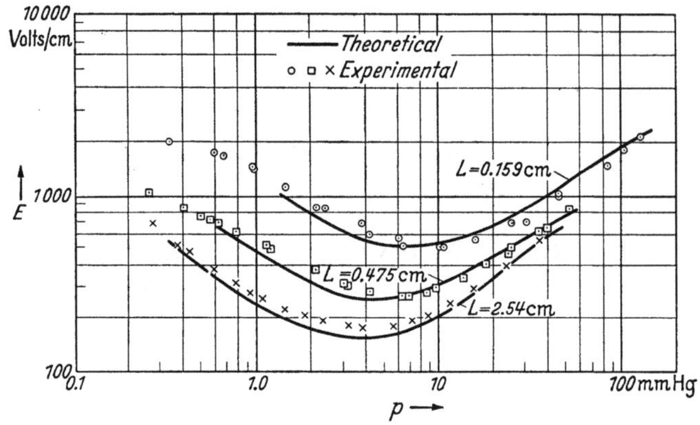

Fig. 5. Comparison of experimental and theoretical curves for breakdown in hydrogen at 3000 Mc/sec. The line is theoretical; the points are experimental.

region $u > u_x$ by ionization and diffusion to the number of electrons lost in the elastic collision range $u < u_x$ by diffusion. This ratio must be equal to unity in a steady state.

The rigorous treatment for constant collision frequency has been applied to hydrogen[1] and helium[2]. For hydrogen, use of the numerical values given under Eqs. (12.2) and (12.3) yielded the agreement shown in Fig. 5. For helium[3], $h_0 = 0.0175$ volt^{-1}, $h_1 = 0.665$, and $h_2 = -6.6$ volts in Eq. (12.2); and $h_i = 5.48 \times 10^{-3}$ volt^{-1}, with $u_i = 24.5$ volts, in Eq. (12.3). The excitation potential u_x was

[1] A. D. MACDONALD and S. C. BROWN: Phys. Rev. **76**, 1634 (1949).
[2] F. H. REDER and S. C. BROWN: Phys. Rev. **95**, 885 (1954).
[3] H. MAIER-LEIBNITZ: Z. Physik **95**, 499 (1935).

taken as 19.9 volts to allow a simple step approximation for the initial part of the excitation function. These values lead to the theoretical line, which is compared with experiment, in Fig. 6.

14. Impure helium. Considerable simplification of the theory presented here results if no excitation need be considered. Such a case is realizable if small

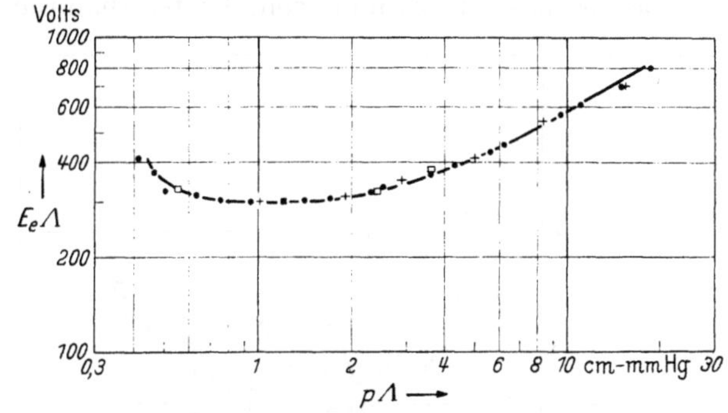

Fig. 6. High-frequency gas-discharge breakdown in pure helium.

amounts of mercury vapor are added to helium, the gas we named Heg in Sect. 3. This synthetic gas adds the simplification of no excitation to the simplification of a constant collision frequency, and one may use it to test the validity of some of the mathematical assumptions necessary in obtaining the solutions previously discussed. The magnitudes of two effects have been thus tested: (a) the result

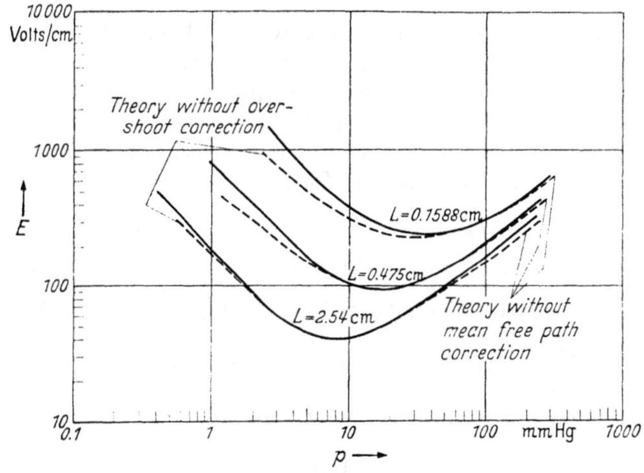

Fig. 7. Theoretical $E-p$ curves illustrating the effect of various correction factors.

of the overshoot in energy due to the fact that the most probable energy at excitation is higher than the excitation energy; and (b) the magnitude of the error introduced by the assumption of constant ν_m, where even for helium this is not true at low energies. For the mathematical details involved see a paper by MacDonald and Brown[1]; the magnitudes of these effects are illustrated in Fig. 7. The excellence of the agreement between the theory and experiment for this simple case is demonstrated in Fig. 8. It should be emphasized that

[1] A. D. MacDonald and S. C. Brown: Phys. Rev. 75, 411 (1949).

although various experimental conditions have been set up to render mathematical manipulation easy, no adjustable parameters appear in the theory.

15. Neon. The high-frequency gas-discharge breakdown in neon has also been studied in detail[1]. In this gas, instead of the collision frequency being considered constant, a constant mean free path is the approximate assumption. The constant mean free path case is considerably more complicated than those previously

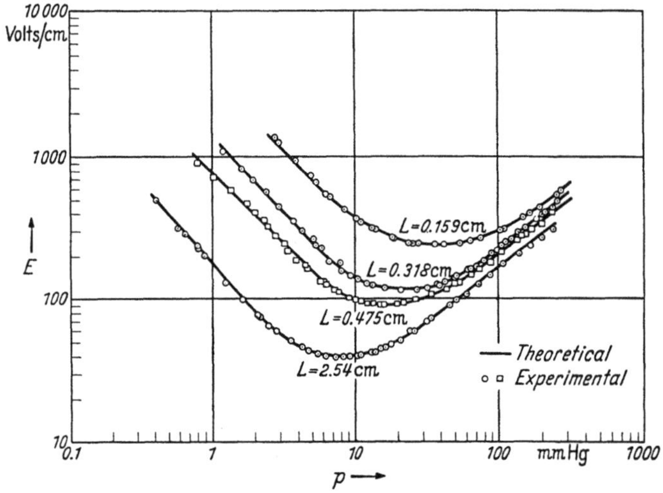

Fig. 8. Experimental breakdown electric fields compared with those theoretically predicted for Heg gas.

described, and the distribution functions were calculated separately for the regions of low $p\lambda$ and high $p\lambda$, corresponding to $(v/l)^2 \ll \omega$ and $(v/l)^2 \gg \omega$. The reader is referred to the original paper[1] for the derivations of the appropriate equations the results of which are as follows.

In the low $p\lambda$ region, the breakdown condition gives an equation for calculating the breakdown electric field in the form

$$\frac{1}{\Lambda^2} = \frac{3s^2}{r^3 l^2} \frac{A h_{i0}[e^{-y_i}(1+D) + D y_i E_i(-y_i)]}{\{2x_x[C I_0(x) - K_0(x)] - 4(x_x)^{\frac{1}{2}} Z_1(x_x) + 4 + A(s^2/r^2)[e^{-y_x} - D E_i(-y_x)]\}}. \quad (15.1)$$

Here, h_i is the ionization efficiency fitted to MAIER-LEIBNITZ's data[2] as

$$\frac{\nu_i}{\nu_m} = h_i(u - u_0) = 4.48 \times 10^{-3}(u - 21.5).$$

The mean free path assumption is written as

$$l = 1/10p;$$

the other quantities are defined as follows:

$$x = us = u\left(1 - \frac{(p\Lambda)^2}{30.8}\right)\left(\frac{636}{E\Lambda p\lambda}\right)^2,$$

$$y = ur = u\left(\frac{412}{E\lambda}\right),$$

$$D = -\frac{\beta y_i^2}{(3+\beta)\, y_i + 4}; \quad \beta = \frac{412}{E\lambda}\left[16 - \frac{0.595}{(p\Lambda)^2}\right].$$

[1] A. D. MACDONALD and D. D. BETTS: Canad. J. Phys. **30**, 565 (1952); **32**, 812 (1954).
[2] H. MAIER-LEIBNITZ: Z. Physik **95**, 499 (1935).

$Z_1(x) = K_1(x)^{\frac{1}{2}} + C I_1(x)^{\frac{1}{2}}$, where K_1 and I_1 are first-order BESSEL functions of imaginary argument.

$$C = \frac{K_2(x)^{\frac{1}{2}} - \varphi K_1(x)^{\frac{1}{2}}}{I_2(x)^{\frac{1}{2}} + \varphi I_1(x)^{\frac{1}{2}}},$$

$$A = \frac{K_1(x)^{\frac{1}{2}} + C I_1(x)^{\frac{1}{2}}}{(x)^{\frac{1}{2}} e^{-y}(1/y + D/y^2)},$$

$$\varphi = \frac{2r(x)^{\frac{1}{2}}}{s} \frac{[y^2 + (1+D)y + 2D]}{D_y + y^2},$$

and $E_i(-y_i)$ is the exponential integral

$$\int_{y_i}^{\infty} e^{-t}/t\, dt,$$

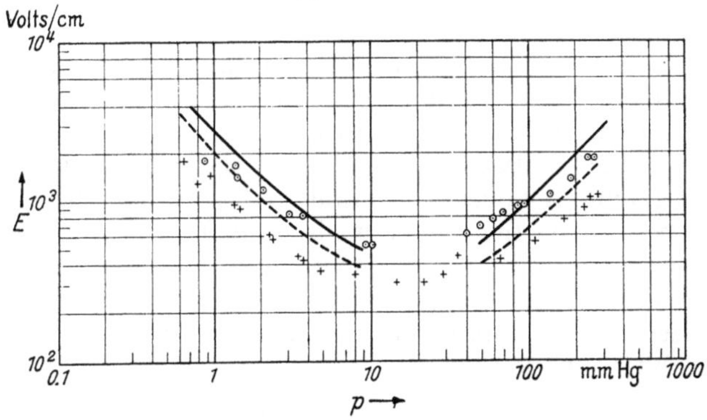

Fig. 9. Comparison of measured breakdown fields in neon with theory. ○ $\Lambda = 0.101$ cm.; + $\Lambda = 0.202$ cm.

tables of which are available in JAHNKE and EMDE[1]. The subscripts i and x refer to ionization and excitation.

In the high $p\lambda$ region, the breakdown is predicted by

$$\frac{1}{\Lambda^2} = \frac{6B}{a^2 \mu l^2} h_i \frac{e^{-a z_i}}{\{z_x e^{-z_x}[M_2(z_x) + F W_2(z_x)] - F/\delta + (B/a) e^{-a z_x}\}}, \tag{15.2}$$

$$h_i = 0.90 \times 10^{-4}; \qquad \mu = 1.62\left(\frac{p}{10E}\right)^2,$$

$$z = u^2\left(\frac{p}{E}\right)^2 / 123.4; \qquad \delta = \frac{30.8}{(p\Lambda)^2},$$

$$\varrho = 4.2\left(\frac{p}{10E}\right)^2; \qquad a = \frac{1}{2}\left(1 + \sqrt{1 + 4\varrho/\mu^2}\right),$$

$$F = -\frac{\frac{d}{dz} M_1(z) + (a-1) M_1(z)}{\frac{d}{dz} W_1(z) + (a-1) W_1(z)},$$

$$B = e^{(a-1)_z}[M_1(z) + F W_1(z)],$$

where the notation $M_1(z)$ holds for the confluent hypergeometric function $M(\delta, 1; z)$, with a similar notation for the second solution.

Eqs. (15.1) and (15.2) are implicit expressions for the breakdown electric field as a function of pressure, container size, and frequency of applied field. The breakdown fields are computed in practice by successive approximations. Breakdown curves for neon, both experimental and theoretical, are shown in Fig. 9.

[1] E. JAHNKE and F. EMDE: Functionentafeln. Leipzig 1933.

b) The effects of nonuniform fields.

16. Cylinders of arbitrary length[1]. The differential equation and boundary condition that lead to the breakdown field strength are obtained from the continuity requirement on the electrons. The equation resulting from Eqs. (2.2) and (7.9) is

$$\nabla^2 \psi + \zeta E^2 \psi = 0, \tag{16.1}$$

where the electron diffusion current density potential ψ is given by $\psi = D n$, and ζ is the high-frequency ionization coefficient. The boundary condition that ψ be zero on the walls of the discharge cavity is sufficiently accurate. The electric field appears explicitly in Eq. (16.1) because it varies with position in the cavity. On the other hand, $p\lambda$ is constant throughout the cavity. The electric field in an arbitrary cavity is given in the form $E = E_0 f(x, y, z)$, where E_0 is the maximum value of the electric field, and f is a geometrical factor obtained from a solution for the field distribution within the cavity as an electromagnetic boundary value problem. The value of f is unity at the maximum field point. The degree of excitation of the cavity is expressed by E_0, and the relative field distribution through the cavity is independent of the excitation. The boundary value problem of finding a nonzero solution to Eq. (16.1), with the boundary condition that ψ be zero on the cavity walls, leads to a characteristic value of E_0, which is the breakdown field at the maximum field point.

Integration of Eq. (16.1) is simplified by the use of an approximation to the ionization coefficient. This approximation expresses the ionization coefficient as the simple analytic function

$$\zeta = \zeta_0 \left(\frac{E}{E_0}\right)^{\beta-2} = \left(\frac{k}{E_0}\right)^2 \left(\frac{E}{E_0}\right)^{\beta-2}, \tag{16.2}$$

where ζ_0 is the value of the ionization coefficient at the maximum field point. The quantity k is introduced for mathematical convenience in the following equations. It has the dimensions of reciprocal length, and appears multiplied into the radius variable below. The quantity $(\beta - 2)$ is obtained as the slope of the ζ versus E/p plot on a logarithmic scale at the point E_0/p. This approximation gives accurate results because it is correct where the ionization is high, is inaccurate only where the ionization is low, and therefore has little effect on the solution of the equation.

The electric field in the TM_{010}-mode cylindrical cavity is given by the expression

$$E = E_0 J_0(2.405\, r/R). \tag{16.3}$$

It depends on the radial coordinate only, which allows the differential equation, Eq. (16.1), to separate. Separation results in $\psi = A \sin(\pi z/L)\, \varphi(r)$, where A is a constant, L is the length of the cylindrical cavity, z is the axial coordinate, and φ is a function only of r, determined from the differential equation

$$\frac{1}{r}\frac{d}{dr}\left(r \frac{d\varphi}{dr}\right) + [\zeta E^2 - (\pi^2)/(L^2)]\, \varphi = 0. \tag{16.4}$$

The approximation of Eq. (16.2) and the electric field of Eq. (16.3) substituted in Eq. (16.4) lead to the equation

$$\frac{1}{r}\frac{d}{dr}\left(r \frac{d\varphi}{dr}\right) + [k^2 J_0^\beta (2.405\, r/R) - (\pi^2)/(L^2)]\, \varphi = 0. \tag{16.5}$$

[1] M. A. HERLIN and S. C. BROWN: Phys. Rev. **74**, 1650 (1948).

It is difficult to find an analytic solution to this equation. A good approximation is obtained by expressing the BESSEL function as the first two terms of its power series. This approximation is also accurate where the ionization is high and fails only where it is low. Eq. (16.5) then becomes

$$\frac{1}{r}\frac{d}{dr}\left(r\frac{d\varphi}{dr}\right)+\left\{k^2\left[1-\frac{r^2}{b^2}\right]-\frac{\pi^2}{L^2}\right\}\varphi=0, \tag{16.6}$$

where

$$b=0.831\,R\,\beta^{-\frac{1}{2}} \tag{16.7}$$

is the radius at which the ionization goes to zero under the assumptions given above.

The ionization function in Eq. (16.6) is negative beyond $r=b$, which is not physically correct. Accordingly, the ionization is set equal to zero from $r=b$ to $r=R$. Eq. (16.6) is used in the range $0\leq r\leq b$, and in the range $b\leq r\leq R$ the equation

$$\frac{1}{r}\frac{d}{dr}\left(r\frac{d\varphi}{dr}\right)-\frac{\pi^2}{L^2}\varphi=0 \tag{16.8}$$

is applied. The ionization function employed here is compared with the actual ionization function in Fig. 10. They are identical near $r=0$, where the ionization is high. The error in the approximation becomes positive as r increases; negative, as it approaches the radius b. Beyond $r=b$, the ionization drops rapidly to zero and is approximated by the value zero. The boundary conditions on φ are that it be zero at $r=R$ and that its derivatives and value match at $r=b$.

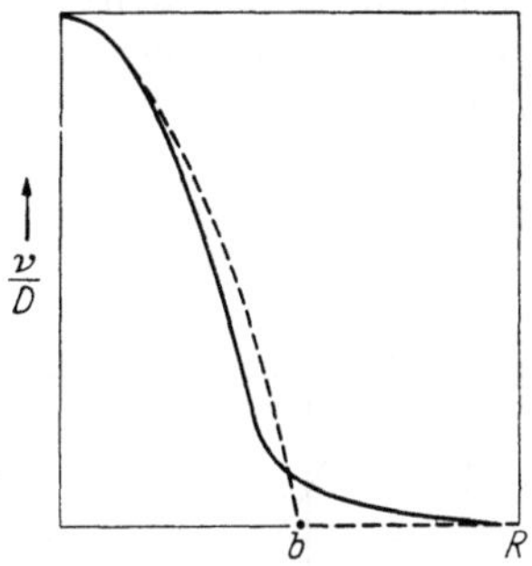

Fig. 10. Comparison of actual ionization function (solid curve) with its approximation (dotted curve).

The solution of Eq. (16.6) is

$$\varphi=\exp\left(-\frac{\sigma x^2}{2}\right)F\left(\frac{2\sigma-1}{4\sigma},\,1;\,\sigma x^2\right), \tag{16.9}$$

where

$$\sigma=\frac{1}{k\,b}\left(1-\frac{\pi^2}{k^2L^2}\right),$$

$$x=\sqrt{1-\frac{\pi^2}{k^2L^2}}\,k r,$$

and F is the confluent hypergeometric function. The second solution is omitted because it has a singularity at the origin. The solution of Eq. (16.8) is

$$\varphi=i\,H_0^{(1)}(i\pi r/L)-\varkappa J_0(i\pi r/L), \tag{16.10}$$

where $\varkappa$ is a constant of integration. It is chosen to make φ equal to zero at the point $r=R$, and it is thus a function of the ratio R/L.

The BESSEL function in Eq. (16.10) is an exponentially increasing function of r; the HANKEL function is an exponentially decreasing function of r. Therefore, $\varkappa$ is zero when R/L is infinity. The exponential decrease in electron density in the region where the ionization rate is very small, and assumed to be zero, is a result of diffusion to the end plates of the cavity. If R/L is not infinite, the negative exponentially increasing BESSEL function term provides the extra decrease in electron density which causes it to go to zero at the finite cavity radius. Numerical computation of the relative magnitudes of these two terms shows, however, that for a wide range of the ratio R/L, the value of $\varkappa$ is very small and the contribution from the BESSEL function term is correspondingly negligible. Thus, for R/L greater than 0.5, the BESSEL term may be neglected. The range below 0.5 may be computed if the BESSEL term is retained. The numerical computations were performed only for $R/L > 0.5$, so that the results presented here are applicable to cavities whose heigths are smaller than their diameters. This increase in the coverage of the range of R/L is a substantial gain over the coverage of the parallel-plate treatment, for which R/L should be greater than 15.

The solutions given in Eqs. (16.9) and (16.10) should each be written with another constant of integration which appears as a multiplying constant. This constant is not written because the matching condition may be satisfied by making the ratio φ'/φ equal on both sides of the matching point $r=b$. The multiplying constant cancels in the ratio. The resulting equation is a transcendental equation for the breakdown field:

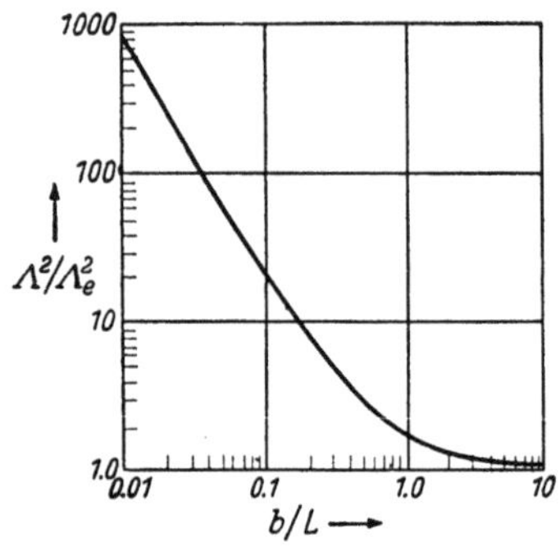

Fig. 11.

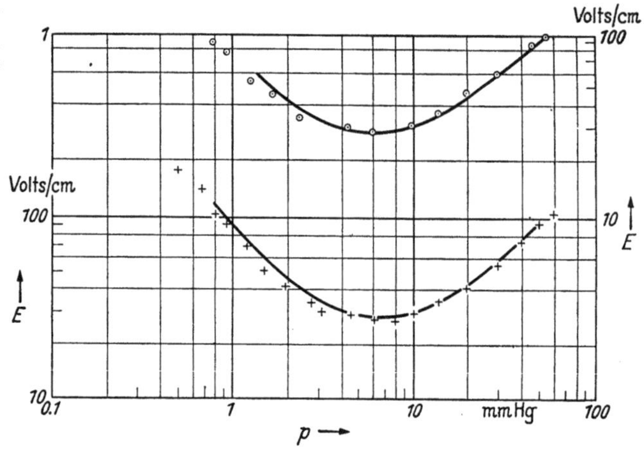

Fig. 12.

Fig. 11. The ratio of the effective diffusion length to real diffusion length for cylinders of arbitrary height as a function of the geometrical dimensions.

Fig. 12. Comparison of experimental points and theoretical curves for a cylindrical 3-inch cavity (+) and a spherical cavity (0) for Heg gas.

$$x_0 \frac{H_1^{(1)}(i x_0)}{i H_0^{(1)}(i x_0)} = y_0 \left[\frac{2\sigma - 1}{2\sigma} \frac{F\left(\frac{6\sigma - 1}{4\sigma}, 2; y_0\right)}{F\left(\frac{2\sigma - 1}{4\sigma}, 1; y_0\right)} - 1 \right], \tag{16.11}$$

where $x_0 = \pi b/L$, and $y_0 = k b$.

Eq. (16.11) may be solved for kb as a function of kL. The results are most conveniently written by expressing $(\pi/kL)^2 = \Lambda_e^2$ as a function of b/L. This plot is shown in Fig. 11. Parallel-plate breakdown requires that $k = \pi/L$, so that the ordinate approaches unity for large b/L. If the tube is long or the slope of the ionization coefficient curve is large, b/L is small; and a larger value of k, and therefore electric field, is required for breakdown, relative to the value that would be required with a uniform field. The procedure used in finding E begins with a plot of ζ versus E/p. Assume a value of E_e/p, read β and ζ_e from the plot, and compute Λ_e. From this we determine $E_e = (\zeta_e)^{\frac{1}{2}}/\Lambda_e$ and compute the corresponding $p = E_e/(E_e/p)$. We can then determine E from E_e, p, and λ. Agreement between the calculated electric field in a TM_{010}-mode cavity and measured values for Heg gas is shown in Fig. 12.

17. Spherical container[1]. The electric field in the lowest electric mode in a spherical cavity may be given by

$$E_r = E_0 \left(\frac{a}{2.75 r}\right) \cos\vartheta\, j_1(2.75 r/a), \tag{17.1}$$

$$E_\vartheta = E_0 \left(\frac{a}{2.75 r}\right) \sin\vartheta \frac{d}{dr} \left[(2.75 r/a)\, j_1(2.75 r/a)\right], \tag{17.2}$$

$$E_\varphi = 0,$$

where r, ϑ, and φ are the spherical coordinate, a is the radius of the sphere, and j_1 is the first-order spherical Bessel function[2].

It is seen that the electric field depends on both r and ϑ the introduction of which makes Eq. (16.1) inseparable. However, in breakdown, we are interested only in the energy transfer from the field to the electrons so that we need take into account only the magnitude of the field at a given point. Near the center of the cavity, where the field, and therefore the ionization, is high, the magnitude of the electric field is approximately spherically symmetric. If we assume that the electric field may be expressed as the average of these values over the whole of the cavity, we may write

$$E = E_0 \left[1 - (r/a)^2\right]. \tag{17.3}$$

This is a good approximation to the average electric field except near the boundaries, where it does not matter.

The assumption that the electric field is independent of ϑ and φ makes ψ independent of these variables. Therefore, with the assumption of Eq. (16.2), we may write Eq. (16.1) as

$$\frac{1}{r^2} \frac{d}{dr} \left(r^2 \frac{d\psi}{dr}\right) + k^2 \left(\frac{E}{E_0}\right)^\beta \psi = 0, \tag{17.4}$$

and introducing the value of E from Eq. (17.3), we have

$$\frac{1}{r^2} \frac{d}{dr} \left(r^2 \frac{d\psi}{dr}\right) + k^2 \left[1 - (r/a)^2\right]^\beta \psi = 0. \tag{17.5}$$

We expand the term in r/a by the binomial theorem and drop powers of (r/a) greater than 2. This makes an appreciable error only near the boundaries, where again the accuracy of the method is unimportant. Then

$$\frac{d^2\psi}{dr^2} + \frac{2}{r} \frac{d\psi}{dr} + k^2 \left(1 - \mu^2 r^2\right) \psi = 0, \tag{17.6}$$

where $\mu^2 = \beta/a^2$; $1/\mu$ is the radius at which the ionization goes to zero under the assumptions given above. Beyond $1/\mu$ these assumptions lead to a negative ζ, which is not physically correct, so we set $\zeta = 0$ for $r > 1/\mu$.

For mathematical convenience, we transform to a dimensionless independent variable and let $k\mu r^2 = x$; the Eq. (17.6) becomes

$$\frac{d^2\psi}{dx^2} + \frac{3}{2x} \frac{d\psi}{dx} + \frac{k^2}{x} \left(\frac{1}{4k\mu} - \frac{x}{4k^2}\right) \psi = 0; \quad x < \frac{k}{\mu}. \tag{17.7}$$

We transform the dependent variable by letting

$$\psi = e^{-x/2} g,$$

[1] A. D. MacDonald and S. C. Brown: Canad. J. Res. **28**, 168 (1950).

[2] For definition see J. Meixner, vol. I of this Encyclopedia, p. 154.

and

$$\frac{d^2 g}{d x^2} + \frac{d g}{d x}\left(\frac{2}{3x} - 1\right) - \frac{\alpha g}{x} = 0, \tag{17.8}$$

where

$$\alpha = \frac{3}{4} - \frac{k}{4\mu}.$$

Eq. (17.8) is the equation for the confluent hypergeometric function in parameters $\frac{3}{2}$ and α. The second solution is not allowed by the boundary condition, and therefore

$$\psi_1 = e^{-x/2} F(\alpha, \tfrac{3}{2}; x), \tag{17.9}$$

where we designate by ψ_1 that part of ψ for which r is less than $1/\mu$ or x is less than k/μ. When x is less than k/μ, ζ is zero, and the differential Eq. (16.1) becomes

$$\frac{d}{dx}\left(x^{\frac{3}{2}} \frac{d\psi}{dx}\right) = 0,$$

of which the solution is

$$\psi_2 = C\left[1 - \left(\frac{x_0}{x}\right)^{\frac{1}{2}}\right], \tag{17.10}$$

where $x_0 = k\mu a^2$ and is determined by the condition that ψ_2 shall be zero on the boundary; C is an arbitrary constant. We must match the solutions of Eqs. (17.9) and (17.10) at the point where $r = 1/\mu$, and therefore

$$\frac{\psi_1'}{\psi_1} = \frac{\psi_2'}{\psi_2},$$

which gives us

$$\frac{\frac{2}{3}\alpha F(\alpha + 1, \frac{5}{2}; x_1)}{F(\alpha; \frac{3}{2}; x_i)} - \frac{1}{2} = \frac{1}{2x_1\left[\left(\frac{x_1}{x_0}\right)^{\frac{1}{2}} - 1\right]}, \tag{17.11}$$

where

$$x_1 = \frac{k}{\mu} = \frac{k a}{\beta^{\frac{1}{2}}}.$$

Eq. (17.11) relates a (the radius of the cavity), β (determined from the slope of the ζ curve), and k, which is inversely proportional to the characteristic diffusion length. The equation may be written in the form

$$\left[\frac{\frac{2}{3}\alpha F(\alpha + 1, \frac{5}{2}; y)}{F(\alpha, \frac{3}{2}; y)} - \frac{1}{2}\right] 2y = \frac{x}{1 - x}, \tag{17.12}$$

where

$$\alpha = \frac{3 - y}{4}, \quad y = \frac{k a}{\beta^{\frac{1}{2}}}, \quad \text{and} \quad x = \beta^{\frac{1}{2}}.$$

Eq. (17.12) is an equation in which the left side is a function of $ka/\beta^{\frac{1}{2}}$ only, and the right side is a function of $\beta^{\frac{1}{2}}$ only. Therefore it is a simple matter to find ka as a function of β. For the case of a uniform field, the characteristic diffusion length in a sphere is a/π, and inspection of Eq. (16.1) indicates that for a uniform field, $k = \pi/a$. We can now consider k as a measure of the effective radius of the discharge for diffusion; ka/π is then Λ/Λ_e, where Λ is the characteristic diffusion length as determined by the geometry of the container, and Λ_e is the effective characteristic diffusion length. Eq. (17.21) is solved numerically and plotted in a fashion similar to that of Fig. 11.

In Fig. 12 are plotted the theoretical values of E as a function of p for a spherical cavity for Heg gas, using the calculated values of Λ_e. Agreement with experiment is also shown.

18. Coaxial cylinders[1]. The value of ζE^2 as a function of position in the discharge cavity is obtained, and from the electric field as a function of position, we have

$$E = [V/r \ln(b/a)] \sin \pi z/L,$$

where V is the "voltage" at the center of the coacial cavity, a and b are the inner and outer conductor radii, L is the length of the cavity, and r and z are the radial and axial space coordinates. This expression applies to a cavity a half-wavelength in length, which was used expreimentally.

The expression given above for the electric field is a function of two space coordinates and leads to a nonseparable differential equation. This difficulty can be avoided by choosing an inner conductor radius which is small compared to the length L, which is a half-wavelength. Since the electric field in the region near the center conductor does not vary much with distance along the conductor, the conditions of infinite length are approached. We may consider the field to be given by the formula

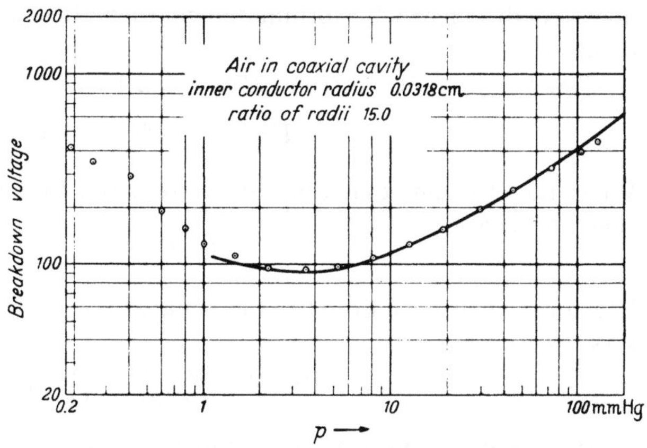

Fig. 13. Comparison of experimental points and theoretical curve for coaxial breakdown.

$$E = V/r \ln(b/a), \qquad (18.1)$$

from which ζE^2 is computed as a function of position. To this approximation Eq. (16.1) is a second-order linear differential equation in r only.

Eq. (16.1), (18.1), and (16.2) now combine into

$$\frac{d^2\psi}{dr^2} + \frac{1}{r}\frac{d\psi}{dr} + k^2\left(\frac{a}{r}\right)^{\beta}\psi = 0, \qquad (18.2)$$

whose solution is

$$\psi = Z_0\left[\frac{2}{\beta-2}\, ka\left(\frac{a}{r}\right)^{1/2(\beta-2)}\right], \qquad (18.3)$$

where Z_0 is a Bessel function of zero order. The boundary condition that $\psi = 0$ at $r = a$ and $r = b$ leads to the following transcendental equation expressing the breakdown condition

$$J_0(x)\, N_0[(b/a)^{1/2(\beta-2)}\, x] - J_0[(b/a)^{1/2(\beta-2)}\, x]\, N_0(x) = 0, \qquad (18.4)$$

where

$$x = \frac{2}{\beta-2}\, ka\left(\frac{a}{b}\right)^{1/2(\beta-2)}. \qquad (18.5)$$

Roots of this equation are tabulated, giving x as a function of $(b/a)^{1/2(\beta-2)}$. Multiplying x by $(b/a)^{1/2(\beta-2)}$, gives $[2/(\beta-2)]\, ka$ as a function of $(b/a)^{1/2(\beta-2)}$.

The agreement between theory and experiment is illustrated in Fig. 13 for the breakdown between coaxial cylinders in air.

[1] M. A. Herlin and S. C. Brown: Phys. Rev. 74, 910 (1948).

c) Effects of a superimposed d. c. field.

19. Breakdown in hydrogen. Solutions for the energy distribution function of electrons in a gas show a very close similarity between the distribution functions under the action of a.c. and of d.c. fields[1,2]. This similarity makes it possible to modify the a.c. distribution function theory for breakdown to take account of the superposition of a d.c. field and thus to predict the behavior of breakdown with both fields acting.

The gas in a cavity will break down when the losses of electrons to the walls of the cavity are replaced by ionization in the body of the gas. When an a.c. field alone is applied, electrons are lost by diffusion. When a small d.c. sweeping field is applied, electrons are lost both by diffusion and mobility. The breakdown condition can be formulated mathematically by a consideration of these processes.

The d.c. flow of electrons $\boldsymbol{\Gamma}$ is given by

$$\boldsymbol{\Gamma} = -n\mu \boldsymbol{E}_{\text{d.c.}} - D\nabla n. \tag{19.1}$$

When the electrons that are lost are replaced by new one resulting from ionization, we have

$$\nabla\cdot\boldsymbol{\Gamma} = \nu_i n. \tag{19.2}$$

If the divergence of Eq. (19.1) is equated to $\nu_i n$, and $\boldsymbol{E}_{\text{d.c.}}$ is directed along the z-axis we obtain

$$\nabla^2 n + \frac{E_{\text{d.c.}}}{D/\mu}\frac{\partial n}{\partial z} + \left(\frac{\nu_i}{D}\right) n = 0. \tag{19.3}$$

Eq. (19.3) can be solved readily for a cylinder of axial heigth L and axial coordinate z, radius R, and radial coordinate r.

By separation of variables,

$$n = M(r)\,N(z),$$

we obtain two equations

$$\nabla_r^2 M + k_1^2 M = 0,$$

$$\frac{d^2N}{dz^2} + \frac{E_{\text{d.c.}}}{D/\mu}\frac{dN}{dz} + \left(\frac{\nu_i}{D} - k_1^2\right) N = 0, \tag{19.4}$$

where k_1^2 is the separation constant, and ∇_r^2 is the two-dimensional LAPLACian in the plane perpendicular to z. The solutions are

$$M = \text{const}\, J_0(k_1 r), \tag{19.5}$$

$$N = \text{const}\,\sin(\pi/L)\,z\,\exp(-\mu E_{\text{d.c.}}\,z/2D), \tag{19.6}$$

where $k_1 = 2.404/R$, and J_0 is the zero-order BESSEL function. The exponential represents the deformation of the sine caused by the sweeping of electrons. This solution is subject to the condition

$$\nu_i/D = 1/\Lambda_{\text{d.c.}}^2, \tag{19.7}$$

where $\Lambda_{\text{d.c.}}$ defines a modified diffusion length given by the relation

$$1/\Lambda_{\text{d.c.}}^2 = 1/\Lambda^2 + [E_{\text{d.c.}}/(2D/\mu)]^2. \tag{19.8}$$

---

[1] T. HOLSTEIN: Phys. Rev. **69**, 50 (1946).

[2] L. J. VARNERIN and S. C. BROWN: Phys. Rev. **79**, 946 (1950).

For this case, the characteristic diffusion legnth is given by

$$1/\Lambda^2 = (\pi/L)^2 + (2.404/R)^2. \tag{19.9}$$

The only difference between the breakdown condition (19.7) in the a.c.-d.c. case and the pure a.c. case is the substitution of a modified diffusion length $\Lambda_{\text{d.c.}}$ for the characteristic diffusion length Λ. Note that the modified diffusion length of a cavity is smaller than the characteristic diffusion length. A cavity whose electron losses are increased by a d.c. sweeping field is equivalent to a smaller cavity without a sweeping field.

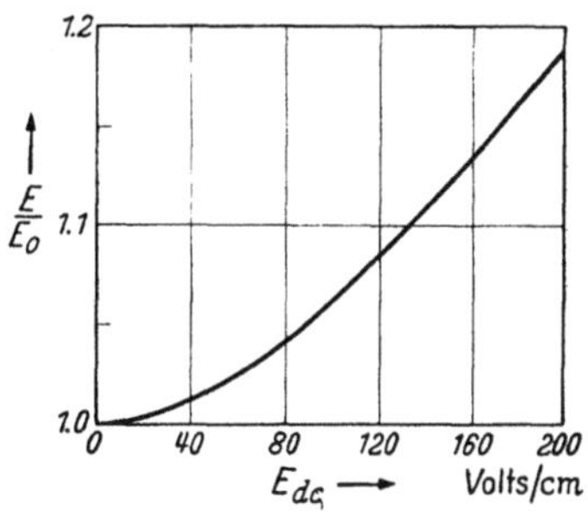

Fig. 14. Relative increase of a.c. breakdown field with superimposed d.c. field for air at a pressure of 38 mm. Hg.

Using the proper distribution function theory to calculate breakdown[1] and the modified diffusion length presented here, a theoretical breakdown curve for an $(E/p)_{\text{d.c.}}$ of 12 volts/cm. has been obtained. The effect of superimposing a d.c. field on an a.c. field is illustrated in Fig. 14.

d) Effects of a superimposed magnetic field.

The breakdown of gases by high-frequency electric fields in the presence of a constant magnetic field has been studied by Townsend and Gill[2]; by A. E. Brown[3]; and by Lax, Allis, and S. C. Brown[4]. As in the case of strictly a.c. field breakdown, two approaches are available: the average electron theory and the distribution function theory. As there are advantages to each, both will be used. In Sect. 20, the average electron theory is given. In this method the orbit of a free electron in the assumed fields is computed first, and from this one computes the displacement and the energy gain in the time $\tau = t - t_0$ elapsed since a collision. These quantities are then averaged over the phase of the a.c. field at the time t_0 of the last collision, over the direction in space of the velocity after the collision, and over the free time up to the next collision. The result is the mean-square displacement and energy gain of the average electron between collisions. One can then discuss an average electron from its initial low energy until it ionizes a gas atom or diffuses out of the tube. The condition for breakdown is that these two final achievements shall be equally probable. This method has the advantage that each step in the analysis has a direct physical meaning.

In Sect. 21 the Boltzmann transport equation is expanded in spherical harmonics in space and in Fourier series in time. There results a differential equation for the distribution function, which is integrated. Most of the properties of a discharge follow directly from a knowledge of the distribution function.

20. Average electron theory. Consider the motion of an electron between collisions under the influence of a uniform electroc field along the x-axis, $E = E_p \exp(j\omega t)$, and a uniform and constant magnetic field B along the z-axis. The equation of motion is then

$$\boldsymbol{F} = -e\boldsymbol{E} - e\boldsymbol{v} \times \boldsymbol{B} = m\dot{\boldsymbol{v}}. \tag{20.1}$$

The solution of this equation is the sum of a general and a particular integral, which correspond to the superposition of a random circular helical motion and

[1] L. J. Varnerin and S. C. Brown: Phys. Rev. **79**, 946 (1950).
[2] J. S. Townsend and E. W. B. Gill: Phil. Mag. **26**, 290 (1938).
[3] A. E. Brown: Phil. Mag. **29**, 302 (1940).
[4] B. Lax, W. P. Allis and S. C. Brown: J. Appl. Phys. **21**, 1297 (1950).

a plane elliptical motion. For the helical motion, whose axis is along the magnetic field, the velocity is

$$\left.\begin{aligned} v_{1x} &= (a + j b) \exp(j \omega_b \tau), \\ v_{1y} &= (b - j a) \exp(j \omega_b \tau), \\ v_{1z} &= c \end{aligned}\right\} \tag{20.2}$$

and oscillates at the cyclotron frequency $\omega_b = e B/m$. The helical motion contains the three arbitrary constants; the energy of this motion is constant and is given in electron-volts by

$$u_1 = \frac{m v_1^2}{2e} = \frac{m}{e} \frac{a^2 + b^2 + c^2}{2}. \tag{20.3}$$

For the elliptical motion, the velocity is

$$\left.\begin{aligned} v_{2x} &= \frac{e E_p}{m} \frac{j\omega}{\omega^2 - \omega_b^2} \exp(j \omega t), \\ v_{2y} &= \frac{e E_p}{m} \frac{\omega_b}{\omega^2 - \omega_b^2} \exp(j \omega t) \end{aligned}\right\} \tag{20.4}$$

and oscillates at the frequency of the applied field[1]. The kinetic energy of this motion is uniquely determined by the magnitude and frequency of the applied field and is given by

$$u_2 = \frac{e E_p^2}{8m} \left[\frac{1}{(\omega - \omega_b)^2} + \frac{1}{(\omega + \omega_b)^2} + \frac{2 \cos 2\omega t}{\omega_b^2 - \omega^2}\right]. \tag{20.5}$$

The total energy $u = (m/2e)(\boldsymbol{v}_1 + \boldsymbol{v}_2)^2$ will contain cross-product terms u_{12} which are important but rather lengthy to write down. Their value will be given later. The three constants, a, b, c, of the helical motion are determined by the velocity $\boldsymbol{v}_0 = \boldsymbol{v}_1(\tau = 0) + \boldsymbol{v}_2(t = t_0)$ immediately after a collision. Since the time τ has been used in Eq. (20.2), we have simply

$$\left.\begin{aligned} a &= v_{0x} + \frac{e E_p}{m} \frac{\omega}{\omega^2 - \omega_b^2} \sin(j \omega t_0), \\ b &= v_{0y} - \frac{e E_p}{m} \frac{\omega_b}{\omega^2 - \omega_b^2} \cos(j \omega t_0), \\ c &= v_{0z}. \end{aligned}\right\} \tag{20.6}$$

It is noted that the elliptical motion exhibits a resonance at frequencies near the cyclotron frequency. Exactly at this frequency, Eqs. (20.4) no longer hold, and the solution corresponds to a spiral; as collisions interrupt the motion it will not be necessary to use this singular solution.

From the velocities one obtains the displacements x_1, y_1, z_1, x_2, y_2 by integration. From these the mean displacements $\langle x_1 \rangle$, etc., can be calculated, but these all vanish on averaging over orientations of v_0, which is assumed isotropic, and over collision times t_0. In the average, an electron stays where it is in a high-frequency discharge.

We are interested, however, in the mean-square displacements $\langle x_1^2 \rangle, \langle y_1^2 \rangle, \langle z_1^2 \rangle$ because these lead to the diffusion coefficient. One finds that the cross-product terms such as $\langle x_1 y_1 \rangle$ all vanish when averaged.

$$\begin{aligned} x_1^2 + y_1^2 &= 2 \frac{a^2 + b^2}{\omega_b^2} (1 - \cos \omega_b \tau), \\ z_1^2 &= c^2 \tau^2. \end{aligned} \tag{20.7}$$

[1] ALLIS: Eqs. (3.2) and (3.4).

Averaging over orientations and times t_0, one finds that

$$\langle a^2\rangle = \langle b^2\rangle = \langle c^2\rangle = \frac{v_1^2}{3}; \quad \langle a b\rangle = 0. \tag{20.8}$$

The cross terms between the helical and elliptical motions also vanish, but terms $\langle x_2^2\rangle$ and $\langle y_2^2\rangle$ do not vanish. However, these terms represent the mean-square displacements from mobility in the applied a.c. field and are not wanted in calculating the diffusion.

The average of a quantity X over the free times $\tau_m = 1/\nu_m$ between collisions ist, by definition,

$$X = \int_0^\infty X \exp(-\nu_m \tau)\, \nu_m\, d\tau. \tag{20.9}$$

Applying this to the quantities in Eqs. (20.7) and defining the diffusion coefficients in terms of the mean-square displacements, we obtain[1]

$$D_{yy} = D_{xx} = \frac{\langle x_1^2\rangle}{2\tau_m} = \frac{v_1^2 \nu_m}{3(\omega_b^2 + \nu_m^2)},$$
$$D_{zz} = v_1^2/3\nu_m. \tag{20.10}$$

This definition of the diffusion tensor is of necessity symmetric. We shall see later that there are skew-symmetric terms which the random-walk definition cannot give. Diffusion along the z-axis is not altered by the magnetic field, but in the plane at right angles to the field it is reduced in the ratio $\nu_m^2/(\omega_b^2 + \nu_m^2)$. For a given collision frequency, the diffusion coefficient is proportional to the energy of the electrons in their helical motion[2].

The mean energy gain between collision is best obtained by considering the power input to an electron $P = -e\boldsymbol{E}\cdot\boldsymbol{v}$. Since the velocity v_{2x} is out of phase with the field, the corresponding power P_2 into this motion is zero in the average, and only the power P_1 need be considered, where

$$P_1 = -eE_p \cos\omega t\,(a\cos\omega_b\tau - b\sin\omega_b\tau), \tag{20.11}$$

and the constants a and b are given by Eqs. (20.6). In averaging over orientations of the initial velocity the terms in v_0 drop out. Averaging over t_0 yields

$$\overline{P} = \frac{e^2 E_p^2}{4m}\left[\frac{\sin(\omega+\omega_b)\tau}{\omega+\omega_b} + \frac{\sin(\omega-\omega_b)\tau}{\omega-\omega_b}\right] \tag{20.12}$$

from which the average energy is obtained by integrating with respect to τ from 0 to τ.

$$\langle u_{12}\rangle = \frac{eE_p^2}{4m}\left[\frac{1-\cos(\omega+\omega_b)\tau}{(\omega-\omega_b)^2} + \frac{1-\cos(\omega-\omega_b)\tau}{(\omega-\omega_b)^2}\right]. \tag{20.13}$$

Averaging this quantity over collision times gives the mean energy gain between collisions:

$$u_c = \frac{eE_p^2}{4m}\left[\frac{1}{(\omega+\omega_b)^2+\nu_m^2} + \frac{1}{(\omega-\omega_b)^2+\nu_m^2}\right] \equiv \frac{eE_e^2}{m\nu_m^2}. \tag{20.14}$$

This is a fundamental quantity in this theory. At low pressures ($\nu_m \to 0$) we see that it approaches twice the mean energy $\overline{u}_2$ of the elliptical motion of an electron, and in no case does it exceed this. At higher pressures, many collisions per oscillation, the energy u_2 loses its meaning, and the collision energy becomes

[1] ALLIS: Eq. (13.8).

[2] A more complete discussion of this is given in ALLIS, Sect. 13.

$eE_p^2/2mv_m^2$. One can use Eq. (20.14) to define an effective field E_e which is the root-mean-square field at high pressure. This concept is useful when the collision frequency ν_m is independent of velocity because this single function takes into account the effects of frequency and magnetic field on the energy.

At low pressures the effective field has a maximum at resonance with the cyclotron frequency, as shown in Fig. 15.

The electrons produced by ionization have initially very little energy, but this increases by steps of u_c until the energy reaches a value u_i, at which ionization occurs. This is above the ionization potential V_i by an amount which we shall neglect. Excitations are disregarded in the following simple theory. The number N of free times for ionizing is $N = u_i/u_c$ when ν_m is constant.

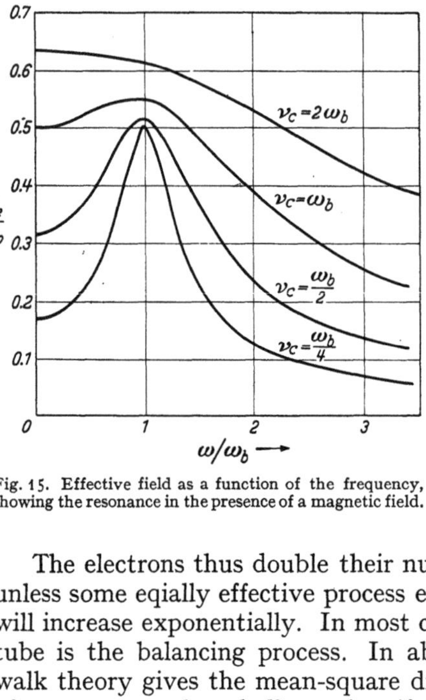

Fig. 15. Effective field as a function of the frequency, showing the resonance in the presence of a magnetic field.

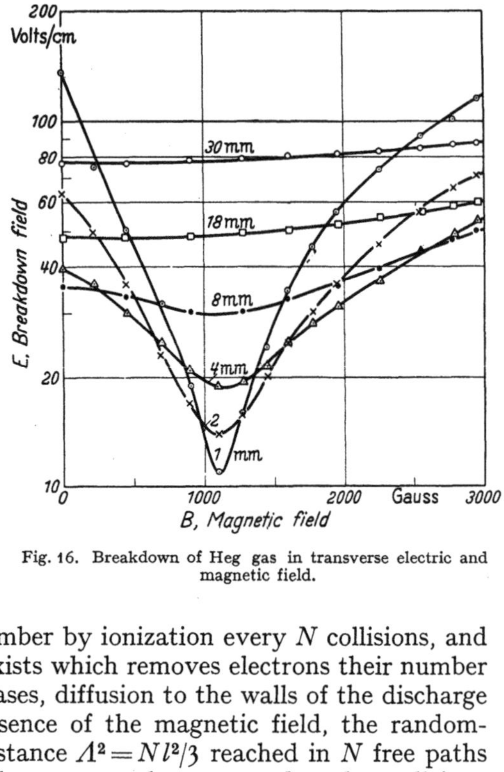

Fig. 16. Breakdown of Heg gas in transverse electric and magnetic field.

The electrons thus double their number by ionization every N collisions, and unless some eqially effective process exists which removes electrons their number will increase exponentially. In most cases, diffusion to the walls of the discharge tube is the balancing process. In absence of the magnetic field, the random-walk theory gives the mean-square distance $\Lambda^2 = N l^2/3$ reached in N free paths of mean-square length l^2, so that if the average electron reaches the wall in a distance Λ the diffusion process will just balance ionization. This is the condition for breakdown, and we can write it

$$\frac{u_c}{u_i} = \frac{\langle l^2 \rangle}{3\Lambda^2} = \frac{\langle v^2 \rangle}{3\Lambda^2 \nu_m^2}, \tag{20.15}$$

where Λ is now a length characteristic of the discharge tube.

If there is a magnetic field, u_c will be altered according to Eq. (20.14). At the same time the random-walk theory must be altered to take into account the curved paths between collisions. This may be done by appropriately decreasing l or increasing Λ. We shall adopt the latter procedure and denote the new length by Λ_e. Its value will be given later.

When the mean free path is much smaller than Λ_e, the intercollision energy gain u_c is correspondingly smaller than the ionization potential. From Eq. (20.15) we see that breakdown should occur at the same effective field if the ratio of the mean free path to the effective diffusion length is the same; that is, the effective field for breakdown is a function of $p\Lambda_e$ only.

Combining Eqs. (20.14) and (20.15) we get an expression similar to Eq. (3.3):

$$E_e^2 = \frac{2}{3}\frac{\bar{u}\,u_i}{\Lambda_e^2}. \tag{20.16}$$

Experimental data for breakdown in Heg are shown in Fig. 16. This resonance effect of the magnetic field and the high frequency are removed by using the effective field.

In order to test Eq. (20.10) for the diffusion coefficient, breakdown was studied in a flat cylindrical cavity whose length was very short compared to the radius. With the magnetic field placed transverse to the axis most of the diffusion has to take place perpendicular to the magnetic field and hence will show the full reduction. By Eq. (20.10), the mean square of the distance traveled by an electron is proportional to the diffusion coefficient D, and therefore the effective diffusion length Λ_e appropriate to infinite parallel plates is

$$\Lambda_e^2 = \frac{\omega_b^2 + \nu_c^2}{\nu_m^2}\Lambda^2. \tag{20.17}$$

The effect of a magnetic field is to make the dimensions of the cavity at right angles to the field appear larger to an electron. By Eq. (20.16) this should reduce the effective field for breakdown in the same proportion.

However, Eq. (20.16) does not correspond with experiment except when used for comparative purposes. Eq. (20.14) and the random-walk theory give the number of free times for the average electron to ionize and reach the wall, respectively. But in a discharge it is the faster-than-average electron that ionizes and the more-mobile-than-average electron that leaves the tube, and these are not the same electron. The mean free path method must therefore fail in predicting quantitative breakdown, and the failure should be worst when there are many collisions and therefore the greatest deviations from the mean.

21. BOLTZMANN Theory. Introducing the magnetic field into the BOLTZMANN transport equation, Eq. (6.1), gives

$$C = \frac{\partial F}{\partial t} + \nabla\cdot(\boldsymbol{v}\,F) - \nabla_v\cdot\frac{e\,\boldsymbol{E}\,F}{m} - \nabla_v\cdot\frac{e}{m}\,\boldsymbol{v}\times\boldsymbol{B}\,F. \tag{21.1}$$

Using Eq. (6.2), we obtain

$$\left.\begin{aligned}
&(\nu_x + \nu_i - q)\,F_0^0 + \frac{m}{M}\frac{1}{v^2}\frac{\partial}{\partial v}\left(\nu_m v^3 F_0^0\right) - \frac{v}{3}\nabla\cdot\boldsymbol{F}_1^0 + \\
&\qquad + \frac{e}{m}\frac{1}{3v^2}\frac{\partial}{\partial v}\left[\frac{v^2}{2}\boldsymbol{E}_p\cdot(\boldsymbol{F}_1^1)_{\mathrm{real}}\right] = 0,\\
&\nu_m\boldsymbol{F}_1^0 + v\,\nabla F_0^0 - \frac{e}{m}\boldsymbol{B}\times\boldsymbol{F}_1^0 = 0,\\
&(\nu_m + j\,\omega)\,\boldsymbol{F}_1^1 - \frac{e}{m}\boldsymbol{E}_p\frac{\partial F_0^0}{\partial v} - \frac{e}{m}\boldsymbol{B}\times\boldsymbol{F}_1^1 = 0.
\end{aligned}\right\} \tag{21.2}$$

These are the necessary equations[1] for handling breakdown problems, which represent steady state conditions for the electrons. Eq. (2.2) are applicable for any orientation of $\boldsymbol{E}$ and $\boldsymbol{B}$. We shall consider only those cases in which they are perpendicular or parallel to each other.

[1] ALLIS: Sect. 24, expecially Eqs. (24.5) to (24.8).

Integrating the first expression of Eq. (21.2) over velocity space in spherical coordinates, the second and fourth terms vanish at the limits. The first term gives the total production rate of electrons, $\nu_i n$, from ionization. This is because excitations merely withdraw fast electrons to replace them by slow ones, whereas ionizations add an extra electron. The third term gives the divergence of the flow vector $\boldsymbol{\Gamma}$

$$\boldsymbol{\Gamma} = \int_0^\infty \frac{4\pi}{3} \boldsymbol{F}_1^0 v^3 \, dv \tag{21.3}$$

so that

$$n \nu_i = \nabla \cdot \boldsymbol{\Gamma}. \tag{21.4}$$

Solving the second expression of Eq. (21.2) for $\boldsymbol{F}_1^0$, we find

$$\boldsymbol{F}_1^0 = -\frac{v}{\nu_m} \frac{\nu_m^2 + \boldsymbol{\omega}_b \boldsymbol{\omega}_b \cdot + \nu_m \boldsymbol{\omega}_b \times}{\nu_m^2 + \omega_b^2} \nabla F_0^0, \qquad \omega_b = (e/m) \boldsymbol{B} \tag{21.5}$$

which we substitute in Eq. (21.3) and, assuming that F_0^0 can be written as a product $n(x, y, z) f_0(v)$, we find that Γ is proportional to ∇n but not necessarily in the same direction. Accordingly, it is possible to define a diffusion coefficient which is a tensor:

$$\Gamma_i = -\Sigma_j D_{ij} \frac{\partial n}{\partial j}; \qquad i, j = x, y, z. \tag{21.6}$$

If the magnetic field is taken along the z-axis, D_{ij} is given[1] by

$$D_{ij} = \int_0^\infty f_0 \begin{bmatrix} \frac{\nu_m}{\nu_m^2 + \omega_b^2} & \frac{-\omega_b}{\nu_m^2 + \omega_b^2} & 0 \\ \frac{\omega_b}{\nu_m^2 + \omega_b^2} & \frac{\nu_m}{\nu_m^2 + \omega_b^2} & 0 \\ 0 & 0 & \frac{1}{\nu_m} \end{bmatrix} \frac{4}{3} \pi v^4 \, dv. \tag{21.7}$$

This expression reduces to the ordinary coefficient $D = \langle v^2/3\nu_m \rangle$ with no magnetic field, $\omega_b = 0$. It is equivalent to Eq. (20.10) except that the present tensor has skewsymmetric terms that were not obtained by the random-walk definition.

Substituting in Eq. (21.4), we obtain the diffusion equation

$$D_{xx} \frac{\partial^2 n}{\partial x^2} + D_{yy} \frac{\partial^2 n}{\partial y^2} + D_{zz} \frac{\partial^2 n}{\partial z^2} + \nu_i n = 0 \tag{21.8}$$

which determines the spatial distribution of the electrons. One can use the normal diffusion coefficient D if lengths are expanded at right angles to the magnetic field in the ratio $(\nu_m^2 + \omega_b^2)^{\frac{1}{2}}/\nu_m$.

The solution of this equation depends on the boundary conditions. One must define an effective diffusion length Λ_e for the whole cavity which takes into account these expansions:

$$\frac{1}{\Lambda_e^2} = \left(\frac{1}{\Lambda_x^2} + \frac{1}{\Lambda_y^2}\right) \frac{\nu_m^2}{\nu_m^2 + \omega_b^2} + \frac{1}{\Lambda_z^2}. \tag{21.9}$$

The effect of the magnetic field is equivalent to expanding the cavity in the ratio $(\nu_m^2 + \omega_b^2)^{\frac{1}{2}}/\nu_m$ in all directions perpendicular to the magnetic field.

The distribution function F_0^0 is obtained by eliminating $\boldsymbol{F}_1^0$ and $\boldsymbol{F}_1^1$ from Eqs. (21.2). In this way one obtains Eq. (8.1), in which Λ is replaced by Λ_e, and E by E_e [defined in Eq. (20.14)]. It follows that all of the breakdown data should plot on a single curve when $E_e \Lambda_e$ is plotted against E_e/p. This is shown in Fig. 17.

[1] ALLIS: Eqs. (30.3) to (30.7).

This exact law is, however, of very limited applicability. The effective quantities E_e and Λ_e depend on ν_m, which is a function of the electron's energy. It is therefore impossible, in general, to make the effective values the same at all of the energies, as the law requires. For helium and hydrogen the collision frequency ν_m is very nearly constant, so that the effective values are significant.

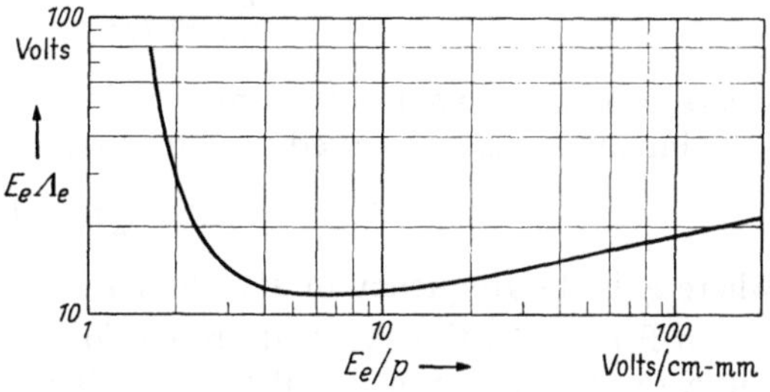

Fig. 17. Theoretical curve for the effective breakdown in Heg gas.

To study the effect of the magnetic field on diffusion alone, the electric and magnetic fields are oriented in the same direction; and in order to reduce diffusion along the magnetic field it is nece ssary to perform the experiment in a cavity whose heigth is greater than the radius. In such a cavity the electric field may no longer be considered uniform, and a correction to the computations must be made in a manner shown in Sect. 16. In the presence of the longitudinal magnetic field the equivalent diffusion length of the cylinder, from Eq. (21.9), is

$$\frac{1}{\Lambda_e^2} = \frac{\nu_m^2}{\nu_m^2+\omega_b^2}\left(\frac{2.405}{R}\right)^2 + \left(\frac{\pi}{L}\right)^2. \tag{21.10}$$

Using this and the nonuniform field correction to the BOLTZMANN theory one obtains the agreement with experiment shown in Fig. 18. This result confirms the predicted effect of the magnetic field upon diffusion.

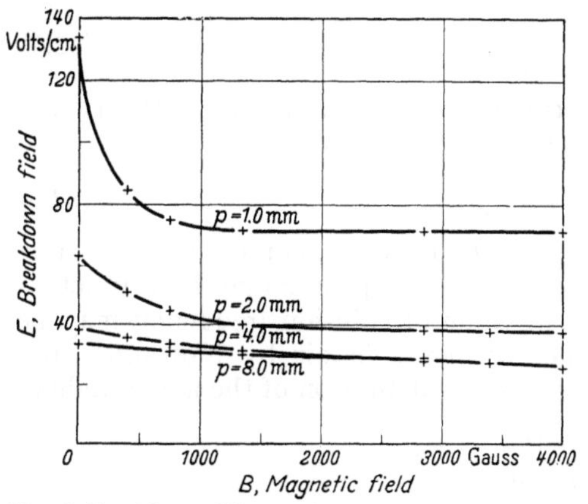

Fig. 18. Breakdown of Heg gas in parallel electric and magnetic fields in a cylindrical cavity. The solid curves are theoretical; the points are experimental.

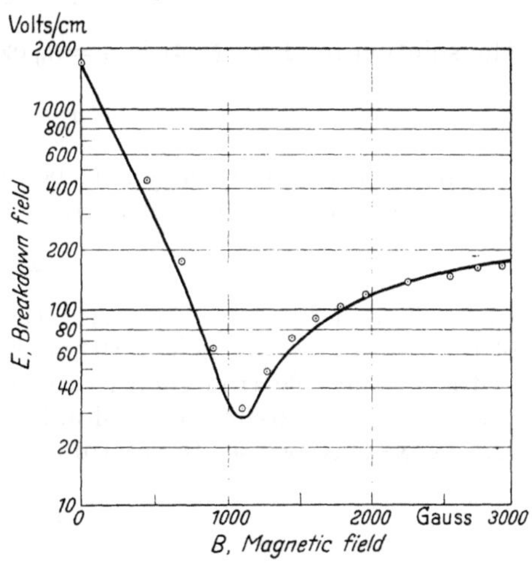

Fig. 19. Breakdown of helium at a pressure of 1 mm. Hg in a cylindrical cavity. Solid line is obtained from the BOLTZMANN theory; points are experimental.

The breakdown measurements shown in Fig. 16 were made in a flat cavity with the magnetic field transverse to the axis. The effect of the latter is to require the solution of the diffusion equation in an elliptical cylinder whose diffusion length is then given by

$$\frac{1}{\Lambda_e^2} = \frac{\nu_m^2}{\nu_m^2+\omega_b^2}\left[\left(\frac{\pi}{L}\right)^2 + \frac{1}{2}\left(\frac{2.405}{R}\right)^2\right] + \frac{1}{2}\left(\frac{2.405}{R}\right)^2. \tag{21.11}$$

Using this and the nonuniform field correction gives the theoretical curves shown in Fig. 19.

B. Electron attachment controlled breakdown.

22. Breakdown in air. The continuity equation for electrons introduced in Eq. (1.8) can be written in a more general form:

$$\frac{\partial n}{\partial t} = \nu_i n - \nu_e n + S \tag{22.1}$$

where ν_e is any loss mechanism. If, in addition to the diffusion loss, electron attachment is also observed, ν_e, the frequency of loss, may be given as

$$\nu_e n = \nu_a n - D \nabla^2 n, \tag{22.2}$$

where ν_a is the frequency of attachment.

In Eq. (22.1), the external production rate, S, will be considered as introducing a residual electron density n_0, so that the continuity equation becomes

$$\frac{\partial n}{\partial t} = \nu_i n - (\nu_a n - D \nabla^2 n) \tag{22.3}$$

with the initial condition that at $t=0$, $n=n_0$.

For those experiments that are available for testing a simple average electron theory of high-frequency breakdown in air, it is a good assumption to take the diffusion as occurring between two parallel plates separated by a distance L. The density configuration is then of the form $\cos(\pi x/L)$, and to a first approximation the term $\nabla^2 n$ can be replaced by $(-\pi^2 n/L)$. Eq. (22.3) becomes

$$\frac{\partial n}{\partial t} = \left(\nu_i - \nu_a - \frac{D\pi^2}{L^2}\right) n = \gamma n. \tag{22.4}$$

The solution for Eq. (22.4) is $n = n_0 \exp(\gamma t)$. For breakdown, $\gamma=0$, and

$$\nu_i = \nu_a + \frac{D\pi^2}{L^2}. \tag{22.5}$$

The ionization frequency may be determined from a knowledge of the TOWNSEND first ionization coefficient α, and the a.c. drift velocity v:

$$\nu_i = \alpha v = \alpha \mu_{\text{d.c.}} E_e, \tag{22.6}$$

where E_e is defined by Eq. (1.3). If $\nu_m^2 \gg \omega^2$, E_e is equivalent to the rms value of the high-frequency field. Although the collision frequency may be a function of the average electron energy, an average value of collision frequency may be used as an approximation. Mobility measurements in air by NIELSEN[1], and RIEMANN[2] may be used to determine μ through the definition of the a.c. mobility:

$$\mu = \frac{(e/m)\,\nu_m}{\nu_m^2 + \omega^2}.$$

These data yield a value of ν_m for air of 4.3×10^9 p. The pressure must be greater than 20 mm. Hg at a 10 cm. wavelength and greater than 70 mm. Hg at a 3 cm. wavelength in order for the effective field to be equivalent to the rms field.

The attachment frequency may be determined from a knowledge of the number of electrons attached per electron in a path of one centimeter, which is represented by β, and the a.c. drift velocity. The ratio of the diffusion coefficient to the mobility is given by $(D/\mu) = 2u_{av}/3$ where u_{av} is the average electron energy in electron volts, and the numerical constant depends on the electron distribution function. The value of $\frac{2}{3}$ is correct for a MAXWELLian velocity distribution.

[1] R. A. NIELSEN: Phys. Rev. **50**, 950 (1936).
[2] W. RIEMANN: Z. Physik **122**, 216 (1944).

The condition for breakdown becomes:

$$\frac{\alpha}{p} = \frac{\beta}{p} + \frac{2}{3} \frac{u_{av}\,\pi^2}{(E_e/p)\,(pL)^2}\,. \tag{22.7}$$

The quantities α/p, β/p, and u_{av} are all functions of E_e/p and depend upon the energy distribution function. Measurements of HARRISON and GEBALLE[1] in air yield experimental values of α/p and β/p as a function of E/p. The average electron energy as a function of E/p is given by HEALEY and REED[2].

The solid line in Fig. 20 shows E_e/p as a function of pL as predicted by Eq. (22.7). Measurements of HERLIN and BROWN[3] at 3000 Mc./sec. with the distance varying from 0.635 cm. to 0.158 cm. and the pressure varying from 70 mm. Hg to 2 mm. Hg are shown. Measurements of PIM[4] at 2000 Mc./sec.

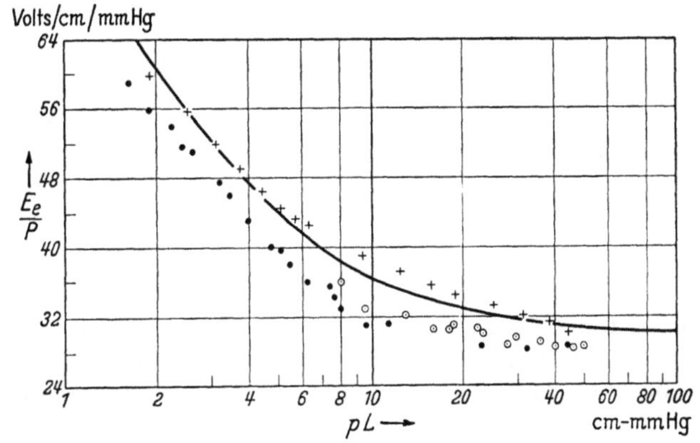

Fig. 20. Breakdown in air. Solid line, theory; ● data of HERLIN and BROWN; ⊙ data of PIM; + "pure" air.

with the distance varying from 0.08 cm. to 0.06 cm. and the pressure varying from 760 mm. Hg to 160 mm. Hg are also given. The experimental results of these observers agree well among themselves but differ from the theoretical curve by as much as 10%. These data were taken by breaking down the gas at the highest pressure possible, for a given set of experimental conditions, reducing the pressure incrementally by pumping, breaking it down again, and so on. Thus, any breakdown products formed at one stage might be present at any subsequent breakdown. The data marked "pure" were taken by introducing fresh air for each measurement. The air was pumped to the desired pressure and only one breakdown initiated. The system was then pumped to 10 μ Hg or below, and the procedure was repeated for the next pressure. Breakdown impurities (such as the oxides of nitrogen) remaining in the gas were thus reduced between readings by a factor of about 10^4. Data collected in this way are shown in Fig. 20 and agree much better with the theory of Eq. (22.7).

23. Breakdown in oxygen. The breakdown condition of Eq. (22.7) may also be solved for oxygen, since we have the measurement of α/p and β/p for oxygen from the work of HARRISON and GEBALLE[5]. A theoretical plot of E_e/p as a function of pL for O_2 is shown in Fig. 21. Microwave measurements in oxygen at

[1] M. A. HARRISON and R. GEBALLE: Phys. Rev. **50**, 1 (1953).
[2] R. H. HEALEY and F. W. REED: The Behaviour of Slow Electrons in Gases. Sydney 1941.
[3] M. A. HERLIN and S. C. BROWN: Phys. Rev. **74**, 291 (1948).
[4] J. A. PIM: Proc. Inst. Electr. Engr., Part. III **96**, 117 (1949).
[5] M. A. HARRISON and R. GEBALLE: Phys. Rev. **91**, 1 (1953).

3000 Mc./sec. are also shown, with $L = 0.635$ cm. and over a range of pressures from 70 to 2 mm. Hg. In calculating the value of E_e from Eq. (1.5), the value of ν_m obtained from the mobility measurements of NIELSEN and BRADBURY[1] and the relation for the a.c. mobility. This value of ν_m is 3.5×10^9 p.

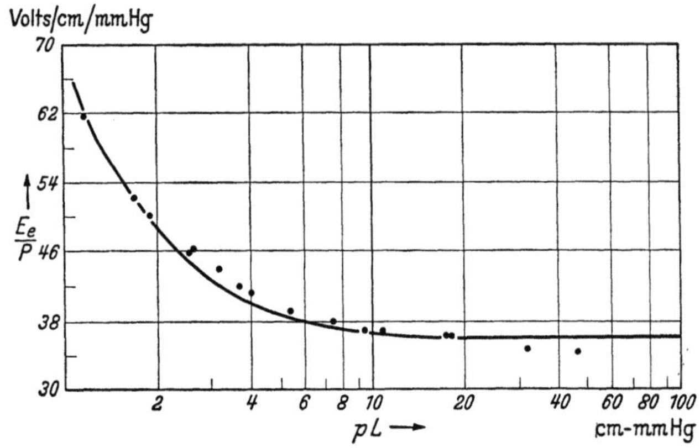

Fig. 21. Breakdown in oxygen. The solid curve is the theory, the points are experimental.

C. Mobility-controlled breakdown.

24. Behavior of the average electron. When the amplitude of the electron oscillation in an electric field is sufficiently high, the electrons can travel completely across the tube and collide with the walls on every half-cycle. If this

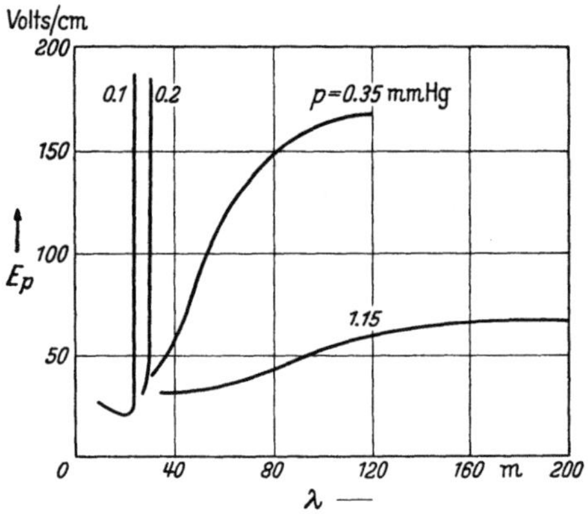

Fig. 22. Breakdown in neon for a flat-ended cylindrical tube 3.55 cm. in length.

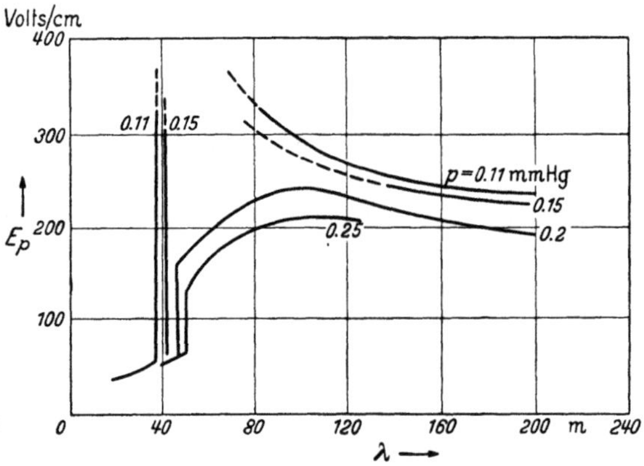

Fig. 23. Breakdown in nitrogen for a flat-ended cylindrical tube 3.55 cm. in length.

can occur, the loss of electrons will be controlled by this mobility motion, and the breakdown field will depend upon the amplitude of the electron oscillation[2,3] as given by Eq. (5.8). This equation gave the oscillation amplitude limit for diffusion. On crossing this limit from the diffusion side, a new loss mechanism occurs. The electrons collide with the walls so that the electric field necessary for causing breakdown rises abruptly. This is illustrated[3] in Figs. 22, 23, and 24. Referring to these figures, breakdown to the left of the sharp discontinuities

[1] R. A. NIELSEN and N. E. BRADBURY: Phys. Rev. **51**, 69 (1937).
[2] J. THOMPSON: Phil. Mag. **23**, 1 (1937). — S. GITHENS jr.: Phys. Rev. **57**, 822 (1940).
[3] E. W. B. GILL and A. VON ENGEL: Proc. Roy. Soc. Lond., Ser. A **197**, 107 (1949).

is diffusion-controlled. At that wavelength of the a.c. field corresponding to the amplitude of oscillation limit for diffusion, electrons are lost by colliding with the walls, and the breakdown field must be sharply increased to make up for this new loss. The breakdown field in the mobility-controlled region is a function of the type of material of which the walls are composed, as we would expect.

In contradistinction to the low-pressure type of secondary emission-controlled breakdown (the electron resonance breakdown, discussed in Chapter D) this high-pressure case is very much a function of the type of gas. This is evident from Eq. (1.2), since the velocity of the oscillating electrons is mobility-controlled, and the mobility is a function of the type of gas. The dependence of the amplitude of oscillation on the electron mobility provides a good substantiation of the factors controlling the critical conditions for crossing the amplitude of oscillation transition. Let us illustrate by taking the case in hydrogen where the drift velocity of electrons is approximately linearly proportional to E/p. This should predict that the critical electric field at the transition value should be proportional to the gas pressure. This is found to be approximately the case by Gill and von Engel, as illustrated in Fig. 25.

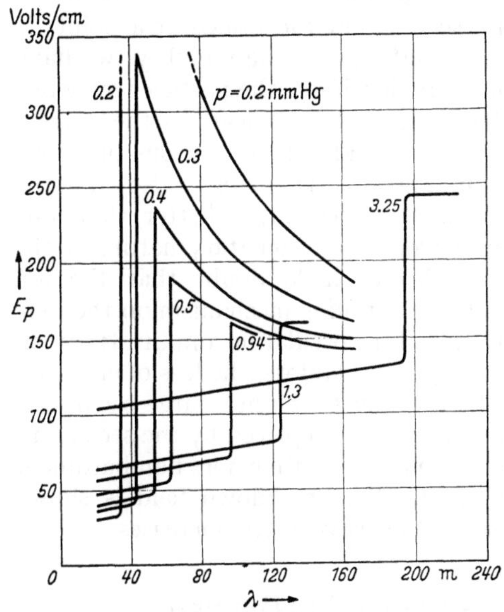

Fig. 24. Breakdown in hydrogen for a flat-ended cylindrical tube 3.55 cm. in length.

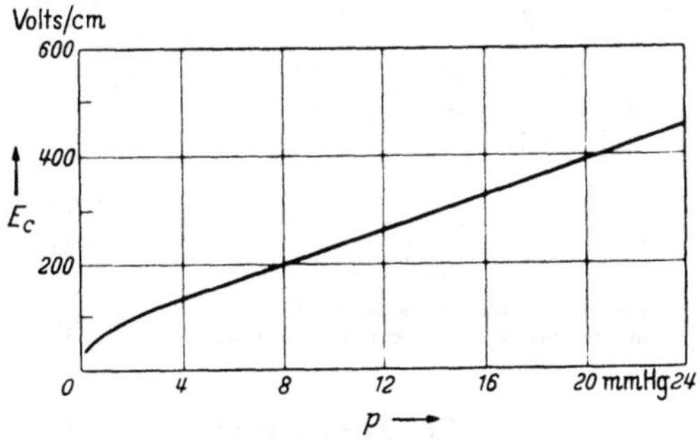

Fig. 25. Cut-off field strength as a function of pressure for hydrogen.

One of the most striking characteristics of the mobility-controlled breakdown is the fact that after the electron oscillation amplitude is greater than the tube dimensions—that is, after all the electrons are lost to the walls—breakdown is still observed. This is illustrated in Figs. 22 to 24 by the measured breakdown fields on the longer wavelength side of the oscillation amplitude transition. These data also show that the longer the wavelength of the applied a.c. field, the more easily the discharge starts. The observation that it is such a strong function of the frequency of the electric field requires the effect to be associated with charged particles. The fact that the electrons are being lost every half-cycle to the walls suggests that the only particles left to account for this increase in ionization must result from the positive ions.

Gill and von Engel used their data in hydrogen to prove that the breakdown beyond the amplitude of oscillation transition was controlled by the positive ion mobility. The amplitude of the oscillation of the velocity obtained from Eq. (1.2) is

$$v_{\max} = \frac{e E_p}{m^* (\omega^2 + \nu_m^2)^{\frac{1}{2}}} = \frac{\sqrt{2}\, e E_e}{m^* \nu_m} \tag{24.1}$$

where m^* is the mass of whatever particle is effective in controlling the discharge. If breakdown occurs when the maximum velocity has reached a constant value v, which is a reasonable assumption when the electrons are produced by secondary emission from the walls, we may write

$$e^2 E^2 = (m^*)^2 v^2 (\omega^2 + \nu_m^2), \tag{24.2}$$

and E^2 plotted against ω^2 should be a straight line. This is illustrated for the data of GILL and VON ENGEL for hydrogen in Fig. 26.

For any pressure, the straight lines are seen to cut the E-axis at a point E_0 where $\omega = 0$. Here $eE_0 = m^* v \nu_m$, but since $(e/m^* \nu_m) = \mu$, the mobility, we have the possibility of identifying the particles by determining what values of e/m^* lead to the correct relation between E and ω at the various pressures measured in Fig. 26. The e/m^* ratio for the molecular hydrogen ion agrees better than one would expect from the elementary nature of the analysis, and leaves little doubt that the ion mobility is the controlling mechanism on the long wavelength side of the oscillation amplitude transition. In this region, therefore, the loss of electrons at the walls resulting from their large oscillation amplitude is more than replaced by the secondary electron emission by positive ion bombardment of the walls, and the breakdown field is shown to decrease as the wavelength increases.

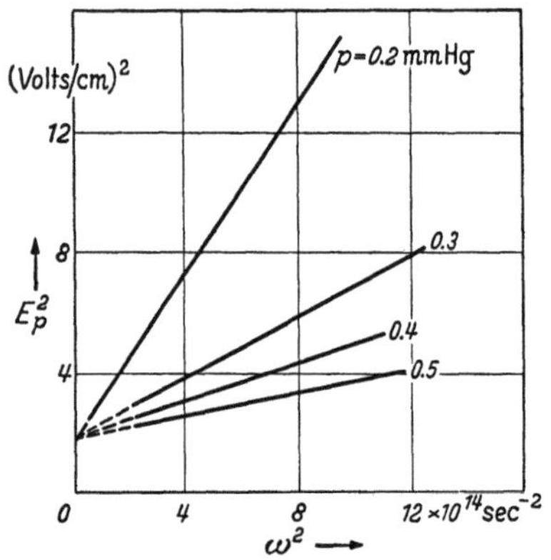

Fig. 26. Relation between field strength and frequency beyond cut-off in hydrogen.

D. Secondary electron resonance breakdown.

25. Behavior of the average electron. Many studies have been made of high-frequency breakdown in regions outside that controlled by diffusion. If the pressure is low, the mean free path becomes long compared with the containing vessel, and ionization in the gas becomes highly unlikely. A number of workers have studied this case and have shown that the secondary emission of electrons by direct bombardment of the walls can cause a breakdown to occur. Not only is the magnitude of the electric field important, but the phase of the electron motion with respect to the field has a governing effect. Under optimum conditions, the electron motion must be in phase with the field. Thus, an electron starting across the gap between the walls should collide with the walls and release secondary electrons just as the electric field passes through zero. The reversed electric field accelerates the secondary electrons back across the gap. The electric field must be of such a value that the transit time across the gap shall be equal to one-half cycle of the a.c. field. In this way the secondary electrons formed by the initial electron become primary electrons for the next half-cycle to form another group of secondary electrons, with the optimum conditions again requiring that the secondaries be formed just as the field reverses its direction.

It is obvious that a breakdown does not require the optimum conditions to occur, and there is a fairly broad region of fields and frequencies over which such a phenomenon may be observed. It should be apparent that for any one frequency, breakdown should be possible in a bounded region between two values of the field corresponding to too little or too much acceleration of the electrons to maintain the proper phase relations.

Because this type of breakdown relies for its electron multiplication on the secondary emission of electrons from the walls, the breakdown field is independent of the type of gas but very dependent on the nature of the walls of the vessel in which the discharge takes place.

With this introduction, let us turn to a more quantitative description of the problem. The motion of an electron acted on by a sinusoidally varying electric field of peak value E_p and radian frequency ω is described by the equation

$$m \frac{dv}{dt} = e E_p \sin(\omega t + \varphi), \tag{25.1}$$

where φ is the phase angle of the secondary-emission electrons, and we here neglect any collision effects between the electrons and gas atoms. For simplicity, let us assume that all electrons have one-half-cycle transit times between the walls and integrate Eq. (25.1) over this half-cycle transit time. We get for the arrival velocity

$$v = v_0 + (2 e E_p / m \omega) \cos \varphi, \tag{25.2}$$

where v_0 is the component in the direction of the electric field of the velocity of emission of the secondary electrons. Integrating again, evaluating over the one-half-cycle transit time, and setting the distance of electron travel equal to the tube wall separation L gives the transit equation

$$L = \pi v_0/\omega + (e E_p / m \omega^2)(\pi \cos \varphi + 2 \sin \varphi). \tag{25.3}$$

Solving this equation for the electric field, we obtain

$$E_p = \frac{\omega^2 L - \pi v_0 \omega}{(e/m)(\pi \cos \varphi + 2 \sin \varphi)}. \tag{25.4}$$

Let us assume that the ratio between the electron arrival velocity and the electron emission velocity is a constant, $(v/v_0) = k$, so that we may combine Eqs. (25.2) and (25.4) to give

$$v = \left(\frac{k}{k-1}\right) \frac{2 e E_p \cos \varphi}{m \omega}, \tag{25.5}$$

and

$$E_p = \omega^2 L \Phi^{-1} m/e, \tag{25.6}$$

where

$$\Phi = \frac{k+1}{k-1} \pi \cos \varphi + 2 \sin \varphi. \tag{25.7}$$

If we combine Eqs. (25.5) and (25.6), writing the velocity in terms of the electron arrival energy u given by $eu = mv^2/2$, we obtain a frequency

$$f = \frac{(k-1)\Phi}{k \pi L \cos \varphi} \left(\frac{eu}{8m}\right)^{\frac{1}{2}}. \tag{25.8}$$

Furthermore, for a given experiment at fixed ω and fixed d the minimum electric field in Eq. (25.6) occurs for a maximum Φ. Maximizing Φ with respect to the phase of the emitted electrons, φ, gives

$$\varphi = \arctan\left[\left(\frac{k-1}{k+1}\right)\left(\frac{2}{\pi}\right)\right]. \tag{25.9}$$

It is an unfortunate fact that we do not have available the necessary information on the fundamental processes involved in these considerations to substitute values of the parameters in Eqs. (25.6), (25.7), (25.8), and (25.9) to obtain a numerical solution. We therefore have to use them as semiempirical relations to calculate this breakdown region.

HATCH and WILLIAMS[1] have made fairly extensive measurements to which we may fit these equations. Their data are shown in Fig. 27. Data on the lower

[1] A. J. HATCH and H. B. WILLIAMS: J. Appl. Phys. **25**, 417 (1954).

branch of the curve were taken by increasing the electric field strength until breakdown occurred. By suddenly applying a high field and then lowering it slowly, the upper breakdown curve was observed. Eq. (25.6) fits the nearly straight line segment $a-b$ in Fig. 27 when $\varphi=6.6$. Using this empirically fitted value for φ in Eqs. (25.7) and (25.9) we find that for this case $k=3$, $\varphi=18°$. If we continue to assume that k is constant, we may obtain a family of straight lines of slope 2 on a log-log plot of E_p versus frequency for different values of φ. Such a plot is shown in Fig. 28. The resonance line $A-B$ for $\varphi=18°$ corresponds to the segment $a-b$ of Fig. 28. The resonance line $C-D$ in Fig. 28 for $\varphi=-56°$ corresponds to the segment $c-d$ of Fig. 27.

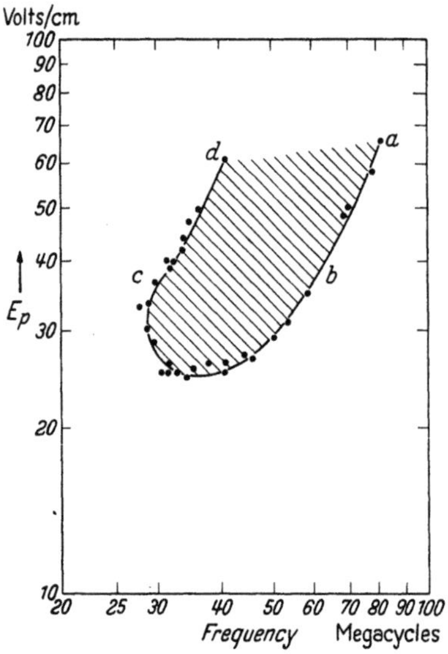

Fig. 27.

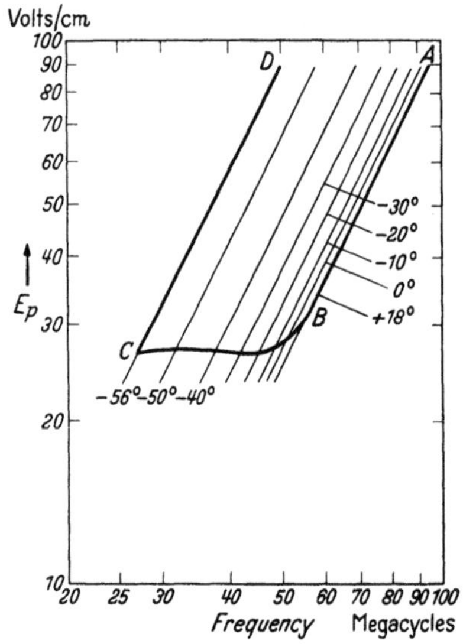

Fig. 28.

Fig. 27. Breakdown in hydrogen at a pressure less than 0.1 μ Hg.

Fig. 28. Theoretical calculation of breakdown for 3-cm. electrode separation, minimum electron arrival energy of 60 ev. and $k=3$. Various lines represent condition for half-cycle electron transit times at indicated phase angles.

From Eq. (25.8) a frequency may be determined for each value of φ corresponding to a particular electron arrival energy u. The value of the arrival energy which most nearly corresponds to the experimental curve is $u=60$ ev. This is the section $B-C$ of Fig. 28. In Fig. 29 are compared the results of this theory with the experimental measurements of HATCH and WILLIAMS.

A fair amount of self-consistency is evidenced by the various studies of this type of secondary electron resonance breakdown. HATCH and WILLIAMS found $k=3$, $-56° \leqq \varphi \leqq 18°$, and $u=60$ ev. DANIELSSON[1] assumed $k=\infty$ ($v_0=0$), $0° \leqq \varphi \leqq 90°$, and found $u=80$ ev, HENNEBERG, ORTHUBER, and STENDEL[2] calculated $0° \leqq \varphi \leqq 32.5°$, assuming zero electron-emission energy. GILL and VON ENGEL[3] found $k=4$, $u=90$ ev, and the cut-off occurred at $\varphi-58°$. Their treatment did not include a second limiting value of φ. These values of k and u are not inconsistent with known secondary electron-emission energies and yields.

Most observers have not taken their data in such a way as to measure the upper branch of the breakdown curve. The common experimental method is

[1] U. DANIELSSON: Diplomarbeit K. Techn. Hochschule, Stockholm 1943.
[2] W. HENNEBERG, R. ORTHUBER and E. STENDEL: Z. techn. Phys. **17**, 115 (1936).
[3] E. W. B. GILL and A. v. ENGEL: Proc. Roy. Soc. Lond., Ser. A **192**, 446 (1948).

to raise the voltage until the tube breaks down at a given frequency. In this case the data take the form shown in Fig. 30, in which the vertical segment of the curve is termed a cut-off. Such a condition corresponds to point C of Fig. 28. This cut-off frequency may be calculated from Eq. (25.8) by using the upper breakdown phase angle Φ_u:

$$f_{\text{co}} = \frac{(k-1)\,\Phi_u}{k\,\pi\,L\cos\varphi_u}\left(\frac{e\,u}{8\,m}\right)^{\frac{1}{2}}. \qquad (25.10)$$

For a constant k, φ_u, and u, this takes the form

$$f_{\text{co}} = (\text{constant})/L, \qquad (25.11)$$

where for any particular case the constant may be obtained by fitting to the data.

A collection of measured values for the cut-off frequency, f_{co}, as a function of the reciprocal electrode separations, $(1/L)$ is plotted[1] in Fig. 31. It seems obvious that Eq. (25.11) is verified by these data.

The breakdown voltage at cut-off is

$$V_{\text{co}} = \frac{u\,(k-1)^2\,\Phi_u}{2\,k^2\cos^2\varphi_u}. \qquad (25.12)$$

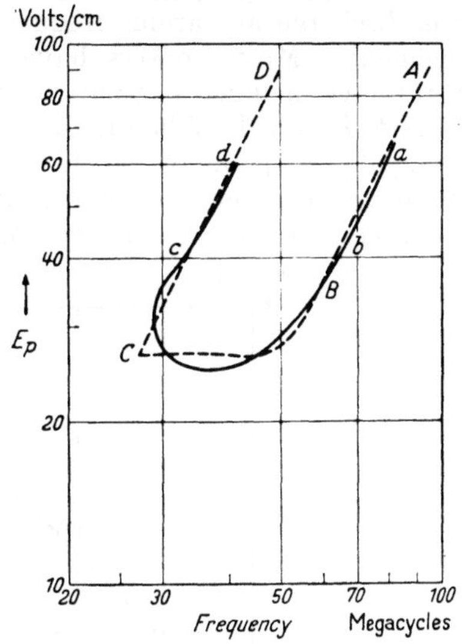

Fig. 29. Comparison of experimental curve of Fig. 27 and theoretical curve of Fig. 28.

Thus the breakdown voltage at cut-off is independent of electrode separation and applied frequency. For fixed k and φ_u, Eq. (25.12) may be written as

$$V_{\text{co}} = (\text{constant})\,u \qquad (25.13)$$

where the constant may be determined empirically.

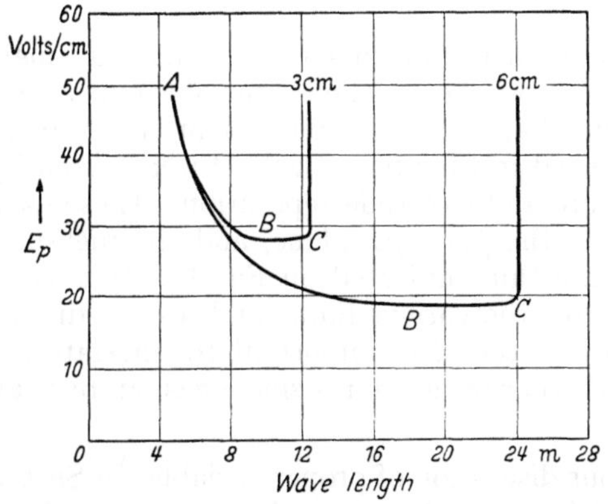

Fig. 30.

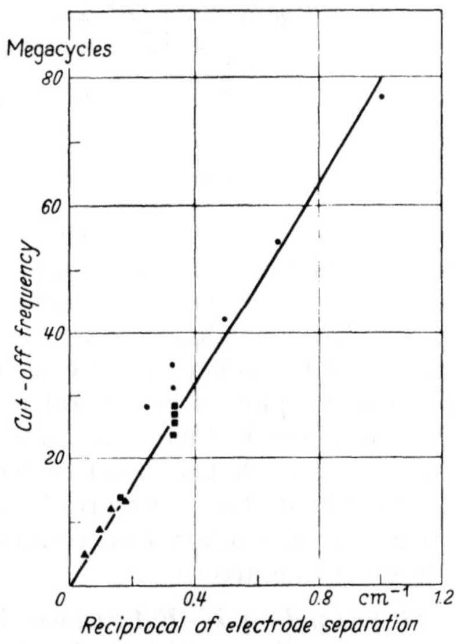

Fig. 31.

Fig. 30. Breakdown field strength as a function of wavelength for flat-ended cylindrical tubes 3 cm. and 6 cm. long with axes parallel to the field for hydrogen at a pressure of 1 μ Hg.

Fig. 31. Summary of cut-off frequencies versus reciprocal electron separation from data of several observers. ▲ GUTTON; ■ GILL and VON ENGEL; ● HATCH and WILLIAMS: Solid line, cut-off law, $f_{\text{co}} = 79/L$.

[1] Other workers cited in this section and C. GUTTON: C. R. Acad. Sci. Paris **178**, 467 (1924). — C. GUTTON and H. GUTTON: C. R. Acad. Sci. Paris **186**, 303 (1928). — H. GUTTON: Ann. Phys. **13**, 62 (1930).

E. Summary of high-frequency breakdown mechanisms.

26. The $p\Lambda - p\lambda$ plane. Many experimenters have studied the breakdown of a gas discharge at various frequencies and in various different geometrical arrangements. Most workers have obtained breakdown data in hydrogen and several have included in their reports sufficient detail for determining the parameters p, λ, and Λ. When these parameters are known, the data may be plotted in the $p\Lambda - p\lambda$ plane of Fig. 3. Two general experimental methods of collecting data have been used. The breakdown measurements reported in Chapter A were taken with the frequency constant with a given electrode separation and

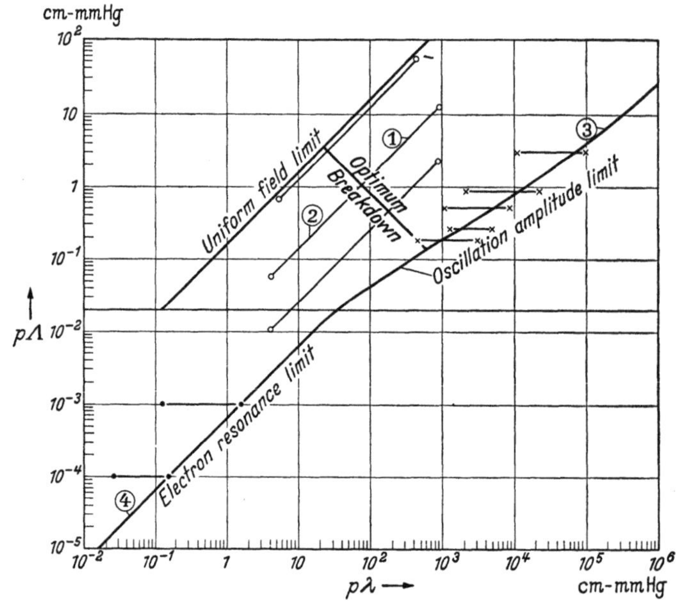

Fig. 32. Summary of the regions of high-frequency breakdown on the $p\Lambda - p\lambda$ plane.

with a varying pressure. Such a technique results in data in which a single run would plot as a line at 45° on the $p\Lambda - p\lambda$ plane. The extent of the measurements taken in hydrogen, presented in Chapter A, are plotted in this way in Fig. 32. The breakdown curves reproduced in Chapters C and D were taken by varying the frequency at a given pressure and electrode separation. These data will therefore plot as horizontal lines on the $p\Lambda - p\lambda$ plane, and the measurements discussed in Chapters C and D are thus indicated on Fig. 32. It is clear from this figure that data exist which cross the various limits and regions discussed throughout this entire article, and it is therefore important to correlate the data of various workers to determine whether or not, for a single gas, an over-all picture may be arrived at.

27. The $p\Lambda - p\lambda - E\Lambda$ surface. In our discussion of proper variables in Sect. 4 we arrived at the variables $p\Lambda$, $p\lambda$, and $E\Lambda$ from dimensional analysis considerations, and therefore these variables should not be restricted to any particular breakdown mechanism. All breakdown data taken for a given gas should fall in a single $p\Lambda - p\lambda - E\Lambda$ surface. This is shown for hydrogen in Fig. 33, constructed from the actual data reported in the literature. Not only do all of the data fall on a single surface, but insofar as the work of different experimenters overlaps on the illustrated surface, their measurements agree within the accuracy of drawing such a model.

Fig. 33 has been drawn to summarize the sections of this article. Regions 1 and 2 were discussed in Chapter A, particularly in Sect. 3 and 12. Region 1 corresponds to the high-pressure data for hydrogen illustrated in Fig. 2. In Region 2 the effective field defined by Ep. 1.5 is significant. The rising curve along the optimum breakdown line for increasing Λ results, for example, from

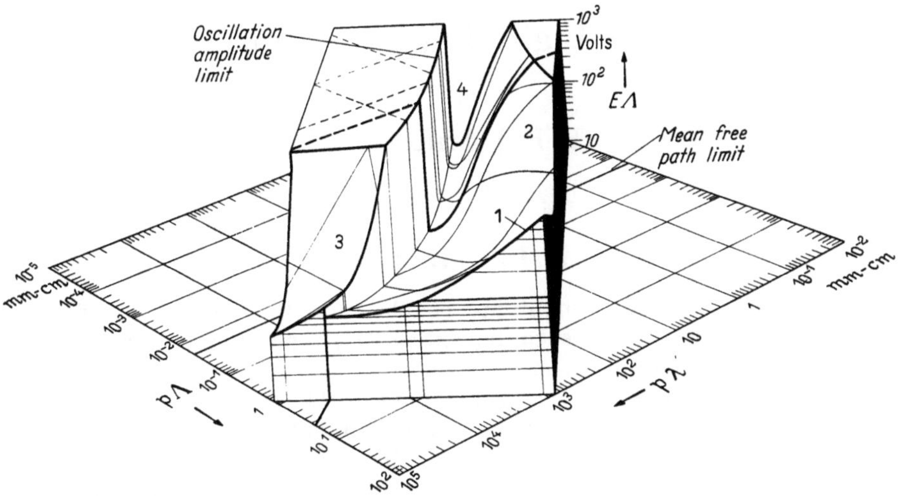

Fig. 33. Summary of the regions of high-frequency breakdown on the $p\Lambda - p\lambda - E\Lambda$ surface.

the nonuniform field effects discussed in Sect. 16, and illustrated in Fig. 11. Region 3 is described in Chapter C; Region 4, in Chapter D.

A similar $p\Lambda - p\lambda - E\Lambda$ surface could be drawn for different gases and for the superimposed d.c. and magnetic field of Sect. c and d by substituting for the geometrical Λ the appropriate effective Λ discussed in these sections.

The Lightning Discharge.

By

B. F. J. Schonland.

With 43 Figures.

A. Introduction.

1. Scope. The study of lightning enables investigations of the transient electrical breakdown of gases to be extended to potentials of the order of 5×10^8 volts, two orders of magnitude higher than can at present be attained in the laboratory. With electrode spacings of several kilometres the time occupied by the breakdown process is very much longer than that involved in the laboratory spark. Though neither the onset of the discharge nor the conditions in the air along its path can be controlled, these great potential differences and path lengths have made lightning studies of value in the understanding of the processes of breakdown in the controlled laboratory spark. It is with these aspects of the lightning discharge that the present article deals.

2. Historical. That the lightning discharge possessed many of the characteristics of the electric spark occurred to a number of electrical experimenters in the 17th and 18th centuries, but it was Benjamin Franklin who first proposed to put the identity to the test of experiment. In 1750 he suggested a crucial point-discharge experiment; this was successfully performed in France under the direction of d'Alibard on May 10th, 1752 and independently by Franklin himself in Philadelphia in June of that year. In September, 1752, Franklin, by collecting some of the charges developed on an insulated rod on a number of occasions when storms were overhead, found that the main lower thundercloud charge responsible for the electric field at the ground was of negative polarity, a conclusion now known to be generally true [*1*].

Little further information about the lightning discharge was obtained for a hundred years, until the photographic camera was employed by Hansel (1883), Kayser (1885), Hoffert (1890), and others, particularly Walter (1902 onwards) to reveal by displacement of the recorded image such general features of the discharge process as its separate partial discharges and the first leader [*2*]. The photographic method reached its most productive form with the development by Boys of his high-speed double-lens camera. This has proved of great value in elucidating the various fastmoving streamer processes in the discharge (Schonland, Malan, Hodges and Collens, 1934 onwards).

No very important new information was provided by electrical studies of lightning for a century and a half after Franklin, though useful estimates of the polarity and magnitude of currents in flashes to ground were obtained by a number of workers, particularly Pockels (1898) and Toepler (1901) [*2*]. Advance in this direction awaited the classic studies of thunderstorm electric fields and field-changes made by C. T. R. Wilson (1916→) and his school. These were followed by the introduction of the cathode ray oscillograph for the recording of the fine-structure of the field-changes due to lightning by APPLETON, WATSON-WATT and HERD (1923) and NORINDER (1923).

3. Forms taken by the discharge. Three main types of lightning can be distinguished:

a) Streak or forked lightning or a flash to ground (foudre, Zickzackblitz, Linienblitz), popularly called a thunderbolt.

A discharge in the air passing between cloud and ground adopting a tortuous course and usually branched downwards from a main channel or trunk.

b) Cloud discharges (éclairs dans le nuage) which take place within a thundercloud and are popularly called sheet lightning. These give a diffuse illumination and no distinct channel is usually seen.

c) Air discharges (éclairs dans l'air, décharges atmosphériques). Sinuous discharges, often with a long horizontal section, passing from a thundercloud into the air without striking the ground.

Subsidiary forms of some physical interest are:

a 1) Ribbon lightning (éclair en ruban) a flash to ground in which the channels of component successive strokes are separated by the wind to form a broad ribbon.

a 2) Bead or chain lightning (éclair en chapelet, Perlschnurblitz). A phenomenon which may follow forked lightning, the channel to ground breaking up into fragments some 50 metres in length which become roughly globular and may persist for an appreciable time.

a 3) Ball lightning (éclair en boule, Kugelblitz), a luminous globe usually reported to appear soon after a discharge to ground. Its diameter has been reported to lie between 10 and 20 cm. and occasionally to reach one metre. It moves slowly in the air or on the ground and usually disappears with a violent explosion. Well authenticated cases of this phenomenon are rare and it will not be discussed here.

c 1) Rocket lightning (éclair en fusée), an air discharge giving the impression of fairly slow visible progression along its channel and branches.

B. General features of ground and air discharges.

The general features of ground and air discharges which were established early by simple camera and electrical studies are outlined below.

4. The component strokes of a flash to ground. Many lightning flashes to ground are seen to flicker and camera studies show that the flicker arises from the fact that the discharge consists of a number of intermittent partial discharges or strokes, which generally travel along the same path. Their luminosity lasts for a few milliseconds only, the intervals between them being variable but most frequently about 40 millisecs.

A stroke is to be distinguished from a simple fluctuation in the luminous intensity of the channel by the fact that it begins with a downward-moving leader process, which on reaching the ground, gives rise to a return streamer (see Sect. 14).

5. The number of separate strokes per flash. Observations of the number of strokes in a flash to ground have been made by a number of workers, using methods of differing reliability. An analysis of the available data made by BRUCE and GOLDE in 1941 led them to conclude [*3*] that the stroke number does not vary greatly in different parts of the world and is most frequently three. This quantity has been determined in South Africa by the writer and MALAN from BOYS camera and field-change records of 1800 discharges over a long period of years. The results, which are shown in Fig. 1, indicate that in this region the most frequent number of strokes is four and that 10% of discharges to ground consist of more than nine strokes. For English storms WORMELL [9] has reported from a similar long series of observations that the majority of discharges (65%) consist of one stroke only. The number evidently depends upon the nature of the storms sampled and increases in general with their severity.

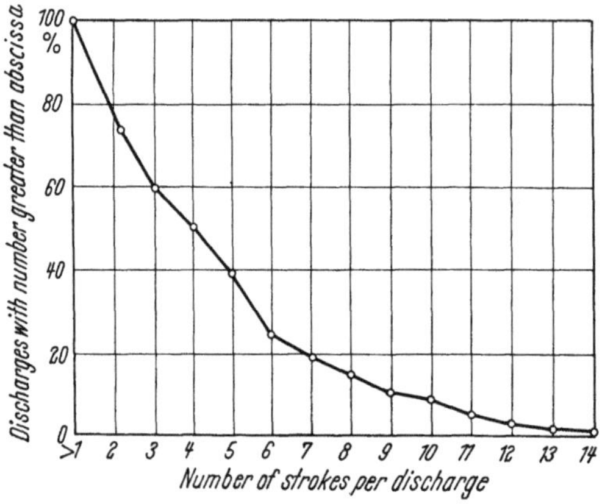

Fig. 1.

Very large numbers of apparently separate strokes, up to 42, in a single flash have been noted on camera records but it would appear that these are not all separate partial discharges, involving a leader-return stroke combination in each case. Many of them are subsidiary fluctuations in the luminosity, called M components (see Sect. 23) which are associated with a long-continuing stroke.

Experience with large frontal thunderstorms in South Africa indicates that flashes of many separate strokes are more common from these storms than from minor convective storms.

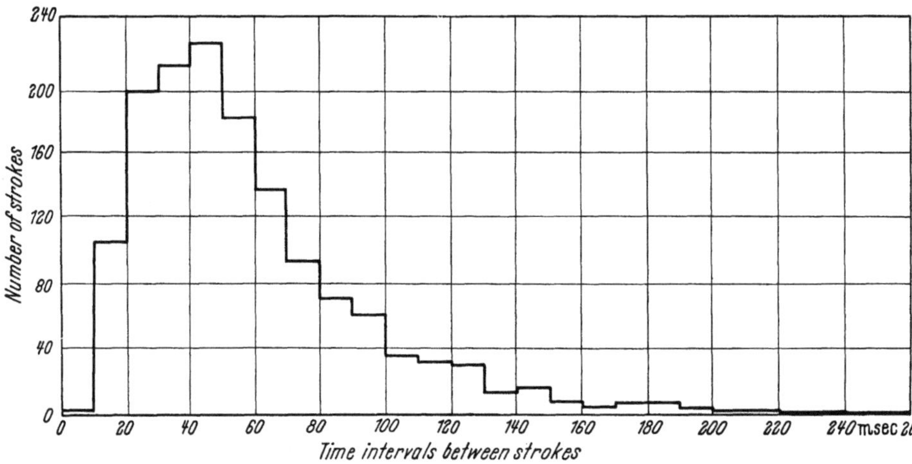

Fig. 2. Frequency analysis of strokes, short intervals.

6. Time-intervals between strokes. BRUCE and GOLDE have also analysed information from various parts of the world on the distribution of time-intervals between strokes [3]. The results obtained depend upon the method used to record the strokes, the less sensitive instruments giving modal values of 0.07 to 0.11 secs. and more sensitive ones, such as the cathode ray oscillograph, modal values close to 0.04 secs. The results of a frequency analysis of 1482 separate strokes by the writer and D. J. MALAN using the latter method on flashes from frontal storms are given in Fig. 2. They indicate that there are very few separate strokes with time-intervals less than 10 m-secs. or greater than 150 m-secs. The range 10 to 100 m-secs. includes 87% of the intervals observed and the most frequent interval is 40 m-secs. Fig. 3 shows the distribution of the long time-intervals in the range 150 to 400 m-secs., the maximum interval observed by these authors.

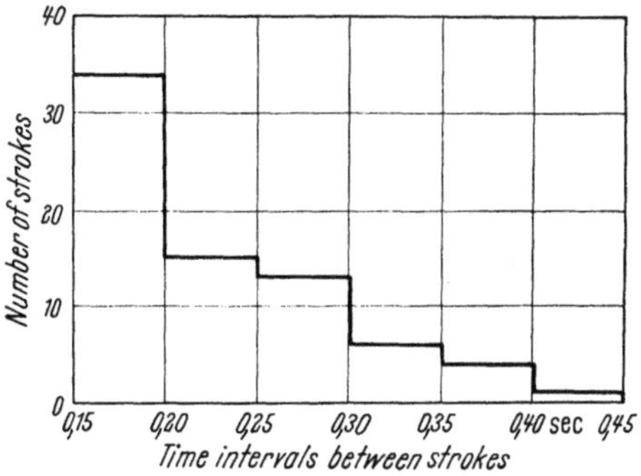

Fig. 3. Frequency analysis, long intervals.

7. The total duration of a discharge to ground. The analysis by BRUCE and GOLDE quoted above concludes from the product of the number of strokes after the first and the average time interval between them that flashes to ground have a most frequent duration of 0.25 secs., durations exceeding one second being unusual. Direct measurements of the total durations of a large number of South African discharges yield a most frequent value of 0.20 secs.

8. The distribution of time-intervals between strokes and M-components. The time-intervals between successive strokes of a discharge to ground are irregularly distributed and show no tendency either towards equality or progressive change with the order of the stroke in the series.

The M-components referred to in Sect. 5 do, however, occasionally show fairly regular separations in time, suggesting some type of relaxation oscillation within the cloud.

Two discharges showing a number of apparently separate strokes at nearly equal time-intervals have been reported by MCCANN[1]. Since none of these showed the leader processes characteristic of true strokes, they are to be considered as M components of long-continuing discharges.

9. The branching of lightning. *α) Direction.* In a flash to ground and in an air-discharge the branches are almost always directed away from the cloud. Since laboratory sparks between two electrodes, neither of which is earthed, branch away from the anode, it was suggested by G. C. SIMPSON[2] that lightning branching, in contradiction to the early experimental results of FRANKLIN, indicated that the polarity of the lower portion of a thundercloud was positive. This question was directly tested by HALLIDAY[3] in South Africa and JENSEN[4] in the United States by photographing lightning discharges whose polarity was simultaneously determined from records of the field-change caused by the transfer of charge to ground. Their results showed that the SIMPSON branching criterion was not valid, for flashes bringing negative charge to ground exhibited considerable downward branching, though JENSEN reported that occasional flashes from positively-charged regions were the more extensively branched.

SCHONLAND and ALLIBONE have carried out experiments with a surge generator to show that branching directed away from the cathode of the laboratory spark can easily be obtained if the anode is an earthed plate provided with sharp points to ensure a flow of positive ions to the inter-electrode space before breakdown occurs[5]. The arrangement was considered by them to correspond closely to natural conditions of point discharge from the ground.

[1] G. D. MCCANN: Trans. Amer. Inst. Electr. Engrs. **63**, 1157 (1944).
[2] G. C. SIMPSON: Proc. Roy. Soc. Lond., Ser. A **114**, 376 (1927).
[3] E. C. HALLIDAY: Proc. Roy. Soc., Lond., Ser A **138**, 205 (1932).
[4] J. C. JENSEN: J. Franklin Inst. **216**, 707 (1933).
[5] B. F. J. SCHONLAND and T. E. ALLIBONE: Nature, Lond. **128**, 794 (1931).

This experiment showed that branching in the lightning flash, though not a criterion of polarity, is an important indication of space-charge between cloud and ground, giving clear evidence of the positive ion emission from elevated pointed objects on the ground whose importance was first emphasised by C. T. R. WILSON [4]. It follows that unusually high space-charge fields must be presumed to exist in regions between cloud and ground when prolific branching is observed. This conclusion is confirmed by measurements of streamer velocities and step-lengths in these regions (see Sect. 16).

A discharge to ground is almost always initiated by a negative downward-moving streamer process, the first leader (see Sect. 14). It is this streamer which creats the branched channel of the discharge and the branching observed indicates its direction of movement. It also gives evidence of the guiding action of local electric fields which have their origin in positive-ion space-charge.

In conformity with these observations it has been found by McEACHRON and his associates in the study of flashes to the Empire State Building in New York that these discharges frequently begin with an upward-moving and extensive upwardly-branched leader streamer.

β) Root-branching. This term is used to describe the branching which is often observed at the lower end of the lightning channel, where it may divide into two or more forks, each of which connects up with the ground. The phenomenon suggests that as the downward-moving leader streamer approaches the earth it creates electric fields at projecting points which are strong enough to cause the development of several separate upward streamers which meet it above the ground.

GOLDE[1] has estimated that for the average lightning conductor these ground streamers should be about 16 m. long and that when the leader approaches open ground their length should be about 5 m.

A photograph of these streamers, one of which is 3 m. long, produced by a flash striking a flat beach, has been published by McEACHRON and MORRIS. Another camera record by MALAN of a flash to open country which shows root branching from a height of about 50 m. has been discussed by GOLDE.

γ) Branching and stroke-order. The branching of a flash to ground is usually confined to the first stroke only. When it is shown by a second or third stroke it is almost always the continuation of a very pronounced branch which was formed, but not completed, by the first stroke, indicating that the considerable space-charge in the air which led to the development of branching in the first stroke was not completely removed by this stroke.

10. Continuing strokes. The most frequent duration of the luminosity in the channel of a stroke (and the corresponding change of electric field) is one or two m-sec., but about 20% of all flashes to open country show at least one "continuing" stroke of long duration which may extend to 200 m-secs.

Such strokes are more frequent in the special case of discharges to tall structures like the Empire State Building (see Sect. 25).

C. Photographic studies of luminosity in the discharge.

I. Methods.

11. Types of camera. An ordinary single-lens camera left open at night and given a translationary movement during the occurrence of a flash to ground produces sufficient displacement of the images of the successive strokes in the flash to separate them from one another. The photographic camera was first

[1] R. H. GOLDE: Nature, Lond. **160**, 395 (1947).

employed for this purpose by H. HOFFERT (1890). More comprehensive studies were made from 1902 onwards by B. WALTER of Hamburg[1], who used a camera which was moved regularly back and forth through a small circular arc [2]. Though he was able by this means to discover several of the main features of the intermittent discharge, including the existence of the first leader (Vorentladung) the method had insufficient time-resolution for study of the fine-structure of the breakdown process and of the properties of the streamers involved in the discharge. Information on these questions was first obtainable when C. V. BOYS[2]

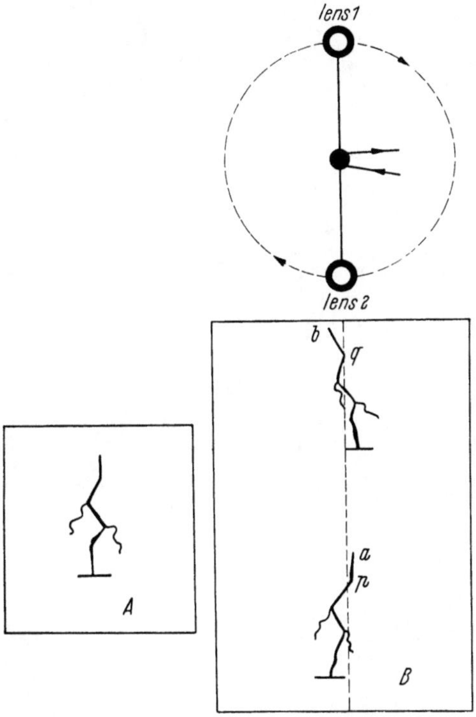

Fig. 4. Boys Camera.

devised a camera which was based upon the relative distortions of the images of the luminous channel which are produced by two separate lenses moving in opposite directions.

In its original and most widely-used form this camera (Fig. 4) employed two lenses rotating in a circle in front of a fixed plate or film. A single stroke, as shown in Fig. 4a, if it happened to occur at the moment when the line joining the two lenses was vertical, would produce two oppositely distorted images, as shown in Fig. 4b, the distortions corresponding in this case to a movement of luminosity in the downward direction bq (or ap). The relative distortion at any point in the channel can be measured by utilising a diameter qp drawn through two corresponding points and mounting the prints of the two images as shown in Fig. 4c. The interval of time between the appearance of luminosity at p (or q) and at a (or b) is given by $(pq - ab)/2v$ when v is the tangential velocity of each lens. The value of v found most suitable by the author and his colleagues D. J. MALAN and H. COLLENS is 1350 cm./sec., the radius of the circle being 10 cm.

[1] B. WALTER: Ann. Phys. **10**, 393 (1902). — Jb. Hamburg. Wiss. Anstalten **20** (1903).
[2] C. V. BOYS: Nature, Lond. **118**, 749 (1926).

The distance $pq - ab$ can under the best conditions be measured to ± 0.002 cm., consequently the best time-resolution obtainable with this camera is $0.7\,\mu$-secs. The use of higher speeds of lens or film has been found to require more sensitive emulsions to obtain satisfactory details of the more feebly luminous processes and since these emulsions are subject to disturbing halation from the brighter processes, the above figure probably represents the practical limit to the time-resolution obtainable with any form of fast camera for lightning studies.

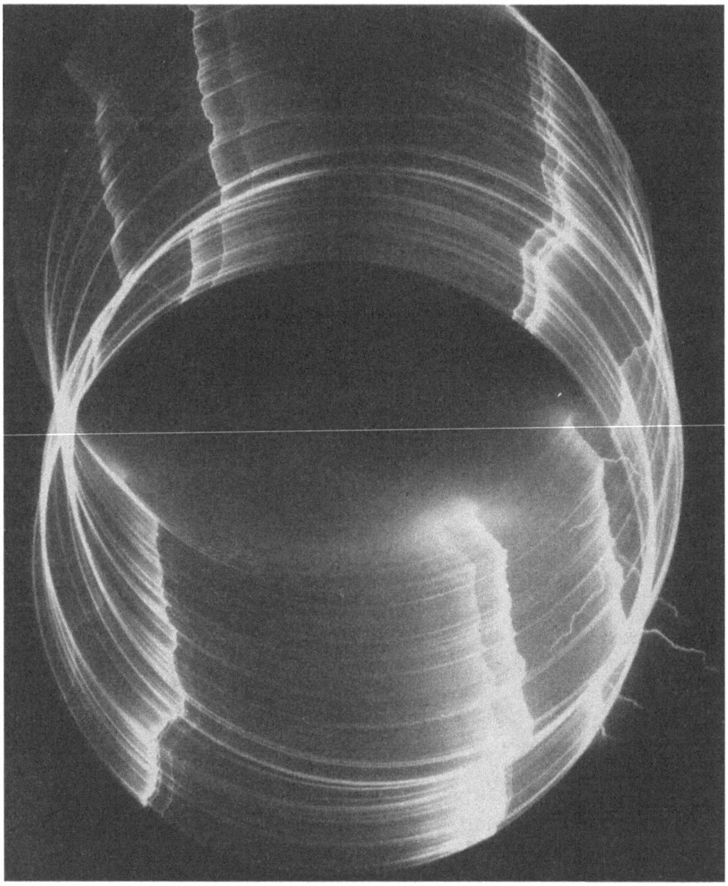

Fig. 5. BOYS Camera record of a flash of seven strokes.

Referring to Fig. 4 it will be clear that if a stroke occurs at a time when the camera lenses are not vertically above each other the same method of mounting and measurement of distortion can be employed. When for example the lenses are momentarily moving vertically the vertical portions of one image will be lengthened and of the other will be shortened. Drag of luminosity, however. makes such measurements a good deal less accurate.

The rate of rotation of the lenses is most easily checked and kept constant in field-work by turning them at a speed just below a noticeable resonance vibration of the camera body. In making measurements from prints care must be taken to avoid unequal shrinkage in drying; for this reason glazed prints are not employed. A BOYS camera record of a flash of seven strokes is reproduced as Fig. 5.

This first form of BOYS camera suffers from several disadvantages, one of which is the reduced accuracy of measurement when the distortion to be measured has a significant component parallel to the direction of movement of the luminous process. A second is that the images from a discharge of many component strokes are distributed over a circle of fairly small diameter and tend to overlap if the distortion is large. This effect is particularly troublesome in the case of the first leader streamer, which is comparatively slow and may occupy more than half the circumference of the circle. As a result weakly luminous processes such as this leader tend to be obscured by halation from overlapping brighter ones.

These difficulties were to some extent removed in a second form of camera developed by BOYS (Fig. 6) in which a moving film was mounted in the interior of a hollow drum rotated about a vertical axis[1]. Two images, one at each end of a vertical diameter of the drum are projected on to the film by separate optical systems, each consisting of a lens with its axis horizontal and a totally reflecting prism. This camera has a resolving power of the same order as the previous one, with the advantage of continuous horizontal distortion of the two images and no serious overlapping of the images of separate processes. It suffers, however, from two defects, a restricted lens aperture which reduces its range and a system of mounting the film which involves much manipulation and loss of valuable time when it has to be changed in order to photograph another flash.

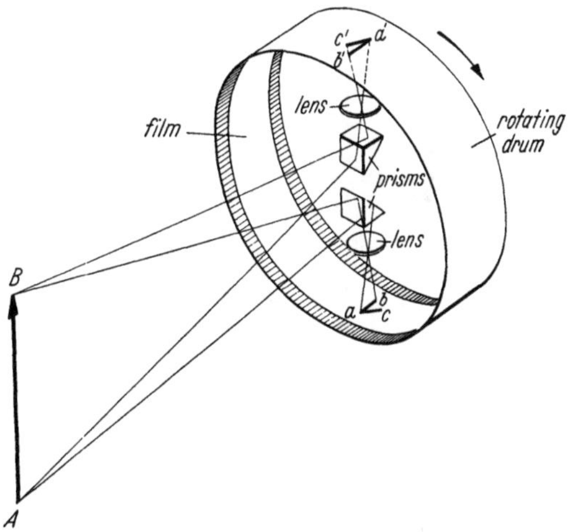

Fig. 6. Boys Drum Camera

Studies with the BOYS camera have shown (see Sect. 14) that discharges to ground involve three separate streamer processes, the most brilliant and easily recorded of which, the return streamer, travels at such a high velocity that it passes from ground to cloud in less than 100 μ-secs. Although the double-lens BOYS camera is still essential for studies of the return streamer, it is now more convenient to use fast single- lens cameras for the photographic recording of the distortions of the images of the slower streamer processes. This can be done if the small distortion of the channel of the return streamer is neglected and this channel is used as a reference from which to measure the much larger distortions of the slower processes.

Various improvements on the original single-lens cameras have been developed for this purpose. Japanese workers have employed a number of conventional cameras mounted on the circumference of a rotating wheel. The writer has used a single lens rotating in a circle in front of a fixed film and also a fixed lens focussed on a fast-moving film mounted on the circumference of a drum rotated about a vertical axis, with special arrangements for keeping the film tight. The most useful high-speed single lens camera has, however, been developed by MALAN[2]. This employs (Fig. 7, not drawn to scale) a fixed film (a portion of a separately mounted 35 mm. film reel) mounted in guides in the inside of a fixed drum of diameter 26 cm. The moving optical system consists of a right-angle prism P, a Wollaston prism W and a lens L, all rotated rapidly by an electric motor about

[1] C. V. BOYS: Nature, Lond. **124**, 54 (1929).
[2] D. J. MALAN: Rev. Opt. **29**, 513 (1950).

an axis coinciding with that of the film drum. Unwanted rotation of the final image about *PWL* is compensated for by an automatic rotation of *W* about the same axis at half the angular velocity of the optical system. This rotation is produced by the epicyclic conical gearing *C*. Additional gearing *T*, *V* and *S* serves to wind up the exposed portion of the film reel and to replace it in a few onds secby a fresh portion, one of the guides *O* having teeth which engage every tenth perforation on one side of the film.

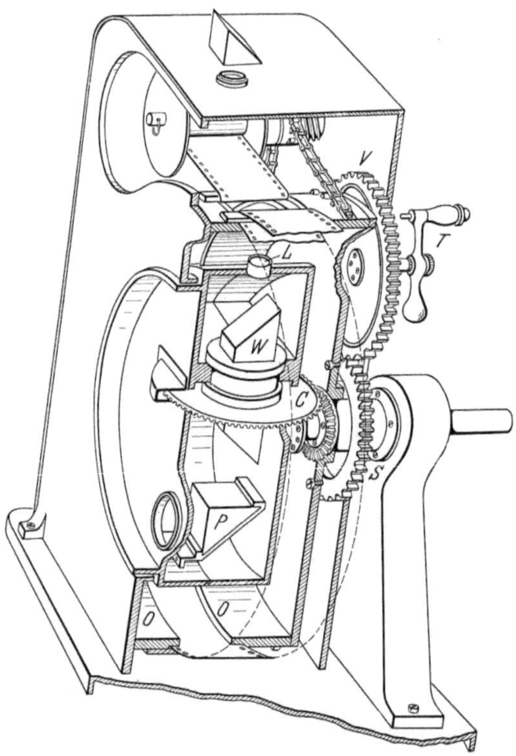

Fig. 7. MALAN Camera.

For the study of very feebly luminous processes, which require a large aperture long-focus lens, MALAN has devised a camera in which the film takes the form of a fast moving belt passing over two pulleys and kept flat behind the lens by a set of metal guides.

12. Photography of lightning in daylight. Photographic studies of lightning in the daytime have been made by the writer by placing an ultraviolet filter, in the form of a flat-faced quartz cell filled with a mixture of chlorine and bromine in front of a quartz-fluorite lens, but a more useful device employs an anticipatory trigger device to open the shutter of a single-lens camera by means of an electromagnet actuated by a thyratron[1]. This is switched on in $\sim$0.002 secs. by a radio receiver tuned to the 20 to 30 kc./sec. band. The early stages of the first breakdown process produce strong electromagnetic radiation in this band of frequencies and the shutter can thus be opened by the discharge itself before the first leader has progressed very far in its downward course. When this device is applied to a fast moving drum camera carrying 180 cm. of film on its circumference, the shutter being allowed to remain open for 0.5 secs., the degree of daylight fogging produced is quite small if the sky is overcast at the time and the fainter leader processes in the discharge can be recorded from flashes up to 15 km. away. An example of such a daylight record obtained at a distance of 3.5 km. is given in Fig. 8. The second stroke shows a dart-leader preceding the return streamer but the first leader is too faint for reproduction.

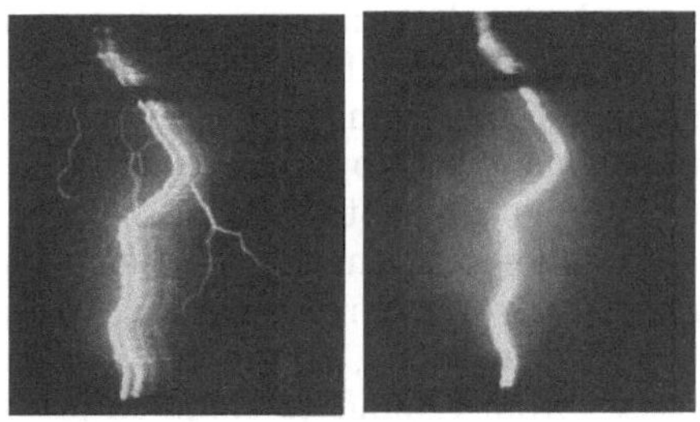

Fig. 8. Flash of two strokes photographed in daylight.

13. Photography of luminosity within the cloud. Flashes within the cloud generally give rise to weak and diffuse illumination with no indication of discharge

[1] B. F. J. SCHONLAND and J. S. ELDER: J. Franklin Inst. **231**, 39 (1941).

channels and are therefore not suitable for detailed study by the photographic method. It is, however, of importance to determine whether any cloud luminosity occurs in the apparently quiescent intervals between successive strokes to ground. This has been done by MALAN [5], making use of a double lens camera which combines a simple rotating lens in front of a fixed film to record the occurrence of strokes beneath the cloud, with a rotating combination of a circular aperture focussed on the film by a second lens. Luminosity falling on this aperture is recorded as a spot of light and the relation between such spots and the images of the strokes produced by the first lens provides the information required (see Sect. 27).

II. Results obtained by the photographic method.

Although the BOYS camera was devised in 1900, a description of it was not published until 1926. It was first put to extensive use by SCHONLAND and COLLENS in 1933 [6] following upon a successful trial of its possibilities by HALLIDAY[1]. The results obtained in South Africa with this and with newer types of single lens cameras have been confirmed and extended by investigators in the United States, Russia, France and Japan. The picture which has emerged from this work is summarised in the following sections [6], [7].

a) Flashes to ground.

14. General. Each separate stroke of a flash to ground is of a dual nature. A downward-moving weakly luminous process, the leader streamer, is followed,

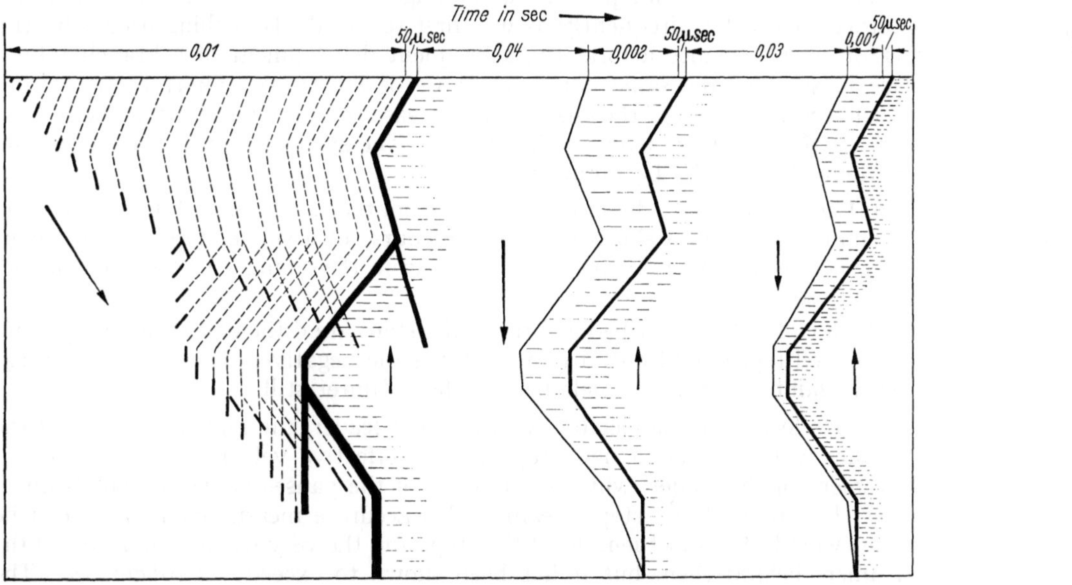

Fig. 9. Streamer processes in a flash to ground.

upon arrival near the ground, by an intensely luminous and much more rapid main or return streamer which traverses the leader channel in the reverse direction. Fig. 9 in which the horizontal time-scale has been distorted for clarity shows what is observed in the case of a discharge consisting of three strokes

[1] E. C. HALLIDAY: Phil. Mag. **15**, 409 (1933).

photographed by a single lens camera in which the film has been moved in a straight line from right to left. The distortion of the image of the return-stroke channel, from which a time-delay of 40 μ-secs. between bottom and top has been derived is of course determined in practice by means of the double lens BOYS camera.

Most of the characteristic downward branching of the flash arises during the downward movement of the first leader process which blazes the trail usually followed by all subsequent processes.

15. The stepped leader. The first leader appears to advance in a series of very rapid discontinuous steps each about 50 m. long (range 10 to 200 m.) and separated by pauses ranging from 40 to 100 μ-secs. These steps have the following characteristics:

α) Each appears as the termination of a fainter streamer extending the whole way down from the point of first appearance of the process, the streamer images increasing in brightness and apparently decreasing in width at their ends.

β) A step generally starts a little way back on part of the track formed by the previous step, so that the portion of the step which is entirely new is about 90% of the whole.

γ) The track apparently blazed by a new step is not always followed in detail by the channels of subsequent processes, the tendency being for the smaller twists and kinks to be smoothed out. In a few cases the form of the subsequent channel is changed so completely as to suggest that a fork previously existed, the fainter branch of which was adopted in later stages. A beaded effect is sometimes observed, but not as frequently as was first reported. Branching occurs by the division of a step into a fork, the subsequent development of a branch taking place in steps of the same nature as those of the trunk. Short branches and branches near the top of the channel often cease to develop some time before the main trunk steps have reached the ground. The start of a branch never occurs elsewhere than at the beginning of a step.

δ) Each step takes a direction different from that of the end of its predecessor; it is unusual to find a sharp bend in the course of a single step. The tortuous course of the lightning channel thus arises from the changes in direction of successive steps.

ε) Though often described as bright, the steps are difficult to photograph and can be well reproduced only by intensifying the negative. They are usually extremely faint and the channels behind them fainter still.

Two velocities are associated with the first leader, the high velocity V of the streamer which creates a new step and the effective velocity, v, of downward movement of the leader, which on account of the pauses between steps is much lower than that of the step streamer. No accurate measurement of V can be made owing to the short lengths of the steps and the very feeble luminosity of the streamers behind them but it has been shown to exceed 5×10^9 cm./sec. The effective velocity v ranges from 0.8 to 2.4×10^8 cm./sec. most of the observed values being close to 1.0×10^7 cm./sec. (see Sect. 19).

Both the lengths of the steps and their brightness are found to increase as v increases.

16. Alpha and Beta type first leaders. SCHONLAND, MALAN, COLLENS [7 b] have distinguished between two types of first leader. Type α leaders show a low and fairly uniform value of v ($\sim 1.0 \times 10^7$ cm./sec.) throughout their passage between cloud-base and ground, and make up 55% of those which have been photographed in South Africa. The steps are short and vary little in length

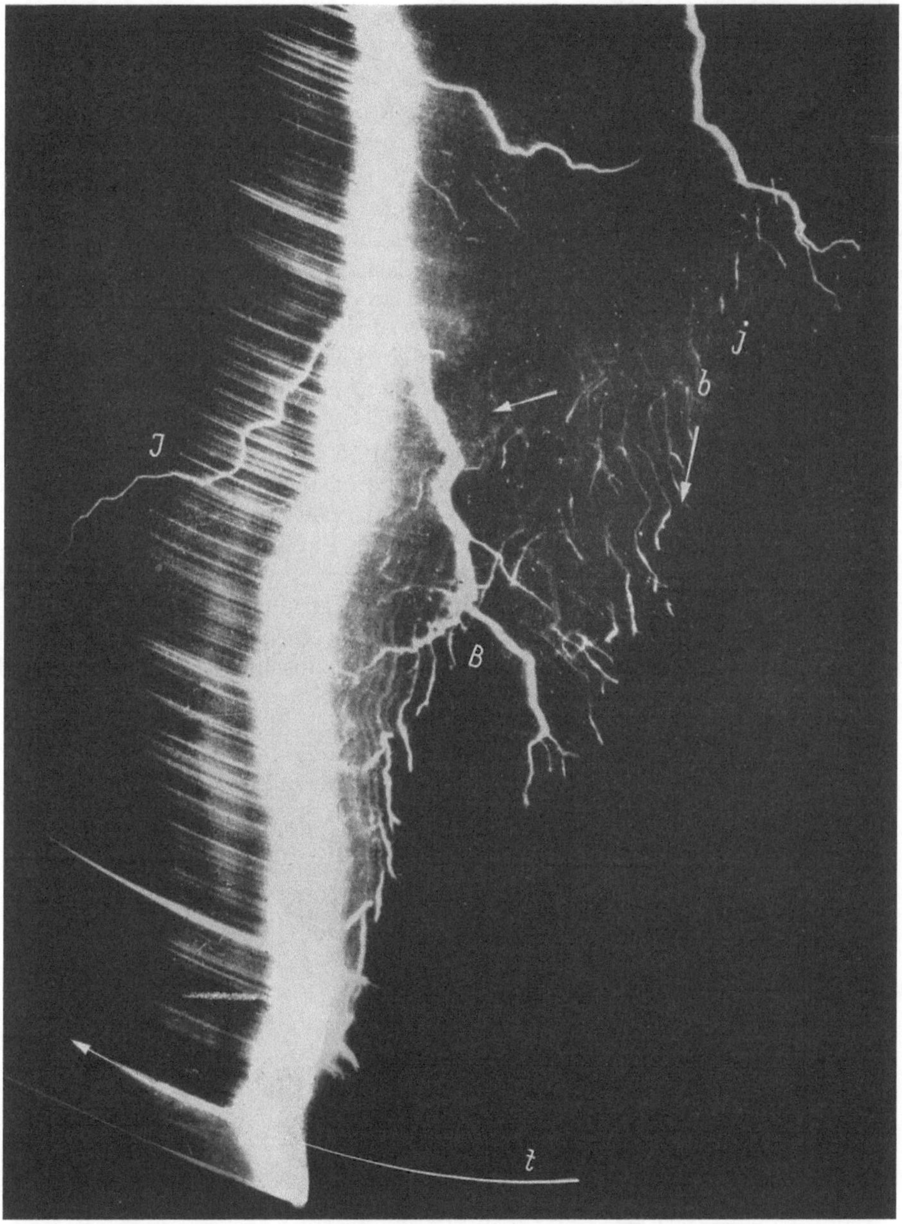

Fig. 10. Type α leader.

and brightness but are so weakly luminous that they are not easy to photograph unless the discharge is very close or the camera has a large aperture lens.

In type β leaders the top of the channel is extensively branched and v is high (8×10^7 to 2.4×10^8 cm./sec.), the steps being long and fairly bright. In its lower and later portion v drops to lower values, similar to those for type α leaders. The high velocities and extensive branching observed in the early stages point to the existence of strong space-charge fields in the air just below the cloud-base (see Sect. 9α).

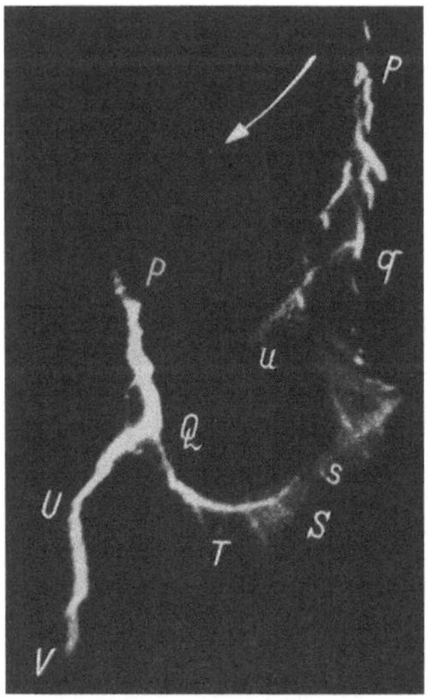

Fig. 11. Type β leader.

Both types of leader show an increase of v and of step brightness as the ground is approached. This can be attributed to the stronger fields between the tip of the advancing channel and the earth.

An example of a type α leader (flash 76) is given in Fig. 10 and of a type β in Fig. 11 (flash 40). Flash 76, details of which are discussed in Sect. 18 was recorded by Dr. D. J. MALAN at a distance of about 2 km., the BOYS camera lens moving clockwise along a circular arc. The leader branched at j to form branch J and at b to form branch B [7d].

The lens motion for flash 40 was similar. The rapid first stage of this type β leader extended from p to q, corresponding to PQ on the return streamer channel. Extensive branching to points around those corresponding to U, S and T occurred and was succeeded by a very faint slow stepped portion reaching the ground at V [7b].

Fig. 12 gives the timetables of the leaders and return streamers of flashes 69 and 70. These were probably branches of a single discharge within the cloud.

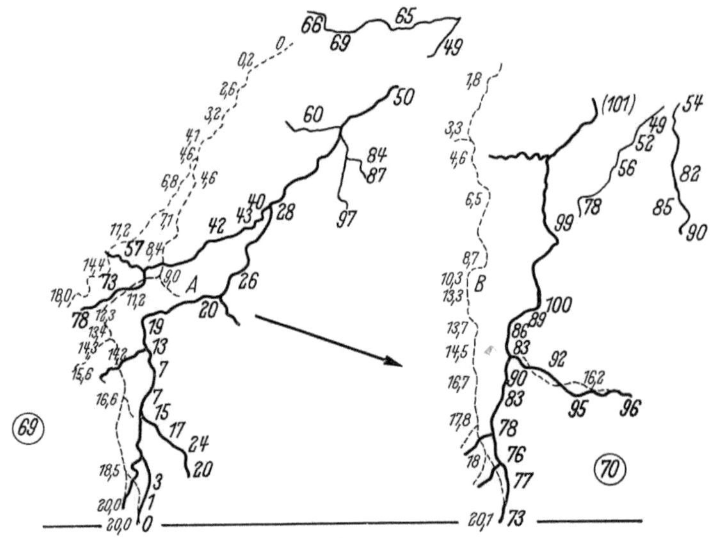

Fig. 12.

The leader times are in milliseconds measured from the moment of appearance of flash 69. The top portion of flash 70 appears 1.8 m-secs. later but this flash, by taking a shorter path to ground arrived there only 73 μ-secs. later than flash 69. The path-lengths and effective velocities of these leaders are 2.6 and 2.2 km.,

1.30×10^7 and 1.22×10^7 cm./sec. respectively. The velocities remained unusually uniform along different parts of the trunk and branches. The return streamer timetables are shown in bold figures and are measured in μ-secs. from the start upwards of flash 69.

The return streamer record of these two discharges is reproduced in Fig. 22.

17. Type β_1 and β_2, first leaders. SCHONLAND, MALAN and COLLENS have reported a number of type β first leaders which show the effects of space-charge in the air below the cloud in an extreme form [*7b*]. They fall into two classes of which the first (type β_1) differs from a normal β leader only in that it shows a very marked discontinuity in the effective velocity of its downward movement. In the heavily branched upper portion, v is high (0.8 to 2.6×10^8 cm./sec.) but it suddenly changes to a low value (0.7 to 3.2×10^7 cm./sec.) in the lower portion with steps of feeble luminosity and no branches.

In the second class, β_2, the movement of the slow second stage is interrupted by the sudden appearance of an overtaking fast dart streamer from the cloud to the leader tip. This appears to enable the leader to proceed onwards to ground at a very low effective velocity ($\sim 1.0 \times 10^7$ cm./sec.). The rejuvenating dart process may recur several times before the leader is able to reach the ground, giving the impression, which has been the conclusion of some workers, that the leader actually advances in a series of three or more very long steps at considerable intervals of time.

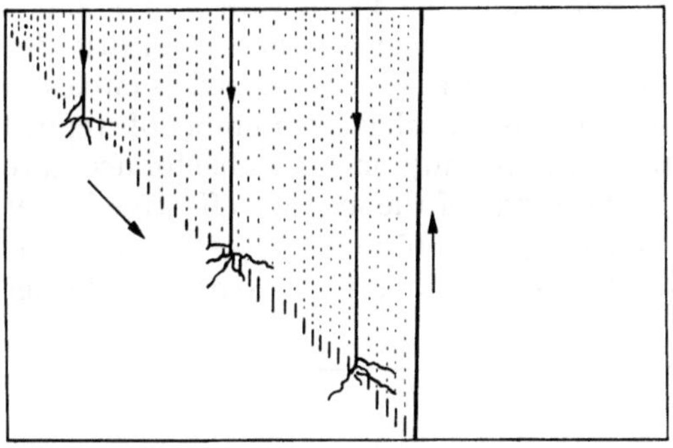

Fig. 13. Type β_2 leader.

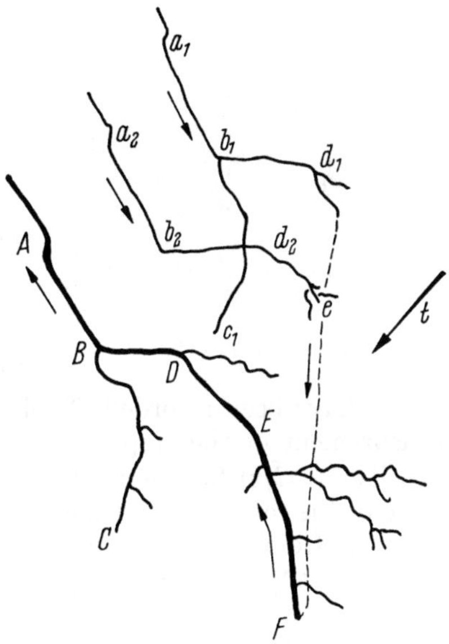

Fig. 14. Type β_2 leader.

The appearance of these darts is observed to be associated with profuse leader branching at their ends (Fig. 13). This suggests that they are called into being when the weak stepped leader process reaches regions of fairly dense positive space-charge.

The β_2 leader is of interest because it is the slowest and weakest form of leader process by means of which the cloud charge can make contact with the ground and call forth a return streamer. If it fails to do so the step leader process, with or without intermittent darts, ends in the air as an air-discharge.

An example of a β_2 leader is shown in Fig. 14. It began with a bright high velocity section $a_1\, b_1\, d_1$ with a prominent branch to c_1, after which it moved slowly from d_1 to e at 9.2×10^6 cm./sec. At e the leader was caught up by a fast dart leader $a_2\, b_2\, d_2\, e$ and continued onwards at 7×10^6 cm./sec. (velocities are given in two dimensions only) to reach the ground at F and initiate the return streamer $FEDBA$. A leader with two such dart components, D_1 and D_2 is illustrated in Fig. 15.

18. Step-lengths and pause-times. Except in the sudden transitions of velocity described above for type β leaders, the step-length, l, does not vary greatly in the course of the first leader, despite inevitable fluctuations due to components of motion of the steps along the line of sight. The pause-time, τ, also remains remarkably constant.

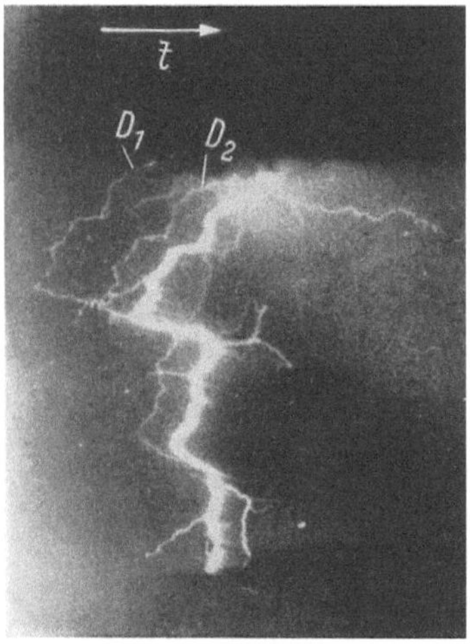

Fig. 15. Type β_2 leader.

Flash 76 (Fig. 10) which shows 24 easily measurable trunk steps, serves as an illustration of this regularity in the step process. The mean step length, $\bar{l}$, is 24.6 m. and the mean pause-time, $\bar{\tau}$, is 52 μ-secs. The root mean square deviation of l is $0.21\,\bar{l}$ and of τ is $0.45\,\bar{\tau}$. The variations are somewhat less if the lowest steps are excluded, for these show increased l's and reduced τ's as the ground is approached. The same regularity is observed in other discharges, smaller variations than those above being usual.

To show how the mean values of l and τ vary in different discharges, these quantities for 21 different first leaders are plotted as circles in Fig. 16. Though a much larger number of stepped leaders have been recorded (Table 1) these are the only ones suitable for accurate measurement of the step parameters. The range of the average effective leader velocity $v\,(=\bar{l}/\bar{\tau})$ for the circled data is from 2.6×10^7 to 2.1×10^8 cm./sec. The range of the mean pause time is 37 to 124 μ-secs. and for all except one leader (20/21) is from 37 to 92 μ-secs.

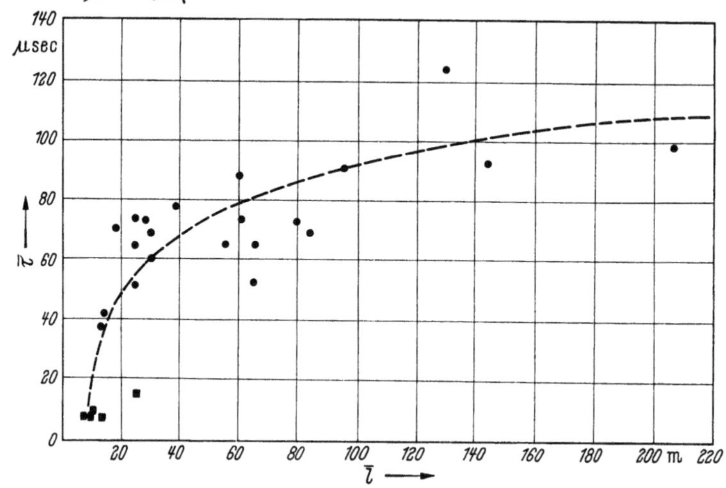

Fig. 16. Mean step parameters.

The figure also gives as squares the step parameters for five dart-stepped leaders to subsequent strokes. The dart-stepped leader is described in Sect. 21.

19. The pilot streamer. As already indicated all high velocity streamers traverse previously blazed and hence previously ionized paths. The high velocity of the step streamers in the first leader suggests that these too travel along a pre-ionized path and follow upon a pilot streamer of weak luminosity moving with the effective velocity v of the leader.

Such a pilot streamer advancing into un-ionised air would have to create the necessary conductivity by means of electron avalanche processes at its tip and should exhibit a minimum velocity v_c equal to the minimum drift velocity at which an electron can ionize. It is known from theoretical considerations (see Sect. 50) that v_c must lie between 1.4×10^6 and 1.8×10^7 cm./sec. The cloud-chamber observations of RAETHER[1] indicate that in ordinary air it is 1.25×10^7 cm./sec.

The argument for the pilot-streamer would thus be strengthened if it could be shown that first leaders exhibit a minimum value of effective velocity, [7e].

Table 1 shows the values of v for 102 first leaders including 19 observed from the Empire State Building by MCEACHRON *et al.* and a number of previously unpublished results obtained by MALAN and the writer.

Column 2 of the Table is concerned with type α leaders which have the smallest values of v, the velocities most frequently lying between 1.0×10^7 and 2.0×10^7 cm./sec. as do those for the upward-moving and usually unbranched leaders from the Empire State Building in Column 5. Included in the range 1.0×10^7 and 2.0×10^7 cm./sec. are two values of 8.0×10^6 and two of 9.0×10^6 cm./sec. in Column 2 and two between 6.0×10^6 and 1.0×10^7 cm./sec. in Column 5. These are in two dimensions.

Table 1. *Distribution of effective velocities of stepped leaders.*

1 (× 10⁷ cm./sec.)	2 Type α leaders	3 Type β leaders and air discharges	4 Dart-stepped leaders	5 Empire State leaders	6 Total	7 Fraction of total with observed steps
1.0–2.0	27	—	—	9	36	0.45
2.0–3.0	9	—	—	4	13	0.46
3.0–4.0	13	—	—	2	15	1.0
4.0–5.0	4	—	—	2	6	1.0
5.0–6.0	3	1	—	1	5	0.8
6.0–7.0	2	1	1	—	4	0.8
7.0–8.0	2	1	—	1	4	0.6
8.0–9.0	—	3	—	—	3	1.0
9.0–10.0	—	2	—	—	2	1.0
10.0–11.0	—	2	1	—	3	0.66
11.0–12.0	—	2	1	—	3	1.0
12.0–13.0	—	1	1	—	2	1.0
16.0–17.0	—	—	2	—	2	1.0
17.0–18.0	—	1	—	—	1	1.0
20.0–21.0	—	1	—	—	1	1.0
23.0–24.0	—	1	—	—	1	1.0
25.0–26.0	—	1	—	—	1	1.0
Totals	60	17	6	19	102	

The observations thus show a minimum value of v in the expected range. Allowing for movement in three dimensions by an average increase of path-length of 30%, v_c is estimated to be 1.3×10^7 cm./sec.

The heavily branched air discharges and type β leaders of Column 3, both of which travel in enhanced fields, show high effective velocities with a maximum value of 2.6×10^8 cm./sec.

Dart-stepped leaders (see Sect. 21) as shown in Column 4 exhibit values of v between 6.0 and 17.0×10^7 cm./sec.

The values of v have in many cases been determined from slow-moving cameras, either through lack of fast camera information or because the leaders were too faint to show on these cameras. Column 6 shows how the fraction of the observations on which steps have been recorded rises with leader velocity and consequent step brightness.

Recent unpublished observations by MALAN [5] with a camera of large aperture have established that all low velocity first leaders without exception show the step process.

[1] H. RAETHER: Z. Physik **107**, 91 (1937).

20. Dart leaders. Leaders to strokes after the first are usually neither stepped nor branched but travel as continuously moving streamers to ground along the main channel formed by the first stroke. They appear on camera records as faint streamers with bright tips about 40 m. long and for this reason are called dart leaders. An example is shown in Fig. 17.

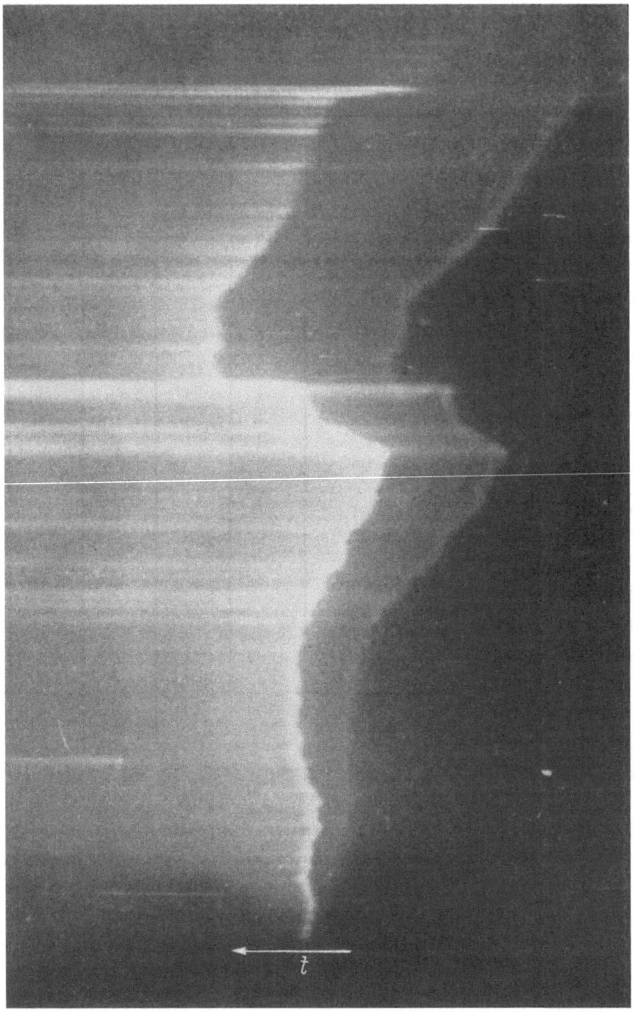

Fig. 17. Dart leader.

Their velocities range from 1.0×10^8 to 2.1×10^9 cm./sec. with a most frequent value of 2.0×10^8 cm./sec. It may be inferred that their high velocity and their continuous movement are associated with the fact that they travel along a previously ionized path. It is in fact found that there is a marked tendency for dart leader velocities to be related to the interval of time which has elapsed since the preceding stroke. SCHONLAND, MALAN and COLLENS [7a] noted that high velocities were generally associated with short intervals, and low velocities with long intervals, a result confirmed by ALIZADE [7f] from BOYS camera observations in the Ural mountains. The results of SCHONLAND, MALAN and COLLENS

Table 2.

Number of dart leaders selected	Interval since previous stroke (secs.)	Mean interval (secs.)	Velocities (cm./sec. $\times 10^8$)	Mean velocity cm./sec. $\times 10^8$)	Intensity of return stroke	Mean Intensity
5	0.005–0.12	0.044	15 to 22	19	0.3 to 0.8	0.46
5	0.07–0.48	0.17	1.7 to 2.8	2.0	1.2 to 5.0	2.2

are illustrated by the data in Table 2 above, which includes information on the luminous intensity of the following return stroke. This quantity, which is estimated in terms of the intensity of the first return stroke, is seen to be larger the greater the time which has elapsed since the preceding stroke.

21. Dart-stepped leaders. An important variant of the subsequent stroke leader is sometimes found when the interval since the preceding stroke is exceptionally long. The dart leader, though following the same channel as before, may alter from a continuously moving streamer of velocity v_1 to a stepped streamer of high velocity v_2 ($\sim 10^8$ cm./sec.) with very short steps and short time intervals between steps. It thus behaves as if the pre-existing ionization had decayed to such an extent that a fast pilot streamer was required to revive it. In some parts of the channel the pre-existing ionization would seem to have effectively disappeared and the leader, after traversing the original channel for some distance in the fast dart-stepped form, is observed to adopt a new direction and travel at a low velocity v_3 ($\sim 2\times 10^7$ cm./sec.) in the normal stepped manner.

Table 3. *Dart-stepped leaders.*

Flash No.	Stroke order	v_2 (cm./sec. $\times 10^7$)	$\bar{l}$ (metres)	$\bar{\tau}$ (μ-sec.)	Length dart-stepped (kms.)	Initial dart velocity, v_1 (cm./sec. $\times 10^7$)	Final step-leader velocity (cm./sec. $\times 10^7$)
67	2	12.2	9.0	7.4	0.35	57	None
64	2	11.1	10.0	9.0	1.0	100	None
75	2	10.0	7.4	7.4	0.4	62	None
130	2	16.6	25.0	15.0	2.1	None	3.8
C 57	2	16.6	13.0	7.8	3.0	None	None
X 7	2	4.8	12.0	25.0	0.6	None	1.3

Details of the cases observed are given in Table 3. An error in the published values of v_2 for the first three [*7b*] has been corrected and three previously unpublished cases have been added.

The significance of these figures is discussed in Sect. 52. An example of a dart-stepped leader is shown in Fig. 18 in which the leader b_2 to the second stroke shows short steps from p to s, corresponding to P and S on the return portion of the first stroke (to the left). These have an average length of 25 m. and an average pause-time of 15 μ-sec. and follow the same channel as the first stroke at 1.7×10^8 cm./sec. At s_1 two-thirds of the way down from the cloud, the leader adopts a completely new path with steps of the same length as before but with $\bar{\tau} = 65$ μ-sec. and an effective velocity of 3.8×10^7 cm./sec. These can be seen as widely separated spots above the dotted arrow. Owing to their low velocity they are recorded as reaching the ground to the right of the section of the Boys camera photograph reproduced in the figure.

22. Return streamers. These streamers travel with high velocity and are very brilliant. In strokes subsequent to the first their velocity ranges from 2.4×10^9 to 1.1×10^{10} cm./sec.; for individual streamers in subsequent strokes it is remarkably

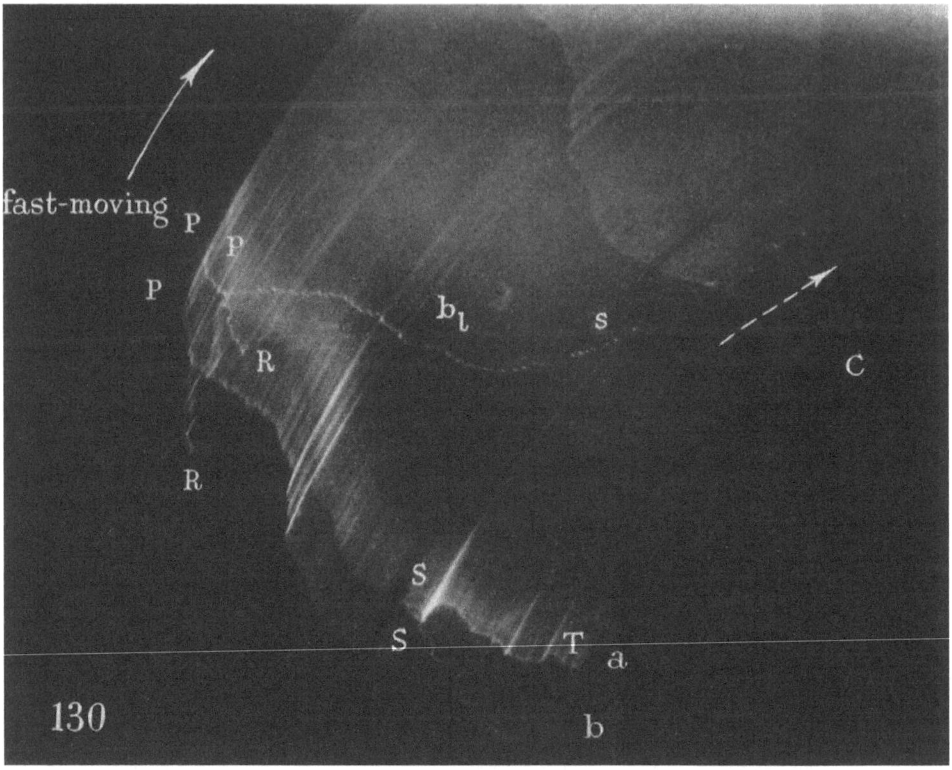

Fig. 18. Dart-stepped leader.

uniform throughout their passage from ground to cloud[1]. This is shown by the timetables in Fig. 19 in which the ordinates are distances along the channel in two dimensions, from the ground up. The luminosity of the streamer tends to decrease as it travels upwards but the effect is not very marked.

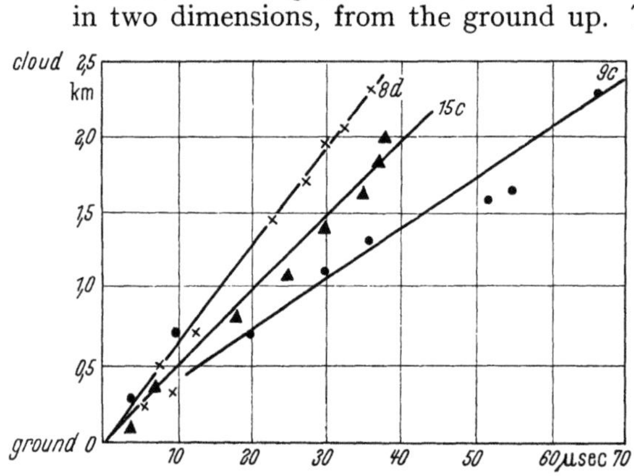

Fig. 19. Time-tables for subsequent strokes.

Branched return streamers to first strokes behave differently. The luminosity of the main channel streamer decreases abruptly and markedly in intensity after each branching-point from the bottom upwards and becomes quite low after the last branch has been reached. The luminous changes are accompanied by changes in the velocity along the main channel, which is initially very high ($\sim 1.0 \times 10^{10}$ cm./sec.) and decreases as each branch point is reached. Final values near the top of the channel are most frequently 4×10^9 cm./sec. The velocities measured outwards along the branches range from 1.5×10^9 to 1.2×10^{10} cm./sec. and show a tendency

[1] The contrary statement in major reference 3 (Sect. 3.1) is incorrect for subsequent strokes.

to be higher than the velocity of the main channel streamer at the point of bifurcation. A time-table for a first return streamer, taken from a BOYS camera record, is shown in Fig. 20; owing to the fall in luminosity near the top of the channel, the measurements there are less accurate than near the base, the resolutions at top and bottom being 2 μ-sec. and 1 μ-sec. respectively. The upward two-dimensional velocities along the trunk measured in cm./sec. $\times 10^9$ were:—base to A, 16.1; A to B, 21.0; B to C, 9.7; C to D, 5.5. There is evidence of a pause lasting about 5 μ-sec. at the moment the trunk streamer arrived at the point B, the start of the third and longest branch. After B the bright luminosity is much reduced and the streamer obviously proceeds upwards with much less development of energy. A photograph of this return streamer is reproduced in Fig. 22.

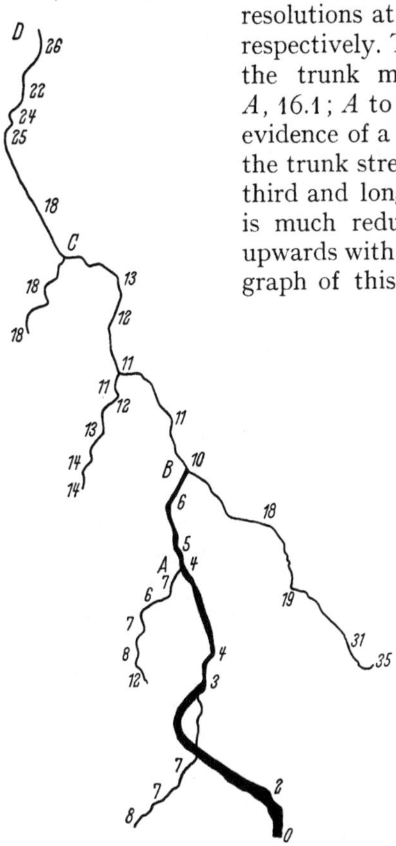

Fig. 20. Time-table of return streamer (μ-secs).

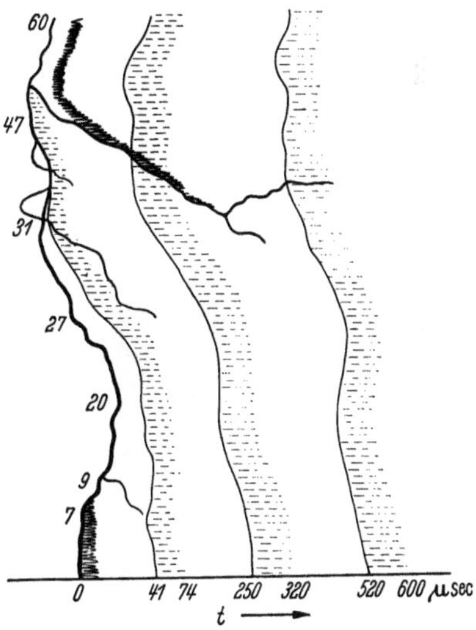

Fig. 21. Branch and M components.

In all such first strokes the streamer suddenly brightens up along its whole length to ground on arrival at each new branching point. It then divides; one part moves outwards along the branch and often continues to do so long after the part which travels up the main trunk of the leader has reached the cloud base. The observations as a whole indicate clearly that the return streamer is engaged in removing charge which has been brought down into the air by the leader and is held chiefly on its branches. It is very rare to find branches blazed by the leader which are not followed up by the return streamer.

23. Branch and *M* components. Two types of fluctuation in the luminosity of the return streamer at its base are observed. The first, which are described as "branch" components, are, as their name indicates, called into being at the moments the streamer reaches a branch. The direct relation between branch components and branches has been established by MALAN and COLLENS [7c]. The second type occurs after the streamer has entered the cloud. These M components (see Sect. 5 and 42) are observed during the period of continuing luminosity which may sometimes last for as long as 200 m-sec. but is usually much less. They are also found at the base of apparently unbranched subsequent strokes and must be associated with processes of an unknown nature, probably branching, inside the cloud. The distribution of time intervals between M components has already been discussed (see Sect. 8). It is generally quite irregular.

The duration of the period of enhanced luminosity in an M component on the other hand increases regularly with the order of the component. This is confirmed by measurements of the duration of the associated field-change (see Sect. 42). Fig. 21 reproduces the nature of the branch and M components in an actual discharge. Photographs of three return streamers which illustrate the pronounced changes in luminosity at branching points are reproduced in Fig. 22.

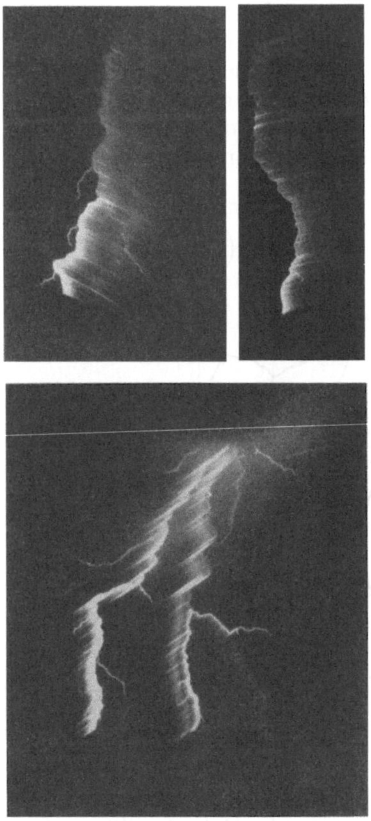

Fig. 22. Return streamers.

24. Recoil streamers. If the leader is still actively developing a branch when the return streamer starts from the ground, it is not infrequently observed that after the return streamer has reached the end of this branch, a slow inwardmoving "recoil" streamer retraverses it from its tip to the main channel. Recoil streamers have velocities of the order of 6×10^7 cm./sec. and may continue to provide luminosity in the branches concerned long after all other illumination has fallen to very low values. They appear to be due to concentrations of space-charge which are slowly redistributed along part of the conducting channel left by the return stroke. An example of a recoil streamer is shown in Fig. 21.

25. Discharges to very high buildings. It has been shown by MCEACHRON and his associates [*8*] in the case of the Empire State Building and verified by MCCANN[1] elsewhere that discharges to very high structures in the North-Eastern United States frequently begin with an upward moving stepped leader process. This has the same characteristics (*vide* Table 1) as the downward leader to open country and to less elevated structures. On entry into the cloud the leader is not followed by a downward return streamer but the channel remains luminous as a continuing discharge. The continuing discharge may be of short duration ($\sim$10 m-sec.) in which case it is often followed by an intermittent succession of downward dart and upward return streamers, or it may last up to 0.65 secs., with no subsequent strokes at all (Fig. 30).

b) Air discharges.

26. It follows from Sect. 17 that discharges which fail to reach the ground should be identical with type β leaders and this is found to be the case.

Most air discharges consist of stepped leader streamers moving at high velocity ($\sim 10^8$ cm./sec.) and profusely branched. A less frequent type has a single

[1] G. D. MCCANN: Trans. Amer. Inst. Electr. Engrs. **63**, 1157 (1944).

dart streamer, or an intermittent series of dart leaders, superimposed on intermediate steps and is identical in appearance with the type β_2 leader.

A number of air-discharges has been studied with the BOYS camera by SOURDILLON[1] who has shown clearly how their development is connected with space-charge concentrations in the air. He has also pointed out that the β_2 type of air discharge is likely to give rise to reports of rocket lightning (see Sect. 3), the successive darts occurring at such long intervals ($\sim$0.045 secs.) as to make their progress perceptible to the eye.

Recoil streamers (see Sect. 24) are not infrequently observed at the branched ends of air discharges, as would be expected if local fields due to space-charge concentrations were responsible for the directions of these discharges.

c) Cloud discharges.

27. MALAN has found by means of a special camera (Sect. 13) that cloud discharges, though giving rise to continuous luminosity, show intermittent bursts of intensity at intervals similar to those between separate strokes to ground. The most frequent duration of the cloud discharges studied by him is 250 milliseconds [5].

By the same means he has found definite evidence that luminous discharges within the cloud occur during the intervals between separate strokes to ground. This provides direct proof of the existence of the junction streamer process which has been deduced from field-change (see Sect. 43) and radar studies (see Sect. 45).

d) The apparent diameters of streamer channels.

28. Stepped leader channels. For the channel diameter of a stepped leader to be found from measurements of its photographic image, the flash must be near enough for the camera lens to give a true and not a diffraction image of its width and the leader channel must be well separated on the photograph from that of the return streamer, to avoid halation effects. If separation is achieved by camera motion, this must be slow enough to exclude artificial widening of the channel width of the image.

A simple method of satisfying these requirements is to photograph a very close air-discharge on a fixed or slowly moving camera. Such photographs are unfortunately extremely rare; a remarkable one taken by LARSEN[2] which from internal evidence was free from widening due to camera motion, indicated a diameter of six metres.

Recent measurements by the writer of BOYS camera photographs of steps in the first leader indicate diameters of between one and ten metres.

29. Return streamer channels. The measurement of the diameters of return streamer channels from fixed camera photographs of a very near flash is possible if the paths of successive strokes have been displaced by the action of the wind, giving rise to ribbon lightning (see Sect. 3). One such photograph has been found to give diameters of 11 return streamer channels ranging from 15 to 23 cms., the average value being 16 cm. Branches had diameters ranging from 11 to 17 cm.[3]

---

[1] M. SOURDILLON: Ann. Géophys. **8**, 349 (1952).
[2] C. G. ABBOT: Smithson. Miscell. Coll. **92**, No. 12 (1934).
[3] B. F. J. SCHONLAND: Phil. Mag. **23**, 503 (1937).

D. Electrical Observations; methods and results.

I. Methods of low time resolution.

30. The capillary electrometer method of C. T. R. WILSON. Much valuable information about the electrical structure of the thundercloud and the quantities of electricity involved in discharges has been obtained from the work of WILSON and his school. The detailed discussion of these matters is more appropriately dealt with in connection with atmospheric electricity [9] and they will be but briefly considered here.

In the method of C. T. R. WILSON [*10*] an exposed conductor, such as an insulated sphere at a height of 5 m. above the ground or an insulated plate flush with the surface, is connected to earth through a mercury-sulphuric acid capillary electrometer. Changes in the electric field F due to lightning discharges produce corresponding movements of the mercury threads and these are recorded, through a slit and low-power microscope objective, on a photographic plate moving at right angles to the slit. The inertia of the system limits the time resolution to $\sim$0.1 sec. The range of the equipment can be increased from 40 to 150 km. by introducing a simple amplifier between the exposed conductor and the electrometer.

The magnitude of a field-change δF is given by

$$\delta F = k_1 k_2 \delta x \qquad (30.1)$$

where k_1 is an instrumental constant obtained by discharging a known quantity of electricity through the electrometer, k_2 is a constant which can be calculated from the geometry of the exposed conductor and δx is the movement of the mercury threads.

The field-strength F at any instant can also be determined by shielding the exposed conductor and observing the resultant δx, when in (30.1), $\delta F = F$.

31. Point-discharge methods; the alti-electrograph. In sufficiently strong fields a fine needle-point connected to earth through a galvanometer provides a point discharge current I whose magnitude is related to F by the equation

$$I = a(F^2 - M^2) \qquad (31.1)$$

where a and M are constants for the arrangement employed.

An instrument known as an alti-electrograph has been devised by SIMPSON and SCRASE[1] to make use of this principle in free balloon observations of the electric fields at points near and within thunderclouds. Two separate iron pins rest upon a sheet of pole-finding paper and are connected to vertical wires, the upper one attached to a bamboo frame-work and the lower one hanging freely downwards. In a sufficiently strong field, the difference of potential appearing at the pins causes a small current to flow between them and a stain to appear on the paper near the positive point. The position of the stain indicates the direction of the prevailing field and its width can be related roughly to the field-strength. The disc carrying the pole-finding paper is rotated by clock-work to give a record which can be correlated with simultaneous pressure, humidity and temperature records.

32. The distribution of thundercloud charges; types of discharge. From these investigations of fields and field-changes at the earth's surface and above it full confirmation has been found for the original conclusion of C. T. R. WILSON [*4*]

[1] G. C. SIMPSON and F. J. SCRASE: Proc. Roy. Soc. Lond., Ser. A **161**, 309 (1937). — G. C. SIMPSON and G. D. ROBINSON: Proc. Roy. Soc. Lond., Ser. A **177**, 281 (1940).

that the main charges in a mature thundercloud cell constitute a dipole with a positive charge P (Fig. 23) nearly vertically above a negative charge N. These charges are not necessarily equal in magnitude and must be expected to be subject to different mechanisms, and therefore different rates, of dissipation. The dipole structure has been confirmed by measurements made in aircraft flying above active thunderstorms[1].

There is evidence both from alti-electrograph balloon observations and from measurements at the ground that a small subsidiary positive charge, p, often

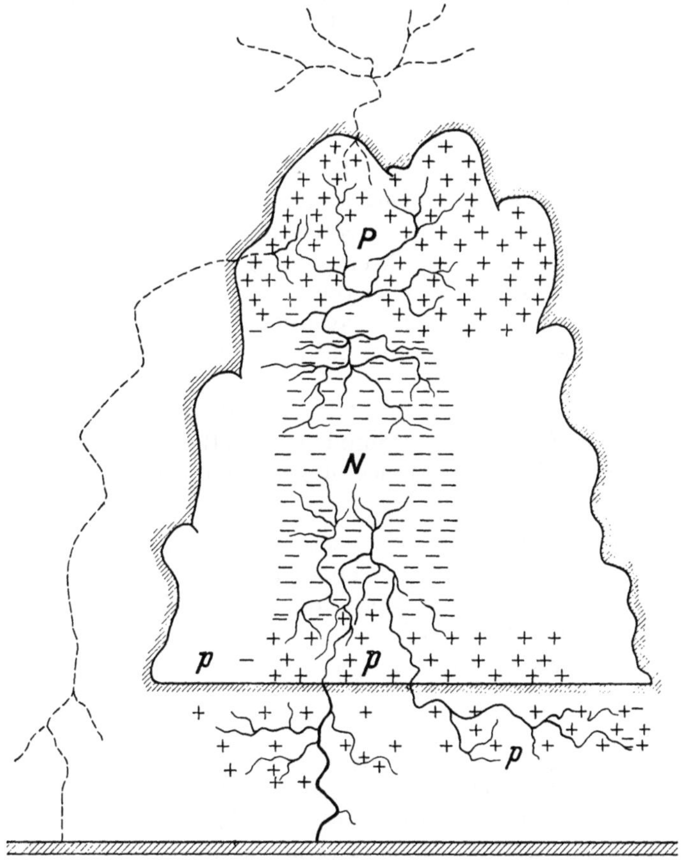

Fig. 23. Distribution of thundercloud charges.

exists below N[2]. It is carried both on the water-drops forming the base of the cloud and as a positive space-charge on ions in the air immediately below the cloud-base. The p charge, which is considered to be produced by a mechanism different from that responsible for the P-N dipole has been found to be not infrequently displaced by as much as several kms. from the dipole axis[3].

Two chief types of cloud discharge may be expected from this distribution; P-N and N-p, as shown in the figure. The former can be identified with the type of cloud-discharge most frequently studied by WILSON and others and the latter with a cloud discharge taking place at a lower height (and hence producing

[1] O. H. GISH and G. R. WAIT: J. Geophys. Res. **55**, 473 (1950). — G. R. WAIT: In Chapt. X, Thunderstorm Electricity. Univ. of Chicago Press 1953.
[2] D. J. MALAN: Ann. Géophys. **8**, 385 (1952).
[3] C. A. HACKING: J. Geophys. Res. **59**, 449 (1954).

a field change reversing in sign at a shorter distance) found by WORMELL [11]. When p is concentrated in the air beneath the cloud base, the N-p discharge becomes an air-discharge which, if conditions are favourable, (in particular if the cloud-base is low) proceeds to earth as a flash to ground. It has been suggested[1] that the N-p discharge is an essential preliminary to the starting of all flashes to ground. The sign of the currents measured in direct discharges to structures such as towers and moored balloons is predominantly negative and indicates that the great majority of these discharges come from the N region. Out of 2503 such measurements in Europe, the USA., Russia and Japan, only 266 have been found to involve a nett transfer of positive electricity [3]. The predominance of N discharges in storms at the mature stage is likely to be even greater than these figures indicate since many of the positive discharges must have come from storms in the final disintegrating stage, when most or all of the N charge has been removed by precipitation.

In the case of very tall structures such as the Empire State Building, most discharges begin in the enhanced fields at their summits and travel upwards (see Sect. 25). These discharges also involve a net transfer of negative (N) charge to ground. It has, however, been pointed out that the initial upward leader in this case may be produced by the field of the lower p charge, whose proximity to the top of the Building is exceptional[2].

Other types of discharge are possible but rare; these are indicated in Fig. 23 by dashed lines.

Studies by MALAN and SCHONLAND of the heights of the points of origin of discharges show that in South African thunderstorms the N charge is of columnar form and extends upwards for a distance of 5 to 6 km., from the -5° C to the -40°C isotherms (see Sect. 47).

33. Changes in electric moment and quantities of charge in flashes to ground. The movement of a charge Q between heights H_1 and H_2 (where H_2 may be zero) produces a change of field δF at a point on the surface of the earth at a horizontal distance D ($D \gg H_1$ or H_2) given by[3]

$$\delta F = 2Q(H_1 - H_2) = \delta F \times D^3. \tag{33.1}$$

Hence,

$$M = 2Q(H_1 - H_2) = \delta F \times D^3. \tag{33.2}$$

M, the change in the electric dipole moment, called by C. T. R. WILSON [4], [10] the electric moment of the discharge, can thus be determined by observations of δF and D on discharges at distances exceeding 10 km. The mean values found for M by various observers are summarised in Table 4.

These results show though M has a very wide range its mean value is close to 100 C-km. and differs little between cloud and ground flashes. If $H_1 - H_2$ in equation (33.2), following C. T. R. WILSON, is $\sim$2.5 km. for both types of discharge, Q is 20 C. A more detailed examination by WORMELL suggests a value of 14.4 C with a probable range of 13 to 34 C [11].

PIERCE and WORMELL [12] have recently determined M for the separate strokes of a number of discharges to ground, making use of amplifier and cathode-ray oscillograph equipment of long time-constant (see Sect. 34). The first stroke has been found to give the largest M (64 C-km.), subsequent strokes having only half this electric moment. The slow leader field-change initiating the first stroke gives $\overline{M} = 35$ C-km. and the return process 29 C-km.

[1] D. J. MALAN and B. F. J. SCHONLAND: Proc. Roy. Soc. Lond., Ser. A **209**, 158 (1951).
[2] B. F. J. SCHONLAND and D. J. MALAN: J. Franklin Inst. **258**, 271 (1954).
[3] B. F. J. SCHONLAND and J. CRAIB: Proc. Roy. Soc. Lond., Ser. A **114**, 229 (1927).

Table 4. *Mean electric moments of discharges* (*Coulomb-kms.*).

Observer	Country	$\overline{M}$ (cloud)	$\overline{M}$ (ground)	Range of M	Number of strokes involved
WILSON [*10*]	England	100		5–500	?
WORMELL [*11*]	England	110	130	22–1200	?
SCHONLAND and CRAIB[3] . .	South Africa	94	94	9–580	?
PIERCE and WORMELL [*12*] .	England		89		1
			125		2
			161		3
			196		4
			107		1 (positive)

Observations of the total quantity of electricity conveyed to ground in flashes to structures fitted with current measuring equipment (see Sect. 37) give values of Q for the whole discharge which agree with those deduced from changes of electric moment. MCEACHRON *et al* [*8*] report Q for flashes to the Empire State Building to be 19C (most frequent value) with a range of 0.5 to 164 C and MCCANN for other flashes reports a modal value of 10C [*13*].

Studies of the pre-discharge field by MALAN[1] suggest that the total p charge is of the order of 10 C. Only a fraction of this ($< \frac{1}{2}$) takes part in the discharge.

It should be noted that equations (33.1) and (33.2) require alteration when applied to the first stroke of a discharge if this stroke involves both N (Q_- at H_1) and p (Q_+ at H_2). Equation (33.2) then becomes

$$2Q_- H_1 - 2Q_+ H_2 = \delta F \times D^3. \qquad (33.3)$$

II. Methods of high and intermediate time-resolution.

34. The amplifier-oscillograph method. This method, which can record electrical field changes taking place in less than a microsecond, is shown diagrammatically in Fig. 24. C_0 is the capacity of an exposed conductor (which for nearby storms, to avoid point-discharge effects, is usually a metal sphere and for far storms is a horizontal wire aerial) and R_1 is a resistance ($\sim 50000\,\omega$) chosen to make the aerial circuit aperiodic. C_1 is a condenser variable in steps to adjust the sensitivity of the equipment and R_2 is a high resistance such that the time-constant $C_1 R_2$ is large in comparison with the duration of the field-change to be measured.

If the electric field alters by δF as a result of a movement of charge in the cloud, this will produce a change $\delta V = h \cdot \delta F$ in the potential of the exposed conductor, where h is its effective height. A change $-\delta q$ must then take place in the charge on the conductor and that on the upper plate of C_1 must increase by $+\delta q$, creating a change of potential δv across C_1 given by

$$\delta v = \frac{\delta q}{C_1} = \frac{\delta V - \delta q}{C_0} \qquad (34.1)$$

since both expressions give the change of potential at X. Hence

$$\delta v = \delta V \cdot \frac{C_0}{C_0 + C_1} = \delta F \cdot h \frac{C_0}{C_0 + C_1}. \qquad (34.2)$$

The arrangement thus operates as a condenser potentiometer. An amplifier with suitably long time-constants is attached to C_1 and its output produces a deflection of the spot of a cathode-ray oscillograph which is proportional to δv and hence

[1] D. J. MALAN: Ann. Géophys. **8**, 385 (1952).

to δF. The oscillograph spot holds this deflection only momentarily unless all the time-constants, including $C_1 R_2$ are very large.

Two difficulties interfere with the use of long time-constants in the study of the effects of near storms; the first is the tendency of the summation of the successive changes in field to reach magnitudes so great as to overload the oscillograph and cause some of the record to be missed. The second is that, when storms are very near, point discharge leakage from the exposed conductor affects its insulation.

For these reasons it is preferable to record the slower field-changes accompanying near lightning discharges by means of a device of intermediate time resolution described in the next section.

If C_1 and R_1 are removed from the circuit of Fig. 24, the arrangement can be used to record the rate of change of the field, dF/dt. From equation (34.1)

$$\left.\begin{aligned} h\frac{dF}{dt} &= \frac{dV}{dt} = \frac{dv}{dt} + \frac{dq/dt}{C_0} \\ &= \frac{dv}{dt} + \frac{v}{C_0 R_2} \quad \text{since} \quad \frac{dq}{dt} = \frac{v}{R_2}. \end{aligned}\right\} \tag{34.3}$$

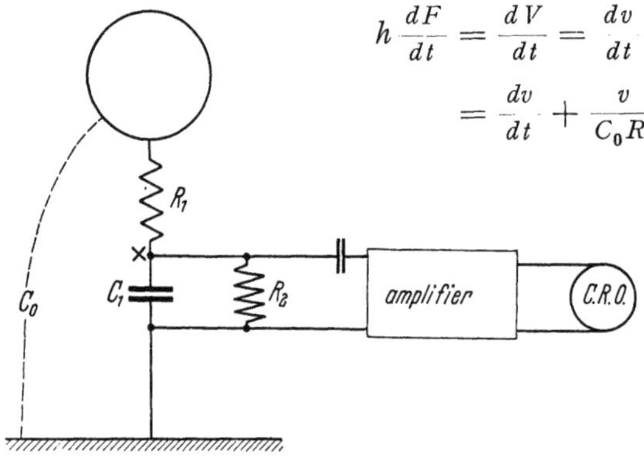

Fig. 24.

If $C_0 R_2$ is small, the second term is much greater than the first and

$$v \approx h\, C_0 R_2 \cdot \frac{dF}{dt}. \tag{34.4}$$

In this form the method has been used by NORINDER[1] and by CHAPMAN [*14*]. Various devices have been developed for recording the transient spot movements involved in the amplifier-oscillograph method. The simplest is to black out the zero position of the spot and record its movement on a film carried on a high speed drum camera, whose axis is regularly advanced by a screw. Since the traces from neighbouring turns then overlap and the whole duration of a discharge may occupy several turns of the drum, a second but slow-moving drum camera is focussed on the spot to enable the field-changes from successive strokes to be placed in their correct order.

In another form of recording device the spot is moved at right angles to the main deflecting plates by a conventional time base and successive traces are separated from one another by recording them on a slow-moving drum camera with film-movement at right angles to the time base.

CHAPMAN [*14*] has used a C. R. O. screen of long after-glow and a conventional time-base to record transient field-changes with a fixed camera whose shutter was triggered by the transient. A very rapid trigger device operated by the transient field-change itself was first used by NORINDER[1]. SCHONLAND and ELDER have employed the anticipatory trigger device described in Sect. 12 and originally developed for daylight photography.

35. The electrostatic fluxmeter. This device has been developed by MALAN and SCHONLAND from the "field-mills" of RUSSELVEDT and of GUNN to provide a recording instrument of intermediate resolving power ($\sim$0.001 secs.) to supplement the amplifier-oscillograph method in the study of the fields and field-changes of near storms[2].

[1] H. A. NORINDER: J. Franklin Inst. **218**, 717 (1934).
[2] D. J. MALAN and B. F. J. SCHONLAND: Proc. Phys. Soc. Lond. B **43**, 402 (1950).

If a conductor, connected to earth through a resistance R, is periodically exposed to an electric field F, an alternating potential difference of amplitude proportional to F will appear across R. This can be amplified and applied to a C.R.O. to record F and its changes. For rapid response the capacity of the conductor must be small and the rate of screening and exposure as fast as possible.

Fig. 25 shows one form of the instrument. Eighteen metal studs (diam. 6.5 mm.) are set in an insulating disc B on a circle of radius 10 cm. The studs are joined together to the control grid of a low capacity triode mounted inside the box C and provided with a grid leak of 2 $M\omega$. The triode output is taken from a cathode follower circuit by cable to a C.R.O. A screening disc A with 18 equally spaced holes is rotated above the studs at 4000 r.p.m. by the steel shaft E which is belt-driven by an electric motor. The mounting G enables the whole instrument

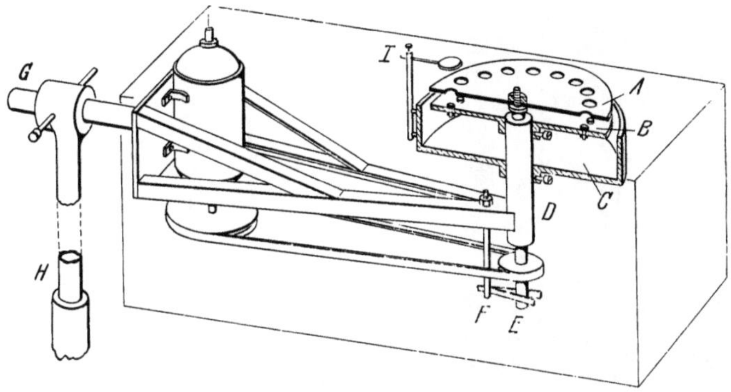

Fig. 25. Electric fluxmeter.

to be inverted to avoid disturbances from rain and hail, a procedure which has been found to halve its sensitivity. An essential part of the equipment is the earthing brush F which serves to remove frictional charges collected by A. The instrument in Fig. 25 has a frequency of 1200 c/s and a measured response-time of half a cycle (0.004 sec.). As described, it does not directly indicate the polarity of the field; this information is obtained by adding an additional set of four studs placed in pairs opposite each other and nearer the centre of B than the main studs, to which they are connected. The disc A is provided with an additional pair of holes to expose these studs, with the result that each ninth cycle of the output has its crest slightly increased if the field is positive and its trough increased if the field is negative.

The instrument can best be used in the observation of transient field-changes by storing its output on moving magnetic tape for subsequent "playback" and photographic recording. The 1200 c/s information yielded by it is suitable for transmission by radio link from distant stations. Two examples of tape-recorded field-changes obtained by this method are shown in Fig. 26. The upper record (Fig. 26a) refers to a flash, consisting of a single stroke, which struck the ground about 200 m. away (Hacking). The negative hook-shaped change L due to the downward-moving leader and the rapid subsequent R change (see Sect. 40) are clearly shown, as is the subsequent recovery of the field to its original negative value after a long period during which it remains close to zero.

The lower record (Fig. 26b) shows the R field-changes due to four strokes of a discharge at a greater distance. It illustrates the negative-going interstroke J field changes discussed in Sect. 43.

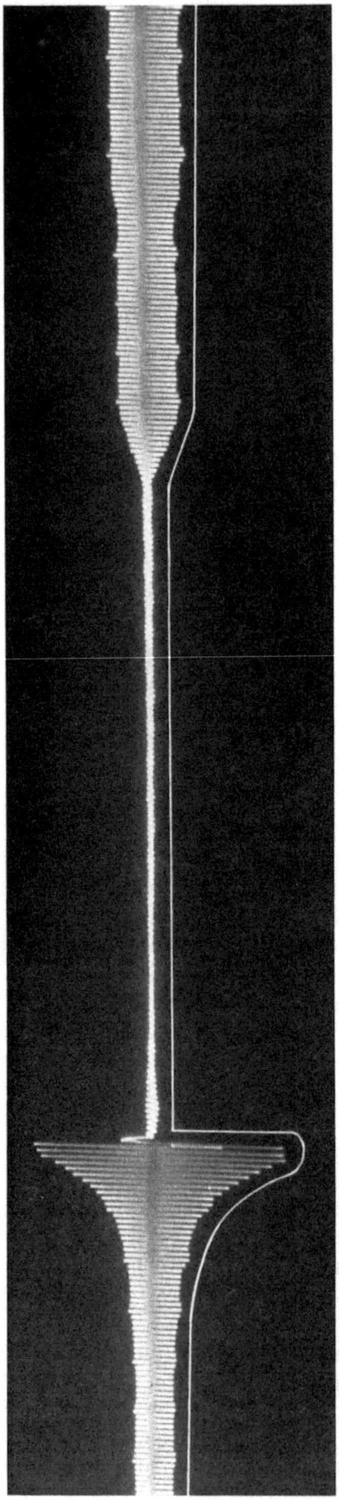

Fig. 26a. Field-change due to a single near stroke.

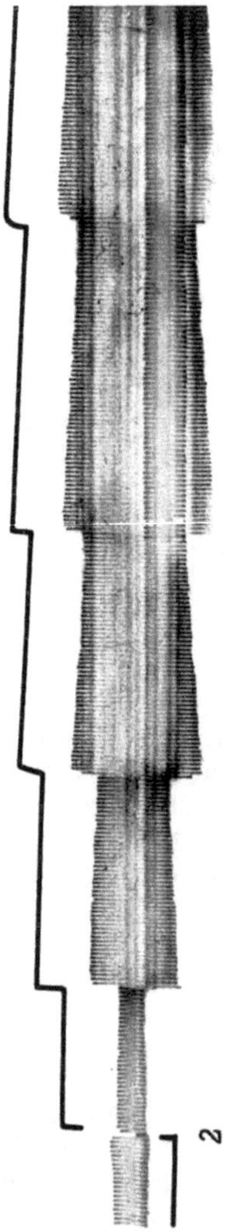

Fig. 26b. Field-change due to a multiple discharge.

III. Currents and current-time variations.

36. Peak current values. Many measurements of the peak values of currents to structures such as the towers of power-lines have been made by means of magnetic links, devices in which the remanent magnetism of suitably placed steel rods is used as the current indicator. The results are plotted in Fig. 27 (after BRUCE and GOLDE) [3] and show most frequent values of ~20 kA. Currents exceeding 100 kA are rare.

37. Current-time variations. Most of the not very abundant information on the variation of current with time has been obtained by attaching cathode-ray oscillograph equipment to a non-inductive resistance network through which flows the current from a direct stroke to a power line, tower, or captive balloon, [3]. An arrangement of magnetic links on the surface of a rapidly rotating drum called a fulchronograph[1] has also given valuable

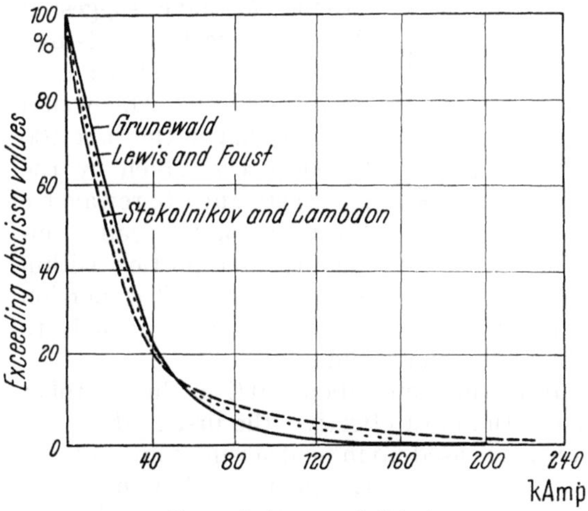

Fig. 27. Peak currents in lightning.

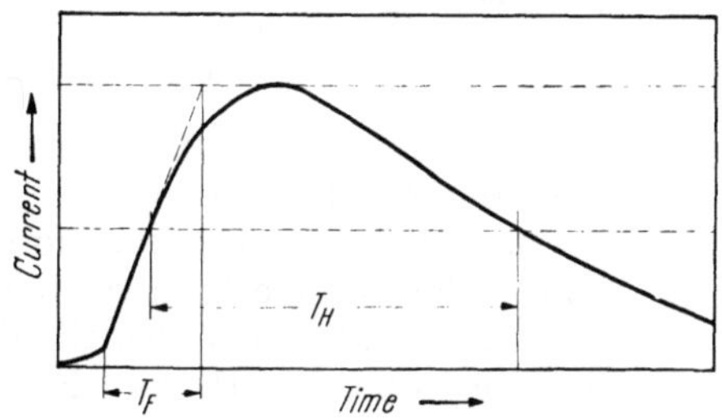

Fig. 28.

information of lower time resolution. By recording the current induced in loop aerials, which is proportional to the rate of change of current in the distant discharge channel, and subsequent integration, NORINDER and MATHIAS have deduced the current-time variation. BERGER has similarly used the magnetic field due to a direct stroke to a transmission-line tower to derive the maximum rate of change of current in the discharge.

The information available from all these sources is unfortunately incomplete as it rarely refers to strokes subsequent to the first of the series making up a discharge. In the case of flashes to power lines it is somewhat ambiguous for it requires correction for corona loss and for distortion of the wave-form of the resulting travelling wave in its passage to the point of observation. Most of the power line records refer to the weaker discharges which do not flash over the tower insulators and so miss the more powerful discharges.

These observations suggest that the current I in the return streamer varies with time according to the equation

$$I = I_0 (e^{-\alpha t} - e^{-\beta t}) \tag{37.1}$$

where I_0, α and β are constants, giving the curve shown in Fig. 28. The range of variation of the "wavefront" time T_F is from 1 to 15 μ-secs. and its most frequent value 2.5 μ-secs. The figure shows that the actual time taken to reach peak value may be as much as twice T_F[2]. These times are comparable with the very short

[1] C. F. WAGNER and G. D. McCANN: Electr. Engng. **60**, 439 (1940).
[2] D. MÜLLER-HILLEBRAND: Elektrotechn. u. Masch.-Bau **54**, 361 (1936).

duration of the first branch-component of luminosity in the first return streamer (see Sect. 23) which has been found by MALAN and COLLENS to be over in less than 10 μ-secs. Thus the peak value of current may well be reached at the first branch from the ground upwards.

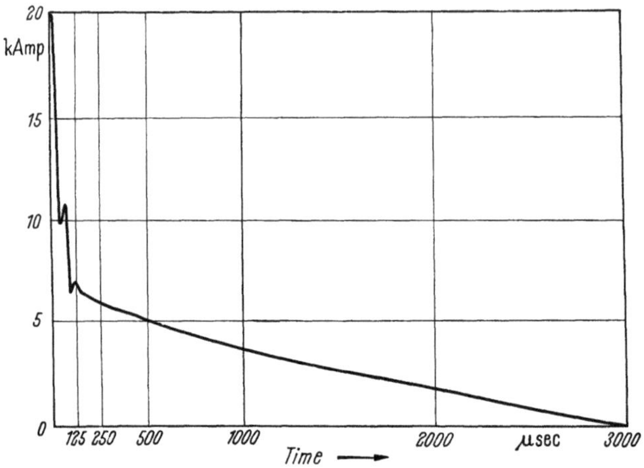

Fig. 29.

The most frequent value of the time T_H between half peak-value points is 15 μ-secs. and its range is from 2 to 100 μ-secs.

From an analysis of the available data, BRUCE and GOLDE [*3*] have suggested that the typical current-time curve for a first stroke is as shown in Fig. 29. This would indicate that in the time 100 to 600 μ-secs. taken by the return streamer to retraverse the whole of the leader, including that portion within the cloud, the quantity transferred to earth is between 1 and 4 C. The portion of the total charge residing on the leader system which is easily accessible to the return streamer must therefore be of this magnitude.

Since the average flash removes from the cloud about 20 C in three strokes (see Sect. 33 and 5) of which the first is the most effective, the first leader must carry at least 9 C into the air. Part of this is used up in neutralising some of the *p* charge in or below the cloud-base and part provides the current in the tail of the curve of Fig. 29 which usually lasts for 1 to 2 m-secs. during which (except in the case of very tall structures) it falls from ~6 kA at 100 to 600 μ-secs. after the start, to very small or zero values. It thus passes a further 3 to 6 C to earth.

Table 5. *Rate of change of current in return streamers.*

Object struck	Reported by	Range (kA/μ-sec.)	Most frequent values (kA/μ-sec.)
Empire State	HAGENGUTH and ANDERSON	0–50	12.5
Tall structures	MCCANN	0–45	9.0
Transmission line towers	BERGER	0–31	11.0
Ground (from frame aerial observations)	NORINDER	0–30	4.5

The most probable value is ~ 10 kA/μ-sec.

The source of this current may be the "mopping-up" by the return streamer of charge either in the cloud or, as BRUCE [*17*] has suggested, situated in a corona envelope on the fringes of the leader channel throughout its length. As it is not yet known whether the considerable "tail" of Fig. 29 exists in all first strokes and to what extent it is present in subsequent strokes, a detailed examination of these alternatives is not profitable at this stage.

From the above discussion, however, it would seem reasonable to estimate the total charge distributed over the channel of the more powerful first leaders at 4 C.

38. Maximum rate of change of currents. Direct observations of the maximum rate of change of current in flashes to ground are summarised in Table 5. This rate occurs at the start of the return streamer.

IV. Discharges to very tall structures.

39. For the study of the time-variation of current in discharges to the Empire State Building, McEachron, Hagenguth and their associates [8] employed equipment which in its final form consisted of two cathode-ray oscillographs; one, with a slow sweep, covered a period of about a second while the other gave details on a 100 μ-sec. time-base of the form of the current variation in individual strokes. Discharges to (or from) the antenna passed through a shunt consisting of a constant non-inductive resistance of 0.01 ω and a non-linear "thyrite" resistance in parallel. Both oscillographs were connected across this shunt, current ranges being 1000 to 20000 amps for the high speed instrument and 50 to

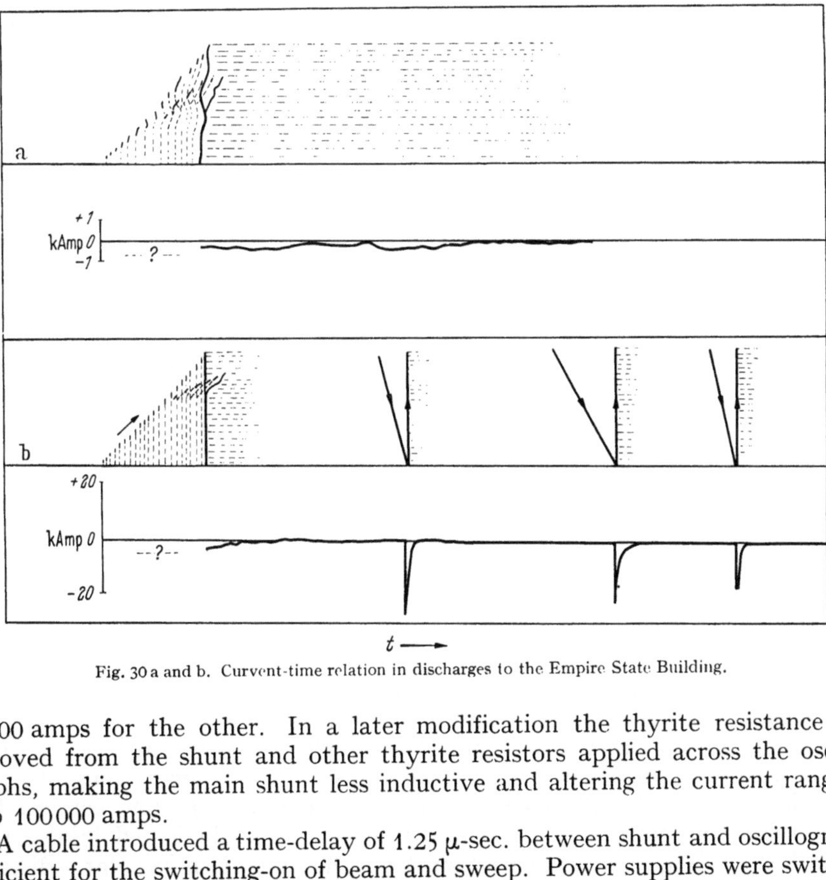

Fig. 30 a and b. Current-time relation in discharges to the Empire State Building.

24000 amps for the other. In a later modification the thyrite resistance was removed from the shunt and other thyrite resistors applied across the oscillographs, making the main shunt less inductive and altering the current range to 0 to 100000 amps.

A cable introduced a time-delay of 1.25 μ-sec. between shunt and oscillograph, sufficient for the switching-on of beam and sweep. Power supplies were switched on when the corona current from a separate point-discharger exceeded a certain value.

These observations have shown that discharges to the Building fall into two classes corresponding to the Boys camera observations discussed in Sect. 25. For the first, recorded in half of the 135 flashes studied, current flow was continuous throughout with a most frequent duration of 0.27 secs. and did not exceed 250 amps. Its maximum observed duration was 0.625 secs. In the second it fell to small or zero values after about 10 to 20 m-secs. and was followed by a series of transients of the type found in flashes to lower structures (Fig. 29) The maximum duration of this form of discharge was 1.5 secs. All flashes produced initial current flow which indicated a negative cloud charge, three changing to positive at or near the end of the discharge. It was not usually found possible to record the sense or the magnitude of the current at the very beginning i.e. in most of the initial upward-moving leaders. These may well have shown a reverse polarity corresponding to movement into the p charge (see Sect. 32).

Examples of the continuous and the intermittent type of current-time relation are shown in Fig. 30a and b.

V. Fast electrical field-change studies.

40. General correlations with camera observations. The amplifier-oscillograph method described in Sect. 34, which was first introduced by APPLETON and CHAPMAN, has been employed by a number of workers in detailed studies of the variations in the electric field in successive stages of the discharge [*15*], [*7d*]. It has

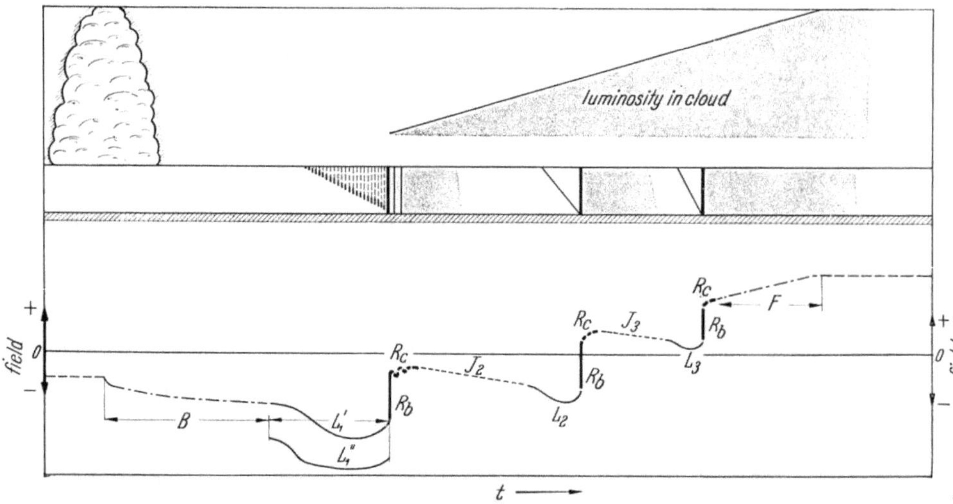

Fig. 31. Distance 3 km.

been found that practically all the observations can be interpreted in terms of the results of camera work. Direct correlations between simultaneously operated electrical and photographic equipment have been made.

A general statement of the results obtained will be given in this section, details being left for later discussion. To provide a background Fig. 31 shows diagrammatically, with time-scale suitably distorted, the variations observed in the electric field at a distance of 3 km. from a flash to ground consisting of three strokes, the corresponding luminous effects being indicated immediately above. Fig. 32 shows the field variations from the same flash when at a distance of 15 km.

The leader (L) field change is shown by the thin continuous line. Its duration corresponds roughly with that of the photographed leader process but whenever a direct correlation is obtained the L change lasts from 1.14 to 4.5 times as long, the difference increasing with the order of the stroke in the series. This indicates that a considerable length of the leader channel is hidden in the cloud and that successive strokes come from progressively higher regions.

As shown in Fig. 31 and 32, L changes are negative and hook-shaped when near and positive and approximately parabolic when far. The change in sign with distance shows that the leader lowers negative charge from the cloud and distributes this charge along its channel.

The return field-change R is made up of two parts. The rapid portion R_b shown by the thick line has a duration of 50 to 250 μ-secs. (most frequent value 165 μ-secs). and evidently corresponds to the rapid upward movement of the return streamer. Correlations between camera and oscillograph again show that its duration is between 1.5 and 2.0 times as great as the time taken for the return streamer to reach the cloud-base. R_b is followed by a slower field-change R_c (indicated by a thick dashed line) which lasts from 70 to 900 μ-secs., corresponding roughly to the duration of continuing luminosity in the return streamer channel. Both R_b

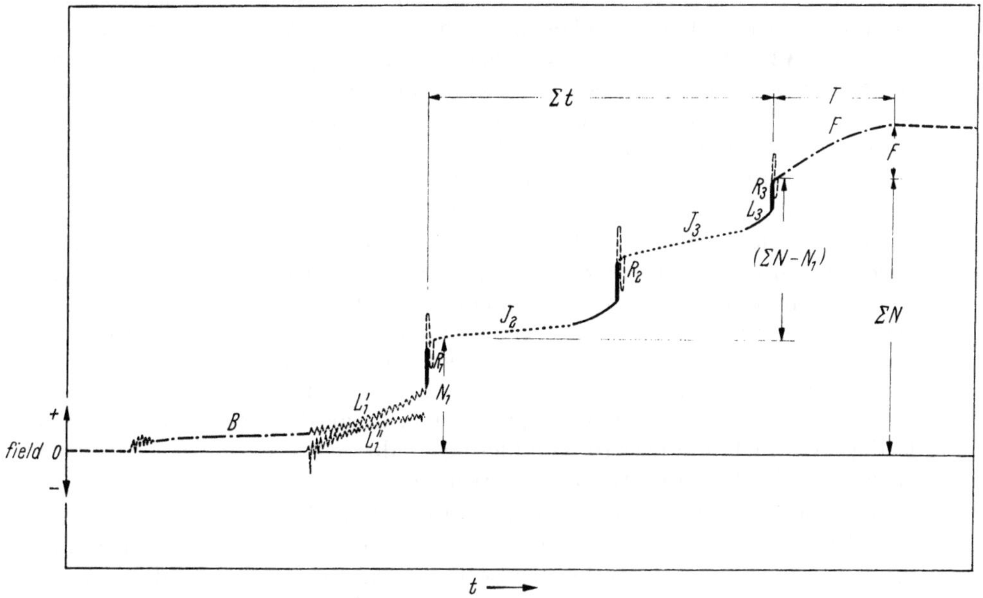

Fig. 32. Distance 15 km.

and R_c are of positive sign at all distances as would be expected if they were due to the removal to earth of negative charge from the leader channel and the cloud.

An observation of some importance, first made by CHAPMAN [*14*] and confirmed by MALAN and SCHONLAND [*7d*], is that the L_1 change shows no appreciable sudden pulsations corresponding to step leader processes when the discharge is very near and the field is predominantly electrostatic in nature. At greater distances, however, (Fig. 32) when induction and radiation components contribute significantly to the field, pulsations which can be associated with steps are prominent. This observation indicates that in spite of its intermittent luminosity, the lowering of charge by the first leader is essentially a continuous process, the brighter steps being due to a secondary mechanism involving a rapid change of current and a comparatively small movement of charge in a very short time ($\ll 1$ μ-sec., see Sect. 15).

The radiation effects of the rapid rate of change of current in the return streamer are prominent on records from distant discharges and are shown in Fig. 32 like those on L_1 by thin dashed lines.

Two distinct types of L_1 change are shown in the figures. These can be interpreted as corresponding to type α and type β first leaders (see Sect. 16) and are marked L_1' and L_1''. L_1' in Fig. 32 shows radiation pulses of fairly regular amplitude over its whole length. Their mean amplitude is ~1/15 of the pulses from the subsequent return streamer[1]. L_1'' in Fig. 32 begins with a considerable change of field lasting from 1 to 2 m-secs. with radiation pulses of amplitude ~1/3.5 of that of the return streamer. This phase is succeeded by a slower change of field with smaller pulses. MALAN has found that L_1'' types constitute 65% of all L_1 changes and has deduced that since only 35% of photographs show type β leaders, a considerable number of so-called α leaders have their initial β portions hidden in the cloud.

It is rare to find any significant radiation pulses on the field records during the intervals between strokes or on the L portions of subsequent strokes but the fields in the intervals between strokes shows the significant slow changes marked J in Fig. 31 and 32. These are negative when the discharge is very near, predominantly positive or zero when it is at a distance of 12 to 20 kms., and exhibit interesting changes at intermediate and at very great distances (see Sect. 43). They have been found to be associated with upward-moving "junction" streamer processes within the cloud, which have been studied by MALAN with the cloud camera (see Sect. 27) and by HEWITT by radar methods (see Sect. 45).

The two portions of the field record marked B and F in Figs. 31 and 32 have been the subject of recent studies. The B change which precedes L_1 in the figure is small in amplitude but of considerable duration (5 to 18 m-secs.). It has been suggested by WORMELL[2] that it represents a precursor to the leader process which travels from cloud to ground at a low velocity (~1.5×10^6 cm./sec.) and with luminosity too feeble to be photographed. MALAN and the writer, however, consider that the B change is due to a preliminary break-down process between N and p (Fig. 23) which produces the conducting conditions necessary in the lower N region before a leader can develop (see Sect. 56).

The F change, which is most prominent at the close of a discharge with a small number of strokes (<4) has been shown to be caused by a flow of continuing current after the last return stroke (see Sect. 44) and thus with a final upward J process mopping up what remains of the charge on the N column.

41. The form of the leader field-change in subsequent strokes. A comparison of the relative magnitudes of the L_1, R_b and R_c portions of records of distant field-changes from subsequent strokes led SCHONLAND, HODGES and COLLENS [*15b*] to the conclusion that the lowering of charge by the dart leader involves an approximately uniform distribution along the whole length of the leader. This conclusion for obvious reasons does not apply to the multiply-branched first leader. They also showed that a uniform charge q per unit length and a uniform velocity v of the dart leader along a vertical channel would produce a change of field δF varying with time t according to the equation

$$\delta F = q\, v^2\, t^2 / D^3 \tag{41.1}$$

when D, the horizontal distance of the flash is large compared with the height from which it starts. This gives the observed parabolic relation between F and t for distant discharges.

A more general expression [*7d*], applicable to dart leader field-changes at all distances from a discharge from height H to ground along a vertical channel is

$$\delta F = 2q\,[\{(H - v\,t)^2 + D^2\}^{-\frac{1}{2}} - \{H^2 + D^2\}^{-\frac{1}{2}} - H\,v\,t\,\{H^2 + D^2\}^{-\frac{3}{2}}]\,. \tag{41.2}$$

[1] D. B. HODGES: Proc. Phys. Soc. Lond. B. **67**, 582 (1954).
[2] T. W. WORMELL: Quart. J. Roy. Met. Soc. **79**, 474 (1953).

Curves showing the way in which the L change given by this expression alters in form with different values of D/H at constant v are given in Fig. 33. If H is constant the shape alters progressively with decreasing D from the positive parabolic form at a to the negative hook form shown at f, [7d].

If D is constant and H increases progressively from stroke to stroke of the series making up a discharge, Fig. 33 shows that the L changes of the strokes should alter progressively in form from a to f. Examples of this phenomenon, which owing to the existence of upper and lower limits to H only occur in practice for values of D ranging from 4 to 10 kms., are shown in Fig. 34. This was drawn from recordings by the oscillograph method, with interstroke fields omitted. The arrows mark the transition from positive to negative L changes. Though there are occasional variations, each discharge channel departing from the vertical in its own way, the examples reproduce the curves of Fig. 33 reasonably well. This correspondence not only indicates that H increases progressively from

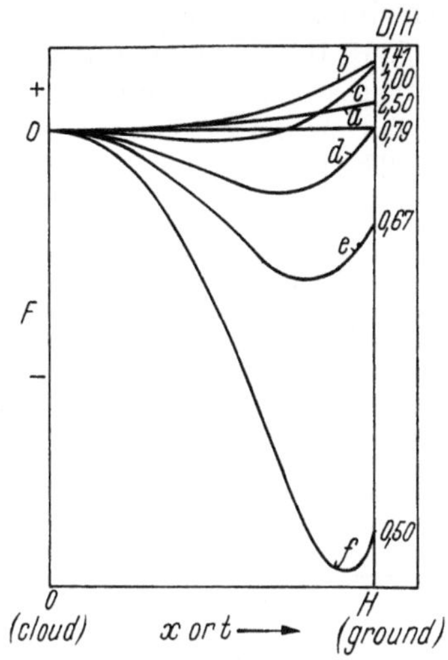

Fig. 33. Calculated L changes.

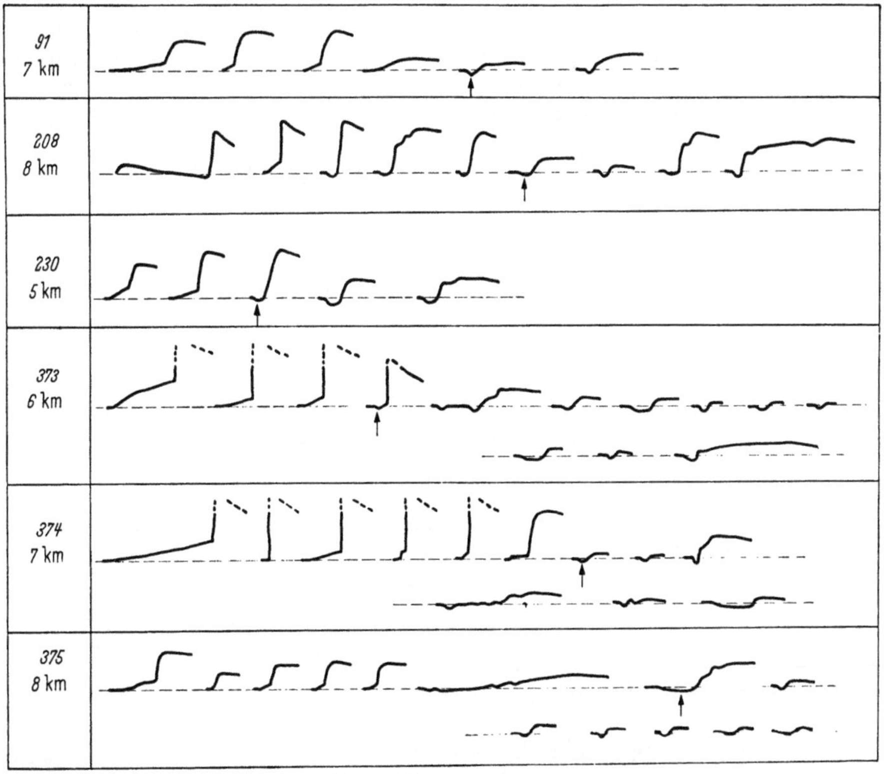

Fig. 34. Observed L changes.

stroke to stroke but also enables the height of the charged region responsible for the stroke after the arrow to be determined, for as Fig. 33 shows the nett L change is zero when

$$H = D/0.8. \tag{41.3}$$

42. *M* changes. Fig. 31 indicates that the R_c portion of the field-change from a near flash to ground frequently shows small hook-shaped field-changes [*7d*], which occur for a period up to 6 m-secs. after R_b. As exemplified by the flash in Fig. 35 these can be directly related to the M components of luminosity in the return streamer channel described in Sect. 23. At a distance they show as minor radiation pulses only.

The hook changes are always similar in general form. Each begins with a negative field-change followed, without any pause, by a larger positive change. The net field-change is always positive and usually from one-fifth to one-hundredth of the main R_b change preceding it.

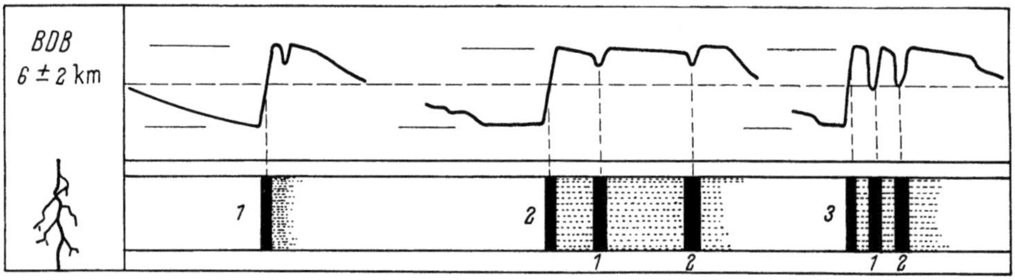

Fig. 35. *M* component field-changes.

The whole hook process generally occupies between 200 and 800 μ-sec., though durations up to 1600 μ-sec. have been noted when a long time-interval has elapsed since the R_b change. When there is more than one hook component to a stroke it is found that the durations T of successive hooks increase with the time which has elapsed since the R_b change, as shown in Table 6.

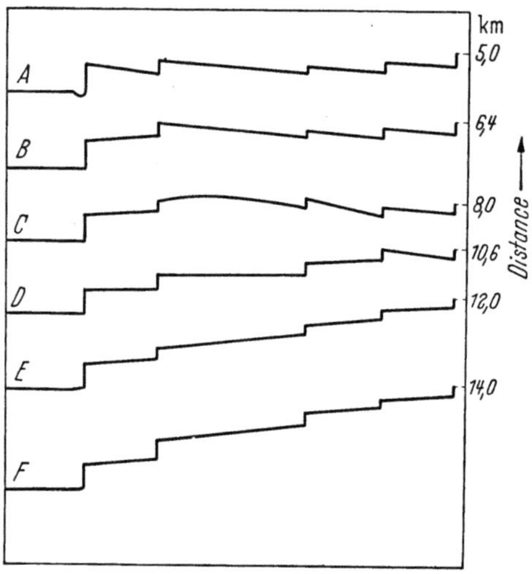

Fig. 36. *J* field-changes at various distances.

Their correlation with M components of luminosity and the dependence of their duration upon the age of the channel indicate that hook changes arise from a slow and minor form of discharge between cloud and ground. This would explain their general resemblance to the L-R change in the main discharge. It is estimated that the charge involved is from 1/5 to 1/100 of that in the main stroke [*7d*].

Table 6.

Interval since R_b (μ-sec.)	Most frequent value of T (μ-sec.)
500	275
1000	320
2000	400
(3000)	(600)

43. The *J* field-change. This field-change in the interval between separate strokes of the discharge to ground is found to be negative for near flashes and positive, or zero, for flashes at distances of the order of 15 kms. The reversal in sign with distance and the continuance of luminosity within the cloud (see Sect. 27) during the same interval show that the J change is due to a discharge process taking place inside the cloud [*15c*].

When the flash is at an intermediate distance (5 to 12 kms.) successive strokes show J changes which alter progressively in slope from positive to negative, as illustrated by the examples in Fig. 36. This and certain other properties of the J change have led MALAN and SCHONLAND to conclude that it is due to a positive streamer process passing progressively upwards from the base to the top of the N column.

The basis for this conclusion is illustrated in Fig. 37 on which the horizontal distance to a supposedly vertical J streamer has values, marked from A to F,

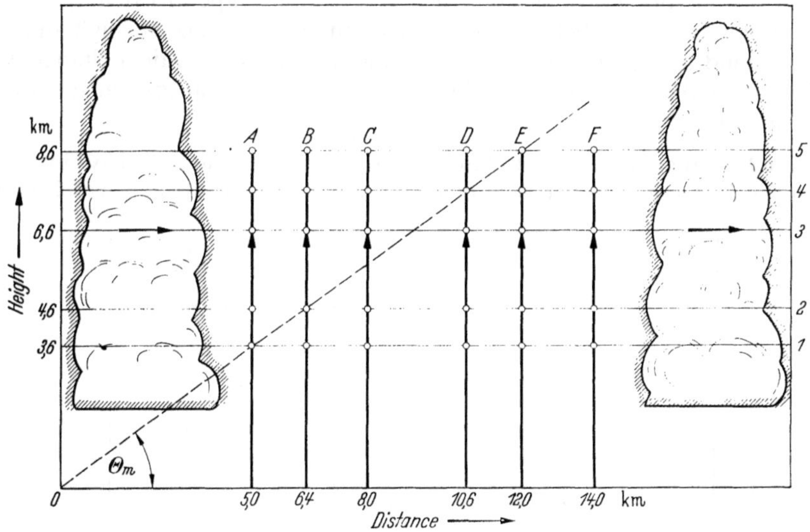

Fig. 37.

ranging from 5 to 14 km. The horizontal lines 1 to 5 represent the mean heights of the charge centres tapped successively by a discharge of five strokes, corresponding to Fig. 36.

The field at O due to a charge Q at the point H, D is

$$F = 2\,Q \sin\vartheta \cos^2\vartheta / D^2 \tag{43.1}$$

where $\tan\vartheta = H/D$. As ϑ increases, F passes through a maximum at $\tan\vartheta_m = 1/\sqrt{2}$ and the slope of the corresponding J change should alter from positive to negative if Q is positive. The discharge under consideration should therefore show the following effects in the intervals between successive strokes if either positive streamers are moving upwards or negative streamers downwards between one charged region and the next.

Table 7.

Distance (km.)	Slope of J change			
	1 to 2	2 to 3	3 to 4	4 to 5
5.0 (A)	—	—	—	—
6.4 (B)	+	—	—	—
10.6 (D)	+	+	+	—
12.0 (E)	+	+	+	+
14.0 (F)	+	+	+	+
8.0 (C)	+	?	—	—

It will be seen that these results are in accord with the observations. The J change in case C between levels 2 and 3 will show an upward curvature if it is caused by positive streamers moving upwards. This has been found in eight out of nine observations of such curved J changes.

The existence of these streamers has been directly demonstrated by the radar method of HEWITT (see Sect. 45) who finds the average vertical component of their velocity to be 2×10^6 cm./sec., in satisfactory agreement with that obtained by dividing the average distance between the regions responsible for successive strokes (see Sect. 46) by the average time-interval between strokes.

Observations of the J change by WORMELL and PIERCE [*12*] in England have been made on more distant discharges (20 to 100 km.). They have found only 25% instead of 64% of the distant J changes to be positive, the remainder being reported as zero, and consider that this indicates a difference between the distribution of the N charge in England and South Africa, a nearly vertical column being much less frequent in the clouds studied by them. It is possible, however, that such distant observations are affected by a new type of discharge process producing slow negative field changes. This may take place so high up, in or above the cloud top, that at moderate distances it produces a field-change which is small compared with that caused by the J process while at greater distances it is of comparable size.

44. The F-change. Significant connections have been established by MALAN [16] between the F change and the field changes in the separate strokes which precede

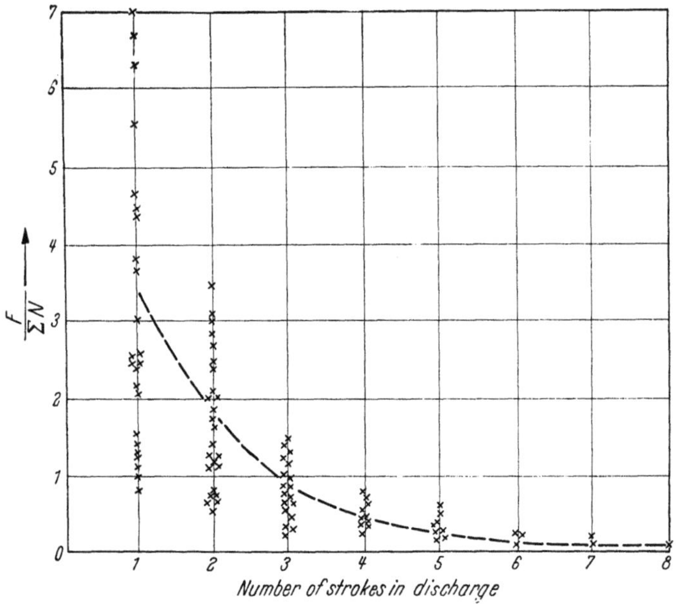

Fig. 38.

it. Table 8 shows that the probability of an F change being observed at all depends upon the number of strokes in the discharge and becomes very small when this number is large.

Table 8.

Number of separate stroker in the discharge .	1	2	3	4	5	6—12
Percentage of flashes giving a final F change .	50	43	43	21	15	6
No. of discharges examined.	71	92	89	107	61	110

Next it is found that the ratio $F/\Sigma N$ of the magnitude of the F change to the total field change preceding it (Fig. 32) decreases, as shown in Fig. 38, as the number of strokes in the discharge increases.

Thirdly, (Fig. 39) the total duration T of the F change, is related to that of all J processes Σt (neglecting the much smaller time occupied by L and R processes after the first stroke) by the equation

$$\frac{F}{T} = \frac{\Sigma N - N_1}{\Sigma t} \tag{44.1}$$

where $\Sigma N - N_1$ is the total preceding field change less L_1 and R_1 (Fig. 32). From this equation and the two preceding results MALAN concludes that the F change is due to a final continuous form of the J streamer process, travelling upwards at about the same velocity ($\sim 2.5 \times 10^6$ cm./sec.). When it occurs it drains the remaining portion of the N column, a portion which is large and frequently able to maintain such a continuous process when the number of preceding strokes is small and *vice versa*. Equation (44.1) shows that both F and J processes remove, or make available for removal, the same average quantity of charge per unit time.

The F process is thus considered to show the frequent existence of an alternative and continuous method of discharging the N column, one which is already familiar in discharges to the Empire State Building but which in flashes to less elevated objects follows upon the intermittent type of discharge.

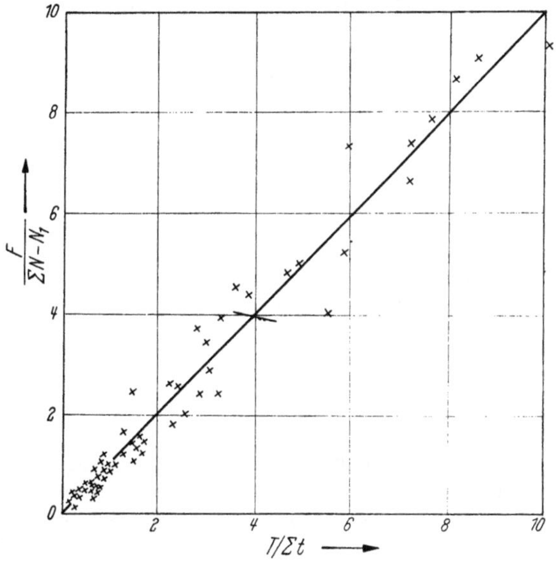

Fig. 39.

VI. Radar studies of interstroke streamers in the cloud.

45. HEWITT[1] has employed a radar equipment on a wavelength of 50 cm. with a pulse-recurrence frequency of 1000 per sec. for the study of streamer movement within thunderclouds in the intervals between strokes to ground. On this wavelength echoes from waterdrops and ice-particles are weak but fairly strong echoes are observed at the time of the occurrence of a discharge. These appear to come from free electrons in the ionized channels of streamers.

The observations show that ascending echoes occur at increasing ranges and angles of elevation in the interstroke intervals, the heights increasing with the order of the intervals in the series. This is in accord with what would be expected if the echoes came from J streamers and as stated in Sect. 43 the observed vertical velocity agrees with that found for the J streamer.

The J echoes are found to be strongest at the start of the interstroke interval and to decrease markedly in amplitude with time, becoming very weak just before the occurrence of the next stroke. This is held by HEWITT to indicate that the intensity of the J process decreases as the interstroke period proceeds and that the free electron density decreases with time by capture and recombination processes (see Sect. 56).

A further observation is that echoes at lower heights ($\sim$4 km.) persist throughout the series of ascending streamers and often show considerable horizontal movement.

[1] F. J. HEWITT: Proc. Phys. Soc. Lond. B **66**, 895 (1953) and Ph. D. Thesis, University of the Witwatersrand 1955.

At the close of each interstroke period the top portion of the J streamer is found to become the source of bursts of considerable radiation noise on the frequency (600 Mc./sec.) used. The bursts of noise start from progressively higher regions in the cloud but are observed to travel downwards with a velocity of $\sim 1.5 \times 10^9$ cm./sec. and to last for 1 m-sec. or less. These characteristics suggest that they are produced by the downward-moving dart leader.

VII. Heights of the regions initiating successive strokes.

46. Methods of determination. Four methods have been used by MALAN and SCHONLAND [*15d*] to estimate the heights of the charged regions responsible for successive strokes.

(a) A comparison is made of the duration, t_e, of the electrical field-change L_m due to the leader to the m'th stroke, with the time, t_p, determined from a simultaneous fast camera record, taken by the leader to pass from cloud-base to ground.

As noted in Sect. 40, t_e is always found to be greater than t_p.

Assuming the velocity of the leader to be constant throughout its length, the height H_m from which it started is given by

$$H_m = H_b \frac{t_e}{t_p} \tag{46.1}$$

where H_b is the measured height of the cloud-base..

(b) As shown in Sect. 41, the net field-change due to the leader, if it carries an approximately uniformly distributed charge along a vertical channel, alters from negative to positive at a horizontal distance D given by

$$H_m = \frac{D}{0.8}. \tag{46.2}$$

The assumption of uniform distribution of charge cannot apply when $m = 1$ but for higher orders the equation enables the most frequent value of H_m to be determined for various values of m from field-change records of a large number of discharges at different distances.

(c) On the same assumptions as used in (b), if ϑ is the angle of elevation of the upper end of the channel of the leader, the L and R_b field-changes must satisfy

$$\frac{L}{R_b} = (1 + \cos\vartheta)\cos\vartheta - 1. \tag{46.3}$$

Since for the m'th stroke, $\vartheta_m = \operatorname{arc\,tan}\,[H_m/D]$, measurements of L, R_b and D yield H_m for $m \gtrless 2$.

(d) An examination of equation (43.1) shows that a complete reversal of the slope of the J field-changes between strokes occurs for a vertical J streamer when

$$H_m = \frac{D}{\sqrt{2}}. \tag{46.4}$$

From the records of the J changes in a large number of flashes at different distances, the most frequent value of H_m for values of $m \gtrless 2$ can thus be determined.

Although methods (b) and (c) do not apply to the first stroke, which is heavily branched, it has been found that a special statistical application of (b) yields a useable modal value of H_1.

Table 9. *Mean values of H_m (km.).*

Method	Order, m 1	2	3	4	5	6	7	8	9	10	11	12	12
a	2.9 (20)	3.5 (4)											6.7 (3)
b and c	4.4	5.6 (17)	6.3 (13)	7.1 (11)	8.0 (8)	8.4 (10)		8.8 (10)			9.1 (3)		
d		3.5 (2)	4.5 (6)	4.7 (5)	6.8 (5)								
Unweighted mean	3.7	4.2 (23)	5.4 (19)	5.9 (16)	7.7 (23)			8.8 (10)			9.1 (3)		

47. Results of height determinations. The mean values of H_m found by these methods are given in Table 9, the numbers of separate flashes used to obtain them being shown in brackets.

Since methods b and c use the same L data, they have been combined in the Table.

All four methods agree in placing successive charge centres at progressively increasing mean heights. The same progressive increase is shown by individual strokes, for which successive values of H_m have been obtained by methods a and c.

This conclusion has been confirmed by HACKING[1] using a fifth method in which the total $(L+R)$ field-changes from successive strokes were separately and simultaneously recorded by electrostatic fluxmeters [*35*] at three different points on the ground. From these he was able to show a general regular increase in H_m with m, though (at the short distances involved) departures of the leader channel from the vertical made direct determinations of H_m impossible. The radar measurements of HEWITT (in Sect. 45) give similar confirmation.

All these measurements have been made in South Africa at 1.2 km. above sea-level. They indicate that in the powerful frontal storms studied the whole discharge to ground comes from a single region of N charge (Fig. 23) in which the portions contributing to separate strokes are contiguous and approximately vertically above one another.

The separate strokes of the discharge process thus do not arise (as was previously supposed by the writer and others) from the separation of the successively discharged regions in space but from a mechanism causing intermittent discharge in time, the nature of which is further discussed in Sect. 56.

VIII. Cloud and air discharges.

48. The field-changes of cloud and air discharges can readily be distinguished from those of flashes to ground by the absence of any sudden R_b change, because such discharges show no return process. The usual electrostatic field-change is a slow rise to a final maximum. At distances exceeding 10 to 15 km. this carries superimposed radiation pulses similar in appearance and time-separation ($\sim 50\,\mu$-secs.) to those from stepped leaders.

The observations of electrical effects are thus exactly what would be expected from BOYS camera studies and indicate that both cloud and air discharges involve extensive stepped streamer processes.

[1] C. A. HACKING: J. Geophys. Res. **59**, 449 (1954).

E. The mechanism of breakdown and of streamer propagation.

49. Initial breakdown in the discharge to ground. It has been shown that the great majority of discharges to ground involve the transfer of negative charge from the N region of the cloud. The long duration of the B field-change and the amplitude of its initial radiation pulses (Fig. 32) indicate that some preliminary breakdown process takes place before the starting of the first leader from the N region.

WORMELL and PIERCE[1] have suggested that this process may be the downward passage into the air below the cloud-base of a very slow ($v \sim 2 \times 10^6$ cm./sec.) preliminary pilot streamer of such feeble luminosity as to be undetectable in camera studies. The evidence examined in Sect. 19, however, indicates that the minimum possible velocity for such a streamer is $\sim 1.0 \times 10^7$ cm./sec.

The conclusion that an ionizing junction process (Sect. 43) precedes the emergence of all dart leaders to subsequent strokes suggests that a similar process is responsible for the first part of the L_1 change. It would indeed be expected that breakdown first occurs in the strong field between the N and p charges of Fig. 23. As in the J process it would lead to the creation of a good conducting region at the lower end of the N column. Such conducting properties of the cloud region tapped by the first leader would seem to be essential if it is to draw the considerable current (see Sect. 51) required for its advance.

50. The tip-field and velocity of the advancing leader. A negative streamer can advance into virgin air as a result of ionization produced either by electrons in its tip or by photoelectrons generated in front of its tip. In either case a lower limit to streamer velocity is set by a critical field-strength X_c in front of the tip, below which ionization by electron impact cannot recur. The corresponding critical electron drift velocity, v_c, must be the minimum velocity of advance of a negative streamer, since the streamer as a whole cannot move more slowly than the ionizing agents which produce its extension.

For inelastic collisions v_c is given by

$$v_c = \sqrt{\frac{2 X_c e \lambda}{\pi m}}\,. \tag{50.1}$$

X_c is known to be 30000 volts/cm. when λ, the gas-kinetic mean free path at 20° C and 76 cm. Hg. is 3.8×10^{-5} cm. and since $X_c \lambda$ is sensibly independent of pressure between 54 cm. Hg. (the mean pressure in these discharges) and 76 cm., v_c is found to be 3.6×10^7 cm./sec. for purely inelastic collisions.

Two factors combine to make this result too high[2]. In the first place it is known that for nitrogen the value chosen for λ is too great by a factor ranging from 0.8 to 0.3 according to the velocity with which the electrons are moving. Secondly, all the collisions will not be inelastic when v_c is approached. For purely elastic collisions the value of v_c is less than that given above by the factor $\sqrt[4]{\pi^2 m/4M}$ or 0.8, m being the mass of the electron and M that of an "air" molecule. Applying the first correction v_c cannot exceed 1.8×10^7 cm./sec. while from the second it cannot be less than 1.4×10^6 cm./sec. As shown in Sect. 19 the minimum effective velocity of the first leader (pilot) streamer is found to be close to 1.0×10^7 cm./sec.

Observations have shown that the average field between cloud and ground is much less than 30000 volts/cm. The requirement that the streamer tip field should

[1] T. W. WORMELL: Quart. J. Roy. Met. Soc. **79**, 474 (1953).
[2] References are to be found in [7e].

exceed this value can, however, be met if the channel of the leader is a sufficiently good conductor and is connected to a conducting region of the cloud capable of supplying the current necessary to maintain the potential V of the streamer-tip with reference to its surroundings at a sufficiently high value.

If r is the tip-radius a rough estimate of the tip-field X can be obtained by taking the capacity of the tip as $r/2$ and its charge as $Vr/2$ from which

$$X \simeq \frac{V}{2r}. \tag{50.2}$$

If $r = 250$ cm. (see Sect. 28) and X is 30000 volts/cm., V is $\sim 1.5 \times 10^7$ volts. For smaller values of r it would be correspondingly reduced. Since, as shown by C. T. R. WILSON [4], the initial potential of the cloud charge is between 10^8 and 10^9 volts, a streamer of this nature would only be prevented from advancing if it had a very considerable potential drop along its length, of the order of 5×10^8 volts in 5 kms. or 1000 volts/cm. If the leader channel behind the tip has been highly ionized it may be expected to pass currents of several hundred amperes (Sect. 51) without giving rise to potential drops of this order, indeed KOMELKOV has reported drops of only 60 volts/cm.[1] in the leader channels of spark discharges. The position is different in the case of corona-like streamers such as the pilot streamer (see Sect. 55) which cannot proceed indefinitely without the intervention of some more highly ionizing streamer to reduce the resistance of the channel behind it.

51. The current in a streamer. The current which is required by an advancing streamer is made up of three parts: I_1, that needed to charge up freshly formed sections of the channel; I_2, to neutralise any space-charge of opposite sign encountered; I_3, to make up loss from any corona discharges from the surface of the channel. I_2 is not usually required in the case of leaders to subsequent strokes. I_3, which is not required if the radius of the channel is sufficiently great for the surface field to be below 30000 volts/cm., is further discussed in Sect. 54.

Two approximate expressions have been derived for I_1. The first, due to RÜDENBERG[2] and in another form to JEHLE[3], equates I_1 to the displacement current in front of the leader tip or to the convection current carried by the tip charge and gives when corrected for an error,

$$I_1 = \frac{X v r}{2} \tag{51.1}$$

where r and v are the radius and velocity of the hemispherical tip and X is the (presumed radial) field in front of it.

The second expression is derived from the consideration that a streamer of velocity v carrying a uniform charge q per unit length requires for the charging of its stem that

$$I_1 = q v. \tag{51.2}$$

Estimates of the current carried by a first leader streamer can be made from both these expressions. By substituting $v_c = 1.0 \times 10^7$ cm./sec., $X = 30000$ volts per cm. $= 100$ e.s.u. and $r = 2.5$ m. (see Sect. 28) equation (51.1) gives for the minimum current in the pilot streamer (or the average current in the first leader at its minimum effective velocity) $I_1 = 75$ A. For the second estimate, taking q

[1] V. S. KOMELKOV: Bull. Acad. Sci. USSR **8**, 955 (1947). — C. R. Acad. Sci. USSR. **68**, 57 (1947).

[2] R. RÜDENBERG: Wiss. Veröff. Siemens-Werk **9**, pt. 1 (1930).

[3] H. JEHLE: Z. Physik **82**, 784 (1933).

to be the result of the distribution of 3 C of charge along 5 km. of leader or 6×10^{-6} C/cm., I_1 is found from equation (51.2) to have a value of 60A.

In spite of the crudeness of the assumptions involved in these estimates they are of value in showing, as was first pointed out by BRUCE [*17*], that the average current in the first leader streamer may be of the order of several hundred amperes, when values of v above the minimum and the contribution of I_2 are considered.

The current in the return streamer can also be estimated from equation (51.2). Since q is the same as for the leader and the velocity of the streamer is from 100 to 1000 times v_c, I_1 at its maximum should be between 7500 and 75000 A, in accord with the results of direct observation (see Sect. 36).

The currents in dart leaders, for which v is 10 to 200 times v_c, may be expected to lie between 750 and 15000 A, since the R_b field-changes in subsequent strokes and hence the values of q, though smaller, are not very far different from those in first strokes.

The considerable difference between the radius of the first leader streamer (~2.5 m., see Sect. 28) and that of a return streamer (16 cm., see Sect. 29) may be ascribed to the well-known "pinch" effect of the magnetic fields due to the currents in the two cases. Since the constricting mechanical forces vary as the square of the current in the streamer, they would differ in these two extreme cases by a factor of 10^6.

The conclusion above that the dart streamer carries a current intermediate between these extremes, a conclusion which is supported by its intermediate luminosity, suggests that it has a radius of the order of tens of cm.

52. The mechanism of the dart leader. The dart leader (see Sect. 20) does not find its way through virgin air but travels along a pre-existing and therefore pre-ionized trail. Its velocity decreases with the age of this trail and when a long interval has elapsed since the previous stroke it either becomes dart-stepped (see Sect. 21) or seeks a new path as a normal stepped leader. It follows that it owes its apparently continuous movement and its high velocity to the existence of free electrons (or easily broken-down ions) in front of its tip. The manner in which it makes use of these has been discussed by CRAVATH and LOEB[1] and by the writer [*7e*] and is summarised below.

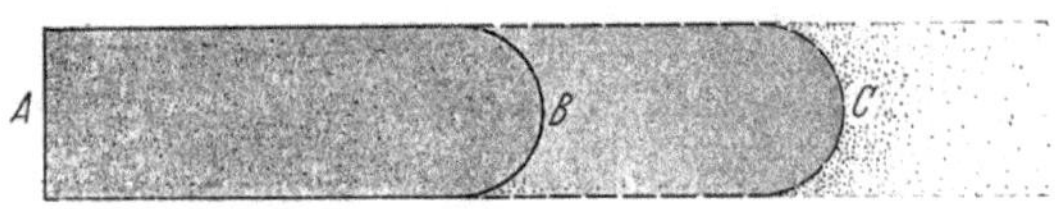

Fig. 40.

In Fig. 40, AB is the negative dart streamer stem and the shaded portion in front of the tip B shows the region BC over which the field exceeds the critical values for impact ionization by electrons. Electron avalanche processes take place throughout BC and as soon as they have rendered it sufficiently conducting the streamer has effectively extended to C. It is thus unnecessary for any electrons to travel the full distance BC and the streamer velocity U can considerably exceed the mean electron drift velocity $\bar{v}$ in BC.

The time t necessary for the ionization of BC may be taken as the time required for it to be effectively covered by a network of small filaments, one from each original electron. The simplest possible assumption is that t is the time required for an electron to cover the mean distance from one electron to the next. If n is the electron density in BC, $t=1/n^{\frac{1}{3}}\,\bar{v}$ and

$$U=\frac{BC}{t}=n^{\frac{1}{3}}\,\bar{v}\times BC. \qquad (52.1)$$

[1] A. CRAVATH and L. B. LOEB: Physics **6**, 125 (1935).

As an illustration a dart streamer with a tip radius of 10 cm. and potential V of 1.0×10^7 volts would in this way be able to ionize for a distance BC of 30 cm. in front of its tip. At a distance of 15 cm., X would be 7.2×10^4 volts/cm. and v about 1.6×10^7 cm./sec. If $n = 10^3$ electrons/cm.^{3}, U from equation (52.1) is 4.8×10^9 cm./sec. more than 100 times v, the drift velocity of electrons at B.

53. The return streamer. Unlike the streamers which move downwards from the cloud, the channel and tip of the return streamer are positively charged, so that in its advance electron avalanches move inwards into its tip and are neutralised by the positive charge thereon. Apart from this the rapid motion of the streamer must arise from a mechanism identical with that discussed above, for it moves forward into the negatively charged and highly ionized channel formed by the preceding leader. This accounts for its exceptionally high velocity at the base of the channel. Decreases in its velocity as it moves upwards can be interpreted in terms of the increased age and hence reduced free electron density n of the channel in front of it. This interpretation is supported by the observation that in the case of subsequent strokes for which the "age" of the top of the leader channel is of the order of 1 m-sec. the return streamer (see Sect. 22) shows little diminution in velocity whereas for first strokes in which the corresponding quantity is 10 m-sec. the falling off of velocity with height is considerable.

It also affords one explanation of the fact that the return streamer travels more rapidly and more brightly along branches (see Sect. 22) than along the portions of the main channel which are being traversed by it at the same time as branches, since as the streamer travels along a branch it encounters portions of the leader of increasingly recent age and of higher n. The resulting increase in velocity and hence of current as the streamer travels along a branch would be responsible for the increases in luminosity at the base of the channel described in Sect. 22. These observations can, however, also be interpreted as showing that the branched system of the first leader carries a higher charge than the main trunk.

54. The pilot streamer. Reasons for postulating that the first leader step process follows upon the track of a pilot streamer have been discussed in Sect. 19 and 50. The small current (see Sect. 51) and feeble luminosity of the pilot explain why it has not as yet been recorded on Boys camera photographs but direct evidence for its existence in the lightning discharge is afforded by the continuous nature of the L field change from near-by leaders (see Sect. 40).

The pilot streamer has been clearly photographed by Liao and Anderson in the stepped spark discharge in oil by making use of a fast camera fitted with a lens transmitting ultra-violet light[1]. They have also reported studies of the luminous and electrical effects of this streamer in the case of stepped spark discharges along the surface of an oiled insulating plate[2]. The observed velocity of the pilot is 5.9×10^5 cm./sec., while that roughly estimated for the minimum ionising velocity of electrons in oil is about 3×10^5 cm./sec.

Mitchell and Snoddy have observed an interesting parallel to the lightning pilot in the transient breakdown of low-pressure discharges in long tubes[3]. The velocity of movement of a negative potential-wave, impulsively applied to one end of a tube, is of the order of magnitude of the drift speed of electrons in the tip field.

When travelling at its most frequent near-minimum velocity the advance of the pilot-streamer can be adequately explained by electron-drift ionization at

[1] T. W. Liao and J. G. Anderson: Electr. Engng. **72**, 641 (1941).
[2] J. G. Anderson and T. W. Liao: Electr. Engng. **73**, 637 (1954).
[3] F. H Mitchell and L. B. Snoddy: Phys. Rev. **72**, 1202 (1947).

its tip. Higher velocities—up to 2×10^8 cm./sec.—are undoubtedly due to movement in enhanced space-charge fields but cannot involve electron drift alone, for if this were the case fields of the order of 3×10^6 volts/cm. would be required. It is probable that with the larger current densities involved, excitation processes in the channel behind the tip lead to forward photoionization which creates a pre-ionized trail for the pilot as for other fast streamers (see Sect. 52). If this is so there should be a maximum as well as a minimum velocity of advance for this streamer given by

$$v_m = \frac{R}{t} \tag{54.1}$$

where R is the maximum range of the ionizing photons and t the average duration of the excited states of molecules in the tip. Approximate values of R and t are 1 cm. and 10^{-8} sec. and yield $v_m \approx 10^8$ cm./sec. not far from the maxima recorded in Table 1 (see Sect. 19). The dart-stepped streamer (see Sect. 20) which, though stepped, travels at an effective velocity of 1 to 2×10^8 cm./sec. affords some confirmation of these ideas. Travelling along a path of considerable age and hence of low electron density, it nevertheless provides for the pilot streamer a small measure of pre-ionization. With the high tip fields made possible by a small tip radius and high channel conductivity this pre-ionization enables the pilot to advance rapidly by the CRAVATH-LOEB mechanism, photo-ionization aiding in the process.

55. The step process in first leaders. *α) General.* The small range of variation in the pause-time between the steps of the first leader (see Sect. 18) makes it improbable that the step process derives from a relaxation oscillation either in the cloud-charge supplying the current to the leader channel or in the channel behind the advancing pilot streamer. Both these effects would be expected to show a much wider range of τ and to operate in all leaders, not in the first leader only. It must be concluded that the step-process involves a primary gas-discharge mechanism associated with streamer advance into previously unionized air.

The mean amplitude of the electrostatic field-changes associated with the steps themselves (see Sect. 40) is less than one-tenth of the change of field in the intervals between them. The negative charge q_1, continuously carried forward by the pilot in its movement over a step-distance l must therefore considerably exceed the additional charge q_2 involved in the subsequent transformation of the pilot channel by the fast step streamer.

For a fairly fast leader ($v = 4 \times 10^7$ cm./sec.) which ultimately distributes about 4 C over its fully extended length of 5 km., $q_1 = 0.016$ C for $l = 20$ m. It has been shown[1] by comparing the average amplitudes of the radiation pulses from distant step streamers with those from distant return streamers, for which the maximum rate of change of current is 10 kA/μ-sec. (see Sect. 38) that the maximum rate of change of current in a step is about 1 kA/μ-sec. The duration of the formation of a step being about 1 μ-sec., the current in the transformation is ~1000 A and q_2 is ~0.001 C, less than 1/16 of q_1, in agreement with observation.

β) Explanations of the step process. Some explanations which have been offered for the step process involve rejection of the early evidence for the existence of the pilot streamer and seek for a cause for the halting of the main leader streamer which ignores the pilot. The theory of SZPOR[2] attributes this halt to a gradual increase of tip radius, caused by outward electron diffusion, which makes further

[1] D. B. HODGES; Proc. Phys. Soc. Lond. B. **67**, 582 (1954).
[2] S. SZPOR: Bull. Ass. Suisse El **1**, 1 (1942).

advance impossible for a time owing to the consequent decrease of the field in front of the tip. That of KOMELKOV[1] attributes the halt to a fall of tip potential in the process of moving forward, which necessitates a pause for this potential to be restored by current flow. The further evidence for the existence of the pilot streamer which has been discussed in Sect. 54 would seem to rule out both these explanations.

The evidence indeed suggests strongly that the primary mechanism is connected with the instability of the pilot and is one in which the pilot stem is rapidly transformed from a state of weak corona-like ionization to one of arc-like conductivity. The halting of the main streamer is due to the need for it to await this transformation.

An explanation originally proposed by the writer and given quantitative form by LOEB and MEEK[2] ascribes the change to electron capture and recombination processes in the pilot stem as a result of which an increased field is required to maintain the current. When this field reaches a critical value a step streamer

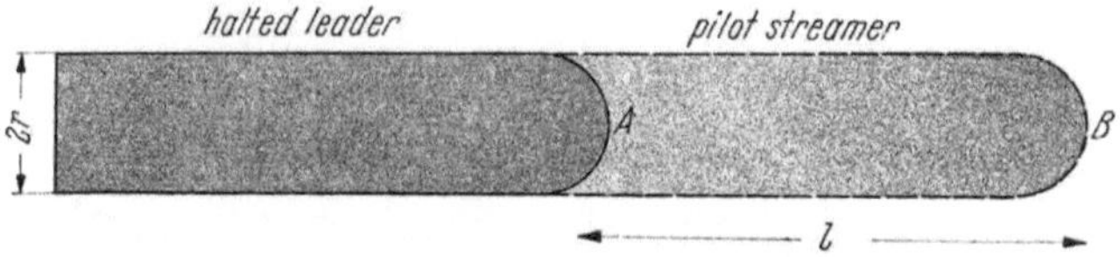

Fig. 41. The step mechanism (SCHONLAND).

starts from the tip of the halted main leader behind the pilot and ionizes it heavily. The decrease in ionization was shown by MEEK to be insensitive to the initial ion density and from appropriate data it was found that the pause-time was close to the observed value of 50 μ-sec. on the supposition that the current in the leader and pilot was ~0.1 A and the streamer radius ~0.3 cm. These quantities are now known to be 1000 times larger and for this and other reasons the theory in this form is untenable.

The writer [*18*] has applied these ideas in another form which starts from the argument that since there is now evidence that all very long sparks in air are preceded by a corona or pilot streamer which crosses the whole gap. The average value of the field-strength, $\overline{X}$, for such breakdown, ~5000 volts/cm., gives the approximate value of the field-strength in the stem of such a pilot streamer below which it cannot advance for lack of free electrons to carry the current. Thus in Fig. 41, since the negatively charged stem of the pilot creates a reverse field behind it and the field at its starting point A must decrease steadily as the pilot advances, a stage will be reached when this field falls below $\overline{X}$. The consequent cessation of normal current at A causes a layer of positive space-charge to form in front of it. This will proceed until the field between it and A reaches 30000 volts per cm., when a fast streamer can travel from A to B along a pre-ionized path and transform the pilot stem. The time required for the formation of the necessary space-charge is found to be 3.1 μ-sec.

From these considerations it can be shown that the step-length l is related to the charge q per unit length of the pilot and leader by the equation

$$\frac{q r}{l^2} + \frac{q}{l} + r = \overline{X} \tag{55.1}$$

[1] V. S. KOMELKOV: Bull. Acad. Sci. USSR. **8**, 955 (1947). — C. R. Acad. Sci. USSR. **68**, 57 (1947).

[2] J. M. MEEK: Phys. Rev. **55**, 972 (1939). — L. B. LOEB and J. MEEK: The Mechanism of the Electric Spark. Stanford University Press 1941.

where r is the streamer radius. For $q = 8.2 \times 10^{-6}$ C/cm. and $r = 2.5$ m. equation (55.1) gives $l = 12.5$ m.

Various additional considerations raise this to 16.2 m., in reasonable agreement with the value of 20 m. to be expected. Applied to the oil gap spark discharges of ANDERSON and LIAO (see Sect. 54), using appropriate values of $\overline{X}$ and of q ($2q/r \approx X_c$, the breakdown field strength in short gaps) equation (55.1) gives values of l/r in agreement with observation, in spite of the enormous difference in scale between oil sparks and lightning.

From (55.1), since $r \ll l$, it follows that leaders with small values of r should show small values of l. This offers an explanation of the small steplengths and correspondingly small pause-times observed in dart-stepped leaders (see Sect. 21).

The very long step lengths observed in regions of high space-charge field (see Sect. 16) are explained on this theory as due to the large values of l needed

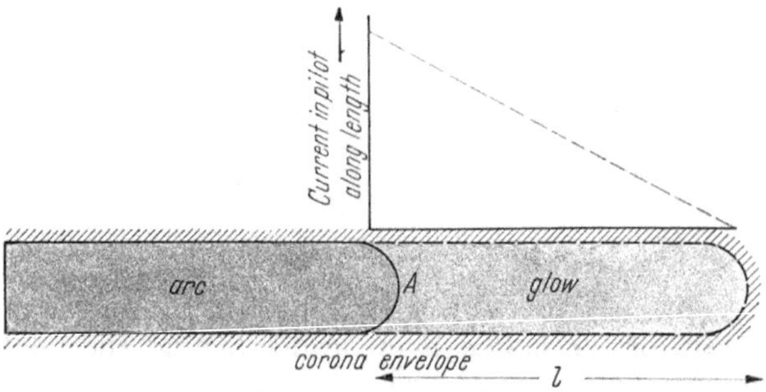

Fig. 42. The step mechanism (BRUCE).

to reduce the field at A (Fig. 41) below $\overline{X}$. The value of $\tau = l/v$ in such cases would not be expected to be much altered since the pilot streamer would move with higher velocity in the enhanced field.

Another and a more phenomenological theory of the step process has been advanced by BRUCE [*17*]. This is based upon the view (see Sect. 37) that the electric field at the surface of the pilot streamer channel is always sufficient to give rise to corona currents to surrounding space. The current entering the pilot from the halted leader at A (Fig. 41) increases as the pilot advances since it has to supply an increasing length of corona-emitting channel. BRUCE considers that a change from glow to arc conditions will take place in the pilot stem at A (Fig. 42) when this current reaches a critical value which he deduces from laboratory observations of glow-arc transitions to be $\sim 1\,A$. His criterion for transformation can be written

$$k V^2 l = 1 A. \tag{55.2}$$

where k is the corona constant per cm. length, V is the difference of potential between the streamer and its surroundings and l is the length of a step.

Bruce has applied equation (55.2) to the case of the upward stepped leader from the Empire State Building for which he estimates V to be 2×10^6 volts at the start and 5.6×10^6 volts at the upper end of the channel. Using for k a value of 4×10^{-17} found from D.C. corona measurements on cylinders, the equation yields values of l from 5.9 to 54 m., in accord with observation. The calculation has not been applied to downward moving stepped leaders owing to uncertainty as to the value of V. In this case $V = V_0 + (X_1 - X_a)\, L$ where V_0 is the initial potential of the cloud charge, X_1 the average pre-discharge field-strength between cloud and ground, X_a the potential gradient in the main leader channel and L

the total length of this channel. Reference to the Empire State case shows that unless V_0 is $< 5 \times 10^6$ volts and $X_1 - X_a$ is very small, equation (55.2) will give values of l which are much smaller than those observed. BRUCE holds that both V_0 (see Sect. 50) and X_1 have previously been overestimated by an order of magnitude.

On this theory not only the pilot streamer but also the channel behind it is surrounded by a corona envelope. The total current required to maintain this is

$$I_3 \approx k L (V_0^2 + V_1 X L + X_1^2 L^2/3) \qquad (55.3)$$

which for the Empire State leaders reaches very high values, of the order of 5000 A. Unless the value of k is $\sim 10^{-10}$, this would seem to be an objection to the theory, for the luminosity of the first leader suggests currents which are very much smaller. Such a change in k would, however, disturb the agreement with observed values of l given by equation (55.2).

The writer's theory of the step process envisages a pilot and leader channel of such large radius that lateral corona currents do not occur while BRUCE's theory would appear to consider a channel consisting of a conducting core of smaller radius surrounded by a corona envelope. Such cored leader channels have been observed by KOMELKOV in the spark discharge but it has not been established that they are of the same nature as pilot streamers in either spark or lightning discharges.

As noted in Sect. 37 BRUCE has suggested that the considerable quantity of electricity removed to earth in the R_c field-change period ($\sim 4C$) is not the result of mopping up by the return streamer of residual charge in the N region but of a corona process passing from the narrow positive return streamer channel and slowly removing the negative corona envelope of the preceding leader to earth.

Further observations on pilot and leader discharges in long sparks and on glow-arc transitions are needed to decide between these two theories of the step process.

56. The intermittence of the discharge process. MALAN and SCHONLAND [*15c*] have given an explanation of the intermittent manner in which separate strokes can take place from a continuous and approximately vertical column of preponderantly negative charge in a non-conducting cloud. The evidence given in Sect. 47 shows that this column may on occasion extend upwards from 3 km. to 8 or 9 km. above the ground.

The series of processes which it is suggested is involved in the n-th stroke is depicted in Fig. 43a to e. Fig. 43a shows the position at the close of the final portion of the $(n-1)$th stroke. Here $+q$, at an effective height H_1, is the positive charge carried in the main channel and on the lower portion of the original column which has already been discharged and rendered conductive by the filamentary branches of preceding strokes.

The remaining negative charge on the column lies above this branched system. A portion $-Q$, at an effective height H_2, will later be rendered conducting and thus available for the production of the n-th stroke, but at the instant shown in Fig. 43a this region is not conducting; an internal streamer process is required before it can take part in the discharge.

Immediately after the stage depicted in Fig. 43a junction streamers proceed upwards in the strong field between $+q$ and the negative column above it. These streamers penetrate into the column, carrying positive charge upwards, until most of $+q$ has been neutralized. In this way a new conducting region is formed in the space previously occupied by $-Q$; the charge remaining in this region will be $-(Q-q)$, as shown in Fig. 43b.

The junction process is a slow one and occupies a time of the order of 0.03 sec. Its connection with the ground is effectively broken not more than a few milliseconds after the end of the previous return stroke, since the main channel rapidly loses its high conductivity as a result of electron capture and recombination and diffusion of ions. The increased resistance of the channel thus prevents any appreciable quantity of positive charge from flowing upwards from the ground as a "continuing current" to add to the charge $+q$ in the cloud. Evidence for this cut-off effect is afforded by the observations of McCANN [13] on currents in lightning conductors. These show that continuous currents of magnitude exceeding

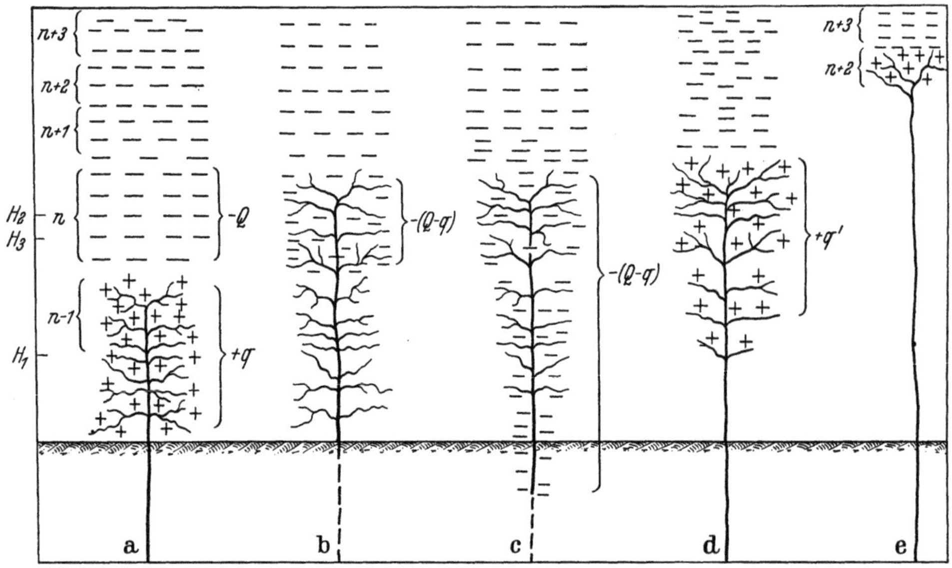

Fig. 43 a—e. The intermittent discharge.

10 A rarely flow in the channel in the intervals between strokes, once the R_c portion of the discharge is over. For much of the interval the current is found by him not to be measurable (less than 0.1 A).

The third stage, shown in Fig. 43 c is the downward movement of $-(Q-q)$ as a leader streamer, the whole of the conducting portion of the column and the original channel being filled with negative charge.

This is followed by the rapid return process, which after removing $-(Q-q)$ to earth leaves a charge of $+q'$ at effective height H_3, greater than H_1, on the conducting part of the column in the cloud, Fig. 43 d. This positive charge then prepares the way for the next stroke by a new upward J process.

Fig. 43 e is a repetition of Fig. 43 a at a later stage in the discharge, when the J process of the last stroke is about to start. It illustrates the fact, for which MALAN and SCHONLAND have cited other evidence, that when the greater portion of the negative charge on the column has been removed, the positive induced charge on the conducting channel, q (or q') diminishes from stroke to stroke.

The interval between separate strokes in thus considered to be the time required for the J streamers, isolated from the ground by the high resistance of the channel, to penetrate sufficiently far and wide into the N column to neutralise the positive charge on the top of the originally ground-connected channel. Each fresh stroke is preceded by a form of cloud discharge (similar to the $N-P$

discharge preceding the first stroke) which renders a new region of the column conducting and which must cease before the leader to the stroke can develop downwards.

This interpretation of the intermittent mechanism obtains support from recent radar observations described in Sect. 45 which indicate that the ionizing processes involving the neutralisation of q within the cloud become very weak just before the start of a dart leader.

57. The mechanism of the continuing discharge. Discharges to the Empire State Building (see Sect. 25 and 39) and the final F change observed in many discharges to lower structures show that the intermittent mode of discharge can on occasion be replaced by a continuous one. MALAN [*16*] has shown that in such cases, as evidenced by the continuing luminosity of the channel below the cloud, the channel remains a conducting link between the J streamers and the ground. He estimates the current flowing in the channel to be between 500 and 800 A, of the same order as that directly measured by MCCANN [*13*] for continuing strokes ($\sim$1000 A max.). That such continuing discharges should occur frequently in flashes to the Empire State Building is to be expected. The cloud bases are here exceptionally close to the ground, as represented by the top of the Building (separation $\sim$600 m.). Fields and consequently channel currents must be expected to be exceptionally high and the channel can retain its conductivity for a longer period than usual.

The occurrence of continuing discharges in other cases MALAN ascribes to the existence, particularly in the upper part of the column, of a charge density (though not necessarily a total charge per unit length) sufficiently great for abnormally heavy J currents to flow and to keep the channel to ground conducting.

A special combination of the intermittent and the continuous method of discharge occasionally occurs. In 6% of the discharges studied by MALAN large slow changes of field, very like the final F change, occupy about half of an abnormally long interval between two consecutive strokes and last from 100 to 400 m-secs. Photographic observations show occasional continuing interstroke luminosity of the same nature. These flashes do not cease with the continuing discharge but resume the intermittent form after a pause or after a further preliminary J process. They are interpreted as due to abnormally high concentrations of charge in the N column, which give rise to J processes drawing sufficient current from the channel to ground to keep it conducting for an abnormally long time.

Major references and notes.

[*1*] FRANKLIN, BENJAMIN: Experiments and Observations on Electricity (1774), edit. by J. B. COHEN. Harvard Univ. Press 1941.

[*2*] KÄHLER, K.: Die Elektrizität der Gewitter. Berlin: Gebrüder Borntraeger 1924.

[*3*] BRUCE, C. E. R., and R. H. GOLDE: J. Inst. Electr. Engrs. **88**, Pt. II, 487 (1941).

[*4*] WILSON, C. T. R.: Phil. Trans. Roy. Soc. Lond., Ser. A **221**, 73 (1920). (With reference [*10*], the classical paper on the subject of the lightning discharge.)

[*5*] MALAN, D. J.: Unpublished work. I am grateful to Dr. MALAN for permission to use this information.

[*6*] SCHONLAND, B. F. J., and H. COLLENS: Proc. Roy. Soc. Lond., Ser. A **143**, 654 (1934).

[*7a*] SCHONLAND, B. F. J., D. J. MALAN and H. COLLENS: Proc. Roy. Soc. Lond., Ser. A **152**, 595 (1935).

[*7b*] SCHONLAND, B. F. J., D. J. MALAN and H. COLLENS: Proc. Roy. Soc. Lond. **168**, 455 (1938).

[*7c*] MALAN, D. J., and H. COLLENS: Proc. Roy. Soc. Lond., Ser. A, **162**, 175 (1937).

[*7d*] MALAN, D. J., and B. F. J. SCHONLAND: Proc. Roy. Soc. Lond., Ser. A **191**, 481 (1947).

[*7e*] SCHONLAND, B. F. J.: Proc. Roy. Soc. Lond., Ser. A **164**, 132 (1938).

[*7f*] ALIZADE, A. S.: Zh. Tekh. Fiz. **21**, 652 (1951).

[8a] MCEACHRON, K. B.: J. Franklin Inst. **227**, 149 (1939). — Trans. Amer. Inst. Electr. Engrs. **60**, 885 (1941).
[8b] HAGENGUTH, J. H., and J. G. ANDERSON: Power Applied Systems No. 1, pp. 641—649. 1952.
[9] A very complete bibliography of general thunderstorm electrification papers up to 1949 is given by J. A. CHALMERS, in "Atmospheric Electricity", Clarendon Press, Oxford 1949 and in a chapter on the same subject in Rep. Progr. Phys. **17**, 101 (1954). "Atmospheric Electricity" by B. F. J. SCHONLAND, Methuen, London 1953 gives a more general account of the subject. A valuable critical study, including some newer data and ideas on the lightning discharge has been given by T. W. WORMELL: Quart. J. Roy. Met. Soc. **59**, 449 (1954).
[10] WILSON, C. T. R.: Proc. Roy. Soc. Lond., Ser. A. **92**, 555 (1916).
[11] WORMELL, T. W.: Phil. Trans. Roy. Soc. Lond., Ser. A **238**, 249 (1939).
[12] PIERCE, E. T., and T. W. WORMELL: Thunderstorm Electricity edit H. R. BYERS. Univ. of Chicago Press 1953. Also PIERCE, E. T.: Quart. J. Roy. Met. Soc. **81**, 21, 229 (1955).
[13] MCCANN, G. D.: Trans. Amer. Inst. Electr. Engrs. **63**, 1157 (1944).
[14] CHAPMAN, F. W.: Proc. Phys. Soc. Lond. **51**, 876 (1939).
[15a] APPLETON, E. V., and F. W. CHAPMAN: Proc. Roy. Soc. Lond., Ser. A **158**, 1 (1937).
[15b] SCHONLAND, B. F. J., D. B. HODGES and H. COLLENS: Proc. Roy. Soc. Lond., Ser. A **166**, 56 (1938).
[15c] MALAN, D. J., and B. F. J. SCHONLAND: Proc. Roy. Soc. Lond., Ser. A **206**, 145 (1951).
[15d] MALAN, D. J., and B. F. J. SCHONLAND: Proc. Roy. Soc. Lond., Ser. A **209**. 158 (1951).
[16] MALAN, D. J.: Ann. Géophys. **10**, 71 (1954).
[17] BRUCE, C. E. R.: Proc. Roy. Soc. Lond., Ser. A **183**, 228 (1944).
[18] SCHONLAND, B. F. J.: Proc. Roy. Soc. Lond., Ser. A **220**, 25 (1953).

The following general surveys of lightning studies are recommended:

(a) MEEK, J. M., and F. R. PERRY: The Lightning Discharge. Rep. Progr. Phys. **10**, 314—354 (1946). This article is used and extended in MEEK, J. M., and J. D. CRAGGS: Electrical Breakdown of Gases, Chapt. V. Oxford: Clarendon Press 1953.

(b) WICHMANN, H.: Grundprobleme der Physik des Gewitters, Wulfenbütteler Verlagsanstalt 1948, gives an up-to-date critical account of the subject with many original ideas.

(c) Thunderstorm Electricity, editor: H. R. BYERS, University of Chicago Press 1953, contains papers on lightning discharge studies by most of the leading workers in this field.

Sachverzeichnis.

(Deutsch-Englisch.)

Bei gleicher Schreibweise in beiden Sprachen sind die Stichwörter nur einmal aufgeführt.

Absorption, *absorption* 38, 39.
—, gesamte, *total absorption* 360.
Absorptionskoeffizient, *absorption coefficient* 356, 358.
adsorbiertes Gas, *adsorbed gas* 85.
Ähnlichkeit, *similarity* 17, 20, 82, 138, 153, 193.
Ähnlichkeitsgesetze, *similarity laws* 70, 97, 89, 117, 166.
Ähnlichkeitsprinzip, physikalische Bedeutung, *physical significance of the similarity principle* 18.
Alti-Elektrograph, *alti-electrograph* 598.
Anfangsdurchschlag, *initial breakdown* 618.
Anfangspotential, *starting potential* 55, 85, 165, 176, 202.
angeregte Atome, *excited atoms* 6.
— Molekülzustände des Argons, *excited molecular states of argon* 505.
Anlagerung, *attachment* 178.
Anode, Stromdichte an der, *current density at the anode* 425.
—, Ausdehnung und Form, *size and shape of the anode* 146.
—, bewegliche, *movable anode* 115.
—, kugelförmige, *spherical anode* 146.
—, Wärme an der, *heat at the anode* 150.
Anodendunkelraum, *anode dark space* 57.
Anodenfall, *anode fall* 61, 83, 145, 146, 148, 165, 279, 422.
—, negativer, *negative anode fall* 146.
—, normaler, *normal anode fall* 436.
—, positiver, *positive anode fall* 146.
Anodenfalltheorien, *anode fall, theories* 434.
Anodenflecken, *anode spots* 146, 151, 165.
Anodengebiet, *anode region* 145, 152.
Anodenglimmlicht, *anode glow* 57, 146, 149, 247.
—, Schwingungen, *oscillations of the anode glow* 144.
Anodenstrahlen, *anode rays* 147.
Anoden- und Anodenplasmatemperatur, *anode and anode-plasma temperature* 424.
Anodenverdampfung, *anode evaporation* 431.
anodische Kontraktion, *contraction at the anode* 425.
Anregung, *ecxitation* 14, 209, 218.
— durch Teilchenstoß, *excitation by particle impact* 219.
—, stufenweise, *stepwise excitation* 225.
Anregungsfunktion, *excitation function* 60, 135.
Anregungstemperatur, *excitation temperature* 309.
Anregungswahrscheinlichkeit, *probability of excitation* 127.
Anwachs-Kurve, Form der, *shape of growth curve* 8.
Äquipotentialflächen, *equipotential surfaces* 121.
ASTONscher oder erster Dunkelraum, ASTON *or primary dark space* 56, 96, 97.
asymmetrische Felder, *asymmetrical fields* 449.
Aufbauzeit, *formative time lag* 45, 49, 471, 494.
—, beobachtete und wahre, *observed and true formative lag* 472.
—, lange, *long formative time lag* 526.
—, Schwankungen in der, *fluctuation in the formative lag* 474.
—, stark anwachsende, mit fallenden Überspannungen, *formative time lag increased rapidly as overvoltages decreased* 492.
Aureole 57, 110, 112, 372.
Auslöseelektronen, *triggering electrons* 506.
Auslösungsursache, *triggering agency* 449.
Austrittsarbeit, *work function* 14, 15.

Bandblitz, *ribbon lightning* 577.
Bandenspektren, *band spectra* 98, 100, 184.
B-Änderung, *B change* 610.
BECK-Bogen, BECK-*arc* 293, 428, 429.
Belag, *tarnish layer* 15, 30.
— auf Elektroden-Oberflächen, *tarnished electrode surface* 21.
Bestrahlung, äußere, *external irradiation* 202.
— innere, *internal irradiation* 110.
Beweglichkeit, *mobility* 103, 178.
Bildungsdauer (s. Aufbauzeit), *formative time* 41, 50.
Blitze, Aufnahmen bei Tageslicht, *photography of lightning in daylight* 584.
Blitzentladungen, *flashing discharge* 191, 577.
Blitzentladung, Historisches, *lightning discharge, historical* 574.
Bögen, abgeschaltete, *switched off arcs* 382.
—, elektroden-stabilisierte, *electrode stabilized arcs* 276, 391.
—, nichtstationäre, *non-stationary arcs* 290.
—, Wiederzünden gelöschter, *reignition of arcs after extinguishing* 291.

Bogencharakteristik, dynamische, *dynamical arc characteristic* 286.
Bogendurchmesser, *arc diameter* 402.
Bogenentladung, *arc discharge* 55.
Bogenkathode, Temperaturen der, *temperatures of the arc-cathode* 407.
Bogenlinien, *arc lines* 112.
Bogensäule, *arc column* 366.
Bolometer 69.
BOLTZMANNsche Gleichung, BOLTZMANN *equation* 121.
—, Verteilung, BOLTZMANN *distribution* 70, 106, 107, 126, 306.
Bombardement der Kathode, *bombardment of the cathode* 13.
BOYS-Kamera, BOYS *camera* 581.
BOYS-Trommelkamera, BOYS *drum camera* 583.
Brechungsindex, *refraction index* 376.

Charakteristik, *characteristic* 10, 15, 16, 147.
—, negative, *negative characteristic* 506, 513.
Charakteristiken von zweiter Art, *characteristics of the second kind* 176.
chemischesPotential, *chemical potential* 317.
COLLI-FACCHINIscher Effekt, COLLI-FACCHINI *effect* 463.
corona 506.
CROOKESscher (oder HITTORFscher) Dunkelraum, CROOKES (*or* HITTORF) *dark space* 56.
—, (oder Kathoden-) Dunkelraum, CROOKES (*or cathode*) *dark space* 246.

Dämpfe, organische, *organic vapours* 181.
Dämpfung, *damping* 360.
dart leaders 592.
dart leader, Mechanismus des, *mechanism of the dart leader* 620.
dart-stepped leaders 590, 591, 593.
Dauer des Anwachsens, *time of growth* 41.
Dauerentladung, *steady discharge* 513.
Dauer-Koronadurchbruch, Einsetzen des, *onset of steady corona* 522.
DAVIDSONsche Gleichung für das Anwachsen, DAVIDSON's *growth equation* 42, 44.
dead time 455, 511, 522.
Dichtemessung, *measurement of density* 376.
Diffusion, *diffusion* 16, 77, 80, 169, 323.
—, ambipolare, *ambipolar diffusion* 122, 129, 170, 174, 332.
—, laterale, *lateral diffusion* 169.
—, radiale, *radial diffusion* 169.
— von Elektronen, *diffusion of electrons* 143.
Diffusionslänge, *diffusion length* 533.
Diffusionstheorie, *diffusion theory* 128, 537.
Dispersionsbeziehungen, *dispersion relations* 193.
Dispersionstheorie des Plasmas, *dispersion theory of the plasma* 356.
Dissoziationsrekombination, *dissociative recombination* 232.
divergentes Feld, *divergent field* 21.
Doppelschichten, *double striations* 139.
—, *double layers* 147.
Doppelschicht, dünne, *thin double layer* 143.
DOPPLER-Effekt, DOPPLER *effect* 70, 98, 136.
DOPPLER-Verbreiterung, DOPPLER *broadening* 215, 361.
DOPPLER-Verteilung, DOPPLER *contribution* 372.
Druck des Elektronengases, *pressure of the electron gas* 174.
Druckverbreiterung, *pressure broadening* 362.
Druckwelle, *shock wave* 253.
Dunkelentladung, *dark discharge* 202.
Dunkelraum, Energieverteilung von Ionen und Elektronen, *dark space, energy distribution of ions and electrons* 95.
—, Länge des, *length of the dark space* 105.
—, optische Eigenschaften des, *optical properties of the dark space* 96.
Dunkelstrom, *dark current* 189.
Durchschlag, *breakdown* 453, 525.
—, Anwachsen der Ionisation vor dem, *growth of pre-breakdown ionization* 30.
—, Bedingung für einen lokalisierten, *threshold for a localized burst* 484.
—, beweglichkeitsgeregelter, *mobility-controlled breakdown* 568.
—, durch Elektroneneinfang geregelter, *electron attachment controlled breakdown* 566.
—, einmaliger, *burst pulse* 515, 518.
— in asymmetrischen Feldern, *breakdown in asymmetrical fields* 506.
— infolge Feldemission, *breakdown by field emission* 450.
— in funkenverhindernden Gasen, *breakdown in spark suppressing gases* 522.
— in inhomogenen Feldern, *breakdown in nonuniform fields* 480.
— in Luft, *breakdown in air* 524, 566.
— in reinem Argon, *breakdown in pure argon* 509, 513.
— in Sauerstoff, *breakdown in oxygen* 567.
—, sekundärer, *secondary pulse* 522.
—, Sekundärelektronen-Resonanz, *secondary electron resonance breakdown* 570.
—, zeitlicher Ablauf des, *time sequence of breakdown events* 2.
Durchschlagsbedingung, *breakdown condition* 6.
Durchschlagskriterium, *breakdown criterion* 5, 9, 16, 32, 33, 35, 37, 45.
—, physikalische Bedeutung des, *physical significance of the breakdown criterion* 9.
Durchschlagsspannung, *breakdown potential* 1, 5, 10, 12, 15, 26, 451.
—, *sparking potential* 11, 12, 20, 45.
—, minimale, *minimum sparking potential* 1, 5, 10, 12, 15.
—, Kurve für die PASCHENsche, *sparking potential curve of* PASCHEN 161.
Durchschlagsstrom, *spark current* 2.
Durchschlagsmechanimus, *mechanism of breakdown* 39.
Durchschlagswahrscheinlichkeit, *breakdown probability* 469.
Durchschlagszeit, *breakdown time* 28, 454.
—, wahre, *true breakdown time* 472.

ECCLEsche Beziehung, ECCLE's *relation* 357.
Edelgasbögen, *arcs in rare gases* 296.
EGGERT-SAHAsche Gleichung, EGGERT-SAHA *equation* 308.
Eigengrößen, *proper variables* 536.
Einkristalloberflächen, *single crystal surfaces* 87.
Eigenstabilisierung, *self-stabilization* 276.
Einfangfrequenz, *attachment frequency* 566.
Einschnürungen, enge, *narrow constrictions* 147, 163.
EINSTEINsche Konstante der spontanen Emission, EINSTEIN *constant of spontaneous emission* 127.
Elatation 209, 218.
Elatationswirkungsquerschnitt, totaler, *total elatation cross-section* 245.
elektrisches Feld, radiales, *radial electric field* 171.
Elektrode, mechanische Kraft darauf, *mechanical force on an electrode* 62.
Elektroden, Polarisation der, *polarisation of the electrodes* 190.
—, Reinheit der, *cleanliness of electrodes* 507.
Elektrodenmaterialien, *electrode materials* 11.
elektrodynamische Kraft, *electrodynamic force* 169.
Elektronen, auslösende, *triggering electrons* 473.
—, Energieverteilung, *energy distribution of electrons* 96, 170.
—, schnelle, *fast electrons* 143, 173.
— und Ionen in Gasen, mittlere Energien, *mean energies of electrons and ions in gases* 2, 14.
Elektronenanlagerung, *electron attachment* 77, 80, 85, 129, 130.
— an neutrale Moleküle, *electron attachment to neutral molecules* 19.
Elektronenbeweglichkeit, *electron mobility* 121.
Elektronendichte, *electron density* 374.
Elektronendruck, *electron pressure* 192.
Elektronenemission, *electron emission* 9.
Elektronenentladungen, monoenergetische, *monoenergetic electron discharges* 242.
Elektronenkonzentration, Mikrowellenmessung, *microwave measurements of electron concentration* 68.
Elektronenlawine, *electron avalanche* 9, 33, 35, 48.
Elektronenschwingungen, *electronic oscillations* 191.
Elektronenstrahl, Ablenkung eines, *deflection of an electron beam* 91.
—, monoenergetischer, *monoenergetic electron beam* 243.
Elektronentemperatur, *electron temperature* 63, 117, 120, 121, 127 ,130, 132, 171, 174.
— in Gasgemischen, theoretisch, *electron temperature in gas mixtures, theoretical* 184.
Elektrophorese, *electrophoresis* 195.
ELENBAAS-HELLERsche Gleichung, ELENBAAS-HELLER *equation* 337.
Emission 527.
—, erzwungene, *enforced emission* 210, 359.
— infolge Einfalls von Photonen, *emission due to incidence of photons* 6.
— von Elektronen durch positive Ionen, *emission of electrons by positive ions* 447.
Emissionseigenschaften, stabile, *stable emissive properties* 34.
Emissionskoeffizient, *emission coefficient* 353, 354.
Empire State Building 591, 607.
Endeffekte, *end effects* 172, 174.
Energiebilanz, *energy balance* 122, 126, 132, 133, 381, 392, 401.
Energie-Erhaltung, *energy conservation* 317.
Energiestrom, *flux of energy* 322.
enge Röhren, *narrow tubes* 62.
Entartung, *degeneracy* 38.
Enthalpie, *enthalpy* 342.
Entionisierungszeit, *clearing time*, 455 519, 520.
Entladung, abnorme, *abnormal discharge* 55.
—, Aufbauzeit der, *time of formation of the discharge* 190.
—, behinderte, *obstructed discharge* 56, 57, 161, 168.
—, charakteristische Farbe, *characteristic colour of the discharge* 58, 114.
—, charakteristische Kurve, *characteristic curve of the discharge* 201.
—, effektiver Widerstand der, *effective resistance of the discharge* 189.
—, eingeschnürte, *constricted discharge* 163, 175.
—, elektrodenlose, *electrodeless discharge* 144, 195, 251.
—, Entwickungszeit der, *time of development of the discharge* 188.
—, gleichmäßig über die Elektroden verteilte, *discharge spread uniformly over the electrodes* 505.
—, intermittierende, *intermittent discharge* 187, 625.
—, magnetisches Feld der, *magnetic field of discharge* 73.
—, Mechanismus der sich fortsetzenden, *mechanism of the continuing discharge* 627.
—, normale, *normal discharge* 55, 82.
—, photographische Erforschung der Leuchtstärke der, *photographic studies of luminosity in the discharge* 580.
—, pulsierende, *throbbing discharge* 190.
—, Selbstkapazität der, *self capacity of the discharge* 188.
Entladung, spezielle Typen der, *special types of discharge* 161, 598.
—, Stabilität von, *stability of discharge* 201.
—, subnormale, *subnormal discharge* 55, 82.
Entladungen an kalter Kathode, *cold cathode discharge* 246.
— (Blitz) zu sehr hohen Gebäuden, *discharges (lightning) to very high buildings* 596, 607.
— durch thermische Elektronen, *thermal electrons discharges* 247.

Entladungen im magnetischen Feld, *discharges in magnetic fields* 164.
— in elektronegativen Gasen, *discharges in electronegative gases* 177.
—, instabile, *unstable discharges* 506.
—, Meßmethoden der, *methods of measurement discharges* 61.
Entladungsfolge, Unterbrechung, *interruption of the discharge sequence* 455.
Entropie, *entropy* 342.
Entropiebilanz, *entropy balance* 317.
Entropieproduktion, *entropy production* 383.
Entropiequelle, *entropy source* 318.
Entropiestrom, *entropy current (flux)* 318.
Erdentladung, Gesamtdauer einer, *total duration of a discharge to ground* 579.
Erd- und Luftentladungen, allgemeine Züge, der, *general features of ground and air discharges* 577.
Erhaltungs-Bedingung, *maintenance condition* 22.
Erhaltungsgleichung, *equation of maintenance* 103.
Erhaltungs-Prozeß, *maintenance process* 23.
erlaubte und verbotene Prozesse, *permitted and forbidden processes* 80.
Erzeugungsfunktion, *production function* 217.

FARADAYscher Dunkelraum, FARADAY *dark space* 56, 57, 60, 110, 224.
Feinstruktur, *fine structure* 143.
Feld, asymmetrisches, *asymmetrical field* 455.
—, effektives, *effective field* 532.
—, Grenze des homogenen, *uniform field limit* 537.
—, homogenes, *uniform field* 455.
—, inhomogenes, *nonuniform field* 552.
—, longitudinales, *longitudinal field* 124, 125.
Feldemission, *field emission* 79, 80, 163, 449, 450.
—, thermische, *thermal field emission* 414.
Feldverzerrung, *field distortion* 8, 29, 33.
FERMI-Entartung, FERMI *degeneration* 314.
Film, dünner, *thin film* 13.
—, metallischer, *metallic film* 13, 14.
Flammenbögen, *flaming arcs* 292.
Flecke, *spots* 507, 511, 514, 521.
Fluoreszenz, *fluorescence* 37f.
flüssige Oberflächen, *liquid surfaces* 158.
freie Elektronen als Gas, *free electron gas* 512.
frei-frei-Strahlung, *free-free-radiation* 354.
Fremdatome, *impurities* 85, 197.
freon 523.
Frequenzgrenze, obere, *upper frequency limit* 192.
Funken, *spark flash* 1, 51.
—, *flash* 1, 51, 506.
—, *spark* 1, 51, 506.
— in sehr kurzen Zündstrecken, *spark in very short gaps* 527.
—, zeitliche Schwankungen des Auftretens von, *fluctuation of appearance of sparks in time* 471.
Funkendurchschlag, s. Durchschlag, *breakdown* 1, 8, 17, 20, 27, 28, 35, 38, 39, 40, 49.
Funkenentladung, *spark breakdown* 451, 453.
—, *spark discharge* 252.
Funkenlinien, *spark lines* 112.
Funkenquerschnitt, *spark diameter* 47.
Funkenspannung, *sparking potential* 1, 5, 10, 12, 15, 82.

Gasabsorption, *gas absorption* 198.
Gasdichte, *gas density* 70, 72.
Gasdruck, *gas pressure* 80.
—, niedriger, *low gas pressure* 2.
Gasdynamik, *gas dynamics* 402.
Gasreinheit, *gaseous purity* 507.
Gasentladung bei niedrigem Druck, *gas discharge at low pressure* 216.
Gasfluß, *gas flow* 200.
Gasgemisch, Entladungen in, *discharges in gas mixtures* 181.
—, normale Stromdichte in, *normal current density in gas mixtures* 183.
Gasinterferometer, *gas interferometer* 196.
Gasprozesse, erlaubte, *permitted gas processes* 75.
Gasschicht, adsorbierte, *adsorbed gas layer* 155, 157.
—, *gas layer* 190.
Gasstrom, longitudinaler, *longitudinal gas stream* 202.
Gastemperatur, *gas temperature* 69, 100, 136.
Gaszähigkeit, *gas viscosity* 200.
Gase, elektronegative (negative Ionen), *electronegative gases, (negative ions)* 129, 146.
Gazekathode, *gauze cathode* 110.
GEIGER-Impuls-Ausbreitung, Schwelle für, *threshold for a* GEIGER *pulse spreading* 484.
GEIGER-Zähler, GEIGER *counter* 180, 240, 483, 515, 522.
GEISSLERsche oder Glimmentladung, GEISSLER *or glow discharge* 452.
GERDIEN-Bogen, GERDIEN-*arc* 300, 371.
Gewitterwolke, Ladungsverteilung, *distribution of thundercloud charges* 598.
GIBBSsche Gleichung, GIBBS *equation* 317.
— Potential, GIBB's *potential* 342.
Gleichgewicht, thermisches, *thermal equilibrium* 305.
Glimmentladung, *glow discharge* 242, 506.
—, alternierende, *alternating glow discharge* 194.
—, behinderte, *obstructed glow discharge* 190.
—, Entstehung einer, *formation of a glow discharge* 53.
—, bei hohem Druck, *high pressure glow discharge* 194.
Glimmentladung in bewegtem Gas, *glow discharge in moving gas* 202.
— in Metallröhren, *glow discharge in metal tubes* 55.
Glimmfunke, *spark flash in form of a glow* 1, 2, 9, 10, 33.
Glimmlicht, diffuses, *diffuse glow* 28.
—, Länge des negativen, *length of the negative glow* 113.
—, negatives, *negative glow* 56, 110, 143.

Glimmlicht, negatives: Theorie, *theory of the negative glow* 113.
—, Saum des, *boundary of the glow* 102.
Glühemission, *thermionic emission* 449, 450.
Glühkathode, *thermionic cathode* 243.
Glühkathodenentladung, *hot cathode discharge* 90.
Glühsonde, *hot probe* 62, 67.
GOLDSTEINsches Gesetz, GOLDSTEIN'S *law* 138, 139.
Grenzamplitude, *oscillation amplitude limit* 539.
Gyrofrequenz, *gyro-frequency* 177.

HALE oder MOHLER Prozesse, HALE *or* MOHLER *processes* 446.
Halogene, *halogens* 522, 523.
Heg-Gas, *Heg gas* 535.
HEHLsches Gesetz, HEHL'S *law* 82.
Helium, unreines, *impure helium* 549.
He_2-Spektrum, He_2 *spectrum* 166.
Heizfaden mit Oxydschicht, *oxide coated filament* 90.
Hg-Bogen, *Hg-arc* 390.
Hilfsanode, *auxiliary anode* 152.
Hochdruckbogen, wasserstabilisierter, *water stabilized high pressure arc* 379.
—, Xenon-, *Xenon high pressure arc* 390.
Hochdruckglimmentladung, *high pressure glow discharge* 193.
Hochfrequenzentladung, *high frequency discharge* 177, 247, 251.
Hochstromkohlebogen, *high current carbon arc* 372, 427, 432.
Hochvakuumprozesse, *high vacuum processes* 79.
hoher Druck, *high pressure* 27, 527.
Höhenbestimmung (im Blitz), *height determination (in lightning)* 617.
Hohlkathodenentladung, *hollow cathode discharge* 91, 97, 108.
HOLST-OOSTERHUIS Entladung, HOLST-OOSTERHUIS *discharge* 241.
HOLTSMARKsche Wahrscheinlichkeitsfunktion, HOLTSMARK *probability function* 213.
HORNBECK-MOLNARscher Effekt, HORNBECK-MOLNAR *effect* 446.
homogenes statisches Feld, *uniform static field* 32.
homologe Bereiche, *homologous regions* 18.
— System, *homologous system* 17.
— Volumina, *homologous volumes* 19.
Hysteresis, *hysteresis* 189, 194.

Impuls- und Energieaustausch der Elektronen mit Plasmawellen, *momentum and energy exchange of electrons with plasma waves* 192.
Impuls-Felder, *impulse fields* 45.
Impuls, scharf begrenzter, *sharply peaked impulse* 27.
Induktivität, *inductance* 201.
—, effektive, *effective inductance* 201.
Infrarot-Emissionsspektra, *infra red emission spectra* 250.
inhomogene Felder, *non-uniform fields* 40, 526.
— symmetrische Felder, *non-uniform symmetrical fields* 525.
innere Energie, *internal energy* 341.
instabiles Gebiet, *unstable region* 452.
Instabilität, *instability* 453.
Intensitäten, Absolutmessungen von, *absolute measurement of intensities* 368, 369.
Interferenzaufnahme, *interference photograph* 396.
Interferenzspektrogramm, *interference spectrogram* 373.
Ionen, *ions* 9, 33.
—, Beschuß mit positiven, *ions: positive ion bombardment* 31.
—, Bewegung der, *motion of ions* 135.
—, Energieverteilung der, *energy distribution of the ions* 64.
—, Erhaltung der Anzahl der, *number balance for ions* 121.
—, langsame, *slow ions* 14.
—, negative, *negative ions* 93, 130, 143, 154, 178.
—, radiale Verteilung der, *radial distribution of ions* 171.
—, Vernichtung negativer, *ions: distruction of negative ions* 179.
Ionenbombardement, *ion bombardment* 81.
Ionendichte, *ion density* 28, 44.
Ionenerzeugung, *ion production* 121.
Ionenquerschnitt, *ion cross-section* 349, 380.
Ionen, positive Raumladungen, *ions, positive space charges* 523.
Ionenstrom zu den Wänden, *ion current to the walls* 126.
Ionenwanderungsgeschwindigkeit, *ion drift velocity* 136.
Ionisationsgleichgewicht, *ionization equilibrium* 308.
Ionisationskammer, *ionization chamber* 28, 29, 113.
Ionisierung, *ionization* 1.
—, Anwachsen der, *ionization growth* 1, 2, 14, 30, 43, 47, 49.
—, äußere Quelle der, *external source of ionization* 54.
— durch Einzelstöße, *ionization by single collisions* 75, 80.
— durch positive Ionen, *ionization by positive ions* 6, 448.
— durch Stoß, *ionization by collision* 446.
—, räumliche Zunahme in statischen Feldern bei hohen Drucken, *spatial growth of ionization, in static fields at high pressures* 27.
—, räumliches Anwachsen in statischen konstanten Feldern, *spatial growth of ionization in static uniform fields* 2, 5.
—, räumliches Anwachsen, *spatial ionization growth* 32.
—, sekundäre, *secondary ionization* 33, 34.
—, stufenweise, *stepwise ionization* 7, 19, 75, 80, 128, 146.
—, zeitliches Anwachsen, *temporal growth of ionization* 40.

Ionisierung, Zunahme in inhomogenen Feldern, *growth of ionization in non-uniform fields* 21.
Ionisierungskoeffizient, *ionization coefficient* 533.
Ionisierungsfunktion, *ionization function* 135, 162.
Ionisierungsspannung, *ionization potential* 89, 196, 526.
Ionisierungs- und Anregungswahrscheinlichkeit, *probabilities of ionization and excitation* 18, 89, 196.
Ionisierungswelle, *ionizing wave* 526.
Ionisierungszone, *ionizing zone* 513.
irreversible Prozesse (Thermodynamik), *irreversible processes (thermodynamics)* 315.

JOULEsche Wärme, JOULE's *heat* 339.

Kameratypen, *types of camera* 580.
Kamerabeobachtungen, Zusammenhang mit, *correlations with camera observations* 608.
Kanal, s. streamer (= leuchtender Faden) und Leitkanal, *canal, cf. streamer or leader.*
Kanalstrahlen, *canal rays* 57, 59, 96.
Kanalstrahlentladungen, *canal ray discharges* 162.
Kapillarelektrometer, *capillary electrometer* 598.
Kapillarröhren, *capillar tubes* 91, 118, 163, 174, 199.
Kaskaden-Konstanten, *cascading constants* 234.
Kataphorese, *cataphoresis* 195.
Kathode, bestrahlte, *irradiated cathode* 1.
—, bewegliche, *movable cathode* 120.
—, Energiebilanz an der, *energy balance at the cathode* 99.
—, Erhitzung der, *heating of the cathode* 99, 100.
—, Feld an der, *field at the cathode* 105.
—, hochgespannte, *highly stressed cathode* 526.
—, lichtempfindliche, *photo-sensitive cathode* 202.
—, Oxydation der, *oxidation of the cathode* 33.
—, positive Ionen auf der, *positive ions on the cathode* 6.
—, verdampfende, *evaporating cathode* 420.
—, virtuelle, *virtual cathode* 59.
Kathoden, konkave und konvexe, *concave and convex cathodes* 97, 99.
Kathodenausdehnung, *area of the cathode* 94.
Kathodendunkelraum, *cathode dark space* 56, 62, 81, 165.
—, Dicke, *thickness of cathode dark space* 106.
—, Länge des, *length of the cathode dark space* 92, 165.
—, Länge in Gasgemischen, *length of cathode dark space in gas mixtures* 183.
Kathodeneffekte, *cathode effects* 10.
Kathodenemission, *cathode emission* 33.
Kathodenfall, *cathode fall* 81, 82, 83, 87, 89, 101, 103, 106, 108, 165, 193, 279, 405, 412, 420.
—, abnormaler, *abnormal cathode fall* 84.
—, Fehler der Theorien, *cathode fall, defects of the theories* 107.
— in Gasgemischen, *cathode fall in gas mixtures* 183.
—, normaler, *normal cathode fall* 84.
—, Schichtungen im, *striations in the cathode fall* 100.
Kathodengebiet, Einflüsse magnetischer Felder, *magnetic field influences on the cathode region* 165.
Kathodengeometrie, *cathode geometry* 97.
Kathodenglimmlicht, *cathode glow* 96, 246.
—, Ionenstrom am, *ion current at the cathode glow* 103.
Kathodeninstabilität, *cathode instability* 506.
Kathodenlicht, *cathode light* 59, 96.
Kathodenmaterial, *cathode material* 12, 59, 88, 156.
Kathodenmechanismus von Lichtbögen, *cathode-mechanism of arcs* 403.
Kathodenmetall, Zerstäubung des, *sputtering of the cathode metal* 86, 87.
Kathodenoberfläche, *cathode surface* 15, 108, 408.
Kathodenprozesse, *cathode processes* 14.
Kathodenschicht, erste, *first cathode layer* 56, 60, 96, 146.
Kathodenstrahlen, *cathode rays* 59, 147.
Kathodentemperatur, *cathode temperature* 90, 157.
Kathodenzerstäubung, *cathode sputtering* 154.
KIRCHHOFFscher Satz, KIRCHHOFF's *law* 358, 359.
koaxiale Zylinder, *coaxial cylinders* 516.
— —, Durchschlag in Luft und reinem Sauerstoff, *coaxial cylinders, breakdown in air and pure oxygen* 516.
— —, Durchschlag in reinen Gasen, *coaxial cylinders, breakdown in pure gases* 514.
Kohlebogen, *carbon arc* 291.
—, zischender, *hissing carbon arc* 425, 427.
Kompression, eigenmagnetische, *self-magnetic compression* 396.
Kontaktpotential, *contact potential* 65.
Kontinuitätsgleichung, *continuity equation* 36, 42.
Kontinuum, *continuum* 229.
konzentrische Ringe, *concentric rings* 151.
Konzentration, *concentration* 209.
Korona-Entladung, *corona discharge* 252.
— zwischen positiver Spitze und Platte, *positive point-to-plane corona* 452, 521.
—, selbständige, *self-sustaining corona* 512.
Koronadurchschlag, einmaliger, *burst pulse corona* 518.
—, Zündschwelle für einmaligen, *burst pulse corona threshold* 482.
Kraftgleichung, *force equation* 317.
Krümmung der Begrenzungen, *curvature of the boundaries* 57.

Kugelblitz, *ball lightning* 577.
KUNZMANsche Quelle, KUNZMAN *source* 145.

Ladungen (im Blitz), Größenordnungen der, *quantities of charge (in lightning)* 600.
Ladungsverformung, *charge distortion* 503.
Ladungstransport, *charge transfer* 77, 79, 80, 99, 106, 109, 163.
LANGEVINsche Beweglichkeitsgleichung, LANGEVIN'S *mobility equation* 122, 131.
LANGMUIRscher Dunkelraum, LANGMUIR *dark space* 224.
LANGMUIR-Sonden, LANGMUIR *probes* 61, 62, 110, 115, 138.
LARENZ und BARTELS, Methode von, *method of* LARENZ *and* BARTELS 371.
LAVAL-Düse, LAVAL *jet* 403.
Lawine, *avalanche* 4, 28, 29, 47, 446.
Lawinenschwankungen, statistische, *avalanche fluctuation* 447.
LECHER-Drähte, LECHER *wires* 68, 137.
Leitfähigkeit, *conductivity* 159.
—, elektrische, *electric conductivity* 326, 352, 378.
—, thermische, *thermal conductivity* 134.
Leitkanal, Feldänderung, *leader field change* 608, 610.
—, Feld und Geschwindigkeit am Kopf, *tipfield and velocity while advancing leader* 618.
—, stufenweise gebildeter, *stepped leader* 586, 597.
Leitkanäle, erste, α- und β-Typ, *α and β type first leaders* 586, 587, 588, 589.
—, stufenweise gebildete, Verteilung der effektiven Geschwindigkeiten, *distribution of effective velocities of stepped leaders* 591.
Leuchten, *luminosity* 44.
—, Fortpflanzung des —s, *propagation of luminosity* 51.
Lichtabsorption, *absorption of light* 113.
Licht, teilweise polarisiertes, *partially polarized light* 81.
—, ultraviolettes. *ultra-violet light* 198.
Lichtbogen, *arc* 1, 2, 9.
—, Charakteristik eines, *characteristic of an arc* 276.
—, chemische Vorgänge in, *chemical reactions in arc* 284.
—, Materialtransport, *matter transport in an arc* 283.
—, selbständiger, *self sustained arc* 255, 304.
—, Stabilisierung eines, *stabilization of an arc* 265.
—, Strahlungseigenschaften von, *radiation properties of arc* 281.
—, thermischer, *thermal arc* 257.
—, Zündung von, *ignition of arc* 262.
Lichtbogen-Mechanismus, *arc-mechanism* 257.
Lichtbogenöfen, *arc furnaces* 303.
Lichtbogenphysik, *arc physics* 255.
Lichtbogentemperaturen, *temperatures of arc* 282.
Lichtbogentypen, *types of arcs* 291.
Lichtstärke, schwankende, *fluctuating luminosity* 191.
Linien- oder Gabelblitz, *streak or forked lightning* 577.
Linienintensitäten, *line intensities* 353.
Linienspektrum, *line spectrum* 135.
Linienverschiebung, *line shift* 375.
LORENTZ-Kräfte, LORENTZ-*forces* 326, 396.
Löschspannung, *extinction potential* 188.
Luftentladungen, *air discharges* 577, 596.

MACH-Zahl, MACH-*number* 403.
magnetische Ablenkung, *magnetic deflection* 95.
magnetisches Feld, longitudinales, *longitudinal magnetic field* 166.
— —, transversales, *transverse magnetic field* 166, 170.
— —, überlagertes, *superimposed magnetic field* 559.
magnetische Glieder, *magnetic links* 605.
— Linsen, *magnetic lenses* 173.
— Strahlen, *magnetic rays* 177.
MALAN-Kamera, MALAN *Camera* 584.
MALTER-Effekt, MALTER *effect* 164.
Massenerhaltung, *conservation of mass* 316.
Massenkonzentration, *mass concentration* 316.
MAXWELL-Verteilung, MAXWELL*ian distribution* 63, 65, 111, 121, 135.
M-Komponenten, *M-components* 579.
Membranschwingungen, *membrane vibrations* 128.
MERKsches Kriterium, MERK'S *criterion* 29.
Metalldampf, *metal vapour* 185.
Metalldampfbögen, *metallic vapour arcs* 294.
metastabile Atome, *metastable atoms* 6, 15, 20f., 76, 79f., 80, 112, 448.
— — im Gas, Ionisation durch, *ionization by metastable atoms in the gas* 448.
— —, Zerfall von, *decay of metastable atoms* 202, 448.
— A_2-Moleküle, *metastable* A_2 *molecules* 509.
— Gasmoleküle, *metastable gas molecules* 65, 509.
metastabiles Neon, *metastable Neon* 87.
Meteorlicht, Erzeugung des, *meteoric light production* 224.
Mikrowellenentladungen, *microwave discharges* 251.
Minimumprinzip, *minimum principle* 104, 108, 151.
mittlere freie Weglänge, *mean free path* 80.
— — — für photoionisierende Strahlung, *mean free path for photoionizing radiation* 485.

Nachglimmen, *after-glows* 228.
natürliche Linienbreite, *natural line breadth* 211.
Nebenfunktionen, *tributary functions* 234.
negative Zonen, *negative zones* 57.
Neon, *neon* 550.
Neutralisierung, wechselseitige, *natural neutralization* 233.

Niederdrucksäule, *low pressure column* 174.
Niederspannungsbögen, *low voltage arc* 242.
Niedervoltbögen, *low-voltage arc* 257, 304.
Nordlicht, *polar aurora* 224, 242.

Oberfläche, Reinigung der, *cleaning of the surface* 520.
OHMsches Gesetz, OHM's *law* 326.
— —, Gültigkeitsbereich, OHM's *law regime* 513, 520.
ONSAGERs Reziprozitätssätze, ONSAGER's *reciprocity laws* 319.
optisch dicke Schicht, *optical thick layer* 373.
optische Schichtdicke, *optic thickness of layer* 367.
— Tiefe, *optic depth* 358.
Oszillatorstärke, *oscillator strength* 211, 360.
Oszillographentechnik, *oscillographic techniques* 140.
Oxydfilm, *oxide film* 13.
Oxydschicht, *oxide layer* 25, 156, 163, 164.
O_2-Banden, O_2 *bands* 119.
O_2, Einfluß von kleinen Mengen, O_2, *effect of small amounts* 510.

PASCHENsche Kurven, PASCHEN *curve* 23.
PASCHENsches Gesetz, PASCHEN's *law* 10, 17, 462.
— —, Versagen für Trägerdichte, PASCHEN's *law, failure of carrier density* 463.
Pausen, *pause times* 590.
PENNING-Effekt, PENNING *effect* 75, 76, 80, 85, 446, 465.
PENNING-Gas, PENNING *gas* 466.
Perlschnurblitz, *bead (or chain) lightning* 577.
photoelektrische Ausbeute, *photoelectric efficiency* 33.
— Emission, *photoelectric emission* 3, 20, 25, 98, 447.
— Empfindlichkeit, *photoelectric sensitivity* 30.
— Wirkung im Gas, *photoelectric action in the gas* 517.
photoelektrischer Effekt, *photoelectric effect* 14, 26, 44, 80, 81, 225.
— Strom, *photoelectric current* 2, 202.
photographische Messungen, *photographic measurements* 198, 585.
Photoionisation, *photo-ionization* 7, 14, 20, 28, 35, 36, 37, 39, 77, 80.
— in Gasen, *photo-ionization in gases* 449, 480, 481, 511, 515.
— in Luft, *photo-ionization in air* 481.
Photomultiplier 140.
Photonen, *photons* 9, 14, 33, 35, 37.
Photonenabsoption, *photon absorption* 225.
Photozellen, *photocells* 140.
pilot streamer 590, 621.
pinch effect 175.
Plasma 256.
—, Kontraktionstheorie des, *theory of plasma contraction* 416.
—, magnetischeEigenschaften des, *magnetic properties of a plasma* 175.
—, neutrales, *neutral plasma* 114.
Plasma, quasi-neutrales, *quasi-neutral plasma* 120.
—, thermisches, *thermal plasma* 305.
Plasmadynamik, *plasma dynamics* 402.
Plasmakugel, *plasma sphere* 390.
Plasmaschwingungen, *plasma oscillations* 66, 132, 142, 144, 191, 356.
Plasmastrahl, *plasma beam* 400.
—, kathodischer, *plasma beam, cathodic* 409.
Plasmaströmung, *plasma flow* 396, 401.
— mit Überschallgeschwindigkeit, *plasma flow with supersonic speed* 403.
Plasmawechselwirkung, *plasma interaction* 314.
Polarisation, *polarization* 223.
POISSONsche Gleichung, POISSON's *equation* 327.
Population 209.
positive Ionen, *positive ions* 13.
Potentialemission, *potential liberation* 447.
Potentialverteilung, radiale, *radial potential distribution* 172.
Primäreffekt, *primary action* 446.
Primärelektronen, *primary electrons* 111.
—, schnelle, *fast primary electrons* 147, 161.
Primär-Impuls, *primary pulse* 522.
Primärionisation, *primary ionization* 4.

Quadrupolstrahlung, *quadripole radiation* 210.
Quasineutralität, *quasi-neutrality* 312.
Quecksilberbogen, *mercury arc* 297.
Quecksilberdampf-Gleichrichter, *mercury vapor rectifier* 304.

Radiowellen, *radio waves* 243.
RAETHERsches Kriterium, RAETHER's *criterion* 29.
Raketenblitz, *rocket lightning* 577, 597.
RAMSAUER-Effekt, RAMSAUER *effect* 72, 73, 132.
RAMSAUER-Querschnitt, RAMSAUER *cross-section* 348, 378.
Raumladung, *space-charge* 2, 8, 15, 16, 33, 34, 35, 43, 45, 47, 50, 471, 499, 505, 526.
—, negative, *negative space charge* 523.
—, positive, *positive space charge* 29, 523.
Raumladungsanhäufung, *space charge accumulation* 481.
raumladungsbegrenzte Ströme, *space charge limited currents* 513.
Raumladungseinfluß, *space charge effect* 454, 460.
Raumladungsfeld, *space charge field* 486.
Raumladungsverzerrung, *space charge distortion* 30.
Rauschen, *noise* 193.
Rauschmessungen, *noise measurements* 69.
Rauschspannung, *noise voltage* 373.
Rauschtemperatur, *noise temperature* 69.
RAYLEIGH-Effekt, RAYLEIGH *effect* 224, 252.
Reibungskräfte, *frictional forces* 318.
Rekombination, *recombination* 19, 36, 78, 80, 107, 112, 136, 139, 245.

Rekombination, Drei-Elektronen-, *three body electron recombination* 232.
—, Kriterium für, *criterion for recombination* 227.
— auf Oberflächen, *surface recombination* 227.
—, Volumen-, *volume recombination* 227.
—, Zwei-Elektronen-, *dielectronic recombination* 231.
Rekombinationskoeffizient, *recombination coefficient* 253
relativistische Theorie, *relativistic theory* 130.
Relaxationszeit, *relaxation time* 209.
Resonanzlinien, *resonance lines* 133.
Resonanzphotonen, *resonance photons* 79.
Resonanzstrahlung, *resonance radiation* 76, 112, 209, 225.
Ringelektroden, *ring electrodes* 94.
Ringentladung, *ring discharge* 195.
Röhren, enge: s. Kapillarröhren, *tubes, narrow: cf. capillar tubes* 118.
—, lange, *long tubes* 140.
—, sehr weite, *very wide tubes* 120.
Röhrenradius, *tube radius* 117f.
Rohrentladungen, *pipe discharge* 181.
Röntgenstrahlen, *X-rays* 202.
—, Absorption, *absorption of X-rays* 200.
—, weiche, *soft x-rays* 247.
Rotationsbande, *rotational band* 370.
Rotationsbandenspektra, *rotational band spectra* 70.
Rückdiffusion, *back diffusion* 154, 512.
Rückdiffusionsverluste, *back diffusion losses* 447.
rückkoppelnd, *regenerative* 450.
Rückstreuung von Elektronen, *back-scattering of electrons* 4, 15.

Säule, ebene positive, *plane positive column* 169.
—, eingeschnürte, *constricted column* 178.
—, gleichförmige, in Atomgasen, *uniform columns in atomic gases* 114, 115, 116.
—, gleichförmige, in Molekülgasen, *uniform columns in molecular gases* 118.
—, geschichtete, *striated column* 57, 137, 142, 252.
—, instabile, *unstable column* 194.
—, Länge der, *length of the column* 114.
—, lineare, *linear column* 128, 129.
—, magnetisches Feld in der positiven, *magnetic field on the positive column* 168.
— mit geringem Strom (subnormal), *low current (subnormal) column* 130.
— mit rechteckigem Querschnitt, *column of rectangular section* 128.
—, positive, *positive column* 57, 114, 120, 165, 174, 247.
—, Zusammenziehung der, *contraction of the column* 57, 175, 193.
Säulengradient, *column gradient* 279.
Schallgeschwindigkeit, *speed of sound* 344, 377.
Schallwellen, *sound waves* 192.
Schaltbögen, *switch-arcs* 302.
Schatteneffekte, *shadow effects* 59.
Schichten, inhomogene, *inhomogeneous layers* 365.
Schichtung, *striations* 92, 141, 142, 143, 169, 192, 252.
—, Abstand, *distance striations* 138.
— (CH_2), blaue, *blue striations* (CH_2) 139.
— in Kapillarröhren, *striations in capillary tubes* 145.
—, Krümmung der, *curvature of the striations* 137.
—, stationäre, *stationary striations* 115, 137.
—, verbreiterte, in Wasserstoff, *broadened striations in hydrogen* 138.
—, wandernde, *moving striations* 115, 140.
Schläge eines Blitzes, *strokes of a flash* 577, 580.
—, Höhenbestimmung für Einsetzen sukzessiver, *height determination of initiation of successive strokes* 616.
—, Zeitabstand, *time-intervals between strokes* 578, 579.
Schmelzpunkt, *melting point* 87.
SCHOTTKYsche Theorie, SCHOTTKY'S *theory* 120, 122, 124.
Schwankungen, statistische, *statistical fluctuations* 454, 474.
Schwellenpotential, Variation des, *variation of threshold potential* 517.
Schweißbögen, *welding arc* 294.
SEELIGERsche Regel, SEELIGER'S *rule* 60.
Sekundärdurchbruch, *secondary burst* 522.
Sekundäreffekt, *secondary action* 446.
Sekundärlektronen, *secondary electrons* 23, 24, 78, 80.
sekundärer Koeffizient, *secondary coefficient* 9.
Sekundärionisation, *secondary ionization* 5, 11.
Sekundärprozesse, *secondary processes* 6, 7, 14, 23, 28, 39, 41, 49, 78, 447, 503.
Selbstabsorption, *self-absorption* 225, 358.
selbständige Entladung, Schwellenbedingung, *threshold condition of a self sustaining discharge* 450.
Selbstabstoßung, *self repulsion* 77, 80.
selbständiger Strom, *self sustaining current* 450.
selbständige TOWNSEND-Entladung, *self sustaining* TOWNSEND *discharge* 54, 449, 452.
Selbstumkehr, *self inversion* 373.
Sonde in Brückenschaltung, *probe in bridge circuit* 66.
Sonden, *probes* 65, 82, 150, 381.
—, abgeschirmte, *screened probes* 67, 132.
—, spezielle Typen, *special types of probes* 67.
—, zylindrische, *cylindrical probes* 66.
Sondenausdehnung, *area of the probe* 63, 64.
Sondencharakteristiken in einem magnetischen Feld, *probe characteristics in a magnetic field* 173.
Sondenkurven, *probe curves* 65.
Spannungsschwankungen, *fluctuations in voltage* 141.

Spannungszusammenbruch, *voltage collapse* 27.
Spektren, kontinuierliche, *continuous spectra* 251.
Spektrallinien geringster Energie, *spectral lines of lowest energy* 60.
Spektralmessungen, *spectral measurements* 112.
Spektrum der angeregten Ionen, *spectrum of excited ions* 136.
spezifische Wärme, *specific heat* 159, 193.
Spitzenentladung, negative, *negative pointlike discharge* 526.
Spitzenentladungsmethoden, *point-discharge methods* 598.
Spitze-Platte-Entladung in N_2, H_2 and A mit 1%igem O_2, *point-to-plane discharge in* N_2, H_2 *and* A *with* 1% O_2 508, 510.
—, negative, in Luft, *negative point-to-plane discharge in air* 518.
—, positive, *positive point-to-plane discharge* 524.
Spitze-Spitze-Zündstrecke, *point-to-point gap* 526.
Spitzenstromwerte, *peak current values* 605.
Sprühentladung, *spray discharge* 163.
Spurenverunreinigung, *traces of impurity* 518.
Stabilität, Formen der, *forms of stability* 201.
Stabilitätsbedingung der Entladung, *stability condition of discharge* 187, 189.
STARK-Effekt, STARK *effect* 62, 101, 363.
STARK-Effekt-Verbreiterung, STARK *effect broadening* 374.
STEENBECKsches Minimumprinzip, STEENBECK's *minimum principle* 383.
Stickstoff, *nitrogen* 507.
STOLETOW 463.
Stoß durch schwere Teilchen, *massive particle impact* 223.
Stoßdurchschlag, *impulse breakdown* 526.
Stöße, elastische, *elastic collisions* 125.
—, unelastische, *inelastic collisions* 125, 133, 194.
— erster Art, *collisions of the first kind* 218.
— zweiter Art, *collisions of the second kind* 75, 85, 112, 136, 183, 184, 236.
Stoßdämpfung, *collision damping* 361.
Stoßhäufigkeit, *collision frequency* 80, 538.
Stoßquerschnitt, *collision cross-section* 344, 348.
Stoßverbreiterung, *impact-broadening* 212.
Strahlung, blockierte, *blockaded radiation* 209, 217, 236.
—, kontinuierliche, *continuous radiation* 354.
Strahlungsdiffusion, *radiation diffusion* 365.
Strahlungseffekte, *radiation effects* 609.
Strahlungsleistung, *radiation efficiency* 388.
Strahlungsnormal, *radiation standard* 368.
Strahlungsrekombination von Elektronen, *radiative recombination of electrons* 229.
Strahlungstheorie, *radiation theory* 9, 202, 210.
Streamer 1, 28, 30, 477, 480, 481, 489, 517, 523.
Streamer, Anodenkorona, *anode streamer corona* 515.
—, Bildung, *streamer formation* 449, 456.
—, Durchschlag, *breakdown streamer* 518, 524, 526.
—, Entladung, Zündbedingung für, *threshold relation for streamer discharge* 489.
—, Fortpflanzung, Mechanismus, *mechanism of streamer propagation* 618.
—, Funken, *streamer spark* 454, 496, 513, 517.
—, Haupt- oder Rückkehr, *main or return streamer* 585, 593, 596.
—, in der Entladung im homogenen Feld, *streamer in uniform field breakdown* 490.
—, Kanaldurchmesser, *diameter of streamer channel* 597.
—, Kathoden, *cathode streamer* 526.
—, Korona, der Zündung, *pre-onset corona streamer* 518.
—, negativer, *negative streamer* 526.
—, —, auf die Anode gerichteter, *anode directed negative streamer* 481.
—, positiver Korona, *positive corona streamer* 481.
— zwischen Schlägen, Radarerforschung der, *radar studies of interstroke streamers* 615.
—, Rückkehr, *return streamer* 597, 621.
—, Rückstoß, *recoil streamer* 596.
—, Strom, *current in a streamer* 619.
—, Übergang zum TOWNSEND-Funken, *transition from streamer to* TOWNSEND *spark* 491.
— vor der Zündung, *pre-onset streamer* 484, 516, 524.
Strom, Anwachsen, *current growth* 4, 7, 33, 499.
—, durch das Feld verstärkter, *field intensified current* 446, 450, 452.
—, raumladungsbegrenzter, *space charge limited current* 93.
—, Stabilität, *stability of the current* 10.
Stromänderung, ungerichtete irreversible, *undirectional irreversible current change* 453.
Stromdichte, *current density* 2, 3, 16, 17, 18, 20, 21, 49, 326, 330.
— an der Kathode, *current density at the cathode* 407.
Stromfaden, s. streamer.
Strommesser, elektrostatischer, *electrostatic fluxmeter* 602, 605.
Strom-Spannungscharakteristiken, Einfluß magnetischer Felder auf, *influence of magnetic field on current-voltage characteristics* 175.
Stufenlängen, *step-lengths* 590.
Stufenprozeß, *step process* 622.
Suszeptibilität, diamagnetische, *diamagnetic susceptibility* 175.
SUTHERLAND-Faktor, SUTHERLAND-*factor* 348.
symmetrische Felder, *symmetrical fields* 525.

Temperatur, Abhängigkeit vom Radius in einem zylindersymmetrischen Bogen, *dependence of temperature, upon radius in a cylindrical arc* 387.

Temperaturmessung, *measurement of temperature* 369.
Temperaturverteilung, *temperature distribution* 100, 134.
TONKS und LANGMUIR, Theorie von, TONKS *and* LANGMUIR *theory* 125.
Thermodiffusion, *thermo-diffusion* 322.
Thoriumoberflächen, *thorium surfaces* 158.
TOWNSEND-Entladung, TOWNSEND (dark) discharge 54, 240, 477.
—, Anwachsen der, *growth of the* TOWNSEND *discharge* 457.
— bei großer Überspannung, TOWNSEND *breakdown in highly overvolted gaps* 497.
—, vorhergehende, *antecedent* TOWNSEND *discharge* 492.
TOWNSEND-Entladungen, statistische Schwankungen, *statistical fluctuations of* TOWNSEND *breakdowns* 469.
TOWNSENDsche Beziehungen für α, TOWNSEND *relations for* α 4, 103, 104.
TOWNSENDscher β-Effekt, TOWNSEND'S β *effect* 162.
— Ionisierungskoeffizient, TOWNSEND's *ionization coefficient* 81.
— Koeffizient, TOWNSEND *coefficient* 446.
— zweiter Koeffizient, TOWNSEND's *second coefficient* 5.
— Typ der Vorentladung, TOWNSEND *type of predischarge* 505.
TOWNSENDsches Gesetz, TOWNSEND's *law* 513.
— Integral, TOWNSEND *integral* 446.
— Kriterium, TOWNSEND *criterion* 5, 10, 29, 30.
Transportkoeffizienten, *transport coefficients* 328, 333.
TRICHEL-Impulse, TRICHEL *pulses* 511.
TRICHELscher Korona-Durchbruch, TRICHEL *pulse corona* 518.

Überführungswärme, *heat of transfer* 322.
Übergang von Glimmentladung zu Lichtbogen, *transition from glow to arc* 16.
Übergangswahrscheinlichkeit, *transition probability* 210, 309, 353.
Übergangszeit, *transit time* 27, 448.
Übergangszustände, *transient states* 195.
Überspannung, *overvoltage* 42, 44f., 47, 50, 51, 470.
—, *overshoot in potential* 508, 511, 513.
Ultra-Hochfrequenzentladung, *ultra-high-frequency breakdown* 535.
ultraviolettes Licht, *ultra violet light* 53.

Vakuumfunken, *vacuum spark* 527.
Verbreiterung, statistische, *statistical broadening* 213.
Verdampfung, *evaporation* 154, 160.
Verdampfungsenergie, *evaporation energy* 158.
Verdichtungsstoß, *shock* 377.
Verstärkungsfaktor, *amplification factor* 446.
Verteilung, *distribution* 4.
Vervielfachung, *multiplication* 81, 102, 103, 197.
Verzweigung des Blitzes, *branching of lightning* 579.
— und Schlagreihenfolge, *branching and stroke-order* 580.
Volumenrekombination, *volume recombination* 128, 133, 194.

Wachstumskurve, *growth curve* 364.
Wälzbogen, *rotating arc* 299.
Wandeffekte, *wall effects* 92, 199.
Wanderungsgeschwindigkeit, *drift velocity* 1, 169.
— von Elektronen, *drift velocity of electrons* 18, 19.
— der positiven Ionen, *drift velocity of positive ions* 18, 19.
Wandladungen, *wall charges* 54, 59, 62, 123, 190.
Wandladungsdichte, *wall charge density* 80.
Wandstrom, *wall current* 64, 124, 174, 202.
Wandtemperatur, *wall temperature* 90, 134, 185.
Wandverlust, *wall loss* 133.
Wärmefluß, *heat flow* 159.
Wärmeleitung, *heat conduction* 325, 338, 346, 351, 381.
Wasserdampf, *water vapour* 190.
Wasserstoff, *hydrogen* 545.
—, reiner, Entladung zwischen koaxialen Zylindern, *pure hydrogen, breakdown betweeen coaxial cylinders* 508.
Wechselstrombögen, *alternating current arcs* 286.
Widerstand, effektiver, *effective resistance* 201.
—, negativer, *negative resistance* 452, 453.
Widerstandslinien, *resistance line* 189, 201.
WIEDEMANN-FRANZsches Gesetz, WIEDEMANN-FRANZ *law* 329.
WIENsches Verschiebungsgesetz, WIEN's *displacement law* 69.
WIGNERsche Regel, WIGNER's *rule* 224.
Wirbelstabilisierung, *whirl-stabilization* 300.
Wirkungsquerschnitt, *cross-section* 5, 219.
—, elektrischer, *electrical cross section* 219.
— für Absorption, *cross section for absorption* 37.
— für Ionisierung, *cross section for ionization* 221.
— für Stöße, *cross section in collisions* 219.
—, optischer, *optical cross section* 220.
Wirkungsquerschnittsangaben, *cross-sectional data* 37.
Wolke, Leuchtstärke der, *luminosity of the cloud* 584.
Wolkenentladung, *cloud discharge* 577, 597.
Wolken- und Luftentladungen, *cloud and air discharges* 617.
Wurzel-Verzweigung, *root-branching* 580.

Zeit, charakteristische, *characteristic time* 471.
Zeitkonstante, *time: unit-constant* 471.
zeitliches Anwachsen, *temporal growth* 41f.
Zeitlupenanordnung, *slow motion arrangement* 383, 398.

Zeitverzögerungen, *time lags* 469.
Zeitverzögerungen, beobachtbare, *observational time lags* 471, 474.
—, experimentelle Untersuchung, *experimental study of time lags* 473
—, lange, Interpretation, *interpretation of long time lags* 493.
—, —, in Nähe der Zündspannung, *long time lags, near threshold* 492.
—, statistische, *statistical time lag* 471.
zerstäubte Atome, *sputtered atoms* 96.
zerstäubtes Material, *sputtered material* 157.
Zerstäubung, chemische, *chemical sputtering* 509.
— der Kathode, *sputtering of cathode* 161.
—, Oberflächenänderung, *surface changes by sputtering* 158.
—, radiale Änderung der, *radial variation of sputtering* 160.
—, Theorien der, *theories of sputtering* 158.
Zerstäubungserscheinungen, *sputterung phenomena* 509.
Zerstäubungsverhältnis, *sputtering rate* 165.
Zischmechanismus, *hissing mechanism* 425.
Zündbedingung, *threshold condition* 459.
Zündschwelle bei Elektroneneinfang, *breakdown threshold with electron attachment* 468.
Zündschwellen-Erniedrigung durch positive Raumladung, *threshold lowering by positive space charge* 461.
Zündspannung, *breakdown threshold* 450f.
—, genaue Formulierung, *breakdown threshold, accurate formulation* 456.
—, kleinste, *minimum sparking potential* 464.
Zündung durch α-Strahlen, *triggering single discharge sequences by α-rays* 512.
Zusammenbruch der Spannung, *collaps of voltage* 41, 45, 50.
Zusammenstoß mit Molekülen, *impact with molecules* 448.
Zustandssumme, *partition function* 321,334, 350.
Zweig- und M-Komponenten, *branch and M components* 595.
Zweistufenionisierung, *two stages ionization* 194.
Zylinder, koaxiale, *coacial cylinders* 510, 557.
—, —, in Luft, *coacial cylinders in air* 511.
—, —, in reinem Stickstoff, *coaxial cylinders in pure nitrogen* 514.
— beliebiger Länge, *cylinders of arbitrary length* 552.

Subject Index.

(English-German.)

Where English and German spelling of a word is identical the German version is omitted.

Absorption, *Absorption* 38, 39.
Absorption coefficient, *Absorptionskoeffizient* 356, 358.
—, total, *gesamte Absorption* 360.
Adsorbed gas, *adsorbiertes Gas* 85.
After-glows, *Nachglimmen* 228.
Air discharges, *Luftentladungen* 577, 596.
Alti-electrograph, *Alti-Elektrograph* 598.
Amplification factor, *Verstärkungsfaktor* 446.
Anode and anode-plasma temperature, *Anoden- und Anodenplasmatemperatur* 424.
—, contraction at the, *anodische Kontraktion* 425.
—, current density at the, *Stromdichte an der Anode* 425.
— dark space, *Anodendunkelraum* 57.
— evaporation, *Anodenverdampfung* 431.
— fall, *Anodenfall* 61, 83, 145, 146, 148, 165, 279, 422.
— —, negative, *negativer Anodenfall* 146.
— —, normal, *normaler Anodenfall* 436.
— —, positive, *positiver Anodenfall* 146.
— —, theories, *Anodenfalltheorien* 434.
— glow, *Anodenglimmlicht* 57, 146, 149, 247.
— —, oscillations of the, *Schwingungen des Anodenglimmlichtes* 144.
—, heat at the, *Wärme an der Anode* 150.
—, movable, *bewegliche Anode* 115.
— rays, *Anodenstrahlen* 147.
— region, *Anodengebiet* 145, 152.
—, size and shape of the, *Ausdehnung und Form der Anode* 146.
—, spherical, *kugelförmige Anode* 146.
— spots, *Anodenflecken* 146, 151, 165.
Arc, *Lichtbogen* 1, 2, 9.
—, carbon, *Kohlebogen* 291.
— cathode, temperature of the, *Temperaturen der Bogenkathode* 407.
—, characteristic of an, *Charakteristiken eines Lichtbogens* 276.
— characteristic, dynamical, *dynamische Bogencharakteristik* 286.
—, chemical reactions in, *chemische Vorgänge im Lichtbogen* 284.
— column, *Bogensäule* 366.
— diameter, *Bogendurchmesser* 402.
— discharge, *Bogenentladung* 55.
— furnaces, *Lichtbogenöfen* 303.
—, GERDIEN-, GERDIEN-*Bogen* 300, 371.
—, ignition of, *Zündung von Lichtbögen* 262.
— lines, *Bogenlinien* 112.
Arc, matter transport, *Materialtransport-Lichtbogen* 283.
— mechanism, *Lichtbogen-Mechanismus* 257.
— physics, *Lichtbogenphysik* 255.
—, radiation properties of, *Strahlungseigenschaften des Lichtbogens* 281.
—, rotating, *Wälzbogen* 299.
—, self sustained, *selbständiger Lichtbogen* 255, 304.
—, stabilization of an, *Stabilisierung eines Lichtbogens* 265.
—, temperatures, *Lichtbogentemperaturen* 282.
—, thermal, *thermischer Lichtbogen* 257.
Arcs, alternating current, *Wechselstrombögen* 286.
—, electrode stabilized, *elektroden-stabilisierte Bögen* 276, 391.
—, metallic vapour, *Metalldampfbögen* 294.
—, non-stationary, *nichtstationäre Bögen* 290.
— in rare gases, *Edelgasbögen* 296.
—, reignition of, after extinguishing, *Wiederzünden gelöschter Bögen* 291.
—, switch, *Schaltbögen* 302.
—, switched off, *abgeschaltete Bögen* 382.
—, types of, *Lichtbogentypen* 291.
ASTON or primary dark space, ASTON*scher oder erster Dunkelraum* 56, 96, 97.
Asymmetrical fields, *asymmetrische Felder* 449.
Attachment, *Anlagerung* 178.
— frequency, *Einfangfrequenz* 566.
Aureole 57, 110, 112, 372.
Auxiliary anode, *Hilfsanode* 152.
Avalanche, *Lawine* 4, 28, 29, 47, 446.
— fluctuation, statistical, *statistische Lawinenschwankungen* 447.

Back diffusion, *Rückdiffusion* 154, 512.
— — losses, *Rückdiffusionsverluste* 447.
Back-scattering of electrons, *Rückstreuung von Elektronen* 4, 15.
Ball lightning, *Kugelblitz* 577.
Band spectra, *Bandenspektren* 98, 100, 184.
B change, *B-Änderung* 610.
Bead (or chain) lightning, *Perlschnurblitz* 577.
BECK-arc, BECK-*Bogen* 293, 428, 429.
Bolometer 69.
BOLTZMANN distribution, BOLTZMANN*sche Verteilung* 70, 106, 107, 126, 306.
— equation, BOLTZMANN*sche Gleichung* 121.

Bombardment of the cathode, *Bombardement der Kathode* 13.
Boys camera, Boys-Kamera 581.
— drum camera, Boys-*Trommelkamera* 583.
Branch and *M* components, *Zweig- und M-Komponenten* 595.
Branching of lightning, *Verzweigung des Blitzes* 579.
— and stroke-order, *Verzweigung und Schlagreihenfolge* 580.
Breakdown, *Funkendurchschlag, s. Durchschlag* 1, 8, 17, 20, 27, 28, 35, 38, 39, 40, 49.
—, *Durchschlag* 453, 525.
— in air, *Durchschlag in Luft* 524, 566.
— in asymmetrical fields, *Durchschlag in asymmetrischen Feldern* 506.
— condition, *Durchschlagsbedingung* 6.
— criterion, *Durchschlags-Kriterium* 5, 9, 16, 32, 33, 35, 37, 45.
— —, physical significance of the, *physikalische Bedeutung des Durchschlags* 9.
—, electron attachment controlled, *durch Elektroneneinfang geregelter Durchschlag* 566.
— events, time sequence of, *zeitlicher Ablauf des Durchschlags* 2.
— by field emission, *Durchschlag infolge Feldemission* 450.
—, mechanism, *Durchschlagsmechanismus* 39.
—, mobility-controlled, *beweglichkeitsgeregelter Durchschlag* 568.
— in non-uniform fields, *Durchschlag in inhomogenen Feldern* 480.
— in oxygen, *Durchschlag in Sauerstoff* 567.
— potential, *Durchschlagsspannung* 1, 5, 10, 12, 15, 26, 451.
— probability, *Durchschlagswahrscheinlichkeit* 469.
— in pure argon, *Durchschlag in reinem Argon* 509, 513.
—, secondary electron resonance, *Sekundärelektronen-Resonanz-Durchschlag* 570.
— in spark suppressing gases, *Durchschlag in funkenverhindernden Gasen* 522.
— threshold, *Zündspannung* 450 f.
— —, accurate formulation, *genaue Fromulierung der Zündspannung* 456.
— — with electron attachment, *Zündschwelle bei Elektroneneinfang* 468.
— time, *Durchschlagszeit* 28, 454.
— —, true, *wahre Durchschlagszeit* 472.
Broadening, statistical, *statistische Verbreiterung* 213.
Burst pulse, *einmaliger Durchschlag* 515, 518.
—, threshold for a localized, *Bedingung für einen lokalisierten Durchschlag* 484.

Camera observations, correlations with, *Zusammenhang mit Kamerabeobachtungen* 608.
—, types of, *Kameratypen* 580.
Canal, cf. streamer, or leader, *Kanal, s. Streamer (=leuchtender Faden) und Leitkanal.*
Canal ray discharges, *Kanalstrahlentladungen* 162.
— rays, *Kanalstrahlen* 57, 59, 96.
Capillar tubes, *Kapillarröhren* 91, 118, 163, 174, 199.
Capillary electrometer, *Kapillarelektrometer* 598.
Carbon arc, hissing, *zischender Kohlebogen* 425, 427.
Cascading constants, *Kaskaden-Konstanten* 234.
Cataphoresis, *Kataphorese* 195.
Cathode, area of the, *Kathodenausdehnung* 94.
— dark space, *Kathodendunkelraum* 56, 62, 81, 165.
— — —, length, *Länge des Kathodendunkelraumes* 92, 165.
— — —, length in gas mixtures, *Länge des Kathodendunkelraumes in Gasgemischen* 183.
— — —, thickness of, *Kathodendunkelraumdicke* 106.
— effects, *Kathodeneffekte* 10.
— emission, *Kathodenemission* 33.
—, energy balance at the, *Energiebilanz an der Kathode* 99.
—, evaporating, *verdampfende Kathode* 420.
— fall, *Kathodenfall,* 81, 82, 83, 87, 89, 101, 103, 106, 108, 165, 193, 279, 405, 412, 420.
— —, abnormal, *abnormaler Kathodenfall* 84.
— —, defects of the theories, *Kathodenfall, Fehler der Theorien* 107.
— — in gas mixtures, *Kathodenfall in Gasgemischen* 183.
— —, normal, *normaler Kathodenfall* 84.
— —, striations, *Schichtung im Kathodenfall* 100.
—, field at the, *Feld an der Kathode* 105.
— geometry, *Kathodengeometrie* 97.
— glow, *Kathodenglimmlicht* 96, 246.
— —, ion current at the, *Ionenstrom am Kathodenglimmlicht* 103.
—, heating of the, *Erhitzung der Kathode* 99, 100.
—, highly stressed, *hochgespannte Kathode* 526.
— instability, *Kathodeninstabilität* 506.
— irradiated, *bestrahlte Kathode* 1.
— layer, first, *erste Kathodenschicht* 56, 60, 96, 146.
— light, *Kathodenlicht* 59, 96.
— material, *Kathodenmaterial* 12, 59, 88, 156.
— mechanism of arcs, *Kathodenmechanismus von Lichtbögen* 403.
— metal, sputtering of the, *Zerstäubung des Kathodenmetalls* 86, 87.
—, movable, *bewegliche Kathode* 120.
—, oxidation of the, *Oxydation der Kathode* 33.
—, photo-sensitive, *lichtempfindliche Kathode* 202.
—, positive ions on the, *positive Ionen auf der Kathode* 6.
— processes, *Kathodenprozesse* 14.

Cathode rays, *Kathodenstrahlen* 59, 147.
— region, magnetic field influences, *Kathodengebiet, Einflüsse magnetischer Felder* 165.
— sputtering, *Kathodenzerstäubung* 154.
— surface, *Kathodenoberfläche* 15, 108, 408.
— temperature, *Kathodentemperatur* 90, 157.
—, thermionic, *Glühkathode* 243.
—, virtual, *virtuelle Kathode* 59.
Cathodes, concave and convex, *konkave und konvexe Kathoden* 97, 99.
Characteristic, *Charakteristik* 10, 15, 16, 147.
—, negative, *negative Charakteristik* 506, 513.
Characteristics of the second kind, *Charakteristiken von zweiter Art* 176.
Charge distortion, *Ladungsverformung* 503.
— distortion, space, *Raumladungsverzerrung* 30.
— (in ligthning), quantities of, *Größenordnungen der Ladungen (im Blitz)* 600.
— transfer, *Ladungstransport* 77, 79, 80, 99, 106, 109, 163.
Chemical potential, *chemisches Potential* 317.
Clearing time, *Entionisierungszeit* 455, 519, 520.
Cloud and air discharges, *Wolken- und Luftentladungen* 617.
— discharge, *Wolkenentladung* 577, 597.
— luminosity, *Leuchtstärke einer Wolke* 584.
Coaxial cylinders, *koaxiale Zylinder* 517.
— —, breakdown in air and pure oxygen, *koaxiale Zylinder, Durchschlag in Luft und reinem Sauerstoff* 516.
— —, — in pure gases, *koaxiale Zylinder, Durchschlag in reinen Gasen* 514.
Collapse of voltage, *Zusammenbruch der Spannung* 41, 45, 50.
COLLI-FACCHINI effect, COLLI-FACCHINI*scher Effekt* 463.
Collision cross-section, *Stoßquerschnitt* 344, 348.
— damping, *Stoßdämpfung* 361.
— frequency, *Stoßhäufigkeit* 80, 538.
Collisions, elastic, *elastische Stöße* 125.
—, inelastic, *unelastische Stöße* 125, 133, 194.
— of the first kind, *Stöße erster Art* 218.
— of the second kind, *Stöße zweiter Art* 75, 85, 112, 136, 183, 184, 236.
Column, constricted, *eingeschnürte Säule* 178.
—, contraction of the, *Zusammenziehung der Säule* 57, 175, 193.
— gradient, *Säulengradient* 279.
—, length of the, *Länge der Säule* 114.
—, linear, *lineare Säule* 128, 129.
—, low current (subnormal), *Säule mit geringem Strom (subnormal)* 130.
—, magnetic field on the positive, *magnetisches Feld in der positiven Säule* 168.
—, plane positive, *ebene positive Säule* 169.
—, positive, *positive Säule* 57, 114, 120, 165, 174, 247.
— of rectangular section, *Säule mit rechteckigem Querschnitt* 128.
—, striated, *geschichtete Säule* 57, 137, 142, 252.
Column, unstable, *instabile Säule* 194.
—, uniform, in atomic gases, *gleichförmige Säulen in Atomgasen* 114, 115, 116.
—, uniform, in molecular gases, *gleichförmige Säulen in Molekülgasen* 118.
Compression, self-magnetic, *eigenmagnetische Kompression* 396.
Concentration, *Konzentration* 209.
Concentric rings, *konzentrische Ringe* 151.
Conductivity, *Leitfähigkeit* 159.
—, electric, *elektrische Leitfähigkeit* 326, 352, 378.
—, thermal, *thermische Leitfähigkeit* 134.
Conservation of mass, *Massenerhaltung* 316.
Constrictions, narrow, *enge Einschnürungen* 147, 163.
Contact potential, *Kontaktpotential* 65.
Continuity equation, *Kontinuitätsgleichung* 36, 42.
Continuum, *Kontinuum* 229.
Corona 506.
—, burst pulse, *einmaliger Koronadurchschlag* 518.
— discharge, *Korona-Entladung* 252.
—, positive point-to-plane, *Koronaentladung zwischen positiver Spitze und Platte* 452, 521.
—, self-sustaining, *selbständige Koronaentladung* 512.
— threshold, burst pulse, *Zündschwelle für einmaligen Koronadurchschlag* 482.
CROOKES (or cathode) dark space, CROOKES*scher (oder Kathoden-) Dunkelraum* 246.
— (or HITTORF) dark space, CROOKES*scher (oder* HITTORF*scher) Dunkelraum* 56.
Cross section, *Wirkungsquerschnitt* 5, 219.
— —, for absorption, *Wirkungsquerschnitt für Absorption* 37.
— — in collisions, *Wirkungsquerschnitt für Stöße* 219.
— —, electrical, *elektrischer Wirkungsquerschnitt* 219.
— — for ionization, *Wirkungsquerschnitt für Ionisierung* 221.
— —, optical, *optischer Wirkungsquerschnitt* 220.
Cross-sectional data, *Wirkungsquerschnittsangaben* 37.
Current change, undirectional irreversible, *ungerichtete irreversible Stromänderung* 453.
— density, *Stromdichte* 2, 3, 16, 17, 18, 20, 21, 49, 326, 330.
— — at cathode, *Stromdichte an der Kathode* 407.
—, field intensified, *durch das Feld verstärkter Strom* 446, 450, 452.
— growth, *Anwachsen des Stroms* 4, 7, 33, 499.
—, space charge limited, *raumladungsbegrenzter Strom* 93.
—, stability of the, *Stabilität des Stromes* 10.
— voltage characteristics, influence of magnetic field, *Strom-Spannungscharakteristiken, Einfluß magnetischer Felder* 175.

Curvature of the boundaries, *Krümmung der Begrenzungen* 57.
Cylinders of arbitrary length, *Zylinder von beliebiger Länge* 552.
—, coaxial, *koaxiale Zylinder* 510, 557.
—, —, in air, *koaxiale Zylinder in Luft* 511.
—, —, in pure nitrogen, *koaxiale Zylinder in reinem Stickstoff* 514.

Damping, *Dämpfung* 360.
Dark current, *Dunkelstrom* 189.
— discharge, *Dunkelentladung* 202.
— space, energy distribution of ions and electrons, *Dunkelraum, Energieverteilung von Ionen und Elektronen* 95.
— —, length of the, *Länge des Dunkelraumes* 105.
— —, optical properties, *optische Eigenschaften des Dunkelraumes* 96.
Dart leader, mechanism of the, *Mechanismus des dart leader* 620.
— leaders 592.
Dart-stepped leaders 590, 591, 593.
DAVIDSON's growth equation, DAVIDSON*sche Gleichung für das Anwachsen* 42, 44.
Dead time 455, 511, 522.
Degeneracy, *Entartung* 38.
Density, measurement of, *Dichtemessung* 376.
Diffusion, *Diffusion* 16, 77, 80, 169, 323.
—, ambipolar, *ambipolare Diffusion* 122, 129, 170, 174, 332.
— of electrons, *Diffusion von Elektronen* 143.
—, lateral, *laterale Diffusion* 169.
— length, *Diffusionslänge* 533.
—, radial, *radiale Diffusion* 169.
— theory, *Diffusionstheorie* 128, 537.
Discharge, abnormal, *abnorme Entladung* 55.
—, characteristic colour, *Entladung, charakteristische Farbe* 58, 114.
—, characteristic curve of the, *charakteristische Kurve der Entladung* 201.
—, cold cathode, *Entladungen an kalter Kathode* 246.
—, constricted, *eingeschnürte Entladung* 163, 175.
—, effective resistance of the, *effektiver Widerstand der Entladung* 189.
—, electrodeless, *elektrodenlose Entladung* 144, 195, 251.
—, flashing, *Blitzentladungen* 191, 577.
— to ground, total duration of a, *Gesamtdauer einer Erdentladung* 579.
—, hot cathode, *Glühkathodenentladung* 90.
—, intermittent, *intermittierende Entladung* 187, 625.
—, magnetic fields, *magnetisches Feld der Entladung* 73.
—, mechanism of the continuing, *Mechanismus der sich fortsetzenden Entladung* 627.
—, normal, *normale Entladung* 55, 82.
—, obstructed, *behinderte Entladung* 56, 57, 161, 168.
—, photographic studies of luminosity in the, *photographische Erforschung der Leuchtstärke bei der Entladung* 580.
Discharge, self capacity of the, *Selbstkapazität der Entladung* 188.
— sequence, interruption of it, *Unterbrechung der Entladungsfolge* 455.
—, special types of, *spezielle Typen der Entladung* 161, 598.
— spread uniformly over the electrodes, *gleichmäßig über die Elektroden verteilte Entladung* 505.
—, stability of, *Stabilität von Entladung* 201.
—, subnormal, *subnormale Entladung* 55, 82.
— by thermal electrons, *Entladungen durch thermische Elektronen* 247.
—, throbbing, *pulsierende Entladung* 190.
—, time of development of the, *Entwicklungszeit der Entladung* 188.
—, time of formation of the, *Aufbauzeit der Entladung* 190.
Discharges in electronegative gases, *Entladungen in elektronegativen Gasen* 177.
— (lightning) to very high buildings, *Entladungen (Blitz) zu sehr hohen Gebäuden* 596, 607.
— in magnetic fields, *Entladungen im magnetischen Feld* 164.
—, methods of measurement, *Entladungen, Meßmethoden* 61.
—, monoenergetic electron, *monoenergetische Elektronenentladungen* 242.
—, unstable, *instabile Entladungen* 506.
Dispersion relations, *Dispersionsbeziehungen* 193.
— theory of the plasma, *Dispersionstheorie des Plasmas* 356.
Dissociative recombination, *Dissoziationsrekombination* 232.
Distribution, *Verteilung* 4.
Divergent field, *divergentes Feld* 21.
DOPPLER effect, DOPPLER-*Effekt* 70, 98, 136.
— broadening, DOPPLER-*Verbreiterung* 215, 361.
— contribution, DOPPLER-*Verteilung* 372.
Double layer, thin, *dünne Doppelschicht* 143.
— layers, *Doppelschichten* 147.
Drift velocity, *Wanderungsgeschwindigkeit* 1, 169.
— — of electrons, *Wanderungs-Geschwindigkeit von Elektronen* 18, 19.
— — of positive ions, *Wanderungsgeschwindigkeit der positiven Ionen* 18, 19.

ECCLE's relation, ECCLE*sche Beziehung* 357.
EGGERT-SAHA equation, EGGERT-SAHA*sche Gleichung* 308.
EINSTEIN constant of spontaneous emission, EINSTEIN*sche Konstante der spontanen Emission* 127.
Elatation 209, 218.
— cross-section, total, *totaler Elatationswirkungsquerschnitt* 245.
Electric field, radial, *radiales elektrisches Feld* 171.
Electrode materials, *Elektrodenmaterialien* 11.
—, mechanical force on it, *mechanische Kraft auf eine Elektrode* 62.

Electrodes, cleanliness of, *Reinheit der Elektroden* 507.
—, polarisation of the, *Polarisation der Elektroden* 190.
Electrodynamic force, *elektrodynamische Kraft* 169.
Electron attachment, *Elektronenanlagerung* 77, 80, 85, 129, 130.
— attachment to neutral molecules, *Elektronenanlagerung an neutrale Moleküle* 19.
— avalanche, *Elektronenlawine* 9, 33, 35, 48.
— beam, deflection of an, *Ablenkung eines Elektronenstrahls* 91.
— concentration, microwave measurement, *Mikrowellenmessung der Elektronenkonzentration* 68.
— density, *Elektronendichte* 374.
— emission, *Elektronenemission* 9.
— mobility, *Elektronenbeweglichkeit* 121.
— beam, monoenergetic, *monoenergetischer Elektronenstrahl* 243.
— pressure, *Elektronendruck* 192.
— temperature, *Elektronentemperatur* 63, 117, 120, 121, 124, 130, 132, 171, 174.
— temperature in gas mixtures, theoretical, *Elektronentemperatur in Gasgemischen, theoretische* 184.
Electronic oscillations, *Elektronenschwingungen* 191.
Electrons, energy distribution of, *Energieverteilung der Elektronen* 96, 170.
—, fast, *schnelle Elektronen* 143, 173.
— and ions in gases, mean energies, *mittlere Energien der Elektronen und Ionen in Gasen* 2, 14.
—, triggering, *auslösende Elektronen* 473.
Electrophoresis, *Elektrophorese* 195.
ELENBAAS-HELLER *equation*, ELENBAAS-HELLER*sche Gleichung* 337.
Emission 527.
— coefficient, *Emissionskoeffizient* 353, 354.
— due to incidence of photons, *Emission infolge Einfalls von Photonen* 6.
—, enforced, *erzwungene Emission* 210, 359.
— of electrons by positive ions, *Emission von Elektronen durch positive Ionen* 447.
Emissive properties, stable, *stabile Emissionseigenschaften* 34.
Empire State Building 591, 607.
End effects, *Endeffekte* 172, 174.
Energy balance, *Energiebilanz* 122, 126, 132, 133, 381, 392, 401.
— conservation, *Energieerhaltung* 317.
—, flux of, *Energiestrom* 322.
Enthalpy, *Enthalpie* 342.
Entropy, *Entropie* 342.
— balance, *Entropiebilanz* 317.
— current (flux), *Entropiestrom* 318.
— production, *Entropieproduktion* 383.
— source, *Entropiequelle* 318.
Equilibrium, thermal, *thermisches Gleichgewicht* 305.
Equipotential surfaces, *Äquipotentialflächen* 121.
Evaporation, *Verdampfung* 154, 160.
— energy, *Verdampfungsenergie* 158.
Excitation, *Anregung* 17, 209, 218.
— function, *Anregungsfunktion* 60, 135.
— by particle impact, *Anregung durch Teilchenstoß* 219.
— probability of, *Anregungswahrscheinlichkeit* 127.
—, stepwise, *stufenweise Anregung* 225.
— temperature, *Anregungstemperatur* 309.
Excited atoms, *angeregte Atome* 6.
— molecular states of argon, *angeregte Molekülzustände des Argons* 505.
Extinction potential, *Löschspannung* 188.

FARADAY dark space, FARADAY*scher Dunkelraum* 56, 57, 60, 110, 224.
FERMI degeneration, FERMI-*Enartung* 314.
Field, asymmetrical, *asymmetrisches Feld* 455.
— distortion, *Feldverzerrung* 8, 29, 33.
—, effective, *effektives Feld* 532.
— emission, *Feldemission* 79, 80, 163, 449, 450.
— —, thermal, *thermische Feldemission* 414.
— limit, uniform, *Grenze des homogenen Feldes* 537.
—, longitudinal, *longitudinales Feld* 124, 125.
—, nonuniform, *inhomogenes Feld* 552.
—, uniform, *homogenes Feld* 455.
Filament, oxide coated, *Heizfaden mit Oxydschicht* 90.
Film, metallic, *metallischer Film* 13, 14.
—, thin, *dünner Film* 13.
Fine structure, *Feinstruktur* 143.
Flaming arcs, *Flammenbögen* 292.
Flash, *Funke* 1, 51, 506.
Fluctuations, statistical, *statistische Schwankungen* 454, 474.
Fluorescence, *Fluoreszenz* 37f.
Fluxmeter, electrostatic, *elektrostatischer Strommesser* 602, 605.
Force equation, *Kraftgleichung* 317.
Formative lag, fluctuation in the, *Schwankungen in der Aufbauzeit* 474.
— —, observed and true, *beobachtete und wahre Aufbauzeit* 472.
— time, *Bildungsdauer (s. Aufbauzeit)* 41, 50.
— — lag, *Aufbauzeit* 45, 49, 471, 494.
— — — increased rapidly as overvoltages decreased, *mit fallenden Überspannungen stark anwachsende Aufbauzeit* 492.
— — —, long, *lange Aufbauzeit* 526.
Free electron gas, *freie Elektronen als Gas* 512.
Free-free- radiation, *frei-frei-Strahlung* 354.
Freon 523.
Frequency limit, upper, *obere Frequenzgrenze* 192.
Frictional forces, *Reibungskräfte* 318.

Gas absorption, *Gasabsorption* 198.
— density, *Gasdichte* 70, 72.
— discharge at low pressure, *Gasentladung bei niedrigem Druck* 216.
— dynamics, *Gasdynamik* 402.

Gas flow, *Gasfluß* 200.
— interferometer, *Gasinterferometer* 196.
— layer, *Gasschicht* 190.
— —, adsorbed, *adsorbierte Gasschicht* 155, 157.
— mixtures, discharges in, *Entladungen in Gasgemische* 181.
— —, normal current density, *normale Stromdichte in Gasgemischen* 183.
— pressure, *Gasdruck* 80.
— —, low, *niedriger Gasdruck* 2.
— processes, permitted, *erlaubte Gasprozesse* 75.
— stream, longitudinal, *longitudinaler Gasstrom* 202.
— temperature, *Gastemperatur* 69, 100, 136.
— viscosity, *Gaszähigkeit* 200.
Gaseous purity, *Gasreinheit* 507.
Gases, electronegative (negative ions), *elektronegative Gase (negative Ionen)* 129, 146.
Gauze cathode, *Gazekathode* 110.
GEIGER counter, GEIGER-*Zähler* 180, 240, 483, 515, 522.
— pulse spreading, threshold for a, *Schwelle für* GEIGER-*Impuls-Ausbreitung* 484.
GEISSLER or glow discharge, GEISSLER*sche oder Glimmentladung* 452.
GIBBS equation, GIBBS*sche Gleichung* 317.
— potential, GIBBS*sches Potential* 342.
Glow, boundary of the, *Saum des Glimmlichtes* 102.
—, diffuse, *diffuses Glimmlicht* 28.
— discharge, *Glimmentladung* 242, 506.
— —, alternating, *alternierende Glimmentladung* 194.
— —, formation of a, *Entstehung einer Glimmentladung* 53.
— —, high pressure, *Glimmentladung bei hohem Druck* 194.
— — in metal tubes, *Glimmentladung in Metallröhren* 55.
— — in moving gas, *Glimmentladung in bewegtem Gas* 202.
— —, obstructed, *behinderte Glimmentladung* 190.
—, length of the negative, *Länge des negativen Glimmlichts* 113.
—, negative, *negatives Glimmlicht* 56, 110, 143.
—, —, theory, *negatives Glimmlicht, Theorie* 113.
—, spark flash in form of a, *Glimmfunke* 1, 2, 9, 10, 33.
GOLDSTEIN's law, GOLDSTEIN*sches Gesetz* 138, 139.
Growth curve, *Wachstumskurve* 364.
— —, shape, *Form der Anwachs-Kurve* 8.
Ground and air discharges, general features of, *allgemeine Züge der Erd- und Luftentladungen* 577.
Gyro-frequency, *Gyrofrequenz* 177.

HALE or MOHLER processes, HALE *oder* MOHLER *Prozesse* 446.
Halogens, *Halogene* 522, 523.
Heat conduction, *Wärmeleitung* 325, 338, 346, 351, 381.
— flow, *Wärmefluß* 159.
— of transfer, *Überführungswärme* 322.
He_2 spectrum, He_2-*Spektrum* 166.
Heg gas, *Heg-Gas* 535.
Hg-arc, *Hg-Bogen* 390.
Height determination (in lightning), *Höhenbestimmung (im Blitz)* 617.
HEHL's law, HEHL*sches Gesetz* 82.
Helium, impure, *unreines Helium* 549.
High current carbon arc, *Hochstromkohlebogen* 372, 427, 432.
— frequency discharge, *Hochfrequenzentladung* 177, 247, 251.
— pressure, *hoher Druck* 27, 527.
— — arc, water stabilized, *wasserstabilisierter Hochdruckbogen* 379.
— — arc, Xenon, *Xenon-Hochdruckbogen* 390.
— — glow discharge, *Hochdruckglimmentladung* 193.
— vacuum processes, *Hochvakuumprozesse* 79.
Hissing mechanism, *Zischmechanismus* 425.
Hollow cathode discharge, *Hohlkathodenentladung* 91, 97, 108.
HOLST-OOSTERHUIS discharge, HOLST-OOSTERHUIS *Entladung* 241.
HOLTSMARK probability function, HOLTSMARK*sche Wahrscheinlichkeitsfunktion* 213.
Homologous regions, *homologe Bereiche* 18.
— system, *homologes System* 17.
— volumes, *homologe Volumina* 19.
HORNBECK-MOLNAR effect, HORNBECK-MOLNAR*scher Effekt* 446.
Hydrogen, *Wasserstoff* 545.
—, pure, breakdown between coaxial cylinders, *Wasserstoff, reiner, Entladung zwischen koaxialen Zylindern* 508.
Hysteresis, *Hysteresis* 189, 194.

Impact broadening, *Stoßverbreiterung* 212.
—, massive particle, *Stoß durch schwere Teilchen* 223.
— with molecules, *Zusammenstoß mit Molekülen* 448.
Impulse breakdown, *Stoßdurchschlag* 526.
— fields, *Impuls-Felder* 45.
—, sharply peaked, *scharf begrenzter Impuls* 27.
Impurities, *Fremdatome* 85, 197.
Inductance, *Induktivität* 201.
—, effective, *effektive Induktivität* 201.
Infra red emission spectra, *Infrarot-Emissionsspektra* 250.
Initial breakdown, *Anfangsdurchschlag* 618.
Instability, *Instabilität* 453.
Intensities, absolute measurement of, *Absolutmessungen von Intensitäten* 368, 369.
Interference photograph, *Interferenzaufnahme* 396.
— spectrogram, *Interferenzspektrogramm* 373.

Internal energy, *innere Energie* 341.
Ion bombardment, *Ionenbombardement* 81.
— cross-section, *Ionenquerschnitt* 349, 380.
— current to the walls, *Ionenstrom zu den Wänden* 126.
— density, *Ionendichte* 28, 44.
— drift velocity, *Ionenwanderungsgeschwindigkeit* 136.
— production, *Ionenerzeugung* 121.
Ions, *Ionen* 9, 33.
—, destruction of negative, *Vernichtung negativer Ionen* 179.
—, energy distribution of the, *Energieverteilung der Ionen* 64.
—, motion of, *Bewegung der Ionen* 135.
—, negative, *negative Ionen* 93, 130, 143, 154, 178.
—, number balance for, *Erhaltung der Anzahl der Ionen* 121.
—, positive ion bombardment, *Beschuß mit positiven Ionen* 31.
—, —: space charges, *Ionen, positive: Raumladungen* 523.
—, radial distribution of, *radiale Verteilung der Ionen* 171.
—, slow, *langsame Ionen* 14.
Ionization, *Ionisierung* 1.
— chamber, *Ionisationskammer* 28, 29, 113.
— coefficient, *Ionisierungskoeffizient* 533.
— by collision, *Ionisierung durch Stoß* 446.
— equilibrium, *Ionisationsgleichgewicht* 308.
— and excitation probabilities, *Ionisierungs- und Anregungswahrscheinlichkeit* 18, 89, 196.
—, external source of, *äußere Quelle der Ionisierung* 54.
— function, *Ionisierungsfunktion* 135, 162.
— growth, *Anwachsen der Ionisierung* 1, 2, 14, 30, 43, 47, 49.
—, growth of, in nonuniform fields, *Zunahme der Ionisierung in inhomogenen Feldern* 21.
— —, spatial, *räumliches Anwachsen der Ionisierung* 32.
— by positive ions, *Ionisierung durch positive Ionen* 6, 448.
— potential, *Ionisierungsspannung* 89, 196, 526.
—, secondary, *sekundäre Ionisierung* 33, 34.
— by single collisions, *Ionisierung durch Einzelstöße* 75, 80.
—, spatial growth of, in static fields at high pressure, *räumliche Zunahme der Ionisierung in statischen Feldern bei hohen Drucken* 27.
—, — — of, in static uniform fields, *räumliches Anwachsen der Ionisierung in statischen konstanten Feldern* 2, 5.
—, stepwise, *stufenweise Ionisierung* 7, 19, 75, 80, 128, 146.
—, temporal growth of, *zeitliches Anwachsen der Ionisierung* 40.
— in two stages, *Zweistufenionisierung* 194.
Ionizing wave, *Ionisierungswelle* 526.
— zone, *Ionisierungszone* 513.
Irradiation, external, *äußere Bestrahlung* 202.
—, internal, *innere Bestrahlung* 110.
Irreversible processes (thermodynamics), *irreversible Prozesse (Thermodynamik)* 315.

JOULE's heat, JOULE*sche Wärme* 339.

KIRCHHOFF's law, KIRCHHOFF*scher Satz* 358, 359.
KUNZMAN source, KUNZMAN*sche Quelle* 145.

LANGEVIN's mobility equation, LANGEVIN*sche Beweglichkeitsgleichung* 122, 131.
LANGMUIR dark space, LANGMUIR*scher Dunkelraum* 224.
— probes, LANGMUIR-*Sonden* 61, 62, 110, 115, 138.
LARENZ and BARTELS, method of, *Methode von* LARENZ *und* BARTELS 371.
LAVAL jet, LAVAL-*Düse* 403.
Layers, inhomogeneous, *inhomogene Schichten* 365.
Leader field change, *Leitkanal, Feldänderung* 608, 610.
—, stepped, *stufenweise gebildeter Leitkanal* 586, 597.
—, tip-field and velocity while advancing, *Leitkanal, Feld und Geschwindigkeit am Kopf* 618.
Leaders, α and β type, first, *erste α- und β-Typ-Leitkanäle* 586, 587, 588, 589.
—, distribution of effective velocities of stepped, *stufenweise gebildete Leitkanäle, Verteilung der effektiven Geschwindigkeiten* 591.
LECHER wires, LECHER-*Drähte* 68, 137.
Light, absorption of, *Lichtabsorption* 113.
—, partially polarized, *teilweise polarisiertes Licht* 81.
—, ultraviolet, *ultraviolettes Licht* 198.
Lightning, daylight photography of, *Aufnahmen von Blitzen bei Tageslicht* 584.
— discharge, historical, *Blitzentladung, Historisches* 574.
Line intensities, *Linienintensitäten* 353.
— shift, *Linienverschiebung* 375.
— spectrum, *Linienspektrum* 135.
Liquid surfaces, *flüssige Oberflächen* 158.
Long time lags, interpretation of, *Interpretation der langen Zeitverzögerungen* 493.
LORENTZ-forces, LORENTZ-*Kräfte* 326, 396.
Low pressure column, *Niederdrucksäule* 174.
— voltage arc, *Niederspannungsbogen* 242.
— — —, *Niedervoltbogen* 257, 304.
Luminosity, *Leuchten* 44.
—, fluctuating, *schwankende Lichtstärke* 191.
—, propagation of, *Fortpflanzung des Leuchtens* 51.

MACH-number, MACH-*Zahl* 403.
Magnetic deflection, *magnetische Ablenkung* 95.
— field, longitudinal, *longitudinales magnetisches Feld* 166.

Magnetic field, superimposed, *überlagertes magnetisches Feld* 559.
— —, transverse, *transversales magnetisches Feld* 166, 170.
— lenses, *magnetische Linsen* 173.
— links, *magnetische Glieder* 605.
— rays, *magnetische Strahlen* 177.
Maintenance condition, *Erhaltungs-Bedingung* 22.
—, equation of, *Erhaltungsgleichung* 103.
— process, *Erhaltungs-Prozeß* 23.
MALAN Camera, MALAN-*Kamera* 584.
MALTER effect, MALTER *Effekt* 164.
Mass concentration, *Massenkonzentration* 316.
MAXWELLian distribution, MAXWELL-*Verteilung* 63, 65, 121, 135, 111.
M-components, *M-Komponenten* 579.
Mean free path, *mittlere frei Weglänge* 80.
— — — for photoionizing radiation, *mittlere freie Weglänge für photoionisierende Strahlung* 485.
Melting point, *Schmelzpunkt* 87.
Membrane, vibrations, *Membranschwingungen* 128.
Mercury arc, *Quecksilberbogen* 297.
— vapor rectifier, *Quecksilberdampf-Gleichrichter* 304.
MERK's criterion, MERK*sches Kriterium* 29.
Metal vapour, *Metalldampf* 185.
Metastable A_2 molecules, *metastabile* A_2-*Moleküle* 509.
— atoms, *metastabile Atome* 6, 15, 20f., 76, 79f., 80, 112, 448.
— —, decay of, *Zerfall von metastabilen Atomen* 202, 448.
— —, ionization in the gas by, *Ionisation durch metastabile Atome im Gas* 448.
— gas molecules, *metastabile Gasmoleküle* 65, 509.
— Neon, *metastabiles Neon* 87.
Meteoric light production, *Erzeugung des Meteorlichtes* 224.
Microwave discharges, *Mikrowellenentladungen* 251.
Minimum principle, *Minimumprinzip* 104, 108, 151.
Mobility, *Beweglichkeit* 103, 178.
Momentum and energy exchange of electrons with plasma waves, *Impuls- und Energieaustausch der Elektronen mit Plasmawellen* 192.
Multiplication, *Vervielfachung* 81, 102, 103, 197.

Narrow tubes, *enge Röhren* 62.
Natural line breadth, *natürliche Linienbreite* 211.
Negative zones, *negative Zonen* 57.
Neon 550.
Neutralization, structural, *wechselseitige Neutralisierung* 233.
Nitrogen, *Stickstoff* 507.
Noise, *Rauschen* 193.
Noise measurements, *Rauschmessungen* 69.
— temperature, *Rauschtemperaturen* 69.
— voltage *Rauschspannung* 373.
Non-uniform fields, *inhomogene Felder* 40, 526.
— symmetrical fields, *inhomogene symmetrische Felder* 525.

O_2 bands, O_2-*Banden* 119.
O_2, effect of small amounts, *Einfluß von kleinen Mengen* O_2 510.
OHM's law, OHM*sches Gesetz* 326.
— — regime, OHM*sches Gesetz, Gültigkeitsbereich* 513, 520.
ONSAGER's reciprocity laws, ONSAGERS *Reziprozitätssätze* 319.
Optical thick layer, *optisch dicke Schicht* 373.
— depth, *optische Tiefe* 358.
— thickness of layer, *optische Schichtdicke* 367.
Oscillation amplitude limit, *Grenzamplitude* 539.
Oscillator strength, *Oszillatorstärke* 211, 360.
Oscillographic techniques, *Oszillographentechnik* 140.
Overshoot in potential, *Überspannung* 508, 511, 513.
Overvoltage, *Überspannung* 42, 44f., 47, 50, 51, 470.
Oxide film, *Oxydfilm* 13.
— layer, *Oxydschicht* 25, 156, 163, 164.

Partition function, *Zustandssumme* 321, 334, 350.
PASCHEN curves, PASCHEN*sche Kurven* 23.
PASCHEN's law, PASCHEN*sches Gesetz* 10, 17. 462.
— —, failure of carrier density, *Versagen des* PASCHEN*schen Gesetzes für Trägerdichte* 463.
Pause times, *Pausen* 590.
Peak current values, *Spitzenstromwerte* 605.
PENNING effect, PENNING-*Effekt* 75, 76, 80, 85, 446, 465.
— gas, PENNING-*Gas* 466.
Permitted and forbidden processes, *erlaubte und verbotene Prozesse* 80.
Pilot streamer 590, 621.
Pinch effect 175.
Pipe discharge, *Rohrentladungen* 181.
Photocells, *Photozellen* 140.
Photoelectric action in the gas, *photoelektrische Wirkung im Gas* 517.
— current, *photoelektrischer Strom* 2, 202.
— effect, *photoelektrischer Effekt* 14, 26, 44, 80, 81, 225.
— efficiency, *photoelektrische Ausbeute* 33.
— emission, *photoelektrische Emission* 3, 20, 25, 98, 447.
— sensitivity, *photoelektrische Empfindlichkeit* 30.
Photographic measurements, *photographische Messungen* 198, 585.

Photo-ionization, *Photoionisation* 7, 14, 20, 28, 35, 36, 37, 39, 77, 80.
— in air, *Photoionisation in Luft* 481.
— in gases, *Photoionisation in Gasen* 449, 480, 481, 511, 515.
Photomultiplier 140.
Photon absorption, *Photonenabsorption* 225.
Photons, *Photonen* 9, 14, 33, 35, 37.
Plasma 256.
— beam, *Plasmastrahl* 400.
— —, cathodic, *kathodischer Plasmastrahl* 409.
— contraction, theory, *Kontraktionstheorie des Plasmas* 416.
— dynamics, *Plasmadynamik* 402.
— flow, *Plasmaströmung* 396, 401.
— flow with supersonic speed, *Plasmaströmung mit Überschallgeschwindigkeit* 403.
— interaction, *Plasmawechselwirkung* 314.
—, magnetic properties of a, *magnetische Eigenschaften des Plasmas* 175.
—, neutral, *neutrales Plasma* 114.
— oscillations, *Plasmaschwingungen* 66, 132, 142, 144, 191, 356.
—, quasi-neutral, *quasi-neutrales Plasma* 120.
— sphere, *Plasmakugel* 390.
—, thermal, *thermisches Plasma* 305.
Point-discharge methods, *Spitzenentladungsmethoden* 598.
Pointlike discharge, negative, *negative Spitzenentladung* 526.
Point-to-plane discharge in N_2, H_2 and A with 1% O_2, *Spitze-Platte-Entladung in* N_2, H_2 *und* A *mit* 1% O_2 508, 510.
— — negative, in air, *negative Spitze-Platte-Entladung in Luft* 518.
— — —, positive, *positive Spitze-Platte-Entladung* 524.
Point-to-point gap, *Spitze-Spitze-Zündstrecke* 526.
POISSON's equation, POISSON*sche Gleichung* 327.
Polar aurora, *Nordlicht* 224, 242.
Polarization, *Polarisation* 223.
Population 209.
Positive ions, *positive Ionen* 13.
Potential distribution, radial, *radiale Potentialverteilung* 172.
— liberation, *Potentialemission* 447.
Pre-breakdown ionization, growth of, *Anwachsen der Ionisation vor dem Durchschlag* 30.
Pressure broadening, *Druckverbreiterung* 362.
— of the electron gas, *Druck des Elektronengases* 174.
Primary action, *Primäreffekt* 446.
— electrons, *Primärelektronen* 111.
— electrons, fast, *schnelle Primärelektronen* 147, 161.
— ionization, *Primärionisation* 4.
— pulse, *Primär-Impuls* 522.
Probe, area of the, *Sondenausdehnung* 63, 64.
— in bridge circuit, *Sonde in Brückenschaltung* 66.
Probe characteristics in a magnetic field, *Sondencharakteristiken in einem magnetischen Feld* 173.
— curves, *Sondenkurven* 65.
—, hot, *Glühsonden* 62, 67.
Probes, *Sonden* 65, 82, 150, 381.
—, cylindrical, *zylindrische Sonden* 66.
—, screened, *abgeschirmte Sonden* 67, 132.
—, special types of, *Sonden, spezielle Typen* 67.
Production function, *Erzeugungsfunktion* 217.
Proper variables, *Eigengrößen* 536.
Pulse, secondary, *sekundärer Durchschlag* 522.

Quadripole radiation, *Quadrupolstrahlung* 210.
Quasi-neutrality, *Quasineutralität* 312.

Radio waves, *Radiowellen* 243.
Radiation, blockaded, *blockierte Strahlung* 209, 217, 236.
—, continuous, *kontinuierliche Strahlung* 354.
— diffusion, *Strahlungsdiffusion* 365.
— effects, *Strahlungseffekte* 609.
— efficiency, *Strahlungsleistung* 388.
— standard, *Strahlungsnormal* 368.
— theory, *Strahlungstheorie* 9, 202, 210.
Radiative recombination of electrons, *Strahlungsrekombination von Elektronen* 229.
RAETHER's criterion, RAETHER*sches Kriterium* 29.
RAMSAUER effect, RAMSAUER-*Effekt* 72, 73, 132.
— cross-section, RAMSAUER-*Querschnitt* 348, 378.
RAYLEIGH effect, RAYLEIGH-*Effekt* 224, 252.
Recombination, *Rekombination* 19, 36, 78, 80, 107, 112, 136, 139, 245.
— coefficient, *Rekombinationskoeffizient* 253.
—, criterion for, *Kriterium für Rekombination* 227.
—, dielectronic, *Zwei-Elektronen-Rekombination* 231.
— surface, *Rekombination auf Oberflächen* 227.
—, three body electron, *Drei-Elektronen-Rekombination* 232.
—, volume, *Volumenrekombination* 227.
Refraction index, *Brechungsindex* 376.
Regenerative, *rückkoppelnd* 450.
Relativistic theory, *relativistische Theorie* 130.
Relaxation time, *Relaxationszeit* 209.
Resistance, effective, *effektiver Widerstand* 201.
— line, *Widerstandslinie* 189, 201.
—, negative, *negativer Widerstand* 452, 453.
Resonance lines, *Resonanzlinien* 133.
— photons, *Resonanzphotonen* 79.
— radiation, *Resonanzstrahlung* 76, 112, 209, 225.
Ribbon lightning, *Bandblitz* 577.
Ring discharge, *Ringentladung* 195.
— electrodes, *Ringelektroden* 94.

Rocket lightning, *Raketenblitz* 577, 597.
Root-branching, *Wurzel-Verzweigung* 580.
Rotational band, *Rotationsbande* 370.
Rotational band spectra, *Rotationsbandenspektren* 70.

SCHOTTKY's theory, SCHOTTKY*sche Theorie* 120, 122, 124.
Secondary action, *Sekundäreffekt* 446.
— burst, *Sekundärdurchbruch* 522.
— coefficient, *sekundärer Koeffizient* 9.
— electrons, *Sekundärelektronen* 23, 24, 78, 80.
— ionization, *Sekundärionisation* 5, 11.
— processes, *Sekundärprozesse* 6, 7, 14, 23, 28, 39, 41, 49, 78, 447, 503.
SEELIGER's rule, SEELIGER*sche Regel* 60.
Self absorption, *Selbstabsorption* 225, 358.
— inversion, *Selbstumkehr* 373.
— repulsion, *Selbstabstoßung* 77, 80.
— stabilization, *Eigenstabilisierung* 276.
— sustaining current, *selbständiger Strom* 450.
— — threshold condition of a discharge, *Schwellenbedingung einer selbständigen Entladung* 450.
— — TOWNSEND discharge, *selbständige* TOWNSEND-*Entladung* 54, 449, 452.
Shadow effects, *Schatteneffekte* 59.
Shock, *Verdichtungsstoß* 377.
— wave, *Druckwelle* 253.
Similarity, *Ähnlichkeit* 17, 20, 82, 138, 153, 193.
— laws, *Ähnlichkeitsgesetze* 70, 97, 99, 117, 166.
— principle, physical significance of the, *physikalische Bedeutung des Ähnlichkeitsprinzips* 18.
Single crystal surfaces, *Einkristalloberflächen* 87.
Slow motion arrangement, *Zeitlupenanordnung* 383, 398.
Sound waves, *Schallwellen* 192.
Space-charge, *Raumladung* 2, 8, 15, 16, 33, 34, 35, 43, 45, 47, 50, 471, 499, 505, 526.
— accumulation, *Raumladungsanhäufung* 481.
— effect, *Raumladungseinfluß* 454, 460.
— field, *Raumladungsfeld* 486.
— limited currents, *Raumladungsbegrenzte Ströme* 513.
—, negative, *negative Raumladung* 523.
—, positive, *positive Raumladung* 29, 523.
Spark, *Funke* 1, 51, 506.
— breakdown, *Funkenentladung* 451, 453.
— current, *Durchschlagsstrom* 2.
— diameter, *Funkenquerschnitt* 47.
— discharge, *Funkenentladung* 252.
— flash, *Funken* 1, 51.
— lines, *Funkenlinien* 112.
— in very short gaps, *Funken in sehr kurzen Zündstrecken* 527.
Sparks, time fluctuation of appearance of, *zeitliche Schwankungen des Auftretens von Funken* 471.
Sparking potential, *Durchschlagsspannung* 1, 5, 10, 11, 12, 15, 20, 45, 82.
— —, curve of PASCHEN, *Kurve für die* PASCHEN*sche Durchschlagsspannung* 161.
— —, minimum, *minimale Durchschlagsspannung* 1, 5, 10, 12, 15, 464.
Specific heat, *spezifische Wärme* 159, 193.
Spectra, continuous, *kontinuierliche Spektren* 251.
Spectral lines of lowest energy, *Spektrallinien geringster Energie* 60.
— measurements, *Spektralmessungen* 112.
Spectrum of excited ions, *Spektrum der angeregten Ionen* 136.
Speed of sound, *Schallgeschwindigkeit* 344, 377.
Spots, *Flecke* 507, 511, 514, 521.
Spray discharge, *Sprühentladung* 163.
Sputtered atoms, *zerstäubte Atome* 96.
— material, *zerstäubtes Material* 157.
Sputtering of cathode, *Zerstäubung der Kathode* 161.
—, chemical, *chemische Zerstäubung* 509.
— phenomena, *Zerstäubungserscheinungen* 509.
—, radial variation of, *radiale Änderung der Zerstäubung* 160.
— rate, *Zerstäubungsverhältnis* 165.
—, surface changes, *Oberflächenänderung durch Zerstäubung* 158.
—, theories of, *Theorien der Zerstäubung* 158.
Stability condition of discharge, *Stabilitätsbedingung der Entladung* 187, 189.
—, forms of, *Formen der Stabilität* 201.
STARK effect, STARK-Effekt 62, 101, 363.
— — broadening, STARK-*Effekt-Verbreiterung* 374.
Starting potential, *Anfangspotential* 55, 85, 165, 176, 202.
Steady corona, onset of, *Einsetzen des Dauer-Koronadurchbruchs* 522.
— discharge, *Dauerentladung* 513.
STEENBECK's minimum principle, STEENBECK*sches Minimumprinzip* 383.
Step-lengths, *Stufenlängen* 590.
Step process, *Stufenprozeß* 622.
STOLETOW 463.
Streak or forked lightning, *Linien oder Gabelblitz* 577.
Streamer 1, 28, 30, 477, 480, 481, 489, 517, 523.
—, anode directed negative, *auf die Anode gerichteter negativer Streamer* 481.
—, breakdown, *Durchschlag-Streamer* 518, 524, 526.
—, cathode, *Kathoden-Streamer* 526.
— channel, diameter of, *Streamer-Kanaldurchmesser* 597.
— corona, anode, *Streamer-Anodenkorona* 515.
—, current in a, *Streamer-Strom* 619.

Streamer, discharge, threshold, *Streamer-Entladung, Zündbedingung* 489.
— formation, *Streamer-Bildung* 449, 456.
—, main or return, *Haupt- oder Rückkehr-Streamer* 585, 593, 596.
—, negative, *negativer Streamer* 526.
—, positive corona, *positiver Korona-Streamer* 481.
—, pre-onset, *Streamer vor der Zündung* 484, 516, 524.
—, pre-onset corona, *Korona-Streamer vor der Zündung* 518.
— propagation, mechanism of breakdown and of, *Mechanismus des Durchschlags und der Streamer-Fortpflanzung* 618.
—, radar studies of interstroke, *Radarerforschung der Streamer zwischen Schlägen* 615.
—, recoil, *Rückstoß-Streamer* 596.
—, return, *Rückkehr-Streamer* 597, 621.
— spark, *Streamer-Funken* 454, 496, 513, 517.
—, transition from, to TOWNSEND spark, *Übergang vom Streamer zum* TOWNSEND-*Funken* 491.
— in uniform field breakdown, *Streamer in der Entladung im homogenen Feld* 490.
Striations, *Schichtung* 92, 141, 142, 143, 169, 192, 252.
—, blue (CH_2), *blaue Schichtung* (CH_2) 139.
—, broadened, in hydrogen, *verbreiterte Schichtung in Wasserstoff* 138.
— in capillary tubes, *Schichtung in Kapillarröhren* 145.
—, curvature of the, *Krümmung der Schichtung* 137.
—, distance between, *Abstand zwischen Schichtung* 138.
—, double, *Doppelschichten* 139.
—, moving, *wandernde Schichtung* 115, 140.
—, stationary, *stationäre Schichtung* 115, 137.
Strokes, time-intervals between, *Zeitabstand zwischen zwei Schlägen* 578, 579.
— of a flash, *Schläge eines Blitzes* 577, 580.
—, height determination of initiation of successive, *Höhenbestimmung für Einsetzen sukzessiver Schläge* 616.
Surface, cleaning of the, *Reinigung der Oberfläche* 520.
Susceptibility, diamagnetic, *diamagnetische Suszeptibilität* 175.
SUTHERLAND-factor, SUTHERLAND-*Faktor* 348.
Symmetrical fields, *symmetrische Felder* 525.

Tarnish layer, *Belag* 15, 30.
Tarnished electrode surface, *Belag auf Elektroden-Oberflächen* 21.
Temperature, dependence upon radius in a cylindrical arc, *Temperatur, Abhängigkeit vom Radius in einem zylindersymmetrischen Bogen* 387.
— distribution, *Temperaturverteilung* 100, 134.
Temperature, measurement of, *Temperaturmessung* 369.
Temporal growth, *zeitliches Anwachsen* 41f.
Thermionic emission, *Glühemission* 449, 450.
Thermo-diffusion, *Thermodiffusion* 322.
Thorium surfaces, *Thoriumoberflächen* 158.
Threshold condition, *Zündbedingung* 459.
— lowering by positive space charge, *Zündschwellen-Erniedrigung durch positive Raumladung* 461.
— potential, variation of, *Variation des Schwellenpotentials* 517.
Thundercloud charges, distribution of, *Ladungsverteilung in der Gewitterwolke* 598.
Time, characteristic, *charakteristische Zeit* 471.
— of growth, *Dauer des Anwachsens* 41.
— lags, *Zeitverzögerungen* 469.
— —, experimental study, *experimentelle Untersuchung der Zeitverzögerungen* 473.
— —, long, near threshold, *lange Zeitverzögerungen in der Nähe der Zündspannung* 492.
— —, observational, *beobachtbare Zeitverzögerungen* 471, 474.
— —, statistical, *statistische Zeitverzögerungen* 471.
—, unit, constant, *Zeitkonstante* 471.
TONKS and LANGMUIR theory, *Theorie von* TONKS *und* LANGMUIR 125.
TOWNSEND breakdown in highly overvolted gaps, TOWNSEND-*Entladung bei großer Überspannung* 497.
— —, statistical fluctuations of, *statistische Schwankungen der* TOWNSEND-*Entladungen* 469.
— coefficient, TOWNSEND*scher Koeffizient* 446.
— criterion, TOWNSEND*sches Kriterium* 5, 10, 29, 30.
— (dark) discharge, TOWNSEND-*Entladung* 54, 240, 477.
— discharge, antecedent, *vorhergehende* TOWNSEND-*Entladung* 492.
— —, growth of the, *Anwachsen der* TOWNSEND-*Entladung* 457.
— integral, TOWNSEND*sches Integral* 446.
— type of predischarge, TOWNSEND*scher Typ der Vorentladung* 505.
— relation for α, TOWNSEND*sche Beziehung für* α 4, 103, 104.
TOWNSEND's β effect, TOWNSEND*scher* β-*Effekt* 162.
— ionization coefficient, TOWNSEND*scher Ionisierungskoeffizient* 81.
— law, TOWNSEND*sches Gesetz* 513.
— second coefficient, TOWNSEND*scher zweiter Koeffizient* 5.
Traces of impurity, *Spurenverunreinigung* 518.
Transient states, *Übergangszustände* 195.
— time, *Übergangszeit* 27, 448.
Transition from glow to arc, *Übergang von Glimmentladung zu Lichtbogen* 16.
— probability, *Übergangswahrscheinlichkeit* 210, 309, 353.

Transport coefficients, *Transportkoeffizienten* 328, 333.
Tributary functions, *Nebenfunktionen* 234.
TRICHEL pulses, TRICHEL-*Impulse* 511.
— pulse corona, TRICHEL*scher Korona-Durchbruch* 518.
Triggering agency, *Auslösungsursache* 449.
— electrons, *Auslöseelektronen* 506.
— single discharge sequences by α-rays, *Zündung durch α-Strahlen* 512.
Tube radius, *Röhrenradius* 117f.
Tubes, long, *lange Röhren* 140.
—, narrow, cf. capillar tubes, *enge Röhren, s. Kapillarröhren* 118.
—, very wide, *sehr weite Röhren* 120.

Ultra-high-frequency breakdown, *Ultra-Hochfrequenzentladung* 535.
Ultra violet light, *ultraviolettes Licht* 53.
Uniform static field, *homogenes statisches Feld* 32.
Unstable region, *instabiles Gebiet* 452.

Vacuum spark, *Vakuumfunken* 527.
Vapours, organic, *organische Dämpfe* 181.
Voltage, fluctuations *Spannungsschwankungen* 141.
Voltage collapse, *Spannungszusammenbruch* 27.
Volume recombination, *Volumenrekombination* 128, 133, 194.

Wall charge density, *Wandladungsdichte* 80.
— charges, *Wandladungen* 54, 59, 62, 123, 190.
— current, *Wandstrom* 64, 124, 174, 202.
— effects, *Wandeffekte* 92, 199.
— loss, *Wandverlust* 133.
— temperature, *Wandtemperatur* 90, 134, 185.
Water vapour, *Wasserdampf* 190.
Welding arc, *Schweißbogen* 294.
Whirl-stabilization, *Wirbelstabilisierung* 300.
WIEDEMANN-FRANZ law, WIEDEMANN-FRANZ*sches Gesetz* 329.
WIEN's displacement law, WIEN*sches Verschiebungsgesetz* 69.
WIGNER's rule, WIGNER*sche Regel* 224.
Work function, *Austrittsarbeit* 14, 15.

X-rays, *Röntgenstrahlen* 202.
—, absorption, *Absorption der Röntgenstrahlen* 200.
—, soft, *weiche Röntgenstrahlen* 247.

MIX
Papier aus verantwortungsvollen Quellen
Paper from responsible sources
FSC® C105338

If you have any concerns about our products,
you can contact us on
ProductSafety@springernature.com

In case Publisher is established outside the EU,
the EU authorized representative is:
Springer Nature Customer Service Center GmbH
Europaplatz 3, 69115 Heidelberg, Germany

Printed by Libri Plureos GmbH
in Hamburg, Germany